TECHNOLOGY OVERVIEW

For Teachers

- Big Ideas **Exam***View*® Assessment Suit
- **Lesson***View*® Dynamic Planning Tool
- **Puzzle***View*® Vocabulary Puzzle Builder
- **Mind***Point*® Quiz*Show*
- Interactive Glossary:
  English, Spanish, Haitian Creole
- Dynamic Classroom
- Answer Presentation Tool
- Lesson Tutorials
- *BigIdeasMath.com*
  - **Student Companion Website**
  - **Teacher Companion Website**

Exam*View*® Assessment Suite

Lesson*View*®

Puzzle*View*®

Mind*Point*® **Quiz***Show*

BIG IDEAS MATH®

TEACHING EDITION

BLUE

Ron Larson
Laurie Boswell

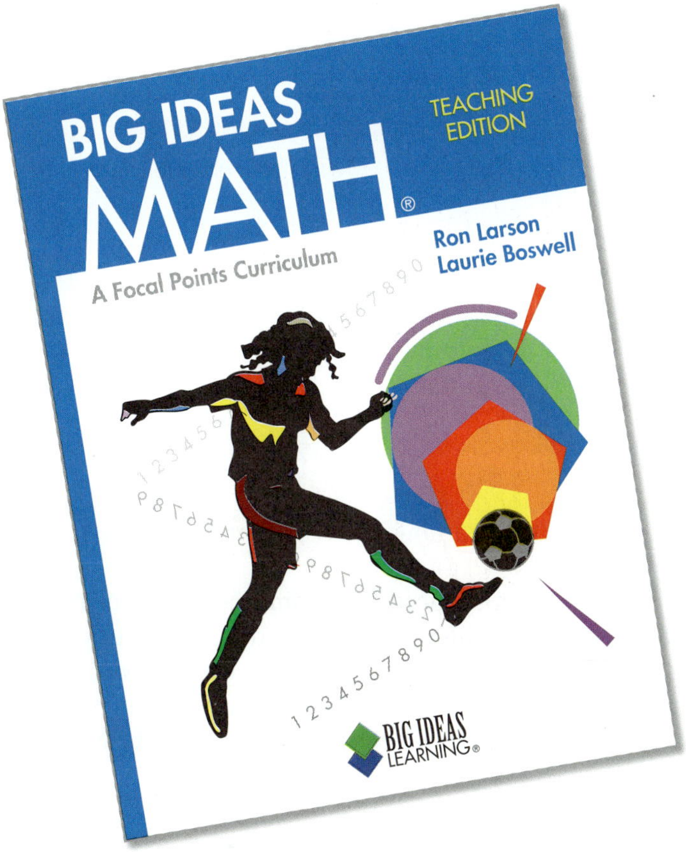

BIG IDEAS LEARNING®

Erie, Pennsylvania
BigIdeasLearning.com

Big Ideas Learning, LLC
1762 Norcross Road
Erie, PA 16510-3838
USA

For product information and customer support, contact Big Ideas Learning at **1-877-552-7766** or visit us at ***BigIdeasLearning.com***.

Printed in the U.S.A.

ISBN 13: 978-1-60840-018-8
ISBN 10: 1-60840-018-2

3 4 5 6 7 8 9 10 WEB 13 12 11 10

AUTHORS

Ron Larson is a professor of mathematics at Penn State University in Erie, Pennsylvania where he has taught since receiving his Ph.D. in mathematics from the University of Colorado in 1970. Dr. Larson is well known as the lead author of a comprehensive program for mathematics that spans middle school, high school, and college courses. His high school and Advanced Placement books are published by Holt McDougal. Dr. Larson's numerous professional activities keep him in constant touch with the needs of students, teachers, and supervisors. Dr. Larson and Laurie Boswell began writing together in 1992. Since that time, they have authored over two dozen textbooks. In their collaboration, Dr. Larson is primarily responsible for the pupil edition and Laurie is primarily responsible for the teaching edition of the text.

Laurie Boswell is a mathematics teacher at the Riverside School in Lyndonville, Vermont. She is a recipient of the Presidential Award for Excellence in Mathematics Teaching. Laurie has taught math to students at all levels, elementary through college. In addition, Laurie was a Tandy Technology Scholar, and served on the NCTM Board of Directors from 2002 to 2005. She currently serves on the board of NCSM, and is a popular national speaker. Along with Dr. Larson, Laurie has co-authored numerous math programs.

ABOUT THE BOOK

This book is brand new! It is not a revision of a previously published book.

When the NCTM released its new *Curriculum Focal Points* for mathematics for grades 6–8, we were delighted. The traditional mile-wide and inch-deep programs that have been followed for years have clearly not worked. Middle school students need something new . . . fewer topics with deeper coverage.

- **DEEPER** Each section is designed for 2–3 day coverage.
- **DYNAMIC** Each section begins with a full class period of active learning.
- **DOABLE** Each section is accompanied by full student and teacher support.
- **DAZZLING** How else can we say this? This book puts the dazzle back in math!

Ron Larson   Laurie Boswell

TEACHER REVIEWERS

- Aaron Eisberg
  Napa Valley Unified School District
  Napa, CA
- Gail Englert
  Norfolk Public Schools
  Norfolk, VA
- Alexis Kaplan
  Lindenwold Public Schools
  Lindenwold, NJ
- Lou Kwiatkowski
  Millcreek Township School District
  Erie, PA
- Marcela Mansur
  Broward County Public Schools
  Fort Lauderdale, FL
- Bonnie Pendergast
  Tolleson Union High School District
  Tolleson, AZ
- Tammy Rush
  Hillsborough County Public Schools
  Tampa, FL
- Patricia D. Seger
  Polk County Public Schools
  Bartow, FL
- Elizabeth Wright
  Pinellas County Schools
  Largo, FL

STUDENT REVIEWERS

- Ashley Benovic
- Vanessa Bowser
- Sara Chinsky
- Kaitlyn Grimm
- Lakota Noble
- Norhan Omar
- Jack Puckett
- Abby Quinn
- Victoria Royal
- Madeline Su
- Lance Williams

CONSULTANTS

- Patsy Davis
  Educational Consultant
  Knoxville, Tennessee

- Deb Johnson
  Differentiated Instruction Consultant
  Missoula, Montana

- Mark Johnson
  Mathematics Assessment Consultant
  Raymond, New Hampshire

- Ryan Keating
  Special Education Advisor
  Gilbert, Arizona

- Michael McDowell
  Project-Based Instruction Specialist
  Scottsdale, Arizona

- Sean McKeighan
  Interdisciplinary Advisor
  Midland, Texas

- Bill Setzer
  Retired Mathematics Supervisor
  Roanoke County, Virginia

- Bonnie Spence
  Differentiated Instruction Consultant
  Missoula, Montana

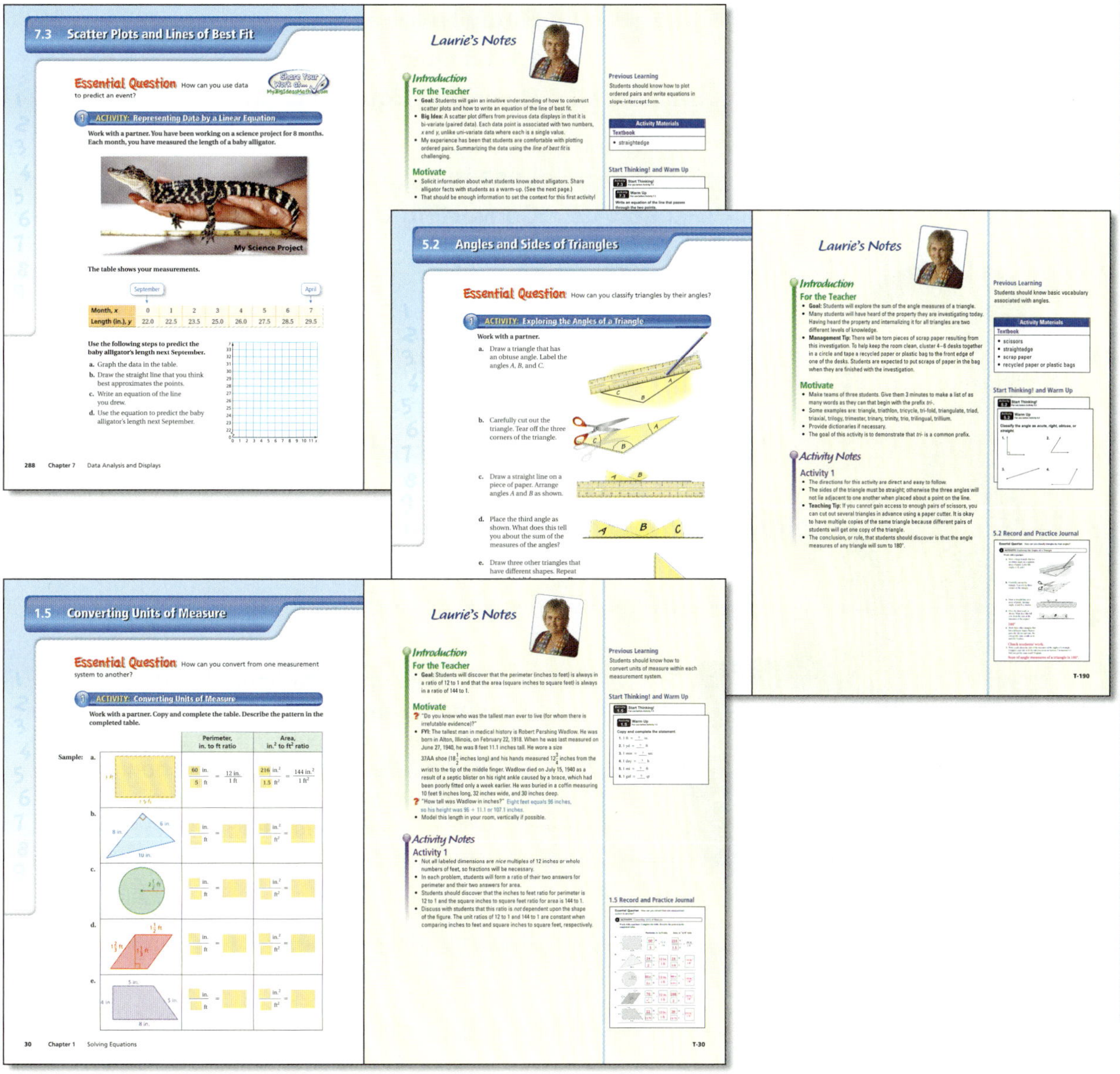

Unit 1
Chapters 1–4

Linear Functions and Systems of Linear Equations

Solving Equations

"I love my math book. It has so many interesting examples and homework problems. I have always liked math, but I didn't know how it could be used. Now I have lots of ideas."

Graphing Linear Equations and Linear Systems

"I like starting each new lesson with a partner activity. I just moved to this school and the activities helped me make friends."

Writing Linear Equations and Linear Systems

"I like having the book on the Internet. The online tutorials help me with my homework when I get stuck on a problem."

4

Functions

"I love the cartoons. They are funny and they help me remember the math. I want to be a cartoonist some day."

Angles and Similarity

"I like how I can click on the words in the book that is online and hear them read to me. I like to pronouce words correctly, but sometimes I don't know how to do that by just reading the words."

Square Roots and the Pythagorean Theorem

"I really liked the projects at the end of the book. The history project on ancient Egypt was my favorite. Someday I would like to visit Egypt and go to the pyramids."

Unit 3 Chapter 7 Analyzing and Summarizing Data Sets

Data Analysis and Displays

"I like how the glossary in the book is part of the index. When I couldn't remember how a vocabulary word was defined, I could go to the index and find where the word was defined in the book."

Unit 4 Inequalities and Exponents
Chapters 8 and 9

Linear Inequalities

"I like the practice tests in the book. I get really nervous on tests. So, having a practice test to work on at home helped me to chill out when the real test came."

Exponents and Scientific Notation

"I like the review at the beginning of each chapter. This book has examples to help me remember things from last year. I don't like it when the review is just a list of questions."

Appendix A: My Big Ideas Projects

Appendix B: Financial Literacy

"I like the workbook (Record and Practice Journal). It saved me a lot of work to not have to copy all the questions and graphs."

PROGRAM OVERVIEW

Print

Available in print, online, and in digital format

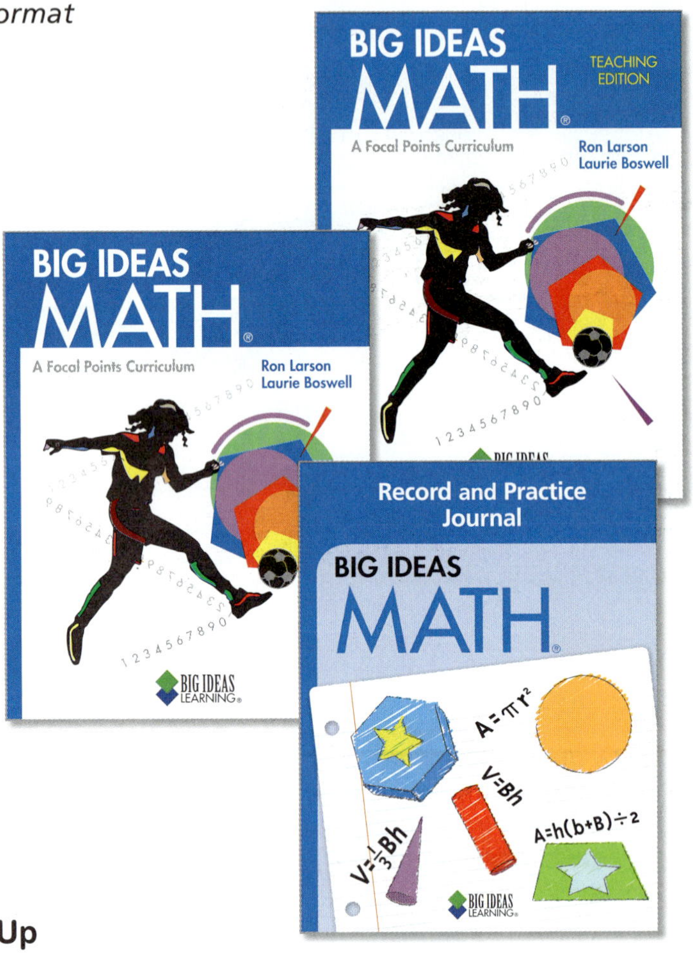

- **Pupil Edition**
- **Teaching Edition**
- **Record and Practice Journal**
- **Assessment Book**
  - Pre-Course Test
  - Quizzes
  - Chapter Tests
  - Standardized Test Practice
  - Alternative Assessment
  - End-of-Course Tests
- **Resources by Chapter**
  - Start Thinking! and Warm Up
  - Family and Community Involvement: English, Spanish, and Haitian Creole
  - School-to-Work
  - Graphic Organizers/Study Help
  - Financial Literacy
  - Technology Connection
  - Life Connections
  - Stories in History
  - Extra Practice
  - Enrichment and Extension
  - Puzzle Time
  - Unit Project with Rubric
  - Cumulative Practice

- Differentiating the Lesson
- Skills Review Handbook
- Basic Skills Handbook
- Worked-Out Solutions
- Lesson Plans
- Teacher Tools

Skills Review Handbook

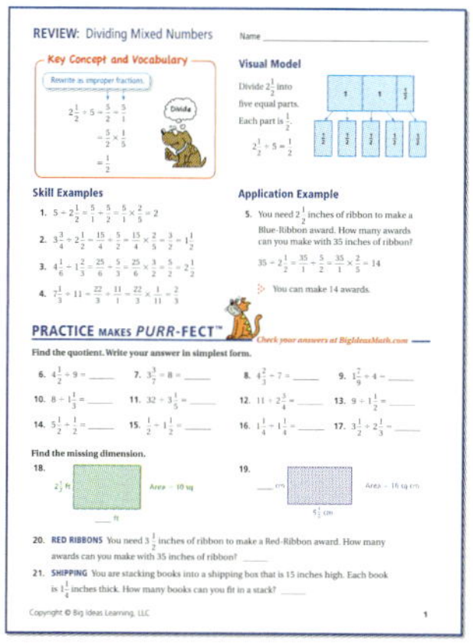
REVIEW: Dividing Mixed Numbers
Key Concept and Vocabulary
Visual Model
Skill Examples
Application Example
PRACTICE MAKES PURR-FECT™

Basic Skills Handbook

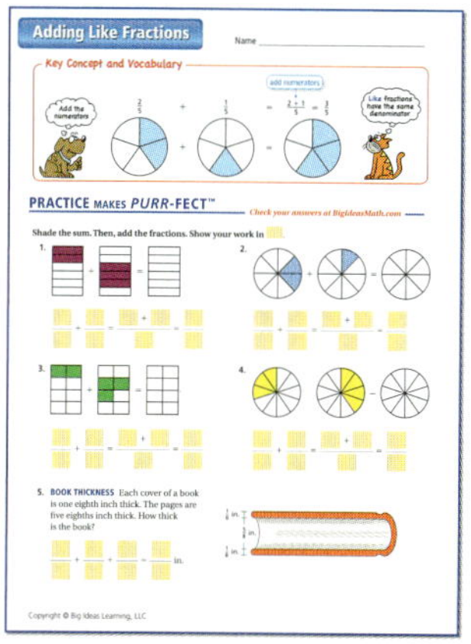
Adding Like Fractions
Key Concept and Vocabulary
PRACTICE MAKES PURR-FECT™

ExamView Assessment Suite | LessonView | PuzzleView | MindPoint QuizShow

Technology

- Big Ideas **Exam***View*® Assessment Suite
  Includes
  - Test Generator
  - Test Player
  - Test Manager
- **Lesson***View*® Dynamic Planning Tool
- **Puzzle***View*® Vocabulary Puzzle Builder
- **Mind***Point*® Quiz*Show*
- Interactive Glossary:
  English, Spanish, Haitian Creole
- Dynamic Classroom
- Answer Presentation Tool
- *BigIdeasMath.com*
  - Student Companion Website
  - Teacher Companion Website
- Lesson Tutorials
- Online Pupil Edition

FOCAL POINT TO BOOK CORRELATION

After a topic is introduced, it is revisited many times in subsequent activities, lessons, and exercises.

FOCAL POINT: Analyze and represent linear functions, and solve linear equations and systems of linear equations.

Topics

- Create and interpret tables, graphs, and models to represent, analyze, and solve problems related to linear equations, including analysis of domain, range, and the difference between discrete and continuous data.
- Interpret the slope and the x- and y-intercepts when graphing a linear equation for a real-world problem.
- Use tables, graphs, and models to represent, analyze, and solve real-world problems related to systems of linear equations.
- Identify the solution to a system of linear equations using graphs.
- Translate among verbal, tabular, graphical, and algebraic representations of linear functions.
- Compare the graphs of linear and non-linear functions for real-world situations.

Section 1.1 **Solving Simple Equations**
Essential Question: How can you use inductive reasoning to discover rules in mathematics? How can you test a rule?

Section 1.2 **Solving Multi-Step Equations**
Essential Question: How can you solve a multi-step equation? How can you check the reasonableness of your solution?

Section 1.3 **Solving Equations with Variables on Both Sides**
Essential Question: How can you solve an equation that has variables on both sides?

Section 2.1 **Graphing Linear Equations**
Essential Question: How can you recognize a linear equation? How can you draw its graph?

Section 2.2 **Slope of a Line**
Essential Question: How can the slope of a line be used to describe the line?

Section 2.3 **Graphing Linear Equations in Slope-Intercept Form**
Essential Question: How can you describe the graph of the equation $y = mx + b$?

Section 2.4 Graphing Linear Equations in Standard Form
Essential Question: How can you describe the graph of the equation $ax + by = c$?

Section 2.5 Systems of Linear Equations
Essential Question: How can you solve a system of linear equations?

Section 2.6 Special Systems of Linear Equations
Essential Question: Can a system of linear equations have no solution? Can a system of linear equations have many solutions?

Section 2.7 Solving Equations by Graphing
Essential Question: How can you use a system of linear equations to solve an equation that has variables on both sides?

Section 3.1 Writing Equations in Slope-Intercept Form
Essential Question: How can you write an equation of a line when you are given the slope and y-intercept of the line?

Section 3.2 Writing Equations Using a Slope and a Point
Essential Question: How can you write an equation of a line when you are given the slope and a point on the line?

Section 3.3 Writing Equations Using Two Points
Essential Question: How can you write an equation of a line when you are given two points on the line?

Section 3.4 Solving Real-Life Problems
Essential Question: How can you use a linear equation in two variables to model and solve a real-life problem?

Section 3.5 Writing Systems of Linear Equations
Essential Question: How can you use a system of linear equations to model and solve a real-life problem?

Section 4.1 Domain and Range of a Function
Essential Question: How can you find the domain and range of a function?

Section 4.2 Discrete and Continuous Domains
Essential Question: How can you decide whether the domain of a function is discrete or continuous?

Section 4.3 Linear Function Patterns
Essential Question: How can you use a linear function to describe a linear pattern?

Section 4.4 Comparing Linear and Nonlinear Functions
Essential Question: How can you recognize when a pattern in real life is linear or nonlinear?

FOCAL POINT: Analyze two- and three-dimensional figures by using distance and angle.

Topics

- Use similar triangles to solve problems that include height and distances.
- Classify and determine the measure of angles, including angles created when parallel lines are cut by transversals.
- Demonstrate that the sum of the angles in a triangle is 180-degrees and apply this fact to find unknown measure of angles and the sum of angles in polygons.
- Validate and apply Pythagorean Theorem to find distances in real world situations or between points in the coordinate plane.

Section 5.1 Classifying Angles
Essential Question: How can you classify two angles as complementary or supplementary?

Section 5.2 Angles and Sides of Triangles
Essential Question: How can you classify triangles by their angles?

Section 5.3 Angles of Polygons
Essential Question: How can you find a formula for the sum of the angle measures of any polygon?

Section 5.4 Using Similar Triangles
Essential Question: Which properties of triangles make them special among all other types of polygons?

Section 5.5 Parallel Lines and Transversals
Essential Question: How can you use properties of parallel lines to solve real-life problems?

Section 6.2 The Pythagorean Theorem
Essential Question: How are the lengths of the sides of a right triangle related?

Section 6.5 Using the Pythagorean Theorem
Essential Question: How can you use the Pythagorean Theorem to solve real-life problems?

FOCAL POINT: Analyze and summarize data sets.

Topics

- Select, organize and construct appropriate data displays, including box and whisker plots, scatter plots, and lines of best fit to convey information and make conjectures about possible relationships.
- Determine and describe how changes in data values impact measures of central tendency.

Section 7.1 Measures of Central Tendency
Essential Question: How can you use measures of central tendency to distribute an amount evenly among a group of people?

Section 7.2 Box-and-Whisker Plots
Essential Question: How can you use a box-and-whisker plot to describe a population?

Section 7.3 Scatter Plots and Lines of Best Fit
Essential Question: How can you use data to predict an event?

Section 7.4 Choosing a Data Display
Essential Question: How can you display data in a way that helps you make decisions?

NARROWER AND DEEPER™

"It's not just what you teach. It's how you teach it. This program begins each lesson with an in-depth, hands-on dynamic full-class period activity."

Ron Larson
Laurie Boswell

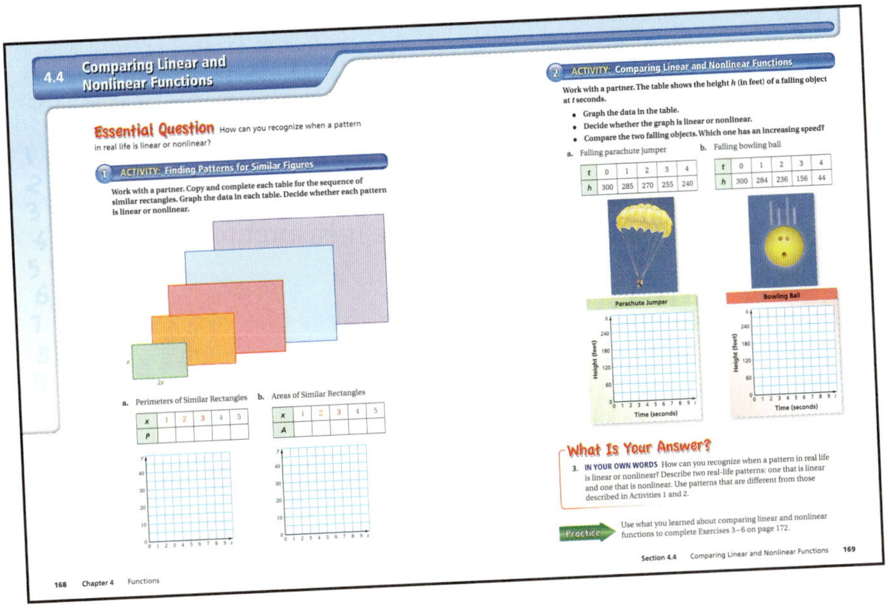

4.4 Comparing Linear and Nonlinear Functions

Essential Question How can you recognize when a pattern in real life is linear or nonlinear?

1 ACTIVITY: Finding Patterns for Similar Figures

Work with a partner. Copy and complete each table for the sequence of similar rectangles. Graph the data in each table. Decide whether each pattern is linear or nonlinear.

a. Perimeters of Similar Rectangles

x	1	2	3	4	5
P					

b. Areas of Similar Rectangles

x	1	2	3	4	5
A					

168 Chapter 4 Functions

2 ACTIVITY: Comparing Linear and Nonlinear Functions

Work with a partner. The table shows the height h (in feet) of a falling object at t seconds.

- Graph the data in the table.
- Decide whether the graph is linear or nonlinear.
- Compare the two falling objects. Which one has an increasing speed?

a. Falling parachute jumper

t	0	1	2	3	4
h	300	285	270	255	240

b. Falling bowling ball

t	0	1	2	3	4
h	300	284	236	156	44

What Is Your Answer?

3. IN YOUR OWN WORDS How can you recognize when a pattern in real life is linear or nonlinear? Describe two real-life patterns: one that is linear and one that is nonlinear. Use patterns that are different from those described in Activities 1 and 2.

Practice Use what you learned about comparing linear and nonlinear functions to complete Exercises 3–6 on page 172.

Section 4.4 Comparing Linear and Nonlinear Functions 169

CONNECTION: Algebra

Topics

- Solve literal equations for a specified variable.
- Solve and graph one- and two-step inequalities in one variable.

Section 1.4 Rewriting Equations and Formulas
Essential Question: How can you use a formula for one measurement to write a formula for a different measurement?

Section 8.1 Writing and Graphing Inequalities
Essential Question: How can you use an inequality to describe a real-life statement?

Section 8.2 Solving Inequalities Using Addition or Subtraction
Essential Question: How can you use addition or subtraction to solve an inequality?

Section 8.3 Solving Inequalities Using Multiplication or Division
Essential Question: How can you use multiplication or division to solve an inequality?

Section 8.4 Solving Multi-Step Inequalities
Essential Question: How can you use an inequality to describe the area and perimeter of a composite figure?

CONNECTION: Geometry and Measurement

Topics

- Compare, contrast, and convert units of measure between different measurement systems (US customary or metric (SI)) and dimensions including temperature, area, volume, and derived units to solve problems.

Section 1.4 Rewriting Equations and Formulas
Essential Question: How can you use a formula for one measurement to write a formula for a different measurement?

Section 1.5 Converting Units of Measure
Essential Question: How can you convert from one measurement system to another?

CONNECTION: Number and Operations

Topics

- Use exponents and scientific notation to write large and small numbers and vice versa and to solve problems.
- Make reasonable approximations of square roots and mathematical expressions that include square roots, and use them to estimate solutions to problems and to compare mathematical expressions involving real numbers and radical expressions.
- Simplify real number expressions using the laws of exponents.
- Perform operations on real numbers (including integer exponents, radicals, percents, scientific notation, absolute value, rational numbers, and irrational numbers) using multi-step and real world problems.

Section 6.1 Finding Square Roots
Essential Question: How can you find the side length of a square when you are given the area of the square?

Section 6.2 The Pythagorean Theorem
Essential Question: How are the lengths of the sides of a right triangle related?

Section 6.3 Approximating Square Roots
Essential Question: How can you find decimal approximations of square roots that are irrational?

Section 6.4 Simplifying Square Roots
Essential Question: How can you use a square root to describe the golden ratio?

Section 6.5 Using the Pythagorean Theorem
Essential Question: How can you use the Pythagorean Theorem to solve real-life problems?

Section 9.1 Exponents
Essential Question: How can you use exponents to write numbers?

Section 9.2 Product of Powers Property
Essential Question: How can you multiply two powers that have the same base?

Section 9.3 Quotient of Powers Property
Essential Question: How can you divide two powers that have the same base?

Section 9.4 Zero and Negative Exponents
Essential Question: How can you define zero and negative exponents?

Section 9.5 Reading Scientific Notation
Essential Question: How can you read numbers that are written in scientific notation?

Section 9.6 Writing Scientific Notation
Essential Question: How can you write a number in scientific notation?

BOOK TO STANDARDS CORRELATION

Chapter 1 Solving Equations

1.1 Solving Simple Equations

- Create and interpret tables, graphs, and models to represent, analyze, and solve problems related to linear equations, including analysis of domain, range, and the difference between discrete and continuous data.

1.2 Solving Multi-Step Equations

- Create and interpret tables, graphs, and models to represent, analyze, and solve problems related to linear equations, including analysis of domain, range, and the difference between discrete and continuous data.

1.3 Solving Equations with Variables on Both Sides

- Create and interpret tables, graphs, and models to represent, analyze, and solve problems related to linear equations, including analysis of domain, range, and the difference between discrete and continuous data.

1.4 Rewriting Equations and Formulas

- Solve literal equations for a specified variable.
- Compare, contrast, and convert units of measure between different measurement systems (US customary or metric (SI)) and dimensions including temperature, area, volume, and derived units to solve problems.

1.5 Converting Units of Measure

- Compare, contrast, and convert units of measure between different measurement systems (US customary or metric (SI)) and dimensions including temperature, area, volume, and derived units to solve problems.

Chapter 2 Graphing Linear Equations and Linear Systems

2.1 Graphing Linear Equations

- Create and interpret tables, graphs, and models to represent, analyze, and solve problems related to linear equations, including analysis of domain, range, and the difference between discrete and continuous data.

2.2 Slope of a Line

- Interpret the slope and the x- and y-intercepts when graphing a linear equation for a real-world problem.

2.3 Graphing Linear Equations in Slope-Intercept Form

- Create and interpret tables, graphs, and models to represent, analyze, and solve problems related to linear equations, including analysis of domain, range, and the difference between discrete and continuous data.
- Interpret the slope and the x- and y-intercepts when graphing a linear equation for a real-world problem.

2.4 Graphing Linear Equations in Standard Form

- Create and interpret tables, graphs, and models to represent, analyze, and solve problems related to linear equations, including analysis of domain, range, and the difference between discrete and continuous data.
- Interpret the slope and the x- and y-intercepts when graphing a linear equation for a real-world problem.

2.5 Systems of Linear Equations

- Use tables, graphs, and models to represent, analyze, and solve real-world problems related to systems of linear equations.
- Identify the solution to a system of linear equations using graphs.

2.6 Special Systems of Linear Equations

- Use tables, graphs, and models to represent, analyze, and solve real-world problems related to systems of linear equations.
- Identify the solution to a system of linear equations using graphs.

2.7 Solving Equations by Graphing

- Create and interpret tables, graphs, and models to represent, analyze, and solve problems related to linear equations, including analysis of domain, range, and the difference between discrete and continuous data.
- Use tables, graphs, and models to represent, analyze, and solve real-world problems related to systems of linear equations.
- Identify the solution to a system of linear equations using graphs.

Chapter 3 Writing Linear Equations and Linear Systems

3.1 Writing Equations in Slope-Intercept Form

- Create and interpret tables, graphs, and models to represent, analyze, and solve problems related to linear equations, including analysis of domain, range, and the difference between discrete and continuous data.
- Interpret the slope and the x- and y-intercepts when graphing a linear equation for a real-world problem.

3.2 Writing Equations Using a Slope and a Point

- Create and interpret tables, graphs, and models to represent, analyze, and solve problems related to linear equations, including analysis of domain, range, and the difference between discrete and continuous data.
- Interpret the slope and the x- and y-intercepts when graphing a linear equation for a real-world problem.

3.3 Writing Equations Using Two Points

- Create and interpret tables, graphs, and models to represent, analyze, and solve problems related to linear equations, including analysis of domain, range, and the difference between discrete and continuous data.
- Interpret the slope and the x- and y-intercepts when graphing a linear equation for a real-world problem.

3.4 Solving Real-Life Problems

- Create and interpret tables, graphs, and models to represent, analyze, and solve problems related to linear equations, including analysis of domain, range, and the difference between discrete and continuous data.
- Interpret the slope and the x- and y-intercepts when graphing a linear equation for a real-world problem.

3.5 Writing Systems of Linear Equations

- Use tables, graphs, and models to represent, analyze, and solve real-world problems related to systems of linear equations.
- Identify the solution to a system of linear equations using graphs.

Chapter 4 Functions

4.1 Domain and Range of a Function

- Create and interpret tables, graphs, and models to represent, analyze, and solve problems related to linear equations, including analysis of domain, range, and the difference between discrete and continuous data.
- Translate among verbal, tabular, graphical, and algebraic representations of linear functions.

4.2 Discrete and Continuous Domains

- Create and interpret tables, graphs, and models to represent, analyze, and solve problems related to linear equations, including analysis of domain, range, and the difference between discrete and continuous data.
- Translate among verbal, tabular, graphical, and algebraic representations of linear functions.

4.3 Linear Function Patterns

- Create and interpret tables, graphs, and models to represent, analyze, and solve problems related to linear equations, including analysis of domain, range, and the difference between discrete and continuous data.
- Translate among verbal, tabular, graphical, and algebraic representations of linear functions.

4.4 Comparing Linear and Nonlinear Functions

- Compare the graphs of linear and non-linear functions for real-world situations.

Chapter 5 Angles and Similarity

5.1 Classifying Angles

- Classify and determine the measure of angles, including angles created when parallel lines are cut by transversals.

5.2 Angles and Sides of Triangles

- Demonstrate that the sum of the angles in a triangle is 180-degrees and apply this fact to find unknown measure of angles and the sum of angles in polygons.

5.3 Angles of Polygons

- Demonstrate that the sum of the angles in a triangle is 180-degrees and apply this fact to find unknown measure of angles and the sum of angles in polygons.

5.4 Using Similar Triangles

- Use similar triangles to solve problems that include height and distances.

5.5 Parallel Lines and Transversals

- Classify and determine the measure of angles, including angles created when parallel lines are cut by transversals.

Chapter 6 Square Roots and the Pythagorean Theorem

6.1 Finding Square Roots

- Perform operations on real numbers (including integer exponents, radicals, percents, scientific notation, absolute value, rational numbers, and irrational numbers) using multi-step and real world problems.

6.2 The Pythagorean Theorem

- Validate and apply Pythagorean Theorem to find distances in real world situations or between points in the coordinate plane.
- Perform operations on real numbers (including integer exponents, radicals, percents, scientific notation, absolute value, rational numbers, and irrational numbers) using multi-step and real world problems.

6.3 Approximating Square Roots

- Make reasonable approximations of square roots and mathematical expressions that include square roots, and use them to estimate solutions to problems and to compare mathematical expressions involving real numbers and radical expressions.

6.4 Simplifying Square Roots

- Make reasonable approximations of square roots and mathematical expressions that include square roots, and use them to estimate solutions to problems and to compare mathematical expressions involving real numbers and radical expressions.
- Perform operations on real numbers (including integer exponents, radicals, percents, scientific notation, absolute value, rational numbers, and irrational numbers) using multi-step and real world problems.

6.5 Using the Pythagorean Theorem

- Validate and apply Pythagorean Theorem to find distances in real world situations or between points in the coordinate plane.
- Make reasonable approximations of square roots and mathematical expressions that include square roots, and use them to estimate solutions to problems and to compare mathematical expressions involving real numbers and radical expressions.
- Perform operations on real numbers (including integer exponents, radicals, percents, scientific notation, absolute value, rational numbers, and irrational numbers) using multi-step and real world problems.

Chapter 7 Data Analysis and Displays

7.1 Measures of Central Tendency

- Determine and describe how changes in data values impact measures of central tendency.

7.2 Box-and-Whisker Plots

- Select, organize and construct appropriate data displays, including box and whisker plots, scatter plots, and lines of best fit to convey information and make conjectures about possible relationships.

7.3 Scatter Plots and Lines of Best Fit

- Select, organize and construct appropriate data displays, including box and whisker plots, scatter plots, and lines of best fit to convey information and make conjectures about possible relationships.

7.4 Choosing a Data Display

- Select, organize and construct appropriate data displays, including box and whisker plots, scatter plots, and lines of best fit to convey information and make conjectures about possible relationships.

Chapter 8 Linear Inequalities

8.1 Writing and Graphing Inequalities

- Solve and graph one- and two-step inequalities in one variable.

8.2 Solving Inequalities Using Addition or Subtraction

- Solve and graph one- and two-step inequalities in one variable.

8.3 Solving Inequalities Using Multiplication or Division

- Solve and graph one- and two-step inequalities in one variable.

8.4 Solving Multi-Step Inequalities

- Solve and graph one- and two-step inequalities in one variable.

Chapter 9 Exponents and Scientific Notation

9.1 Exponents

- Use exponents and scientific notation to write large and small numbers and vice versa and to solve problems.
- Perform operations on real numbers (including integer exponents, radicals, percents, scientific notation, absolute value, rational numbers, and irrational numbers) using multi-step and real world problems.

9.2 Product of Powers Property

- Use exponents and scientific notation to write large and small numbers and vice versa and to solve problems.
- Simplify real number expressions using the laws of exponents.
- Perform operations on real numbers (including integer exponents, radicals, percents, scientific notation, absolute value, rational numbers, and irrational numbers) using multi-step and real world problems.

9.3 Quotient of Powers Property

- Use exponents and scientific notation to write large and small numbers and vice versa and to solve problems.
- Simplify real number expressions using the laws of exponents.
- Perform operations on real numbers (including integer exponents, radicals, percents, scientific notation, absolute value, rational numbers, and irrational numbers) using multi-step and real world problems.

9.4 Zero and Negative Exponents

- Use exponents and scientific notation to write large and small numbers and vice versa and to solve problems.
- Simplify real number expressions using the laws of exponents.
- Perform operations on real numbers (including integer exponents, radicals, percents, scientific notation, absolute value, rational numbers, and irrational numbers) using multi-step and real world problems.

9.5 Reading Scientific Notation

- Use exponents and scientific notation to write large and small numbers and vice versa and to solve problems.
- Perform operations on real numbers (including integer exponents, radicals, percents, scientific notation, absolute value, rational numbers, and irrational numbers) using multi-step and real world problems.

9.6 Writing Scientific Notation

- Use exponents and scientific notation to write large and small numbers and vice versa and to solve problems.
- Perform operations on real numbers (including integer exponents, radicals, percents, scientific notation, absolute value, rational numbers, and irrational numbers) using multi-step and real world problems.

The Focal Points are completely covered in Chapters 1–9. Appendix B is included as an end-of-the-year option.

Appendix B Financial Literacy

B.1 Simple and Compound Interest

- Create and interpret tables, graphs, and models to represent, analyze, and solve problems related to linear equations, including analysis of domain, range, and the difference between discrete and continuous data.
- Compare the graphs of linear and non-linear functions for real-world situations.

B.2 Formula for Compound Interest

- Simplify real number expressions using the laws of exponents.
- Perform operations on real numbers (including integer exponents, radicals, percents, scientific notation, absolute value, rational numbers, and irrational numbers) using multi-step and real world problems.

B.3 Installment Loans

- Solve percent problems, including problems involving discounts, simple interest, taxes, tips, and percents of increase or decrease.
- Perform operations on real numbers (including integer exponents, radicals, percents, scientific notation, absolute value, rational numbers, and irrational numbers) using multi-step and real world problems.

B.4 Checking Accounts

- Solve percent problems, including problems involving discounts, simple interest, taxes, tips, and percents of increase or decrease.
- Perform operations on real numbers (including integer exponents, radicals, percents, scientific notation, absolute value, rational numbers, and irrational numbers) using multi-step and real world problems.

B.5 Credit Cards

- Solve percent problems, including problems involving discounts, simple interest, taxes, tips, and percents of increase or decrease.
- Perform operations on real numbers (including integer exponents, radicals, percents, scientific notation, absolute value, rational numbers, and irrational numbers) using multi-step and real world problems.

B.6 Payroll Deductions

- Solve percent problems, including problems involving discounts, simple interest, taxes, tips, and percents of increase or decrease.
- Perform operations on real numbers (including integer exponents, radicals, percents, scientific notation, absolute value, rational numbers, and irrational numbers) using multi-step and real world problems.

SKILLS REVIEW HANDBOOK

The Skills Review Handbook provides complete coverage of pre-course skills. Each of the 80 lessons includes Key Concepts and Vocabulary, Visual Models, Skill Examples, Application Examples, and practice.

This complete, full color handbook was authored and designed by Ron Larson, ensuring consistency and thoroughness in coverage of the content.

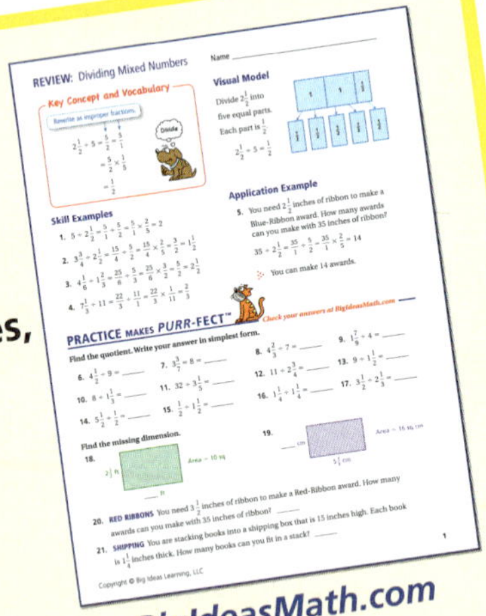

BigIdeasMath.com

BASIC SKILLS HANDBOOK

The Basic Skills Handbook covers basic skills needed for success in middle school mathematics. This handbook is designed for students who are working two or more grades below grade level. This colorful, highly visual handbook uses few words, making it ideal for students with reading difficulties or students learning English.

The Basic Skills Handbook was also authored and designed by Ron Larson.

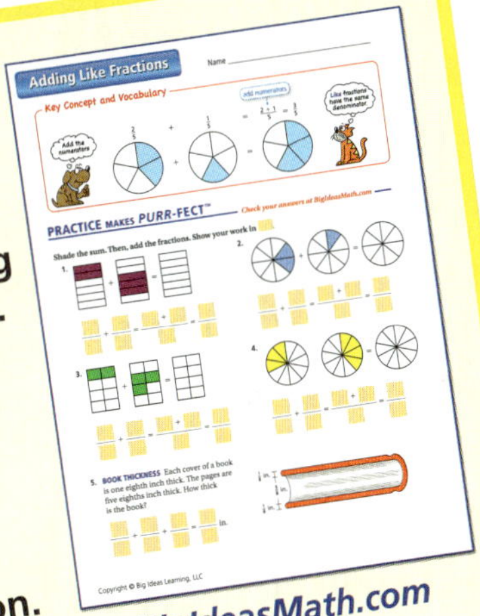

BigIdeasMath.com

ENGAGING PUPIL BOOKS

1 2 3

Activity - Lesson - Exercises

1 ACTIVITY

Each section begins with a 2-page ***Activity*** *that is introduced by an Essential Question.*

- ***Deeper***
- ***Dynamic***
- ***Doable***
- ***Dazzling***

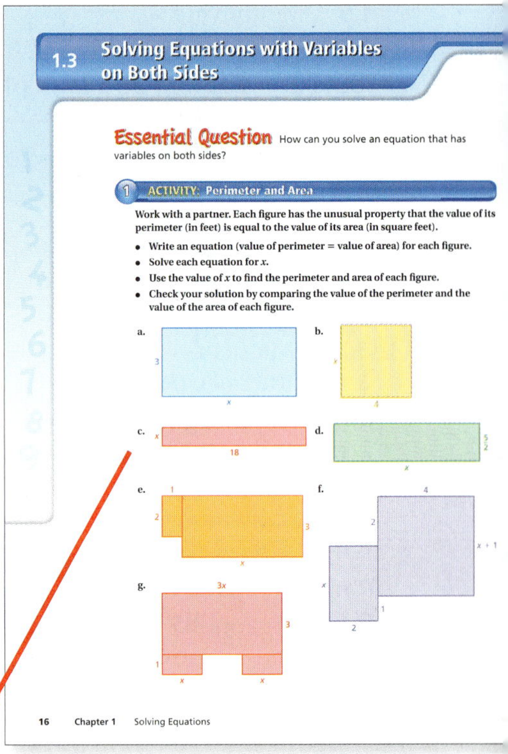

Students gain a deeper understanding of topics through inductive reasoning and exploration.

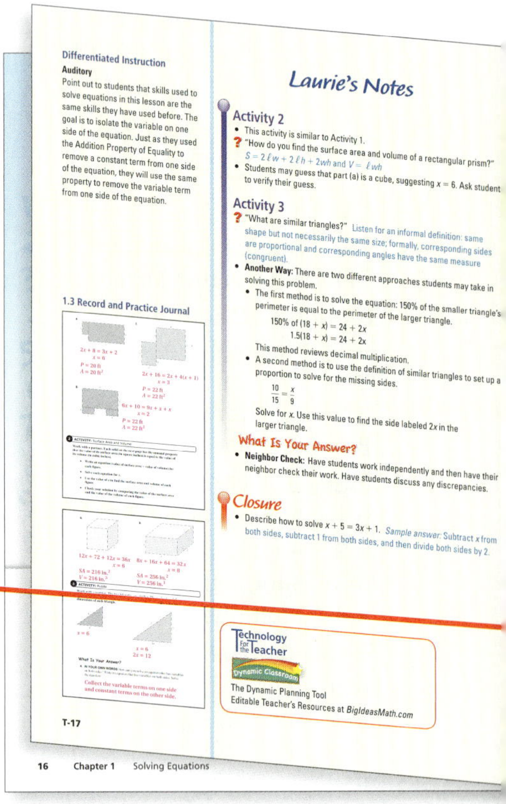

Students develop communication and problem-solving skills by answering Essential Questions.

INSPIRING TEACHER BOOKS

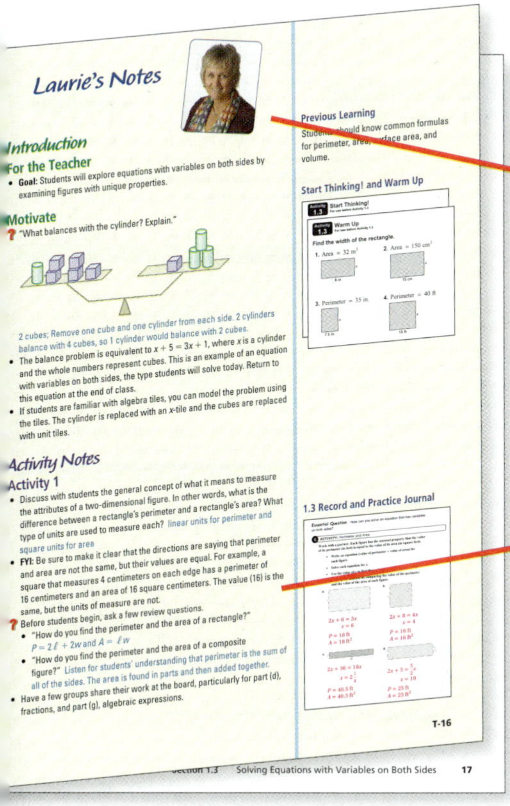

Laurie's Notes

Introduction

For the Teacher

- **Goal:** Students will explore equations with variables on both sides by examining figures with unique properties.

Motivate

"What balances with the cylinder? Explain."

2 cubes; Remove one cube and one cylinder from each side. 2 cylinders balance with 4 cubes, so 1 cylinder would balance with 2 cubes.

- The balance problem is equivalent to $x + 5 = 3x + 1$, where x is a cylinder and the whole numbers represent cubes. This is an example of an equation with variables on both sides, the type students will solve today. Return to this equation at the end of class.
- If students are familiar with algebra tiles, you can model the problem using the tiles. The cylinder is replaced with an x-tile and the cubes are replaced with unit tiles.

Activity Notes

Activity 1

- Discuss with students the general concept of what it means to measure the attributes of a two-dimensional figure. In other words, what is the difference between a rectangle's perimeter and a rectangle's area? What type of units are used to measure each? linear units for perimeter and square units for area
- **FYI:** Be sure to make it clear that the directions are saying that perimeter and area are not the same, but their values are equal. For example, a square that measures 4 centimeters on each edge has a perimeter of 16 centimeters and an area of 16 square centimeters. The value (16) is the same, but the units of measure are not.

Before students begin, ask a few review questions.

- "How do you find the perimeter and the area of a rectangle?" $P = 2\ell + 2w$ and $A = \ell w$
- "How do you find the perimeter and the area of a composite figure?" Listen for students' understanding that perimeter is the sum of all of the sides. The area is found in parts and then added together.

- Have a few groups share their work at the board, particularly for part (d), fractions, and part (g), algebraic expressions.

Previous Learning

Students should know common formulas for perimeter, area, surface area, and volume.

Start Thinking! and Warm Up

Activity 1.3 Start Thinking!

Activity 1.3 Warm Up

Find the width of the rectangle.

1. Area = 32 m²
2. Area = 150 cm²
3. Perimeter = 35 in.
4. Perimeter = 40 ft

1.3 Record and Practice Journal

T-16

Section 1.3 Solving Equations with Variables on Both Sides 17

Teachers have the benefit of **Laurie Boswell's** 20-plus years of classroom experience reflected in her lively and informative notes.

Teachers are offered suggestions for questioning that help guide students toward better understanding.

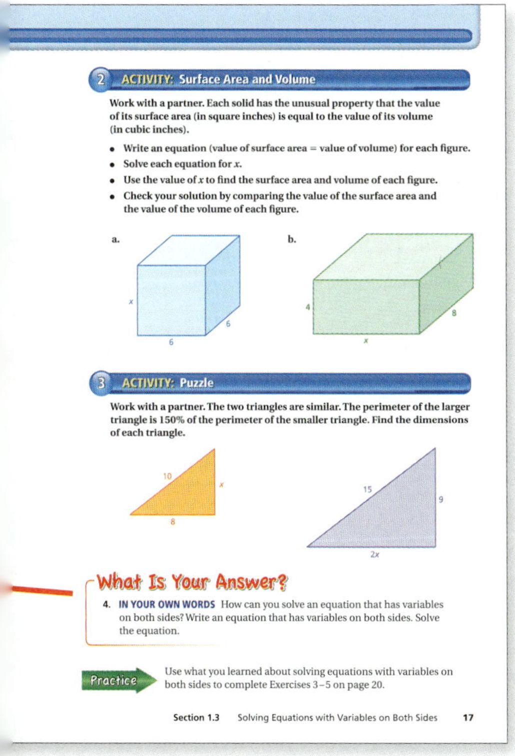

2 ACTIVITY: Surface Area and Volume

Work with a partner. Each solid has the unusual property that the value of its surface area (in square inches) is equal to the value of its volume (in cubic inches).

- Write an equation (value of surface area = value of volume) for each figure.
- Solve each equation for x.
- Use the value of x to find the surface area and volume of each figure.
- Check your solution by comparing the value of the surface area and the value of the volume of each figure.

3 ACTIVITY: Puzzle

Work with a partner. The two triangles are similar. The perimeter of the larger triangle is 150% of the perimeter of the smaller triangle. Find the dimensions of each triangle.

What Is Your Answer?

4. **IN YOUR OWN WORDS** How can you solve an equation that has variables on both sides? Write an equation that has variables on both sides. Solve the equation.

Practice Use what you learned about solving equations with variables on both sides to complete Exercises 3–5 on page 20.

Section 1.3 Solving Equations with Variables on Both Sides 17

"Laurie's Notes"

COMPLETE ACTIVITY AND TIME MANAGEMENT SUPPORT FROM A MASTER CLASSROOM TEACHER

CLEAR PUPIL BOOKS

Activity - **Lesson** - **Exercises**

2 LESSON

After the concept has been introduced with a full-class period ***Activity****, it is extended the following day through the* ***Lesson****.*

- ***Key Ideas***
- ***Examples***
- ***"On Your Own" Questions***

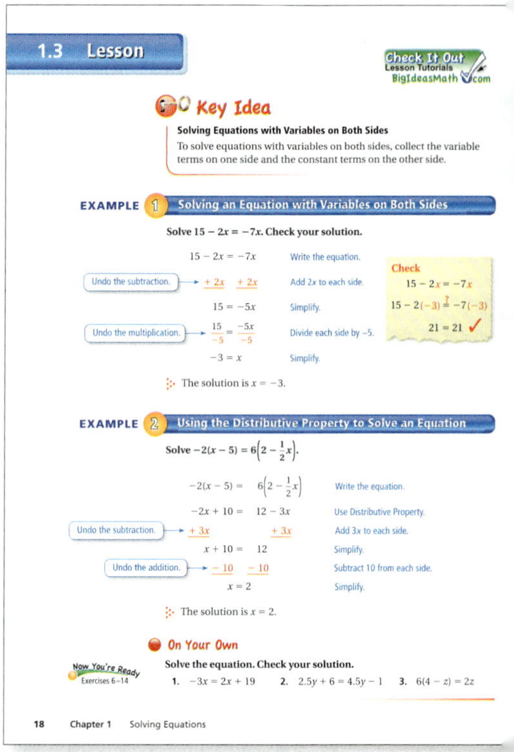

1.3 Lesson

Check It Out
Lesson Tutorials
BigIdeasMath.com

Key Idea

Solving Equations with Variables on Both Sides

To solve equations with variables on both sides, collect the variable terms on one side and the constant terms on the other side.

EXAMPLE 1 Solving an Equation with Variables on Both Sides

Solve $15 - 2x = -7x$. Check your solution.

	$15 - 2x = -7x$	Write the equation.
Undo the subtraction.	$+2x \quad +2x$	Add $2x$ to each side.
	$15 = -5x$	Simplify.
Undo the multiplication.	$\frac{15}{-5} = \frac{-5x}{-5}$	Divide each side by -5.
	$-3 = x$	Simplify.

Check
$15 - 2x = -7x$
$15 - 2(-3) \stackrel{?}{=} -7(-3)$
$21 = 21$ ✓

The solution is $x = -3$.

EXAMPLE 2 Using the Distributive Property to Solve an Equation

Solve $-2(x - 5) = 6\left(2 - \frac{1}{2}x\right)$.

	$-2(x - 5) = 6\left(2 - \frac{1}{2}x\right)$	Write the equation.
	$-2x + 10 = 12 - 3x$	Use Distributive Property.
Undo the subtraction.	$+3x \quad +3x$	Add $3x$ to each side.
	$x + 10 = 12$	Simplify.
Undo the addition.	$-10 \quad -10$	Subtract 10 from each side.
	$x = 2$	Simplify.

The solution is $x = 2$.

On Your Own

Now You're Ready
Exercises 6–14

Solve the equation. Check your solution.

1. $-3x = 2x + 19$
2. $2.5y + 6 = 4.5y - 1$
3. $6(4 - z) = 2z$

18 Chapter 1 Solving Equations

Each example in the pupil edition is accompanied by teaching suggestions from Laurie.

Each concept is accompanied by clear, stepped-out examples.

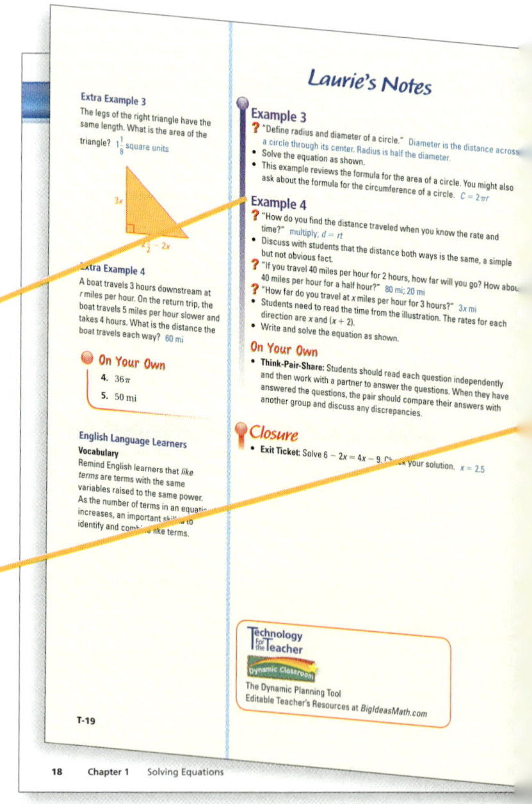

Extra Example 3

The legs of the right triangle have the same length. What is the area of the triangle? $1\frac{1}{8}$ square units

Extra Example 4

A boat travels 3 hours downstream at r miles per hour. On the return trip, the boat travels 5 miles per hour slower and takes 4 hours. What is the distance the boat travels each way? 60 mi

On Your Own

4. 36π
5. 50 mi

English Language Learners

Vocabulary

Remind English learners that *like terms* are terms with the same variables raised to the same power. As the number of terms in an equation increases, an important skill is to identify and combine like terms.

T-19

Laurie's Notes

Example 3

- ? "Define radius and diameter of a circle." Diameter is the distance across a circle through its center. Radius is half the diameter.
- Solve the equation as shown.
- This example reviews the formula for the area of a circle. You might also ask about the formula for the circumference of a circle. $C = 2\pi r$

Example 4

- ? "How do you find the distance traveled when you know the rate and time?" multiply; $d = rt$
- Discuss with students that the distance both ways is the same, a simple but not obvious fact.
- ? "If you travel 40 miles per hour for 2 hours, how far will you go? How about 40 miles per hour for a half hour?" 80 mi; 20 mi
- ? "How far do you travel at x miles per hour for 3 hours?" $3x$ mi
- Students need to read the time from the illustration. The rates for each direction are x and $(x + 2)$.
- Write and solve the equation as shown.

On Your Own

- **Think-Pair-Share:** Students should read each question independently and then work with a partner to answer the questions. When they have answered the questions, the pair should compare their answers with another group and discuss any discrepancies.

Closure

- **Exit Ticket:** Solve $6 - 2x = 4x - 9$. Check your solution. $x = 2.5$

Technology for the Teacher
Dynamic Classroom
The Dynamic Planning Tool
Editable Teacher's Resources at *BigIdeasMath.com*

18 Chapter 1 Solving Equations

INSIGHTFUL TEACHER BOOKS

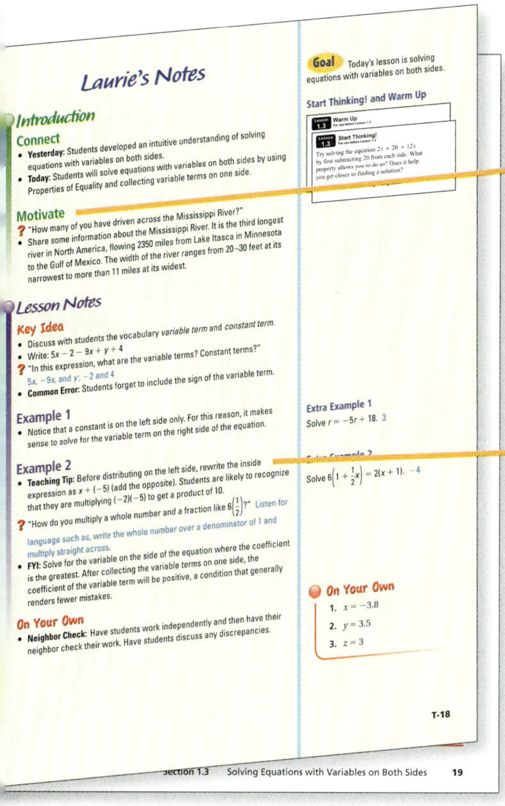

Laurie's Notes

Introduction

Connect
- Yesterday: Students developed an intuitive understanding of solving equations with variables on both sides.
- Today: Students will solve equations with variables on both sides by using Properties of Equality and collecting variable terms on one side.

Motivate
- ? "How many of you have driven across the Mississippi River?"
- Share some information about the Mississippi River. It is the third longest river in North America, flowing 2350 miles from Lake Itasca in Minnesota to the Gulf of Mexico. The width of the river ranges from 20–30 feet at its narrowest to more than 11 miles at its widest.

Lesson Notes

Key Idea
- Discuss with students the vocabulary *variable term* and *constant term*.
- Write: $5x - 2 - 9x + y + 4$
- ? "In this expression, what are the variable terms? Constant terms?" $5x$, $-9x$, and y; -2 and 4
- Common Error: Students forget to include the sign of the variable term.

Example 1
- Notice that a constant is on the left side only. For this reason, it makes sense to solve for the variable term on the right side of the equation.

Example 2
- Teaching Tip: Before distributing on the left side, rewrite the inside expression as $x + (-5)$ (add the opposite). Students are likely to recognize that they are multiplying $(-2)(-5)$ to get a product of 10.
- ? "How do you multiply a whole number and a fraction like $6\left(\frac{1}{2}\right)$?" Listen for language such as, write the whole number over a denominator of 1 and multiply straight across.
- FYI: Solve for the variable on the side of the equation where the coefficient is the greatest. After collecting the variable terms on one side, the coefficient of the variable term will be positive, a condition that generally renders fewer mistakes.

On Your Own
- Neighbor Check: Have students work independently and then have their neighbor check their work. Have students discuss any discrepancies.

Goal Today's lesson is solving equations with variables on both sides.

Start Thinking! and Warm Up

Lesson 1.3 Warm Up

Lesson 1.3 Start Thinking!

Try solving the equation $2x + 20 = 12x$ by first subtracting 20 from each side. What property allows you to do so? Does it help you get closer to finding a solution?

Extra Example 1

Solve $r = -5r + 18$. 3

Extra Example 2

Solve $6\left(1 + \frac{1}{2}x\right) = 2(x + 1)$. -4

On Your Own
1. $x = -3.8$
2. $y = 3.5$
3. $z = 3$

T-18

Section 1.3 Solving Equations with Variables on Both Sides 19

Student-friendly and teacher-tested motivation activities start each lesson.

Laurie shares insights she has gained through years of teaching experience.

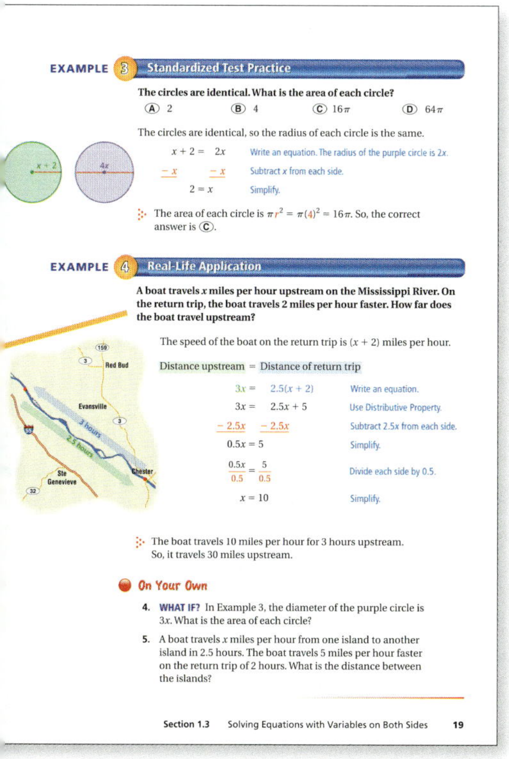

EXAMPLE 3 Standardized Test Practice

The circles are identical. What is the area of each circle?

Ⓐ 2 Ⓑ 4 Ⓒ 16π Ⓓ 64π

The circles are identical, so the radius of each circle is the same.

$x + 2 = 2x$ Write an equation. The radius of the purple circle is $2x$.

$-x \quad -x$ Subtract x from each side.

$2 = x$ Simplify.

The area of each circle is $\pi r^2 = \pi(4)^2 = 16\pi$. So, the correct answer is Ⓒ.

EXAMPLE 4 Real-Life Application

A boat travels x miles per hour upstream on the Mississippi River. On the return trip, the boat travels 2 miles per hour faster. How far does the boat travel upstream?

The speed of the boat on the return trip is $(x + 2)$ miles per hour.

Distance upstream = Distance of return trip

$3x = 2.5(x + 2)$ Write an equation.

$3x = 2.5x + 5$ Use Distributive Property.

$-2.5x \quad -2.5x$ Subtract $2.5x$ from each side.

$0.5x = 5$ Simplify.

$\frac{0.5x}{0.5} = \frac{5}{0.5}$ Divide each side by 0.5.

$x = 10$ Simplify.

The boat travels 10 miles per hour for 3 hours upstream. So, it travels 30 miles upstream.

On Your Own

4. WHAT IF? In Example 3, the diameter of the purple circle is $3x$. What is the area of each circle?
5. A boat travels x miles per hour from one island to another island in 2.5 hours. The boat travels 5 miles per hour faster on the return trip of 2 hours. What is the distance between the islands?

Section 1.3 Solving Equations with Variables on Both Sides 19

"Laurie's Notes"

COMPREHENSIVE TEACHING SUPPORT FROM A MASTER CLASSROOM TEACHER

DEEPER PUPIL BOOKS

1 2 3

Activity - Lesson - Exercises

3 EXERCISES

*After students explore the concept in the **Activity** and extend it in the **Lesson**, they can practice it with skills and applications in the **Exercises**.*

- ***Vocabulary and Concept Check***
- ***Practice and Problem Solving***
- ***Fair Game (Cumulative) Review***

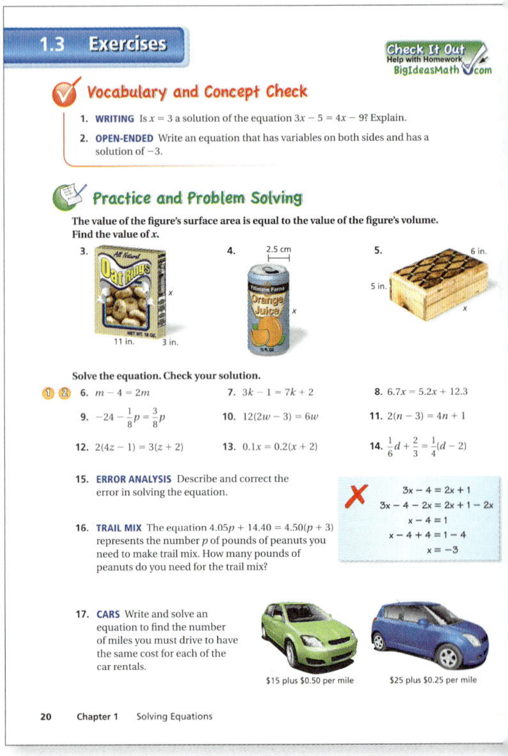

1.3 Exercises

Check It Out
Help with Homework
BigIdeasMath.com

Vocabulary and Concept Check

1. WRITING Is $x = 3$ a solution of the equation $3x - 5 = 4x - 9$? Explain.
2. OPEN-ENDED Write an equation that has variables on both sides and has a solution of -3.

Practice and Problem Solving

The value of the figure's surface area is equal to the value of the figure's volume. Find the value of x.

3. 4. 5.

Solve the equation. Check your solution.

6. $m - 4 = 2m$ 7. $3k - 1 = 7k + 2$ 8. $6.7x = 5.2x + 12.3$
9. $-24 - \frac{1}{8}p = \frac{3}{8}p$ 10. $12(2w - 3) = 6w$ 11. $2(n - 3) = 4n + 1$
12. $2(4z - 1) = 3(z + 2)$ 13. $0.1x = 0.2(x + 2)$ 14. $\frac{1}{6}d + \frac{2}{3} = \frac{1}{4}(d - 2)$

15. ERROR ANALYSIS Describe and correct the error in solving the equation.

$3x - 4 = 2x + 1$
$3x - 4 - 2x = 2x + 1 - 2x$
$x - 4 = 1$
$x - 4 + 4 = 1 - 4$
$x = -3$

16. TRAIL MIX The equation $4.05p + 14.40 = 4.50(p + 3)$ represents the number p of pounds of peanuts you need to make trail mix. How many pounds of peanuts do you need for the trail mix?

17. CARS Write and solve an equation to find the number of miles you must drive to have the same cost for each of the car rentals.

\$15 plus \$0.50 per mile   \$25 plus \$0.25 per mile

20 Chapter 1 Solving Equations

In each section, one of the problem-solving exercises is accompanied by detailed suggestions for taking the mathematics deeper.

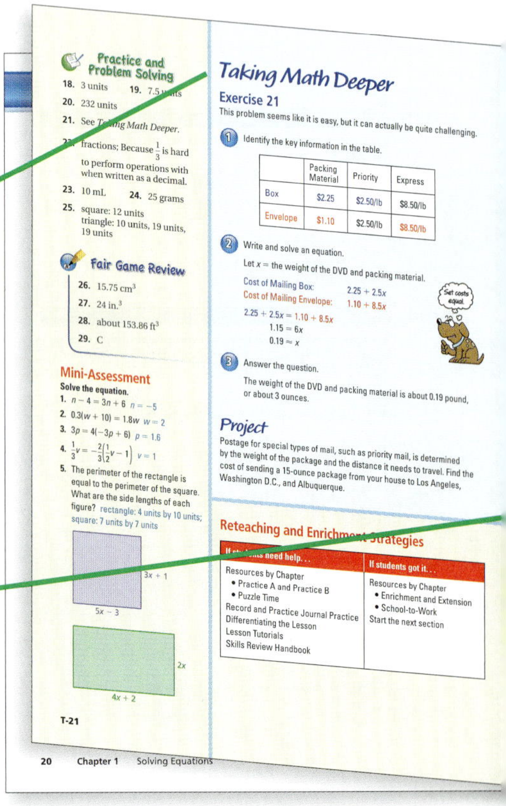

Practice and Problem Solving
18. 3 units 19. 7.5 units
20. 232 units
21. See Taking Math Deeper.
22. fractions; Because $\frac{1}{3}$ is hard to perform operations with when written as a decimal.
23. 10 mL 24. 25 grams
25. square: 12 units triangle: 10 units, 19 units, 19 units

Fair Game Review
26. 15.75 cm³
27. 24 in.³
28. about 153.86 ft³
29. C

Mini-Assessment
Solve the equation.
1. $n - 4 = 3n + 6$ $n = -5$
2. $0.3(w + 10) = 1.8w$ $w = 2$
3. $3p = 4(-3p + 6)$ $p = 1.6$
4. $\frac{1}{3}v = -\frac{2}{3}\left(\frac{1}{2}v - 1\right)$ $v = 1$
5. The perimeter of the rectangle is equal to the perimeter of the square. What are the side lengths of each figure? rectangle: 4 units by 10 units; square: 7 units by 7 units

T-21

Taking Math Deeper
Exercise 21
This problem seems like it is easy, but it can actually be quite challenging.
1 Identify the key information in the table.

	Packing Material	Priority	Express
Box	\$2.25	\$2.50/lb	\$8.50/lb
Envelope	\$1.10	\$2.50/lb	\$8.50/lb

2 Write and solve an equation.
Let x = the weight of the DVD and packing material.
Cost of Mailing Box: $2.25 + 2.5x$
Cost of Mailing Envelope: $1.10 + 8.5x$
$2.25 + 2.5x = 1.10 + 8.5x$
$1.15 = 6x$
$0.19 \approx x$
Set costs equal.
3 Answer the question.
The weight of the DVD and packing material is about 0.19 pound, or about 3 ounces.

Project
Postage for special types of mail, such as priority mail, is determined by the weight of the package and the distance it needs to travel. Find the cost of sending a 15-ounce package from your house to Los Angeles, Washington D.C., and Albuquerque.

Reteaching and Enrichment Strategies

If students need help...	If students got it...
Resources by Chapter • Practice A and Practice B • Puzzle Time Record and Practice Journal Practice Differentiating the Lesson Lesson Tutorials Skills Review Handbook	Resources by Chapter • Enrichment and Extension • School-to-Work Start the next section

20 Chapter 1 Solving Equations

Deeper problem-solving exercises follow each lesson.

DEEPER TEACHER BOOKS

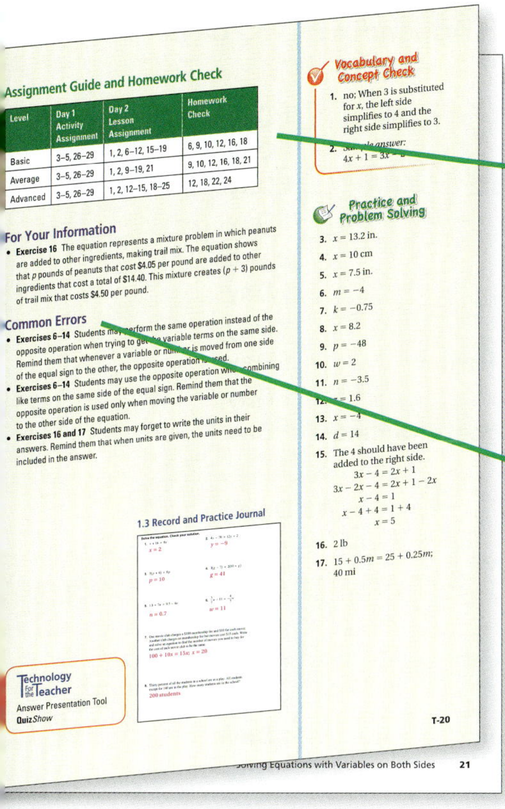

Assignment Guide and Homework Check

Level	Day 1 Activity Assignment	Day 2 Lesson Assignment	Homework Check
Basic	3–5, 26–29	1, 2, 6–12, 15–19	6, 9, 10, 12, 16, 18
Average	3–5, 26–29	1, 2, 9–19, 21	9, 10, 12, 16, 18, 21
Advanced	3–5, 26–29	1, 2, 12–15, 18–25	12, 18, 22, 24

For Your Information

- **Exercise 16** The equation represents a mixture problem in which peanuts are added to other ingredients, making trail mix. The equation shows that p pounds of peanuts that cost \$4.05 per pound are added to other ingredients that cost a total of \$14.40. This mixture creates $(p + 3)$ pounds of trail mix that costs \$4.50 per pound.

Common Errors

- **Exercises 6–14** Students may perform the same operation instead of the opposite operation when trying to get the variable terms on the same side. Remind them that whenever a variable or number is moved from one side of the equal sign to the other, the opposite operation is used.
- **Exercises 6–14** Students may use the opposite operation when combining like terms on the same side of the equal sign. Remind them that the opposite operation is used only when moving the variable or number to the other side of the equation.
- **Exercises 16 and 17** Students may forget to write the units in their answers. Remind them that when units are given, the units need to be included in the answer.

1.3 Record and Practice Journal

Technology for the Teacher
Answer Presentation Tool
QuizShow

Vocabulary and Concept Check

1. no; When 3 is substituted for x, the left side simplifies to 4 and the right side simplifies to 3.
2. Sample answer: $4x + 1 = 3x - 2$

Practice and Problem Solving

3. $x = 13.2$ in.
4. $x = 10$ cm
5. $x = 7.5$ in.
6. $m = -4$
7. $k = -0.75$
8. $x = 8.2$
9. $p = -48$
10. $w = 2$
11. $n = -3.5$
12. $z = 1.6$
13. $x = -4$
14. $d = 14$
15. The 4 should have been added to the right side.
$3x - 4 = 2x + 1$
$3x - 2x - 4 = 2x + 1 - 2x$
$x - 4 = 1$
$x - 4 + 4 = 1 + 4$
$x = 5$
16. 2 lb
17. $15 + 0.5m = 25 + 0.25m$; 40 mi

T-20

Solving Equations with Variables on Both Sides 21

The Assignment Guide and Homework Check provides suggestions for basic, average, and advanced assignments.

Teachers are provided proven strategies that help students avoid common errors.

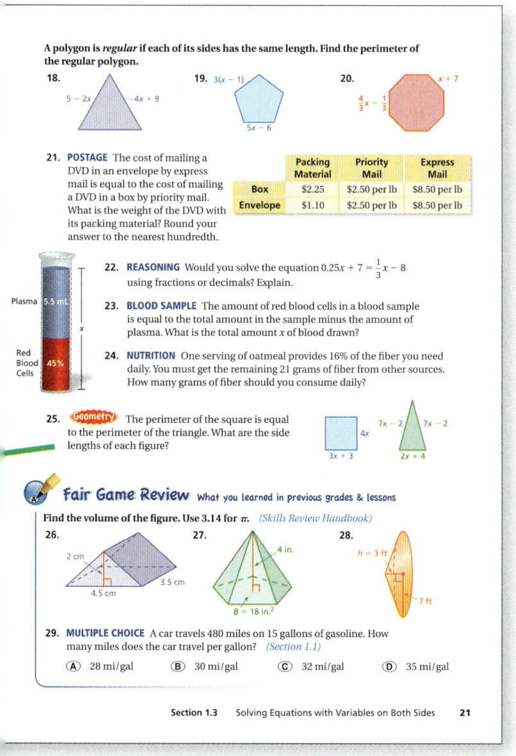

A polygon is *regular* if each of its sides has the same length. Find the perimeter of the regular polygon.

18. $5 - 2x$, $4x + 9$
19. $3(x - 1)$, $5x - 6$
20. $\frac{4}{3}x - \frac{1}{3}$, $x + 7$

21. **POSTAGE** The cost of mailing a DVD in an envelope by express mail is equal to the cost of mailing a DVD in a box by priority mail. What is the weight of the DVD with its packing material? Round your answer to the nearest hundredth.

	Packing Material	Priority Mail	Express Mail
Box	\$2.25	\$2.50 per lb	\$8.50 per lb
Envelope	\$1.10	\$2.50 per lb	\$8.50 per lb

Plasma 5.5 mL
Red Blood Cells 45%

22. **REASONING** Would you solve the equation $0.25x + 7 = \frac{1}{3}x - 8$ using fractions or decimals? Explain.
23. **BLOOD SAMPLE** The amount of red blood cells in a blood sample is equal to the total amount in the sample minus the amount of plasma. What is the total amount x of blood drawn?
24. **NUTRITION** One serving of oatmeal provides 16% of the fiber you need daily. You must get the remaining 21 grams of fiber from other sources. How many grams of fiber should you consume daily?
25. **Geometry** The perimeter of the square is equal to the perimeter of the triangle. What are the side lengths of each figure?
$4x$, $3x + 3$; $7x - 2$, $7x - 2$, $2x + 4$

Fair Game Review What you learned in previous grades & lessons

Find the volume of the figure. Use 3.14 for π. *(Skills Review Handbook)*

26. 2 cm, 3.5 cm, 4.5 cm
27. 4 in., $B = 18$ in.^{2}
28. $h = 3$ ft, 7 ft

29. **MULTIPLE CHOICE** A car travels 480 miles on 15 gallons of gasoline. How many miles does the car travel per gallon? *(Section 1.1)*
Ⓐ 28 mi/gal Ⓑ 30 mi/gal Ⓒ 32 mi/gal Ⓓ 35 mi/gal

Section 1.3 Solving Equations with Variables on Both Sides 21

PACING GUIDE

Each page in the book is in the ***Pacing Guide***.

- Chapters 1–9: 150 days
- With Appendix B: 165 days

Chapter 1 (18 Days)

Chapter Opener	1 Day
Activity 1.1	1 Day
Lesson 1.1	2 Days
Activity 1.2	1 Day
Lesson 1.2	1 Day
Activity 1.3	1 Day
Lesson 1.3	1 Day
Study Help/Quiz	1 Day
Activity 1.4	2 Days
Lesson 1.4	1 Day
Activity 1.5	1 Day
Lesson 1.5	2 Days
Quiz/Chapter Review	1 Day
Chapter Test	1 Day
Standardized Test Practice	1 Day

Chapter 2 (21 Days)

Chapter Opener	1 Day
Activity 2.1	1 Day
Lesson 2.1	1 Day
Activity 2.2	1 Day
Lesson 2.2	2 Days
Activity 2.3	1 Day
Lesson 2.3	1 Day
Activity 2.4	1 Day
Lesson 2.4	1 Day
Study Help/Quiz	1 Day
Activity 2.5	2 Days
Lesson 2.5	1 Day
Activity 2.6	1 Day
Lesson 2.6	1 Day
Activity 2.7	1 Day
Lesson 2.7	1 Day
Quiz/Chapter Review	1 Day
Chapter Test	1 Day
Standardized Test Practice	1 Day

Chapter 3 (17 Days)

Chapter Opener	1 Day
Activity 3.1	1 Day
Lesson 3.1	1 Day
Activity 3.2	1 Day
Lesson 3.2	2 Days
Activity 3.3	1 Day
Lesson 3.3	1 Day
Study Help/Quiz	1 Day
Activity 3.4	1 Day
Lesson 3.4	2 Days
Activity 3.5	1 Day
Lesson 3.5	1 Day
Quiz/Chapter Review	1 Day
Chapter Test	1 Day
Standardized Test Practice	1 Day

Chapter 4 (13 Days)

Chapter Opener	1 Day
Activity 4.1	1 Day
Lesson 4.1	1 Day
Activity 4.2	1 Day
Lesson 4.2	1 Day
Study Help/Quiz	1 Day
Activity 4.3	1 Day
Lesson 4.3	1 Day
Activity 4.4	1 Day
Lesson 4.4	1 Day
Quiz/Chapter Review	1 Day
Chapter Test	1 Day
Standardized Test Practice	1 Day

Chapter 5 (17 Days)

Chapter Opener	1 Day
Activity 5.1	1 Day
Lesson 5.1	1 Day
Activity 5.2	1 Day
Lesson 5.2	1 Day
Activity 5.3	1 Day
Lesson 5.3	2 Days
Study Help/Quiz	1 Day
Activity 5.4	1 Day
Lesson 5.4	1 Day
Activity 5.5	1 Day
Lesson 5.5	2 Days
Quiz/Chapter Review	1 Day
Chapter Test	1 Day
Standardized Test Practice	1 Day

Chapter 6 (17 Days)

Chapter Opener	1 Day
Activity 6.1	1 Day
Lesson 6.1	1 Day
Activity 6.2	1 Day
Lesson 6.2	1 Day
Study Help/Quiz	1 Day
Activity 6.3	1 Day
Lesson 6.3	2 Days
Activity 6.4	2 Days
Lesson 6.4	1 Day
Activity 6.5	1 Day
Lesson 6.5	1 Day
Quiz/Chapter Review	1 Day
Chapter Test	1 Day
Standardized Test Practice	1 Day

Chapter 7 (15 Days)

Chapter Opener	1 Day
Activity 7.1	1 Day
Lesson 7.1	1 Day
Activity 7.2	1 Day
Lesson 7.2	1 Day
Study Help/Quiz	1 Day
Activity 7.3	1 Day
Lesson 7.3	2 Days
Activity 7.4	1 Day
Lesson 7.4	2 Days
Quiz/Chapter Review	1 Day
Chapter Test	1 Day
Standardized Test Practice	1 Day

Chapter 8 (15 Days)

Chapter Opener	1 Day
Activity 8.1	1 Day
Lesson 8.1	1 Day
Activity 8.2	1 Day
Lesson 8.2	1 Day
Study Help/Quiz	1 Day
Activity 8.3	1 Day
Lesson 8.3	2 Days
Activity 8.4	2 Days
Lesson 8.4	1 Day
Quiz/Chapter Review	1 Day
Chapter Test	1 Day
Standardized Test Practice	1 Day

Chapter 9 (17 Days)

Chapter Opener	1 Day
Activity 9.1	1 Day
Lesson 9.1	1 Day
Activity 9.2	1 Day
Lesson 9.2	1 Day
Activity 9.3	1 Day
Lesson 9.3	1 Day
Study Help/Quiz	1 Day
Activity 9.4	1 Day
Lesson 9.4	1 Day
Activity 9.5	1 Day
Lesson 9.5	1 Day
Activity 9.6	1 Day
Lesson 9.6	1 Day
Quiz/Chapter Review	1 Day
Chapter Test	1 Day
Standardized Test Practice	1 Day

Appendix B (15 Days)

Activity B.1	1 Day
Lesson B.1	1 Day
Activity B.2	1 Day
Lesson B.2	1 Day
Activity B.3	1 Day
Lesson B.3	1 Day
Activity B.4	1 Day
Lesson B.4	1 Day
Study Help/Quiz	1 Day
Activity B.5	1 Day
Lesson B.5	1 Day
Activity B.6	1 Day
Lesson B.6	1 Day
Quiz/Chapter Review	1 Day
Chapter Test	1 Day

PROFESSIONAL DEVELOPMENT

Big Ideas Learning, LLC is a professional development and publishing company founded by Dr. Ron Larson. We are dedicated to providing 21st century teaching and learning in the area of mathematics. We work with middle schools across the country as they implement world-class standards for mathematics.

As teachers and school districts move forward in implementing new standards, the fundamental ideas of rigor and relevance and big ideas that are deep and focused take on new meaning. Big Ideas Learning provides a rich, hands-on experience that allows for astute understanding and practice, not only of the challenging world-class standards, but of the underlying mathematics pedagogy as well.

WORKSHOPS

- Creating Highly Motivating Classrooms
- Implementing the NCTM Curriculum Focal Points

Activities for the Mathematics Classroom:

- Teaching More with Less
- Best Practices in the Mathematics Classroom
- Questioning in the Mathematics Classroom
- The Three R's: Rigor, Relevance and Reality
- Reading in the Content Areas
- Games for Numerical Fluency
- Engaging the Tech Natives

Our professional staff of experienced instructors can also assist you in creating customized training sessions tailored to achieving your desired outcome.

"Do you think the stripes in this shirt make me look too linear?"

A Focal Points Curriculum

Chapter Coverage for Focal Points and Connections

1 2 3 4 5 6 7 8 9

Focal Point Analyze and represent linear functions and solve linear equations and systems of linear equations.

- Create and interpret tables, graphs, and models to represent, analyze, and solve problems related to linear equations, including analysis of domain, range and the difference between discrete and continuous data.
- Interpret the slope and the x- and y-intercepts when graphing a linear equation for a real-world problem.
- Use tables, graphs, and models to represent, analyze, and solve real-world problems related to systems of linear equations.
- Identify the solution to a system of linear equations using graphs.
- Translate among verbal, tabular, graphical and algebraic representations of linear functions.
- Compare the graphs of linear and non-linear functions for real-world situations.

1 2 3 4 5 6 7 8 9

Focal Point Analyze two- and three-dimensional figures by using distance and angle.

- Use similar triangles to solve problems that include height and distances.
- Classify and determine the measure of angles, including angles created when parallel lines are cut by transversals.
- Demonstrate that the sum of the angles in a triangle is 180-degrees and apply this fact to find unknown measures of angles, and the sum of angles in polygons.
- Validate and apply the Pythagorean Theorem to find distances in real-world situations or between points in the coordinate plane.

1 2 3 4 5 6 7 8 9

Focal Point Analyze and summarize data sets.

- Select, organize, and construct appropriate data displays, including box and whisker plots, scatter plots, and lines of best fit to convey information and make conjectures about possible relationships.
- Determine and describe how changes in data values impact measures of central tendency.

1 2 3 4 5 6 7 8 9

Connection Algebra

- Solve literal equations for a specified variable.
- Solve and graph one- and two-step inequalities in one variable.

1 2 3 4 5 6 7 8 9

Connection Geometry and Measurement

- Compare, contrast, and convert units of measure between different measurement systems (US customary or metric (SI)) and dimensions including temperature, area, volume, and derived units to solve problems.

1 2 3 4 5 6 7 8 9

Connection Number and Operations

- Use exponents and scientific notation to write large and small numbers and vice versa and to solve problems.
- Make reasonable approximations of square roots and mathematical expressions that include square roots, and use them to estimate solutions to problems and to compare mathematical expressions involving real numbers and radical expressions.
- Simplify real number expressions using the laws of exponents.
- Perform operations on real numbers (including integer exponents, radicals, percents, scientific notation, absolute value, rational numbers, and irrational numbers) using multi-step and real-world problems.

How to Use Your Math Book

- Read the **Essential Question** in the activity.

  Work with a partner to decide **What Is Your Answer?**

  Now you are ready to do the **Practice** problems.

- Find the **Key Vocabulary** words, **highlighted in yellow**.

  Read their definitions. Study the concepts in each **Key Idea**.

  If you forget a definition, you can look it up online in the

  Multi-Language Glossary at BigIdeasMath.com.

- After you study each **EXAMPLE**, do the exercises in the **On Your Own**.

  Now You're Ready to do the exercises that correspond to the example.

  As you study, look for a **Study Tip** or a **Common Error**.

- The exercises are divided into 3 parts.

  Vocabulary and Concept Check

  Practice and Problem Solving

  Fair Game Review

  If an exercise has a 1 next to it, look back at Example 1 for help with that exercise.

  More help is available at **Check It Out Lesson Tutorials BigIdeasMath.com**.

- To help study for your test, use the following.

  Quiz **Study Help**

  Chapter Review **Chapter Test**

SCAVENGER HUNT

Use this *Scavenger Hunt* to find where things are in **Chapter 1**.

Newton

1. Find an **Essential Question**.
2. Find a **Practice**.
3. Find the definition of a **conversion factor**.
4. Use the index in the back of the book. Find the (Key Idea) Division Property of Equality.
5. Find an **EXAMPLE** of solving a two-step equation.
6. Find an **On Your Own** about $-3x = 2x + 19$.
7. Find **Now You're Ready** Exercises 18–23.
8. Find a **Remember** about formulas.
9. Find a **Vocabulary** question about a literal equation.
10. Find a **Problem Solving** question about **ANTS**. Find the answer in the back of the book.
11. Which examples (?) help with Exercise 8 on page 14?
12. Find 2 **Quiz**zes.
13. How many questions are on the **Chapter Test**?

Descartes

1 Solving Equations

Connections to Previous Learning

- Write, solve, and graph one-step and two-step linear equations.
- Solve problems using a formula.

- Formulate and use different strategies to solve one-step and two-step linear equations, including equations with rational coefficients.
- Use properties of equality to rewrite an equation and to show two equations are equivalent.
- Compare, contrast, and convert between different measurement systems.

- Create models to represent, analyze, and solve problems related to linear equations.
- Solve literal equations for a variable.
- Compare, contrast, and convert between different measurement systems including dimensions for temperature, area, volume, and rates.

Math in History

There are two uses of the number 0 in mathematics.

★ Zero can be used as a place holder in a number system. For instance, the numbers 27 and 207 are different. The Mayans used zero in this way.

★ Zero can also be used to represent a number on the number line. The properties of 0, such as "the sum of zero and a number is that number" were described by Indian mathematicians over 3000 years ago.

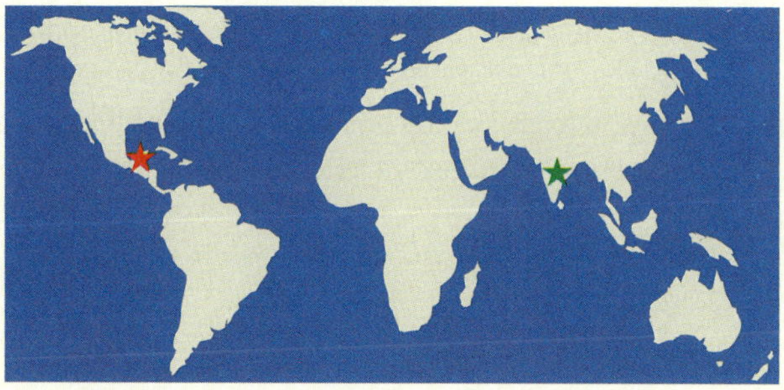

Pacing Guide for Chapter 1

Chapter Opener	1 Day
Section 1 Activity Lesson	 1 Day 2 Days
Section 2 Activity Lesson	 1 Day 1 Day
Section 3 Activity Lesson	 1 Day 1 Day
Study Help / Quiz	1 Day
Section 4 Activity Lesson	 2 Days 1 Day
Section 5 Activity Lesson	 1 Day 2 Days
Quiz / Chapter Review	1 Day
Chapter Test	1 Day
Standardized Test Practice	1 Day
Total Chapter 1	18 Days
Year-to-Date	18 Days

Check Your Resources

- Record and Practice Journal
- Resources by Chapter
- Skills Review Handbook
- Assessment Book
- Worked-Out Solutions

Technology For the Teacher

Dynamic Classroom

The Dynamic Planning Tool
Editable Teacher's Resources at
BigIdeasMath.com

Additional Topics for Review

- Compare, contrast, and convert measures within the same dimension
- Converting mixed numbers and improper fractions

Try It Yourself

1. No. *Sample answer:* One-quarter cup is about 59.25 milliliters and you need 60 milliliters.
2. $2\frac{5}{8}$ ounces

Record and Practice Journal

1. 4.75 L
2. 9.84 in.
3. 0.8454 cup
4. 21.3 oz
5. about 337.5 g
6. about 0.42 cup
7. about 2.25 kg
8. $1\frac{1}{6}$
9. $1\frac{31}{40}$
10. $\frac{3}{10}$
11. $1\frac{1}{4}$
12. $\frac{5}{8}$ tsp
13. $2\frac{7}{12}$ cups

Math Background Notes

Vocabulary Review

- Mixed Number
- Common Denominator
- Least Common Multiple

Converting Measures

- Students should know how to convert measures.
- **Management Tip:** It may be helpful to photocopy and distribute a list of commonly used conversion factors prior to completing the examples. This way, information is easily accessible to students.
- Remind students that converting measures does not change the value of the measurement. The purpose of converting measures is to write an equivalent form of the measure using different units.
- Remind students that they must multiply the measure by a conversion factor equal to one.
- **Common Error:** Students may use the conversion factor upside-down. Remind them to always write the given measure first. Align the units you want to divide out diagonally from one another.

Adding and Subtracting Fractions

- Students should know how to add and subtract fractions.
- Remind students that adding and subtracting fractions always requires a common denominator.
- If two fractions do not already share a common denominator, students should find one by determining the least common multiple of the denominators.
- **Common Error:** After finding the common denominator and renaming each fraction, students may try to combine the numerators and then combine the denominators. Remind them that combining the numerators is fine but that the common denominator should be carried through to the answer.
- You may want to review converting mixed numbers to improper fractions prior to completing Example 4.
- Encourage students to express their answers in simplest (reduced) form.
- Remind students that in the context of a recipe, they must include units with their answers.

Reteaching and Enrichment Strategies

If students need help. . .	If students got it. . .
Record and Practice Journal • Fair Game Review Skills Review Handbook Lesson Tutorials	Game Closet at *BigIdeasMath.com* Start the next section

What You Learned Before

"Once upon a time, there lived the most handsome dog who just happened to be a genius at math. He..."

Converting Measures

You find a recipe for wheat germ bread.

Example 1 **How many cups of water do you need to make the recipe?**

$$340 \text{ mL} \times \frac{1 \text{ cup}}{237 \text{ mL}} \approx 1.4 \text{ cups}$$

You need about 1.4 cups of water to make the recipe.

Example 2 **Do you need more whole wheat grains or more whole wheat flour?**

$$400 \text{ grams} \times \frac{1 \text{ oz}}{28 \text{ grams}} \approx 14.3 \text{ oz of flour}$$

Because 14.3 ounces > $2\frac{3}{4}$ ounces, you need more whole wheat flour.

WHEAT GERM BREAD

Preheat Oven to 220°C

Ingredients:

- $2\frac{3}{4}$ oz whole wheat grains
- 100 g wheat germ
- 400 g whole wheat flour
- $\frac{3}{4}$ tsp salt
- 340 mL water
- 60 mL orange juice
- 40 g honey
- $1\frac{2}{3}$ tsp yeast
- $\frac{1}{8}$ tsp cinnamon

Adding and Subtracting Fractions

Example 3 **How many teaspoons of spice (salt and cinnamon) are in the recipe?**

$$\frac{3}{4} + \frac{1}{8} = \frac{6}{8} + \frac{1}{8} = \frac{7}{8}$$

So, there is $\frac{7}{8}$ teaspoon of spice.

Example 4 **How many more teaspoons of yeast than salt are in the recipe?**

$$1\frac{2}{3} - \frac{3}{4} = \frac{5}{3} - \frac{3}{4} = \frac{20}{12} - \frac{9}{12} = \frac{11}{12}$$

So, there is $\frac{11}{12}$ teaspoon more yeast than salt.

Try It Yourself

Use the recipe to answer the questions.

1. You have one-quarter cup of orange juice. Do you have enough to make the recipe? Explain.
2. You have $\frac{1}{8}$ ounce of whole wheat grains. How many more ounces do you need to make the recipe?

1.1 Solving Simple Equations

Essential Question How can you use inductive reasoning to discover rules in mathematics? How can you test a rule?

1 ACTIVITY: Sum of the Angles of a Triangle

Work with a partner. Copy the triangles. Use a protractor to measure the angles of each triangle. Copy and complete the table to organize your results.

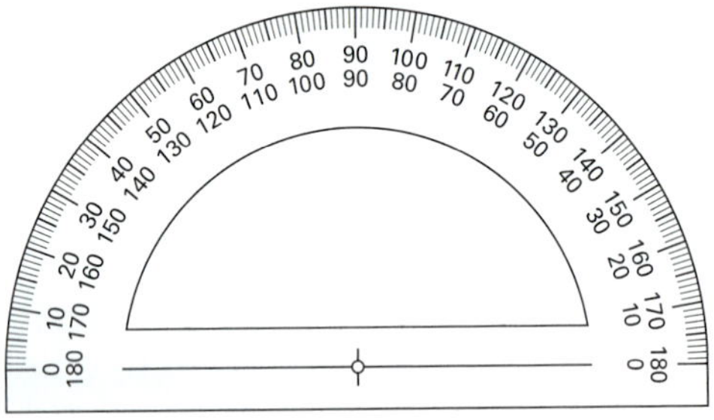

a.

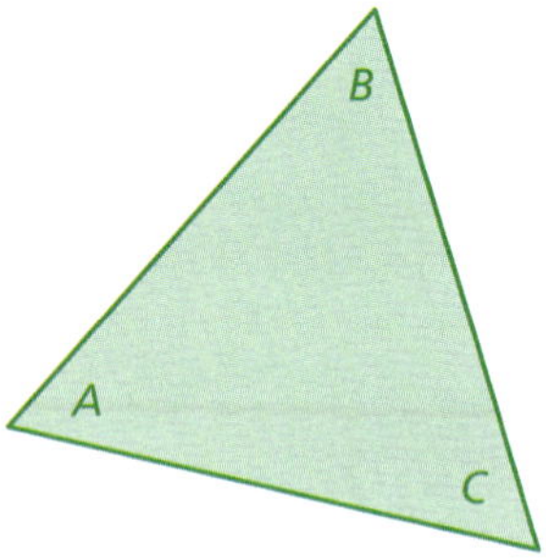

b.

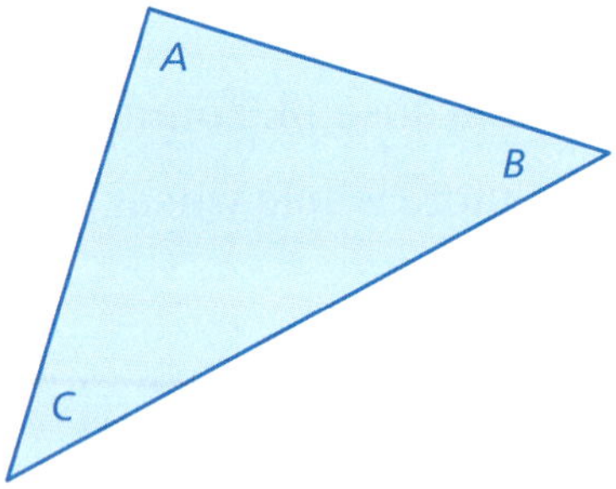

c.

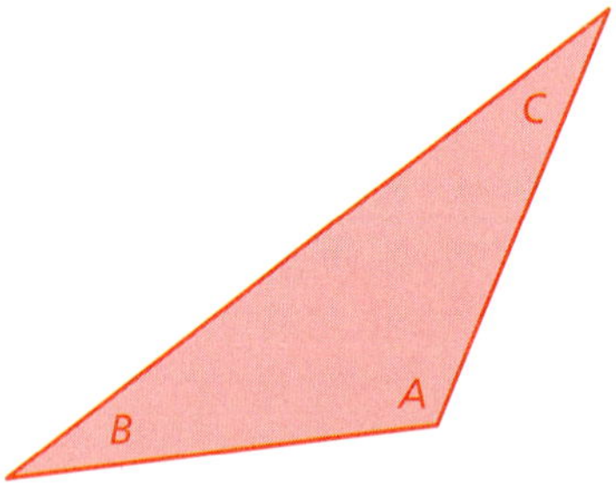

d.

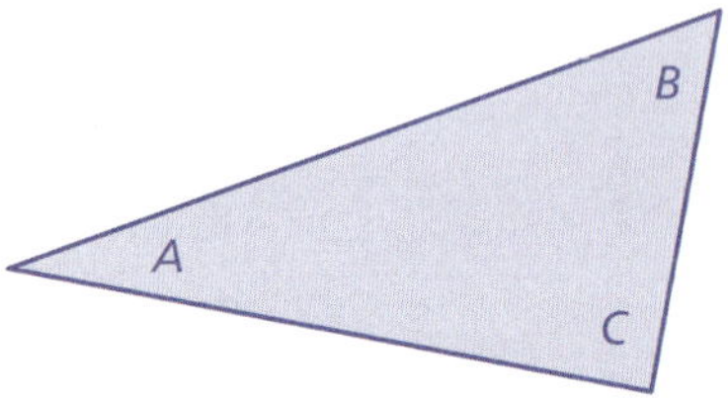

Triangle	Angle A (degrees)	Angle B (degrees)	Angle C (degrees)	$A + B + C$
a.				
b.				
c.				
d.				

Laurie's Notes

Introduction

For the Teacher

- **Goal:** Students will explore the sum of the angle measures of a triangle to develop an understanding of writing and solving simple equations.
- Note that the triangles drawn in the activities are not shown in the standard orientation, with a side parallel to the horizontal edge. You do not want students to believe, by repeated example, that triangles must have a horizontal base.

Motivate

? "What do Tony Hawk, Shaun White, and Rodney Mullen have in common?" All are famous skateboarders. Shaun White is also a snowboarder.

- Today's activity is about angle measures. Boarders know a lot about angle measure, in particular the multiples of 180°, because of the different tricks they perform.

Activity Notes

Words of Wisdom

? "What does it mean to measure an angle?" Listen for an understanding of the rotation from one ray to a second ray. Both angles shown have the same measure, although some students would say the angle on the left is greater.

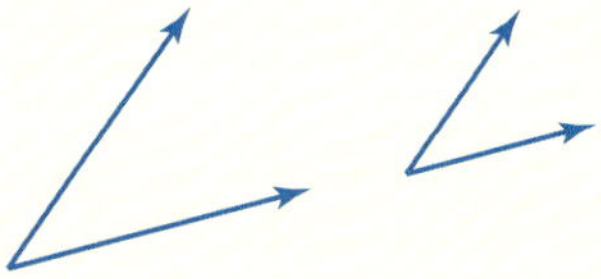

- Review with students how to place the protractor on the angle, and how to read the protractor.
- **Common Error:** Notice that 0° does not always align with the bottom edge of some protractors, nor does the vertex of the angle always align with the bottom edge. It is common for students to align the bottom edge of the protractor with one ray of the angle, producing an error of more than 5°.

Activity 1

? "Do you see any pattern(s) in the table? Describe the pattern(s)." The sum of the angle measures is 180°, or close to 180°. Students might also mention that in part (a), all of the angles are congruent (same measure) and in parts (b) and (c), two of the angles are congruent.

- **FYI:** If a sum is significantly different from 180°, the student may have read the protractor incorrectly (i.e., they recorded 150° instead of 30°).

Previous Learning

Students should know the vocabulary of angles, such as ray, vertex, acute, obtuse, right, and straight.

Activity Materials
Textbook
• protractors

Start Thinking! and Warm Up

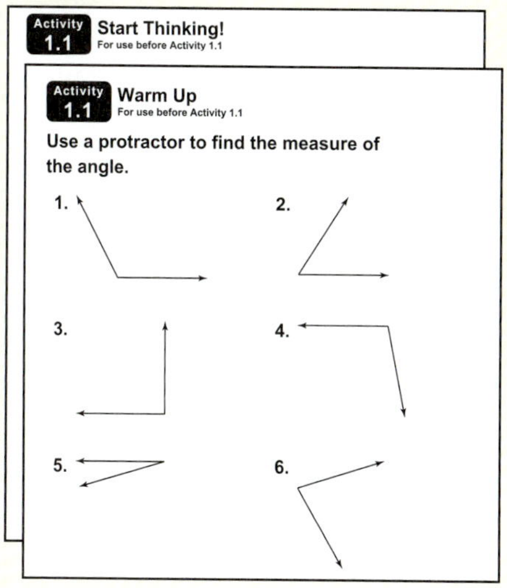

Activity 1.1 Start Thinking!
For use before Activity 1.1

Activity 1.1 Warm Up
For use before Activity 1.1

Use a protractor to find the measure of the angle.

1. 2. 3. 4. 5. 6.

1.1 Record and Practice Journal

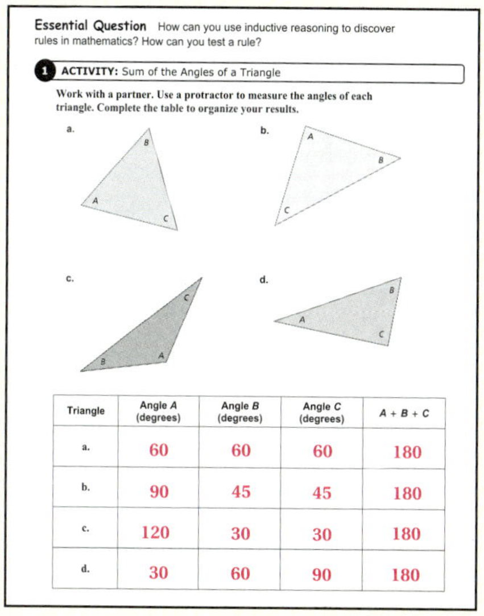

Essential Question How can you use inductive reasoning to discover rules in mathematics? How can you test a rule?

1 ACTIVITY: Sum of the Angles of a Triangle

Work with a partner. Use a protractor to measure the angles of each triangle. Complete the table to organize your results.

a. b. c. d.

Triangle	Angle A (degrees)	Angle B (degrees)	Angle C (degrees)	A + B + C
a.	60	60	60	180
b.	90	45	45	180
c.	120	30	30	180
d.	30	60	90	180

Differentiated Instruction

Kinesthetic

Ask two students to assist you at the board or overhead when solving equations. Assign one student to the left side of the equation and the other student to the right side. Each student is responsible for performing the operations on his or her side. Emphasize that to keep the equality, both students must perform the same operation to solve the equation.

1.1 Record and Practice Journal

2 ACTIVITY: Writing a Rule

Work with a partner. Use inductive reasoning to write and test a rule.

a. Use the completed table in Activity 1 to write a rule about the sum of the angle measures of a triangle.

The sum of the angle measures of a triangle is 180°.

b. **TEST YOUR RULE** Draw four triangles that are different from those in Activity 1. Measure the angles of each triangle. Organize your results in a table. Find the sum of the angle measures of each triangle.

Check students' work.

3 ACTIVITY: Applying Your Rule

Work with a partner. Use the rule you wrote in Activity 2 to write an equation for each triangle. Then, solve the equation to find the value of x. Use a protractor to check the reasonableness of your answer.

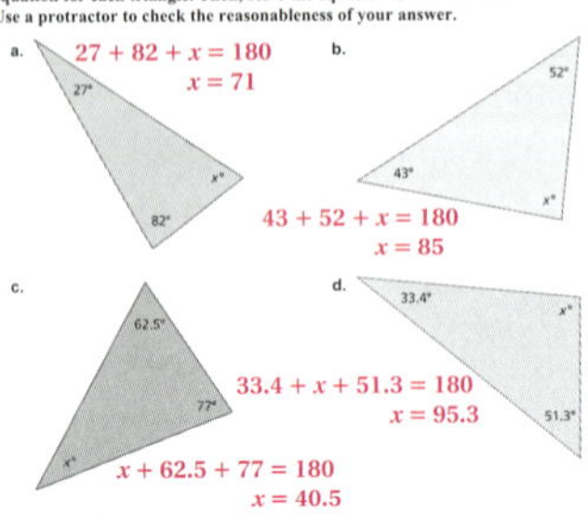

a. $27 + 82 + x = 180$, $x = 71$

b. $43 + 52 + x = 180$, $x = 85$

c. $x + 62.5 + 77 = 180$, $x = 40.5$

d. $33.4 + x + 51.3 = 180$, $x = 95.3$

What Is Your Answer?

4. **IN YOUR OWN WORDS** How can you use inductive reasoning to discover rules in mathematics? How can you test a rule? How can you use a rule to solve problems in mathematics?

Check students' work.

Laurie's Notes

Activity 2

? "What rule did you write for the sum of the angle measures of a triangle?" Students should write that the sum of the angle measures of a triangle equals 180°.

- Suggest to students that they make their triangles larger so that it is easier to measure the angles. They should also use a straight edge to make straight lines.

? "Did you measure all three angles for each triangle?" Some students will not measure all three angles! They will only measure two and do a quick computation to find the third.

Activity 3

- You should model the first problem and write the equation. Otherwise, students will do a computation to find the missing angle.
- **Write:** $27 + 82 + x = 180$. Students have solved equations previously and may simply write the answer as the second step: $x = 71$. Focus on the representation of equation solving instead of the intuitive sense of how to solve this addition equation. Model the second step by showing 109 subtracted from each side of the equation.
- Note that parts (c) and (d) integrate decimal review. Their answers should be exact.
- Have students share their answers.

What Is Your Answer?

? "What is inductive reasoning and how was it used in the activities today?" *Sample answer:* Inductive reasoning is writing a general rule based on examples. Today I found that the sum of the angle measures of several triangles equals 180°, so I wrote a rule for triangles in general.

- Have students discuss their ideas to the questions posed in Question 4.

Closure

- **Exit Ticket:** Two angles of a triangle measure 48.2° and 63.8°. Make a reasonable sketch of the triangle. Write and solve an equation to find the measure of the third angle. $48.2 + 63.8 + x = 180$; $x = 68$

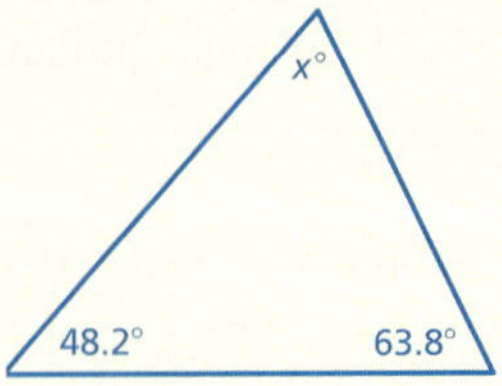

Technology For the Teacher

Dynamic Classroom

The Dynamic Planning Tool

Editable Teacher's Resources at *BigIdeasMath.com*

2 ACTIVITY: Writing a Rule

Work with a partner. Use inductive reasoning to write and test a rule.

a. Use the completed table in Activity 1 to write a rule about the sum of the angle measures of a triangle.

b. **TEST YOUR RULE** Draw four triangles that are different from those in Activity 1. Measure the angles of each triangle. Organize your results in a table. Find the sum of the angle measures of each triangle.

3 ACTIVITY: Applying Your Rule

Work with a partner. Use the rule you wrote in Activity 2 to write an equation for each triangle. Then, solve the equation to find the value of *x*. Use a protractor to check the reasonableness of your answer.

a.

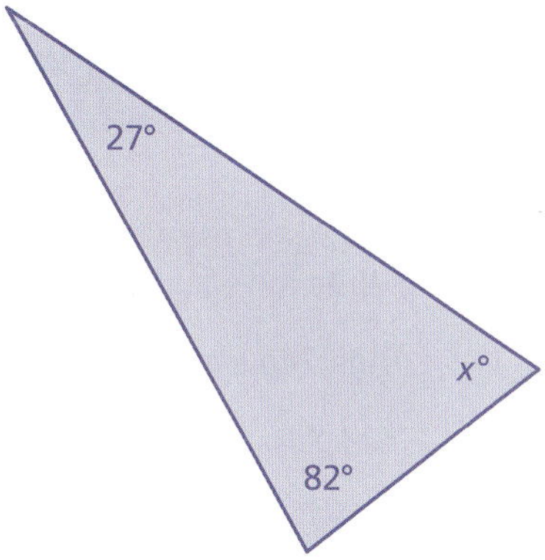

b.

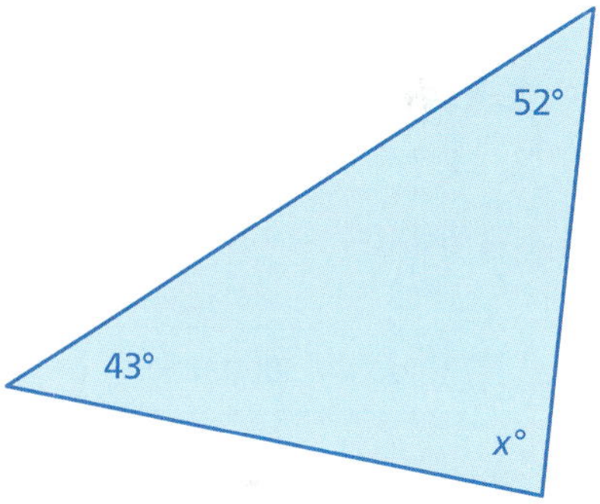

c.

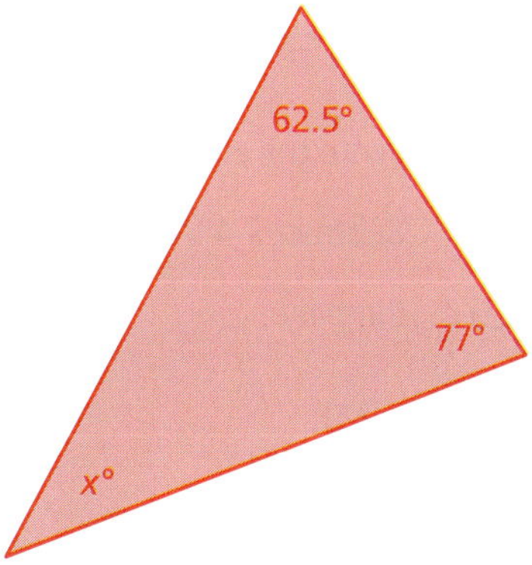

d.

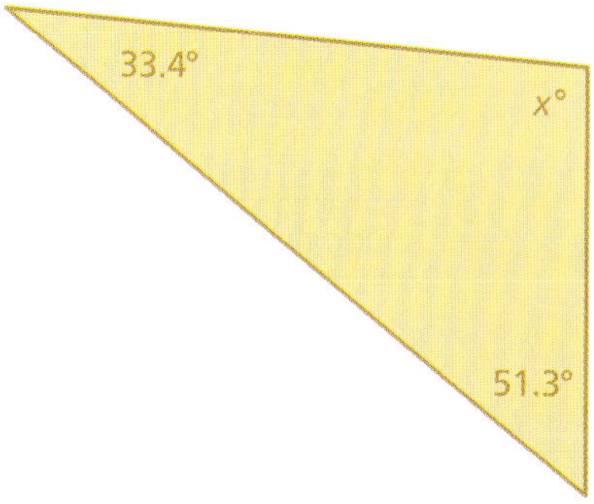

What Is Your Answer?

4. **IN YOUR OWN WORDS** How can you use inductive reasoning to discover rules in mathematics? How can you test a rule? How can you use a rule to solve problems in mathematics?

Practice Use what you learned about solving simple equations to complete Exercises 4–6 on page 7.

1.1 Lesson

Check It Out
Lesson Tutorials
BigIdeasMath.com

Remember

Addition and subtraction are inverse operations.

Key Ideas

Addition Property of Equality

Words Adding the same number to each side of an equation produces an equivalent equation.

Algebra If $a = b$, then $a + c = b + c$.

Subtraction Property of Equality

Words Subtracting the same number from each side of an equation produces an equivalent equation.

Algebra If $a = b$, then $a - c = b - c$.

EXAMPLE 1 Solving Equations Using Addition or Subtraction

a. Solve $x - 7 = -6$.

	$x - 7 = -6$	Write the equation.
Undo the subtraction. →	$+7 \quad +7$	Add 7 to each side.
	$x = 1$	Simplify.

The solution is $x = 1$.

Check

$x - 7 = -6$

$1 - 7 \stackrel{?}{=} -6$

$-6 = -6$ ✓

b. Solve $y + 3.4 = 0.5$.

	$y + 3.4 = 0.5$	Write the equation.
Undo the addition. →	$-3.4 \quad -3.4$	Subtract 3.4 from each side.
	$y = -2.9$	Simplify.

The solution is $y = -2.9$.

Check

$y + 3.4 = 0.5$

$-2.9 + 3.4 \stackrel{?}{=} 0.5$

$0.5 = 0.5$ ✓

c. Solve $h + 2\pi = 3\pi$.

	$h + 2\pi = 3\pi$	Write the equation.
Undo the addition. →	$-2\pi \quad -2\pi$	Subtract 2π from each side.
	$h = \pi$	Simplify.

The solution is $h = \pi$.

Laurie's Notes

Introduction

Connect

- **Yesterday:** Students used the sum of the angle measures of a triangle to explore simple equation solving.
- **Today:** Students will use Properties of Equality to solve one-step equations.

Motivate

- Tell students that you are going to play a quick game of *REVERSO*. The directions are simple: you give a command to a student and your opponent must give the reverse (inverse) command to undo your command. For example, you say, "take 3 steps forward " and your opponent would say "take 3 steps backward."
- Sample commands: turn lights on; step up on a chair; turn to your right; fold 2 sheets of paper; draw a square; open the door
- The goal is for students to think about inverse operations.

Lesson Notes

Key Ideas

- Write the Key Ideas.
- Redefine *equivalent equations*. Two equations that have the same solution are *equivalent equations*.
- **Teaching Tip:** Use an alternate color to show adding (subtracting) c to (from) each side of the equation.
- Remind students of the big idea. Whatever you do to one side of the equation, you must do to the other side of the equation.

Example 1

- Work through each part. Note that the vertical format of equation solving is used. The number being added to or subtracted from each side of the equation is written vertically below the number with which it will be combined.

? "What is the approximate value of π?" 3.14 "of 2π?" 6.28

- Remind students that 2π and 3π are (irrational) numbers, so you can treat these numbers as you would integers. It is common for students to think of π as a variable and they will say that there are two variables in $h + 2\pi$.

Goal Today's lesson is solving one-step equations.

Start Thinking! and Warm Up

Lesson 1.1 Warm Up
For use before Lesson 1.1

Lesson 1.1 Start Thinking!
For use before Lesson 1.1

The Addition Property of Equality states that adding the same number to each side of an equation produces an equivalent equation. What do you think the Subtraction, Multiplication, and Division Properties of Equality state?

Describe a real-life situation that you can relate to one of the properties of equality.

Extra Example 1

a. Solve $d - \frac{1}{4} = -\frac{1}{2}$. $-\frac{1}{4}$

b. Solve $m + 4.8 = 9.2$ 4.4

c. Solve $r - 6\pi = 2\pi$. 8π

On Your Own

1. $b = -7$ 2. $g = 0.8$
3. $k = -6$ 4. $r = 2\pi$
5. $t = -\frac{1}{2}$ 6. $z = -13.6$

Extra Example 2

a. Solve $\frac{2}{5}m = -4$. -10

b. Solve $3p = -\frac{2}{3}$. $-\frac{2}{9}$

On Your Own

7. $y = -28$ 8. $x = 6$
9. $w = 20$

English Language Learners

Vocabulary

In this section, students will learn to use inverse (or opposite) operations to solve equations. Students will use addition to solve a subtraction equation and use subtraction to solve an addition equation. Review these pairs of words that are essential to understanding mathematics. Give students one word of a pair and ask them to provide the opposite.

Examples:

odd, even	positive, negative
add, subtract	sum, difference
multiply, divide	product, quotient
plus, minus	

Laurie's Notes

On Your Own

- Circulate as students work on these six questions. Remind students that it is the practice of *representing* their work that is important in these questions.

Key Ideas

- Write the Key Ideas.
- **Representation:** Review different ways in which multiplication is represented.

$$a(c) = b(c) \qquad ac = bc \qquad a \times c = b \times c \qquad a \cdot c = b \cdot c$$

- Generally, when there are variables in equations, you do not want to use $\times$ to represent multiplication because it can be mistaken for a variable.
- **Representation**: Review different ways in which division is represented.

$$a \div c = b \div c \qquad \frac{a}{c} = \frac{b}{c} \qquad a/c = b/c$$

Example 2

- Work through each part.
- Remind students that the goal is to solve for the variable so that it has a coefficient of 1.

? "What is the coefficient of *n*?" $-\frac{3}{4}$ "What operation is represented?" multiplication "How do you undo multiplication?" divide

? "What is equivalent to dividing by $-\frac{3}{4}$?" multiplying by $-\frac{4}{3}$

- When the coefficient is a fraction, it is more efficient to multiply by the reciprocal.
- The purpose of part (b) is to practice working with π in an algebraic expression.

On Your Own

- Circulate as students work on these three questions. Stress representation with students. Do not let them short-cut the process and simply record the answer. They should show what operation is being performed on each side of the equation.
- **Common Error:** Students may try to subtract 6π from πx or subtract πx from 6π. Remind them that the variable they are solving for is x.

On Your Own

Now You're Ready
Exercises 7–15

Solve the equation. Check your solution.

1. $b + 2 = -5$
2. $g - 1.7 = -0.9$
3. $-3 = k + 3$
4. $r - \pi = \pi$
5. $t - \frac{1}{4} = -\frac{3}{4}$
6. $5.6 + z = -8$

Remember

Multiplication and division are inverse operations.

Key Ideas

Multiplication Property of Equality

Words Multiplying each side of an equation by the same number produces an equivalent equation.

Algebra If $a = b$, then $a \cdot c = b \cdot c$.

Division Property of Equality

Words Dividing each side of an equation by the same number produces an equivalent equation.

Algebra If $a = b$, then $a \div c = b \div c$, $c \neq 0$.

EXAMPLE 2 Solving Equations Using Multiplication or Division

a. Solve $-\frac{3}{4}n = -2$.

$-\frac{3}{4}n = -2$ Write the equation.

Use the reciprocal. → $-\frac{4}{3} \cdot \left(-\frac{3}{4}n\right) = -\frac{4}{3} \cdot -2$ Multiply each side by $-\frac{4}{3}$, the reciprocal of $-\frac{3}{4}$.

$n = \frac{8}{3}$ Simplify.

The solution is $n = \frac{8}{3}$.

b. Solve $\pi x = 3\pi$.

$\pi x = 3\pi$ Write the equation.

Undo the multiplication. → $\frac{\pi x}{\pi} = \frac{3\pi}{\pi}$ Divide each side by π.

$x = 3$ Simplify.

The solution is $x = 3$.

Check

$\pi x = 3\pi$

$\pi(3) \stackrel{?}{=} 3\pi$

$3\pi = 3\pi$ ✓

On Your Own

Now You're Ready
Exercises 18–26

Solve the equation. Check your solution.

7. $\frac{y}{4} = -7$
8. $6\pi = \pi x$
9. $0.09w = 1.8$

EXAMPLE 3 Standardized Test Practice

What value of k makes the equation $k + 4 \div 0.2 = 5$ true?

Ⓐ -15 Ⓑ -5 Ⓒ -3 Ⓓ 1.5

$k + 4 \div 0.2 = 5$	Write the equation.
$k + 20 = 5$	Divide 4 by 0.2.
$\underline{-20} \quad \underline{-20}$	Subtract 20 from each side.
$k = -15$	Simplify.

The correct answer is Ⓐ.

EXAMPLE 4 Real-Life Application

The melting point of bromine is $-7°C$.

The *melting point* of a solid is the temperature at which the solid becomes a liquid. The melting point of bromine is $\frac{1}{30}$ of the melting point of nitrogen. Write and solve an equation to find the melting point of nitrogen.

Words	The melting point of bromine	is	$\frac{1}{30}$	of	the melting point of nitrogen.
Variable	Let n be the melting point of nitrogen.				
Equation	-7	$=$	$\frac{1}{30}$	$\cdot$	n

$-7 = \frac{1}{30}n$	Write the equation.
$30 \cdot (-7) = 30 \cdot \left(\frac{1}{30}n\right)$	Multiply each side by 30.
$-210 = n$	Simplify.

The melting point of nitrogen is $-210°C$.

On Your Own

Now You're Ready
Exercises 33–38

10. Solve $p - 8 \div \frac{1}{2} = -3$.

11. Solve $q + |-10| = 2$.

12. The melting point of mercury is about $\frac{1}{4}$ of the melting point of krypton. The melting point of mercury is $-39°C$. Write and solve an equation to find the melting point of krypton.

Laurie's Notes

Lesson Notes

Example 3

? "What is $10 + 4 \div 2$?" listen for order of operations; Answer is 12, *not* 7.

- Students could use *Guess, Check, and Revise*. However, it is more efficient to use order of operations and then solve the equation.

Example 4

- Note the color-coding of the words and symbols. Discuss this feature with students. Students find it difficult to read a word problem and translate it into symbols. This skill is practiced throughout the text.
- **Representation:** The term $\frac{1}{30}n$ could also have been written as $\frac{n}{30}$. Make sure students understand why. It is how a fraction and number are multiplied.
- **FYI:** The final answer $-210 = n$ can also be written as $n = -210$.

On Your Own

- Remind students to perform the operations following the order of operations.
- **Common Error:** $8 \div \frac{1}{2} \neq 4$; $8 \div \frac{1}{2} = 16$

? "For Question 11, what does $|-10|$ mean?" absolute value of -10, which equals 10

Closure

- Describe in words how to solve a one-step equation.
- Write and solve a one-step equation.

Technology For the Teacher

Dynamic Classroom

The Dynamic Planning Tool
Editable Teacher's Resources at *BigIdeasMath.com*

Extra Example 3

Solve $w - 4 \div \frac{1}{2} = 5$. $w = 13$

Extra Example 4

The melting point of ice is $\frac{2}{9}$ of the melting point of candle wax. The melting point of ice is 32°F. Write and solve an equation to find the melting point of candle wax. $\frac{2}{9}x = 32$; 144°F

On Your Own

10. $p = 1$

11. $q = -8$

12. $-39 = \frac{1}{4}k$; -156°C

Vocabulary and Concept Check

1. + and − are inverses. × and ÷ are inverses.
2. yes; The solution of each equation is $x = -3$.
3. $x - 3 = 6$; It is the only equation that does not have $x = 6$ as a solution.

Practice and Problem Solving

4. $x = 32$
5. $x = 57$
6. $x = 111$
7. $x = -5$
8. $g = 24$
9. $p = 21$
10. $y = -2.04$
11. $x = 9\pi$
12. $w = 10\pi$
13. $d = \frac{1}{2}$
14. $r = -\frac{7}{24}$
15. $n = -4.9$
16. $p - 14.50 = 53$; $67.50
17. a. $105 = x + 14$; $x = 91$

    b. no; Because $82 + 9 = 91$, you did not knock down the last pin with the second ball of the frame.

Assignment Guide and Homework Check

Level	Day 1 Activity Assignment	Day 2 Lesson Assignment	Homework Check
Basic	4–6, 45–48	1–3, 7–15 odd, 16, 19–33 odd, 28, 37	7, 11, 16, 19, 21, 33
Average	4–6, 45–48	1–3, 10–14, 17, 21–24, 27–37 odd, 41	10, 11, 17, 21, 22, 33
Advanced	4–6, 45–48	1–3, 17, 24–27, 30–38 even, 39–44	17, 24, 30, 34, 39, 42

For Your Information

- **Exercise 17** Students may not know what a spare is in bowling. A spare means that all of the pins were knocked down after the second ball of a frame was thrown. To calculate the score after a spare, you add the number of pins knocked down on your next ball to 10. For example, if you got a spare in the first frame and then knocked down 6 pins on your next ball, your score for the first frame would be 16.

Common Errors

- **Exercises 4–6** Students may struggle using the protractor to find the missing angle. Encourage them to trace the triangle and extend the sides so they can get a more accurate reading.
- **Exercises 7–15** Students may perform the same operation on both sides instead of the opposite operation. Remind them that to solve for the variable, they must *undo* the operation by using the opposite (or inverse) operation.
- **Exercise 16** Students may write the wrong equation for the problem. Encourage them to rewrite the problem so that it is clear what equation they should write. Remind them that subtraction is not commutative.

1.1 Record and Practice Journal

Solve the equation. Check your solution.

1. $x + 5 = 16$ — $x = 11$
2. $11 = w - 12$ — $w = 23$
3. $\frac{3}{4} + z = \frac{5}{6}$ — $z = \frac{1}{12}$
4. $3y = 18$ — $y = 6$
5. $\frac{k}{7} = 10$ — $k = 70$
6. $\frac{4}{5}n = \frac{9}{10}$ — $n = \frac{9}{8}$
7. $x - 12 + 6 = 9$ — $x = 11$
8. $h + |-8| = 15$ — $h = 7$
9. $1.3(2) + p = 7.9$ — $p = 5.3$
10. A coupon subtracts $5.16 from the price p of a shirt. You pay $15.48 for the shirt after using the coupon. Write and solve an equation to find the original price of the shirt.

    $p - 5.16 = 15.48$; $p = \$20.64$
11. After a party, you have $\frac{1}{6}$ of the cookies you made left over. There are a dozen cookies left. How many cookies did you make for the party?

    $c = 72$

Technology For the Teacher

Answer Presentation Tool

Quiz*Show*

1.1 Exercises

Check It Out
Help with Homework
BigIdeasMath.com

Vocabulary and Concept Check

1. **VOCABULARY** Which of the operations $+$, $-$, $\times$, and $\div$ are inverses of each other?
2. **VOCABULARY** Are the equations $3x = -9$ and $4x = -12$ equivalent? Explain.
3. **WHICH ONE DOESN'T BELONG?** Which equation does *not* belong with the other three? Explain your reasoning.

| $x - 2 = 4$ | $x - 3 = 6$ | $x - 5 = 1$ | $x - 6 = 0$ |

Practice and Problem Solving

Find the value of x. Use a protractor to check the reasonableness of your answer.

4. 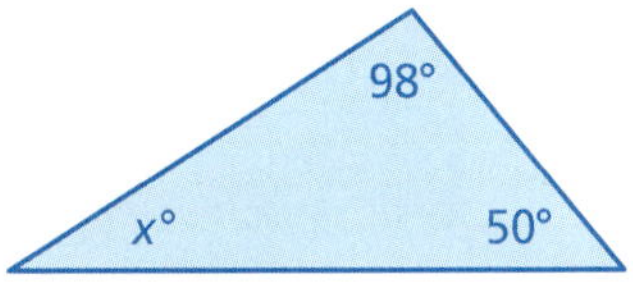

5. 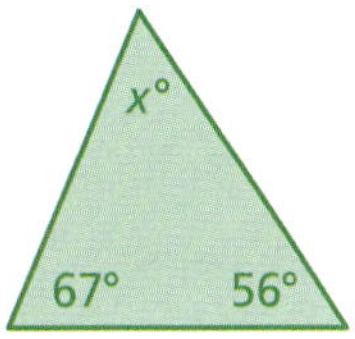

6. 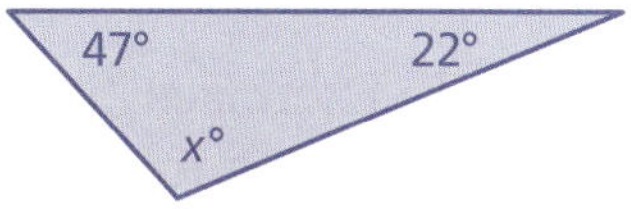

Solve the equation. Check your solution.

1 7. $x + 12 = 7$
8. $g - 16 = 8$
9. $-9 + p = 12$
10. $0.7 + y = -1.34$
11. $x - 8\pi = \pi$
12. $4\pi = w - 6\pi$
13. $\frac{5}{6} = \frac{1}{3} + d$
14. $\frac{3}{8} = r + \frac{2}{3}$
15. $n - 1.4 = -6.3$

16. **CONCERT** A discounted concert ticket is \$14.50 less than the original price p. You pay \$53 for a discounted ticket. Write and solve an equation to find the original price.

17. **BOWLING** Your friend's final bowling score is 105. Your final bowling score is 14 pins less than your friend's final score.

    a. Write and solve an equation to find your final score.

    b. Your friend made a spare in the tenth frame. Did you? Explain.

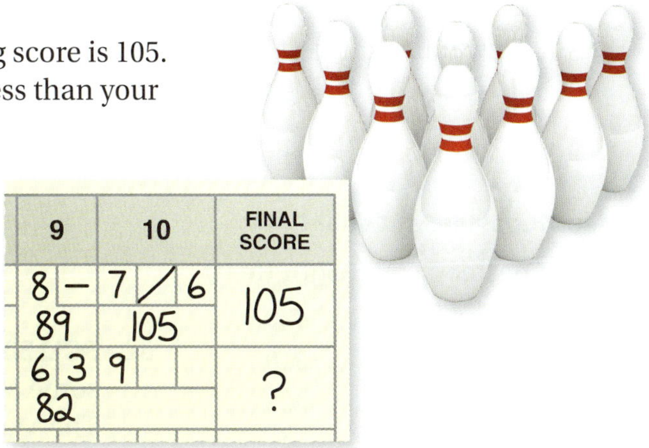

Solve the equation. Check your solution.

(2) **18.** $7x = 35$ **19.** $4 = -0.8n$ **20.** $6 = -\frac{w}{8}$

21. $\frac{m}{\pi} = 7.3$ **22.** $-4.3g = 25.8$ **23.** $\frac{3}{2} = \frac{9}{10}k$

24. $-7.8x = -1.56$ **25.** $-2 = \frac{6}{7}p$ **26.** $3\pi d = 12\pi$

27. ERROR ANALYSIS Describe and correct the error in solving the equation.

✗
$-1.5 + k = 8.2$
$k = 8.2 + (-1.5)$
$k = 6.7$

28. TENNIS A gym teacher orders 42 tennis balls. Each package contains 3 tennis balls. Which of the following equations represents the number x of packages?

$x + 3 = 42$ $3x = 42$ $\frac{x}{3} = 42$ $x = \frac{3}{42}$

In Exercises 29–32, write and solve an equation to answer the question.

29. PARK You clean a community park for 6.5 hours. You earn \$42.25. How much do you earn per hour?

30. SPACE SHUTTLE A space shuttle is scheduled to launch from Kennedy Space Center in 3.75 hours. What time is it now?

31. BANKING After earning interest, the balance of an account is \$420. The new balance is $\frac{7}{6}$ of the original balance. How much interest was earned?

Tallest Coasters at Cedar Point

Roller Coaster	Height (feet)
Top Thrill Dragster	420
Millennium Force	310
Magnum XL-200	205
Mantis	?

32. ROLLER COASTER Cedar Point amusement park has some of the tallest roller coasters in the United States. The Mantis is 165 feet shorter than the Millennium Force. What is the height of the Mantis?

Common Errors

- **Exercises 18–26** Students may use the same operation instead of the opposite operation to get the variable by itself. Remind them that to *undo* the operation, they must use the opposite (or inverse) operation. Demonstrate that using the same operation will not work. For example:

Incorrect	Correct
$7x = 35$	$7x = 35$
$7 \cdot 7x = 35 \cdot 7$	$\frac{7x}{7} = \frac{35}{7}$
$49x = 245$	$x = 5$

- **Exercise 32** Students may skip the step of writing the equation and just subtract the difference in height from the height of the Millennium Force. Encourage them to develop the problem solving technique of writing the equation before solving. This skill will be useful later in mathematics.
- **Exercises 33–38** Students may forget to use the order of operations when solving for the variable. Remind them of the order of operations and encourage them to simplify both sides of the equation before solving.

English Language Learners

Vocabulary

Have students create a table in their notebooks of the common words used to indicate addition, subtraction, multiplication, and division.

For instance,

Addition	Subtraction
added to	subtracted from
plus	minus
sum of	difference of
more than	less than
increased by	decreased by
total of	fewer than
and	take away
Multiplication	**Division**
multiplied by	divided by
times	quotient of
product of	
twice	
of	

Practice and Problem Solving

18. $x = 5$ **19.** $n = -5$

20. $w = -48$ **21.** $m = 7.3\pi$

22. $g = -6$ **23.** $k = 1\frac{2}{3}$

24. $x = 0.2$ **25.** $p = -2\frac{1}{3}$

26. $d = 4$

27. They should have added 1.5 to each side.

$$-1.5 + k = 8.2$$
$$k = 8.2 + 1.5$$
$$k = 9.7$$

28. $3x = 42$

29. $6.5x = 42.25$; \$6.50 per hour

30. $x + 3\frac{3}{4} = 11\frac{1}{3}$; 7:35 A.M.

31. $420 = \frac{7}{6}b, b = 360$; \$60

32. $x + 165 = 310$; 145 ft

Practice and Problem Solving

33. $h = -7$ **34.** $w = 19$

35. $q = 3.2$ **36** $d = 0$

37. $x = -1\frac{4}{9}$ **38.** $p = -\frac{1}{12}$

39. greater than; Because a negative number divided by a negative number is a positive number.

40. *Sample answer:* $x - 2 = -4$, $\frac{x}{2} = -1$

41. 3 mg

42. See *Taking Math Deeper.*

43. 8 in.

44. **a.** \$18, \$27, \$45

b. *Sample answer:* Everyone did not do an equal amount of painting.

Fair Game Review

45. $7x - 4$ **46.** $1.6b - 3.2$

47. $\frac{25}{4}g - \frac{2}{3}$

48. A

Mini-Assessment

Solve the equation.

1. $t + 17 = 3$ $t = -14$
2. $-2\pi + d = -3\pi$ $d = -\pi$
3. $-13.5 = 2.7s$ $s = -5$
4. $\frac{2}{3}j = 8$ $j = 12$
5. You earn \$9.65 per hour. This week, you earned \$308.80 before taxes. Write and solve an equation to find the number of hours you worked this week. $9.65x = 308.8$; You worked 32 hours this week.

Taking Math Deeper

Exercise 42

A nice way to organize the given information is to put it into a table.

1 Use a table to organize the information.

	Total	Retake
Girls	x	$\frac{1}{4}x = 16$
Boys	y	$\frac{1}{8}y = 7$

2 Use the equations to solve for x and y.

Girls: $\frac{1}{4}x = 16$

$x = 64$

Boys: $\frac{1}{8}y = 7$

$y = 56$

3 Answer the question.

There are $64 + 56 = 120$ students in the eighth grade.

Project

Find out how many retakes were done at your school last year. Do the given ratios work for your school? What do you think are some of the reasons students have retakes?

Reteaching and Enrichment Strategies

If students need help. . .	If students got it. . .
Resources by Chapter • Practice A and Practice B • Puzzle Time Record and Practice Journal Practice Differentiating the Lesson Lesson Tutorials Skills Review Handbook	Resources by Chapter • Enrichment and Extension Start the next section

Solve the equation. Check your solution.

3 **33.** $-3 = h + 8 \div 2$

34. $12 = w - |-7|$

35. $q + |6.4| = 9.6$

36. $d - 2.8 \div 0.2 = -14$

37. $\frac{8}{9} = x + \frac{1}{3}(7)$

38. $p - \frac{1}{4} \cdot 3 = -\frac{5}{6}$

39. CRITICAL THINKING Is the solution of $-2x = -15$ *greater than* or *less than* -15? Explain.

40. OPEN-ENDED Write a subtraction equation and a division equation that each has a solution of -2.

41. ANTS Some ant species can carry 50 times their body weight. It takes 32 ants to carry the cherry. About how much does each ant weigh?

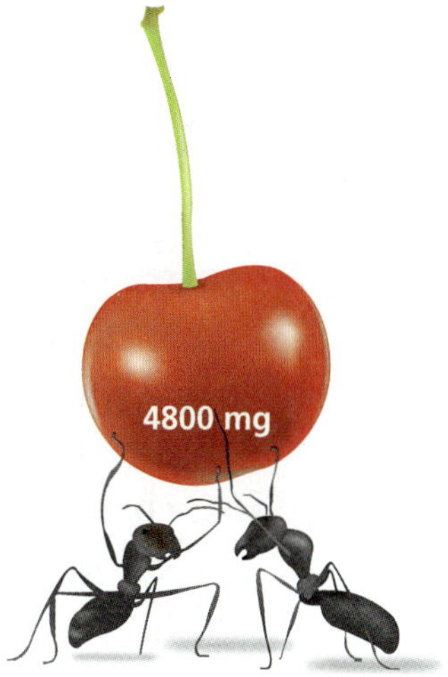

42. PICTURES One-fourth of the girls and one-eighth of the boys in an eighth grade retake their school pictures. The photographer retakes pictures for 16 girls and 7 boys. How many students are in the eighth grade?

43. VOLUME The volume V of the cylinder is 72π cubic inches. Use the formula $V = Bh$ to find the height h of the cylinder.

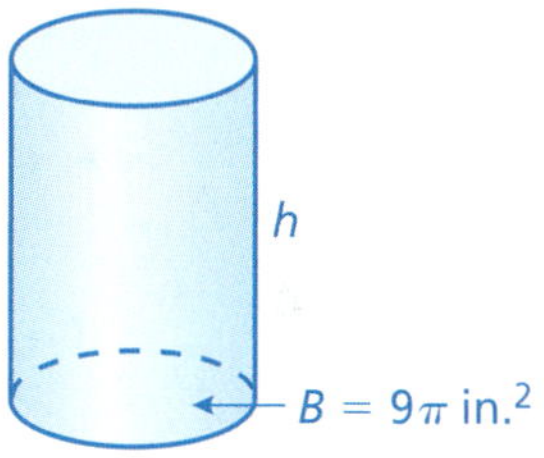

44. Critical Thinking A neighbor pays you and two friends \$90 to paint her garage. The money is divided three ways in the ratio 2 : 3 : 5.

a. How much is each share?

b. What is one possible reason the money is not divided evenly?

Fair Game Review What you learned in previous grades & lessons

Simplify the expression. *(Skills Review Handbook)*

45. $2(x - 2) + 5x$

46. $0.4b - 3.2 + 1.2b$

47. $\frac{1}{4}g + 6g - \frac{2}{3}$

48. MULTIPLE CHOICE The temperature at 4 P.M. was -12°C. By 11 P.M. the temperature had dropped 14 degrees. What was the temperature at 11 P.M.? *(Skills Review Handbook)*

Ⓐ -26°C Ⓑ -2°C Ⓒ 2°C Ⓓ 26°C

1.2 Solving Multi-Step Equations

Essential Question How can you solve a multi-step equation? How can you check the reasonableness of your solution?

1 ACTIVITY: Solving for the Angles of a Triangle

Work with a partner. Write an equation for each triangle. Solve the equation to find the value of the variable. Then find the angle measures of each triangle. Use a protractor to check the reasonableness of your answer.

a.

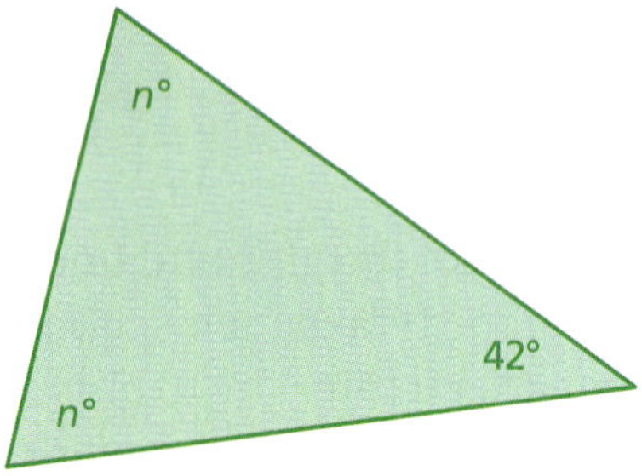

b.

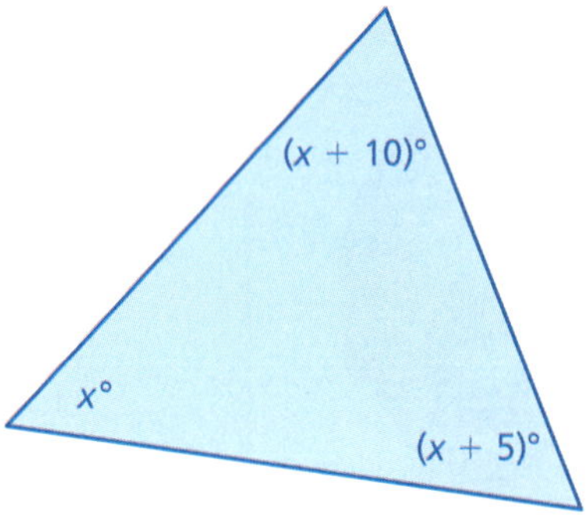

c.

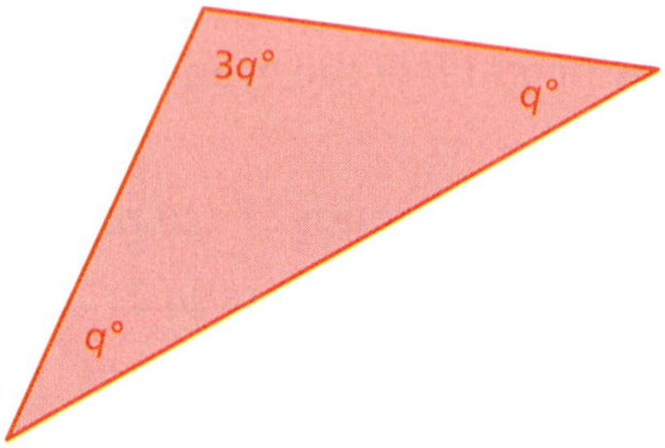

d.

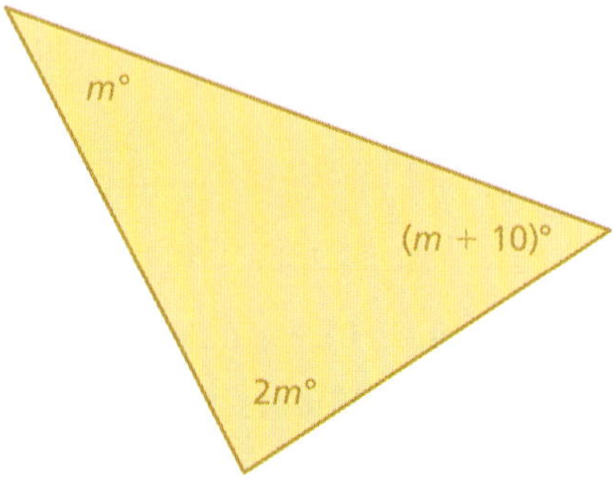

e.

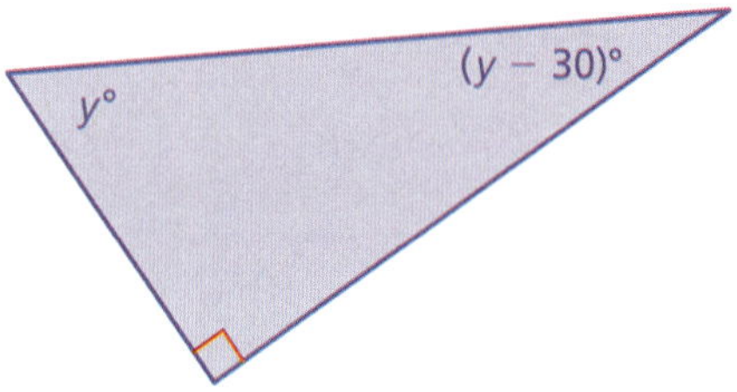

f.

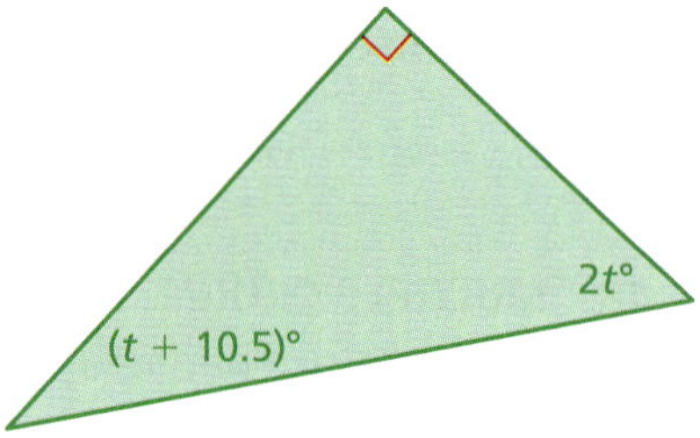

Laurie's Notes

Introduction

For the Teacher

- **Goal:** Students will develop an understanding of solving multi-step equations using the sum of the angle measures of a triangle.
- Have an informal discussion of the vocabulary associated with the triangles shown.

Motivate

- Make a card for each student in your class. Write a variable term on each card. Students will walk around to find others with a card containing a *like term* to the one they are holding.

  Samples: $5x$, $-13x$, $5y$, $6xy$, x, $3.8x$, $\frac{1}{2}y$, $-3.8y$
- Ask students to explain what it means for terms to be *like* terms.

Activity Notes

Activity 1

? Ask a few questions to prepare students for the activity.
  - "In the previous lesson, what did you conclude about the sum of the angle measures of a triangle?" sum = 180°
  - "So if two angles measure 65° and 75°, what does the third angle measure?" 40°
  - "If the angles of a triangle measure $x°$, $2x°$, and $3x°$, could you determine the measure of each angle?" Students should say yes.
- Model how to write and solve the equation $x + 2x + 3x = 180$. Be sure to mention like terms when solving. Ask about the coefficient of x.
- **Common Error:** After solving the equation, you still need to substitute the value into each angle expression to solve for each angle measure. Students sometimes forget this step.
- **FYI:** The triangles are drawn to scale, so the angle measures can be checked using a protractor.
- Ask for volunteers to show a few of the solutions at the board.

? "Why are there only two angles with variable expressions written in parts (e) and (f)?" The third angle in each is a right angle.

Previous Learning

Students should know how to use inverse operations to solve one-step equations.

Activity Materials
Introduction
• index cards

Start Thinking! and Warm Up

Activity 1.2 **Start Thinking!** For use before Activity 1.2

Activity 1.2 **Warm Up** For use before Activity 1.2

Simplify the expression.

1. $2n + 5 + 3n$	2. $x - 7 - 4x$
3. $4f + f + 6f$	4. $(9 - m) + 4m + 7$
5. $17 + 2t - 9 + 2t$	6. $(y + 7) + (2y - 5)$

1.2 Record and Practice Journal

Essential Question How can you solve a multi-step equation? How can you check the reasonableness of your solution?

1 ACTIVITY: Solving for the Angles of a Triangle

Work with a partner. Write an equation for each triangle. Solve the equation to find the value of the variable. Then find the angle measures of each triangle. Use a protractor to check the reasonableness of your answer.

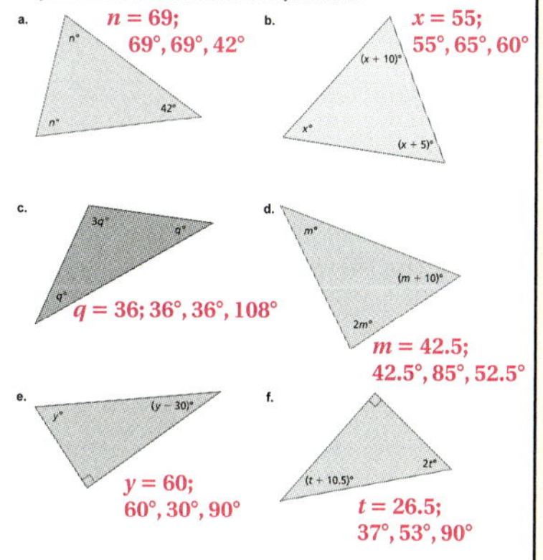

a. $n = 69$; 69°, 69°, 42°

b. $x = 55$; 55°, 65°, 60°

c. $q = 36$; 36°, 36°, 108°

d. $m = 42.5$; 42.5°, 85°, 52.5°

e. $y = 60$; 60°, 30°, 90°

f. $t = 26.5$; 37°, 53°, 90°

Differentiated Instruction

Auditory

Remind students that in order to solve an equation, the variable must be isolated on one side of the equation. The operations on the same side as the variable are those that need to be undone.

1.2 Record and Practice Journal

2 ACTIVITY: Problem Solving Strategy

Work with a partner.

The six triangles form a rectangle.

Find the angle measures of each triangle. Use a protractor to check the reasonableness of your answers.

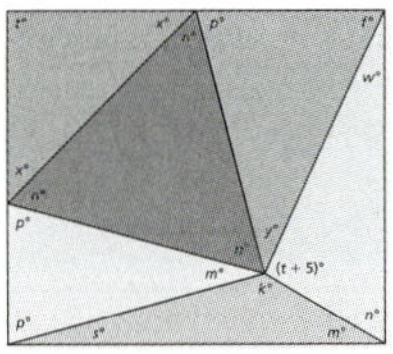

$f = 65$ $s = 15$

$k = 135$ $t = 90$

$m = 30$ $w = 25$

$n = 60$ $x = 45$

$p = 75$ $y = 40$

See Additional Answers.

3 ACTIVITY: Puzzle

Work with a partner. A survey asked 200 people to name their favorite weekday. The results are shown in the circle graph.

Favorite Weekday

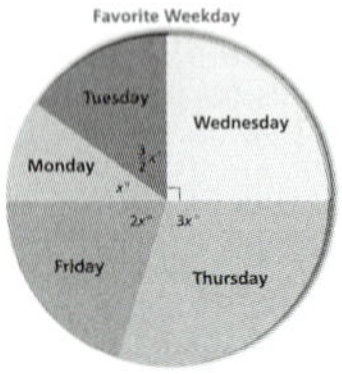

a. How many degrees are in each part of the circle graph?

b. What percent of the people chose each day?

c. How many people chose each day?

d. Organize your results in a table.

a–d.

	Mon.	Tues.	Wed.	Thurs.	Fri.
Degrees	36°	54°	90°	108°	72°
Percent	10%	15%	25%	30%	20%
People	20	30	50	60	40

What Is Your Answer?

4. **IN YOUR OWN WORDS** How can you solve a multi-step equation? How can you check the reasonableness of your solution?

Use inverse operations.

Check by substituting solution back into original equation.

Laurie's Notes

Activity 2

- Students will have different strategies for solving this puzzle.
- ? "Define a straight angle." An angle that measures 180°.
- Remind students to look for a variety of ways to check their answers.
- Discuss results and strategies for finding the angle measures. Listen for: right angles at the vertices of the rectangle; sum of angle measures forming a straight angle equals 180°; sum of angle measures about a point equals 360°; sum of the angle measures of a triangle equals 180°.
- Most students will not write a formal equation, but the thinking involved is an equation. For example, $k + m + s = 180$. If you know k and m, you can use mental math to solve for s.

Activity 3

- This example reviews fraction addition, mixed numbers, fraction division, and percents.
- ? Ask a few questions to help students begin the activity.
  - "How many people were surveyed?" 200
  - "What is the sum of the five central angle measures?" 360°
  - "What is the angle measure of the sector labeled Wednesday?" 90°
- Some students may use all five angles and set the expression equal to 360, while other students may only consider the four angles represented by a variable expression and set it equal to 270.
- ? "How do you find the percent each angle measure represents?"

  Convert $\frac{\text{angle measure}}{360}$ to a percent.

Words of Wisdom

- There are many steps in Activity 3, but it is possible to solve. This problem takes time and students will feel a sense of accomplishment when they finish.

Closure

- Find the angle measures in the right triangle. $x = 60$, $y = 30$, $z = 120$

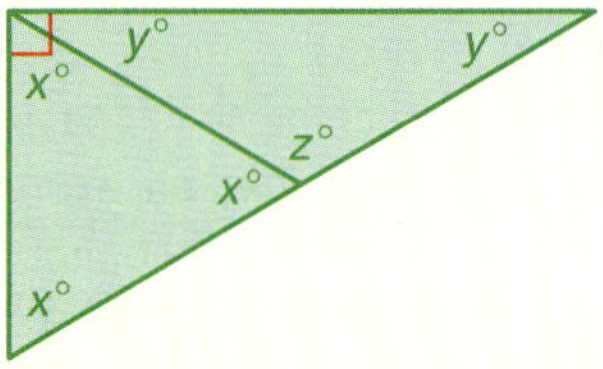

Technology For the Teacher

Dynamic Classroom

The Dynamic Planning Tool

Editable Teacher's Resources at *BigIdeasMath.com*

2 ACTIVITY: Problem-Solving Strategy

Work with a partner.

The six triangles form a rectangle.

Find the angle measures of each triangle. Use a protractor to check the reasonableness of your answers.

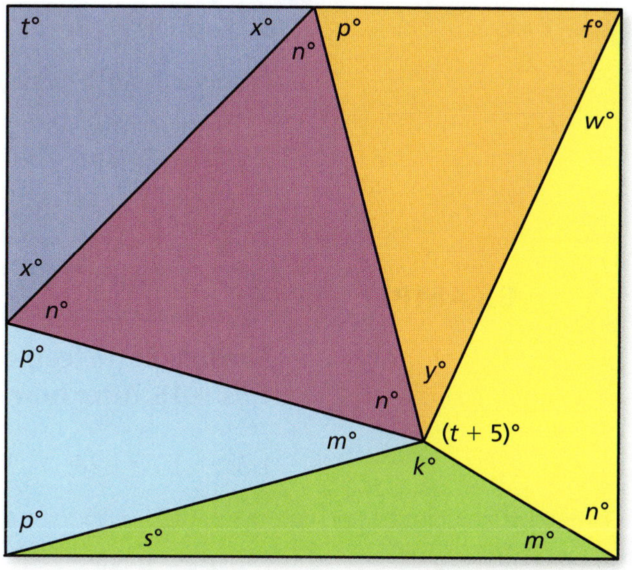

3 ACTIVITY: Puzzle

Work with a partner. A survey asked 200 people to name their favorite weekday. The results are shown in the circle graph.

a. How many degrees are in each part of the circle graph?

b. What percent of the people chose each day?

c. How many people chose each day?

d. Organize your results in a table.

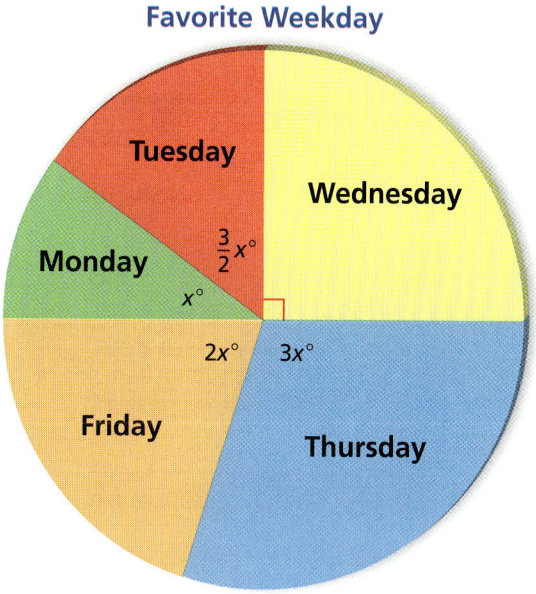

What Is Your Answer?

4. **IN YOUR OWN WORDS** How can you solve a multi-step equation? How can you check the reasonableness of your solution?

Practice Use what you learned about solving multi-step equations to complete Exercises 3–5 on page 14.

1.2 Lesson

Check It Out
Lesson Tutorials
BigIdeasMath.com

Key Idea

Solving Multi-Step Equations

To solve multi-step equations, use inverse operations to isolate the variable.

EXAMPLE 1 Solving a Two-Step Equation

The height (in feet) of a tree after x years is $1.5x + 15$. After how many years is the tree 24 feet tall?

	$1.5x + 15 = 24$	Write an equation.
Undo the addition. →	$\underline{-15} \quad \underline{-15}$	Subtract 15 from each side.
	$1.5x = 9$	Simplify.
Undo the multiplication. →	$\frac{1.5x}{1.5} = \frac{9}{1.5}$	Divide each side by 1.5.
	$x = 6$	Simplify.

∴ The tree is 24 feet tall after 6 years.

EXAMPLE 2 Combining Like Terms to Solve an Equation

Solve $8x - 6x - 25 = -35$.

	$8x - 6x - 25 = -35$	Write the equation.
	$2x - 25 = -35$	Combine like terms.
Undo the subtraction. →	$\underline{+25} \quad \underline{+25}$	Add 25 to each side.
	$2x = -10$	Simplify.
Undo the multiplication. →	$\frac{2x}{2} = \frac{-10}{2}$	Divide each side by 2.
	$x = -5$	Simplify.

∴ The solution is $x = -5$.

On Your Own

Now You're Ready
Exercises 6–9

Solve the equation. Check your solution.

1. $-3z + 1 = 7$ **2.** $\frac{1}{2}x - 9 = -25$ **3.** $-4n - 8n + 17 = 23$

Laurie's Notes

Introduction

Connect

- **Yesterday:** Students developed an intuitive understanding about solving multi-step equations.
- **Today:** Students will solve multi-step equations by using inverse operations to isolate the variable.

Motivate

- Share information with students about the three man-made, palm-shaped islands built in Dubai, the self-proclaimed "Eighth Wonder of the World!"
  - Each of the islands is being built in the shape of a palm tree consisting of a trunk and a crown with fronds. Each of the palm-shaped islands is surrounded by a crescent island that acts as a breakwater.
  - There will be over 100 luxury hotels, exclusive residential beachside villas and apartments, marinas, water theme parks, restaurants, shopping malls, sports facilities, and health spas on the islands.

Lesson Notes

Key Idea

- **Connection:** When you evaluate an expression, you follow the order of operations. Solving an equation undoes the evaluating, in reverse order. The goal is to isolate the variable term and then solve for the variable.

Example 1

- One way to explain the equation is to think of the tree as being 15 feet tall when being planted. It then grows 1.5 feet each year.
- **Extension:** Make a table to show the height of the tree from the first year to the sixth year.

Example 2

? "Why is $8x - 6x = 2x$?" Use the Distributive Property to subtract the terms; $8x - 6x = (8 - 6)x = 2x$.

On Your Own

- In Question 2, students may divide both sides by $\frac{1}{2}$ and get $x = -8$. Remind students that dividing by $\frac{1}{2}$ is the same as multiplying by 2.

Goal Today's lesson is solving multi-step equations.

Start Thinking! and Warm Up

Lesson 1.2 Warm Up
For use before Lesson 1.2

Lesson 1.2 Start Thinking!
For use before Lesson 1.2

A multi-step equation requires two or more operations to solve the equation. Explain why the following situation can be modeled by a multi-step equation.

A plumber charges \$80 per hour for labor plus \$60 for parts.

Come up with your own scenario that can be modeled by a multi-step equation.

Extra Example 1

The height (in inches) of a plant after t days is $\frac{1}{2}t + 6$. After how many days is the plant 21 inches tall? 30 days

Extra Example 2

Solve $-2m + 4m + 5 = -3$. $m = -4$

On Your Own

1. $z = -2$
2. $x = -32$
3. $n = -0.5$

Extra Example 3

Solve $-4(3g - 5) + 10g = 19$. 0.5

Extra Example 4

You have scored 7, 10, 8, and 9 on four quizzes. Write and solve an equation to find the score you need on the fifth quiz so that your mean score is 8.

$\frac{x + 7 + 10 + 8 + 9}{5} = 8$; 6

On Your Own

4. $x = -1.5$ 5. $d = -1$

6. $\frac{88 + 92 + 87 + x}{4} = 90$; $x = 93$

English Language Learners

Vocabulary

English learners will benefit from understanding that a *term* is a number, a variable, or the product of a number and variable. *Like terms* are terms that have identical variable parts.

3 and 16 are like terms because they contain no variable.

$4x$ and $7x$ are like terms because they have the same variable x.

$5a$ and $5b$ are *not* like terms because they have different variables.

Laurie's Notes

Example 3

- Ask students to identify the operations involved in this equation. from left to right: multiplication (by 2), subtraction, multiplication ($5x$), addition
- **Note:** Combining like terms in the third step is not obvious to students. When the like terms are not adjacent, students are unsure of how to combine them. Rewrite the left side of the equation as $2 + (-10)x + 4$.

Words of Wisdom

- Take time to work through the Study Tip and discuss the steps. Instead of using the Distributive Property, both sides of the equation are divided by 2 in the third step. This will not be obvious to students, nor will they know why it is okay to do this.
- Explain to students that the left side of the equation is 2 times an expression. When the expression $2(1 - 5x)$ is divided by 2, it leaves the expression $1 - 5x$. In the next step, students want to add 1 to each side because of the subtraction operation shown. Again, it is helpful to write $1 - 5x$ as $1 + (-5)x$ so that it makes sense to students why 1 is subtracted from each side.

Example 4

- You may need to review *mean* with the students.
- Discuss the information displayed in the table and write the equation.

? "Is it equivalent to write $\frac{x + 3.5}{5} = 1.5$ instead of $\frac{3.5 + x}{5} = 1.5$? Explain." yes; Commutative Property of Addition

- **FYI:** It may be helpful to write the third step with parentheses: $5\left(\frac{3.5 + x}{5}\right)$.
- **Note:** This is a classic question. When all of the data are known except for one, what is needed in order to achieve a particular average? Students often ask this in the context of wanting to know what they have to score on a test in order to achieve a certain average.

On Your Own

- Encourage students to work with a partner. Students need to be careful with multi-step equations and it is helpful to have a partner check each step.

Closure

- **Exit Ticket:** Solve $8x + 9 - 4x = 25$. Check your solution. $x = 4$

Technology For the Teacher

Dynamic Classroom

The Dynamic Planning Tool

Editable Teacher's Resources at *BigIdeasMath.com*

EXAMPLE 3 Using the Distributive Property to Solve an Equation

Solve $2(1 - 5x) + 4 = -8$.

$2(1 - 5x) + 4 = -8$ Write the equation.

$2(1) - 2(5x) + 4 = -8$ Use Distributive Property.

$2 - 10x + 4 = -8$ Multiply.

$-10x + 6 = -8$ Combine like terms.

$\underline{-6} \quad \underline{-6}$ Subtract 6 from each side.

$-10x = -14$ Simplify.

$\frac{-10x}{-10} = \frac{-14}{-10}$ Divide each side by –10.

$x = 1.4$ Simplify.

> **Study Tip**
>
> Here is another way to solve the equation in Example 3.
>
> $2(1 - 5x) + 4 = -8$
>
> $2(1 - 5x) = -12$
>
> $1 - 5x = -6$
>
> $-5x = -7$
>
> $x = 1.4$

EXAMPLE 4 Real-Life Application

Use the table to find the number of miles x you need to run on Friday so that the mean number of miles run per day is 1.5.

Day	Miles
Monday	2
Tuesday	0
Wednesday	1.5
Thursday	0
Friday	x

Write an equation using the definition of mean.

sum of the data → number of values →

$\frac{2 + 0 + 1.5 + 0 + x}{5} = 1.5$ Write the equation.

$\frac{3.5 + x}{5} = 1.5$ Combine like terms.

Undo the division. → $5 \cdot \frac{3.5 + x}{5} = 5 \cdot 1.5$ Multiply each side by 5.

$3.5 + x = 7.5$ Simplify.

Undo the addition. → $\underline{-3.5} \quad \underline{-3.5}$ Subtract 3.5 from each side.

$x = 4$ Simplify.

∴ You need to run 4 miles on Friday.

On Your Own

Now You're Ready
Exercises 10 and 11

Solve the equation. Check your solution.

4. $-3(x + 2) + 5x = -9$

5. $5 + 1.5(2d - 1) = 0.5$

6. You scored 88, 92, and 87 on three tests. Write and solve an equation to find the score you need on the fourth test so that your mean test score is 90.

1.2 Exercises

Check It Out
Help with Homework
BigIdeasMath.com

Vocabulary and Concept Check

1. **WRITING** Write the verbal statement as an equation. Then solve.

   2 more than 3 times a number is 17.

2. **OPEN-ENDED** Explain how to solve the equation $2(4x - 11) + 9 = 19$.

Practice and Problem Solving

Find the value of the variable. Then find the angle measures of the polygon. Use a protractor to check the reasonableness of your answer.

3.

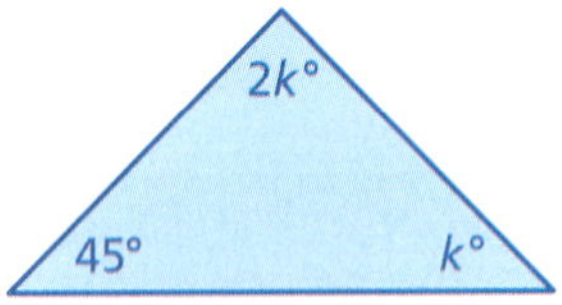

Sum of angle measures: 180°

4.

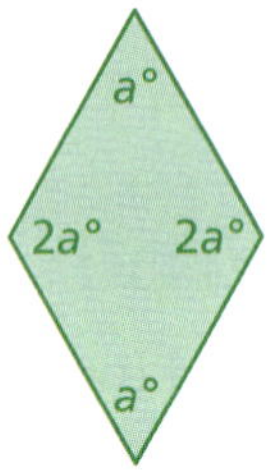

Sum of angle measures: 360°

5.

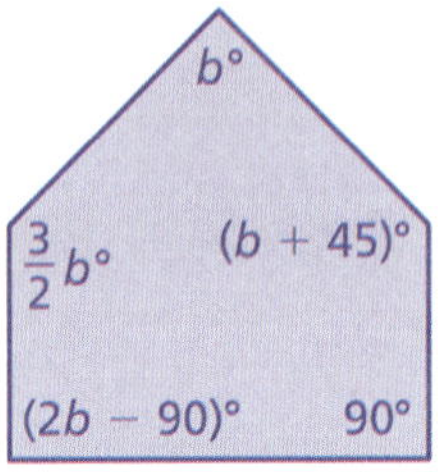

Sum of angle measures: 540°

Solve the equation. Check your solution.

① ② 6. $10x + 2 = 32$

7. $19 - 4c = 17$

8. $1.1x + 1.2x - 5.4 = -10$

9. $\frac{2}{3}h - \frac{1}{3}h + 11 = 8$

③ 10. $6(5 - 8v) + 12 = -54$

11. $21(2 - x) + 12x = 44$

12. **ERROR ANALYSIS** Describe and correct the error in solving the equation.

    ✗
    $$-2(7 - y) + 4 = -4$$
    $$-14 - 2y + 4 = -4$$
    $$-10 - 2y = -4$$
    $$-2y = 6$$
    $$y = -3$$

13. **WATCHES** The cost (in dollars) of making n watches is represented by $C = 15n + 85$. How many watches are made when the cost is \$385?

14. **HOUSE** The height of the house is 26 feet. What is the height x of each story?

Assignment Guide and Homework Check

Level	Day 1 Activity Assignment	Day 2 Lesson Assignment	Homework Check
Basic	3–5, 19–22	1, 2, 6–14	6, 8, 10, 12, 14
Average	3–5, 19–22	1, 2, 7, 9, 11, 12–17	7, 9, 11, 14, 16
Advanced	3–5, 19–22	1, 2, 6–14 even, 15–18	10, 12, 14, 16

Common Errors

- **Exercises 8 and 9** When combining like terms, students may square the variable. Remind them that $x^2 = x \cdot x$, and in these exercises they are not multiplying the variables. Remind them that when adding and subtracting variables, they perform the addition or subtraction on the coefficient of the variable.
- **Exercises 10 and 11** When using the Distributive Property, students may forget to distribute to all the values within the parentheses. Remind them that they need to distribute to all the values and encourage them to draw arrows showing the distribution, if needed.
- **Exercise 16** Students may struggle with writing the equation for this problem because of the tip that is added to the total. Encourage them to write an expression for the cost of the food and then add on the tip.

1.2 Record and Practice Journal

Solve the equation. Check your solution.

1. $3x - 11 = 22$
   $x = 11$
2. $24 - 10b = 9$
   $b = 1.5$
3. $2.4z + 1.2z - 6.5 = 0.7$
   $z = 2$
4. $\frac{3}{4}w - \frac{1}{2}w - 4 = 12$
   $w = 64$
5. $2(a + 7) - 7 = 9$
   $a = 1$
6. $20 + 8(q - 11) = -12$
   $q = 7$
7. Find the width of the rectangular prism when the surface area is 208 square centimeters.
   $w = 4$ cm

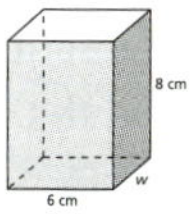

8. The amount of money in your savings account after m months is represented by $A = 135m + 225$. After how many months do you have $765 in your savings account?
   $m = 4$ months

Technology For the Teacher

Answer Presentation Tool
Quiz*Show*

Vocabulary and Concept Check

1. $2 + 3x = 17; x = 5$
2. *Sample answer:* Subtract 9 from each side. Divide each side by 2. Add 11 to each side. Divide each side by 4.

Practice and Problem Solving

3. $k = 45$; 45°, 45°, 90°
4. $a = 60$; 60°, 120°, 60°, 120°
5. $b = 90$; 90°, 135°, 90°, 90°, 135°
6. $x = 3$
7. $c = 0.5$
8. $x = -2$
9. $h = -9$
10. $v = 2$
11. $x = -\frac{2}{9}$
12. They did not distribute the -2 properly.

$$\begin{aligned} -2(7 - y) + 4 &= -4 \\ -14 + 2y + 4 &= -4 \\ 2y - 10 &= -4 \\ 2y &= 6 \\ y &= 3 \end{aligned}$$

13. 20 watches
14. 10 ft

Practice and Problem Solving

15. $4(b + 3) = 24$; 3 in.

16. $1.15(2p + 1.5) = 11.5$; \$4.25

17. $\frac{2580 + 2920 + x}{3} = 3000$; 3500 people

18. See *Taking Math Deeper.*

Fair Game Review

19. $<$ **20.** $=$

21. $>$ **22.** D

Mini-Assessment

Solve the equation.

1. $18 = 5a - 2a + 3$ $a = 5$
2. $2(4 - 2w) - 8 = -4$ $w = 1$
3. $2.3y + 4.4y - 3.7 = 16.4$ $y = 3$
4. $\frac{3}{4}z + \frac{1}{4}z - 6 = -5$ $z = 1$
5. The perimeter of the picture is 36 inches. What is the height of the picture? 10 in.

Taking Math Deeper

Exercise 18

This problem points out that mathematics and algebra are used in many different fields. This is a nice example using Olympic scoring.

1 Begin by translating the scoring system into a mathematical formula.

Score = 0.6(degree of difficulty)(sum of countries' scores) — minus the highest and lowest

2 Substitute the given information.

Let x = the degree of difficulty.

$$77.7 = 0.6(x)(7.5 + 8.0 + 7.0 + 7.5 + 7.0)$$
$$77.7 = 0.6x(37)$$
$$77.7 = 22.2x$$
$$3.5 = x$$

a. The degree of difficulty is 3.5.

3 This question has many answers.

Let x = sum of the five countries' scores.

$$97.2 = 0.6(4)(x)$$
$$97.2 = 2.4x$$
$$40.5 = x$$

One possibility is the following:

b. 8.0 + 8.0 + 8.0 + 8.0 + 8.5 with a low score of 7.5 and a high score of 9.0

Project

Use the Internet or school library to find all the different dives that are scored in a diving competition. Find the degree of difficulty that goes with each dive.

Reteaching and Enrichment Strategies

If students need help. . .	If students got it. . .
Resources by Chapter • Practice A and Practice B • Puzzle Time Record and Practice Journal Practice Differentiating the Lesson Lesson Tutorials Skills Review Handbook	Resources by Chapter • Enrichment and Extension Start the next section

In Exercises 15–17, write and solve an equation to answer the question.

15. POSTCARD The area of the postcard is 24 square inches. What is the width b of the message (in inches)?

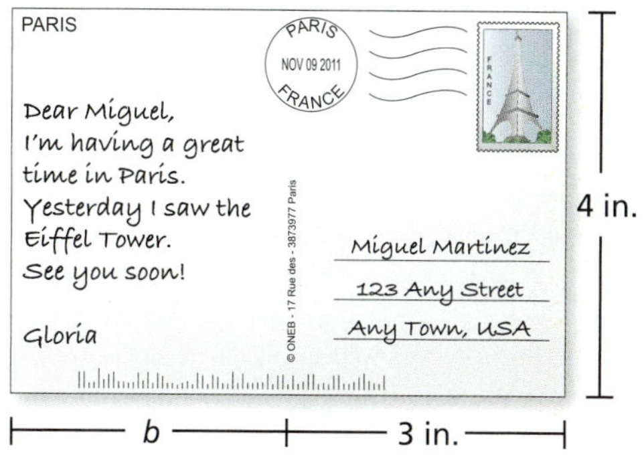

16. BREAKFAST You order two servings of pancakes and a fruit cup. The cost of the fruit cup is \$1.50. You leave a 15% tip. Your total bill is \$11.50. How much does one serving of pancakes cost?

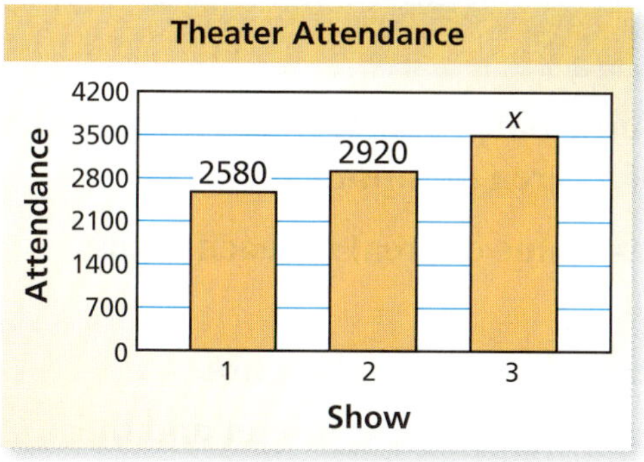

17. THEATER How many people must attend the third show so that the average attendance for the three shows is 3000?

18. DIVING Olympic divers are scored by an international panel of judges. The highest and lowest scores are dropped. The total of the remaining scores is multiplied by the degree of difficulty of the dive. This product is multiplied by 0.6 to determine the final score.

a. A diver's final score is 77.7. What is the degree of difficulty of the dive?

Judge	Russia	China	Mexico	Germany	Italy	Japan	Brazil
Score	7.5	8.0	6.5	8.5	7.0	7.5	7.0

b. Critical Thinking The degree of difficulty of a dive is 4.0. The diver's final score is 97.2. Judges award half or whole points from 0 to 10. What scores could the judges have given the diver?

© Paul Slaughter, www.slaughterphoto.com
Greg Louganis diving at the 1984 Olympics

Fair Game Review What you learned in previous grades & lessons

Let $a = 3$ and $b = -2$. Copy and complete the statement using <, >, or =.
(Skills Review Handbook)

19. $-5a$ ▢ 4 **20.** 5 ▢ $b + 7$ **21.** $a - 4$ ▢ $10b + 8$

22. MULTIPLE CHOICE What value of x makes the equation $x + 5 = 2x$ true?
(Skills Review Handbook)

Ⓐ -1 Ⓑ 0 Ⓒ 3 Ⓓ 5

1.3 Solving Equations with Variables on Both Sides

Essential Question How can you solve an equation that has variables on both sides?

1 ACTIVITY: Perimeter and Area

Work with a partner. Each figure has the unusual property that the value of its perimeter (in feet) is equal to the value of its area (in square feet).

- **Write an equation (value of perimeter = value of area) for each figure.**
- **Solve each equation for x.**
- **Use the value of x to find the perimeter and area of each figure.**
- **Check your solution by comparing the value of the perimeter and the value of the area of each figure.**

a.

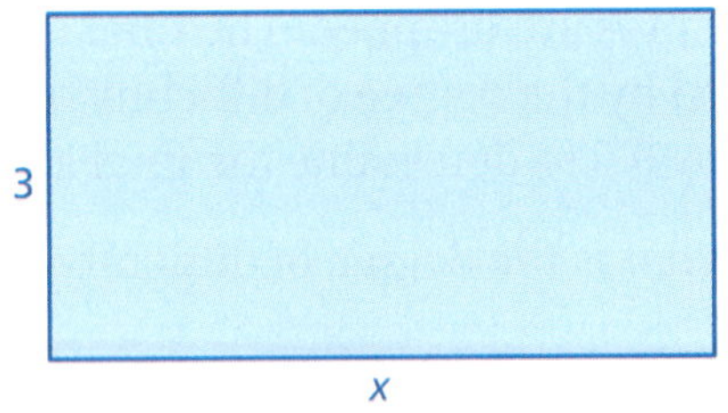

b.

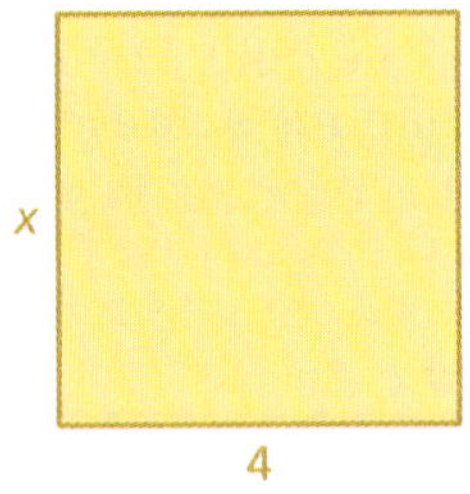

c.

d.

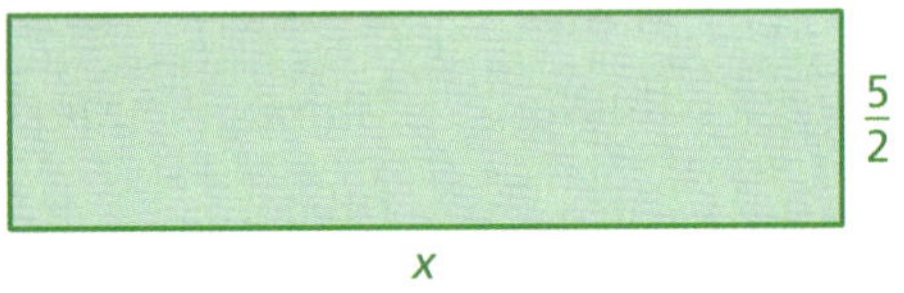

e.

f.

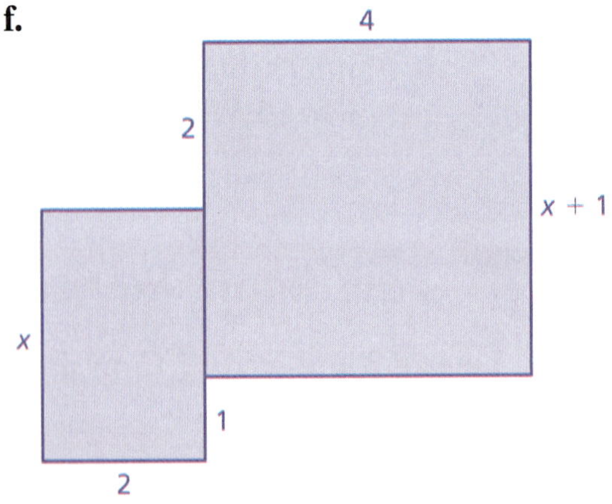

g.

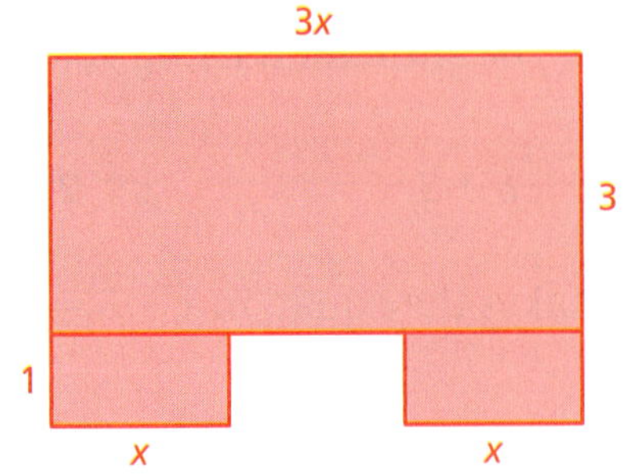

Laurie's Notes

Introduction

For the Teacher

- **Goal:** Students will explore equations with variables on both sides by examining figures with unique properties.

Motivate

? "What balances with the cylinder? Explain."

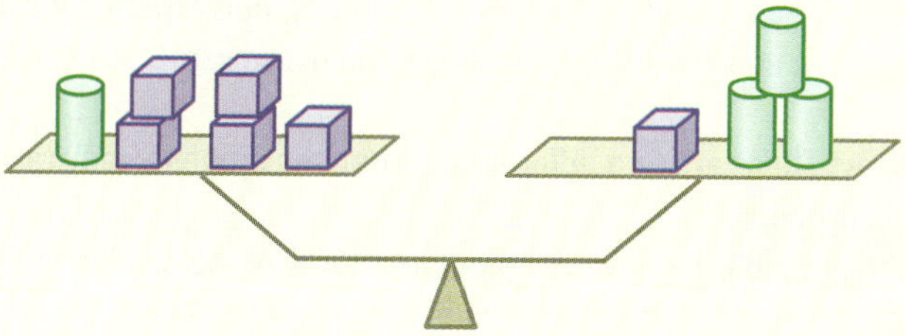

2 cubes; Remove one cube and one cylinder from each side. 2 cylinders balance with 4 cubes, so 1 cylinder would balance with 2 cubes.

- The balance problem is equivalent to $x + 5 = 3x + 1$, where x is a cylinder and the whole numbers represent cubes. This is an example of an equation with variables on both sides, the type students will solve today. Return to this equation at the end of class.
- If students are familiar with algebra tiles, you can model the problem using the tiles. The cylinder is replaced with an x-tile and the cubes are replaced with unit tiles.

Activity Notes

Activity 1

- Discuss with students the general concept of what it means to measure the attributes of a two-dimensional figure. In other words, what is the difference between a rectangle's perimeter and a rectangle's area? What type of units are used to measure each? linear units for perimeter and square units for area
- **FYI:** Be sure to make it clear that the directions are saying that perimeter and area are not the same, but their values are equal. For example, a square that measures 4 centimeters on each edge has a perimeter of 16 centimeters and an area of 16 square centimeters. The value (16) is the same, but the units of measure are not.

? Before students begin, ask a few review questions.

- "How do you find the perimeter and the area of a rectangle?" $P = 2\ell + 2w$ and $A = \ell w$
- "How do you find the perimeter and the area of a composite figure?" Listen for students' understanding that perimeter is the sum of all of the sides. The area is found in parts and then added together.

- Have a few groups share their work at the board, particularly for part (d), fractions, and part (g), algebraic expressions.

Previous Learning

Students should know common formulas for perimeter, area, surface area, and volume.

Start Thinking! and Warm Up

Activity 1.3 Start Thinking!
For use before Activity 1.3

Activity 1.3 Warm Up
For use before Activity 1.3

Find the width of the rectangle.

1. Area = 32 m^2 (length 8 m, width x)
2. Area = 150 cm^2 (length 15 cm, width x)
3. Perimeter = 35 in. (7.5 in., x)
4. Perimeter = 40 ft (10 ft, x)

1.3 Record and Practice Journal

Essential Question How can you solve an equation that has variables on both sides?

1 ACTIVITY: Perimeter and Area

Work with a partner. Each figure has the unusual property that the value of its perimeter (in feet) is equal to the value of its area (in square feet).

- Write an equation (value of perimeter = value of area) for each figure.
- Solve each equation for x.
- Use the value of x to find the perimeter and area of each figure.
- Check your solution by comparing the value of the perimeter and the value of the area of each figure.

a. $2x + 6 = 3x$; $x = 6$; $P = 18$ ft; $A = 18$ ft^2

b. $2x + 8 = 4x$; $x = 4$; $P = 16$ ft; $A = 16$ ft^2

c. $2x + 36 = 18x$; $x = 2\frac{1}{4}$; $P = 40.5$ ft; $A = 40.5$ ft^2

d. $2x + 5 = \frac{5}{2}x$; $x = 10$; $P = 25$ ft; $A = 25$ ft^2

Differentiated Instruction

Auditory

Point out to students that skills used to solve equations in this lesson are the same skills they have used before. The goal is to isolate the variable on one side of the equation. Just as they used the Addition Property of Equality to remove a constant term from one side of the equation, they will use the same property to remove the variable term from one side of the equation.

1.3 Record and Practice Journal

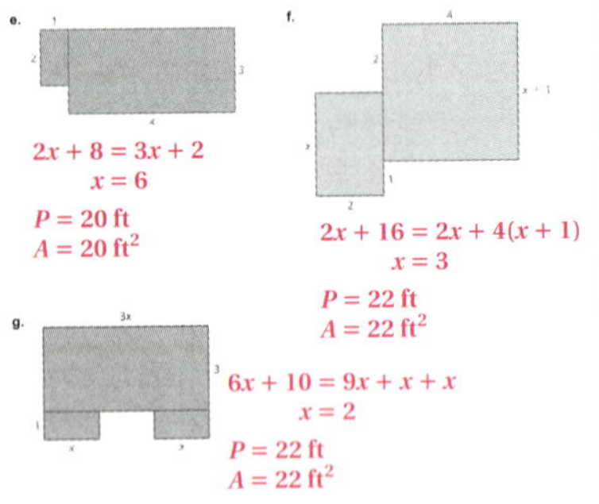

e. $2x + 8 = 3x + 2$
$x = 6$
$P = 20$ ft
$A = 20$ ft^2

f. $2x + 16 = 2x + 4(x + 1)$
$x = 3$
$P = 22$ ft
$A = 22$ ft^2

g. $6x + 10 = 9x + x + x$
$x = 2$
$P = 22$ ft
$A = 22$ ft^2

2 ACTIVITY: Surface Area and Volume

Work with a partner. Each solid on the next page has the unusual property that the value of its surface area (in square inches) is equal to the value of its volume (in cubic inches).

- Write an equation (value of surface area = value of volume) for each figure.
- Solve each equation for x.
- Use the value of x to find the surface area and volume of each figure.
- Check your solution by comparing the value of the surface area and the value of the volume of each figure.

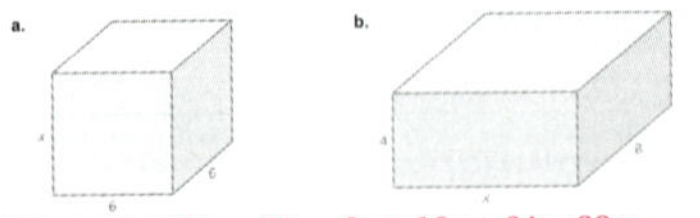

a. $12x + 72 + 12x = 36x$
$x = 6$
$SA = 216$ in.^{2}
$V = 216$ in.^{3}

b. $8x + 16x + 64 = 32x$
$x = 8$
$SA = 256$ in.^{2}
$V = 256$ in.^{3}

3 ACTIVITY: Puzzle

Work with a partner. The two triangles are similar. The perimeter of the larger triangle is 150% of the perimeter of the smaller triangle. Find the dimensions of each triangle.

$x = 6$

$x = 6$
$2x = 12$

What Is Your Answer?

4. IN YOUR OWN WORDS How can you solve an equation that has variables on both sides? Write an equation that has variables on both sides. Solve the equation.

Collect the variable terms on one side and constant terms on the other side.

Laurie's Notes

Activity 2

- This activity is similar to Activity 1.

? "How do you find the surface area and volume of a rectangular prism?" $S = 2\ell w + 2\ell h + 2wh$ and $V = \ell wh$

- Students may guess that part (a) is a cube, suggesting $x = 6$. Ask students to verify their guess.

Activity 3

? "What are similar triangles?" Listen for an informal definition: same shape but not necessarily the same size; formally, corresponding sides are proportional and corresponding angles have the same measure (congruent).

- **Another Way:** There are two different approaches students may take in solving this problem.
  - The first method is to solve the equation: 150% of the smaller triangle's perimeter is equal to the perimeter of the larger triangle.

    $$150\% \text{ of } (18 + x) = 24 + 2x$$
    $$1.5(18 + x) = 24 + 2x$$

    This method reviews decimal multiplication.
  - A second method is to use the definition of similar triangles to set up a proportion to solve for the missing sides.

    $$\frac{10}{15} = \frac{x}{9}$$

    Solve for x. Use this value to find the side labeled $2x$ in the larger triangle.

What Is Your Answer?

- **Neighbor Check:** Have students work independently and then have their neighbor check their work. Have students discuss any discrepancies.

Closure

- Describe how to solve $x + 5 = 3x + 1$. *Sample answer:* Subtract x from both sides, subtract 1 from both sides, and then divide both sides by 2.

Technology For the Teacher

Dynamic Classroom

The Dynamic Planning Tool
Editable Teacher's Resources at *BigIdeasMath.com*

2 ACTIVITY: Surface Area and Volume

Work with a partner. Each solid has the unusual property that the value of its surface area (in square inches) is equal to the value of its volume (in cubic inches).

- **Write an equation (value of surface area = value of volume) for each figure.**
- **Solve each equation for x.**
- **Use the value of x to find the surface area and volume of each figure.**
- **Check your solution by comparing the value of the surface area and the value of the volume of each figure.**

a.

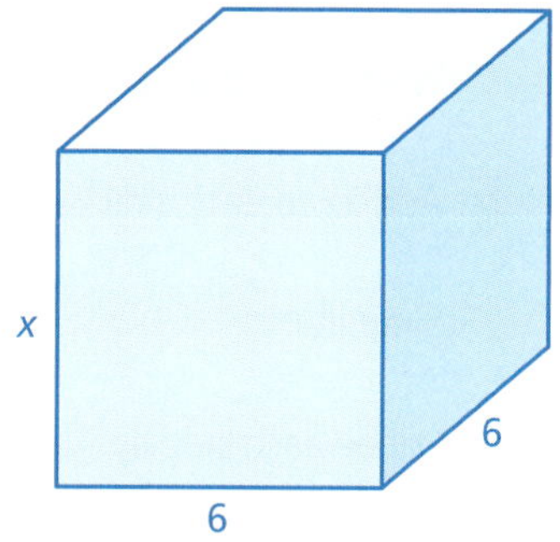

b.

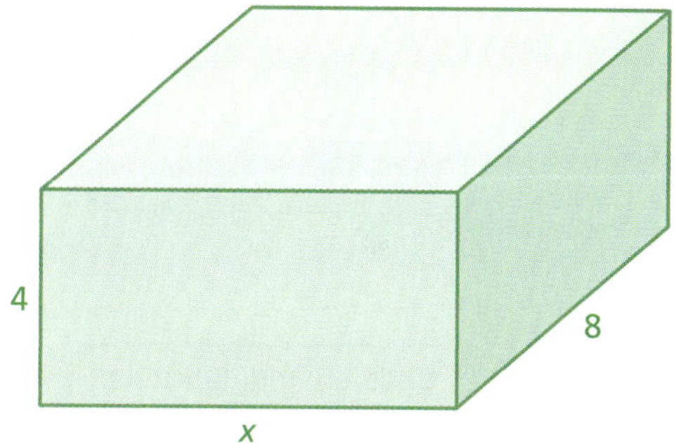

3 ACTIVITY: Puzzle

Work with a partner. The two triangles are similar. The perimeter of the larger triangle is 150% of the perimeter of the smaller triangle. Find the dimensions of each triangle.

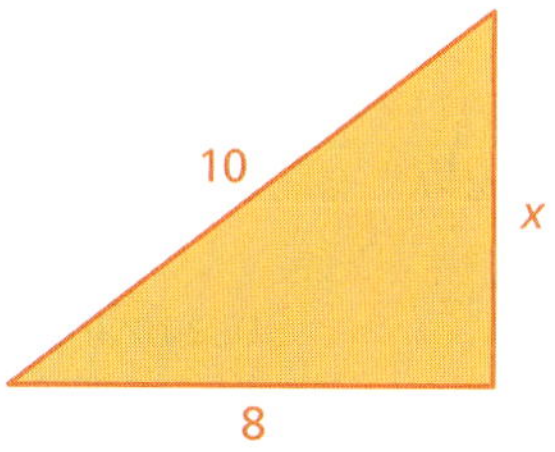

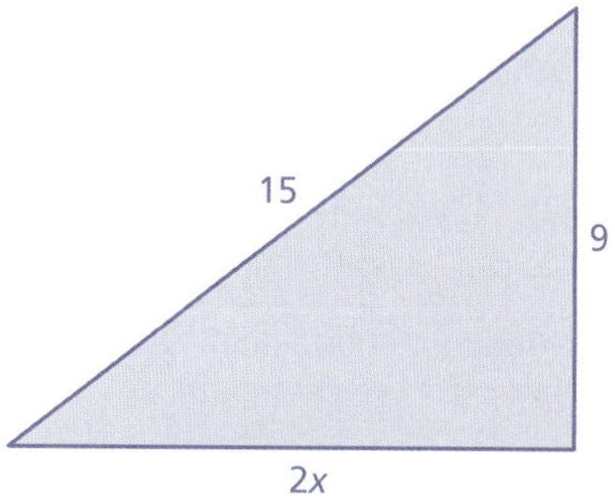

What Is Your Answer?

4. **IN YOUR OWN WORDS** How can you solve an equation that has variables on both sides? Write an equation that has variables on both sides. Solve the equation.

Practice Use what you learned about solving equations with variables on both sides to complete Exercises 3–5 on page 20.

1.3 Lesson

Check It Out Lesson Tutorials BigIdeasMath.com

Key Idea

Solving Equations with Variables on Both Sides

To solve equations with variables on both sides, collect the variable terms on one side and the constant terms on the other side.

EXAMPLE 1 Solving an Equation with Variables on Both Sides

Solve $15 - 2x = -7x$. Check your solution.

	Equation	Step
	$15 - 2x = -7x$	Write the equation.
Undo the subtraction. →	$+2x \quad +2x$	Add $2x$ to each side.
	$15 = -5x$	Simplify.
Undo the multiplication. →	$\frac{15}{-5} = \frac{-5x}{-5}$	Divide each side by -5.
	$-3 = x$	Simplify.

Check

$15 - 2x = -7x$

$15 - 2(-3) \stackrel{?}{=} -7(-3)$

$21 = 21$ ✓

The solution is $x = -3$.

EXAMPLE 2 Using the Distributive Property to Solve an Equation

Solve $-2(x - 5) = 6\left(2 - \frac{1}{2}x\right)$.

	Equation	Step
	$-2(x - 5) = 6\left(2 - \frac{1}{2}x\right)$	Write the equation.
	$-2x + 10 = 12 - 3x$	Use Distributive Property.
Undo the subtraction. →	$+3x \qquad +3x$	Add $3x$ to each side.
	$x + 10 = 12$	Simplify.
Undo the addition. →	$-10 \quad -10$	Subtract 10 from each side.
	$x = 2$	Simplify.

The solution is $x = 2$.

On Your Own

Now You're Ready
Exercises 6–14

Solve the equation. Check your solution.

1. $-3x = 2x + 19$
2. $2.5y + 6 = 4.5y - 1$
3. $6(4 - z) = 2z$

Laurie's Notes

Introduction

Connect

- **Yesterday:** Students developed an intuitive understanding of solving equations with variables on both sides.
- **Today:** Students will solve equations with variables on both sides by using Properties of Equality and collecting variable terms on one side.

Motivate

? "How many of you have driven across the Mississippi River?"

- Share some information about the Mississippi River. It is the third longest river in North America, flowing 2350 miles from Lake Itasca in Minnesota to the Gulf of Mexico. The width of the river ranges from 20–30 feet at its narrowest to more than 11 miles at its widest.

Lesson Notes

Key Idea

- Discuss with students the vocabulary *variable term* and *constant term.*
- Write: $5x - 2 - 9x + y + 4$

? "In this expression, what are the variable terms? Constant terms?" $5x$, $-9x$, and y; -2 and 4

- **Common Error:** Students forget to include the sign of the variable term.

Example 1

- Notice that a constant is on the left side only. For this reason, it makes sense to solve for the variable term on the right side of the equation.

Example 2

- **Teaching Tip:** Before distributing on the left side, rewrite the inside expression as $x + (-5)$ (add the opposite). Students are likely to recognize that they are multiplying $(-2)(-5)$ to get a product of 10.

? "How do you multiply a whole number and a fraction like $6\left(\frac{1}{2}\right)$?" Listen for language such as, write the whole number over a denominator of 1 and multiply straight across.

- **FYI:** Solve for the variable on the side of the equation where the coefficient is the greatest. After collecting the variable terms on one side, the coefficient of the variable term will be positive, a condition that generally renders fewer mistakes.

On Your Own

- **Neighbor Check:** Have students work independently and then have their neighbor check their work. Have students discuss any discrepancies.

Goal Today's lesson is solving equations with variables on both sides.

Start Thinking! and Warm Up

Lesson 1.3 Warm Up
For use before Lesson 1.3

Lesson 1.3 Start Thinking!
For use before Lesson 1.3

Try solving the equation $2x + 20 = 12x$ by first subtracting 20 from each side. What property allows you to do so? Does it help you get closer to finding a solution?

What is a better first step? Explain.

Extra Example 1

Solve $r = -5r + 18$. 3

Extra Example 2

Solve $6\left(1 + \frac{1}{2}x\right) = 2(x + 1)$. -4

On Your Own

1. $x = -3.8$
2. $y = 3.5$
3. $z = 3$

Extra Example 3

The legs of the right triangle have the same length. What is the area of the triangle? $1\frac{1}{8}$ square units

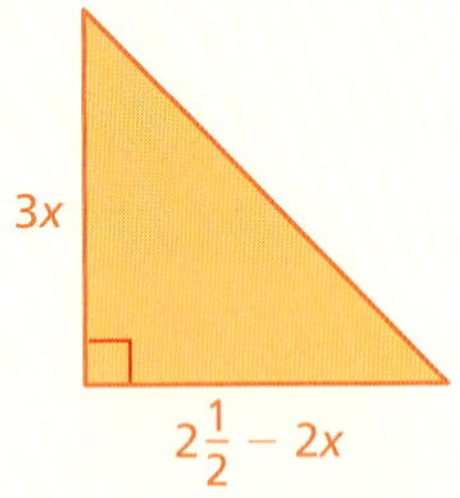

Extra Example 4

A boat travels 3 hours downstream at r miles per hour. On the return trip, the boat travels 5 miles per hour slower and takes 4 hours. What is the distance the boat travels each way? 60 mi

On Your Own

4. 36π

5. 50 mi

English Language Learners

Vocabulary

Remind English learners that *like terms* are terms with the same variables raised to the same power. As the number of terms in an equation increases, an important skill is to identify and combine like terms.

Laurie's Notes

Example 3

- ? "Define radius and diameter of a circle." Diameter is the distance across a circle through its center. Radius is half the diameter.
- Solve the equation as shown.
- This example reviews the formula for the area of a circle. You might also ask about the formula for the circumference of a circle. $C = 2\pi r$

Example 4

- ? "How do you find the distance traveled when you know the rate and time?" multiply; $d = rt$
- Discuss with students that the distance both ways is the same, a simple but not obvious fact.
- ? "If you travel 40 miles per hour for 2 hours, how far will you go? How about 40 miles per hour for a half hour?" 80 mi; 20 mi
- ? "How far do you travel at x miles per hour for 3 hours?" $3x$ mi
- Students need to read the time from the illustration. The rates for each direction are x and $(x + 2)$.
- Write and solve the equation as shown.

On Your Own

- **Think-Pair-Share:** Students should read each question independently and then work with a partner to answer the questions. When they have answered the questions, the pair should compare their answers with another group and discuss any discrepancies.

Closure

- **Exit Ticket:** Solve $6 - 2x = 4x - 9$. Check your solution. $x = 2.5$

Technology For the Teacher

Dynamic Classroom

The Dynamic Planning Tool
Editable Teacher's Resources at *BigIdeasMath.com*

EXAMPLE 3 Standardized Test Practice

The circles are identical. What is the area of each circle?

Ⓐ 2 Ⓑ 4 Ⓒ 16π Ⓓ 64π

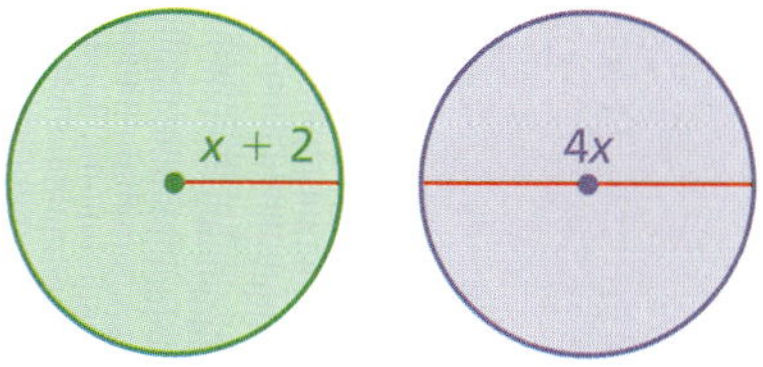

The circles are identical, so the radius of each circle is the same.

$x + 2 = 2x$	Write an equation. The radius of the purple circle is $2x$.
$-x \quad -x$	Subtract x from each side.
$2 = x$	Simplify.

∴ The area of each circle is $\pi r^2 = \pi(4)^2 = 16\pi$. So, the correct answer is Ⓒ.

EXAMPLE 4 Real-Life Application

A boat travels x miles per hour upstream on the Mississippi River. On the return trip, the boat travels 2 miles per hour faster. How far does the boat travel upstream?

The speed of the boat on the return trip is $(x + 2)$ miles per hour.

Distance upstream = Distance of return trip

$3x = 2.5(x + 2)$	Write an equation.
$3x = 2.5x + 5$	Use Distributive Property.
$-2.5x \quad -2.5x$	Subtract $2.5x$ from each side.
$0.5x = 5$	Simplify.
$\frac{0.5x}{0.5} = \frac{5}{0.5}$	Divide each side by 0.5.
$x = 10$	Simplify.

∴ The boat travels 10 miles per hour for 3 hours upstream. So, it travels 30 miles upstream.

On Your Own

4. **WHAT IF?** In Example 3, the diameter of the purple circle is $3x$. What is the area of each circle?

5. A boat travels x miles per hour from one island to another island in 2.5 hours. The boat travels 5 miles per hour faster on the return trip of 2 hours. What is the distance between the islands?

1.3 Exercises

Check It Out
Help with Homework
BigIdeasMath.com

Vocabulary and Concept Check

1. **WRITING** Is $x = 3$ a solution of the equation $3x - 5 = 4x - 9$? Explain.
2. **OPEN-ENDED** Write an equation that has variables on both sides and has a solution of -3.

Practice and Problem Solving

The value of the figure's surface area is equal to the value of the figure's volume. Find the value of *x*.

3.

4.

5.

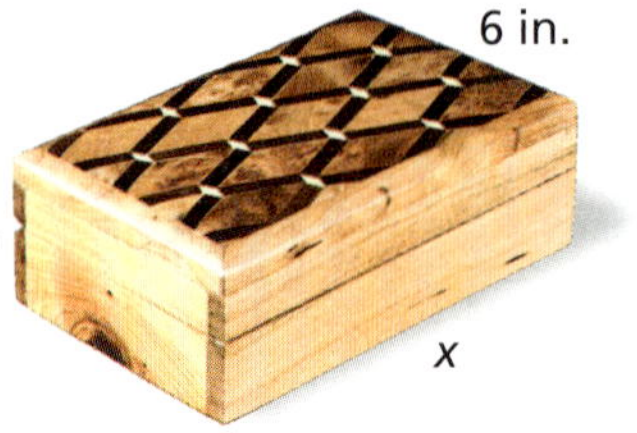

Solve the equation. Check your solution.

1 2

6. $m - 4 = 2m$
7. $3k - 1 = 7k + 2$
8. $6.7x = 5.2x + 12.3$
9. $-24 - \frac{1}{8}p = \frac{3}{8}p$
10. $12(2w - 3) = 6w$
11. $2(n - 3) = 4n + 1$
12. $2(4z - 1) = 3(z + 2)$
13. $0.1x = 0.2(x + 2)$
14. $\frac{1}{6}d + \frac{2}{3} = \frac{1}{4}(d - 2)$
15. **ERROR ANALYSIS** Describe and correct the error in solving the equation.

✗
$$3x - 4 = 2x + 1$$
$$3x - 4 - 2x = 2x + 1 - 2x$$
$$x - 4 = 1$$
$$x - 4 + 4 = 1 - 4$$
$$x = -3$$

16. **TRAIL MIX** The equation $4.05p + 14.40 = 4.50(p + 3)$ represents the number p of pounds of peanuts you need to make trail mix. How many pounds of peanuts do you need for the trail mix?
17. **CARS** Write and solve an equation to find the number of miles you must drive to have the same cost for each of the car rentals.

$15 plus $0.50 per mile

$25 plus $0.25 per mile

Assignment Guide and Homework Check

Level	Day 1 Activity Assignment	Day 2 Lesson Assignment	Homework Check
Basic	3–5, 26–29	1, 2, 6–12, 15–19	6, 9, 10, 12, 16, 18
Average	3–5, 26–29	1, 2, 9–19, 21	9, 10, 12, 16, 18, 21
Advanced	3–5, 26–29	1, 2, 12–15, 18–25	12, 18, 22, 24

For Your Information

- **Exercise 16** The equation represents a mixture problem in which peanuts are added to other ingredients, making trail mix. The equation shows that p pounds of peanuts that cost \$4.05 per pound are added to other ingredients that cost a total of \$14.40. This mixture creates $(p + 3)$ pounds of trail mix that costs \$4.50 per pound.

Common Errors

- **Exercises 6–14** Students may perform the same operation instead of the opposite operation when trying to get the variable terms on the same side. Remind them that whenever a variable or number is moved from one side of the equal sign to the other, the opposite operation is used.
- **Exercises 6–14** Students may use the opposite operation when combining like terms on the same side of the equal sign. Remind them that the opposite operation is used only when moving the variable or number to the other side of the equation.
- **Exercises 16 and 17** Students may forget to write the units in their answers. Remind them that when units are given, the units need to be included in the answer.

1.3 Record and Practice Journal

Solve the equation. Check your solution.

1. $x + 16 = 9x$
   $x = 2$
2. $4y - 70 = 12y + 2$
   $y = -9$
3. $5(p + 6) = 8p$
   $p = 10$
4. $3(g - 7) = 2(10 + g)$
   $g = 41$
5. $1.8 + 7n = 9.5 - 4n$
   $n = 0.7$
6. $\frac{3}{7}w - 11 = -\frac{4}{7}w$
   $w = 11$
7. One movie club charges a \$100 membership fee and \$10 for each movie. Another club charges no membership fee but movies cost \$15 each. Write and solve an equation to find the number of movies you need to buy for the cost of each movie club to be the same.
   $100 + 10x = 15x$; $x = 20$
8. Thirty percent of all the students in a school are in a play. All students except for 140 are in the play. How many students are in the school?
   200 students

Technology For the Teacher

Answer Presentation Tool
Quiz*Show*

Vocabulary and Concept Check

1. no; When 3 is substituted for x, the left side simplifies to 4 and the right side simplifies to 3.
2. *Sample answer:* $4x + 1 = 3x - 2$

Practice and Problem Solving

3. $x = 13.2$ in.
4. $x = 10$ cm
5. $x = 7.5$ in.
6. $m = -4$
7. $k = -0.75$
8. $x = 8.2$
9. $p = -48$
10. $w = 2$
11. $n = -3.5$
12. $z = 1.6$
13. $x = -4$
14. $d = 14$
15. The 4 should have been added to the right side.

$$3x - 4 = 2x + 1$$
$$3x - 2x - 4 = 2x + 1 - 2x$$
$$x - 4 = 1$$
$$x - 4 + 4 = 1 + 4$$
$$x = 5$$

16. 2 lb
17. $15 + 0.5m = 25 + 0.25m$; 40 mi

Practice and Problem Solving

18. 3 units **19.** 7.5 units

20. 232 units

21. See *Taking Math Deeper.*

22. fractions; Because $\frac{1}{3}$ is hard to perform operations with when written as a decimal.

23. 10 mL **24.** 25 grams

25. square: 12 units
triangle: 10 units, 19 units, 19 units

Fair Game Review

26. 15.75 cm^3

27. 24 $in.^3$

28. about 153.86 ft^3

29. C

Mini-Assessment

Solve the equation.

1. $n - 4 = 3n + 6$ $n = -5$
2. $0.3(w + 10) = 1.8w$ $w = 2$
3. $3p = 4(-3p + 6)$ $p = 1.6$
4. $\frac{1}{3}v = -\frac{2}{3}\left(\frac{1}{2}v - 1\right)$ $v = 1$
5. The perimeter of the rectangle is equal to the perimeter of the square. What are the side lengths of each figure? rectangle: 4 units by 10 units; square: 7 units by 7 units

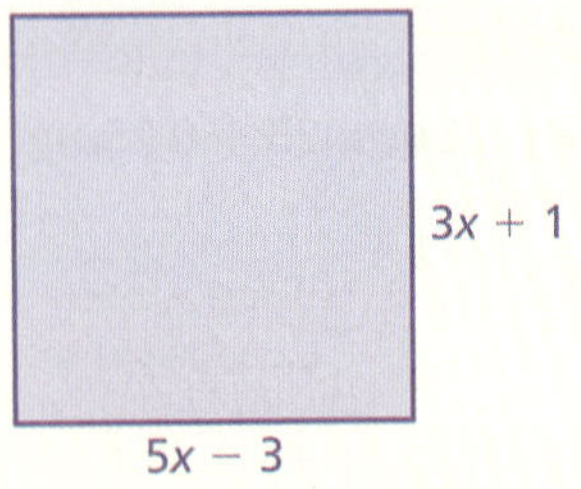

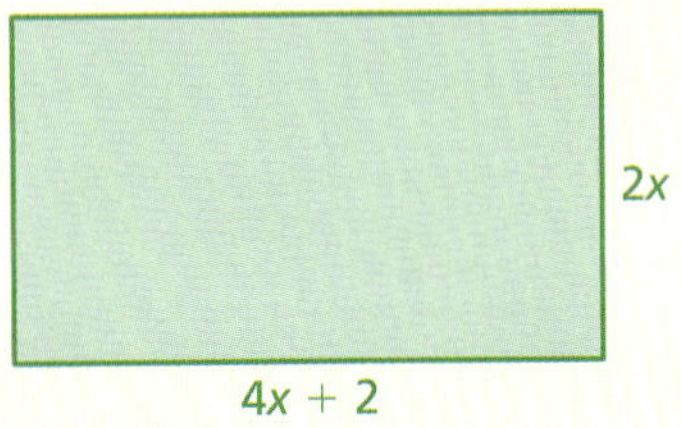

Taking Math Deeper

Exercise 21

This problem seems like it is easy, but it can actually be quite challenging.

1 Identify the key information in the table.

	Packing Material	Priority	Express
Box	\$2.25	\$2.50/lb	\$8.50/lb
Envelope	\$1.10	\$2.50/lb	\$8.50/lb

2 Write and solve an equation.

Let $x =$ the weight of the DVD and packing material.

Cost of Mailing Box: $2.25 + 2.5x$

Cost of Mailing Envelope: $1.10 + 8.5x$

$$2.25 + 2.5x = 1.10 + 8.5x$$
$$1.15 = 6x$$
$$0.19 \approx x$$

3 Answer the question.

The weight of the DVD and packing material is about 0.19 pound, or about 3 ounces.

Project

Postage for special types of mail, such as priority mail, is determined by the weight of the package and the distance it needs to travel. Find the cost of sending a 15-ounce package from your house to Los Angeles, Washington D.C., and Albuquerque.

Reteaching and Enrichment Strategies

If students need help. . .	If students got it. . .
Resources by Chapter • Practice A and Practice B • Puzzle Time Record and Practice Journal Practice Differentiating the Lesson Lesson Tutorials Skills Review Handbook	Resources by Chapter • Enrichment and Extension • School-to-Work Start the next section

A polygon is *regular* if each of its sides has the same length. Find the perimeter of the regular polygon.

18. 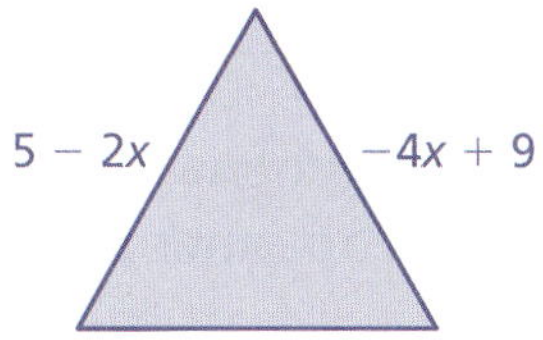

19. 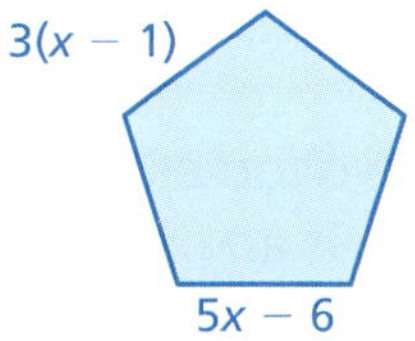

20. 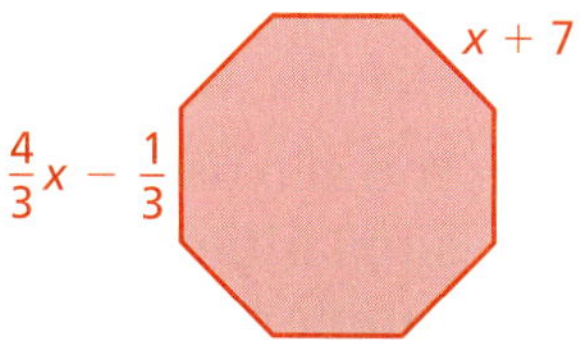

21. **POSTAGE** The cost of mailing a DVD in an envelope by express mail is equal to the cost of mailing a DVD in a box by priority mail. What is the weight of the DVD with its packing material? Round your answer to the nearest hundredth.

	Packing Material	Priority Mail	Express Mail
Box	$2.25	$2.50 per lb	$8.50 per lb
Envelope	$1.10	$2.50 per lb	$8.50 per lb

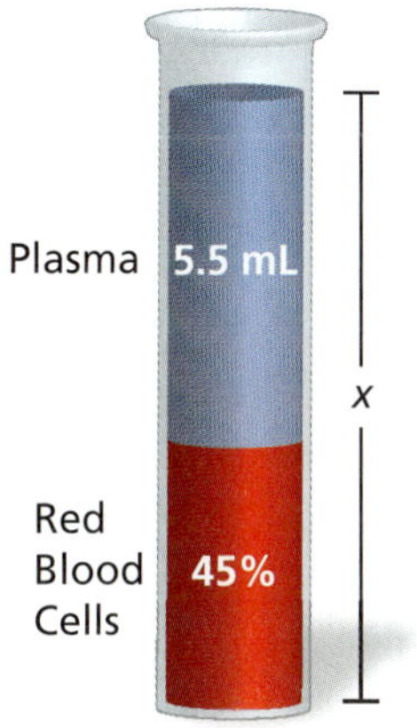

22. **REASONING** Would you solve the equation $0.25x + 7 = \frac{1}{3}x - 8$ using fractions or decimals? Explain.

23. **BLOOD SAMPLE** The amount of red blood cells in a blood sample is equal to the total amount in the sample minus the amount of plasma. What is the total amount x of blood drawn?

24. **NUTRITION** One serving of oatmeal provides 16% of the fiber you need daily. You must get the remaining 21 grams of fiber from other sources. How many grams of fiber should you consume daily?

25. **Geometry** The perimeter of the square is equal to the perimeter of the triangle. What are the side lengths of each figure?

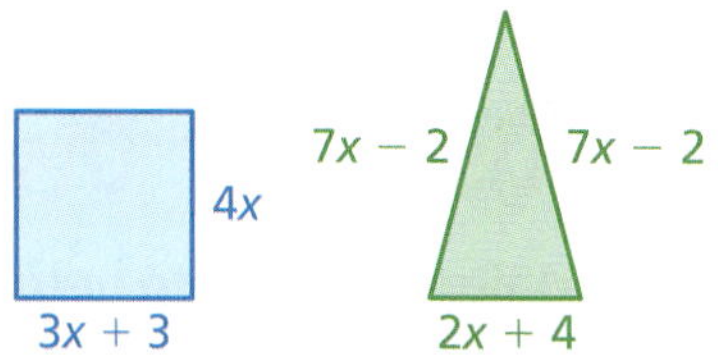

Fair Game Review *What you learned in previous grades & lessons*

Find the volume of the figure. Use 3.14 for π. *(Skills Review Handbook)*

26. 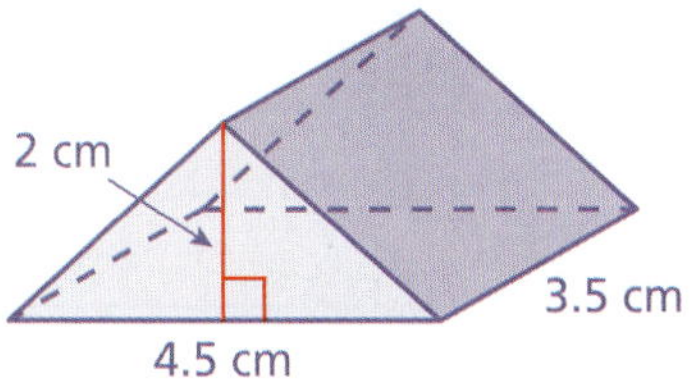

27. 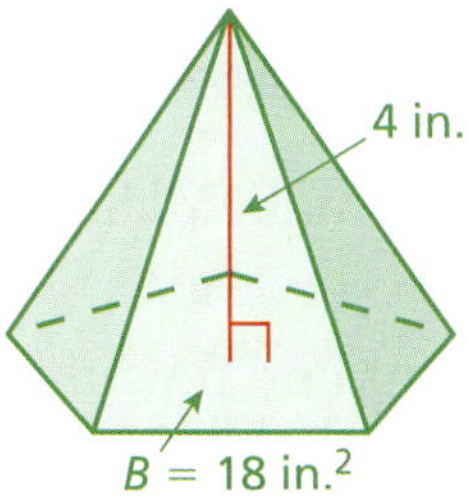

28. 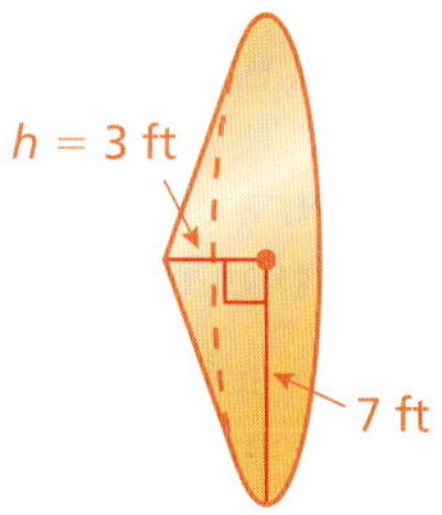

29. **MULTIPLE CHOICE** A car travels 480 miles on 15 gallons of gasoline. How many miles does the car travel per gallon? *(Section 1.1)*

Ⓐ 28 mi/gal Ⓑ 30 mi/gal Ⓒ 32 mi/gal Ⓓ 35 mi/gal

1 Study Help

Check It Out
Graphic Organizer
BigIdeasMath.com

You can use a **Y chart** to compare two topics. List differences in the branches and similarities in the base of the Y. Here is an example of a Y chart that compares solving simple equations to solving multi-step equations.

Solving Simple Equations

- You can solve the equation in one step.

Solving Multi-Step Equations

- You must use more than one step to solve the equation.
- Undo the operations in the reverse order of the order of operations.

Both

- As necessary, use the Addition, Subtraction, Multiplication, and Division Properties of Equality to solve for the variable.
- The variable can end up on either side of the equation.
- It is always a good idea to check your solution.

On Your Own

Make a Y chart to help you study and compare these topics.

1. solving equations with the variable on one side and solving equations with variables on both sides
2. solving multi-step equations and solving equations with variables on both sides

After you complete this chapter, make Y charts for the following topics.

3. solving multi-step equations and rewriting literal equations
4. converting meters to feet and converting feet to meters
5. converting one unit to another and converting one rate to another

"I made a Y chart to compare and contrast yours and Fluffy's characteristics."

Sample Answers

1.

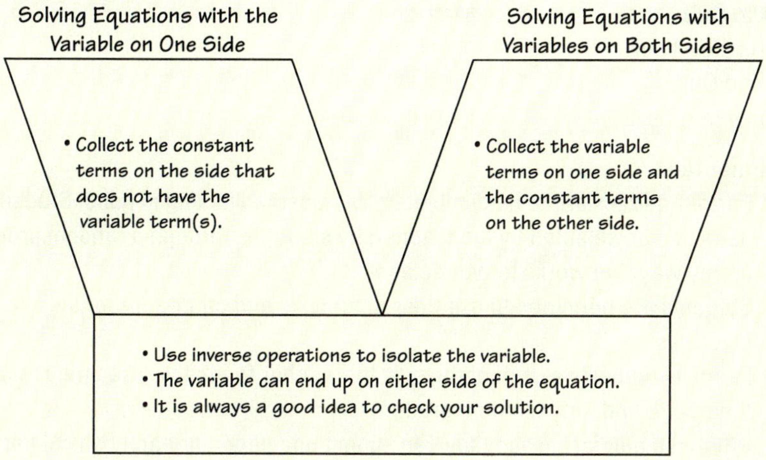

2.

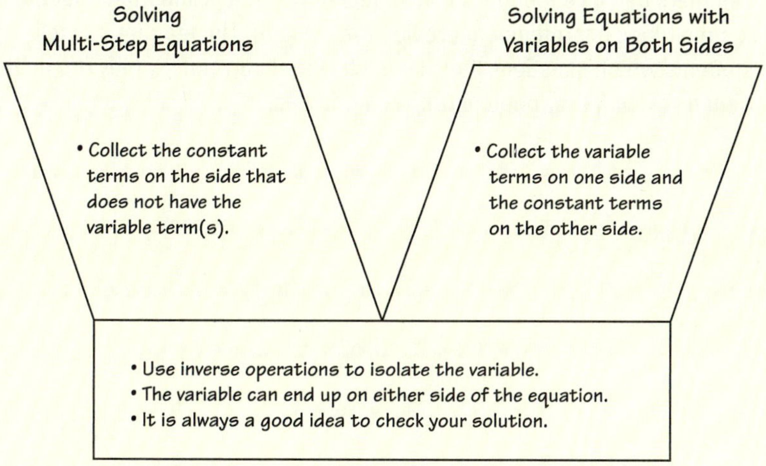

List of Organizers

Available at *BigIdeasMath.com*

Comparison Chart
Concept Circle
Definition (Idea) and Example Chart
Example and Non-Example Chart
Formula Triangle
Four Square
Information Frame
Information Wheel
Notetaking Organizer
Process Diagram
Summary Triangle
Word Magnet
Y Chart

About this Organizer

A **Y Chart** can be used to compare two topics. Students list differences between the two topics in the branches of the Y and similarities in the base of the Y. A Y chart serves as a good tool for assessing students' knowledge of a pair of topics that have subtle but important differences. You can include blank Y charts on tests or quizzes for this purpose.

Technology For the Teacher

Vocabulary Puzzle Builder

Answers

1. $y = \frac{1}{2}$
2. $w = 5\pi$
3. $m = 0.5$
4. $k = 4$
5. $z = 16$
6. $x = 60$; 55°, 60°, 65°
7. $x = 126$; 63°, 80°, 126°, 91°
8. $x = -1$
9. $s = 6$
10. $32
11. 50 ft, 150 ft, 75 ft, 180 ft
12. $230x = 1265$; 5.5 hours
13. passing beach: 13 miles
 passing park: 15 miles

Assessment Book

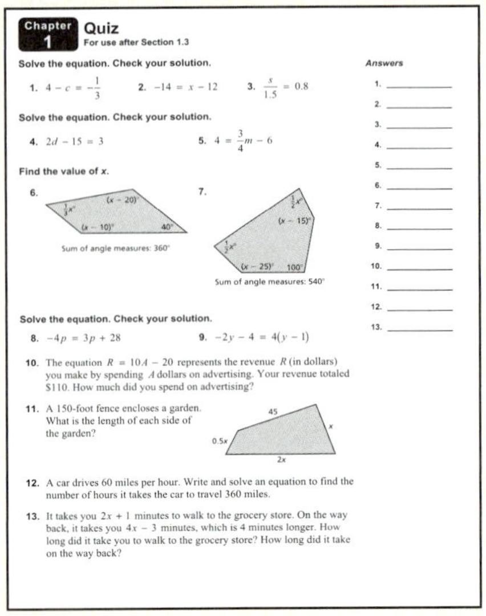

Chapter 1 Quiz
For use after Section 1.3

Solve the equation. Check your solution.

1. $4 - c = -\frac{1}{3}$ 2. $-14 = x - 12$ 3. $\frac{s}{1.5} = 0.8$

Solve the equation. Check your solution.

4. $2d - 15 = 3$ 5. $4 = \frac{3}{4}m - 6$

Find the value of x.

6. Sum of angle measures: 360°

7. Sum of angle measures: 540°

Solve the equation. Check your solution.

8. $-4p = 3p + 28$ 9. $-2y - 4 = 4(y - 1)$

10. The equation $R = 10A - 20$ represents the revenue R (in dollars) you make by spending A dollars on advertising. Your revenue totaled $110. How much did you spend on advertising?

11. A 150-foot fence encloses a garden. What is the length of each side of the garden?

12. A car drives 60 miles per hour. Write and solve an equation to find the number of hours it takes the car to travel 360 miles.

13. It takes you $2x + 1$ minutes to walk to the grocery store. On the way back, it takes you $4x - 3$ minutes, which is 4 minutes longer. How long did it take you to walk to the grocery store? How long did it take on the way back?

Answers

1. ______
2. ______
3. ______
4. ______
5. ______
6. ______
7. ______
8. ______
9. ______
10. ______
11. ______
12. ______
13. ______

Alternative Quiz Ideas

100% Quiz
Error Notebook
Group Quiz
Homework Quiz
Math Log
Notebook Quiz
Partner Quiz
Pass the Paper

Partner Quiz

- Partner quizzes are to be completed by students working in pairs. Student pairs can be selected by the teacher, by students, through a random process, or any way that works for your class.
- Students are permitted to use their notebooks and other appropriate materials.
- Each pair submits a draft of the quiz for teacher feedback. Then they revise their work and turn it in for a grade.
- When the pair is finished they can submit one paper, or each can submit their own.
- Teachers can give feedback in a variety of ways. It is important that the teacher does not reteach or provide the solution. The teacher can tell students which questions they have answered correctly, if they are on the right track, or if they need to rethink a problem.

Reteaching and Enrichment Strategies

If students need help. . .	If students got it. . .
Resources by Chapter • Study Help • Practice A and Practice B • Puzzle Time Lesson Tutorials *BigIdeasMath.com* Practice Quiz Practice from the Test Generator	Resources by Chapter • Enrichment and Extension • School-to-Work Game Closet at *BigIdeasMath.com* Start the next section

Technology For the Teacher

Answer Presentation Tool
Big Ideas Test Generator

1.1–1.3 Quiz

Check It Out
Progress Check
BigIdeasMath.com

Solve the equation. Check your solution. *(Section 1.1)*

1. $-\frac{1}{2} = y - 1$

2. $-3\pi + w = 2\pi$

3. $1.2m = 0.6$

Solve the equation. Check your solution. *(Section 1.2)*

4. $-4k + 17 = 1$

5. $\frac{1}{4}z + 8 = 12$

Find the value of *x*. Then find the angle measures of the polygon. *(Section 1.2)*

6.

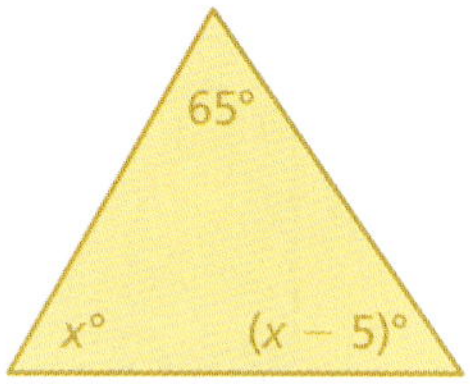

Sum of angle measures: 180°

7.

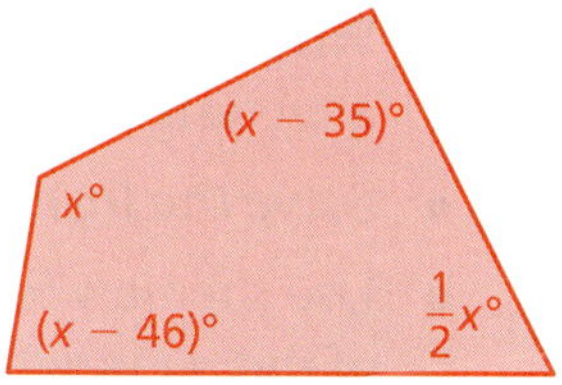

Sum of angle measures: 360°

Solve the equation. Check your solution. *(Section 1.3)*

8. $2(x + 4) = -5x + 1$

9. $\frac{1}{2}s = 4s - 21$

10. **JEWELER** The equation $P = 2.5m + 35$ represents the price P (in dollars) of a bracelet, where m is the cost of the materials (in dollars). The price of a bracelet is \$115. What is the cost of the materials? *(Section 1.2)*

11. **PASTURE** A 455-foot fence encloses a pasture. What is the length of each side of the pasture? *(Section 1.2)*

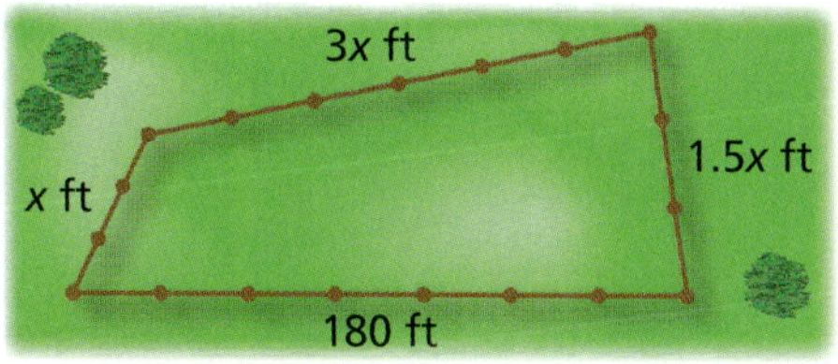

12. **POSTERS** A machine prints 230 movie posters each hour. Write and solve an equation to find the number of hours it takes the machine to print 1265 posters. *(Section 1.1)*

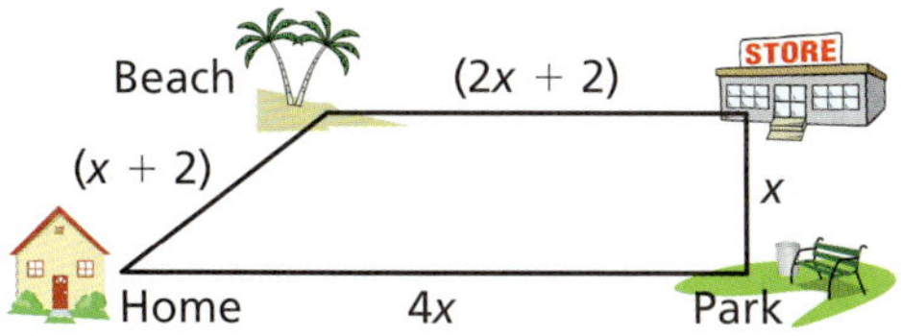

13. **ROUTES** From your home, the route to the store that passes the beach is 2 miles shorter than the route to the store that passes the park. What is the length of each route? *(Section 1.3)*

1.4 Rewriting Equations and Formulas

Essential Question How can you use a formula for one measurement to write a formula for a different measurement?

1 ACTIVITY: Using Perimeter and Area Formulas

Work with a partner.

a. • Write a formula for the perimeter P of a rectangle.
• Solve the formula for w.
• Use the new formula to find the width of the rectangle.

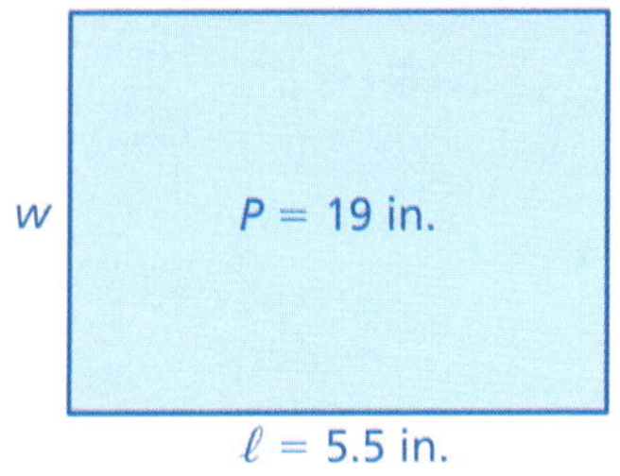

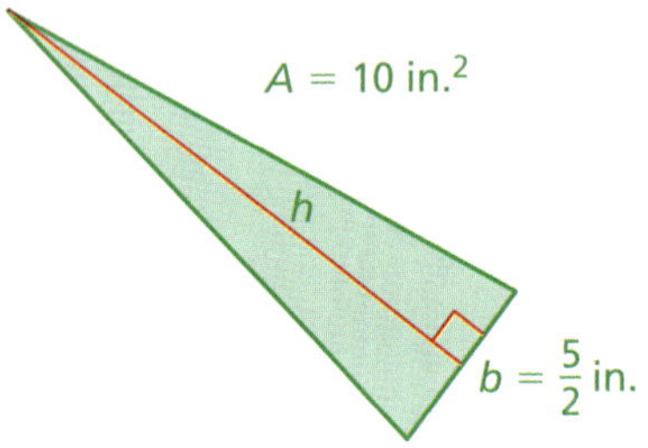

b. • Write a formula for the area A of a triangle.
• Solve the formula for h.
• Use the new formula to find the height of the triangle.

c. • Write a formula for the circumference C of a circle.
• Solve the formula for r.
• Use the new formula to find the radius of the circle.

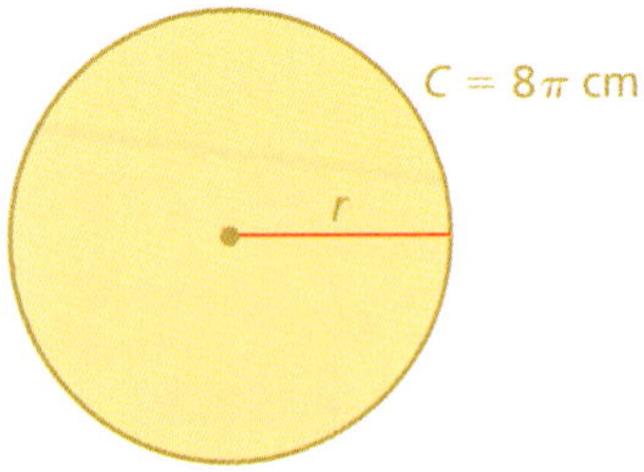

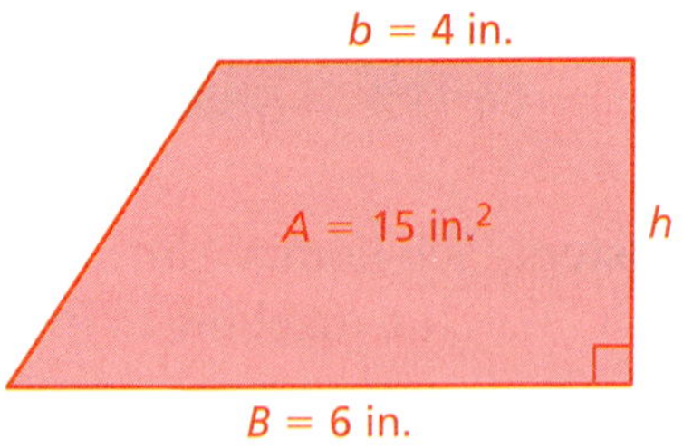

d. • Write a formula for the area A of a trapezoid.
• Solve the formula for h.
• Use the new formula to find the height of the trapezoid.

e. • Write a formula for the area A of a parallelogram.
• Solve the formula for h.
• Use the new formula to find the height of the parallelogram.

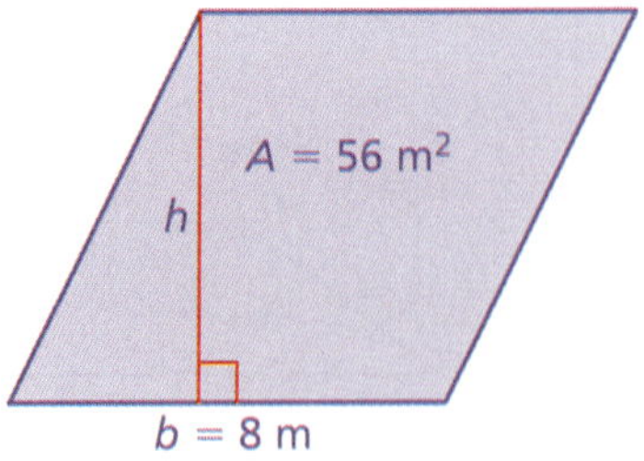

Laurie's Notes

Introduction

For the Teacher

- **Goal:** Students will use common formulas for perimeter, area, and volume to explore rewriting equations.
- **Teaching Tip:** Continually ask students what operations are being performed in each formula.

Motivate

- **Preparation:** Make a set of formula cards. My set is a collection of five cards for each shape: the labeled diagram, the two measurements, and the two formulas being found.

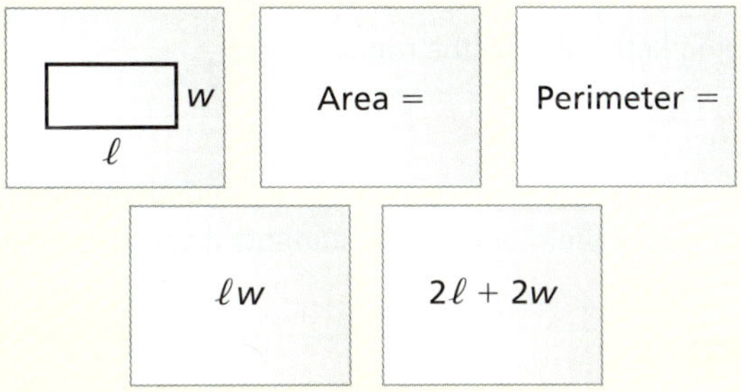

- Depending upon the number of students in your class, use some or all of the cards. Pass out the cards and have students form groups matching all 5 cards for the shape.
- When all of the matches have been made, ask each group to read their formulas aloud.

Activity Notes

Activity 1

- Solving literal equations can be one of the most challenging skills for students. Model a problem, such as solving $A = \ell w$ for width.
- Fractional coefficients can also be a challenge, so model an additional problem, such as solving $A = \frac{1}{2}xy$ for y. First, multiply both sides by 2, then divide both sides by x.
- **Teaching Tip:** You may find that students are substituting the known values of the variables and then solving the equation, instead of solving the equation and then substituting.
- **Connection:** The reason for solving for the variable is that the equation can be used for the width of any rectangle given the perimeter and length, not just the specific example shown. It is a general solution that can be reused.
- **Teaching Tip:** After 2 or more groups have correctly solved part (a), have a volunteer write the solution on the board for the other groups to see.
- For parts (b) and (d), suggest students start by multiplying both sides by the reciprocal of $\frac{1}{2}$. In part (c), 2π is a number and can be manipulated as such, so divide both sides by 2π.

Previous Learning

Students should know the common formulas for area, perimeter, and volume.

Activity Materials
Introduction
• formula cards (index cards)

Start Thinking! and Warm Up

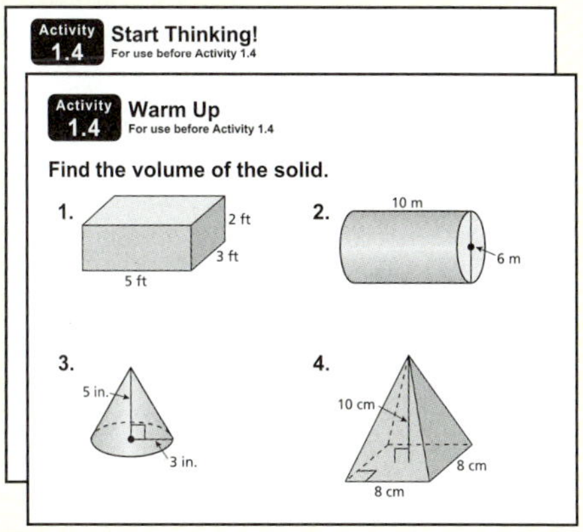

Activity 1.4 Start Thinking!
For use before Activity 1.4

Activity 1.4 Warm Up
For use before Activity 1.4

Find the volume of the solid.

1. 2 ft, 3 ft, 5 ft
2. 10 m, 6 m
3. 5 in., 3 in.
4. 10 cm, 8 cm, 8 cm

1.4 Record and Practice Journal

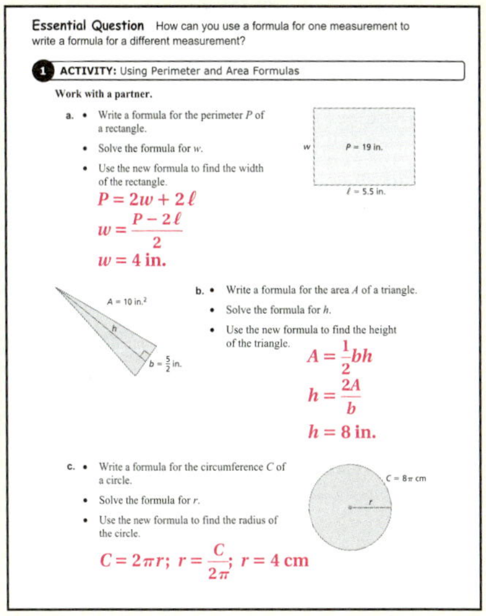

Essential Question How can you use a formula for one measurement to write a formula for a different measurement?

1 ACTIVITY: Using Perimeter and Area Formulas

Work with a partner.

a. • Write a formula for the perimeter P of a rectangle.
• Solve the formula for w.
• Use the new formula to find the width of the rectangle.

$P = 2w + 2\ell$
$w = \frac{P - 2\ell}{2}$
$w = 4$ in.

b. • Write a formula for the area A of a triangle.
• Solve the formula for h.
• Use the new formula to find the height of the triangle.

$A = \frac{1}{2}bh$
$h = \frac{2A}{b}$
$h = 8$ in.

c. • Write a formula for the circumference C of a circle.
• Solve the formula for r.
• Use the new formula to find the radius of the circle.

$C = 2\pi r$; $r = \frac{C}{2\pi}$; $r = 4$ cm

Differentiated Instruction

Kinesthetic

Have kinesthetic learners model the areas of the polygons on grid paper. Then compare their answers with the answers found using the area formulas.

1.4 Record and Practice Journal

- **Write a formula for the area *A*.**
- **Solve the formula for *h*.**
- **Use the new formula to find the height.**

d.

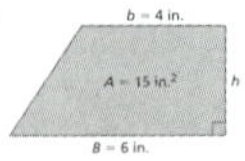

$A = \frac{1}{2}h(b + B)$

$h = \frac{2A}{b + B}$

$h = 3$ in.

e.

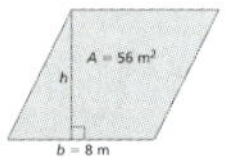

$A = bh$

$h = \frac{A}{b}$

$h = 7$ m

2 ACTIVITY: Using Volume Formulas

Work with a partner.

a. • Write a formula for the volume V of a prism.

- Solve the formula for h.
- Use the new formula to find the height of the prism.

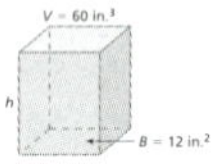

$V = Bh$

$h = \frac{V}{B}$

$h = 5$ in.

- **Write a formula for the volume *V*.**
- **Solve the formula for *B*.**
- **Use the new formula to find the area of the base.**

b.

$V = \frac{1}{3}Bh$

$B = \frac{3V}{h}$

$B = 16 \text{ ft}^2$

c.

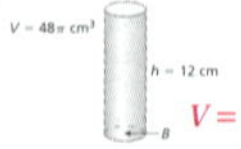

$V = Bh$

$B = \frac{V}{h}$

$B = 4\pi \text{ cm}^2$

d. • Write a formula for the volume V of a cone.

- Solve the formula for h.
- Use the new formula to find the height of the cone.

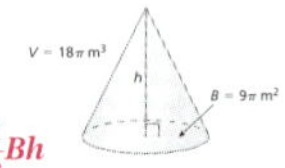

$V = \frac{1}{3}Bh$

$h = \frac{3V}{B}$

$h = 6$ m

What Is Your Answer?

3. **IN YOUR OWN WORDS** How can you use a formula for one measurement to write a formula for a different measurement? Give an example that is different from the examples on these three pages.

Solve a given formula for a different variable.

Laurie's Notes

Activity 2

- This activity is similar to Activity 1, where students worked with perimeter and area formulas. In Activity 2, students will work with volume formulas.
- Note that all of the diagrams use B for the area of the base instead of having students use specific area formulas. Using this approach, the volume formulas for parts (a) and (c) are the same ($V = Bh$) and the volume formulas for parts (b) and (d) are the same $\left(V = \frac{1}{3}Bh\right)$. This helps students to recall that structurally the prism and the cylinder are the same, and that structurally the pyramid and the cone are the same.
- Use the *Teaching Tips* from Activity 1. Have students work in groups of 3 or 4 and post a correct solution on the board after 2 or more groups have been successful.
- **Common Error:** For parts (b) and (d), suggest that students start by multiplying both sides by the reciprocal of $\frac{1}{3}$.

What Is Your Answer?

- **Neighbor Check:** Have students work independently and then have their neighbor check their work. Have students discuss any discrepancies.

Closure

- Describe how to solve $d = rt$ for t. *Sample answer:* Divide both sides of the equation by r.

Technology For the Teacher

Dynamic Classroom

The Dynamic Planning Tool

Editable Teacher's Resources at *BigIdeasMath.com*

2 ACTIVITY: Using Volume Formulas

Work with a partner.

a.
- Write a formula for the volume V of a prism.
- Solve the formula for h.
- Use the new formula to find the height of the prism.

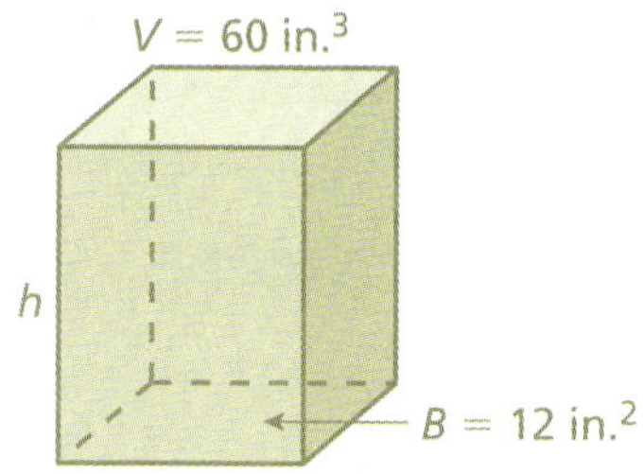

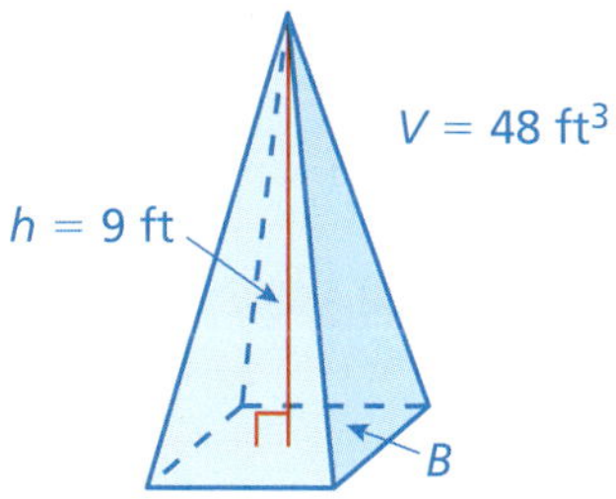

b.
- Write a formula for the volume V of a pyramid.
- Solve the formula for B.
- Use the new formula to find the area of the base of the pyramid.

c.
- Write a formula for the volume V of a cylinder.
- Solve the formula for B.
- Use the new formula to find the area of the base of the cylinder.

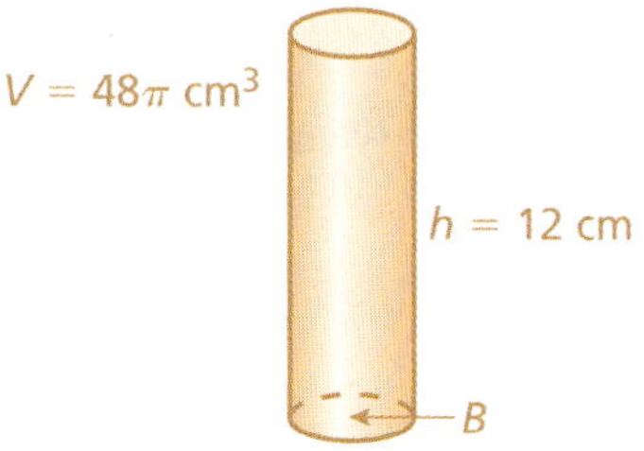

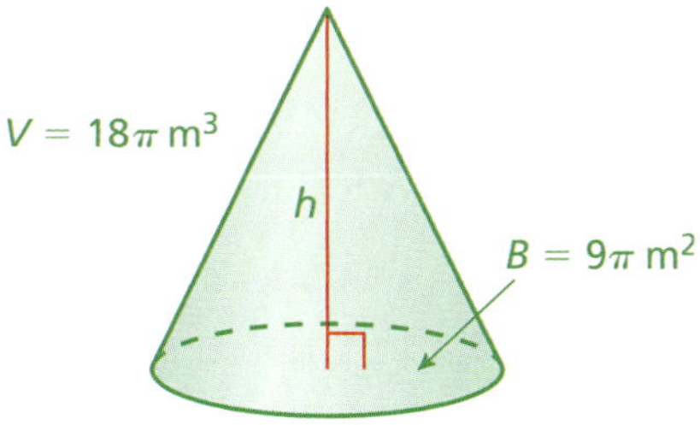

d.
- Write a formula for the volume V of a cone.
- Solve the formula for h.
- Use the new formula to find the height of the cone.

What Is Your Answer?

3. **IN YOUR OWN WORDS** How can you use a formula for one measurement to write a formula for a different measurement? Give an example that is different from the examples on these two pages.

Practice Use what you learned about rewriting equations and formulas to complete Exercises 3 and 4 on page 28.

1.4 Lesson

Check It Out
Lesson Tutorials
BigIdeasMath.com

Key Vocabulary
literal equation, p. 26

An equation that has two or more variables is called a **literal equation**. To rewrite a literal equation, solve for one variable in terms of the other variable(s).

EXAMPLE 1 Rewriting an Equation

Solve the equation $2y + 5x = 6$ for y.

	$2y + 5x = 6$	Write the equation.
Undo the addition. →	$2y + 5x - 5x = 6 - 5x$	Subtract $5x$ from each side.
	$2y = 6 - 5x$	Simplify.
Undo the multiplication. →	$\frac{2y}{2} = \frac{6 - 5x}{2}$	Divide each side by 2.
	$y = 3 - \frac{5}{2}x$	Simplify.

On Your Own

Now You're Ready
Exercises 5–10

Solve the equation for y.

1. $5y - x = 10$
2. $4x - 4y = 1$
3. $12 = 6x + 3y$

EXAMPLE 2 Rewriting a Formula

Remember
A *formula* shows how one variable is related to one or more other variables. A formula is a type of literal equation.

The formula for the surface area S of a cone is $S = \pi r^2 + \pi r\ell$. Solve the formula for the slant height ℓ.

$S = \pi r^2 + \pi r\ell$	Write the equation.
$S - \pi r^2 = \pi r^2 - \pi r^2 + \pi r\ell$	Subtract πr^2 from each side.
$S - \pi r^2 = \pi r\ell$	Simplify.
$\frac{S - \pi r^2}{\pi r} = \frac{\pi r\ell}{\pi r}$	Divide each side by πr.
$\frac{S - \pi r^2}{\pi r} = \ell$	Simplify.

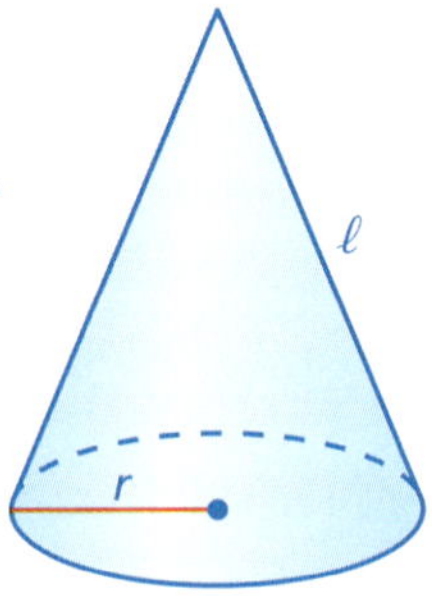

On Your Own

Now You're Ready
Exercises 14–19

Solve the formula for the red variable.

4. Area of rectangle: $A = bh$
5. Simple interest: $I = Prt$
6. Surface area of cylinder: $S = 2\pi r^2 + 2\pi rh$

Laurie's Notes

Introduction

Connect

- **Yesterday:** Students practiced rewriting common geometric formulas.
- **Today:** Students will use the techniques explored yesterday to solve literal equations.

Motivate

- Share with students the following highest and lowest recorded temperatures.

State	Highest Recorded Temperatures		State	Lowest Recorded Temperatures	
NM	122°F	50°C	NM	−50°F	−46°C
NH	106°F	41°C	NH	−47°F	−44°C
PA	111°F	44°C	PA	−42°F	−41°C

- The purpose is to pique interest and have students observe that the temperatures are measured in degrees Fahrenheit or degrees Celsius.

Lesson Notes

Example 1

- Write the definition of literal equation.
- ? "Can 6 and $5x$ be combined? Explain." no; They are not like terms.
- Simplifying the last step is not obvious to all students. Relate it to fractions. You subtract the numerators and keep the same denominator. For example:

$$\frac{5-3}{7} = \frac{5}{7} - \frac{3}{7} \quad \text{and} \quad \frac{6-5x}{2} = \frac{6}{2} - \frac{5x}{2} = 3 - \frac{5}{2}x.$$

On Your Own

- Notice in Question 2 that the coefficient of y is -4. Suggest students rewrite the equation as $4x + (-4)y = 1$.

Example 2

- **Teaching Tip:** Highlight the variable ℓ in red as shown in the textbook. Discuss the idea that everything except the variable ℓ must be moved to the left side of the equation using Properties of Equality.
- ? "The term πr^2 is added to the term $\pi r \ell$. How do you move it to the left side of the equation?" Subtract πr^2 from each side of the equation.
- Discuss the technique of dividing by πr in one step, instead of dividing by π and then dividing by r.

On Your Own

- **Think-Pair-Share:** Students should read each question independently and then work with a partner to answer the questions. When they have answered the questions, the pair should compare their answers with another group and discuss any discrepancies.

Goal Today's lesson is solving **literal equations**.

Start Thinking! and Warm Up

Lesson 1.4 Warm Up
For use before Lesson 1.4

Lesson 1.4 Start Thinking!
For use before Lesson 1.4

How does solving the equation $5x + 4y = 14$ for x compare to solving the equation $5x + 20 = 14$ for x? Describe the steps involved in each solution.

Extra Example 1

Solve the equation $-2x - 3y = 6$ for y.

$y = -\frac{2}{3}x - 2$

On Your Own

1. $y = 2 + \frac{1}{5}x$
2. $y = x - \frac{1}{4}$
3. $y = 4 - 2x$

Extra Example 2

The formula for the surface area of a square pyramid is $S = x^2 + 2x\ell$. Solve the formula for the slant height ℓ.

$\ell = \frac{S - x^2}{2x}$

On Your Own

4. $b = \frac{A}{h}$
5. $P = \frac{I}{rt}$
6. $h = \frac{S - 2\pi r^2}{2\pi r}$

English Language Learners

Vocabulary

Have students start a *Formula* page in their notebooks with the formulas used in this section. Each formula should be accompanied by a description of what each of the variables represents and an example. In the case of area formulas, units of measure should be included with the description (e.g., units and square units). As students progress throughout the year, additional formulas can be added to the *Formula* notebook page.

Extra Example 3

Solve the temperature formula $F = \frac{9}{5}C + 32$ for C. $C = \frac{5}{9}(F - 32)$

Extra Example 4

Which temperature is greater, 400°F or 200°C? 400°F

On Your Own

7. greater than

Laurie's Notes

Key Idea

- Write the formula for converting from degrees Fahrenheit to degrees Celsius.
- Use this formula if you know the temperature in degrees Fahrenheit and you want to find the temperature in degrees Celsius.
- ? "You are traveling abroad and the temperature is always stated in degrees Celsius. How can you figure out the temperature in degrees Fahrenheit, with which you are more familiar?" Students may recognize that you will want to have a different conversion formula that allows you to substitute for C and calculate F.

Example 3

- ? "What is the reciprocal of $\frac{5}{9}$?" $\frac{9}{5}$
- Remind students that multiplying by the reciprocal $\frac{9}{5}$ is more efficient than dividing by the fraction $\frac{5}{9}$.

Example 4

- **FYI:** The graphic on the left provides information about the temperature of a lightning bolt and the temperature of the surface of the sun. The two temperatures use different scales.
- ? "How can you compare two temperatures that are in different scales?" Listen for understanding that one of the temperatures must be converted.
- ? "How do you multiply $\frac{9}{5}$ times 30,000?" Students may recall that you can simplify before multiplying. Five divides into 30,000 six thousand times, so $6000 \times 9 = 54{,}000$.
- ? "Approximately how many times hotter is a lightning bolt than the surface of the sun?" 5 times This is a *cool* fact for students to know!

On Your Own

- **Neighbor Check:** Have students work independently and then have their neighbor check their work. Have students discuss any discrepancies.

Closure

- **Exit Ticket:** Solve $2x + 4y = 11$ for y. Check your solution. $y = -\frac{1}{2}x + \frac{11}{4}$

Technology For the Teacher

Dynamic Classroom

The Dynamic Planning Tool

Editable Teacher's Resources at *BigIdeasMath.com*

Key Idea

Temperature Conversion

A formula for converting from degrees Fahrenheit F to degrees Celsius C is

$$C = \frac{5}{9}(F - 32).$$

EXAMPLE 3 Rewriting the Temperature Formula

Solve the temperature formula for F.

	$C = \frac{5}{9}(F - 32)$	Write the temperature formula.
Use the reciprocal. →	$\frac{9}{5} \cdot C = \frac{9}{5} \cdot \frac{5}{9}(F - 32)$	Multiply each side by $\frac{9}{5}$, the reciprocal of $\frac{5}{9}$.
	$\frac{9}{5}C = F - 32$	Simplify.
Undo the subtraction. →	$\frac{9}{5}C + 32 = F - 32 + 32$	Add 32 to each side.
	$\frac{9}{5}C + 32 = F$	Simplify.

∴ The rewritten formula is $F = \frac{9}{5}C + 32$.

EXAMPLE 4 Real-Life Application

Which has the greater temperature?

Convert the Celsius temperature of lightning to Fahrenheit.

$F = \frac{9}{5}C + 32$	Write the rewritten formula from Example 3.
$= \frac{9}{5}(30{,}000) + 32$	Substitute 30,000 for C.
$= 54{,}032$	Simplify.

∴ Because 54,032 °F is greater than 11,000 °F, lightning has the greater temperature.

On Your Own

7. Room temperature is considered to be 70 °F. Suppose the temperature is 23 °C. Is this greater than or less than room temperature?

1.4 Exercises

Check It Out
Help with Homework
BigIdeasMath.com

Vocabulary and Concept Check

1. **VOCABULARY** Is $-2x = \frac{3}{8}$ a literal equation? Explain.

2. **DIFFERENT WORDS, SAME QUESTION** Which is different? Find "both" answers.

Solve $4x - 2y = 6$ for y.	Solve $6 = 4x - 2y$ for y.
Solve $4x - 2y = 6$ for y in terms of x.	Solve $4x - 2y = 6$ for x in terms of y.

Practice and Problem Solving

3. a. Write a formula for the area A of a triangle.
   b. Solve the formula for b.
   c. Use the new formula to find the base of the triangle.

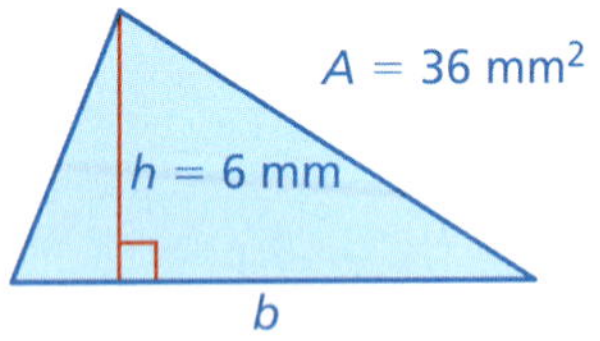

4. a. Write a formula for the volume V of a prism.
   b. Solve the formula for B.
   c. Use the new formula to find the area of the base of the prism.

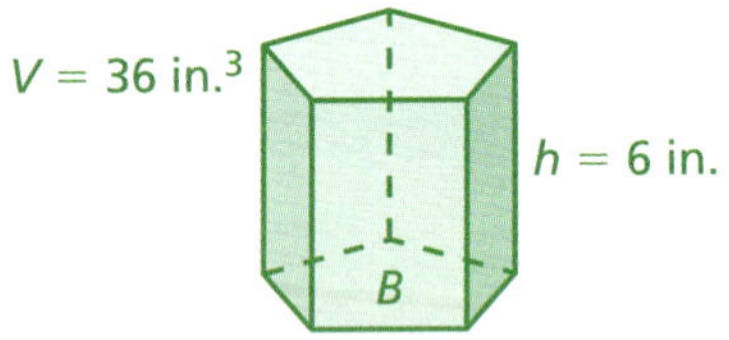

Solve the equation for *y*.

1 5. $\frac{1}{3}x + y = 4$

6. $3x + \frac{1}{5}y = 7$

7. $6 = 4x + 9y$

8. $\pi = 7x - 2y$

9. $4.2x - 1.4y = 2.1$

10. $6y - 1.5x = 8$

11. **ERROR ANALYSIS** Describe and correct the error in rewriting the equation.

✗
$2x - y = 5$
$y = -2x + 5$

12. **TEMPERATURE** The formula $K = C + 273.15$ converts temperatures from Celsius C to Kelvin K.
    a. Solve the formula for C.
    b. Convert 300 K to Celsius.

13. **INTEREST** The formula for simple interest is $I = Prt$.
    a. Solve the formula for t.
    b. Use the new formula to find the value of t in the table.

I	\$75
P	\$500
r	5%
t	

Assignment Guide and Homework Check

Level	Day 1 Activity Assignment	Day 2 Lesson Assignment	Homework Check
Basic	3, 4, 24–28	1, 2, 5–19 odd, 12	7, 12, 13, 17
Average	3, 4, 24–28	1, 2, 8–11, 15–21 odd, 20	8, 10, 17, 20
Advanced	3, 4, 24–28	1, 2, 10, 11, 17–23	10, 17, 20, 22

For Your Information

- **Exercise 2** *Different Words, Same Question* is a new type of exercise. Three of the four choices pose the same question using different words. The remaining choice poses a different question. So there are two answers.

Common Errors

- **Exercises 5–10** Students may solve the equation for the wrong variable. Remind them that they are solving the equation for *y*. Encourage them to make *y* a different color when solving so that it is easy to remember that they are solving for *y*.
- **Exercises 14–19** Each equation has a different step that could confuse students. Remind them to take their time when solving for the red variable. Remind them of the process for solving for a variable. They should start away from the variable and move toward it.

1.4 Record and Practice Journal

Solve the equation for *y*.

1. $2x + y = -9$
   $y = -2x - 9$
2. $4x - 10y = 12$
   $y = \frac{2}{5}x - \frac{6}{5}$
3. $13 = \frac{1}{6}y + 2x$
   $y = -12x + 78$

Solve the equation for the bold variable.

4. $V = \ell wh$
   $w = \frac{V}{\ell h}$
5. $f = \frac{1}{2}(r + 6.5)$
   $r = 2f - 6.5$
6. $S = 2\pi r^2 + 2\pi rh$
   $h = \frac{S - 2\pi r^2}{2\pi r}$

7. The formula for the area of a triangle is $A = \frac{1}{2}bh$.
   a. Solve the formula for h.
   $h = \frac{2A}{b}$
   b. Use the new formula to find the value of h.
   $h = 9$ in.

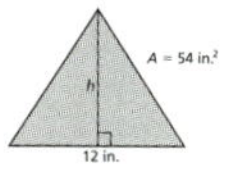

Technology For the Teacher

Answer Presentation Tool
Quiz*Show*

Vocabulary and Concept Check

1. no; The equation only contains one variable.
2. Solve $4x - 2y = 6$ for x in terms of y.; $x = \frac{3}{2} + \frac{1}{2}y$; $y = -3 + 2x$

Practice and Problem Solving

3. a. $A = \frac{1}{2}bh$
   b. $b = \frac{2A}{h}$
   c. $b = 12$ mm
4. a. $V = Bh$
   b. $B = \frac{V}{h}$
   c. $B = 6$ in.^{2}
5. $y = 4 - \frac{1}{3}x$
6. $y = 35 - 15x$
7. $y = \frac{2}{3} - \frac{4}{9}x$
8. $y = \frac{7}{2}x - \frac{\pi}{2}$
9. $y = 3x - 1.5$
10. $y = \frac{4}{3} + \frac{1}{4}x$
11. The y should have a negative sign in front of it.
    $2x - y = 5$
    $-y = -2x + 5$
    $y = 2x - 5$
12. a. $C = K - 273.15$
    b. 26.85°C
13. a. $t = \frac{I}{Pr}$
    b. $t = 3$ yr

Practice and Problem Solving

14. $t = \frac{d}{r}$

15. $m = \frac{e}{c^2}$

16. $C = R - P$

17. $\ell = \frac{A - \frac{1}{2}\pi w^2}{2w}$

18 $V = \frac{Bh}{3}$

19. $w = 6g - 40$

20. The rewritten formula is a general solution that can be reused.

21. a. $F = 32 + \frac{9}{5}(K - 273.15)$

 b. 32°F

 c. liquid nitrogen

22. See *Taking Math Deeper*.

23. $r^3 = \frac{3V}{4\pi}$; $r = 4.5$ in.

Fair Game Review

24. $3\frac{3}{4}$

25. $6\frac{2}{5}$

26. $\frac{1}{3}$

27. $1\frac{1}{4}$

28. D

Mini-Assessment

Solve the formula for the red variable.

1. Distance Formula: $d = rt$ $r = \frac{d}{t}$
2. Area of a triangle: $A = \frac{1}{2}bh$ $h = \frac{2A}{b}$
3. Circumference of a circle: $C = 2\pi r$ $r = \frac{C}{2\pi}$
4. The temperature in Portland, Oregon is 37°F. The temperature in Mobile, Alabama is 22°C. In which city is the temperature higher? Mobile, Alabama

Taking Math Deeper

Exercise 22

This problem is a nice review of circles and percents, as well as distance, rate, and time. It also has a bit of history related to George Ferris, who designed the first Ferris wheel for the 1893 World's Fair in Chicago.

1 Organize the given information.

Circumference (Navy Pier Ferris Wheel): $C = 439.6$ ft

Circumference (first Ferris wheel): x ft

Relationship: $439.6 = 0.56x$

2 Find the radius of each wheel.

a. Radius (Navy Pier Ferris Wheel):

$$C = 2\pi r$$
$$439.6 \approx 2(3.14)r$$
$$70 = r$$

Circumference (first Ferris wheel):

$$439.6 = 0.56x$$
$$785 = x$$

b. Radius (first Ferris wheel):

$$785 \approx 2(3.14)R$$
$$125 = R$$

3 **c.** The first Ferris wheel made 1 revolution in 9 minutes. How fast was the wheel moving?

$$\text{rate} = \frac{785 \text{ ft}}{9 \text{ min}} \approx 87.2 \text{ ft per min}$$

It might be interesting for students to know that the first Ferris wheel had 36 cars, each of which held 60 people!

Project

Use your school's library or the Internet to find how long one revolution takes for the Ferris wheel on the Navy Pier in Chicago and the one in London, England. Which one has the greater circumference? Which one travels faster? How do you know?

Reteaching and Enrichment Strategies

If students need help. . .	If students got it. . .
Resources by Chapter • Practice A and Practice B • Puzzle Time Record and Practice Journal Practice Differentiating the Lesson Lesson Tutorials Skills Review Handbook	Resources by Chapter • Enrichment and Extension • School-to-Work Start the next section

Solve the equation for the red variable.

2 **14.** $d = rt$

15. $e = mc^2$

16. $R - C = P$

17. $A = \frac{1}{2}\pi w^2 + 2\ell w$

18. $B = 3\frac{V}{h}$

19. $g = \frac{1}{6}(w + 40)$

20. WRITING Why is it useful to rewrite a formula in terms of another variable?

21. TEMPERATURE The formula $K = \frac{5}{9}(F - 32) + 273.15$ converts temperatures from Fahrenheit F to Kelvin K.

a. Solve the formula for F.

b. The freezing point of water is 273.15 Kelvin. What is this temperature in Fahrenheit?

c. The temperature of dry ice is −78.5 °C. Which is colder, dry ice or liquid nitrogen?

Navy Pier Ferris Wheel

22. FERRIS WHEEL The Navy Pier Ferris Wheel in Chicago has a circumference that is 56% of the circumference of the first Ferris wheel built in 1893.

a. What is the radius of the Navy Pier Ferris Wheel?

b. What was the radius of the first Ferris wheel?

c. The first Ferris wheel took 9 minutes to make a complete revolution. How fast was the wheel moving?

23. Geometry The formula for the volume of a sphere is $V = \frac{4}{3}\pi r^3$. Solve the formula for r^3. Use guess, check, and revise to find the radius of the sphere.

$V = 381.51$ in.^{3} r

Fair Game Review What you learned in previous grades & lessons

Multiply. *(Skills Review Handbook)*

24. $5 \times \frac{3}{4}$

25. $2.4 \times \frac{8}{3}$

26. $\frac{1}{4} \times \frac{3}{2} \times \frac{8}{9}$

27. $25 \times \frac{3}{5} \times \frac{1}{12}$

28. MULTIPLE CHOICE Which of the following is not equivalent to $\frac{3}{4}$? *(Skills Review Handbook)*

Ⓐ 0.75 Ⓑ 3:4 Ⓒ 75% Ⓓ 4:3

1.5 Converting Units of Measure

Essential Question How can you convert from one measurement system to another?

1 ACTIVITY: Converting Units of Measure

Work with a partner. Copy and complete the table. Describe the pattern in the completed table.

		Perimeter, in. to ft ratio	Area, in.^{2} to ft^2 ratio
Sample: a.	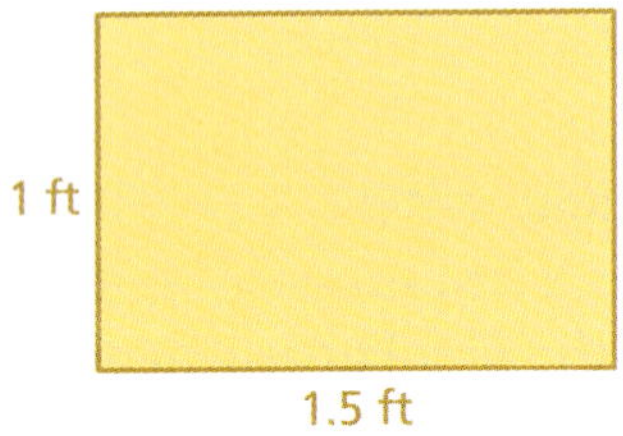	$\frac{60 \text{ in.}}{5 \text{ ft}} = \frac{12 \text{ in.}}{1 \text{ ft}}$	$\frac{216 \text{ in.}^2}{1.5 \text{ ft}^2} = \frac{144 \text{ in.}^2}{1 \text{ ft}^2}$
b.	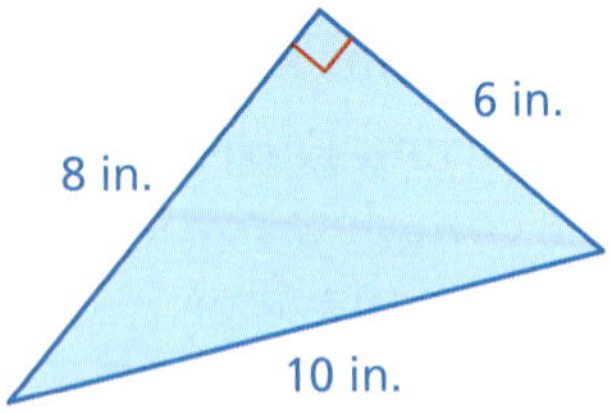	$\frac{\square \text{ in.}}{\square \text{ ft}} = \square$	$\frac{\square \text{ in.}^2}{\square \text{ ft}^2} = \square$
c.	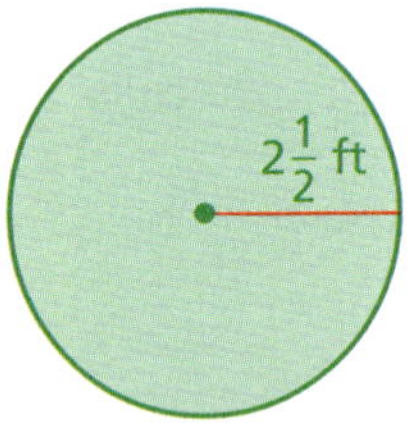	$\frac{\square \text{ in.}}{\square \text{ ft}} = \square$	$\frac{\square \text{ in.}^2}{\square \text{ ft}^2} = \square$
d.	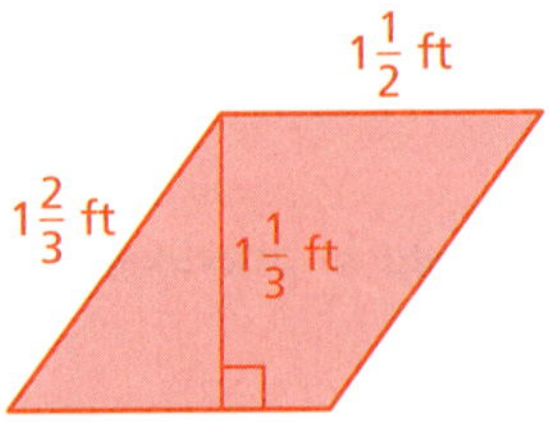	$\frac{\square \text{ in.}}{\square \text{ ft}} = \square$	$\frac{\square \text{ in.}^2}{\square \text{ ft}^2} = \square$
e.	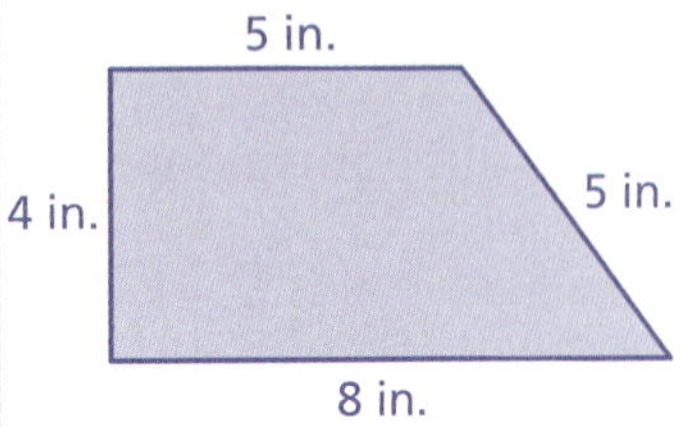	$\frac{\square \text{ in.}}{\square \text{ ft}} = \square$	$\frac{\square \text{ in.}^2}{\square \text{ ft}^2} = \square$

Laurie's Notes

Introduction

For the Teacher

- **Goal:** Students will discover that the perimeter (inches to feet) is always in a ratio of 12 to 1 and that the area (square inches to square feet) is always in a ratio of 144 to 1.

Motivate

? "Do you know who was the tallest man ever to live (for whom there is irrefutable evidence)?"

- **FYI:** The tallest man in medical history is Robert Pershing Wadlow. He was born in Alton, Illinois, on February 22, 1918. When he was last measured on June 27, 1940, he was 8 feet 11.1 inches tall. He wore a size 37AA shoe ($18\frac{1}{2}$ inches long) and his hands measured $12\frac{3}{4}$ inches from the wrist to the tip of the middle finger. Wadlow died on July 15, 1940 as a result of a septic blister on his right ankle caused by a brace, which had been poorly fitted only a week earlier. He was buried in a coffin measuring 10 feet 9 inches long, 32 inches wide, and 30 inches deep.

? "How tall was Wadlow in inches?" Eight feet equals 96 inches, so his height was 96 + 11.1 or 107.1 inches.

- Model this length in your room, vertically if possible.

Activity Notes

Activity 1

- Not all labeled dimensions are *nice* multiples of 12 inches or whole numbers of feet, so fractions will be necessary.
- In each problem, students will form a ratio of their two answers for perimeter and their two answers for area.
- Students should discover that the inches to feet ratio for perimeter is 12 to 1 and the square inches to square feet ratio for area is 144 to 1.
- Discuss with students that this ratio is *not* dependent upon the shape of the figure. The unit ratios of 12 to 1 and 144 to 1 are constant when comparing inches to feet and square inches to square feet, respectively.

Previous Learning

Students should know how to convert units of measure within each measurement system.

Start Thinking! and Warm Up

Activity 1.5 Start Thinking!
For use before Activity 1.5

Activity 1.5 Warm Up
For use before Activity 1.5

Copy and complete the statement.

1. 1 ft = __?__ in.
2. 1 yd = __?__ ft
3. 1 min = __?__ sec
4. 1 day = __?__ h
5. 1 mi = __?__ ft
6. 1 gal = __?__ qt

1.5 Record and Practice Journal

Essential Question How can you convert from one measurement system to another?

1 ACTIVITY: Converting Units of Measure

Work with a partner. Complete the table. Describe the pattern in the completed table.

		Perimeter, in. to ft ratio	Area, $in.^2$ to ft^2 ratio
a.	1 ft, 1.5 ft	$\frac{60 \text{ in.}}{5 \text{ ft}} = \frac{12 \text{ in.}}{1 \text{ ft}}$	$\frac{216 \text{ in.}^2}{1.5 \text{ ft}^2} = \frac{144 \text{ in.}^2}{1 \text{ ft}^2}$
b.	8 in., 6 in., 10 in.	$\frac{24 \text{ in.}}{2 \text{ ft}} = \frac{12 \text{ in.}}{1 \text{ ft}}$	$\frac{24 \text{ in.}^2}{1/6 \text{ ft}^2} = \frac{144 \text{ in.}^2}{1 \text{ ft}^2}$
c.	$2\frac{1}{2}$ ft	$\frac{60\pi \text{ in.}}{5\pi \text{ ft}} = \frac{12 \text{ in.}}{1 \text{ ft}}$	$\frac{900\pi \text{ in.}^2}{6.25\pi \text{ ft}^2} = \frac{144 \text{ in.}^2}{1 \text{ ft}^2}$
d.	$1\frac{1}{2}$ ft, $1\frac{2}{3}$ ft, $1\frac{1}{3}$ ft	$\frac{76 \text{ in.}}{6\frac{1}{3} \text{ ft}} = \frac{12 \text{ in.}}{1 \text{ ft}}$	$\frac{288 \text{ in.}^2}{2 \text{ ft}^2} = \frac{144 \text{ in.}^2}{1 \text{ ft}^2}$
e.	5 in., 4 in., 5 in., 8 in.	$\frac{22 \text{ in.}}{11/6 \text{ ft}} = \frac{12 \text{ in.}}{1 \text{ ft}}$	$\frac{26 \text{ in.}^2}{13/72 \text{ ft}^2} = \frac{144 \text{ in.}^2}{1 \text{ ft}^2}$

English Language Learners

Words and Abbreviations

Make a poster to display both customary and metric units of measure. Show the full word, both singular and plural, and the abbreviation.

Word		Abbreviation
Singular	Plural	
inch	inches	in.
foot	feet	ft
meter	meters	m

1.5 Record and Practice Journal

2 ACTIVITY: Comparing Units of Measure

Work with a partner. Name the units for each pair of "rulers."

a. 

inches and centimeters

b. 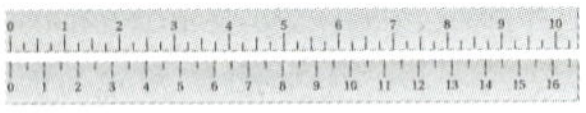

miles and kilometers

c. 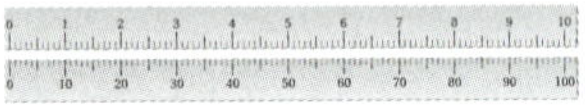

centimeters and millimeters

3 ACTIVITY: Puzzle

Who is correct, Fred or Sam? Explain your reasoning.

John said, "We left camp this morning, and walked 1 mile due south. Then, we saw a polar bear and turned due east and ran 1 kilometer. Finally ,we turned due north and walked 1 mile and ended back at camp."

Fred said, "That is not possible!"

Sam explained, "Yes it is. And I know exactly where the camp was."

Sam; Camp is at the North Pole.

What Is Your Answer?

4. **IN YOUR OWN WORDS** How can you convert from one measurement system to another? The examples on these three pages are measurements of length and area. Describe a conversion between two types of temperature units.

Use a formula, usually involving a ratio.

Laurie's Notes

Activity 2

- Many rulers have both *customary* and *metric* measures on the same side of the ruler, although in opposite directions. If you could put the two scales adjacent and in the same direction, would you recognize the ratio of inches to centimeters? That is what is happening in this activity. One scaled ruler is above the other. The puzzle is to figure out what the units are in each part.
- This activity requires some familiarity with customary and metric linear measures.
- One strategy is to look for nice benchmarks between the two rulers. For example, in part (a), 2 on the top scale is about 5 on the bottom scale. For part (b), 5 on the top scale is about 8 on the bottom scale. For part (c), 1 on the top scale is 10 on the bottom scale.

Activity 3

- This is a classic puzzle that some students may have heard before.
- Sam is correct. The only place on Earth that this is possible is the North Pole.

What Is Your Answer?

- **Extension:** Check to see if students have a sense about conversions with cubic measurements. For instance, 1 $\text{ft}^3 = 1728 \text{ in.}^3$

Closure

- Find the perimeter and area of the rectangle shown. Describe your method. $P = 60$ in. or 5 ft and $A = 189 \text{ in.}^2$ or 1.3125 ft^2

9 in.

$1\frac{3}{4}$ ft

Technology For the Teacher

Dynamic Classroom

The Dynamic Planning Tool
Editable Teacher's Resources at *BigIdeasMath.com*

2 ACTIVITY: Comparing Units of Measure

Work with a partner. Name the units for each pair of "rulers".

a.

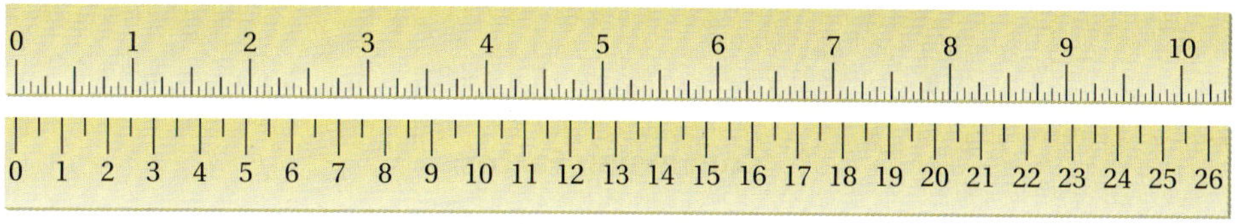

b.

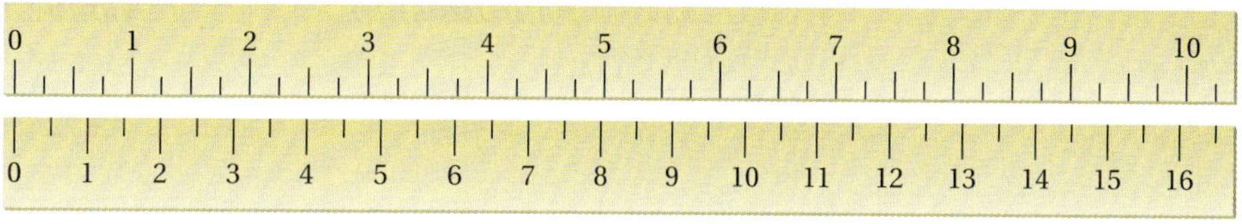

c.

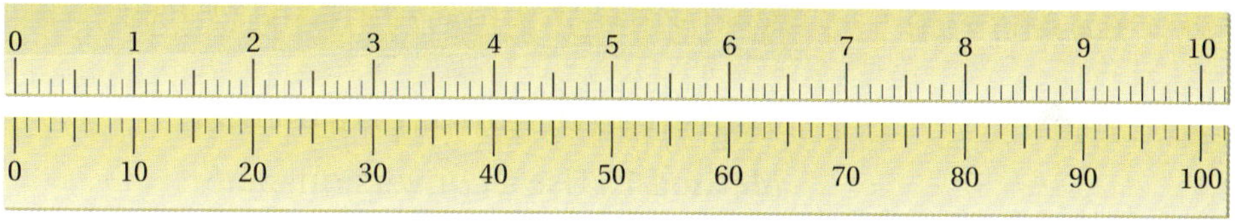

3 ACTIVITY: Puzzle

Who is correct, Fred or Sam? Explain your reasoning.

John said, "We left camp this morning, and walked 1 mile due south. Then, we saw a polar bear and turned due east and ran 1 kilometer. Finally, we turned due north and walked 1 mile and ended back at camp."

Fred said, "That is not possible!"

Sam explained, "Yes it is. And I know exactly where the camp was."

What Is Your Answer?

4. **IN YOUR OWN WORDS** How can you convert from one measurement system to another? The examples on these two pages are measurements of length and area. Describe a conversion between two types of temperature units.

Practice Use what you learned about converting units of measure to complete Exercises 4–6 on page 35.

1.5 Lesson

Check It Out
Lesson Tutorials
BigIdeasMath.com

Key Vocabulary
conversion factor, *p. 32*

To convert between customary and metric units, multiply by one or more *conversion factors*.

Key Idea

Conversion Factor

A **conversion factor** is a rate that equals 1.

	Relationship	*Conversion factors*
Example	$1 \text{ m} \approx 3.28 \text{ ft}$	$\frac{1 \text{ m}}{3.28 \text{ ft}}$ and $\frac{3.28 \text{ ft}}{1 \text{ m}}$

EXAMPLE 1 Converting Between Systems

Convert 20 centimeters to inches.

Method 1: Use a conversion factor.

1 in. ≈ 2.54 cm

$$20 \text{ cm} \cdot \frac{1 \text{ in.}}{2.54 \text{ cm}} \approx 7.87 \text{ in.}$$

So, 20 centimeters is about 7.87 inches.

Method 2: Use a proportion.

Let x be the number of inches equivalent to 20 centimeters.

inches → $\frac{1}{2.54} \approx \frac{x}{20}$ ← inches
centimeters → ← centimeters

$$\frac{1}{2.54} \approx \frac{x}{20} \quad \text{Write a proportion.}$$

$$20 \cdot \frac{1}{2.54} \approx 20 \cdot \frac{x}{20} \quad \text{Multiply each side by 20.}$$

$$7.87 \approx x \quad \text{Simplify.}$$

So, 20 centimeters is about 7.87 inches.

On Your Own

Now You're Ready
Exercises 7–15

Copy and complete the statement.

1. $10 \text{ qt} \approx$ ___ L

2. $4 \text{ km} \approx$ ___ mi

3. $18 \text{ in.} \approx$ ___ cm

4. $84 \text{ lb} \approx$ ___ kg

Multi-Language Glossary at BigIdeasMath.com.

Laurie's Notes

Introduction

Connect

- **Yesterday:** Students explored conversions with customary units.
- **Today:** Students will convert *between* customary and metric units.

Motivate

- **History of Customary Length:**
  - The Egyptian cubit was developed about 3000 B.C. The unit was based on the length of the arm (from the elbow to the extended finger).
  - The Romans adopted the foot from the Greeks and divided it into 12 sections called *unicae*, which came to be known as an inch.
- **History of Metric:**
  - The word *meter* is from the Greek word *metron*, which means a *measure*. In 1793, the meter was defined to be one ten-millionth of *the length of the earth's meridian along a quadrant*. This is the same as one ten-millionth of the distance from the North Pole to the equator, along the meridian running near Dunkirk in France.
  - The United States is the only industrialized country that does not use the metric system as its predominant system of measurement.

Lesson Notes

Key Idea

- Write the Key Idea. Define *conversion factor* and give several examples. Point out that each conversion factor can be written two ways, which are reciprocals of each other.
- **Big Idea:** The technique that is used today is called unit analysis. You multiply the known amount by the conversion factor, written in the form that allows the known units to divide out and the desired units to remain.
- There are conversion factors for converting from U.S. customary to Metric (i.e., 1 inch $\approx$ 2.54 centimeters) and from Metric to U.S. customary (i.e., 1 centimeter $\approx$ 0.39 inch). Depending upon which conversion factor students use, their answers may vary due to rounding.

Example 1

- Model this problem with a ruler.
- The first method uses the technique of unit analysis.
- Write and model the second method which uses a proportion.

On Your Own

- **Think-Pair-Share:** Students should read each question independently and then work with a partner to answer the questions. When they have answered the questions, the pair should compare their answers with another group and discuss any discrepancies.

Goal Today's lesson is converting between customary and metric units.

Lesson Materials
Textbook
• ruler

Start Thinking! and Warm Up

Lesson 1.5 Warm Up
For use before Lesson 1.5

Lesson 1.5 Start Thinking!
For use before Lesson 1.5

If you know that a door is 30 inches wide and 7 feet tall, how can you find the area of the door?

Is there more than one correct answer?

Extra Example 1

Convert 56 kilograms to pounds.
about 124.4 lb

On Your Own

1. 9.5
2. 2.5
3. 45.72
4. 37.8

Extra Example 2

Convert the flow rate of 3 liters per minute to quarts per minute. about 3.16 quarts per minute

Extra Example 3

Convert the speed 3 kilometers per hour to meters per minute. 50 meters per minute

On Your Own

5. 5 gallons per second
6. 176 feet per second
7. 0.6 kilometer per minute

English Language Learners

Labels

English learners may recognize the fraction bar as division, but may not be familiar with its use in the concept of rate. The following unit rates are equivalent.

$\frac{3\text{ m}}{1\text{ h}}$ 3 m/h 3 meters per hour

Each of these rates can be read as "three meters for every hour."

Laurie's Notes

Example 2

- Write the equation as shown.
- ? "What are the units for the rate given?" quarts per minute "What units are desired for the rate?" liters per minute
- The strategy is to look at the units given (qt/min) and the units desired (L/min). Because you need to eliminate quarts and introduce liters, write the conversion factor with liters in the numerator and quarts in the denominator as shown. The common units of quarts divide out, leaving L/min.
- **FYI:** The average adult has about 6 quarts of blood in his or her body. In one minute, almost all of it will be pumped through the heart.

Example 3

- ? "What is a zip line? How does it work?" A zip line is a wire, usually hung between trees high off of the ground. It is hung at a diagonal, so you use gravity to slide down. You wear a harness and climb to the top of a tower, push off and slide either to the ground or to another platform.
- ? "What are the units for the known rate and what units do you want?" known rate: mi/h; want rate: ft/sec
- In this example, the linear measure must be converted to feet and the time must be converted to seconds. So, you must multiply by two conversion factors.
- Discuss the alternate method of converting the units given in the *Study Tip*.

On Your Own

- **Question 5 Strategy:** You have gallons per minute and want gallons per second. You need to eliminate minutes and introduce seconds.
- Students should work with a partner on these three questions.

Key Idea

- Discuss the Key Idea. Ask students to explain how they would convert 5 square feet to square inches.

$$5\text{ ft}^2 = 5\text{ ft}^2 \bullet \left(\frac{12\text{ in.}}{1\text{ ft}}\right)^2$$
$$= 5\text{ ft}^2 \bullet \frac{144\text{ in.}^2}{1\text{ ft}^2}$$
$$= 720\text{ in.}^2$$

EXAMPLE 2 Converting a Rate: Changing One Unit

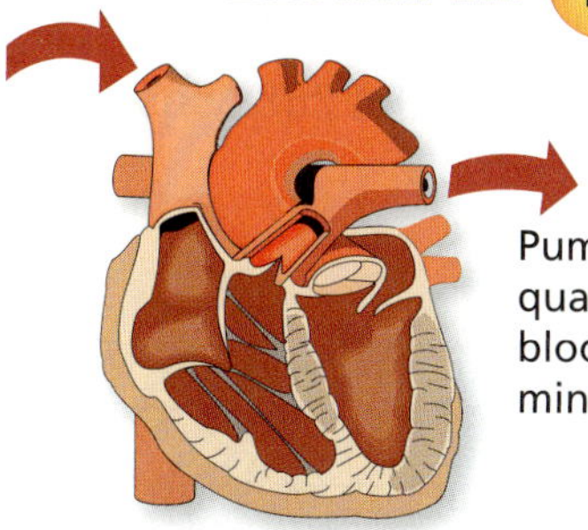

Convert the pumping rate of the human heart to liters per minute.

1 qt ≈ 0.95 L

$$\frac{5 \text{ qt}}{1 \text{ min}} \cdot \frac{0.95 \text{ L}}{1 \text{ qt}} \approx \frac{4.75 \text{ L}}{1 \text{ min}}$$

∴ The rate of 5 quarts per minute is about 4.75 liters per minute.

EXAMPLE 3 Converting a Speed: Changing Both Units

Convert the speed of the zip liner to feet per second.

$$\frac{15 \text{ mi}}{1 \text{ h}}\left(\frac{5280 \text{ ft}}{1 \text{ mi}}\right)\left(\frac{1 \text{ h}}{3600 \text{ sec}}\right) = \frac{15 \cdot 5280 \text{ ft}}{3600 \text{ sec}}$$

1 mi = 5280 ft

1 h = 3600 sec

$$= \frac{79{,}200 \text{ ft}}{3600 \text{ sec}}$$

$$= \frac{22 \text{ ft}}{1 \text{ sec}}$$

∴ The speed of the zip liner is 22 feet per second.

Study Tip

Here is another way to convert the rate in Example 3.

- Write the rate as $15 \frac{\text{miles}}{\text{hour}}$.
- Substitute 5280 feet for miles and 3600 seconds for hour.

On Your Own

Now You're Ready
Exercises 18–23

5. An oil tanker is leaking oil at a rate of 300 gallons per minute. Convert this rate to gallons per second.
6. A tennis ball travels at a speed of 120 miles per hour. Convert this rate to feet per second.
7. A kite boarder travels at a speed of 10 meters per second. Convert this rate to kilometers per minute.

Key Idea

Converting Units for Area or Volume

To convert units for area, multiply the area by the *square* of the conversion factor.

To convert units for volume, multiply the volume by the *cube* of the conversion factor.

EXAMPLE 4 Converting Units for Area

Remember

Area is measured in *square units*. Volume is measured in *cubic units*.

The painting *Fracture* by Benedict Gibson has an area of 2880 square inches. What is the area of the painting in square feet?

1 ft = 12 in.

$$2880 \text{ in.}^2 = 2880 \text{ in.}^2 \cdot \left(\frac{1 \text{ ft}}{12 \text{ in.}}\right)^2$$

$$= 2880 \cancel{\text{in.}^2} \cdot \frac{1 \text{ ft}^2}{144 \cancel{\text{in.}^2}}$$

$$= \frac{2880}{144} \text{ ft}^2$$

$$= 20 \text{ ft}^2$$

The area of the painting is 20 square feet.

EXAMPLE 5 Converting Units for Volume

What is the volume of the cylinder in cubic centimeters?

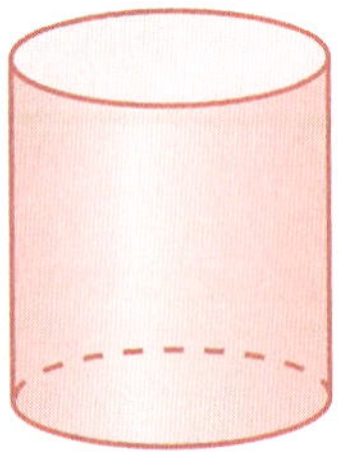

Volume = 80 m^3

1 m = 100 cm

$$80 \text{ m}^3 = 80 \text{ m}^3 \cdot \left(\frac{100 \text{ cm}}{1 \text{ m}}\right)^3$$

$$= 80 \cancel{\text{m}^3} \cdot \frac{1{,}000{,}000 \text{ cm}^3}{1 \cancel{\text{m}^3}}$$

$$= 80{,}000{,}000 \text{ cm}^3$$

The volume is 80,000,000 cubic centimeters.

On Your Own

Now You're Ready
Exercises 30–35

8. The painting *Busy Market* by Haitian painter Frantz Petion has an area of 6 square feet. What is the area of the painting in square inches?

9. The volume of a pyramid is 50 cubic centimeters. What is the volume of the pyramid in cubic millimeters?

Laurie's Notes

Example 4

- "What is the conversion factor for feet to inches?" $\frac{1 \text{ ft}}{12 \text{ in.}}$ "square feet to square inches?" $\frac{1 \text{ ft}^2}{144 \text{ in.}^2}$
- **Strategy:** The measurement is given in square inches and you want to convert this to square feet.

Example 5

- Make sure your students read carefully. They are now working with cubic units of measure because the example is converting units for volume.
- If you have small cubes that are cubic centimeters, use them to help students visualize the problem. Explain that a cubic meter is a cube with an edge length of 1 meter. A visual model of a cubic meter is approximately a baby's play pen. Given these visual models, it should make sense that $1 \text{ m}^3 = 1{,}000{,}000 \text{ cm}^3$.

On Your Own

- **Neighbor Check:** Have students work independently and then have their neighbor check their work. Have students discuss any discrepancies.

Closure

- A major league pitcher throws a 90 mile per hour fastball. How many feet per second is this? 132 feet per second

Extra Example 4

An acre has an area of 4840 square yards. What is the area of an acre in square feet? $43{,}560 \text{ ft}^2$

Extra Example 5

The volume of a cube is 216 cubic feet. What is the volume of the cube in cubic yards? 8 yd^3

On Your Own

8. 864 in.^2

9. $50{,}000 \text{ mm}^3$

Technology For the Teacher

Dynamic Classroom

The Dynamic Planning Tool

Editable Teacher's Resources at *BigIdeasMath.com*

Vocabulary and Concept Check

1. yes; Because 1 centimeter is equal to 10 millimeters, the conversion factor equals 1.
2. First convert liters to milliliters by multiplying by 1000. Then convert hours to seconds by dividing by 3600.
3. 6.25 ft; The other three represent the same length.

Practice and Problem Solving

4. 36 ft, 12 yd
5. 11 yd, 33 ft
6. about 25.12 yd, about 75.36 ft
7. 12.63
8. 45.92
9. 1.22
10. 28.8
11. 0.19
12. 190.5
13. 37.78
14. 5.91
15. 14.4
16. The conversion factor is wrong.

$$8\text{ L} \approx 8\,\cancel{\text{L}} \cdot \frac{1\text{ qt}}{0.95\,\cancel{\text{L}}}$$
$$\approx 8.42\text{ qt}$$

17. **a.** about 60.67 m

    b. about 8.04 km

Assignment Guide and Homework Check

Level	Day 1 Activity Assignment	Day 2 Lesson Assignment	Homework Check
Basic	4–6, 41–45	1–3, 7–25 odd, 16, 24, 31–35 odd	7, 16, 19, 21, 31, 33
Average	4–6, 41–45	1–3, 13–17, 19–25 odd, 29–37 odd, 36	13, 19, 21, 31, 33, 36
Advanced	4–6, 41–45	1–3, 14–16, 22–24, 29, 30–36 even, 37, 39, 40	14, 22, 30, 34, 36

Common Errors

- **Exercises 7–15** Students may use the wrong conversion factor or may write the units they are trying to remove in the numerator. For example, a student may write $14\text{ m} \cdot \frac{1\text{ m}}{3.28\text{ ft}}$ instead of $14\text{ m} \cdot \frac{3.28\text{ ft}}{1\text{ m}}$. Remind them to put the given unit in the denominator so that the units will divide out. Also caution students to make sure that they use the correct conversion factor.
- **Exercise 17** Students may convert the unit of measure but may not understand the context of what they are finding. Ask them to write a sentence explaining what the new measurement means in the context of the problem.

1.5 Record and Practice Journal

Complete the statement.

1. 3 m ≈ 9.84 ft
2. 32 cm ≈ 12.6 in.
3. 16 qt ≈ 15.2 L
4. $\frac{50\text{ mi}}{\text{h}} \approx \frac{80\text{ km}}{\text{h}}$
5. $\frac{25\text{ gal}}{\text{min}} = \frac{1.67\text{ qt}}{\text{sec}}$
6. $\frac{1000\text{ m}}{\text{sec}} = \frac{60\text{ km}}{\text{min}}$
7. 20 in.^{2} ≈ 0.14 ft^2
8. 50 ft^2 ≈ 5.56 yd^2
9. 50 m^3 = ____ cm^3 50,000,000
10. Your doctor prescribes you to take 400 milligrams of medicine every 8 hours. How many ounces of medicine do you take in a day?

    about 0.043 ounce
11. In Canada, a speed limit is 100 kilometers per hour. What is the speed limit in miles per hour?

    about 62.5 mi/h

Technology For the Teacher

Answer Presentation Tool

Quiz*Show*

1.5 Exercises

Check It Out
Help with Homework
BigIdeasMath.com

Vocabulary and Concept Check

1. **VOCABULARY** Is $\frac{10 \text{ mm}}{1 \text{ cm}}$ a conversion factor? Explain.

2. **WRITING** Describe how to convert 2 liters per hour to milliliters per second.

3. **WHICH ONE DOESN'T BELONG?** Which measurement does *not* belong with the other three? Explain your reasoning.

100 in.	254 cm	6.25 ft	2.54 m

Practice and Problem Solving

Find the perimeter in feet and in yards.

4.
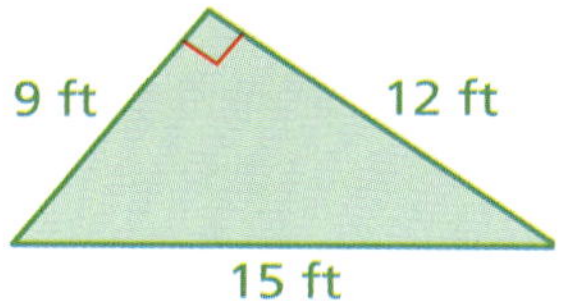

5.
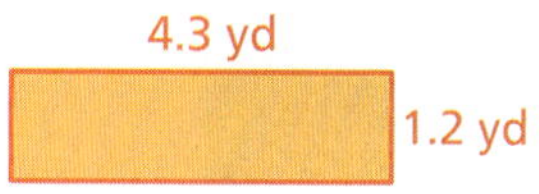

6.
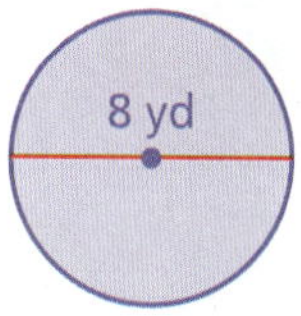

Copy and complete the statement.

1 7. 12 L ≈ ___ qt

8. 14 m ≈ ___ ft

9. 4 ft ≈ ___ m

10. 64 lb ≈ ___ kg

11. 0.3 km ≈ ___ mi

12. 75 in. ≈ ___ cm

13. 17 kg ≈ ___ lb

14. 15 cm ≈ ___ in.

15. 9 mi ≈ ___ km

16. **ERROR ANALYSIS** Describe and correct the error in converting the units.

✗
$$8 \text{ L} \approx 8 \text{ L} \cdot \frac{0.95 \text{ qt}}{1 \text{ L}}$$
$$= 8 \cancel{\text{L}} \cdot \frac{0.95 \text{ qt}}{1 \cancel{\text{L}}}$$
$$= 7.6 \text{ qt}$$

17. **BRIDGE** The Mackinac Bridge, in Michigan, is the third longest suspension bridge in the United States.

a. How high above the water is the roadway in meters?

b. The bridge has a length of 26,372 feet. What is the length in kilometers?

Copy and complete the statement.

(2) (3) 18. $\dfrac{13\text{ km}}{\text{h}} \approx \dfrac{\square\text{ mi}}{\text{h}}$

19. $\dfrac{22\text{ L}}{\text{min}} = \dfrac{\square\text{ L}}{\text{h}}$

20. $\dfrac{63\text{ mi}}{\text{h}} = \dfrac{\square\text{ mi}}{\text{sec}}$

21. $\dfrac{3\text{ km}}{\text{min}} \approx \dfrac{\square\text{ mi}}{\text{h}}$

22. $\dfrac{17\text{ gal}}{\text{h}} \approx \dfrac{\square\text{ qt}}{\text{min}}$

23. $\dfrac{6\text{ cm}}{\text{min}} = \dfrac{\square\text{ m}}{\text{sec}}$

24. **SNAIL** What is the speed of the snail in kilometers per hour?

25. **BLOOD DRIVE** A donor gives blood at a rate of 0.125 pint per minute. What is the rate in milliliters per second?

26. **POSTER** A poster of your favorite band has a width of 15 inches. You have a space on your wall that has a width of 1.2 feet. Will the poster fit? Explain.

27. **ROME** Ancient Romans used the *talent* and the *mina* as measures of weight. How many minas are in 100 pounds?

28. **FUEL EFFICIENCY** The fuel efficiency standard for cars in Japan is 20 kilometers per liter. The fuel efficiency standard for cars in the United States is 28 miles per gallon. Which country has a greater fuel efficiency standard?

29. **BIRDS** The table shows the flying speeds of several birds.

a. Which bird is the fastest? Which is the slowest?

b. The peregrine falcon has a dive speed of 322 kilometers per hour. Is the dive speed of the peregrine falcon faster than the flying speed of any of the birds? Explain.

Bird	Speed
Spine-tailed swift	2843.2 m/min
Spur-winged goose	129.1 ft/sec
Eider duck	31.3 m/sec
Mallard	65 mi/h

Copy and complete the statement.

(4) 30. $4\text{ yd}^2 = \square\text{ ft}^2$

31. $0.00125\text{ mi}^2 = \square\text{ ft}^2$

32. $30\text{ mm}^2 = \square\text{ cm}^2$

(5) 33. $3\text{ km}^3 = \square\text{ m}^3$

34. $2\text{ ft}^3 = \square\text{ in.}^3$

35. $420\text{ cm}^3 = \square\text{ m}^3$

Common Errors

- **Exercises 18–20** Students may change the wrong units. For example, a student may change from hours to minutes instead of kilometers to miles. Remind them to read the units carefully and encourage them to write down what unit is changing before converting.
- **Exercises 21–23** Students may write the units they are trying to remove in the numerator. This is common because they are converting both units. Remind them to put the given unit in the denominator so that the units will divide out. Some students may need to convert one unit, and then convert the other unit as a separate problem.
- **Exercises 30–35** Students may forget to square or cube the conversion factor, or may square it when they should have cubed it or vice versa. Remind them to carefully examine the units. When the units are cubed, the conversion factor must also be cubed. When the units are squared, the conversion factor must also be squared.

Differentiated Instruction

Visual

Create a poster to display the units of 1 (or ratios) that are used in converting units of measure between customary and metric.

Length

$\frac{1 \text{ in.}}{2.54 \text{ cm}}, \frac{2.54 \text{ cm}}{1 \text{ in.}}, \frac{1 \text{ mi}}{1.6 \text{ km}}, \frac{1.6 \text{ km}}{1 \text{ mi}}$

Capacity

$\frac{1 \text{ qt}}{0.95 \text{ L}}, \frac{0.95 \text{ L}}{1 \text{ qt}}$

Weight and Mass

$\frac{1 \text{ lb}}{0.45 \text{ kg}}, \frac{0.45 \text{ kg}}{1 \text{ lb}}$

Practice and Problem Solving

18. 8.125 **19.** 1320

20. 0.0175 **21.** 112.5

22. 1.13 **23.** 0.001

24. 0.0468 km/h

25. about 0.99 mL/sec

26. no; The space on the wall is only 14.4 inches wide.

27. 80

28. Japan

29. **a.** spine-tailed swift; mallard

b. yes, It is faster than all of the other birds in the table. Its dive speed is about 201.25 miles per hour.

30. 36

31. 34,848

32. 0.3

33. 3,000,000,000

34. 3456

35. 0.00042

Practice and Problem Solving

36. 4.74 yd^3

37. **a.** 120 $in.^3$

b. 138 tissues

38. **a.** *Sample answer:*

Country	Currency	Value in dollars
United States	Dollar	\$1
Japan	Yen	\$0.001
Spain	Euro	\$1.27
Great Britain	Pound	\$1.54

b. *Sample answer:* 2000 Yen, 15.75 Euros, 12.99 Pounds

39. 113,000 mm^3

40. See *Taking Math Deeper*.

Fair Game Review

41–44.

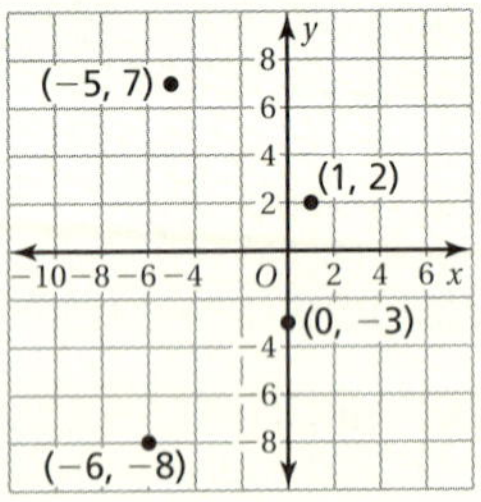

45. B

Mini-Assessment

Copy and complete the statement.

1. 2 km ≈ ☐ mi 1.25
2. 60 in. ≈ ☐ cm 152.4
3. 30 lb ≈ ☐ kg 13.5
4. 4 L ≈ ☐ qt 4.21
5. Your cell phone is 4 inches long. How long is the cell phone in centimeters? about 10.16 cm

Taking Math Deeper

Exercise 40

This problem provides students with practice in converting between metric and standard measures.

1 Organize the capacities in a table.
Note that 4 cups = 2 pints = 1 quart. Also, as a reasonable approximation, students can use 1 quart ≈ 1 liter. (The actual is 0.94635.)

Ingredient	Standard	Metric
Water	2 c	0.5 L
Sugar	2 c	0.5 L
Ginger ale		1 L
Carbonated water		1 L
Sherbet	1 pt = 2 c	0.5 L
Ice	4 c	1 L
Lemonade	1.5 c	0.4 L
Orange juice	1.5 c	0.4 L
	Total	5.3 L

2 **a.** The recipe makes about 5.3 liters (about 5.08 liters using the actual conversion instead of the approximation.) In either case, it will fit into a bowl that holds 6 liters.

3 **b.** The recipe makes about 5.3 liters, which is 5300 milliliters = 5300 cubic centimeters. So, the punch will not fit into a container that has a volume of only 3000 cubic centimeters.

Reteaching and Enrichment Strategies

If students need help. . .	If students got it. . .
Resources by Chapter • Practice A and Practice B • Puzzle Time Record and Practice Journal Practice Differentiating the Lesson Lesson Tutorials Skills Review Handbook	Resources by Chapter • Enrichment and Extension • School-to-Work • Financial Literacy Start the Chapter Review

36. FIREWOOD The volume of a cord of firewood is 128 cubic feet. What is the volume of a cord of firewood in cubic yards? Round your answer to the nearest hundredth.

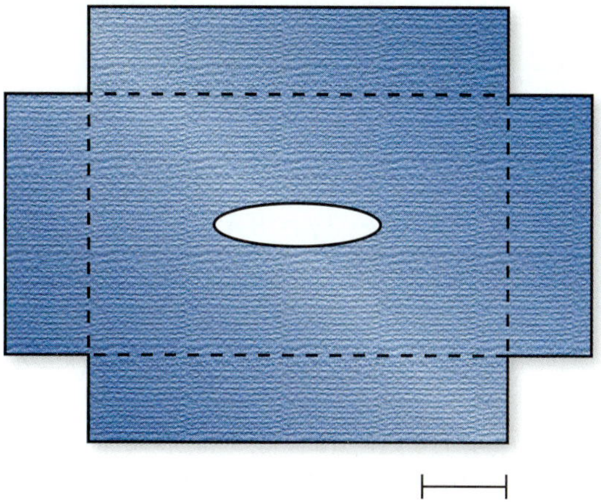

1 cm : 2 in.

37. FABRIC COVER The pattern shows the dimensions of a fabric cover for a tissue box.

a. Use the pattern and a ruler to estimate the volume of a tissue box.

b. The volume of a tissue is about 0.864 cubic inch. About how many tissues are in a box?

38. PROJECT The table shows the currencies of four countries.

a. RESEARCH Use the Internet to find the exchange rates for the currencies listed in the table.

b. How much of each currency would you receive in exchange for $20?

Country	Currency	Value in Dollars
United States	Dollar	$1
Japan	Yen	
Spain	Euro	
Great Britain	Pound	

39. SHAMPOO Your shampoo bottle is 80% full. The total volume of the bottle is 565 cubic centimeters. How much shampoo have you used? Write your answer in cubic millimeters.

40. Critical Thinking You make Floating Island Punch for a party.

a. Your punch bowl holds 6 liters. Will the punch fit into the bowl? Explain.

b. One milliliter is equal to 1 cubic centimeter. Can you store the punch in a container with a capacity of 3000 cubic centimeters?

Recipe for: Floating Island Punch
From: Mom

2 cups water	2 cups sugar
1 L ginger ale	1 L carbonated water
1 pt orange sherbet	4 cups ice
$1\frac{1}{2}$ cups frozen lemonade concentrate	$1\frac{1}{2}$ cups frozen orange juice concentrate

Fair Game Review What you learned in previous grades & lessons

Plot the ordered pair in a coordinate plane. *(Skills Review Handbook)*

41. (1, 2) **42.** (0, −3) **43.** (−6, −8) **44.** (−5, 7)

45. MULTIPLE CHOICE Which equation shows direct variation? *(Skills Review Handbook)*

Ⓐ $y = 2x + 1$ Ⓑ $y = \frac{1}{3}x$ Ⓒ $4 = xy$ Ⓓ $y = 2x - 1$

1.4–1.5 Quiz

Check It Out
Progress Check
BigIdeasMath.com

Solve the equation for *y*. *(Section 1.4)*

1. $6x - 3y = 9$

2. $8 = 2y - 10x$

Solve the formula for the red variable. *(Section 1.4)*

3. Volume of a cylinder: $V = \pi r^2 h$

4. Area of a trapezoid: $A = \frac{1}{2}h(b + B)$

Copy and complete the statement. *(Section 1.5)*

5. 30 cm $\approx$ ____ in.

6. 0.7 km $\approx$ ____ mi

7. 15 L $\approx$ ____ qt

8. $3000\text{ cm}^2 =$ ____ m^2

9. $45\text{ in.}^3 =$ ____ ft^3

10. $50\text{ yd}^3 =$ ____ ft^3

11. **TEMPERATURE** In which city is the water temperature higher? *(Section 1.4)*

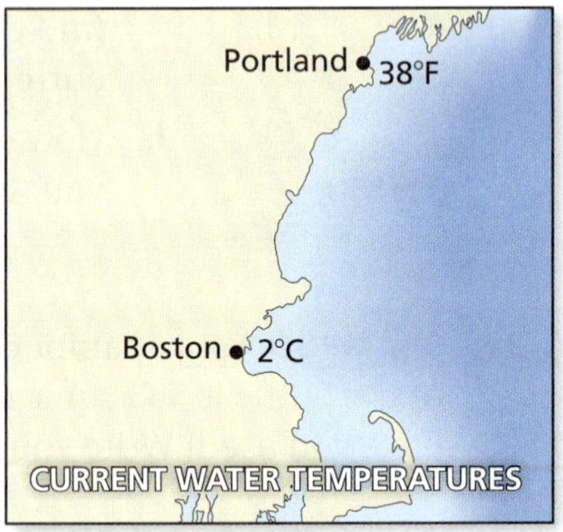

12. **INTEREST** The formula for simple interest I is $I = Prt$. Solve the formula for the interest rate r. What is the interest rate r if the principal P is \$1500, the time t is 2 years, and the interest earned I is \$900? *(Section 1.4)*

13. **HIKING** The Black Mountain Loop, a hiking trail near Lake George, NY, is 7 miles long. How long is the trail in kilometers? *(Section 1.5)*

14. **MEDICINE** A grain is a measure of weight equal to 0.06 gram. An aspirin tablet weighs 5 grains. How many milligrams does the tablet weigh? *(Section 1.5)*

15. **HAWK** What is the speed of the hawk in kilometers per hour? *(Section 1.5)*

Alternative Assessment Options

Math Chat | Student Reflective Focus Question
Structured Interview | Writing Prompt

Math Chat

- Have students work in pairs. One student describes how to convert units of measure, giving examples. The other student probes for more information. Students then switch roles and repeat the process for how to rewrite equations and formulas.
- The teacher should walk around the classroom listening to the pairs and asking questions to ensure understanding.

Study Help Sample Answers

Remind students to complete Graphic Organizers for the rest of the chapter.

3. Solving Multi-Step Equations | Rewriting Literal Equations

Solving Multi-Step Equations:
- Isolate the variable.
- Solution is numerical.

Rewriting Literal Equations:
- Solve for one variable in terms of the other variable(s).

Both:
- Use the Addition, Subtraction, Multiplication, and Division Properties of Equality as necessary.
- The variable being isolated or solved for can end up on either side of the equation.

4. Converting Meters to Feet | Converting Feet to Meters

Converting Meters to Feet:
- You can use the conversion factor $\frac{3.28 \text{ ft}}{1 \text{ m}}$.
- You can use a proportion in which one side is the ratio of feet to meters.

Converting Feet to Meters:
- You can use the conversion factor $\frac{1 \text{ m}}{3.28 \text{ ft}}$.
- You can use a proportion in which one side is the ratio of meters to feet.

Both:
- Use the relationship 1 m ≈ 3.28 ft.
- You can multiply by a conversion factor or use a proportion.
- When using a conversion factor, make sure the units you are converting divide out and the units you are converting to remain.
- It is always a good idea to check your answer for reasonableness.

5. Available at *BigIdeasMath.com.*

Reteaching and Enrichment Strategies

If students need help. . .	If students got it. . .
Resources by Chapter • Study Help • Practice A and Practice B • Puzzle Time Lesson Tutorials *BigIdeasMath.com* Practice Quiz Practice from the Test Generator	Resources by Chapter • Enrichment and Extension • School-to-Work Game Closet at *BigIdeasMath.com* Start the Chapter Review

Answers

1. $y = 2x - 3$
2. $y = 5x + 4$
3. $h = \dfrac{V}{\pi r^2}$
4. $b = \dfrac{2A}{h} - B$
5. 11.81
6. 0.4375
7. 15.79
8. 0.3
9. 0.03
10. 1350
11. Portland
12. 30%
13. about 11.2 km
14. 300 mg
15. 46.8 km/h

Technology For the Teacher
Answer Presentation Tool

Assessment Book

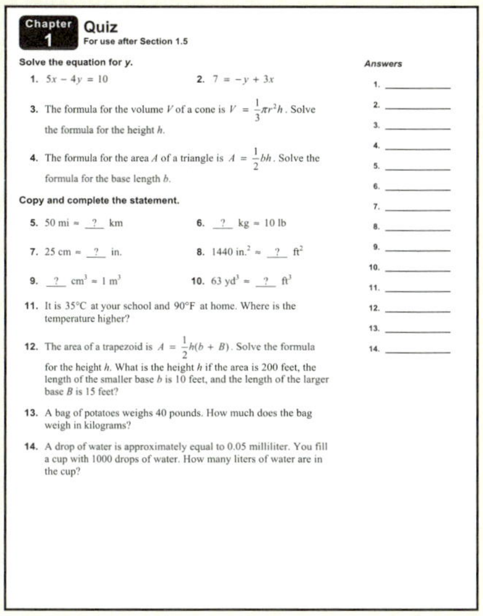
Chapter 1 Quiz
For use after Section 1.5

Solve the equation for *y*.

1. $5x - 4y = 10$
2. $7 = -y + 3x$
3. The formula for the volume V of a cone is $V = \frac{1}{3}\pi r^2 h$. Solve the formula for the height h.
4. The formula for the area A of a triangle is $A = \frac{1}{2}bh$. Solve the formula for the base length b.

Copy and complete the statement.

5. 50 mi ≈ __?__ km
6. __?__ kg ≈ 10 lb
7. 25 cm ≈ __?__ in.
8. 1440 in.^2 ≈ __?__ ft^2
9. __?__ cm^3 = 1 m^3
10. 63 yd^3 = __?__ ft^3
11. It is 35°C at your school and 90°F at home. Where is the temperature higher?
12. The area of a trapezoid is $A = \frac{1}{2}h(b + B)$. Solve the formula for the height h. What is the height h if the area is 200 feet, the length of the smaller base b is 10 feet, and the length of the larger base B is 15 feet?
13. A bag of potatoes weighs 40 pounds. How much does the bag weigh in kilograms?
14. A drop of water is approximately equal to 0.05 milliliter. You fill a cup with 1000 drops of water. How many liters of water are in the cup?

Answers

1. ______
2. ______
3. ______
4. ______
5. ______
6. ______
7. ______
8. ______
9. ______
10. ______
11. ______
12. ______
13. ______
14. ______

For the Teacher
Additional Review Options

- **Quiz***Show*
- Big Ideas Test Generator
- Game Closet at *BigIdeasMath.com*
- Vocabulary Puzzle Builder
- Resources by Chapter
  Puzzle Time
  Study Help

Answers

1. $y = -19$
2. $n = -8$
3. $t = 12\pi$

Review of Common Errors

Exercises 1–3

- Students may perform the same operation that is in the equation instead of the inverse operation. Remind them that they must use an inverse operation to undo an operation. Also, remind them to check their solution in the original equation.

Exercises 4–6

- Students may change the exponent of the variable when combining like terms. For example, they may write the sum $x + x + \frac{1}{2}x + \frac{1}{2}x$ as $3x^3$. Remind them how to correctly combine like terms that have variables.

Exercise 7

- Students may make mistakes when collecting the variable terms on one side and the constant terms on the other side. Remind them that when a term is moved from one side of an equation to the other, the inverse operation is used. Also, remind them to check their solution in the original equation.

Exercises 8 and 9

- Students may multiply only one of the terms in parentheses by the factor outside the parentheses. Remind them how to correctly use the Distributive Property.

Exercises 10 and 11

- Students may be unsure about how to solve the formula for the specified variable. Point out that they should work through the order of operations *backwards,* using inverse operations to isolate the variable.

Exercises 12–16

- When using conversion factor(s), students may write them inverted. Point out that a conversion factor should be written so that they can divide out the units in the given measure.

Exercise 17

- Students may forget to cube the conversion factor. Point out that when they work with cubic units, they must use the cube of the linear conversion factor.

1 Chapter Review

Check It Out
Vocabulary Help
BigIdeasMath.com

Review Key Vocabulary

literal equation, *p. 26*

conversion factor, *p. 32*

Review Examples and Exercises

1.1 Solving Simple Equations *(pp. 2–9)*

The *boiling point* of a liquid is the temperature at which the liquid becomes a gas. The boiling point of mercury is about $\frac{41}{200}$ of the boiling point of lead. Write and solve an equation to find the boiling point of lead.

Let x be the boiling point of lead.

$$\frac{41}{200}x = 357 \qquad \text{Write the equation.}$$

$$\frac{200}{41} \cdot \left(\frac{41}{200}x\right) = \frac{200}{41} \cdot 357 \qquad \text{Multiply each side by } \frac{200}{41}.$$

$$x \approx 1741 \qquad \text{Simplify.}$$

The boiling point of lead is about 1741°C.

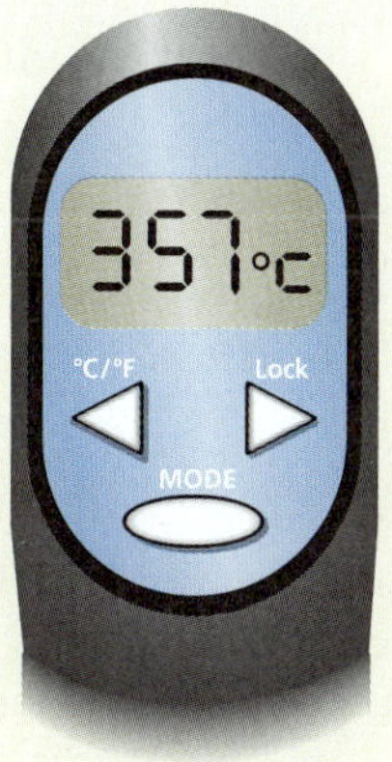

Mercury 357°C

Exercises

Solve the equation. Check your solution.

1. $y + 8 = -11$

2. $3.2 = -0.4n$

3. $-\frac{t}{4} = -3\pi$

1.2 Solving Multi-Step Equations *(pp. 10–15)*

Solve $-14x + 28 + 6x = -44$.

$$-14x + 28 + 6x = -44 \qquad \text{Write the equation.}$$

$$-8x + 28 = -44 \qquad \text{Combine like terms.}$$

$$-28 \quad -28 \qquad \text{Subtract 28 from each side.}$$

$$-8x = -72 \qquad \text{Simplify.}$$

$$\frac{-8x}{-8} = \frac{-72}{-8} \qquad \text{Divide each side by } -8.$$

$$x = 9 \qquad \text{Simplify.}$$

The solution is $x = 9$.

Exercises

Find the value of x. Then find the angle measures of the polygon.

4.

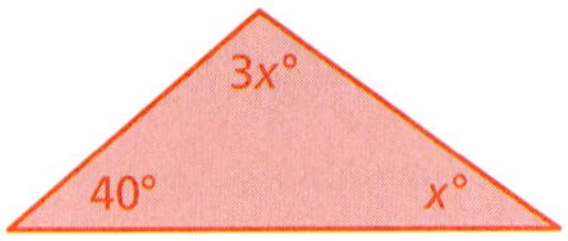

Sum of angle measures: 180°

5.

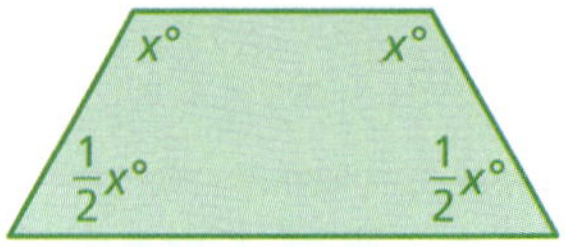

Sum of angle measures: 360°

6.

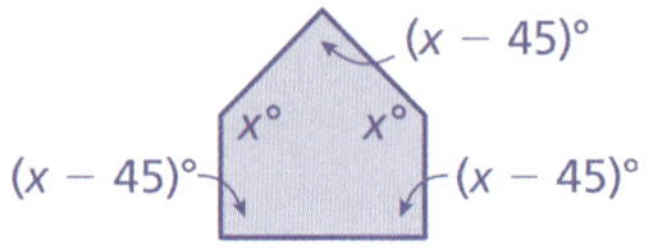

Sum of angle measures: 540°

1.3 Solving Equations with Variables on Both Sides *(pp. 16–21)*

Solve $3(x - 4) = -2(4 - x)$.

$3(x - 4) = -2(4 - x)$	Write the equation.
$3x - 12 = -8 + 2x$	Use Distributive Property.
$-2x \qquad\qquad -2x$	Subtract $2x$ from each side.
$x - 12 = -8$	Simplify.
$+12 \qquad +12$	Add 12 to each side.
$x = 4$	Simplify.

∴ The solution is $x = 4$.

Exercises

Solve the equation. Check your solution.

7. $5m - 1 = 4m + 5$ **8.** $3(5p - 3) = 5(p - 1)$ **9.** $\frac{2}{5}n + \frac{1}{10} = \frac{1}{2}(n + 4)$

1.4 Rewriting Equations and Formulas *(pp. 24–29)*

The equation for a line in slope-intercept form is $y = mx + b$. Solve the equation for x.

$y = mx + b$	Write the equation.
$y - b = mx + b - b$	Subtract b from each side.
$y - b = mx$	Simplify.
$\frac{y - b}{m} = \frac{mx}{m}$	Divide each side by m.
$\frac{y - b}{m} = x$	Simplify.

∴ So, $x = \frac{y - b}{m}$.

Review Game

Scavenger Hunt

Big Ideas Game Closet

Materials per Group:

- paper
- pencil

Directions:

Divide the class into groups. Each group finds 5 different printed measurements in the classroom and converts them. If the measurement is in metric units, students convert it to customary units. If the measurement is in customary units, students convert it to metric units.

Sources for measurements could include water bottle labels, glue bottle labels, rulers, yardsticks, or similar items. If possible, allow students to look for measurements outside of the classroom, such as in the cafeteria or the gym.

Who Wins?

The first group to find and correctly convert 5 measurements wins.

For the Student Additional Practice

- Lesson Tutorials
- Study Help (textbook)
- Student Website
  - Multi-Language Glossary
  - Practice Assessments

Answers

4. $x = 35$; 40°, 105°, 35°

5. $x = 120$; 60°, 120°, 120°, 60°

6. $x = 135$; 90°, 135°, 90°, 135°, 90°

7. $m = 6$

8. $p = 0.4$

9. $n = -19$

10. **a.** $K = \frac{5}{9}(F - 32) + 273.15$

b. about 388.71 K

11. **a.** $A = \frac{1}{2}h(b + B)$

b. $h = \frac{2A}{b + B}$

c. $h = 6$ cm

12. 43.18 **13.** 84.21

14. 1.45 **15.** 136.4 ft/sec

16. about 624 gal/h

17. 0.0343 m^3

My Thoughts on the Chapter

Teacher Tip

Not allowed to write in your teaching edition? Use sticky notes to record your thoughts.

What worked. . .

What did not work. . .

What I would do differently. . .

Exercises

10. **a.** The formula $F = \frac{9}{5}(K - 273.15) + 32$ converts a temperature from Kelvin K to Fahrenheit F. Solve the formula for K.

b. Convert 240 °F to Kelvin K. Round your answer to the nearest hundredth.

11. **a.** Write the formula for the area A of a trapezoid.

b. Solve the formula for h.

c. Use the new formula to find the height h of the trapezoid.

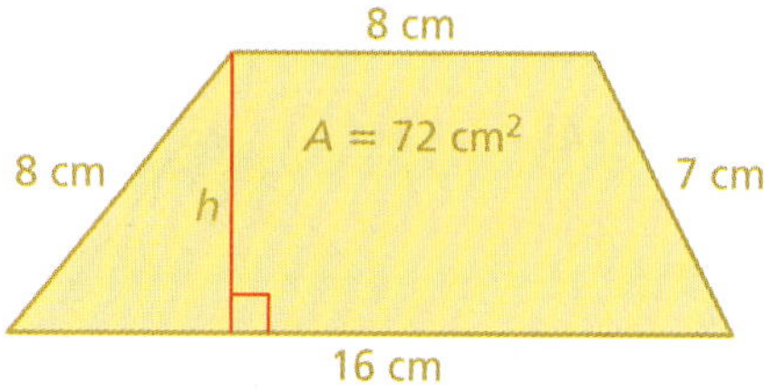

1.5 Converting Units of Measure (pp. 30–37)

Convert the speed of the train to miles per hour.

$$\frac{5\text{ km}}{1\text{ min}}\left(\frac{60\text{ min}}{1\text{ h}}\right)\left(\frac{0.6\text{ mi}}{1\text{ km}}\right)$$

60 min = 1 h

0.6 mi ≈ 1 km

$$\frac{5\text{ km}}{1\text{ min}}\left(\frac{60\text{ min}}{1\text{ h}}\right)\left(\frac{0.6\text{ mi}}{1\text{ km}}\right) = \frac{5 \cdot 60 \cdot 0.6\text{ mi}}{1\text{ h}}$$

$$= \frac{180\text{ mi}}{1\text{ h}}$$

∴ The speed of the train is 180 miles per hour.

Exercises

Copy and complete the statement.

12. 17 in. ≈ ___ cm

13. 80 L ≈ ___ qt

14. 4800 ft ≈ ___ km

15. **BASEBALL** A baseball pitch is clocked at 93 miles per hour. Convert this rate to feet per second.

16. **POOL** A community pool is filling at a rate of 40 liters per minute. Convert this rate to gallons per hour.

17. **GEOMETRY** What is the volume of the cube in cubic meters?

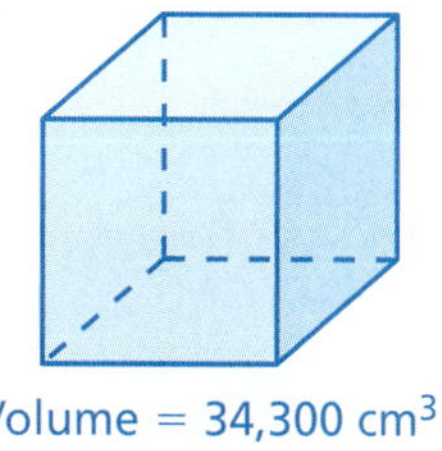

1 Chapter Test

Check It Out
Test Practice
BigIdeasMath.com

Solve the equation. Check your solution.

1. $4 + y = 9.5$

2. $x - 3\pi = 5\pi$

3. $3.8n - 13 = 1.4n + 5$

Find the value of *x*. Then find the angle measures of the polygon.

4.

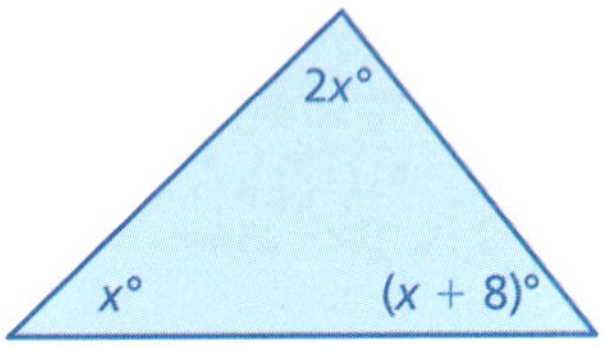

Sum of angle measures: 180°

5.

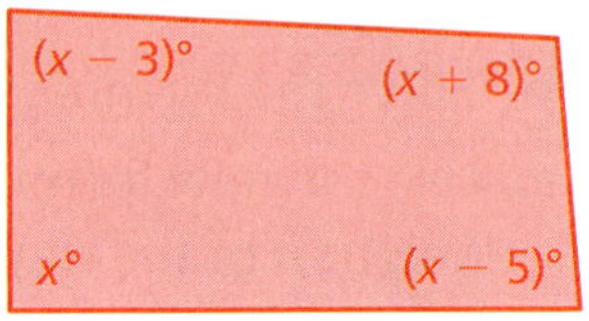

Sum of angle measures: 360°

Solve the equation for *y*.

6. $1.2x - 4y = 28$

7. $0.5 = 0.4y - 0.25x$

Solve the formula for the red variable.

8. Perimeter of a rectangle: $P = 2\ell + 2w$

9. Distance formula: $d = rt$

Copy and complete the statement.

10. 27 in. ≈ ___ cm

11. 14 mi ≈ ___ km

12. 12 L ≈ ___ qt

13. **BASKETBALL** Your basketball team wins a game by 13 points. The opposing team scores 72 points. Write and solve an equation to find your team's score.

14. **AQUARIUM** You want to buy the aquarium that has the greater volume. Which of the aquariums should you buy? Explain.

20 quarts

19 liters

15. **AIRPORT** Runway 17 at an airport is 5020 feet long. What is the length of the runway in meters? Round your answer to the nearest whole number.

16. **JOBS** Your profit for mowing lawns this week is \$24. You are paid \$8 per hour and you paid \$40 for gas for the lawnmower. How many hours did you work this week?

Test Item References

Chapter Test Questions	Section to Review
1, 2, 13	1.1
4, 5, 16	1.2
3	1.3
6–9	1.4
10–12, 14, 15	1.5

Test-Taking Strategies

Remind students to quickly look over the entire test before they start so that they can budget their time. When working with equations, students need to write all numbers and variables clearly, line up terms in each step, and not crowd their work. Also, there are conversions on the test, so remind students to jot down conversions on the back of their test before they start. Have students use the **Stop** and **Think** strategy.

Common Assessment Errors

- **Exercises 1–3** Students may perform the same operation that is in the equation instead of the inverse operation. Remind them that they must use an inverse operation to undo an operation. Also, remind students to check their solution in the original equation.
- **Exercises 4 and 5** Students may change the exponent of the variable when combining like terms that have variables. For example, they may write the sum $2x + x + (x + 8)$ as $4x^3 + 8$. Remind them how to correctly combine like terms that have variables.
- **Exercises 6–9** Students may be unsure about how to solve for the specified variable. Point out that they should work through the order of operations *backwards,* using inverse operations to isolate the variable.
- **Exercises 10–12, 14, and 15** When using a conversion factor, students may write it inverted. Point out that a conversion factor should be written so that they can divide out the units in the given measure.

Reteaching and Enrichment Strategies

If students need help. . .	If students got it. . .
Resources by Chapter • Practice A and Practice B • Puzzle Time Record and Practice Journal Practice Differentiating the Lesson Lesson Tutorials Practice from the Test Generator Skills Review Handbook	Resources by Chapter • Enrichment and Extension • School-to-Work • Financial Literacy Game Closet at *BigIdeasMath.com* Start Standardized Test Practice

Answers

1. $y = 5.5$
2. $x = 8\pi$
3. $n = 7.5$
4. $x = 43; 43°, 86°, 51°$
5. $x = 90; 90°, 87°, 98°, 85°$
6. $y = 0.3x - 7$
7. $y = 0.625x + 1.25$
8. $w = \frac{P}{2} - \ell$
9. $r = \frac{d}{t}$
10. 68.58
11. 22.4
12. 12.63
13. $x - 13 = 72; x = 85$ points
14. It doesn't matter. The aquariums have about the same volume.
15. 1506 m
16. 8 hours

Assessment Book

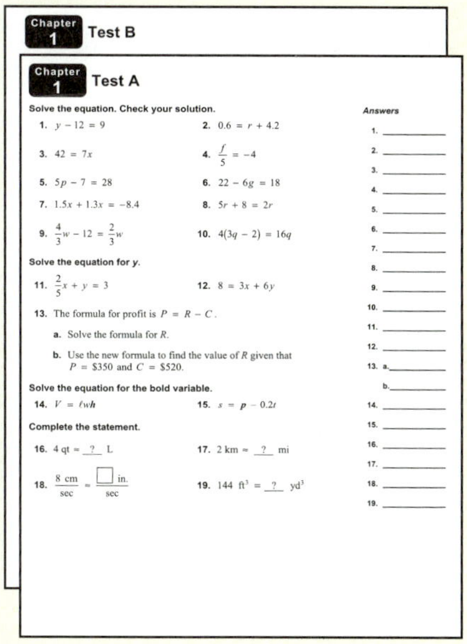
Chapter 1 Test B

Chapter 1 Test A

Solve the equation. Check your solution.

1. $y - 12 = 9$
2. $0.6 = r + 4.2$
3. $42 = 7x$
4. $\frac{f}{5} = -4$
5. $5p - 7 = 28$
6. $22 - 6g = 18$
7. $1.5x + 1.3x = -8.4$
8. $5r + 8 = 2r$
9. $\frac{4}{3}w - 12 = \frac{2}{3}w$
10. $4(3q - 2) = 16q$

Solve the equation for *y*.

11. $\frac{2}{5}x + y = 3$
12. $8 = 3x + 6y$
13. The formula for profit is $P = R - C$.
    a. Solve the formula for R.
    b. Use the new formula to find the value of R given that $P = \$350$ and $C = \$520$.

Solve the equation for the bold variable.

14. $V = \ell w \mathbf{h}$
15. $s = \mathbf{p} - 0.2t$

Complete the statement.

16. 4 qt ≈ ? L
17. 2 km ≈ ? mi
18. $\frac{8 \text{ cm}}{\text{sec}} \approx \frac{\square \text{ in.}}{\text{sec}}$
19. 144 ft^3 = ? yd^3

Answers

1. ____ 2. ____ 3. ____ 4. ____ 5. ____ 6. ____ 7. ____ 8. ____ 9. ____ 10. ____ 11. ____ 12. ____ 13. a. ____ b. ____ 14. ____ 15. ____ 16. ____ 17. ____ 18. ____ 19. ____

Test-Taking Strategies

Available at *BigIdeasMath.com*

After Answering Easy Questions, Relax
Answer Easy Questions First
Estimate the Answer
Read All Choices before Answering
Read Question before Answering
Solve Directly or Eliminate Choices
Solve Problem before Looking at Choices
Use Intelligent Guessing
Work Backwards

About this Strategy

When taking a multiple choice test, be sure to read each question carefully and thoroughly. Before answering a question, determine exactly what is being asked, then eliminate the wrong answers and select the best choice.

Answers

1. A
2. I
3. D
4. 5 yd
5. G

Technology For the Teacher

Big Ideas Test Generator

Item Analysis

1. **A.** Correct answer
   B. The student subtracts 4 from 32 instead of dividing 32 by 4.
   C. The student adds 4 to 32 instead of dividing 32 by 4.
   D. The student multiplies 4 and 32 instead of dividing 32 by 4.

2. **F.** The student correctly subtracts 3 from 39, but then multiplies instead of dividing.
   G. The student correctly subtracts 3 from 39, but then subtracts 2 instead of dividing.
   H. The student incorrectly adds 3 and 39 instead of subtracting, then performs division correctly.
   I. Correct answer

3. **A.** The student mishandles the fraction, multiplying by $\frac{1}{4}$ instead of dividing.
   B. The student finds the number of teaspoons instead of quarter-teaspoons in the bottle.
   C. The student adds 6 and 4 instead of multiplying, then works correctly.
   D. Correct answer

4. **Gridded Response:** Correct answer: 5 yd

   Common Error: The student correctly calculates 15 feet as the distance traveled in 1 second, but fails to convert feet to yards.

5. **F.** The student moves r over in the equation instead of dividing both sides by r.
   G. Correct answer
   H. The student subtracts r instead of dividing by r.
   I. The student divides r by d and moves t to the other side of the equal sign instead of dividing d by r.

6. **A.** The student misunderstands that $3x$ and 5 are not like terms.
   B. Correct answer
   C. The student does not realize that the Distributive Property must first be used to multiply 2 by $x + 7$.
   D. The student does not realize that the Distributive Property must first be used to multiply 2 by $x + 7$.

Standardized Test Practice Icons

Gridded Response

Short Response (2-point rubric)

Extended Response (4-point rubric)

1 Standardized Test Practice

1. Which value of x makes the equation true?

$$4x = 32$$

A. 8 **C.** 36

B. 28 **D.** 128

2. A taxi ride costs \$3 plus \$2 for each mile driven. When you rode in a taxi, the total cost was \$39. This can be modeled by the equation below, where m represents the number of miles driven.

$$2m + 3 = 39$$

How long was your taxi ride?

F. 72 mi **H.** 21 mi

G. 34 mi **I.** 18 mi

Test-Taking Strategy

Solve Directly or Eliminate Choices

"You can eliminate A and D. Then, solve directly to determine that the correct answer is B."

3. One fluid ounce (fl oz) contains 6 teaspoons. You add $\frac{1}{4}$ teaspoon of vanilla each time you make hot chocolate. How many times can you make hot chocolate using the bottle of vanilla shown?

A. 6 **C.** 40

B. 24 **D.** 96

4. A bicyclist is riding at a speed of 900 feet per minute. How many yards does the bicyclist ride in 1 second?

5. The formula below relates distance, rate, and time.

$$d = rt$$

Solve this formula for t.

F. $t = dr$ **H.** $t = d - r$

G. $t = \frac{d}{r}$ **I.** $t = \frac{r}{d}$

6. What could be the first step to solve the equation shown below?

$$3x + 5 = 2(x + 7)$$

A. Combine $3x$ and 5.

B. Multiply x by 2 and 7 by 2.

C. Subtract x from $3x$.

D. Subtract 5 from 7.

7. You work as a sales representative. You earn $400 per week plus 5% of your total sales for the week.

Think Solve Explain

Part A Last week, you had total sales of $5000. Find your total earnings. Show your work.

Part B One week, you earned $1350. Let s represent your total sales that week. Write an equation that could be used to find s.

Part C Using your equation from Part B, find s. Show all steps clearly.

8. In ten years, Maria will be 39 years old. Let m represent Maria's age today. Which equation can be used to find m?

F. $m = 39 + 10$

G. $m - 10 = 39$

H. $m + 10 = 39$

I. $10m = 39$

9. Which value of y makes the equation below true?

$$3y + 8 = 7y + 11$$

A. -4.75

B. -0.75

C. 0.75

D. 4.75

10. The equation below is used to convert a Fahrenheit temperature F to its equivalent Celsius temperature C.

$$C = \frac{5}{9}(F - 32)$$

Which formula can be used to convert a Celsius temperature to its equivalent Fahrenheit temperature?

F. $F = \frac{5}{9}(C - 32)$

G. $F = \frac{9}{5}(C + 32)$

H. $F = \frac{9}{5}C + \frac{32}{5}$

I. $F = \frac{9}{5}C + 32$

Item Analysis (continued)

7. **4 points** The student demonstrates a thorough understanding of evaluating expressions, writing equations, and solving equations, and presents his or her steps clearly. The following answers should be obtained: Part A: \$650; Part B: $0.05s + 400 = 1350$; Part C: \$19,000.

   3 points The student demonstrates an essential but less than thorough understanding. In particular, the correct equation or its equivalent should be given in Part B, but an arithmetic error may have been performed in Part C.

   2 points The student demonstrates a partial understanding of the processes of writing and solving equations. Part A should be correctly completed, but the equation in Part B may be written incorrectly. Alternatively, the correct equation could be written in Part B, but Part C might display misunderstanding of how to proceed.

   1 point The student demonstrates a limited understanding of equation writing and solving, as well as working with percents. The student's response is incomplete and exhibits many flaws.

   0 points The student provided no response, a completely incorrect or incomprehensible response, or a response that demonstrates insufficient understanding of percents and equations.

8. **F.** The student misunderstands the problem and decides to add the two numbers together.

   G. The student gets the idea that *m* and 39 are 10 apart, but chose subtraction instead of addition to relate them.

   H. Correct answer

   I. The student mistakes $10m$ for $10 + m$.

9. **A.** The student adds 8 and 11 instead of subtracting, and either misplaces a negative sign when subtracting $3x$ and $7x$ or performs a sign error when dividing.

   B. Correct answer

   C. The student subtracts 8 and 11 correctly, but then misplaces a negative sign when subtracting $3x$ and $7x$ or performs a sign error when dividing.

   D. The student adds 8 and 11 instead of subtracting.

10. **F.** The student simply interchanges the variables.

    G. The student "inverts" the variables, the fraction, and the subtraction.

    H. The student correctly multiplies both sides of the equation by $\frac{9}{5}$, but also incorrectly applies operations to 32.

    I. Correct answer

11. **Gridded Response:** Correct answer: 14 weeks

    Common Error: The student adds 35 and 175 to get 210, then divides by 10 to get 21 as an answer.

Answers

6. B
7. *Part A* \$650

   Part B $0.05s + 400 = 1350$

   Part C \$19,000
8. H
9. B
10. I

Answers

11. 14 weeks

12. D

13. I

14. A

Answer for Extra Example

1. **A.** The student multiplies 4 by 0.95 but does not carry out the multiplication correctly.

   B. Correct answer

   C. The student divides 4 by 0.95 instead of multiplying.

   D. The student adds 4 and 0.95.

Item Analysis (continued)

12. **A.** The student multiplies by 12 because there are 12 inches in 1 foot.

B. The student realizes that volume requires accounting for "12 inches = 1 foot" three times, but adds the factors instead of multiplying them.

C. The student finds the number of square inches in 1 square foot.

D. Correct answer

13. **F.** The student distributes correctly but then makes a mistake combining the constant terms, yielding $2x = -11$.

G. The student does not distribute the left side of the equation correctly, yielding $6x - 3$.

H. The student combines $6x$ and $4x$ incorrectly, yielding $10x$ instead of $2x$.

I. Correct answer

14. **A.** Correct answer

B. The student incorrectly uses the fact that there are 4 items on one side and 2 on the other to get the ratio $\frac{2}{4} = \frac{1}{2}$.

C. The student incorrectly uses the fact that there are 4 items on one side and 2 on the other to get the ratio $\frac{2}{4} = \frac{1}{2}$, and then misuses the order of the ratio.

D. The student gets the correct ratio of $\frac{1}{3}$, but misuses it.

Extra Example for Standardized Test Practice

1. Which of the following best completes the table below?

1 quart ≈ 0.95 liters
1 gallon = 4 quarts
1 gallon ≈ ☐ liters

A. 3.6 **C.** 4.21

B. 3.8 **D.** 4.95

11. You have already saved \$35 for a new cell phone. You need \$175 in all. You think you can save \$10 per week. At this rate, how many more weeks will you need to save money before you can buy the new cell phone?

12. The cube shown below has edge lengths of 1 foot. What is the volume of the cube?

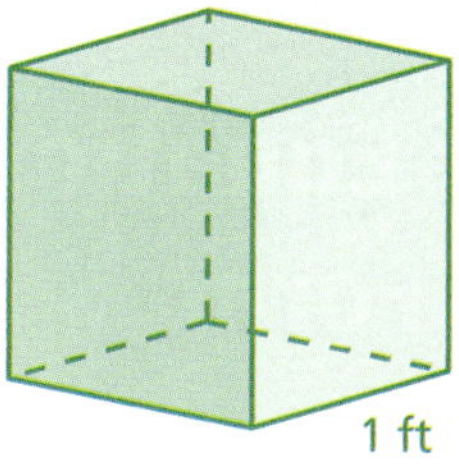

A. 12 in.^{3}

B. 36 in.^{3}

C. 144 in.^{3}

D. 1728 in.^{3}

13. Which value of x makes the equation below true?

$$6(x - 3) = 4x - 7$$

F. -5.5

G. -2

H. 1.1

I. 5.5

14. The drawing below shows equal weights on two sides of a balance scale.

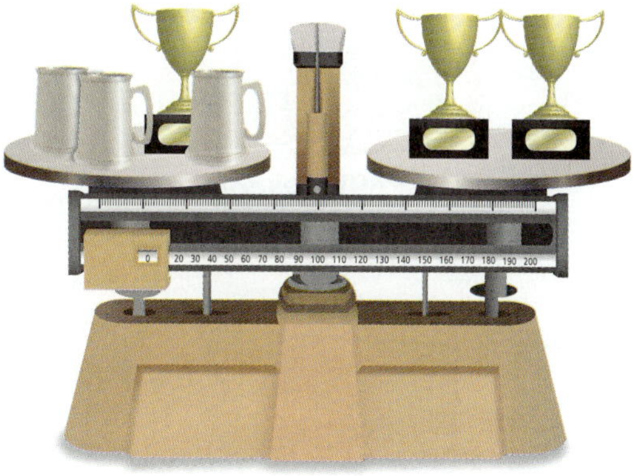

What can you conclude from the drawing?

A. A mug weighs one-third as much as a trophy.

B. A mug weighs one-half as much as a trophy.

C. A mug weighs twice as much as a trophy.

D. A mug weighs three times as much as a trophy.

2 Graphing Linear Equations and Linear Systems

"Okay Descartes, stand on the *y*-axis and try to intercept the pass when I throw."

"Here's an easy example of a line with a slope of 1."

"You eat one mouse treat the first day. Two treats the second day. And so on. Get it?"

Connections to Previous Learning

- Write, solve, and graph one-step and two-step linear equations.
- Construct and analyze tables, graphs, and equations to describe linear functions and other simple relations.

- Graph proportional relationships and identify the unit rate as slope of linear functions.

- Construct and analyze tables, graphs, and models to describe linear equations.
- Interpret slope and *x*- and *y*-intercepts when graphing a linear equation for a real world problem.
- Use tables, graphs, and models to represent, analyze, and solve real world problems related to systems of linear equations.
- Identify the solution to a system of linear equations using graphs.

Math in History

The concept of writing all numbers by using only ten different symbols appears to have originated in India.

★ Here are some of the symbols that were used in the Brahmi system in India until around 400 A.D.

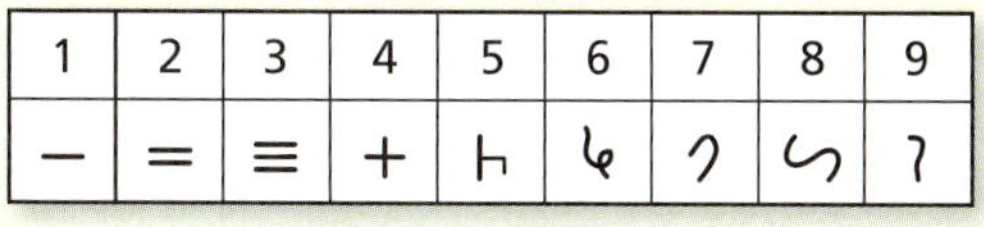

Notice that there is no symbol for 0. That concept was not yet devised. Also notice that the symbols for 2 and 3 are related to our modern symbols for 2 and 3.

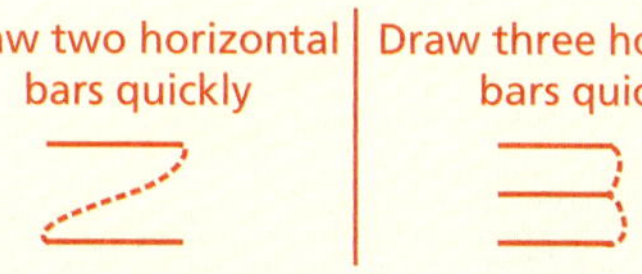

★ By comparing the Brahmi symbols to another culture's symbols (such as the Chinese), you can see that our modern symbols are more closely related to the ancient Brahmi symbols.

1	2	3	4	5	6	7	8	9
一	二	三	四	五	六	七	八	九

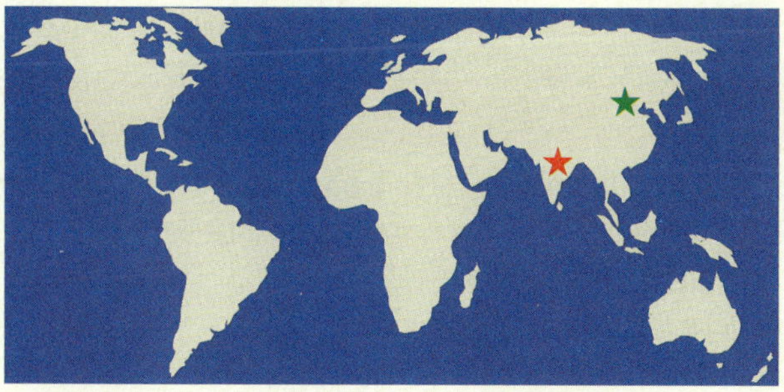

Pacing Guide for Chapter 2

Chapter Opener	1 Day
Section 1 Activity Lesson	 1 Day 1 Day
Section 2 Activity Lesson	 1 Day 2 Days
Section 3 Activity Lesson	 1 Day 1 Day
Section 4 Activity Lesson	 1 Day 1 Day
Study Help / Quiz	1 Day
Section 5 Activity Lesson	 2 Days 1 Day
Section 6 Activity Lesson	 1 Day 1 Day
Section 7 Activity Lesson	 1 Day 1 Day
Quiz / Chapter Review	1 Day
Chapter Test	1 Day
Standardized Test Practice	1 Day
Total Chapter 2	21 Days
Year-to-Date	39 Days

Check Your Resources

- Record and Practice Journal
- Resources by Chapter
- Skills Review Handbook
- Assessment Book
- Worked-Out Solutions

Technology For the Teacher

Dynamic Classroom

The Dynamic Planning Tool
Editable Teacher's Resources at *BigIdeasMath.com*

Additional Topics for Review

- Order of Operations
- Exponents
- Plotting points in Quadrant I

Try It Yourself

1. -12 **2.** -23

3. 15 **4.** $4\frac{3}{4}$

5. $(0, 4)$ **6.** $(4, 2)$

7. Point R **8.** Point N

Record and Practice Journal

1. 5 **2.** 16

3. -5 **4.** $-38\frac{1}{2}$

5. 108 **6.** 65

7. $-3\frac{7}{19}$ **8.** 262

9. $50.00

10. $(-5, 0)$ **11.** $(3, -5)$

12. Point F **13.** Point G

14. Point B, Point H

15. Point C, Point E

16–20.

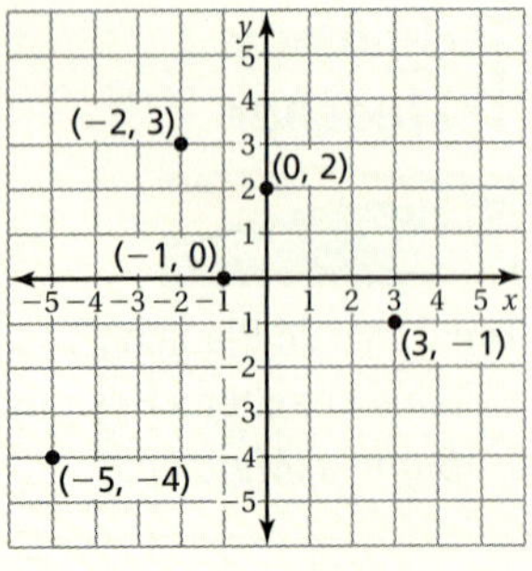

Math Background Notes

Vocabulary Review

- Evaluate
- Expression
- Order of Operations
- Substitute
- Coordinates

Evaluating Expressions Using Order of Operations

- Students should know how to substitute values into algebraic expressions and evaluate the results using order of operations.
- **Teaching Tip:** Sometimes color coding substitutions can help students to evaluate expressions. Each time you want to substitute a number in place of a variable, you must substitute your lead pencil for a colored pencil.
- Remind students that after they substitute values in for x and y, they must use the correct order of operations to continue simplifying the expression.
- **Common Error:** Encourage students to use a set of parentheses whenever they do a substitution. This will help students distinguish between subtracting 7 and multiplying by -7.

Plotting Points

- Students should know how to plot points in all four quadrants.
- **Common Error:** Students may write the coordinates backwards. Remind them that coordinates are written in alphabetical order with the x move (horizontal) written before the y move (vertical).
- **Common Error:** Students may also have difficulty with the negative numbers associated with plotting outside Quadrant I. Remind them that the negatives are directional. A negative x-value communicates a move to the left of the origin and a negative y-coordinate communicates a move downward from the origin.

Reteaching and Enrichment Strategies

If students need help. . .	If students got it. . .
Record and Practice Journal • Fair Game Review Skills Review Handbook Lesson Tutorials	Game Closet at *BigIdeasMath.com* Start the next section

What You Learned Before

"I estimate that we are on a slope of about −0.625. What do you think?"

Evaluating Expressions Using Order of Operations

Example 1 **Evaluate $2xy + 3(x + y)$ when $x = 4$ and $y = 7$.**

$2xy + 3(x + y) = 2(4)(7) + 3(4 + 7)$	Substitute 4 for x and 7 for y.
$= 8(7) + 3(4 + 7)$	Use order of operations.
$= 56 + 3(11)$	Simplify.
$= 56 + 33$	Multiply.
$= 89$	Add.

Try It Yourself

Evaluate the expression when $a = \frac{1}{4}$ and $b = 6$.

1. $-8ab$
2. $16a^2 - 4b$
3. $\frac{5b}{32a^2}$
4. $12a + (b - a - 4)$

Plotting Points

Example 2 **Write the ordered pair that corresponds to Point U.**

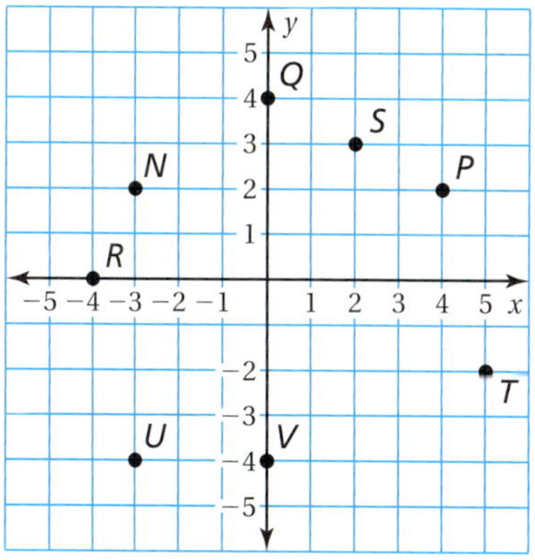

Point U is 3 units to the left of the origin and 4 units down. So, the x-coordinate is −3 and the y-coordinate is −4.

∴ The ordered pair (−3, −4) corresponds to Point U.

Example 3 **Which point is located at (5, −2)?**

Start at the origin. Move 5 units right and 2 units down.

∴ Point T is located at (5, −2).

Try It Yourself

Use the graph to answer the question.

5. Write the ordered pair that corresponds to Point Q.
6. Write the ordered pair that corresponds to Point P.
7. Which point is located at (−4, 0)?
8. Which point is located in Quadrant II?

2.1 Graphing Linear Equations

Essential Question How can you recognize a linear equation? How can you draw its graph?

1 ACTIVITY: Graphing a Linear Equation

Work with a partner.

a. Use the equation $y = \frac{1}{2}x + 1$ to complete the table. (Choose any two x-values and find the y-values.)

	Solution Points	
x		
$y = \frac{1}{2}x + 1$		

b. Write the two ordered pairs given by the table. These are called **solution points** of the equation.

c. Plot the two solution points. Draw a line *exactly* through the two points.

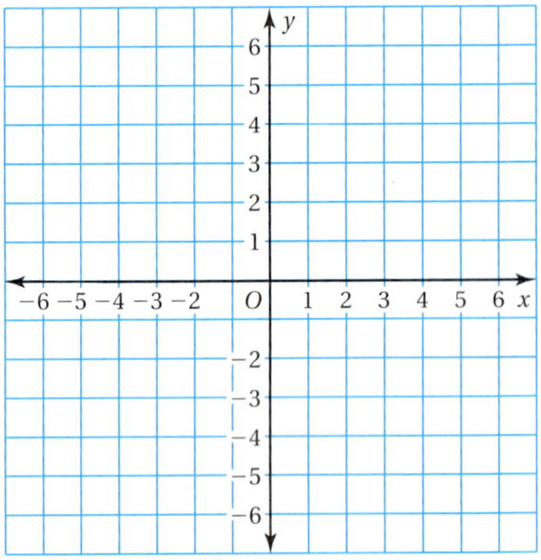

d. Find a different point on the line. Check that this point is a solution point of the equation $y = \frac{1}{2}x + 1$.

e. **GENERALIZE** Do you think it is true that *any* point on the line is a solution point of the equation $y = \frac{1}{2}x + 1$? Explain.

f. Choose five additional x-values for the table. (Choose positive and negative x-values.) Plot the five corresponding solution points. Does each point lie on the line?

	Solution Points				
x					
$y = \frac{1}{2}x + 1$					

g. **GENERALIZE** Do you think it is true that *any* solution point of the equation $y = \frac{1}{2}x + 1$ is a point on the line? Explain.

h. **THE MEANING OF A WORD** Why is $y = ax + b$ called a *linear equation*?

Laurie's Notes

Introduction

For the Teacher

- **Goal:** Students will use solution points to graph linear equations.
- In honor of Rene Descartes, the coordinate plane is often called the *Cartesian plane.*
- Throughout this chapter, you may encounter applications that show a graph of discrete data with a line through the points. At this point in the text, we do not think it is necessary for students to distinguish between discrete data (plotting points only) and continuous data (plotting points along with the line). Students can draw a line through discrete points to help them solve an exercise. They will learn more about discrete and continuous data at a later time.

Motivate

- Play a game of coordinate BINGO.
- Distribute small coordinate grids to students. They should plot ten ordered pairs, where the *x*- and *y*-coordinates are integers between −4 and 4.
- Generate a random ordered pair in the grid. Write the integers from −4 to 4 on slips of paper and place them in a bag. Draw and replace an integer twice to generate the ordered pair, then write it on the board.
- Each time you record a new ordered pair, the students check to see if it is one of their 10 ordered pairs. If it is, they put an X there. The goal is to be the first person with three X's. If a student thinks they have won, they read their ordered pairs for you to check against the master list.

? "Are there ordered pairs that are not on lattice points, meaning the *x*- or *y*-coordinate is not an integer? Explain." yes; It's possible for the ordered pair to be $\left(3.5, \frac{1}{2}\right)$. Plot whatever example students give.

- Remind students that the ordered pairs are always (*x*, *y*), where *x* is the horizontal direction and *y* is the vertical direction.

Activity Notes

Activity 1

- Some students will recognize right away that if they substitute an even number for *x*, the *y*-coordinate will not be a fraction. It is likely that students will only try positive *x*-values. Encourage them to try negative values for *x*.
- In part (d), suggest that students consider only those ordered pairs that appear to be lattice points.
- Listen and discuss student responses to the generalizations in parts (e) and (g).
- **Big Idea:** The goal of this activity is for students to recognize and understand two related, but different, ideas. 1) *All* solution points of a linear equation lie on the same line. 2) *All* points on the line are solution points of the equation.

Previous Learning

Students should know about slope as a ratio. Students should know how to plot ordered pairs.

Activity Materials	
Introduction	**Textbook**
• small coordinate grid	• straightedge

Start Thinking! and Warm Up

Activity 2.1 Start Thinking!
For use before Activity 2.1

Activity 2.1 Warm Up
For use before Activity 2.1

Copy and complete the table using the given equation.

1. $y = 2x - 3$

x	−1	0	1	2
y				

2. $y = -x + 1$

x	−1	0	1	2
y				

3. $y = x + 5$

x	−1	0	1	2
y				

4. $y = \frac{1}{2}x + 2$

x	−1	0	1	2
y				

2.1 Record and Practice Journal

Essential Question How can you recognize a linear equation? How can you draw its graph?

1 ACTIVITY: Graphing a Linear Equation

Work with a partner.

Sample answer:

a. Use the equation $y = \frac{1}{2}x + 1$ to complete the table. (Choose any two *x*-values and find the *y*-values).

	Solution Points	
x	−2	2
$y = \frac{1}{2}x + 1$	0	2

b. Write the two ordered pairs given by the table. These are called **solution points** of the equation.

(−2, 0), (2, 2)

c. Plot the two solution points. Draw a line *exactly* through the two points.

d. Find a different point on the line. Check that this point is a solution point of the equation $y = \frac{1}{2}x + 1$.

(0, 1)

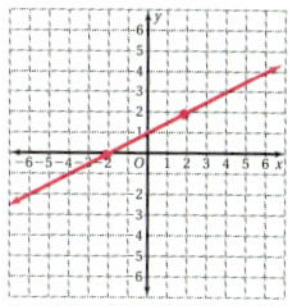

e. GENERALIZE Do you think it is true that *any* point on the line is a solution point of the equation $y = \frac{1}{2}x + 1$? Explain.

yes

Differentiated Instruction

Kinesthetic

For students that are kinesthetic learners and have difficulty in plotting points in the coordinate plane, suggest they use a finger for tracing. Have the student place their finger at the origin and trace left or right along the x-axis to the first coordinate, then trace up or down to the second coordinate. Students should also practice writing the ordered pair of a plotted point. Guide students with questions such as, "Should you move left or right? How far? Should you move up or down? How far?"

2.1 Record and Practice Journal

f. Choose five additional x-values for the table. (Choose positive and negative x-values.) Plot the five corresponding solution points on the previous page. Does each point lie on the line? *Sample answer:*

	Solution Points				
x	-6	-4	1	4	6
$y = \frac{1}{2}x + 1$	-2	-1	$1\frac{1}{2}$	3	4

g. GENERALIZE Do you think it is true that *any* solution point of the equation $y = \frac{1}{2}x + 1$ is a point on the line? Explain.

yes

h. THE MEANING OF A WORD Why is $y = ax + b$ called a *linear equation*?

The graph is a line.

2 HISTORY: Analytic Geometry

Rene Descartes was a French philosopher, scientist, and mathematician.

Up until the time of Descartes, *algebra* and *geometry* were separate fields of mathematics. Descartes's invention of the coordinate plane was of huge importance to mathematics. For the first time, people could "see" solutions of equations. No longer did people have to work with algebra from a purely symbolic point of view.

René Descartes (1596–1650)

Descartes's combination of geometry and algebra is called *analytic* (or algebraic) *geometry*. One of the main discoveries in analytic geometry is that all of the important types of graphs (lines, parabolas, circles, ellipses, and so on) can be represented by simple algebraic equations.

Within a few dozen years, other mathematicians were able to discover all of *calculus*, a field of mathematics that is of great value in business, science, and engineering.

In this book, you will study lines. In Algebra 1 and Algebra 2, you will study many other types of equations.

Line: $y = ax + b$ Parabola: $y = ax^2 + b$ Circle: $x^2 + y^2 = r^2$

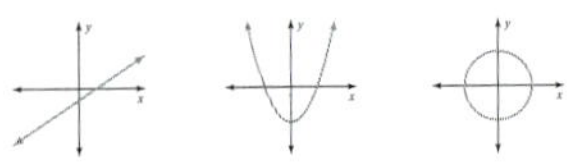

What Is Your Answer?

3. IN YOUR OWN WORDS How can you recognize a linear equation? How can you draw its graph? Write an equation that is linear. Write an equation that is *not* linear.

The form $y = ax + b$ and its graph is a line; find solution points and draw a line through them.

4. Are you a visual learner? Most people can learn mathematics more easily when they see "pictures" of the mathematics. Why do you think Descartes's invention was important to mathematics?

Check students' work.

Laurie's Notes

History 2

- This activity gives students a broader perspective about graphing in the Cartesian plane. Ask volunteers to read this information aloud.
- **FYI:** Many articles have been written about Descartes and in most of them it is reported that Descartes was not healthy in his youth. Because of his delicate health, he was permitted to lie in bed until late morning. This was a custom which he followed until adulthood.
  It is reported that Descartes' invention of the coordinate plane, as a means of connecting algebra and geometry, came to him during a dream on November 10, 1619 while he was serving as a soldier in the army.
- **Extension:** Have students write a short report on a famous mathematician.

What Is Your Answer?

- Discuss students' responses to each question.

Closure

- Find three ordered pairs that are solutions of the equation $y = 2x - 3$. Draw the graph of the line. *Sample answer:* $(-1, -5)$, $(0, -3)$, and $(1, -1)$

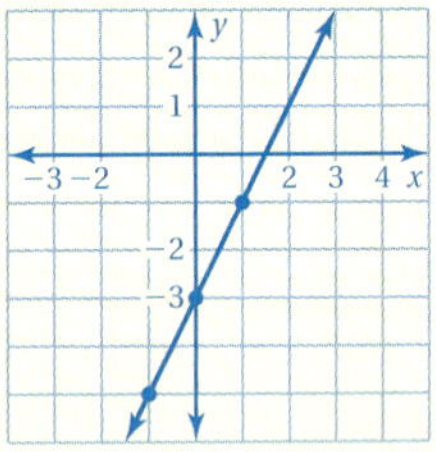

2 HISTORY: Analytic Geometry

René Descartes (1596–1650)

René Descartes was a French philosopher, scientist, and mathematician.

Up until the time of Descartes, *algebra* and *geometry* were separate fields of mathematics. Descartes's invention of the coordinate plane was of huge importance to mathematics. For the first time, people could "see" solutions of equations. No longer did people have to work with algebra from a purely symbolic point of view.

Descartes's combination of geometry and algebra is called *analytic* (or algebraic) *geometry*. One of the main discoveries in analytic geometry is that all of the important types of graphs (lines, parabolas, circles, ellipses, and so on) can be represented by simple algebraic equations.

Within a few dozen years, other mathematicians were able to discover all of *calculus*, a field of mathematics that is of great value in business, science, and engineering.

In this book, you will study lines. In Algebra 1 and Algebra 2, you will study many other types of equations.

Line: $y = ax + b$ Parabola: $y = ax^2 + b$ Circle: $x^2 + y^2 = r^2$

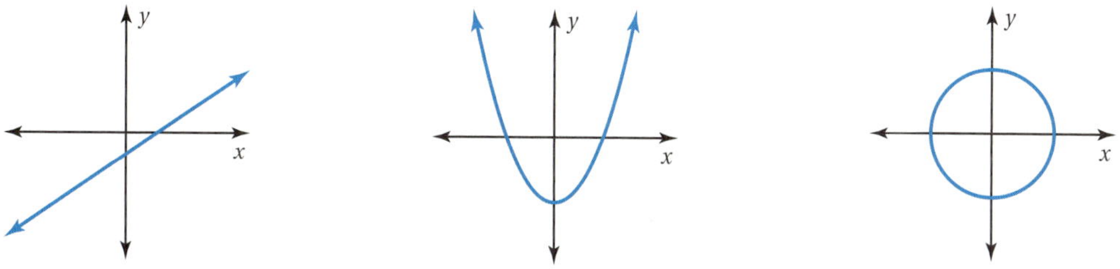

What Is Your Answer?

3. **IN YOUR OWN WORDS** How can you recognize a linear equation? How can you draw its graph? Write an equation that is linear. Write an equation that is *not* linear.

4. Are you a visual learner? Most people can learn mathematics more easily when they see "pictures" of the mathematics. Why do you think Descartes's invention was important to mathematics?

Practice Use what you learned about graphing linear equations to complete Exercises 3 and 4 on page 52.

2.1 Lesson

Check It Out
Lesson Tutorials
BigIdeasMath.com

Key Vocabulary
linear equation, *p. 50*
solution of a linear equation, *p. 50*

Remember
An ordered pair (x, y) is used to locate a point in a coordinate plane.

Key Idea

Linear Equations

A **linear equation** is an equation whose graph is a line. The points on the line are **solutions** of the equation.

You can use a graph to show the solutions of a linear equation. The graph below is for the equation $y = x + 1$.

x	y	(x, y)
-1	0	$(-1, 0)$
0	1	$(0, 1)$
2	3	$(2, 3)$

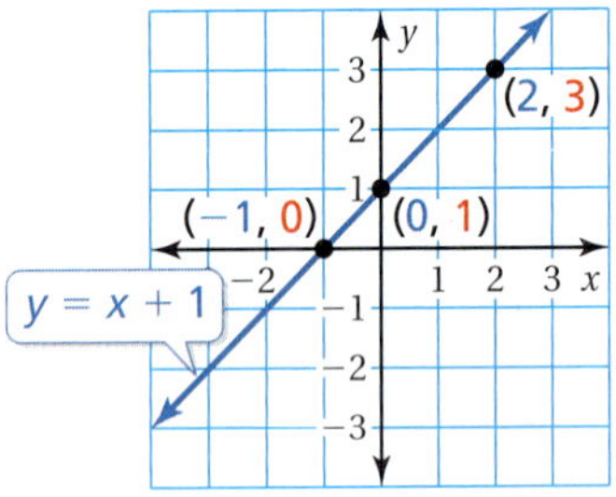

EXAMPLE 1 Graphing a Linear Equation

Graph $y = -2x + 1$.

Step 1: Make a table of values.

x	$y = -2x + 1$	y	(x, y)
-1	$y = -2(-1) + 1$	3	$(-1, 3)$
0	$y = -2(0) + 1$	1	$(0, 1)$
2	$y = -2(2) + 1$	-3	$(2, -3)$

Step 2: Plot the ordered pairs.

Step 3: Draw a line through the points.

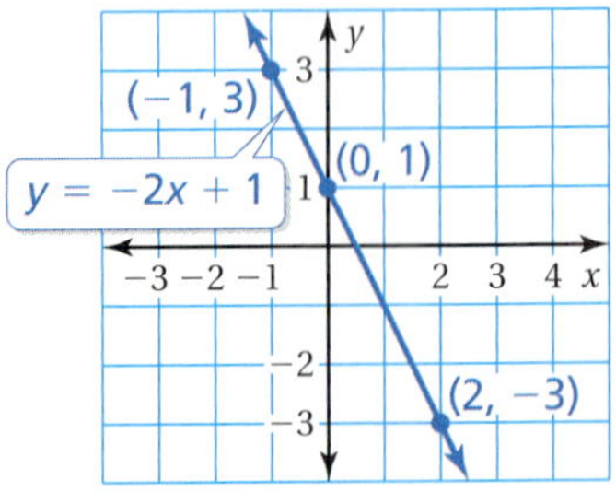

Key Idea

Graphing a Horizontal Line

The graph of $y = a$ is a horizontal line passing through $(0, a)$.

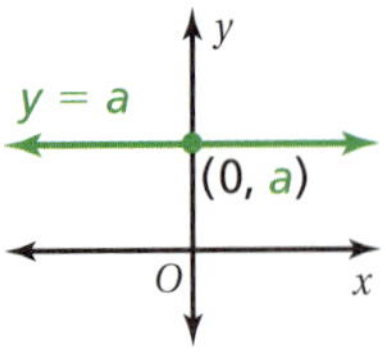

Multi-Language Glossary at BigIdeasMath.com.

Laurie's Notes

Introduction

Connect

- **Yesterday:** Students explored the graphs of linear equations.
- **Today:** Students will graph linear equations using a table of values.

Motivate

- Discuss a fact about wind speeds related to Example 3. During a wild April storm in 1934, a wind gust of 231 miles per hour (372 kilometers per hour) pushed across the summit of Mt. Washington in New Hampshire. This wind speed still stands as the all-time surface wind speed record.

Lesson Notes

Key Idea

- Define *linear equation* and *solutions* of the equation.
- Note the use of color in the input-output table. The equation used is a simple equation that helps students focus on the representation of the solutions as ordered pairs. The y-coordinate is always 1 greater than the x-coordinate, just as the equation states.

Example 1

? As a quick review, ask a volunteer to review the rules for integer multiplication. If the factors have the same signs, the product is positive. If the factors have different signs, the product is negative.

- Write the 4-column table. Take the time to show how the x-coordinate is being substituted in the second column. The number in blue is the only quantity that varies (variable); the other quantities are always the same (constant). Values from the first and third columns form the ordered pair.

? "From the graph, can you estimate the solution when $x = \frac{1}{2}$? Verify your answer by evaluating the equation when $x = \frac{1}{2}$." yes; $\left(\frac{1}{2}, 0\right)$

Key Idea

- Students are sometimes confused by the equation $y = a$. Explain to students that a is a variable. It can equal any number.
- **Teaching Tip:** Another way to discuss the equation $y = a$ is to say that "y always equals a certain number, while x can equal anything." For example, if $y = -4$, the table of values will look like this:

x	−1	0	1	2
y	−4	−4	−4	−4

Goal Today's lesson is graphing **linear equations**.

Start Thinking! and Warm Up

Lesson 2.1 Warm Up
For use before Lesson 2.1

Lesson 2.1 Start Thinking!
For use before Lesson 2.1

Think about how much energy you have on an average day.

Graph your energy level (on a scale of 0 to 10) throughout an average day.

Are any sections of your graph linear?

Extra Example 1

Graph $y = \frac{1}{2}x - 3$.

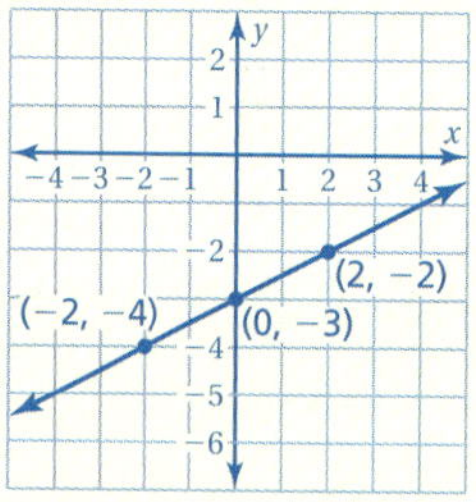

Extra Example 2

Graph $y = 4$.

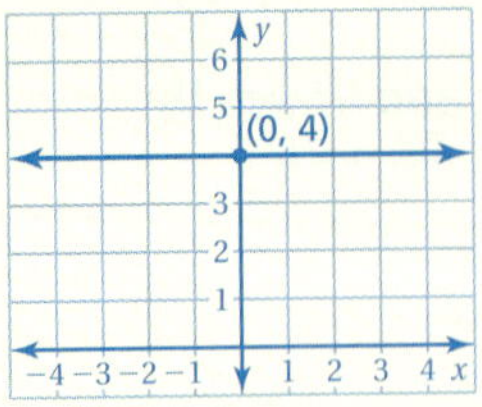

On Your Own

1–4. See Additional Answers.

Extra Example 3

The cost y (in dollars) for making friendship bracelets is $y = 0.5x + 2$, where x is the number of bracelets.

a. Graph the equation.

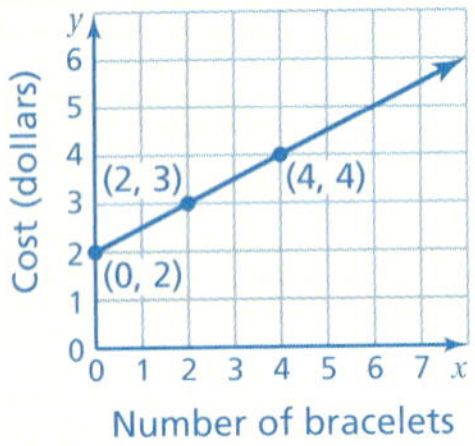

b. How many bracelets can be made for $10? 16

On Your Own

5. 8 hours after it enters the Gulf of Mexico

English Language Learners

Vocabulary

Make sure students understand that the graph of a *linear* equation is a *line*. Only two points are needed to graph a line, but if one of the points is incorrect the wrong line will be graphed. Plotting three points for a line in the coordinate plane and making sure that the points form a line provides students with a check when graphing.

Laurie's Notes

Example 2

? "What are other points on the line $y = -3$?" *Sample answer:* (5, −3), or anything of the form $(x, -3)$

On Your Own

- Ask volunteers to share their graphs at the board.

Example 3

? "What does x represent in the problem? What does y represent?" x = number of hours after the storm enters the Gulf of Mexico; y = wind speed

- Work through the problem using the 4-column table to generate solutions of the equation.
- Note that the y-coordinate is much greater than the x-coordinate. For this reason, a broken vertical axis is used. Students should *not* scale the y-axis beginning at 0.

? "Why are only non-negative numbers substituted for x?" Because x equals the number of hours after the storm enters the Gulf of Mexico, you do not know if the equation makes sense for x-values before that.

- Note that the ordered pairs are all located in Quadrant I because x is a non-negative number. Even though this restriction was not stated explicitly, you know from reading the description of x that it needs to be non-negative.
- In part (b), help students read the graph. Starting with a y-value of 74 on the y-axis, trace horizontally until you reach the graph of the line, and then trace straight down (vertically) to the x-axis. The x-coordinate is 4.

On Your Own

- **Neighbor Check:** Have students work independently and then have their neighbor check their work. Have students discuss any discrepancies.

Closure

- Explain how you know if an equation is linear. *Sample answer:* The graph of the equation is a line.

Technology For the Teacher

Dynamic Classroom

The Dynamic Planning Tool
Editable Teacher's Resources at *BigIdeasMath.com*

EXAMPLE 2 Graphing a Horizontal Line

Graph $y = -3$.

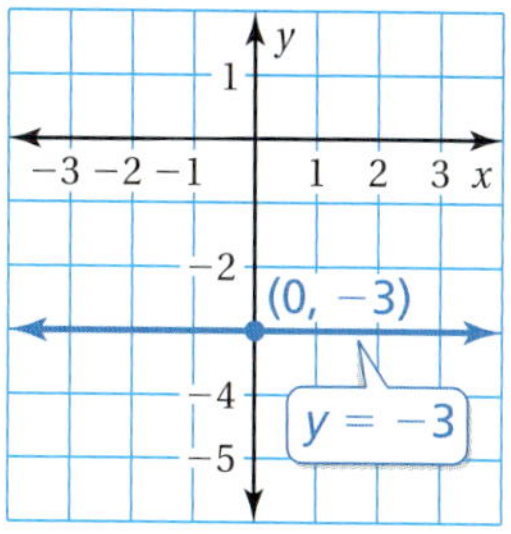

The graph of $y = -3$ is a horizontal line passing through $(0, -3)$.

Plot $(0, -3)$. Draw a horizontal line through the point.

On Your Own

Now You're Ready
Exercises 5–13

Graph the linear equation.

1. $y = 3x$
2. $y = -\frac{1}{2}x + 2$
3. $y = \pi$
4. $y = -1.5$

EXAMPLE 3 Real-Life Application

The wind speed y (in miles per hour) of a tropical storm is $y = 2x + 66$, where x is the number of hours after the storm enters the Gulf of Mexico.

a. Graph the equation.

b. When does the storm become a hurricane?

A tropical storm becomes a hurricane when wind speeds are at least 74 miles per hour.

a. Make a table of values.

x	$y = 2x + 66$	y	(x, y)
0	$y = 2(0) + 66$	66	$(0, 66)$
1	$y = 2(1) + 66$	68	$(1, 68)$
2	$y = 2(2) + 66$	70	$(2, 70)$
3	$y = 2(3) + 66$	72	$(3, 72)$

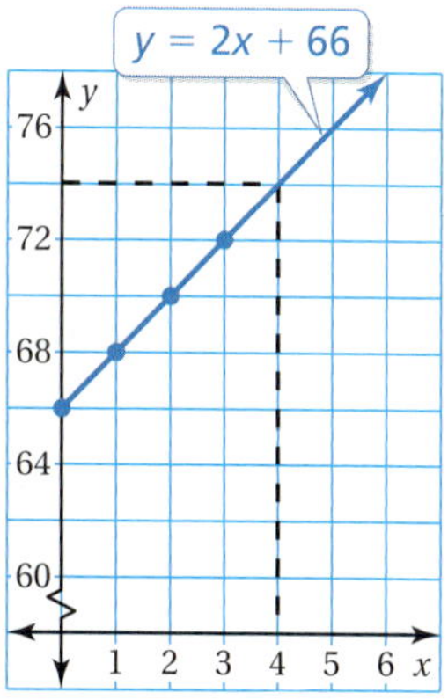

Plot the ordered pairs and draw a line through the points.

b. From the graph, you can see that $y = 74$ when $x = 4$.

So, the storm becomes a hurricane 4 hours after it enters the Gulf of Mexico.

On Your Own

5. **WHAT IF?** In Example 3, the wind speed of the storm is $y = 1.5x + 62$. When does the storm become a hurricane?

2.1 Exercises

Check It Out
Help with Homework
BigIdeasMath.com

Vocabulary and Concept Check

1. **VOCABULARY** What type of graph represents the solutions of the equation $y = 2x + 3$?

2. **WHICH ONE DOESN'T BELONG?** Which equation does *not* belong with the other three? Explain your reasoning.

   $y = 0.5x - 0.2$ $\quad$ $4x + 3 = y$ $\quad$ $y = x^2 + 6$ $\quad$ $\frac{3}{4}x + \frac{1}{3} = y$

Practice and Problem Solving

Copy and complete the table. Plot the two solution points and draw a line *exactly* through the two points. Find a different solution point on the line.

3.

x		
$y = 3x - 1$		

4.

x		
$y = \frac{1}{3}x + 2$		

Graph the linear equation.

1 2

5. $y = -5x$
6. $y = \frac{1}{4}x$
7. $y = 5$
8. $y = x - 3$
9. $y = -7x - 1$
10. $y = -\frac{x}{3} + 4$
11. $y = \frac{3}{4}x - \frac{1}{2}$
12. $y = -\frac{2}{3}$
13. $y = 6.75$

14. **ERROR ANALYSIS** Describe and correct the error in graphing the equation.

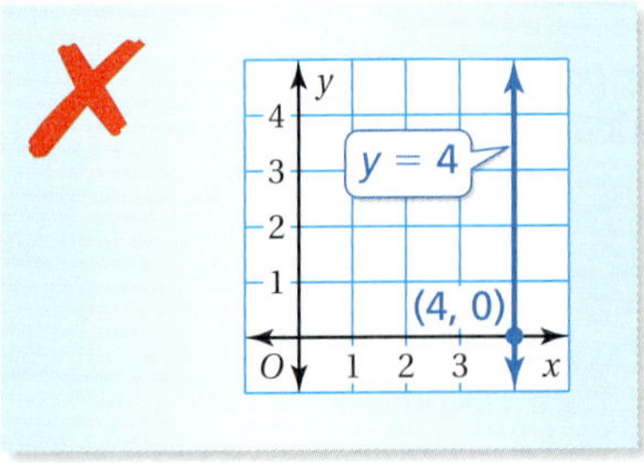

15. **MESSAGING** You sign up for an unlimited text messaging plan for your cell phone. The equation $y = 20$ represents the cost y (in dollars) for sending x text messages. Graph the equation.

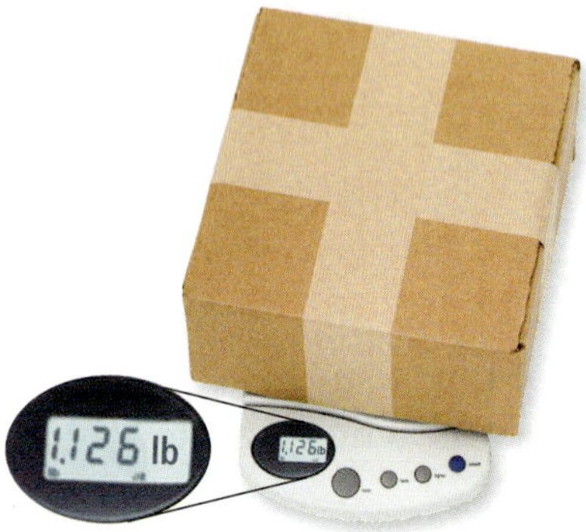

16. **MAIL** The equation $y = 2x + 3$ represents the cost y (in dollars) of mailing a package that weighs x pounds.

    a. Graph the equation.

    b. Use the graph to estimate how much it costs to mail the package.

    c. Use the equation to find exactly how much it costs to mail the package.

Assignment Guide and Homework Check

Level	Day 1 Activity Assignment	Day 2 Lesson Assignment	Homework Check
Basic	3, 4, 25–29	1, 2, 5–19 odd, 14, 16	5, 7, 11, 16, 19
Average	3, 4, 25–29	1, 2, 9–14, 17, 19, 21, 22	11, 13, 19, 22
Advanced	3, 4, 25–29	1, 2, 10–24 even, 21, 23	10, 12, 18, 22

Common Errors

- **Exercises 5–13** Students may make a calculation error for one of the ordered pairs in a table of values. If they only find two ordered pairs for the graph, they may not recognize their mistake. Encourage them to find at least three ordered pairs when drawing a graph.
- **Exercises 7, 12, and 13** Students may draw a vertical line through a point on the x-axis instead of through the corresponding point on the y-axis. Remind them that the equation is a horizontal line. Ask them to identify the y-coordinate for several x-coordinates. For example, what is the y-coordinate for $x = 5$? $x = 6$? $x = -4$? Students should answer with the same y-coordinate each time.
- **Exercises 17–20** Students may make a mistake in solving for y, such as using the same operation instead of the opposite operation.

2.1 Record and Practice Journal

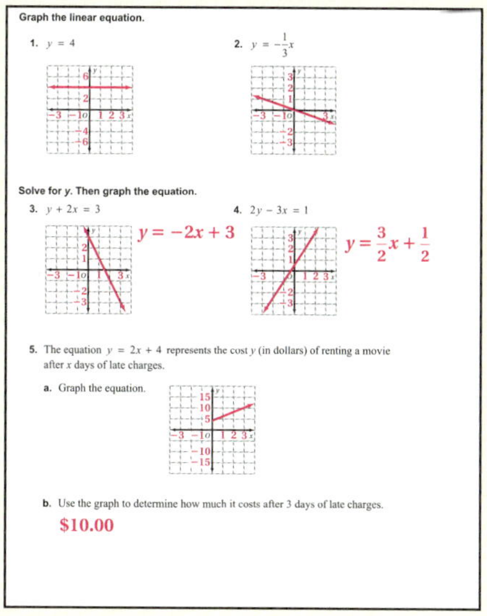
Graph the linear equation.

1. $y = 4$

2. $y = -\frac{1}{3}x$

Solve for y. Then graph the equation.

3. $y + 2x = 3$ $\quad y = -2x + 3$

4. $2y - 3x = 1$ $\quad y = \frac{3}{2}x + \frac{1}{2}$

5. The equation $y = 2x + 4$ represents the cost y (in dollars) of renting a movie after x days of late charges.

   a. Graph the equation.

   b. Use the graph to determine how much it costs after 3 days of late charges.

   $10.00

Technology For the Teacher

Answer Presentation Tool
Quiz*Show*

Vocabulary and Concept Check

1. a line
2. $y = x^2 + 6$ does not belong because it is not a linear equation.

Practice and Problem Solving

3. *Sample answer:*

x	0	1
$y = 3x - 1$	-1	2

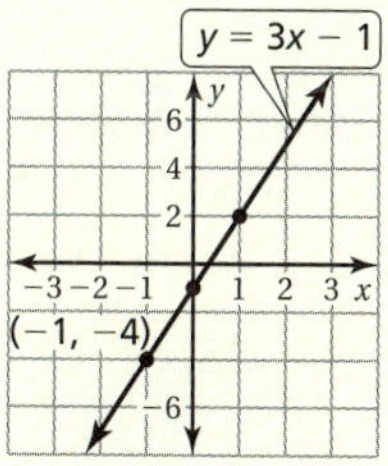

4. *Sample answer:*

x	0	3
$y = \frac{1}{3}x + 2$	2	3

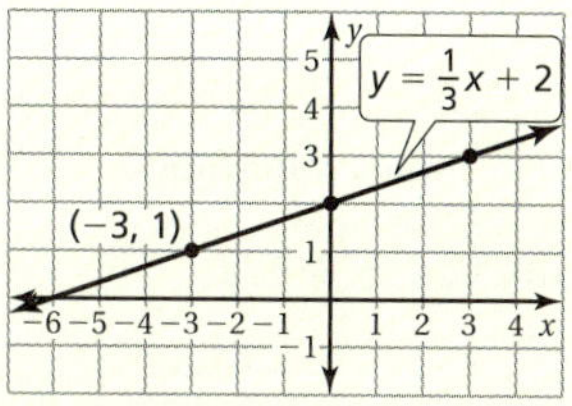

5.

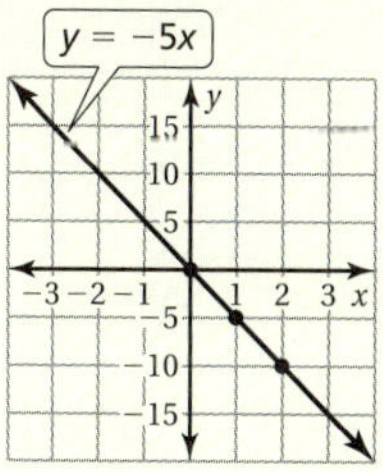

6–16. See Additional Answers.

Practice and Problem Solving

17. $y = 3x + 1$

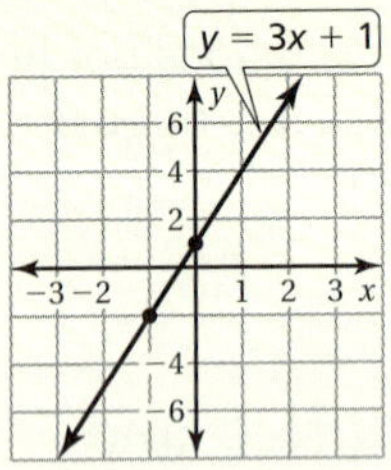

18–21. See Additional Answers.

22. See *Taking Math Deeper.*

23–24. See Additional Answers.

Fair Game Review

25. (5, 3) **26.** (−6, 6)

27. (2, −2) **28.** (−4, −3)

29. B

Mini-Assessment

Graph the linear equation.

1. Graph $y = -\frac{1}{2}x + 2$.

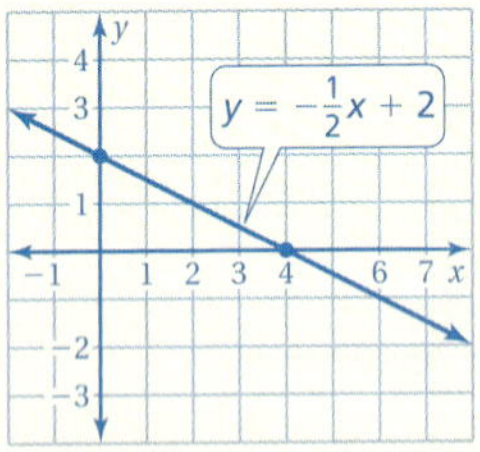

2. You have $100 in your savings account and plan to deposit $20 each month. Write and graph a linear equation that represents the balance in your account. $y = 20x + 100$

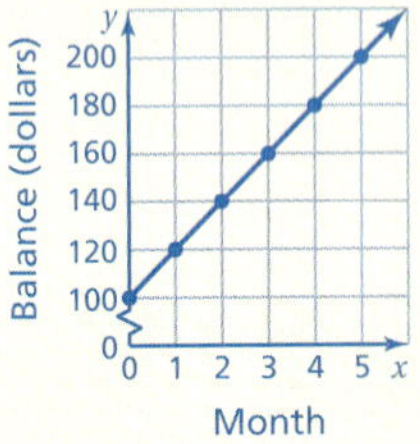

Taking Math Deeper

Exercise 22

Some of the information for this exercise is given in the photo and some is given in the text. It is a good idea to start by listing all of the given information.

1 List the given information.

- The camera can store 250 pictures.
- 1 second of video = 2 pictures
- Video time used = 90 seconds
- Let y = number of pictures
- Let x = number of seconds of video

2 **a.** Write and graph an equation for x and y.

$$y + 2x = 250$$
$$y = -2x + 250$$

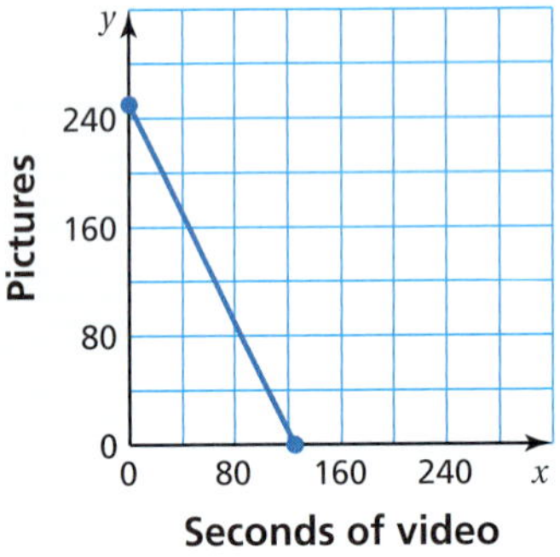

3 **b.** Answer the question.

When $x = 90$, the value of y is as follows.

$$y = -2x + 250$$
$$= -2(90) + 250$$
$$= 70$$

Your camera can store 70 pictures.

Project

Research digital cameras. Find the number of pictures that can be stored on five different cameras. Compare the prices of the cameras. What do you consider to be the better buy? Why?

Reteaching and Enrichment Strategies

If students need help...	If students got it...
Resources by Chapter • Practice A and Practice B • Puzzle Time Record and Practice Journal Practice Differentiating the Lesson Lesson Tutorials Skills Review Handbook	Resources by Chapter • Enrichment and Extension Start the next section

Solve for *y*. Then graph the equation.

17. $y - 3x = 1$

18. $5x + 2y = 4$

19. $-\frac{1}{3}y + 4x = 3$

20. $x + 0.5y = 1.5$

21. **SAVINGS** You have $100 in your savings account and plan to deposit $12.50 each month.

a. Write and graph a linear equation that represents the balance in your account.

b. How many months will it take you to save enough money to buy 10 acres of land on Mars?

22. **CAMERA** One second of video on your digital camera uses the same amount of memory as two pictures. Your camera can store 250 pictures.

a. Write and graph a linear equation that represents the number *y* of pictures your camera can store if you take *x* seconds of video.

b. How many pictures can your camera store after you take the video shown?

23. **SEA LEVEL** Along the U.S. Atlantic Coast, the sea level is rising about 2 millimeters per year.

a. Write and graph a linear equation that represents how much sea level rises over a period of time.

b. How many millimeters has sea level risen since you were born?

24. **Geometry** The sum *S* of the measures of the angles of a polygon is $S = (n - 2) \cdot 180°$, where *n* is the number of sides of the polygon. Plot four points (n, S) that satisfy the equation. Do the points lie on a line? Explain your reasoning.

Fair Game Review What you learned in previous grades & lessons

Write the ordered pair corresponding to the point. *(Skills Review Handbook)*

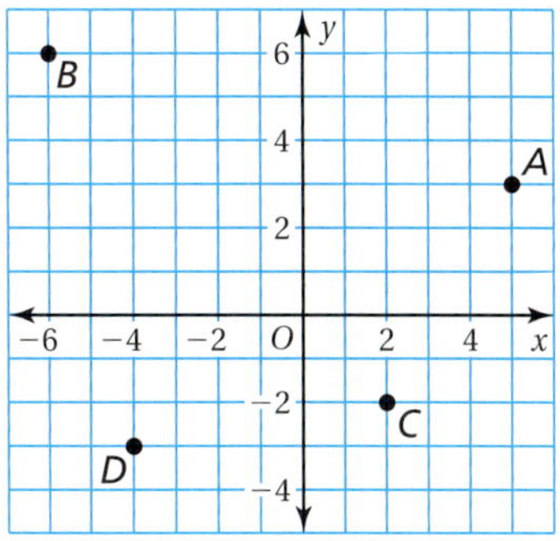

25. Point *A*

26. Point *B*

27. Point *C*

28. Point *D*

29. **MULTIPLE CHOICE** A debate team has 15 female members. The ratio of females to males is 3 : 2. How many males are on the debate team? *(Skills Review Handbook)*

Ⓐ 6 Ⓑ 10 Ⓒ 22 Ⓓ 25

2.2 Slope of a Line

Essential Question How can the slope of a line be used to describe the line?

Slope is the rate of change between any two points on a line. It is the measure of the *steepness* of the line.

To find the slope of a line, find the ratio of the change in y (vertical change) to the change in x (horizontal change).

$$\text{slope} = \frac{\text{change in } y}{\text{change in } x}$$

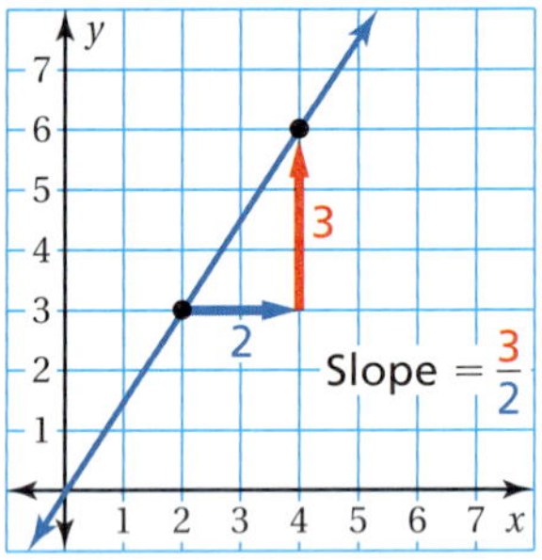

1 ACTIVITY: Finding the Slope of a Line

Work with a partner. Find the slope of each line using two methods.

Method 1: Use the two black points. •

Method 2: Use the two pink points. •

Do you get the same slope using each method?

a.

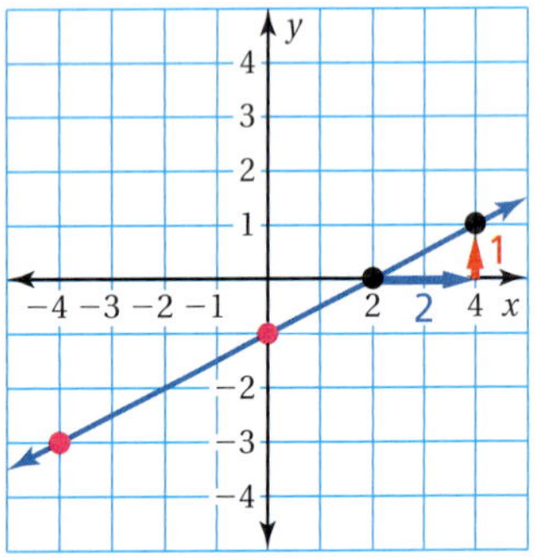

b.

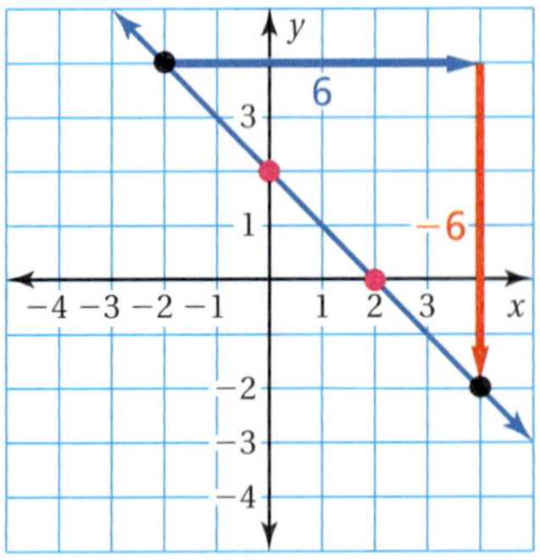

c.

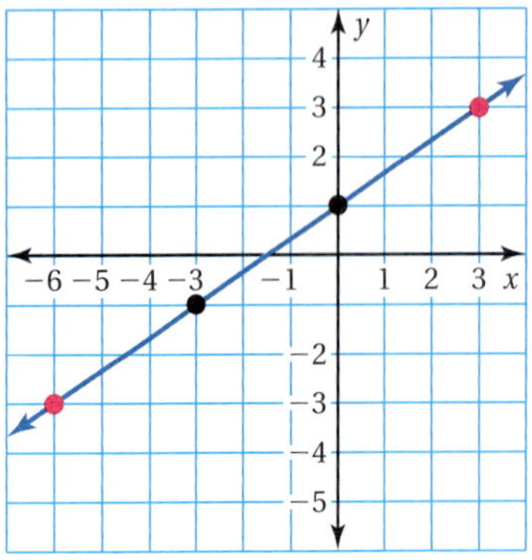

d.

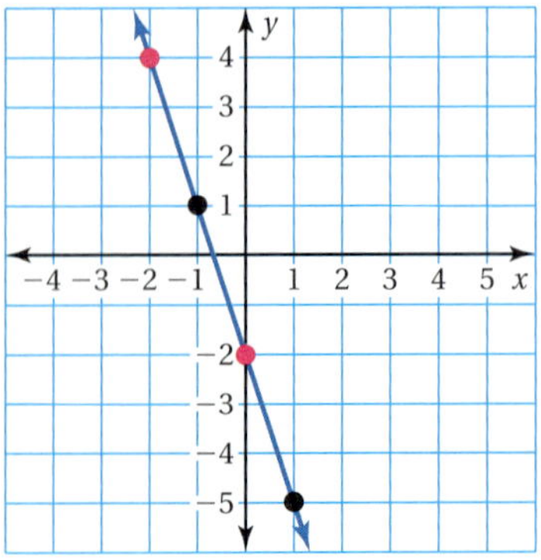

Laurie's Notes

Introduction

For the Teacher

- **Goal:** Students will learn that they can use any two points on a line to calculate the slope of that line.
- The formal definition of slope, when computed, uses subscripts $(y_2 - y_1)$. The subscript notation will be taught in a future course.

Motivate

? "How many of you have been on a roller coaster?"

- Discuss with students what makes one roller coaster more thrilling than another. Students will usually describe how quickly the coaster drops or the steepness of the hill. This is similar to the *change in y* of a line when finding the slope.

Activity Notes

Discuss

? "Does anyone remember what is meant by slope of a line?" At least one student should recall that it measures the steepness of a line.

- Write the definition for slope. Sketch the graph shown to demonstrate what is meant by change in y (red vertical arrow) and change in x (blue horizontal arrow).
- Remind students that slope is always the change in y in the numerator and the change in x in the denominator. This can be confusing for students because graphs are read from left to right, and we have a tendency to move in the x-direction first. For this reason, students want to write the change in x in the numerator.
- **Note:** In this book, the change in x will always be positive. However, you should show students that the change in x can be negative.

? "Can the change in x be negative? Explain." Yes; moving to the left horizontally is negative.

? "Can the change in y be negative? Explain." Yes; moving down vertically is negative.

Activity 1

- Encourage students to draw the change arrows for each pair of points. Label the change in x (or y) next to the arrow.
- **Big Idea:** The slope of a line is always the same regardless of what two ordered pairs are selected.
- **Common Error:** Students may forget to make the change negative when moving downward in the y-direction.

Previous Learning

Students should know that slope is the rate of change between two points on a line.

Start Thinking! and Warm Up

Activity 2.2 Start Thinking!
For use before Activity 2.2

Activity 2.2 Warm Up
For use before Activity 2.2

Write the fraction in simplest form.

1. $\frac{6}{2}$
2. $\frac{8}{28}$
3. $\frac{10}{25}$
4. $\frac{10}{8}$
5. $\frac{6}{9}$
6. $\frac{16}{12}$

2.2 Record and Practice Journal

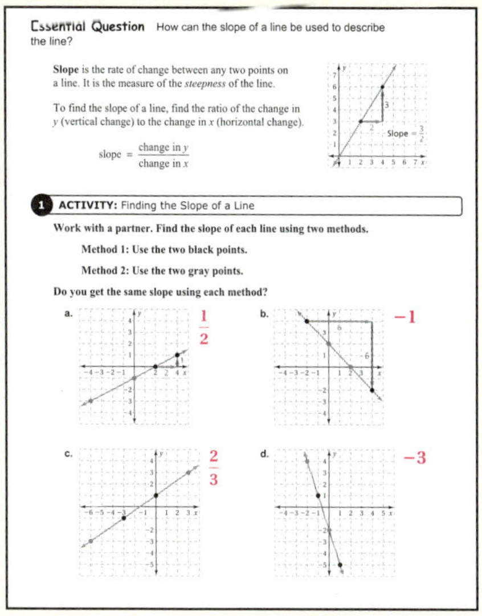

Essential Question How can the slope of a line be used to describe the line?

Slope is the rate of change between any two points on a line. It is the measure of the *steepness* of the line.

To find the slope of a line, find the ratio of the change in y (vertical change) to the change in x (horizontal change).

$$\text{slope} = \frac{\text{change in } y}{\text{change in } x}$$

Slope $= \frac{3}{2}$

1 ACTIVITY: Finding the Slope of a Line

Work with a partner. Find the slope of each line using two methods.

Method 1: Use the two black points.

Method 2: Use the two gray points.

Do you get the same slope using each method?

a. $\frac{1}{2}$

b. -1

c. $\frac{2}{3}$

d. -3

Differentiated Instruction

Kinesthetic

Help students develop number sense about slope. Have them draw lines in the coordinate plane through the following pairs of points.

(0, 0) and (3, 5) (0, 0) and (3, 4)
(0, 0) and (3, 3) (0, 0) and (3, 2)
(0, 0) and (3, 1)

Next have students find the slope of each line. Point out that the line passing through (3, 3) has a slope of 1. The lines with y-coordinates greater than 3 have a slope greater than 1. The lines with y-coordinates less than 3 have a slope less than 1. For positive slopes, the steeper lines will have a greater slope.

2.2 Record and Practice Journal

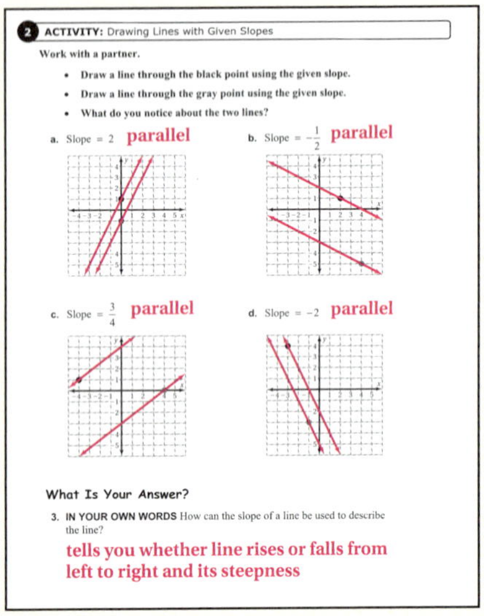

2 ACTIVITY: Drawing Lines with Given Slopes

Work with a partner.

- Draw a line through the black point using the given slope.
- Draw a line through the gray point using the given slope.
- What do you notice about the two lines?

a. Slope = 2 **parallel**

b. Slope = $-\frac{1}{2}$ **parallel**

c. Slope = $\frac{3}{4}$ **parallel**

d. Slope = −2 **parallel**

What Is Your Answer?

3. IN YOUR OWN WORDS How can the slope of a line be used to describe the line?

tells you whether line rises or falls from left to right and its steepness

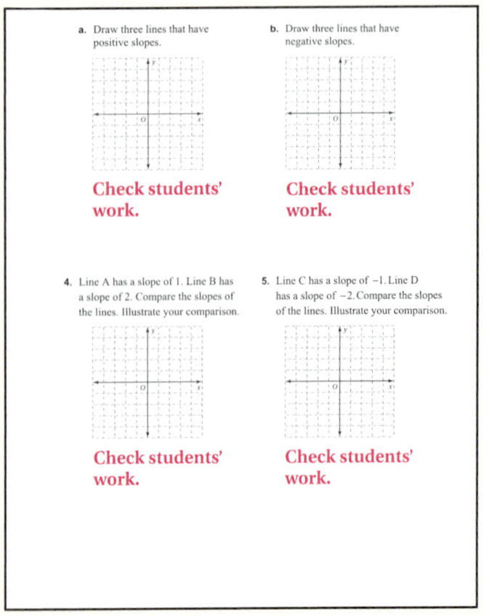

a. Draw three lines that have positive slopes.

Check students' work.

b. Draw three lines that have negative slopes.

Check students' work.

4. Line A has a slope of 1. Line B has a slope of 2. Compare the slopes of the lines. Illustrate your comparison.

Check students' work.

5. Line C has a slope of −1. Line D has a slope of −2. Compare the slopes of the lines. Illustrate your comparison.

Check students' work.

Laurie's Notes

Activity 2

- In addition to being able to determine the slope of a line that has been graphed, you want students to be able to draw a line that has a particular slope. This is the goal of Activity 2.
- ? "What does it mean for a line to have a slope of $\frac{2}{3}$?" For every 2 units of change in the y-direction, there is a change of 3 units in the x-direction.
- ? "What does it mean for a line to have a slope of −3?" For every −3 units of change in the y-direction, there is a change of 1 unit in the x-direction. This is the same as 3 units in the y-direction and −1 unit in the x-direction.
- Explain to students that $-3 = \frac{-3}{1} = \frac{3}{-1}$. Demonstrate what this means in a coordinate plane.
- **Teaching Tip:** If possible, give each pair of students two colored pencils. Have them draw the first line (through the black point) in one color and draw the second line (through the pink point) in the other color.
- ? "What do you notice about the two lines you have drawn?" Students should observe that the lines are parallel. Students may also observe that the positive slopes rise (from left to right) and the negative slopes fall (from left to right).
- **Big Idea:** Slope is a measure of the steepness of a line. Two different lines with the same slope are parallel (they have the same steepness).

What Is Your Answer?

- Discuss student responses to each question. These are significant questions.

Closure

- Plot the point (0, 3). Draw the line through this point that has a slope of $\frac{1}{3}$. Name two points on the line. *Sample answer:* (3, 4), (−3, 2)

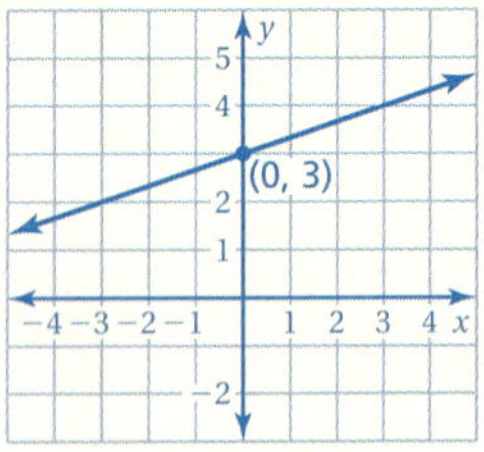

Technology For the Teacher

Dynamic Classroom

The Dynamic Planning Tool
Editable Teacher's Resources at *BigIdeasMath.com*

2 ACTIVITY: Drawing Lines with Given Slopes

Work with a partner.

- **Draw a line through the black point using the given slope.**
- **Draw a line through the pink point using the given slope.**
- **What do you notice about the two lines?**

a. Slope = 2

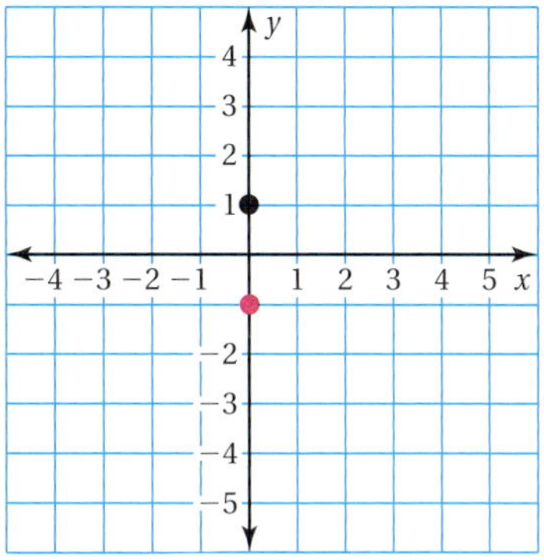

b. Slope = $-\frac{1}{2}$

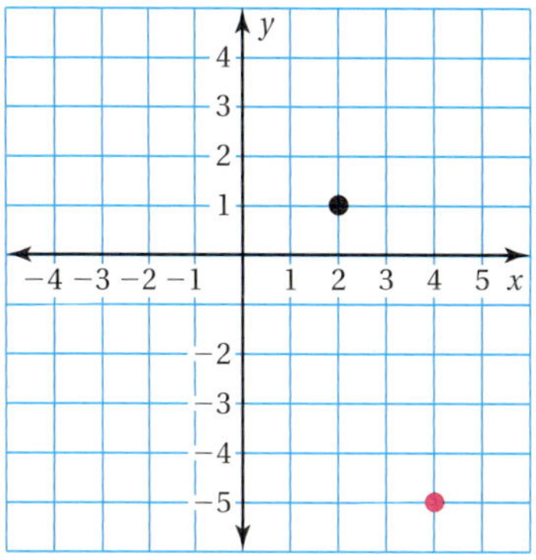

c. Slope = $\frac{3}{4}$

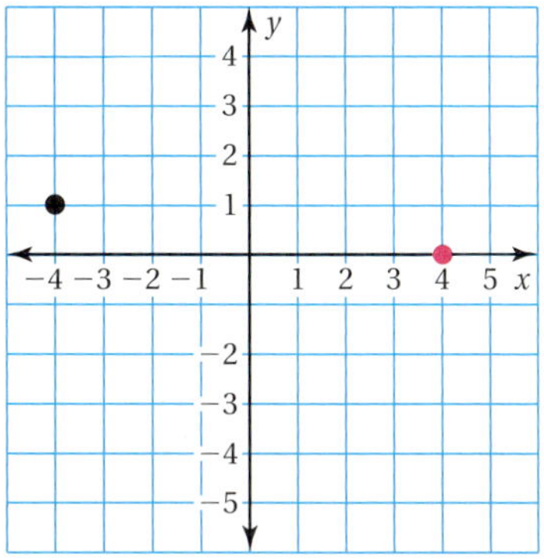

d. Slope = −2

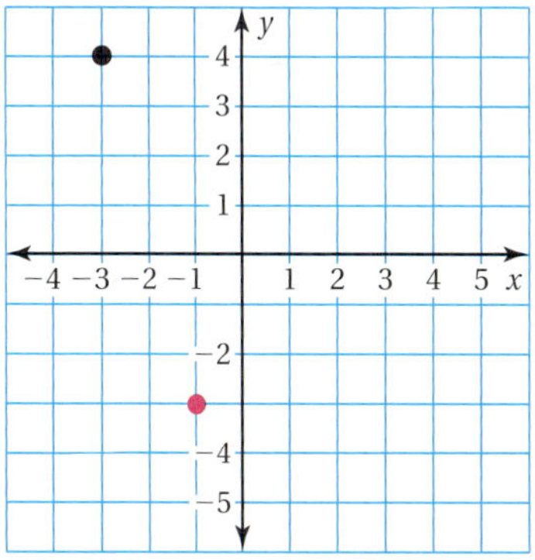

What Is Your Answer?

3. **IN YOUR OWN WORDS** How can the slope of a line be used to describe the line?

a. Draw three lines that have positive slopes.

b. Draw three lines that have negative slopes.

4. Line A has a slope of 1. Line B has a slope of 2. Compare the slopes of the lines. Illustrate your comparison.

5. Line C has a slope of −1. Line D has a slope of −2. Compare the slopes of the lines. Illustrate your comparison.

Practice Use what you learned about the slope of a line to complete Exercises 4–6 on page 59.

2.2 Lesson

Check It Out
Lesson Tutorials
BigIdeasMath.com

Key Vocabulary
slope, *p. 56*
rise, *p. 56*
run, *p. 56*

Key Idea

Slope

The **slope** of a line is a ratio of the change in y (the **rise**) to the change in x (the **run**) between any two points on the line.

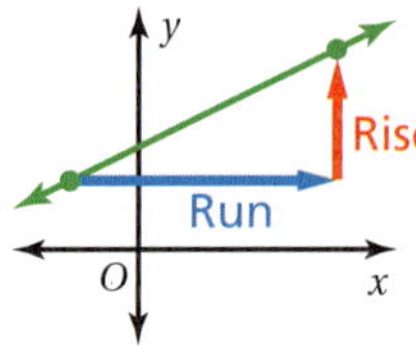

$$\text{slope} = \frac{\text{change in } y}{\text{change in } x} = \frac{\text{rise}}{\text{run}}$$

Positive slope

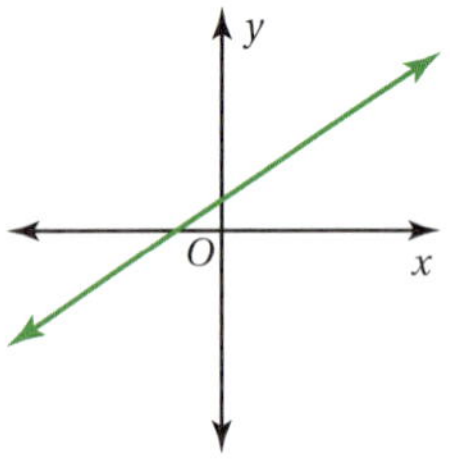

The line rises from left to right.

Negative slope

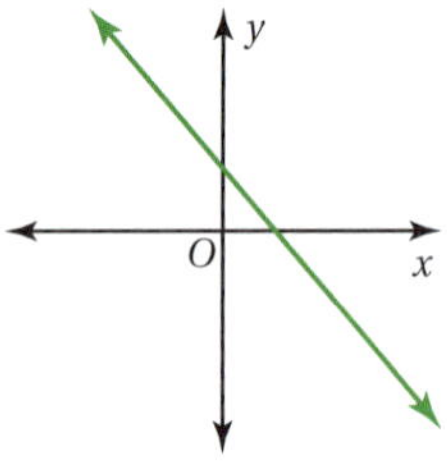

The line falls from left to right.

EXAMPLE 1 Finding the Slope of a Line

Tell whether the slope of the line is *positive* or *negative*. Then find the slope.

a.

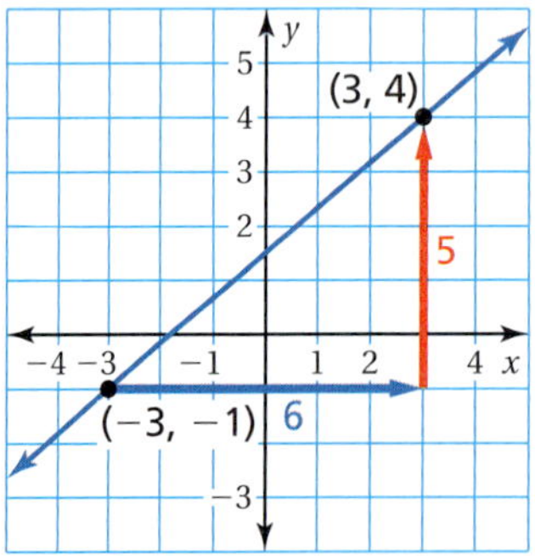

The line rises from left to right. So, the slope is positive.

$$\text{slope} = \frac{\text{rise}}{\text{run}}$$

$$= \frac{5}{6}$$

The slope is $\frac{5}{6}$.

b.

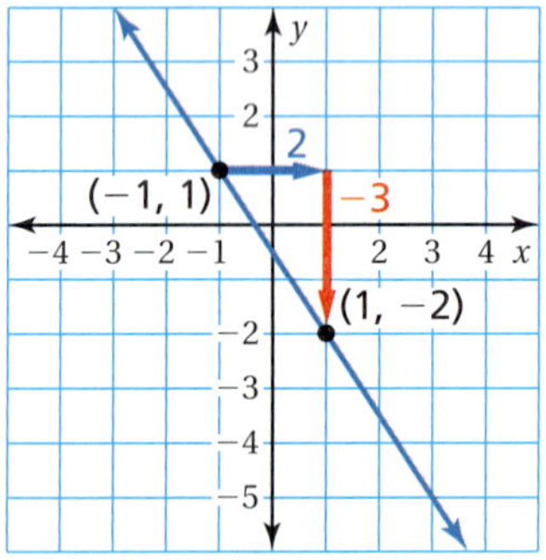

The line falls from left to right. So, the slope is negative.

$$\text{slope} = \frac{\text{rise}}{\text{run}}$$

$$= \frac{-3}{2}, \text{ or } -\frac{3}{2}$$

The slope is $-\frac{3}{2}$.

Laurie's Notes

Introduction

Connect

- **Yesterday:** Students explored slopes of lines.
- **Today:** Students will find the slopes of lines in a variety of contexts.

Motivate

- Have students plot four points: $A(5, 0)$, $B(0, 5)$, $C(-5, 0)$, and $D(0, -5)$. Connect the points to form the quadrilateral $ABCD$.

? "What type of quadrilateral is $ABCD$?" Without proof, students should say square.

? "What is the slope of each side, meaning the slopes of the lines through AB, BC, CD, and DA?" Slopes of AB and CD are both -1. Slopes of BC and DA are both 1.

? "If you drew another square inside $ABCD$ that had one vertex at (0, 3), could you predict what the slopes of the four sides would be?" Students should be able to visualize this and predict that two sides would have a slope of 1 and two sides would have a slope of -1.

Lesson Notes

Key Idea

- Write the Key Idea. Define slope of a line.
- Note the use of color in the definition and on the graph. The *change in y* and the *vertical change arrow* are both red. The *change in x* and the *horizontal change arrow* are both blue.
- Discuss the difference in positive and negative slopes, a concept students explored yesterday.
- Remind students that graphs are read from left to right.

Example 1

- Work through each part. The arrows, words, and numbers are color-coded.
- Students often ask if they can move in the y-direction first, followed by the x-direction. The answer is yes. Demonstrate this on either graph.
  - In part (a), start at $(-3, -1)$ and move up 5 units in the y-direction and then to the right 6 units in the x-direction. You will end at (3, 4).
  - In part (b), start at $(-1, 1)$ and move down 3 units in the y-direction and then to the right 2 units in the x-direction. You will end at $(1, -2)$.

Goal Today's lesson is finding the **slope** of a line.

Start Thinking! and Warm Up

Lesson 2.2 Warm Up
For use before Lesson 2.2

Lesson 2.2 Start Thinking!
For use before Lesson 2.2

1. Each student must choose an ordered pair.
2. Choose a partner and work together to find the slope of the line joining your two points. Use a graph to help you.
3. Repeat the process several times with different partners.

Were any of the slopes positive? negative? zero?

Extra Example 1

Tell whether the slope of the line is *positive* or *negative*. Then find the slope.

a.

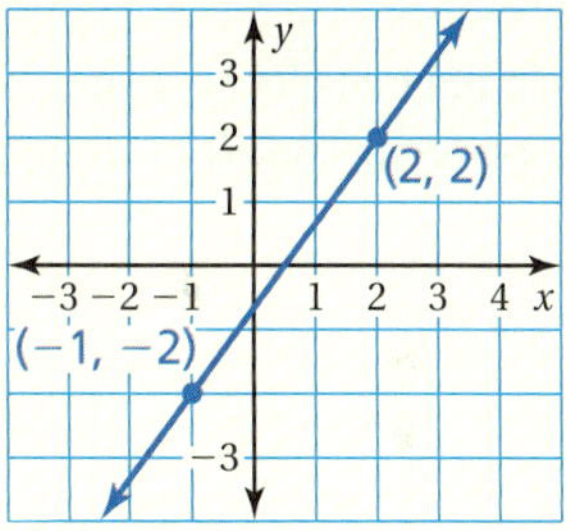

positive; $\frac{4}{3}$

b.

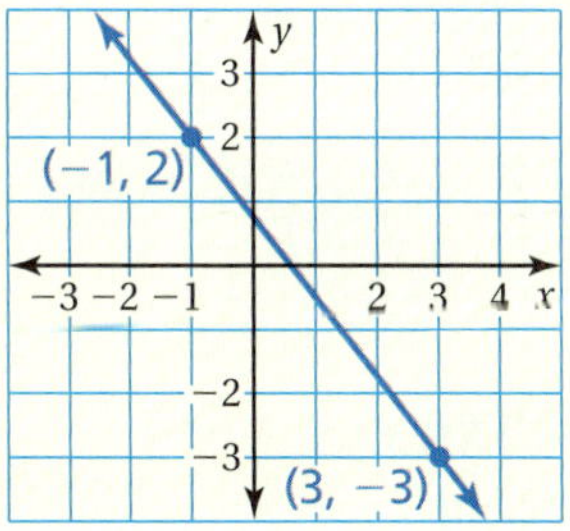

negative; $-\frac{5}{4}$

Extra Example 2

Find the slope of the line.

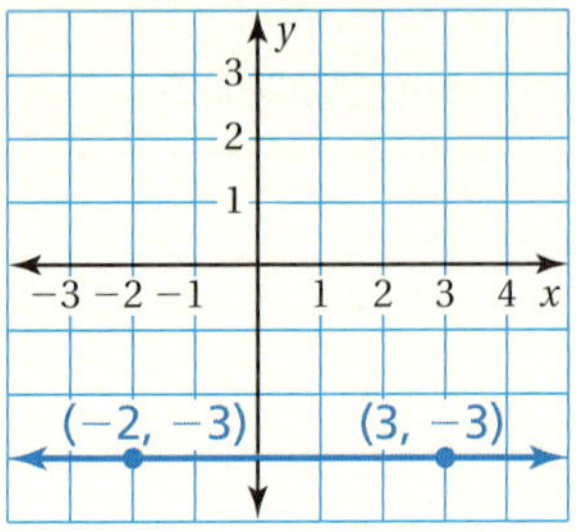

0

On Your Own

1. $-\frac{1}{5}$
2. 0
3. $\frac{5}{2}$

Extra Example 3

The points in the table lie on a line. Find the slope of the line. Then draw its graph.

x	−2	−1	0	1
y	−8	−5	−2	1

3

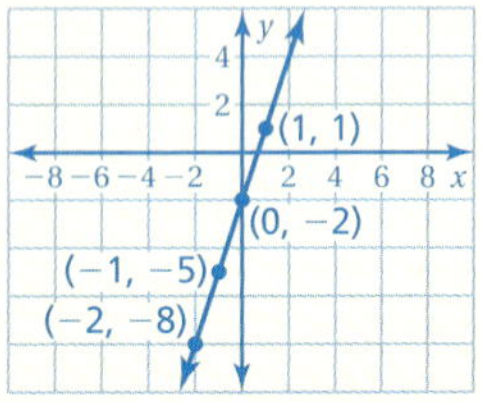

English Language Learners

Comprehension

The Key Idea box states "The slope of a line is a ratio of the change in *y* to the change in *x* between any two points on the line." Have students choose four points on a line. Use two points to find the slope. Then find the slope using the other two points. Students will find that the slopes are the same and should understand that the slope of the line is the same for the entire infinite length of the line.

Laurie's Notes

Example 2

? "How does a slope of $\frac{1}{2}$ compare to a slope of $\frac{1}{5}$? Describe the lines." A slope of $\frac{1}{2}$ runs 2 units for every 1 unit it rises. A slope of $\frac{1}{5}$ runs 5 units for each 1 unit it rises. A slope of $\frac{1}{5}$ is not as steep.

? "What would a slope of $\frac{1}{10}$ look like?" A slope of $\frac{1}{10}$ is less steep than a slope of $\frac{1}{5}$, so it is almost flat.

? "How steep do you think a horizontal line is?" Listen for students to describe a horizontal line as having no rise. In this example, they will see it has a slope of 0.

- Work through the example.
- **Big Idea:** There is a difference between a slope of 0 and no slope. A line with no slope is vertical. A horizontal line has a slope of 0.

Example 3

? "What do you notice about the *x*-values and the *y*-values?" The *x*-values are increasing by 3 and the *y*-values are decreasing by 2.

- Compute the slope between any two points in the table.
- Using (1, 8) and (4, 6): slope $= \frac{\text{change in } y}{\text{change in } x} = \frac{6-8}{4-1} = -\frac{2}{3}$
- Using (1, 8) and (7, 4): slope $= \frac{\text{change in } y}{\text{change in } x} = = \frac{4-8}{7-1} = -\frac{4}{6} = -\frac{2}{3}$
- **Connection:** The slope is the same regardless of which two points are selected. The slope triangles that are formed are similar. (Note: A slope triangle is the triangle formed by the line and the change in *x* and change in *y* arrows.)

? "The line has a negative slope. What do you notice about the line?" The line falls from left to right.

EXAMPLE 2 Finding the Slope of a Horizontal Line

Study Tip

The slope of a horizontal line is always 0.

Find the slope of the line.

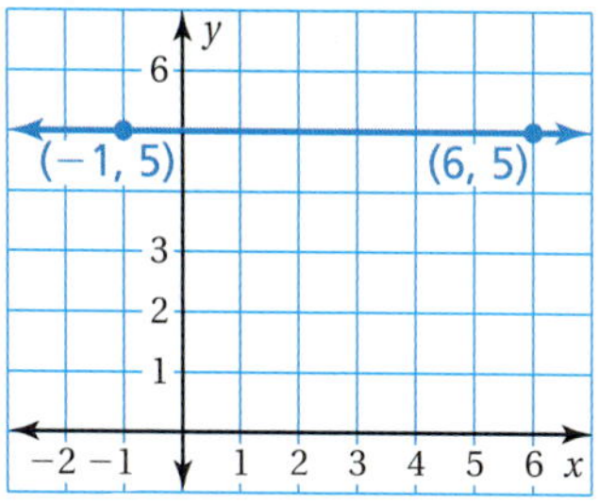

The line is not rising or falling. So, the rise is 0.

$$\text{slope} = \frac{\text{rise}}{\text{run}}$$

$$= \frac{0}{7}$$

∴ The slope is 0.

On Your Own

Now You're Ready
Exercises 7–12

Find the slope of the line.

1.

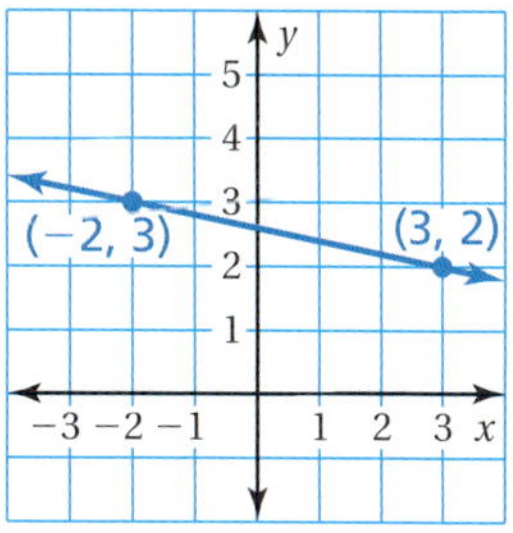

2.

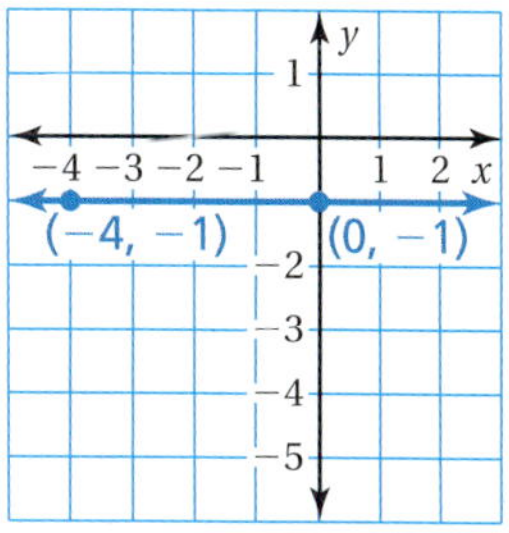

3.

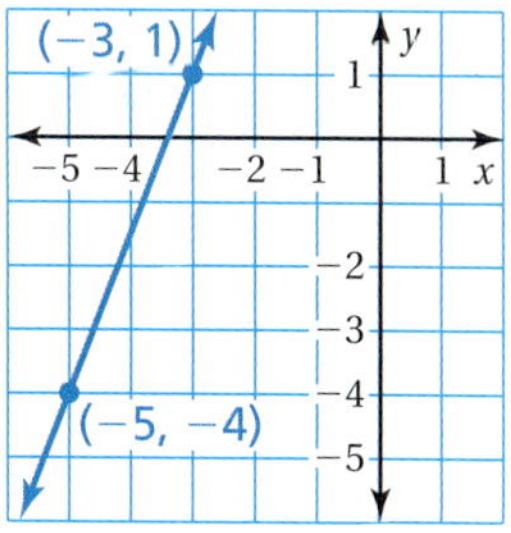

EXAMPLE 3 Finding Slope from a Table

The points in the table lie on a line. Find the slope of the line. Then draw its graph.

x	1	4	7	10
y	8	6	4	2

Choose any two points from the table. Then find the change in y and the change in x.

Use the points (1, 8) and (4, 6).

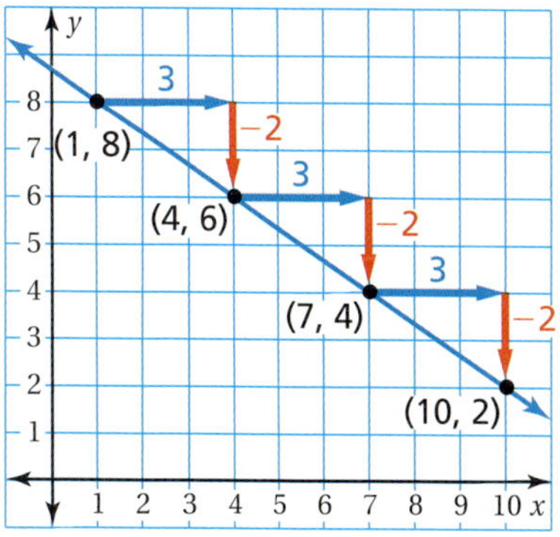

$$\text{slope} = \frac{\text{change in } y}{\text{change in } x}$$

$$= \frac{6 - 8}{4 - 1}$$

$$= \frac{-2}{3}$$

∴ The slope is $-\frac{2}{3}$.

On Your Own

Now You're Ready
Exercises 15–18

The points in the table lie on a line. Find the slope of the line. Then draw its graph.

4.

x	1	3	5	7
y	2	5	8	11

5.

x	−3	−2	−1	0
y	6	4	2	0

Key Idea

Parallel Lines and Slopes

Two lines in the same plane that do not intersect are parallel lines. Two lines with the same slope are parallel.

EXAMPLE 4 Finding Parallel Lines

Which two lines are parallel? Explain.

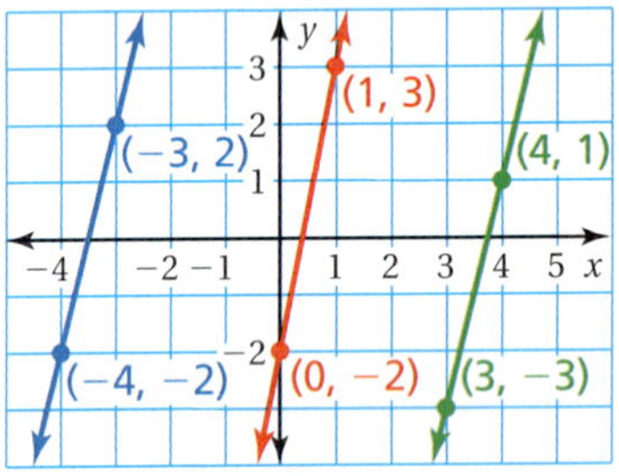

Find the slope of each line.

Blue Line

$$\text{slope} = \frac{\text{rise}}{\text{run}} = \frac{4}{1} = 4$$

Red Line

$$\text{slope} = \frac{\text{rise}}{\text{run}} = \frac{5}{1} = 5$$

Green Line

$$\text{slope} = \frac{\text{rise}}{\text{run}} = \frac{4}{1} = 4$$

The slope of the blue and green lines is 4. The slope of the red line is 5.

∴ The blue and green lines have the same slope, so they are parallel.

On Your Own

Now You're Ready
Exercises 21 and 22

6. Which two lines are parallel? Explain.

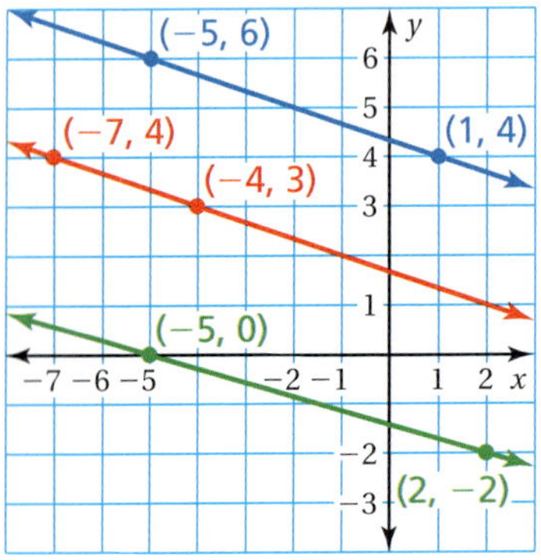

Laurie's Notes

On Your Own

- **Neighbor Check:** Have students work independently and then have their neighbor check their work. Have students discuss any discrepancies.
- **Connection:** In Question 4, students may recognize from the table that both x and y are increasing and the slope is positive. In Question 5, as x increases, the y-values are decreasing and the slope is negative.

Key Idea

- Students explored this idea yesterday. Students generally understand the fact that parallel lines have the same slope. They have the same steepness.

Example 4

- Work through the example. Students may look quickly and believe that all of the lines are parallel. They should compute the slope of each line to prove which lines are parallel.

On Your Own

- **Think-Pair-Share:** Students should read the question independently and then work with a partner to answer the question. When they have answered the question, the pair should compare their answer with another group and discuss any discrepancies.

Closure

- Plot the ordered pairs and find the slope of the line.

x	-1	0	1	2
y	-3	-1	1	3

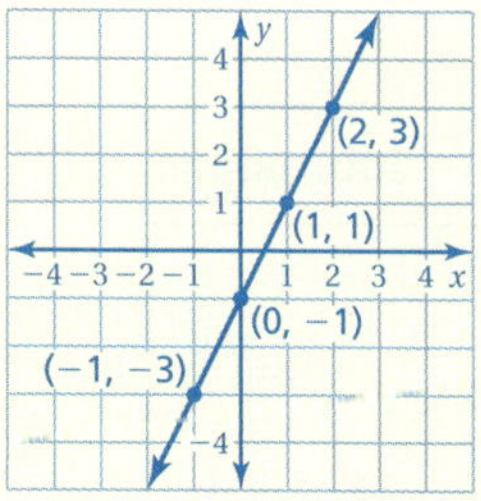

slope = 2

Technology For the Teacher

Dynamic Classroom

The Dynamic Planning Tool
Editable Teacher's Resources at *BigIdeasMath.com*

On Your Own

4. $\frac{3}{2}$

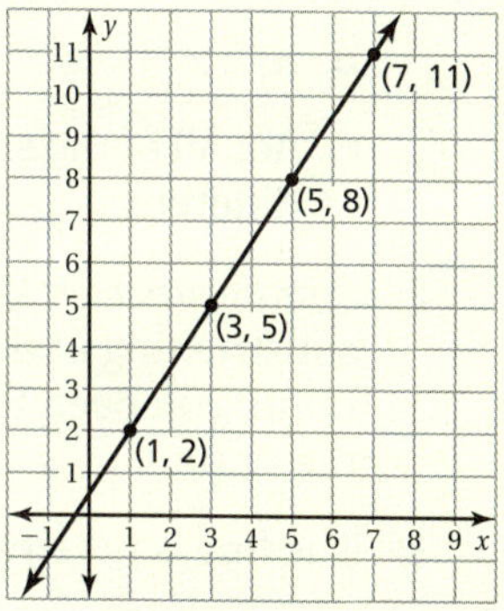

5. -2

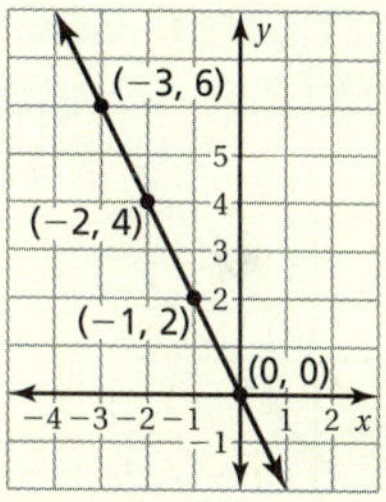

Extra Example 4

Which two lines are parallel? Explain.

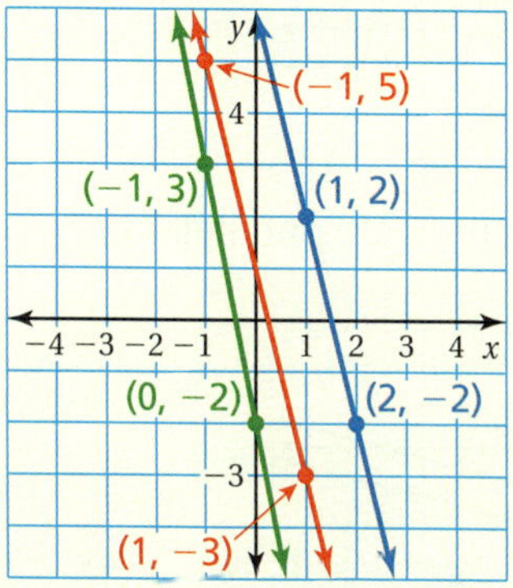

The blue and red lines both have a slope of -4, so they are parallel.

On Your Own

6. The blue and red lines both have a slope of $-\frac{1}{3}$, so they are parallel.

Vocabulary and Concept Check

1. **a.** B and C

   b. A

   c. no; All of the slopes are different.

2. *Sample answer:* When constructing a wheelchair ramp, you need to know the slope.

3. The line is horizontal.

Practice and Problem Solving

4.

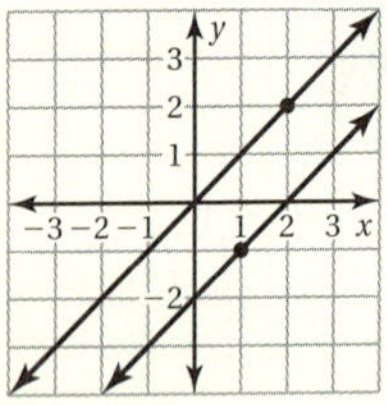

The lines are parallel.

5.

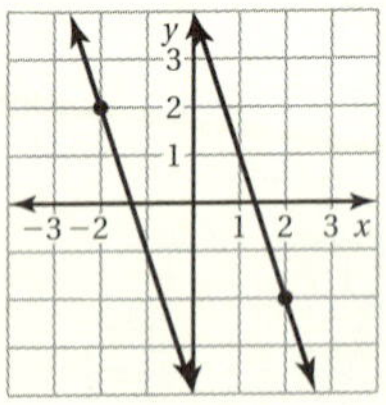

The lines are parallel.

6.

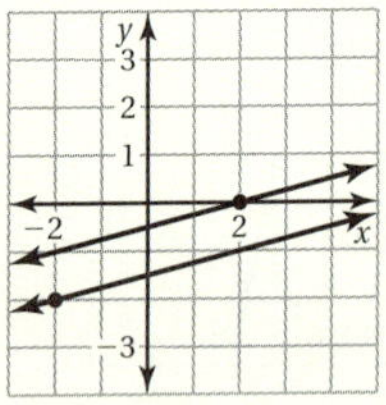

The lines are parallel.

7. $\frac{3}{4}$ **8.** $-\frac{5}{4}$

9. $-\frac{3}{5}$ **10.** $\frac{1}{6}$

11. 0 **12.** $\frac{5}{2}$

Assignment Guide and Homework Check

Level	Day 1 Activity Assignment	Day 2 Lesson Assignment	Homework Check
Basic	4–6, 28–31	1–3, 7–14, 15–21 odd	7, 11, 14, 15, 21
Average	4–6, 28–31	1–3, 7–25 odd, 14, 22	7, 11, 14, 15, 22
Advanced	4–6, 28–31	1–3, 11–13, 16, 18, 21–27	12, 16, 22, 24, 26

Common Errors

- **Exercises 7–12** Students may find the reciprocal of the slope because they mix up rise and run. Remind them that the change in *y* is the numerator and the change in *x* is the denominator.
- **Exercises 7–12** Students may forget negatives, or include them when they are not needed. Remind them that if the line rises from left to right the slope is positive, and if the line falls from left to right the slope is negative.

2.2 Record and Practice Journal

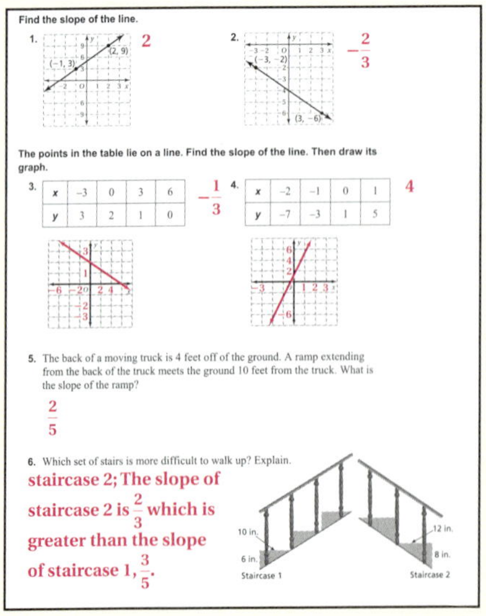

Find the slope of the line.

1. (−1, 3), (2, 9) — 2

2. (−3, −2), (3, −6) — $-\frac{2}{3}$

The points in the table lie on a line. Find the slope of the line. Then draw its graph.

3.

x	−3	0	3	6
y	3	2	1	0

$-\frac{1}{3}$

4.

x	−2	−1	0	1
y	−7	−3	1	5

4

5. The back of a moving truck is 4 feet off of the ground. A ramp extending from the back of the truck meets the ground 10 feet from the truck. What is the slope of the ramp?

$\frac{2}{5}$

6. Which set of stairs is more difficult to walk up? Explain.

staircase 2; The slope of staircase 2 is $\frac{2}{3}$ which is greater than the slope of staircase 1, $\frac{3}{5}$.

Technology For the Teacher

Answer Presentation Tool

Quiz*Show*

2.2 Exercises

Check It Out
Help with Homework
BigIdeasMath.com

Vocabulary and Concept Check

1. **CRITICAL THINKING** Refer to the graph.

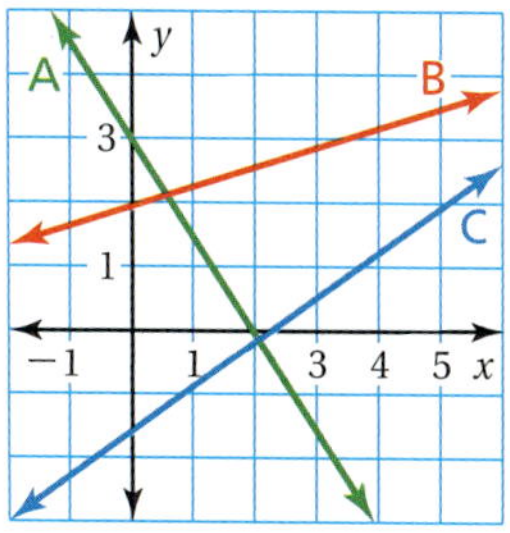

   a. Which lines have positive slopes?
   b. Which line has the steepest slope?
   c. Are any two of the lines parallel? Explain.

2. **OPEN-ENDED** Describe a real-life situation that involves slope.

3. **REASONING** The slope of a line is 0. What do you know about the line?

Practice and Problem Solving

Draw a line through each point using the given slope. What do you notice about the two lines?

4. Slope = 1

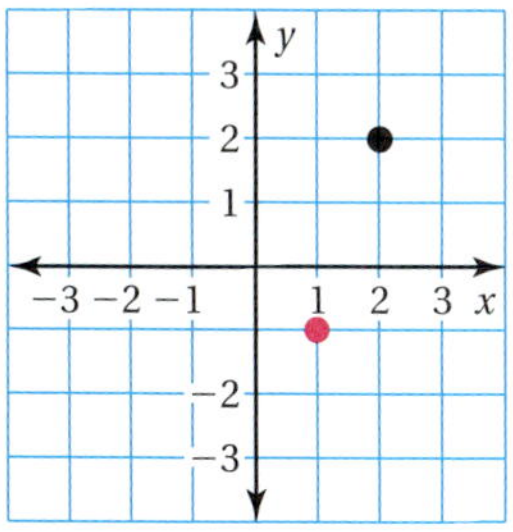

5. Slope = −3

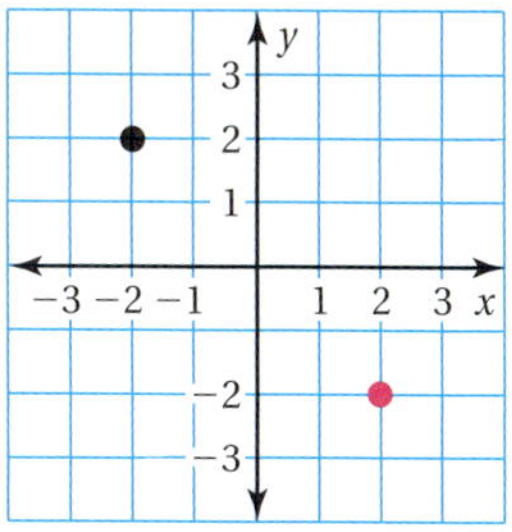

6. Slope = $\frac{1}{4}$

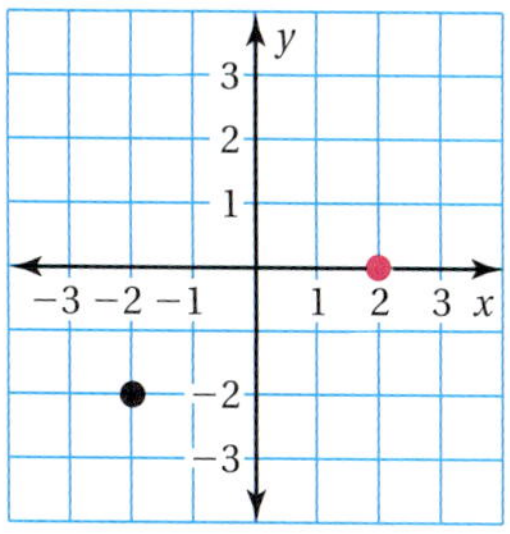

Find the slope of the line.

1 2 7.

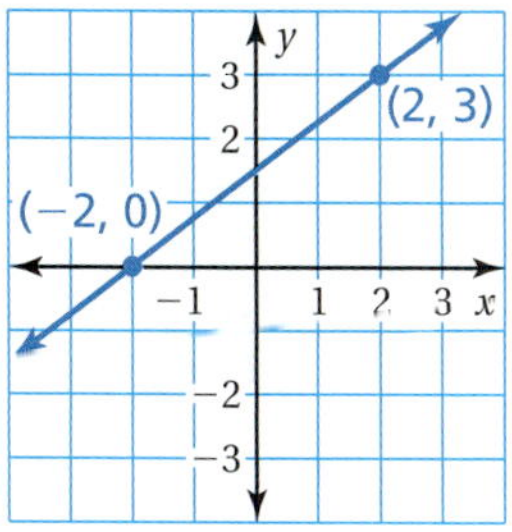

8.

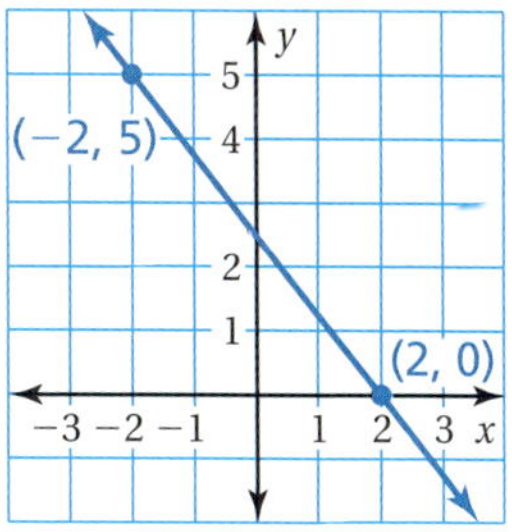

9.

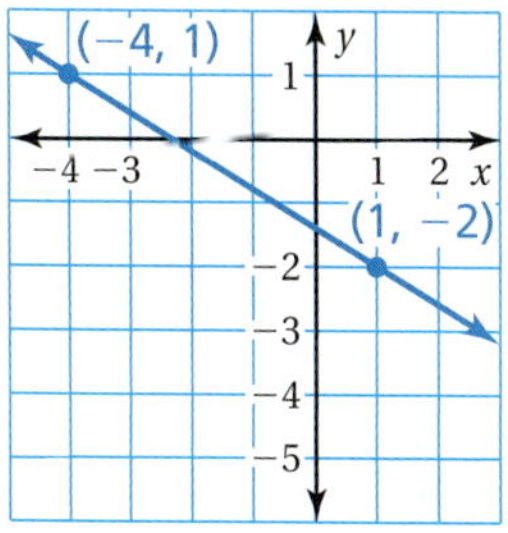

10.

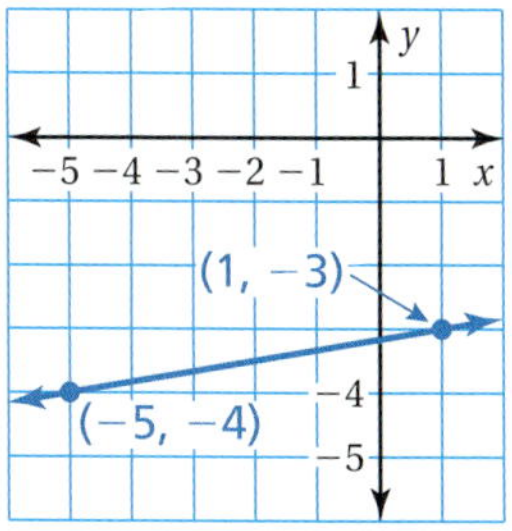

11.

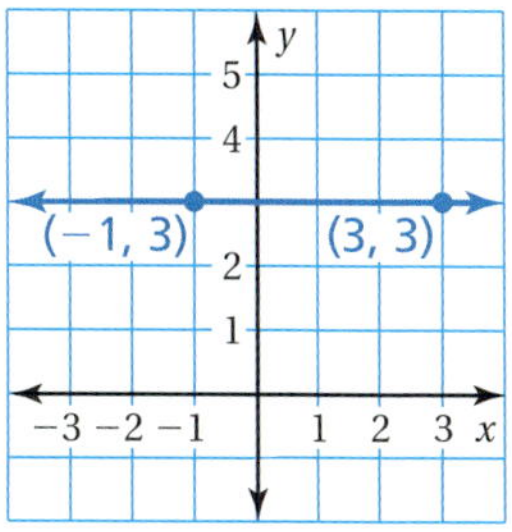

12.

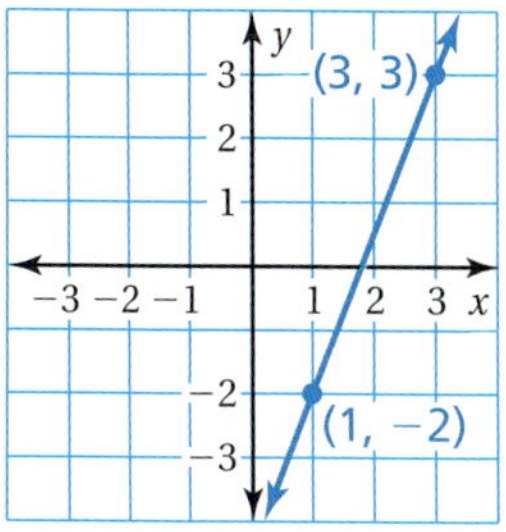

13. **ERROR ANALYSIS** Describe and correct the error in finding the slope of the line.

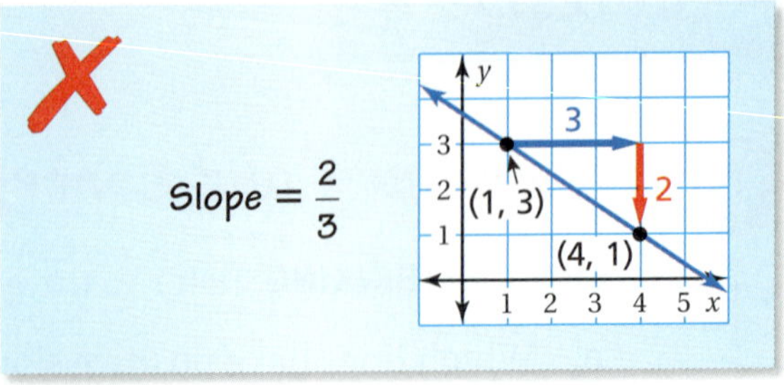

14. **CRITICAL THINKING** Is it more difficult to walk up the ramp or the hill? Explain.

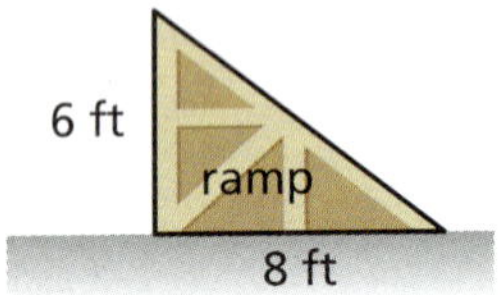

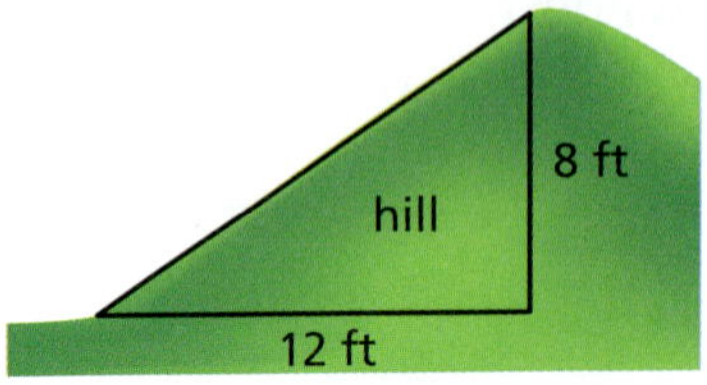

The points in the table lie on a line. Find the slope of the line. Then draw its graph.

③ 15.

x	1	3	5	7
y	2	10	18	26

16.

x	−3	2	7	12
y	0	2	4	6

17.

x	−6	−2	2	6
y	8	5	2	−1

18.

x	−8	−2	4	10
y	8	1	−6	−13

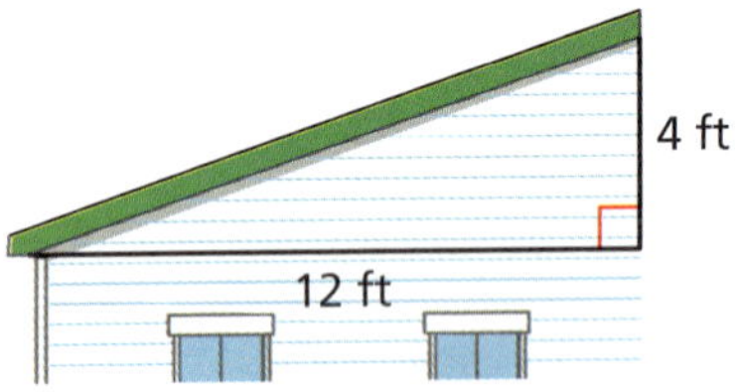

19. **PITCH** Carpenters refer to the slope of a roof as the *pitch* of the roof. Find the pitch of the roof.

20. **PROJECT** The guidelines for a wheelchair ramp suggest that the ratio of the rise to the run be no greater than 1 : 12.

   a. Find a wheelchair ramp in your school or neighborhood. Measure its slope. Does the ramp follow the guidelines?

   b. Design a wheelchair ramp that provides access to a building with a front door that is 2.5 feet higher than the sidewalk. Illustrate your design.

Which two lines are parallel? Explain.

④ 21.

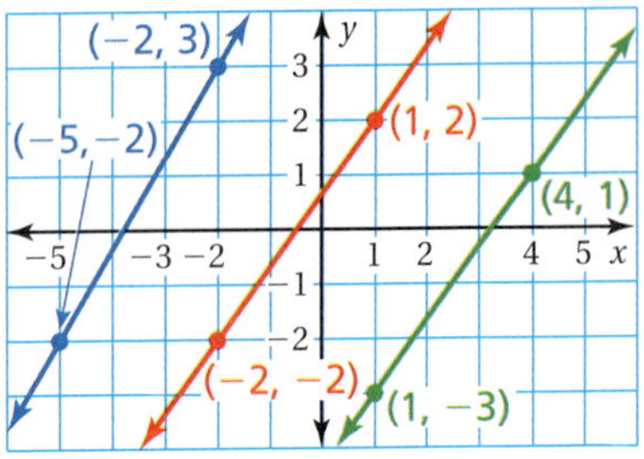

22.

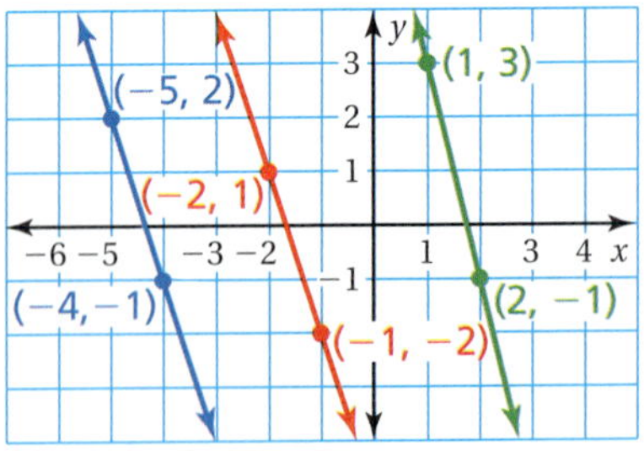

Common Errors

- **Exercise 14** Students may get confused because one of the slopes is negative and the other is positive. Tell them to think of the absolute values of the slopes when comparing. Encourage them to graph the slopes on a number line to check their answer.
- **Exercises 15–18** Students may reverse the *x*- and *y*-coordinates when plotting the ordered pairs given in the table. Remind them that the first row is the *x*-coordinate and the second row is the *y*-coordinate. Encourage students to write the ordered pairs before graphing.
- **Exercises 21 and 22** Students may use their eyes to guess which two lines are parallel without finding the actual slopes. Encourage them to find the slope of each line. The slopes of the lines are close enough that it is difficult to tell the difference visually.
- **Exercises 23 and 24** Students may not remember the definition of a parallelogram. Remind them that they have to compare the slopes of the opposite sides. Even if one pair of sides is parallel, the other pair may not be parallel. Remind students to compare both pairs of opposite sides.

Practice and Problem Solving

13. The 2 should be -2 because it goes down.

Slope $= -\frac{2}{3}$

14. The ramp because its slope is steeper.

15. 4

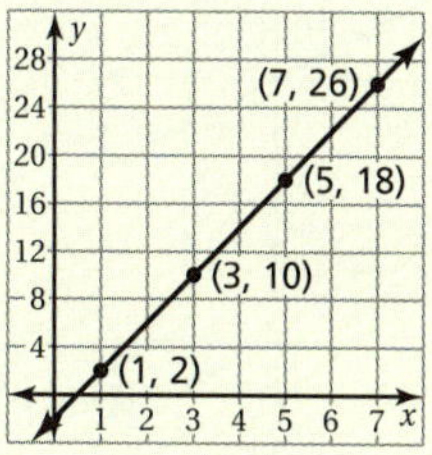

16. $\frac{2}{5}$

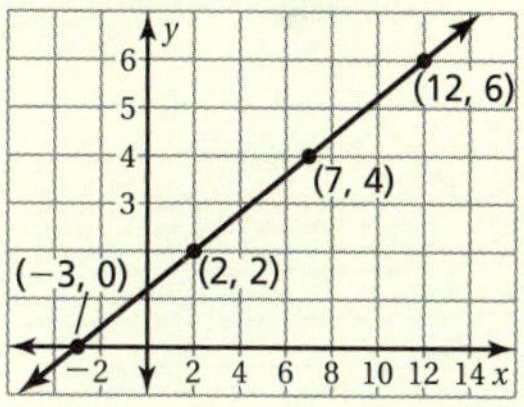

17. $-\frac{3}{4}$

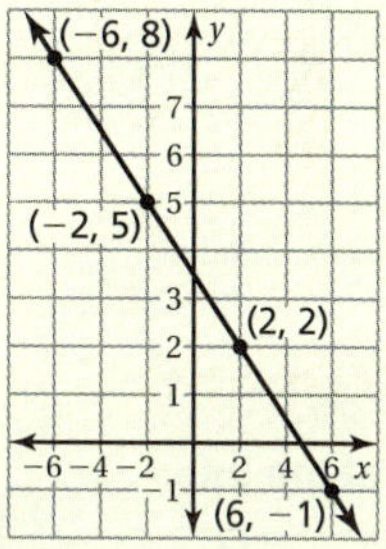

18–22. See Additional Answers.

Differentiated Instruction

Auditory

Discuss how the rate of change in a rate problem is related to slope. For example, the cost to travel on a turnpike (cost per mile) can be expressed as $\frac{\text{cost (in dollars)}}{\text{miles driven}}$, where the cost is the change in *y*-values and the miles driven is the change in *x*-values.

Practice and Problem Solving

23–26. See Additional Answers.

27. See *Taking Math Deeper.*

Fair Game Review

28.

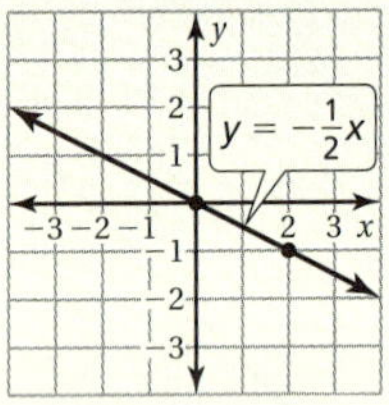

29.

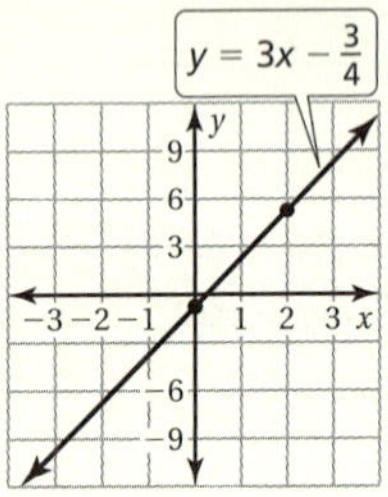

30.

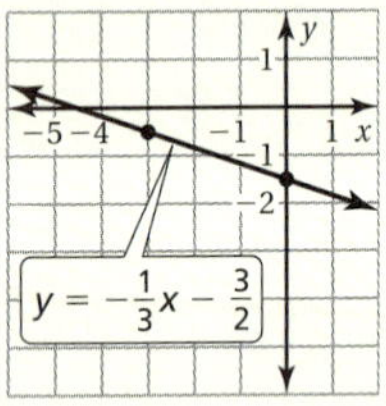

31. B

Mini-Assessment

Find the slope of the line.

1.

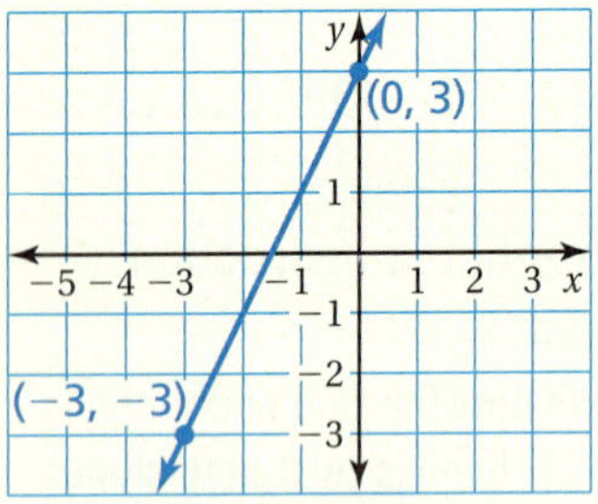

slope = 2

2.

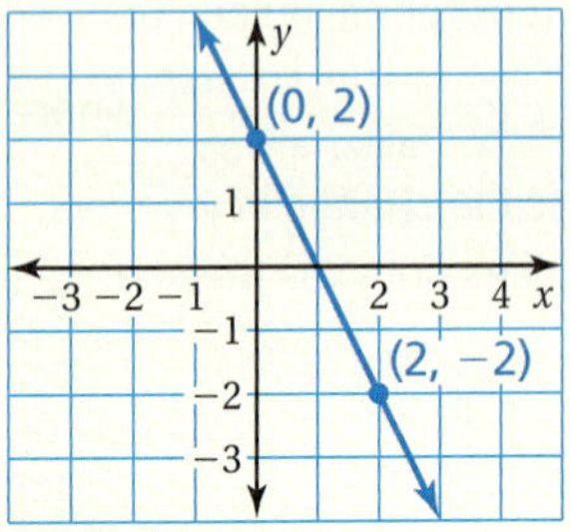

slope = −2

Taking Math Deeper

Exercise 27

This exercise is a nice example of the power of a diagram. Instead of using the drawing of the slide, encourage students to draw the slide in a coordinate plane. Once that is done, the question is easier to answer.

1 Draw a diagram.

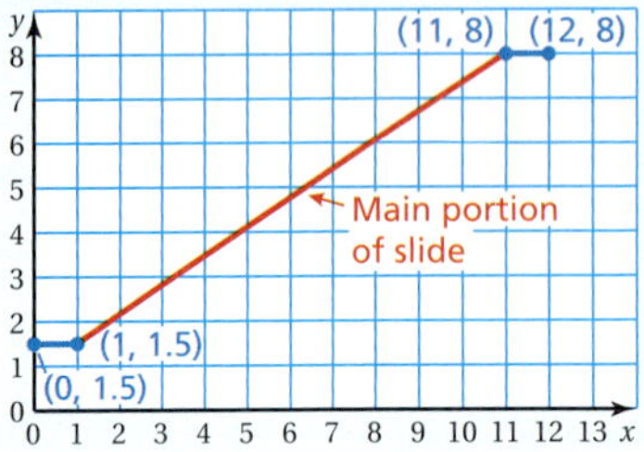

2 **a.** Find the slope of the slide.

$$\text{Slope} = \frac{\text{change in } y}{\text{change in } x} = \frac{8 - 1.5}{11 - 1} = \frac{6.5}{10} = 0.65$$

3 Compare the slopes.

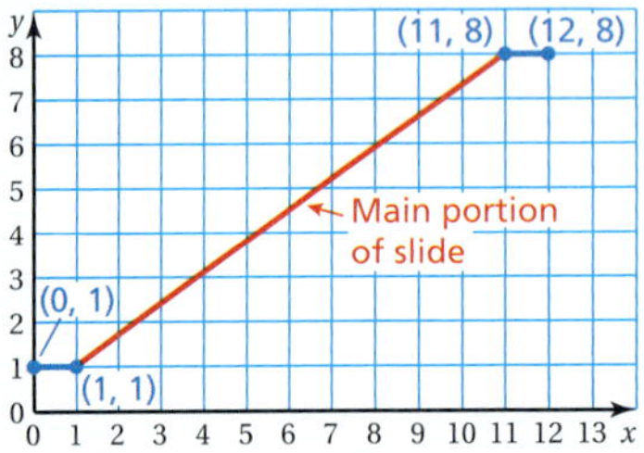

Because 0.7 > 0.65, the slide is steeper.

b.

$$\text{Slope} = \frac{\text{change in } y}{\text{change in } x} = \frac{8 - 1}{11 - 1} = \frac{7}{10} = 0.7$$

Project

Many water parks and amusement parks have water slides. Find the height of a slide and calculate the slope of the main part of the slide.

Reteaching and Enrichment Strategies

If students need help. . .	If students got it. . .
Resources by Chapter • Practice A and Practice B • Puzzle Time Record and Practice Journal Practice Differentiating the Lesson Lesson Tutorials Skills Review Handbook	Resources by Chapter • Enrichment and Extension Start the next section

Tell whether the quadrilateral is a parallelogram. Explain.

23.

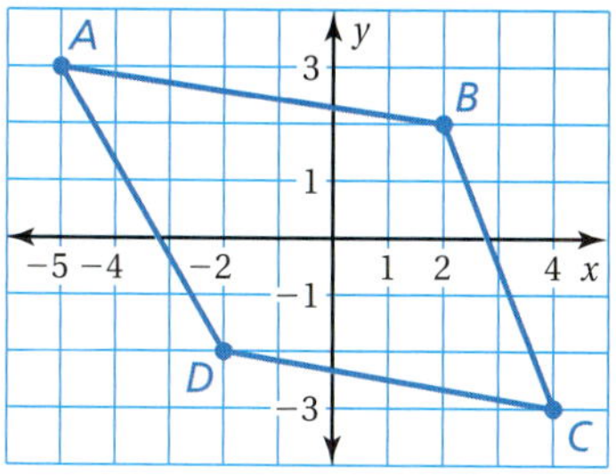

24.

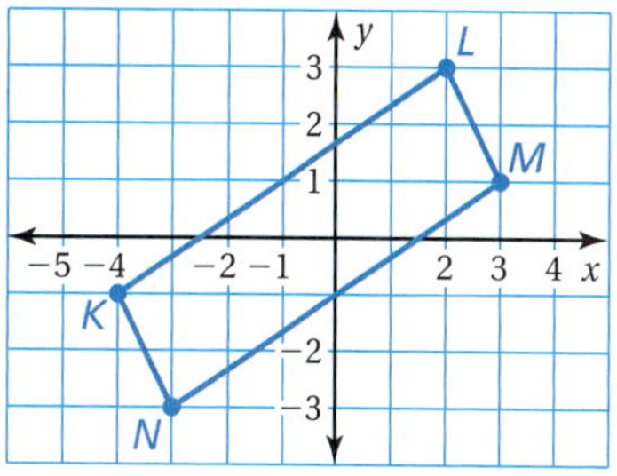

25. **TURNPIKE TRAVEL** The graph shows the cost of traveling by car on a turnpike.

a. Find the slope of the line.

b. Explain the meaning of the slope as a rate of change.

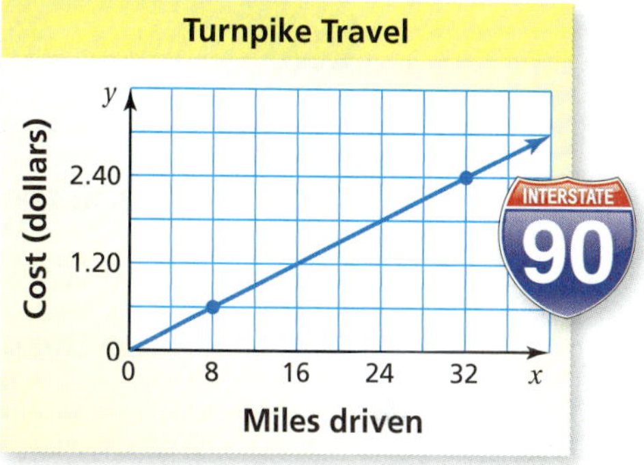

26. **BOAT RAMP** Which is steeper: the boat ramp or a road with a 12% grade? Explain. (*Note:* Road grade is the vertical increase divided by the horizontal distance.)

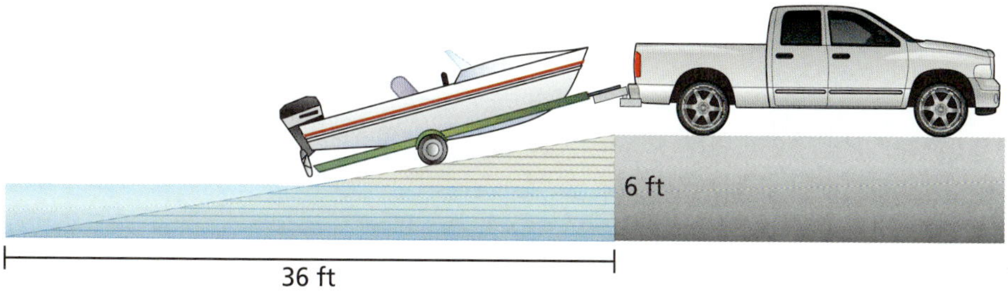

27. Critical Thinking The top and bottom of the slide are parallel to the ground.

a. What is the slope of the main portion of the slide?

b. How does the slope change if the bottom of the slide is only 12 inches above the ground? Is the slide steeper? Explain.

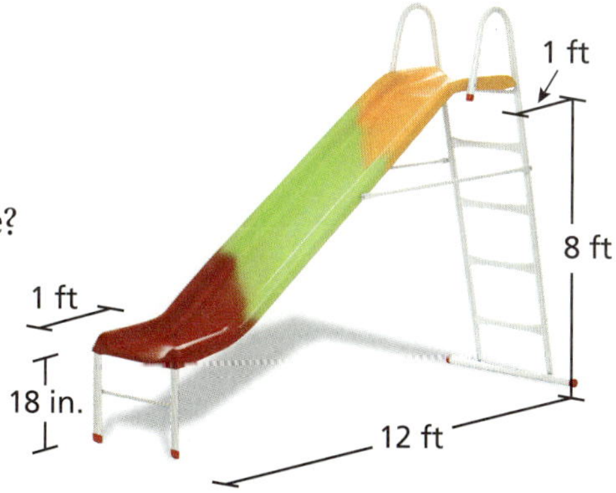

Fair Game Review What you learned in previous grades & lessons

Graph the linear equation. *(Section 2.1)*

28. $y = -\frac{1}{2}x$ **29.** $y = 3x - \frac{3}{4}$ **30.** $y = -\frac{x}{3} - \frac{3}{2}$

31. **MULTIPLE CHOICE** What is the prime factorization of 84? *(Skills Review Handbook)*

Ⓐ $2 \times 3 \times 7$ Ⓑ $2^2 \times 3 \times 7$ Ⓒ $2 \times 3^2 \times 7$ Ⓓ $2^2 \times 21$

2.3 Graphing Linear Equations in Slope-Intercept Form

Essential Question How can you describe the graph of the equation $y = mx + b$?

1 ACTIVITY: Finding Slopes and y-Intercepts

Work with a partner.

- **Graph the equation.**
- **Find the slope of the line.**
- **Find the point where the line crosses the *y*-axis.**

a. $y = -\frac{1}{2}x + 1$

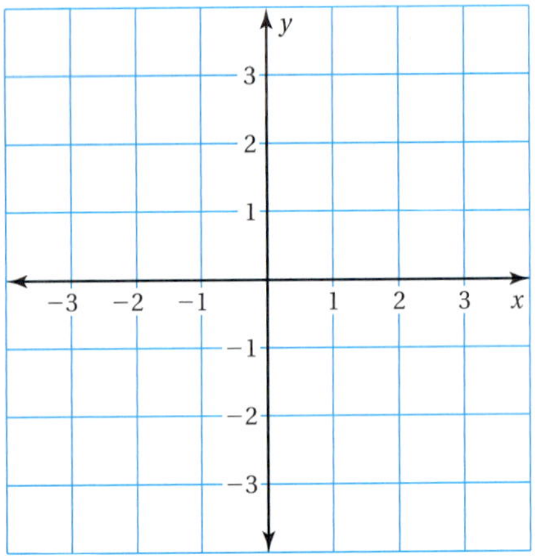

b. $y = -x + 2$

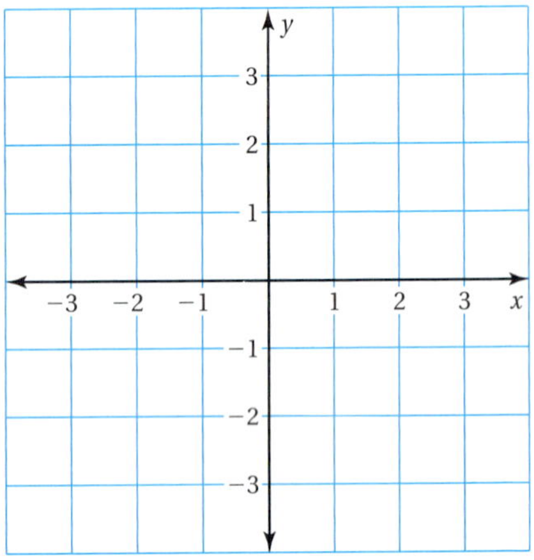

c. $y = -x - 2$

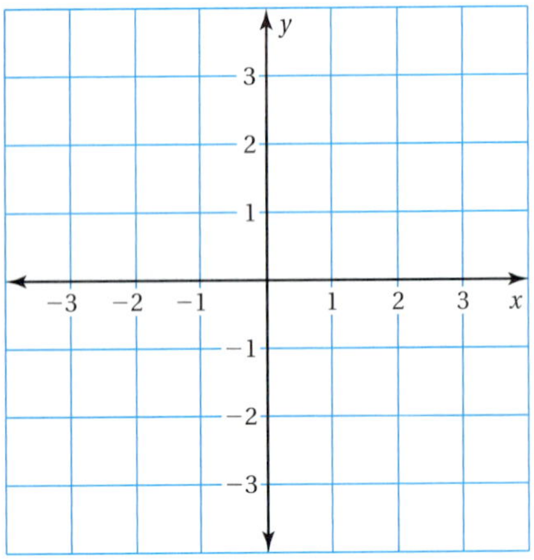

d. $y = \frac{1}{2}x + 1$

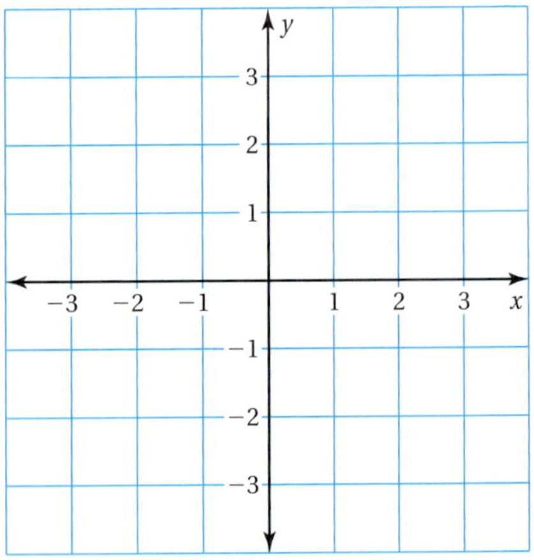

Laurie's Notes

Introduction

For the Teacher

- **Goal:** Students will explore the connection between the equation of a line and its graph.

Motivate

- **Preparation:** Make three demonstration cards on 8.5"x 11" paper. The x-axis is labeled "time" and the y-axis is labeled "distance from home."
- Sample cards A, B, and C are shown.

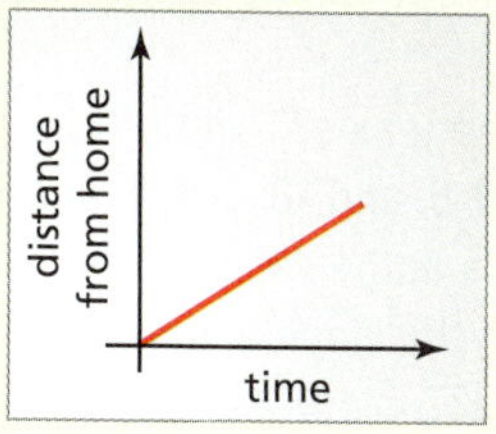

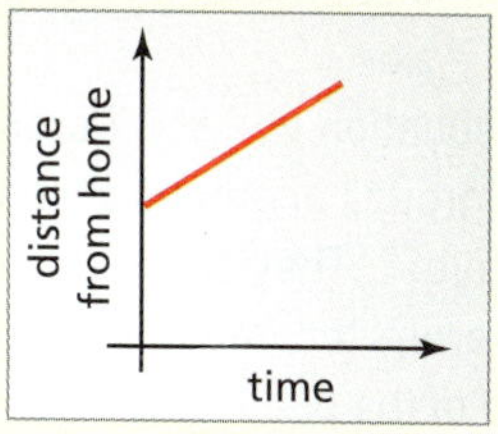

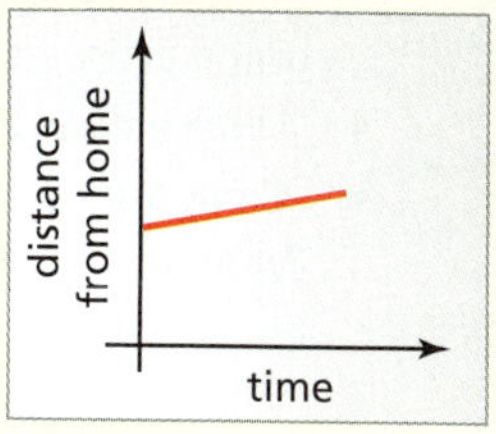

- Ask 3 students to hold the cards for the class to see.
- ? "Consider how the axes are labeled. What does the slope of the line represent?" $\frac{\text{distance}}{\text{time}} = \text{rate}$
- ? "What story does each card tell? How are the stories similar and different?" A: you begin at home; B: you travel at the same rate, but you start away from home; C: you start away from home, but you travel at a slower rate
- **Management Tip:** If you plan to use the demonstration cards again next year, laminate them.

Activity Notes

Activity 1

- ? "How do you graph an equation?" Plot several points, then connect the points with a line.
- ? "Is there a way to organize the points you need to plot?" Use an input-output table.
- Review with students how input-output tables are set up and what values of x they should substitute. When the coefficient of x is a fraction, it is wise to select x-values that are multiples of the denominator. This will eliminate fractional values.
- ? "How many points do you need in order to graph the equation?" Minimum is 2. Plot 3 to be safe.
- Remind students to evaluate the equation for $x = 0$. This will ensure that they find the point where the graph crosses the y-axis.
- The slope and the point where the line crosses the y-axis will be recorded in the table on the next page.
- Check students' work before going on to the Inductive Reasoning.

Previous Learning

Students should know how to find the slopes of lines.

Activity Materials	
Introduction	**Textbook**
• demonstration cards	• straightedge • grid paper
Closure	
• demonstration cards	

Start Thinking! and Warm Up

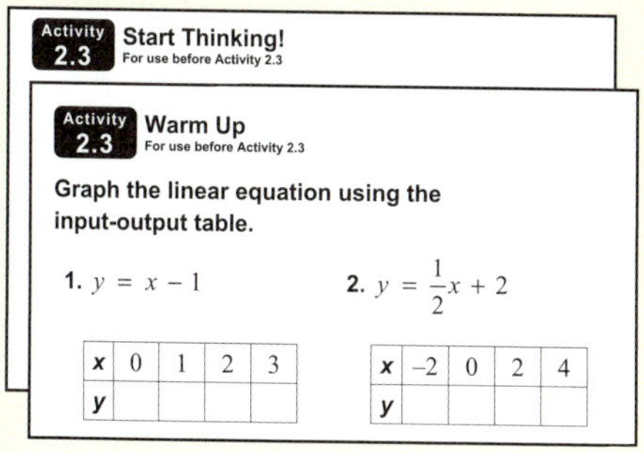

Activity 2.3 Start Thinking!
For use before Activity 2.3

Activity 2.3 Warm Up
For use before Activity 2.3

Graph the linear equation using the input-output table.

1. $y = x - 1$

x	0	1	2	3
y				

2. $y = \frac{1}{2}x + 2$

x	−2	0	2	4
y				

2.3 Record and Practice Journal

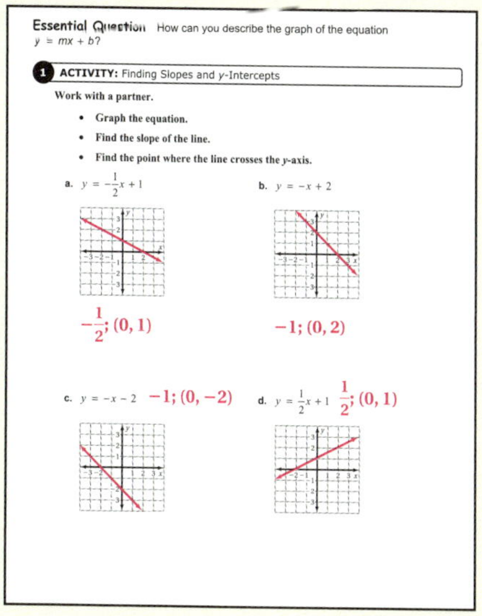

Essential Question How can you describe the graph of the equation $y = mx + b$?

1 ACTIVITY: Finding Slopes and y-Intercepts

Work with a partner.

- Graph the equation.
- Find the slope of the line.
- Find the point where the line crosses the y-axis.

a. $y = -\frac{1}{2}x + 1$ $-\frac{1}{2}$; (0, 1)

b. $y = -x + 2$ −1; (0, 2)

c. $y = -x - 2$ −1; (0, −2)

d. $y = \frac{1}{2}x + 1$ $\frac{1}{2}$; (0, 1)

English Language Learners

Build on Past Knowledge

Remind students from their study of rational numbers that the slope -2 can be written as the fraction $\frac{-2}{1}$. By writing the integer as a fraction, students can see that the slope has a run of 1 and a rise of -2. This will help in graphing the line of a linear function.

2.3 Record and Practice Journal

Inductive Reasoning

Work with a partner. Graph each equation. Then complete the table.

	Equation	Description of Graph	Slope of Graph	Point of Intersection with y-axis
1a	2. $y = -\frac{1}{2}x + 1$	Line	$-\frac{1}{2}$	$(0, 1)$
1b	3. $y = -x + 2$	Line	-1	$(0, 2)$
1c	4. $y = -x - 2$	Line	-1	$(0, -2)$
1d	5. $y = \frac{1}{2}x + 1$	Line	$\frac{1}{2}$	$(0, 1)$
	6. $y = x + 2$	Line	1	$(0, 2)$
	7. $y = x - 2$	Line	1	$(0, -2)$
	8. $y = \frac{1}{2}x - 1$	Line	$\frac{1}{2}$	$(0, -1)$
	9. $y = -\frac{1}{2}x - 1$	Line	$-\frac{1}{2}$	$(0, -1)$
	10. $y = 3x + 2$	Line	3	$(0, 2)$
	11. $y = 3x - 2$	Line	3	$(0, -2)$
	12. $y = -2x + 3$	Line	-2	$(0, 3)$

What Is Your Answer?

13. **IN YOUR OWN WORDS** How can you describe the graph of the equation $y = mx + b$?

a line with slope m and crosses the y-axis at $(0, b)$

a. How does the value of m affect the graph of the equation?

steepness of line

b. How does the value of b affect the graph of the equation?

Moves graph up and down.

c. Check your answers to parts (a) and (b) with three equations that are not in the table.

Check students' work.

14. Why is $y = mx + b$ called the "slope-intercept" form of the equation of a line?

m is the slope and b is the y-intercept.

Laurie's Notes

Inductive Reasoning

- Give students sufficient time to complete the table. Provide grid paper for Questions 6–12.
- Students should begin to observe patterns as they complete the table.
- Circulate to ensure that graphs are drawn correctly.
- Encourage students to draw the directed arrows in order to help them find the slope of the line.
- Have students put a few graphs on the board to help facilitate discussion.

? When students have finished, ask a series of summary questions.

- "Compare certain pairs of graphs such as 6 and 7; or 10 and 11. What do you observe?" They have the same steepness (slope) and the number at the end of the equation is the y-coordinate of where the graph crosses the y-axis.
- "Where does the equation $y = x + 7$ cross the y-axis?" at $(0, 7)$
- "Compare certain groups of graphs such as 3, 6, and 10; or 4, 7, and 11. What do you observe?" They cross the y-axis at the same point, but the slopes are different; the coefficient of x is the slope of the line.
- "What is the slope of the equation $y = 7x + 2$?" slope $= 7$

Words of Wisdom

- Students may not use mathematical language to describe their observations. Listen for the concept, the vocabulary will come later.
- In equations such as $y = x - 2$, students do not always think of the subtraction operation as making the constant negative. You may need to remind students that this is the same as *adding the opposite*. So, $y = x - 2$ is equivalent to $y = x + (-2)$.

What Is Your Answer?

- These answers should follow immediately from discussing student observations.

Closure

- Refer back to the demonstration cards A, B, and C. Have students describe how the equations would be similar and how they would be different. *Sample answer:* A and B have the same slope, but different y-intercepts. B and C have different slopes, but the same y-intercept.

Technology For the Teacher

Dynamic Classroom

The Dynamic Planning Tool

Editable Teacher's Resources at *BigIdeasMath.com*

Inductive Reasoning

Work with a partner. Graph each equation. Then copy and complete the table.

	Equation	Description of Graph	Slope of Graph	Point of Intersection with *y*-axis
1a	**2.** $y = -\frac{1}{2}x + 1$	Line	$-\frac{1}{2}$	(0, 1)
1b	**3.** $y = -x + 2$			
1c	**4.** $y = -x - 2$			
1d	**5.** $y = \frac{1}{2}x + 1$			
	6. $y = x + 2$			
	7. $y = x - 2$			
	8. $y = \frac{1}{2}x - 1$			
	9. $y = -\frac{1}{2}x - 1$			
	10. $y = 3x + 2$			
	11. $y = 3x - 2$			
	12. $y = -2x + 3$			

What Is Your Answer?

13. **IN YOUR OWN WORDS** How can you describe the graph of the equation $y = mx + b$?

a. How does the value of m affect the graph of the equation?

b. How does the value of b affect the graph of the equation?

c. Check your answers to parts (a) and (b) with three equations that are not in the table.

14. Why is $y = mx + b$ called the "slope-intercept" form of the equation of a line?

Practice Use what you learned about graphing linear equations in slope-intercept form to complete Exercises 4–6 on page 66.

2.3 Lesson

Check It Out
Lesson Tutorials
BigIdeasMath.com

Key Vocabulary
x-intercept, *p. 64*
y-intercept, *p. 64*
slope-intercept form, *p. 64*

Key Ideas

Intercepts

The **x-intercept** of a line is the x-coordinate of the point where the line crosses the x-axis. It occurs when $y = 0$.

The **y-intercept** of a line is the y-coordinate of the point where the line crosses the y-axis. It occurs when $x = 0$.

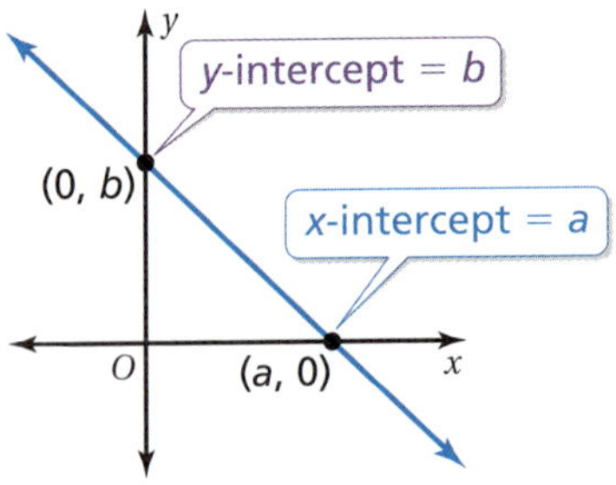

Slope-Intercept Form

Words An equation written in the form $y = mx + b$ is in **slope-intercept form**. The slope of the line is m and the y-intercept of the line is b.

Algebra $y = mx + b$

slope (m) · y-intercept (b)

EXAMPLE 1 Identifying Slopes and y-Intercepts

Find the slope and y-intercept of the graph of each linear equation.

a. $y = -4x - 2$

$y = -4x + (-2)$ Write in slope-intercept form.

∴ The slope is -4 and the y-intercept is -2.

b. $y - 5 = \frac{3}{2}x$

$y = \frac{3}{2}x + 5$ Add 5 to each side.

∴ The slope is $\frac{3}{2}$ and the y-intercept is 5.

On Your Own

Now You're Ready
Exercises 7–15

Find the slope and y-intercept of the graph of the linear equation.

1. $y = 3x - 7$

2. $y - 1 = -\frac{2}{3}x$

Laurie's Notes

Introduction

Connect

- **Yesterday:** Students explored the connection between the equation of a line and its graph.
- **Today:** Students will use the slope-intercept form of a line to graph the line.

Motivate

- Share the following taxi information. All trips start at a convention center.

Destination	Distance	Taxi Fare
Football stadium	18.7 mi	\$39 approx.
Airport	12 mi	\$32 flat fee
Shopping district	9.5 mi	\$20 approx.

? "How do you think taxi fares are determined?" Answers will vary; listen for distance, number of passengers, tolls.

- Discuss why some locations, often involving airports, have flat fees associated with them.

Lesson Notes

Key Ideas

- Write the Key Ideas on the board. Draw the graph and discuss the vocabulary of this lesson: *x*-intercept, *y*-intercept, and slope-intercept form.
- Explain to students that the equation must be written with *y* as a function of *x*. This means that the equation must be solved for *y*.
- **FYI:** Students may ask why the letters *m* and *b* are used. Historically, there is no definitive answer. I tell my students that mathematicians, much older than myself, have used *m* for slope for centuries. Using *b* for the *y*-intercept appears to be an American phenomenon.

Example 1

? "What is a linear equation?" an equation whose graph is a line

- Write part (a). This is written in the form $y = mx + b$, enabling students to quickly identify the slope and *y*-intercept.
- Write part (b).

? "Is $y - 5 = \frac{3}{2}x$ in slope-intercept form?" no "Can you rewrite it so that it is?" yes; Add 5 to each side of the equation.

On Your Own

- **Think-Pair-Share:** Students should read each question independently and then work with a partner to answer the questions. When they have answered the questions, the pair should compare their answers with another group and discuss any discrepancies.

Goal Today's lesson is graphing the equation of a line written in **slope-intercept form**.

Lesson Materials
Textbook
• straightedge

Start Thinking! and Warm Up

Lesson 2.3 Warm Up
For use before Lesson 2.3

Lesson 2.3 Start Thinking!
For use before Lesson 2.3

Describe a situation involving online shopping that can be modeled with a linear equation.

What is the slope?

What is the *y*-intercept?

Extra Example 1

Find the slope and *y*-intercept of the graph of each linear equation.

a. $y = \frac{3}{4}x - 5$

slope: $\frac{3}{4}$; *y*-intercept: -5

b. $y + \frac{1}{2} = -6x$

slope: -6; *y*-intercept: $-\frac{1}{2}$

On Your Own

1. slope: 3; *y*-intercept: -7
2. slope: $-\frac{2}{3}$; *y*-intercept: 1

Extra Example 2

Graph $y = -\frac{2}{3}x - 2$. Identify the x-intercept.

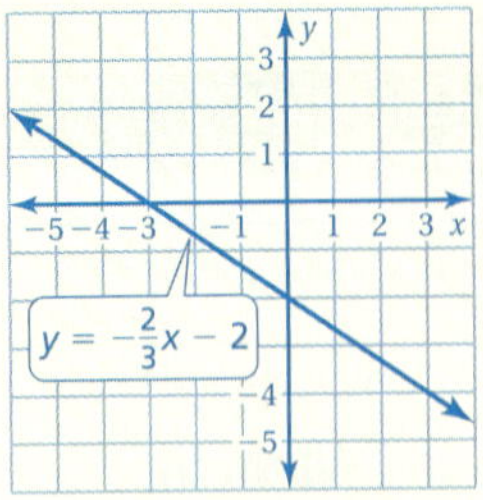

−3

Extra Example 3

The cost y (in dollars) for making friendship bracelets is $y = 0.5x + 2$, where x is the number of bracelets.

a. Graph the equation.

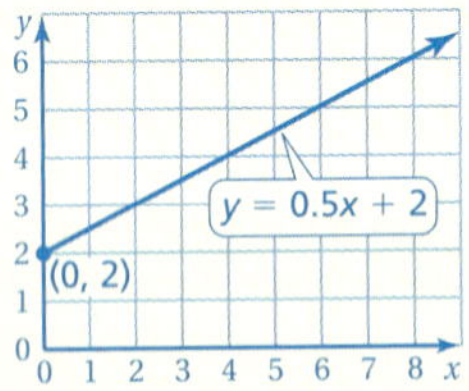

b. Interpret the slope and y-intercept.
The slope is 0.5. So, the cost per bracelet is \$0.50. The y-intercept is 2. So, there is an initial cost of \$2 to make the bracelets.

On Your Own

3–5. See Additional Answers.

Differentiated Instruction

Kinesthetic

When graphing a linear function using the slope-intercept form, students must apply the slope correctly after plotting the point for the y-intercept. Have students plot (0, 3) in the coordinate plane. Then graph the lines $y = 4x + 3$, $y = \frac{1}{4}x + 3$, $y = -4x + 3$, and $y = -\frac{1}{4}x + 3$ in the same coordinate plane using (0, 3) as the starting point. Make sure students identify the correct rise and run for each line.

Laurie's Notes

Example 2

? "How can knowing the slope and the y-intercept help you graph a line?" Listen for student understanding of what slope and y-intercept mean.

- Remind students that a slope of −3 can be interpreted as $\frac{-3}{1} = \frac{3}{-1}$. Starting at the y-intercept, you can move to the right 1 unit and down 3 units, or to the left 1 unit and up 3 units. In both cases, you land on a point which satisfies the equation.

? **Extension:** "In this problem, you found the x-intercept by interpreting the slope and it coincidentally landed on the x-axis. If the line had missed the x-intercept, how would you find the x-intercept?" Set $y = 0$ and solve for x.

Example 3

? "Have any of you taken a taxi that had a meter on the front dashboard?"

- **FYI:** Not all taxis use meters. Where I live, which is rural, taxis charge a flat fee to travel from one region of town to another.
- Write the equation $y = 2.5x + 2$ on the board.

? "What is the slope and y-intercept for this equation?" Slope = 2.5 and y-intercept is 2.

- Suggest to students that because the slope is 2.5, any ratio equivalent to 2.5 can also be used, such as $\frac{2.5}{1} = \frac{5}{2}$. Using whole numbers instead of decimals improves the accuracy of graphing.
- Explain that the graph of this equation will only be in Quadrant I because it does not make sense to have a negative number of miles or a negative cost.
- Interpreting the slope and y-intercept is an important step, particularly for real-life applications.

? "What is the cost for a 2-mile taxi ride? a 10-mile taxi ride?" \$7; \$27

On Your Own

- Have students share results at the board after graphing the equations.

Closure

- **Exit Ticket:** Graph $y - 4 = 2x$ and identify the slope and y-intercept.

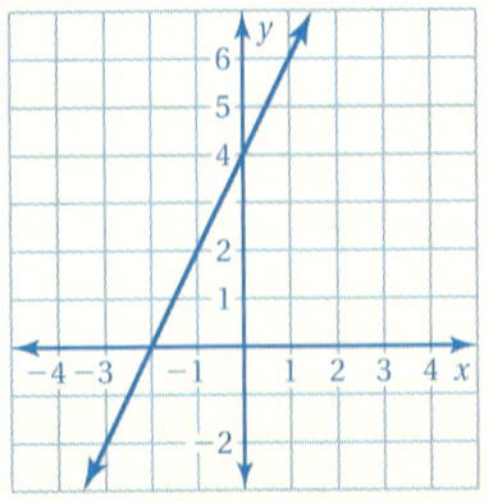

slope = 2
y-intercept = 4

Technology For the Teacher

Dynamic Classroom

The Dynamic Planning Tool
Editable Teacher's Resources at *BigIdeasMath.com*

EXAMPLE 2 Graphing a Linear Equation in Slope-Intercept Form

Graph $y = -3x + 3$. Identify the x-intercept.

Step 1: Find the slope and y-intercept.

$$y = -3x + 3$$

slope → -3 y-intercept → 3

Step 2: The y-intercept is 3. So, plot (0, 3).

Step 3: Use the slope to find another point and draw the line.

$$\text{slope} = \frac{\text{rise}}{\text{run}} = \frac{-3}{1}$$

Plot the point that is 1 unit right and 3 units down from (0, 3). Draw a line through the two points.

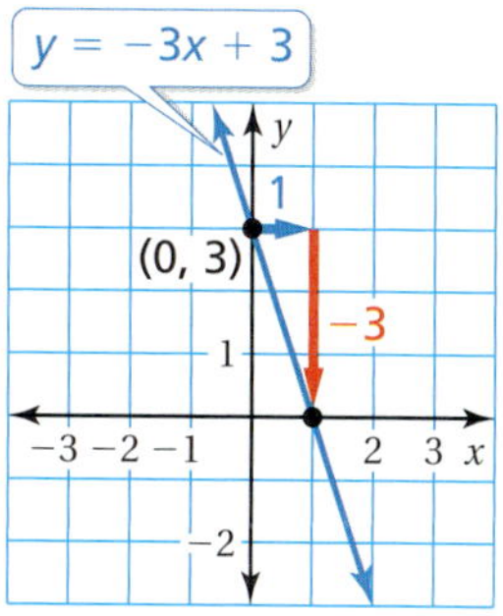

The line crosses the x-axis at (1, 0). So, the x-intercept is 1.

Study Tip

You can check the x-intercept by substituting $y = 0$ in the equation and solving for x.

$y = -3x + 3$

$0 = -3x + 3$

$-3 = -3x$

$1 = x$

EXAMPLE 3 Real-Life Application

The cost y (in dollars) of taking a taxi x miles is $y = 2.5x + 2$. (a) Graph the equation. (b) Interpret the y-intercept and slope.

a. The slope of the line is $2.5 = \frac{5}{2}$. Use the slope and y-intercept to graph the equation.

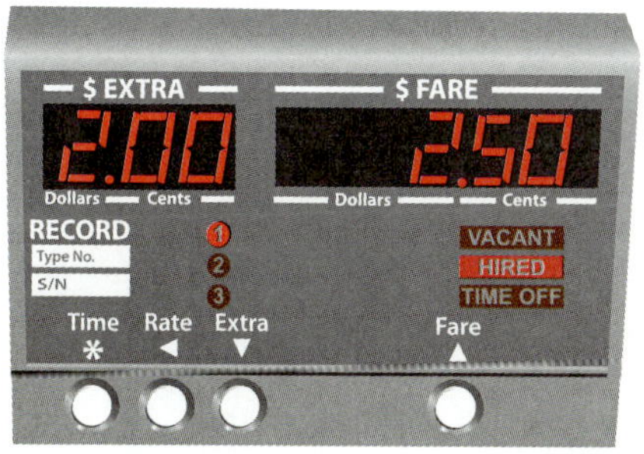

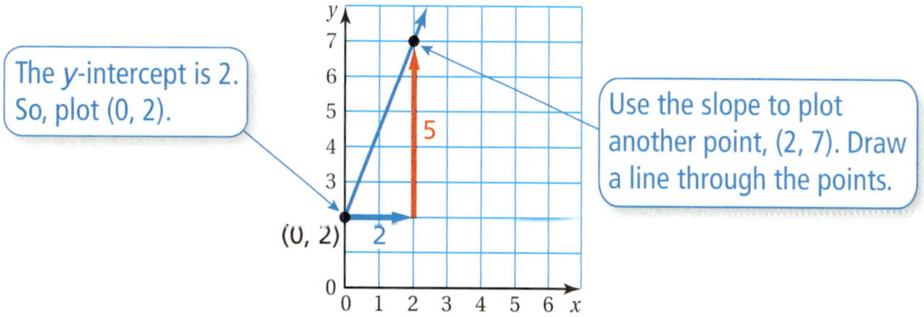

b. The slope is 2.5. So, the cost per mile is \$2.50. The y-intercept is 2. So, there is an initial fee of \$2 to take the taxi.

On Your Own

Now You're Ready
Exercises 18–23

Graph the linear equation. Identify the x-intercept.

3. $y = x - 4$

4. $y = -\frac{1}{2}x + 1$

5. In Example 3, the cost y (in dollars) of taking a different taxi x miles is $y = 2x + 1.5$. Interpret the y-intercept and slope.

2.3 Exercises

Check It Out
Help with Homework
BigIdeasMath.com

Vocabulary and Concept Check

1. **VOCABULARY** How can you find the x-intercept of the graph of $2x + 3y = 6$?
2. **CRITICAL THINKING** Is the equation $y = 3x$ in slope-intercept form? Explain.
3. **OPEN-ENDED** Describe a real-life situation that can be modeled by a linear equation. Write the equation. Interpret the y-intercept and slope.

Practice and Problem Solving

Match the equation with its graph. Identify the slope and y-intercept.

4. $y = 2x + 1$

5. $y = \frac{1}{3}x - 2$

6. $y = -\frac{2}{3}x + 1$

A.

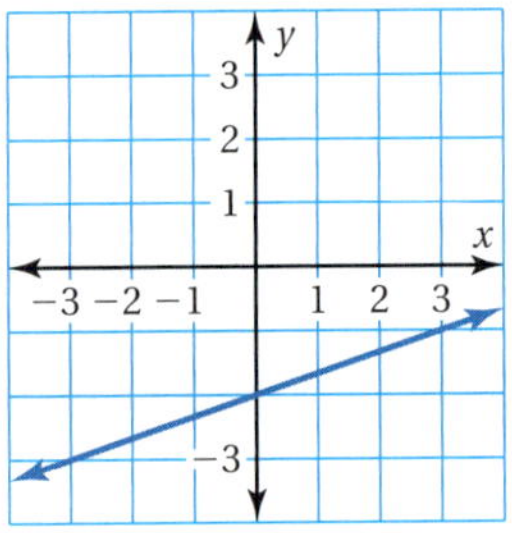

B.

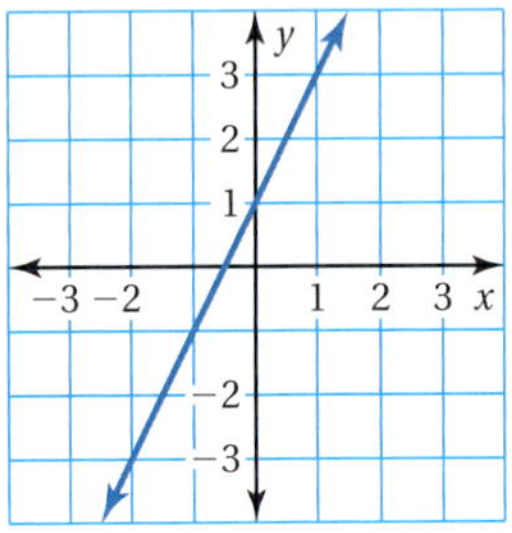

C.

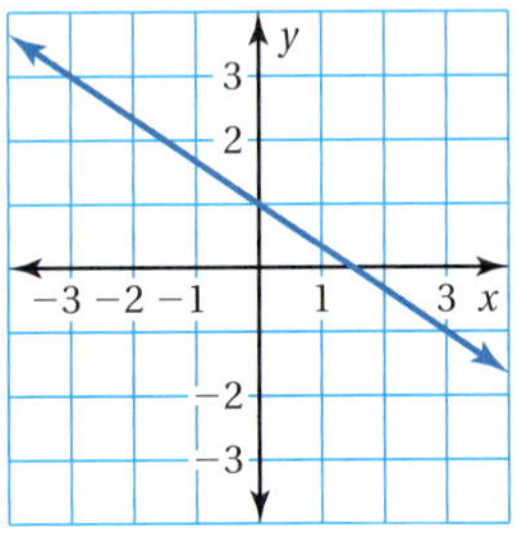

Find the slope and y-intercept of the graph of the linear equation.

7. $y = 4x - 5$
8. $y = -7x + 12$
9. $y = -\frac{4}{5}x - 2$
10. $y = 2.25x + 3$
11. $y + 1 = \frac{4}{3}x$
12. $y - 6 = \frac{3}{8}x$
13. $y - 3.5 = -2x$
14. $y + 5 = -\frac{1}{2}x$
15. $y = 1.5x + 11$

16. **ERROR ANALYSIS** Describe and correct the error in finding the slope and y-intercept of the graph of the linear equation.

> ✗ $y = 4x - 3$
> The slope is 4 and the y-intercept is 3.

17. **SKYDIVING** A skydiver parachutes to the ground. The height y (in feet) of the skydiver after x seconds is $y = -10x + 3000$.

    a. Graph the equation.

    b. Interpret the x-intercept and slope.

Assignment Guide and Homework Check

Level	Day 1 Activity Assignment	Day 2 Lesson Assignment	Homework Check
Basic	4–6, 29–33	1–3, 7–25 odd, 16	7, 15, 16, 19, 23
Average	4–6, 29–33	1–3, 13–16, 19–27 odd, 24, 26	14, 16, 19, 23, 24
Advanced	4–6, 29–33	1–3, 13–16, 18–28 even, 27	14, 16, 18, 22, 24

Common Errors

- **Exercises 7–15** Students may forget to include negatives with the slope and/or y-intercept. Remind them to look at the sign in front of the slope and the y-intercept. Also remind students that the equation is $y = mx + b$. This means that if the linear equation has "minus b," then the y-intercept is negative.
- **Exercises 11–14** Students may identify the opposite y-intercept because they forget to solve for y. Remind them that slope-intercept form has y by itself, so they must solve for y before identifying the slope and y-intercept.
- **Exercises 18–23** Students may use the reciprocal of the slope when graphing and may find an incorrect x-intercept. Remind them that slope is *rise* over *run*, so the numerator represents vertical change, not horizontal.

2.3 Record and Practice Journal

Find the slope and y-intercept of the graph of the linear equation.

1. $y = -3x + 9$
   slope: -3; y-intercept: 9
2. $y = 4 - \frac{2}{5}x$
   slope: $-\frac{2}{5}$; y-intercept: 4
3. $6 + y = 8x$
   slope: 8; y-intercept: -6

Graph the linear equation. Identify the x-intercept.

4. $y = \frac{2}{3}x + 6$
   x-intercept: -9

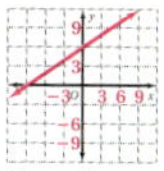

5. $y - 10 = -5x$
   x-intercept: 2

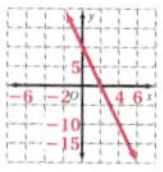

6. The equation $y = -90x + 1440$ represents the time (in minutes) left after x games of a tournament.

   a. Graph the equation.

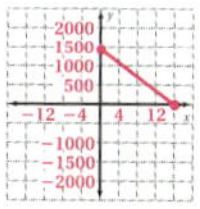

   b. Interpret the x-intercept and slope.

   The slope is -90. So, the length of each game is 90 minutes. The x-intercept is 16. So, there are 16 games in the tournament.

Technology For the Teacher
Answer Presentation Tool
Quiz*Show*

Vocabulary and Concept Check

1. Find the x-coordinate of the point where the graph crosses the x-axis.
2. yes; The slope is 3 and the y-intercept is 0.
3. *Sample answer:* The amount of gasoline y (in gallons) left in your tank after you travel x miles is $y = -\frac{1}{20}x + 20$.
   The slope of $-\frac{1}{20}$ means the car uses 1 gallon of gas for every 20 miles driven. The y-intercept of 20 means there is originally 20 gallons of gas in the tank.

Practice and Problem Solving

4. B; slope: 2; y-intercept: 1
5. A; slope: $\frac{1}{3}$; y-intercept: -2
6. C; slope: $-\frac{2}{3}$; y-intercept: 1
7. slope: 4; y-intercept: -5
8. slope: -7; y-intercept: 12
9. slope: $-\frac{4}{5}$; y-intercept: -2
10. slope: 2.25; y-intercept: 3
11. slope: $\frac{4}{3}$; y-intercept: -1
12. slope: $\frac{3}{8}$; y-intercept: 6
13. slope: -2; y-intercept: 3.5
14. slope: $-\frac{1}{2}$; y-intercept: -5
15. slope: 1.5; y-intercept: 11

16–17. See Additional Answers.

Practice and Problem Solving

18.

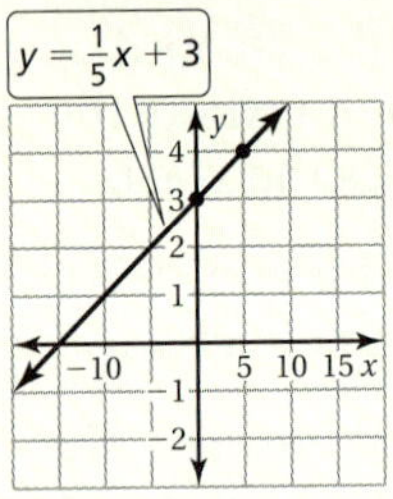

x-intercept: −15

19–27. See Additional Answers.

28. See *Taking Math Deeper.*

Fair Game Review

29. $y = 2x + 3$

30. $y = -\frac{4}{5}x + \frac{13}{5}$

31. $y = \frac{2}{3}x - 2$

32. $y = -\frac{7}{4}x + 2$

33. B

Mini-Assessment

Find the slope and y-intercept of the graph of the equation. Then graph the equation.

1. $y = -5x + 3$

slope $= -5$, y-intercept $= 3$

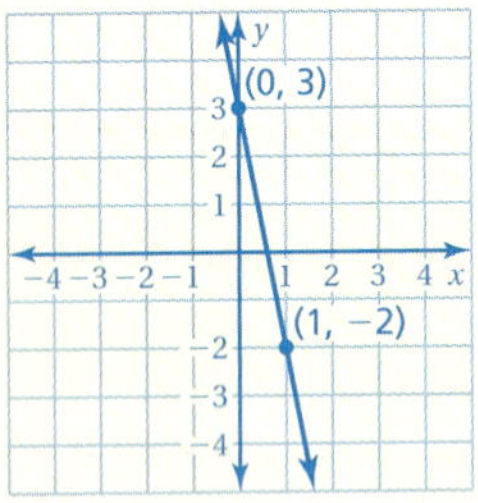

2. $y - 4 = \frac{1}{2}x$

slope $= \frac{1}{2}$, y-intercept $= 4$

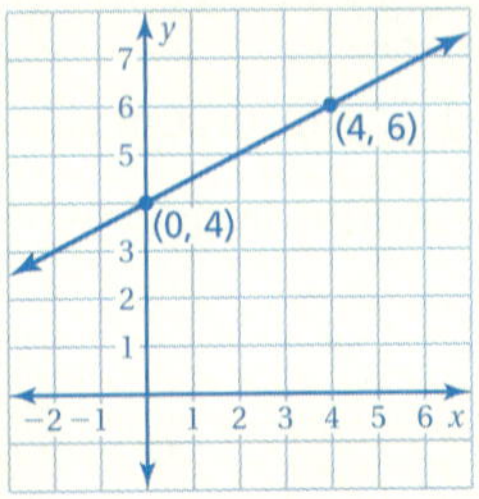

Taking Math Deeper

Exercise 28

This is a classic business problem. You have monthly costs for your business. The question is how much do you have to sell to cover your costs and start making a profit.

1 Organize the given information.

- The site sells 5 banner ads.
- Monthly income is \$0.005 per click.
- It costs \$120 per month to run the site.
- Let y be the monthly income.
- Let x be the number of clicks per month.

2 **a.** Write an equation for the income.

$y = 0.005x$

3 **b.** Graph the equation.

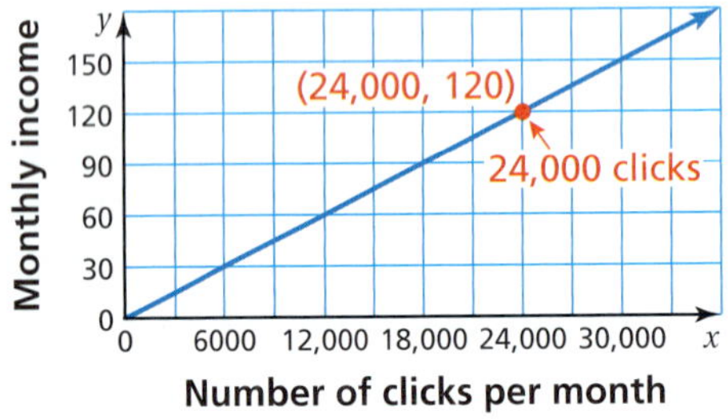

When the ads start to get 24,000 clicks a month, the income will be \$120 per month. Each banner ad needs to average $\frac{24{,}000}{5} = 4800$ clicks. Any additional clicks per month will start earning a profit.

Project

Use the Internet or the school library to research methods for determining the number of clicks on a website.

Reteaching and Enrichment Strategies

If students need help. . .	If students got it. . .
Resources by Chapter • Practice A and Practice B • Puzzle Time Record and Practice Journal Practice Differentiating the Lesson Lesson Tutorials Skills Review Handbook	Resources by Chapter • Enrichment and Extension • School-to-Work Start the next section

Graph the linear equation. Identify the x-intercept.

2 **18.** $y = \frac{1}{5}x + 3$ **19.** $y = 6x - 7$ **20.** $y = -\frac{8}{3}x + 9$

21. $y = -1.4x - 1$ **22.** $y + 9 = -3x$ **23.** $y - 4 = -\frac{3}{5}x$

24. **PHONES** The cost y (in dollars) of making a long distance phone call for x minutes is $y = 0.25x + 2$.

a. Graph the equation.

b. Interpret the slope and y-intercept.

25. **APPLES** Write a linear equation that models the cost y of picking x pounds of apples. Graph the equation.

26. **ELEVATOR** The basement of a building is 40 feet below ground level. The elevator rises at a rate of 5 feet per second. You enter the elevator in the basement. Write an equation that represents the height y (in feet) of the elevator after x seconds. Graph the equation.

27. **BONUS** You work in an electronics store. You earn a fixed amount of $35 per day, plus a 15% bonus on the merchandise you sell. Write an equation that models the amount y (in dollars) you earn for selling x dollars of merchandise in one day. Graph the equation.

28. Critical Thinking Six friends create a website. The website earns money by selling banner ads. The site has five banner ads. It costs $120 a month to operate the website.

a. A banner ad earns $0.005 per click. Write a linear equation that represents the monthly income y (in dollars) for x clicks.

b. Draw a graph of the equation in part (a). On the graph, label the number of clicks needed for the friends to start making a profit.

Fair Game Review What you learned in previous grades & lessons

Solve the equation for y. *(Section 1.4)*

29. $y - 2x = 3$ **30.** $4x + 5y = 13$ **31.** $2x - 3y = 6$ **32.** $7x + 4y = 8$

33. **MULTIPLE CHOICE** Which point is a solution of the equation $3x - 8y = 11$? *(Section 2.1)*

Ⓐ $(1, 1)$ Ⓑ $(1, -1)$ Ⓒ $(-1, 1)$ Ⓓ $(-1, -1)$

2.4 Graphing Linear Equations in Standard Form

Essential Question How can you describe the graph of the equation $ax + by = c$?

1 ACTIVITY: Using a Table to Plot Points

Work with a partner. You sold a total of $16 worth of tickets to a school concert. You lost track of how many of each type of ticket you sold.

$$\frac{\$4}{\text{Adult}} \cdot \text{Number of Adult Tickets} + \frac{\$2}{\text{Child}} \cdot \text{Number of Child Tickets} = \$16$$

a. Let x represent the number of adult tickets.

Let y represent the number of child tickets.

Write an equation that relates x and y.

b. Copy and complete the table showing the different combinations of tickets you might have sold.

Number of Adult Tickets, x					
Number of Child Tickets, y					

c. Plot the points from the table. Describe the pattern formed by the points.

d. If you remember how many adult tickets you sold, can you determine how many child tickets you sold? Explain your reasoning.

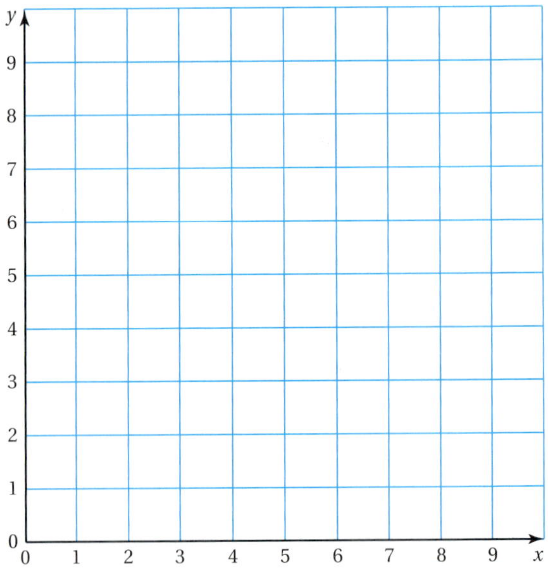

Laurie's Notes

Introduction

For the Teacher

- **Goal:** Students will explore the graph of a function in standard form.

Motivate

- **Preparation:** Make a set of equation cards on strips of paper. The equations are all the same when simplified and need to be written large enough to be read by students sitting at the back of the classroom.
- Here is a sample set of equations: $y = 2x + 1$, $-2x + y = 1$, $2x - y = -1$, $4x - 2y = -2$
- Ask 4 students to stand at the front of the room and hold the cards so only they can see the equations.
- As you state an ordered pair, the students holding the cards determine if it is a solution of the equation they are holding. If it is, they raise their hand. If not, they do nothing. State several ordered pairs, four that are solutions and two that are not. Plot all of the points that you state. Keep the ordered pairs simple.
- The four ordered pairs that are solutions will be in a line.

? "How many lines can pass through any two points?" one "How many lines pass through the four solutions points?" Students will say 4; now is the time to discuss the idea of one line written in different forms.

- Have each student reveal their equation to the class and read it aloud. Write each of the equations on the board.
- Explain to students that equations can be written in different forms. Today they will explore a new form of a linear equation.

Activity Notes

Activity 1

- Read the problem aloud. Discuss what the variables x and y represent.
- Note that a verbal model is shown for the equation $4x + 2y = 16$.

? "Could you have sold 5 adult tickets? Explain." No; 5 adult tickets would be \$20, which is too much.

- Students may say that they do not know how to figure out x and y. Students may not realize that there is more than one solution. Remind students that *Guess, Check, and Revise* would be an appropriate strategy to use.
- Discuss part (c). The points lie on a line.
- Discuss part (d). Students may not recognize that in knowing x, they can substitute and solve for y. This is not an obvious step for students.

? "Could $x = 1.5$? Explain." No, you cannot sell 1.5 tickets.

? "What are the different numbers of adult tickets that are possible to sell?" 0, 1, 2, 3, 4

- **Note:** This is an example of a discrete domain; there are only 5 possible values for the variable x. This will be taught at a later time.

Previous Learning

Students should know how to graph lines in slope-intercept form.

Activity Materials	
Introduction	**Textbook**
• equation cards	• straightedge
Closure	
• equation cards	

Start Thinking! and Warm Up

Activity 2.4 **Start Thinking!** For use before Activity 2.4

Activity 2.4 **Warm Up** For use before Activity 2.4

Solve the equation for y.

1. $x + y = 4$
2. $2x + y = 10$
3. $3x + 4y = 12$
4. $-5x + 10y = 8$
5. $-4x + 2y = 10$
6. $-x + 2y = 4$

2.4 Record and Practice Journal

Essential Question How can you describe the graph of the equation $ax + by = c$?

1 ACTIVITY: Using a Table to Plot Points

Work with a partner. You sold a total of \$16 worth of tickets to a school concert. You lost track of how many of each type of ticket you sold.

$\frac{\$4}{\text{Adult}}$ • Number of Adult Tickets + $\frac{\$2}{\text{Child}}$ • Number of Child Tickets = \$16

a. Let x represent the number of adult tickets. Let y represent the number of child tickets. Write an equation that relates x and y. $4x + 2y = 16$

b. Complete the table showing the different combinations of tickets you might have sold.

Number of Adult Tickets, x	0	1	2	3	4
Number of Child Tickets, y	8	6	4	2	0

c. Plot the points from the table. Describe the pattern formed by the points.

form a line

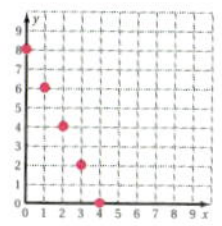

d. If you remember how many adult tickets you sold, can you determine how many child tickets you sold? Explain your reasoning.

yes

Differentiated Instruction

Visual

Have students create a chart in their notebooks of the equation forms and how to graph them.

Slope-intercept form $y = mx + b$	• Plot $(0, b)$. • Use the slope m to plot a second point. • Draw a line through the two points.
Horizontal line $y = c$	• Draw a horizontal line through $(0, c)$.
Standard form $ax + by = c$	• Find the y-intercept. • Find the x-intercept. • Plot the associated points. Draw a line through the two points.

2.4 Record and Practice Journal

2 ACTIVITY: Rewriting an Equation

Work with a partner. You sold a total of \$16 worth of cheese. You forgot how many pounds of each type of cheese you sold.

$\frac{\$4}{\text{lb}}$ • Pounds of Swiss + $\frac{\$2}{\text{lb}}$ • Pounds of Cheddar = \$16

a. Let x represent the number of pounds of Swiss cheese. Let y represent the number of pounds of Cheddar cheese. Write an equation that relates x and y.

$4x + 2y = 16$

b. Write the equation in slope-intercept form. Then graph the equation.

$y = -2x + 8$

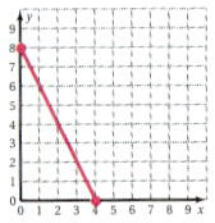

What Is Your Answer?

3. IN YOUR OWN WORDS How can you describe the graph of the equation $ax + by = c$?

a line with slope $-\frac{a}{b}$ and y-intercept of $\frac{c}{b}$

4. Activities 1 and 2 show two different methods for graphing $ax + by = c$. Describe the two methods. Which method do you prefer? Explain.

Check students' work.

5. Write a real-life problem that is similar to those shown in Activities 1 and 2.

Check students' work.

Laurie's Notes

Activity 2

- Read the problem aloud. Discuss what the variables x and y represent.
- Note that a verbal model is shown for the equation $4x + 2y = 16$.

? "Could you have sold 5 pounds of Swiss cheese? Explain." No; 5 pounds of Swiss cheese would be \$20, which is too much.

- Give time for students to work with their partner. While this may be the same equation as Activity 1, the approach is different. Students are asked to write the equation in slope-intercept form. After the equation is in slope-intercept form, students can substitute a value for x, and find y. This is generally not the case for equations written in standard form.

? "Could $x = 1.5$? Explain." yes; You can buy a portion of a pound.

- **Note:** This is an example of a continuous domain; all numbers $0 \le x \le 4$ are possible. This will be taught at a later time.
- Students might observe that both examples have graphs in the first quadrant. This is common for real-life examples.

What Is Your Answer?

- **Question 3:** Students may guess that the graph is linear from Activity 1. However, some students may not be secure with this knowledge yet.

Closure

- Refer back to the equation cards. Rewrite the last three equations in slope-intercept form. $y = 2x + 1$

Technology For the Teacher

Dynamic Classroom

The Dynamic Planning Tool

Editable Teacher's Resources at *BigIdeasMath.com*

2 ACTIVITY: Rewriting an Equation

Work with a partner. You sold a total of $16 worth of cheese. You forgot how many pounds of each type of cheese you sold.

$$\frac{\$4}{\text{lb}} \cdot \text{Pounds of Swiss} + \frac{\$2}{\text{lb}} \cdot \text{Pounds of Cheddar} = \$16$$

a. Let x represent the number of pounds of Swiss cheese.
Let y represent the number of pounds of Cheddar cheese.
Write an equation that relates x and y.

b. Write the equation in slope-intercept form. Then graph the equation.

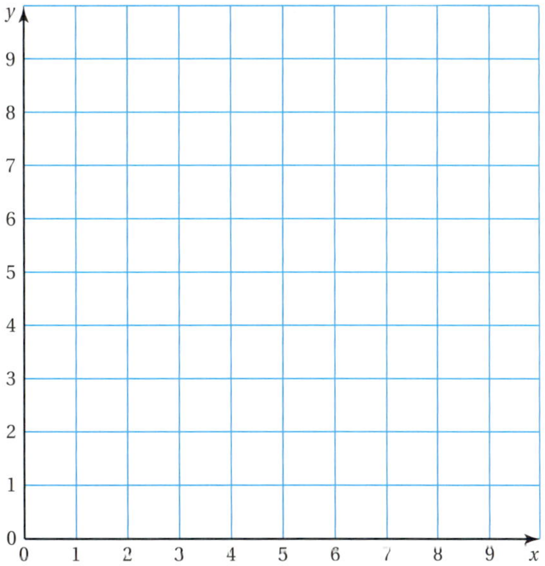

What Is Your Answer?

3. **IN YOUR OWN WORDS** How can you describe the graph of the equation $ax + by = c$?

4. Activities 1 and 2 show two different methods for graphing $ax + by = c$. Describe the two methods. Which method do you prefer? Explain.

5. Write a real-life problem that is similar to those shown in Activities 1 and 2.

Practice Use what you learned about graphing linear equations in standard form to complete Exercises 3 and 4 on page 72.

2.4 Lesson

Check It Out
Lesson Tutorials
BigIdeasMath.com

Key Vocabulary
standard form, *p. 70*

Study Tip
Any linear equation can be written in standard form.

Key Idea

Standard Form of a Linear Equation

The **standard form** of a linear equation is

$$ax + by = c$$

where a and b are not both zero.

EXAMPLE 1 Graphing a Linear Equation in Standard Form

Graph $-2x + 3y = -6$.

Step 1: Write the equation in slope-intercept form.

$$-2x + 3y = -6 \quad \text{Write the equation.}$$

$$3y = 2x - 6 \quad \text{Add } 2x \text{ to each side.}$$

$$y = \frac{2}{3}x - 2 \quad \text{Divide each side by 3.}$$

Step 2: Use the slope and y-intercept to graph the equation.

$$y = \frac{2}{3}x + (-2)$$

slope: $\frac{2}{3}$; y-intercept: -2

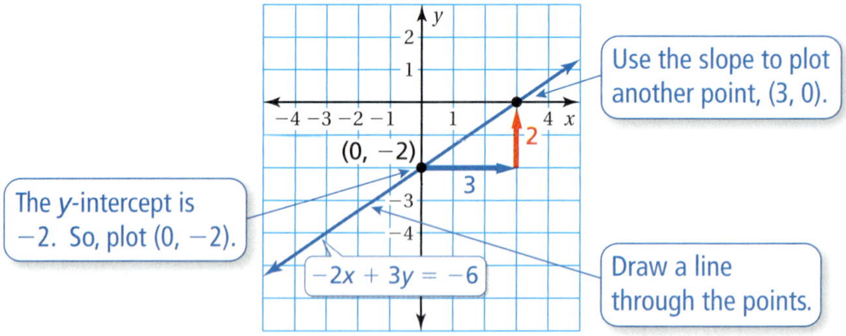

On Your Own

Now You're Ready
Exercises 5–10

Graph the linear equation.

1. $x + y = -2$
2. $-\frac{1}{2}x + 2y = 6$
3. $-\frac{2}{3}x + y = 0$
4. $2x + y = 5$

Multi-Language Glossary at BigIdeasMath.com.

Laurie's Notes

Introduction

Connect

- **Yesterday:** Students explored the graph of an equation written in standard form.
- **Today:** Students will graph equations written in standard form.

Motivate

? "How many pairs of numbers can you think of that have a sum of 5?" Encourage students to write their numbers on paper as ordered pairs. Example, (2, 3)

? "Did any of you include numbers that are not whole numbers?" Check to see if anyone had negative numbers or rational numbers.

- Ask one student to name the x-coordinate in one of their ordered pairs and another student to provide the y-coordinate. Plot the ordered pairs in a coordinate plane.

? "What do you think the equation of this line would be?" $x + y = 5$

Lesson Notes

Key Idea

- Define the standard form of a linear equation.
- Students may ask why both a and b cannot be zero. Explain that if $a = 0$ and $b = 0$, you would not have the equation of a line.
- **Teaching Tip:** Students are often confused when the standard form is written with parameters a, b, and c. Students see 5 variables. Show examples of equations written in standard form and identify a, b, and c.

Example 1

- Have students identify a, b, and c. $a = -2$, $b = 3$, and $c = -6$

? "How do you solve for y?" Add $2x$ to each side, then divide both sides by 3.

- **Common Error:** Students only divide one of the two terms on the right side of the equation by 3. Relate this to fraction operations. You are separating the expression into two terms and then simplifying.

? "Now that the equation is in slope-intercept form, explain how to graph the equation." Plot the ordered pair for the y-intercept. To plot another point, start at (0, −2) and move to the right 3 units and up 2 units. Note that you can also move 3 units to the left and down 2 units. Connect these points with a line.

- Substitute the additional ordered pairs into the original equation to verify that they are solutions of the equation.

On Your Own

- In Questions 2 and 3, the fractional coefficient of x may present a problem.

Goal Today's lesson is graphing a line written in **standard form**.

Lesson Materials
Textbook
• straightedge

Start Thinking! and Warm Up

Lesson 2.4 Warm Up
For use before Lesson 2.4

Lesson 2.4 Start Thinking!
For use before Lesson 2.4

You have \$40 to spend on turkey and cheese for a party. At the deli, turkey is \$10 per pound and cheese is \$6 per pound.

Is it easier to write an equation to represent the situation in *slope-intercept form* or *standard form*? Why?

Extra Example 1

Graph $3x - 2y = 2$.

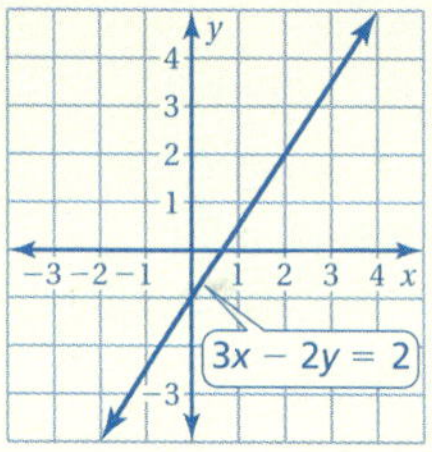

On Your Own

1.
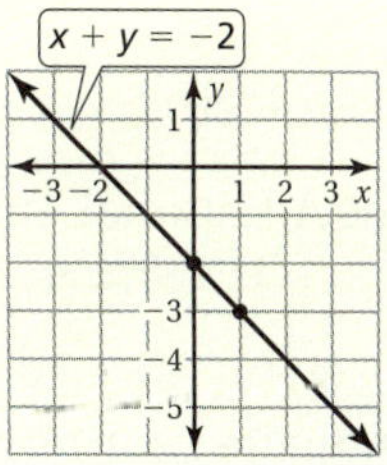

2.
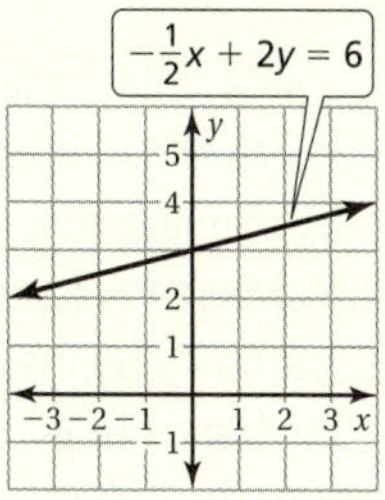

3–4. See Additional Answers.

Extra Example 2

Graph $5x - y = -5$ using intercepts.

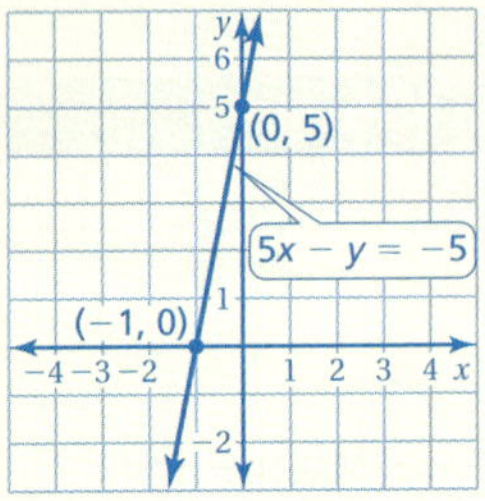

Extra Example 3

You have \$2.40 to spend on grapes and bananas.

a. Graph the equation $1.2x + 0.6y = 2.4$, where x is the number of pounds of grapes and y is the number of pounds of bananas.

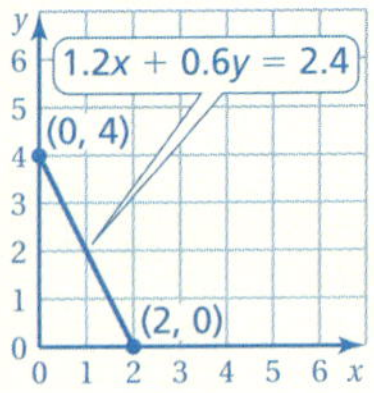

b. Interpret the intercepts. The x-intercept shows that you can buy 2 pounds of grapes, if you do not buy any bananas. The y-intercept shows that you can buy 4 pound of bananas, if you do not buy any grapes.

On Your Own

5–7. See Additional Answers.

English Language Learners

Vocabulary

For English learners, relate the word *intercept* with the football term *interception.* A defensive player on a football team crosses the path of the football to catch it and make an interception. Similarly, the y-intercept is the point where the line crosses the y-axis and the x-intercept is where the line crosses the x-axis.

Laurie's Notes

Example 2

- Start with a simple equation in standard form, such as $x + y = 4$. In this example, $a = 1$, $b = 1$, and $c = 4$. Explain to students that this could be solved for y by subtracting x from each side of the equation. Instead, you want to leave the equation as it was written.
- ? "Another way to think of this equation is *the sum of two numbers is 4.* Can you name some ordered pairs that would satisfy the equation?" Students should give many, including the two intercepts, (0, 4) and (4, 0).
- Explain to students that sometimes an equation in standard form is graphed by using the two intercepts, instead of rewriting the equation in slope-intercept form.
- Write the equation shown: $x + 3y = -3$.
- ? "To find the x-intercept, what is the value of y? To find the y-intercept, what is the value of x?" 0; 0
- Finish the problem as shown.
- **Big Idea:** When the equation is in standard form, you can plot the points for the two intercepts and then draw the line through them.

Example 3

- Read the problem. Write the equation $1.5x + 0.6y = 6$ on the board.
- ? "What are the intercepts for this equation?" The x-intercept is 4 and the y-intercept is 10.
- Interpreting the intercepts in part (b) is an important step, particularly for real-life applications.
- Explain to students that negative values of x and y are not included in the graph because it does not make sense to have negative pounds of apples and bananas.
- ? "What is the cost of 2 pounds of apples and 5 pounds of bananas?" \$6

On Your Own

- Students should work with a partner.

Closure

- **Writing Prompt:** To graph the equation $2x + y = 4$ … *Sample answer:* Find and plot the points for the x- and y-intercepts, then draw a line through these two points.

Technology For the Teacher

Dynamic Classroom

The Dynamic Planning Tool
Editable Teacher's Resources at *BigIdeasMath.com*

EXAMPLE 2 Graphing a Linear Equation in Standard Form

Graph $x + 3y = -3$ using intercepts.

Step 1: To find the x-intercept, substitute 0 for y.

$$x + 3y = -3$$
$$x + 3(0) = -3$$
$$x = -3$$

To find the y-intercept, substitute 0 for x.

$$x + 3y = -3$$
$$0 + 3y = -3$$
$$y = -1$$

Step 2: Graph the equation.

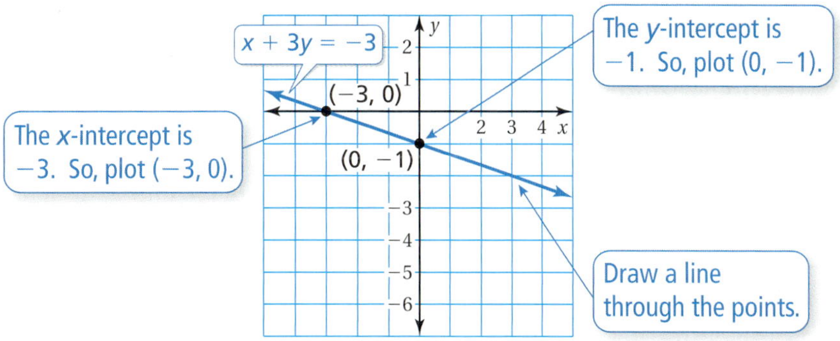

EXAMPLE 3 Real-Life Application

You have \$6 to spend on apples and bananas. (a) Graph the equation $1.5x + 0.6y = 6$, where x is the number of pounds of apples and y is the number of pounds of bananas. (b) Interpret the intercepts.

a. Find the intercepts and graph the equation.

***x*-intercept**

$$1.5x + 0.6y = 6$$
$$1.5x + 0.6(0) = 6$$
$$x = 4$$

***y*-intercept**

$$1.5x + 0.6y = 6$$
$$1.5(0) + 0.6y = 6$$
$$y - 10$$

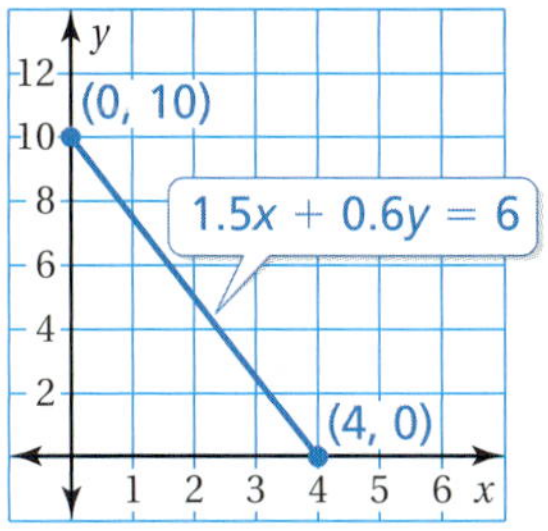

b. The x-intercept shows that you can buy 4 pounds of apples if you don't buy any bananas. The y-intercept shows that you can buy 10 pounds of bananas if you don't buy any apples.

On Your Own

Now You're Ready
Exercises 16–18

Graph the linear equation using intercepts.

5. $2x - y = 8$
6. $x + 3y = 6$
7. **WHAT IF?** In Example 3, you buy y pounds of oranges instead of bananas. Oranges cost \$1.20 per pound. Graph the equation $1.5x + 1.2y = 6$. Interpret the intercepts.

2.4 Exercises

Check It Out
Help with Homework
BigIdeasMath.com

Vocabulary and Concept Check

1. **VOCABULARY** Is the equation $y = -2x + 5$ in standard form? Explain.

2. **REASONING** Does the graph represent a linear equation? Explain.

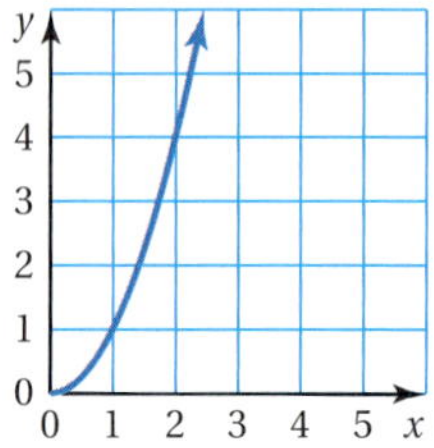

Practice and Problem Solving

Define two variables for the verbal model. Write an equation in slope-intercept form that relates the variables. Graph the equation.

3. $\dfrac{\$2.00}{\text{pound}}$ • Pounds of peaches + $\dfrac{\$1.50}{\text{pound}}$ • Pounds of apples = \$15

4. $\dfrac{16\text{ miles}}{\text{hour}}$ • Hours biked + $\dfrac{2\text{ miles}}{\text{hour}}$ • Hours walked = 32 miles

Write the linear equation in slope-intercept form.

1 5. $2x + y = 17$

6. $5x - y = \dfrac{1}{4}$

7. $-\dfrac{1}{2}x + y = 10$

Graph the linear equation.

8. $-18x + 9y = 72$

9. $16x - 4y = 2$

10. $\dfrac{1}{4}x + \dfrac{3}{4}y = 1$

Use the graph to find the *x*- and *y*-intercepts.

11.

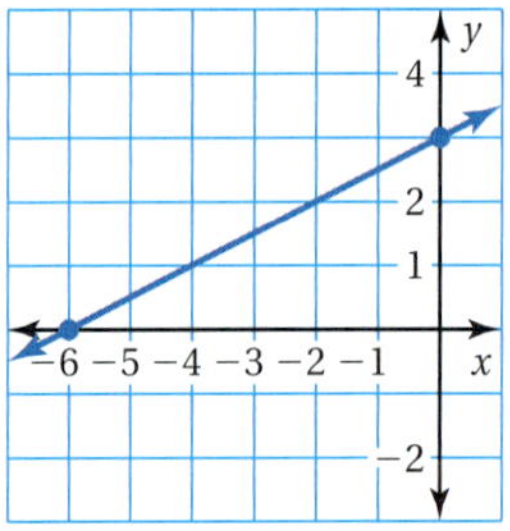

12.

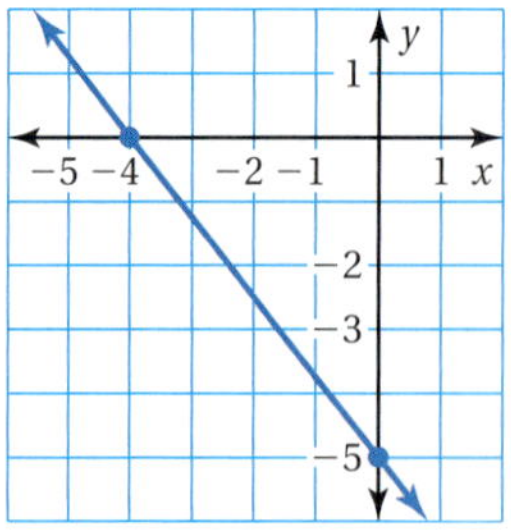

13.

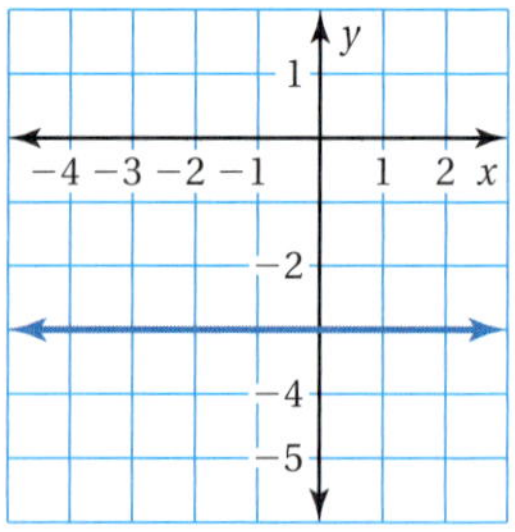

14. **ERROR ANALYSIS** Describe and correct the error in finding the x-intercept.

$$-2x + 3y = 12$$
$$-2(0) + 3y = 12$$
$$3y = 12$$
$$y = 4$$

15. **BRACELET** A charm bracelet costs \$65, plus \$25 for each charm.

 a. Write an equation in standard form that represents the total cost of the bracelet.

 b. How much does the bracelet shown cost?

Assignment Guide and Homework Check

Level	Day 1 Activity Assignment	Day 2 Lesson Assignment	Homework Check
Basic	3, 4, 24–26	1, 2, 5, 6, 8, 9, 11, 12, 14–19	6, 8, 12, 14, 16, 19
Average	3, 4, 24–26	1, 2, 5–13 odd, 14–21	7, 9, 13, 14, 16, 20
Advanced	3, 4, 24–26	1, 2, 8–14 even, 13, 16–23	10, 13, 14, 16, 20, 22

Common Errors

- **Exercises 5–10** Students may use the same operation instead of the opposite operation when rewriting the equation in slope-intercept form.
- **Exercises 11 and 12, 16–18** Students may mix up the *x*- and *y*-intercepts. Remind them that the *x*-intercept is the *x*-coordinate of where the line crosses the *x*-axis and the *y*-intercept is the *y*-coordinate of where the line crosses the *y*-axis.
- **Exercise 13** Because the line is horizontal and there is no *x*-intercept, students may say that the *x*-intercept is zero. Remind them that this would mean that the *x*-intercept is at the origin; however, there is no *x*-intercept.

2.4 Record and Practice Journal

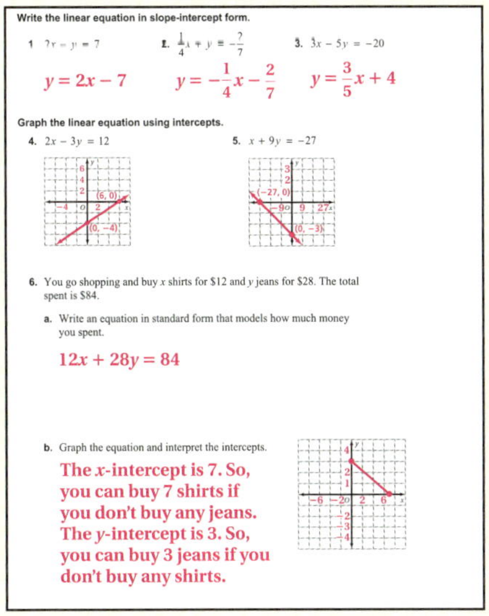
Write the linear equation in slope-intercept form.

1. $2x - y = 7$ $y = 2x - 7$
2. $\frac{1}{4}x + y = -\frac{2}{7}$ $y = -\frac{1}{4}x - \frac{2}{7}$
3. $3x - 5y = -20$ $y = \frac{3}{5}x + 4$

Graph the linear equation using intercepts.

4. $2x - 3y = 12$
5. $x + 9y = -27$

6. You go shopping and buy x shirts for \$12 and y jeans for \$28. The total spent is \$84.

a. Write an equation in standard form that models how much money you spent.

$12x + 28y = 84$

b. Graph the equation and interpret the intercepts.

The x-intercept is 7. So, you can buy 7 shirts if you don't buy any jeans. The y-intercept is 3. So, you can buy 3 jeans if you don't buy any shirts.

Technology For the Teacher

Answer Presentation Tool

Quiz*Show*

Vocabulary and Concept Check

1. no; The equation is in slope-intercept form.
2. no; The graph is not a line.

Practice and Problem Solving

3. x = pounds of peaches
y = pounds of apples
$y = -\frac{4}{3}x + 10$

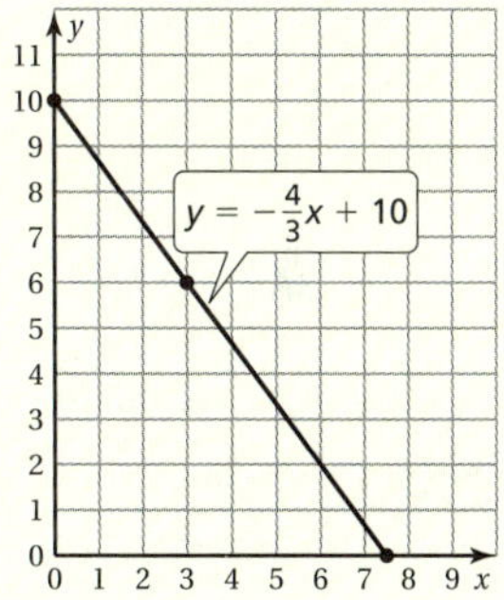

4. x = hours biked
y = hours walked
$y = -8x + 16$

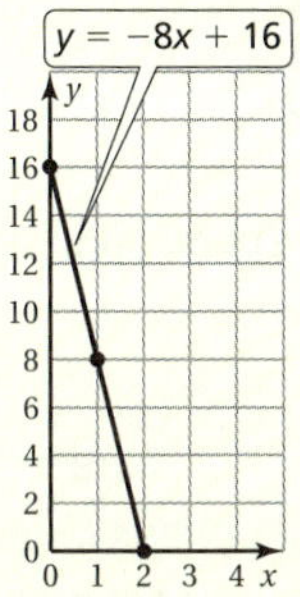

5. $y = -2x + 17$
6. $y = 5x - \frac{1}{4}$
7. $y = \frac{1}{2}x + 10$
8.

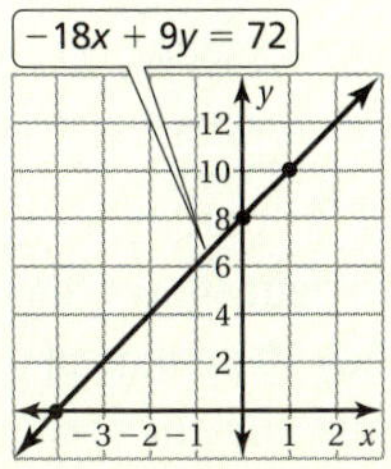

9–15. See Additional Answers.

Practice and Problem Solving

16–19. See Additional Answers.

20. See *Taking Math Deeper.*

21–23. See Additional Answers.

Fair Game Review

24. 1; 3; 5; 7; 9

25. 1; −2; −5; −8; −11

26. D

Mini-Assessment

1. Graph $-2x + 4y = 16$ using intercepts.

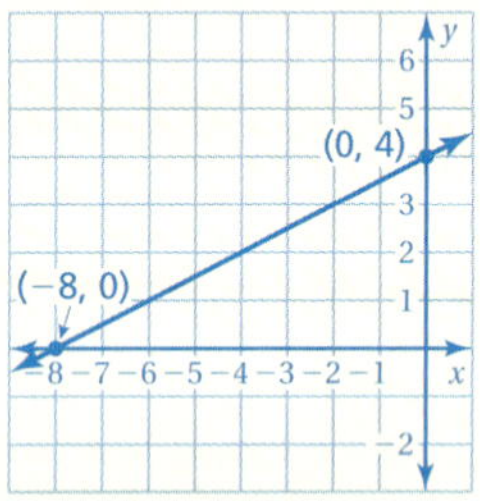

2. **You have $12 to spend on pears and oranges.**

   a. Graph the equation $1.2x + 0.8y = 12$, where x is the number of pounds of pears and y is the number of pounds of oranges.

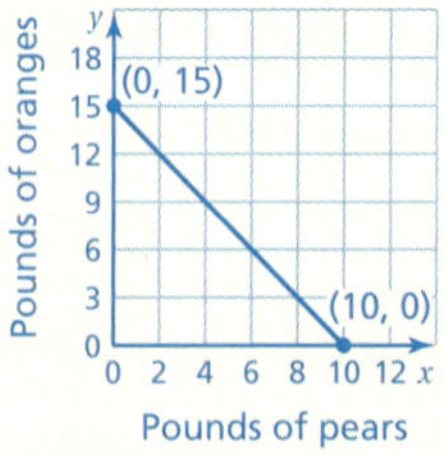

   b. Interpret the intercepts.
   The x-intercept shows that you can buy 10 pounds of pears if you do not buy any oranges. The y-intercept shows that you can buy 15 pounds of oranges if you do not buy any pears.

Taking Math Deeper

Exercise 20

As with many real-life problems, it helps to start by summarizing the given information.

1 Summarize the given information.

- Let x = days for renting boat.
- Let y = days for renting scuba gear.
- Cost of boat = $250 per day.
- Cost of scuba gear = $50 per day.
- Total spent = $1000.

2 **a.** Write an equation.

$$250x + 50y = 1000$$

3 **b.** Graph the equation and interpret the intercepts.

$$y = -5x + 20$$

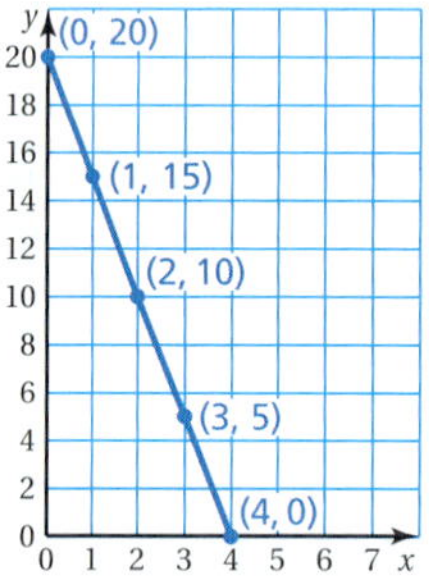

If $x = 0$, the group rented only the scuba gear for 20 days.
If $y = 0$, the group rented only the boat for 4 days.

Project

To go on a professional scuba diving tour, you need to be a certified diver. Use the school library or the Internet to research the requirements to become certified in scuba diving.

Reteaching and Enrichment Strategies

If students need help. . .	If students got it. . .
Resources by Chapter • Practice A and Practice B • Puzzle Time Record and Practice Journal Practice Differentiating the Lesson Lesson Tutorials Skills Review Handbook	Resources by Chapter • Enrichment and Extension • School-to-Work Start the next section

Graph the linear equation using intercepts.

2 **16.** $3x - 4y = -12$ **17.** $2x + y = 8$ **18.** $\frac{1}{3}x - \frac{1}{6}y = -\frac{2}{3}$

19. SHOPPING The amount of money you spend on x CDs and y DVDs is given by the equation $14x + 18y = 126$. Find the intercepts and graph the equation.

20. SCUBA Five friends go scuba diving. They rent a boat for x days and scuba gear for y days. The total spent is $1000.

a. Write an equation in standard form that represents the situation.

b. Graph the equation and interpret the intercepts.

21. WAGES You work at a restaurant as a host and a server. You earn $9.45 for each hour you work as a host and $7.65 for each hour you work as a server.

a. Write an equation in standard form that models your earnings.

b. Graph the equation.

Basic Information
Pay to the Order of:
.... John Doe
# of hours worked as
.... host: x
# of hours worked as
.... server: y
Earnings for this pay
.... period: $160.65

22. REASONING Does the graph of every linear equation have an x-intercept? Explain your reasoning. Include an example.

23. Critical Thinking For a house call, a veterinarian charges $70, plus $40 an hour.

a. Write an equation that represents the total fee y charged by the veterinarian for a visit lasting x hours.

b. Find the x-intercept. Will this point appear on the graph of the equation? Explain your reasoning.

c. Graph the equation.

Fair Game Review What you learned in previous grades & lessons

Copy and complete the table of values. *(Skills Review Handbook)*

24.

x	-2	-1	0	1	2
$2x + 5$					

25.

x	-2	-1	0	1	2
$-5 - 3x$					

26. MULTIPLE CHOICE Which value of x makes the equation $4x - 12 = 3x - 9$ true? *(Section 1.3)*

Ⓐ -1 Ⓑ 0 Ⓒ 1 Ⓓ 3

2 Study Help

Check It Out
Graphic Organizer
BigIdeasMath.com

You can use a **process diagram** to show the steps involved in a procedure. Here is an example of a process diagram for graphing a linear equation.

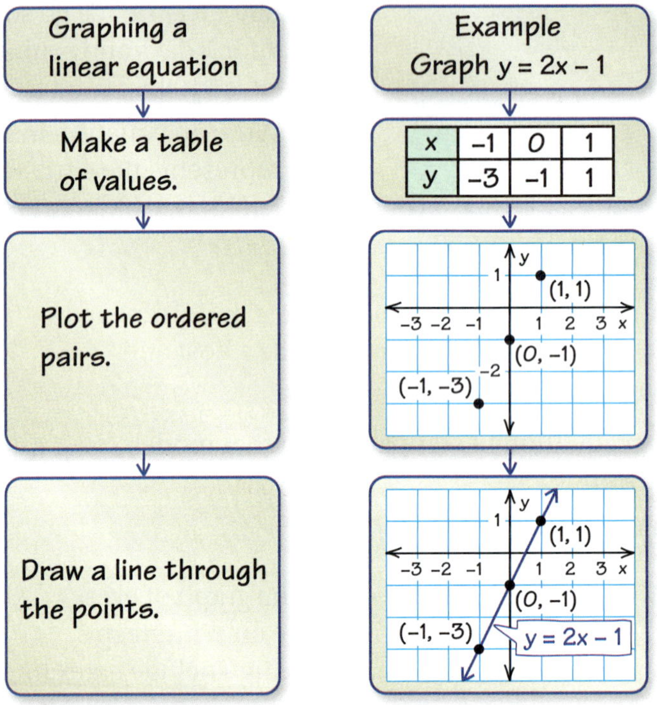

On Your Own

Make a process diagram with an example to help you study these topics.

1. finding the slope of a line
2. graphing a linear equation using
   a. slope and y-intercept
   b. x- and y-intercepts

After you complete this chapter, make process diagrams for the following topics.

3. solving a linear system
   a. using a table
   b. using a graph
   c. algebraically
4. finding the number of solutions of a linear system
5. solving an equation by graphing

"Here is a process diagram with suggestions for what to do if a hyena knocks on your door."

Sample Answers

1.

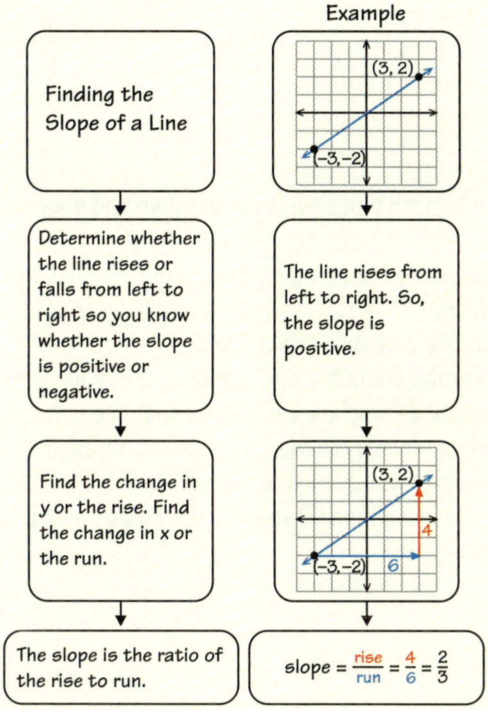

2a.

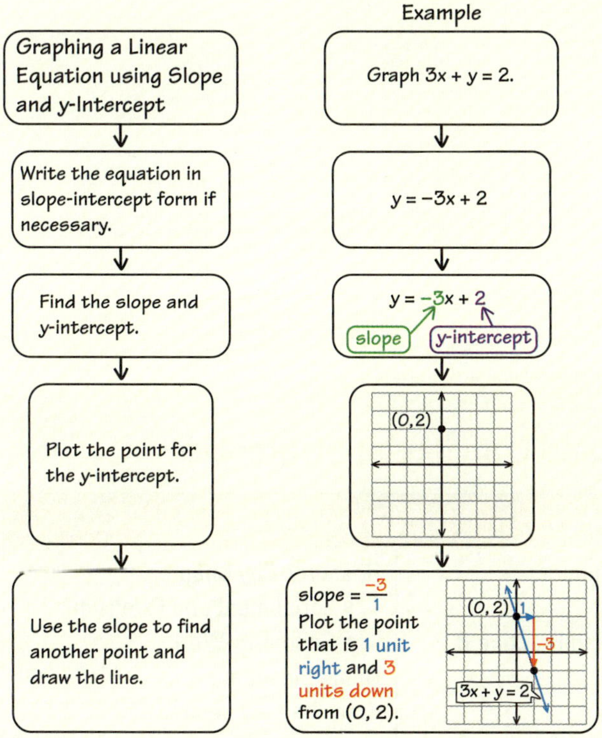

2b. Available at *BigIdeasMath.com*.

List of Organizers

Available at *BigIdeasMath.com*

Comparison Chart
Concept Circle
Definition (Idea) and Example Chart
Example and Non-Example Chart
Formula Triangle
Four Square
Information Frame
Information Wheel
Notetaking Organizer
Process Diagram
Summary Triangle
Word Magnet
Y Chart

About this Organizer

A **Process Diagram** can be used to show the steps involved in a procedure. Process diagrams are particularly useful for illustrating procedures with two or more steps, and they can have one or more branches. As shown, students' process diagrams can have two parallel parts, in which the procedure is stepped out in one part and an example illustrating each step is shown in the other part. Or, the diagram can be made up of just one part, with example(s) included in the last "bubble" to illustrate the steps that precede it.

Technology For the Teacher

Vocabulary Puzzle Builder

Answers

1–4. See Additional Answers.

5. $-\frac{1}{2}$

6. 2

7. slope: $\frac{1}{4}$
y-intercept: -8

8. slope: -1
y-intercept: 3

9. x-intercept: 4
y-intercept: -6

10. x-intercept: 15
y-intercept: 3

11.

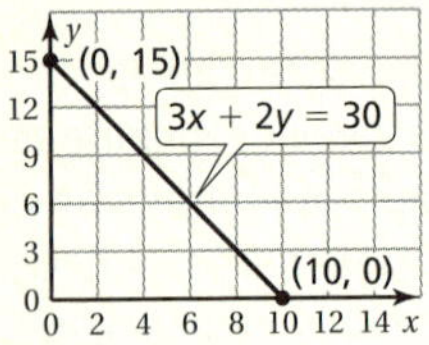

The x-intercept, 10, shows that your family can buy 10 pounds of beef if they do not buy any chicken. The y-intercept, 15, shows that your family can buy 15 pounds of chicken if they do not buy any beef.

12–14. See Additional Answers.

Assessment Book

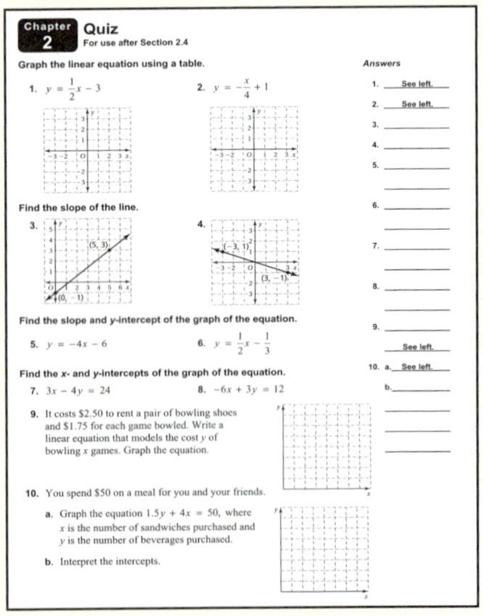

Chapter 2 Quiz
For use after Section 2.4

Graph the linear equation using a table.

1. $y = \frac{1}{2}x - 3$
2. $y = -\frac{x}{4} + 1$

Find the slope of the line.

3.
4.

Find the slope and y-intercept of the graph of the equation.

5. $y = -4x - 6$
6. $y = \frac{1}{2}x - \frac{1}{3}$

Find the x- and y-intercepts of the graph of the equation.

7. $3x - 4y = 24$
8. $-6x + 3y = 12$

9. It costs $2.50 to rent a pair of bowling shoes and $1.75 for each game bowled. Write a linear equation that models the cost y of bowling x games. Graph the equation.

10. You spend $50 on a meal for you and your friends.
  a. Graph the equation $1.5y + 4x = 50$, where x is the number of sandwiches purchased and y is the number of beverages purchased.
  b. Interpret the intercepts.

Answers
1. See left.
2. See left.
3.
4.
5.
6.
7.
8.
9. See left.
10. a. See left.
b.

Alternative Quiz Ideas

100% Quiz
Error Notebook
Group Quiz
Homework Quiz
Math Log
Notebook Quiz
Partner Quiz
Pass the Paper

Pass the Paper

- Work in groups of four. The first student copies the problem and does a step, explaining his or her work.
- The paper is passed and the second student works through the next step, also explaining his or her work.
- This process continues until the problem is completed.
- The second member of the group starts the next problem. Students should be allowed to question and debate as they are working through the quiz.
- Student groups can be selected by the teacher, by students, through a random process, or any way that works for your class.
- The teacher walks around the classroom listening to the groups and asks questions to ensure understanding.

Reteaching and Enrichment Strategies

If students need help. . .	If students got it. . .
Resources by Chapter • Study Help • Practice A and Practice B • Puzzle Time Lesson Tutorials *BigIdeasMath.com* Practice Quiz Practice from the Test Generator	Resources by Chapter • Enrichment and Extension • School-to-Work Game Closet at *BigIdeasMath.com* Start the next section

Technology For the Teacher

Answer Presentation Tool
Big Ideas Test Generator

2.1–2.4 Quiz

Check It Out
Progress Check
BigIdeasMath.com

Graph the linear equation using a table. *(Section 2.1)*

1. $y = -12x$

2. $y = -x + 8$

3. $y = \frac{x}{3} - 4$

4. $y = 3.5$

Find the slope of the line. *(Section 2.2)*

5.

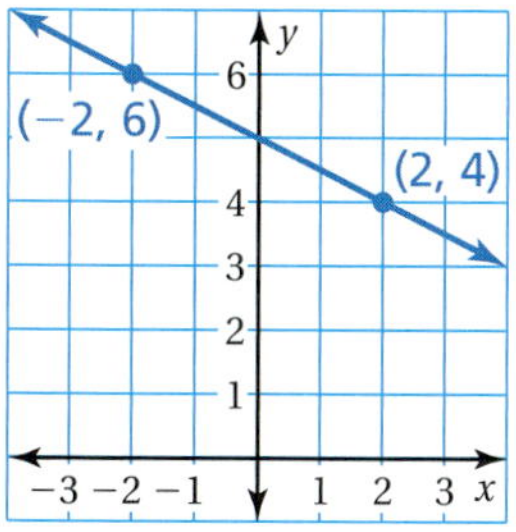

6.

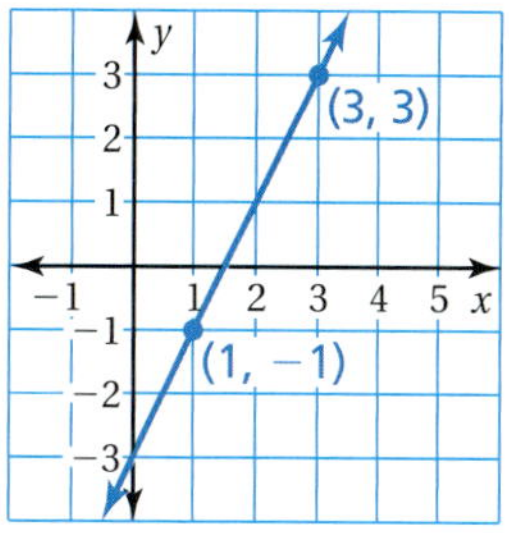

Find the slope and y-intercept of the graph of the equation. *(Section 2.3)*

7. $y = \frac{1}{4}x - 8$

8. $y = -x + 3$

Find the x- and y-intercepts of the graph of the equation. *(Section 2.4)*

9. $3x - 2y = 12$

10. $x + 5y = 15$

11. BARBEQUE The equation $3x + 2y = 30$ represents the amount of money your family spends on x pounds of beef and y pounds of chicken for a barbeque. Graph the equation and interpret the intercepts. *(Section 2.4)*

12. BANKING A bank charges \$3 each time you use an out-of-network ATM. At the beginning of the month, you have \$1500 in your bank account. You withdraw \$60 from your bank account each time you use an out-of-network ATM. Write and graph a linear equation that represents the balance in your account after you use an out-of-network ATM x times. *(Section 2.1)*

13. STATE FAIR Write a linear equation that models the cost y of one person going on x rides at the fair. Graph the equation. *(Section 2.3)*

14. PAINTING You used \$90 worth of paint for a school float. *(Section 2.4)*

a. Graph the equation $18x + 15y = 90$, where x is the number of gallons of blue paint and y is the number of gallons of white paint.

b. Interpret the intercepts.

2.5 Systems of Linear Equations

Essential Question How can you solve a system of linear equations?

1 ACTIVITY: Writing a System of Linear Equations

Work with a partner.

Your family starts a bed-and-breakfast in your home. You spend $500 fixing up a bedroom to rent. Your cost for food and utilities is $10 per night. Your family charges $60 per night to rent the bedroom.

a. Write an equation that represents your costs.

Cost, C (in dollars) = $10 per night • Number of nights, x + $500

b. Write an equation that represents your revenue (income).

Revenue, R (in dollars) = $60 per night • Number of nights, x

c. A set of two (or more) linear equations is called a **system of linear equations**. Write the system of linear equations for this problem.

2 ACTIVITY: Using a Table to Solve a System

Use the cost and revenue equations from Activity 1 to find how many nights you need to rent the bedroom before you recover the cost of fixing up the bedroom. This is the *break-even point* for your business.

a. Copy and complete the table.

x	0	1	2	3	4	5	6	7	8	9	10	11
C												
R												

b. How many nights do you need to rent the bedroom before you break even?

Laurie's Notes

Introduction

For the Teacher

- **Goal:** Students will explore solving a system of linear equations using graphs, tables, and algebra.
- This is a long investigation with many parts. Do not rush students.

Motivate

- Share a story about a trip to Indianapolis, where Market Street intersects Meridian Street at Monument Circle. Draw a sketch.

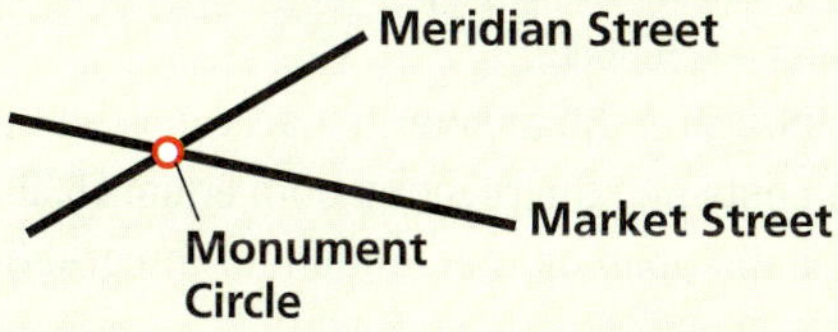

? Ask if students have ever visited a town or city where a monument was located in the middle of two streets.

- **Connection:** The monument is located on both streets. In other words, you will find the monument where the streets intersect.

Activity Notes

Activity 1

- Discuss what is known about your costs and your income.
- **Financial Literacy:** Do not assume that students are knowledgeable about concepts such as *costs* (fixed and variable) and *revenue* (income). Explain these words as you use them.
- Point out to students that the units in the verbal model agree. This means that "dollars per night × nights" is equal to dollars. So, the units in the equation are dollars = dollars + dollars.
- Ask a volunteer to share the equations they wrote with their partner.
- Discuss the definition of a system of linear equations.

Activity 2

- Read the problem aloud. Define and discuss the break-even point.

? "Why would a business want to know the break-even point?" You want to know how many nights it will take before you start to make money.

- Give time for students to work with their partner to fill in the table. Make sure that all students are using correct equations.

? **Extension:** "What patterns do you observe in the table?" The two rows continue to get closer together until they are finally equal at $x = 10$. Then the revenue is greater than the cost.

- Make sure to interpret the answer to part (b). When you rent the room for 10 nights, the costs and the revenue both equal \$600. A solution of each equation is (10, 600).

Previous Learning

Students should know how to solve equations with variables on both sides.

Activity Materials	
Introduction	**Textbook**
• picture of Monument Circle	• straightedge

Start Thinking! and Warm Up

Activity 2.5 **Start Thinking!** For use before Activity 2.5

Activity 2.5 **Warm Up** For use before Activity 2.5

Solve for *y*. Then graph the equation.

1. $y - 2x = 6$
2. $2x + y = 10$
3. $\frac{1}{2}y - x = 4$
4. $x + 3y = 9$
5. $x - 2y = 2$
6. $5x + 3y = 30$

2.5 Record and Practice Journal

Essential Question How can you solve a system of linear equations?

1 ACTIVITY: Writing a System of Linear Equations

Work with a partner.

Your family starts a bed-and-breakfast in your home. You spend \$500 fixing up a bedroom to rent. Your cost for food and utilities is \$10 per night. Your family charges \$60 per night to rent the bedroom.

a. Write an equation that represents your costs.

Cost, *C* (in dollars)	=	\$10 per night	•	Number of nights, *x*	+	\$500

$C = 10x + 500$

b. Write an equation that represents your revenue (income).

Revenue, *R* (in dollars)	=	\$60 per night	•	Number of nights, *x*

$R = 60x$

c. A set of two (or more) linear equations is called a **system of linear equations**. Write the system of linear equations for this problem.

$C = 10x + 500$
$R = 60x$

English Language Learners

Pair Activity

Pair each English learner with an English speaker. Ask one student to solve the system using algebra and the other student to solve the same system by graphing. Students then compare their answers. Partners should alternate solution methods as they continue to solve problems.

2.5 Record and Practice Journal

2 ACTIVITY: Using a Table to Solve a System

Use the cost and revenue equations from Activity 1 to find how many nights you need to rent the bedroom before you recover the cost of fixing up the bedroom. This is the *break-even point* for your business.

a. Complete the table.

x	0	1	2	3	4	5	6	7	8	9	10	11
C	500	510	520	530	540	550	560	570	580	590	600	610
R	0	60	120	180	240	300	360	420	480	540	600	660

b. How many nights do you need to rent the bedroom before you break even?

10 nights

3 ACTIVITY: Using a Graph to Solve a System

a. Graph the cost equation from Activity 1.

b. In the same coordinate plane, graph the revenue equation from Activity 1.

c. Find the point of intersection of the two graphs. The x-value of this point is the number of nights you need to rent the bedroom to break even.

(10, 600); 10 nights

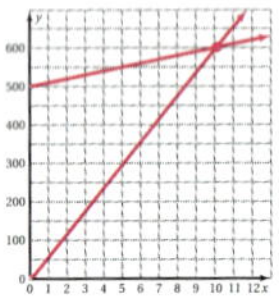

4 ACTIVITY: Using an Equation to Solve a System

a. Write the cost equation from Activity 1.

$C = 10x + 500$

b. Write the revenue equation from Activity 1.

$R = 60x$

c. The break-even point occurs when $C = R$. Set the expression for C equal to the expression for R. You should obtain an equation with x on both sides. Solve this equation for x. The solution is your break-even point.

$$10x + 500 = 60x$$
$$x = 10$$

d. Did you obtain the same break-even point in Activities 2, 3, and 4? If not, check your work. The break-even point should be the same in all three activities.

yes

What Is Your Answer?

5. IN YOUR OWN WORDS How can you solve a system of linear equations?

using a table, a graph, or an equation

6. When solving a system of linear equations, explain why it is a good idea to use two different ways to find the solution.

to check the solution

Laurie's Notes

Activity 3

- ? "Look at the scaling of the axes for this activity. What are the units that will be used to graph each of these equations?" number of nights, dollars
- The cost equation is $C = 10x + 500$. The revenue equation is $R = 60x$.
- ? "In what form is this Cost equation? What strategy can be used to graph the equation?" slope-intercept form; Students might say just plot the points from the table in Activity 2. Others might say plot the point for the y-intercept and then use a slope of 10 (which is equivalent to right 10 units, up 100 units).
- ? "In what form is the Revenue equation? What strategy can be used to graph the equation?" similar response to previous question
- Provided that students have graphed the equations carefully, the lines should intersect at (10, 600).
- Make sure to interpret the answer to part (c). When you rent the room for 10 nights, the costs and the revenue both equal $600. A point on each line is (10, 600). It is the point of intersection for the graphs.

Activity 4

- Students should be able to get started with this problem right away.
- ? "How do you solve this equation with variables on both sides of the equation?" Subtract $10x$ from each side and divide by 50.

What Is Your Answer?

- For Question 5, have students share their thoughts with the whole class.

Closure

- Refer back to Monument Circle in Indianapolis. How is it related to a system of equations? *Sample answer:* Each street is similar to an equation in a system and the monument at the intersection represents the point that satisfies both equations.
- **FYI:** The Soldiers and Sailors Monument located at Monument Circle in Indianapolis was designed to honor Indiana's Civil War veterans. It now honors all of Indiana's men and women who served in the military prior to World War I.

3 ACTIVITY: Using a Graph to Solve a System

a. Graph the cost equation from Activity 1.

b. In the same coordinate plane, graph the revenue equation from Activity 1.

c. Find the point of intersection of the two graphs. The x-value of this point is the number of nights you need to rent the bedroom to break even.

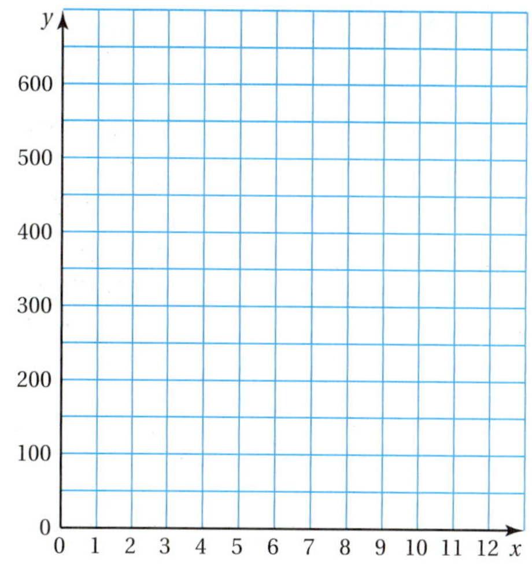

4 ACTIVITY: Using an Equation to Solve a System

a. Write the cost equation from Activity 1.

$C =$ ______

b. Write the revenue equation from Activity 1.

$R =$ ______

c. The break-even point occurs when $C = R$. Set the expression for C equal to the expression for R. You should obtain an equation with x on both sides. Solve this equation for x. The solution is your break-even point.

d. Did you obtain the same break-even point in Activities 2, 3, and 4? If not, check your work. The break-even point should be the same in all three activities.

What Is Your Answer?

5. **IN YOUR OWN WORDS** How can you solve a system of linear equations?

6. When solving a system of linear equations, explain why it is a good idea to use two different ways to find the solution.

Practice Use what you learned about systems of linear equations to complete Exercises 3 and 4 on page 80.

2.5 Lesson

Check It Out
Lesson Tutorials
BigIdeasMath.com

Key Vocabulary
system of linear equations, *p. 78*
solution of a system of linear equations, *p. 78*

A **system of linear equations** is a set of two or more linear equations in the same variables. A **solution of a system of linear equations** in two variables is an ordered pair that makes each equation true.

EXAMPLE 1 Solving a System of Linear Equations Using a Table

Solve the system. $y = x - 5$ Equation 1

$y = -3x + 7$ Equation 2

Reading
A system of linear equations is also called a *linear system*.

Step 1: Make a table of values.

Step 2: Find an x-value that gives the same y-value for both equations.

x	0	1	2	3
$y = x - 5$	−5	−4	−3	−2
$y = -3x + 7$	7	4	1	−2

The solution is $(3, -2)$.

EXAMPLE 2 Solving a System of Linear Equations Using a Graph

Solve the system. $y = 2x + 3$ Equation 1

$y = -x + 6$ Equation 2

Step 1: Graph each equation.

Step 2: Find the point of intersection. The graphs appear to intersect at $(1, 5)$.

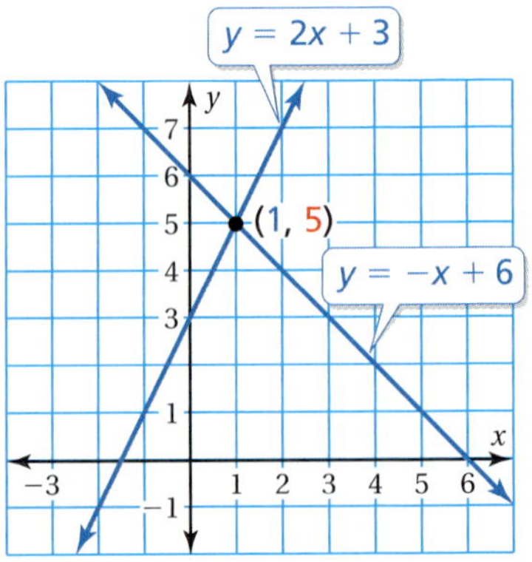

Step 3: Check your solution.

Equation 1

$y = 2x + 3$

$5 \stackrel{?}{=} 2(1) + 3$

$5 = 5$ ✓

Equation 2

$y = -x + 6$

$5 \stackrel{?}{=} -1 + 6$

$5 = 5$ ✓

The solution is $(1, 5)$.

On Your Own

Now You're Ready
Exercises 5–7 and 10–12

Solve the system of linear equations using a table and using a graph.

1. $y = x - 1$
   $y = -x + 3$
2. $y = -5x + 14$
   $y = x - 10$
3. $y = x$
   $y = 2x + 1$

Multi-Language Glossary at BigIdeasMath.com.

Laurie's Notes

Introduction

Connect

- **Yesterday:** Students explored several ways to solve a real-life problem using a system of equations.
- **Today:** Students will solve systems of linear equations using three different techniques.

Motivate

- Share a cooking story about salsa.

	Cilantro	Tomatoes	Onion
Summer Salsa	$\frac{1}{2}$ cup	3 cups	$\frac{3}{4}$ cup
Romero's Salsa	$1\frac{1}{4}$ cups	8 cups	2 cups

? "Do you think algebra can help a cook figure out how much salsa of each type can be made if you have 5 cups of cilantro and 40 cups of crushed tomatoes?" Comments will vary.

- Explain to students that the techniques they will study today are used to solve problems of this type.

Lesson Notes

Example 1

- Define a system of linear equations and the solution of a system of linear equations.
- Discuss with students that the x-value is written only once, in the top row. The next two rows are y-value.

? "When $x = 0$, what is the value of each equation (y-value)?" -5 and 7

- **Common Error:** Students may look at the table and incorrectly read the solution as $(-2, -2)$. Remind students that the y-values for the equations are both -2 when the x-value is 3.

Example 2

- Discuss with students how the graphs are drawn; plotting the point for the y-intercept first and then using the slope (rise over run) to plot 1 or 2 additional points.

? "Where do the lines appear to intersect?" (1, 5)

? "How can you determine if (1, 5) is actually the point of intersection and not (1.1, 4.9)?" Listen for checking the point in the equations.

- Students' understanding of systems of linear equations and their vocabulary are very elementary at this point. They know (1, 5) works, meaning the ordered pair satisfies each equation.
- Reinforce that solving a system of linear equations means that you want to find the value of x that gives the same value of y when substituted into each equation.

Goal Today's lesson is solving a **system of linear equations** using three different methods.

Start Thinking! and Warm Up

Lesson 2.5 Warm Up
For use before Lesson 2.5

Lesson 2.5 Start Thinking!
For use before Lesson 2.5

Write a word problem that can be solved using a system of linear equations.

Exchange problems with a classmate and solve your classmate's problem.

What method did you use to solve the problem?

Extra Example 1

Solve the system using a table.

$y = x - 3$

$y = -2x + 3$

(2, −1)

x	0	1	2
$y = x - 3$	−3	−2	−1
$y = -2x + 3$	3	1	−1

Extra Example 2

Solve the system using a graph.

$y = 2 - x$

$y = 3x - 10$

(3, −1)

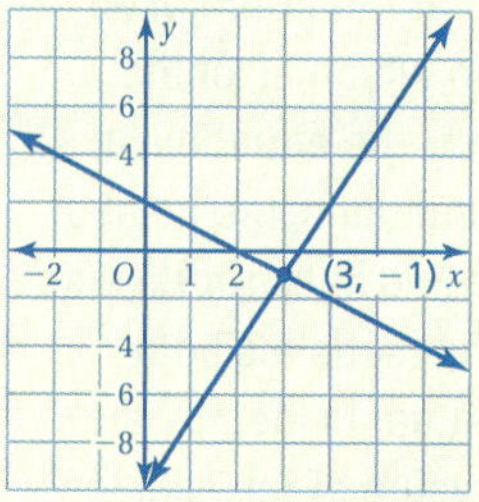

On Your Own

1. (2, 1)
2. (4, −6)
3. (−1, −1)

Laurie's Notes

Key Idea

- Write the Key Idea. Discuss and answer questions about each step. I try to limit the questions and focus on modeling each step in a specific example.

Example 3

- Ask a volunteer to read the problem. Write the verbal model. Reread the problem to demonstrate how the models are a direct translation of the first two sentences.
- Write the equations. Discuss the form of each. One equation is in standard form and the other is in slope-intercept form.
- ? **Step 1:** "Why do you need to rewrite the first equation, solving it for y?" You need both equations solved for y so that the expressions can be set equal to each other.
- **Step 2:** Set the two expressions equal to each other. Repeat to students that "y is equal to $35 - x$ and y is *also* equal to $x + 7$."
- Solve the equation for x. Substitute this value back into either of the original equations to find the value for y.
- **Extension:** You may have a student who recognizes an alternate approach to this problem.

  Let x = # of boys

  Let $x + 7$ = # of girls $\longrightarrow$ $x + x + 7 = 35$

  $$2x + 7 = 35$$

  $$2x = 28$$

  $x = 14$ (boys) and $x + 7 = 21$ (girls)

On Your Own

- **Think-Pair-Share:** Students should read the question independently and then work with a partner to answer the question. When they have answered the question, the pair should compare their answer with another group and discuss any discrepancies.

Closure

- "What does it mean if (2, 4) is the solution of the system $2x + y = 8$ and $y = 2x$?" *Sample answer:* (2, 4) is the point where the graphs of these two equations intersect. It is the point where the two equations are equal.

Technology For the Teacher

Dynamic Classroom

The Dynamic Planning Tool
Editable Teacher's Resources at *BigIdeasMath.com*

Extra Example 3

In Example 3, the yearbook committee has 23 members. There are 5 more girls than boys. Use the models to write a system of linear equations. Then solve the system to find the number of boys x and the number of girls y.

$x + y = 23$

$y = x + 5$

9 boys, 14 girls

On Your Own

4. $x + y = 45$

$y = x + 7$

19 boys, 26 girls

Differentiated Instruction

Visual

Remind students that the graph of an equation is a line in which all of the ordered pairs satisfy the equation. In a system of linear equations, the point of intersection of the two lines represents the ordered pair that satisfies both equations. Have students identify the x-value and the y-value of the solution and what the values represent in the problem.

Key Idea

Solving a System of Linear Equations Algebraically

Step 1 Solve both equations for one of the variables.

Step 2 Set the expressions equal to each other and solve for the variable.

Step 3 Substitute back into one of the original equations and solve for the other variable.

EXAMPLE 3 Solving a System of Linear Equations Algebraically

A middle school yearbook committee has 35 members. There are 7 more girls than boys. Use the models to write a system of linear equations. Then solve the system to find the number of boys *x* and the number of girls *y*.

Number of boys, x + Number of girls, y = 35

Number of girls, y = Number of boys, x + 7

The system is $x + y = 35$ and $y = x + 7$.

Step 1: Solve $x + y = 35$ for y.

$y = 35 - x$ — Subtract x from each side.

Step 2: Set the expressions equal to each other and solve for x.

$35 - x = x + 7$ — Set expressions equal to each other.

$28 = 2x$ — Subtract 7 from each side. Add x to each side.

$14 = x$ — Divide each side by 2.

Step 3: Substitute $x = 14$ into one of the original equations and solve for y.

$y = x + 7$ — Write one of the original equations.

$= 14 + 7$ — Substitute 14 for x.

$= 21$ — Add.

∴ There are 14 boys and 21 girls on the yearbook committee.

Study Tip

Be sure to check your solutions.

On Your Own

Now You're Ready
Exercises 13–15

4. **WHAT IF?** In Example 3, the yearbook committee has 45 members. Use the models to write a system of linear equations. Then solve the system to find the number of boys x and the number of girls y.

2.5 Exercises

Check It Out
Help with Homework
BigIdeasMath.com

Vocabulary and Concept Check

1. **VOCABULARY** Do the equations $4a - 3b = 5$ and $7b + 2a = -8$ form a system of linear equations? Explain.

2. **REASONING** Can a point in Quadrant II be a break-even point for a system? Explain.

Practice and Problem Solving

Use the table to find the break-even point. Check your solution.

3. $C = 15x + 150$

   $R = 45x$

x	0	1	2	3	4	5	6
C							
R							

4. $C = 24x + 80$

   $R = 44x$

x	0	1	2	3	4	5	6
C							
R							

Solve the system of linear equations using a table.

1 5. $y = x + 4$

   $y = 3x - 1$

6. $y = 1.5x - 2$

   $y = -x + 13$

7. $y = \frac{2}{3}x - 3$

   $y = -2x + 5$

8. **ERROR ANALYSIS** Describe and correct the error in solving the system of linear equations.

x	0	1	2	3
$y = -2x - 1$	−1	−3	−5	−7
$y = x - 7$	−7	−6	−5	−4

The solution is (−5, −5).

9. **CARRIAGE RIDES** The cost C (in dollars) for the care and maintenance of a horse and carriage is $C = 15x + 2000$, where x is the number of rides.

   a. Write an equation for the revenue R in terms of the number of rides.

   b. How many rides are needed for the business to break even?

Assignment Guide and Homework Check

Level	Day 1 Activity Assignment	Day 2 Lesson Assignment	Homework Check
Basic	3, 4, 19–22	1, 2, 5, 6, 8–11, 13, 14	6, 8, 9, 10, 14
Average	3, 4, 19–22	1, 2, 6–11, 13, 14, 16	6, 8, 9, 10, 14
Advanced	3, 4, 19–22	1, 2, 7, 8, 12–18	8, 12, 14, 16

Common Errors

- **Exercises 5–7** Students may substitute the *y*-value found in the first equation for the *x*-value in the second equation. Remind them that they need to substitute the *x*-values in the top row in each equation.
- **Exercises 10–12** Students may not show enough of the graph, so the lines will not intersect. Encourage them to extend their lines until they intersect. All of the systems of linear equations in this section have a solution.
- **Exercises 13–15** Students may perform the same operation instead of the opposite operation when trying to get the variables on the same side. Remind them that whenever a variable or number is moved from one side of the equal sign to the other, the opposite operation is used.

Vocabulary and Concept Check

1. yes; The equations are linear and in the same variables.
2. no; Points in Quadrant II have negative *x*-coordinates. You cannot break-even selling a negative number of items.

Practice and Problem Solving

3–4. See Additional Answers.

5. (2.5, 6.5)

6. (6, 7)

7. (3, −1)

8. They used both *y*-values for the point. The solution is (2, −5).

9. a. $R = 35x$

b. 100 rides

2.5 Record and Practice Journal

Solve the system of linear equations using a table.

1. $y = x + 6$
   $y = 4x - 3$
   (3, 9)

2. $y = 0.5x - 3$
   $y = -6x + 10$
   (2, −2)

Solve the system of linear equations using a graph.

3. $y = 3x + 1$ **(−1, −2)**
   $y = -2x - 4$

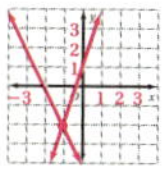

4. $y = -\frac{2}{5}x - 2\frac{3}{5}$ **(1, −3)**
   $y = 5x - 8$

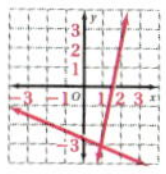

Solve the system of linear equations algebraically.

5. $y = \frac{2}{3}x + 2$
   $x - y = 4$
   (18, 14)

6. $y = x - 7$
   $y + x = 3$
   (5, −2)

7. There are 26 students in your class. There are 4 more girls than boys. Use a system of linear equations to find how many boys are in your class. How many girls are in your class?
   11 boys; 15 girls

Technology For the Teacher

Answer Presentation Tool

Quiz*Show*

Practice and Problem Solving

10. $(-1, 7)$

11. $(-5, 1)$

12. $(-4, -3)$

13. $(12, 15)$

14. $(5, 22)$

15. $(8, 1)$

16. $x + y = 42$
$y = x - 10$
26 math problems,
16 science problems

17. **a.** 6 h

b. 49 mi

18. See *Taking Math Deeper.*

Fair Game Review

19. yes

20. yes

21. no

22. D

Mini-Assessment

1. Solve the linear system algebraically.

$$y = \frac{1}{2}x - 6$$
$$y = -3x + 8$$

$(4, -4)$

2. Solve the linear system using a graph.

$$y = 3x + 5$$
$$y = -2x + 10$$

$(1, 8)$

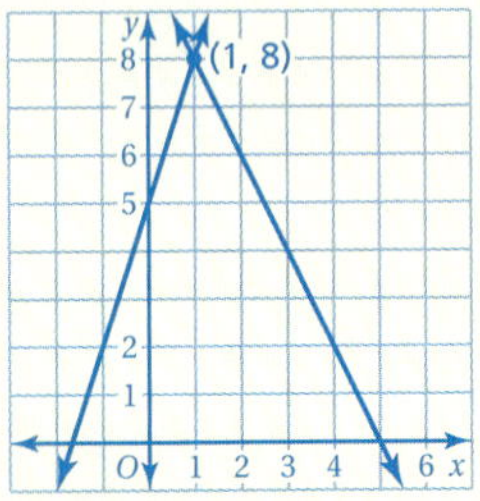

Taking Math Deeper

Exercise 18

Students know how to solve a system of linear equations using three methods; graphs, tables, and algebra.

1 Solve each equation for y.

Let x = number of bottles of face paint.
Let y = number of brushes.

First Store: $10x + 7.5y = 42.5$ (dollars)

$$y = -\frac{4}{3}x + \frac{17}{3}$$

Second Store: $8x + 6y = 34$ (dollars)

$$y = -\frac{4}{3}x + \frac{17}{3}$$

2 Graph each equation.

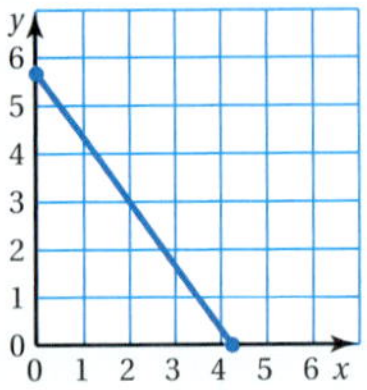

Because both equations are the same line, you cannot tell how many bottles and brushes were purchased.

3 Use a table.

x	0	1	2	3	4
$y = -\frac{4}{3}x + \frac{17}{3}$	$\frac{17}{3}$	$\frac{13}{3}$	3	$\frac{5}{3}$	$\frac{1}{3}$

From the table, you can see that there is only one whole number solution for positive values of x and y. Because you cannot buy parts of bottles of paint or parts of brushes, the solution is $x = 2$ bottles of paint and $y = 3$ brushes.

Reteaching and Enrichment Strategies

If students need help. . .	If students got it. . .
Resources by Chapter • Practice A and Practice B • Puzzle Time Record and Practice Journal Practice Differentiating the Lesson Lesson Tutorials Skills Review Handbook	Resources by Chapter • Enrichment and Extension • School-to-Work • Financial Literacy Start the next section

Solve the system of linear equations using a graph.

2 **10.** $y = 2x + 9$
$y = 6 - x$

11. $y = -x - 4$
$y = \frac{3}{5}x + 4$

12. $y = 2x + 5$
$y = \frac{1}{2}x - 1$

Solve the system of linear equations algebraically.

3 **13.** $x + y = 27$
$y = x + 3$

14. $y - x = 17$
$y = 4x + 2$

15. $x - y = 7$
$0.5x + y = 5$

16. **HOMEWORK** You have 42 math and science problems for homework. You have 10 more math problems than science problems. Use the model to write a system of linear equations. How many problems do you have in each subject?

Number of math problems, x + Number of science problems, y = 42

Number of science problems, y = Number of math problems, x − 10

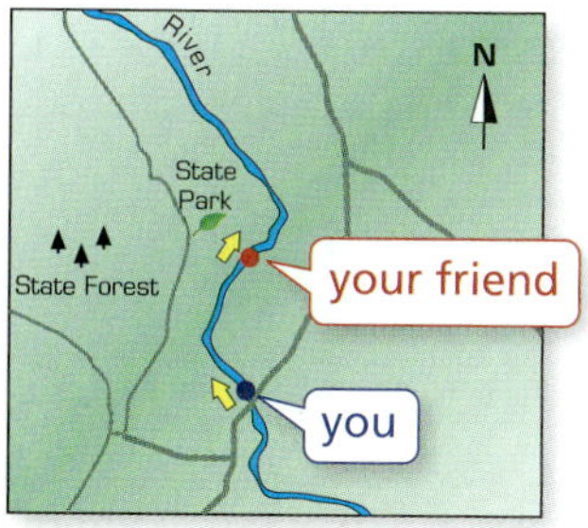

17. **CANOEING** You and your friend are canoeing. Your position on the river y (in miles) is represented by $y = 3.5x + 28$, where x is in hours. Your friend's position is represented by $y = 2x + 37$.

a. How long will it take you to catch up with your friend?

b. How far will you each have traveled when you catch up with your friend?

18. **Critical Thinking** You buy x bottles of face paint and y brushes at two stores. The amounts you spend are represented by $10x + 7.5y = 42.5$ and $8x + 6y = 34$. How many bottles of face paint and brushes did you buy?

Fair Game Review What you learned in previous grades & lessons

Decide whether the two equations are equivalent. *(Section 1.2 and Section 1.3)*

19. $4n + 1 = n - 8$
$3n = -9$

20. $2a + 6 = 12$
$a + 3 = 6$

21. $7v - \frac{3}{2} = 5$
$14v - 3 = 15$

22. **MULTIPLE CHOICE** Which line has the same slope as $y = \frac{1}{2}x - 3$? *(Section 2.3)*

Ⓐ $y = -2x + 4$ Ⓑ $y = 2x + 3$ Ⓒ $y - 2x = 5$ Ⓓ $2y - x = 7$

2.6 Special Systems of Linear Equations

Essential Question Can a system of linear equations have no solution? Can a system of linear equations have many solutions?

1 ACTIVITY: Writing a System of Linear Equations

Work with a partner.

Your cousin is 3 years older than you. Your ages can be represented by two linear equations.

$y = t$ Your age

$y = t + 3$ Your cousin's age

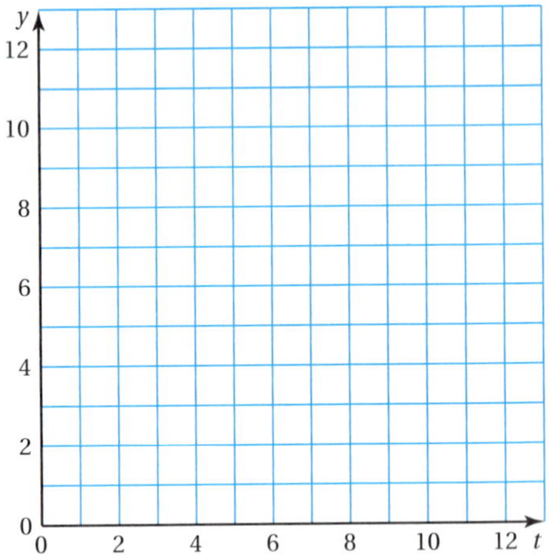

a. Graph both equations in the same coordinate plane.

b. What is the vertical distance between the two graphs? What does this distance represent?

c. Do the two graphs intersect? If not, what does this mean in terms of your age and your cousin's age?

2 ACTIVITY: Using a Table to Solve a System

Work with a partner. You invest \$500 for equipment to make dog backpacks. Each backpack costs you \$15 for materials. You sell each backpack for \$15.

a. Copy and complete the table for your cost C and your revenue R.

x	0	1	2	3	4	5	6	7	8	9	10
C											
R											

b. When will your company break even? What is wrong?

Laurie's Notes

Introduction

For the Teacher

- **Goal:** Students will explore special systems of linear equations that have no solution or an infinite number of solutions.

Motivate

- If you have an overhead projector or document camera, place 2 pieces of spaghetti on display. Say, "These represent two lines. Right now they are intersecting."
- ? "Is there any other relationship they could have?" Listen for parallel (non-intersecting) and a reasonable description of coinciding lines.
- Now place a transparency of a coordinate grid on top of the spaghetti.
- Suggest that lines, when graphed, do not always have to intersect.
- **Management Tip:** I use spaghetti for many models throughout the year, so I keep a box of spaghetti in my desk.

Activity Notes

Activity 1

- ? "How do you measure the vertical distance between two graphs?" Use a vertical line on the graph to count the units from one graph to the other.
- **Part (c):** Students will likely say that the two are not the same age. The solution actually says more: you and your cousin will never be the same age at the same time. Your cousin will always be 3 years older.

Activity 2

- Have students write the cost equation and the revenue equation. Students should turn back to the previous lesson as needed for help.
- Give time for students to work with their partner to fill in the table. Make sure that all students are using correct equations.
- ? **Extension:** "What patterns do you observe in the table?" Both rows of numbers are increasing by the same amount (15). The values are not getting closer together.
- Discuss student answers for part (b). Students should recognize that when you sell an item for the exact price that it costs, you will never make a profit, nor pay off the original investment of $500.
- ? **Extension:** "If you were to graph this system, what would you expect the graph to look like?" parallel lines; same slope

Previous Learning

Students should know how to solve equations with variables on both sides.

Activity Materials	
Introduction	**Textbook**
• spaghetti • transparency grid	• straightedge

Start Thinking! and Warm Up

Activity 2.6 Start Thinking!
For use before Activity 2.6

Activity 2.6 Warm Up
For use before Activity 2.6

Solve the system of linear equations algebraically.

1. $x + y = 19$
   $y = x + 5$
2. $x - y = 10$
   $x + y = 6$
3. $y = 2x + 1$
   $y = 3x$
4. $x + 2y = 17$
   $y = 2x + 6$

2.6 Record and Practice Journal

Essential Question Can a system of linear equations have no solution? Can a system of linear equations have many solutions?

1 ACTIVITY: Writing a System of Linear Equations

Work with a partner.

Your cousin is 3 years older than you. Your ages can be represented by two linear equations.

$y = t$ Your age

$y = t + 3$ Your cousin's age

a. Graph both equations in the same coordinate plane.

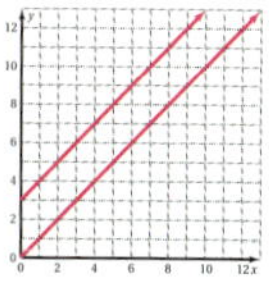

b. What is the vertical distance between the two graphs? What does this distance represent?

3; difference in age

c. Do the two graphs intersect? If not, what does this mean in terms of your age and your cousin's age?

no; You and your cousin will never be the same age at the same time.

2 ACTIVITY: Using a Table to Solve a System

Work with a partner. You invest $500 for equipment to make dog backpacks. Each backpack costs you $15 for materials. You sell each backpack for $15.

a. Complete the table for your cost C and your revenue R.

x	0	1	2	3	4	5	6	7	8	9	10
C	500	515	530	545	560	575	590	605	620	635	650
R	0	15	30	45	60	75	90	105	120	135	150

Differentiated Instruction

Visual

To help students remember that lines with the same slope are parallel write "If the slopes on the board are =, the lines are parallel." Relate that the segments that make up the equal sign are parallel segments.

2.6 Record and Practice Journal

b. When will your company break even? What is wrong?

never

3 ACTIVITY: Using a Graph to Solve a Puzzle

Work with a partner. Let x and y be two numbers. Here are two clues about the values of x and y.

	Words	Equation
Clue 1:	The value of y is 4 more than twice the value of x.	$y = 2x + 4$
Clue 2:	The difference of $3y$ and $6x$ is 12.	$3y - 6x = 12$

a. Graph both equations in the same coordinate plane.

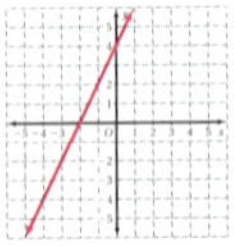

b. Do the two lines intersect? Explain.

yes; same line

c. What is the solution of the puzzle?

any point on the line $y = 2x + 4$

d. Use the equation $y = 2x + 4$ to complete the table.

x	0	1	2	3	4	5	6	7	8	9	10
y	4	6	8	10	12	14	16	18	20	22	24

e. Does each solution in the table satisfy *both* clues?

yes

f. What can you conclude? How many solutions does the puzzle have? How can you describe them?

There are many solutions.;
all points on the line $y = 2x + 4$

What Is Your Answer?

4. **IN YOUR OWN WORDS** Can a system of linear equations have no solution? Can a system of linear equations have many solutions? Give examples to support your answers.

Yes if lines are parallel;
yes if equations are the same.

Laurie's Notes

Activity 3

- Students enjoy puzzles. Present the next activity in this context.
- ? "Look at the words for each clue and then look at the equation. Does the translation make sense? Explain." Listen to student explanations.
- ? "In what form is each equation written? What strategy can be used to graph each equation?" Listen to student explanations.
- Provided students have graphed the equations carefully, the lines should coincide. There is only one graph.
- Discuss answers to part (f). The fact that the two graphs coincide means the equations are equivalent and have the same graph. There are many solutions to the puzzle! Help students recognize this by selecting two or three ordered pairs from the table. Talk through the substitution of the ordered pair into each equation. All of the ordered pairs will be solutions of both equations.
- **Extension:** Students have heard of puzzles where you start with a number, do a few computations, and eventually end up with a number twice the original. Here's an example. (I have annotated the algebraic representation to the right.) Suggest to students that they start with a small number so that the computation is simple.

Step 1: Pick a number, perhaps your age.	x
Step 2: Add 10.	$x + 10$
Step 3: Multiply by 4.	$4(x + 10) = 4x + 40$
Step 4: Divide by 2.	$(4x + 40) \div 2 = 2x + 20$
Step 5: Subtract 20.	$2x + 20 - 20 = 2x$

Announce that you should now have a number that is twice what you started with. In relationship to this lesson, you would have:

$$y = \frac{4(x + 10)}{2} - 20 \text{ and } y = 2x.$$

The graph of both lines is $y = 2x$.

What Is Your Answer?

- This question tries to help students summarize the two additional cases for the solution of a system of linear equations.

Closure

- Sketch a graph of a system of equations that has no solution.
- Sketch a graph of a system of equations that has many solutions.

Technology For the Teacher

Dynamic Classroom

The Dynamic Planning Tool
Editable Teacher's Resources at *BigIdeasMath.com*

3 ACTIVITY: Using a Graph to Solve a Puzzle

Work with a partner. Let x and y be two numbers. Here are two clues about the values of x and y.

	Words	Equation
Clue 1:	y is 4 more than twice the value of x.	$y = 2x + 4$
Clue 2:	The difference of $3y$ and $6x$ is 12.	$3y - 6x = 12$

a. Graph both equations in the same coordinate plane.

b. Do the two lines intersect? Explain.

c. What is the solution of the puzzle?

d. Use the equation $y = 2x + 4$ to complete the table.

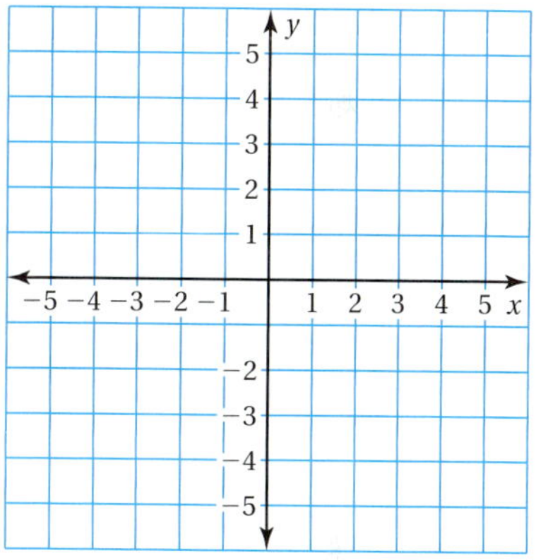

x	0	1	2	3	4	5	6	7	8	9	10
y											

e. Does each solution in the table satisfy *both* clues?

f. What can you conclude? How many solutions does the puzzle have? How can you describe them?

What Is Your Answer?

4. IN YOUR OWN WORDS Can a system of linear equations have no solution? Can a system of linear equations have many solutions? Give examples to support your answers.

Practice Use what you learned about special systems of linear equations to complete Exercises 4 and 5 on page 86.

2.6 Lesson

Check It Out
Lesson Tutorials
BigIdeasMath.com

EXAMPLE 1 Solving a Special System of Linear Equations

Solve the system.

$y = 3x + 1$ Equation 1

$y = 3x - 5$ Equation 2

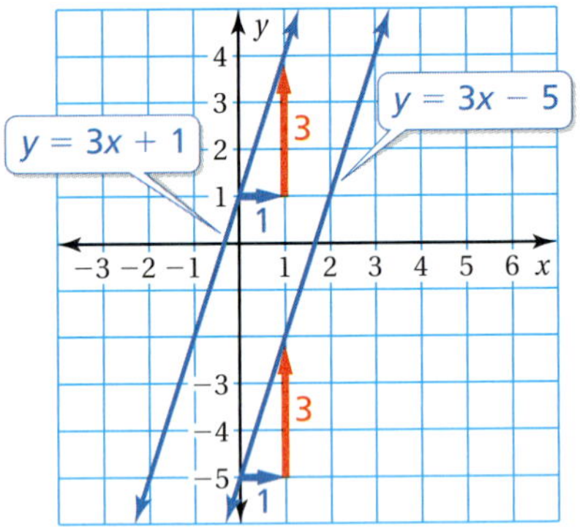

Graph each equation.

The lines have the same slope and different y-intercepts. So, the lines are parallel.

Because parallel lines do not intersect, there is no point that is a solution of both equations.

So, the system of linear equations has no solution.

EXAMPLE 2 Solving a Special System of Linear Equations

Solve the system.

$y = -2x + 4$ Equation 1

$4x + 2y = 8$ Equation 2

Write $4x + 2y = 8$ in slope-intercept form.

$4x + 2y = 8$ Write the equation.

$2y = -4x + 8$ Subtract $4x$ from each side.

$y = -2x + 4$ Divide each side by 2.

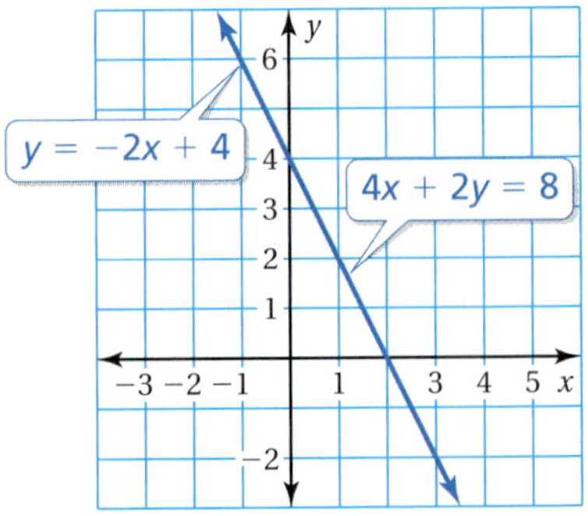

The equations are the same.

The solution of the system is all the points on the line $y = -2x + 4$.

So, the system of linear equations has infinitely many solutions.

On Your Own

Now You're Ready
Exercises 6–11

Solve the system of linear equations.

1. $-4x + 4y = 8$
   $y = x + 2$
2. $y = -5x - 2$
   $5x + y = 3$
3. $x + y = 6$
   $y = -x$

Laurie's Notes

Introduction

Connect

- **Yesterday:** Students explored the graphs of two special systems of linear equations.
- **Today:** Students will solve special systems of linear equations by writing the equations in slope-intercept form before graphing.

Motivate

- Make a set of cards in advance: equation card, slope card, and y-intercept card. Make enough cards so that everyone in the class has a card.

Examples:

$y = 3x - 4$	$m = 3$	$b = -4$
$3x + y = 4$	$m = -3$	$b = 4$
$y = 4x - 2$	$m = 4$	$b = -2$
$y = 2 - 4x$	$m = -4$	$b = 2$

? "What is the general form of an equation written in slope-intercept form?" $y = mx + b$

- Distribute all of the cards and have students form a matching set of 3.
- The goal of this quick activity is to review the slope-intercept form of an equation.

Lesson Notes

Example 1

? "How do you graph equations in slope-intercept form?" Plot the point for the y-intercept and then use the slope to locate two additional points.

- Because both equations are in slope-intercept form, students will observe the same slope for each equation and will not be surprised that the lines are parallel.
- Graphing is a visual check that the lines are parallel.

Example 2

? "What is different about this system of equations than in Example 1?" The second equation is in standard form.

? "Explain the steps needed to rewrite this equation in slope-intercept form." Listen to student directions.

- **Common Question:** When you subtract $4x$ from each side of the equation, students want to write $2y = 8 - 4x$. Explain that this is okay, but discuss the properties that allow you to rewrite it as $2y = -4x + 8$.

$2y = 8 - 4x$	Given equation
$2y = 8 + (-4x)$	Definition of subtraction
$2y = -4x + 8$	Commutative Property of Addition

On Your Own

- Students should check with a neighbor after completing each question.

Goal Today's lesson is solving a system of equations with no solution or infinitely many solutions.

Lesson Materials	
Introduction	**Textbook**
• cards	• straightedge

Start Thinking! and Warm Up

Lesson 2.6 Warm Up
For use before Lesson 2.6

Lesson 2.6 Start Thinking!
For use before Lesson 2.6

A car is traveling at an average rate of 60 miles per hour.

Another car starts from the same location one hour later and travels on the same route at the same average rate.

Will the second car ever catch up?

How does this situation relate to solving a system of equations?

Extra Example 1

Solve the system.

$y = -2x + 5$

$y = -2x + 1$

no solution

Extra Example 2

Solve the system.

$y = x - 3$

$2x - 2y = 6$

infinitely many solutions; all points on the line $y = x - 3$

On Your Own

1. infinitely many solutions; all points on the line $y = x + 2$
2. no solution
3. no solution

Extra Example 3

The perimeter of the trapezoid is 10 units. The perimeter of the triangle is 5 units. Write and solve a linear system to find the values of x and y.

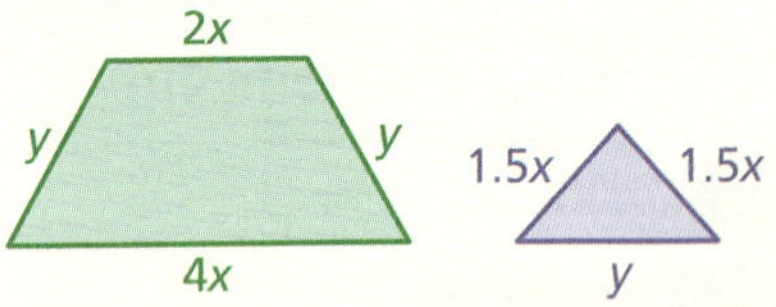

infinitely many solutions

On Your Own

4. There is no solution because the resulting equation $y = -\frac{1}{2}x + \frac{27}{4}$ is parallel to the equation of the perimeter of the triangle.

English Language Learners

Visual

Make a poster of the Summary box on the number of solutions of systems of linear equations. Use color to help English learners make connections between concepts and language.

Laurie's Notes

Example 3

- Ask a volunteer to read the problem. Check to see that students are comfortable with finding the perimeter of a triangle and a rectangle.
- Write the equations. Simplify each.
- Rewrite each equation in slope-intercept form. Note that in each case when the x-term is subtracted from each side of the equation, it is written before the constant term.
- Take time to explain that dividing each side of the equation by the coefficient of y relates back to addition of fractions.

$$\frac{-4x + 36}{8} = \frac{-4x}{8} + \frac{36}{8} = \frac{-1x}{2} + \frac{9}{2} = -\frac{1}{2}x + \frac{9}{2}$$

- When you add fractions, you keep the same denominator and add the numerators.
- Another point of confusion for students is why $\frac{-1x}{2}$ can be rewritten as $-\frac{1}{2}x$. Remind students of how a fraction $\left(-\frac{1}{2}\right)$ is multiplied by a number (x).
- Note that there are infinitely many solutions of this system. However, they are limited to Quadrant I because the perimeter cannot be negative.

On Your Own

- **Think-Pair-Share:** Students should read the question independently and then work with a partner to answer the question. When they have answered the question, the pair should compare their answer with another group and discuss any discrepancies.

Summary

- Write the information from the Summary box. Connect this back to the spaghetti used yesterday to introduce special systems of linear equations.

Closure

- **Exit Ticket:** Write a system of equations that will have no solution.
  Sample answer: $y = 2x + 4$ and $y = 2x - 6$

Technology For the Teacher

Dynamic Classroom

The Dynamic Planning Tool
Editable Teacher's Resources at *BigIdeasMath.com*

EXAMPLE 3 Solving a Special System of Linear Equations

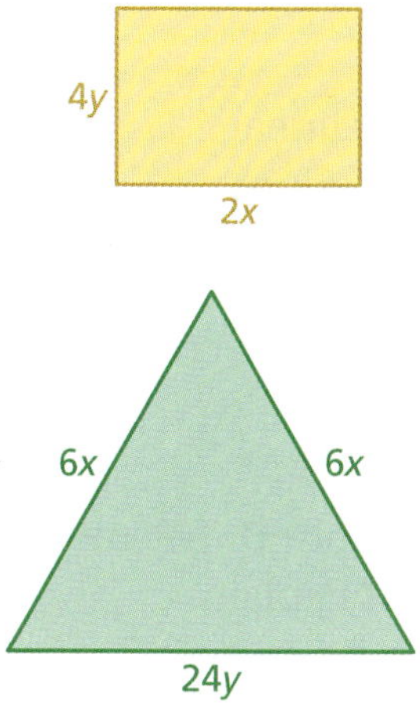

The perimeter of the rectangle is 36 units. The perimeter of the triangle is 108 units. Write and solve a system of linear equations to find the values of x and y.

Perimeter of rectangle

$2(2x) + 2(4y) = 36$

$4x + 8y = 36$ Equation 1

Perimeter of triangle

$6x + 6x + 24y = 108$

$12x + 24y = 108$ Equation 2

The system of linear equations is $4x + 8y = 36$ and $12x + 24y = 108$.

Write both equations in slope-intercept form.

$$4x + 8y = 36$$
$$8y = -4x + 36$$
$$y = -\frac{1}{2}x + \frac{9}{2}$$

$$12x + 24y = 108$$
$$24y = -12x + 108$$
$$y = -\frac{1}{2}x + \frac{9}{2}$$

The equations are the same. So, the solution of the system is all the points on the line $y = -\frac{1}{2}x + \frac{9}{2}$.

∴ The system of linear equations has infinitely many solutions.

On Your Own

4. **WHAT IF?** What happens to the solution in Example 3 if the perimeter of the rectangle is 54 units? Explain.

Summary

Solutions of Systems of Linear Equations

A system of linear equations can have *one solution, no solution*, or *infinitely many solutions*.

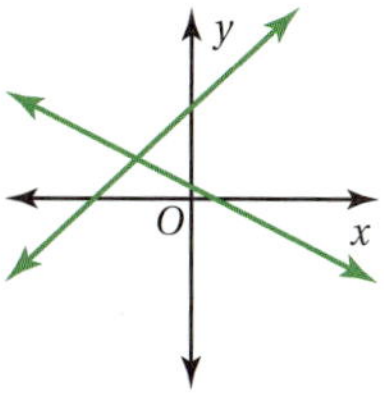

One solution

The lines intersect.

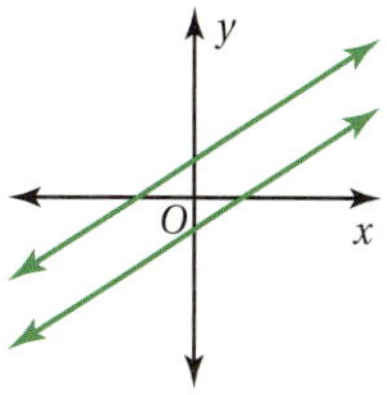

No solution

The lines are parallel.

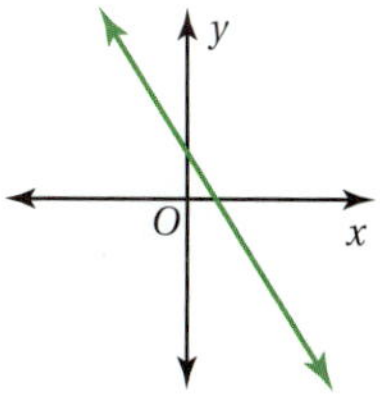

Infinitely many solutions

The lines are the same.

2.6 Exercises

Check It Out
Help with Homework
BigIdeasMath.com

Vocabulary and Concept Check

1. **VOCABULARY** What is the difference between the graph of a system of linear equations that has *no solution* and the graph of a system of linear equations that has *infinitely many solutions*?
2. **NUMBER SENSE** Determine the number of solutions of the system of linear equations without writing the equations in slope-intercept form. Explain your reasoning.

$$2x + y = 5$$
$$4x + 2y = 10$$

3. **REASONING** One equation in a system of linear equations has a slope of -3. The other equation has a slope of 4. How many solutions does the system have? Explain.

Practice and Problem Solving

Let x and y be two numbers. Find the solution of the puzzle.

4. y is $\frac{1}{3}$ more than 4 times the value of x.

   The difference of $3y$ and $12x$ is 1.

5. $\frac{1}{2}$ of x plus 3 is equal to y.

   x is 6 more than twice the value of y.

Solve the system of linear equations.

1 2

6. $-6x + 3y = 18$
   $y = 2x - 2$
7. $y = -\frac{1}{6}x + 5$
   $x + 6y = 30$
8. $-x + 2y = -3$
   $9x - 3y = -3$
9. $3x + 2y = 0$
   $y = x - 5$
10. $y = \frac{4}{9}x + \frac{1}{3}$
    $-4x + 9y = 3$
11. $y = -6x + 8$
    $12x + 2y = -8$

12. **ERROR ANALYSIS** Describe and correct the error in solving the system of linear equations.

$$y = -2x + 6$$
$$8x + 4y = 24$$

✗
$8x + 4y = 24$
$4y = -8x + 24$
$y = -2x + 6$
The lines have the same slope so there is no solution.

13. **PIG RACE** In a pig race, your pig gets a head start of 3 feet running at a rate of 2 feet per second. Your friend's pig is running at a rate of 2 feet per second. A system of linear equations that represents this situation is $y = 2x + 3$ and $y = 2x$. Will your friend's pig catch up to your pig? Explain.

Assignment Guide and Homework Check

Level	Day 1 Activity Assignment	Day 2 Lesson Assignment	Homework Check
Basic	4, 5, 20–23	1–3, 6–13	6, 8, 10, 12
Average	4, 5, 20–23	1–3, 8–14, 17	8, 10, 11, 12, 14
Advanced	4, 5, 20–23	1–3, 10, 12, 14–19	10, 12, 14, 18

Common Errors

- **Exercises 6–11** Students may see that the slope is the same for both equations and immediately say that the system of linear equations has no solution. Remind them that they need to compare the slope *and* y-intercepts when determining the number of solutions. Encourage them to check algebraically and graphically that the system of linear equations has the number of solutions they found.
- **Exercises 14–16** Students may make calculation errors when solving the equations for y. Encourage them to be careful when solving for y. Remind them that they have already mastered the skills of adding, subtracting, multiplying and dividing decimals, so they can focus on solving the equations for y.

2.6 Record and Practice Journal

Solve the system of linear equations.

1. $y = 2x$ [illegible]
   $y = 2x + 7$
   no solution

2. $3x + 4y = -10$
   $y = -\frac{3}{4}x - \frac{5}{2}$
   infinitely many solutions; all points on the line $y = -\frac{3}{4}x - \frac{5}{2}$

3. $y = x + 6$
   $2x + y = 9$
   one solution;(1, 7)

4. $x - y = 8$
   $2y = 2x - 16$
   infinitely many solutions; all points on the line $y = x - 8$

5. You start reading a book for your literature class two days before your friend. You both read 10 pages per night. A system of linear equations that represents this situation is $y = 10x + 20$ and $y = 10x$. Will your friend finish the book before you? Explain.
   no; You have a 20 page head start and you both read at the same pace.

6. You and a friend buy music from different online stores. The system of linear equations represents the total amount spent on x songs and y albums. Both of you buy the same number of albums. Compare the number of songs each of you buy.
   $0.95x + 10y = 39.5$
   $1.9x + 20y = 79$
   You both buy the same number of songs.

Technology For the Teacher

Answer Presentation Tool

Quiz*Show*

Vocabulary and Concept Check

1. The graph of a system with no solution has two parallel lines, and the graph of a system with infinitely many solutions is one line.
2. infinitely many solutions; If you multiply the first equation by 2, the result is the second equation.
3. one solution; because the lines are not parallel and will not be the same equation

Practice and Problem Solving

4. infinitely many solutions; all points on the line $y = 4x + \frac{1}{3}$
5. no solution
6. no solution
7. infinitely many solutions; all points on the line $y = -\frac{1}{6}x + 5$
8. one solution; $(-1, -2)$
9. one solution; $(2, -3)$
10. infinitely many solutions; all points on the line $y = \frac{4}{9}x + \frac{1}{3}$
11. no solution
12. They did not look at the entire equation. Because the two equations are the same, there are inifinitely many solutions. The solution of the system is all the points on the line $y = -2x + 6$.
13. no; because they are running at the same speed and your pig had a head start

Practice and Problem Solving

14. infinitely many solutions; all points on the line $y = 1.2x + 0.4$

15. no solution

16. one solution; $(-2.4, -3.5)$

17. **a.** 6 h

b. You both work the same number of hours.

18. See *Taking Math Deeper.*

19. **a.** *Sample answer:* $y = -7$

b. *Sample answer:* $y = 3x$

c. *Sample answer:* $2y - 6x = -2$

Fair Game Review

20. $x = 6$

21. $x = -3$

22. $x = -8$

23. B

Mini-Assessment

Solve the system of equations using any method.

1. $2x + 3y = 5$
   $2x + 3y = 7$
   no solution
2. $x + 2y = 12$
   $y = -\frac{1}{2}x + 6$
   infinitely many solutions; all points on the line $y = -\frac{1}{2}x + 6$
3. $-3x + 2y = 2$
   $4x + 3y = 20$
   one solution; (2, 4)

Taking Math Deeper

Exercise 18

The interesting thing about this problem is that two boats can tie for a race, even though they cross the finish line at different times.

1 Which boat left first?

a. From the graph, Boat 1 left when $x = 0$ hours and Boat 2 left when $x = 1$ hour. So, Boat 1 left first.

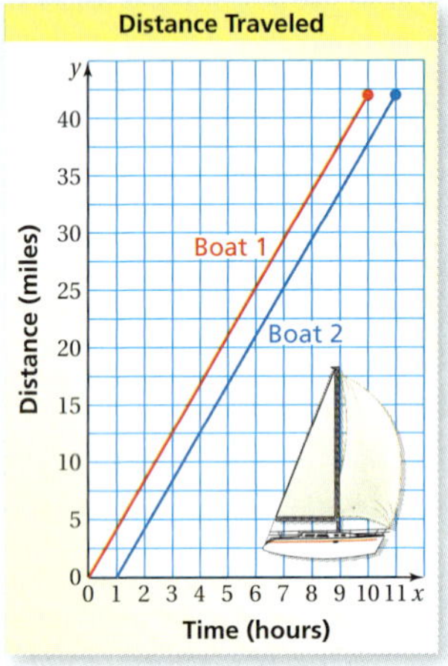

2 Compare the speeds.

b. Boat 1 completed the 42-mile race in 10 hours. So, its speed was 4.2 miles per hour.

The lines are parallel, so Boat 2's speed was the same as Boat 1's speed.

3 **c.** It took each boat 10 hours to complete the race.

Project

Write a report on the Americas Cup Race. Remember to include who, what, when, and where.

Reteaching and Enrichment Strategies

If students need help. . .	If students got it. . .
Resources by Chapter • Practice A and Practice B • Puzzle Time Record and Practice Journal Practice Differentiating the Lesson Lesson Tutorials Skills Review Handbook	Resources by Chapter • Enrichment and Extension • School-to-Work • Financial Literacy • Technology Connection Start the next section

Solve the system of linear equations.

14. $y + 4.6x = 5.8x + 0.4$
$-4.8x + 4y = 1.6$

15. $y = \frac{\pi}{3}x + \pi$
$-\pi x + 3y = -6\pi$

16. $-2x + y = 1.3$
$2(0.5x - y) = 4.6$

$4x + 8y = 64$
$8x + 16y = 128$

17. MONEY You and a friend both work two different jobs. The system of linear equations represents the total earnings for x hours worked at the first job and y hours worked at the second job. Your friend earns twice as much as you.

a. One week, both of you work 4 hours at the first job. How many hours do you and your friend work at the second job?

b. Both of you work the same number of hours at the second job. Compare the number of hours you each work at the first job.

18. BOAT RACE Two sailboats enter a timed race that is 42 miles long.

a. Which boat left the starting point first?

b. Compare the speeds of the boats.

c. Estimate how long it takes each boat to finish the race.

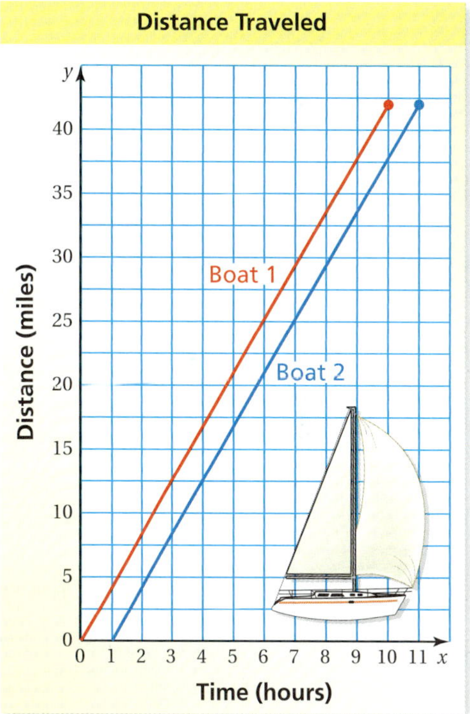

19. Critical Thinking One equation in a system of linear equations is $y = 3x - 1$.

a. Write a second equation so that $(-2, -7)$ is the only solution of the system.

b. Write a second equation so that the system has no solution.

c. Write a second equation so that the system has infinitely many solutions.

Fair Game Review What you learned in previous grades & lessons

Solve the equation. Check your solution. *(Section 1.3)*

20. $3x - 5 = 2x + 1$

21. $-3(x - 1) = -8x - 12$

22. $\frac{1}{2}x + 4 = \frac{3}{4}x + 6$

23. MULTIPLE CHOICE What is the slope of the line represented by the points in the table? *(Section 2.2)*

x	0	4	8	12
y	2	5	8	11

Ⓐ $-\frac{3}{4}$ Ⓑ $\frac{3}{4}$ Ⓒ $\frac{4}{3}$ Ⓓ 2

2.7 Solving Equations by Graphing

Essential Question How can you use a system of linear equations to solve an equation that has variables on both sides?

You learned how to use algebra to solve equations with variables on both sides. Another way is by using a system of linear equations.

1 ACTIVITY: Solving a System of Linear Equations

Work with a partner. Find the solution of $2x - 1 = -\frac{1}{2}x + 4$.

a. Use the left side of the equation to write one linear equation. Then, use the right side to write another linear equation.

$$2x - 1 = -\frac{1}{2}x + 4$$

$y = 2x - 1$ $\qquad$ $y = -\frac{1}{2}x + 4$

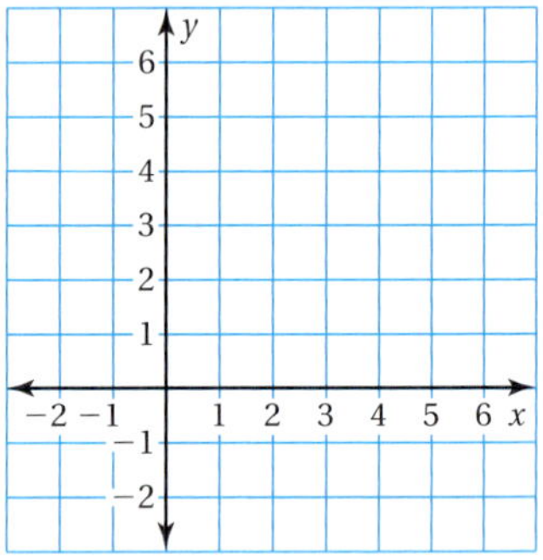

b. Sketch the graphs of the two linear equations. Find the x-value of the point of intersection. The x-value is the solution of

$$2x - 1 = -\frac{1}{2}x + 4.$$

Check the solution.

c. Explain why this "graphical method" works.

2 ACTIVITY: Using a Graphing Calculator

Use a graphing calculator to graph the two linear equations.

$$y = 2x - 1$$

$$y = -\frac{1}{2}x + 4$$

The steps used to enter the equations depend on the calculator model that you have.

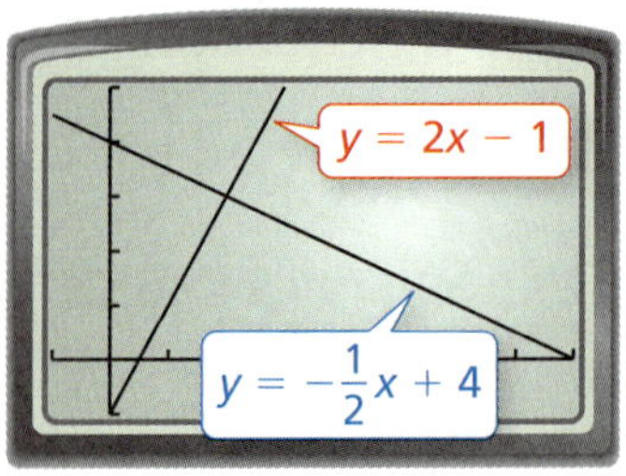

Laurie's Notes

Introduction

For the Teacher

- **Goal:** Students will use two techniques to solve an equation with variables on both sides of the equal sign: graphing a system of equations and solving the equation algebraically.

Motivate

- Use algebra tiles, or two different shapes on a balance scale, to represent $3x + 2 = 2x + 5$.

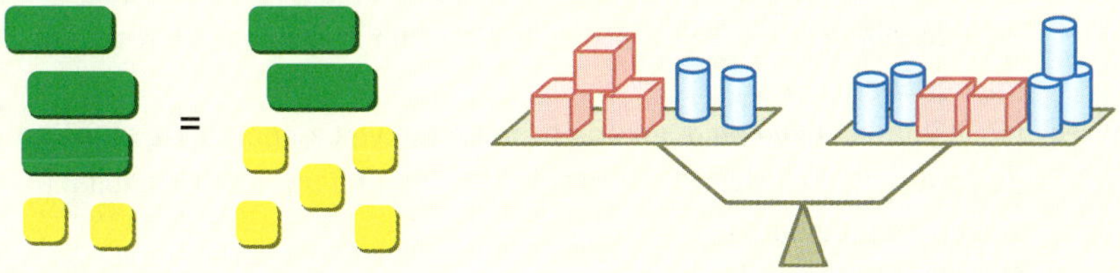

? "What equation is being represented in the problem?" $3x + 2 = 2x + 5$
- Ask a student to talk through the steps necessary to solve for *x*, or to solve for one of the cubes.

Activity Notes

Activity 1

- In this activity, students need to view the equation as two parts. The left side of the equation is set equal to *y* and the right side of the equation is also set equal to *y*. This forms the system of linear equations. One way to think about this is to ask, "What would the graph of each side of the equation look like?"

? "Where do the two graphs intersect?" (2, 3)
- **Part (b):** The ordered pair (2, 3) is a solution of each equation. Help students to understand that when $x = 2$, *both* sides of the equation have a value of 3.
- Ask a student to show algebraically that $x = 2$ is correct. The fractional coefficient makes this a little more challenging.

Activity 2

- Use graphing calculators to solve this equation by graphing each side of the equation as done in Activity 1. If graphing calculators are not available, explain how a graphing calculator can be used to find the solution.

Previous Learning

Students should know how to solve equations with variables on both sides.

Activity Materials	
Introduction	**Textbook**
• algebra tiles	• straightedge

Start Thinking! and Warm Up

Activity 2.7 Start Thinking! For use before Activity 2.7

Activity 2.7 Warm Up For use before Activity 2.7

Solve the equation. Check your solution.

1. $m - 8 = 2m$
2. $3x - 5 = 2x + 1$
3. $12 - 4p = p - 3$
4. $2(x - 8) = -x + 5$
5. $-2r - 9 = -3(r + 2)$
6. $4(x + 8) = -2(x - 1)$

2.7 Record and Practice Journal

Essential Question How can you use a system of linear equations to solve an equation that has variables on both sides?

You learned how to use algebra to solve equations with variables on both sides. Another way is by using a system of linear equations.

1 ACTIVITY: Solving a System of Linear Equations

Work with a partner. Find the solution of $2x - 1 = -\frac{1}{2}x + 4$.

a. Use the left side of the equation to write one linear equation. Then, use the right side to write another linear equation.

$2x - 1 = -\frac{1}{2}x + 4$

$y = 2x - 1$ $\quad$ $y = -\frac{1}{2}x + 4$

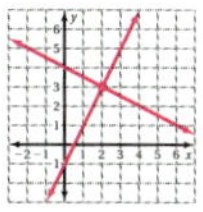

b. Sketch the graphs of the two linear equations. Find the *x*-value of the point of intersection.

The *x*-value is the solution of $2x - 1 = -\frac{1}{2}x + 4$.

Check the solution. 2

c. Explain why this "graphical method" works.

You are not changing either side of the original equation.

2 ACTIVITY: Using a Graphing Calculator

Use a graphing calculator to graph the two linear equations.

$y = 2x - 1$

$y = -\frac{1}{2}x + 4$

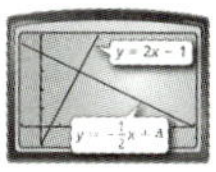

The steps used to enter the equations depend on the calculator model that you have.

Differentiated Instruction

Kinesthetic

Some students benefit by having access to graphing calculators or the graphing software programs available on many computer operating systems. Students can enter the equations, view their graphs, and check the point of intersection.

2.7 Record and Practice Journal

3 ACTIVITY: Using a System of Linear Equations

Work with a partner. Solve the equation using two methods.

- **Method 1:** Use an algebraic method.
- **Method 2:** Use a graphical method.
- Is the solution the same using both methods?

a. $\frac{1}{2}x + 4 = -\frac{1}{4}x + 1$

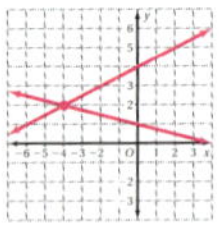

$x = -4$; yes

b. $\frac{2}{3}x + 4 = \frac{1}{3}x + 3$

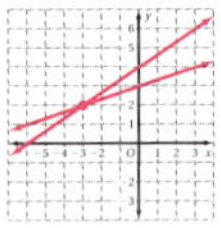

$x = -3$; yes

c. $-\frac{2}{3}x - 1 = \frac{1}{3}x - 4$

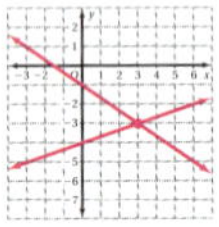

$x = 3$; yes

d. $\frac{4}{5}x + \frac{7}{5} = 3x - 3$

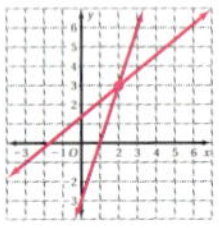

$x = 2$; yes

What Is Your Answer?

4. **IN YOUR OWN WORDS** How can you use a system of linear equations to solve an equation that has variables on both sides? Give an example that is different from those in Activities 1 and 3.

Graph the left side of the equation. Then graph the right side of the equation. The solution is the x-value of the point of intersection.

5. Describe three ways in which Rene Descartes's invention of the coordinate plane allows you to solve algebraic problems graphically.

See Additional Answers.

Laurie's Notes

Activity 3

- These four problems will take some time to solve because each is done using two methods. The problems are a bit more challenging due to the fractional coefficients.

? "What do you need to remember when adding or subtracting fractions?" need a common denominator

- **FYI:** While the fractions may make the algebraic solution a bit messy, the graphical approach is easy.
- This is a good review of fraction computations and equation solving.
- **Teaching Tip:** For graphing the equations, plot the point for the *y*-intercept and then use the slope to plot 2 additional points. For part (d), the *y*-intercept is a fraction. Suggest to students that they make a quick table of values using $x = -3$, 0, and 2 in order to graph the left side of the equation.
- It is helpful to have an overhead transparency of a coordinate plane so that students can share their graphs. It is more accurate than doing a sketch on the board.
- Note that the coordinate planes do not have the axes centered. It should provide sufficient space for students to complete their graphs and be able to see the point of intersection of the two lines.

? "Can you tell from inspection of the equations if the graph of the system will result in parallel lines?" Students should observe that no pair of lines has the same slope.

What Is Your Answer?

- **Question 5:** Students can look in previous sections for examples.

Closure

- Graph the left and right sides of the equation:

$$3x + 2 = 2x + 5$$

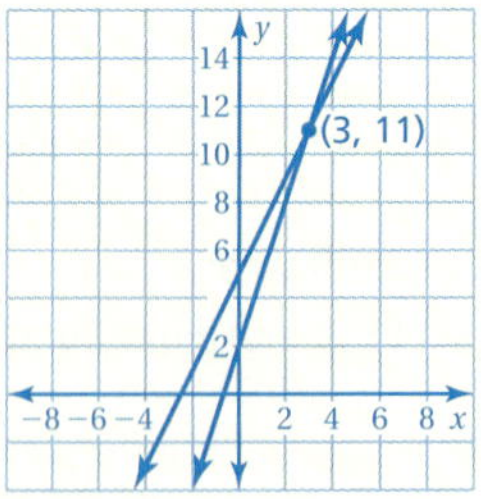

"Is the solution the same as the solution found with algebra tiles?" yes

Technology For the Teacher

Dynamic Classroom

The Dynamic Planning Tool

Editable Teacher's Resources at *BigIdeasMath.com*

3 ACTIVITY: Using a System of Linear Equations

Work with a partner. Solve each equation using two methods.

- **Method 1:** Use an algebraic method.
- **Method 2:** Use a graphical method.
- Is the solution the same using both methods?

a. $\frac{1}{2}x + 4 = -\frac{1}{4}x + 1$

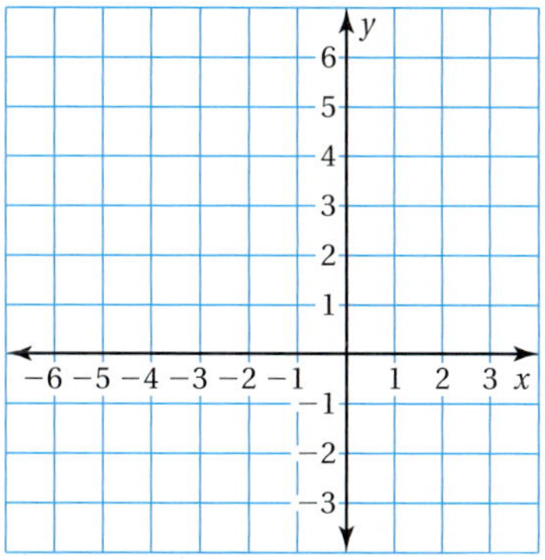

b. $\frac{2}{3}x + 4 = \frac{1}{3}x + 3$

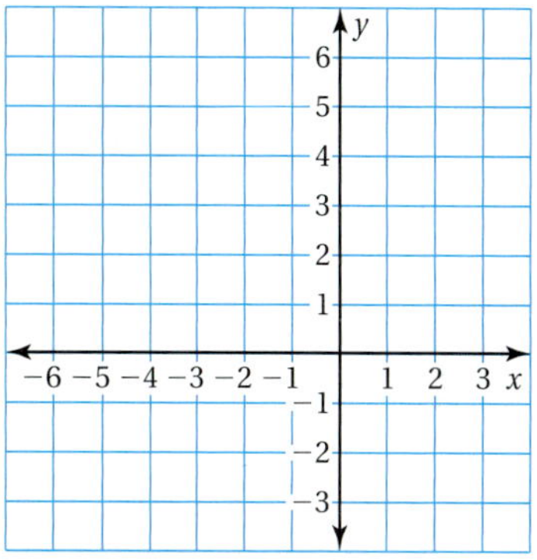

c. $-\frac{2}{3}x - 1 = \frac{1}{3}x - 4$

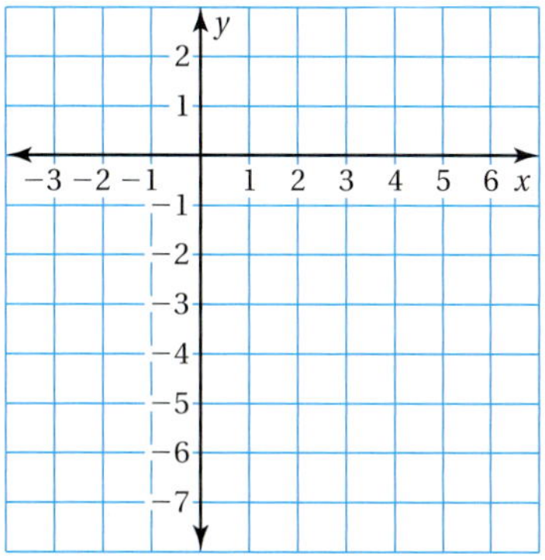

d. $\frac{4}{5}x + \frac{7}{5} = 3x - 3$

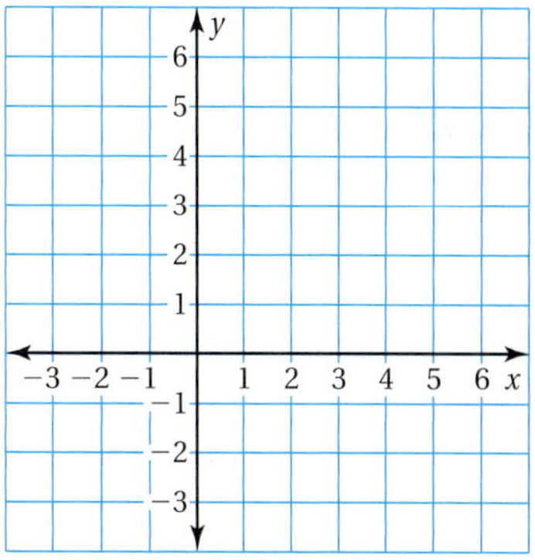

What Is Your Answer?

4. IN YOUR OWN WORDS How can you use a system of linear equations to solve an equation that has variables on both sides? Give an example that is different from those in Activities 1 and 3.

5. Describe three ways in which René Descartes's invention of the coordinate plane allows you to solve algebraic problems graphically.

Practice Use what you learned about solving equations by graphing to complete Exercises 3–5 on page 92.

2.7 Lesson

Key Idea

Solving Equations Using Graphs

Step 1: To solve the equation $ax + b = cx + d$, write two linear equations.

$$ax + b = cx + d$$

$y = ax + b$ and $y = cx + d$

Step 2: Graph the system of linear equations. The x-value of the solution of the system of linear equations is the solution of the equation $ax + b = cx + d$.

EXAMPLE 1 Solving an Equation Using a Graph

Solve $x - 2 = -\frac{1}{2}x + 1$ using a graph. Check your solution.

Step 1: Write a system of linear equations using each side of the equation.

$$x - 2 = -\frac{1}{2}x + 1$$

$y = x - 2$ $\quad$ $y = -\frac{1}{2}x + 1$

Step 2: Graph the system.

$$y = x - 2$$

$$y = -\frac{1}{2}x + 1$$

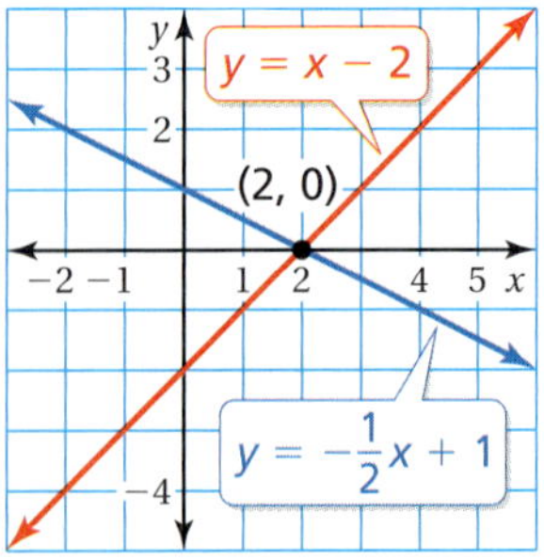

The graphs intersect at (2, 0). So, the solution is $x = 2$.

Check

$$x - 2 = -\frac{1}{2}x + 1$$

$$2 - 2 \stackrel{?}{=} -\frac{1}{2}(2) + 1$$

$$0 = 0 \checkmark$$

On Your Own

Now You're Ready Exercises 6 and 7

Use a graph to solve the equation. Check your solution.

1. $\frac{1}{3}x = x + 8$

2. $1.5x + 2 = 11 - 3x$

Laurie's Notes

Introduction

Connect

- **Yesterday:** Students reviewed how to solve equations with variables on both sides. They compared this technique with graphing a system of equations.
- **Today:** Students will solve an equation with variables on both sides by graphing a system of equations.

Motivate

- ? "Have any of you seen the play *Little Shop of Horrors*? There is a flytrap-like alien plant that lives on human blood and eventually grows large enough to swallow people whole."
- In reality, the Venus flytrap does not grow rapidly. However, the Thuja Giant, a fast growing evergreen tree, grows 3–5 feet per year.

Lesson Notes

Words of Wisdom

- Students often do not see the point of learning a new technique when the old technique worked well. The graphical approach helps to show the connection between the algebraic and geometric approaches. Some equations are easier to graph than they are to manipulate.

Key Idea

- ? "What does it mean to solve an equation with variables on both sides?" Find the value of x that makes both sides equal the same value.
- Write the *Key Idea*.
- ? "For the equation written in red, $y = ax + b$, name the slope and y-intercept." slope $= a$ and y-intercept $= b$
- ? "For the equation written in blue, $y = cx + d$, name the slope and y-intercept." slope $= c$ and y-intercept $= d$

Example 1

- Set each side of the equation equal to y to create the system of equations.
- The check is algebraic. Does the x-coordinate of the point of intersection make both sides of the equation equal the same value, the y-coordinate?

On Your Own

- ? "How do you graph a slope of 1.5?" think of $\frac{\text{rise}}{\text{run}} = \frac{3}{2}$

Goal Today's lesson is solving an equation with variables on both sides by graphing.

Lesson Materials
Textbook
• straightedge

Start Thinking! and Warm Up

Lesson 2.7 Warm Up
For use before Lesson 2.7

Lesson 2.7 Start Thinking!
For use before Lesson 2.7

You have used two ways to solve an equation with variables on both sides: using algebra and using a graph.

Which method do you prefer? Why?

Extra Example 1

Solve $x + 3 = \frac{1}{2}x + 1$ using a graph.

Check your solution. $x = -4$

On Your Own

1. $x = -12$
2. $x = 2$

Laurie's Notes

Extra Example 2

In Example 2, Plant A grows 0.4 inch per month. Plant B grows three times faster per month.

a. Use the model to write an equation. $0.4x + 12 = 1.2x + 9$

b. After how many months x are the plants the same height? 3.75 mo

On Your Own

3. yes; The lines intersect at one point.

4. 6 mo

English Language Learners

Pair Activity

Pair English learners with English speakers. Assign a problem-solving exercise from the exercise set. Have students make a poster illustrating the solution steps to the problem. Students should include defining the variables, writing and solving the system, and graphing the system.

Example 2

- Ask a volunteer to read the problem. Students will also need to read information from the diagram.
- The verbal model helps to focus attention on the goal, which is to determine *when* the two *plants will be the same height.*
- The growth rate is stated in terms of inches per month and is multiplied by months. So, you are adding inches to inches on each side of the equation.

$$\frac{\text{inches}}{\cancel{\text{months}}} \times \cancel{\text{months}} + \text{inches}$$

- Finish working through the problem as shown.

? "How tall are the plants after 5 months?" 15 inches

On Your Own

- **Think-Pair-Share:** Students should read each question independently and then work with a partner to answer the questions. When they have answered the questions, the pair should compare their answers with another group and discuss any discrepancies.

Closure

- **Exit Ticket:** Solve the equation $3x - 4 = \frac{1}{2}x + 1$. $x = 2$

Technology For the Teacher

Dynamic Classroom

The Dynamic Planning Tool
Editable Teacher's Resources at *BigIdeasMath.com*

EXAMPLE 2 Real-Life Application

Plant A grows 0.6 inch per month. Plant B grows twice as fast.

a. **Use the model to write an equation.**

b. **After how many months *x* are the plants the same height?**

Growth rate • Months, x + Original height = Growth rate • Months, x + Original height

a. The equation is $0.6x + 12 = 1.2x + 9$.

b. Write a system of linear equations using each side of the equation. Then graph the system.

$$0.6x + 12 = 1.2x + 9$$

$y = 0.6x + 12$ $y = 1.2x + 9$

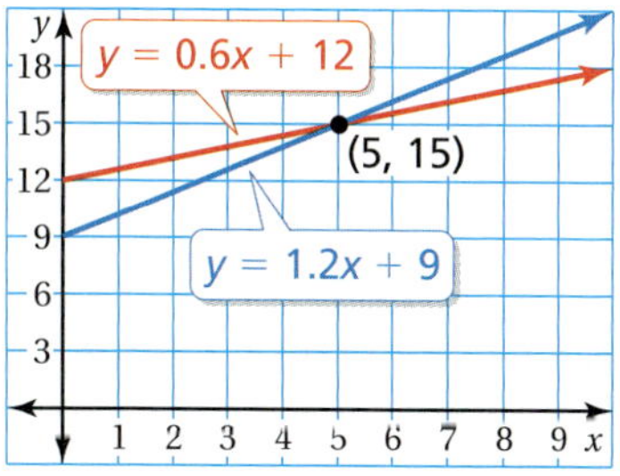

The solution of the system is $(x, y) = (5, 15)$. So, the plants are both 15 inches tall after 5 months.

On Your Own

3. Using the graph in Example 2, is the statement below true? Explain.

The system of linear equations $y = 0.6x + 12$ and $y = 1.2x + 9$ has one solution.

4. **WHAT IF?** In Example 2, the growth rate of Plant A is 0.5 inch per month. After how many months x are the plants the same height?

2.7 Exercises

Check It Out
Help with Homework
BigIdeasMath.com

Vocabulary and Concept Check

1. **CRITICAL THINKING** Would you rather solve the equation $x - \frac{4}{5} = -x + \frac{6}{5}$ using an algebraic method or a graphical method? Explain.

2. **DIFFERENT WORDS, SAME QUESTION** Which is different? Find "both" answers.

What is the solution of the equation $x - 3 = -\frac{1}{3}x + 5$?	What is the x-value of the solution of the linear system $y + 3 = x$ and $y + \frac{1}{3}x = 5$?
What is the y-coordinate of the intersection of $y = x - 3$ and $y = -\frac{1}{3}x + 5$?	What is the x-coordinate of the intersection of $y = x - 3$ and $y = -\frac{1}{3}x + 5$?

Practice and Problem Solving

Solve the equation algebraically and graphically.

3. $\frac{1}{3}x - 2 = -\frac{1}{6}x + 1$

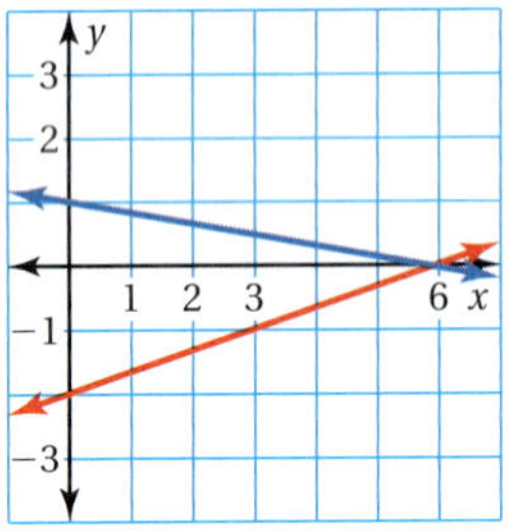

4. $-\frac{3}{4}x + \frac{5}{4} = x + 3$

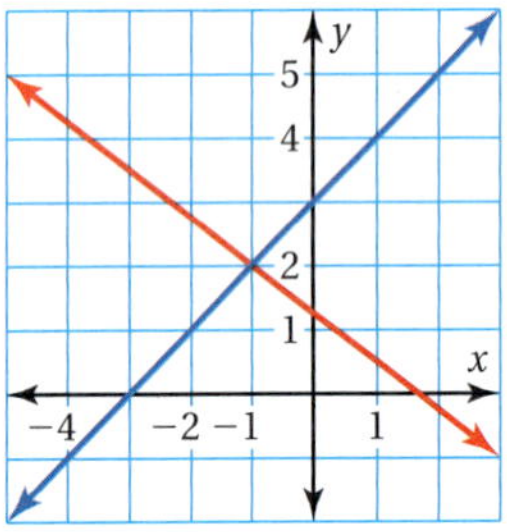

5. $\frac{5}{8}x - \frac{3}{4} = \frac{1}{3}x + 1$

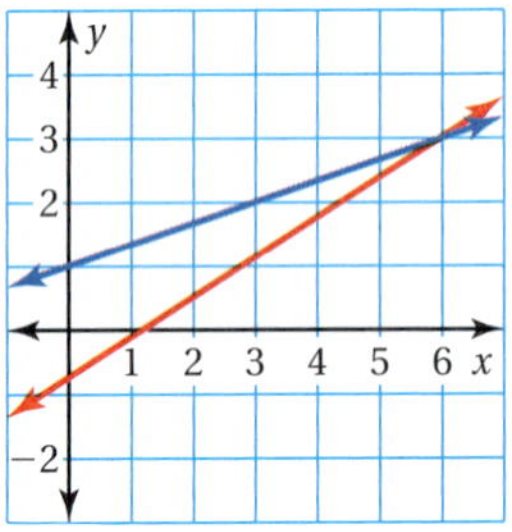

Use a graph to solve the equation. Check your solution.

1 6. $\frac{2}{5}x - 2 = -x + 12$

7. $-\frac{5}{6}x + \frac{1}{2} = -x + 1$

8. **ERROR ANALYSIS** Describe and correct the error in solving the equation $2x + 4 = -\frac{7}{2}x + 11$.

✗

$y = 2x + 4$

$y = -\frac{7}{2}x + 11$

There is no solution.

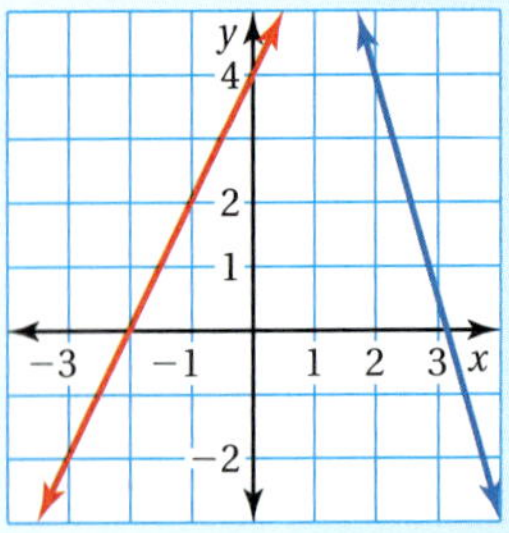

9. **KARAOKE** One night at karaoke, you sang $3x + 2$ songs. The next night, you sang $4x$ songs. Is it possible that you sang the same number of songs each night? Explain.

Assignment Guide and Homework Check

Level	Day 1 Activity Assignment	Day 2 Lesson Assignment	Homework Check
Basic	3–5, 17–21	1, 2, 6–11	6, 8, 9, 10
Average	3–5, 17–21	1, 2, 7–11, 15	7, 8, 10, 15
Advanced	3–5, 17–21	1, 2, 8, 12–16	8, 12, 14, 15

Common Errors

- **Exercises 3–5** Students may make a mistake when solving algebraically or graphically. Encourage them to substitute the *x*-value into the original equation to check that the statement is true.
- **Exercises 6, 7, 10–12** Students may find the wrong intersection point when graphing because they did not draw straight lines. Encourage them to use a straightedge when graphing so that they can draw accurate lines. Remind them to check their solution in the original equation.
- **Exercise 9** Students may find the solution of the equation, but will not answer the question asked. Remind them to examine the context of the problem and to make sure that they are answering the question.

2.7 Record and Practice Journal

Solve the equation algebraically and graphically.

1. $\frac{1}{4}x + 4 = x - 2$

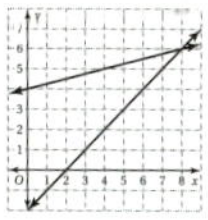

$x = 8$

2. $\frac{1}{2}x + 2 = \frac{7}{8}x + \frac{1}{2}$

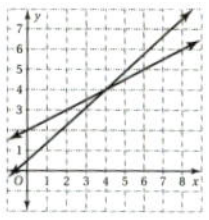

$x = 4$

Use a graph to solve the equation. Check your solution.

3. $\frac{1}{2}x + 4 = -x - 11$

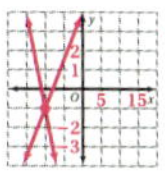

$x = -10$

4. $-x + 1 = -\frac{1}{4}x - \frac{1}{2}$

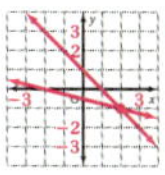

$x = 2$

5. On the first day of your garage sale, you earned $12x + 9$ dollars. The next day you earned $22x$ dollars. Is it possible that you earned the same amount each day? Explain.

yes; You earn the same amount each day if $x = \frac{9}{10}$.

Technology For the Teacher

Answer Presentation Tool

Quiz*Show*

Vocabulary and Concept Check

1. algebraic method; Graphing with fractions is harder than solving the equation algebraically.
2. What is the *y*-coordinate of the intersection of $y = x - 3$ and $y = -\frac{1}{3}x + 5$?; $y = 3$; $x = 6$

Practice and Problem Solving

3. $x = 6$
4. $x = -1$
5. $x = 6$
6. $x = 10$
7. $x = 3$
8. The graph is not big enough. The two lines are not parallel, so there must be a solution. The solution is $x = \frac{14}{11}$.
9. yes; Because a solution of $3x + 2 = 4x$ exists ($x = 2$).

Practice and Problem Solving

10. $x = 4$

11. $x = 2$

12. $x = -3$

13. The two lines are parallel, which means there is no solution. Using an algebraic method, you obtain $-5 = 8$, which is not true and means that there is no solution.

14. 4 min

15. See *Taking Math Deeper*.

16. no; The solution of the system of linear equations describing their volumes gives a negative volume.

Fair Game Review

17. 4 **18.** 6

19. -3 **20.** $-\pi$

21. A

Mini-Assessment

Use a graph to solve the equation.

1. $-\frac{1}{3}x + 5 = 4x - 8$ $x = 3$

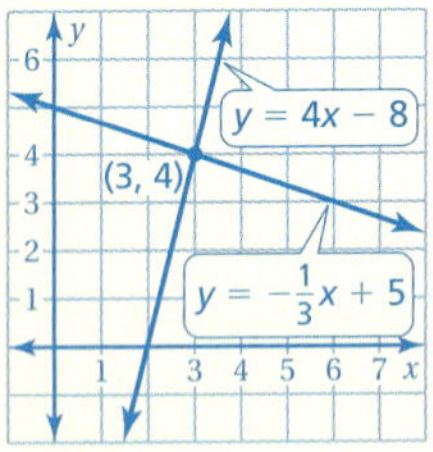

2. $-\frac{3}{2}x - 2 = -2x - 3$ $x = -2$

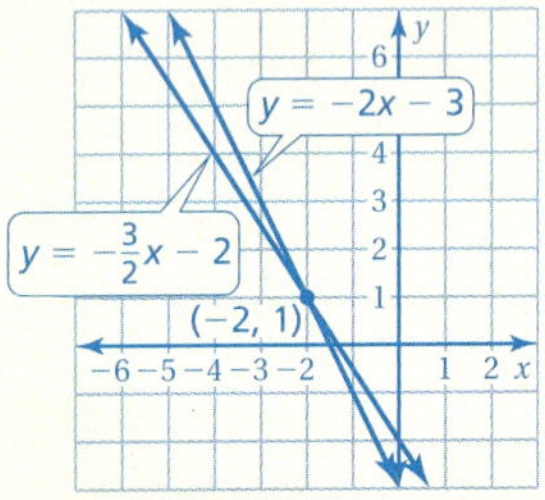

Taking Math Deeper

Exercise 15

For this problem, it helps to start by organizing the information in a table.

1 Use a table to organize the information.

	Home	Away
Last year	11	x
This year	$11 + 4$	$\frac{3}{4}x$

Total number of games played each year is the same.

2 Write and solve an equation.

$$11 + x = 15 + \frac{3}{4}x$$

$$44 + 4x = 60 + 3x$$

$$x = 16$$

3 Answer the question.

The soccer team played 16 away games last year.

Reteaching and Enrichment Strategies

If students need help. . .	If students got it. . .
Resources by Chapter • Practice A and Practice B • Puzzle Time Record and Practice Journal Practice Differentiating the Lesson Lesson Tutorials Skills Review Handbook	Resources by Chapter • Enrichment and Extension • School-to-Work • Financial Literacy • Technology Connection • Life Connections Start the next section

Use a graph to solve the equation. Check your solution.

10. $2.5x + 3 = 4x - 3$

11. $-1.4x + 1 = 1.6x - 5$

12. $0.7x - 1.2 = -1.4x - 7.5$

13. **CRITICAL THINKING** What happens when you use a graphical method to solve $\frac{1}{3}x - 5 = \frac{1}{3}x + 8$? Does an algebraic method give the same result?

14. **HIKING** You hike uphill at a rate of 200 feet per minute. Your friend hikes downhill on the same trail at a rate of 250 feet per minute. How long will it be until you meet?

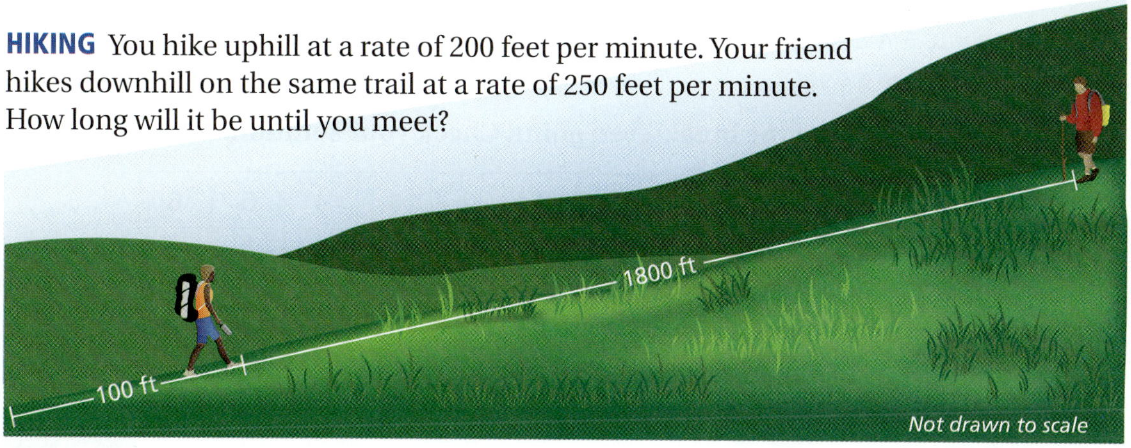

Last Year	
Home	**Away**
11	x

15. **SOCCER** A soccer team played four more home games and three-fourths as many away games this year than last year. The team played the same number of games each season. How many away games did the team play last year?

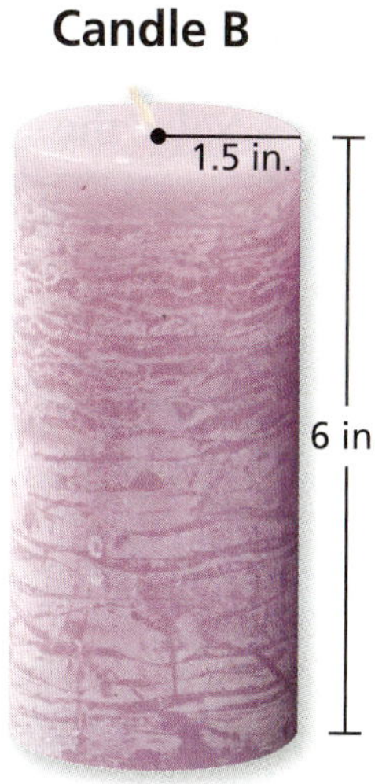

16. **Geometry** Candle A burns at an average rate of 11 cubic centimeters per hour. Candle B burns at an average rate of 18 cubic centimeters per hour. Do the candles ever have the same volume? Explain.

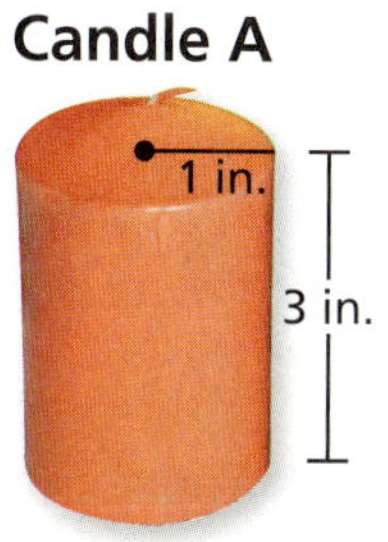

Fair Game Review What you learned in previous grades & lessons

Find the y-intercept of the graph of the linear equation. *(Section 2.3)*

17. $y = 3x + 4$

18. $y = -\frac{2}{3}x + 6$

19. $y = 2.4x - 3$

20. $y = 2x - \pi$

21. **MULTIPLE CHOICE** Which of the following is the slope of the line? *(Section 2.2)*

Ⓐ $-\frac{6}{7}$

Ⓑ $-\frac{7}{6}$

Ⓒ $\frac{6}{7}$

Ⓓ $\frac{7}{6}$

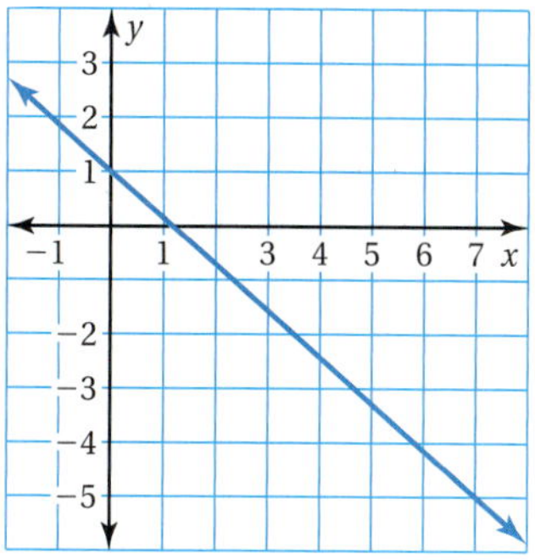

2.5–2.7 Quiz

Check It Out
Progress Check
BigIdeasMath.com

Solve the system of linear equations algebraically. *(Section 2.5)*

1. $y = 4 - x$
$y = x - 4$

2. $y = \frac{x}{2} + 10$
$y = 4x - 4$

Use the table to find the break-even point. Check your solution. *(Section 2.5)*

3. $C = 10x + 180$
$R = 46x$

x	0	1	2	3	4	5	6
C							
R							

Solve the system of linear equations using a graph. *(Section 2.5)*

4. $y = 3x + 2$
$y = x + 4$

5. $y = -3x - 1$
$y = -2x + 5$

Solve the system of linear equations using any method. *(Section 2.6)*

6. $y = 2x - 3$
$y - 6x = -9$

7. $y = 4x + 8$
$2y - 8x = 18$

Use a graph to solve the equation. Check your solution. *(Section 2.7)*

8. $\frac{1}{4}x - 4 = \frac{3}{4}x + 2$

9. $8x - 14 = -2x - 4$

10. BASKETBALL You score 24 points in a basketball game. You make 9 shots. How many three-point shots and two-point shots do you make? *(Section 2.5)*

Number of three-point shots, x + Number of two-point shots, y = 9

Value of a three-point shot • Number of three-point shots, x + Value of a two-point shot • Number of two-point shots, y = 24

11. BICYCLE One day, you ride $2x + 5$ kilometers on your bicycle. The next day, you ride $3x$ kilometers. Is it possible that you rode the same distance each day? Explain. *(Section 2.7)*

12. TEMPERATURE Two students write the expressions $\frac{1}{2}x + 49$ and $2x - 5$ to represent today's high temperature (in degrees Fahrenheit). What is today's high temperature? *(Section 2.7)*

Alternative Assessment Options

Math Chat	Student Reflective Focus Question
Structured Interview	Writing Prompt

Math Chat

- Work in groups of four. Discuss the similarities and differences of solving linear equations compared to solving systems of linear equations. When they are finished, they explain their findings to the other groups in the class.
- The teacher should walk around the classroom listening to the pairs and ask questions to ensure understanding.

Study Help Sample Answers

Remind students to complete Graphic Organizers for the rest of the chapter.

3a.

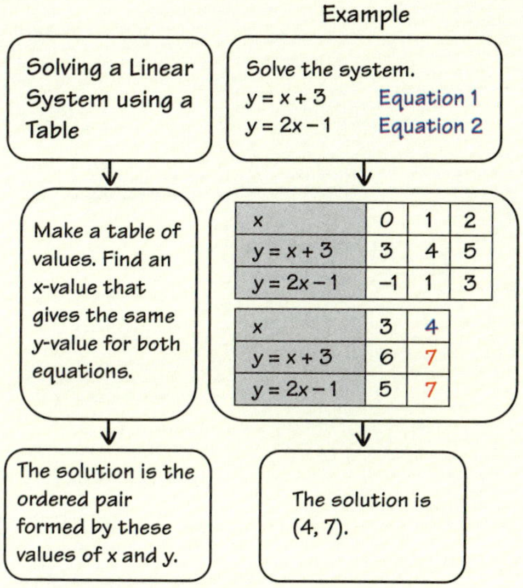

3b, c, 4, 5. Available at *BigIdeasMath.com*.

Reteaching and Enrichment Strategies

If students need help. . .	If students got it. . .
Resources by Chapter • Study Help • Practice A and Practice B • Puzzle Time Lesson Tutorials *BigIdeasMath.com* Practice Quiz Practice from the Test Generator	Resources by Chapter • Enrichment and Extension • School-to-Work Game Closet at *BigIdeasMath.com* Start the Chapter Review

Answers

1. (4, 0)
2. (4, 12)
3. See Additional Answers.
4. (1, 5)
5. (−6, 17)
6. one solution; $\left(\frac{3}{2}, 0\right)$
7. no solution
8. $x = -12$
9. $x = 1$
10. 6 three-point shots and 3 two-point shots
11. yes; Because a solution to $2x + 5 = 3x$ exists ($x = 5$).
12. 67°F

Technology For the Teacher

Answer Presentation Tool

Assessment Book

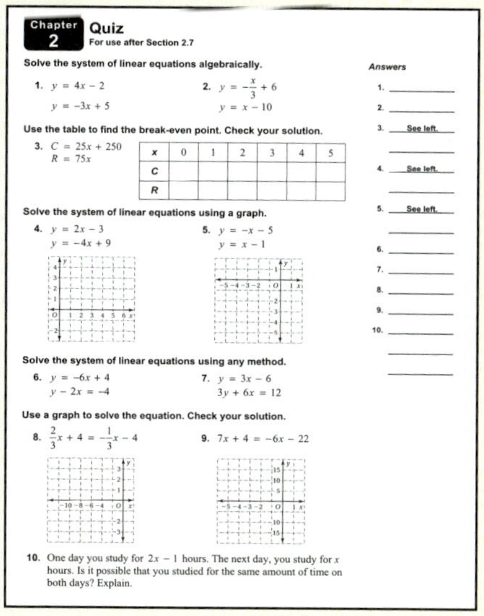

Chapter 2 **Quiz**
For use after Section 2.7

Solve the system of linear equations algebraically.

1. $y = 4x - 2$, $y = -3x + 5$
2. $y = -\frac{x}{3} + 6$, $y = x - 10$

Use the table to find the break-even point. Check your solution.

3. $C = 25x + 250$, $R = 75x$

x	0	1	2	3	4	5
C						
R						

Solve the system of linear equations using a graph.

4. $y = 2x - 3$, $y = -4x + 9$
5. $y = -x - 5$, $y = x - 1$

Solve the system of linear equations using any method.

6. $y = -6x + 4$, $y - 2x = -4$
7. $y = 3x - 6$, $3y + 6x = 12$

Use a graph to solve the equation. Check your solution.

8. $\frac{2}{3}x + 4 = -\frac{1}{3}x - 4$
9. $7x + 4 = -6x - 22$

10. One day you study for $2x - 1$ hours. The next day, you study for x hours. Is it possible that you studied for the same amount of time on both days? Explain.

Answers
1. ______
2. ______
3. See left.
4. See left.
5. See left.
6. ______
7. ______
8. ______
9. ______
10. ______

For the Teacher
Additional Review Options

- **Quiz***Show*
- Big Ideas Test Generator
- Game Closet at *BigIdeasMath.com*
- Vocabulary Puzzle Builder
- Resources by Chapter
  Puzzle Time
  Study Help

Answers

1. 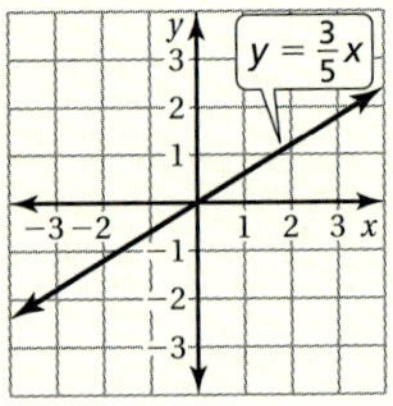

2. 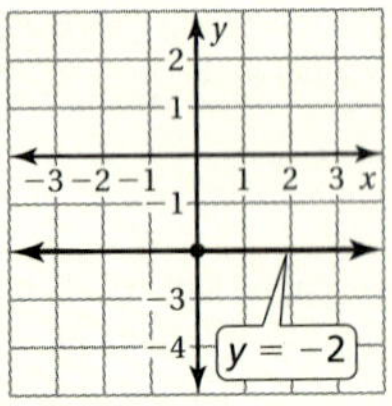

3. 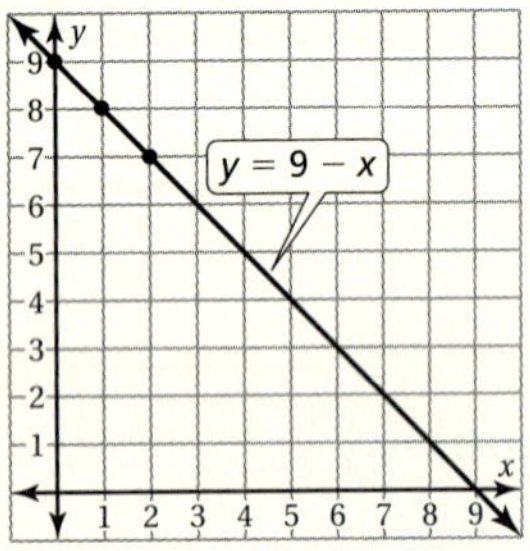

4. 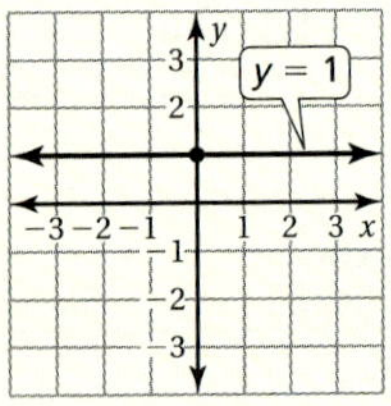

5. 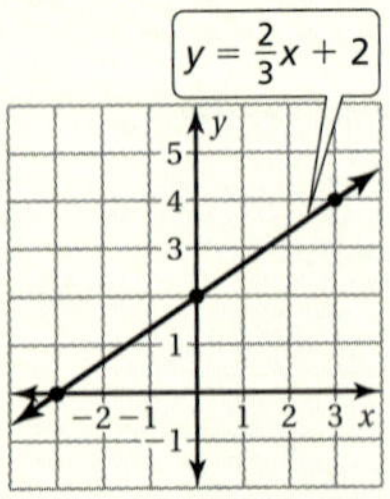

6. 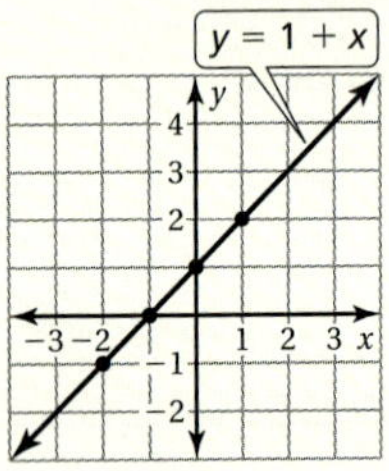

Review of Common Errors

Exercises 1–6

- Students may make a calculation error for one of the ordered pairs in a table of values. If they only find two ordered pairs for the graph, they may not recognize their mistake. Encourage them to find at least three ordered pairs when drawing a graph.

Exercises 2 and 4

- Students may draw a vertical line through a point on the x-axis instead of through the corresponding point on the y-axis. Remind them that the equation is a horizontal line. Ask them to identify the y-coordinate for several x-coordinates. For example, what is the y-coordinate for $x = 5$? $x = 6$? $x = -4$? Students should answer with the same y-coordinate each time.

2 Chapter Review

Check It Out
Vocabulary Help
BigIdeasMath.com

Review Key Vocabulary

linear equation *p. 50*
solution of a linear equation, *p. 50*
slope, *p. 56*
rise, *p. 56*
run, *p. 56*
x-intercept, *p. 64*
y-intercept, *p. 64*
slope-intercept form, *p. 64*
standard form, *p. 70*
system of linear equations, *p. 78*
solution of a system of linear equations, *p. 78*

Review Examples and Exercises

2.1 Graphing Linear Equations *(pp. 48–53)*

Graph $y = 3x - 1$.

Step 1: Make a table of values.

x	$y = 3x - 1$	y	(x, y)
-2	$y = 3(-2) - 1$	-7	$(-2, -7)$
-1	$y = 3(-1) - 1$	-4	$(-1, -4)$
0	$y = 3(0) - 1$	-1	$(0, -1)$
1	$y = 3(1) - 1$	2	$(1, 2)$

Step 2: Plot the ordered pairs.

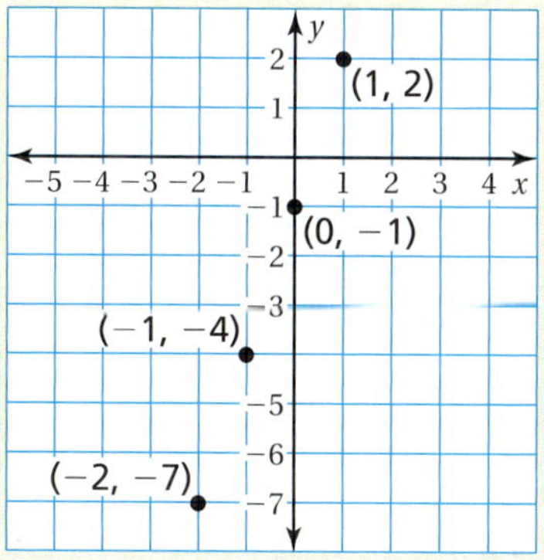

Step 3: Draw a line through the points.

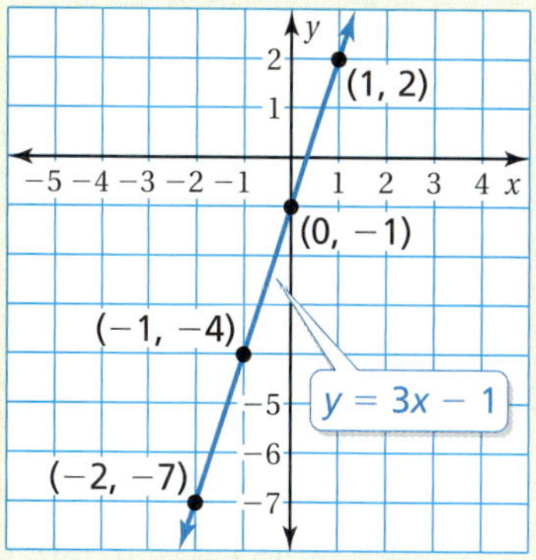

Exercises

Graph the linear equation.

1. $y = \frac{3}{5}x$

2. $y = -2$

3. $y = 9 - x$

4. $y = 1$

5. $y = \frac{2}{3}x + 2$

6. $y = 1 + x$

2.2 Slope of a Line (pp. 54–61)

Which two lines are parallel? Explain.

Find the slope of each line.

Red Line	*Blue Line*	*Green Line*
$\frac{-2}{6} = -\frac{1}{3}$	$\frac{-2}{4} = -\frac{1}{2}$	$\frac{-3}{9} = -\frac{1}{3}$

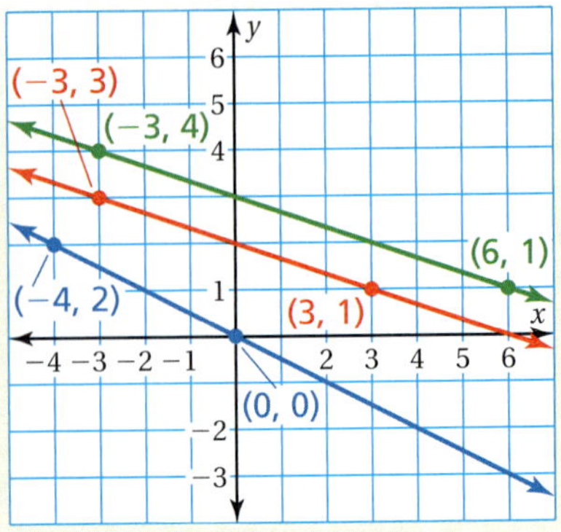

∴ The red and green lines have the same slope, so they are parallel.

Exercises

The points in the table lie on a line. Find the slope of the line. Then draw its graph.

7.

x	0	1	2	3
y	−1	0	1	2

8.

x	−2	0	2	4
y	3	4	5	6

2.3 Graphing Linear Equations in Slope-Intercept Form (pp. 62–67)

Graph $y = 0.5x - 3$. Identify the x-intercept.

Step 1: Find the slope and y-intercept.

$$y = 0.5x + (-3)$$

slope: 0.5; y-intercept: −3

Step 2: The y-intercept is −3. So, plot (0, −3).

Step 3: Use the slope to find another point and draw the line.

$$\text{slope} = \frac{\text{rise}}{\text{run}} = \frac{1}{2}$$

Plot the point that is 2 units right and 1 unit up from (0, –3). Draw a line through the two points.

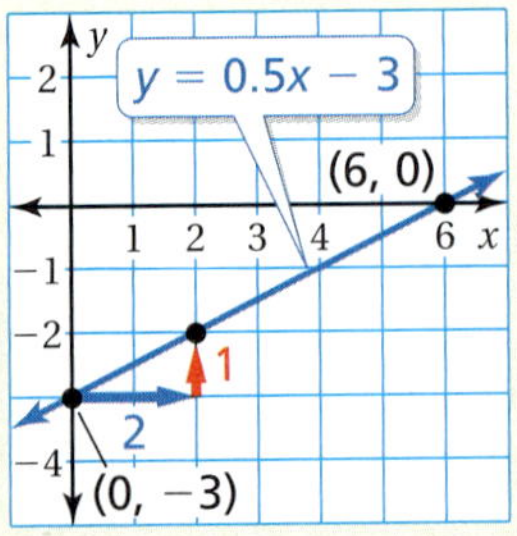

∴ The line crosses the x-axis at (6, 0). So, the x-intercept is 6.

Exercises

Graph the linear equation. Identify the x-intercept.

9. $y = 2x - 6$ **10.** $y = -4x + 8$ **11.** $y = -x - 8$

Review of Common Errors (continued)

Exercises 7 and 8

- Students may reverse the *x*- and *y*-coordinates when plotting the ordered pairs given in the table. Remind them that the first row is the *x*-coordinate and the second row is the *y*-coordinate. Encourage students to write the ordered pairs before graphing.

Exercises 9–11

- Students may forget to include negatives with the slope and/or *y*-intercept. Remind them to look at the sign in front of the slope and the *y*-intercept. Also remind students that the equation is $y = mx + b$. This means that if the linear equation has "minus *b*," then the *y*-intercept is negative.
- Students may use the reciprocal of the slope when graphing and may find an incorrect *x*-intercept. Remind them that slope is *rise* over *run*, so the numerator represents vertical change, not horizontal.

Answers

7. 1

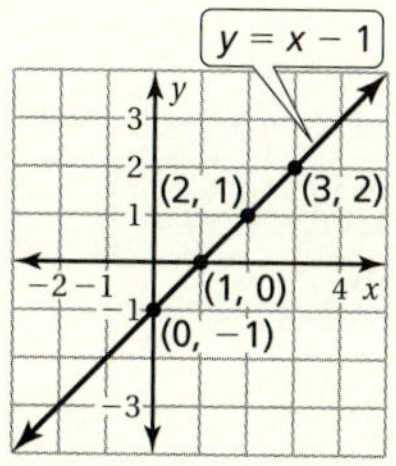

8. $\frac{1}{2}$

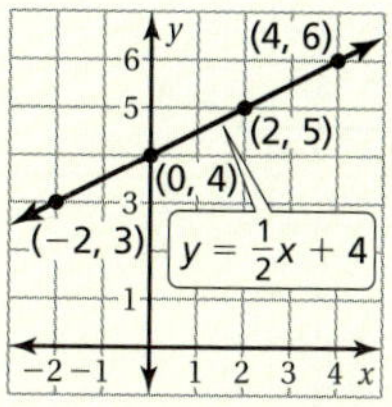

9.

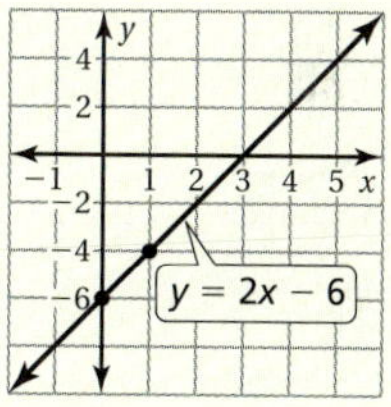

x-intercept: 3

10.

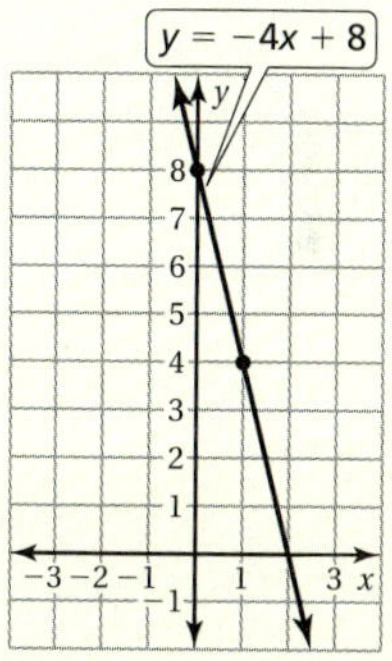

x-intercept: 2

11.

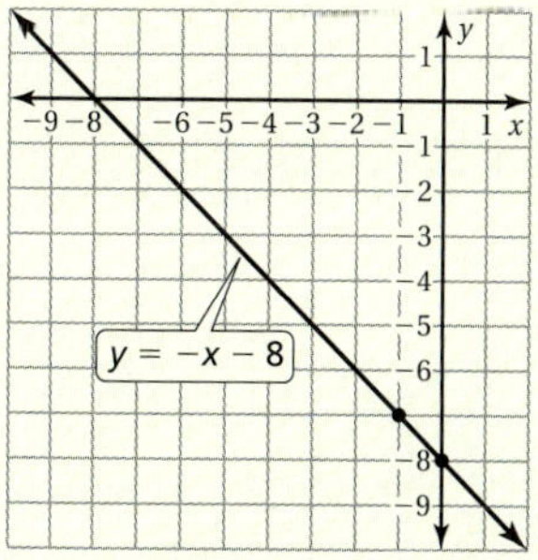

x-intercept: −8

Answers

12. 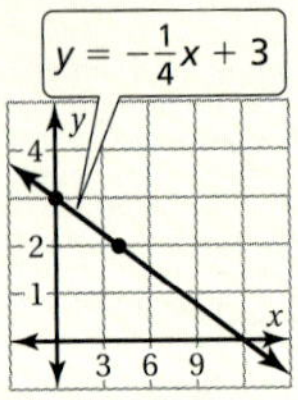

13. 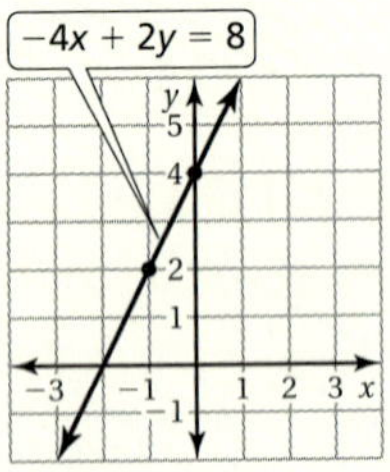

14. 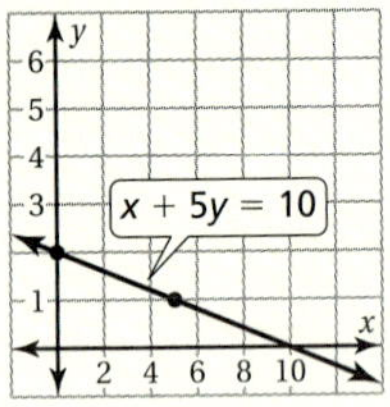

15. 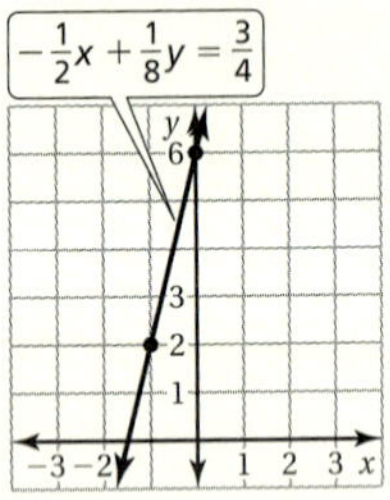

Review of Common Errors (continued)

Exercises 12–15

- Students may use the same operation instead of the inverse operation when rewriting the equation in slope-intercept form. Remind them of the steps to rewrite an equation.
- Students may mix up the *x*- and *y*-intercepts. Remind them that the *x*-intercept is the *x*-coordinate of where the line crosses the *x*-axis and the *y*-intercept is the *y*-coordinate of where the line crosses the *y*-axis.

Exercises 16–18

- Students may perform the same operation instead of the inverse operation when trying to get the variables on the same side. Remind them that whenever a variable or number is moved from one side of the equal sign to the other, the inverse operation is used.

Exercise 19

- Students may see that the slope is the same for both equations and immediately say that the system of linear equations has no solution. Remind them that they need to compare the slope *and* *y*-intercepts when determining the number of solutions. Encourage them to check algebraically and graphically that the system of linear equations has the number of solutions they found.

Exercise 20

- Students may make a mistake when solving algebraically or graphically. Encourage them to substitute the *x*-value into the original equation to check that the statement is true.
- Students may find the wrong intersection point when graphing because they did not draw straight lines. Encourage students to use a straightedge when graphing so that they can draw accurate lines. Remind them to check their solution in the original equation.

2.4 Graphing Linear Equations in Standard Form (pp. 68–73)

Graph $8x + 4y = 16$.

Step 1: Write the equation in slope-intercept form.

$8x + 4y = 16$	Write the equation.
$4y = -8x + 16$	Subtract $8x$ from each side.
$y = -2x + 4$	Divide each side by 4.

Step 2: Use the slope and y-intercept to plot two points.

$$y = -2x + 4$$

slope: -2; y-intercept: 4

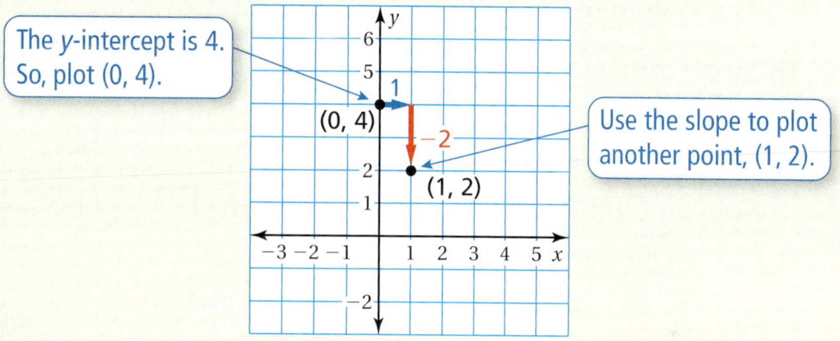

Step 3: Draw a line through the points.

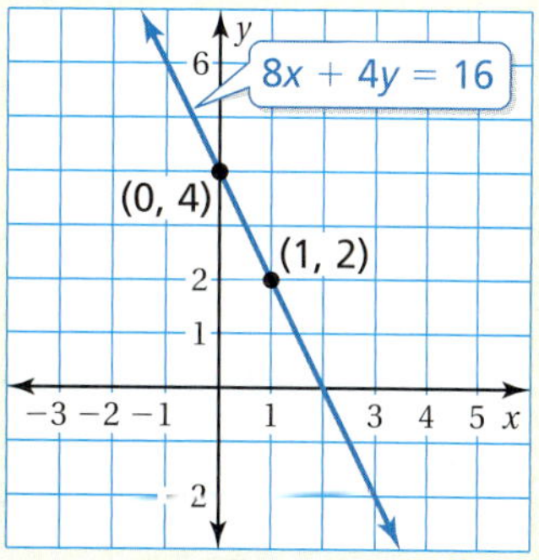

Exercises

Graph the linear equation.

12. $\frac{1}{4}x + y = 3$

13. $-4x + 2y = 8$

14. $x + 5y = 10$

15. $-\frac{1}{2}x + \frac{1}{8}y = \frac{3}{4}$

2.5 Systems of Linear Equations (pp. 76–81)

A middle school science club has 30 members. There are 12 more boys than girls. Use the models to write a system of linear equations. Then solve the system to find the number of boys *x* and the number of girls *y*.

Number of boys, x + Number of girls, y = 30

Number of boys, x = Number of girls, y + 12

The system is $x + y = 30$ and $x = y + 12$.

Step 1: Solve $x + y = 30$ for x.

$x = 30 - y$ — Subtract y from each side.

Step 2: Set the expressions equal to each other and solve for y.

$30 - y = y + 12$ — Set expressions equal to each other.

$18 = 2y$ — Subtract 12 from each side. Add y to each side.

$9 = y$ — Divide each side by 2.

Step 3: Substitute $y = 9$ into one of the original equations and solve for x.

$x = y + 12$ — Write one of the original equations.

$= 9 + 12$ — Substitute 9 for y.

$= 21$ — Add.

There are 21 boys and 9 girls in the science club.

Exercises

Solve the system of linear equations algebraically.

16. $x + y = 20$
$y = 2x - 1$

17. $x - y = 3$
$x + 2y = -6$

18. $2x + y = 8$
$3x + 2y = 30$

Review Game

Graphing Linear Equations

Big Ideas Game Closet

Materials per Group:

- map of the United States
- pencil
- straightedge

Directions:

On a map of the United States, students will place a coordinate plane with the origin located at Wichita, Kansas. The x-axis will go from -1700 miles to 1700 miles and the y-axis will go from -625 miles to 625 miles. These are roughly the dimensions of the United States.

The teacher will write equations and cities, in jumbled order, on the board. Students will work in groups and graph the equations to determine which line goes through which city.

Examples:

Dallas	$y = \frac{156}{50}x - 156$
Denver	$y = \frac{312}{625}x + 312$
Orlando	$y = \frac{100}{200}x + 100$
Chicago	$y = \frac{280}{600}x$
Las Vegas	$y = \frac{625}{1275}x - 625$

Who Wins?

The first group to correctly graph the lines and match the cities wins.

For the Student Additional Practice

- Lesson Tutorials
- Study Help (textbook)
- Student Website
  - Multi-Language Glossary
  - Practice Assessments

Answers

16. $(7, 13)$

17. $(0, -3)$

18. $(-14, 36)$

19. no solution

20. $x = -2$

My Thoughts on the Chapter

Teacher Tip

Not allowed to write in your teaching edition? Use sticky notes to record your thoughts.

What worked. . .

What did not work. . .

What I would do differently. . .

2.6 Special Systems of Linear Equations (pp. 82–87)

Solve the system.

$$y = x - 2 \qquad \text{Equation 1}$$

$$-3x + 3y = -6 \qquad \text{Equation 2}$$

Write $-3x + 3y = -6$ in slope-intercept form.

$-3x + 3y = -6$ Write the equation.

$3y = 3x - 6$ Add $3x$ to both sides.

$y = x - 2$ Divide each side by 3.

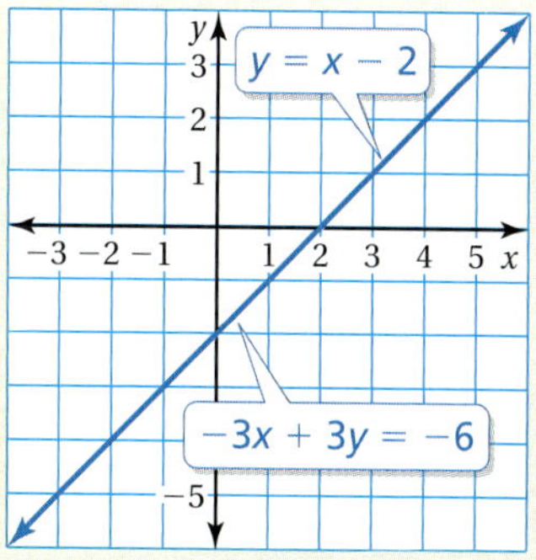

The equations are the same. The solution of the system is all the points on the line $y = x - 2$.

So, the system of linear equations has infinitely many solutions.

Exercises

19. Solve the system $y = -3x + 2$ and $6x + 2y = 10$.

2.7 Solving Equations by Graphing (pp. 88–93)

Solve $-\frac{2}{5}x - 10 = x + 4$ using a graph.

Step 1: Write a system of linear equations using each side of the equation.

$$y = -\frac{2}{5}x - 10 \rightarrow -\frac{2}{5}x - 10 = x + 4 \leftarrow y = x + 4$$

Step 2: Graph the system.

$$y = -\frac{2}{5}x - 10$$

$$y = x + 4$$

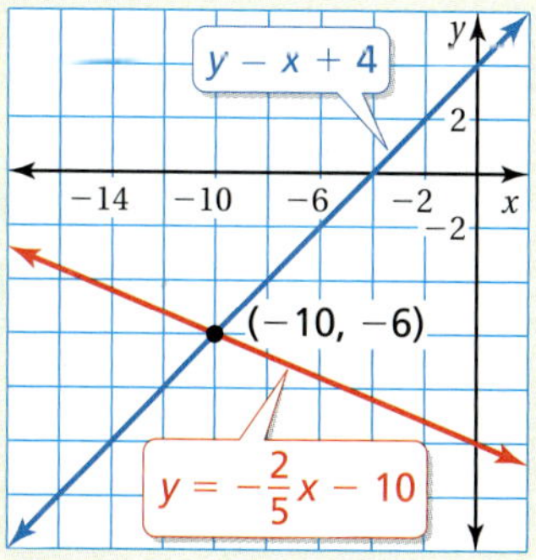

The graphs intersect at $(-10, -6)$.
So, the solution is $x = -10$.

Exercises

20. Use a graph to solve the equation $6x + 3 = 3x - 3$.

2 Chapter Test

Check It Out
Test Practice
BigIdeasMath.com

Find the slope and y-intercept of the graph of the linear equation.

1. $y = 6x - 5$

2. $y = 20x + 15$

3. $y = -5x - 16$

4. $y - 1 = 3x + 8.4$

5. $y + 4.3 = 0.1x$

6. $-\frac{1}{2}x + 2y = 7$

Graph the linear equation.

7. $y = 2x + 4$

8. $y = -\frac{1}{2}x - 5$

9. $-3x + 6y = 12$

10. Which two lines are parallel? Explain.

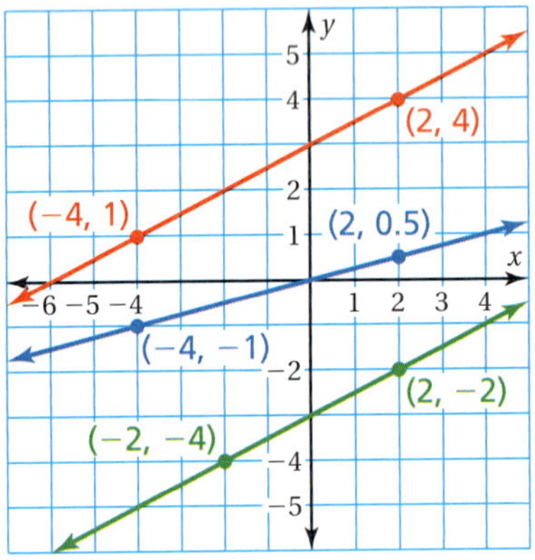

11. The points in the table lie on a line. Find the slope of the line. Then draw its graph.

x	y
−1	−4
0	−1
1	2
2	5

12. Solve the system of linear equations using any method.

$$2x - y = 16$$
$$y = 12 + x$$

13. Use a graph to solve the equation. Check your solution.

$$2x + 6 = 4x - 12$$

14. **MATH TEST** A math class has a test today. There are 30 problems on the test. The test has two types of problems: multiple choice problems and word problems. The multiple choice problems are worth 2 points and the word problems are worth 5 points. The teacher says there are a total of 75 points possible. How many multiple choice and word problems are on the test?

Number of multiple choice problems, x + Number of word problems, y = 30

Point value of multiple choice problems • Number of multiple choice problems, x + Point value of word problems • Number of word problems, y = 75

Test Item References

Chapter Test Questions	Section to Review
7, 8	2.1
10, 11	2.2
1–6	2.3
9	2.4
14	2.5
12	2.6
13	2.7

Test-Taking Strategies

Remind students to quickly look over the entire test before they start so that they can budget their time. Students should jot down the formula for the slope-intercept form of a linear equation on the back of their test before they begin, so they remember the correct sign for the *y*-intercept. Teach students to use the Stop and Think strategy before answer answering. **Stop** and carefully read the question, and **Think** about what the answer should look like.

Common Assessment Errors

- **Exercises 1–6** Students may use the reciprocal of the slope when graphing and may find an incorrect *x*-intercept. Remind them that slope is *rise* over *run*, so the numerator represents vertical change, not horizontal.
- **Exercises 7–9** Students may make a calculation error for one of the ordered pairs in a table of values. If they only find two ordered pairs for the graph, they may not recognize their mistake. Encourage them to find at least three ordered pairs when drawing a graph.
- **Exercises 12 and 13** Students may reverse the *x*- and *y*-coordinates when plotting the ordered pairs given in the table. Remind them that the first row is the *x*-coordinate and the second row is the *y*-coordinate. Encourage students to write the ordered pairs before graphing.
- **Exercise 14** Students may perform the same operation instead of the inverse operation when trying to get the variables on the same side. Remind them that whenever a variable or number is moved from one side of the equal sign to the other, the inverse operation is used.

Reteaching and Enrichment Strategies

If students need help. . .	If students got it. . .
Resources by Chapter • Practice A and Practice B • Puzzle Time Record and Practice Journal Practice Differentiating the Lesson Lesson Tutorials Practice from the Test Generator Skills Review Handbook	Resources by Chapter • Enrichment and Extension • School-to-Work • Financial Literacy • Life Connections Game Closet at *BigIdeasMath.com* Start Standardized Test Practice

Answers

1. slope: 6; *y*-intercept: -5
2. slope: 20; *y*-intercept: 15
3. slope: -5; *y*-intercept: -16
4. slope: 3; *y*-intercept: 9.4
5. slope: 0.1; *y*-intercept: -4.3
6. slope: $\frac{1}{4}$; *y*-intercept: $\frac{7}{2}$

7–9. See Additional Answers.

10. red and green; They both have a slope of $\frac{1}{2}$.
11. 3

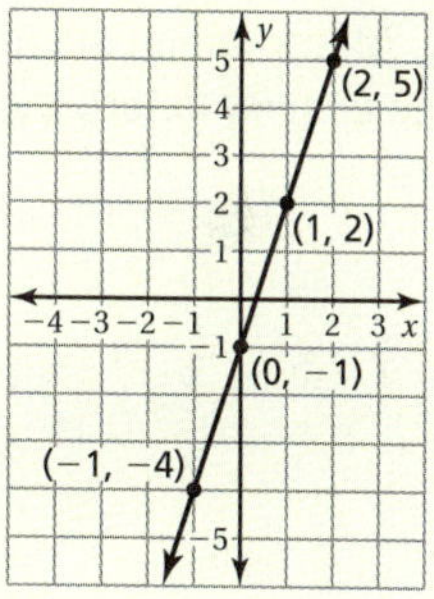

12. (28, 40)
13. $x = 9$
14. 25 multiple choice problems and 5 word problems

Assessment Book

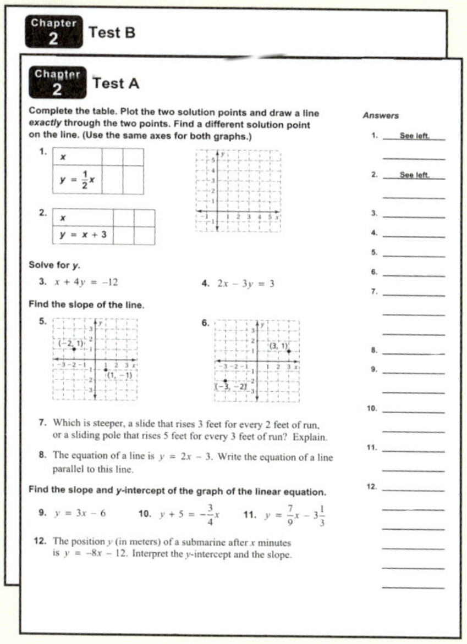

Test-Taking Strategies

Available at *BigIdeasMath.com*

After Answering Easy Questions, Relax
Answer Easy Questions First
Estimate the Answer
Read All Choices before Answering
Read Question before Answering
Solve Directly or Eliminate Choices
Solve Problem before Looking at Choices
Use Intelligent Guessing
Work Backwards

About this Strategy

When taking a multiple choice test, be sure to read each question carefully and thoroughly. After reading the question, estimate the answer before trying to solve.

Answers

1. C
2. F
3. B
4. H

Technology For the Teacher

Big Ideas Test Generator

Item Analysis

1. **A.** The student misreads the graph, either going across from 60 or going down from the corresponding point on the graph.
   B. The student misreads the graph, either going across from 60 or going down from the corresponding point on the graph.
   C. Correct answer
   D. The student misreads the graph, either going across from 60 or going down from the corresponding point on the graph.
2. **F.** Correct answer
   G. The student reads the slope correctly, but uses the wrong point to identify the *y*-intercept.
   H. The student reads the *y*-intercept correctly, but miscalculates the slope.
   I. The student finds the slope and *y*-intercept incorrectly.
3. **A.** The student multiplies correctly by 60, but then divides by 8 instead of 4.
   B. Correct answer
   C. The student multiplies correctly by 60, but then fails to convert to gallons.
   D. The student multiplies correctly by 60, but then multiplies by 4 instead of dividing.
4. **F.** The student interchanges correct values for *x* and *y*.
   G. The student makes two errors: interchanging *x* and *y*, and assigning a negative sign incorrectly.
   H. Correct answer
   I. The student makes a mistake with a correct solution (4, 2) and assigns a negative value to 2, forgetting that there is already a minus sign in the equation.
5. **A.** The student misunderstands the meaning/roles of slope or *y*-intercept, or does not understand what the graphical implications are of a pair of lines having no solution.
   B. Correct answer
   C. The student misunderstands the meaning/roles of slope or *y*-intercept, or does not understand what the graphical implications are of a pair of lines having no solution.
   D. The student misunderstands the meaning/roles of slope or *y*-intercept, or does not understand what the graphical implications are of a pair of lines having no solution.

2 Standardized Test Practice

"Using estimation you can see that there are about 300 hairs. So, it's got to be C."

1. The graph below shows the value of United States dollars compared to Guatemalan quetzals.

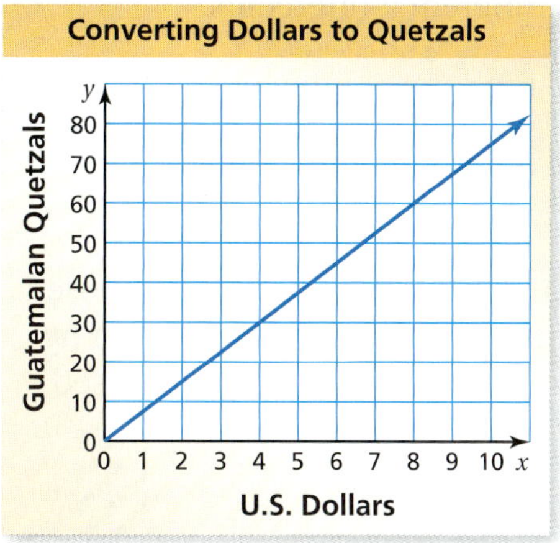

What is the value of 60 quetzals?

A. \$6 **C.** \$8

B. \$7 **D.** \$9

2. Which equation matches the line shown in the graph?

F. $y = 2x - 2$

G. $y = 2x + 1$

H. $y = x - 2$

I. $y = x + 1$

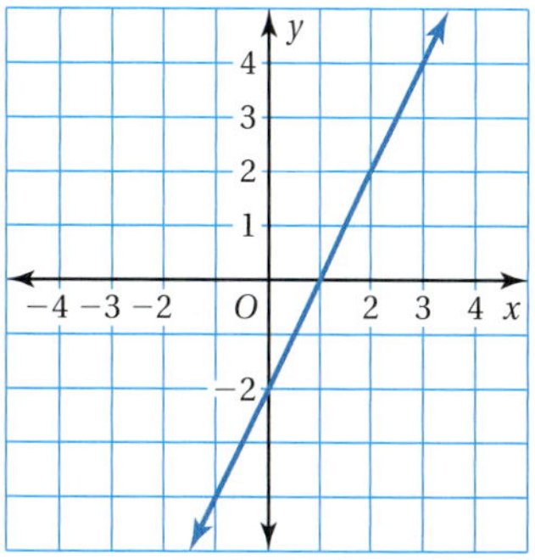

3. A faucet releases 6 quarts of water per minute. How many gallons of water will the faucet release in one hour?

A. 45 gal **C.** 360 gal

B. 90 gal **D.** 1440 gal

4. The equation $6x - 5y = 14$ is written in standard form. Which point lies on the graph of this equation?

F. $(-4, -1)$ **H.** $(-1, -4)$

G. $(-2, 4)$ **I.** $(4, -2)$

5. A system of two linear equations has no solutions. What can you conclude about the graphs of the two equations?

A. The lines have the same slope and the same y-intercept.

B. The lines have the same slope and different y-intercepts.

C. The lines have different slopes and the same y-intercept.

D. The lines have different slopes and different y-intercepts.

6. A cell phone plan costs \$10 per month plus \$0.10 for each minute used. Last month, you spent \$18.50 using this plan. This can be modeled by the equation below, where m represents the number of minutes used.

$$0.1m + 10 = 18.5$$

How many minutes did you use last month?

F. 8.4 min

G. 85 min

H. 185 min

I. 285 min

7. What is the slope of the line that passes through the points $(2, -2)$ and $(8, 1)$?

Think Solve Explain

8. It costs \$40 to rent a car for one day. In addition, the rental agency charges you for each mile driven as shown in the graph.

Part A Determine the slope of the line joining the points on the graph.

Part B Explain what the slope represents.

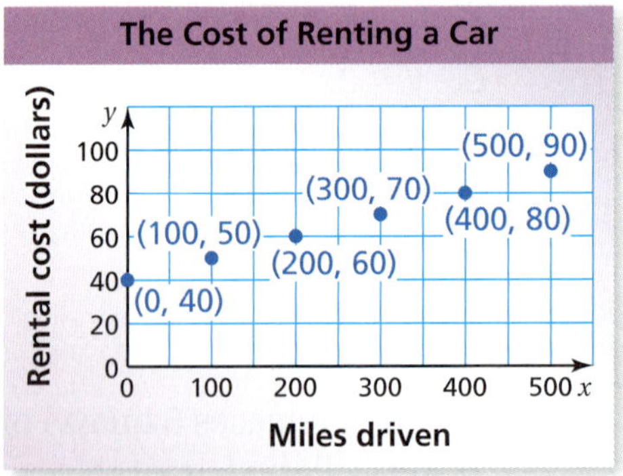

9. Which point is a solution of the system of equations shown below?

$$x + 3y = 10$$

$$x = 2y - 5$$

A. $(1, 3)$

B. $(3, 1)$

C. $(55, -15)$

D. $(-35, -15)$

Item Analysis (continued)

6. **F.** The student correctly subtracts 10 from both sides, but then subtracts 0.1 instead of dividing.

   G. Correct answer

   H. The student ignores 10 and simply divides by 0.1.

   I. The student adds 10 to both sides, and then divides by 0.1.

7. **Gridded Response:** Correct answer: 0.5, or $\frac{1}{2}$

   Common Error: The student performs subtraction incorrectly for the *y*-terms, yielding an answer of $\frac{1}{6}$ or $-\frac{1}{6}$.

8. **2 points** The student demonstrates a thorough understanding of the slope of a line and what it represents, explains the work fully, and calculates the slope accurately. The slope of the line is $\frac{50-40}{100-0} = \frac{10}{100} = \frac{1}{10} = 0.10$.
   The slope represents the rental cost per mile driven, \$0.10 per mile.

   1 point The student's work and explanations demonstrate a lack of essential understanding. The formula for the slope of a line is misstated, or the student incorrectly states what the slope of the line represents.

   0 points The student provides no response, a completely incorrect or incomprehensible response, or a response that demonstates insufficient understanding of the slope of a line and what it represents.

9. **A.** Correct answer

   B. The student interchanges the roles of *x* and *y* when writing the ordered pair.

   C. The student incorrectly sets *x* equal to $3y + 10$ in the first equation, substitutes for *x* in the second equation, and then solves for *y*, yielding -15. An *x*-value of 55 then results from substituting into the original first equation.

   D. The student incorrectly sets *x* equal to $3y + 10$ in the first equation, substitutes for *x* in the second equation, and then solves for *y*, yielding -15. An *x*-value of -35 then results from substituting into the original second equation.

Answers

5. B

6. G

7. $\frac{1}{2}$

8. *Part A* 0.10

   Part B \$0.10 per mile

9. A

Answers

10. I

11. B

12. H

13. 180 kg

14. C

Answer for Extra Example

1. **Gridded Response:** Correct answer: 1.25

   Common Error: The student does not properly distribute the factor 3 in the right hand side of the equation, yielding an incorrect answer of $x = 1$.

Item Analysis (continued)

10. **F.** The student mistakes slope for meaning that a line passes through (0, 0).

G. The student mistakes a vertical line for zero slope.

H. The student mistakes slope for meaning that a line passes through (0, 0).

I. Correct answer

11. **A.** The student divides M by 3, but fails to divide $(K + 7)$ by 3.

B. Correct answer

C. The student divides K by 3, but fails to divide 7 by 3.

D. The student subtracts 7 instead of adding it to both sides.

12. **F.** The student picks the coefficient of x from the equation, failing to convert to slope-intercept form.

G. The student divides properly by 2, but attaches an incorrect sign since x initially has a positive coefficient.

H. Correct answer

I. The student correctly moves $5x$ to the opposite side of the equation, but fails to divide by 2.

13. **Gridded Response:** Correct answer: 180 kg

Common Error: The student does not know the correct conversion rate from grams to kilograms, yielding an answer of 1800 or 18,000.

14. **A.** The student incorrectly assumes that adding 4 to the y-value of a point is the same as saying the slope is 4. Correct point would have 1 added to a as well.

B. The student incorrectly assumes that because the ratio $8 : 2 = 4$, the point must lie on the line of slope 4.

C. Correct answer

D. The student calculates the slope correctly and gets $4 \cdot \frac{b}{a}$, but then assumes this fits the initial problem because its point was (a, b) and its slope was 4.

Extra Example for Standardized Test Practice

1. What value of x makes the equation shown below true?

$$11x - 7 = 3(x + 1)$$

10. Which line has a slope of 0?

F.

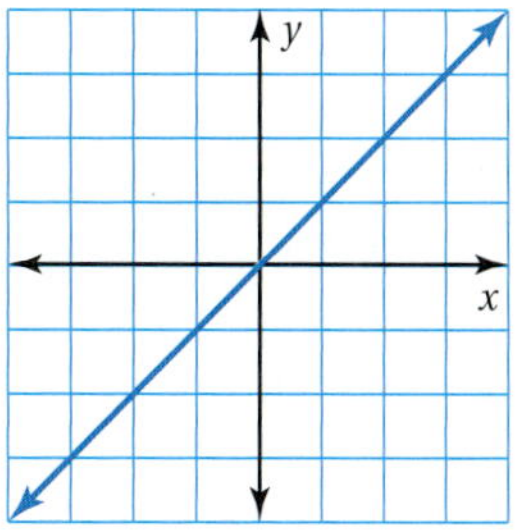

H.

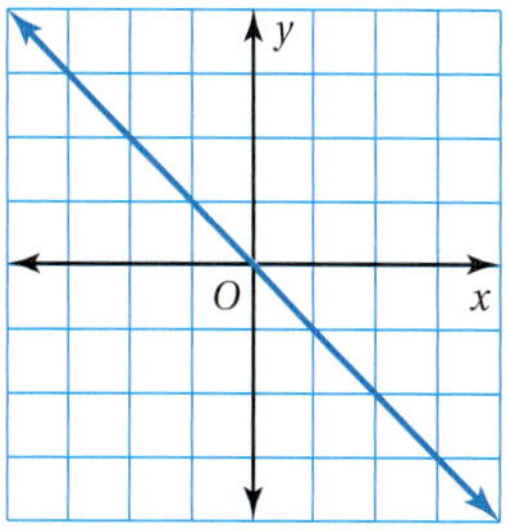

G.

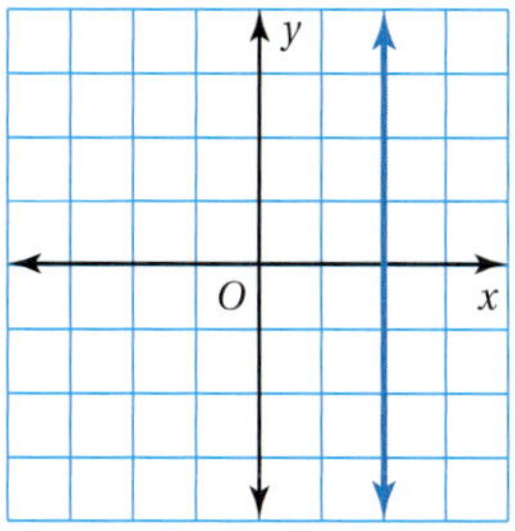

I.

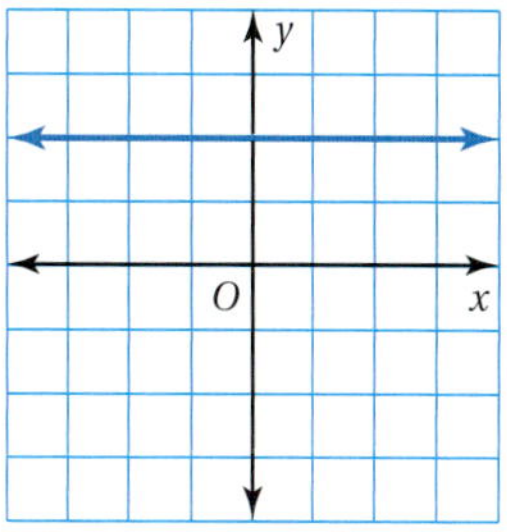

11. Solve the formula $K = 3M - 7$ for M.

A. $M = K + 7$

B. $M = \frac{K + 7}{3}$

C. $M = \frac{K}{3} + 7$

D. $M = \frac{K - 7}{3}$

12. The linear equation $5x + 2y = 10$ is written in standard form. What is the slope of the graph of this equation?

F. 5

G. 2.5

H. -2.5

I. -5

13. A package of breakfast cereal is labeled "750 g." This cereal is shipped in cartons that hold 24 packages. What is the total mass, in kilograms, of the breakfast cereal in 10 cartons?

14. A line has a slope of 4 and passes through the point (a, b). Which point must also lie on this line?

A. $(a, b + 4)$

B. $(2a, 8b)$

C. $(a + 1, b + 4)$

D. $(2a, 5b)$

3 Writing Linear Equations and Linear Systems

Connections to Previous Learning

- Write, solve, and graph one-step and two-step linear equations.
- Construct and analyze tables, graphs, and equations to describe linear functions and other simple relations.

- Use properties of equality to rewrite an equation and to show two equations are equivalent.

- Construct and analyze tables, graphs, and models to describe linear equations.
- Interpret slope and x- and y-intercepts when graphing a linear equation for a real world problem.
- Use tables, graphs, and models to represent, analyze, and solve real-world problems related to systems of linear equations.

Math in History

The number pi was recognized independently by more than one ancient culture, including the Babylonian and Egyptian cultures.

★ The Greek mathematician Archimedes (about 250 B.C.) approximated pi with two regular polygons, one inscribed in a circle and one circumscribed about the circle. His approximation was 3.1429.

★ The Chinese mathematician Liu Hui (about 220 – about 280 A.D.) based his approximation of pi on a regular polygon with 3072 sides. His approximation was 3.14159.

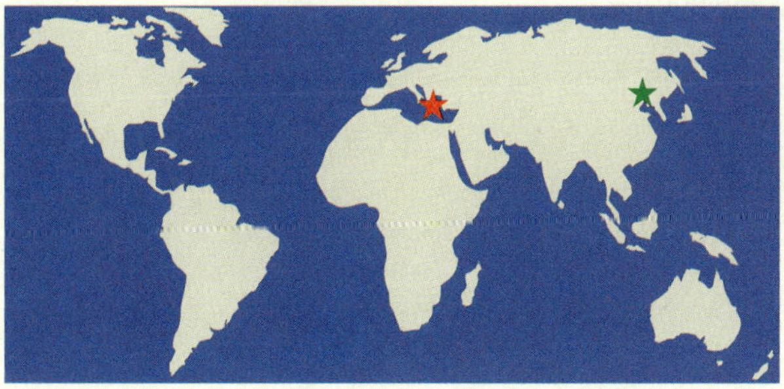

Pacing Guide for Chapter 3

Chapter Opener	1 Day
Section 1 Activity Lesson	 1 Day 1 Day
Section 2 Activity Lesson	 1 Day 2 Days
Section 3 Activity Lesson	 1 Day 1 Day
Study Help / Quiz	1 Day
Section 4 Activity Lesson	 1 Day 2 Days
Section 5 Activity Lesson	 1 Day 1 Day
Quiz / Chapter Review	1 Day
Chapter Test	1 Day
Standardized Test Practice	1 Day
Total Chapter 3	17 Days
Year-to-Date	56 Days

Check Your Resources

- Record and Practice Journal
- Resources by Chapter
- Skills Review Handbook
- Assessment Book
- Worked-Out Solutions

Technology For the Teacher

Dynamic Classroom

The Dynamic Planning Tool
Editable Teacher's Resources at
BigIdeasMath.com

Additional Topics for Review

- Key words (example: of means multiply, is means equals, etc.)
- Simplifying fractions
- Converting between percents and decimals
- Least Common Multiple
- Greatest Common Factor

Try It Yourself

1. $\frac{14}{45}$
2. $\frac{20}{27}$
3. $\frac{3}{8}$
4. $\frac{3}{5}$
5. $3.30
6. $1.85

Record and Practice Journal

1. $\frac{4}{7}$
2. $\frac{2}{15}$
3. $\frac{3}{4}$
4. $\frac{49}{18}$
5. 20 crayons
6. 6 muffins
7. $1.43
8. $1.78
9. 72 questions
10. 6 students

Math Background Notes

Vocabulary Review

- Numerator
- Denominator
- Reciprocal
- Percent

Multiplying and Dividing Fractions

- Students should know how to multiply and divide fractions.
- Remind them that multiplying and dividing fractions does not require the use of a common denominator.
- Remind students that any integer can be made into a fraction. Place the integer in the numerator and place 1 in the denominator.
- To multiply fractions, simply multiply the numerators and multiply the denominators. Simplify the resulting fraction.
- To divide two fractions, keep the first and multiply by the reciprocal of the second.

Using Percents

- Students should be able to work with percents and apply that knowledge to real-life problems.
- Remind students that the word percent means parts per one hundred.
- **Common Error:** Students may forget to convert the given percent into a decimal before multiplying by the total. Remind them that the percent symbol is a convenient notation, but in order to work with percents they should be converted into decimal form.
- **Teaching Tip:** Examples 3 and 4 could encourage discussions about common life skills within your classroom. If time permits, you could discuss sales tax with students. Are all items taxed? Is your state's sales tax higher or lower than neighboring states? What is a fair tip to leave at a restaurant? Are there other situations in which you should tip (at the beauty shop for example)? Such discussions could prove to be very beneficial to students. Some students, especially non-native students, may not have previously encountered this information.

Reteaching and Enrichment Strategies

If students need help. . .	If students got it. . .
Record and Practice Journal • Fair Game Review Skills Review Handbook Lesson Tutorials	Game Closet at *BigIdeasMath.com* Start the next section

What You Learned Before

Multiplying and Dividing Fractions

A container of apple juice contains 16 cups.

Example 1 **You drink $\frac{3}{8}$ of the juice. How many cups of juice did you drink?**

$$16 \cdot \frac{3}{8} = \frac{16}{1} \cdot \frac{3}{8}$$

$$= \frac{\overset{2}{\cancel{16}} \cdot 3}{1 \cdot \underset{1}{\cancel{8}}}$$

$$= 6$$

∴ You drank 6 cups of juice.

Example 2 **A serving of juice is $\frac{4}{5}$ cup. How many servings are in the container?**

$$16 \div \frac{4}{5} = \frac{16}{1} \div \frac{4}{5}$$

$$= \frac{16}{1} \cdot \frac{5}{4}$$

$$= \frac{\overset{4}{\cancel{16}} \cdot 5}{1 \cdot \underset{1}{\cancel{4}}}$$

$$= 20$$

∴ There are 20 servings in the container.

Try It Yourself

Evaluate the expression.

1. $\frac{7}{15} \cdot \frac{2}{3}$

2. $\frac{5}{6} \cdot \frac{8}{9}$

3. $\frac{1}{7} \div \frac{8}{21}$

4. $\frac{9}{20} \div \frac{3}{4}$

Using Percents

Example 3 **Sales tax is 6%. What is the sales tax on an item that costs $43?**

What is 6% of 43?

$43 \cdot 0.06 = 2.58$

∴ The sales tax is $2.58.

Example 4 **Your bill at a restaurant is $21.25. What is the amount of a 20% tip on the bill?**

What is 20% of $21.25?

$21.25 \cdot 0.2 = 4.25$

∴ The amount of the tip is $4.25.

Try It Yourself

5. Sales tax is 6%. What is the sales tax on an item that costs $55?

6. Your bill at a restaurant is $12.30. What is the amount of a 15% tip on the bill?

Writing Equations in Slope-Intercept Form

Essential Question How can you write an equation of a line when you are given the slope and *y*-intercept of the line?

1 ACTIVITY: Writing Equations of Lines

Work with a partner.

- **Find the slope of each line.**
- **Find the *y*-intercept of each line.**
- **Write an equation for each line.**
- **What do the three lines have in common?**

a.

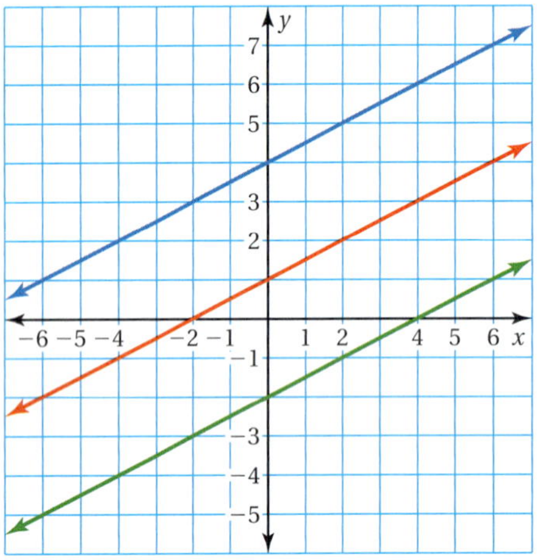

b.

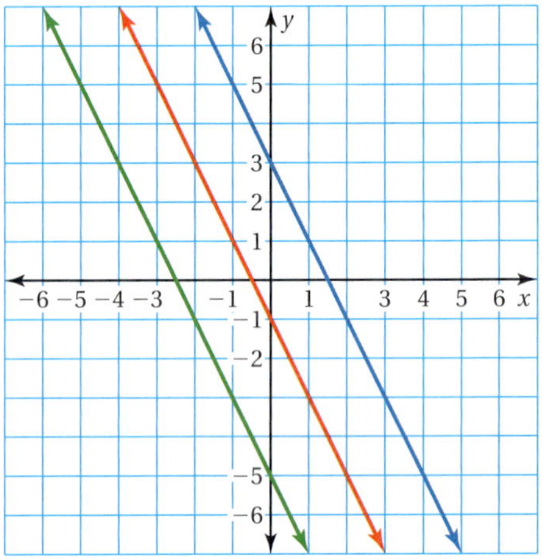

c.

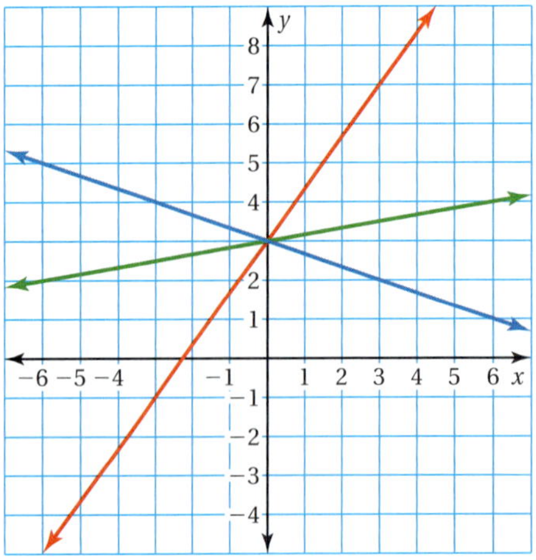

d.

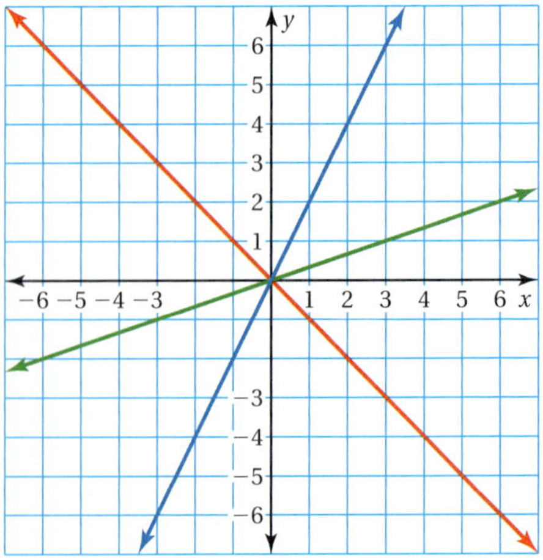

Laurie's Notes

Introduction

For the Teacher

- **Goal:** Students will determine the slope and y-intercept of a line by examining a graph. They will also write an equation in slope-intercept form.

Motivate

- If there is sufficient space in your classroom, hallway, or school foyer, make coordinate axes using masking tape. Use a marker to scale each axis with integers -5 through 5.
- Take turns having two students be the *rope anchors* who then will make a line on the coordinate axes while other students observe.
- Here are a series of directions you can give and some follow-up questions. Remind students that slope is rise over run and that the equation of a line in slope-intercept form is $y = mx + b$.
  - ? Make the line $y = x$. "What is the slope?" 1 "What is the y-intercept?" 0
  - ? Keep the same slope, but make the y-intercept 2. "What is the equation of this line?" $y = x + 2$
  - ? Use the y-intercept 2, but make the slope steeper. "What is the slope of this line?" Answers will vary.
  - ? Keep the same y-intercept, but make the slope $\frac{1}{2}$. "What is the equation?" $y = \frac{1}{2}x + 2$
- **Management Tip:** This activity can also be done by drawing the axes on the board and having the students hold the rope against the board.

Activity Notes

Activity 1

- ? "How do you determine the slope of a line drawn in a coordinate plane?" Use two points that you are sure are on the graph and find the rise and run between the points.
- ? "Does it matter whether you move left-to-right or right-to-left when you're finding the rise and run? Explain." No; Either way the slope will be the same.
- Students may have difficulty writing the equation in slope-intercept form. They think it should be harder to do!
- **FYI:** When the y-intercept is negative, students may leave their equation as $y = 3x + (-4)$ instead of $y = 3x - 4$. Remind students that it is more common to represent the equation as $y = 3x - 4$.
- **Teaching Tip:** If you have a student that is color blind, refer to the lines by a number or letter scheme (1, 2, 3 or A, B, C).
- Ask students to share what they found in common for each trio of lines.

Previous Learning

Students should know how to find the slope of a line. Students should know about parallel lines.

Activity Materials
Introduction
• masking tape • rope or yarn

Start Thinking! and Warm Up

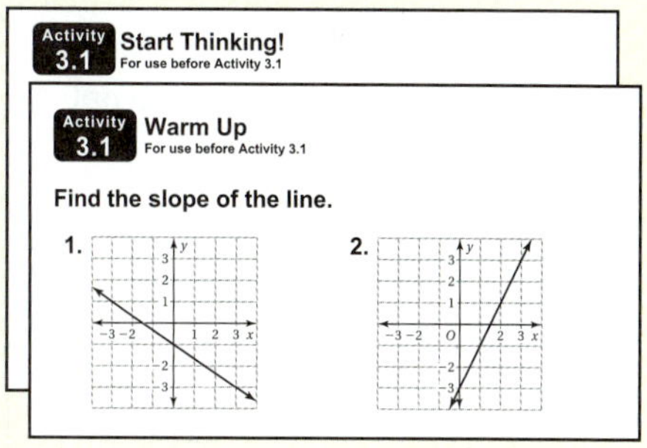

Activity 3.1 Start Thinking! For use before Activity 3.1

Activity 3.1 Warm Up For use before Activity 3.1

Find the slope of the line.

1.

2.

3.1 Record and Practice Journal

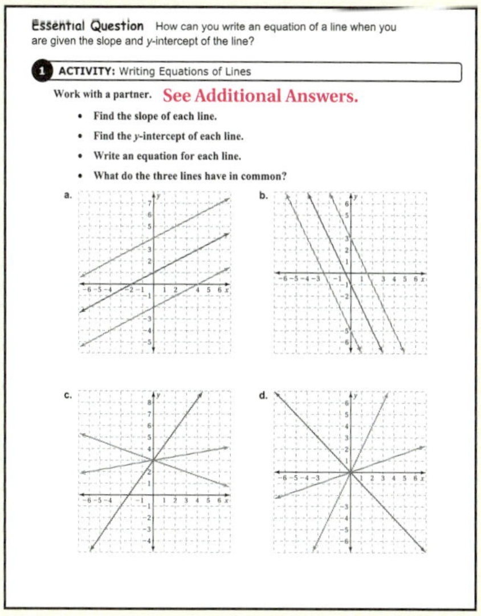

Essential Question How can you write an equation of a line when you are given the slope and y-intercept of the line?

1 ACTIVITY: Writing Equations of Lines

Work with a partner. See Additional Answers.

- Find the slope of each line.
- Find the y-intercept of each line.
- Write an equation for each line.
- What do the three lines have in common?

a. b. c. d.

Differentiated Instruction

Visual

To avoid mistakes when substituting the variables, have students color code the slope and y-intercept of an equation.

slope: 3 y-intercept: 4

$y = mx + b$

$y = 3x + 4$

3.1 Record and Practice Journal

2 ACTIVITY: Describing a Parallelogram

Work with a partner.

- Find the area of each parallelogram.
- Write an equation for each side of each parallelogram.
  See Additional Answers.
- What do you notice about the slopes of the opposite sides of each parallelogram?
  Opposite sides have the same slope.

a.

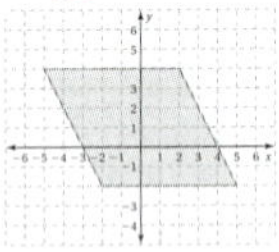

42 square units

b.

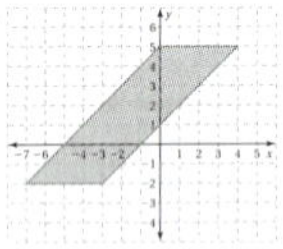

28 square units

3 ACTIVITY: Interpreting the Slope and y-Intercept

Work with a partner. The graph shows a trip taken by a car where t is the time (in hours) and y is the distance (in miles) from Phoenix.

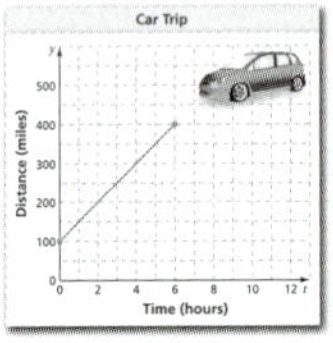

a. How far from Phoenix was the car at the beginning of the trip?
   100 mi

b. What was the car's speed?
   50 mph

c. How long did the trip last?
   6 hours

d. How far from Phoenix was the car at the end of the trip?
   400 mi

What Is Your Answer?

4. **IN YOUR OWN WORDS** How can you write an equation of a line when you are given the slope and y-intercept of the line? Give an example that is different from those in Activities 1, 2, and 3.
   $y = mx + b$ with m as the slope and b as the y-intercept

Laurie's Notes

Activity 2

- ? "How do you find the area of a parallelogram?" area = base × height
- ? "Are the base and height the sides of the parallelogram?" They could be if it's a rectangle. Otherwise, height is the perpendicular distance between the two bases.
- To find the base and height, students simply count the units on the diagram. Note that the height for the parallelogram in part (b) is outside the parallelogram.
- **Common Error:** The slope of the horizontal sides is zero. Students may say that you cannot find the slope *for a flat line.*
- ? "What is the equation of a horizontal line?" $y = b$
- The challenge in this activity is writing the equations for the diagonal sides of the figure in part (a). Suggest that by extending the sides using the slope, the students should be able to determine the y-intercept. For example: the slopes of the two diagonal sides are -2. This means a rise of -2 and a run of 1. Start at a vertex of one of the sides and use the slope to extend the side to the y-axis. Repeat the process for the other side.
- This activity reviews positive, negative, and zero slope.

Activity 3

- If students have difficulty getting started with this activity, remind them to read the labels on the axes. Another hint is to ask them how to interpret the y-intercept. The car was 100 miles from Phoenix at the beginning of the trip.
- Discuss answers to each part of the problem as a class.
- ? **Extension:** Draw the segment from (6, 400) to (12, 0) and explain that this represents the return trip. Ask the following questions.
  - "What is the slope of this line segment? What does the slope mean in the context of the problem?" slope ≈ -67; returning at a rate of about 67 mi/h
  - "What does the point (12, 0) mean in the context of the problem?" You have arrived in Phoenix.
  - "What would the graph look like if the car had stopped for 1 hour?" horizontal segment of length 1 unit

Closure

- **Exit Ticket:** What is the slope and y-intercept of the equation $y = 2x + 4$? slope = 2, y-intercept = 4 Write the equation of a line with a slope of 3 and a y-intercept of 1. $y = 3x + 1$

Technology For the Teacher

Dynamic Classroom

The Dynamic Planning Tool
Editable Teacher's Resources at *BigIdeasMath.com*

2 ACTIVITY: Describing a Parallelogram

Work with a partner.

- **Find the area of each parallelogram.**
- **Write an equation for each side of each parallelogram.**
- **What do you notice about the slopes of the opposite sides of each parallelogram?**

a.

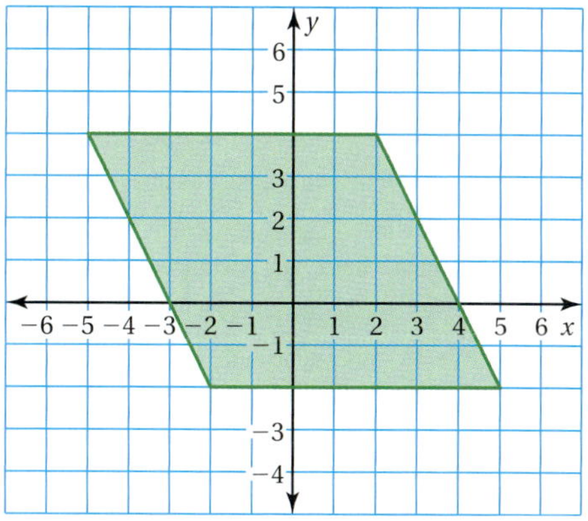

b.

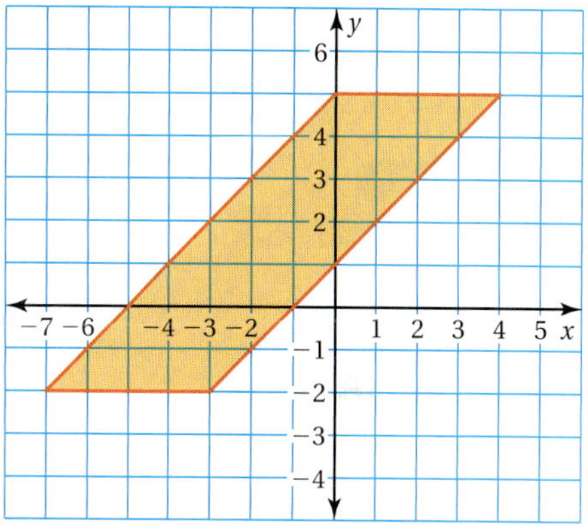

3 ACTIVITY: Interpreting the Slope and *y*-Intercept

Work with a partner. The graph shows a trip taken by a car where *t* is the time (in hours) and *y* is the distance (in miles) from Phoenix.

a. How far from Phoenix was the car at the beginning of the trip?

b. What was the car's speed?

c. How long did the trip last?

d. How far from Phoenix was the car at the end of the trip?

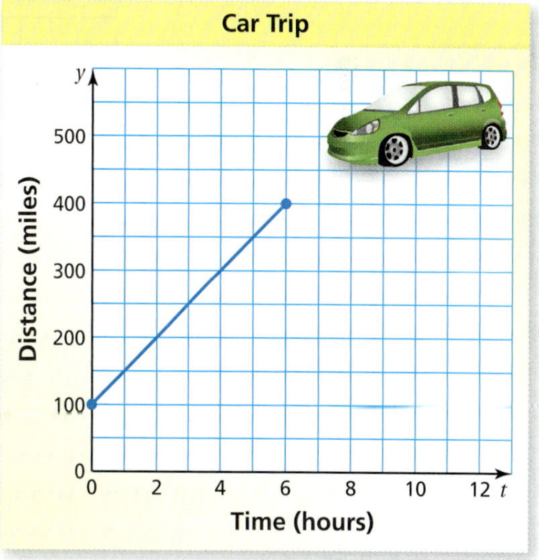

What Is Your Answer?

4. IN YOUR OWN WORDS How can you write an equation of a line when you are given the slope and *y*-intercept of the line? Give an example that is different from those in Activities 1, 2, and 3.

Practice Use what you learned about writing equations in slope-intercept form to complete Exercises 3 and 4 on page 110.

3.1 Lesson

Check It Out
Lesson Tutorials
BigIdeasMath.com

EXAMPLE 1 Writing Equations in Slope-Intercept Form

Write an equation of the line in slope-intercept form.

a.

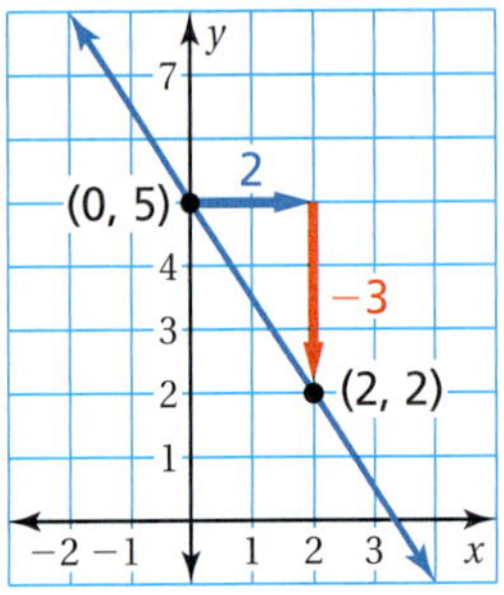

Find the slope and *y*-intercept.

$$\text{slope} = \frac{\text{rise}}{\text{run}} = \frac{-3}{2} = -\frac{3}{2}$$

Because the line crosses the *y*-axis at (0, 5), the *y*-intercept is 5.

slope *y*-intercept

So, the equation is $y = -\frac{3}{2}x + 5$.

Study Tip

After writing an equation, check that the given points are solutions of the equation.

b.

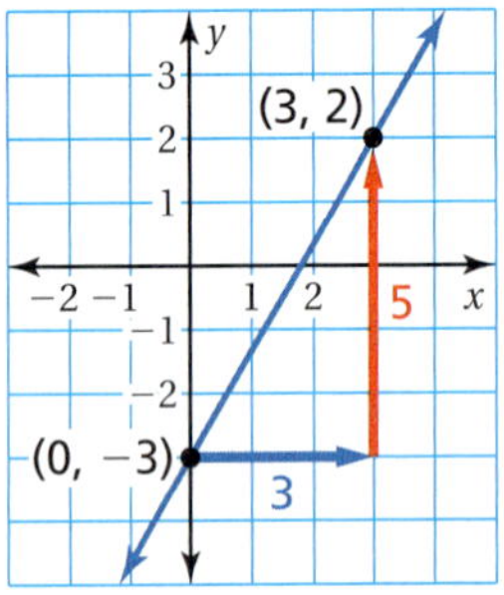

Find the slope and *y*-intercept.

$$\text{slope} = \frac{\text{rise}}{\text{run}} = \frac{5}{3}$$

Because the line crosses the *y*-axis at (0, −3), the *y*-intercept is −3.

slope *y*-intercept

So, the equation is $y = \frac{5}{3}x + (-3)$, or $y = \frac{5}{3}x - 3$.

On Your Own

Now You're Ready
Exercises 5–10

Write an equation of the line in slope-intercept form.

1.

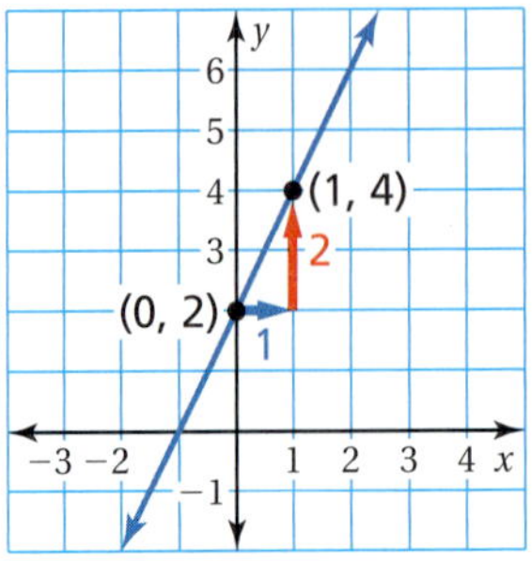

2.

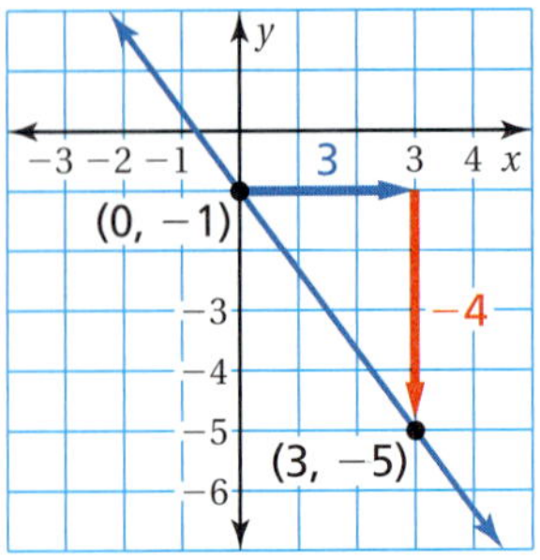

Laurie's Notes

Introduction

Connect

- **Yesterday:** Students developed an intuitive understanding about how to write the equation of a line when you know its slope and *y*-intercept.
- **Today:** Students will write the equation of a line given its slope and *y*-intercept.

Motivate

- **Story Time:** Tell students that as a child you loved to dig tunnels in the sand. Ask if any of them like to dig tunnels or if they have traveled through tunnels. Hold a paper towel tube or other similar model to pique student interest. Share some facts about tunnels.
  - The world's longest overland tunnel is a 21-mile-long rail link under the Alps in Switzerland and was built to ease highway traffic jams in the mountainous country. The tunnel took eight years to build and cost $3.5 billion. It reduces the time trains need to cross between Germany and Italy from 3.5 hours to just under 2 hours.
  - The world's longest underwater tunnel is Seikan Tunnel in Japan. It is 33.49 miles long and runs under the Tsugaru Strait. It opened in 1988 and took 17 years to construct.
  - The Channel Tunnel (Chunnel) connects England and France. It is 31 miles long and travels under the English Channel.

Lesson Notes

Example 1

- Write the slope-intercept form of an equation, $y = mx + b$. Review with students that the coefficient of *x* is the slope, and the constant *b* is the *y*-intercept.
- ? "What do you know about the slope of the line just by inspection? Explain." Slope is negative because the graph falls left to right.
- ? "What are the coordinates of the point where the line crosses the *y*-axis?" (0, 5)
- Use the slope and the *y*-intercept to write the equation.
- Work through part (b). Remind students that you want the more simplified equation $y = \frac{5}{3}x - 3$ instead of $y = \frac{5}{3}x + (-3)$. Stress that while both forms are correct, the simplified version is preferred.

On Your Own

- Before students begin these two problems, they should do a visual inspection. They should make a note of the sign of the slope and *y*-intercept. It is very easy to have the wrong sign(s) when the equation is written.

Goal Today's lesson is writing the equation of a line in slope-intercept form.

Lesson Materials
Introduction
• paper towel tube

Start Thinking! and Warm Up

Lesson 3.1 Warm Up
For use before Lesson 3.1

Lesson 3.1 Start Thinking!
For use before Lesson 3.1

A gym membership has a $20 enrollment fee and costs $40 per month.

Write an equation in slope-intercept form that represents the cost *y* after *x* months of joining the gym.

What does the slope represent?

What does the *y*-intercept represent?

Extra Example 1

Write an equation of the line in slope-intercept form.

a.

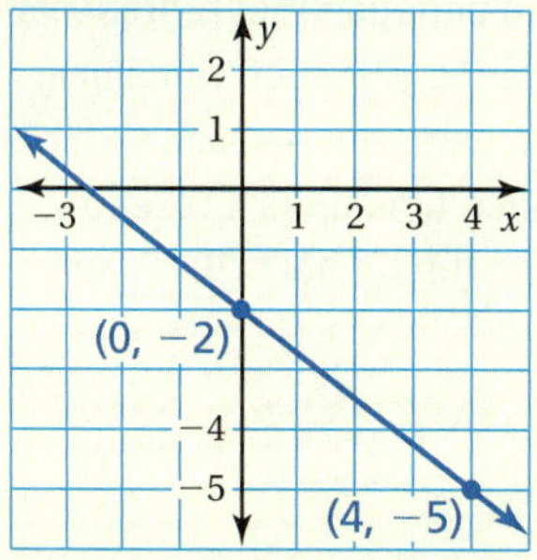

$y = -\frac{3}{4}x - 2$

b.

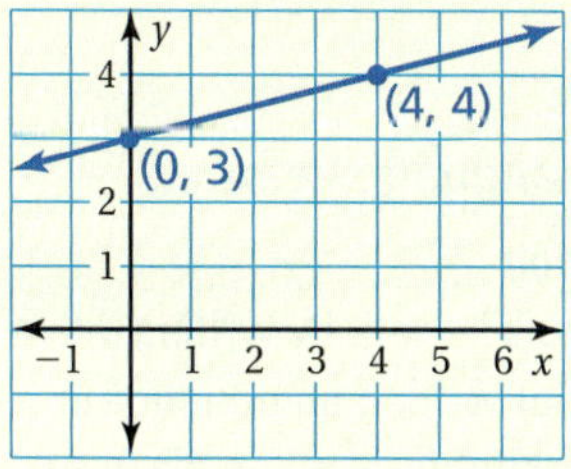

$y = \frac{1}{4}x + 3$

On Your Own

1. $y = 2x + 2$
2. $y = -\frac{4}{3}x - 1$

Laurie's Notes

Example 2

- Make a quick sketch of the graph to reference as you work the problem.
- When finding the slope, students are unsure of how to simplify $\frac{0}{3}$. This is a good time to review the difference between $\frac{0}{3}$ and $\frac{3}{0}$.
- **Teaching Tip:** $\frac{3}{0}$ is undefined. The explanation I give students that seems to resonate with them is to write the problem $8 \div 4 = 2$ on the board. Then I rewrite it as $4\overline{)8}$. To check, multiply the quotient (2) times the divisor (4) and you get the dividend (8). In other words, 2 multiplied by 4 is 8. Do the same thing with $\frac{3}{0}$. Rewrite it using long division, $0\overline{)3}$. What do you multiply 0 by to get 3? There is no quotient, so you say $\frac{3}{0}$ is undefined. You cannot divide by 0.
- Students don't always recognize that $y = -4$ is a linear equation written in slope-intercept form. It helps to write the extra step of $y = (0)x + (-4)$ so students can see that the slope is 0.

Example 3

- Ask a volunteer to read the problem. Discuss information that can be *read* from the graph.
- ? "By visual inspection, what do you know about the sign of the slope and the *y*-intercept in this problem?" The slope is negative. The *y*-intercept is positive.
- Before moving on to part (b), ask students to interpret what a slope of −500 means in the context of this problem. A slope of −500 means that for each additional month of work, the distance left to complete is 500 feet less.
- The *x*-intercept for this graph is 7.
- Note that the graph is in Quadrant I. In the context of this problem, it doesn't make sense for time or distance to be negative.

On Your Own

- For Question 3, encourage students to sketch a graph of the line through the two points to give them a clue as to how to begin. This technique will help students start Question 4.

Closure

- **Writing Prompt:** For a line that has been graphed in a coordinate plane, you can write the equation by … finding the slope and *y*-intercept

Technology For the Teacher

Dynamic Classroom

The Dynamic Planning Tool
Editable Teacher's Resources at *BigIdeasMath.com*

Extra Example 2

Write an equation of the line that passes through the points (0, −1) and (4, −1).
$y = -1$

Extra Example 3

In Example 3, the points are (0, 3500) and (5, 1750).

a. Write an equation that represents the distance *y* (in feet) remaining after *x* months. $y = -350x + 3500$

b. How much time does it take to complete the tunnel? 10 months

On Your Own

3. $y = 5$

4. $8\frac{3}{4}$ mo

English Language Learners

Organization

Students will benefit by writing down the steps for writing an equation in slope-intercept form when given a graph. Have students write the steps in their notebooks. A poster with the steps could be posted in the classroom.

Step 1: Write the slope-intercept form of an equation.

Step 2: Determine the slope of the line.

Step 3: Determine the *y*-intercept of the line.

Step 4: Write the equation in slope-intercept form.

EXAMPLE 2 Standardized Test Practice

Which equation is shown in the graph?

Ⓐ $y = -4$ Ⓑ $y = -3$

Ⓒ $y = 0$ Ⓓ $y = -3x$

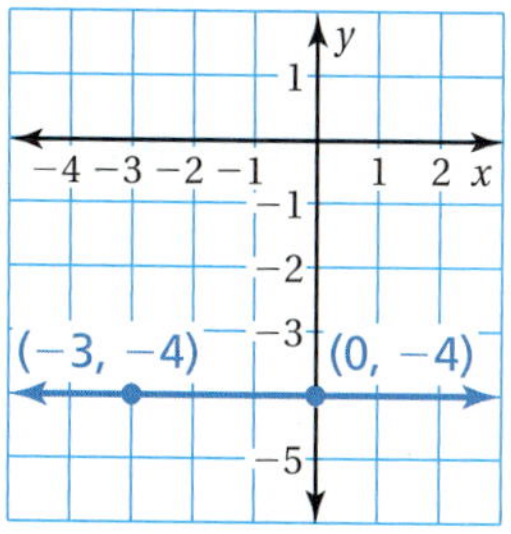

Remember

The graph of $y = a$ is a horizontal line that passes through $(0, a)$.

Find the slope and y-intercept.

The line is horizontal, so the rise is 0.

$$\text{slope} = \frac{\text{rise}}{\text{run}} = \frac{0}{3} = 0$$

Because the line crosses the y-axis at $(0, -4)$, the y-intercept is -4.

∴ So, the equation is $y = 0x + (-4)$, or $y = -4$. The correct answer is Ⓐ.

EXAMPLE 3 Real-Life Application

Engineers used a tunnel boring machine like the one shown above to dig the main tunnel of the Yucca Mountain Repository in Nevada. The main tunnel is 5 miles long and 25 feet wide.

The graph shows the distance remaining to complete a tunnel. (a) Write an equation that represents the distance y (in feet) remaining after x months. (b) How much time does it take to complete the tunnel?

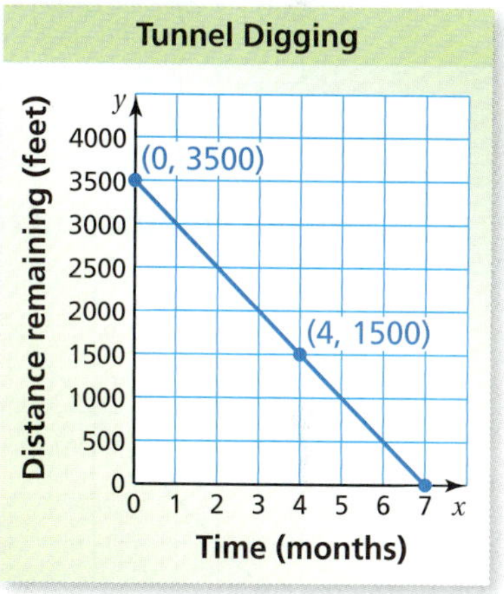

a. Find the slope and y-intercept.

$$\text{slope} = \frac{\text{rise}}{\text{run}} = \frac{-2000}{4} = -500$$

Because the line crosses the y-axis at $(0, 3500)$, the y-intercept is 3500.

∴ So, the equation is $y = -500x + 3500$.

b. The tunnel is complete when the distance remaining is 0 feet. So, find the value of x when $y = 0$.

$y = -500x + 3500$	Write the equation.
$0 = -500x + 3500$	Substitute 0 for y.
$-3500 = -500x$	Subtract 3500 from each side.
$7 = x$	Solve for x.

∴ It takes 7 months to complete the tunnel.

On Your Own

Now You're Ready
Exercises 13–15

3. Write an equation of the line that passes through $(0, 5)$ and $(4, 5)$.
4. **WHAT IF?** In Example 3, the points are $(0, 3500)$ and $(5, 1500)$. How long does it take to complete the tunnel?

3.1 Exercises

Check It Out
Help with Homework
BigIdeasMath.com

Vocabulary and Concept Check

1. **WRITING** Explain how to find the slope of a line given the intercepts of the line.
2. **WRITING** Explain how to write an equation of a line using its graph.

Practice and Problem Solving

Write an equation for each side of the figure.

3. 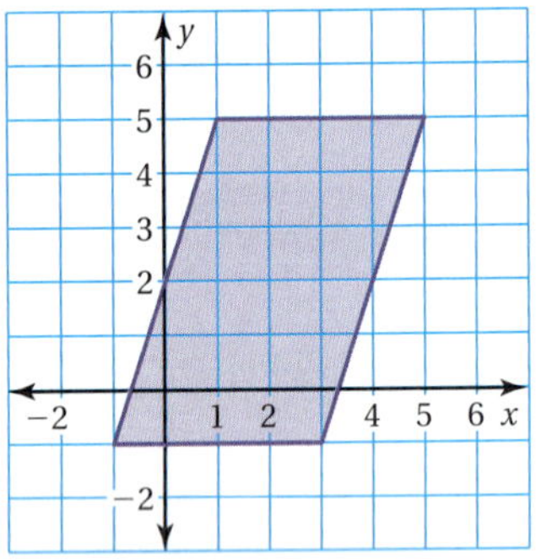

4. 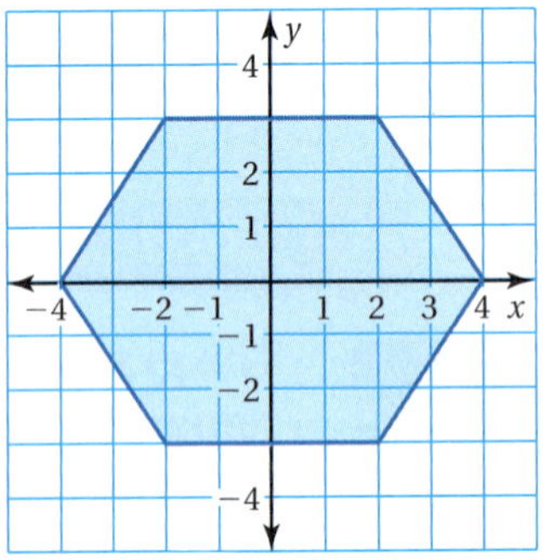

Write an equation of the line in slope-intercept form.

1 5. 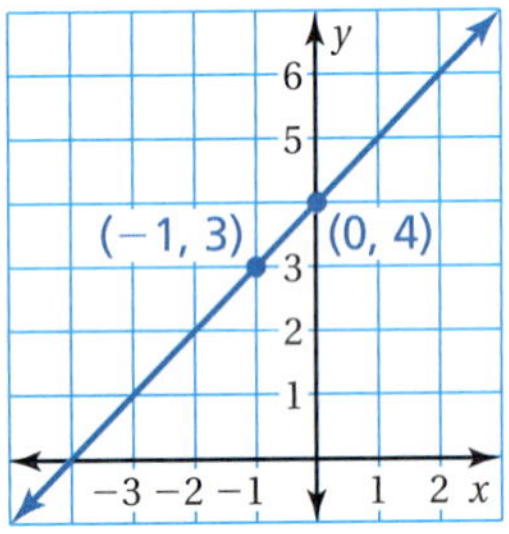

6. 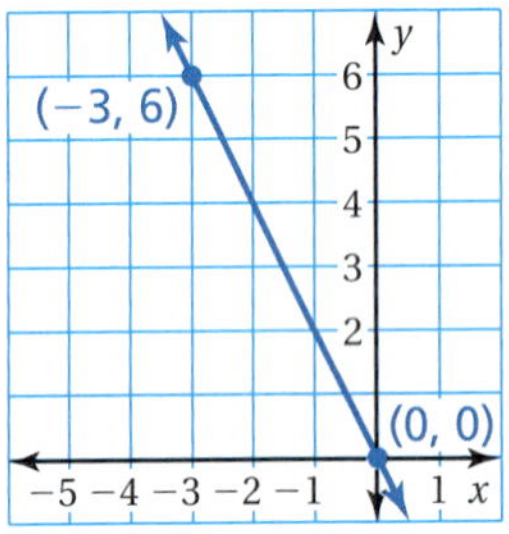

7. 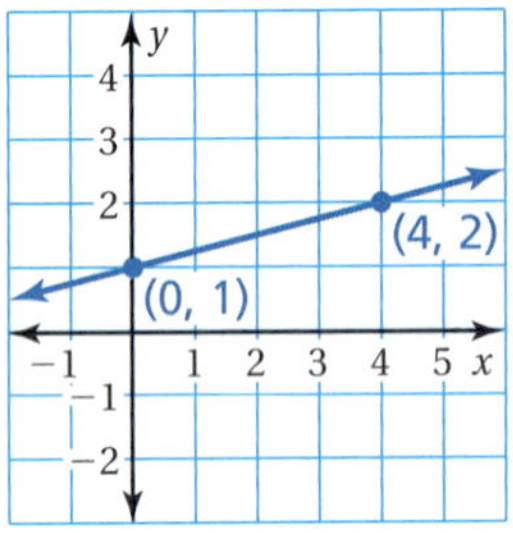

8. 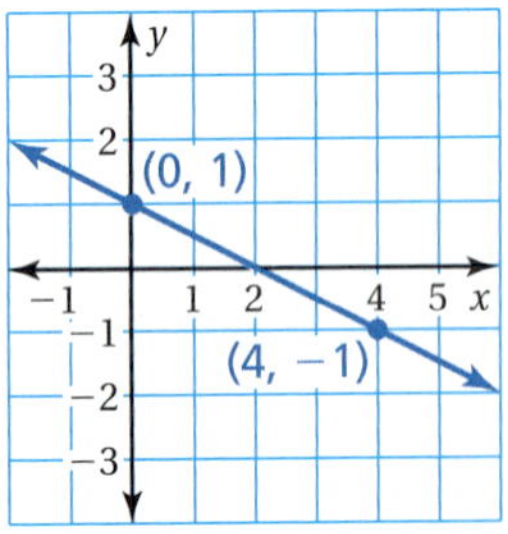

9. 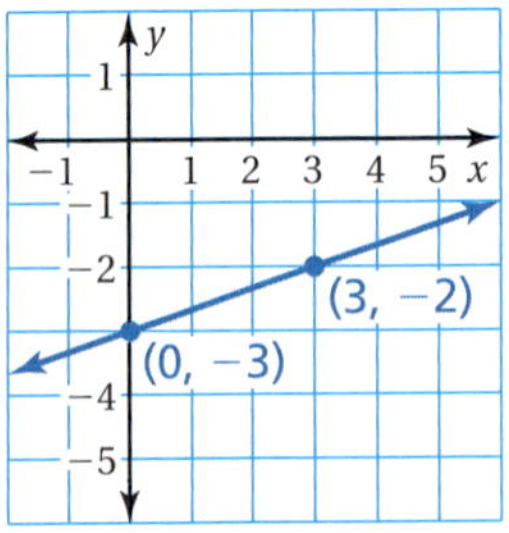

10. 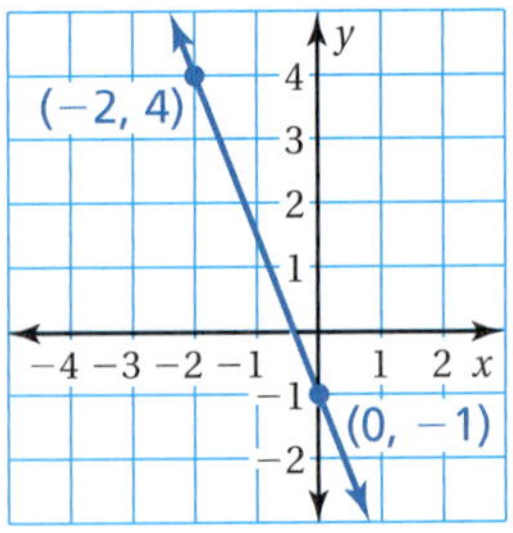

11. **ERROR ANALYSIS** Describe and correct the error in writing the equation of the line.

✗ $y = \frac{1}{2}x + 4$

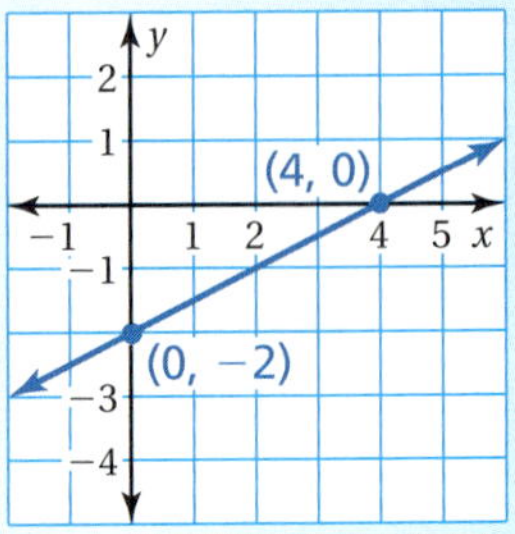

12. **BOA** A boa constrictor is 18 inches long at birth and grows 8 inches per year. Write an equation that represents the length y (in feet) of a boa constrictor that is x years old.

Assignment Guide and Homework Check

Level	Day 1 Activity Assignment	Day 2 Lesson Assignment	Homework Check
Basic	3, 4, 20–24	1, 2, 5–16	6, 9, 12, 13, 16
Average	3, 4, 20–24	1, 2, 6–17	6, 9, 12, 14, 16
Advanced	3, 4, 20–24	1, 2, 8–19	8, 14, 16, 18

Common Errors

- **Exercises 5–10** Students may write the reciprocal of the slope or forget a negative sign. Remind them of the definition of slope. Ask students to predict the sign of the slope based on the rise or fall of the line.
- **Exercises 13–15** Students may write the wrong equation when the slope is zero. For example, instead of $y = 5$, students may write $x = 5$. Ask them what is the rise of the graph (zero) and write this in slope-intercept form with the y-intercept as well, such as $y = 0x + 5$. Then ask students what happens when a variable (or any number) is multiplied by zero. Rewrite the equation as $y = 5$.

3.1 Record and Practice Journal

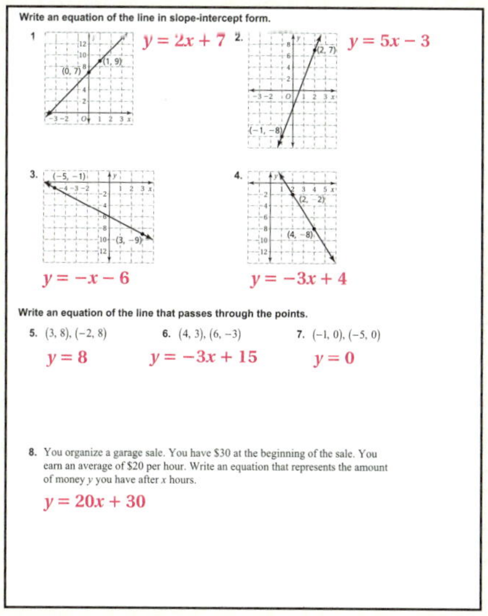

Write an equation of the line in slope-intercept form.

1. $y = 2x + 7$
2. $y = 5x - 3$
3. $y = -x - 6$
4. $y = -3x + 4$

Write an equation of the line that passes through the points.

5. (3, 8), (−2, 8) $y = 8$
6. (4, 3), (6, −3) $y = -3x + 15$
7. (−1, 0), (−5, 0) $y = 0$

8. You organize a garage sale. You have $30 at the beginning of the sale. You earn an average of $20 per hour. Write an equation that represents the amount of money y you have after x hours.

$y = 20x + 30$

Technology For the Teacher

Answer Presentation Tool

Quiz*Show*

Vocabulary and Concept Check

1. *Sample answer:* Find the ratio of the rise to the run between the intercepts.
2. *Sample answer:* Find the slope of the line between any two points. Then find the y-intercept. The equation of the line is $y = mx + b$, where m is the slope and b is the y-intercept.

Practice and Problem Solving

3. $y = 3x + 2$;
   $y = 3x - 10$;
   $y = 5$;
   $y = -1$
4. $y = \frac{3}{2}x + 6$;
   $y = 3$;
   $y = -\frac{3}{2}x + 6$;
   $y = \frac{3}{2}x - 6$;
   $y = -3$;
   $y = -\frac{3}{2}x - 6$
5. $y = x + 4$
6. $y = -2x$
7. $y = \frac{1}{4}x + 1$
8. $y = -\frac{1}{2}x + 1$
9. $y = \frac{1}{3}x - 3$
10. $y = -\frac{5}{2}x - 1$
11. The x-intercept was used instead of the y-intercept.
    $y = \frac{1}{2}x - 2$
12. $y = \frac{2}{3}x + \frac{3}{2}$

Practice and Problem Solving

13. $y = 5$ **14.** $y = 0$

15. $y = -2$ **16.** $y = 0.7x + 10$

17. See Additional Answers.

18. $y = -140x + 500$

19. See *Taking Math Deeper*.

Fair Game Review

20–23.

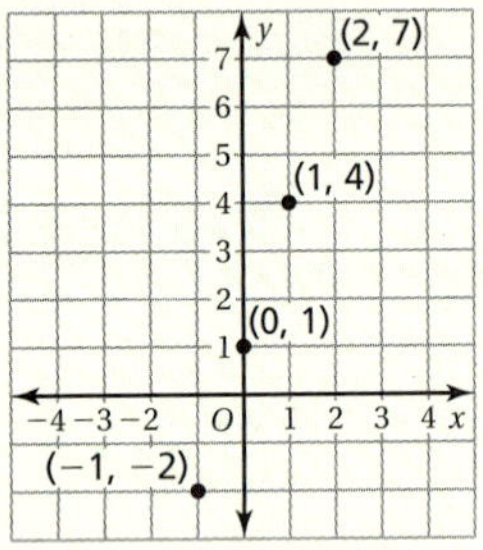

24. C

Mini-Assessment

Write an equation of the line in slope-intercept form.

1. $y = x + 2$

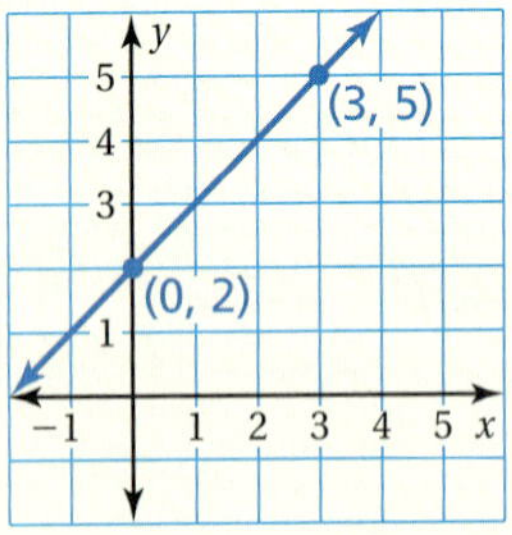

2. $y = -2x - 1$

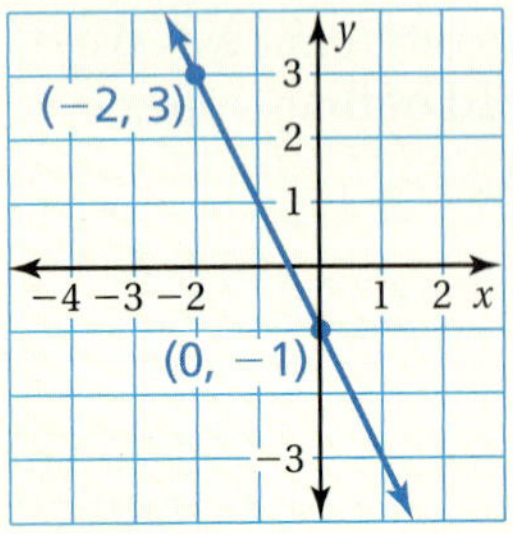

Taking Math Deeper

Exercise 19

This is a nice real-life problem using estimation. For this problem, remember that you are not looking for exact solutions. You want to know *about* how much the trees grow each year so that you can predict their approximate heights.

1 Estimate the heights in the photograph.

a. Height of 10-year-old tree: about 18 ft
Height of 8-year-old tree: about 14 ft

2 **b.** Plot the heights of the two trees.

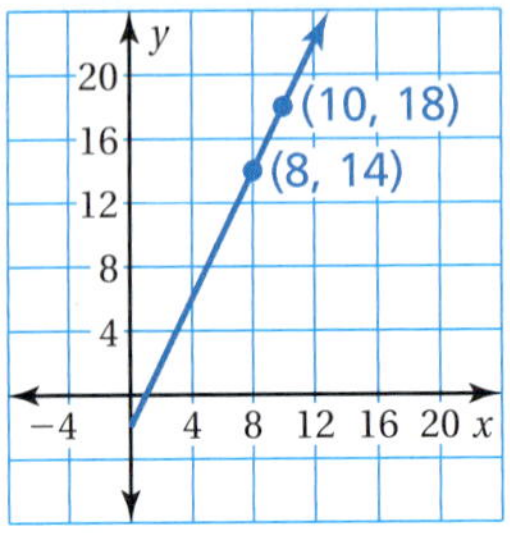

3 **c.** The trees are growing at a rate of about 2 feet per year. Because this would put the height of a 0-year-old tree at −2, it is better to adjust the rate of growth to be about 1.8 feet per year.

d. A possible equation for the growth rate is $y = 1.8x$.

Project

Research information about the palm tree. Pick any kind of palm tree in which you are interested. How old is the longest living palm tree?

Reteaching and Enrichment Strategies

If students need help. . .	If students got it. . .
Resources by Chapter • Practice A and Practice B • Puzzle Time Record and Practice Journal Practice Differentiating the Lesson Lesson Tutorials Skills Review Handbook	Resources by Chapter • Enrichment and Extension Start the next section

Write an equation of the line that passes through the points.

② **13.** (2, 5), (0, 5) **14.** (−3, 0), (0, 0) **15.** (0, −2), (4, −2)

16. WALKATHON One of your friends gives you \$10 for a charity walkathon. Another friend gives you an amount per mile. After 5 miles, you have raised \$13.50 total. Write an equation that represents the amount y of money you have raised after x miles.

17. BRAKING TIME During each second of braking, an automobile slows by about 10 miles per hour.

a. Plot the points (0, 60) and (6, 0). What do the points represent?

b. Draw a line through the points. What does the line represent?

c. Write an equation of the line.

18. PAPER You have 500 sheets of notebook paper. After 1 week, you have 72% of the sheets left. You use the same number of sheets each week. Write an equation that represents the number y of pages remaining after x weeks.

19. Critical Thinking The palm tree on the left is 10 years old. The palm tree on the right is 8 years old. The trees grow at the same rate.

a. Estimate the height y (in feet) of each tree.

b. Plot the two points (x, y), where x is the age of each tree and y is the height of each tree.

c. What is the rate of growth of the trees?

d. Write an equation that represents the height of a palm tree in terms of its age.

Fair Game Review What you learned in previous grades & lessons

Plot the ordered pair in a coordinate plane. *(Skills Review Handbook)*

20. (1, 4) **21.** (−1, −2) **22.** (0, 1) **23.** (2, 7)

24. MULTIPLE CHOICE Which of the following statements is true? *(Section 2.3)*

Ⓐ The x-intercept is 5.

Ⓑ The x-intercept is −2.

Ⓒ The y-intercept is 5.

Ⓓ The y-intercept is −2.

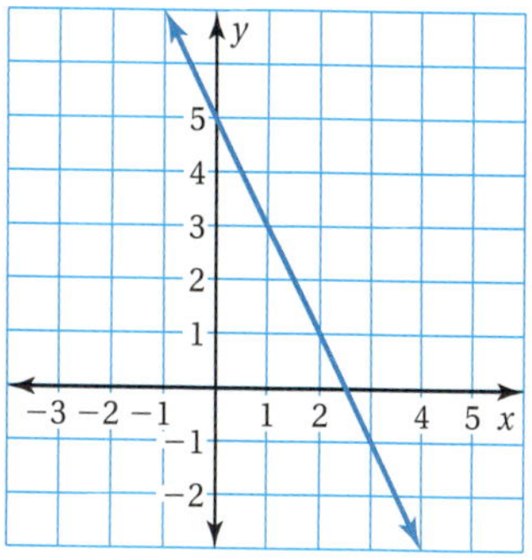

3.2 Writing Equations Using a Slope and a Point

Essential Question How can you write an equation of a line when you are given the slope and a point on the line?

1 ACTIVITY: Writing Equations of Lines

Work with a partner.

- **Sketch the line that has the given slope and passes through the given point.**
- **Find the *y*-intercept of the line.**
- **Write an equation of the line.**

a. $m = -2$

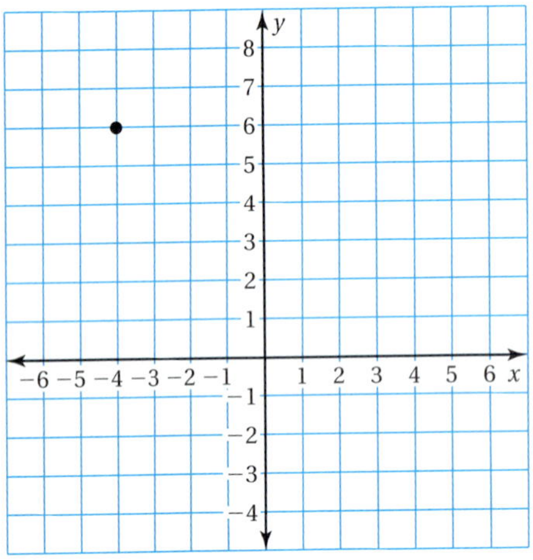

b. $m = \frac{1}{3}$

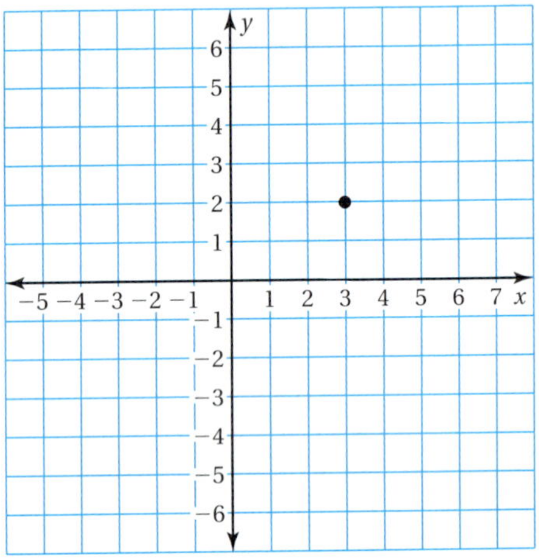

c. $m = -\frac{2}{3}$

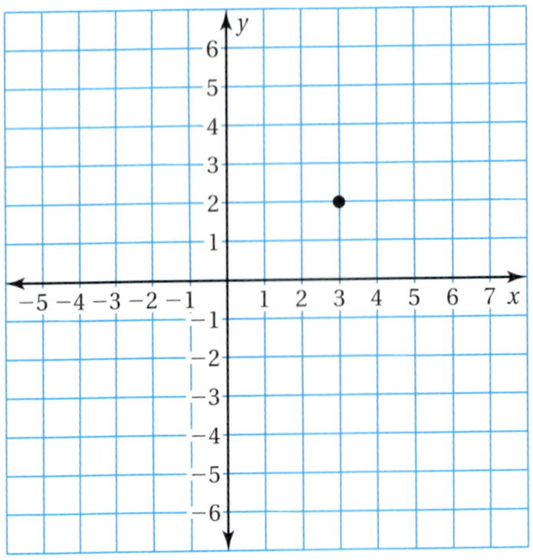

d. $m = \frac{5}{2}$

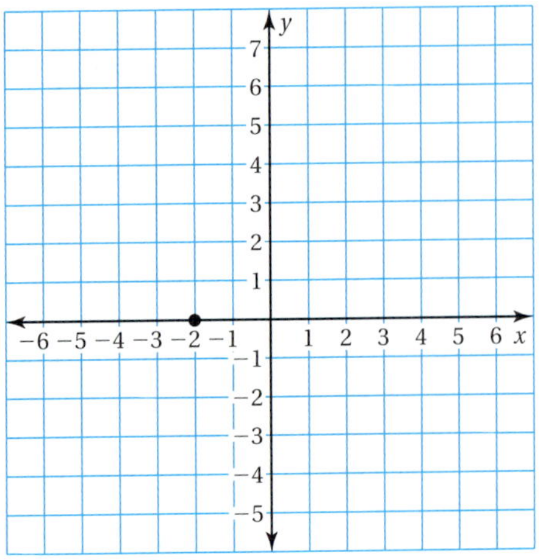

Laurie's Notes

Introduction

For the Teacher

- **Goal:** Students will explore writing an equation of a line given the slope and a point that is not on the *y*-axis.

Motivate

- Hold a piece of ribbon and a pair of scissors in your hands. Snip a one-foot piece of ribbon off. Repeat once or twice more.

? "Do you know how long my ribbon was when I first started?" no

- Your question should prompt students to ask two obvious questions: "How much are you cutting off each time?" and "How many times have you made a cut?" How much you cut off is the slope (−1). How many times you cut the ribbon helps students work backwards to find the length before any cuts were made, which is the *y*-intercept.

Discuss

- Explain that today students will be writing the equation of a line. Some of the problems will be presented on the coordinate grid, while others will be presented as a story, like the ribbon. They need to figure out the slope and the *y*-intercept.

Activity Notes

Activity 1

? "What does it mean for a line to have a slope of −2? A slope of $\frac{1}{3}$?"

For every unit it runs, it falls 2 units. For every 3 units it runs, it rises 1.

- Students may also answer the last question by saying "over 1, down 2" and "over 3, up 1." These geometric answers are fine. Students will need this level of understanding to locate additional points on a line, in order to find the *y*-intercept.
- You cannot sketch the line immediately. You must first find additional points on the line.
- Students should start at the given point and use the slope to find additional points on the line. One of the points will give the *y*-intercept.
- For part (b), it might be helpful to think of the slope of $\frac{1}{3}$ as $\frac{-1}{-3}$. So, start at the point given and move left 3 units and then down 1 unit.
- **Common Error:** Students may interchange the rise and run. Have students look back at their graph to see if the slope looks correct to them.
- **Teaching Tip:** Encourage students to use a pencil and lightly trace the rise and run direction arrows as they locate additional points.
- To share student work, have transparency grids available for the overhead.

? "What made it possible to write the equation of the line?" The slope was given and by using the slope, it was possible to find the *y*-intercept. Then substitute into the formula $y = mx + b$.

Previous Learning

Students should know how to plot ordered pairs and apply the definition of slope.

Activity Materials	
Introduction	**Textbook**
• scissors • yarn, rope, or ribbon	• straightedge • transparency grid

Start Thinking! and Warm Up

Activity 3.2 Start Thinking!
For use before Activity 3.2

Activity 3.2 Warm Up
For use before Activity 3.2

Graph the linear equation.

1. $y = 2x + 3$
2. $y = -x + 2$
3. $y = \frac{2}{3}x - 1$
4. $y = -\frac{1}{2}x + 4$
5. $y = 5x - 10$
6. $y = -\frac{5}{7}x + 14$

3.2 Record and Practice Journal

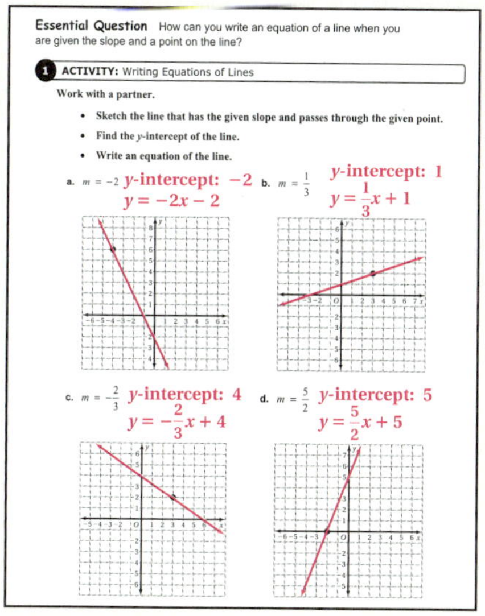
Essential Question How can you write an equation of a line when you are given the slope and a point on the line?

1 ACTIVITY: Writing Equations of Lines

Work with a partner.

- Sketch the line that has the given slope and passes through the given point.
- Find the *y*-intercept of the line.
- Write an equation of the line.

a. $m = -2$ *y*-intercept: −2 $y = -2x - 2$

b. $m = \frac{1}{3}$ *y*-intercept: 1 $y = \frac{1}{3}x + 1$

c. $m = -\frac{2}{3}$ *y*-intercept: 4 $y = -\frac{2}{3}x + 4$

d. $m = \frac{5}{2}$ *y*-intercept: 5 $y = \frac{5}{2}x + 5$

Differentiated Instruction

Kinesthetic

Write a list of linear equations on the board or overhead. Have students copy the equations onto index cards. On the back of each card students are to write the slope and y-intercept of the line. After the cards are completed, students can work in pairs to check each other's work. Finally students can quiz each other with the flash cards they made.

3.2 Record and Practice Journal

2 ACTIVITY: Writing Linear Equations

Work with a partner.

a. For 4 months, you have saved \$25 a month. You now have \$175 in your savings account.

- Draw a graph that shows the balance in your account after t months.
- Write an equation that represents the balance A after t months.

$A = 25t + 75$

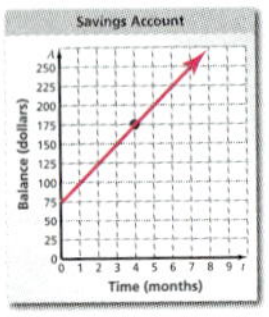

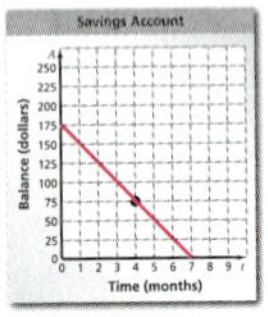

b. For 4 months, you have withdrawn \$25 a month from your savings account. Your account balance is now \$75.

- Draw a graph that shows the balance in your account after t months.
- Write an equation that represents the balance A after t months.

$A = -25t + 175$

c. For 6 years, the population of a town has grown by 5000 people per year. The population is now 70,000.

- Draw a graph that shows the population after t years.
- Write an equation that represents the population P after t years.

$P = 5000t + 40{,}000$

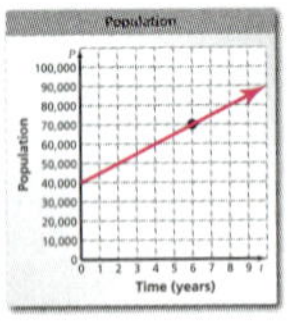

What Is Your Answer?

3. **IN YOUR OWN WORDS** How can you write an equation of a line when you are given the slope and a point on the line? Give an example that is different from those in Activities 1 and 2.

Plot point and use slope to plot additional points to find y-intercept.

Laurie's Notes

Activity 2

- The strategy is the same as for Activity 1. The slope may not be obvious.
- **Part (a):** If students do not understand what the slope is, suggest they work backwards and make a table of values. Students should use the variables t and A instead of x and y.

Month, t	0	1	2	3	4
Balance in Account, A	\$75	\$100	\$125	\$150	\$175

? Ask a few questions to guide students' understanding:

- "What is the slope for this problem? What is the A-intercept?" slope = \$25; A-intercept = \$75
- "Do you have enough information to write the equation?" yes; $A = 25t + 75$
- "Explain why the slope is positive." You are putting money in the bank. Your account is growing.

- **Part (b):** Some students will need to make a table of values, while others will know the slope is -25.

? "What does the A-intercept mean in the context of this problem?" How much money you started with (in your account) 4 months ago.

- **Part (c):** Encourage students to plot points representing the population for prior years.
- **Big Idea:** For each of these real-life problems, the slope is given in words. A point on the line is given so that the P-intercept can be determined using the slope. Each problem is an example of writing an equation of a line, given the slope and a point on the line.

What Is Your Answer?

- **Neighbor Check:** Have students work independently and then have their neighbor check their work. Have students discuss any discrepancies.

Closure

- Refer back to the ribbon and scissors. If the ribbon is now 7 feet and you made 4 equal cuts of 1-foot length, write the equation that gives the length of the ribbon R after n cuts. $R = 11 - n$ or $R = -n + 11$

Technology For the Teacher

Dynamic Classroom

The Dynamic Planning Tool
Editable Teacher's Resources at *BigIdeasMath.com*

2 ACTIVITY: Writing Linear Equations

Work with a partner.

a. For 4 months, you have saved $25 a month. You now have $175 in your savings account.

- Draw a graph that shows the balance in your account after t months.
- Write an equation that represents the balance A after t months.

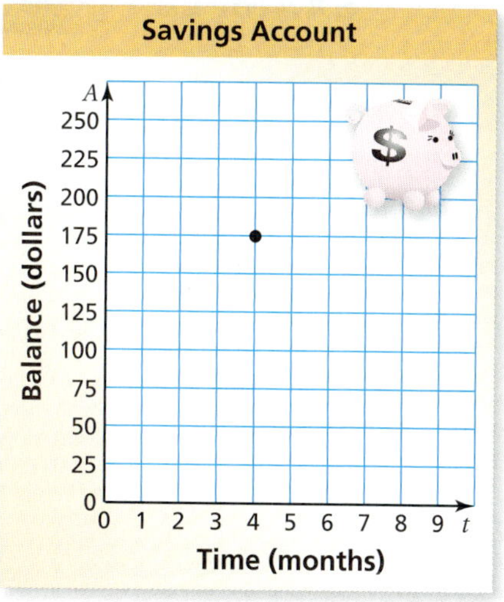

b. For 4 months, you have withdrawn $25 a month from your savings account. Your account balance is now $75.

- Draw a graph that shows the balance in your account after t months.
- Write an equation that represents the balance A after t months.

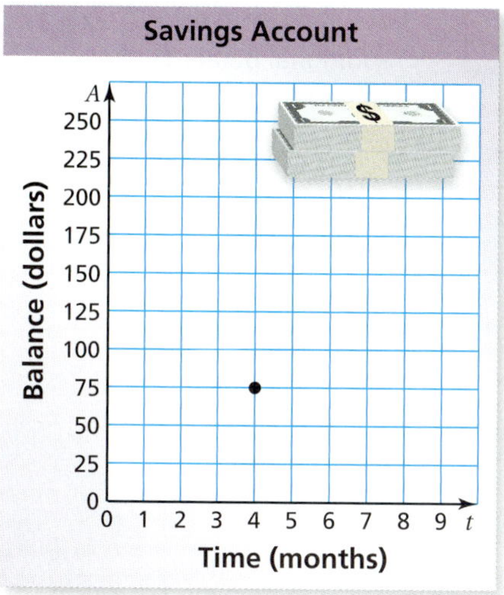

c. For 6 years, the population of a town has grown by 5000 people per year. The population is now 70,000.

- Draw a graph that shows the population after t years.
- Write an equation that represents the population P after t years.

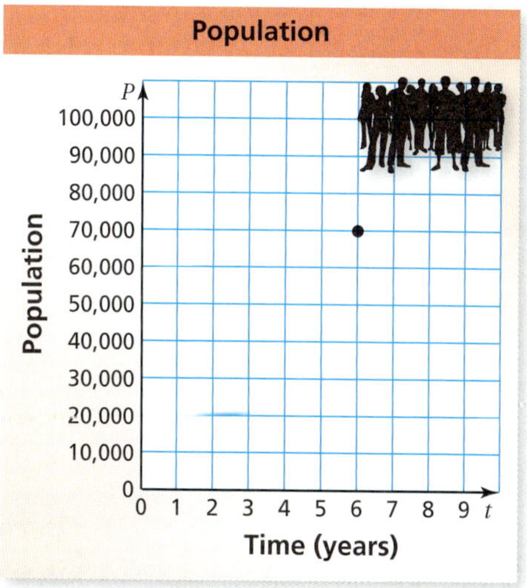

What Is Your Answer?

3. **IN YOUR OWN WORDS** How can you write an equation of a line when you are given the slope and a point on the line? Give an example that is different from those in Activities 1 and 2.

Practice Use what you learned about writing equations using a slope and a point to complete Exercises 3–5 on page 116.

3.2 Lesson

Check It Out
Lesson Tutorials
BigIdeasMath.com

EXAMPLE 1 Writing Equations Using a Slope and a Point

Write an equation of the line with the given slope that passes through the given point.

a. $m = \frac{2}{3}$; $(-6, 1)$

Use a graph to find the y-intercept.

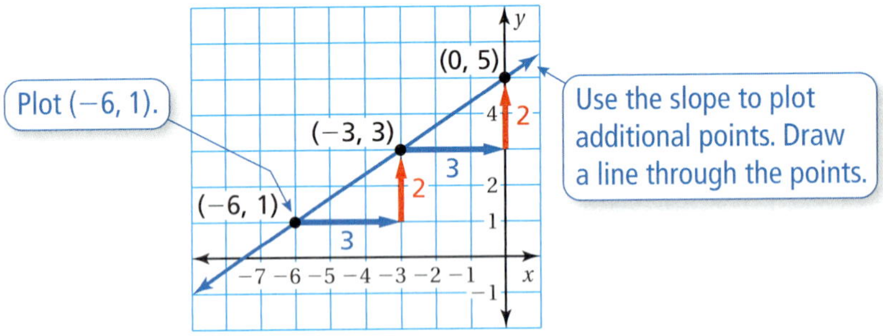

Because the line crosses the y-axis at $(0, 5)$, the y-intercept is 5.

So, the equation is $y = \frac{2}{3}x + 5$.

Check

Check that $(-6, 1)$ is a solution of the equation.

$y = \frac{2}{3}x + 5$

$1 \stackrel{?}{=} \frac{2}{3}(-6) + 5$

$1 = 1$ ✓

b. $m = -3$; $(1, -4)$

Use a graph to find the y-intercept.

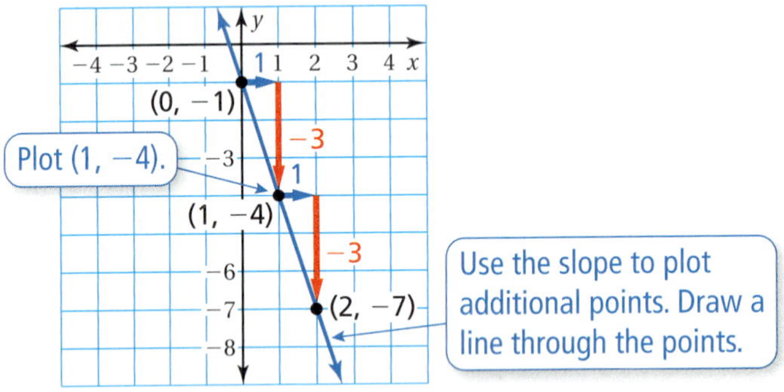

Because the line crosses the y-axis at $(0, -1)$, the y-intercept is -1.

So, the equation is $y = -3x + (-1)$, or $y = -3x - 1$.

Check

Check that $(1, -4)$ is a solution of the equation.

$y = -3x - 1$

$-4 \stackrel{?}{=} -3(1) - 1$

$-4 = -4$ ✓

On Your Own

Now You're Ready
Exercises 6–11

Write an equation of the line with the given slope that passes through the given point.

1. $m = 1$; $(2, 0)$

2. $m = -\frac{1}{2}$; $(2, 3)$

Laurie's Notes

Introduction

Connect

- **Yesterday:** Students developed an intuitive understanding of how to write the equation of a line given the slope and a point.
- **Today:** Students will write the equation of a line given the slope and a point.

Motivate

? "Have you seen an airplane come in for a landing either in real life, on the television, or in movies?" Most will answer yes.

? "Can you describe in words or with a picture what it looks like?" Listen for a smooth approach, meaning a constant rate of descent.

? "If the plane descends 200 feet per second, what is its height 5 seconds before it lands?" 1000 ft

- Make a sketch of this scenario and ask if it's possible to write an equation that models the height h of the airplane, t seconds before it lands.

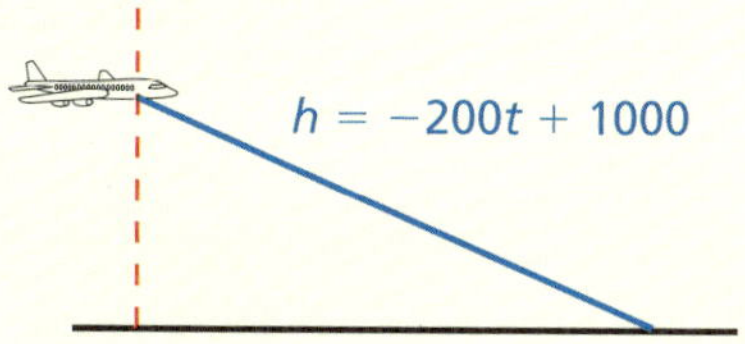

Lesson Notes

Example 1

- Write the slope-intercept form of a linear equation: $y = mx + b$.
- Discuss with students that it is possible to find the y-intercept using the slope and a point, and then plotting additional points on the graph.
- Write the given information: slope $m = \frac{2}{3}$ and the point is $(-6, 1)$.

? After plotting $(-6, 1)$ ask, "Is it possible to find additional points on the line? Explain." Yes; Use the slope of $\frac{2}{3}$ to plot additional points.

- Once additional points are plotted, use a straightedge to draw the line.
- Encourage students to check that the equation contains the given point.
- **Part (b):** The red and blue arrows make it look as though you started at the point $(0, -1)$. Starting at $(1, -4)$, you can also show the arrow going 1 unit to the left (run $= -1$) and up 3 units (rise $= 3$). The slope is -3.

On Your Own

- Students begin by plotting the known point. Ask students to lay their pencil on the ordered pair and angle their pencil so that the given slope is modeled. This gives students a visual image of the line and approximately what the y-intercept is. Model this process at the overhead using a piece of spaghetti and a transparency grid.

Goal Today's lesson is writing the equation of a line given a slope and a point.

Lesson Materials
Textbook
• straightedge

Start Thinking! and Warm Up

Lesson 3.2 Warm Up
For use before Lesson 3.2

Lesson 3.2 Start Thinking!
For use before Lesson 3.2

How is writing the equation of a line given the slope and a point on the line similar to writing the equation of a line given the slope and y-intercept? How is it different?

Extra Example 1

Write an equation of the line with the given slope that passes through the given point.

a. $m = \frac{5}{2}$; $(2, 2)$ $y = \frac{5}{2}x - 3$

b. $m = -\frac{4}{3}$; $(3, 6)$ $y = -\frac{4}{3}x + 10$

On Your Own

1. $y = x - 2$
2. $y = -\frac{1}{2}x + 4$

Laurie's Notes

Extra Example 2

You are pulling down your kite at a rate of 2 feet per second. After 3 seconds, your kite is 54 feet above you.

a. Write an equation that represents the height y (in feet) of the kite above you after x seconds. $y = -2x + 60$

b. At what height was the kite flying? 60 ft

c. How long does it take to pull the kite down? 30 sec

On Your Own

3. after 5.5 sec

English Language Learners

Visual

Encourage English learners to plot the given point in the coordinate plane and then use the slope. The graph will give them a visual reference they can use when writing the equation.

Example 2

- Ask a volunteer to read the problem. Discuss information that can be *read* from the illustration.
- ? "Have any of you parasailed?" Wait for students to respond. Explain that you want a smooth descent, like an airplane.
- ? "What is the slope for this problem? How did you know?" Slope is -10. The arrow pointing down means the slope is negative.
- Students may start at (2, 25) and want to plot additional points to the right of (2, 25). This approach is helpful in answering part (c); however, you still need to determine the y-intercept.
- **Extension:** Discuss the need to restrict this problem to the first quadrant.

Closure

- **Exit Ticket:** Write an equation of the line with a slope of 2 that passes through the point $(-1, 4)$. $y = 2x + 6$

Technology For the Teacher

Dynamic Classroom

The Dynamic Planning Tool

Editable Teacher's Resources at *BigIdeasMath.com*

EXAMPLE 2 Real-Life Application

You finish parasailing and are being pulled back to the boat. After 2 seconds, you are 25 feet above the boat. (a) Write an equation that represents the height y (in feet) above the boat after x seconds. (b) At what height were you parasailing? (c) When do you reach the boat?

a. You are being pulled down at the rate of 10 feet per second. So, the slope is -10. You are 25 feet above the boat after 2 seconds. So, the line passes through (2, 25).

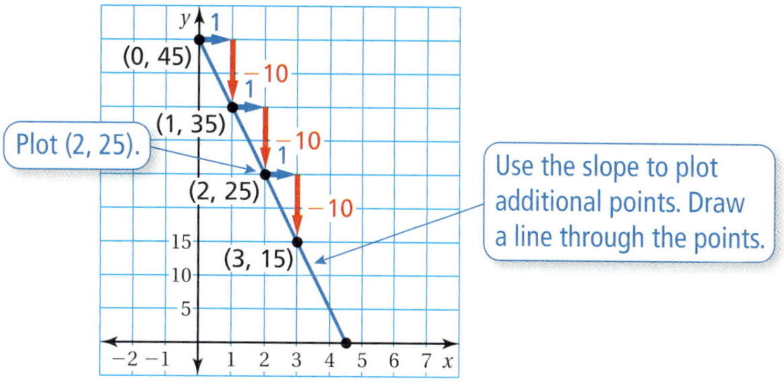

Because the line crosses the y-axis at (0, 45), the y-intercept is 45.

∴ So, the equation is $y = -10x + 45$.

Check

Check that (2, 25) is a solution of the equation.

$y = -10x + 45$	Write the equation.
$25 \stackrel{?}{=} -10(2) + 45$	Substitute.
$25 = 25$ ✓	Simplify.

b. You start descending when $x = 0$. The y-intercept is 45. So, you were parasailing at a height of 45 feet.

c. You reach the boat when $y = 0$.

$y = -10x + 45$	Write the equation.
$0 = -10x + 45$	Substitute 0 for y.
$-45 = -10x$	Subtract 45 from each side.
$4.5 = x$	Solve for x.

∴ You reach the boat after 4.5 seconds.

On Your Own

3. **WHAT IF?** In Example 2, you are 35 feet above the boat after 2 seconds. When do you reach the boat?

3.2 Exercises

Check It Out
Help with Homework
BigIdeasMath.com

Vocabulary and Concept Check

1. **WRITING** What information do you need to write an equation of a line?
2. **WRITING** Describe how to write an equation of a line using its slope and a point on the line.

Practice and Problem Solving

Write an equation of the line with the given slope that passes through the given point.

3. $m = \frac{1}{2}$

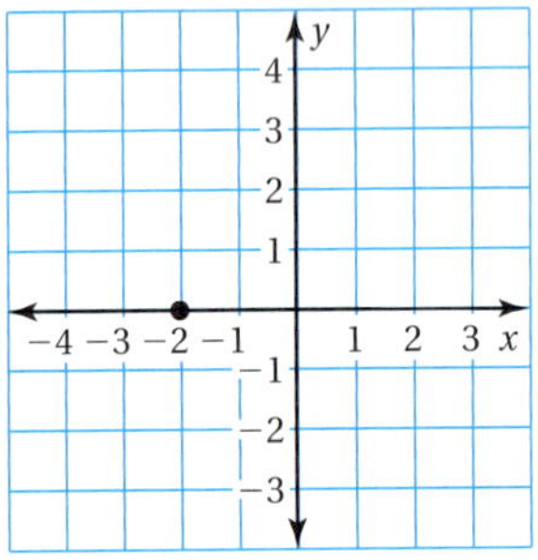

4. $m = -\frac{3}{4}$

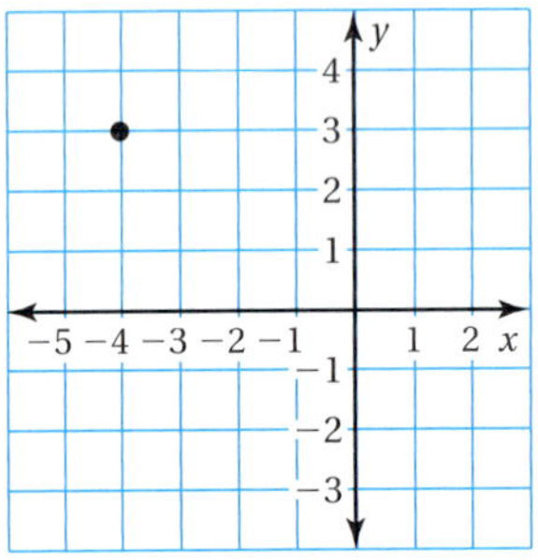

5. $m = -3$

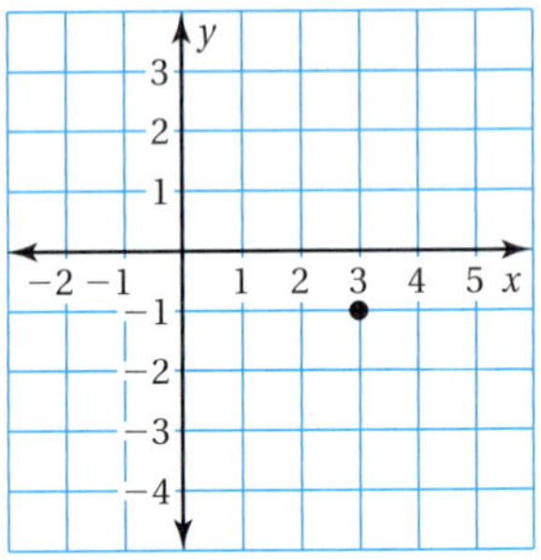

(1) 6. $m = -\frac{2}{3}$; $(3, 0)$

7. $m = \frac{3}{4}$; $(4, 8)$

8. $m = 4$; $(1, -3)$

9. $m = -\frac{1}{7}$; $(7, -5)$

10. $m = \frac{5}{3}$; $(3, 3)$

11. $m = -2$; $(-1, -4)$

12. **ERROR ANALYSIS** Describe and correct the error in writing an equation of the line with a slope of $\frac{1}{3}$ that passes through the point $(6, 4)$.

✗ $y = \frac{1}{3}x + 4$

13. **CHEMISTRY** At 0°C, the volume of a gas is 22 liters. For each degree the temperature T (in degrees Celsius) increases, the volume V (in liters) of the gas increases by $\frac{2}{25}$. Write an equation that represents the volume of the gas in terms of the temperature.

Assignment Guide and Homework Check

Level	Day 1 Activity Assignment	Day 2 Lesson Assignment	Homework Check
Basic	3–5, 18, 19	1, 2, 7–13 odd, 12, 14	7, 9, 12, 14
Average	3–5, 18, 19	1, 2, 7, 9, 11, 12, 14, 15	7, 9, 12, 14
Advanced	3–5, 18, 19	1, 2, 10, 12, 14–17	10, 12, 14, 16

For Your Information

- **Exercise 1** There are many possible answers.
- **Exercise 17** Because of the lack of gravity in space, an astronaut's musculoskeletal system is not used as intensively, resulting in a reduction of bone and muscle strength and size.

Common Errors

- **Exercises 6–11** Students might use the reciprocal of the slope when plotting the second point. Remind students of the definition of slope.
- **Exercises 6–11** Students might say that the *y*-intercept is the *y*-coordinate of the given point. Remind them that the *y*-intercept is represented by *b* in the point (0, *b*), and if the given point is not in this form, they need to use the slope to find where the line crosses the *y*-axis.
- **Exercise 13** Students might have trouble knowing which variable can be compared with *x* and *y* and may write the given point backwards. Review what the words "in terms of" mean when writing an equation. In this problem, *V* could be replaced by *y* and *T* could be replaced by *x*. Remind students to check their equation by substituting the given point and checking that it is a solution of the equation.

3.2 Record and Practice Journal

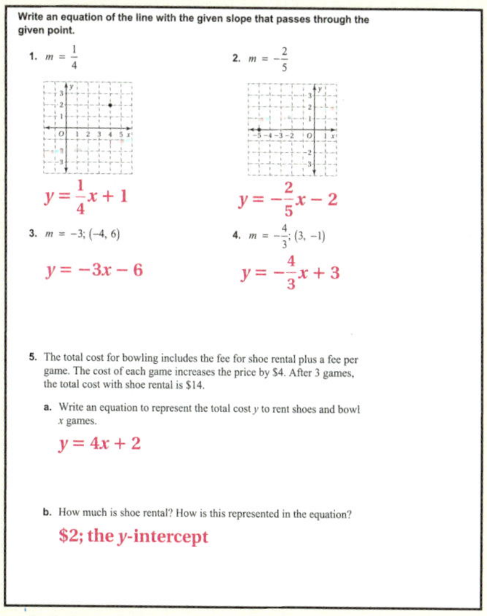

Write an equation of the line with the given slope that passes through the given point.

1. $m = \frac{1}{4}$

$y = \frac{1}{4}x + 1$

2. $m = -\frac{2}{5}$

$y = -\frac{2}{5}x - 2$

3. $m = -3$; $(-4, 6)$

$y = -3x - 6$

4. $m = -\frac{4}{3}$; $(3, -1)$

$y = -\frac{4}{3}x + 3$

5. The total cost for bowling includes the fee for shoe rental plus a fee per game. The cost of each game increases the price by $4. After 3 games, the total cost with shoe rental is $14.

a. Write an equation to represent the total cost *y* to rent shoes and bowl *x* games.

$y = 4x + 2$

b. How much is shoe rental? How is this represented in the equation?

$2; the *y*-intercept

Technology For the Teacher

Answer Presentation Tool

Quiz*Show*

Vocabulary and Concept Check

1. *Sample answer:* slope and a point
2. Plot the given point and use the slope to plot additional points to find the *y*-intercept. Then use the slope *m* and the *y*-intercept *b* to write the equation $y = mx + b$.

Practice and Problem Solving

3. $y = \frac{1}{2}x + 1$
4. $y = -\frac{3}{4}x$
5. $y = -3x + 8$
6. $y = -\frac{2}{3}x + 2$
7. $y = \frac{3}{4}x + 5$
8. $y = 4x - 7$
9. $y = -\frac{1}{7}x - 4$
10. $y = \frac{5}{3}x - 2$
11. $y = -2x - 6$
12. The *y*-intercept is wrong.
    $y = \frac{1}{3}x + 2$
13. $V = \frac{2}{25}T + 22$

Practice and Problem Solving

14. a. $V = -4000x + 30{,}000$

b. \$30,000

15. See *Taking Math Deeper.*

16. a. $y = 30x + 20$

b. the flat fee for renting the airboat

17. a. $y = -0.03x + 2.9$

b. $2\ \text{g/cm}^2$

c. *Sample answer:* Eventually $y = 0$, which means the astronaut's bones will be very weak.

Fair Game Review

18.

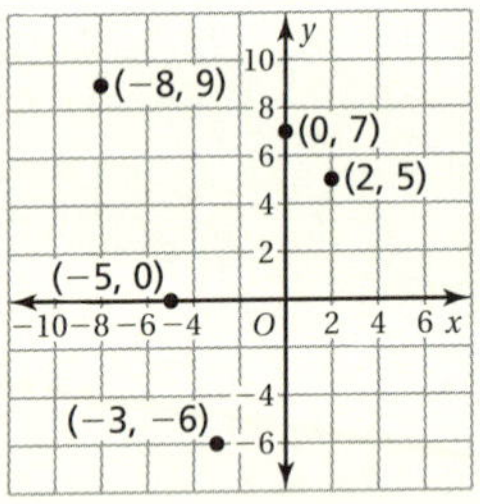

19. B

Mini-Assessment

Write an equation of the line with the given slope that passes through the given point.

1. $m = 3$; $(1, 4)$ $y = 3x + 1$
2. $m = -2$; $(-2, 1)$ $y = -2x - 3$
3. $m = 1$; $(3, 5)$ $y = x + 2$
4. $m = \frac{1}{2}$; $(2, 1)$ $y = \frac{1}{2}x$
5. **You rent a floor sander for \$24 per day. You pay \$82 for 3 days.**
   a. Write an equation that represents your total cost y (in dollars) after x days. $y = 24x + 10$
   b. Interpret the y-intercept. The y-intercept is 10. This means you paid a deposit fee of \$10 to rent the sander.

Taking Math Deeper

Exercise 15

The challenge in this biology problem is to interpret the given information as a rate of change (or slope) and as an ordered pair.

1 Translate the given information into math.

T = temperature (°F)
x = chirps per minute
Rate of change = 0.25 degree per chirp

2 Write an equation.

Given point: $(x, T) = (40, 50)$

With a slope of 0.25, you can determine that the T-intercept of the line is 40. So, the equation is

a. $T = 0.25x + 40.$

3 Use the equation.
If $x = 100$ chirps per minute, then

$T = 0.25(100) + 40$
b. $= 65°\text{F}.$

If $T = 96$, then you can find the number of chirps per minute as follows.

$96 = 0.25x + 40$
$56 = 0.25x$
$224 = x$

c. So, you would expect the cricket to make 224 chirps in one minute.

This relationship between temperature and cricket chirps was first published by Amos Dolbear in 1897 in an article called *The Cricket as a Thermometer.*

Project

Research other plants or animals that predict the temperature or weather.

Reteaching and Enrichment Strategies

If students need help...	If students got it...
Resources by Chapter • Practice A and Practice B • Puzzle Time Record and Practice Journal Practice Differentiating the Lesson Lesson Tutorials Skills Review Handbook	Resources by Chapter • Enrichment and Extension Start the next section

14. CARS After it is purchased, the value of a new car decreases \$4000 each year. After 3 years, the car is worth \$18,000.

a. Write an equation that represents the value V (in dollars) of the car x years after it is purchased.

b. What was the original value of the car?

15. CRICKETS According to Dolbear's Law, you can predict the temperature T (in degrees Fahrenheit) by counting the number x of chirps made by a snowy tree cricket in 1 minute. For each chirp the cricket makes in 1 minute, the temperature rises 0.25 degree.

a. A cricket chirps 40 times in 1 minute when the temperature is 50°F. Write an equation that represents the temperature in terms of the number of chirps in 1 minute.

b. You count 100 chirps in 1 minute. What is the temperature?

c. The temperature is 96°F. How many chirps would you expect the cricket to make?

16. AIRBOATS You rent an airboat. The total cost includes a flat fee plus an hourly fee.

a. After 4 hours the total cost is \$140. Write an equation that represents the total cost y after x hours.

b. Interpret the y-intercept.

17. Critical Thinking Bone mineral density is a measure of the strength of bones. The average bone mineral density of a female astronaut who has never been in space is 2.9 grams per square centimeter. For the first three years she spends in space, her bone density decreases by 0.03 grams per square centimeter per month.

a. Write an equation that represents the bone mineral density y of a female astronaut in terms of the number x of months she spends in space.

b. What is her bone mineral density after 2 years and 6 months in space?

c. Explain why the amount of time an astronaut can spend in space is limited.

Fair Game Review What you learned in previous grades & lessons

18. Plot the ordered pairs in the same coordinate plane. *(Skills Review Handbook)*

(2, 5), (−3, −6), (0, 7), (−5, 0), (−8, 9)

19. MULTIPLE CHOICE What is the y-intercept of the equation $5x - 2y = 28$? *(Section 2.4)*

Ⓐ $-\frac{5}{2}$ Ⓑ -14 Ⓒ $\frac{5}{2}$ Ⓓ 5.6

3.3 Writing Equations Using Two Points

Essential Question How can you write an equation of a line when you are given two points on the line?

1 ACTIVITY: Writing Equations of Lines

Work with a partner.

- **Sketch the line that passes through the given points.**
- **Find the slope and *y*-intercept of the line.**
- **Write an equation of the line.**

a.

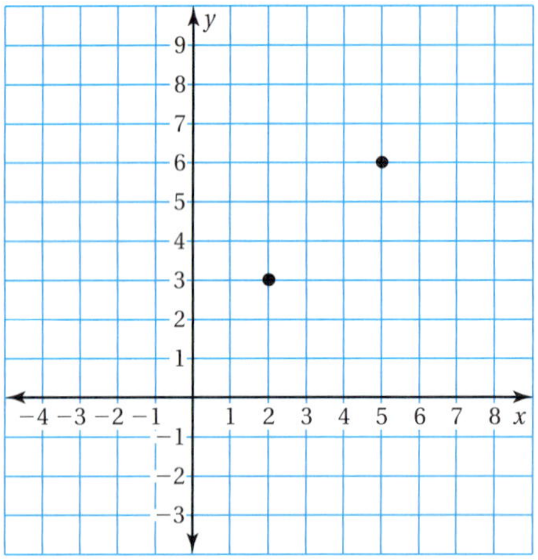

b.

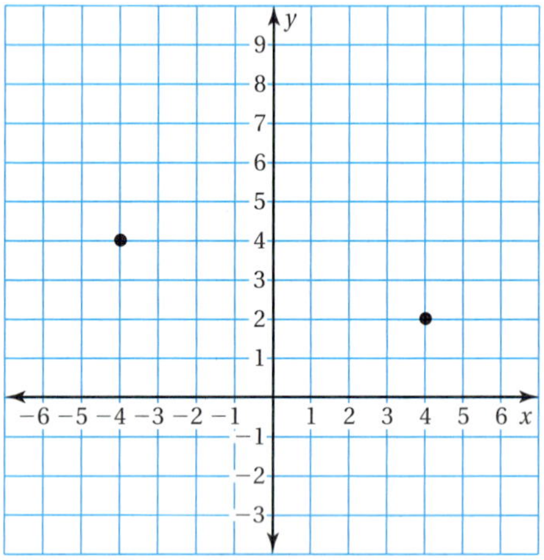

c.

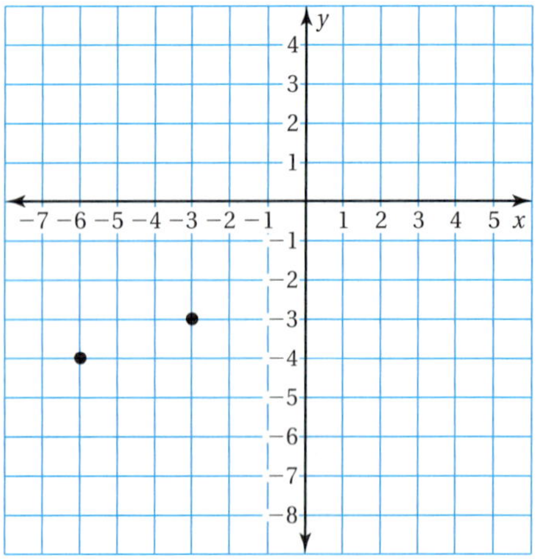

d.

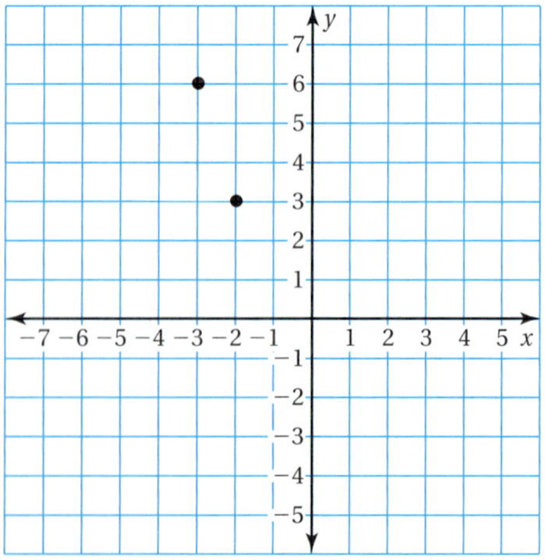

Laurie's Notes

Introduction

For the Teacher

- **Goal:** Students will explore how to write an equation for a line given two points on the line.
- **Big Idea:** The problems in this activity assume that the rate of change (slope) remains constant for a period of time, which in practice is unlikely. The goal is to practice some important mathematical skills in a context that is plausible.

Motivate

? "Have any of you ever taken a hot air balloon ride?"

- If you have taken a hot air balloon ride, share the experience with your students.
- Share a few "math tidbits" followed by posing a question that they will answer later.
  - Traditional balloons look like a sphere at the top and a truncated cone at the bottom.
  - The propane tanks are generally cylindrical.
  - One of the instruments on the balloon is the variometer which keeps track of the rate of climb (vertical speed).

Discuss

- Share a story about a friend who took a hot air balloon ride. Explain that your friend was holding a portable variometer and knew that at 2 minutes the height was 300 feet and at 3 minutes the height was 450 feet. At this rate, how many minutes did it take to reach 900 feet?

Activity Notes

Activity 1

? "How do you find the slope of the line between two points in the coordinate plane?" Draw the horizontal and vertical arrows that represent the run and the rise, and then find the ratio of rise to run.

- Given the similarity of this activity to the previous two, students should be able to get started right away. Once students have found the slope of the line, they will find the y-intercept as they did previously.
- All of the y-intercepts are integer values, so they will be easy for students to find.
- In a formal algebra class, the challenge of this lesson is the symbolism involved in the formulas. Having students draw the lines first, *see* the slope and y-intercept, and then write the equation, has eliminated this challenge.

? "What made it possible to be able to write the equation of the line?" Slope is found by using the definition of slope. Then by using the slope, you can find the y-intercept. Finally, substitute into the formula $y = mx + b$.

Previous Learning

Students should know how to plot points and find the slope between the two points. Students should know how to apply the definition of slope in order to find the y-intercept of a graph.

Start Thinking! and Warm Up

Activity 3.3 Start Thinking!
For use before Activity 3.3

Activity 3.3 Warm Up
For use before Activity 3.3

Write an equation of the line given slope *m* and *y*-intercept *b*.

1. $m = \frac{1}{2}, b = 3$	**2.** $m = -2, b = 0$
3. $m = 1, b = -3$	**4.** $m = \frac{1}{3}, b = -1$
5. $m = -\frac{3}{5}, b = \frac{1}{2}$	**6.** $m = -\frac{7}{5}, b = -\frac{1}{5}$

3.3 Record and Practice Journal

Essential Question How can you write an equation of a line when you are given two points on the line?

1 ACTIVITY: Writing Equations of Lines

Work with a partner.

- Sketch the line that passes through the given points.
- Find the slope and y-intercept of the line.
- Write an equation of the line.

a.

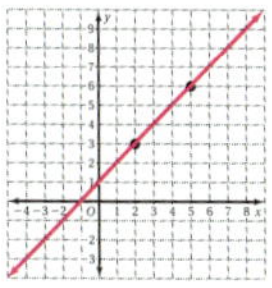

slope: 1 y-intercept: 1
equation: $y = x + 1$

b.

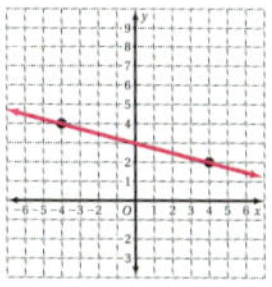

slope: $-1/4$ y-intercept: 3
equation: $y = -\frac{1}{4}x + 3$

c.

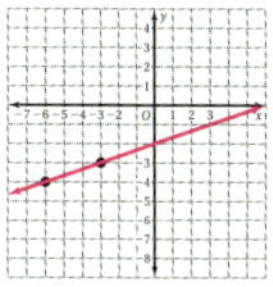

slope: $1/3$ y-intercept: -2
equation: $y = \frac{1}{3}x - 2$

d.

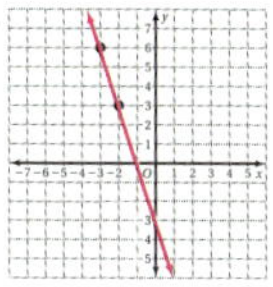

slope: -3 y-intercept: -3
equation: $y = -3x - 3$

English Language Learners

Group Activity

Form small groups of English learners and English speakers. Provide each group with enough information (slope and y-intercept, slope and a point, or two points) to write equations. After writing an equation, the equation should be graphed in the coordinate plane. Students can share their work at the board or overhead.

3.3 Record and Practice Journal

2 ACTIVITY: Writing and Using Linear Equations

Work with a partner.

a. You are rising in a hot air balloon. After 1 minute, you are 200 feet above the ground. After 4 minutes, you are 800 feet above the ground.

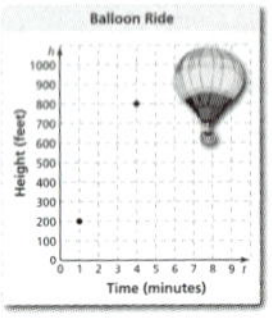

- Write an equation for the height h in terms of the time t.

  $h = 200t$

- Use your equation to find the height of the balloon after 5 minutes.

  1000 ft

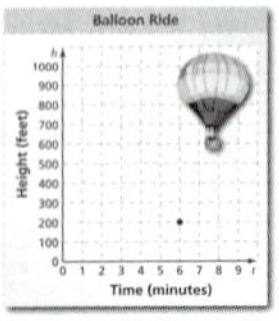

b. After 5 minutes, the hot air balloon starts to descend. After 6 minutes, you are 200 feet above the ground.

- Write an equation for the height h in terms of the time t.

  $h = -800t + 5000$

- Use your equation to estimate when the balloon lands on the ground.

  $6\frac{1}{4}$ minutes

c. You are on a roller coaster. After 3 seconds, you are 190 feet above the ground and have reached maximum speed. One second later, you are 95 feet above the ground.

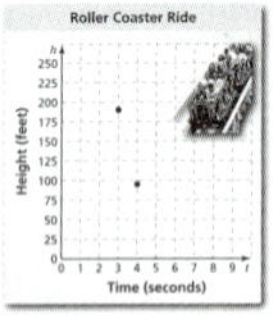

- Write an equation for the height h in terms of the time t.

  $h = -95t + 475$

- When will you reach ground level?

  5 seconds

What Is Your Answer?

3. **IN YOUR OWN WORDS** How can you write an equation of a line when you are given two points on the line? Give an example that is different from those in Activities 1 and 2.

   Plot points and draw the line through them. Use the graph to find slope and y-intercept.

Laurie's Notes

Activity 2

- **Part (a):** Write the two ordered pairs that are described in words and are plotted on the graph, (1, 200) and (4, 800). Ask a few questions.
  - "How did the height of the balloon change in the 3 minutes between these two points?" Height increased 600 feet.
  - "What is the slope between these points?" slope $= \frac{\text{rise}}{\text{run}} = \frac{600}{3} = 200$
  - "What units do you need to describe the slope?" ft per min
- Students should work backwards to find the h-intercept, which is 0.
- It is difficult for students to take the information they know and write the equation for the height h in terms of the time t. Students should be able to write $y = 200x + 0$, or $y = 200x$, but the direction sentence of "write an equation for the height h in terms of the time t" seems to overwhelm students. Have students look at the graph and how the axes are labeled. The time t is on the horizontal axis and the height h is on the vertical axis. In the equation $y = 200x$, replace x with t and replace y with h.
- "What does a slope of 200 mean in the context of this problem?" For every minute that passes, the height of the balloon increases 200 feet.
- "What does an h-intercept of 0 mean in the context of this problem?" The balloon starts at ground level.

- **Part (b):** Students use their answer (5, 1000) from part (a), along with the new ordered pair (6, 200), to find the slope. The balloon is descending (negative slope) at a rate of 800 feet per minute.
- Determining the h-intercept for this problem is challenging for two reasons: the h-intercept is off the graph and the balloon was never at that height. Suggest that students make a table of values.

Time (minutes), t	6	5	4	3	2	1	0
Height (feet), h	200	1000	1800	2600	3400	4200	5000

- "How do you use the equation to find out when the balloon will land on the ground?" Solve $0 = -800t + 5000$.

Closure

- Refer back to the balloon question posed at the beginning and ask students to write an equation for the height of the balloon in terms of the time in minutes. Find the time when the height is 900 feet. $y = 150x$; 6 min

Technology For the Teacher

Dynamic Classroom

The Dynamic Planning Tool

Editable Teacher's Resources at *BigIdeasMath.com*

2 ACTIVITY: Writing and Using Linear Equations

Work with a partner.

a. You are rising in a hot air balloon. After 1 minute, you are 200 feet above the ground. After 4 minutes, you are 800 feet above the ground.

- Write an equation for the height h in terms of the time t.
- Use your equation to find the height of the balloon after 5 minutes.

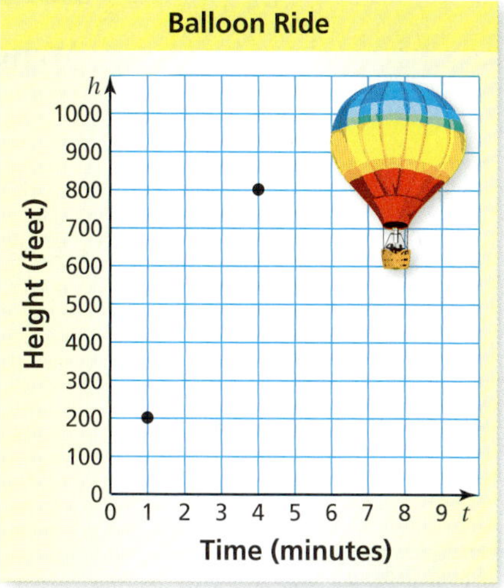

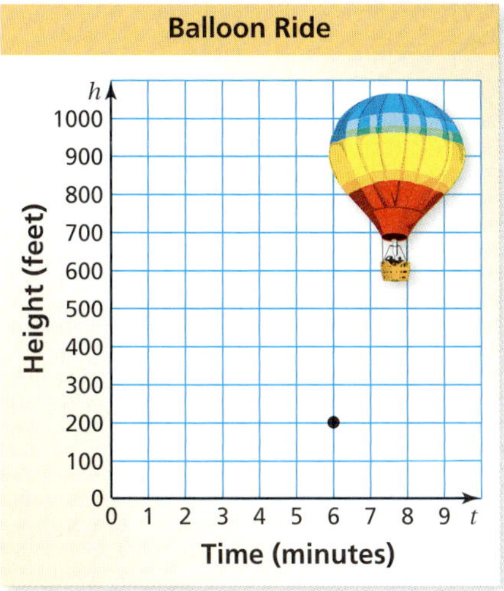

b. After 5 minutes, the hot air balloon starts to descend. After 6 minutes, you are 200 feet above the ground.

- Write an equation for the height h in terms of the time t.
- Use your equation to estimate when the balloon lands on the ground.

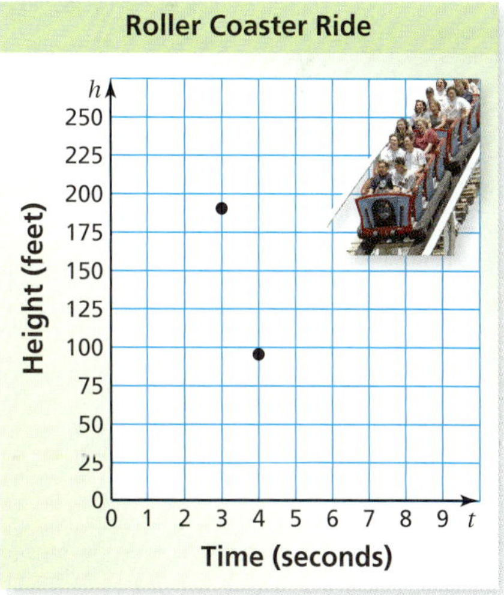

c. You are on a roller coaster. After 3 seconds, you are 190 feet above the ground and have reached maximum speed. One second later, you are 95 feet above the ground.

- Write an equation for the height h in terms of the time t.
- When will you reach ground level?

What Is Your Answer?

3. IN YOUR OWN WORDS How can you write an equation of a line when you are given two points on the line? Give an example that is different from those in Activities 1 and 2.

Practice Use what you learned about writing equations using two points to complete Exercises 3–5 on page 122.

3.3 Lesson

Check It Out
Lesson Tutorials
BigIdeasMath.com

EXAMPLE 1 Writing Equations Using Two Points

Write an equation of the line that passes through the points.

a. $(-6, 6), (-3, 4)$

Use a graph to find the slope and y-intercept.

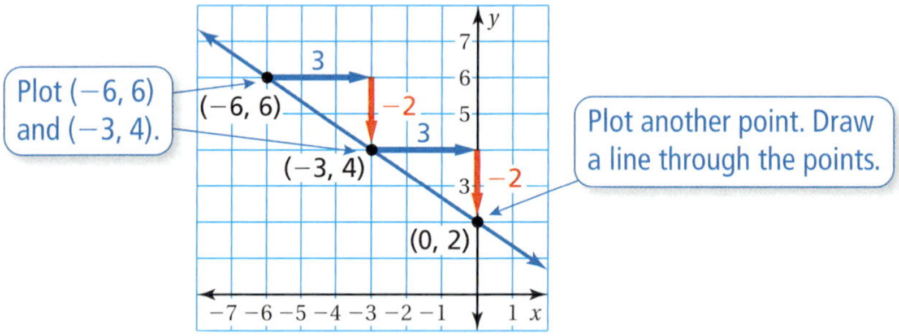

$$\text{slope} = \frac{\text{rise}}{\text{run}} = \frac{-2}{3} = -\frac{2}{3}$$

Because the line crosses the y-axis at $(0, 2)$, the y-intercept is 2.

So, the equation is $y = -\frac{2}{3}x + 2$.

Study Tip

After writing an equation, check that the given points are solutions of the equation.

b. $(-2, -4), (1, -1)$

Use a graph to find the slope and y-intercept.

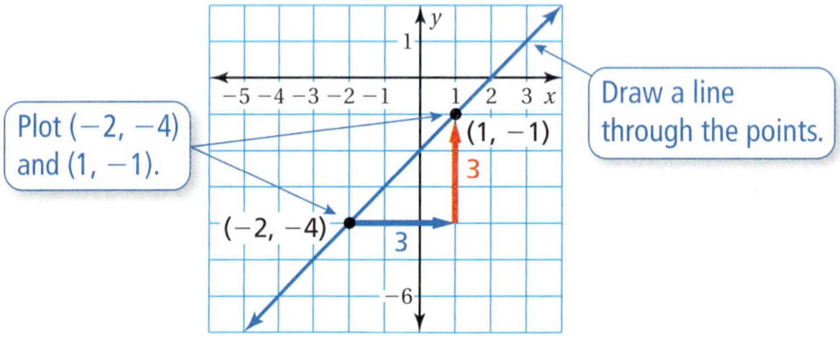

$$\text{slope} = \frac{\text{rise}}{\text{run}} = \frac{3}{3} = 1$$

Because the line crosses the y-axis at $(0, -2)$, the y-intercept is -2.

So, the equation is $y = 1x + (-2)$, or $y = x - 2$.

On Your Own

Now You're Ready
Exercises 6–14

Write an equation of the line that passes through the points.

1. $(2, 3), (4, 4)$

2. $(-1, 2), (1, -4)$

Laurie's Notes

Introduction

Connect

- **Yesterday:** Students developed an intuitive understanding about writing the equation of a line given two points on the line.
- **Today:** Students will write an equation of a line given two points.

Motivate

- **Model:** Model what it looks like when you pour water out of a gallon jug into a bucket at a constant rate. Pour slowly! The bucket should have a uniform shape vertically.
- The variable x is the time in seconds. The variable y is the height of the water in the bucket.
- Write on the board: After 2 seconds, the height of the water in the bucket was 1 inch. Two seconds later, the height was 2 inches.

? "When will the height be 5 inches?" 10 sec

Discuss

- Discuss the water problem. You could work the problem now or wait until the end of the lesson for students to solve it on their own.
- Explain that when two data points are given and they describe a linear equation, it is possible to write the equation of the line.

Lesson Notes

Example 1

- **Part (a):** State and write the given information: points $(-6, 6)$ and $(-3, 4)$.
- Plot both points. Draw the line through the two points.

? "Is the slope positive or negative?" negative

? "What is the slope?" $-\frac{2}{3}$

- Use the slope to plot additional points and to find the point where the graph crosses the y-axis. Write the equation.
- Have students check to make sure that the equation contains the given points.
- **Part (b):** Students must be careful when plotting the points because the line through the given points already crosses the y-axis. If students do not have a sharp pencil, or if the straightedge is not lined up carefully, it may not be obvious to students that the y-intercept is -2.
- **Big Idea:** Before students attempt to write the equation of the line, they should know if the slope and y-intercept are positive or negative.

On Your Own

- Have students share their work at the overhead or at the board.
- Note that Question 2 is similar to part (b). The given ordered pairs are on opposite sides of the y-axis.

Goal Today's lesson is writing the equation of a line given two points.

Lesson Materials	
Introduction	**Textbook**
• bucket • gallon of water	• straightedge

Start Thinking! and Warm Up

Lesson 3.3 **Warm Up**
For use before Lesson 3.3

Lesson 3.3 **Start Thinking!**
For use before Lesson 3.3

The table shows the U.S. Postal Service cost for sending media mail (small and large packages of books, film, manuscripts, sound recordings, video tapes, and computer readable media such as CDs, DVDs, and diskettes).

How can you write an equation to represent the cost y of sending x pounds of media mail?

Weight Not Over (pounds)	Cost
1	$2.23
2	$2.58
3	$2.93
4	$3.28
5	$3.63
6	$3.98

Extra Example 1

Write an equation that passes through the points.

a. $(-2, 3), (-1, 1)$ $y = -2x - 1$

b. $(-2, 2), (2, 4)$ $y = \frac{1}{2}x + 3$

On Your Own

1. $y = \frac{1}{2}x + 2$

2. $y = -3x - 1$

Laurie's Notes

Extra Example 2

Write an equation that passes through $(-4, -3)$ and $(4, 3)$. $y = \frac{3}{4}x$

Extra Example 3

A three-week old puppy weighs 24 ounces. Two weeks later, it weighs 36 ounces.

a. Write an equation to represent the weight y (in ounces) of the puppy x weeks after birth. $y = 6x + 6$

b. How old is the puppy when it weighs 60 ounces? 9 weeks

On Your Own

3. C

4. 8 weeks

Differentiated Instruction

Visual

Students can color code their notes to identify the different parts of the graphs and equations. As in the textbook, use red to show the rise of the graph and the numerator of the slope. Use blue to show the run of the graph and the denominator of the slope. Then green can be used to label the y-intercept on the graph and in the equation.

Example 2

- Plot the two points and draw the line through them.
- ? "Is the slope positive or negative?" negative
- Only two choices have a negative slope, choice A and choice C.
- **Reasoning:** It appears that the y-intercept is 0, so choice A and choice C are still viable choices. It is necessary to actually find the slope.
- ? "What is the slope for this problem? How did you know?" Slope is $-\frac{1}{2}$. The rise to run ratio is $-\frac{1}{2}$.
- **Common Error:** Students may find the reciprocal of the slope.
- **Alternate Approach:** Students can find the answer using a version of Guess, Check, and Revise. Students can substitute the ordered pairs into the equations and check to see which equation is true for *both* points.

Example 3

- Ask a student to read the problem and the information that is given in the photo.
- This example assumes that the kitten's weight is increasing at a constant rate, which may not be exactly true. For the contextual problems, it is assumed that the rate stays constant so that students can practice the algebraic skills for a problem that is plausible.
- ? "In the graph, what do x and y represent?" x is the number of weeks and y is the weight of the kitten in ounces.
- Find the equation of the line. Use the equation to solve part (b).
- ? "Interpret what the slope and y-intercept mean for this problem." Slope of 3 means the kitten's weight is increasing 3 ounces per week. The y-intercept of 3 means the kitten weighed 3 ounces at birth.

On Your Own

- These problems take time, but they provide a good assessment of students' understanding of the lesson.

Closure

- **Exit Ticket:** Write an equation of the line through the points $(0, 4)$ and $(2, 2)$. $y = -x + 4$

Technology For the Teacher

Dynamic Classroom

The Dynamic Planning Tool
Editable Teacher's Resources at *BigIdeasMath.com*

EXAMPLE 2 Standardized Test Practice

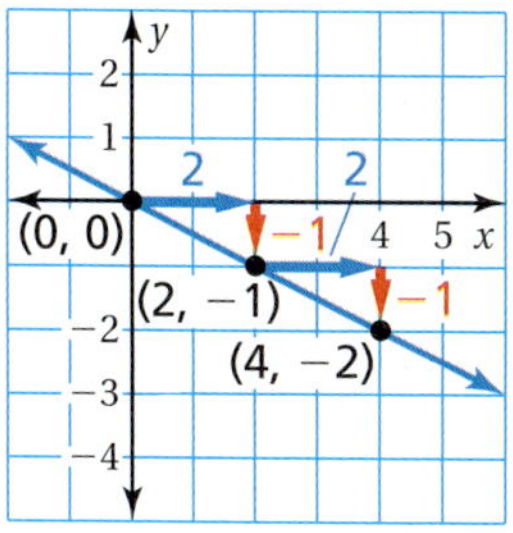

The graph of which equation passes through (2, −1) and (4, −2)?

Ⓐ $y = -\frac{1}{2}x$ Ⓑ $y = \frac{1}{2}x$

Ⓒ $y = -2x$ Ⓓ $y = 2x$

Graph the line through the points. Find the slope and y-intercept.

$$\text{slope} = \frac{\text{rise}}{\text{run}} = \frac{-1}{2} = -\frac{1}{2}$$

Because the line crosses the y-axis at (0, 0), the y-intercept is 0.

∴ So, the equation is $y = -\frac{1}{2}x + 0$, or $y = -\frac{1}{2}x$.

The correct answer is Ⓐ.

EXAMPLE 3 Real-Life Application

22.5 oz

A 2-week old kitten weighs 9 ounces. Two weeks later, it weighs 15 ounces. (a) Write an equation to represent the weight y (in ounces) of the kitten x weeks after birth. (b) How old is the kitten in the photo?

a. The kitten weighs 9 ounces after 2 weeks and 15 ounces after 4 weeks. So, graph the line that passes through (2, 9) and (4, 15).

$$\text{slope} = \frac{\text{rise}}{\text{run}} = \frac{6}{2} = 3$$

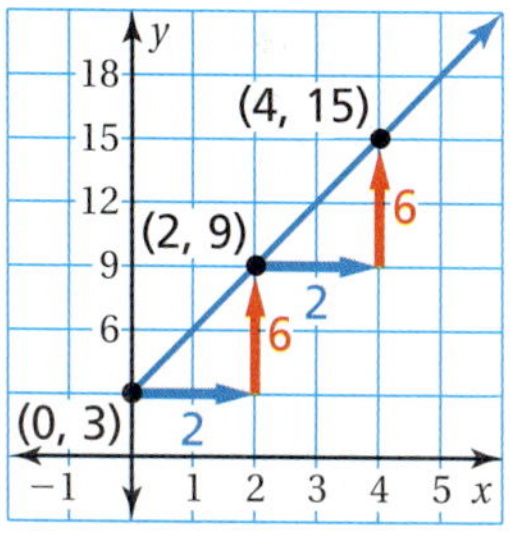

Because the line crosses the y-axis at (0, 3), the y-intercept is 3.

∴ So, the equation is $y = 3x + 3$.

b. Find the value of x when $y = 22.5$.

$y = 3x + 3$	Write the equation.
$22.5 = 3x + 3$	Substitute 22.5 for y.
$19.5 = 3x$	Subtract 3 from each side.
$6.5 = x$	Solve for x.

∴ The kitten in the photo is 6.5 weeks old.

On Your Own

3. The graph of which equation in Example 2 passes through (−2, 4) and (−1, 2)?

4. A 3-week old kitten weighs 12 ounces. Two weeks later, it weighs 18 ounces. How old is the kitten when it weighs 27 ounces?

3.3 Exercises

Check It Out
Help with Homework
BigIdeasMath.com

Vocabulary and Concept Check

1. **WRITING** Describe how to write an equation of a line using two points on the line.

2. **WHICH ONE DOESN'T BELONG?** Which pair of points does *not* belong with the other three? Explain your reasoning.

(0, 1), (2, 3)	(1, 2), (4, 5)	(2, 3), (5, 6)	(1, 2), (4, 6)

Practice and Problem Solving

Find the slope and y-intercept of the line that passes through the points. Then write an equation of the line.

3. 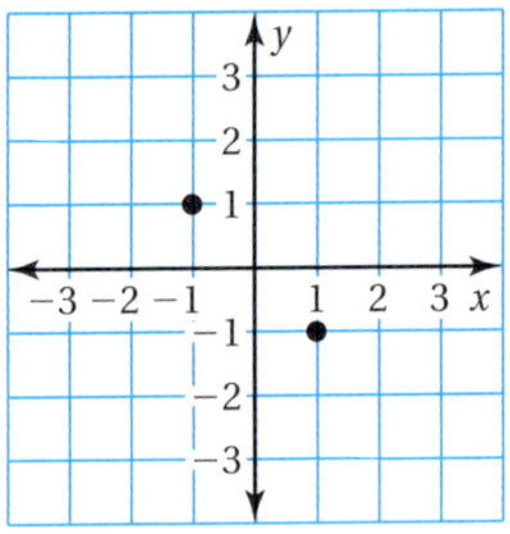

4. 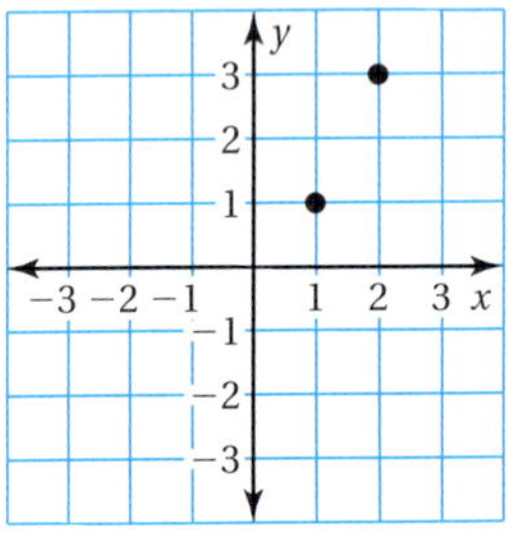

5. 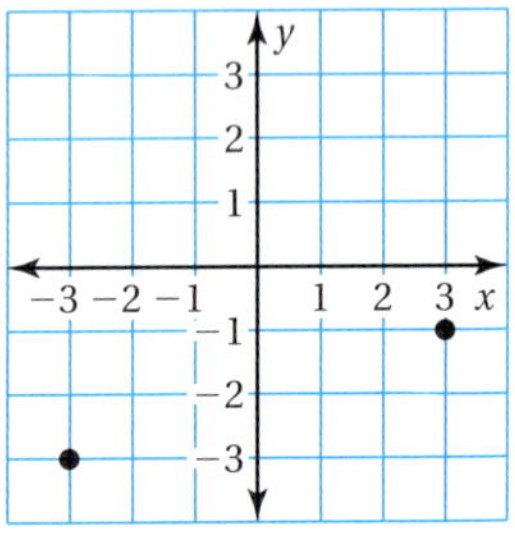

Write an equation of the line that passes through the points.

1 6. $(-1, -1), (1, 5)$

7. $(2, 4), (3, 6)$

8. $(-2, 3), (2, 7)$

9. $(4, 1), (8, 2)$

10. $(-9, 5), (-3, 3)$

11. $(1, 2), (-2, -1)$

12. $(-5, 2), (5, -2)$

13. $(2, -7), (8, 2)$

14. $(1, -2), (3, -8)$

15. **ERROR ANALYSIS** Describe and correct the error in finding the equation of the line that passes through $(-1, -6)$ and $(3, 2)$.

✗ slope $= \dfrac{\text{rise}}{\text{run}} = \dfrac{8}{4} = 2$

The y-intercept is $(0, -4)$.

The equation is $y = -4x + 2$.

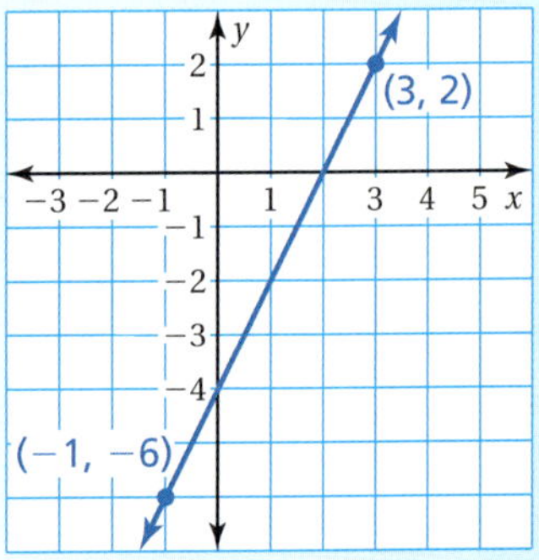

16. **JET SKI** It costs \$175 to rent a jet ski for 2 hours. It costs \$300 to rent a jet ski for 4 hours. Write an equation that represents the cost y (in dollars) of renting a jet ski for x hours.

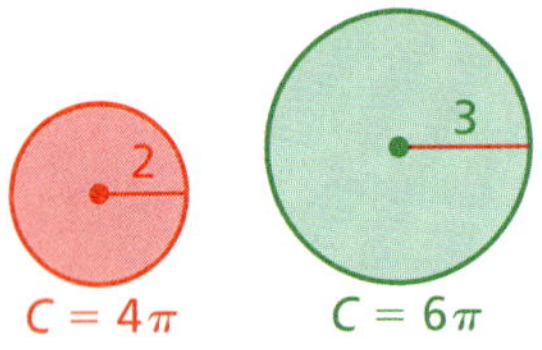

17. **CIRCUMFERENCE** Consider the circles shown.

  a. Plot the points $(2, 4\pi)$ and $(3, 6\pi)$.

  b. Write an equation of the line that passes through the two points.

Assignment Guide and Homework Check

Level	Day 1 Activity Assignment	Day 2 Lesson Assignment	Homework Check
Basic	3–5, 22–25	1, 2, 7–17 odd, 16	2, 7, 16, 17
Average	3–5, 22–25	1, 2, 9–11, 15, 17–19	2, 10, 17, 18
Advanced	3–5, 22–25	1, 2, 12–15, 19–21	2, 12, 19, 20

Common Errors

- **Exercises 6–14** Students may plot the points, draw a line through the points, and then use the point where the line crosses the *y*-axis to find the *y*-intercept. Sometimes this is not the correct point because students may not be able to draw a straight line. Encourage them to use the slope to find the *y*-intercept.
- **Exercise 16** Students may write the first point one way and the second point another way. For example, a student may write (2, 175) and (300, 4). Encourage them to write a description of the ordered pair, for example, (hours, cost).
- **Exercise 17** Students may struggle when plotting the given points because π is used. Encourage them to scale the *y*-axis by increments of π.

3.3 Record and Practice Journal

Write an equation of the line that passes through the points.

1. 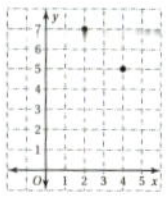

$y = -x + 9$

2. 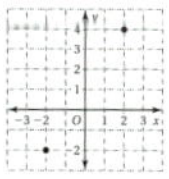

$y = \frac{3}{2}x + 1$

3. (−3, 0), (−2, 3)

$y = 3x + 9$

4. (−6, 10), (6, −10)

$y = -\frac{5}{3}x$

5. It costs \$315 to book a DJ for 3 hours. It costs \$525 to book the same DJ for 5 hours. Write an equation that represents the cost *y* in dollars of booking a DJ for *x* hours.

$y = 105x$

6. Water comes out of a garden hose at a constant rate to fill a pool. After 3 minutes, the pool is filled with 30 gallons of water. After 6.5 minutes, the pool is filled with 65 gallons of water.

   a. Write an equation that represents the number of gallons of water *y* in the pool after *x* minutes.

   $y = 10x$

   b. How long will it take to fill a pool that needs 10,000 gallons of water?

   1000 minutes

Technology For the Teacher

Answer Presentation Tool

Quiz*Show*

Vocabulary and Concept Check

1. Plot both points and draw the line that passes through them. Use the graph to find the slope and *y*-intercept. Then write the equation in slope-intercept form.
2. (1, 2), (4, 6); The slope of the line connecting these two points is $\frac{4}{3}$. The slope of the other pairs of points is 1.

Practice and Problem Solving

3. slope $= -1$; *y*-intercept: 0; $y = -x$
4. slope $= 2$; *y*-intercept: -1; $y = 2x - 1$
5. slope $= \frac{1}{3}$; *y*-intercept: -2; $y = \frac{1}{3}x - 2$
6. $y = 3x + 2$
7. $y = 2x$
8. $y = x + 5$
9. $y = \frac{1}{4}x$
10. $y = -\frac{1}{3}x + 2$
11. $y = x + 1$
12. $y = -\frac{2}{5}x$
13. $y = \frac{3}{2}x - 10$
14. $y = -3x + 1$
15. They switched the slope and *y*-intercept in the equation. $y = 2x - 4$
16. $y = 62.5x + 50$
17. See Additional Answers.

Practice and Problem Solving

18. a. 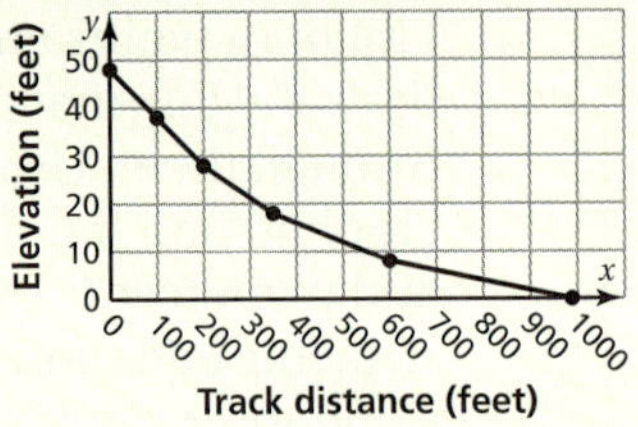

b. no; The graph is not a line.

c. $y = -\frac{1}{10}x + 48$

19. See Additional Answers.

20. See *Taking Math Deeper.*

21. a. $y = 14x - 108.5$

b. 4 m

Fair Game Review

22. 45 **23.** 175

24. −4.5 **25.** D

Mini-Assessment

Write an equation of the line that passes through the points.

1. $(-2, 1), (3, -4)$ $y = -x - 1$
2. $(-4, 0), (2, 3)$ $y = \frac{1}{2}x + 2$
3. $(-3, -8), (4, 6)$ $y = 2x - 2$
4. $(-2, 6), (3, -9)$ $y = -3x$
5. After 2 weeks, you have made \$150 mowing lawns. After 4 weeks, you have made \$300. Write an equation that represents your earnings y (in dollars) after x weeks. $y = 75x$

Taking Math Deeper

Exercise 20

As with many real-life problems, this one has a lot of information. Help students understand that they are not expected to look at this problem and see the solution immediately. Just take a deep breath, relax, and start to organize the given information.

1 Translate the given information into math.

Let y = ounces of water and x = time in seconds.
First given point: (5, 58)
Second given point: (20, 28)

Note that "15 seconds later" means that the second time is $5 + 15 = 20$.

2 Plot the two points.

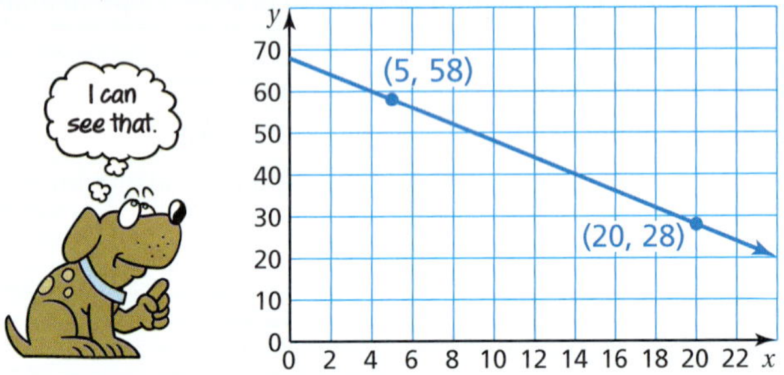

The slope is −2 ounces per second.
The y-intercept is 68. So, the equation is

a. $y = -2x + 68$.

3 Use the equation. When $x = 0$ seconds, the watering can has

$y = -2(0) + 68$
b. $= 68$ ounces.

The watering can will be empty when $y = 0$.

$0 = -2x + 68$
$2x = 68$
c. $x = 34$ seconds

The watering can will be empty after 34 seconds.

Reteaching and Enrichment Strategies

If students need help. . .	If students got it. . .
Resources by Chapter • Practice A and Practice B • Puzzle Time Record and Practice Journal Practice Differentiating the Lesson Lesson Tutorials Skills Review Handbook	Resources by Chapter • Enrichment and Extension • School-to-Work Start the next section

18. SOAP BOX DERBY The table shows the changes in elevation for a Soap Box Derby track.

Track Distance	Elevation
0 ft	48 ft
100 ft	38 ft
200 ft	28 ft
350 ft	18 ft
600 ft	8 ft
989 ft	0 ft

a. Draw a Soap Box Derby track in a coordinate plane.

b. Does each section of the track have the same slope? Explain.

c. Write an equation that represents the elevation y (in feet) of the track between 100 feet and 200 feet.

19. CAR VALUE The value of a car decreases at a constant rate. After 3 years, the value of the car is \$15,000. After 2 more years the value of the car is \$11,000.

a. Write an equation that represents the value y (in dollars) of the car after x years.

b. Graph the equation.

c. What is the y-intercept of the line? Interpret the y-intercept.

Leaning Tower of Pisa

20. WATERING CAN You water the plants in your classroom at a constant rate. After 5 seconds, your watering can contains 58 ounces of water. Fifteen seconds later, the can contains 28 ounces of water.

a. Write an equation that represents the amount y (in ounces) of water in the can after x seconds.

b. How much water was in the can when you started watering the plants?

c. When is the watering can empty?

21. Critical Thinking The Leaning Tower of Pisa in Italy was built between 1173 and 1350.

a. Write an equation for the yellow line.

b. The tower is 56 meters tall. How far off center is the top of the tower?

Fair Game Review What you learned in previous grades & lessons

Find the percent of the number. *(Skills Review Handbook)*

22. 15% of 300 **23.** 140% of 125 **24.** 6% of −75

25. MULTIPLE CHOICE What is the x-intercept of the equation $3x + 5y = 30$? *(Section 2.4)*

Ⓐ −10 Ⓑ −6 Ⓒ 6 Ⓓ 10

3 Study Help

Check It Out
Graphic Organizer
BigIdeasMath.com

You can use a **notetaking organizer** to write notes, vocabulary, and questions about a topic. Here is an example of a notetaking organizer for writing equations using a slope and a point.

Write important vocabulary or formulas in this space. →

Slope-intercept form:

$y = mx + b$

$m = \text{slope} = \frac{\text{rise}}{\text{run}}$

$b = y\text{-intercept}$

Writing equations using a slope and a point

Plot the point. Plot additional points using the slope. Draw a line through the points and find the y-intercept. Write the equation in slope-intercept form.

Example:

slope $= \frac{1}{2}$; $(-4, 0)$

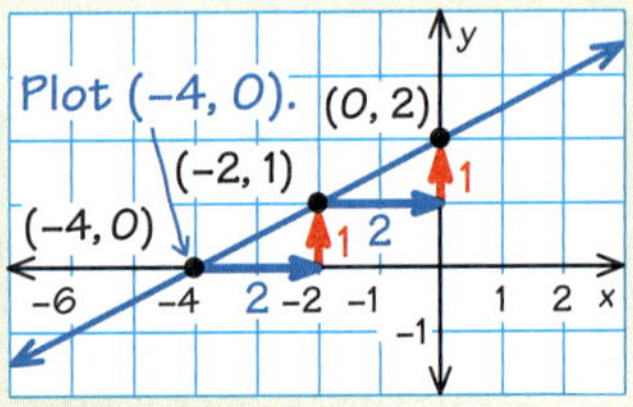

The y-intercept is 2. So, the equation is $y = \frac{1}{2}x + 2$.

← Write your notes about the topic in this space.

How can you write an equation of a line using two points?

← Write your questions about the topic in this space.

On Your Own

Make a notetaking organizer to help you study these topics.

1. writing equations in slope-intercept form
2. writing equations using two points

After you complete this chapter, make notetaking organizers for the following topics.

3. writing systems of linear equations
4. Think of a real-life application that can be modeled using a linear equation. Then make a notetaking organizer to help you study this application.

"My notetaking organizer has me thinking about retirement when I won't have to fetch sticks any more."

Sample Answers

1.

Slope-intercept form:
$y = mx + b$
$m = \text{slope} = \frac{\text{rise}}{\text{run}}$
$b = y\text{-intercept}$

Writing Equations in Slope-Intercept Form

Find the slope m and the y-intercept b. Then subsititute these values into $y = mx + b$.

Example:

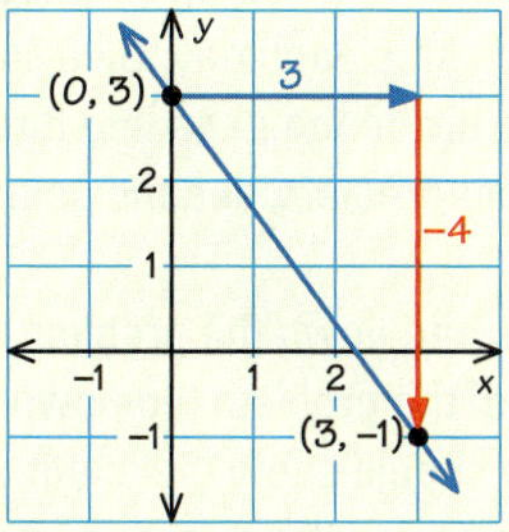

$\text{slope} = \frac{\text{rise}}{\text{run}} = \frac{-4}{3} = -\frac{4}{3}$

The y-intercept is 3. So, the equation is $y = -\frac{4}{3}x + 3$.

How can you write an equation of a horizontal line?

2.

Slope-intercept form:
$y = mx + b$
$m = \text{slope} = \frac{\text{rise}}{\text{run}}$
$b = y\text{-intercept}$

Writing Equations using Two Points

Plot the points and draw a line through them. Use the graph to find the slope and y-intercept.

Example:
(–1, 3), (1, 1)

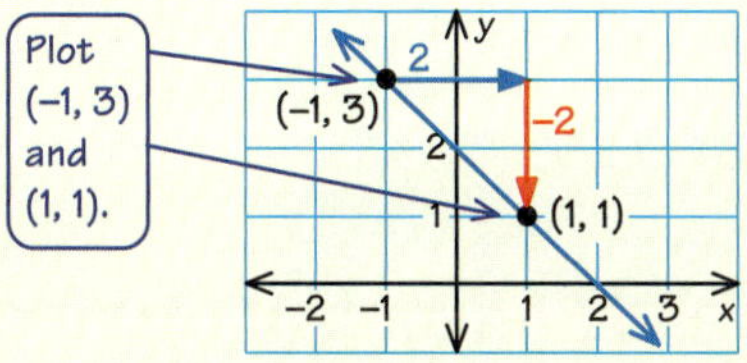

$\text{slope} = \frac{\text{rise}}{\text{run}} = \frac{-2}{2} = -1$

The y-intercept is 2. So, the equation is $y = -1x + 2$, or $y = -x + 2$.

How can you find the y-intercept if the two points are both to the left or both to the right of the y-axis?

List of Organizers

Available at *BigIdeasMath.com*

Comparison Chart
Concept Circle
Definition (Idea) and Example Chart
Example and Non-Example Chart
Formula Triangle
Four Square
Information Frame
Information Wheel
Notetaking Organizer
Process Diagram
Summary Triangle
Word Magnet
Y Chart

About this Organizer

A **Notetaking Organizer** can be used to write notes, vocabulary, and questions about a topic. In the space on the left, students write important vocabulary or formulas. In the space on the right, students write their notes about the topic. In the space at the bottom, students write their questions about the topic. A notetaking organizer can also be used as an assessment tool, in which blanks are left for students to complete.

Technology For the Teacher

Vocabulary Puzzle Builder

Answers

1. $y = -\frac{4}{3}x - 1$
2. $y = x$
3. $y = \frac{1}{2}x - 2$
4. $y = 2x + 1$
5. $y = \frac{1}{3}x - 1$
6. $y = -x + 3$
7. $y = -\frac{1}{8}x - 4$
8. $y = -\frac{2}{3}$
9. $y = -x + 4$
10. $y = \frac{1}{2}x + 13$
11. **a.** $y = 160x + 30$

    b. It is the initial deposit.
12. $y = x - 1$

Assessment Book

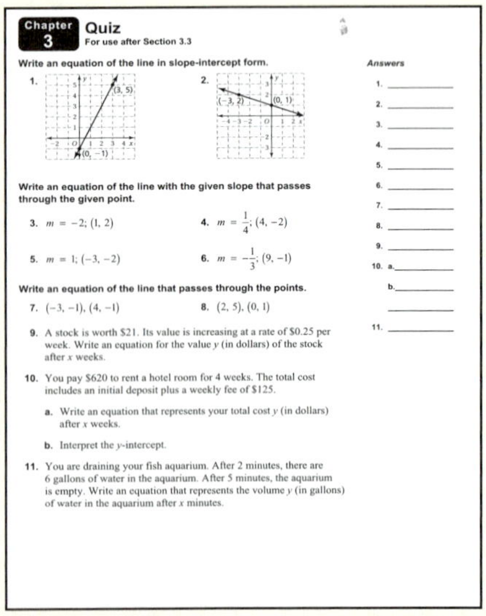

Chapter 3 Quiz

For use after Section 3.3

Write an equation of the line in slope-intercept form.

1. (graph through $(3, 5)$ and $(0, -1)$)

2. (graph through $(-3, 2)$ and $(0, 1)$)

Write an equation of the line with the given slope that passes through the given point.

3. $m = -2$; $(1, 2)$

4. $m = \frac{1}{4}$; $(4, -2)$

5. $m = 1$; $(-3, -2)$

6. $m = -\frac{1}{3}$; $(9, -1)$

Write an equation of the line that passes through the points.

7. $(-3, -1)$, $(4, -1)$

8. $(2, 5)$, $(0, 1)$

9. A stock is worth \$21. Its value is increasing at a rate of \$0.25 per week. Write an equation for the value y (in dollars) of the stock after x weeks.

10. You pay \$620 to rent a hotel room for 4 weeks. The total cost includes an initial deposit plus a weekly fee of \$125.

    a. Write an equation that represents your total cost y (in dollars) after x weeks.

    b. Interpret the y-intercept.

11. You are draining your fish aquarium. After 2 minutes, there are 6 gallons of water in the aquarium. After 5 minutes, the aquarium is empty. Write an equation that represents the volume y (in gallons) of water in the aquarium after x minutes.

Answers

1. ______
2. ______
3. ______
4. ______
5. ______
6. ______
7. ______
8. ______
9. ______
10. a. ______

    b. ______

    ______
11. ______

Alternative Quiz Ideas

100% Quiz	Math Log
Error Notebook	Notebook Quiz
Group Quiz	Partner Quiz
Homework Quiz	Pass the Paper

Error Notebook

An error notebook provides an opportunity for students to analyze and learn from their errors. Have students make an error notebook for this chapter. They should work in their notebook a little each day. Give students the following directions.

- Use a notebook and divide the page into three columns.
- Label the first column *problem*, second column *error*, and third column *correction*.
- In the first column, write down the problem on which the errors were made. Record the source of the problem (homework, quiz, in-class assignment).
- The second column should show the exact error that was made. Include a statement of why you think the error was made. This is where the learning takes place, so it is helpful to use a different color ink for the work in this column.
- The last column contains the corrected problems and comments that will help with future work.
- Separate each problem with horizontal lines.

Reteaching and Enrichment Strategies

If students need help. . .	If students got it. . .
Resources by Chapter • Study Help • Practice A and Practice B • Puzzle Time Lesson Tutorials *BigIdeasMath.com* Practice Quiz Practice from the Test Generator	Resources by Chapter • Enrichment and Extension • School-to-Work Game Closet at *BigIdeasMath.com* Start the next section

Technology For the Teacher

Answer Presentation Tool

Big Ideas Test Generator

3.1–3.3 Quiz

Check It Out
Progress Check
BigIdeasMath.com

Write an equation of the line in slope-intercept form. *(Section 3.1)*

1.

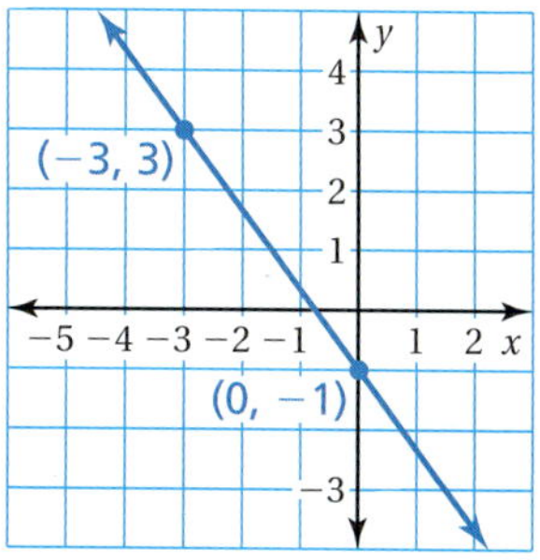

2.

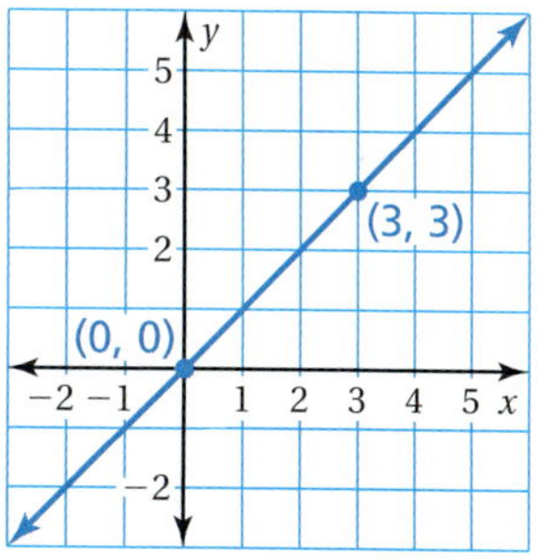

3.

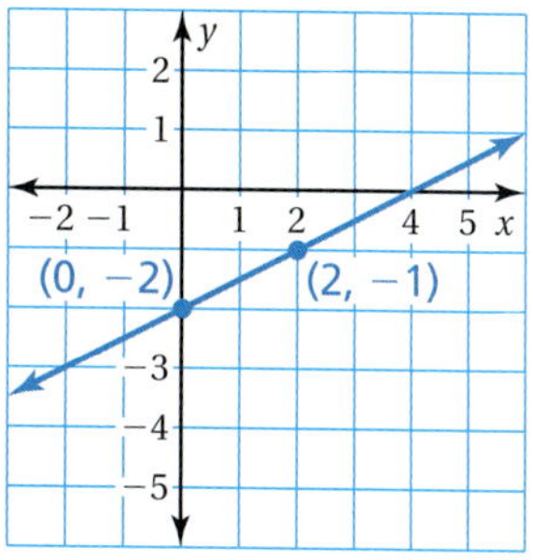

Write an equation of the line with the given slope that passes through the given point. *(Section 3.2)*

4. $m = 2;\ (1, 3)$

5. $m = \frac{1}{3};\ (-3, -2)$

6. $m = -1;\ (-1, 4)$

7. $m = -\frac{1}{8};\ (8, -5)$

Write an equation of the line that passes through the points. *(Section 3.3)*

8. $\left(0, -\frac{2}{3}\right) \left(-3, -\frac{2}{3}\right)$

9. $(4, 0), (0, 4)$

10. CONSTRUCTION A construction crew is extending a highway sound barrier that is 13 miles long. The crew builds $\frac{1}{2}$ mile per week. Write an equation for the length y (in miles) of the barrier after x weeks. *(Section 3.1)*

11. STORAGE You pay $510 to rent a storage unit for 3 months. The total cost includes an initial deposit plus a monthly fee of $160. *(Section 3.2)*

a. Write an equation that represents your total cost y (in dollars) after x months.

b. Interpret the y-intercept.

12. CORN After 3 weeks, a corn plant is 2 feet tall. After 9 weeks, the plant is 8 feet tall. Write an equation that represents the height y (in feet) of the corn plant after x weeks. *(Section 3.3)*

3.4 Solving Real-Life Problems

Share Your Work at... My.BigIdeasMath.com

Essential Question How can you use a linear equation in two variables to model and solve a real-life problem?

1 EXAMPLE: Writing a Story

Write a story that uses the graph at the right.

- **In your story, interpret the slope of the line, the y-intercept, and the x-intercept.**
- **Make a table that shows data from the graph.**
- **Label the axes of the graph with units.**
- **Draw pictures for your story.**

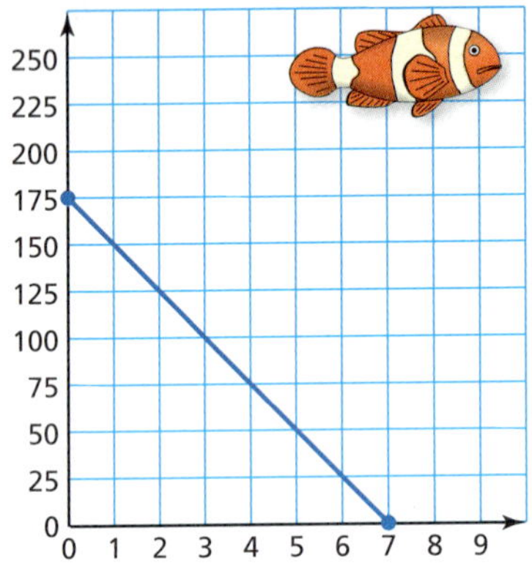

There are many possible stories. Here is one about a reef tank.

Tom works at an aquarium shop on Saturdays. One Saturday, when Tom gets to work, he is asked to clean a 175-gallon reef tank.

His first job is to drain the tank. He puts a hose into the tank and starts a siphon. Tom wonders if the tank will finish draining before he leaves work.

He measures the amount of water that is draining out and finds that 12.5 gallons drain out in 30 minutes. So, he figures that the rate is 25 gallons per hour. To see when the tank will be empty, Tom makes a table and draws a graph.

x-intercept: number of hours to empty the tank

x	0	1	2	3	4	5	6	7
y	175	150	125	100	75	50	25	0

y-intercept: amount of water in full tank

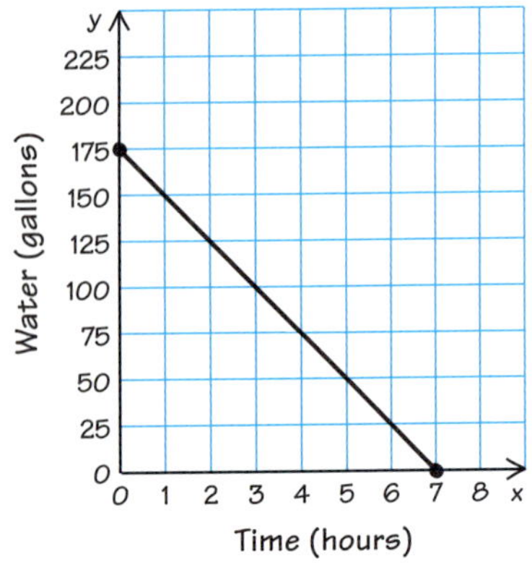

From the table and also from the graph, Tom sees that the tank will be empty after 7 hours. This will give him 1 hour to wash the tank before going home.

Laurie's Notes

Introduction

For the Teacher

- **Goal:** Students will develop an intuitive understanding of solving real-life problems.
- Students have already solved real-life problems involving linear equations. The skills in this lesson are not new. One skill that students have not had much practice with is interpreting the x-intercept.

Motivate

- **Time for a Story!** Tell students that you want to give your friend 10 pounds of chocolate for his birthday. The fancy bars of chocolate come in one-pound and half-pound bars. Hold up two rectangular blocks (or real bars).

? "What are some different ways I could give my friend the 10 pounds of chocolate?" Answers will vary.

- Students should eventually give the two *simple* answers of 10 one-pound bars and no half-pound bars, or 20 half-pound bars and no one-pound bars. If you let x equal the number of one-pound bars and y equal the number of half-pound bars, these two solutions are (10, 0) and (0, 20). Plot the two points.

Discuss

? Depending upon time, you may wish to ask a series of questions and have students answer them now or at the end of the lesson.

- "Are there other combinations besides these two that will equal 10 pounds?" Make a table to show additional solutions.
- "What is the slope of the line that goes through the two points?" -2
- "Could you write an equation for the line through the two points?" $y = -2x + 20$
- "If you bought 5 one-pound bars, how many half-pound bars would you need to purchase?" 10
- "If a one-pound bar costs twice as much as a half-pound bar, what do you know about all of the possible gift combinations?" all cost the same

Activity Notes

Example 1

- Ask one or more students to read through the sample provided.
- The problem does not have to be about fish tanks, but it does need to use the ordered pairs (0, 175) and (7, 0). The table of values will be the same regardless of the context selected. Notice that the equation is not determined, although this could be an extension for some or all students.
- Here are some suggestions for ordered pairs: (# of weeks, $ in the bank); (# of hours biking, kilometers traveled); (# of hours reading, pages to read); (# of wheel barrows, pounds of dirt)

Previous Learning

Students should know how to write equations in slope-intercept form.

Activity Materials	
Introduction	**Textbook**
• 2 rectangular blocks	• straightedge

Start Thinking! and Warm Up

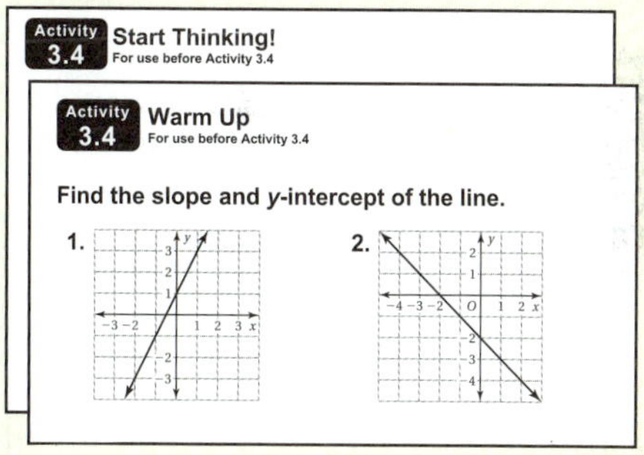

3.4 Record and Practice Journal

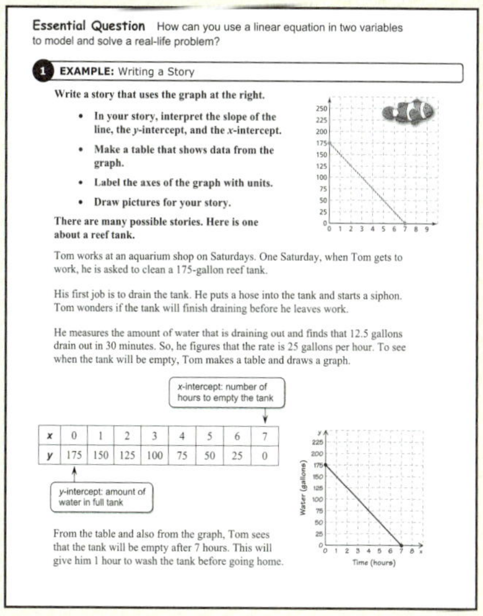

Essential Question How can you use a linear equation in two variables to model and solve a real-life problem?

1 EXAMPLE: Writing a Story

Write a story that uses the graph at the right.

- **In your story, interpret the slope of the line, the y-intercept, and the x-intercept.**
- **Make a table that shows data from the graph.**
- **Label the axes of the graph with units.**
- **Draw pictures for your story.**

There are many possible stories. Here is one about a reef tank.

Tom works at an aquarium shop on Saturdays. One Saturday, when Tom gets to work, he is asked to clean a 175-gallon reef tank.

His first job is to drain the tank. He puts a hose into the tank and starts a siphon. Tom wonders if the tank will finish draining before he leaves work.

He measures the amount of water that is draining out and finds that 12.5 gallons drain out in 30 minutes. So, he figures that the rate is 25 gallons per hour. To see when the tank will be empty, Tom makes a table and draws a graph.

x-intercept: number of hours to empty the tank

x	0	1	2	3	4	5	6	7
y	175	150	125	100	75	50	25	0

y-intercept: amount of water in full tank

From the table and also from the graph, Tom sees that the tank will be empty after 7 hours. This will give him 1 hour to wash the tank before going home.

English Language Learners

English Language Learners
For Activity 2, pair English learners with English speakers to write a story based on the English learners' culture. Encourage students to share how they incorporated the math into the story.

3.4 Record and Practice Journal

2 ACTIVITY: Writing a Story

Work with a partner. Write a story that uses the graph of a line.

- **In your story, interpret the slope of the line, the y-intercept, and the x-intercept.**
- **Make a table that shows data from the graph.**
- **Label the axes of the graph with units.**
- **Draw pictures for your story.**

Check students' work.

3 ACTIVITY: Drawing Graphs

Work with a partner. Describe a real-life problem that has the given rate and intercepts. Draw a line that represents the problem.

a. Rate: -30 feet per second
y-intercept: 150 feet
x-intercept: 5 seconds

Check students' work.

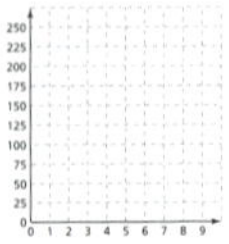

b. Rate: -25 dollars per month
y-intercept: \$200
x-intercept: 8 months

Check students' work.

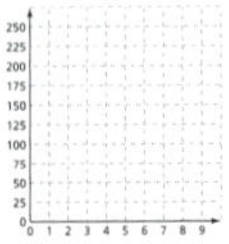

What Is Your Answer?

4. IN YOUR OWN WORDS How can you use a linear equation in two variables to model and solve a real-life problem? List three different rates that can be represented by slopes in real-life problems.

Represent x and y as units and then its slope as a rate.

Laurie's Notes

Activity 2

- **Interdisciplinary:** This activity is open-ended. The goal is to integrate language arts skills while practicing mathematical skills.
- Depending upon time, you may choose not to do this activity.
- You modeled how this problem can be done with the chocolate bars at the beginning of the lesson.
- You may wish to return to this problem *after* students have brainstormed three different rates in Question 4.
- Students should share their stories with the whole class, drawing a sketch of the graph with the axes labeled.

Activity 3

? "What are the ordered pairs for the intercepts?" (0, 150) and (5, 0)

? "Explain how the rate is found." The change is 150 feet in 5 seconds, which simplifies to 30 feet per 1 second.

? "Why is the rate negative?" The amount of feet is decreasing as time increases.

- Students may need to see the graph to understand why the rate is negative.
- **Extension:** Ask students to sketch a graph that represents a rate of 30 feet per second.

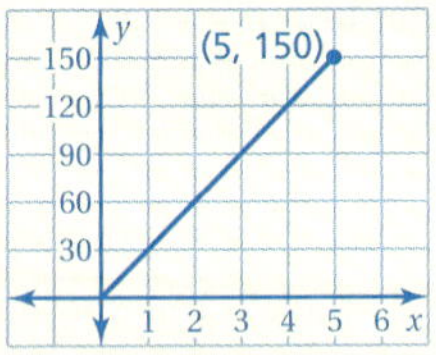

- **Connection:** For most contextual problems, the x- and y-intercepts will be positive numbers, and so the slope of the line between them is negative.

What Is Your Answer?

- **Whole Class Activity:** Have the class brainstorm 1 or 2 rates, and then ask students to think about two or three more on their own. This question will help students think about the variety of contexts in which linear equations can arise.
- **Possible Rates:** miles per hour, feet per second, miles per gallon, outs per inning, points per quarter, people per team, tiles per foot

Closure

- Refer back to the chocolate question given at the beginning of the lesson and ask students to write an equation for this problem. The rate of -2 means that for every 2 half-pound bars that are bought, there is one less one-pound bar. $y = -2x + 20$

Technology For the Teacher

Dynamic Classroom

The Dynamic Planning Tool
Editable Teacher's Resources at *BigIdeasMath.com*

2 ACTIVITY: Writing a Story

Work with a partner. Write a story that uses the graph of a line.

- **In your story, interpret the slope of the line, the *y*-intercept, and the *x*-intercept.**
- **Make a table that shows data from the graph.**
- **Label the axes of the graph with units.**
- **Draw pictures for your story.**

3 ACTIVITY: Drawing Graphs

Work with a partner. Describe a real-life problem that has the given rate and intercepts. Draw a line that represents the problem.

a. Rate: −30 feet per second

y-intercept: 150 feet

x-intercept: 5 seconds

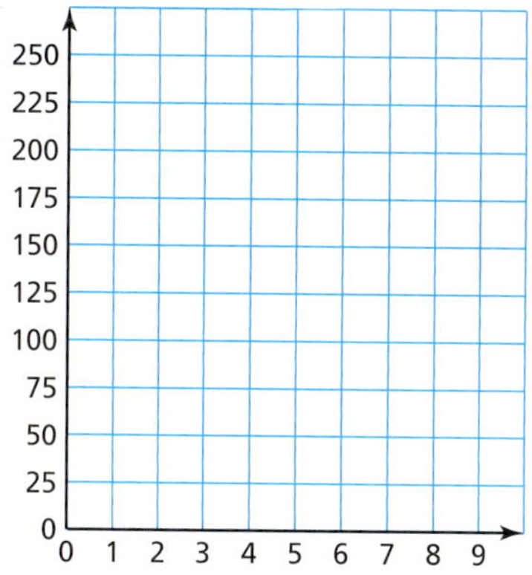

b. Rate: −25 dollars per month

y-intercept: $200

x-intercept: 8 months

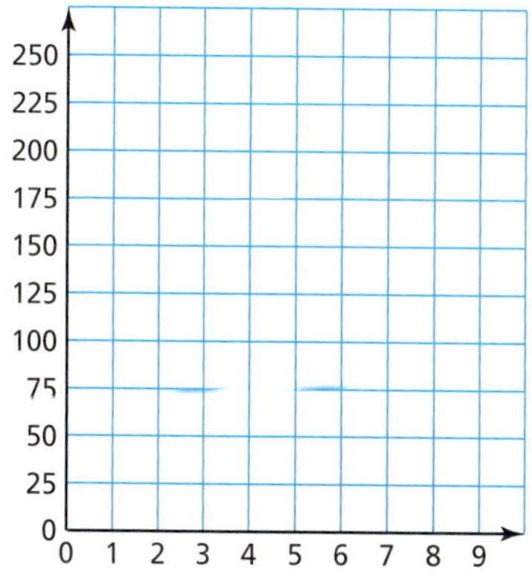

What Is Your Answer?

4. **IN YOUR OWN WORDS** How can you use a linear equation in two variables to model and solve a real-life problem? List three different rates that can be represented by slopes in real-life problems.

Practice Use what you learned about solving real-life problems to complete Exercises 4 and 5 on page 130.

3.4 Lesson

Check It Out
Lesson Tutorials
BigIdeasMath.com

EXAMPLE 1 Real-Life Application

The percent y (in decimal form) of battery power remaining x hours after you turn on a laptop computer is $y = -0.2x + 1$. (a) Graph the equation. (b) Interpret the x- and y-intercepts. (c) After how many hours is the battery power at 75%?

a. Use the slope and the y-intercept to graph the equation.

$$y = -0.2x + 1$$

slope (−0.2) ; y-intercept (1)

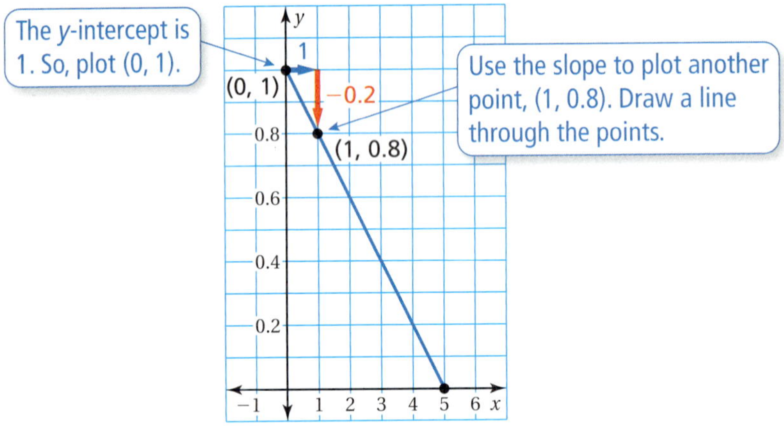

b. To find the x-intercept, substitute 0 for y in the equation.

$y = -0.2x + 1$	Write the equation.
$0 = -0.2x + 1$	Substitute 0 for y.
$5 = x$	Solve for x.

∴ The x-intercept is 5. So, the battery lasts 5 hours. The y-intercept is 1. So, the battery power is at 100% when you turn on the laptop.

75% Remaining

c. Find the value of x when $y = 0.75$.

$y = -0.2x + 1$	Write the equation.
$0.75 = -0.2x + 1$	Substitute 0.75 for y.
$1.25 = x$	Solve for x.

∴ The battery power is at 75% after 1.25 hours.

On Your Own

Now You're Ready
Exercise 6

1. The amount y (in gallons) of gasoline remaining in a gas tank after driving x hours is $y = -2x + 12$. (a) Graph the equation. (b) Interpret the x- and y-intercepts. (c) After how many hours are there 5 gallons left?

Laurie's Notes

Introduction

Connect

- **Yesterday:** Students explored real-life problems involving rates, where the x- and y-intercepts were each positive.
- **Today:** Students will solve real-life problems using equations, graphs, and intercepts.
- **FYI:** The goal is to bring the concepts related to linear equations, graphing, and writing, together into one lesson with more focus on the interpretation of slope and intercepts.

Motivate

- Hold a digital camera and take some pictures of your students. You could also pretend to be taking pictures. Continue to take pictures until someone finally asks how many you're going to take.
- Say, "Until my memory card is full! But don't worry, it's only a 64 megabyte (64 MB) card." If someone asks, tell them that every picture uses about 4 MB.

Discuss

- Discuss how many pictures can be stored on the card. Every time a picture is taken, the megabytes remaining decrease.

Lesson Notes

Example 1

- Discuss the use of a laptop computer and what happens when the computer is running off the battery versus using the AC adapter.
- Read the problem. Write what an ordered pair represents in words: (# of hours computer runs on battery, % of battery remaining as a decimal)

? "If you have just turned your fully charged computer on, how much battery power do you have?" 100%

? "What is the ordered pair associated with turning your computer on?" (0, 1)

- **Common Error:** Students may say (0, 100), but remind them that the percent needs to be in decimal form.
- Refer to the equation and ask about the slope and y-intercept.
- Plot the point for the y-intercept and use the slope to plot additional points on the graph.

? "The y-intercept means you just turned your computer on. What does the x-intercept mean?" The computer ran for 5 hours before the battery died.

? **Big Idea:** Ask students why the graph is contained in Quadrant I only. It does not make sense for the battery power remaining to be greater than 1 (100%) or less than 0.

On Your Own

- This question is modeled after Example 1.

Goal Today's lesson is solving real-life problems.

Lesson Materials	
Introduction	**Textbook**
• SD or memory card of some type • digital camera	• straightedge

Start Thinking! and Warm Up

Lesson 3.4 **Warm Up**
For use before Lesson 3.4

Lesson 3.4 **Start Thinking!**
For use before Lesson 3.4

How can you use linear equations to solve problems about taking a road trip?

Extra Example 1

The percent y (in decimal form) of battery power remaining x hours after you turn on your handheld video game is $y = -0.3x + 1$.

a. Graph the equation.

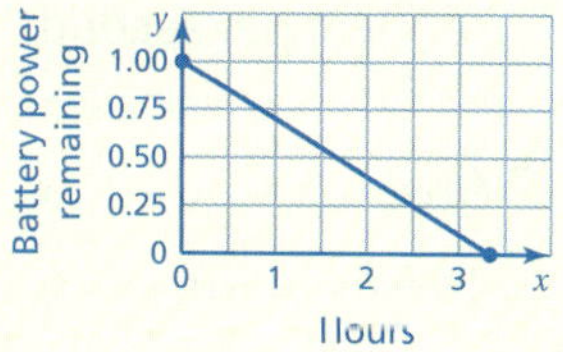

b. Interpret the x- and y-intercepts.

The x-intercept is $3\frac{1}{3}$. So, the battery lasts $3\frac{1}{3}$ hours. The y-intercept is 1. So, the battery power is at 100% when you turn it on.

c. After how many hours is the battery power at 40%? after 2 hours

On Your Own

1. See Additional Answers.

Laurie's Notes

Extra Example 2

The graph shows the cost y (in dollars) of a BMX (Bicycle Motocross) track membership and entry fees for x races at the track.

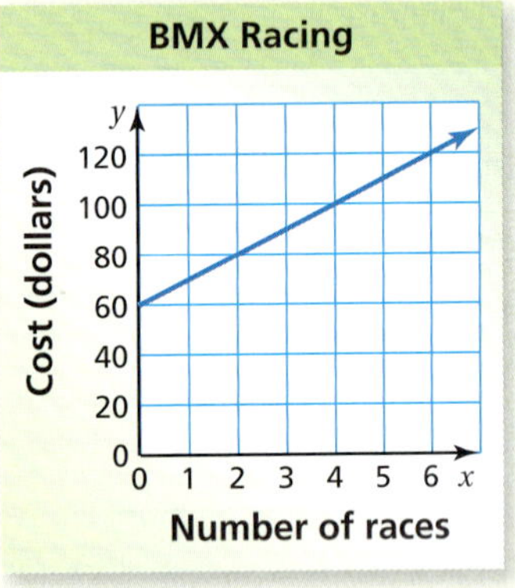

a. Find the slope and y-intercept. slope: 10; y-intercept: 60

b. Write an equation of the line. $y = 10x + 60$

c. How much does it cost to be a member and enter 4 races? $100

On Your Own

2. **a.** The slope is $\frac{3}{2}$. So, the flag is raised at a rate of $\frac{3}{2}$ feet per second.

   b. $y = \frac{3}{2}x + 3$

   c. 16.5 or $16\frac{1}{2}$ ft

Differentiated Instruction

Visual

Encourage students to write down notes when solving word problems, or to underline relevant information and cross out irrelevant information. Allow students to do this on handouts and tests.

Example 2

- Ask a student to read the problem. Write in words what the ordered pairs represent: (°C, °F)

? "Explain in words what the two ordered pairs represent." 0°C is the same as 32°F. 30°C is the same as 86°F.

- Draw a sketch of the graph. In order to find the slope, it is helpful for students to see the arrows representing the change in x (30) and the change in y (54).

? "The slope is $\frac{9}{5}$. What other information is needed to write the equation of this line?" y-intercept

- The y-intercept is shown in the graph. Write the equation.
- When students evaluate the equation for $C = 15$, they may make an error multiplying the fraction and whole number. Remind students of how multiplication of fractions is performed, and to divide out the common factor of 5 before multiplying.
- **Note:** Students have seen the conversion formula $F = \frac{9}{5}C + 32$ before. Because you want to reference the y-intercept, the equation is written in terms of x and y.
- **FYI:** This is a real-life application where it makes sense for the graph to be found in Quadrants I, II and III.

? "What do you think *mean temperature of Earth* means?" Answers will vary.

On Your Own

- In interpreting the slope, the graph is read from left to right, so every 2 seconds the flag's height increases 3 feet.

? "Why does it make sense that the y-intercept is not 0?" Flag's height does not start on the ground.

Closure

- Draw a graph of the memory card problem. Find and interpret the slope. slope $= -4$; Megabytes of free space decrease by 4 for each picture taken. Write the equation of the line. $y = -4x + 64$ Find and interpret the x- and y-intercepts. 16 and 64; When 16 pictures have been taken, there is 0 MB remaining. When 0 pictures have been taken, there is 64 MB of storage.

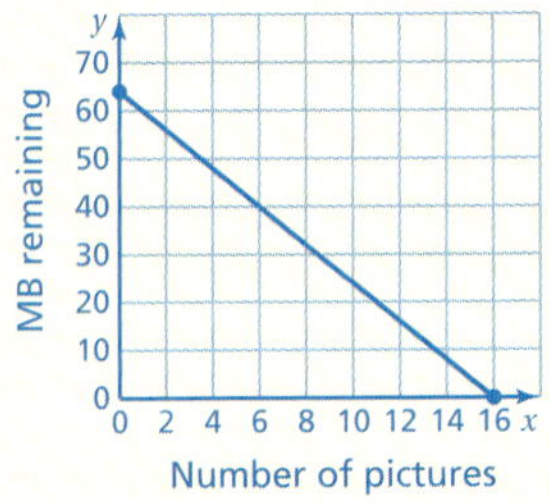

Technology For the Teacher

Dynamic Classroom

The Dynamic Planning Tool
Editable Teacher's Resources at *BigIdeasMath.com*

EXAMPLE 2 Real-Life Application

The graph relates temperatures y (in degrees Fahrenheit) to temperatures x (in degrees Celsius). (a) Find the slope and y-intercept. (b) Write an equation of the line. (c) What is the mean temperature of Earth in degrees Fahrenheit?

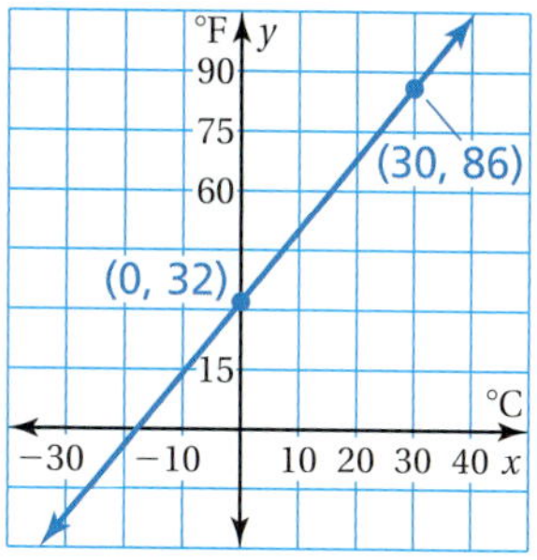

Mean Temperature: 15°C

a. $\text{slope} = \dfrac{\text{change in } y}{\text{change in } x} = \dfrac{54}{30} = \dfrac{9}{5}$

The line crosses the y-axis at (0, 32). So, the y-intercept is 32.

The slope is $\frac{9}{5}$ and the y-intercept is 32.

b. Use the slope and y-intercept to write an equation.

slope

y-intercept

The equation is $y = \frac{9}{5}x + 32$.

c. In degrees Celsius, the mean temperature of Earth is 15°. To find the mean temperature in degrees Fahrenheit, find the value of y when $x = 15$.

$y = \frac{9}{5}x + 32$ — Write the equation.

$= \frac{9}{5}(15) + 32$ — Substitute 15 for x.

$= 59$ — Simplify.

The mean temperature of Earth is 59°F.

On Your Own

Now You're Ready
Exercise 7

2. The graph shows the height y (in feet) of a flag x seconds after you start raising it up a flagpole.

   a. Find and interpret the slope.

   b. Write an equation of the line.

   c. What is the height of the flag after 9 seconds?

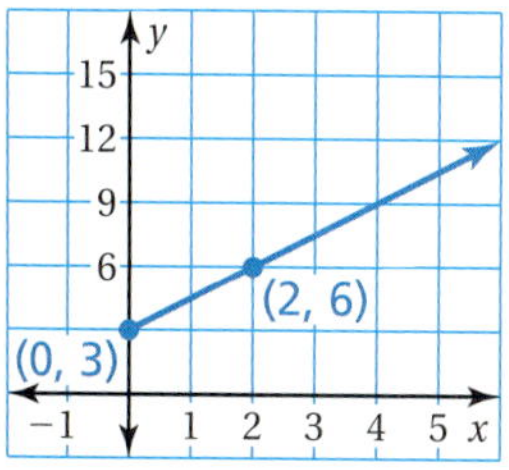

3.4 Exercises

Check It Out
Help with Homework
BigIdeasMath.com

Vocabulary and Concept Check

1. **REASONING** Explain how to find the slope, y-intercept, and x-intercept of the line shown.

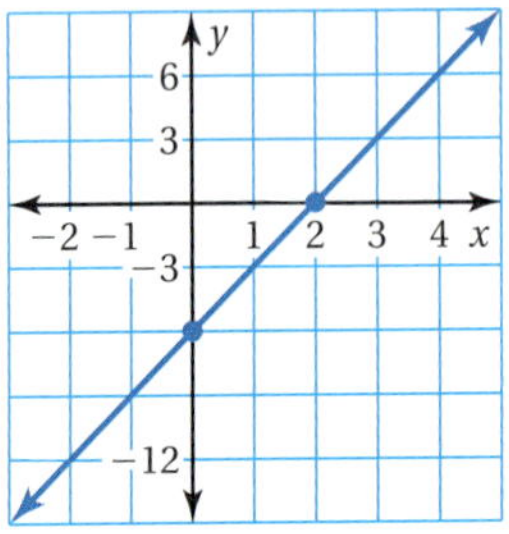

2. **OPEN-ENDED** Describe a real-life situation that uses a negative slope.

3. **REASONING** In a real-life situation, what does the slope of a line represent?

Practice and Problem Solving

Describe a real-life problem that has the given rate and intercepts. Draw a line that represents the problem.

4. Rate: −1.6 gallons per hour
   y-intercept: 16 gallons
   x-intercept: 10 hours

5. Rate: −45 pesos per week
   y-intercept: 180 pesos
   x-intercept: 4 weeks

1 6. **DOWNLOAD** You are downloading a song. The percent y (in decimal form) of megabytes remaining to download after x seconds is $y = -0.1x + 1$.
   a. Graph the equation.
   b. Interpret the x- and y-intercepts.
   c. After how many seconds is the download 50% complete?

2 7. **HIKING** The graph relates temperature y (in degrees Fahrenheit) to altitude x (in thousands of feet).
   a. Find the slope and y-intercept.
   b. Write an equation of the line.
   c. What is the temperature at sea level?

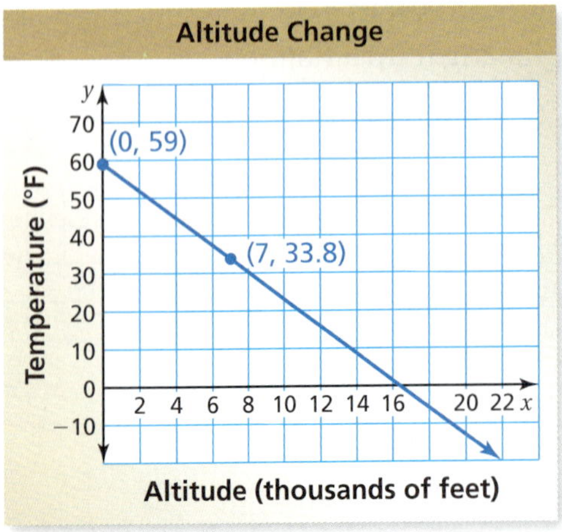

Assignment Guide and Homework Check

Level	Day 1 Activity Assignment	Day 2 Lesson Assignment	Homework Check
Basic	4, 5, 11–14	1–3, 6–8	2, 3, 6, 7, 8
Average	4, 5, 11–14	1–3, 6–9	2, 3, 6, 7, 8
Advanced	4, 5, 11–14	1–3, 7–10	2, 3, 7, 8

For Your Information

- **Exercise 8c** If you were to travel in a straight line, the speed would remain the same but the distance traveled would be less. So, it would take less time to make the trip.
- **Exercise 9** Ask students if their school lies on the line drawn between Denver and Beijing.

Common Errors

- **Exercise 6** Students may forget to convert the percent in part (c) to a decimal before substituting into the equation. Remind them that they need to convert percents to decimals before substituting.
- **Exercise 7** Students may struggle to find the change in *y* for the slope. Encourage them to focus on the *y*-coordinates and to write an expression that represents the change in temperature, 33.8 − 59, to help find the change in *y*.

3.4 Record and Practice Journal

Describe a real-life problem that has the given rate and intercepts. Draw a line that represents the problem.

1. Rate: −6 feet per year
   y-intercept: 24 feet
   x-intercept: 4 years

   Sample answer: the depth of a pond; See Additional Answers.

2. Rate: −2 pages per minute
   y-intercept: 280 pages
   x-intercept: 140 minutes

   Sample answer: recycling paper; See Additional Answers.

3. You are buying shirts for an organization. You have $120 to spend. The number of short-sleeved shirts y you buy after buying x long-sleeved shirts is $y = -1.2x + 24$.

   a. Graph the equation.

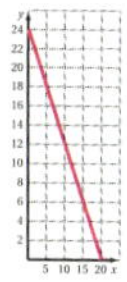

   b. Interpret the x- and y-intercepts.

   y-intercept; you can buy at most 24 short-sleeved shirts if you buy no long-sleeved shirts. x-intercept: you can buy at most 20 long-sleeved shirts if you buy no short-sleeved shirts.

4. The graph relates the percent of battery life y left in an MP3 player to the amount of playing time it has left. Write an equation of the line.

   $y = -5x + 100$

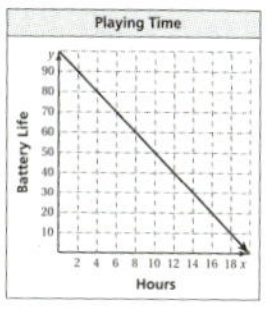

Technology For the Teacher

Answer Presentation Tool

Quiz*Show*

Vocabulary and Concept Check

1. The y-intercept is -6 because the line crosses the y-axis at the point $(0, -6)$. The x-intercept is 2 because the line crosses the x-axis at the point $(2, 0)$. You can use these two points to find the slope.

$$\text{Slope} = \frac{\text{change in } y}{\text{change in } x} = \frac{6}{2} = 3$$

2. *Sample answer:* a balloon descending toward the ground
3. *Sample answer:* the rate at which something is happening

Practice and Problem Solving

4–5. See Additional Answers.

6. a.

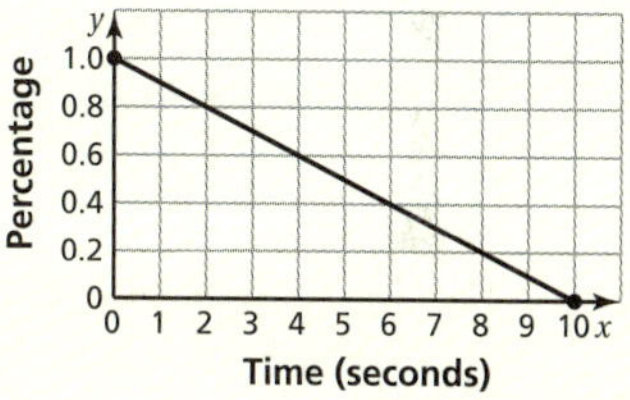

b. The x-intercept is 10. So, it takes 10 seconds to download the song. The y-intercept is 1. So, 100% of the song needs to be downloaded.

c. 5 sec

7. a. slope: -3.6
y-intercept: 59

b. $y = -3.6x + 59$

c. 59°F

Practice and Problem Solving

8. See Additional Answers.

9. a. Antananarivo: 19°S, 47°E
Denver: 39°N, 105°W
Brasilia: 16°S, 48°W
London: 51°N, 0°W
Beijing: 40°N, 116°E

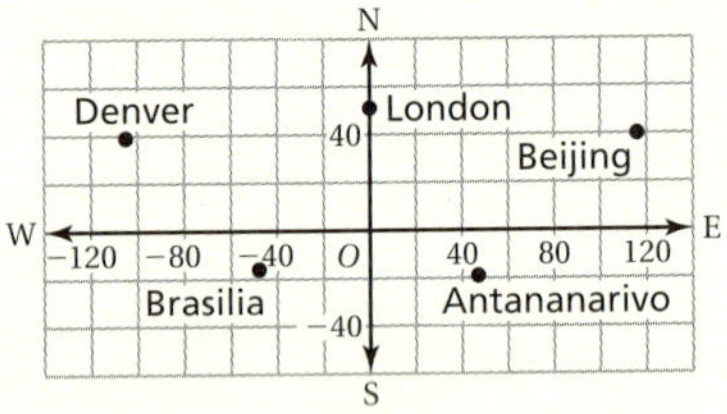

b. $y = \frac{1}{221}x + \frac{8724}{221}$

c. a place that is on the prime meridian

10. See *Taking Math Deeper*.

Fair Game Review

11. infinitely many solutions

12. one solution

13. no solution

14. B

Mini-Assessment

1. **You need \$125 to buy an MP3 player. Your allowance per week is \$5 and you earn \$20 per lawn mowed.**

a. Write an equation that represents your weekly income y (in dollars) for x lawns mowed. $y = 20x + 5$

b. Interpret the y-intercept. The y-intercept is 5. This is your allowance, the amount you started with before mowing lawns.

c. How many lawns do you need to mow to earn enough money in one week to buy the MP3 player? 6 lawns

Taking Math Deeper

Exercise 10

This is a classic "break-even" type of business problem. The band wants to invest \$5000 in new equipment and is trying to project how many tickets need to be sold to pay for the equipment.

1 Organize the given information.

R = band's revenue (income)
x = number of tickets sold
Income = \$1500 + 30% of ticket sales
Price of each ticket = \$20
Maximum capacity = 800

2 Write an equation for the revenue.

$R = 1500 + (30\% \text{ of } \$20 \text{ times } x)$
$= 1500 + 0.3(20)x$
a. $= 1500 + 6x$

In other words, the band receives \$6 per ticket. The organizers of the concert keep the remaining \$14 to cover the expenses of auditorium rental, marketing, and salaries.

3 Use the equation.
To find the number of tickets that need to be sold to earn a revenue of \$5000, substitute 5000 for R and solve for x.

$5000 = 1500 + 6x$
$3500 = 6x$
$583.3 \approx x$

b. So, if the band can sell 584 tickets to the concert, it will earn enough to pay for the new equipment. The capacity of this auditorium is 800, so this is possible.

Project

Draw a poster that could be used to advertise a concert by your favorite band.

Reteaching and Enrichment Strategies

If students need help...	If students got it...
Resources by Chapter • Practice A and Practice B • Puzzle Time Record and Practice Journal Practice Differentiating the Lesson Lesson Tutorials Skills Review Handbook	Resources by Chapter • Enrichment and Extension • School-to-Work Start the next section

8. **TRAVEL** Your family is driving from Cincinnati to St Louis. The graph relates your distance from St Louis y (in miles) and travel time x (in hours).

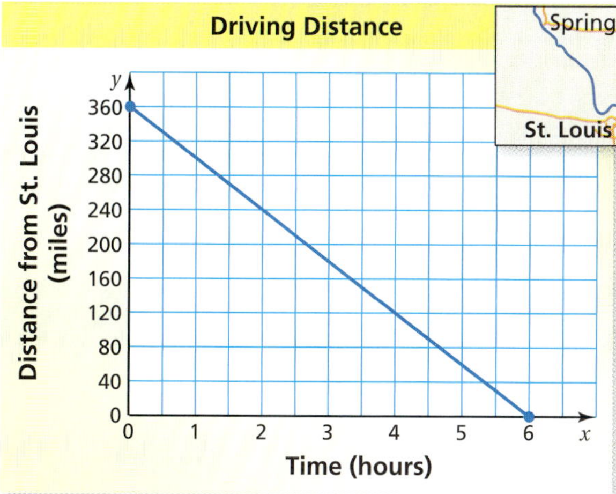

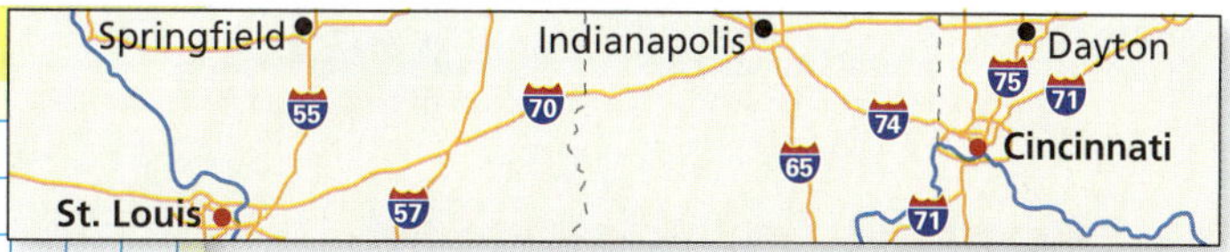

a. Interpret the x- and y-intercepts.

b. What is the slope? What does the slope represent in this situation?

c. Write an equation of the line. How would the graph and the equation change if you were able to travel in a straight line?

9. **PROJECT** Use a map or the Internet to find the latitude and longitude of your school to the nearest whole number. Then find the latitudes and longitudes of: Antananarivo, Madagascar; Denver, Colorado; Brasilia, Brazil; London, England; and Beijing, China.

a. Plot a point for each of the cities in the same coordinate plane. Let the positive y-axis represent north and the positive x-axis represent east.

b. Write an equation of the line that passes through Denver and Beijing.

c. In part (b), what geographic location does the y-intercept represent?

10. **Reasoning** A band is performing at an auditorium for a fee of \$1500. In addition to this fee, the band receives 30% of each \$20 ticket sold. The maximum capacity of the auditorium is 800 people.

a. Write an equation that represents the band's revenue R when x tickets are sold.

b. The band needs \$5000 for new equipment. How many tickets must be sold for the band to earn enough money to buy the new equipment?

Fair Game Review What you learned in previous grades & lessons

Tell whether the system has *one solution, no solution,* or *infinitely many solutions.* *(Section 2.5 and Section 2.6)*

11. $y = -x + 6$
$-4(x + y) = -24$

12. $y = 3x - 2$
$-x + 2y = 11$

13. $-9x + 3y = 12$
$y = 3x - 2$

14. **MULTIPLE CHOICE** Which equation is the slope-intercept form of $24x - 8y = 56$? *(Section 2.3)*

Ⓐ $y = -3x + 7$ Ⓑ $y = 3x - 7$ Ⓒ $y = -3x - 7$ Ⓓ $y = 3x + 7$

3.5 Writing Systems of Linear Equations

Essential Question How can you use a system of linear equations to model and solve a real-life problem?

Share Your Work at... My.BigIdeasMath.com

1 ACTIVITY: Writing a System

Work with a partner.

- **Peak Valley Middle School has 1200 students. Its enrollment is decreasing by 30 students per year.**
- **Southern Tier Middle School has 500 students. Its enrollment is increasing by 40 students per year.**
- **In how many years will the two schools have equal enrollments?**

a. **USE A TABLE** Use a table to answer the question.

Now

Year, x	0	1	2	3	4	5	6	7	8	9	10
Peak Valley MS, P	1200										
Southern Tier MS, S	500										

b. **USE A GRAPH** Write a linear equation that represents each enrollment.

$P =$

$S =$

Then graph each equation and find the point of intersection to answer the question.

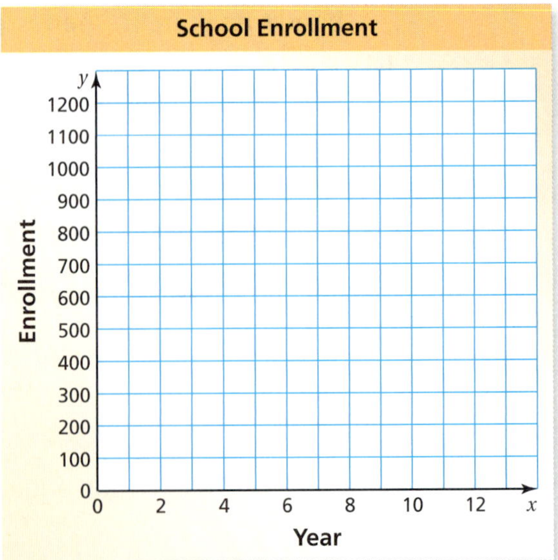

c. **USE ALGEBRA** Answer the question by setting the expressions for P and S equal to each other and solving for x.

Laurie's Notes

Introduction

For the Teacher

- **Goal:** Students will determine when two schools have the same enrollment.
- Students must first write each equation in the system.

Motivate

- Use colored tiles to create a model for a system of equations.
- Make two piles, one with *some* red tiles and one with *a lot* of blue tiles.
- Ask a student to come up and add 3 red tiles to the red pile and remove 5 blue tiles from the blue pile. Repeat this several times.
- ? "Do you think my two piles will ever have the same number of tiles in them?" Answers should express uncertainty and raise some questions about the initial amount (i.e., y-intercept) for each pile.
- ? "What is one way to keep track of the number of tiles in my pile each time?" Make a table of values.

Activity Notes

Activity 1

- Ask one or more students to read the information about each school.
- **Summarize:** The larger school is decreasing in size while the smaller school is increasing in size. When will the schools have equal enrollment?
- **Caution:** In the table of values, the same x-value is used for each school. The two sets of points are (x, P) and (x, S).
- ? When students have had sufficient time to work through the activity, ask questions of the class.
  - "How are the values for P changing in the table? The values for S?" P is decreasing by 30 and S is increasing by 40.
  - "What is the slope of each line?" Peak Valley: slope $= -30$; Southern Tier: slope $= 40$
  - "If one school has decreasing enrollment and the other has increasing enrollment, will the lines always intersect? Explain." No; only if the school with decreasing enrollment started with more students.
- It is natural for students to write the linear equation for Peak Valley as $P = 1200 - 30x$ because they have just finished the table of values and they were subtracting 30 each time. Students should recognize that this is equivalent to $P = -30x + 1200$.
- ? "Interpret the ordered pair where the graphs intersected." (10, 900) means in 10 years the population at each school will be 900.
- Ask a volunteer to show their work for part (c). If there are students who wrote different, yet equivalent equations, have them show their work.
- **Big Idea:** The only new skill in this activity was to write the system.
- ? "You solved a system of equations by making a table, graphing, and solving algebraically. Is one method easier for you than another?" Answers will vary.

Previous Learning

Students should know how to solve systems of equations.

Activity Materials	
Introduction	**Textbook**
• red tiles and blue tiles	• straightedge

Start Thinking! and Warm Up

Activity 3.5 Start Thinking!
For use before Activity 3.5

Activity 3.5 Warm Up
For use before Activity 3.5

Solve the linear system algebraically.

1. $x + y = 10$
   $y = 2x + 1$
2. $x - y = 2$
   $x + y = 18$
3. $y = -2x + 9$
   $y = x - 3$
4. $y = x + 5$
   $y = 6x$

3.5 Record and Practice Journal

Essential Question How can you use a system of linear equations to model and solve a real-life problem?

1 ACTIVITY: Writing a System

Work with a partner.

- Peak Valley Middle School has 1200 students. Its enrollment is decreasing by 30 students per year.
- Southern Tier Middle School has 500 students. Its enrollment is increasing by 40 students per year.
- In how many years will the two schools have equal enrollments?

a. **USE A TABLE** Use a table to answer the question.

Now ↓

Year, x	0	1	2	3	4	5	6	7	8	9	10
Peak Valley MS, P	1200	1170	1140	1110	1080	1050	1020	990	960	930	900
Southern Tier MS, S	500	540	580	620	660	700	740	780	820	860	900

b. **USE A GRAPH** Write a linear equation that represents each enrollment.

$P =$ $-30x + 1200$

$S =$ $40x + 500$

Then graph each equation and find the point of intersection to answer the question.

10 years

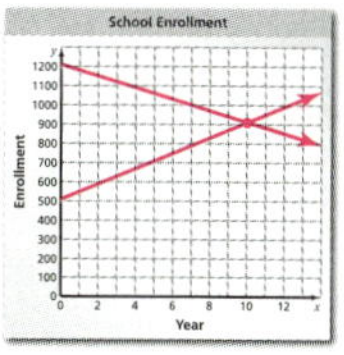

Differentiated Instruction

Organizational

Have students create a chart to describe the methods for writing a linear equation in slope-intercept form.

What I Know	How to Find Slope	How to Find *y*-intercept
slope and *y*-intercept		
slope and a point		
two points		

3.5 Record and Practice Journal

c. **USE ALGEBRA** Answer the question by setting the expressions for P and S equal to each other and solving for x.

$-30x + 1200 = 40x + 500;$
$x = 10$ years

2 ACTIVITY: Writing a System

Work with a partner. The table shows the enrollments of Sizemore Middle School and Wright Middle School for 7 years.

Year, x	0	1	2	3	4	5	6
Sizemore MS, S	1500	1438	1423	1350	1308	1247	1204
Wright MS, W	825	854	872	903	927	946	981

From the enrollment pattern, do you think the two schools will ever have the same enrollment? If so, when?

a. Plot the enrollments of each middle school.

b. Draw a line that approximately fits the points for each middle school.

c. Estimate the year in which the schools will have the same enrollment.

9 years

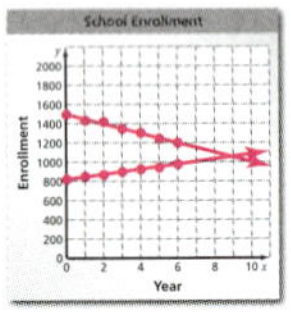

d. Write an equation for each line.

$S = -50x + 1500$

$W = 25x + 825$

e. **USE ALGEBRA** Answer the question by setting the expressions for S and W equal to each other and solving for x.

$-50x + 1500 = 25x + 825;$ about 9 years

What Is Your Answer?

3. **IN YOUR OWN WORDS** How can you use a system of linear equations to model and solve a real-life problem?

Check students' work.

4. **PROJECT** Use the Internet, a newspaper, a magazine, or some other reference to find two sets of real-life data that can be modeled by linear equations.

a. List the data in a table.

Check students' work.

b. Graph the data. Find a line to represent each data set.

Check students' work.

c. If possible, estimate when the two quantities will be equal.

Check students' work.

Laurie's Notes

Activity 2

- This activity is similar to Activity 1. One difference is that the rate of change (slope) is not constant. For this reason, the students are asked to draw a line that *best fits* the points. This is called the line of best fit.

? Ask guiding questions to help students think about how to write the equation of each line in slope-intercept form.

  - "Do you know the *y*-intercept for each line? Explain." Yes; 1500 and 825
  - "For Sizemore MS, by approximately how much did their enrollment change in 6 years? Can you find the approximate change per year?" Enrollment went down about 300. $-300 \div 6$ is -50 students per year.
  - "For Wright MS, by approximately how much did their enrollment change in 6 years? Can you find the approximate change per year?" Enrollment went up about 150. $150 \div 6$ is 25 students per year.

- The best-fit lines should be approximately $S = -50x + 1500$ and $W = 25x + 825$. Students may have other equations that are very similar, meaning their slopes might be -48 and 27 instead of -50 and 25.
- Discuss with students that it would be unusual for a school enrollment to increase or decrease by the same amount for 6 consecutive years, so this activity is more realistic than Activity 1.
- Some students may reason that if one school is losing approximately 50 students each year and the other school is increasing by 25 students each year, the net change is about 75 students each year. The difference in original enrollment is $1500 - 825 = 675$. Divide this by the net change of 75, and it will take 9 years for the schools to have about the same enrollment.

What Is Your Answer?

- Students have solved many real-life problems in this chapter. For Question 3, ask them to think about how many of them might be made into a system of equations. For example, there could be two hot air balloons.

Closure

- Refer back to the red and blue tiles. Start with 5 red and add 3 each time. $y = 3x + 5$ Start with 53 blue and subtract 5 each time. $y = -5x + 53$ When do the piles have the same number of tiles? after 6 trials

Technology For the Teacher

Dynamic Classroom

The Dynamic Planning Tool
Editable Teacher's Resources at *BigIdeasMath.com*

2 ACTIVITY: Writing a System

Work with a partner. The table shows the enrollments of Sizemore Middle School and Wright Middle School for 7 years.

Year, x	0	1	2	3	4	5	6
Sizemore MS, S	1500	1438	1423	1350	1308	1247	1204
Wright MS, W	825	854	872	903	927	946	981

From the enrollment pattern, do you think the two schools will ever have the same enrollment? If so, when?

a. Plot the enrollments of each middle school.

b. Draw a line that approximately fits the points for each middle school.

c. Estimate the year in which the schools will have the same enrollment.

d. Write an equation for each line.

$S =$ ______

$W =$ ______

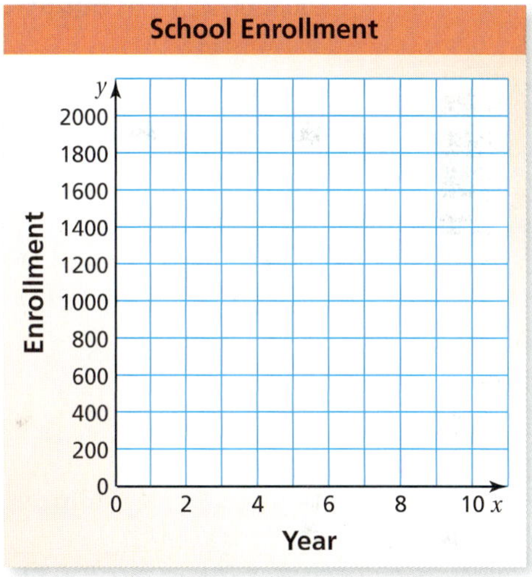

e. **USE ALGEBRA** Answer the question by setting the expressions for S and W equal to each other and solving for x.

What Is Your Answer?

3. **IN YOUR OWN WORDS** How can you use a system of linear equations to model and solve a real-life problem?

4. **PROJECT** Use the Internet, a newspaper, a magazine, or some other reference to find two sets of real-life data that can be modeled by linear equations.

a. List the data in a table.

b. Graph the data. Find a line to represent each data set.

c. If possible, estimate when the two quantities will be equal.

Practice Use what you learned about writing systems of linear equations to complete Exercises 4 and 5 on page 136.

3.5 Lesson

Check It Out
Lesson Tutorials
BigIdeasMath.com

EXAMPLE 1 Writing a System of Linear Equations

A bank teller is counting \$20 bills and \$10 bills. There are 16 bills that total \$200. Write and solve a system of equations to find the number x of \$20 bills and the number y of \$10 bills.

Words	Number of \$20 bills	plus	number of \$10 bills	is	the total number of bills.		
Equation 1 → **Equation**	x	$+$	y	$=$	16		
Words	Twenty times	the number of \$20 bills	plus ten times	the number of \$10 bills	is	the total value.	
Equation 2 → **Equation**	$20\ \cdot$	x	$+\ 10\ \cdot$	y	$=$	200	

The linear system is $x + y = 16$ and $20x + 10y = 200$.

Solve each equation for y. Then make a table of values to find the x-value that gives the same y-value for both equations.

x	0	1	2	3	4
$y = 16 - x$	16	15	14	13	12
$y = 20 - 2x$	20	18	16	14	12

The solution is (4, 12).

So, there are 4 twenty-dollar bills and 12 ten-dollar bills.

Check

Equation 1

$x + y = 16$

$4 + 12 \stackrel{?}{=} 16$

$16 = 16$ ✓

Equation 2

$20x + 10y = 200$

$20(4) + 10(12) \stackrel{?}{=} 200$

$200 = 200$ ✓

On Your Own

1. The length ℓ of the rectangle is 1 more than 3 times the width w. Write and solve a system of linear equations to find the dimensions of the rectangle.

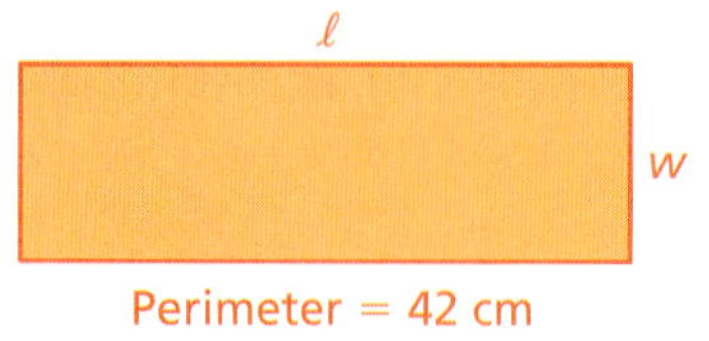

Laurie's Notes

Introduction

Connect

- **Yesterday:** Students explored real-life problems involving school enrollments in order to find out when two schools would have the same enrollment.
- **Today:** Students will write a system of linear equations and solve the system using several methods.

Motivate

- Make a sketch on the board similar to the one shown.

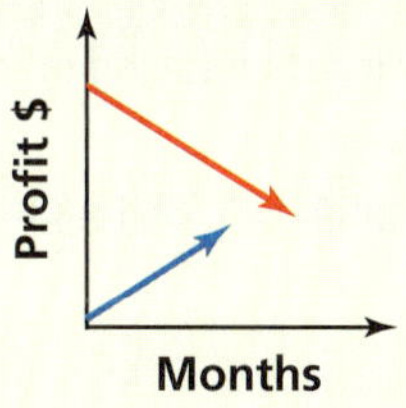

- Ask students to tell a story about the red company and the blue company.

Lesson Notes

Example 1

- **Teaching Tip:** Use play money or pieces of paper with \$20 and \$10 written on them to model this problem.
- ? "If I have 8 of each type of bill, how much money do I have? Explain." $\$240 = 8 \times \$20 + 8 \times \$10$
- ? "Can you use Guess, Check, and Revise to solve Example 1?" yes
- Explain that instead of Guess, Check, and Revise, they are going to write a system of equations to solve the problem.
- ? "What form are the equations written in?" standard form
- Because you do not know how many bills of each type there are, you have to start with some amount. The table begins at $x = 0$, meaning no \$20 bills. For the first equation to be satisfied, there would have to be sixteen \$10 bills. For the second equation to be satisfied, there would need to be twenty \$10 bills.
- Make a point of describing the goal: that you need to find an x-value that gives the same y-value for both equations.
- Try another value for x and repeat the process until the two equations have the same y-value.
- For each trial, you could model the equations using paper money.

On Your Own

- There is information in the diagram that students must consider to write the system of equations.
- Students may use different techniques to solve this problem. Ask volunteers to share their method.

Goal Today's lesson is writing systems of linear equations.

Lesson Materials
Textbook
• straightedge • play paper money

Start Thinking! and Warm Up

Lesson 3.5 **Warm Up** For use before Lesson 3.5

Lesson 3.5 **Start Thinking!** For use before Lesson 3.5

1. Pick any two integers. Use them to copy and complete the statements.
   - The sum of the two numbers is __?__.
   - The difference of the first number and twice the second number is __?__.
2. Exchange your two statements with a classmate. Write two linear equations from your classmate's statements.
3. Solve the system of equations using any method to find your classmate's numbers.

Did you guess the two numbers correctly?

Extra Example 1

A bank teller is counting \$5 bills and \$10 bills. There are 25 bills for a total of \$200. Write and solve a system of equations to find the number x of \$5 bills and the number y of \$10 bills. $x + y = 25$, $5x + 10y = 200$; 10 five dollar bills and 15 ten dollar bills

On Your Own

1. $\ell = 1 + 3w$; $2\ell + 2w = 42$; $\ell = 16$ cm, $w = 5$ cm

Laurie's Notes

Extra Example 2

The sum of two numbers is 24. The second number y is one-fifth of the first number x. Write and solve a system of equations to find the number x and the number y.
$x + y = 24, y = \frac{x}{5}; x = 20, y = 4$

Extra Example 3

In Example 3, the altitude of the Airbus A320 increases 1500 feet each minute. In how many minutes do the jets have the same altitude? 4.5 min

Example 2

- Work through the problem as shown.

Example 3

- **FYI:** Airbus and Boeing are the two largest commercial aircraft manufacturers. Boeing is an American company that moved its headquarters from Seattle to Chicago in 2001. The rival of Boeing is Airbus, a French company that has American headquarters in Virginia.
- Relate this problem to the hot air balloons. If one balloon is lifting off and one is landing, at some point in time the two balloons are at the same height.
- ? "Describe how the altitude of the Airbus is changing. Is it taking off or landing?" increasing by 1000 feet each minute; taking off
- ? "Describe how the altitude of the Boeing is changing. Is it taking off or landing?" decreasing by 500 feet each minute; landing
- ? "What is their difference in altitude to start, at time = 0? What is the difference after 1 minute?" 9000 feet; 7500 feet
- The first method is to plot points. Because data is given for the first 4 minutes only, the lines need to be extended in order to intersect. Notice the language used: the lines *appear* to intersect at (6, 6000).
- In the second method, the equations of the lines are written in slope-intercept form. From the table, students can determine the slope and the y-intercept for each equation.
- ? "How do you solve an equation with variables on both sides?" Gather the variable terms on one side and the constants on the other side.
- ? "Can you think of other ways to solve this problem?" Yes, extend the table, or use the idea that the difference in altitude is changing by 1500 feet every minute and find $9000 \div 1500 = 6$.

On Your Own

- **Think-Pair-Share:** Students should read each question independently and then work with a partner to answer the questions. When they have answered the questions, the pair should compare their answers with another group and discuss any discrepancies.
- Have volunteers share their solution at the board. If different students used different methods, share each method.

On Your Own

2. $x + y = 20; y = 3x;$
$x = 5, y = 15$
3. 5 min

Closure

- **Exit Ticket:** Describe how to solve a system of equations in at least three ways. *Sample answer:* graph both equations and see where they intersect; make a table of values; set the two equations equal to each other.

English Language Learners

Simplified Language
Word problems pose an additional challenge for English learners. You can simplify the statement of word problems so that students can apply their problem-solving skills. Students benefit from comparing the original problem to the simplified statement and learn how to determine which information in the problem is critical to the solution.

Technology For the Teacher

Dynamic Classroom

The Dynamic Planning Tool
Editable Teacher's Resources at *BigIdeasMath.com*

EXAMPLE 2 Standardized Test Practice

The sum of two numbers is 35. The second number y is equal to 4 times the first number x. Which system of linear equations represents the two numbers?

Ⓐ $x + y = 35$, $x = y + 4$ Ⓑ $x + y = 35$, $y = 4x$ Ⓒ $x + y = 35$, $y = -4x$ Ⓓ $x - y = 35$, $y = 4x$

Equation 1 →

Words First number plus second number is 35.

Equation $x + y = 35$

Equation 2 →

Words Second number is equal to 4 times the first number.

Equation $y = 4 \cdot x$

∴ The system is $x + y = 35$ and $y = 4x$. The correct answer is Ⓑ.

EXAMPLE 3 Writing a System of Linear Equations

x	Airbus A320, *A*	Boeing 777, *B*
0	0	9000
1	1000	8500
2	2000	8000
3	3000	7500
4	4000	7000

The table shows the altitudes (in feet) of two jets after x minutes. After how many minutes do the jets have the same altitude?

Method 1: Plot the points and draw each line. The graphs appear to intersect at (6, 6000).

∴ So, the jets have the same altitude after 6 minutes.

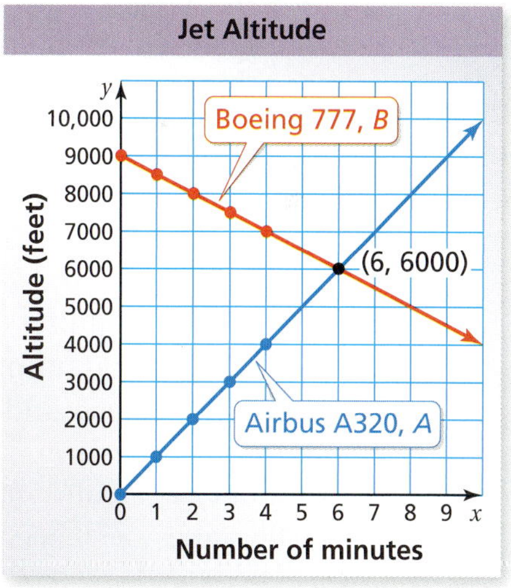

Method 2: Use the slopes and y-intercepts to write equations for A and B. Set the equations equal to each other and solve for x.

$A = 1000x$ $B = -500x + 9000$

$$1000x = -500x + 9000$$
$$1500x = 9000$$
$$x = 6$$

∴ The jets have the same altitude after 6 minutes.

On Your Own

Now You're Ready
Exercises 4–6

2. The sum of two numbers is 20. The second number is 3 times the first number. Write and solve a system of equations to find the two numbers.

3. **WHAT IF?** In Example 3, the altitude of the Boeing 777 decreases 800 feet each minute. After how many minutes do the jets have the same altitude? Solve using both methods.

3.5 Exercises

Check It Out
Help with Homework
BigIdeasMath.com

Vocabulary and Concept Check

1. **VOCABULARY** Why is the equation $2x - y = 4$ called a *linear* equation?
2. **VOCABULARY** What must be true for an ordered pair to be a solution of a system of two linear equations?
3. **WRITING** Describe three ways to solve a system of linear equations.

Practice and Problem Solving

In Exercises 4–6, (a) write a system of linear equations to represent the situation. Then, answer the question using (b) a table, (c) a graph, and (d) algebra.

1 2 3 4. **ATTENDANCE** The first football game has 425 adult fans and 225 student fans. The adult attendance A decreases by 15 each game. The student attendance S increases by 25 each game. After how many games x will the adult attendance equal the student attendance?

Adults:	Attendence each game	is	425	minus 15 times	number of games.
Students:	Attendence each game	is	225	plus 25 times	number of games.

5. **BOUQUET** A bouquet of lilies and tulips has 12 flowers. Lilies cost \$3 each and tulips cost \$2 each. The bouquet costs \$32. How many lilies x and tulips y are in the bouquet?

Number of flowers:	Number of lilies	plus	Number of tulips	is	12.		
Cost of bouquet:	\$3 times	number of lilies	plus \$2 times	number of tulips	is	\$32.	

6. **CHORUS** There are 63 students in a middle school chorus. There are 11 more boys than girls. How many boys x and girls y are in the chorus?

Number of students:	Number of boys	plus	number of girls	is	63.
Boys and girls:	Number of boys	equals	number of girls	plus	11.

Assignment Guide and Homework Check

Level	Day 1 Activity Assignment	Day 2 Lesson Assignment	Homework Check
Basic	4, 5, 14–17	1–3, 6–9	2, 3, 6, 8
Average	4, 5, 14–17	1–3, 7–11	2, 3, 8, 10
Advanced	4, 5, 14–17	1–3, 7, 9–13	2, 3, 10, 12

Common Errors

- **Exercise 4** Students may get confused because there are three variables in this problem, *A*, *S*, and *x*. Tell them to read the word problem carefully and notice that *x* is the variable for which they are solving. After they have written the equations, hint to students who are struggling that they can set the equations equal to each other to answer the question algebraically.
- **Exercises 5 and 6** Students may write the wrong equation because they are not carefully reading the words that tell them what symbol to write. For example, a student may write addition instead of subtraction. Remind them to read carefully.

3.5 Record and Practice Journal

Write a system of linear equations to represent the situation. Then, answer the question using (a) a table, (b) a graph, and (c) algebra.

1. The cost of buying pans and making pies for a bake sale is \$15 plus \$3.50 for each pie. The revenue is \$5.00 for each pie. After how many pies x will the cost equal the revenue?

Cost:	Buying pans and ingredients	is	\$15	plus \$3.50 times	number of pies.
Revenue:	Income from selling pies	is	\$5	times	number of pies.

$C = 15 + 3.5x$
$R = 5x$
$x = 10$

2. You buy 16 candles. Large candles cost \$7 each and small candles cost \$3 each. You spend \$76. How many large candles x and small candles y did you buy?

Number of candles:	Number of large candles	plus	number of small candles	is	16.		
Cost of candles:	\$7 times	number of large candles	plus \$3 times	number of small candles	is	\$76.	

$16 = x + y$
$76 = 7x + 3y$
$y = 9;\ x = 7$

3. There are 62 time slots for cheerleading tryouts on Saturday and Sunday. There are 14 more slots on Saturday than on Sunday. Find the number of Saturday time slots x and the number of Sunday time slots y.

$62 = x + y$
$x = 14 + y$
$x = 38;\ y = 24$

Technology For the Teacher

Answer Presentation Tool
Quiz*Show*

Vocabulary and Concept Check

1. because its graph is a line
2. The ordered pair must be on both lines.
3. You can use a table to see when the two equations are equal. You can use a graph to see whether or not the two lines intersect. You can use algebra and set the equations equal to each other to see when they have the same value.

Practice and Problem Solving

4. a. $A = 425 - 15x$
$S = 225 + 25x$

b.

x	1	2	3	4	5
A	410	395	380	365	350
S	250	275	300	325	350

after 5 games

c.

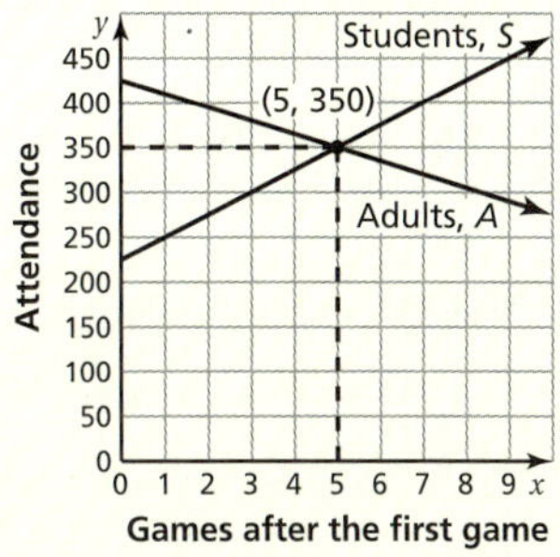

after 5 games

d. $425 - 15x = 225 + 25x$;
$x = 5$; after 5 games

5–6. See Additional Answers.

Practice and Problem Solving

7. **a.** no; You need to know how many more dimes there are than nickels or how many coins there are total.

   b. *Sample answer:* 9 dimes and 1 nickel

8. 15 yr

9. no; A linear system must have either one, none, or infinitely many solutions. Lines cannot intersect at exactly two points.

10. \$16.10

11. Each equation is the same. So, the graph of the system is the same line.

12. See *Taking Math Deeper.*

13. (1, 0), (−2, 3), (−6, 1)

Fair Game Review

14. $y = 2x - 1$

15. $y = \frac{1}{4}x - 2$

16. $y = -2x + 5$

17. B

Mini-Assessment

1. You have \$6.15 worth of quarters and dimes in a jar. There are 30 coins in the jar. Write and solve a system of equations to find the number x of dimes and the number y of quarters. There are 21 quarters and 9 dimes.

2. A bouquet of roses and carnations has 15 flowers. Roses cost \$4 each and carnations cost \$2 each. The bouquet costs \$44. How many roses x and carnations y are in the bouquet? There are 7 roses and 8 carnations.

Taking Math Deeper

Exercise 12

Even though this problem is a straightforward "linear system" problem, students might be thrown off because they don't understand the terminology of World Cup soccer. If they are interested, they might research the topic. However, they should realize that they can still answer the question, even without knowing soccer zones.

1 Organize the information.

Let S = number of teams in the Scottish zone and
P = number of teams in the Puerto Rican zone.
$S + P = 88$
$S = 2P - 17$

Note: The six continental zones are Africa, Asia, North and Central America and Caribbean, South America, Oceana, and Europe. So, the Scottish team is in the European zone and the Puerto Rican team is in the North and Central American and Caribbean zone.

2 Solve the system of equations.

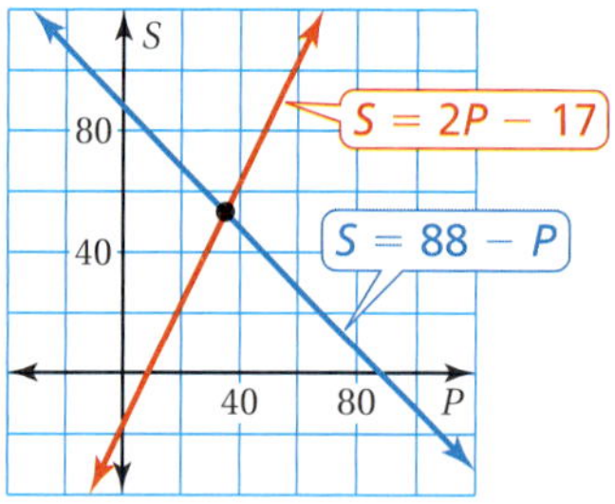

The two lines appear to intersect at about (35, 55). By Guess, Check, and Revise, students can determine that the exact point of intersection is (35, 53).

3 Answer the question.

There are 53 teams in the Scottish zone and
35 teams in the Puerto Rican zone.

Note: You may want to challenge students to see if they can discover a strategy for solving a system algebraically.

Reteaching and Enrichment Strategies

If students need help. . .	If students got it. . .
Resources by Chapter • Practice A and Practice B • Puzzle Time Record and Practice Journal Practice Differentiating the Lesson Lesson Tutorials Skills Review Handbook	Resources by Chapter • Enrichment and Extension • School-to-Work • Financial Literacy Start the next section

7. **WHAT IS MISSING?** You have dimes and nickels in your pocket with a total value of $0.95. There are more dimes than nickels. How many of each coin do you have?

   a. Do you have enough information to write a system of equations to answer the question? If not, what else do you need to know?

   b. Find one possible solution.

x	Account *A*	Account *B*
0	420	465
1	426	468
2	432	471
3	438	474
4	444	477

8. **INTEREST** The table shows the balances (in dollars) of two accounts earning simple interest for x years. After how many years will the accounts have the same balance?

9. **CRITICAL THINKING** Is it possible for a system of two linear equations to have exactly two solutions? Explain.

10. **DINNER** How much does it cost for two specials and two glasses of milk?

11. **REASONING** A system of two linear equations has more than one solution. Describe the graph of the system.

Scottish Team

Puerto Rican Team

12. **WORLD CUP** The global competition for the World Cup is broken up into six continental zones. The number of teams in the Scottish team's zone is 17 less than twice the number of teams in the Puerto Rican team's zone. There is a total of 88 teams in both zones. How many teams are in each zone?

13. **Algebra** The graphs of the three equations form a triangle. Use algebra to find the coordinates of the vertices of the triangle.

$$x + y = 1 \qquad x + 7y = 1 \qquad x - 2y = -8$$

Fair Game Review What you learned in previous grades & lessons

Write an equation of the line that passes through the points. *(Section 3.3)*

14. (0, −1), (1, 1)
15. (−4, −3), (4, −1)
16. (2, 1), (3, −1)

17. **MULTIPLE CHOICE** Which function rule relates x and y for the set of ordered pairs (2, 4), (4, 5), (6, 6)? *(Skills Review Handbook)*

   Ⓐ $y = x - 2$ Ⓑ $y = \frac{1}{2}x + 3$ Ⓒ $y = 2x + 1$ Ⓓ $y = \frac{1}{2}x - 3$

3.4–3.5 Quiz

Check It Out
Progress Check
BigIdeasMath.com

1. **FISH POND** You are draining a fish pond. The amount y (in liters) of water remaining after x hours is $y = -60x + 480$. (a) Graph the equation. (b) Interpret the x- and y-intercepts. *(Section 3.4)*

2. **CABLE CAR** The graph shows the distance y (in meters) that a cable car travels up a mountain in x minutes. *(Section 3.4)*

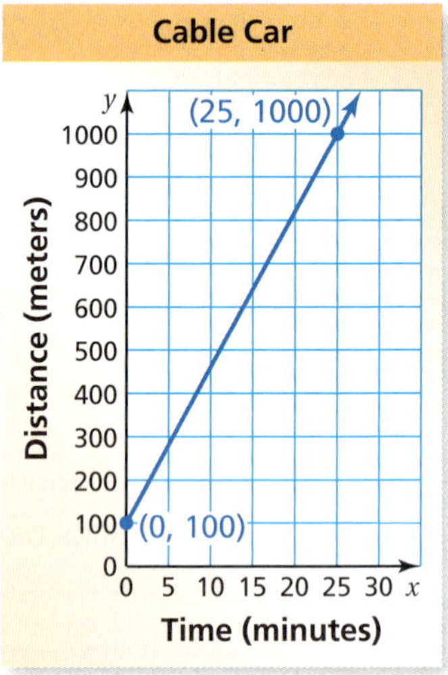

   a. Find and interpret the slope.
   b. Write an equation of the line.
   c. How far does the cable car travel in 15 minutes?

3. **BICYCLE** You need \$160 to buy a mountain bike. You earn \$40 per week for babysitting and \$20 per lawn mowed. *(Section 3.4)*
   a. Write an equation that represents your weekly income y (in dollars) for x lawns mowed.
   b. How many lawns do you need to mow to earn enough money in 1 week to buy the mountain bike?

4. **WATER** A recreation department bought bottled water to sell at a fair. The graph shows the number y of bottles remaining after each hour x. *(Section 3.4)*

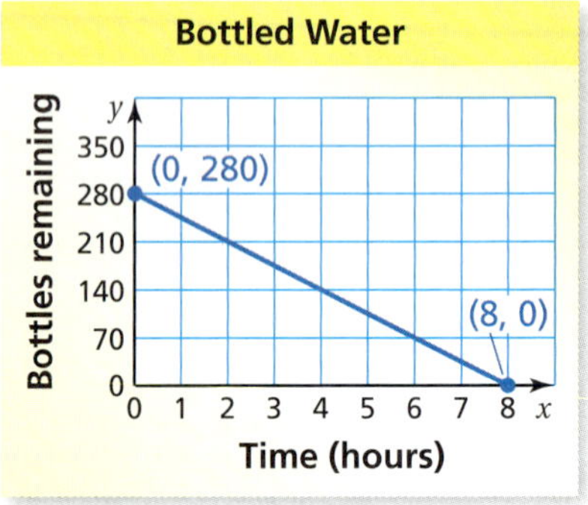

   a. Find the slope and y-intercept.
   b. Write an equation of the line.
   c. The fair started at 10 A.M. When did the recreation department run out of bottled water?

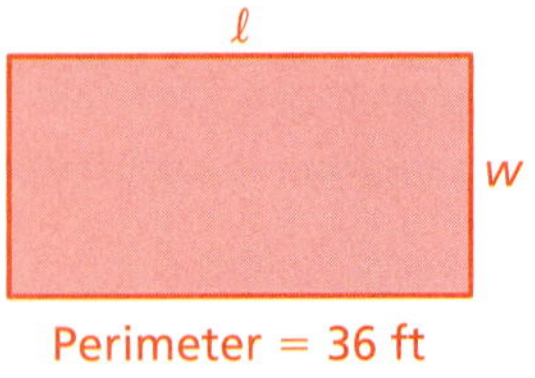

5. **RECTANGLE** The length of the rectangle is twice its width. Write and solve a system of linear equations to find the length ℓ and width w of the rectangle. *(Section 3.5)*

6. **PUZZLE** The difference of two numbers is 8. The first number x is 1 less than twice the second number y. Write and solve a system of linear equations to find the two numbers. *(Section 3.5)*

7. **SHELTER** A cat and dog shelter houses 23 animals. There are 5 more cats x than dogs y. Write and solve a system of linear equations to find the numbers of cats and dogs in the shelter. *(Section 3.5)*

Alternative Assessment Options

Math Chat
Structured Interview
Student Reflective Focus Question
Writing Prompt

Structured Interview

Interviews can occur formally or informally. Ask a student to perform a task and to explain it as they work. Have them describe their thought process. Probe the student for more information. Do not ask leading questions. Keep a rubric or notes.

Teacher Prompts	Student Answers	Teacher Notes
Tell me a story about managing a checking account. Include Rate: −50 dollars/mo y-intercept: \$300 x-intercept: 6 months	I opened a checking account and deposited \$300. Each month, I withdrew \$50. After 6 months, I had \$0 in the checking account.	Student understands the meaning of slope, the y-intercept, and the x-intercept in a real-life problem.

Study Help Sample Answers

Remind students to complete Graphic Organizers for the rest of the chapter.

3.

Linear equation in slope-intercept form:
$y = mx + b$
$m = \text{slope} = \frac{\text{rise}}{\text{run}}$
$b = y$-intercept

Writing Systems of Linear Equations

Write each equation using words. Then translate the words into algebra.

Example:
You have nickels and dimes in your pocket. There are 16 coins that total \$1.40. Write a system of equations to represent this situation.

Words: number of nickels plus number of dimes is 16.
Algebra: $x + y = 16$

Words: value of a nickel times number of nickels plus value of a dime times number of dimes is the total value.
Algebra: $0.05 \cdot x + 0.10 \cdot y = 1.40$

The linear system is $x + y = 16$ and $0.05x + 0.10y = 1.40$.

How can you solve a system of linear equations?

4. Available at *BigIdeasMath.com*

Reteaching and Enrichment Strategies

If students need help. . .	If students got it. . .
Resources by Chapter • Study Help • Practice A and Practice B • Puzzle Time Lesson Tutorials *BigIdeasMath.com* Practice Quiz Practice from the Test Generator	Resources by Chapter • Enrichment and Extension • School-to-Work Game Closet at *BigIdeasMath.com* Start the Chapter Review

Answers

1. a.

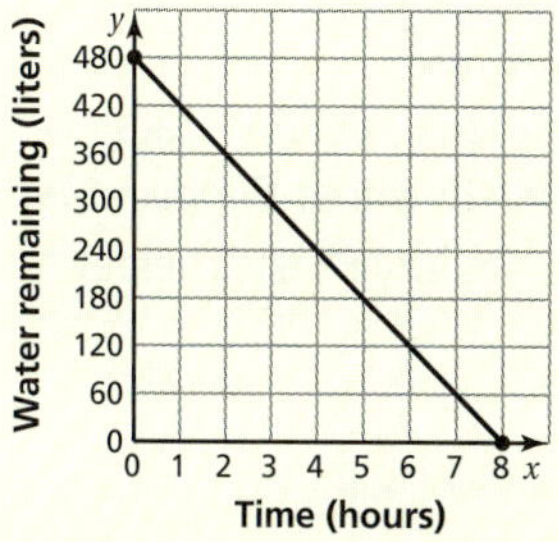

b. The x-intercept is the number of hours it takes to drain the pond. The y-intercept represents the amount of water in the pond initially.

2. a. slope: 36; The slope is the speed at which the cable car is traveling, 36 meters per minute.
   b. $y = 36x + 100$
   c. 640 m
3. a. $y = 20x + 40$
   b. 6 lawns
4. a. slope: -35; y-intercept: 280
   b. $y = -35x + 280$
   c. 6 P.M.

5–7. See Additional Answers.

Technology For the Teacher
Answer Presentation Tool

Assessment Book

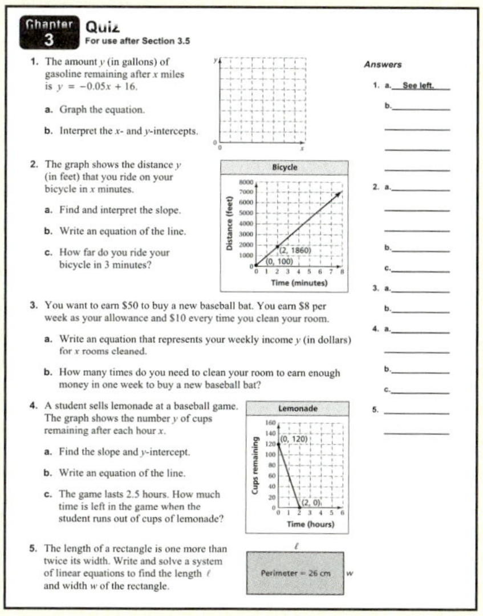
Chapter 3 Quiz
For use after Section 3.5

1. The amount y (in gallons) of gasoline remaining after x miles is $y = -0.05x + 16$.
   a. Graph the equation.
   b. Interpret the x- and y-intercepts.
2. The graph shows the distance y (in feet) that you ride on your bicycle in x minutes.
   a. Find and interpret the slope.
   b. Write an equation of the line.
   c. How far do you ride your bicycle in 3 minutes?
3. You want to earn \$50 to buy a new baseball bat. You earn \$8 per week as your allowance and \$10 every time you clean your room.
   a. Write an equation that represents your weekly income y (in dollars) for x rooms cleaned.
   b. How many times do you need to clean your room to earn enough money in one week to buy a new baseball bat?
4. A student sells lemonade at a baseball game. The graph shows the number y of cups remaining after each hour x.
   a. Find the slope and y-intercept.
   b. Write an equation of the line.
   c. The game lasts 2.5 hours. How much time is left in the game when the student runs out of cups of lemonade?
5. The length of a rectangle is one more than twice its width. Write and solve a system of linear equations to find the length ℓ and width w of the rectangle.

Answers
1. a. See left.

For the Teacher
Additional Review Options

- **Quiz***Show*
- Big Ideas Test Generator
- Game Closet at *BigIdeasMath.com*
- Vocabulary Puzzle Builder
- Resources by Chapter
    - Puzzle Time
    - Study Help

Answers

1. $y = x - 2$
2. $y = -\frac{1}{2}x + 4$
3. $y = 3x - 8$
4. $y = 2x + 2$
5. $y = -0.5x$

Review of Common Errors

Exercises 1 and 2

- Students may write the reciprocal of the slope or forget a negative sign. Remind them of the definition of slope. Ask them to predict the sign of the slope based on the rise or fall of the line.

Exercises 3–5

- Students may use the reciprocal of the slope when finding the second point to plot. Remind them of the definition of slope.
- Students may say that the y-intercept is the y-coordinate of the given point. Remind them that the y-intercept is represented by b in the point $(0, b)$, and if the given point is not in this form, they need to use the slope to find where the line crosses the y-axis.

Exercises 6 and 7

- Students may plot the points, draw a line through the points, and use the point where the line crosses the y-axis to find the y-intercept. Sometimes this is not the correct point because the line may not be perfectly drawn. Encourage students to use the slope to find the y-intercept.

Exercise 8

- Students may forget to include negatives with the slope and/or the y-intercept. Remind them to look at the sign in front of the slope and the y-intercept. Also remind them that the equation is $y = mx + b$. This means that if the linear equation has "minus b," then the y-intercept is negative.

Exercises 9 and 10

- Students may write the wrong equation because they are not carefully reading the words that tell them what symbol to write. For example, a student may write addition instead of subtraction. Remind them to read carefully.

3 Chapter Review

Check It Out
Vocabulary Help
BigIdeasMath.com

Review Examples and Exercises

3.1 Writing Equations in Slope-Intercept Form (pp. 106–111)

Write an equation of the line in slope-intercept form.

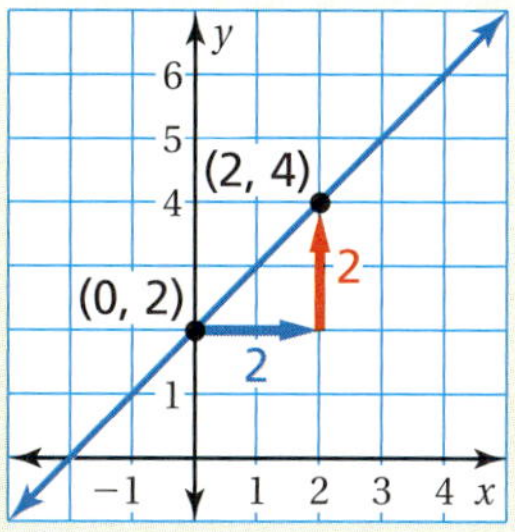

Find the slope and y-intercept.

$$\text{slope} = \frac{\text{rise}}{\text{run}} = \frac{2}{2} = 1$$

Because the line crosses the y-axis at $(0, 2)$, the y-intercept is 2.

slope → y-intercept

So, the equation is $y = 1x + 2$, or $y = x + 2$.

Exercises

Write an equation of the line in slope-intercept form.

1.

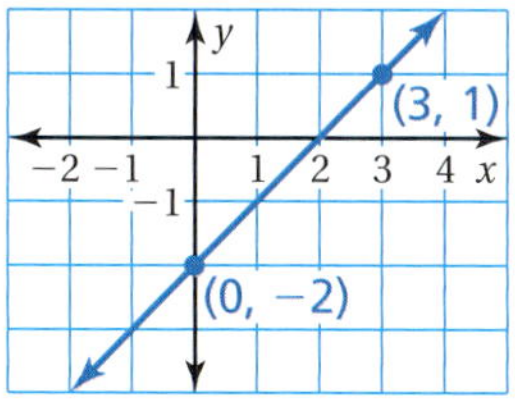

2.

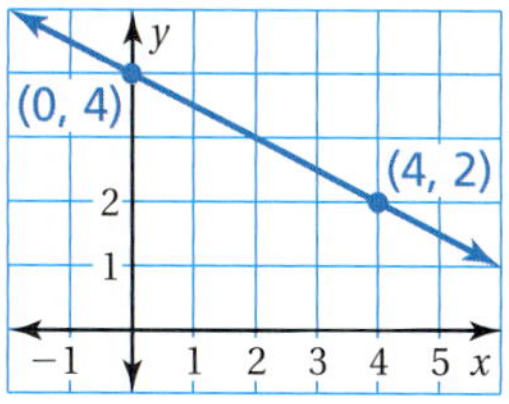

3.2 Writing Equations Using a Slope and a Point (pp. 112–117)

Write an equation of the line with a slope of $\frac{2}{3}$ that passes through the point $(-3, -1)$.

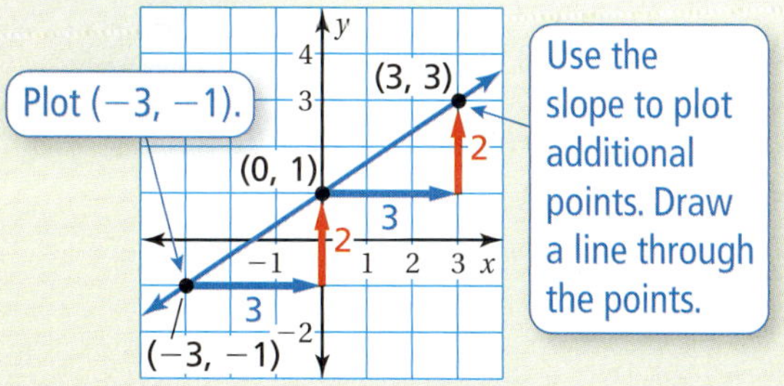

Use a graph to find the y-intercept.

Because the line crosses the y-axis at $(0, 1)$, the y-intercept is 1.

So, the equation is $y = \frac{2}{3}x + 1$.

Exercises

Write an equation of the line with the given slope that passes through the given point.

3. $m = 3;\ (4, 4)$

4. $m = 2;\ (2, 6)$

5. $m = -0.5;\ (-4, 2)$

3.3 Writing Equations Using Two Points (pp. 118–123)

Write an equation of the line that passes through the points (2, 3) and (−2, −3).

Use a graph to find the slope and y-intercept.

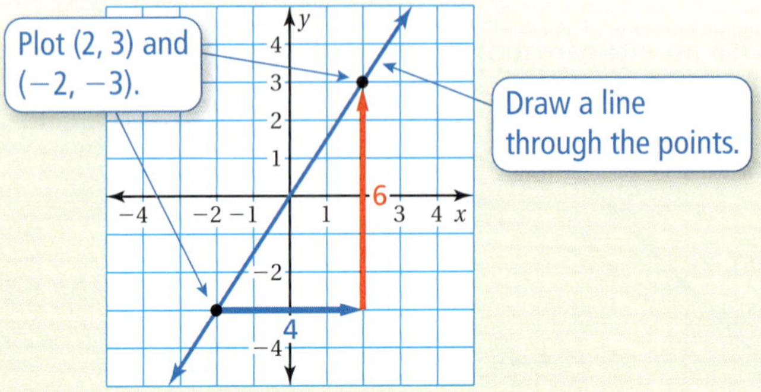

$$\text{slope} = \frac{\text{rise}}{\text{run}} = \frac{6}{4} = \frac{3}{2}$$

Because the line crosses the y-axis at (0, 0), the y-intercept is 0.

∴ So, the equation is $y = \frac{3}{2}x + 0$, or $y = \frac{3}{2}x$.

Exercises

Write an equation of the line that passes through the points.

6. (−2, 0), (2, −4)

7. (−2, −2), (4, 1)

3.4 Solving Real-Life Problems (pp. 126–131)

The amount y (in dollars) of money you have left after playing x games at a carnival is $y = -0.75x + 10$. How much money do you have after playing eight games?

$y = -0.75x + 10$	Write the equation.
$= -0.75(8) + 10$	Substitute 8 for x.
$= 4$	Simplify.

∴ You have $4 left after playing 8 games.

Exercises

8. **HAY** The amount y (in bales) of hay remaining after feeding cows for x days is $y = -3.5x + 105$. (a) Graph the equation. (b) Interpret the x- and y-intercepts. (c) How many bales are left after 10 days?

Review Game

Equations of Lines

Big Ideas Game Closet

Materials per Group:

- map of the city or town where students' homes and school are located
- pencil
- straight edge

Directions:

Students work in groups of four using a local map that includes their home and school locations. Maps can be downloaded from the Internet. The legend for distance on the map must be visible.

The teacher sets an origin, using a popular location like the zoo or library. Placing the origin away from the school will result in having *x*- and *y*-intercepts other than zero.

Each group will plot the points of the school and the point of one of their homes. They should draw a line through the points and write an equation for the line. They will repeat for each of their homes. Students should work with numbers representing actual miles.

Who Wins?

The first group to complete all school to home lines with the correct equations wins.

For the Student Additional Practice

- Lesson Tutorials
- Study Help (textbook)
- Student Website
  - Multi-Language Glossary
  - Practice Assessments

Answers

6. $y = -x - 2$

7. $y = \frac{1}{2}x - 1$

8. a.

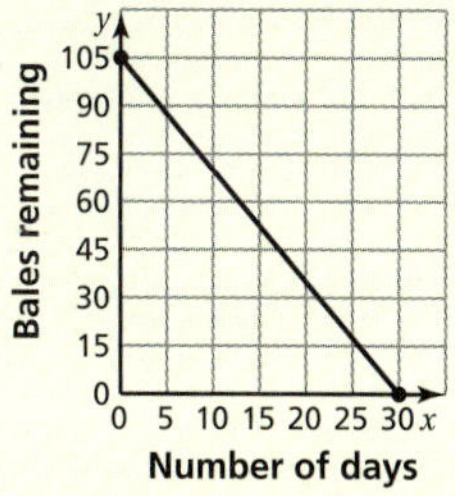

   b. The x-intercept is the number of days it takes to feed all the hay to the cows. The y-intercept represents how many bales of hay there were originally.

   c. 70 bales

9. $j + m = 8$; $6j + 5m = 45$; $j = 5$ jars of jam, $m = 3$ packages of muffin mix

10. $y = \frac{1}{2}p$; $y + p = 36$; $y = 12$ yellow marbles, $p = 24$ purple marbles

My Thoughts on the Chapter

What worked. . .

What did not work. . .

What I would do differently. . .

Teacher Tip

Not allowed to write in your teaching edition? Use sticky notes to record your thoughts.

3.5 Writing Systems of Linear Equations (pp. 132–137)

You have quarters and dimes in your pocket. There are 5 coins that total $0.80. Write and solve a system of equations to find the number x of dimes and the number y of quarters.

Words	Number of dimes	plus	number of quarters	is	five.
Equation 1 → **Equation**	x	$+$	y	$=$	5

Words	Value of a dime	times	number of dimes	plus	value of a quarter	times	number of quarters	is	the total value.
Equation 2 → **Equation**	0.10	$\cdot$	x	$+$	0.25	$\cdot$	y	$=$	0.80

The linear system is $x + y = 5$ and $0.10x + 0.25y = 0.80$.

Solve each equation for y. Then make a table of values to find the x-value that gives the same y-value for both equations.

x	0	1	2	3	4
$y = 5 - x$	5	4	3	2	1
$y = 3.2 - 0.4x$	3.2	2.8	2.4	2	1.6

The solution is (3, 2).

So, you have three dimes and two quarters in your pocket.

Exercises

9. **GIFT BASKET** A gift basket that contains jars of jam and packages of muffin mix costs $45. There are eight items in the basket. Jars of jam cost $6 each and packages of muffin mix cost $5 each. Write and solve a system of equations to find the number of jars of jam and the number of packages of muffin mix.

10. **MARBLES** A bag contains half as many yellow marbles as purple marbles. There are 36 marbles in the bag. Write and solve a system of equations to find the number of marbles of each color.

Check It Out
Test Practice
BigIdeasMath.com

Write an equation of the line in slope-intercept form.

1.

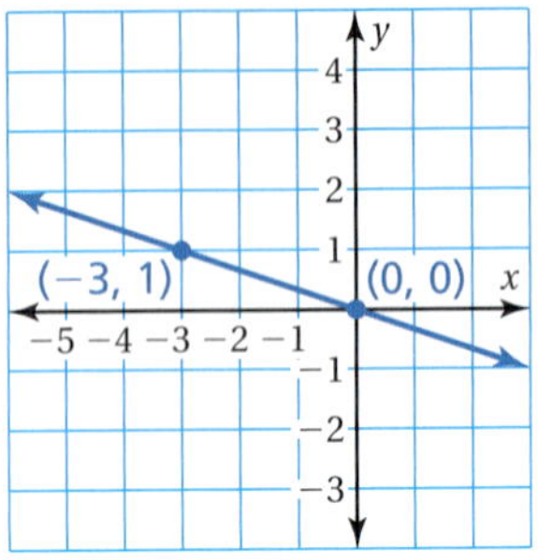

2.

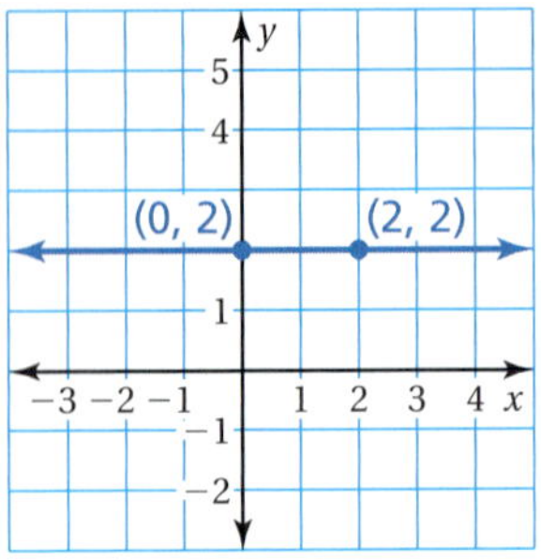

3.

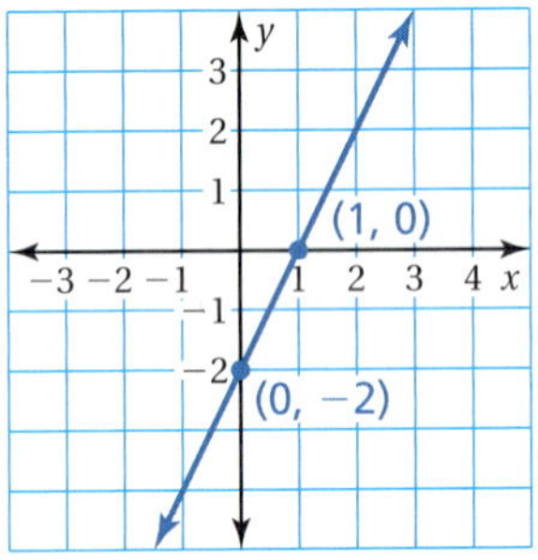

4.

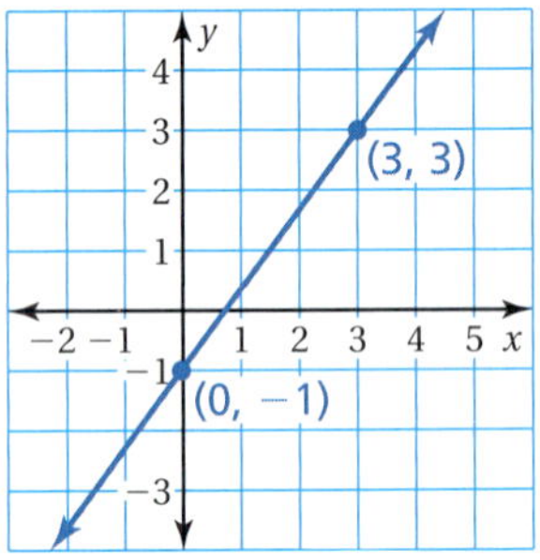

Write an equation of the line with the given slope that passes through the given point.

5. $m = -3$; $(-2, -2)$

6. $m = -\frac{1}{2}$; $(4, -1)$

7. $m = \frac{2}{3}$; $(3, 3)$

8. $m = 1$; $(4, 3)$

Write an equation of the line that passes through the points.

9. $(-1, 5), (3, -3)$

10. $(-4, 1), (4, 3)$

11. $(-2, 5), (-1, 1)$

12. BRAILLE Because of its size and detail, Braille takes longer to read than text. A person reading Braille reads at 25% the rate of a person reading text. Write an equation that represents the average rate y of a Braille reader in terms of the average rate x of a text reader.

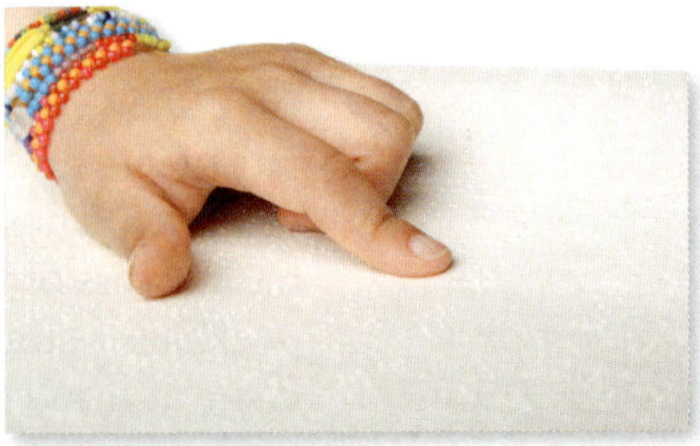

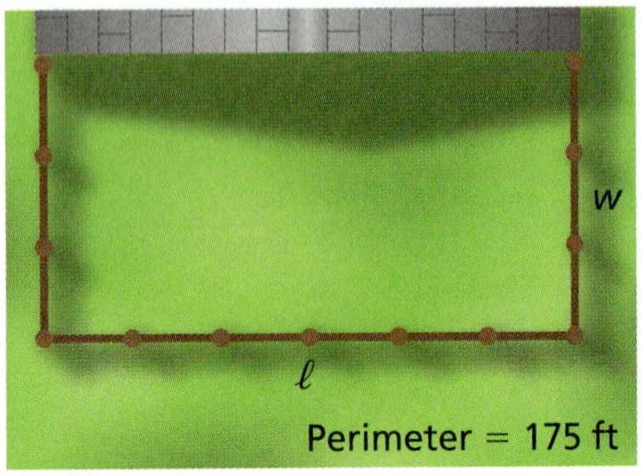

13. PASTURE You are building a fence for a rectangular pasture. The length of the pasture is 1 foot less than twice the width.

a. Write and solve a system of linear equations to find the length and width of the pasture.

b. Find the area of the pasture.

Test Item References

Chapter Test Questions	Section to Review
1–4	3.1
5–8	3.2
9–11	3.3
12	3.4
13	3.5

Test-Taking Strategies

Remind students to quickly look over the entire test before they start so that they can budget their time. When students hurry they make unintentional mistakes, such as writing inverted slopes or making sign errors. Remind them to **Stop** and **Think** before they write their answers.

Common Assessment Errors

- **Exercises 1–4** Students may write the reciprocal of the slope or forget a negative sign. Ask them to predict the sign of the slope based on the rise or fall of the line.
- **Exercises 5–8** Students may use the reciprocal of the slope when finding the second point to plot. Remind them of the definition of slope.
- **Exercises 5–8** Students may think that the *y*-intercept is the *y*-coordinate of the given point. Remind them that the *y*-intercept is represented by *b* in the point (0, *b*), and if the given point is not in this form, they need to use the slope to find where the line crosses the *y*-axis.
- **Exercises 9–11** Students may plot the points, draw a line through the points, and use the point where the line crosses the *y*-axis to find the *y*-intercept. This may not be the correct point because the line may not be perfectly drawn. Encourage students to use the slope to find the *y*-intercept.
- **Exercise 12** Students may have difficulty using the given information to come up with an equation. Point out that the two rates are proportional.
- **Exercise 13** Students may write the equation relating the length ℓ and width w as $\ell = 1 - 2w$.
- **Exercise 13** Students may count the length twice when writing the equation for perimeter. Point out that fencing is only required on 3 sides of the pasture.

Reteaching and Enrichment Strategies

If students need help. . .	If students got it. . .
Resources by Chapter • Practice A and Practice B • Puzzle Time Record and Practice Journal Practice Differentiating the Lesson Lesson Tutorials Practice from the Test Generator Skills Review Handbook	Resources by Chapter • Enrichment and Extension • School-to-Work • Financial Literacy Game Closet at *BigIdeasMath.com* Start Standardized Test Practice

Answers

1. $y = -\frac{1}{3}x$
2. $y = 2$
3. $y = 2x - 2$
4. $y = \frac{4}{3}x - 1$
5. $y = -3x - 8$
6. $y = -\frac{1}{2}x + 1$
7. $y = \frac{2}{3}x + 1$
8. $y = x - 1$
9. $y = -2x + 3$
10. $y = \frac{1}{4}x + 2$
11. $y = -4x - 3$
12. $y = 0.25x$
13. a. $\ell = 2w - 1;\ \ell + 2w = 175;$ $w = 44$ ft, $\ell = 87$ ft

    b. 3828 ft^2

Assessment Book

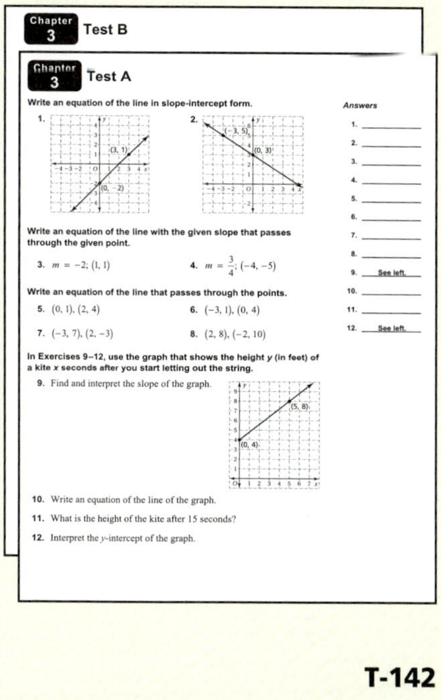
Chapter 3 Test B

Chapter 3 Test A

Write an equation of the line in slope-intercept form.

1. 2.

Answers 1. ___ 2. ___ 3. ___ 4. ___ 5. ___ 6. ___ 7. ___ 8. ___ 9. See left. 10. ___ 11. ___ 12. See left.

Write an equation of the line with the given slope that passes through the given point.

3. $m = -2$; (1, 1)
4. $m = \frac{3}{4}$; (−4, −5)

Write an equation of the line that passes through the points.

5. (0, 1), (2, 4)
6. (−3, 1), (0, 4)
7. (−3, 7), (2, −3)
8. (2, 8), (−2, 10)

In Exercises 9–12, use the graph that shows the height *y* (in feet) of a kite *x* seconds after you start letting out the string.

9. Find and interpret the slope of the graph.
10. Write an equation of the line of the graph.
11. What is the height of the kite after 15 seconds?
12. Interpret the *y*-intercept of the graph.

Test-Taking Strategies

Available at *BigIdeasMath.com*

After Answering Easy Questions, Relax
Answer Easy Questions First
Estimate the Answer
Read All Choices before Answering
Read Question before Answering
Solve Directly or Eliminate Choices
Solve Problem before Looking at Choices
Use Intelligent Guessing
Work Backwards

About this Strategy

When taking a multiple choice test, be sure to read each question carefully and thoroughly. One way to answer the question is to work backwards. Try putting the responses into the question, one at a time, and see if you get a correct solution.

Answers

1. B
2. G
3. 1.75
4. A
5. H
6. D

Technology For the Teacher

Big Ideas Test Generator

Item Analysis

1. **A.** The student finds the slope correctly but then fails to find the *y*-intercept.

B. Correct answer

C. The student finds the reciprocal of the slope and then uses the second point to find the *y*-intercept.

D. The student finds the reciprocal of the slope and then uses the first point to find the *y*-intercept.

2. **F.** The student mistakes the roles of slope and intercept, thinking same intercept means what same slope means.

G. Correct answer

H. The student chooses a conclusion for lines that have the same intercept and the same slope, overlooking that these lines have different slopes.

I. The student is confused by the problem and is grasping at straws.

3. **Gridded Response:** Correct answer: 1.75

Common Error: The student subtracts 11 from -4 instead of adding, yielding an answer of -3.75.

4. **A.** Correct answer

B. The student accounts for the variable term correctly, but fails to realize that the starting point is three years prior to the point at which the value is 21,000.

C. The student accounts for the variable term correctly, but is confused by the roles of 21,000 and three years time. The student decides that subtracting them makes sense because the car is depreciating.

D. The student is badly confused, subtracting numbers out of desperation and fails to account for the proper role of 2,500.

5. **F.** The student finds how much orange juice you drink in 6 days.

G. The student uses 5 days in a week instead of 7.

H. Correct answer

I. The student converts incorrectly, using 16 fluid ounces in a quart instead of 32.

6. **A.** The student finds the point where the line crosses the *y*-axis.

B. The student finds the correct point, but writes its coordinates in reverse.

C. The student finds the point where the line crosses the *y*-axis and then writes its coordinates in reverse.

D. Correct answer

3 Standardized Test Practice

1. A line contains the points $(-3, 5)$ and $(6, 8)$. What is the equation of the line?

   A. $y = \frac{1}{3}x$ **C.** $y = 3x - 10$

   B. $y = \frac{1}{3}x + 6$ **D.** $y = 3x + 14$

2. Two lines have the same y-intercept. The slope of one line is 1 and the slope of the other line is -1. What can you conclude?

   F. The lines are parallel.

   G. The lines meet at exactly one point.

   H. The lines meet at more than one point.

   I. The situation described is impossible.

Test-Taking Strategy
Work Backwards

"**Work backwards** by trying 1, 2, 3, and 4. You will see that 3(4) = 12. So, D is correct."

3. What value of x makes the equation below true?

$$4x - 11 = -4$$

4. A car's value depreciates at a rate of \$2,500 per year. Three years after it was purchased, the car's value was \$21,000. Which equation can be used to find v, its value in dollars, n years after it was purchased?

   A. $v = 28{,}500 - 2{,}500n$ **C.** $v = 18{,}500 - 2{,}500n$

   B. $v = 21{,}000 - 2{,}500n$ **D.** $v = 18{,}500 - n$

5. You drink 8 fluid ounces of orange juice every morning. How many quarts of orange juice do you drink in 6 weeks?

   F. 1.5 qt **H.** 10.5 qt

   G. 7.5 qt **I.** 21 qt

6. The line $4x + 5y = 12$ is written in standard form. At what point does the graph of this line cross the x-axis?

   A. (0, 2.4) **C.** (2.4, 0)

   B. (0, 3) **D.** (3, 0)

7. Water is leaking from a jug at a constant rate. After leaking for two hours, the jug contains 48 fluid ounces of water. After leaking for five hours, the jug contains 42 fluid ounces of water.

Think
Solve
Explain

Part A Find the rate at which water is leaking from the jug.

Part B Find how many fluid ounces of water were in the jug before it started leaking. Show your work and explain your reasoning.

Part C Write an equation that shows how many fluid ounces y of water are left in the jug after it has been leaking for h hours.

Part D Find how many hours it will take the jug to empty entirely. Show your work and explain your reasoning.

8. What is the slope of the line given by $3x - 6y = 33$?

F. -3

G. $-\frac{1}{2}$

H. $\frac{1}{2}$

I. 3

9. You have 40 nickels and dimes. Their total value is \$3.30. Which system of equations could be used to find the number n of nickels and the number d of dimes?

A. $n + d = 40$
$n + d = 3.30$

B. $n + d = 40$
$5n + 10d = 3.30$

C. $5n + 10d = 40$
$n + d = 330$

D. $n + d = 40$
$5n + 10d = 330$

10. You bought a lead pencil for \$5 and three identical markers. You spent \$12.47 in all. Which equation could be used to find the price p of one marker?

F. $5 + p = 12.47$

G. $5 + 3p = 12.47$

H. $3(5 + p) = 12.47$

I. $3p = 12.47 + 5$

Item Analysis (continued)

7. **4 points** The student demonstrates a thorough understanding of how to analyze a problem situation and translate it into a linear equation by building its components step-by-step. In addition, the student is able to set the equation equal to a given value and solve it in clear, complete steps. In Part A, the water is leaking at a rate of 2 fluid ounces per hour. In Part B, there were 52 fluid ounces in the jug before it started leaking. In Part C, the equation is $y = 52 - 2h$, or its equivalent. In Part D, the jug will be empty after 26 hours.

   3 points The student demonstrates an essential but less than thorough understanding of the problem situation and how to attack it. There may be a minor error made along the way, but subsequent work is consistent with the error.

   2 points The student demonstrates a partial understanding of how to interpret the problem and translate it algebraically. The student's work and explanations demonstrate a lack of essential understanding. For example, the student may calculate the rate correctly in Part A, but then overlooks the fact that the leaking process began two hours before the jug contained 48 fluid ounces. Additionally, the student will have trouble translating the work in Parts A and B into an equation.

   1 point The student demonstrates limited understanding. The student's response is incomplete and exhibits many flaws.

   0 points The student provides no response, a completely incorrect or incomprehensible response, or a response that demonstrates insufficient understanding of how to work with a verbal situation that must be translated into algebra.

8. **F.** The student simply moves $3x$ to the right side of the equation and picks up its coefficient.

   G. The student makes a mistake with a negative sign, either when moving terms or when dividing by the coefficient of y.

   H. Correct answer

   I. The student adds $3x$ to the right side of the equation and picks up its coefficient.

9. **A.** The student fails to recognize that the values of the nickels and dimes must be added to get $3.30.

   B. The student adds the values of the nickels and dimes in cents, but writes the total value in dollars.

   C. The student confuses value and number for the two equations.

   D. Correct answer

Answers

7. *Part A* 2 fluid ounces per hour

   Part B 52 fluid ounces

   Part C $y = -2h + 52$

   Part D 26 hours

8. H

9. D

10. G

Answers

11. C

12. F

13. C

14. I

Answer for Extra Example

1. **Gridded Response:**
Correct answer: 5.6

Common Error: The student fails to distribute 10 across the expression on the left side of the equation, yielding the incorrect solution of 0.2.

Item Analysis (continued)

10. **F.** The student adds together the lead pencil and one marker instead of three markers.

G. Correct answer

H. The student overthinks the problem and decides that the factor of 3 applies to the sum.

I. The student accounts for the three markers, but gets muddled on how things add together.

11. **A.** The student misreads the graph.

B. The student finds the value for 20 minutes and adds 5 to that value since 25 minutes is 5 minutes later.

C. Correct answer

D. The student misreads the graph.

12. **F.** Correct answer

G. The student makes an arithmetic error relating c and m.

H. The student picks an equation that fits the first point in the graph, (10, 90).

I. The student interchanges the roles of c and m.

13. **A.** The student uses the slope of 3 instead of -3.

B. The student interchanges the coordinates when finding the equation.

C. Correct answer

D. The student works correctly, but uses the wrong sign on the term 14.

14. **F.** The student performs the first step correctly, but then subtracts P rather than dividing.

G. The student performs the first step correctly, but then divides A by P instead of dividing the entire expression by P.

H. The student performs the first step correctly, but then divides P by P instead of dividing the entire expression by P.

I. Correct answer

Extra Example for Standardized Test Practice

1. What value of x makes this equation true?

$$10(x - 3) = 5x - 2$$

The graph below shows how many calories *c* are burned during *m* minutes of playing basketball. Use the graph for Exercises 11 and 12.

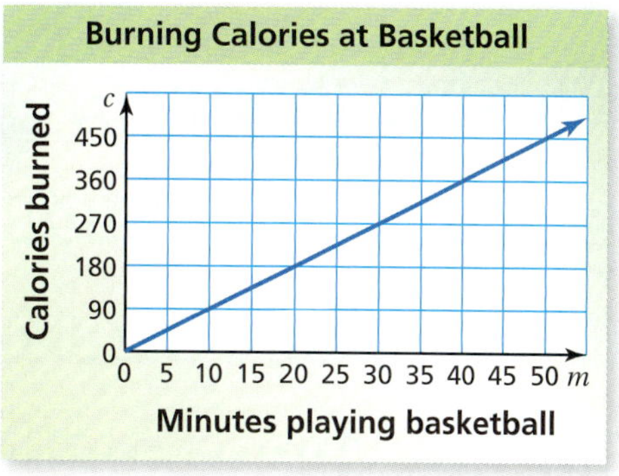

11. How many calories are burned in 25 minutes?

A. 180 cal

B. 185 cal

C. 225 cal

D. 270 cal

12. Which equation fits the data given in the graph?

F. $c = 9m$

G. $c = 90m$

H. $c = m + 80$

I. $m = 9c$

13. The line shown in the graph below has a slope of -3. What is the equation of the line?

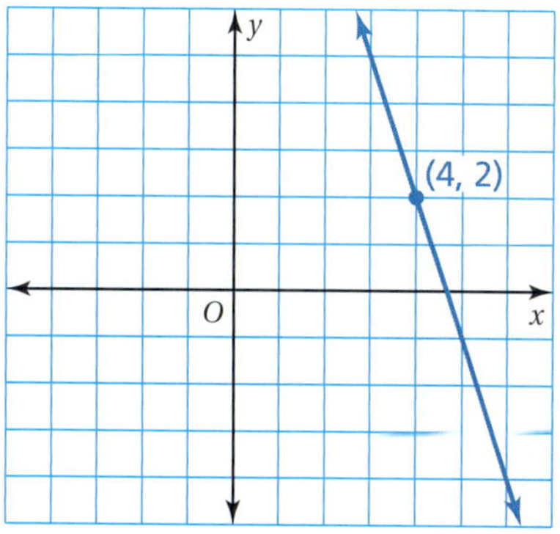

A. $y = 3x - 10$

B. $y = -3x + 10$

C. $y = -3x + 14$

D. $y = -3x - 14$

14. Solve the formula below for I.

$$A = P + PI$$

F. $I = A - 2P$

G. $I = \frac{A}{P} - P$

H. $I = A - \frac{P}{P}$

I. $I = \frac{A - P}{P}$

4 Functions

"Here's how I remember that the range is the y-values."
Where the deer and the antelope play, huh?
"I draw a cabin on the y-axis. Then, I hum 'Home, Home on the range'."

INPUT
OUTPUT
TREAT Converter
I wondered where my cat treats were going.
"It is my treat-converter function machine. However many cat treats I input, the machine outputs TWICE that many dog biscuits. Isn't that cool?"

Connections to Previous Learning

- Differentiate between continuous and discrete data, and choose ways to represent it.
- Graph and describe continuous data, such as a quantity that changes over time.
- Construct and analyze tables, graphs, and equations to describe linear functions and other simple relations.

- Identify and plot ordered pairs in all four quadrants.

- Construct and analyze tables, graphs, and models to represent, analyze, and solve problems related to linear equations, including analysis of domain, range, and the difference between discrete and continuous data.
- Translate among representations of linear functions in words, tables, graphs, and equations.
- Compare graphs of linear and nonlinear functions.

Math in History

Formulas for the volumes of solids have been known and used in many cultures.

★ There are records of Japanese mathematicians using a Chinese text that was written around 200 B.C. The third section of the text contains methods for finding the volumes of prisms, cylinders, pyramids, and cones.

★ Around 628 A.D., a book called Brahmasphutasiddhanta was written by the Indian mathematician Brahmagupta. The book has 25 chapters. In one of them, he describes methods for calculating the volume of a prism and a cone.

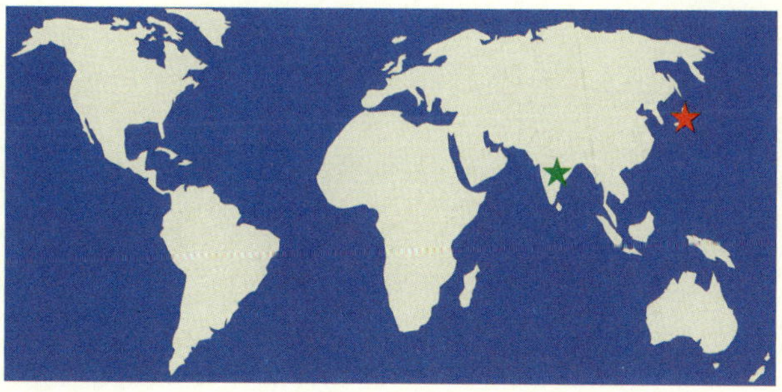

Pacing Guide for Chapter 4

Chapter Opener	1 Day
Section 1	
Activity	1 Day
Lesson	1 Day
Section 2	
Activity	1 Day
Lesson	1 Day
Study Help / Quiz	1 Day
Section 3	
Activity	1 Day
Lesson	1 Day
Section 4	
Activity	1 Day
Lesson	1 Day
Quiz / Chapter Review	1 Day
Chapter Test	1 Day
Standardized Test Practice	1 Day
Total Chapter 4	13 Days
Year-to-Date	69 Days

Check Your Resources

- Record and Practice Journal
- Resources by Chapter
- Skills Review Handbook
- Assessment Book
- Worked-Out Solutions

Technology For the Teacher

Dynamic Classroom

The Dynamic Planning Tool
Editable Teacher's Resources at
BigIdeasMath.com

Additional Topics for Review

- Adding and subtracting decimals and fractions
- Multiplying fractions and decimals
- Plotting points and identifying coordinates in Quadrant I

Try It Yourself

1. As the input decreases by 2, the output increases by 3.
2. As the input increases by 2, the output increases by 1.
3. As the input decreases by 1, the output decreases by 3.5.

Record and Practice Journal

1. As the input increases by 1, the output increases by 2.
2. As the input increases by 2, the output increases by 5.
3. As the input increases by 4, the output increases by 3.
4. As the input increases by 1, the output decreases by 7.
5. As the hours increase by 1, the customers increase by 15.
6. Input Output

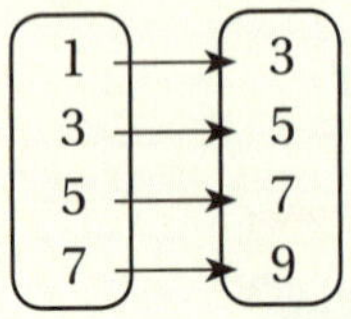

As the input increases by 2, the output increases by 2.

7–10. See Additional Answers.

Math Background Notes

Vocabulary Review

- Input
- Output
- Mapping Diagram

Recognizing Patterns

- Students have been working with patterns since elementary school. They should know how to use mapping diagrams and In and Out tables. They should know how to use graphs to represent and identify patterns.
- Remind students that mapping diagrams, In and Out tables, graphs, and words are four different ways that can be used to express or describe a pattern.
- **Common Error:** Some students may try to find a pattern between the input and output values. Remind them that they are not searching for how the input values relate to output values but rather, how the input values relate to one another and how the output values relate to one another.
- **Teaching Tip:** Some students may have difficulty identifying the pattern. Encourage these students to search for context clues first. For example, are the input values getting progressively greater? If so, the pattern will most likely involve addition or multiplication. If the numbers are all even, try adding or multiplying by even numbers first to try to find the pattern.
- **Teaching Tip:** Rather than trying to identify the pattern using the points on a graph, encourage students to transfer the information contained in the ordered pairs into a mapping diagram as in Example 3. Then use the mapping diagram to find the pattern.

Reteaching and Enrichment Strategies

If students need help. . .	If students got it. . .
Record and Practice Journal • Fair Game Review Skills Review Handbook Lesson Tutorials	Game Closet at *BigIdeasMath.com* Start the next section

What You Learned Before

Recognizing Patterns

Describe the pattern of inputs and outputs.

Example 1

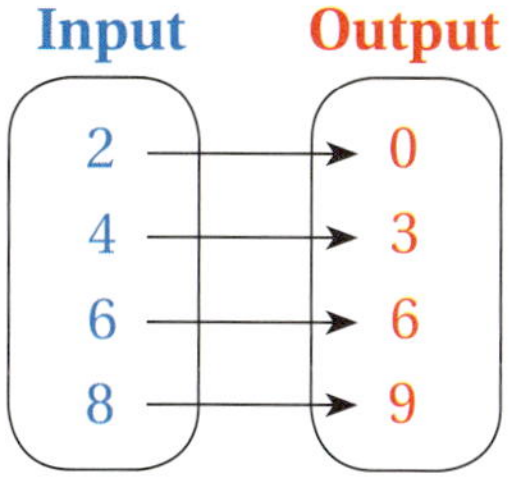

∴ As the input increases by 2, the output increases by 3.

Example 2

Input, *x*	6	1	−4	−9	−14
Output, *y*	7	8	9	10	11

∴ As the input x decreases by 5, the output y increases by 1.

Example 3

Draw a mapping diagram for the graph. Then describe the pattern of inputs and outputs.

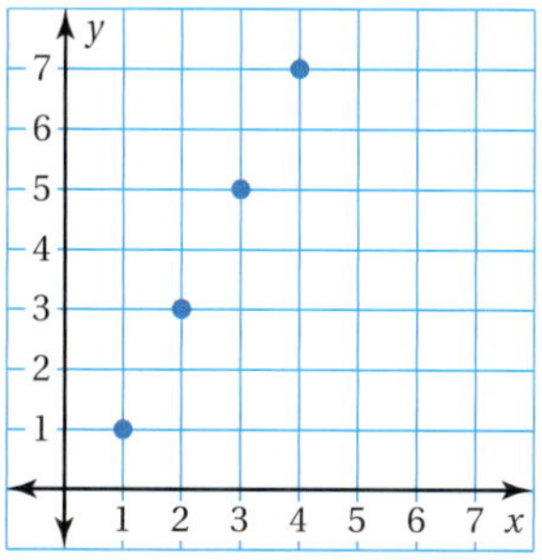

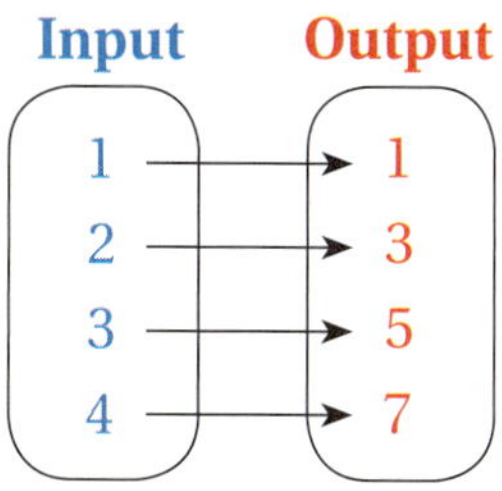

∴ As the input increases by 1, the output increases by 2.

Try It Yourself

Describe the pattern of inputs x and outputs y.

1.

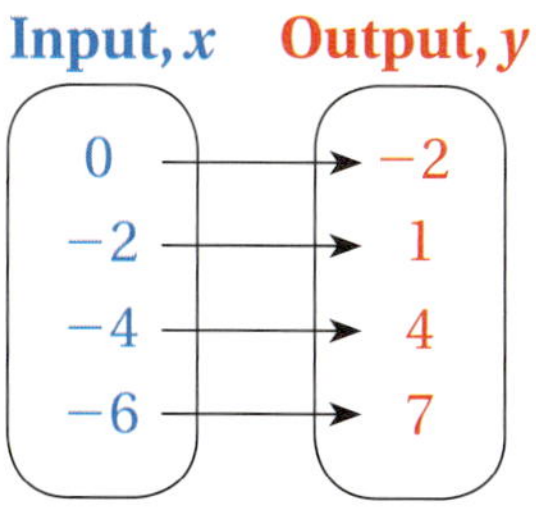

2.

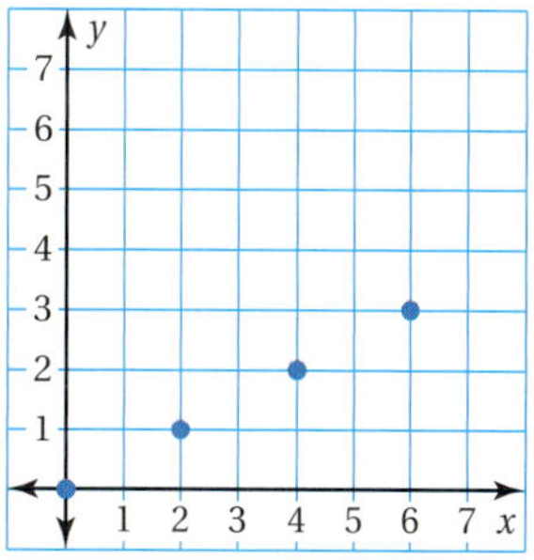

3.

Input, *x*	0	−1	−2	−3	−4
Output, *y*	7	3.5	0	−3.5	−7

4.1 Domain and Range of a Function

Essential Question How can you find the domain and range of a function?

1 ACTIVITY: The Domain and Range of a Function

Work with a partner. The table shows the number of adult and child tickets sold for a school concert.

Input →	Number of Adult Tickets, x	0	1	2	3	4
Output →	Number of Child Tickets, y	8	6	4	2	0

The variables x and y are related by the linear equation $4x + 2y = 16$.

a. Write the equation in **function form** by solving for y.

b. The **domain** of a function is the set of all input values. Find the domain of the function.

Domain = ______

Why is $x = 5$ not in the domain of the function?

Why is $x = \frac{1}{2}$ not in the domain of the function?

c. The **range** of a function is the set of all output values. Find the range of the function.

Range = ______

d. Functions can be described in many ways.

- by an equation
- by an input-output table
- in words
- by a graph
- as a set of ordered pairs

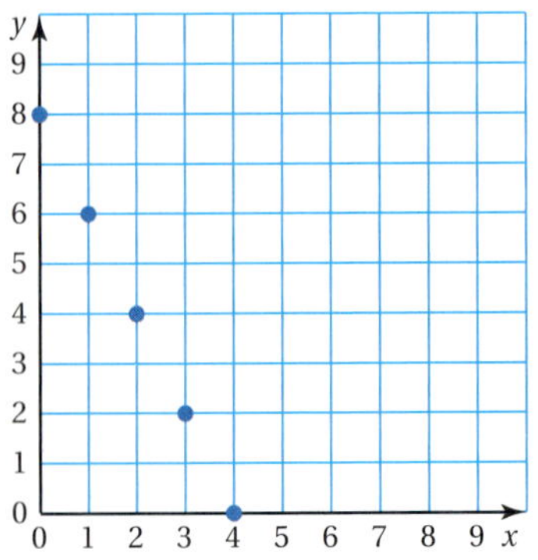

Use the graph to write the function as a set of ordered pairs.

(___ , ___), (___ , ___),
(___ , ___), (___ , ___),
(___ , ___)

Laurie's Notes

Introduction

For the Teacher

- **Goal:** Students will develop an understanding of domain and range by exploring familiar problems.
- The language of functions can be challenging for students, more so than the actual concepts.

Motivate

- Ask for a volunteer who will not mind you measuring his or her head for a hat. See the chart for sizes.

Hat Size	$6\frac{1}{2}$	$6\frac{5}{8}$	$6\frac{3}{4}$	$6\frac{7}{8}$	7	$7\frac{1}{8}$	$7\frac{1}{4}$	$7\frac{3}{8}$	$7\frac{1}{2}$	$7\frac{5}{8}$	$7\frac{3}{4}$	$7\frac{7}{8}$	8
Inches	$20\frac{1}{2}$	$20\frac{7}{8}$	$21\frac{1}{4}$	$21\frac{5}{8}$	22	$22\frac{1}{2}$	$22\frac{7}{8}$	$23\frac{1}{4}$	$23\frac{5}{8}$	24	$24\frac{3}{8}$	$24\frac{3}{4}$	$25\frac{1}{4}$
Centimeters	52	53	54	55	56	57	58	59	60	61	62	63	64
	X-SMALL		SMALL		MEDIUM		LARGE		X-LARGE		XX-LARGE		

- ? "What is the input for determining your hat size?" size of your head in inches or centimeters
- ? "What are the outputs?" hat size as a number $\left(6\frac{1}{2}, 6\frac{5}{8}, \ldots\right)$ or a category (X-small, small, …)
- **Discuss:** Relate the input and output for determining the hat size to the new vocabulary, domain and range.

Activity Notes

Activity 1

- This problem involves reading the language of functions. Students should pay attention to the vocabulary.
- ? "Recall that the linear equation $4x + 2y = 16$ is written in standard form. What other common form have you used to write linear equations?" slope-intercept form
- Writing the equation in slope-intercept form results in an equation in function form. Students may write $y = 8 - 2x$, which is equivalent to $y = -2x + 8$.
- This problem focuses attention on the domain, the set of all input values. We also describe domain as the *permissible values* for x, meaning what numbers can be substituted for x. In the context of this problem, it makes sense that x can only be a whole number between 0 and 4, inclusive.
- Once the domain values are determined, a set of output values, the range, result.
- ? "Why are the ordered pairs of the graph not connected?" The function has whole number solutions. You cannot have fractional or negative numbers of tickets sold.
- ? "How many solutions are there for this function?" five

Previous Learning

Students should know how to determine solutions of a linear equation.

Activity Materials
Introduction
• tape measure

Start Thinking! and Warm Up

Activity 4.1 Start Thinking! For use before Activity 4.1

Activity 4.1 Warm Up For use before Activity 4.1

Copy and complete the input-output table for the function.

1. $y = -x + 2$

x	−2	−1	0	1	2
y					

2. $y = 2x - 7$

x	−2	−1	0	1	2
y					

3. $y = \frac{1}{2}x + 3$

x	0	1	2	3	4
y					

4. $y = \frac{1}{3}x - 1$

x	−6	−3	0	3	6
y					

4.1 Record and Practice Journal

Essential Question How can you find the domain and range of a function?

1 ACTIVITY: The Domain and Range of a Function

Work with a partner. The table shows the number of adult and child tickets sold for a school concert.

Input →	Number of Adult Tickets, x	0	1	2	3	4
Output →	Number of Child Tickets, y	8	6	4	2	0

The variables x and y are related by the linear equation $4x + 2y = 16$.

a. Write the equation in **function form** by solving for y.

$y = 8 - 2x$

b. The **domain** of a function is the set of all input values. Find the domain of the function represented by the table.

Domain = 0, 1, 2, 3, 4

Why is $x = 5$ not in the domain of the function?

The output would be negative.

Why is $x = \frac{1}{2}$ not in the domain of the function?

You cannot sell $\frac{1}{2}$ of a ticket.

c. The **range** of a function is the set of all output values. Find the range of the function represented by the table.

Range = 0, 2, 4, 6, 8

English Language Learners

Vocabulary

Students will find it helpful to relate the words *input* and *output* with the prepositions *in* and *out.*

4.1 Record and Practice Journal

d. Functions can be described in many ways.

- by an equation
- by an input-output table
- in words
- by a graph
- as a set of ordered pairs

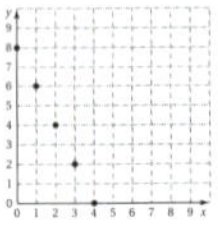

Use the graph to write the function as a set of ordered pairs.

(0, 8), (1, 6), (2, 4), (3, 2), (4, 0)

2 ACTIVITY: Finding Domains and Ranges

Work with a partner.

- **Complete each input-output table.**
- **Find the domain and range of each function represented by the table.**

a. $y = -3x + 4$ D: −2, −1, 0, 1, 2 R: 10, 7, 4, 1, −2

x	−2	−1	0	1	2
y	10	7	4	1	−2

b. $y = \frac{1}{2}x - 6$ D: 0, 1, 2, 3, 4 R: −6, −5.5, −5, −4.5, −4

x	0	1	2	3	4
y	−6	−5.5	−5	−4.5	−4

c.

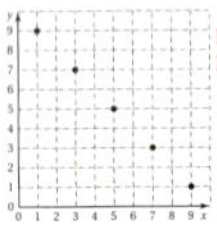

D: 1, 3, 5, 7, 9
R: 9, 7, 5, 3, 1

x	1	3	5	7	9
y	9	7	5	3	1

d.

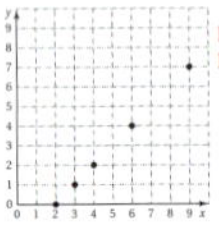

D: 2, 3, 4, 6, 9
R: 0, 1, 2, 4, 7

x	2	3	4	6	9
y	0	1	2	4	7

What Is Your Answer?

3. **IN YOUR OWN WORDS** How can you find the domain and range of a function?

domain is the set of all possible input values; range is the set of all possible output values

4. **The following are general rules for finding a person's foot length.**

To find the length y (in inches) of a woman's foot, divide her shoe size x by 3 and add 7.

To find the length y (in inches) of a man's foot, divide his shoe size x by 3 and add 7.3.

a. Write an equation for one of the statements. women: $y = \frac{1}{3}x + 7$

men: $y = \frac{1}{3}x + 7.3$

b. Make an input-output table for the function in part (a).

Use shoe sizes $5\frac{1}{2}$ to 12.

See Additional Answers.

c. Label the domain and range of the function represented by the table.

See Additional Answers.

Laurie's Notes

Activity 2

- **Connection:** Students have graphed linear equations whose domain and range were the set of real numbers. In fact, students plotted 3 to 4 points that satisfied the equation and then connected the points to graph the linear equation. If the domain is restricted to only a finite set of values, the range becomes restricted to a finite set of values, and the number of solutions is finite.
- Students may ask why they are only using five values for x in parts (a) and (b), and why they are not connecting the ordered pairs in the graph in parts (c) and (d). Again, the focus is on domain and range. Remind students that if you only use certain domain values (inputs for x), the range values (outputs for y) are determined by using the function rule (equation).

? "What is the resulting range for the function $y = -3x + 4$ with a domain of $-2, -1, 0, 1, 2$?" 10, 7, 4, 1, −2

- Students often make these problems more difficult than they are. Remind them to use substitution to evaluate the functions (equations) for parts (a) and (b), then use their eyesight to read the ordered pair solutions for parts (c) and (d). Then students record their answers in the input-output table.

? To focus attention on the language of functions, ask students to describe the function rule for each problem. For instance, the function rule for part (a) is *multiply the input by −3 and then add 4.* Students will need to find the function rule for parts (c) and (d). $y = 10 - x$; $y = x - 2$

What Is Your Answer?

- **Question 4 Extension:** Gather data from students who are willing to share their shoe size and to have their foot measured. How well does the rule fit the data?

Closure

- **Writing Prompt:** Describe what is meant by the domain and range of a function.

Technology For the Teacher

Dynamic Classroom

The Dynamic Planning Tool
Editable Teacher's Resources at *BigIdeasMath.com*

2 ACTIVITY: Finding Domains and Ranges

Work with a partner.

- **Copy and complete each input-output table.**
- **Find the domain and range of the function represented by the table.**

a. $y = -3x + 4$

x	−2	−1	0	1	2
y					

b. $y = \frac{1}{2}x - 6$

x	0	1	2	3	4
y					

c.

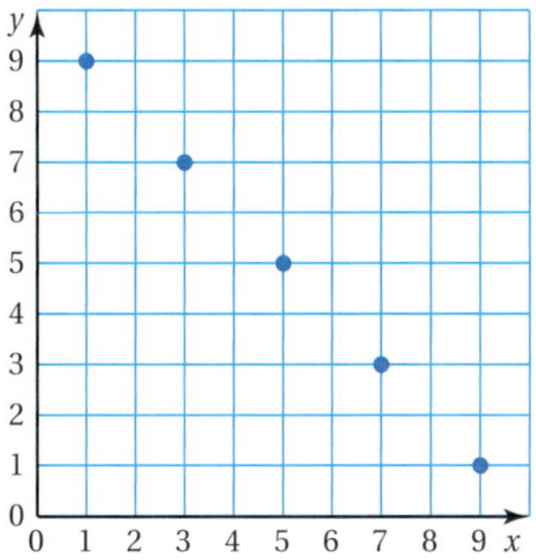

x					
y					

d.

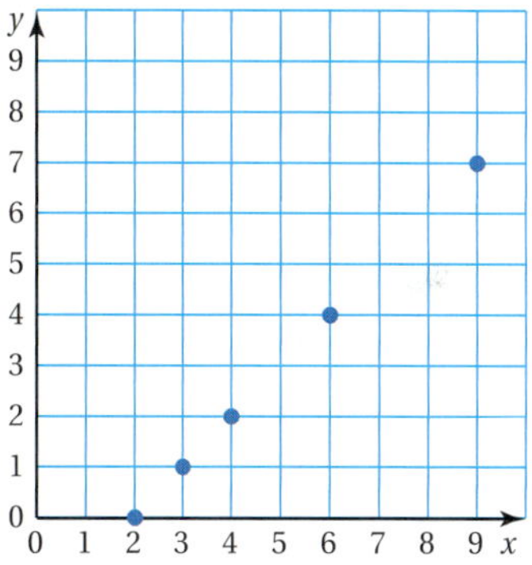

x					
y					

What Is Your Answer?

3. **IN YOUR OWN WORDS** How can you find the domain and range of a function?

4. **The following are general rules for finding a person's foot length.**

To find the length y (in inches) of a woman's foot, divide her shoe size x by 3 and add 7.

To find the length y (in inches) of a man's foot, divide his shoe size x by 3 and add 7.3.

a. Write an equation for one of the statements.

b. Make an input-output table for the function in part (a). Use shoe sizes $5\frac{1}{2}$ to 12.

c. Label the domain and range of the function on the table.

Practice Use what you learned about the domain and range of a function to complete Exercise 3 on page 152.

4.1 Lesson

Check It Out
Lesson Tutorials
BigIdeasMath.com

Key Vocabulary
function, *p. 150*
domain, *p. 150*
range, *p. 150*
function form, *p. 150*

Key Idea

Functions

A **function** is a relationship that pairs each *input* with exactly one *output*. The **domain** is the set of all possible input values. The **range** is the set of all possible output values.

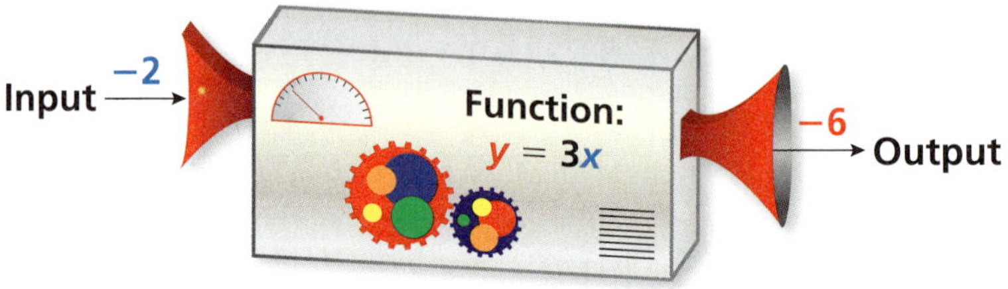

Remember

The ordered pair (x, y) shows the output y for an input x.

EXAMPLE 1 Finding Domain and Range from a Graph

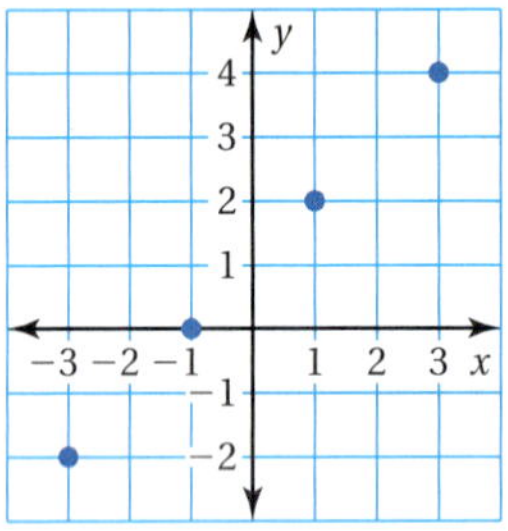

Find the domain and range of the function represented by the graph.

Write the ordered pairs. Identify the inputs and outputs.

inputs

$(-3, -2), (-1, 0), (1, 2), (3, 4)$

outputs

The domain is −3, −1, 1, and 3. The range is −2, 0, 2, and 4.

On Your Own

Now You're Ready
Exercises 4–6

Find the domain and range of the function represented by the graph.

1.

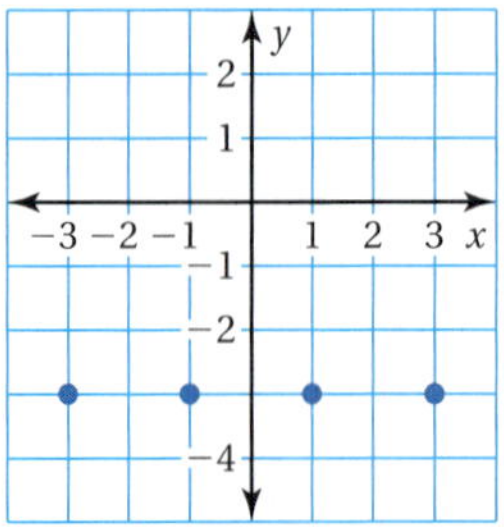

2.

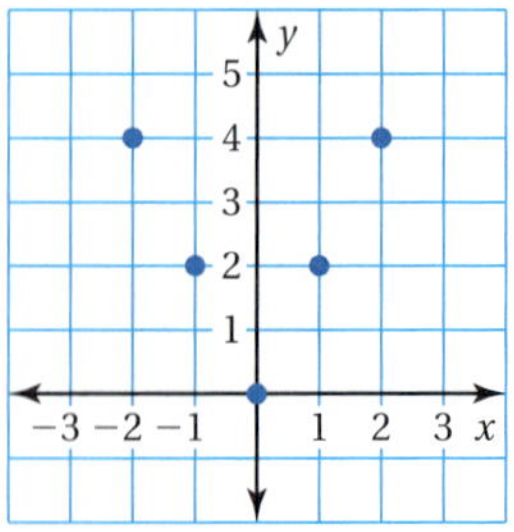

An equation is in **function form** if it is solved for y.

$x + y = 1$	$y = -x + 1$
not in function form	in function form

Multi-Language Glossary at BigIdeasMath.com.

Laurie's Notes

Introduction

Connect

- **Yesterday:** Students explored the domain and range of functions by revisiting familiar problems.
- **Today:** Students will identify the domain and range of a function from a graph and table of values.

Motivate

? "Could someone describe how a vending machine works?" Listen for: put in money, make a selection, and item comes out.

- Explain that a vending machine is like a function. You make a selection (the input) and a specific item comes out (the output).
- Sometimes there are several inputs that give the same output (3 different buttons for the same bottled water), but there are never several different outputs of the same input (if the vending machine is working properly).

Lesson Notes

Key Idea

- Write the Key Idea. The graphic of a *function machine* should help students conceptualize the idea of entering an input value, applying a rule, and obtaining the output value.
- In discussing the definition of a function, describe what is meant by *each input is paired with exactly one output*. This means:
  - one unique input yields one unique output, or
  - two or more inputs yields the same output.
  - The equation is *not* a function if one unique input yields more than one output.

Example 1

- Point out to students that the inputs, or x-coordinates, are the domain, and the outputs, or y-coordinates, are the range.

? "Is each input paired with exactly one output?" yes

On Your Own

- **Question 1:** Students may say this is not a function because there is only one number in the range (students should not list it 4 times). Every domain value is paired with the same range value, and that is okay. Remind students it is only the repeat of x-values they need to consider.

Write

- Write the definition of function form. In previous lessons, students solved equations for y. This is the same as writing the equation in function form.
- Ask students for examples of linear equations in function form and not in function form. *Sample answers:* function form: $y = 5x + 2$, not function form: $-10x + 2y = 4$

Goal Today's lesson is identifying the **domain** and **range** of a **function**.

Start Thinking! and Warm Up

Lesson 4.1 Warm Up
For use before Lesson 4.1

Lesson 4.1 Start Thinking!
For use before Lesson 4.1

The installation and set-up fees for cable internet come to \$150. The monthly cost for internet access is \$40 per month.

Write a function for the cost y of x months of internet service.

What are the domain and range of the function if you only have internet service for 6 months?

Extra Example 1

Find the domain and range of the function represented by the graph.

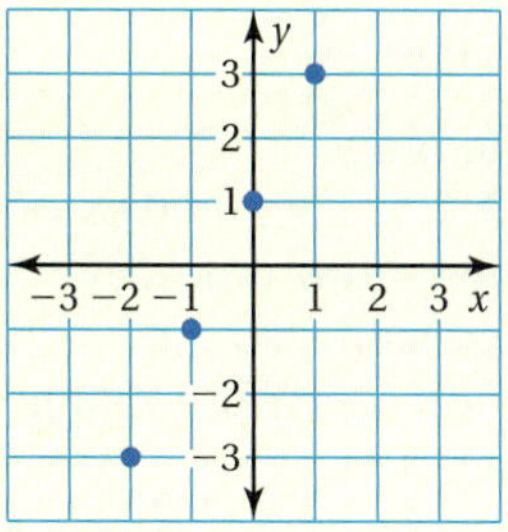

domain: −2, −1, 0, 1; range: −3, −1, 1, 3

On Your Own

1. domain: −3, −1, 1, 3
   range: −3
2. domain: −2, −1, 0, 1, 2
   range: 0, 2, 4

Extra Example 2

The domain of the function represented by $x + y = 5$ is $-3, -2, -1, 0$, and 1. What is the range of the function?
8, 7, 6, 5, 4

Extra Example 3

x	0	1	2	3	4
y	0.11	0.18	0.27	0.37	0.47

a. The table shows the percent y (in decimal form) of the moon that was visible at midnight x days after August 10, 2013. Interpret the domain and range. The domain is 0, 1, 2, 3, and 4. So, the table shows the data for August 10, 11, 12, 13, and 14. The range is 0.11, 0.18, 0.27, 0.37, and 0.47. So, the table shows that the moon was more visible each day.

b. What percent of the moon was visible on August 12, 2013? 27%

On Your Own

3–4. See Additional Answers.

5. a. The domain is 0, 1, 2, 3, and 4. It represents December 17, 18, 19, 20, and 21.

The range is 0.2, 0.3, 0.4, 0.5, and 0.6. These amounts are increasing, so the moon was more visible each day.

b. 60%

Differentiated Instruction

Kinesthetic

Have students build their own function machine using a shoe box, index cards, and sticky notes. Write the function on the sticky note and put it on the box. Write numbers on the index cards to use as the input and output values of the function.

Laurie's Notes

Example 2

- Write the equation. Note that the equation is in standard form.
- Write the equation in function form and complete the input-output table.
- ? "Do you need to write the equation in function form in order to find the output y?" no
- ? "Why is it helpful to write the equation in function form before finding the y-values?" All of the computations are on one side of the equal sign.
- **Extension:** Ask students to describe the patterns in the input-output table. The x-values increase by 2 and the y-values decrease by 4.

Example 3

- **FYI:** During the full moon, the moon's illuminated side is facing Earth and appears to be completely illuminated by direct sunlight. During a new moon, the moon's unilluminated side is facing Earth and the moon is not visible.
- ? Ask a few questions to help students understand the problem.
  - "How do you interpret the x-values in the table?" x is the number of days after January 24, 2011.
  - "What day does $x = 3$ represent?" January 27, 2011
  - "How do you interpret the y-values in the table?" percent of the moon visible at midnight, written as a decimal
  - "Is the moon becoming more or less visible in the week following January 24, 2011? Explain." less; The percents are decreasing.
  - "If this problem were continued, what values would make sense for the domain? What values make sense for the range?" domain: whole numbers; range: decimals between 0 and 1 inclusive

On Your Own

- **Question 4:** Students may rewrite this equation in function form, or they may reason that the question is asking, "What numbers sum to -3?"
- **Question 5:** Ask students if the data suggests that they are moving towards a full moon or away from a full moon.

Closure

- **Exit Ticket:** Make an input-output table for the function $2x + y = 3$ using the inputs $-2, 0, 2, 4$.

x	−2	0	2	4
y	7	3	−1	−5

Technology For the Teacher

Dynamic Classroom

The Dynamic Planning Tool
Editable Teacher's Resources at *BigIdeasMath.com*

EXAMPLE 2 Finding the Range of a Function

Input, x	$-2x + 8$	Output, y
-2	$-2(-2) + 8$	12
0	$-2(0) + 8$	8
2	$-2(2) + 8$	4
4	$-2(4) + 8$	0
6	$-2(6) + 8$	-4

The domain of the function represented by $2x + y = 8$ is -2, 0, 2, 4, and 6. What is the range of the function represented by the table?

Write the function in function form.

$$2x + y = 8$$

$$y = -2x + 8$$

Use this form to make an input-output table.

∴ The range is 12, 8, 4, 0, and -4.

EXAMPLE 3 Real-Life Application

The table shows the percent y (in decimal form) of the moon that was visible at midnight x days after January 24, 2011. (a) Interpret the domain and range. (b) What percent of the moon was visible on January 26, 2011?

x	y
0	0.76
1	0.65
2	0.54
3	0.43
4	0.32

a. Zero days after January 24 is January 24. One day after January 24 is January 25. So, the domain of 0, 1, 2, 3, and 4 represents January 24, 25, 26, 27, and 28.

The range is 0.76, 0.65, 0.54, 0.43, and 0.32. These amounts are decreasing, so the moon was less visible each day.

b. January 26, 2011 corresponds to the input $x = 2$. When $x = 2$, $y = 0.54$. So, 0.54, or 54% of the moon was visible on January 26, 2011.

On Your Own

Now You're Ready
Exercises 9–11

Copy and complete the input-output table for the function. Then find the domain and range of the function represented by the table.

3. $y = 2x - 3$

x	-1	0	1	2
y				

4. $x + y = -3$

x	0	1	2	3
y				

5. The table shows the percent y (in decimal form) of the moon that was visible at midnight x days after December 17, 2012.
(a) Interpret the domain and range.
(b) What percent of the moon was visible on December 21, 2012?

x	0	1	2	3	4
y	0.2	0.3	0.4	0.5	0.6

4.1 Exercises

Vocabulary and Concept Check

1. **VOCABULARY** Is the equation $2x - 3y = 4$ in function form? Explain.

2. **DIFFERENT WORDS, SAME QUESTION** Which is different? Find "both" answers.

Find the range of the function represented by the table.

Find the inputs of the function represented by the table.

Find the x-values of the function represented by (2, 7), (4, 5), and (6, −1).

Find the domain of the function represented by (2, 7), (4, 5), and (6, −1).

x	2	4	6
y	7	5	−1

Practice and Problem Solving

3. The number of earrings and headbands you can buy with \$24 is represented by the equation $8x + 4y = 24$. The table shows the number of earrings and headbands.

   a. Write the equation in function form.

   b. Find the domain and range.

   c. Why is $x = 6$ not in the domain of the function?

Earrings, x	0	1	2	3
Headbands, y	6	4	2	0

Find the domain and range of the function represented by the graph.

1 4.

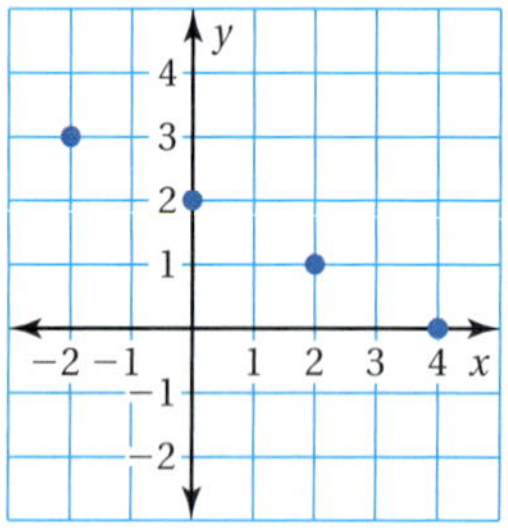

5.

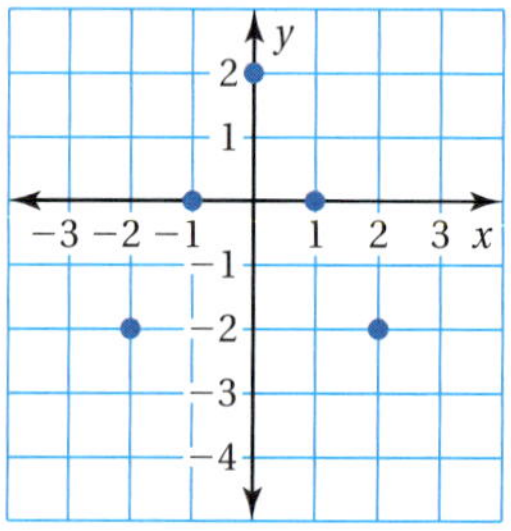

6.

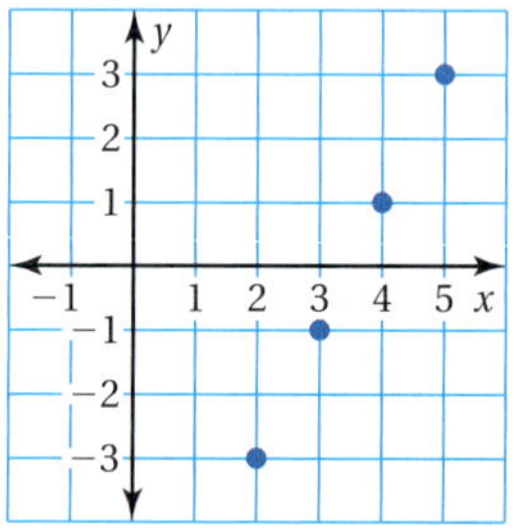

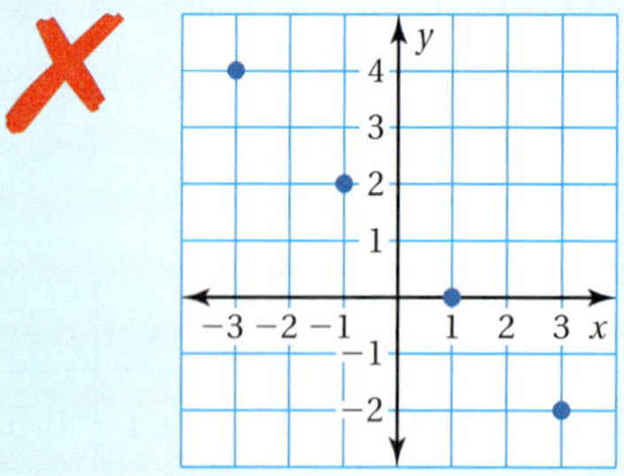

7. **ERROR ANALYSIS** Describe and correct the error in finding the domain and range of the function represented by the graph.

8. **REASONING** Find the domain and range of the function represented by the table.

Tickets, x	2	3	5	8
Cost, y	\$14	\$21	\$35	\$56

Assignment Guide and Homework Check

Level	Day 1 Activity Assignment	Day 2 Lesson Assignment	Homework Check
Basic	3, 15–19	1, 2, 4–11	2, 4, 8, 10
Average	3, 15–19	1, 2, 4–7, 9–12	2, 4, 10, 12
Advanced	3, 15–19	1, 2, 6, 7, 9–14	2, 6, 10, 12

Common Errors

- **Exercises 4–6** Students may mix up the domain and range. For example, a student may give all the y-values of the coordinates as the domain. This can happen with all the ordered pairs, or only one or two. Remind students that the x-coordinates are the domain and the y-coordinates are the range, as shown in Example 1.
- **Exercises 9–11** Students may make mistakes when substituting the values of x and solving for y. For example, a student may write $y = 6(2) + 2 = 6(4) = 24$ instead of $y = 6(2) + 2 = 12 + 2 = 14$. Remind them of the order of operations.

4.1 Record and Practice Journal

Find the domain and range of the function represented by the graph.

1.

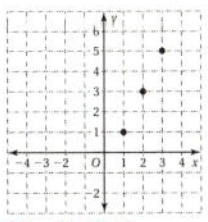

D: 1, 2, 3
R: 1, 3, 5

2.

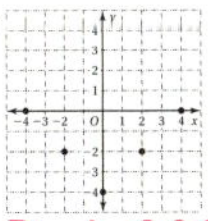

D: −4, −2, 0, 2, 4
R: −4, −2, 0

Complete the input-output table for the function. Then find the domain and range of the function represented by the table.

3. $y = 2x - 5$

x	−2	−1	0	1
y	−9	−7	−5	−3

D: −2, −1, 0, 1
R: −9, −7, −5, −3

4. $y = -\frac{1}{2}x + 3$

x	0	2	4	6
y	3	2	1	0

D: 0, 2, 4, 6
R: 0, 1, 2, 3

5. A factory that makes frozen pies produces 300 pies per minute.
   a. Write an equation in function form that represents the number y of pies made each hour x.
   $y = 18{,}000x$
   b. Create an input-output table for the equation in part (a). Use inputs 1, 2, 4, 8, and 10.
   See Additional Answers.
   c. Find the domain and range of the function represented by the table.
   D: 1, 2, 4, 8, 10
   R: 18,000, 36,000, 72,000, 144,000, 180,000

Technology For the Teacher
Answer Presentation Tool
Quiz*Show*

Vocabulary and Concept Check

1. no; The equation is not solved for y.
2. Find the range of the function represented by the table.; 7, 5, −1; 2, 4, 6

Practice and Problem Solving

3. a. $y = 6 - 2x$
   b. domain: 0, 1, 2, 3
   range: 6, 4, 2, 0
   c. $x = 6$ is not in the domain because it would make y negative, and it is not possible to buy a negative number of headbands.
4. domain: −2, 0, 2, 4
   range: 3, 2, 1, 0
5. domain: −2, −1, 0, 1, 2
   range: −2, 0, 2
6. domain: 2, 3, 4, 5
   range: −3, −1, 1, 3
7. The domain and range are switched. The domain is −3, −1, 1, and 3. The range is −2, 0, 2, and 4.
8. domain: 2, 3, 5, 8
   range: 14, 21, 35, 56
9.

x	−1	0	1	2
y	−4	2	8	14

domain: −1, 0, 1, 2
range: −4, 2, 8, 14

10.

x	0	4	8	12
y	−2	−3	−4	−5

domain: 0, 4, 8, 12
range: −2, −3, −4, −5

Practice and Problem Solving

11.

x	-1	0	1	2
y	1.5	3	4.5	6

domain: $-1, 0, 1, 2$
range: 1.5, 3, 4.5, 6

12. **a.** domain: 1, 2, 3
range: 6.856, 7.923, 8.135

b. The domain represents the round of competition. The range represents the scores received by the vaulter.

c. 7.638

13. See *Taking Math Deeper.*

14. See Additional Answers.

Fair Game Review

15–18. See Additional Answers.

19. D

Mini-Assessment

Find the domain and range of the function represented by the graph.

1.

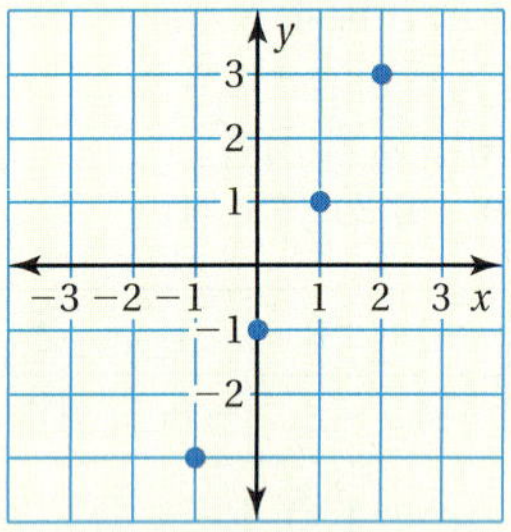

domain: $-1, 0, 1, 2$;
range: $-3, -1, 1, 3$

2.

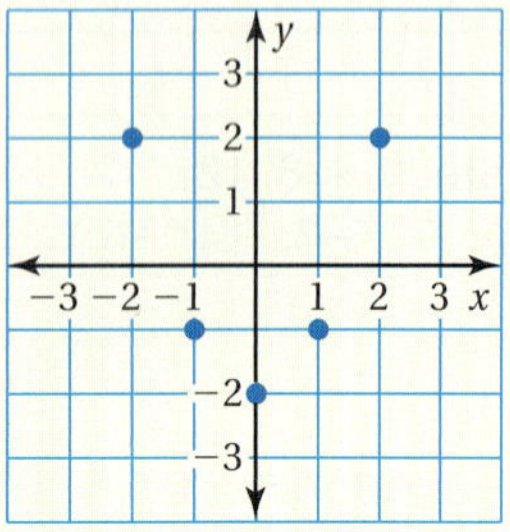

domain: $-2, -1, 0, 1, 2$;
range: $-2, -1, 2$

Taking Math Deeper

Exercise 13

The manatee is a gray, waterplant-eating mammal that reaches up to 15 feet in length and can weigh more than 2000 pounds. Humans (especially motor boats) are the cause of about half of all manatee deaths.

1 Find a function.

Let x = manatee's weight in pounds.
Let y = weight of food per day.

a. $y = 12\%$ of x
$= 0.12x$

2 Describe the function.

Create an input-output table for the function. Then, identify the domain and range.

b.

Input, x	150	300	450	600	750	900
Output, y	18	36	54	72	90	108

For the function, the domain is the set of positive numbers that are possible weights of manatees. An approximation would be the positive numbers up to 2000.

The range would be the positive numbers up to about 240.

c. *For the input-output table shown*, the domain is the x-values and the range is the y-values.

3 Use the function.

The manatees weigh a total of
$300 + 750 + 1050 = 2100$ pounds.

d. In a day, these manatees would eat
$y = 0.12(2100) = 252$ pounds of food.
In a week, they would eat
$y = 7(252) = 1764$ pounds of food.

Reteaching and Enrichment Strategies

If students need help. . .	If students got it. . .
Resources by Chapter • Practice A and Practice B • Puzzle Time Record and Practice Journal Practice Differentiating the Lesson Lesson Tutorials Skills Review Handbook	Resources by Chapter • Enrichment and Extension Start the next section

Copy and complete the input-output table for the function. Then find the domain and range of the function represented by the table.

2 **9.** $y = 6x + 2$

x	−1	0	1	2
y				

10. $y = -\frac{1}{4}x - 2$

x	0	4	8	12
y				

11. $y = 1.5x + 3$

x	−1	0	1	2
y				

12. VAULTING In the sport of vaulting, a vaulter performs a routine while on a moving horse. For each round x of competition, the vaulter receives a score y from 1 to 10.

a. Find the domain and range of the function represented by the table.

b. Interpret the domain and range.

c. What is the mean score of the vaulter?

x	y
1	6.856
2	7.923
3	8.135

13. MANATEE A manatee eats about 12% of its body weight each day.

a. Write an equation in function form that represents the amount y (in pounds) of food a manatee eats each day for its weight x.

b. Create an input-output table for the equation in part (a). Use the inputs 150, 300, 450, 600, 750, and 900.

c. Find the domain and range of the function represented by the table.

d. An aquatic center has manatees that weigh 300 pounds, 750 pounds, and 1050 pounds. How many pounds of food do all three manatees eat in a day? in a week?

14. Critical Thinking Describe the domain and range of the function.

a. $y = |x|$ **b.** $y = -|x|$ **c.** $y = |x| - 6$ **d.** $y = -|x| + 4$

Fair Game Review What you learned in previous grades & lessons

Graph the linear equation. *(Section 2.1)*

15. $y = 2x + 8$ **16.** $5x + 6y = 12$ **17.** $-x - 3y = 2$ **18.** $y = 7x - 5$

19. MULTIPLE CHOICE The minimum number of people needed for a group rate at an amusement park is 8. Which inequality represents the number of people needed to get the group rate? *(Skills Review Handbook)*

Ⓐ $x \leq 8$ Ⓑ $x > 8$ Ⓒ $x < 8$ Ⓓ $x \geq 8$

4.2 Discrete and Continuous Domains

Essential Question How can you decide whether the domain of a function is discrete or continuous?

1 EXAMPLE: Discrete and Continuous Domains

In Activities 1 and 2 in Section 2.4, you studied two real-life problems represented by the same equation.

$$4x + 2y = 16 \quad \text{or} \quad y = -2x + 8$$

a.

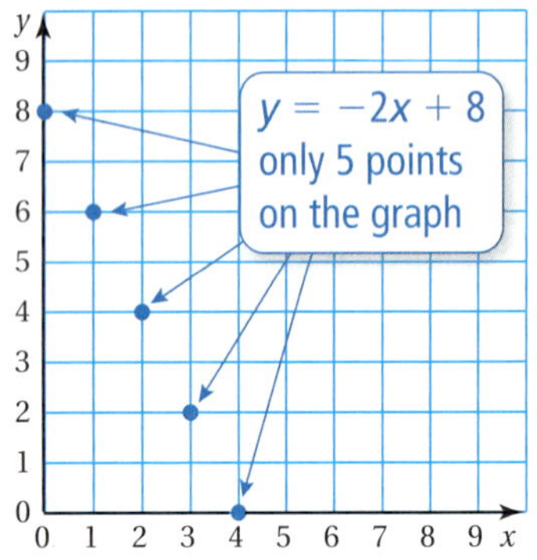

Domain (x-values): 0, 1, 2, 3, 4

Range (y-values): 8, 6, 4, 2, 0

The domain is **discrete** because it consists of only the numbers 0, 1, 2, 3, and 4.

b.

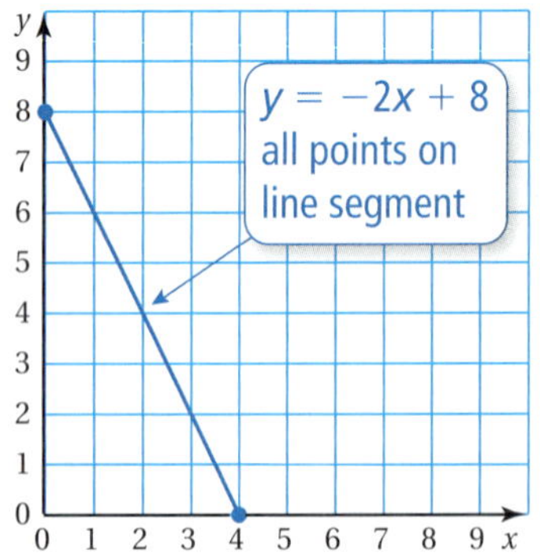

Domain (x-values): $x \geq 0$ and $x \leq 4$
(All numbers from 0 to 4)

Range (y-values): $y \geq 0$ and $y \leq 8$
(All numbers from 0 to 8)

The domain is **continuous** because it consists of all numbers from 0 to 4 on the number line.

Laurie's Notes

Introduction

For the Teacher

- **Goal:** Students will develop an understanding of discrete and continuous domains by exploring familiar problems.

Motivate

- Pass out strips of paper, perhaps 2 inches by 8 inches. Have students fold the paper in half.
- On one half of the paper, have students describe a variable that must be an integer (i.e., 4, -5, 20). On the other half, have them describe a variable that could be a fraction or decimal (i.e., $2\frac{1}{2}$, 4.8, -3.1). Students should write a description of the variable instead of giving a numerical example. Samples:

People at a movie (130)	Hours you work (2.2)
Problems assigned (18)	Length of fingernails (1.6 cm)
Buses on the road (15)	Yards lost on the play $\left(8\frac{1}{2}\right)$

- Have students tear the paper in half and place in two piles.
- Read and discuss examples in each pile.

Activity Notes

Example 1

- Discuss the two problems on this page. Make note of the vocabulary, discrete and continuous, used to describe the domain of each problem.
- ? "Why are the ordered pairs of the graph not connected in part (a)?" The function has whole number solutions. You cannot have fractional or negative numbers of tickets sold.
- ? "What do the intercepts of each graph represent?" Part (a): (0, 8) represents 8 child tickets sold and no adult tickets sold; (4, 0) represents 4 adult tickets sold and no child tickets sold; Part (b): (0, 8) represents 8 pounds of Cheddar sold and no pounds of Swiss sold; (4, 0) represents 4 pounds of Swiss sold and no pounds of Cheddar sold.
- Review the inequality notation used to describe the domain and range for part (b). These two inequalities describe all of the numbers between 0 and 4, including 0 and 4.
- Ask students to describe the range in words.
- Refer to the variables described by students on the strips of paper. The integer variables described are examples of discrete domains. The fraction or decimal variables described are examples of continuous domains.
- **Common Error:** It is incorrect to say that discrete domains are finite and continuous domains are infinite. A domain can be discrete and infinite, such as the set of counting numbers.

Previous Learning

Students should know how to find solutions of a linear equation.

Activity Materials
Introduction
• paper strips

Start Thinking! and Warm Up

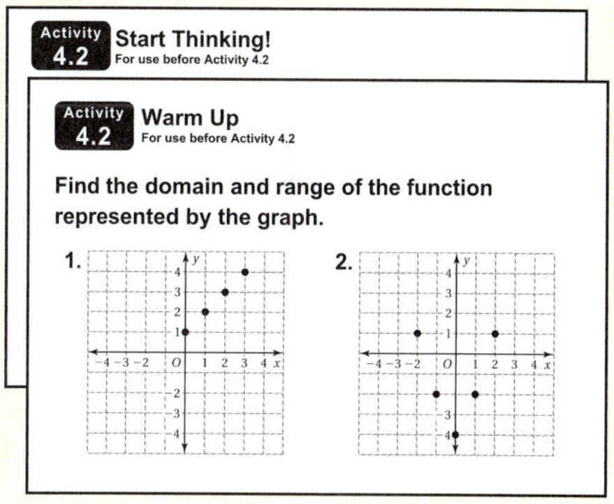

Activity 4.2 Start Thinking! For use before Activity 4.2

Activity 4.2 Warm Up For use before Activity 4.2

Find the domain and range of the function represented by the graph.

1.

2.

4.2 Record and Practice Journal

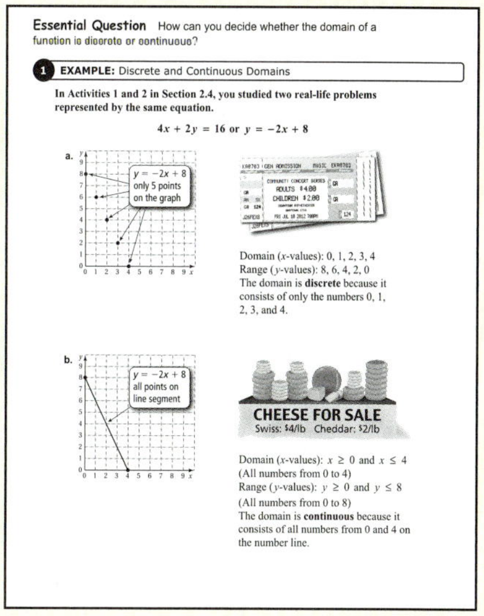

Essential Question How can you decide whether the domain of a function is discrete or continuous?

1 EXAMPLE: Discrete and Continuous Domains

In Activities 1 and 2 in Section 2.4, you studied two real-life problems represented by the same equation.

$$4x + 2y = 16 \text{ or } y = -2x + 8$$

a.

Domain (x-values): 0, 1, 2, 3, 4
Range (y-values): 8, 6, 4, 2, 0
The domain is **discrete** because it consists of only the numbers 0, 1, 2, 3, and 4.

b.

Domain (x-values): $x \ge 0$ and $x \le 4$
(All numbers from 0 to 4)
Range (y-values): $y \ge 0$ and $y \le 8$
(All numbers from 0 to 8)
The domain is **continuous** because it consists of all numbers from 0 and 4 on the number line.

Differentiated Instruction

Auditory

Help students understand the concept of a function by showing them how it is used in everyday life. For example, the number of plates to set on the dinner table is a function of the number of people expected to eat. A person earning an hourly wage has an income that is a function of the number of hours worked. Discuss with students other instances of functions in life.

4.2 Record and Practice Journal

2 ACTIVITY: Discrete and Continuous Domains

Work with a partner.

- **Write a linear function to represent each problem.**
- **Graph each function.**
- **Describe the domain and range of each function. Is the domain discrete or continuous?**

a. You are in charge of reserving hotel rooms for a youth soccer team. Each room costs \$69, plus \$6 tax, per night. You need each room for two nights. You need 10 to 16 rooms. Write a function for the total hotel cost.

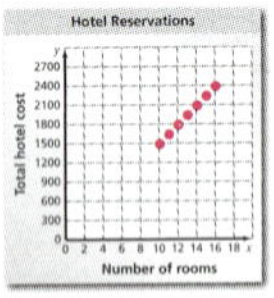

$y = 150x$

Domain: 10, 11, 12, 13, 14, 15, 16

Range: 1500, 1650, 1800, 1950, 2100, 2250, 2400

The domain is discrete.

b. The airline you are using for the soccer trip needs an estimate of the total weight of the team's luggage. You determine that there will be 36 pieces of luggage and each piece will weigh from 25 to 45 pounds. Write a function for the total weight of the luggage.

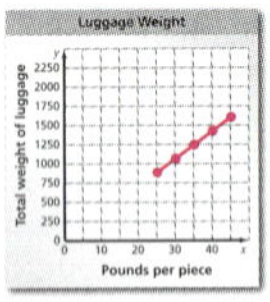

$y = 36x$

Domain: $x \geq 25$ and $x \leq 45$

Range: $y \geq 900$ and $y \leq 1620$

The domain is continuous.

What Is Your Answer?

3. **IN YOUR OWN WORDS** How can you decide whether the domain of a function is discrete or continuous? Describe two real-life examples of functions: one with a discrete domain and one with a continuous domain.

discrete consists of only certain numbers in an interval; continuous consists of all numbers in an interval

Laurie's Notes

Words of Wisdom

- Students may ask you to explain each of the two problems, meaning, show them how to do the problems! Resist the tendency to jump in and solve the problems for them. Students should work with their partner and work through the problem. They need to read and think!

Activity 2

- **Part (a):** Rooms are \$75 a night (\$69 + \$6). If 10 rooms are rented for 2 nights, the cost is \$75 × 10 × 2 = \$1500. Write a verbal model to help you find an equation: Cost = cost per night × number of rooms × 2 nights; if x = the number of rooms rented, the equation is $y = (75)2x = 150x$.
- **Part (b):** There are 36 pieces of luggage that will vary in weight between 25 and 45 pounds. The approximate total weight equals the number of pieces of luggage times the average weight of each piece. If x = average weight of each piece of luggage, the equation is $y = 36x$.

? "Is the domain of part (a) discrete or continuous? Explain." discrete; The number of rooms must be a whole number.

? "Is the domain of part (b) discrete or continuous? Explain." continuous; The weight of each piece of luggage is between 25 and 45 pounds, but could be a fraction of a pound.

What Is Your Answer?

- **Think-Pair-Share:** Students should read the question independently and then work with a partner to answer the question. When they have answered the question, the pair should compare their answer with another group and discuss any discrepancies.

Closure

- **Exit Ticket:** Sketch a graph of the two examples suggested in Question 3. Students should scale the axes with reasonable numbers.

Technology For the Teacher

Dynamic Classroom

The Dynamic Planning Tool

Editable Teacher's Resources at *BigIdeasMath.com*

2 ACTIVITY: Discrete and Continuous Domains

Work with a partner.

- **Write a function to represent each problem.**
- **Graph each function.**
- **Describe the domain and range of each function. Is the domain discrete or continuous?**

a. You are in charge of reserving hotel rooms for a youth soccer team. Each room costs \$69, plus \$6 tax, per night. You need each room for two nights. You need 10 to 16 rooms. Write a function for the total hotel cost.

b. The airline you are using for the soccer trip needs an estimate of the total weight of the team's luggage. You determine that there will be 36 pieces of luggage and each piece will weigh from 25 to 45 pounds. Write a function for the total weight of the luggage.

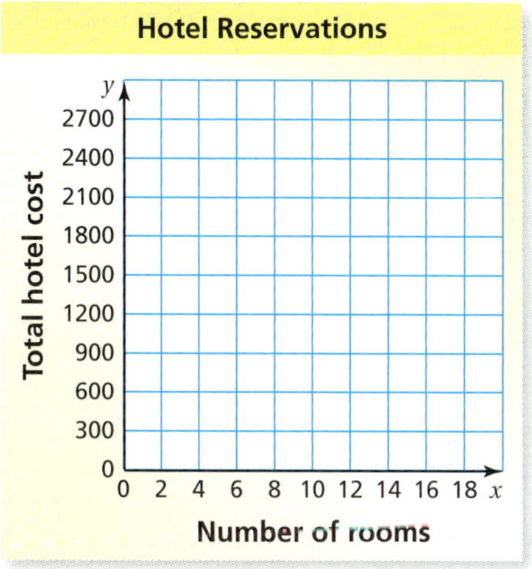

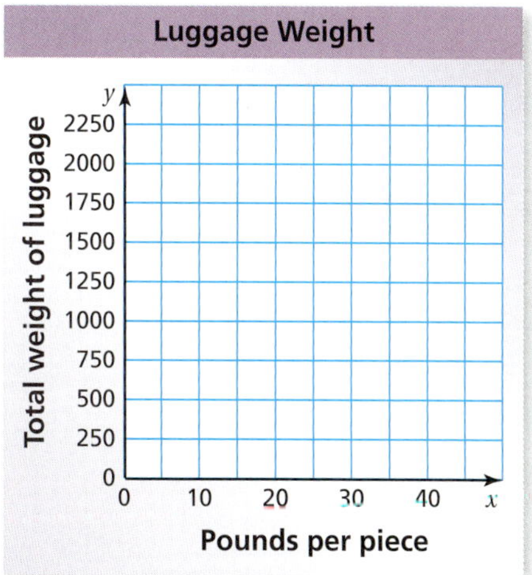

What Is Your Answer?

3. **IN YOUR OWN WORDS** How can you decide whether the domain of a function is discrete or continuous? Describe two real-life examples of functions: one with a discrete domain and one with a continuous domain.

Practice Use what you learned about discrete and continuous domains to complete Exercises 3 and 4 on page 158.

4.2 Lesson

Check It Out
Lesson Tutorials
BigIdeasMath.com

Key Vocabulary
discrete domain, *p. 156*
continuous domain, *p. 156*

Key Idea

Discrete and Continuous Domains

A **discrete domain** is a set of input values that consists of only certain numbers in an interval.

Example: Integers from 1 to 5

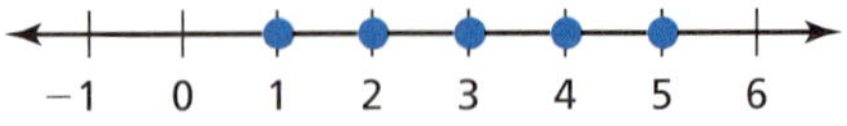

A **continuous domain** is a set of input values that consists of all numbers in an interval.

Example: All numbers from 1 to 5.

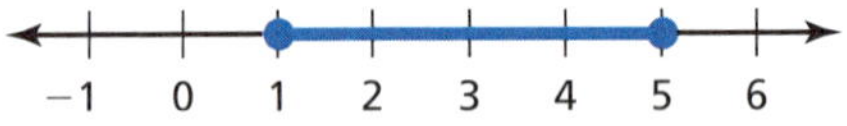

EXAMPLE 1 Graphing Discrete Data

The function $y = 15.95x$ represents the cost y (in dollars) of x tickets for a museum. Graph the function using a domain of 0, 1, 2, 3, and 4. Is the domain of the graph discrete or continuous? Explain.

Make an input-output table.

Input, x	$15.95x$	Output, y	Ordered Pair, (x, y)
0	15.95(0)	0	(0, 0)
1	15.95(1)	15.95	(1, 15.95)
2	15.95(2)	31.9	(2, 31.9)
3	15.95(3)	47.85	(3, 47.85)
4	15.95(4)	63.8	(4, 63.8)

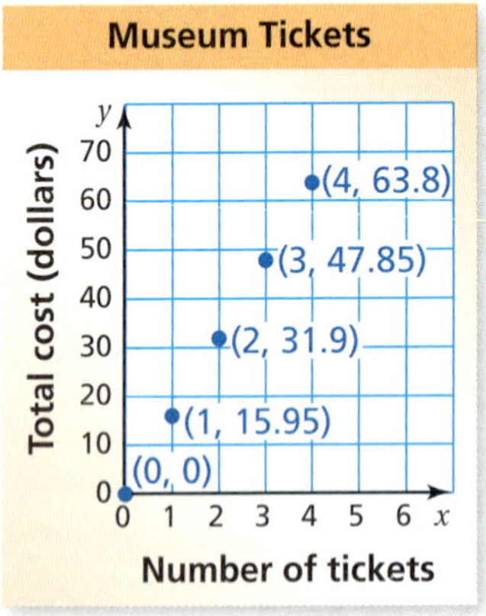

Plot the ordered pairs. Because you cannot buy part of a ticket, the graph consists of individual points.

So, the domain is discrete.

On Your Own

1. The function $m = 50 - 9d$ represents the amount of money m (in dollars) you have after buying d DVDs. Graph the function. Is the domain discrete or continuous? Explain.

Multi-Language Glossary at BigIdeasMath.com.

Laurie's Notes

Introduction

Connect

- **Yesterday:** Students explored problems with discrete and continuous domains.
- **Today:** Students will graph functions and determine if the domain is discrete or continuous.

Motivate

- **FYI:** Share some trivia about the words *continuous* and *discrete.*
- The word *continuous* derives from a Latin root meaning *to hang together* or *to cohere.* This same root gives us the noun *continent* (an expanse of land unbroken by sea).
- The word *discrete* derives from a Latin root meaning *to separate.* This same root yields the verb *discern* (to recognize as distinct or separate) and the cognate *discreet* (to show discernment).

Lesson Notes

Key Idea

- The number line graphs of each example should help students visualize the difference between these two types of domains.
- **Common Error:** Discrete functions do not need to exclude fractions or decimals. A discrete domain might be shoe sizes from 6 to 9, including the half sizes such as $6\frac{1}{2}$.

Example 1

- The table displayed is a good reminder of how equations are evaluated and how solutions can be recorded as ordered pairs.

? "Is this data set a function? Explain." yes; Each input is paired with exactly one output.

- Students may ask why the outputs are not written with the $ symbol and a digit in the hundredths position. If describing the answer to a contextual problem, such as "what is the cost for 2 people to visit the museum?", the answer would be stated with the units ($31.90). Otherwise, the y-coordinate is stated as a real number.

? "What do you notice about the range of this discrete function?" The range is discrete also.

On Your Own

? "Could you buy 0 DVDs? 1 DVD? 2 DVDs? What is the greatest number of DVDs you have money to purchase?" yes; yes; yes; 5

? "Is it possible to spend all of your money on DVDs? Explain." no; if you buy 5 DVDs, you have $5 remaining which is not enough to buy a sixth DVD.

Goal Today's lesson is determining if a domain is **discrete** or **continuous**.

Start Thinking! and Warm Up

Lesson 4.2 Warm Up
For use before Lesson 4.2

Lesson 4.2 Start Thinking!
For use before Lesson 4.2

Discuss whether the following functions have discrete or continuous domains.

A. The air temperature over the course of a day

B. The cost of hot dogs

C. The distance traveled on a road trip

D. The weight of a baby over his first month

E. The cost of parking for a certain number of hours at a parking garage

Extra Example 1

The function $y = 12.90x$ represents the cost y (in dollars) of x admission tickets for a museum. Graph the function using the domain of 0, 1, 2, 3, and 4. Is the domain discrete or continuous? Explain.

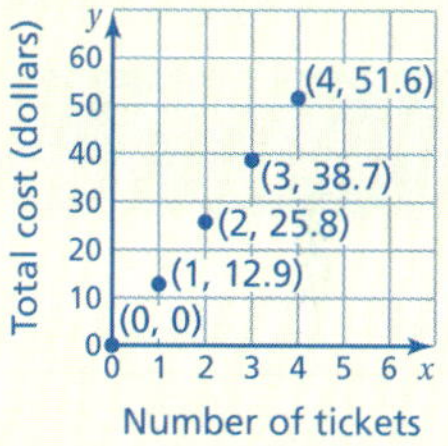

The domain is discrete because you cannot buy part of a ticket.

On Your Own

1.

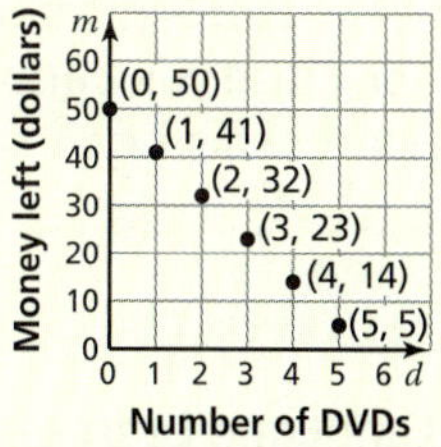

The domain is discrete because you cannot buy part of a DVD.

Extra Example 2

A cereal bar contains 155 calories. The number c of calories consumed is a function of the number b of bars eaten. Graph the function. Is the domain discrete or continuous?

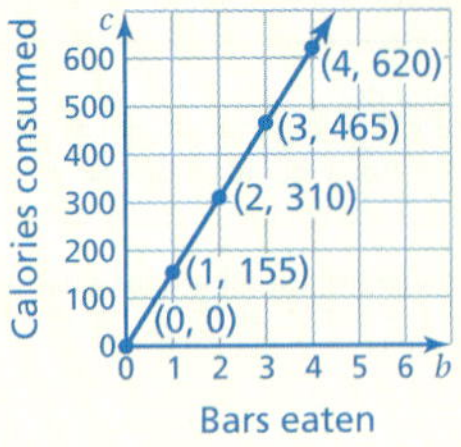

The domain is continuous because you can eat part of a cereal bar.

Extra Example 3

You conduct an experiment on the distance traveled at 55 miles per hour. (a) What is the domain of the function? $t \geq 1$ and $t \leq 4$ (b) Is the domain discrete or continuous? The domain is continuous because the time can be any value between 1 and 4, inclusive.

Input Time, t (hours)	Output Distance, d (miles)
1	55
2	110
3	165
4	220

On Your Own

2. See Additional Answers.
3. The data is discrete because you cannot have part of a story.

English Language Learners

Vocabulary

Have students add the key vocabulary words *function*, *domain*, *range*, *function form*, *discrete domain*, and *continuous domain* to their notebooks. Definitions, examples, and pictures should accompany the words.

Laurie's Notes

Example 2

- Read the problem. This context will be familiar to students. Students will recognize that it is possible to eat some portion of a cereal bar, meaning this is a continuous domain.
- Discuss the graph of this function. Although the table of values stops at the ordered pair (4, 520), it is possible to consume more than 4 cereal bars.
- The domain is *restricted* in the sense that the number of cereal bars must be non-negative.

? **Extension:** "What is the slope and c-intercept of this function?" 130; 0

Example 3

- Read the problem and discuss the table.

? "What are the values of the domain?" Listen for: time (in seconds) between 2 and 10, inclusive.

? "What are the values of the range?" Listen for: distance (in miles) between 0.434 and 2.17, inclusive.

- **Note:** The two inequalities $t \geq 2$ and $t \leq 10$ can also be written as a compound inequality, $2 \leq t \leq 10$. This is true for each of the multiple choice answers. Students have not learned how to read or write compound inequalities, but they will often recognize that the variable is between two numbers.

On Your Own

- These are nice questions that provide another context for understanding the difference between continuous and discrete domains.

Closure

- **Exit Ticket:** Explain how to determine if a graph has a continuous or discrete domain. Listen for: if the points on the graph are connected then it is continuous, but if the points are separated then it is discrete.

Technology For the Teacher

Dynamic Classroom

The Dynamic Planning Tool
Editable Teacher's Resources at *BigIdeasMath.com*

EXAMPLE 2 Graphing Continuous Data

A cereal bar contains 130 calories. The number c of calories consumed is a function of the number b of bars eaten. Graph the function. Is the domain of the graph discrete or continuous?

Make an input-output table.

Input, b	Output, c	Ordered Pair, (b, c)
0	0	(0, 0)
1	130	(1, 130)
2	260	(2, 260)
3	390	(3, 390)
4	520	(4, 520)

Cereal Bar Calories

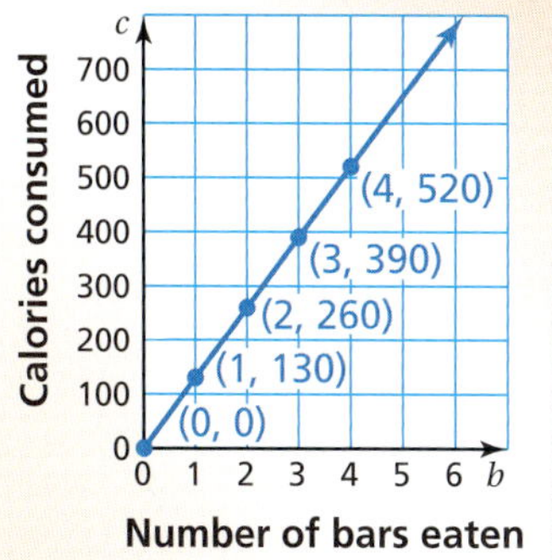

Plot the ordered pairs. Because you can eat part of a cereal bar, b can be any value greater than or equal to 0. Draw a line through the points.

So, the domain is continuous.

EXAMPLE 3 Standardized Test Practice

You conduct an experiment on the speed of sound waves in dry air at 86 °F. You record your data in a table. Which of the following is true?

Input Time, t (seconds)	Output Distance, d (miles)
2	0.434
4	0.868
6	1.302
8	1.736
10	2.170

(A) The domain is $t \geq 2$ and $t \leq 10$ and it is discrete.

(B) The domain is $t \geq 2$ and $t \leq 10$ and it is continuous.

(C) The domain is $d \geq 0.434$ and $d \leq 2.17$ and it is discrete.

(D) The domain is $d \geq 0.434$ and $d \leq 2.17$ and it is continuous.

The domain is the set of possible input values, or the time t. The time t can be any value from 2 to 10. So, the domain is continuous.

The correct answer is (B).

On Your Own

Now You're Ready
Exercises 5–8

2. A 20-gallon bathtub is draining at a rate of 2.5 gallons per minute. The number g of gallons remaining is a function of the number m of minutes. Graph the function. Is the domain discrete or continuous?

3. Are the data shown in the table discrete or continuous? Explain.

Number of Stories	1	2	3	4	5
Height of Building (feet)	12	24	36	48	60

4.2 Exercises

Check It Out
Help with Homework
BigIdeasMath.com

Vocabulary and Concept Check

1. **VOCABULARY** Describe the difference between a discrete domain and a continuous domain.

2. **WRITING** Describe how you can use a graph to determine whether a domain is discrete or continuous.

Practice and Problem Solving

Describe the domain and range of the function. Is the domain discrete or continuous?

3.

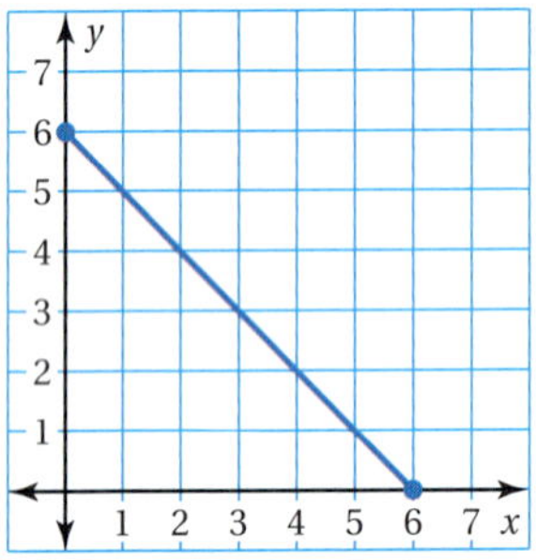

4.

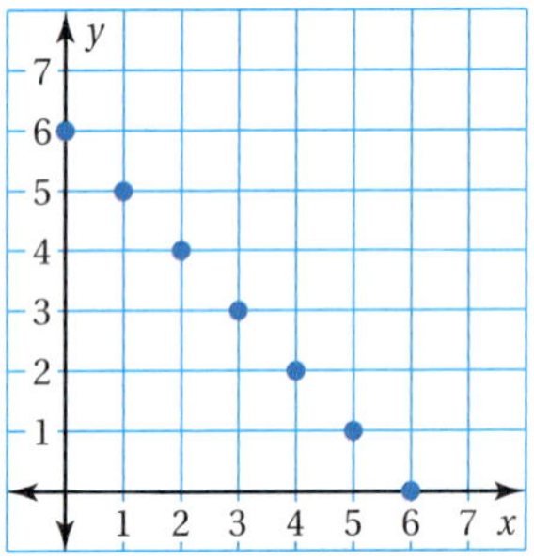

Graph the function. Is the domain of the graph discrete or continuous?

① ② 5.

Input Bags, *x*	Output Marbles, *y*
2	20
4	40
6	60

6.

Input Years, *x*	Output Height of a Tree, *y* (feet)
0	3
1	6
2	9

7.

Input Width, *x* (inches)	Output Volume, *y* (cubic inches)
5	50
10	100
15	150

8.

Input Hats, *x*	Output Cost, *y* (dollars)
0	0
1	8.45
2	16.9

9. **ERROR ANALYSIS** Describe and correct the error in classifying the domain.

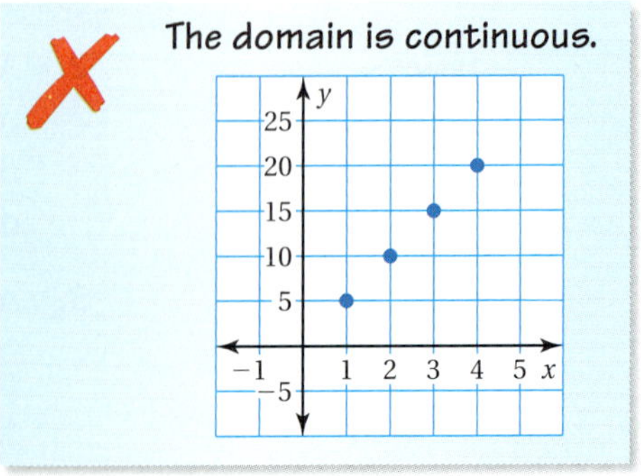

10. **YARN** The function $m = 40 - 8.5b$ represents the amount m of money (in dollars) that you have after buying b balls of yarn. Graph the function using a domain of 0, 1, 2, and 3. Is the domain discrete or continuous?

Assignment Guide and Homework Check

Level	Day 1 Activity Assignment	Day 2 Lesson Assignment	Homework Check
Basic	3, 4, 16–19	1, 2, 5–11	2, 6, 8, 10
Average	3, 4, 16–19	1, 2, 7–13	2, 8, 10, 12
Advanced	3, 4, 16–19	1, 2, 8–15	2, 8, 12, 14

Common Errors

- **Exercises 5–8** Students may mistake the output for the domain and say that the domain is continuous when it is actually discrete. Encourage them to think about the context of the problem. For example, you cannot buy part of a hat, so the domain is discrete.
- **Exercises 5–8** When graphing the function, students may connect the points without considering if the data are discrete or continuous. Remind them that the graph displays discrete and continuous data differently, so the graphs should be different.
- **Exercise 11** Students may say that both functions will have a discrete domain because length is often given as a whole number. Encourage them to think about a context for the two functions, for example, the length of a snake and the cost for several shirts.

4.2 Record and Practice Journal

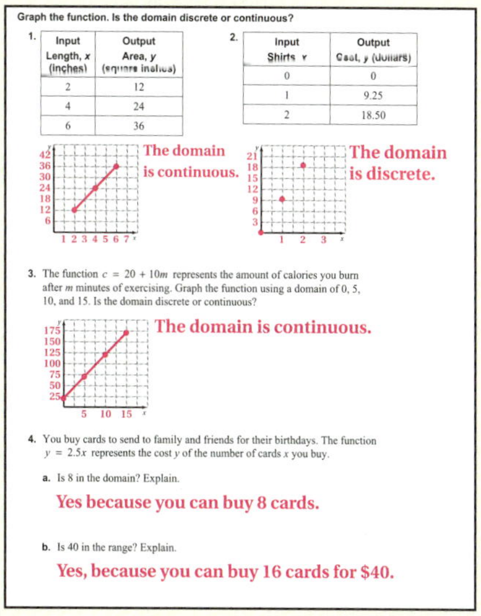
Graph the function. Is the domain discrete or continuous?

1.

Input Length, x (inches)	Output Area, y (square inches)
2	12
4	24
6	36

The domain is continuous.

2.

Input Shirts, x	Output Cost, y (dollars)
0	0
1	9.25
2	18.50

The domain is discrete.

3. The function $c = 20 + 10m$ represents the amount of calories you burn after m minutes of exercising. Graph the function using a domain of 0, 5, 10, and 15. Is the domain discrete or continuous?

The domain is continuous.

4. You buy cards to send to family and friends for their birthdays. The function $y = 2.5x$ represents the cost y of the number of cards x you buy.

a. Is 8 in the domain? Explain.

Yes because you can buy 8 cards.

b. Is 40 in the range? Explain.

Yes, because you can buy 16 cards for $40.

Technology For the Teacher

Answer Presentation Tool

Quiz*Show*

Vocabulary and Concept Check

1. A discrete domain consists of only certain numbers in an interval, whereas a continuous domain consists of all numbers in an interval.
2. If the graph is a line covering all inputs on an interval, then it is a continuous domain. If a graph consists of just points, then it is a discrete domain.

Practice and Problem Solving

3. domain: $x \geq 0$ and $x \leq 6$
range: $y \geq 0$ and $y \leq 6$;
continuous

4. domain: 0, 1, 2, 3, 4, 5, 6
range: 0, 1, 2, 3, 4, 5, 6;
discrete

5.

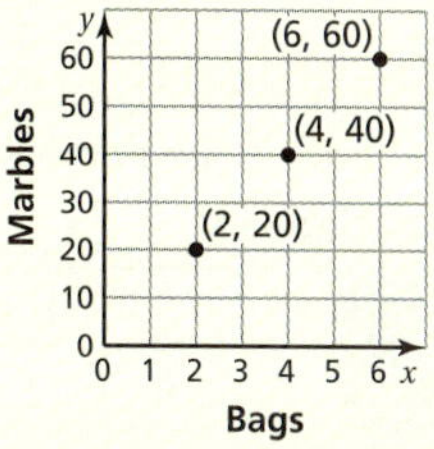

discrete

6–8. See Additional Answers.

9. The domain is discrete because only certain numbers are inputs.

10.

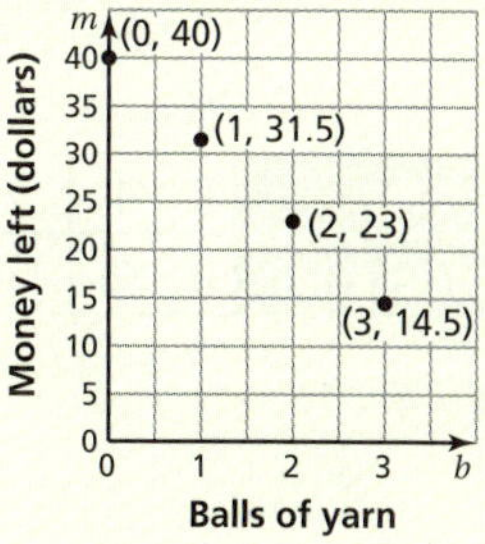

discrete

Practice and Problem Solving

11. The function with an input of length has a continuous domain because you can use any length, but you cannot have half a shirt.

12–13. See Additional Answers.

14. a. yes; You can have 4 boxes.

b. yes; If you have 3 boxes, then 60 books will fit in them.

15. See *Taking Math Deeper*.

Fair Game Review

16. 1

17. $-\frac{5}{2}$

18. $\frac{1}{3}$

19. C

Mini-Assessment

Graph the function. Is the domain discrete or continuous?

1.

Cups, x	0	3	6	9
Cost, y ($)	0	6	12	18

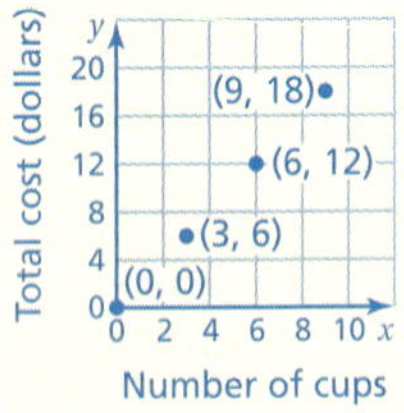

discrete

2.

Time, x (h)	0	1	2	3
Distance, y (mi)	0	55	110	165

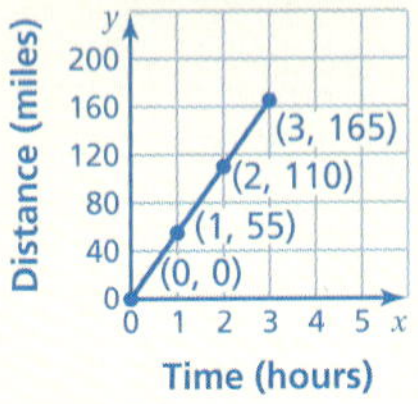

continuous

Taking Math Deeper

Exercise 15

A nice way to start this problem is to draw one possible arrangement. This confirms that everyone understands the question.

1 Draw one possible arrangement.

You might also ask students to cut out six lengths of grid paper that are 4 units long and three lengths that are 8 units long. With these, they can form and record each possible arrangement.

2 Write a function.

Let x = number of pictures in 4-inch frames.

Let y = number of pictures in 8-inch frames.

$4x + 8y = 24$ — Sum is 24 inches.

$8y = 24 - 4x$ — Subtract $4x$ from each side.

a. $y = 3 - 0.5x$ — Divide each side by 8.

3 Graph the function.

Input, x	0	~~1~~	2	~~3~~	4	~~5~~	6
Output, y	3	~~2.5~~	2	~~1.5~~	1	~~0.5~~	0

Note that 1, 3, and 5 are not in the domain because they create "half" pictures.

b.

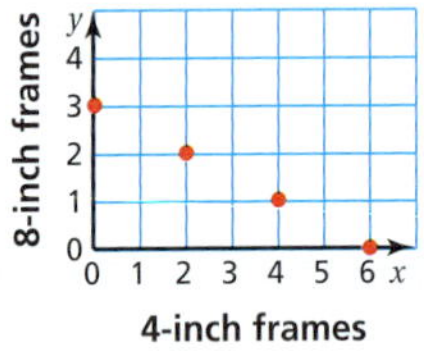

c. The domain is discrete.

Reteaching and Enrichment Strategies

If students need help. . .	If students got it. . .
Resources by Chapter • Practice A and Practice B • Puzzle Time Record and Practice Journal Practice Differentiating the Lesson Lesson Tutorials Skills Review Handbook	Resources by Chapter • Enrichment and Extension Start the next section

11. REASONING The input of one function is *length*. The input of another function is *number of shirts*. Which function has a continuous domain? Explain.

12. DISTANCE The function $y = 3.28x$ converts length from x meters to y feet. Graph the function. Is the domain discrete or continuous?

13. AREA The area A of the triangle is a function of the height h. Graph the function. Is the domain discrete or continuous?

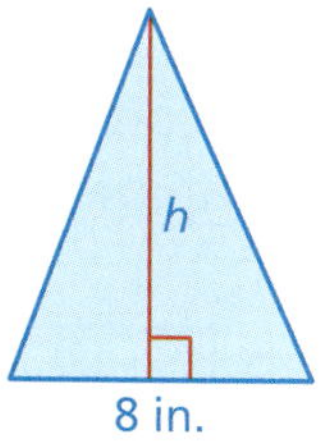

14. PACKING You are packing books into boxes. The function $y = 20x$ represents the number y of books that will fit into x boxes.

a. Is 4 in the domain? Explain.

b. Is 60 in the range? Explain.

15. Reasoning You want to fill a 2-foot shelf with framed pictures. There are x pictures in 4-inch frames and y pictures in 8-inch frames.

a. Write a function for this situation.

b. Graph the function.

c. Is the domain discrete or continuous?

Fair Game Review What you learned in previous grades & lessons

Find the slope of the line. *(Section 2.2)*

16.

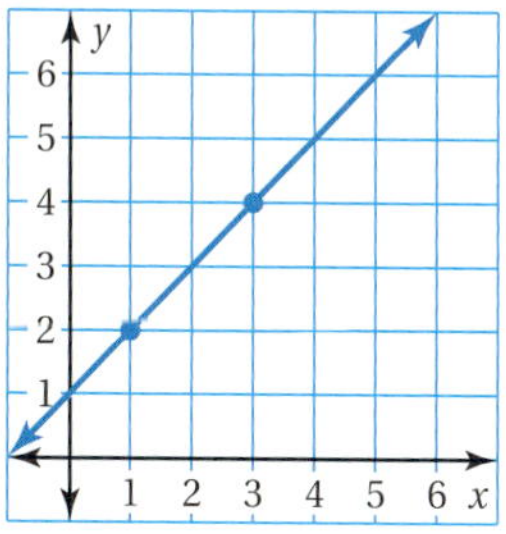

17.

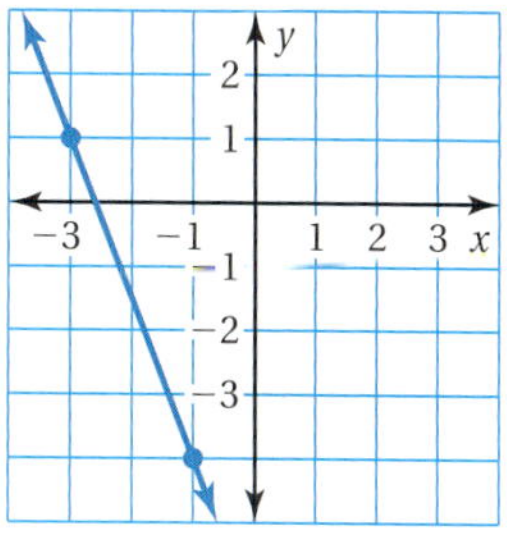

18.

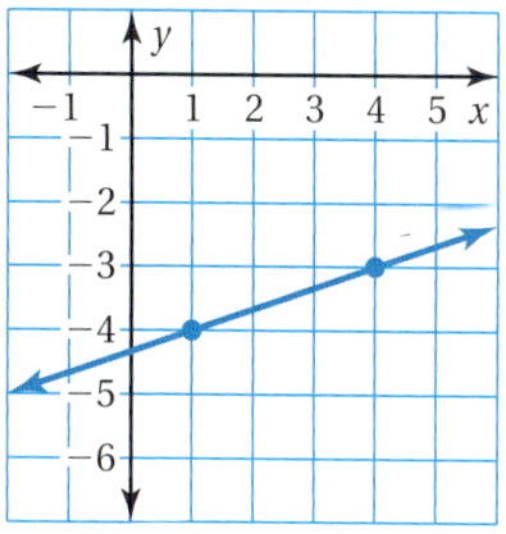

19. MULTIPLE CHOICE What is the y-intercept of the graph of the linear equation? *(Section 2.3)*

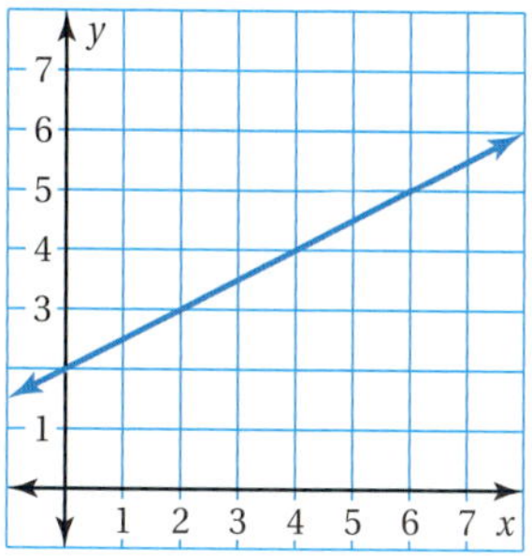

Ⓐ -4 Ⓑ -2

Ⓒ 2 Ⓓ 4

4 Study Help

Check It Out
Graphic Organizer
BigIdeasMath.com

You can use a **comparison chart** to compare two topics. Here is an example of a comparison chart for domain and range.

<table>
<tr><th></th><th>Domain</th><th>Range</th></tr>
<tr><td>Definition</td><td>the set of all possible input values</td><td>the set of all possible output values</td></tr>
<tr><td>Algebra
Example: $y = mx + b$</td><td>x-values</td><td>corresponding y-values</td></tr>
<tr><td>Ordered pairs
Example: (−4, 0), (−3, 1), (−2, 2), (−1, 3)</td><td>−4, −3, −2, −1</td><td>0, 1, 2, 3</td></tr>
<tr><td>Table
Example:<table><tr><td>x</td><td>−1</td><td>0</td><td>2</td><td>3</td></tr><tr><td>y</td><td>1</td><td>0</td><td>4</td><td>9</td></tr></table></td><td>−1, 0, 2, 3</td><td>0, 1, 4, 9</td></tr>
<tr><td>Graph
Example:
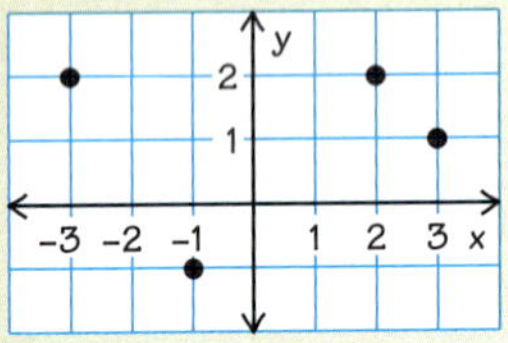</td><td>−3, −1, 2, 3</td><td>−1, 1, 2</td></tr>
</table>

On Your Own

Make a comparison chart to help you study and compare these topics.

1. discrete data and continuous data

After you complete this chapter, make comparison charts for the following topics.

2. linear functions with positive slopes and linear functions with negative slopes
3. linear functions and nonlinear functions

"Creating a comparison chart causes canines to crystalize concepts."

Sample Answers

1.

<table>
<tr><th></th><th>Discrete Data</th><th>Continuous Data</th></tr>
<tr><td>Definition</td><td>Data that consist of only certain numbers in an interval</td><td>Data that consist of all numbers in an interval</td></tr>
<tr><td>Words</td><td>• integers from 0 through 4
• the number of books in a library</td><td>• all numbers from 0 through 4
• gallons of water in a puddle</td></tr>
<tr><td>Table</td><td><table><tr><th>Input
Number in group, x</th><th>Output
Total cost of tickets, y (dollars)</th></tr><tr><td>2</td><td>15</td></tr><tr><td>3</td><td>22.5</td></tr><tr><td>4</td><td>30</td></tr><tr><td>5</td><td>37.5</td></tr></table></td><td><table><tr><th>Input
Years, x</th><th>Output
Height of a tree, y (feet)</th></tr><tr><td>0</td><td>1</td></tr><tr><td>1</td><td>3</td></tr><tr><td>2</td><td>5</td></tr><tr><td>3</td><td>7</td></tr></table></td></tr>
<tr><td>Graphs
Number line

Coordinate axes</td><td>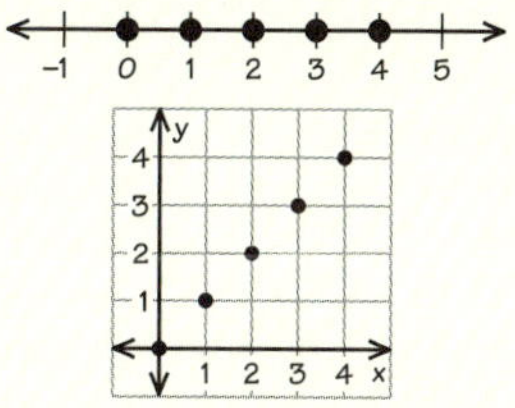</td><td>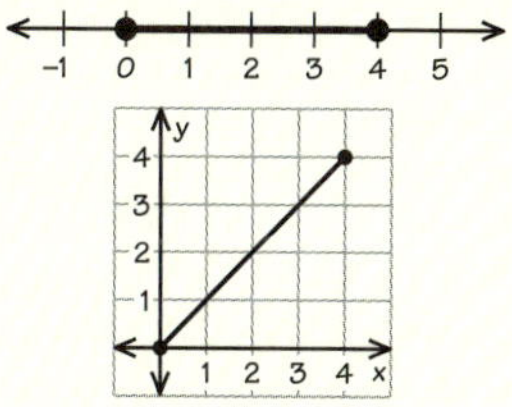</td></tr>
</table>

List of Organizers

Available at *BigIdeasMath.com*

Comparison Chart
Concept Circle
Definition (Idea) and Example Chart
Example and Non-Example Chart
Formula Triangle
Four Square
Information Frame
Information Wheel
Notetaking Organizer
Process Diagram
Summary Triangle
Word Magnet
Y Chart

About this Organizer

A **Comparison Chart** can be used to compare two topics. Students list different aspects of the two topics in the left column. These can include *algebra, definition, description, equation(s), graph(s), table(s),* and *words.* Students write about or give examples illustrating these aspects in the other two columns for the topics being compared. Comparison charts are particularly useful with topics that are related but that have distinct differences. Students can place their comparison charts on note cards to use as a quick study reference.

Technology For the Teacher

Vocabulary Puzzle Builder

Answers

1. domain: −4, −1, 2, 5
   range: 2, 1, 0, −1
2. domain: −2, −1, 0, 1, 2
   range: −1, 1, 3
3. domain: −1, 1, 3, 5
   range: −1
4.

x	0	1	2	3
y	−6	−1	4	9

   domain: 0, 1, 2, 3
   range: −6, −1, 4, 9

5.

x	−1	0	1	2
y	4	2	0	−2

   domain: −1, 0, 1, 2
   range: 4, 2, 0, −2

6. 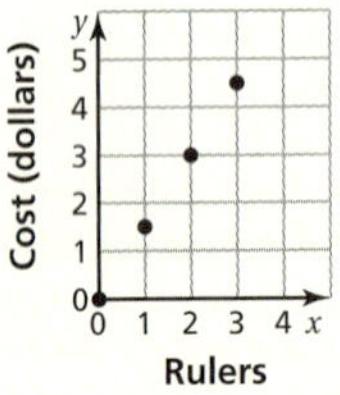

   discrete

7–11. See Additional Answers.

Assessment Book

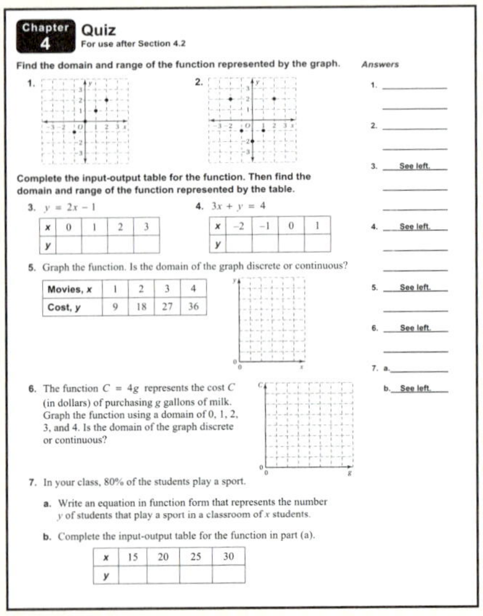

Chapter 4 Quiz
For use after Section 4.2

Find the domain and range of the function represented by the graph. *Answers*

1. 2.

Complete the input-output table for the function. Then find the domain and range of the function represented by the table.

3. $y = 2x - 1$

x	0	1	2	3
y				

4. $3x + y = 4$

x	−2	−1	0	1
y				

5. Graph the function. Is the domain of the graph discrete or continuous?

Movies, *x*	1	2	3	4
Cost, *y*	9	18	27	36

6. The function $C = 4g$ represents the cost C (in dollars) of purchasing g gallons of milk. Graph the function using a domain of 0, 1, 2, 3, and 4. Is the domain of the graph discrete or continuous?

7. In your class, 80% of the students play a sport.
   a. Write an equation in function form that represents the number y of students that play a sport in a classroom of x students.
   b. Complete the input-output table for the function in part (a).

x	15	20	25	30
y				

Answers: 1. ____ 2. ____ 3. See left. 4. See left. 5. See left. 6. See left. 7. a. ____ b. See left.

Alternative Quiz Ideas

100% Quiz
Error Notebook
Group Quiz
Homework Quiz
Math Log
Notebook Quiz
Partner Quiz
Pass the Paper

Partner Quiz

- Students should work in pairs. Each pair should have a small white board.
- The teacher selects certain problems from the quiz and writes one on the board.
- The pairs work together to solve the problem and write their answer on the white board.
- Students show their answers and, as a class, discuss any differences.
- Repeat for as many problems as the teacher chooses.
- For the word problems, teachers may choose to have students read them out of the book.

Reteaching and Enrichment Strategies

If students need help. . .	If students got it. . .
Resources by Chapter • Study Help • Practice A and Practice B • Puzzle Time Lesson Tutorials *BigIdeasMath.com* Practice Quiz Practice from the Test Generator	Resources by Chapter • Enrichment and Extension • School-to-Work Game Closet at *BigIdeasMath.com* Start the next section

Technology For the Teacher

Answer Presentation Tool
Big Ideas Test Generator

4.1–4.2 Quiz

Check It Out
Progress Check
BigIdeasMath.com

Find the domain and range of the function represented by the graph. *(Section 4.1)*

1.

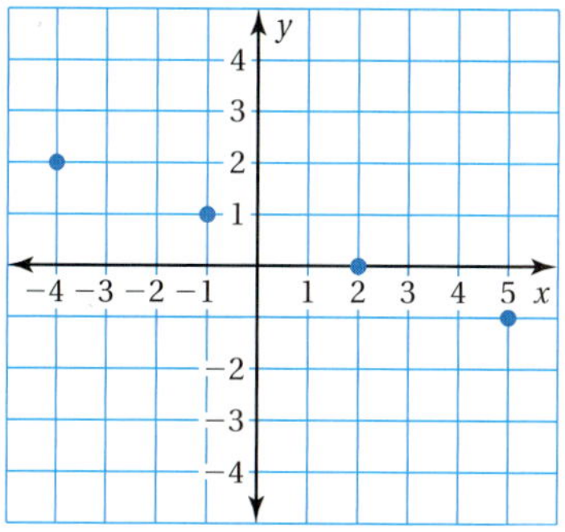

2.

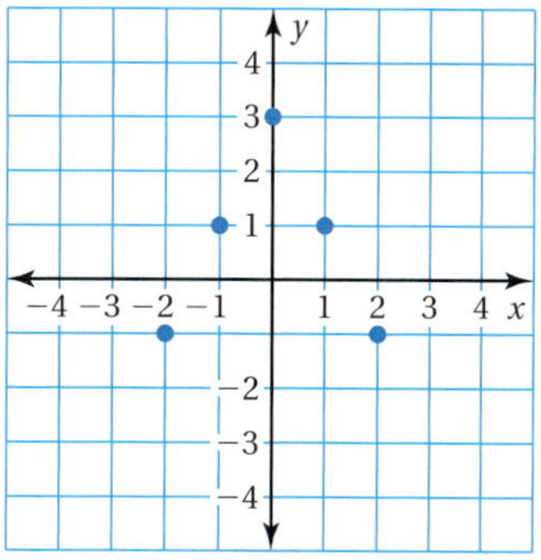

3.

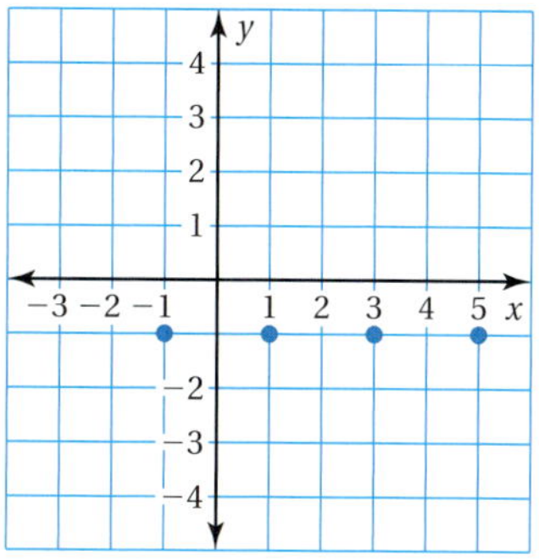

Copy and complete the input-output table for the function. Then find the domain and range of the function represented by the table. *(Section 4.1)*

4. $y = 5x - 6$

x	0	1	2	3
y				

5. $2x + y = 2$

x	−1	0	1	2
y				

Graph the function. Is the domain of the graph discrete or continuous? *(Section 4.2)*

6.

Rulers, *x*	Cost, *y*
0	0
1	1.5
2	3
3	4.5

7.

Gallons, *x*	Miles Remaining, *y*
0	300
1	265
2	230
3	195

8.

Minutes, *x*	0	10	20	30
Height, *y*	40	35	30	25

9.

Relay Teams, *x*	2	4	6	8
Athletes, *y*	8	16	24	32

10. VIDEO GAME The function $m = 30 - 3r$ represents the amount m (in dollars) of money you have after renting r video games. Graph the function using a domain of 0, 1, 2, 3, and 4. Is the domain of the graph discrete or continuous? *(Section 4.2)*

11. WATER Water accounts for about 60% of a person's body weight. *(Section 4.1)*

a. Write an equation in function form that represents the water weight y of a person that weighs x pounds.

b. Make an input-output table for the function in part (a). Use the inputs 100, 120, 140, and 160.

4.3 Linear Function Patterns

Essential Question How can you use a linear function to describe a linear pattern?

1 ACTIVITY: Finding Linear Patterns

Work with a partner.

- **Plot the points from the table in a coordinate plane.**
- **Write a linear equation for the function represented by the graph.**

a.

x	0	2	4	6	8
y	150	125	100	75	50

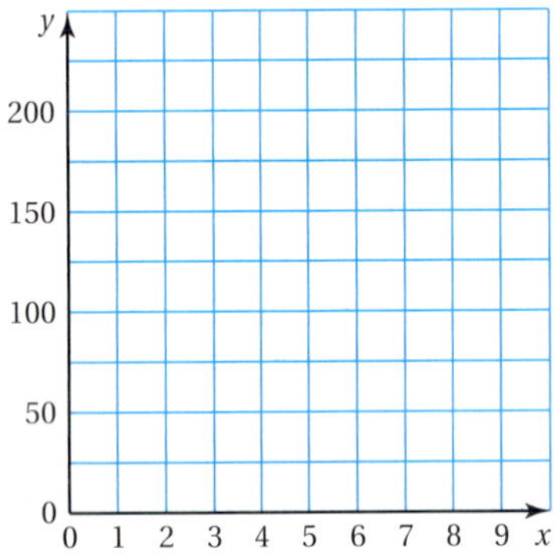

b.

x	4	6	8	10	12
y	15	20	25	30	35

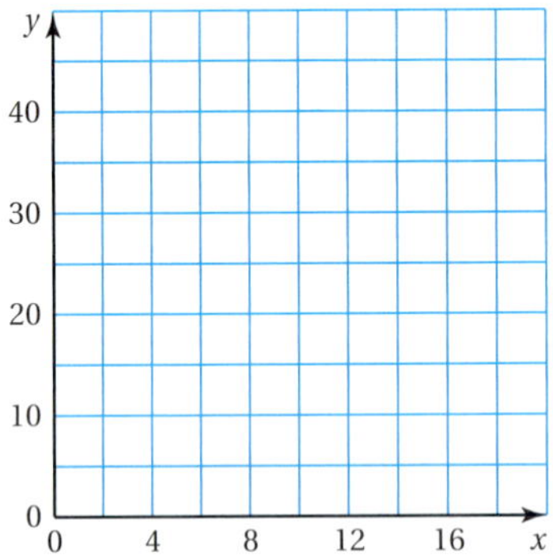

c.

x	−4	−2	0	2	4
y	4	6	8	10	12

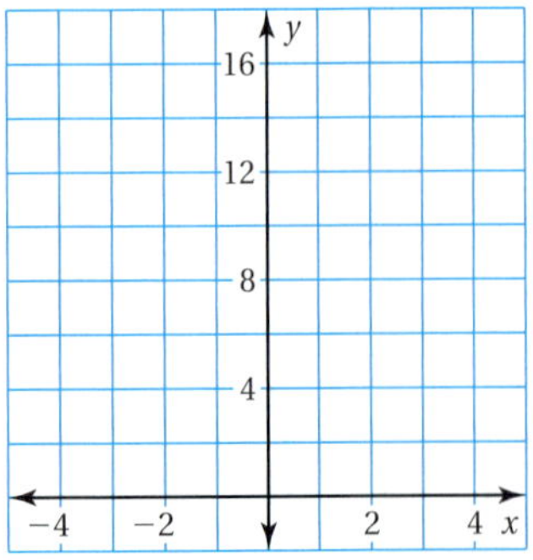

d.

x	−4	−2	0	2	4
y	1	0	−1	−2	−3

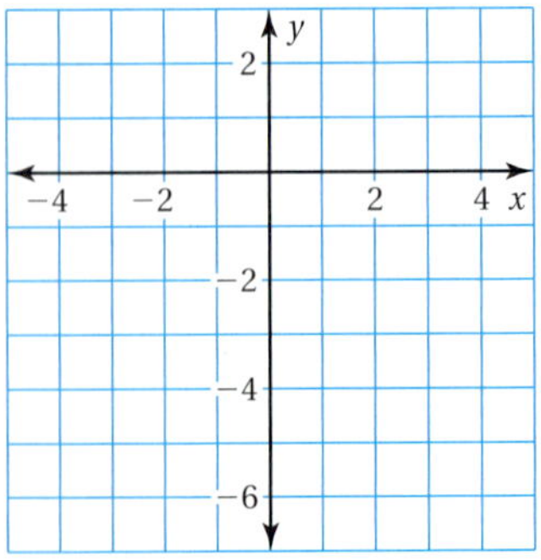

Laurie's Notes

Introduction

For the Teacher

- **Goal:** Students will explore linear patterns in tables and graphs to write linear equations.

Motivate

- Do a quick matching game with students. Have 4–5 graphs on the board with slopes and y-intercepts that are different enough so that students can distinguish between them. Write the equations in a list. Have students work with a partner to match the correct equation with each graph.
- Make sure that students are still focusing on key information from the graph. Is it increasing or decreasing from left to right? Is the slope steeper than 1 or close to 0? Is the y-intercept positive or negative?

Activity Notes

Activity 1

? "What do you notice about the scaling on the axes for each problem?" Answers will vary. Students should recognize the difference of how the x- and y-axes are scaled in each problem.

? "For each problem, you are asked to write a linear equation for the function. How will you do this?" Find the slope and y-intercept.

- Give sufficient time for students to work through the four problems.
- From the graphs, students should be able to determine the slope. It is important that students pay attention to how the axes are scaled when they record values for rise and run.
- **Another Way:** From the table of values, the y-intercept is given for 3 of the 4 problems. In part (a), the ordered pair (0, 150) gives the y-intercept, $b = 150$. To find the slope from the table, notice that every time x increases by 2, y decreases by 25. This means that the run is 2 and the rise is -25.

  So, $m = \frac{\text{rise}}{\text{run}} = \frac{-25}{2} = -12.5$. Now write the equation in slope-intercept form, $y = -12.5x + 150$.

- When students have finished, check their equations.

? "What numeric patterns do you see in the table?" Listen for how the x- and y-values are changing.

- Make sure students recognize the connection between the numeric patterns in the table and the slope of the line.
- **FYI:** For students, recognizing a pattern in the table is the easy part. Helping students translate the pattern into a slope, and then into an equation, is the challenging part. This takes practice.

Previous Learning

Students should know how to write a linear equation in slope-intercept form. Students should know common geometric formulas, such as area and perimeter.

Start Thinking! and Warm Up

Activity 4.3 Start Thinking!
For use before Activity 4.3

Activity 4.3 Warm Up
For use before Activity 4.3

Write an equation of the line in slope-intercept form.

1.

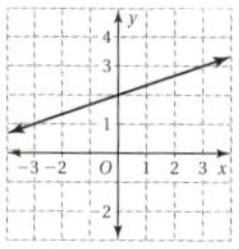

2.

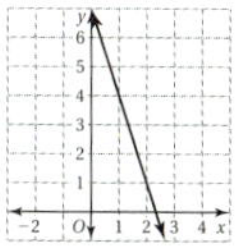

4.3 Record and Practice Journal

Essential Question How can you use a linear function to describe a linear pattern?

1 ACTIVITY: Finding Linear Patterns

Work with a partner.

- Plot the points from the table in a coordinate plane.
- Write a linear equation for the function.

a.

x	0	2	4	6	8
y	150	125	100	75	50

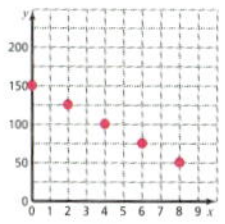

$y = -\frac{25}{2}x + 150$

b.

x	4	6	8	10	12
y	15	20	25	30	35

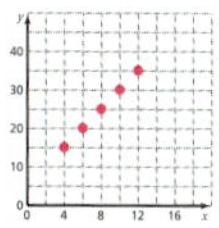

$y = \frac{5}{2}x + 5$

c.

x	−4	−2	0	2	4
y	4	6	8	10	12

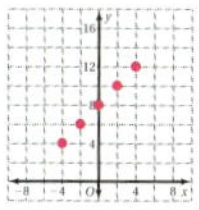

$y = x + 8$

d.

x	−4	−2	0	2	4
y	1	0	−1	−2	−3

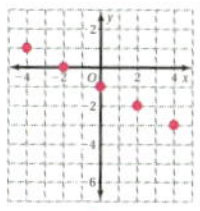

$y = -\frac{1}{2}x - 1$

Differentiated Instruction

Visual

Explain to students that representing a function table as a list of ordered pairs is for convenience. Once the function is represented by ordered pairs, it can be graphed in a coordinate plane. This is a visual representation of the function and is an excellent way to show students the connection between algebra and geometry.

4.3 Record and Practice Journal

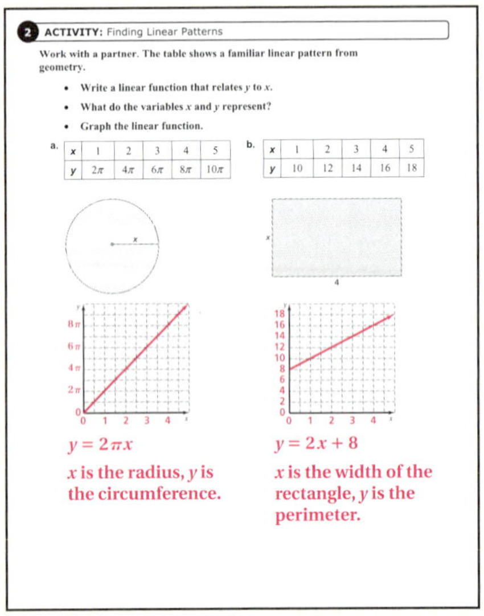

2 ACTIVITY: Finding Linear Patterns

Work with a partner. The table shows a familiar linear pattern from geometry.

- Write a linear function that relates y to x.
- What do the variables x and y represent?
- Graph the linear function.

a.

x	1	2	3	4	5
y	2π	4π	6π	8π	10π

$y = 2\pi x$

x is the radius, y is the circumference.

b.

x	1	2	3	4	5
y	10	12	14	16	18

$y = 2x + 8$

x is the width of the rectangle, y is the perimeter.

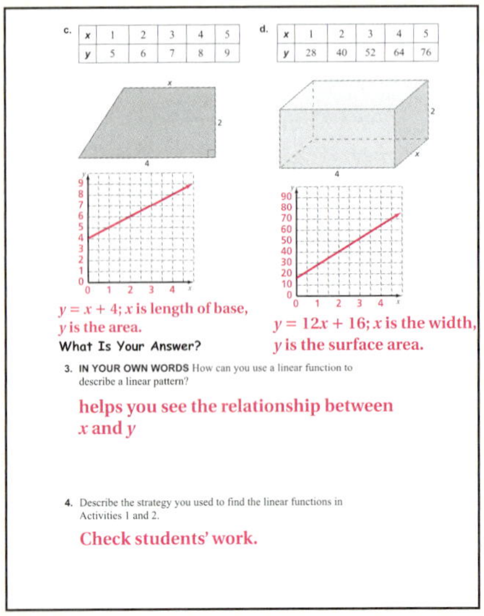

c.

x	1	2	3	4	5
y	5	6	7	8	9

$y = x + 4$; x is length of base, y is the area.

d.

x	1	2	3	4	5
y	28	40	52	64	76

$y = 12x + 16$; x is the width, y is the surface area.

What Is Your Answer?

3. IN YOUR OWN WORDS How can you use a linear function to describe a linear pattern?

helps you see the relationship between x and y

4. Describe the strategy you used to find the linear functions in Activities 1 and 2.

Check students' work.

Laurie's Notes

Activity 2

- The challenge in these problems is that the equation relates to a geometric formula. The figure shown for each problem should provide a hint as to what the variables x and y represent in the problem.
- **Part (a):** Two formulas involving π and circles are circumference ($C = 2\pi r$) and area ($A = \pi r^2$). Substitute the value of x for the radius in each formula. The value of the circumference will match the y-values in the table.
- **Part (b):** Two formulas involving rectangles are perimeter ($P = 2\ell + 2w$) and area ($A = \ell w$). Substitute 4 for the length and the value of x for the width in each formula. The value of the perimeter will match the y-values in the table.

? "Could y represent the perimeter for part (c)? Explain." no; You only know 3 of the 4 side lengths.

- **Part (c):** The formula for the area of a trapezoid is $A = (b + B)h \div 2$. Substitute 4 for B, 2 for h, and the value of x for the length of the shorter base. The value of the area will match the y-values in the table.
- **Part (d):** Two formulas involving a rectangular prism are surface area ($S = 2\ell w + 2wh + 2\ell h$) and volume ($V = \ell wh$). Substitute 4 for the length, 2 for the height, and the value of x for the width in each formula. The value of the surface area will match the y-values in the table.

? Extension: "In part (c), how does the diagram of the trapezoid change as the value of x increases?" When $x = 4$, the trapezoid has the shape of a rectangle. When $x > 4$, the upper base becomes the longer of the two bases.

- Note that in each of these problems, there is a numeric pattern in the table. Have students describe the numeric pattern.

What Is Your Answer?

- **Think-Pair-Share:** Students should read each question independently and then work with a partner to answer the questions. When they have answered the questions, the pair should compare their answers with another group and discuss any discrepancies.

Closure

- **Exit Ticket:** Plot the points given in the table and write a linear equation for the function.

x	−2	0	2	4	6
y	2	3	4	5	6

$y = \frac{1}{2}x + 3$

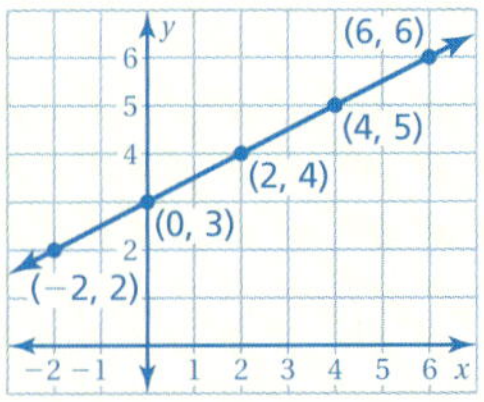

Technology For the Teacher

Dynamic Classroom

The Dynamic Planning Tool

Editable Teacher's Resources at *BigIdeasMath.com*

2 ACTIVITY: Finding Linear Patterns

Work with a partner. The table shows a familiar linear pattern from geometry.

- **Write a linear function that relates *y* to *x*.**
- **What do the variables *x* and *y* represent?**
- **Graph the linear function.**

a.

x	1	2	3	4	5
y	2π	4π	6π	8π	10π

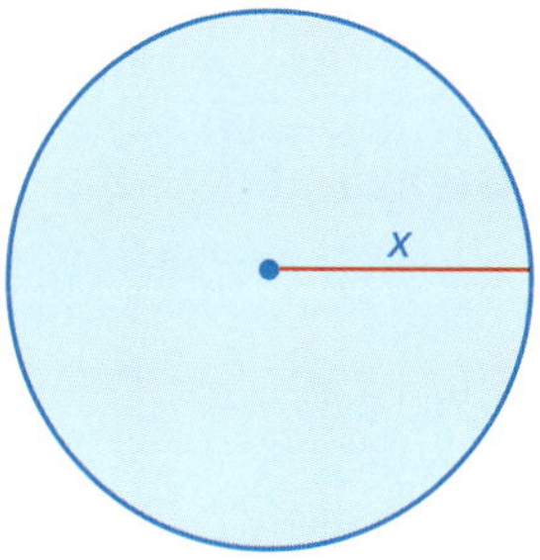

b.

x	1	2	3	4	5
y	10	12	14	16	18

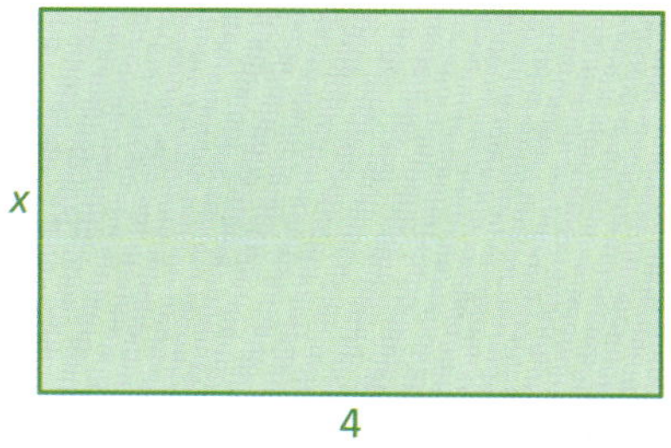

c.

x	1	2	3	4	5
y	5	6	7	8	9

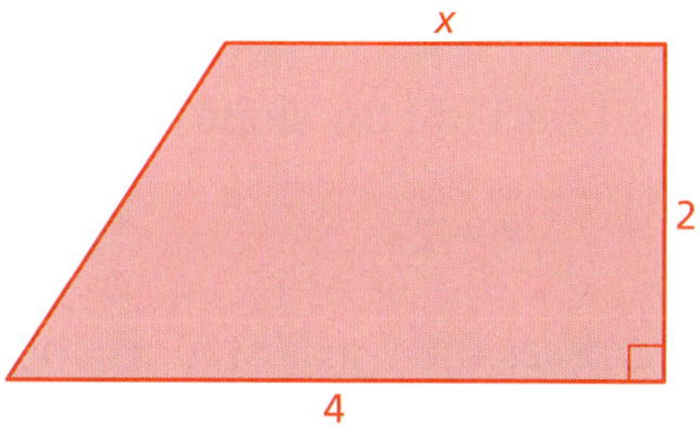

d.

x	1	2	3	4	5
y	28	40	52	64	76

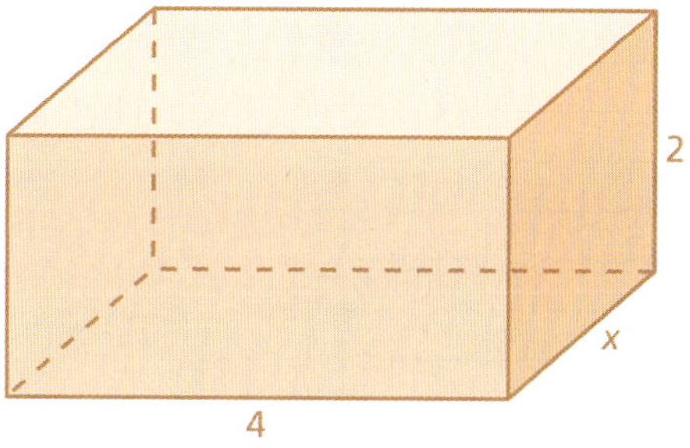

What Is Your Answer?

3. **IN YOUR OWN WORDS** How can you use a linear function to describe a linear pattern?
4. Describe the strategy you used to find the linear functions in Activities 1 and 2.

Practice Use what you learned about linear function patterns to complete Exercises 3 and 4 on page 166.

4.3 Lesson

Check It Out
Lesson Tutorials
BigIdeasMath.com

Key Vocabulary
linear function, *p. 164*

A **linear function** is a function whose graph is a line.

EXAMPLE 1 Finding a Linear Function Using a Graph

Use the graph to write a linear function that relates y to x.

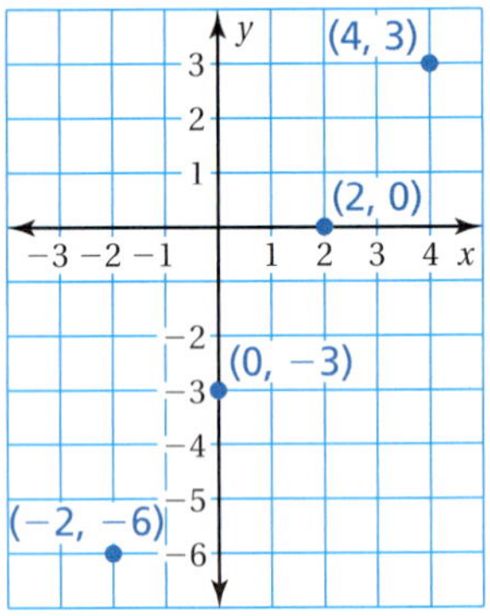

The points lie on a line. Find the slope and y-intercept of the line.

$$\text{slope} = \frac{\text{rise}}{\text{run}} = \frac{3}{2}$$

Because the line crosses the y-axis at $(0, -3)$, the y-intercept is -3.

So, the linear function is $y = \frac{3}{2}x - 3$.

EXAMPLE 2 Finding a Linear Function Using a Table

Use the table to write a linear function that relates y to x.

x	−3	−2	−1	0
y	9	7	5	3

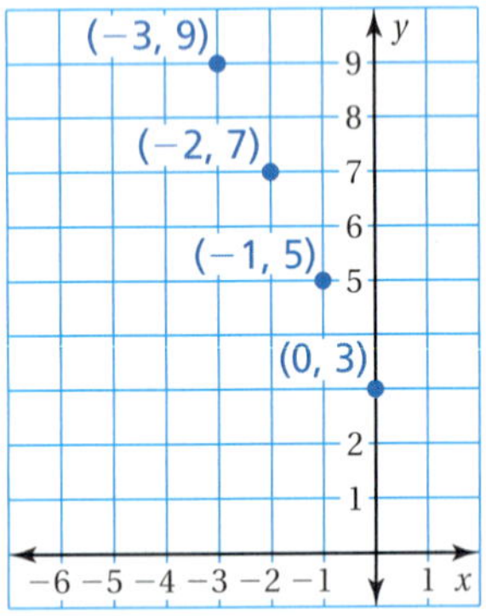

Plot the points in the table.

The points lie on a line. Find the slope and y-intercept of the line.

$$\text{slope} = \frac{\text{rise}}{\text{run}} = \frac{-2}{1} = -2$$

Because the line crosses the y-axis at $(0, 3)$, the y-intercept is 3.

So, the linear function is $y = -2x + 3$.

On Your Own

Now You're Ready
Exercises 5–10

Use the graph or table to write a linear function that relates y to x.

1.

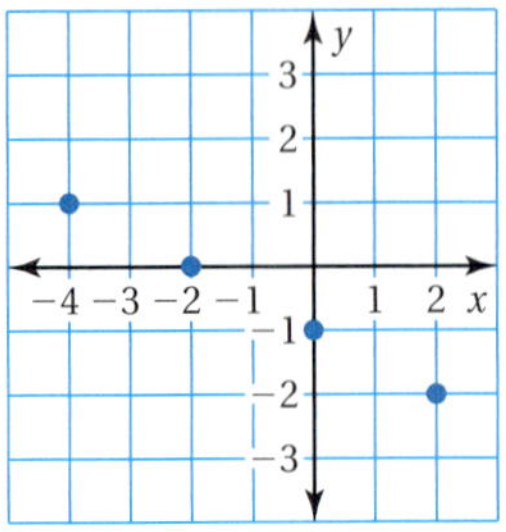

2.

x	−2	−1	0	1
y	2	2	2	2

Multi-Language Glossary at BigIdeasMath.com.

Laurie's Notes

Introduction

Connect

- **Yesterday:** Students gained additional practice in writing linear equations.
- **Today:** Students will write linear functions by recognizing patterns in graphical and tabular information.

Motivate

- Tell the story of Amos Dolbear, who in 1898 noticed that warmer crickets seemed to chirp faster. Dolbear made a detailed study of cricket chirp rates based on the temperature of the crickets' environment and came up with the cricket chirping temperature formula known as Dolbear's Law. Remember that the formula is actually a linear function with a slope and *y*-intercept!

Lesson Notes

Example 1

- Review the definition of a function and the vocabulary associated with functions: domain and range.
- Plot the ordered pairs of the function.
- **Teaching Tip:** To find the slope, draw a right triangle with the hypotenuse between two of the points. Label the legs of the triangle to represent the rise and run. Then compute the slope.

? "Does it matter what two points you select to find the slope? Explain." no; The ratio of rise to run will be the same because the slope triangles are actually similar. It is unlikely students will say this; however, it is the case.

- **Big Idea:** Demonstrate that it does not matter what two points are selected to compute the slope. The slope between (0, −3) and (2, 0) is $\frac{3}{2}$. The slope between (0, −3) and (4, 3) is $\frac{6}{4} = \frac{3}{2}$.

Example 2

- Plot the ordered pairs and repeat the steps from Example 1.

? "Can you tell anything about the slope without plotting the points? Explain." yes; As *x* increases by 1 (run), *y* decreases by 2 (rise).

On Your Own

- **Common Error:** Students may say the slope for Question 1 is −2 instead of $-\frac{1}{2}$. It is very easy to state the reciprocal of the slope.
- **Question 2:** Students may need to graph this function. Once graphed, they will recognize this as a horizontal line whose equation is $y = 2$.

Goal Today's lesson is writing a **linear function** from a graph or a table of values.

Start Thinking! and Warm Up

Lesson 4.3 Warm Up
For use before Lesson 4.3

Lesson 4.3 Start Thinking!
For use before Lesson 4.3

Think about the term *linear function*.

What do you think it means?

Give an example.

Extra Example 1

Use the graph to write a linear function that relates *y* to *x*.

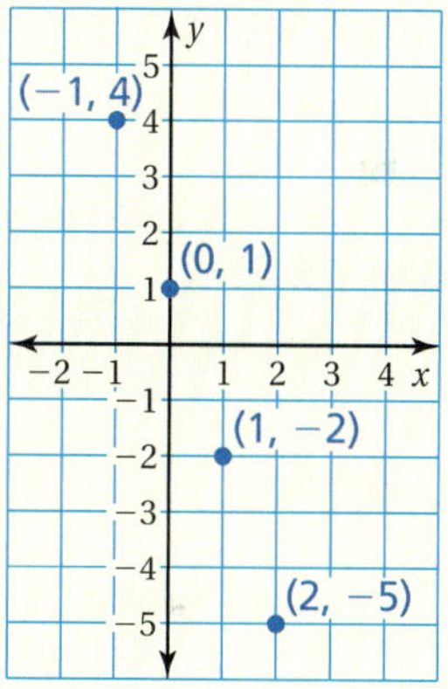

$y = -3x + 1$

Extra Example 2

Use the table to write a linear function that relates *y* to *x*.

x	−2	0	2	4
y	−2	−1	0	1

$y = \frac{1}{2}x - 1$

On Your Own

1. $y = -\frac{1}{2}x - 1$
2. $y = 2$

Extra Example 3

Graph the data.

Hours Jogging, x	Calories Burned, y
2	800
4	1600
6	2400
8	3200

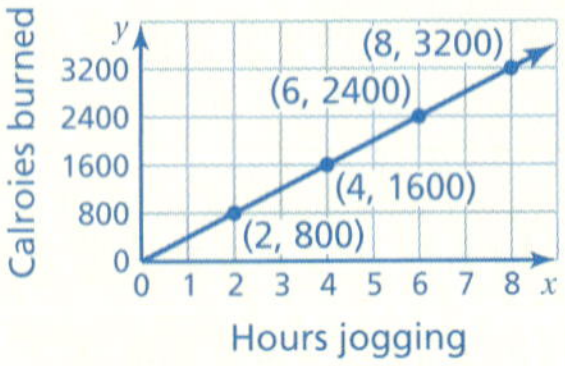

a. Is the domain discrete or continuous? continuous

b. Write a linear function that relates y to x. $y = 400x$

c. How many calories do you burn in 2.5 hours? 1000 calories

On Your Own

3.

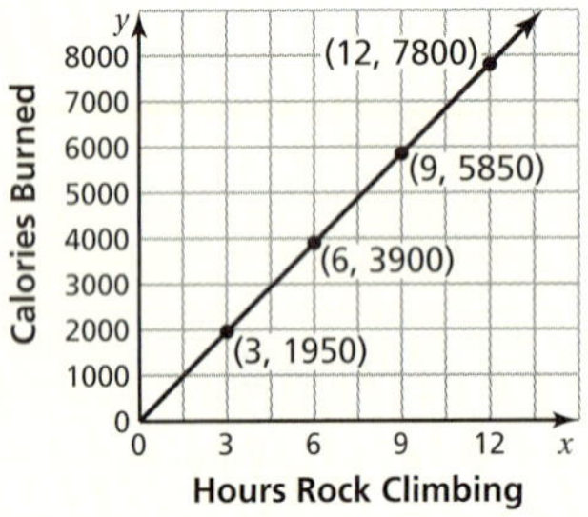

a. continuous

b. $y = 650x$

c. 3575 calories

English Language Learners

Classroom

This chapter gives English learners a chance to share with the rest of the class and the opportunity to build their confidence. Many examples and exercises use tables and graphs giving English learners a rest from interpreting sentences.

Laurie's Notes

Example 3

- Read the problem and discuss the ordered pairs in the table.
- ? Ask questions to check understanding.
  - "Is the slope positive or negative? How do you know?" Listen for positive slope and for recognition that the x- and y-values in the table are both increasing.
  - "What is the domain? Is it continuous?" hours kayaking ≥ 0; yes
  - "What is the range?" calories burned ≥ 2
  - "Explain what a slope of 300 means in the context of this problem." The person is burning 300 calories per hour kayaking.
  - "Explain why a y-intercept of 0 makes sense." If you haven't kayaked yet, you haven't burned any calories.
- **Extension:** Ask students to determine how long the person would have to kayak in order to burn 1000 calories (i.e., given y, solve for x).

On Your Own

- Compare the slope of this line to the slope of the line in Example 3. Which slope is greater? Which line is steeper? Question 3 has the greater slope and the line is steeper.

Summary

- Discuss the Summary. Students should be able to describe how a table of values and a graph can represent a function, and how information is *read* from the graph and table in order to write the linear function.

Closure

- **Exit Ticket:** Write the table of values on the board and ask students to write the equation that relates the temperature to the number of chirps. Acknowledge that this is an approximation and not every Snowy Tree cricket will chirp exactly the same.

Chirps per minute	0	16	32	48	64
Temperature (°F)	40	44	48	52	56

$T = 0.25N + 40$ (Have students check this equation with the one they wrote for Exercise 15 in Section 3.2. It is the same equation.)

Technology For the Teacher

Dynamic Classroom

The Dynamic Planning Tool
Editable Teacher's Resources at *BigIdeasMath.com*

EXAMPLE 3 Real-Life Application

Hours Kayaking, *x*	Calories Burned, *y*
2	600
4	1200
6	1800
8	2400

Graph the data in the table. (a) Is the domain discrete or continuous? (b) Write a linear function that relates *y* to *x*. (c) How many calories do you burn in 4.5 hours?

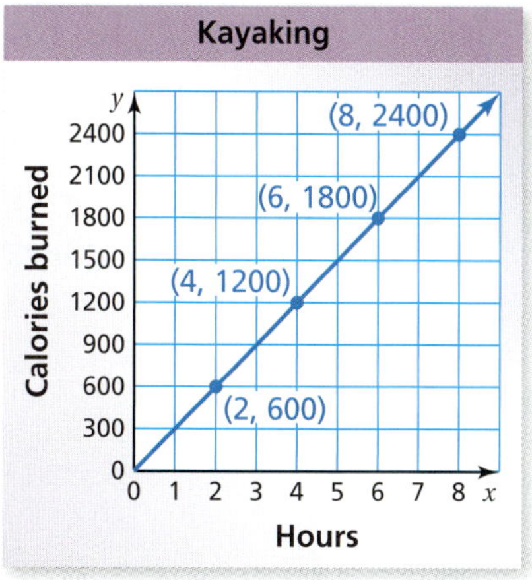

a. Plot the points. Time can represent any value greater than or equal to 0, so the domain is continuous. Draw a line through the points.

b. The slope is $\frac{600}{2} = 300$ and the *y*-intercept is 0.

∴ So, the linear function is $y = 300x$.

c. Find the value of *y* when $x = 4.5$.

$y = 300x$	Write the equation.
$= 300(4.5)$	Substitute 4.5 for *x*.
$= 1350$	Multiply.

∴ You burn 1350 calories in 4.5 hours of kayaking.

On Your Own

Hours Rock Climbing, *x*	Calories Burned, *y*
3	1950
6	3900
9	5850
12	7800

3. Graph the data in the table.

a. Is the domain discrete or continuous?

b. Write a linear function that relates *y* to *x*.

c. How many calories do you burn in 5.5 hours?

Summary

Representing a Function

Words An output is 2 more than the input.

Equation $y = x + 2$

Input-Output Table

Input, *x*	−1	0	1	2
Output, *y*	1	2	3	4

Graph

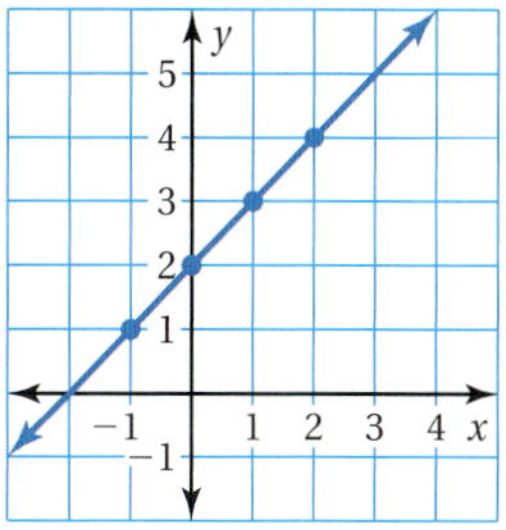

4.3 Exercises

Check It Out
Help with Homework
BigIdeasMath.com

Vocabulary and Concept Check

1. **VOCABULARY** Describe four ways to represent a function.

2. **VOCABULARY** Is the function represented by the graph a linear function? Explain.

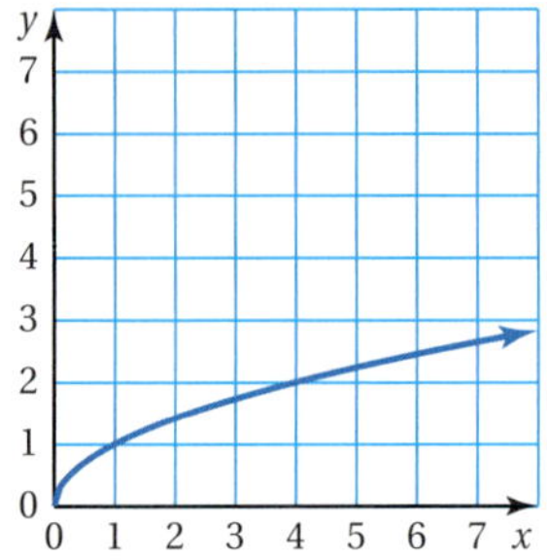

Practice and Problem Solving

The table shows a familiar linear pattern from geometry. Write a linear function that relates *y* to *x*. What do the variables *x* and *y* represent? Graph the linear function.

3.

x	1	2	3	4	5
y	π	2π	3π	4π	5π

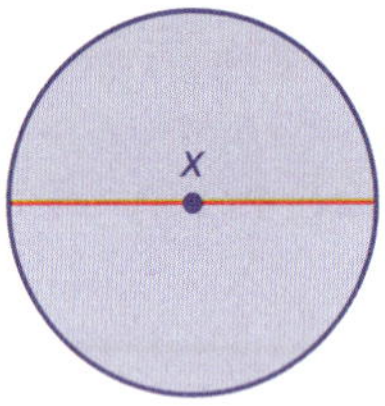

4.

x	1	2	3	4	5
y	2	4	6	8	10

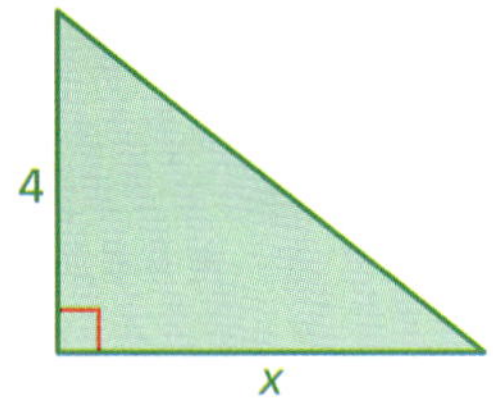

Use the graph or table to write a linear function that relates *y* to *x*.

1 2 5.

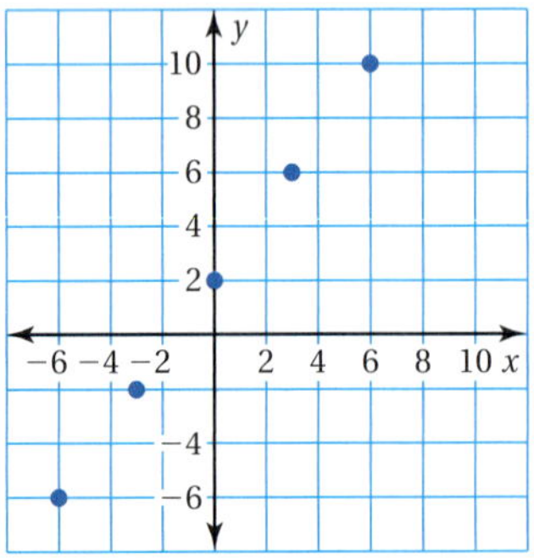

6.

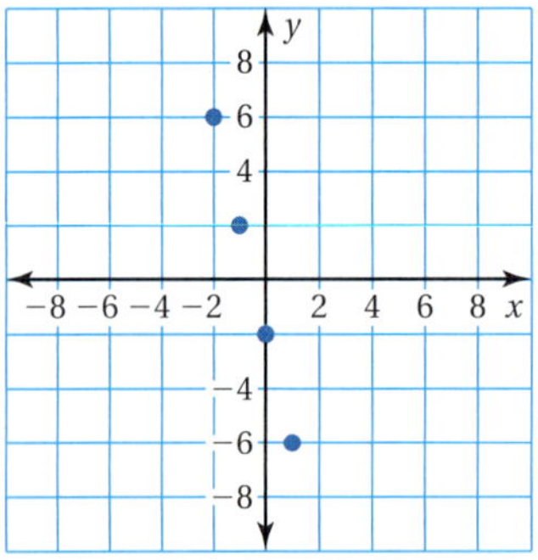

7.

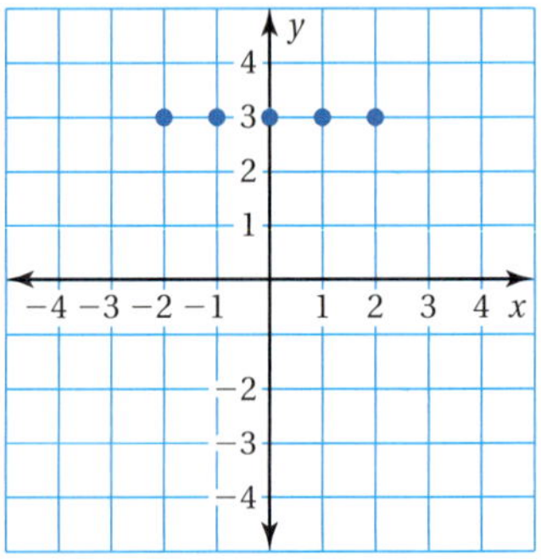

8.

x	−2	−1	0	1
y	−4	−2	0	2

9.

x	−8	−4	0	4
y	2	1	0	−1

10.

x	−3	0	3	6
y	3	5	7	9

11. **MOVIES** The table shows the cost *y* (in dollars) of renting *x* movies.

   a. Graph the data. Is the domain of the graph discrete or continuous?

   b. Write a linear function that relates *y* to *x*.

   c. How much does it cost to rent three movies?

Number of Movies, *x*	0	1	2	4
Cost, *y*	0	3	6	12

Assignment Guide and Homework Check

Level	Day 1 Activity Assignment	Day 2 Lesson Assignment	Homework Check
Basic	3, 4, 15–17	1, 2, 5–11	2, 6, 8, 11
Average	3, 4, 15–17	1, 2, 5–10, 12	2, 6, 8, 12
Advanced	3, 4, 15–17	1, 2, 6–14 even, 13	2, 6, 8, 12

Common Errors

- **Exercises 5 and 6** Students may find the wrong slope because they may misread the scale on an axis. Encourage them to label the points and to use the points they know to write the slope.
- **Exercise 7** Students may not remember how to write the equation for a horizontal line. They may write $x = 3$ instead of $y = 3$. Encourage them to think about the slope-intercept form of an equation.
- **Exercises 8–10** Students may write the reciprocal of the slope when writing the equation from the table. Encourage them to substitute a point into the equation and check to make sure that the equation is true for that point.

4.3 Record and Practice Journal

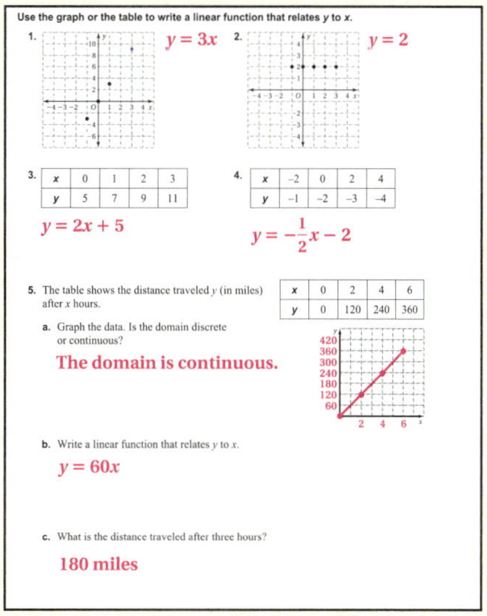
Use the graph or the table to write a linear function that relates y to x.

1. $y = 3x$

2. $y = 2$

3.

x	0	1	2	3
y	5	7	9	11

$y = 2x + 5$

4.

x	−2	0	2	4
y	−1	−2	−3	−4

$y = -\frac{1}{2}x - 2$

5. The table shows the distance traveled y (in miles) after x hours.

x	0	2	4	6
y	0	120	240	360

a. Graph the data. Is the domain discrete or continuous?

The domain is continuous.

b. Write a linear function that relates y to x.

$y = 60x$

c. What is the distance traveled after three hours?

180 miles

Technology For the Teacher

Answer Presentation Tool

Quiz*Show*

Vocabulary and Concept Check

1. words, equation, table, graph
2. no; The graph is not a line.

Practice and Problem Solving

3. $y = \pi x$; x is the diameter; y is the circumference.

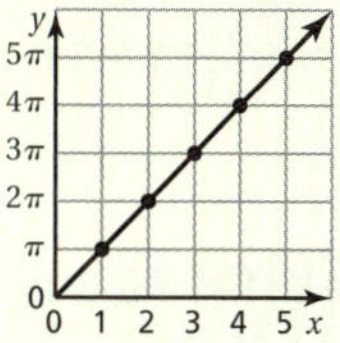

4. $y = 2x$; x is the base of the triangle; y is the area of the triangle.

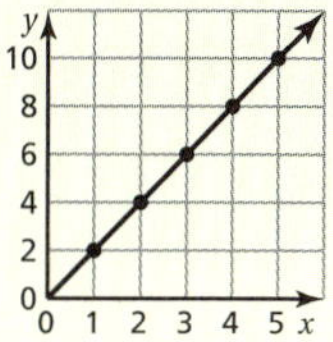

5. $y = \frac{4}{3}x + 2$
6. $y = -4x - 2$
7. $y = 3$
8. $y = 2x$
9. $y = -\frac{1}{4}x$
10. $y = \frac{2}{3}x + 5$
11. a.

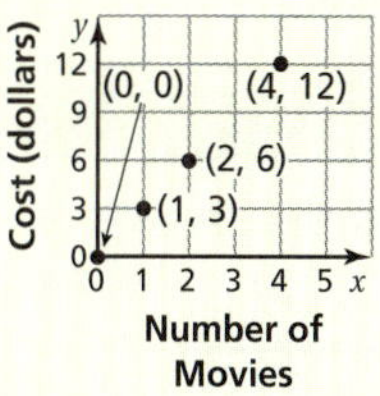

discrete

b. $y = 3x$ c. $9

Practice and Problem Solving

12. **a.**

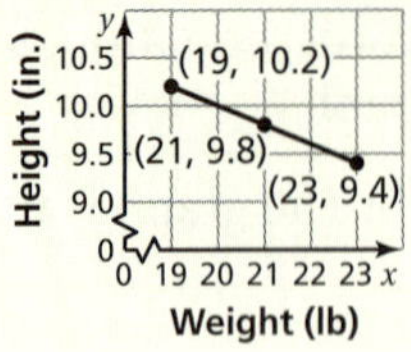

linear

b. $y = -0.2x + 14$

c. 9.7 in.

13. See *Taking Math Deeper.*

14. See Additional Answers.

Fair Game Review

15. 5% **16.** 3%

17. B

Mini-Assessment

Use the graph or table to write a linear function that relates *y* to *x*.

1.

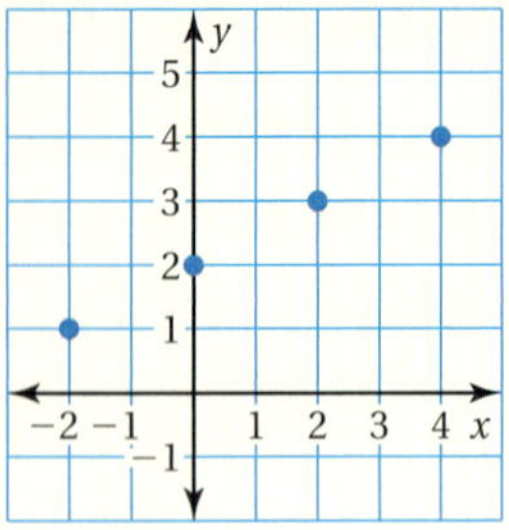

$y = \frac{1}{2}x + 2$

2.

x	−2	−1	0	1
y	9	4	−1	−6

$y = -5x - 1$

Taking Math Deeper

Exercise 13

Students might find it interesting to discover that there is a correlation between years of education and salary. Of course, the correlation only relates annual salaries. There are many examples of people with no years of education beyond high school who have big salaries.

1 Graph the data. Describe the pattern.

a.

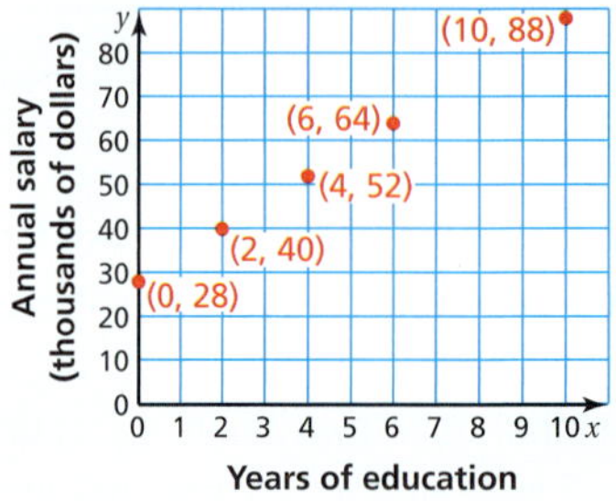

The pattern is that for every 2 years of additional education, the annual salary increases by $12,000.

2 Write a function.

Let x = years of education beyond high school.
Let y = annual salary.

y-intercept = 28
slope = 6

b. $y = 6x + 28$

3 Use the function.

For 8 years of education beyond high school, the annual salary is

c. $y = 6(8) + 28 = 76$, or $76,000.

Check this on the graph to see that it fits the pattern.

Project

Select four careers in which you might be interested. List the annual salary for each. Also list the amount of education required for each career.

Reteaching and Enrichment Strategies

If students need help. . .	If students got it. . .
Resources by Chapter • Practice A and Practice B • Puzzle Time Record and Practice Journal Practice Differentiating the Lesson Lesson Tutorials Skills Review Handbook	Resources by Chapter • Enrichment and Extension • School-to-Work Start the next section

12. **BIKE JUMPS** A bunny hop is a bike trick in which the rider brings both tires off the ground without using a ramp. The table shows the height y (in inches) of a bunny hop on a bike that weighs x pounds.

Weight, x	19	21	23
Height, y	10.2	9.8	9.4

a. Graph the data. Then describe the pattern.

b. Write a linear function that relates the height of a bunny hop to the weight of the bike.

c. What is the height of a bunny hop on a bike that weighs 21.5 pounds?

Years of Education, x	Annual Salary, y
0	28
2	40
4	52
6	64
10	88

13. **SALARY** The table shows a person's annual salary y (in thousands of dollars) after x years of education beyond high school.

a. Graph the data.

b. Write a linear function that relates the person's annual salary to the number of years of education beyond high school.

c. What is the annual salary of the person after 8 years of education beyond high school?

14. Critical Thinking The Heat Index is calculated using the relative humidity and the temperature. For every 1 degree increase in the temperature from 94°F to 98°F at 75% relative humidity, the Heat Index rises 4°F.

a. On a summer day, the relative humidity is 75%, the temperature is 94°F, and the Heat Index is 122°F. Construct a table that relates the temperature t to the Heat Index H. Start the table at 94°F and end it at 98°F.

b. Write a linear function that represents this situation.

c. Estimate the Heat Index when the temperature is 100°F.

Fair Game Review What you learned in previous grades & lessons

Find the annual simple interest rate. *(Skills Review Handbook)*

15. $I = \$60, P = \$400, t = 3$ years

16. $I = \$45, P = \$1000, t = 18$ months

17. **MULTIPLE CHOICE** You buy a pair of gardening gloves for \$2.25 and x packets of seeds for \$0.88 each. Which equation represents the total cost y? *(Skills Review Handbook)*

Ⓐ $y = 0.88x - 2.25$

Ⓑ $y = 0.88x + 2.25$

Ⓒ $y = 2.25x - 0.88$

Ⓓ $y = 2.25x + 0.88$

4.4 Comparing Linear and Nonlinear Functions

Essential Question How can you recognize when a pattern in real life is linear or nonlinear?

1 ACTIVITY: Finding Patterns for Similar Figures

Work with a partner. Copy and complete each table for the sequence of similar rectangles. Graph the data in each table. Decide whether each pattern is linear or nonlinear.

a. Perimeters of Similar Rectangles

x	1	2	3	4	5
P					

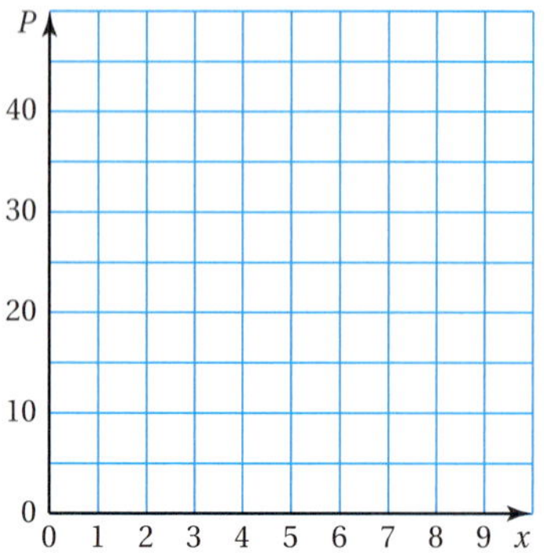

b. Areas of Similar Rectangles

x	1	2	3	4	5
A					

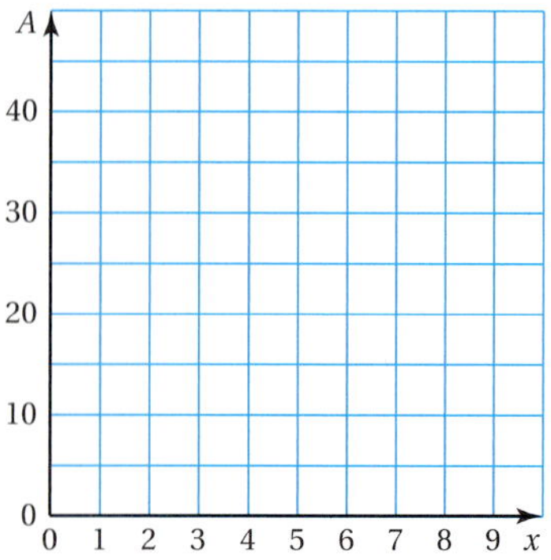

Laurie's Notes

Introduction

For the Teacher

- **Goal:** Students will compare tables and graphs of linear and nonlinear functions.
- Many students get to this point and believe that all equations are linear because that is all they have seen. It is important for students to recognize that not all equations are linear, and therefore not all graphs are linear.
- In future math courses, students will study specific nonlinear functions such as quadratics, exponentials, and rational, just to name a few. For now it is fine to simply refer to them collectively as nonlinear.

Motivate

? "How many of you would like to try skydiving? Why?"

- Share with students that the first successful parachute jump made from a moving airplane was made by Captain Albert Berry in St. Louis, in 1912.
- The first parachute jump from a balloon was completed by André-Jacques Garnerin in 1797 over Monceau Park in Paris.

Activity Notes

Activity 1

? "What does it mean for two rectangles to be similar?" Corresponding sides are proportional and corresponding angles have the same measure.

? "What is the relationship between the length and the width of the green rectangle?" The length is twice the width.

? "What is the relationship between the length and the width of the yellow rectangle? How do you know?" The length is twice the width. Because the rectangles are similar, the lengths of all the rectangles will be twice the widths.

- Explain to students that they will find the perimeter and area of each rectangle for the side lengths given, and then plot the results.
- **Teaching Tip:** It may be helpful to set up a table that includes a row for the second dimension as shown.

Width	x	1	2	3	4	5
Length	$2x$	2	4	6	8	10
Perimeter	P					

- Encourage students to be accurate with their graphing. Because only 5 points are being plotted for each graph, it is possible that students will not see the curvature of the area graph. Students should recognize, however, that the numeric data for area does not have a constant difference between y-values.

Previous Learning

Students should know common geometric formulas, such as area and perimeter.

Activity Materials
Textbook
• handkerchief • floss • tape • small figure

Start Thinking! and Warm Up

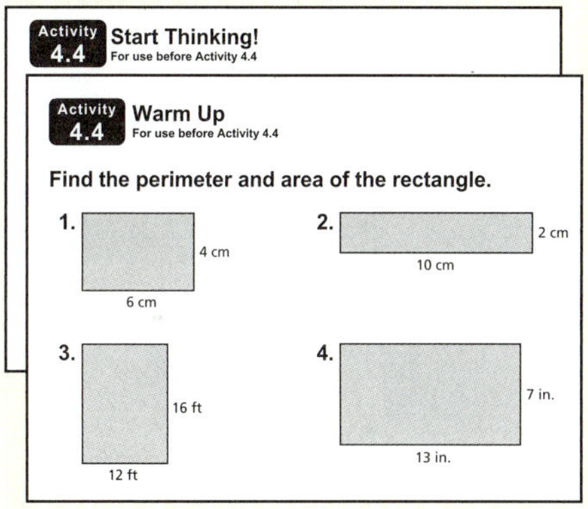

Activity 4.4 Start Thinking!
For use before Activity 4.4

Activity 4.4 Warm Up
For use before Activity 4.4

Find the perimeter and area of the rectangle.

1. 4 cm, 6 cm
2. 2 cm, 10 cm
3. 16 ft, 12 ft
4. 7 in., 13 in.

4.4 Record and Practice Journal

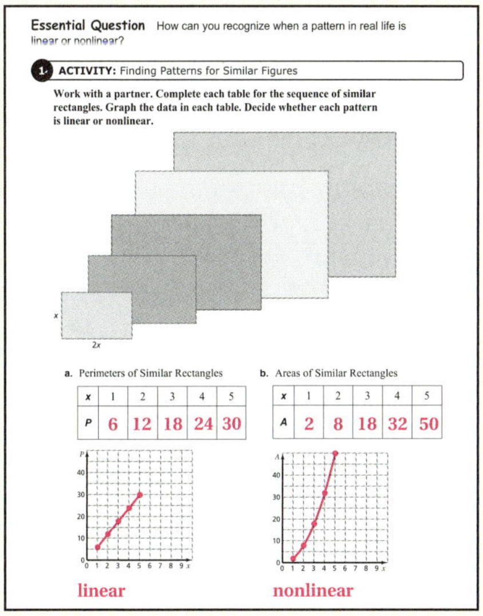

Essential Question How can you recognize when a pattern in real life is linear or nonlinear?

1 ACTIVITY: Finding Patterns for Similar Figures

Work with a partner. Complete each table for the sequence of similar rectangles. Graph the data in each table. Decide whether each pattern is linear or nonlinear.

a. Perimeters of Similar Rectangles

x	1	2	3	4	5
P	6	12	18	24	30

linear

b. Areas of Similar Rectangles

x	1	2	3	4	5
A	2	8	18	32	50

nonlinear

Differentiated Instruction

Visual

Students may be able to describe how the sequence of output numbers is changing, for example, *start with 2 and add 3*, but they may find it difficult to write a function rule for changing an input value to an output value. If students determine that the output increases or decreases by a constant value as the input increases, the function will have an ax term. Have students create function tables for equations such as $y = x + 3$, $y = 4x - 1$, and $y = 0.5x$ to see this pattern.

4.4 Record and Practice Journal

2 ACTIVITY: Comparing Linear and Nonlinear Functions

Work with a partner. The table shows the height h (in feet) of a falling object at t seconds.

- **Graph the data in the table.**
- **Decide whether the graph is linear or nonlinear.**
- **Compare the two falling objects. Which one has an increasing speed?**

a. Falling parachute jumper

t	0	1	2	3	4
h	300	285	270	255	240

b. Falling bowling ball

t	0	1	2	3	4
h	300	284	236	156	44

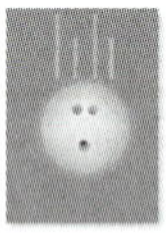

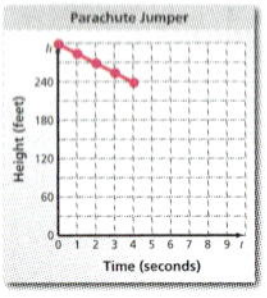

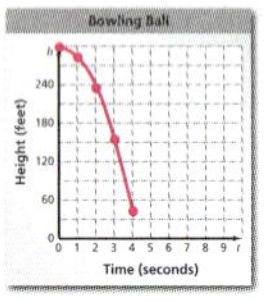

linear

nonlinear

The bowling ball has increasing speed.

What Is Your Answer?

3. **IN YOUR OWN WORDS** How can you recognize when a pattern in real life is linear or nonlinear? Describe two real-life patterns: one that is linear and one that is nonlinear. Use patterns that are different from those described in Activities 1 and 2.

If the rate of change is constant, the pattern is linear.

Laurie's Notes

Activity 2

- This activity is similar to Activity 1, except the ordered pairs are already given. Discuss the two falling objects—one with a parachute and one that is free falling.

? "Do you think there is a difference in the rate at which two objects fall when one is attached to a parachute and the other is left to free fall? Explain." Listen for discussion of rate. It is unlikely students will bring up acceleration.

- If you have the means to make a small parachute (handkerchief, tape, dental floss, small figurine), you could model the difference in a parachute-controlled fall versus a free fall.
- Again, it is necessary for students to be accurate when plotting the ordered pairs given the scale on the y-axis.

? After students have plotted the points, ask about the two graphs. First note that the two graphs begin at the same height (y-intercept), 300 feet.

  - "How far has the jumper fallen after 4 seconds? 60 ft "How far has the bowling ball fallen after 4 seconds?" 256 ft
  - "Describe the flight of the jumper." falling at a constant rate of 15 ft/sec
  - "Describe the flight of the bowling ball." Listen for students to describe that the bowling ball is picking up speed as it falls.

- **Extension:** Students could write a linear equation for the jumper, but not the bowling ball.

What Is Your Answer?

- Students may need help thinking of real-life patterns that are nonlinear. You might suggest area or volume relationships, or even simple story graphs about time and distance.

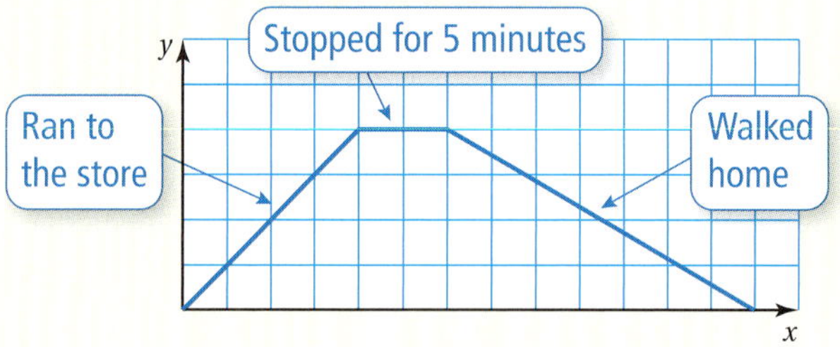

Closure

- Draw two functions with a domain of $x \geq 0$. Have one that is linear and one that is nonlinear. Describe how the graphs are alike and how they are different. Answers will vary.

Technology For the Teacher

Dynamic Classroom

The Dynamic Planning Tool
Editable Teacher's Resources at *BigIdeasMath.com*

2 ACTIVITY: Comparing Linear and Nonlinear Functions

Work with a partner. The table shows the height *h* (in feet) of a falling object at *t* seconds.

- **Graph the data in the table.**
- **Decide whether the graph is linear or nonlinear.**
- **Compare the two falling objects. Which one has an increasing speed?**

a. Falling parachute jumper

t	0	1	2	3	4
h	300	285	270	255	240

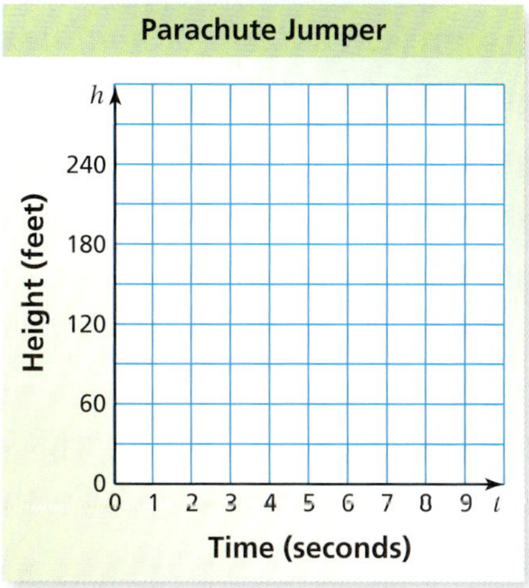

b. Falling bowling ball

t	0	1	2	3	4
h	300	284	236	156	44

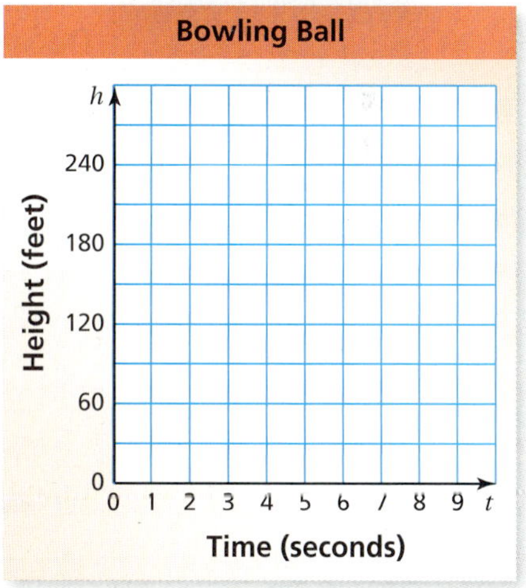

What Is Your Answer?

3. IN YOUR OWN WORDS How can you recognize when a pattern in real life is linear or nonlinear? Describe two real-life patterns: one that is linear and one that is nonlinear. Use patterns that are different from those described in Activities 1 and 2.

Practice

Use what you learned about comparing linear and nonlinear functions to complete Exercises 3–6 on page 172.

4.4 Lesson

Check It Out
Lesson Tutorials
BigIdeasMath.com

Key Vocabulary
nonlinear function, *p. 170*

The graph of a linear function shows a constant rate of change. A **nonlinear function** does not have a constant rate of change. So, its graph is *not* a line.

EXAMPLE 1 Identifying Functions from Tables

Does the table represent a *linear* or *nonlinear* function? Explain.

a.

+3 +3 +3

x	3	6	9	12
y	40	32	24	16

−8 −8 −8

As x increases by 3, y decreases by 8. The rate of change is constant. So, the function is linear.

b.

+2 +2 +2

x	1	3	5	7
y	2	11	33	88

+9 +22 +55

As x increases by 2, y increases by different amounts. The rate of change is *not* constant. So, the function is nonlinear.

EXAMPLE 2 Identifying Functions from Graphs

Does the graph represent a *linear* or *nonlinear* function? Explain.

a.

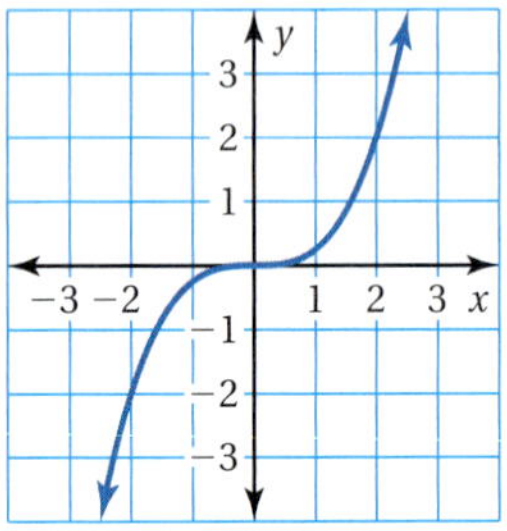

The graph is *not* a line.
So, the function is nonlinear.

b.

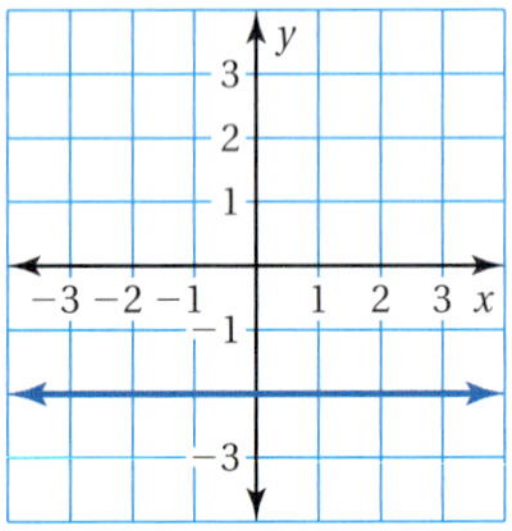

The graph is a line.
So, the function is linear.

On Your Own

Now You're Ready
Exercises 3–11

Does the table or graph represent a *linear* or *nonlinear* function? Explain.

1.

x	*y*
0	25
7	20
14	15
21	10

2.

x	*y*
2	8
4	4
6	0
8	−4

3.

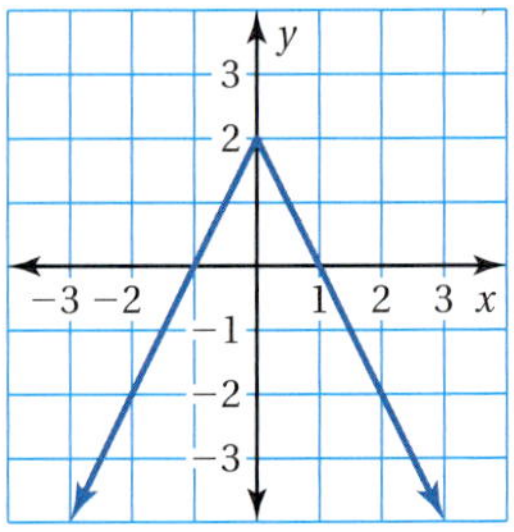

Laurie's Notes

Introduction

Connect

- **Yesterday:** Students explored the graphs of functions that were linear and nonlinear.
- **Today:** Students will compare linear and nonlinear functions.

Motivate

- Ask 5 students to complete a table of values, where the domain is the same for 5 functions.

	−3	−2	−1	0	1	2	3
$y = x + 2$	−1	0	1	2	3	4	5
$y = x - 2$	−5	−4	−3	−2	−1	0	1
$y = 2x$	−6	−4	−2	0	2	4	6
$y = \frac{x}{2}$	$-\frac{3}{2}$	−1	$-\frac{1}{2}$	0	$\frac{1}{2}$	1	$\frac{3}{2}$
$y = x^2$	9	4	1	0	1	4	9

- Spend time discussing the many patterns in the table. Discuss one function at a time. Ask students for their observations about patterns, changes in *y*-values, slope, and *y*-intercept.
- For $y = x^2$, students want it to have a constant slope. Draw a quick plot of the points and show it is not a linear function.

Lesson Notes

Example 1

- Copy the first table of values. Draw attention to the change in *x* (increasing by 3 each time) and the change in *y* (decreasing by 8 each time). Because the rate of change is constant, the function is linear.
- Copy the second table of values. Draw attention to the change in *x* (increasing by 2 each time) and the change in *y* (increasing by different amounts each time). This is a nonlinear function.

Example 2

- Part (b) may seem obvious, but the horizontal line seems like a special case to students. They may not be sure it is a linear function.

? "What is the slope of this line?" 0 "What is the constant rate of change?" Each time *x* increases by 1, *y* stays the same.

On Your Own

? "What are the constant rates of change for Questions 1 and 2?" $-\frac{5}{7}$; −2

? "Why is Question 3 not a linear function?" There are two parts of this function. The rate of change is positive, then negative.

Goal Today's lesson is comparing linear and **nonlinear functions**.

Start Thinking! and Warm Up

Lesson 4.4 **Warm Up** For use before Lesson 4.4

Lesson 4.4 **Start Thinking!** For use before Lesson 4.4

Describe two real-life situations at an amusement park: one that can be represented by a linear function and one that cannot.

Extra Example 1

Does the table represent a *linear* or *nonlinear* function? Explain.

x	3	4	5	6
y	1	2	3	4

linear; As *x* increases by 1, *y* increases by 1.

Extra Example 2

Does the graph represent a *linear* or *nonlinear* function? Explain.

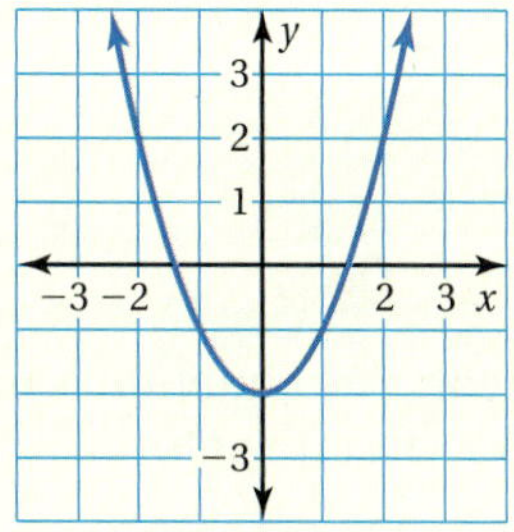

nonlinear; The graph is not a line.

On Your Own

1. linear; As *x* increases by 7, *y* decreases by 5.
2. linear; As *x* increases by 2, *y* decreases by 4.
3. nonlinear; The graph is not a line.

Extra Example 3

Does $y = 6x - 3$ represent a *linear* function? Yes, the equation is written in slope-intercept form.

Extra Example 4

Account A earns simple interest. Account B earns compound interest. The table shows the balances for 5 years. Graph the data and compare the graphs.

Year, t	Account A Balance	Account B Balance
0	\$50	\$50
1	\$55	\$55
2	\$60	\$60.50
3	\$65	\$66.55
4	\$70	\$73.21
5	\$75	\$80.53

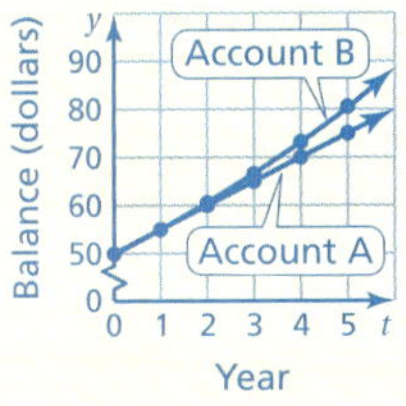

The function representing the balance of Account A is linear. The function representing the balance of Account B is nonlinear.

On Your Own

4. linear; The equation is in slope-intercept form.
5. linear; The equation can be written in slope-intercept form.
6. nonlinear; The equation cannot be written in slope-intercept form.

English Language Learners

Vocabulary

Begin the lesson by reviewing the terms *function* and *linear function.* Define *nonlinear function* and compare it to linear function.

Laurie's Notes

Example 3

- Discuss each equation. Remind students that all linear functions can be written in slope-intercept form.
- Students often see $y = \frac{4}{x}$ and $y = \frac{x}{4}$ as *the same kind of function.* So, many students think this will be a linear function. Remind students of how fractions are multiplied, and use the examples $\frac{4}{x}$ and $\frac{x}{4}$.

  $\frac{4}{x} = \frac{4}{1} \bullet \frac{1}{x} = 4 \bullet \frac{1}{x}$ $\qquad$ $\frac{x}{4} = \frac{x}{1} \bullet \frac{1}{4} = x \bullet \frac{1}{4} = \frac{1}{4} \bullet x$

  So, $y = \frac{x}{4}$ is linear with a slope of $\frac{1}{4}$. The equation $y = \frac{4}{x}$ cannot be written as a linear equation.
- **Note:** The equation $y = \frac{4}{x}$ shows inverse variation.

Example 4

- **Financial Literacy:** Ask a volunteer to read the problem. In addition to looking at linear and nonlinear functions, you also want to integrate financial literacy skills when appropriate.
- ? "Each time the year increases by 1, what happens to the balance of Account A?" It increases by \$10.
- ? "Each time the year increases by 1, what happens to the balance of Account B?" It increases by a greater amount each year.
- The graph of Account B's balance is starting to curve a bit, while the graph of Account A's balance is a line.
- **Extension:** Show students how to calculate the values in the table. Account A's balance is found using $I = Prt$, where $P = 100$, $r = 0.1$, and $t =$ year. Account B's balance can also be found using $I = Prt$, but the principal is changing each year. For the first few years, the calculations are:

Interest	Balance
$I = 100(0.1) = 10$	$100 + 10 = 110$
$I = 110(0.1) = 11$	$110 + 11 = 121$

On Your Own

- Students should see the exponent of 2 in Question 6 and quickly decide that the function is nonlinear.
- ? "What are the slopes of the equations in Questions 4 and 5?" 1; $\frac{4}{3}$

Closure

- **Exit Ticket:** Describe how to determine if a function is linear or nonlinear from (a) the equation, (b) a table of values, and (c) a graph.

Technology For the Teacher

Dynamic Classroom

The Dynamic Planning Tool
Editable Teacher's Resources at *BigIdeasMath.com*

EXAMPLE 3 Standardized Test Practice

Which equation represents a *nonlinear* function?

Ⓐ $y = 4.7$ Ⓑ $y = \pi x$

Ⓒ $y = \frac{4}{x}$ Ⓓ $y = 4(x - 1)$

The equations $y = 4.7$, $y = \pi x$, and $y = 4(x - 1)$ can be rewritten in slope-intercept form. So, they are linear functions.

The equation $y = \frac{4}{x}$ cannot be rewritten in slope-intercept form. So, it is a nonlinear function.

∴ The correct answer is Ⓒ.

EXAMPLE 4 Real-Life Application

Account A earns simple interest. Account B earns compound interest. The table shows the balances for 5 years. Graph the data and compare the graphs.

Remember

The simple interest formula is given by $I = Prt$.

- I is the simple interest
- P is the principal
- r is the annual interest rate
- t is the time in years

Year, t	Account A Balance	Account B Balance
0	\$100	\$100
1	\$110	\$110
2	\$120	\$121
3	\$130	\$133.10
4	\$140	\$146.41
5	\$150	\$161.05

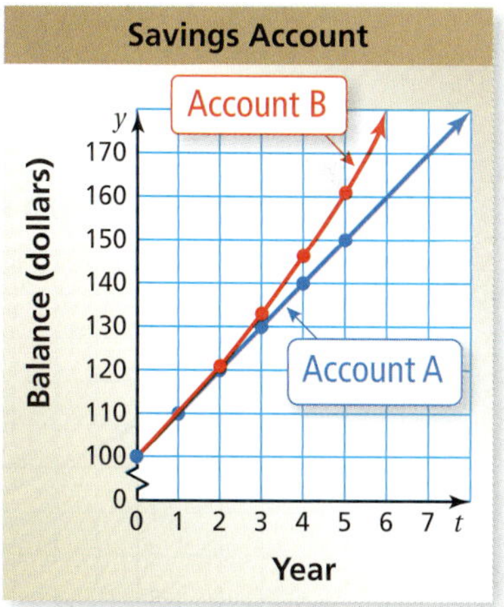

The balance of Account A has a constant rate of change of \$10. So, the function representing the balance of Account A is linear.

The balance of Account B increases by different amounts each year. Because the rate of change is not constant, the function representing the balance of Account B is nonlinear.

On Your Own

Now You're Ready
Exercises 12–14

Does the equation represent a *linear* or *nonlinear* function? Explain.

4. $y = x + 5$ **5.** $y = \frac{4x}{3}$ **6.** $y = 1 - x^2$

4.4 Exercises

Check It Out
Help with Homework
BigIdeasMath.com

Vocabulary and Concept Check

1. **VOCABULARY** Describe the difference between a linear function and a nonlinear function.

2. **WHICH ONE DOESN'T BELONG?** Which equation does *not* belong with the other three? Explain your reasoning.

$5y = 2x$ $\qquad y = \frac{2}{5}x$ $\qquad 10y = 4x$ $\qquad 5xy = 2$

Practice and Problem Solving

Graph the data in the table. Decide whether the function is *linear* or *nonlinear*.

1 3.

x	0	1	2	3
y	4	8	12	16

4.

x	1	2	3	4
y	1	2	6	24

5.

x	6	5	4	3
y	21	15	10	6

6.

x	−1	0	1	2
y	−7	−3	1	5

Does the table or graph represent a *linear* or *nonlinear* function? Explain.

2 7.

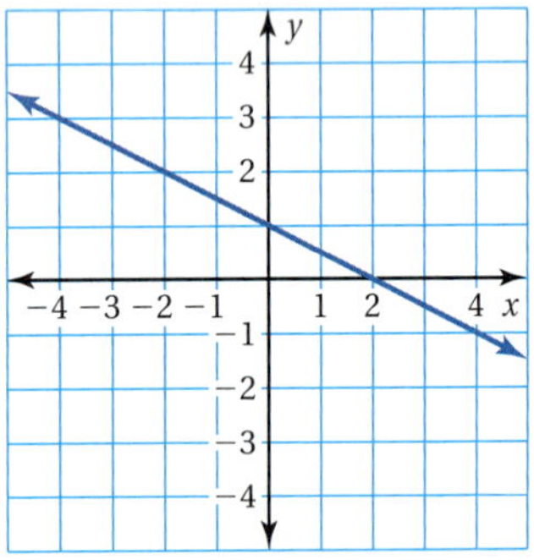

8.

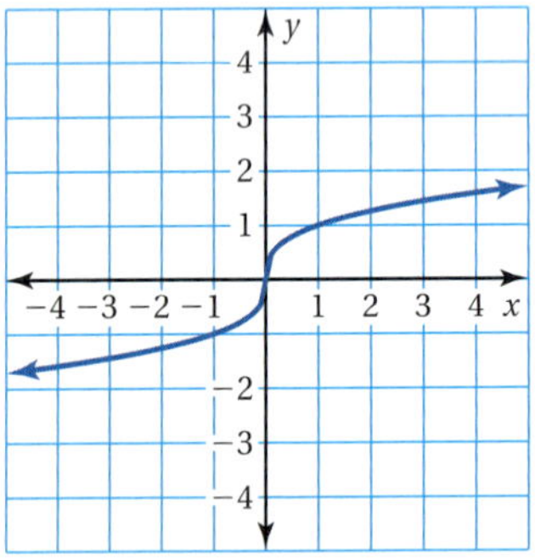

9.

x	5	11	17	23
y	7	11	15	19

10.

x	−3	−1	1	3
y	9	1	1	9

11. **VOLUME** The table shows the volume V (in cubic feet) of a cube with a side length of x feet. Does the table represent a linear or nonlinear function? Explain.

Side Length, x	1	2	3	4	5	6	7	8
Volume, V	1	8	27	64	125	216	343	512

Assignment Guide and Homework Check

Level	Day 1 Activity Assignment	Day 2 Lesson Assignment	Homework Check
Basic	3–6, 20–24	1, 2, 7–15	2, 8, 9, 14
Average	3–6, 20–24	1, 2, 7–10, 13–17	8, 9, 14, 16
Advanced	3–6, 20–24	1, 2, 8–14 even, 15–19	8, 14, 16, 17

Common Errors

- **Exercises 3–6, 9, and 10** Students may say that the function is linear because the *x*-values are increasing or decreasing by the same amount each time. Encourage them to examine the *y*-values to see if the graph represents a line.
- **Exercises 12–14** Students may not rewrite the equation in slope-intercept form and will guess if the equation is linear. Remind them to attempt to write the equation in slope-intercept form as a check.
- **Exercise 16** Students may try to graph the function to determine if it is linear and make an incorrect assumption depending upon how they scale their axes. Encourage them to examine the change in *y* for each *x*-value.

4.4 Record and Practice Journal

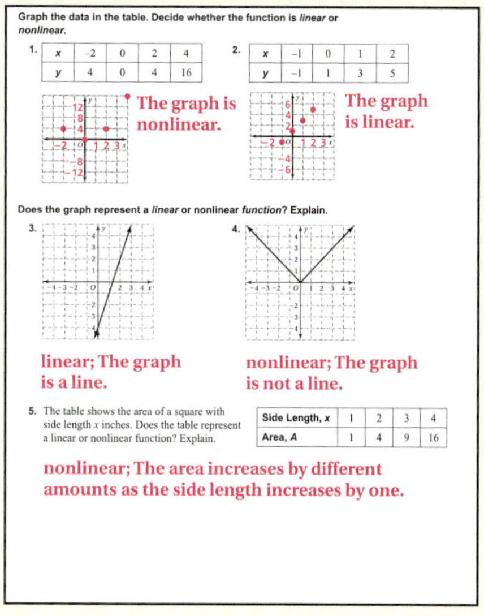
Graph the data in the table. Decide whether the function is *linear* or *nonlinear*.

1.

x	−2	0	2	4
y	4	0	4	16

The graph is nonlinear.

2.

x	−1	0	1	2
y	−1	1	3	5

The graph is linear.

Does the graph represent a *linear* or nonlinear *function*? Explain.

3. linear; The graph is a line.

4. nonlinear; The graph is not a line.

5. The table shows the area of a square with side length *x* inches. Does the table represent a linear or nonlinear function? Explain.

Side Length, *x*	1	2	3	4
Area, *A*	1	4	9	16

nonlinear; The area increases by different amounts as the side length increases by one.

Technology For the Teacher

Answer Presentation Tool

Quiz*Show*

Vocabulary and Concept Check

1. A linear function has a constant rate of change. A nonlinear function does not have a constant rate of change.
2. $5xy = 2$; It cannot be written in slope-intercept form.

Practice and Problem Solving

3. 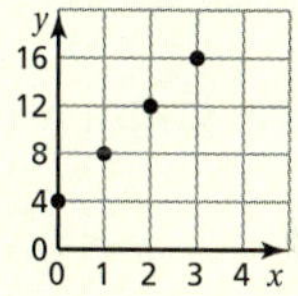
linear

4. 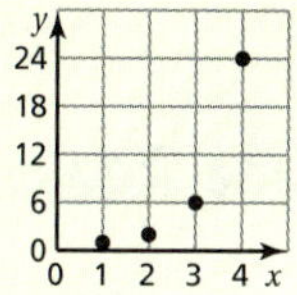
nonlinear

5. 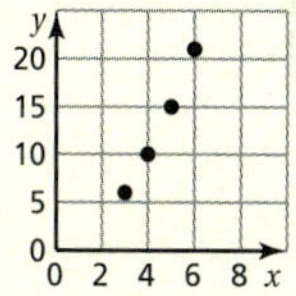
nonlinear

6. 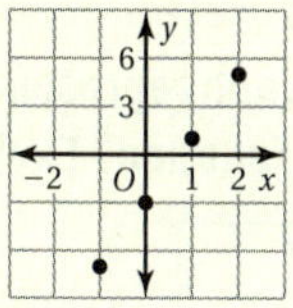
linear

7. linear; The graph is a line.
8. nonlinear; The graph is not a line.
9. linear; As *x* increases by 6, *y* increases by 4.
10. nonlinear; As *x* increases by 2, *y* changes by different amounts.
11. nonlinear; As *x* increases by 1, *V* increases by different amounts.

Practice and Problem Solving

12. linear; The equation can be written in slope-intercept form.
13. linear; The equation can be written in slope-intercept form.
14. nonlinear; The equation cannot be written in slope-intercept form.
15. See *Taking Math Deeper*.
16. nonlinear; As x decreases by 65, y increases by different amounts.
17. nonlinear; The graph is not a line.
18. nonlinear
19. linear

Fair Game Review

20. obtuse
21. straight
22. acute
23. right
24. B

Mini-Assessment

Does the table or graph represent a *linear* or *nonlinear* function? Explain.

1. 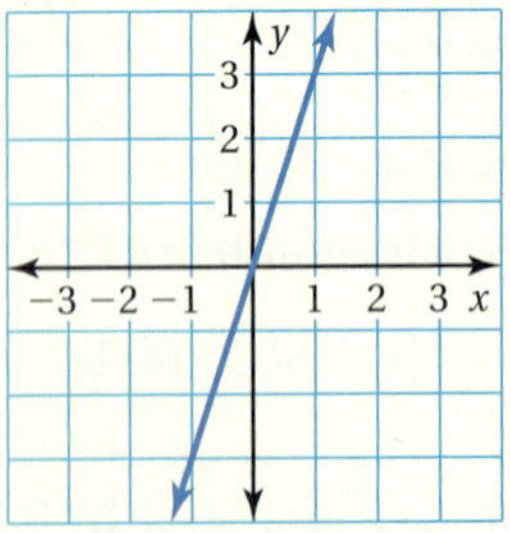

linear; The graph is a line.

2.

x	−2	0	2	4
y	8	0	8	64

nonlinear; The rate of change is not constant.

Taking Math Deeper

Exercise 15

Students can learn a valuable lesson about mathematics from this problem. Even though the problem does not specifically ask them to draw a graph, it is still a good idea. *Seeing* the relationship between pounds and cost is easier than simply finding the relationship using algebra.

1. Plot the two given points.

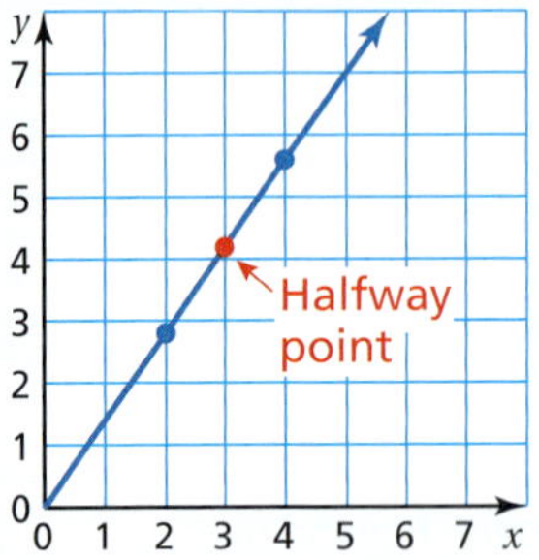

2. Find the halfway point.

Because you want the table to represent a linear function and 3 is halfway between 2 and 4, you need to find the number that is halfway between \$2.80 and \$5.60. This number is the mean of \$2.80 and \$5.60.

a. $\text{Mean} = \dfrac{2.80 + 5.60}{2} = 4.20$

3. Write a function.

Let x = pounds of seeds.

Let y = cost.

y-intercept = 0

$\text{slope} = \dfrac{5.60 - 2.80}{4 - 2} = 1.4$

b. $y = 1.4x$

Project

Plant some sunflower seeds. Keep track of the progress of the plants until they bloom.

Reteaching and Enrichment Strategies

If students need help. . .	If students got it. . .
Resources by Chapter • Practice A and Practice B • Puzzle Time Record and Practice Journal Practice Differentiating the Lesson Lesson Tutorials Skills Review Handbook	Resources by Chapter • Enrichment and Extension • School-to-Work Start the next section

Does the equation represent a *linear* or *nonlinear* function? Explain.

③ **12.** $2x + 3y = 7$

13. $y + x = 4x + 5$

14. $y = \frac{8}{x^2}$

15. SUNFLOWER SEEDS The table shows the cost y (in dollars) of x pounds of sunflower seeds.

Pounds, x	Cost, y
2	2.80
3	?
4	5.60

a. What is the missing y-value that makes the table represent a linear function?

b. Write a linear function that represents the cost y of x pounds of seeds.

16. LIGHT The frequency y (in terahertz) of a light wave is a function of its wavelength x (in nanometers). Does the table represent a linear or nonlinear function? Explain.

Color	Red	Yellow	Green	Blue	Violet
Wavelength, x	660	595	530	465	400
Frequency, y	454	504	566	645	749

17. LIGHTHOUSES The table shows the heights x (in feet) of four Florida lighthouses and the number y of steps in each. Does the table represent a linear or nonlinear function? Explain.

Lighthouse	Height, x	Steps, y
Ponce de Leon Inlet	175	213
St. Augustine	167	219
Cape Canaveral	145	179
Key West	86	98

18. PROJECT The wooden bars of a xylophone produce different musical notes when struck. The pitch of a note is determined by the length of the bar. Use the Internet or some other reference to decide whether the pitch of a note is a linear function of the length of the bar.

19. *Geometry* The radius of the base of a cylinder is 3 feet. Is the volume of the cylinder a linear or nonlinear function of the height of the cylinder?

Fair Game Review What you learned in previous grades & lessons

Classify the angle as *acute, obtuse, right,* or *straight.* *(Skills Review Handbook)*

20.

21.

22.

23.

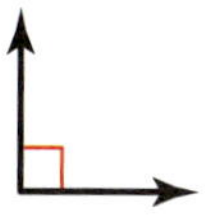

24. MULTIPLE CHOICE What is the value of x? *(Section 1.1)*

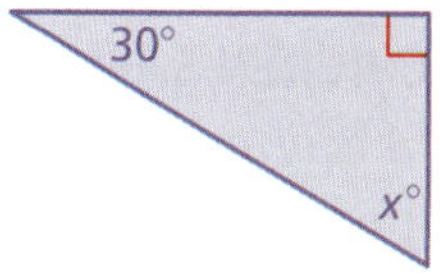

Ⓐ 30 Ⓑ 60 Ⓒ 90 Ⓓ 180

4.3–4.4 Quiz

Use the graph or table to write a linear function that relates *y* to *x*. *(Section 4.3)*

1. 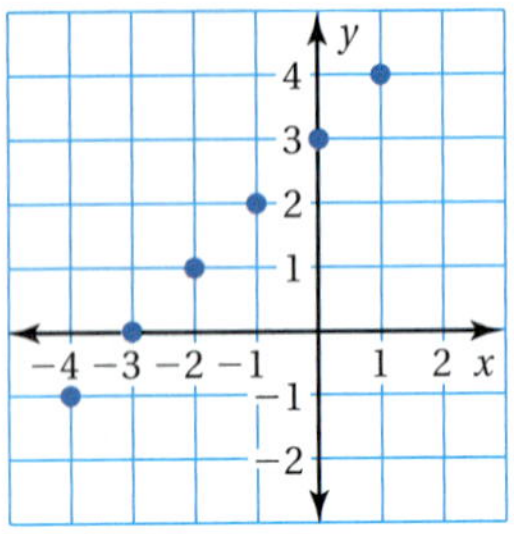

2. 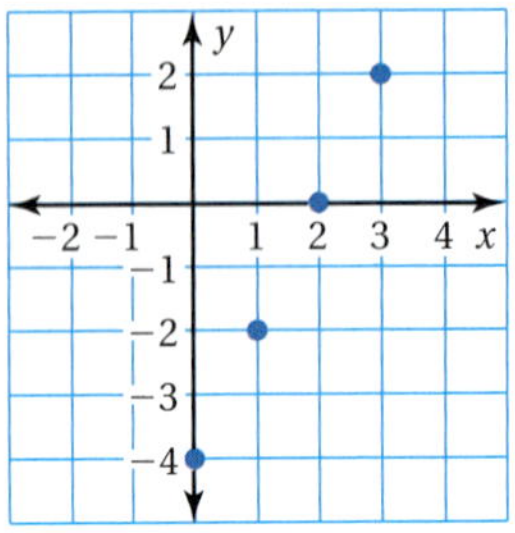

3. 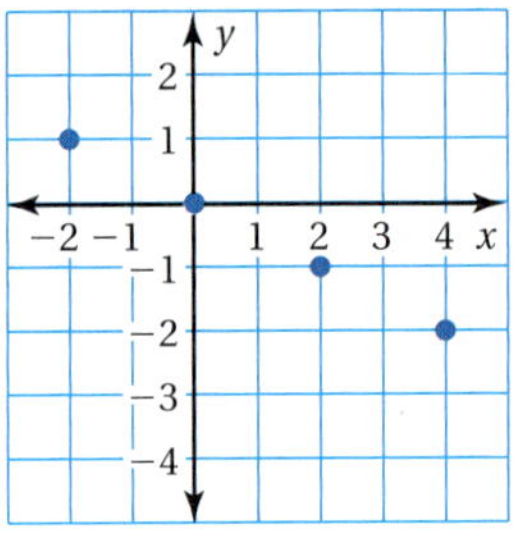

4.

x	0	1	2	3
y	2	1	0	−1

5.

x	−3	0	3	6
y	−3	−1	1	3

Does the table or graph represent a *linear* or *nonlinear* function? Explain. *(Section 4.4)*

6. 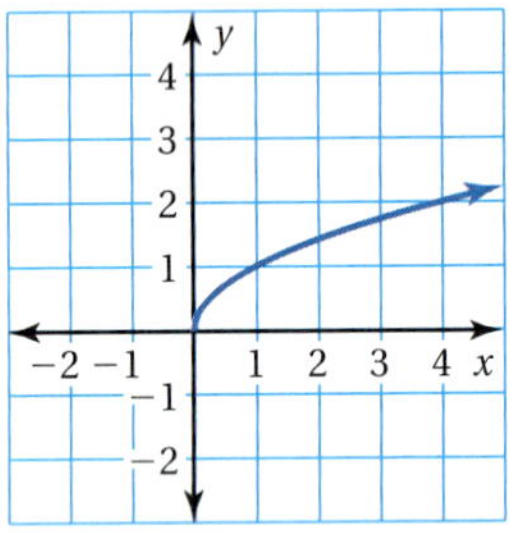

7. 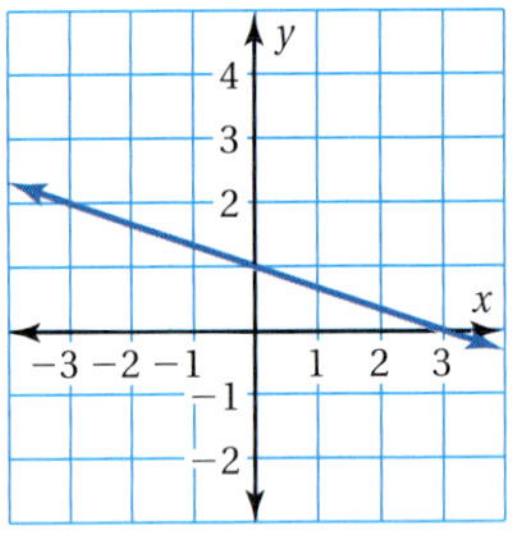

8. 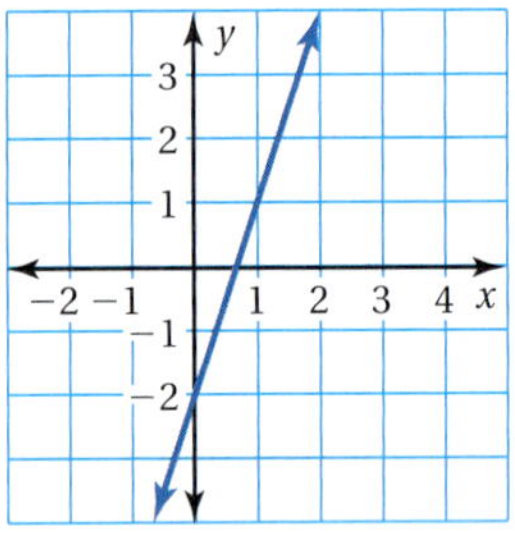

9.

x	y
0	0
2	−2
4	−4
6	−6

10.

x	y
1	−2
3	7
5	23
7	47

11.

x	y
0	3
3	0
6	3
9	6

12. ADVERTISING The table shows the revenue R (in millions of dollars) of a company when it spends A (in millions of dollars) on advertising. *(Section 4.3)*

Advertising, A	Revenue, R
0	2
2	6
4	10
6	14
8	18

a. Write a linear function that relates the revenue to the advertising cost.

b. What is the revenue of the company when it spends $10 million on advertising?

13. CHICKEN SALAD The equation $y = 7.9x$ represents the cost y (in dollars) of buying x pounds of chicken salad. Does this equation represent a linear or nonlinear function? Explain. *(Section 4.4)*

Alternative Assessment Options

Math Chat
Structured Interview
Student Reflective Focus Question
Writing Prompt

Writing Prompt

Ask students to write two different stories. One story should involve data whose domain is continuous and the other story should involve data whose domain is discrete. Both sets of data should be linear. Students should graph their data and write linear functions that relate *y* to *x*. They should include a summary in each story describing how they know whether the domains are discrete or continuous. Students can share their stories and summaries with the class.

Study Help Sample Answers

Remind students to complete Graphic Organizers for the rest of the chapter.

2.

	Linear Functions with Positive Slopes	Linear Functions with Negative Slopes
Algebra slope-intercept form: $y = mx + b$	m is positive	m is negative
Description	Graph is a line that rises from left to right (as x increases, y increases).	Graph is a line that falls from left to right (as x increases, y decreases).
Equations In slope-intercept form	$y = \frac{1}{2}x - 3$ $y = 3x$	$y = -\frac{1}{2}x - 3$ $y = -x + 1$
Not in slope-intercept form	$x - 3y = 0$ $x + 3 = y - 3$	$3x + 2y = 10$ $2x = 8 - \frac{2}{3}y$
Table	x: 0, 1, 2, 3 y: 0, 3, 6, 9	x: 0, 1, 2, 3 y: 0, −2, −4, −6
Graph	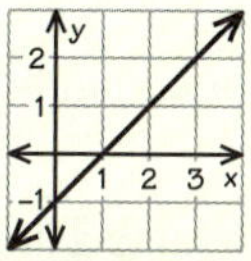	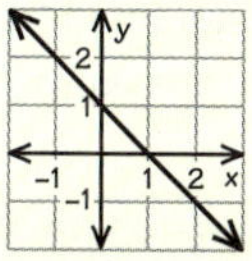

3. Available at *BigIdeasMath.com.*

Reteaching and Enrichment Strategies

If students need help. . .	If students got it. . .
Resources by Chapter • Study Help • Practice A and Practice B • Puzzle Time Lesson Tutorials *BigIdeasMath.com* Practice Quiz Practice from the Test Generator	Resources by Chapter • Enrichment and Extension • School-to-Work Game Closet at *BigIdeasMath.com* Start the Chapter Review

Answers

1. $y = x + 3$
2. $y = 2x - 4$
3. $y = -\frac{1}{2}x$
4. $y = -x + 2$
5. $y = \frac{2}{3}x - 1$
6. nonlinear; The graph is not a line.
7. linear; The graph is a line.
8. linear; The graph is a line.
9. linear; As x increases by 2, y decreases by 2.
10. nonlinear; As x increases by 2, y increases by different amounts.
11. nonlinear; As x increases by 3, y changes by different amounts.
12. **a.** $R = 2A + 2$
    b. $22 million
13. linear; The equation is in slope-intercept form.

Technology For the Teacher
Answer Presentation Tool

Assessment Book

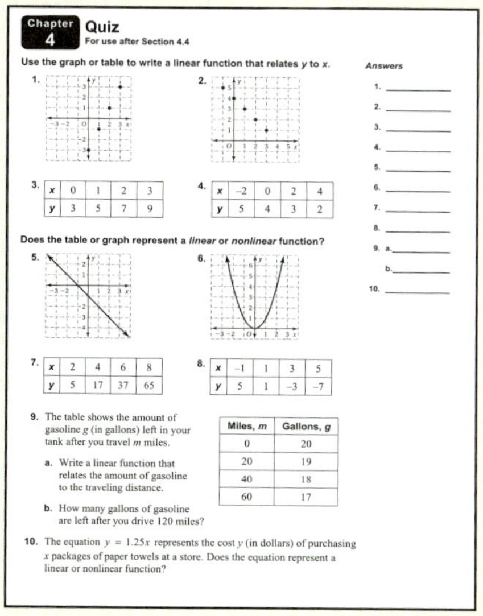
Chapter 4 Quiz
For use after Section 4.4

Use the graph or table to write a linear function that relates *y* to *x*.

1.
2.

3.

x	0	1	2	3
y	3	5	7	9

4.

x	−2	0	2	4
y	5	4	3	2

Does the table or graph represent a *linear* or *nonlinear* function?

5.
6.

7.

x	2	4	6	8
y	5	17	37	65

8.

x	−1	1	3	5
y	5	1	−3	−7

9. The table shows the amount of gasoline g (in gallons) left in your tank after you travel m miles.

Miles, m	Gallons, g
0	20
20	19
40	18
60	17

a. Write a linear function that relates the amount of gasoline to the traveling distance.
b. How many gallons of gasoline are left after you drive 120 miles?

10. The equation $y = 1.25x$ represents the cost y (in dollars) of purchasing x packages of paper towels at a store. Does the equation represent a linear or nonlinear function?

Answers
1. ______
2. ______
3. ______
4. ______
5. ______
6. ______
7. ______
8. ______
9. a. ______
b. ______
10. ______

For the Teacher
Additional Review Options

- **Quiz***Show*
- Big Ideas Test Generator
- Game Closet at *BigIdeasMath.com*
- Vocabulary Puzzle Builder
- Resources by Chapter
  Puzzle Time
  Study Help

Answers

1. domain: −5, −4, −3, −2
   range: 3, 1, −1, −3

2. domain: −2, −1, 0, 1, 2
   range: −4, −2, 0

3.

x	*y*
−1	−4
0	−1
1	2
2	5

domain: −1, 0, 1, 2
range: −4, −1, 2, 5

4.

x	*y*
0	2
1	−2
2	−6
3	−10

domain: 0, 1, 2, 3
range: 2, −2, −6, −10

Review of Common Errors

Exercises 1 and 2

- Students may confuse the domain and range. Remind them that the domain is the set of all possible input values (the *x*-coordinates) and the range is the set of all possible output values (the *y*-coordinates).

Exercises 3 and 4

- Students may make errors when finding or solving for *y*. Remind them how to use the order of operations, and (in Exercise 4) how to solve for *y*.

Exercises 5 and 6

- Students may see decimal numbers in a table and think that the function is continuous, or see whole numbers in a table and think that the function is discrete. Encourage them to think about the context of the problem. Point out that you can drive for part of an hour or part of a mile, but you cannot buy part of a stamp.
- Students may graph the function incorrectly. Remind them that discrete data points are not connected, but continuous data points are.

Exercises 7–9

- Students may try to write the linear function without first finding the slope and *y*-intercept, or they may use the reciprocal of the slope in their function. Encourage them to check their work by making sure that all of the given points are solutions.

Exercise 10

- Students may notice that all of the values of *y* are the same and not be able to write the linear function. Encourage them to plot the points, and if necessary, remind them how to write the equation of a horizontal line.

Exercises 11 and 12

- Students may guess their answer, or they may think that because the *x*-values are increasing by the same amount, the function is linear. Encourage them to examine the *y*-values or to plot the given points so that they can tell if the table represents a linear or nonlinear function.

Exercise 13

- Students may think that a graph represents a linear function if part(s) of the graph appear to be straight. Point out that the graph of a linear function is not curved and does not change direction.

4 Chapter Review

Check It Out
Vocabulary Help
BigIdeasMath.com

Review Key Vocabulary

function, *p. 150*
domain, *p. 150*
range, *p. 150*
function form, *p. 150*
discrete domain, *p. 156*
continuous domain, *p. 156*
linear function, *p. 164*
nonlinear function, *p. 170*

Review Examples and Exercises

4.1 Domain and Range of a Function *(pp. 148–153)*

Find the domain and range of the function represented by the graph.

Write the ordered pairs. Identify the inputs and outputs.

inputs

$(-2, -3), (0, -1), (2, 1), (4, 3)$

outputs

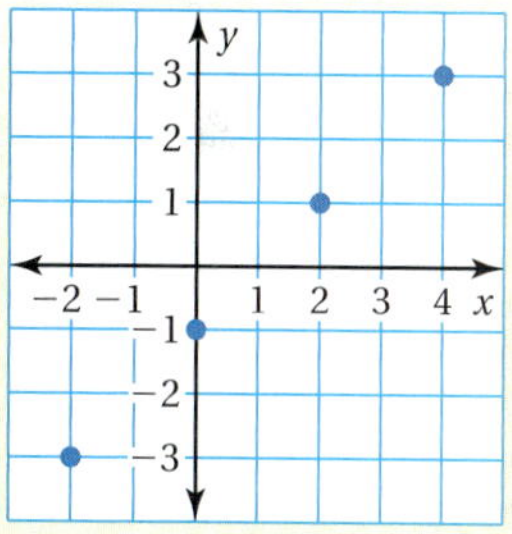

The domain is -2, 0, 2, and 4. The range is -3, -1, 1, and 3.

Exercises

Find the domain and range of the function represented by the graph.

1.

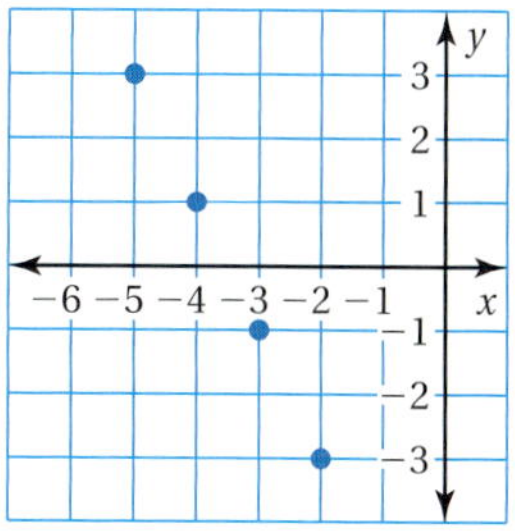

2.

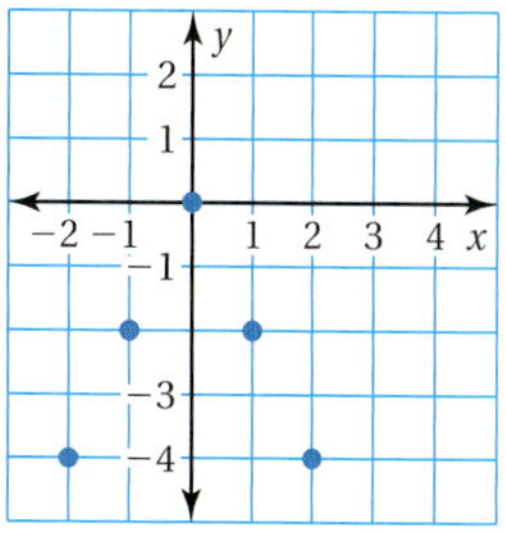

Copy and complete the input-output table for the function. Then find the domain and range of the function represented by the table.

3. $y = 3x - 1$

x	*y*
−1	
0	
1	
2	

4. $4x + y = 2$

x	*y*
0	
1	
2	
3	

4.2 Discrete and Continuous Domains (pp. 154–159)

The function $y = 19.5x$ represents the cost y (in dollars) of x yearbooks. Graph the function. Is the domain of the graph discrete or continuous?

Make an input-output table.

Input, x	$19.5x$	Output, y	Ordered Pair, (x, y)
0	19.5(0)	0	(0, 0)
1	19.5(1)	19.5	(1, 19.5)
2	19.5(2)	39	(2, 39)
3	19.5(3)	58.5	(3, 58.5)
4	19.5(4)	78	(4, 78)

Plot the ordered pairs.

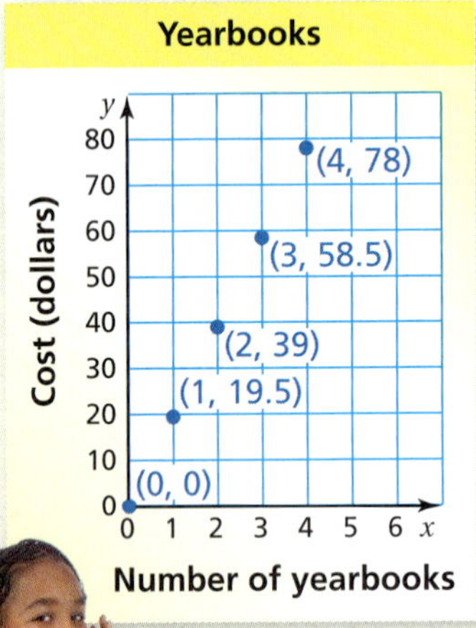

Because you cannot buy part of a yearbook, the graph consists of individual points.

So, the domain is discrete.

Exercises

Graph the function. Is the domain of the graph discrete or continuous?

5.

Hours, x	Miles, y
0	0
1	4
2	8
3	12
4	16

6.

Stamps, x	Cost, y
20	8.4
40	16.8
60	25.2
80	33.6
100	42

4.3 Linear Function Patterns (pp. 162–167)

Use the graph to write a linear function that relates y to x.

The points lie on a line. Find the slope and y-intercept of the line.

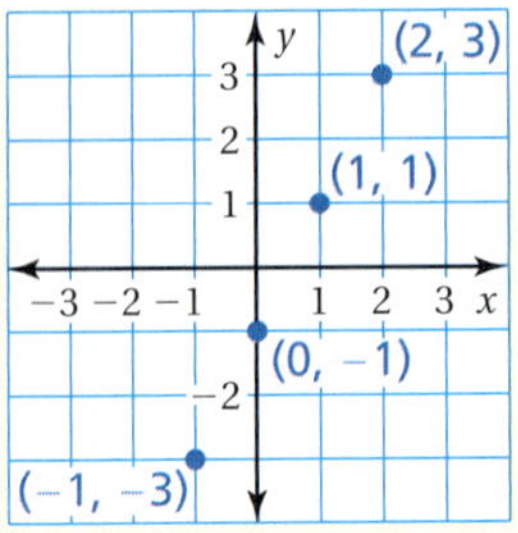

$$\text{slope} = \frac{\text{rise}}{\text{run}} = \frac{2}{1} = 2$$

Because the line crosses the y-axis at $(0, -1)$, the y-intercept is -1.

So, the linear function is $y = 2x - 1$.

Review Game

Writing Linear Functions

Big Ideas Game Closet

Materials per Group:

- paper
- two yard sticks
- pencils

Directions:

- Divide the class into an even number of groups.
- Groups pair up to compete against each other.
- Each pair of groups makes a paper football.
- Students in each pair of groups take turns flicking the football with their fingers as high and as far as they can. Students from the other group measure and record the length and height that the football travels. Both groups have to agree on the measurements. Length can be measured after the football has come to rest, but height must be measured while it is moving.
- Each student writes the domain and range of both the ascent and descent of the football when they took their turn. The domain is the length traveled and the range is the height traveled. (See figure below.) Students write one linear function to approximate the ascent and another linear function to approximate the descent.

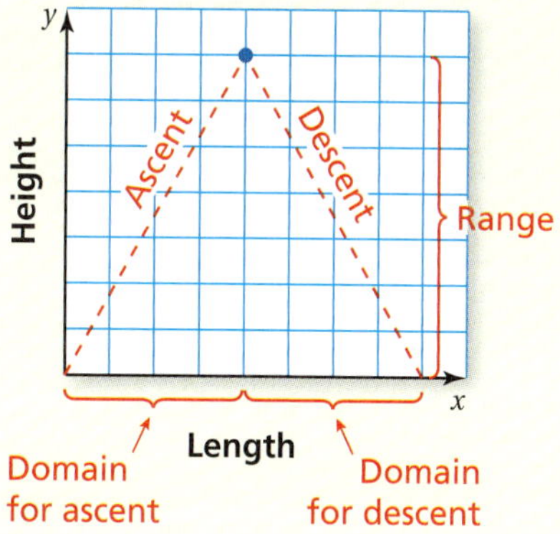

Who Wins?

Each student earns their group a point for each inch achieved in length and height. Points only count if the linear functions correctly model the motion of the football and the domains and ranges are clearly identified. The group with the most points wins.

For the Student
Additional Practice

- Lesson Tutorials
- Study Help (textbook)
- Student Website
  Multi-Language Glossary
  Practice Assessments

Answers

5.

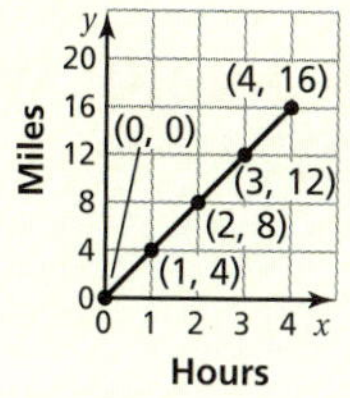

continuous

6.

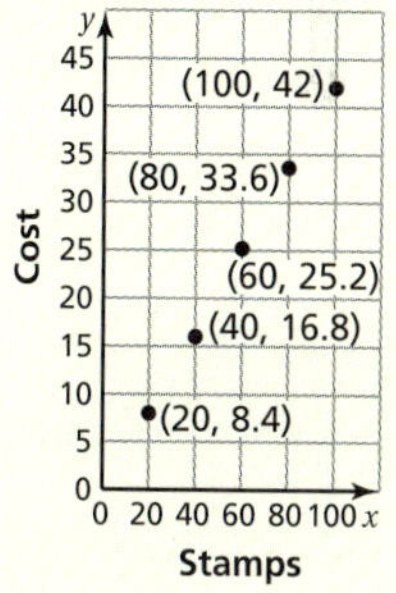

discrete

7. $y = -x - 2$

8. $y = \frac{1}{3}x + 3$

9. $y = 3x + 1$

10. $y = -7$

11. linear, As x increases by 3, y increases by 9.

12. nonlinear; As x increases by 2, y changes by different amounts.

13. nonlinear; The graph is not a line.

My Thoughts on the Chapter

Teacher Tip

Not allowed to write in your teaching edition? Use sticky notes to record your thoughts.

What worked. . .

What did not work. . .

What I would do differently. . .

Exercises

Use the graph or table to write a linear function that relates *y* to *x*.

7.

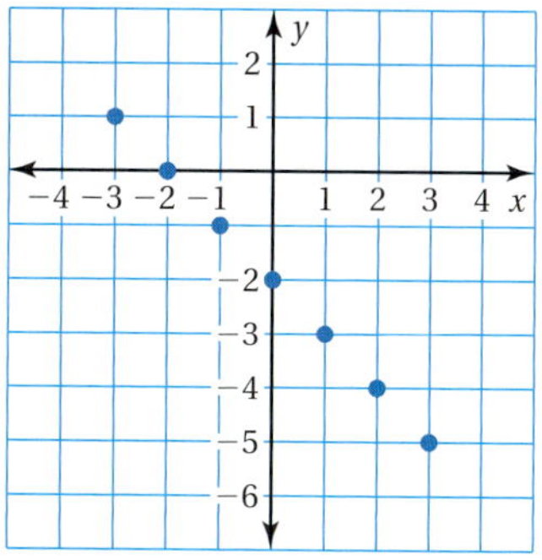

8.

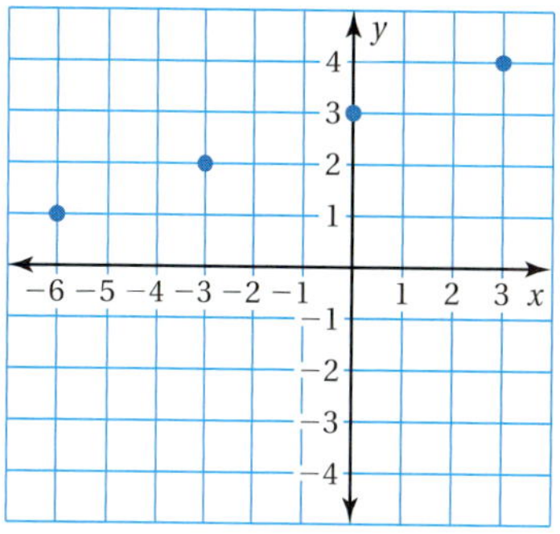

9.

x	−2	−1	0	1
y	−5	−2	1	4

10.

x	−2	0	2	4
y	−7	−7	−7	−7

4.4 Comparing Linear and Nonlinear Functions (pp. 168–173)

Does the table represent a *linear* or *nonlinear* function? Explain.

a.

+2 +2 +2

x	0	2	4	6
y	0	1	4	9

+1 +3 +5

As x increases by 2, y increases by different amounts. The rate of change is *not* constant. So, the function is nonlinear.

b.

x	y
0	50
5	40
10	30
15	20

x: +5, +5, +5 ; y: −10, −10, −10

As x increases by 5, y decreases by 10. The rate of change is constant. So, the function is linear.

Exercises

Does the table or graph represent a *linear* or *nonlinear* function? Explain.

11.

x	y
3	1
6	10
9	19
12	28

12.

x	y
1	3
3	1
5	1
7	3

13.

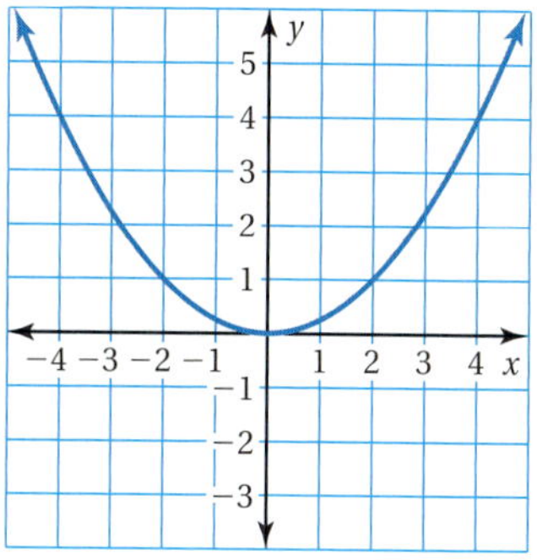

4 Chapter Test

Check It Out
Test Practice
BigIdeasMath.com

1. Find the domain and range of the function represented by the graph.

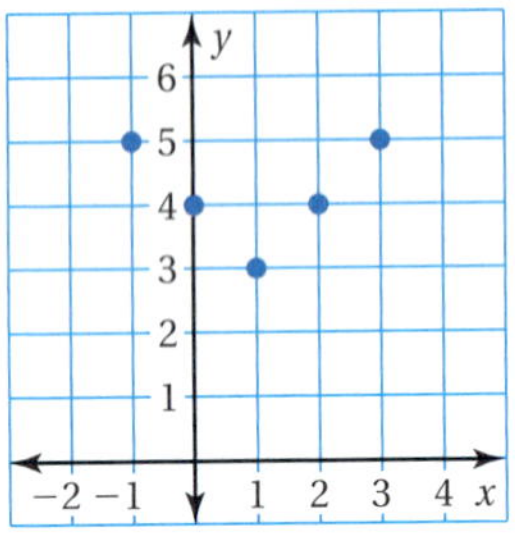

2. Copy and complete the input-output table for the function $y = 7x - 3$. Then find the domain and range of the function represented by the table.

x	−1	0	1	2
y				

Graph the function. Is the domain of the graph discrete or continuous?

3.

Hair Clips, *x*	Cost, *y*
0	0
1	1.5
2	3
3	4.5

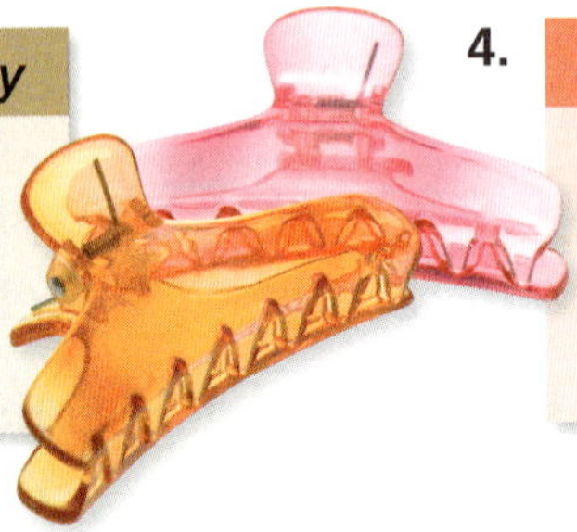

4.

Minutes, *x*	Gallons, *y*
0	60
5	45
10	30
15	15

5. Write a linear function that relates *y* to *x*.

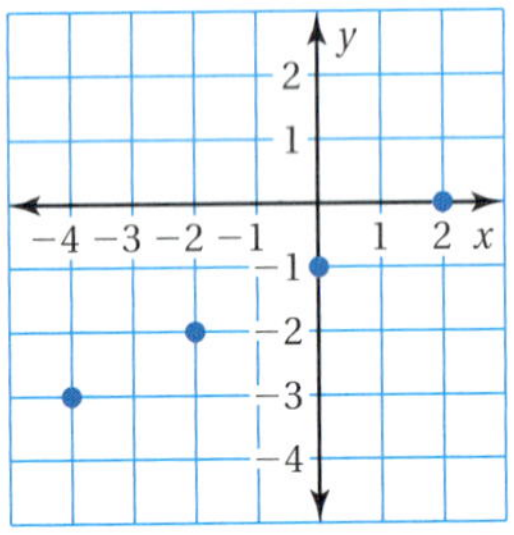

6. Does the table represent a *linear* or *nonlinear* function? Explain.

x	0	2	4	6
y	8	0	−8	−16

7. **SAVINGS** You save 15% of your monthly earnings *x* (in dollars).

   a. Write an equation in function form that represents the amount *y* (in dollars) you save each month.

   b. Create an input-output table for the equation in part (a). Use the inputs 25, 30, 35, and 40.

   c. What is the total amount saved during those 4 months?

8. **FOOD DRIVE** You are putting cans of food into boxes for a food drive. One box holds 30 cans of food. Write a linear function that represents the number *y* of cans of food that will fit in *x* boxes.

9. **SURFACE AREA** The table shows the surface area *S* (in square inches) of a cube with a side length of *x* feet. Does the table represent a linear or nonlinear function? Explain.

Side Length, *x*	1	2	3	4
Surface Area, *S*	6	24	54	96

Test Item References

Chapter Test Questions	Section to Review
1, 2, 7	4.1
3, 4	4.2
5, 8	4.3
6, 9	4.4

Test-Taking Strategies

Remind students to quickly look over the entire test before they start so that they can budget their time. This test involves analyzing pairs of concepts that students can easily confuse, such as domain and range, input and output, rise and run, discrete and continuous, and linear and nonlinear. So, it is important that students use the **Stop** and **Think** strategy before they answer a question.

Common Assessment Errors

- **Exercise 1** Students may confuse the domain and range. Remind them that the domain is the set of all possible input values (the x-coordinates) and the range is the set of all possible output values (the y-coordinates).
- **Exercise 2** Students may make order of operations errors when finding y. Remind them how to use the order of operations.
- **Exercises 3 and 4** Students may see decimal numbers in a table and think that the function is continuous, or see whole numbers in a table and think that the function is discrete. Encourage them to think about the context of the problem. For example, you cannot buy part of a hair clip, so the domain of the function in Exercise 3 is discrete.
- **Exercises 3 and 4** Students may graph the function incorrectly. Remind them that discrete data points are not connected, but continuous data points are.
- **Exercise 5** Students may try to write the linear function without first finding the slope and y-intercept, or they may use the reciprocal of the slope in their function. Encourage them to check their work by making sure that all of the given points are solutions.
- **Exercise 6** Students may guess their answer without providing an explanation. Encourage them to examine the y-values or to plot the given points so that they can tell if the table represents a linear or nonlinear function.

Reteaching and Enrichment Strategies

If students need help. . .	If students got it. . .
Resources by Chapter • Practice A and Practice B • Puzzle Time Record and Practice Journal Practice Differentiating the Lesson Lesson Tutorials Practice from the Test Generator Skills Review Handbook	Resources by Chapter • Enrichment and Extension • School-to-Work Game Closet at *BigIdeasMath.com* Start Standardized Test Practice

Answers

1. domain: $-1, 0, 1, 2, 3$
   range: $5, 4, 3$

2.

x	-1	0	1	2
y	-10	-3	4	11

domain: $-1, 0, 1, 2$
range: $-10, -3, 4, 11$

3–4. See Additional Answers.

5. $y = \frac{1}{2}x - 1$

6. linear; As x increases by 2, y decreases by 8.

7. a. $y = 0.15x$

   b.

x	25	30	35	40
y	3.75	4.5	5.25	6

   c. \$19.50

8. $y = 30x$

9. nonlinear; As x increases by 1, S increases by different amounts.

Assessment Book

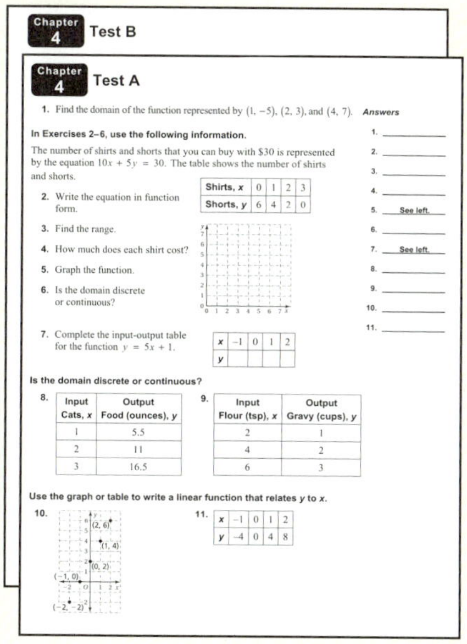
Chapter 4 Test B

Chapter 4 Test A

1. Find the domain of the function represented by $(1, -5)$, $(2, 3)$, and $(4, 7)$.

Answers
1. ______
2. ______
3. ______
4. ______
5. See left.
6. ______
7. See left.
8. ______
9. ______
10. ______
11. ______

In Exercises 2–6, use the following information.

The number of shirts and shorts that you can buy with \$30 is represented by the equation $10x + 5y = 30$. The table shows the number of shirts and shorts.

Shirts, x	0	1	2	3
Shorts, y	6	4	2	0

2. Write the equation in function form.
3. Find the range.
4. How much does each shirt cost?
5. Graph the function.
6. Is the domain discrete or continuous?
7. Complete the input-output table for the function $y = 5x + 1$.

x	-1	0	1	2
y				

Is the domain discrete or continuous?

8.

Input Cats, x	Output Food (ounces), y
1	5.5
2	11
3	16.5

9.

Input Flour (tsp), x	Output Gravy (cups), y
2	1
4	2
6	3

Use the graph or table to write a linear function that relates y to x.

10. (graph with points $(2, 6)$, $(1, 4)$, $(0, 2)$, $(-1, 0)$, $(-2, -2)$)

11.

x	-1	0	1	2
y	-4	0	4	8

Test-Taking Strategies

Available at *BigIdeasMath.com*

After Answering Easy Questions, Relax
Answer Easy Questions First
Estimate the Answer
Read All Choices before Answering
Read Question before Answering
Solve Directly or Eliminate Choices
Solve Problem before Looking at Choices
Use Intelligent Guessing
Work Backwards

About this Strategy

When taking a multiple choice test, be sure to read each question carefully and thoroughly. Look closely for words that change the meaning of the question like not, never, all, every, and always.

Answers

1. D
2. F
3. \$200
4. B

Technology For the Teacher

Big Ideas Test Generator

Item Analysis

1. **A.** The student thinks that reversing the domain gets you the range.

   B. The student takes 5 away from each domain element, ignoring the coefficient 0.2.

   C. The student performs an arithmetic error subtracting integers.

   D. Correct answer

2. **F.** Correct answer

   G. The student picks the number 1 from the formula.

   H. The student uses the numbers in the formula: $1 - 0.25 = 0.75$.

   I. The student picks the number 0.25 from the formula.

3. **Gridded Response:** Correct answer: \$200

   Common Error: The student divides \$500 by 5, or \$800 by 10.

4. **A.** The student thinks that a line with a negative slope is nonlinear.

   B. Correct answer

   C. The student thinks that a horizontal line is nonlinear.

   D. The student thinks that a steep line with a positive slope is nonlinear.

5. **F.** The student associates 0 with a solution.

   G. The student selects this choice because the first equation equals 0 when $x = 4$.

   H. Correct answer

   I. The student has no idea how to interpret the table, or fails to see that the functions are equal at $x = 8$.

6. **Gridded Response:** Correct answer: 3°F

   Common Error: The student subtracts 36 from 54, but fails to divide the result by 6.

7. **A.** The student only includes the domain values greater than or equal to 0.

   B. The student finds the range.

   C. Correct answer

   D. The student puts the domain and range together.

4 Standardized Test Practice

Test-Taking Strategy
Read Question Before Answering

"Take your time and read the question carefully before choosing your answer."

1. The domain of the function $y = 0.2x - 5$ is 5, 10, 15, 20. What is the range of this function?

   A. 20, 15, 10, 5 **C.** 4, 3, 2, 1

   B. 0, 5, 10, 15 **D.** $-4, -3, -2, -1$

2. A toy runs on a rechargeable battery. During use, the battery loses power at a constant rate. The percent P of total power left in the battery x hours after being fully charged, can be found using the equation shown below. When will the battery be fully discharged?

$$P = -0.25x + 1$$

   F. After 4 hours of use

   G. After 1 hour of use

   H. After 0.75 hour of use

   I. After 0.25 hour of use

3. A limousine company charges a fixed cost for a limousine and an hourly rate for its driver. It costs \$500 to rent the limousine for 5 hours and \$800 to rent the limousine for 10 hours. What is the fixed cost, in dollars, to rent the limousine?

4. Which graph shows a nonlinear function?

A.

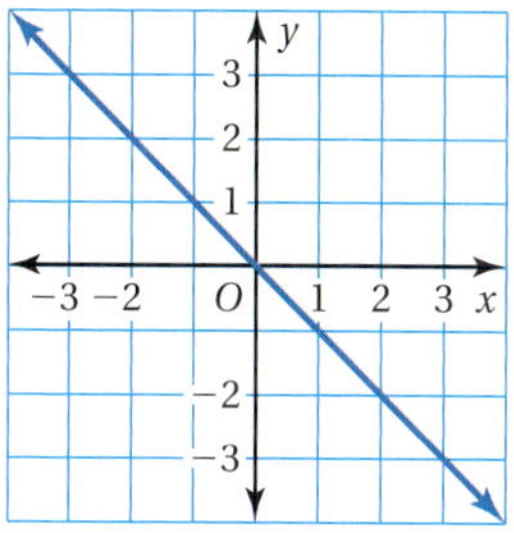

C.

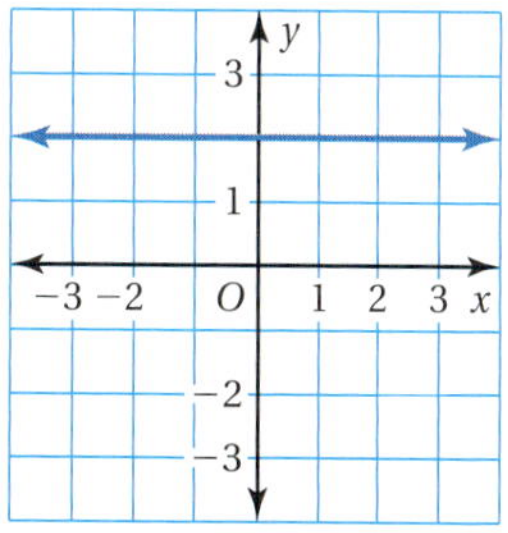

B.

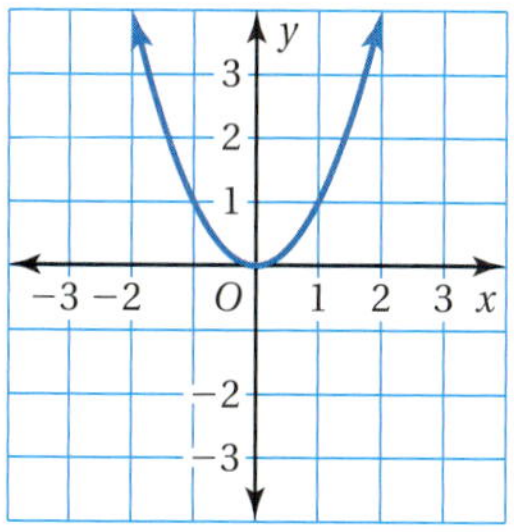

D.

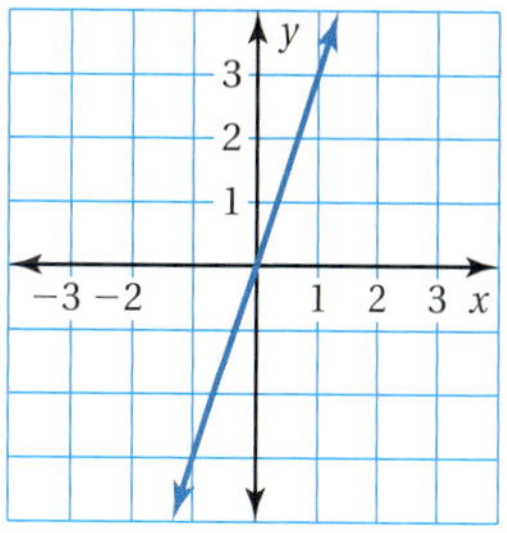

5. The equations $y = -x + 4$ and $y = \frac{1}{2}x - 8$ form a system of linear equations. The table below shows the (x, y) values for these equations at six different values of x.

x	0	2	4	6	8	10
$y = -x + 4$	4	2	0	−2	−4	−6
$y = \frac{1}{2}x - 8$	−8	−7	−6	−5	−4	−3

What can you conclude from the table?

F. The system has one solution, when $x = 0$.

G. The system has one solution, when $x = 4$.

H. The system has one solution, when $x = 8$.

I. The system has no solution.

6. The temperature fell from 54 degrees Fahrenheit to 36 degrees Fahrenheit over a six-hour period. The temperature fell by the same number of degrees each hour. How many degrees Fahrenheit did the temperature fall each hour?

7. What is the domain of the function graphed in the coordinate plane below?

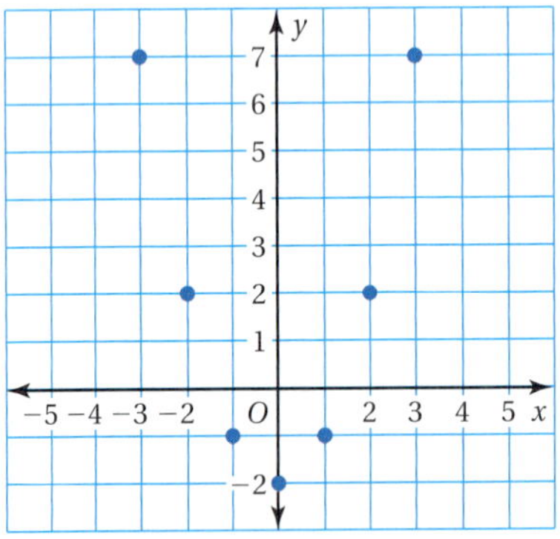

A. 0, 1, 2, 3

B. −2, −1, 2, 7

C. −3, −2, −1, 0, 1, 2, 3

D. −2, −1, 0, 1, 2, 3, 7

8. What value of w makes the equation below true?

$$\frac{w}{3} = 3(w - 1) - 1$$

F. $\frac{3}{2}$

G. $\frac{5}{4}$

H. $\frac{3}{4}$

I. $\frac{1}{2}$

Item Analysis (continued)

8. **F.** Correct answer

   G. The student multiplies both sides by 3, but fails to distribute the 3 correctly on the right side.

   H. The student starts by distributing the 3 on the right side incorrectly.

   I. The student multiplies both sides by 3, distributes the 3 correctly, but does not distribute the 9 correctly. Alternatively, the student starts by distributing the 3 on the right side correctly. But when multiplying both sides by 3, the student does not distribute it across the expression $3w - 4$ correctly.

9. **A.** The student subtracts 5 and -3 incorrectly.

   B. The student subtracts 5 and -3 incorrectly, and subtracts -4 and 1 incorrectly.

   C. Correct answer

   D. The student subtracts -4 and 1 incorrectly.

10. **F.** The student picks the slope of the problem and ignores the need to find a y-intercept.

    G. The student picks the slope of the problem and the y-value from the point (6, 1) for the y-intercept.

    H. Correct answer

    I. The student approaches the problem properly, but misplaces a negative sign for y.

11. **2 points** The student demonstrates a thorough understanding of how to determine whether data show a linear function or a nonlinear function, explains the work fully, and relates perimeter to a linear function and area to a nonlinear function. The first table shows a linear function. The second table shows a nonlinear function.

    1 point The student's work and explanations demonstrate a lack of essential understanding. The slope formula is used incorrectly or a graph of the data is incomplete.

    0 points The student provides no response, a completely incorrect or incomprehensible response, or a response that demonstates insufficient understanding of linear functions and nonlinear functions.

Answers

5. H
6. 3°F
7. C
8. F

Answers

9. C

10. H

11. *Part A* yes

Part B no

12. B

Answers for Extra Examples

1. **A.** The student ignores the starting amount of $300.

   B. The student combines the starting and weekly amounts incorrectly.

   C. The student reverses the roles of the starting and weekly amounts.

   D. Correct answer

2. **F.** The student identifies the domain incorrectly.

   G. The student identifies the domain incorrectly.

   H. Correct answer

   I. The student identifies the domain correctly but thinks it is discrete.

Item Analysis (continued)

12. **A.** The student multiplies by 1000 because there are 1000 mL in 1 L.

B. Correct answer

C. The student has the right idea, but uses the wrong relationship of 1 L = 100 mL.

D. The student divides by 1000 because there are 1000 mL in 1 L.

Extra Examples for Standardized Test Practice

1. Deanna started a savings account with $300 and added $20 per month to the account. Let *n* be the number of weeks that Deanna added money to the account and let *a* be the total amount in her account. Which equation describes the relationship between *a* and *n*?

   A. $a = 20n$

   B. $a = 320n$

   C. $a = 300n + 20$

   D. $a = 20n + 300$

2. Julia is studying how attendance at an amusement park is related to temperature. To build her study, Julia uses the temperature each day as an input value and amusement park attendance as a daily output value. Which statement best describes this situation?

   F. The domain is daily attendance and it is continuous.

   G. The domain is daily attendance and it is discrete.

   H. The domain is temperature and it is continuous.

   I. The domain is temperature and it is discrete.

9. What is the slope of the line shown in the graph below?

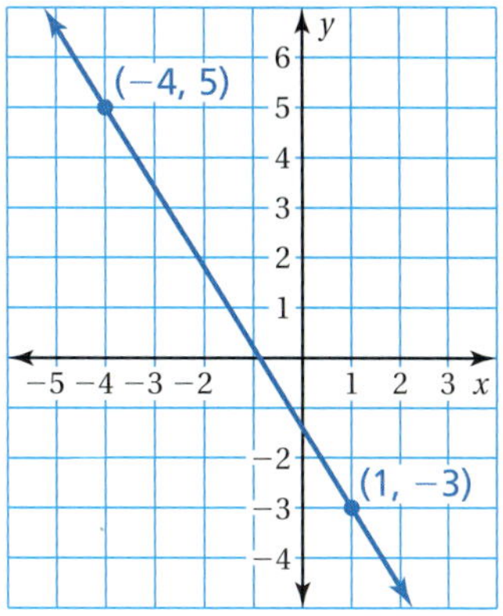

A. $-\frac{2}{5}$

B. $-\frac{2}{3}$

C. $-\frac{8}{5}$

D. $-\frac{8}{3}$

10. A line with slope of $\frac{1}{3}$ contains the point (6, 1). What is the equation of the line?

F. $y = \frac{1}{3}x$

G. $y = \frac{1}{3}x + 1$

H. $x - 3y = 3$

I. $x + 3y = 3$

11. The tables show how the perimeter and area of a square are related to its side length. Examine the data in the table.

Think Solve Explain

Side Length	1	2	3	4	5	6
Perimeter	4	8	12	16	20	24

Side Length	1	2	3	4	5	6
Area	1	4	9	16	25	36

Part A Does the first table show a linear function? Explain your reasoning.

Part B Does the second table show a linear function? Explain your reasoning.

12. A bottle of orange extract marked 25 mL costs $2.49. What is the cost per liter?

A. $2490.00 per L

B. $99.60 per L

C. $9.96 per L

D. $0.00249 per L

5 Angles and Similarity

5.1 Classifying Angles
5.2 Angles and Sides of Triangles
5.3 Angles of Polygons
5.4 Using Similar Triangles
5.5 Parallel Lines and Transversals

"Start with any triangle."

"Tear off the angles. You can always rearrange the angles so that they form a straight line."

"What does that prove?"

"Let's use shadows and similar triangles to indirectly measure the height of the giant hyena standing right behind you."

Connections to Previous Learning

• Use reasoning about multiplication and division to solve ratio and rate problems.
• Solve problems with similar figures. • Use proportions to solve problems.
• Classify and determine the measures of angles, including angles created when parallel lines are cut by transversals. • Demonstrate that the sum of the angles in a triangle is 180-degrees and apply this fact to find unknown measures of angles, and the sum of the angles in polygons. • Use similar triangles to solve problems that include height and distances.

Math in History

Geometry and special ratios have been used in the building industry for thousands of years.

★ The geometry used in the Indus Valley Civilization of North India and Pakistan from around 3000 B.C. was developed mostly as a result of building cities. The geometry originated from practical things, such as designing bricks. Brick sizes were in a perfect ratio of 4 : 2 : 1. Even today, the ratio for brick dimensions 4 : 2 : 1 is considered optimal for effective bonding.

★ Omar Khayyám (born 1048 A.D.) was a Persian mathematician, philosopher, and poet who described his philosophy through poems known as quatrains in the Rubaiyat of Omar Khayyám. Omar Khayyám is known for using geometry to solve algebraic equations.

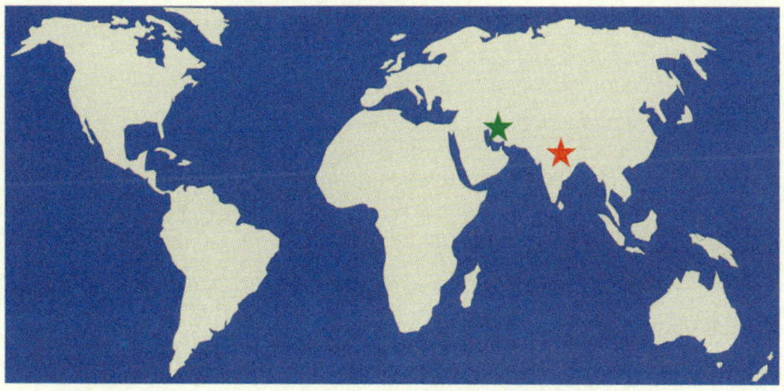

Pacing Guide for Chapter 5

Chapter Opener	1 Day
Section 1 Activity Lesson	 1 Day 1 Day
Section 2 Activity Lesson	 1 Day 1 Day
Section 3 Activity Lesson	 1 Day 2 Days
Study Help / Quiz	1 Day
Section 4 Activity Lesson	 1 Day 1 Day
Section 5 Activity Lesson	 1 Day 2 Days
Quiz / Chapter Review	1 Day
Chapter Test	1 Day
Standardized Test Practice	1 Day
Total Chapter 5	17 Days
Year-to-Date	86 Days

Check Your Resources

- Record and Practice Journal
- Resources by Chapter
- Skills Review Handbook
- Assessment Book
- Worked-Out Solutions

Technology For the Teacher

Dynamic Classroom

The Dynamic Planning Tool
Editable Teacher's Resources at
BigIdeasMath.com

Additional Topics for Review

- Identifying similar figures
- Cross Products Property
- Solving simple equations
- Obtuse, acute, and right angles
- Naming polygons

Try It Yourself

1. 10 in.
2. 12.5 mm

Record and Practice Journal

1. $x = 3$
2. $x = 24$
3. $x = 4.5$
4. $x = 18$
5. $x = 8$
6. $x = 8$
7. $x = 30$
8. $x = 4$
9. $x = 9$
10. $x = 9$
11. 60 feet

Math Background Notes

Vocabulary Review

- Similar Figures
- Proportion
- Ratio
- Cross Products Property
- Corresponding Parts
- Polygon

Finding Unknown Measures in Similar Figures

- Students should be able to write proportions and use them to solve problems.
- Remind students that a complete answer includes a measure and the appropriate units for that measurement.
- **Common Student Question:** "Is that the only proportion that works?" The answer to this question is *no*. In Example 1, 16 and 12 appear in the numerators of the ratios because they both describe the red triangle. Similarly, 18 and x are in the denominators of the ratios because they describe the blue triangle. The position of the measurements in the ratios is irrelevant as long as students are consistent. For example, writing the proportion $\frac{18}{16} = \frac{x}{12}$ is also correct. In this proportion, the measurements describing the blue triangle appear in the numerators.
- **Common Student Question:** "Does the order in which I cross multiply matter?" The answer to this question is *no*. It makes no difference whether you multiply downward to the right or downward to the left first. Encourage students to develop a method and use that method every time for consistency.
- **Common Error:** Some students struggle when given the ratio of the perimeters. Some will be unsure of how to incorporate this ratio with the given side lengths of the polygons. Remind students that the smaller polygon will have the smaller perimeter.

Reteaching and Enrichment Strategies

If students need help. . .	If students got it. . .
Record and Practice Journal • Fair Game Review Skills Review Handbook Lesson Tutorials	Game Closet at *BigIdeasMath.com* Start the next section

What You Learned Before

Finding Unknown Measures in Similar Triangles

Example 1 The two triangles are similar. Find the value of *x*.

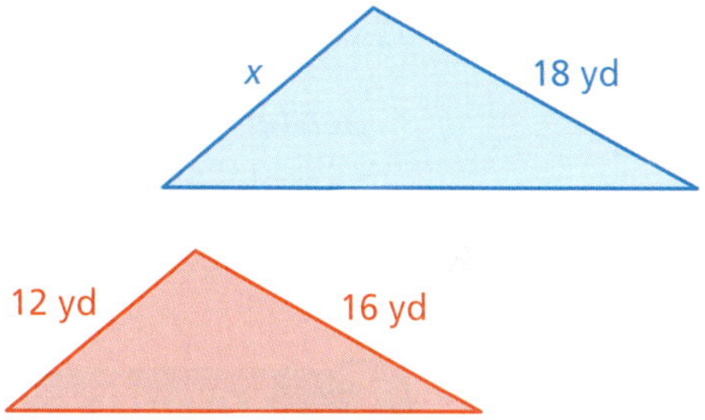

$\frac{16}{18} = \frac{12}{x}$ Write a proportion.

$16x = 216$ Use Cross Products Property.

$x = 13.5$ Divide each side by 16.

So, x is 13.5 yards.

Example 2 The two quadrilaterals are similar. The ratio of their perimeters is 4 : 5. Find the value of *x*.

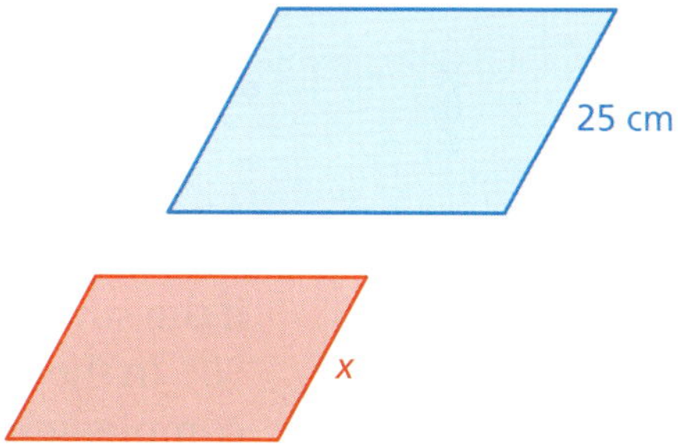

$\frac{4}{5} = \frac{x}{25}$ Write a proportion.

$100 = 5x$ Use Cross Products Property.

$20 = x$ Divide each side by 5.

So, x is 20 centimeters.

Try It Yourself

The polygons are similar. Find the value of *x*.

1. 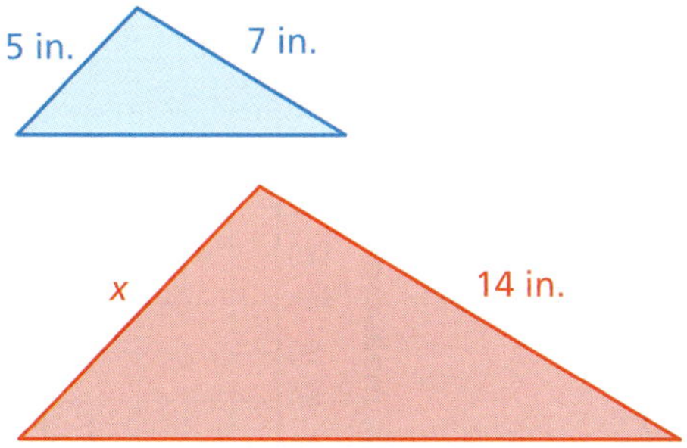

2. The ratio of the perimeters is 2 : 1.

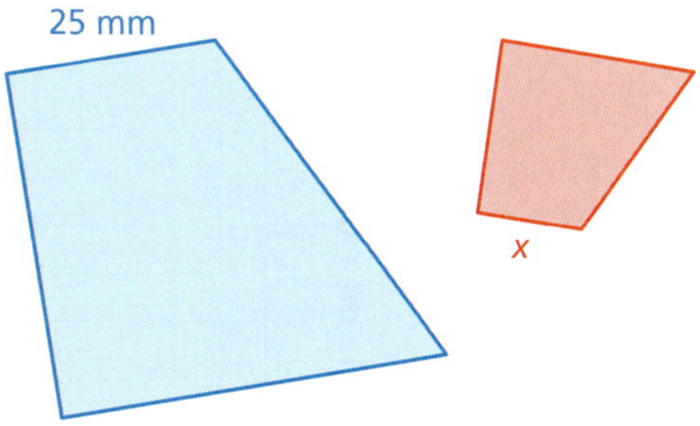

5.1 Classifying Angles

Essential Question How can you classify two angles as complementary or supplementary?

Classification of Angles

Acute: Less than 90°	***Right:*** Equal to 90°	***Obtuse:*** Greater than 90° and less than 180°	***Straight:*** Equal to 180°

1 ACTIVITY: Complementary and Supplementary Angles

Work with a partner.

- **Copy and complete each table.**
- **Graph each function. Is the function linear?**
- **Write an equation for *y* as a function of *x*.**
- **Describe the domain of each function.**

a. Two angles are **complementary** if the sum of their measures is 90°. In the table, x and y are complementary.

x	15°	30°	45°	60°	75°
y					

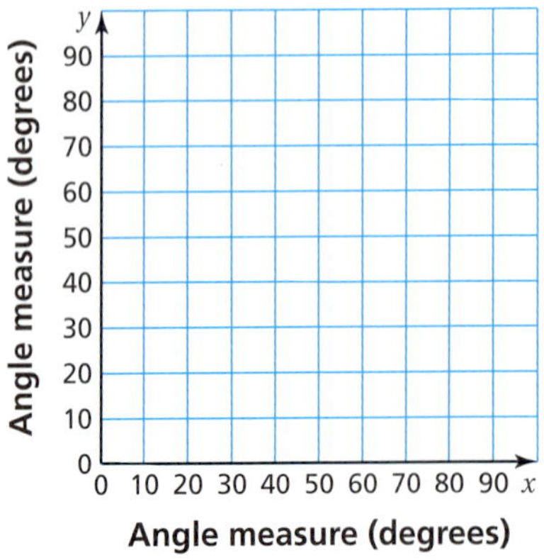

b. Two angles are **supplementary** if the sum of their measures is 180°. In the table, x and y are supplementary.

x	30°	60°	90°	120°	150°
y					

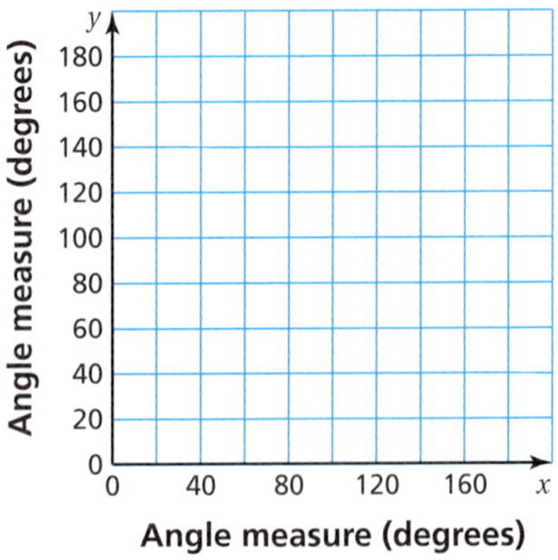

Laurie's Notes

Introduction

For the Teacher

- **Goal:** Students will explore the properties of complementary and supplementary angles.
- Do not assume that all students will recall the vocabulary of angles, know how to use a protractor, or know what an angle measure means.

Motivate

- **Preparation:** Make a model to practice estimation skills with angle measures. Cut two circles (6-inch diameter) out of file folders. Cut a slit in each. On one circle, label every 10°. The second circle is shaded. Insert one circle into the other so the angle measure faces you and the shaded angle faces the students.
- Ask students to estimate the measure of the shaded angle. You can read the answer from your side of the model. Repeat several times.

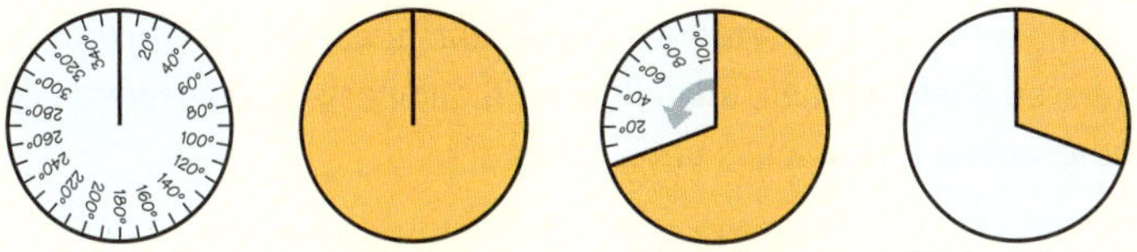

Labeled   Shaded   Your view   Students' view

Activity Notes

Discuss

? "What names do you use to classify angles, and what does each mean?" acute: less than 90°, right: 90°, obtuse: greater than 90°, straight: 180°

- **Caution:** Do not draw every angle in this chapter with the initial ray horizontal and extending rightward. Use varied orientation to gauge students' understanding of reading angle measures.

Activity 1

- Students should read the definition of complementary and supplementary angles, complete the table, and then plot the ordered pairs.

? "What is the slope and y-intercept for part (a)?" slope $= -1$; y-intercept $= 90$

? "What makes sense for the domain of this function?" angle measures between 0 and 90, but not including 0° and 90° (Note: In higher-level math courses, zero degree angles are allowed.)

- This may be the first time students have seen a domain where the endpoints of the interval are not included. The endpoints are not included because a 90° angle does not have a complement.
- **Big Idea:** As an angle increases, its complement decreases. This should make sense because the sum of the two angle measures must always be 90°. A similar pattern occurs with supplementary angles.

? "What is the range of each function?" part (a): $0° < y < 90°$ and part (b): $0° < y < 180°$

Previous Learning

Students should know basic vocabulary associated with angles.

Activity Materials
Introduction
• two circles cut out of file folders

Start Thinking! and Warm Up

Activity 5.1 Start Thinking!
For use before Activity 5.1

Activity 5.1 Warm Up
For use before Activity 5.1

Graph the linear function.

1. $y = -x + 3$
2. $y = 2x - 2$
3. $y = \frac{1}{2}x - 1$
4. $y = \frac{1}{3}x + 2$
5. $y = -\frac{3}{4}x + 2$
6. $y = \frac{2}{3}x + 4$

5.1 Record and Practice Journal

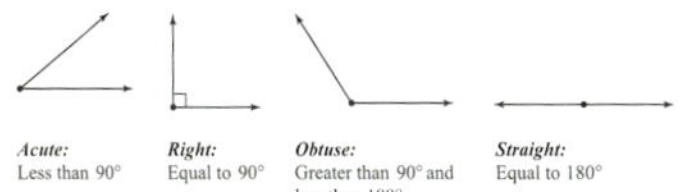

Essential Question How can you classify two angles as complementary or supplementary?

Classification of Angles

Acute: Less than 90°   Right: Equal to 90°   Obtuse: Greater than 90° and less than 180°   Straight: Equal to 180°

1 ACTIVITY: Complementary and Supplementary Angles

Work with a partner.

- Complete each table.
- Graph each function. Is the function linear?
- Write an equation for y as a function of x.
- Describe the domain of each function.

a. Two angles are **complementary** if the sum of their measures is 90°. In the table, x and y are complementary.

x	15°	30°	45°	60°	75°
y	75°	60°	45°	30°	15°

b. Two angles are **supplementary** if the sum of their measures is 180°. In the table, x and y are supplementary.

x	30°	60°	90°	120°	150°
y	150°	120°	90°	60°	30°

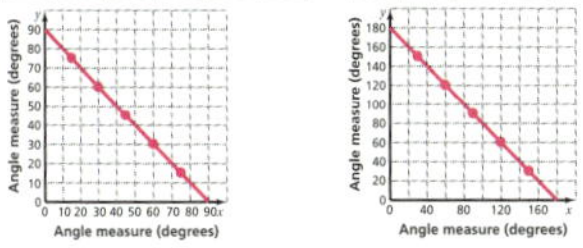

See Additional Answers.

English Language Learners

Illustrate

Explain to English learners that the name *right angle* does not come from the orientation of the angle opening to the right, as shown in the activity. Students might think that if the angle opens to the left, it is called a *left angle*. Point out that any angle that measures 90° is a *right angle*.

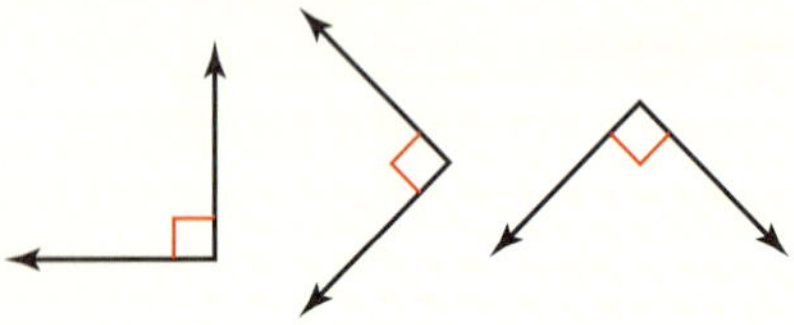

5.1 Record and Practice Journal

2 ACTIVITY: Exploring Rules About Angles

Work with a partner. Complete each sentence with *always*, *sometimes*, or *never*.

a. If x and y are complementary angles, then both x and y are **always** acute.

b. If x and y are supplementary angles, then x is **sometimes** acute.

c. If x is a right angle, then x is **never** acute.

3 ACTIVITY: Naming Angles

Some angles, such as $\angle A$, can be named by a single letter. When this does not clearly identify an angle, you should use three letters, as follows.

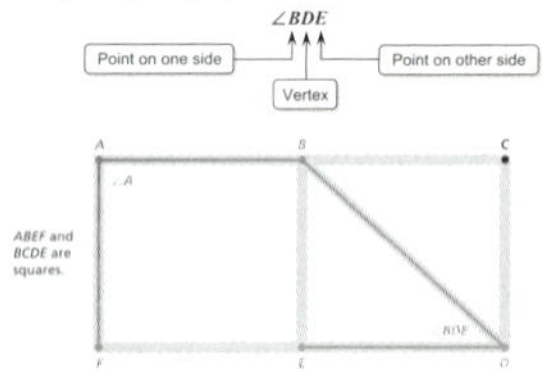

Work with a partner.

a. Name all pairs of complementary angles in the diagram above.

$\angle BDE$ and $\angle DBE$, $\angle CBD$ and $\angle CDB$, $\angle CBD$ and $\angle DBE$, $\angle CDB$ and $\angle BDE$

b. Name all pairs of supplementary angles in the diagram above.

See Additional Answers.

What Is Your Answer?

4. **IN YOUR OWN WORDS** How can you classify two angles as complementary or supplementary? Give examples of each type.

complementary: sum of angle measures is 90°
supplementary: sum of angle measures is 180°

5. Find examples of real-life objects that use complementary and supplementary angles. Make a drawing of each object and approximate the degree measure of each angle.

Check students' work.

Laurie's Notes

Activity 2

- Give time for partners to discuss the problems. The graphs and tables of values from Activity 1 should help students think through their answers.
- **Teaching Tip:** When answers are *sometimes* true, it is important to give students a sample of when the statement is true and when the statement is false. For example, in part (b), $x = 75°$ and $y = 105°$ which makes x acute, or $x = 105°$ and $y = 75°$ which makes x obtuse.

Activity 3

- This activity reviews how angles are named. Discuss when there is a need for three letters instead of one. Also discuss that $\angle EBD$ and $\angle DBE$ name the same angle (because the vertex position is the same).
- ? "What other angles could be named using just one letter?" $\angle C$, $\angle F$
- Although the right angles are not labeled in this diagram, it is assumed that those which appear to be right angles are right angles.

What Is Your Answer?

- **Think-Pair-Share:** Students should read each question independently and then work with a partner to answer the questions. When they have answered the questions, the pair should compare their answers with another group and discuss any discrepancies.

Closure

- Look around the room. Name angles that appear to be acute, right, obtuse, or straight. Name angles that appear to be complementary or supplementary.

Technology For the Teacher

Dynamic Classroom

The Dynamic Planning Tool
Editable Teacher's Resources at *BigIdeasMath.com*

2 ACTIVITY: Exploring Rules About Angles

Work with a partner. Copy and complete each sentence with *always*, *sometimes*, or *never*.

a. If x and y are complementary angles, then both x and y are __________ acute.

b. If x and y are supplementary angles, then x is __________ acute.

c. If x is a right angle, then x is __________ acute.

3 ACTIVITY: Naming Angles

Some angles, such as $\angle A$, can be named by a single letter. When this does not clearly identify an angle, you should use three letters, as follows.

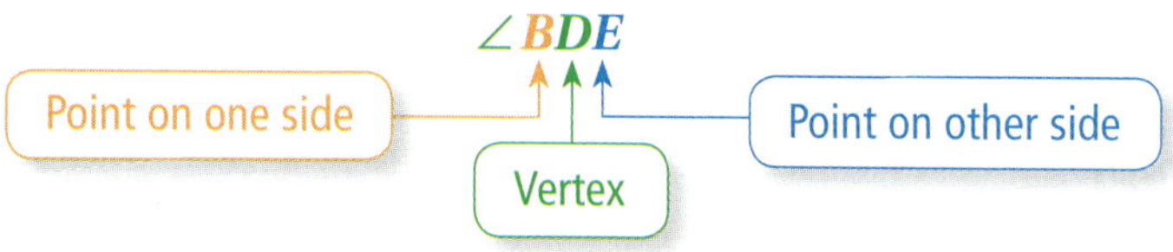

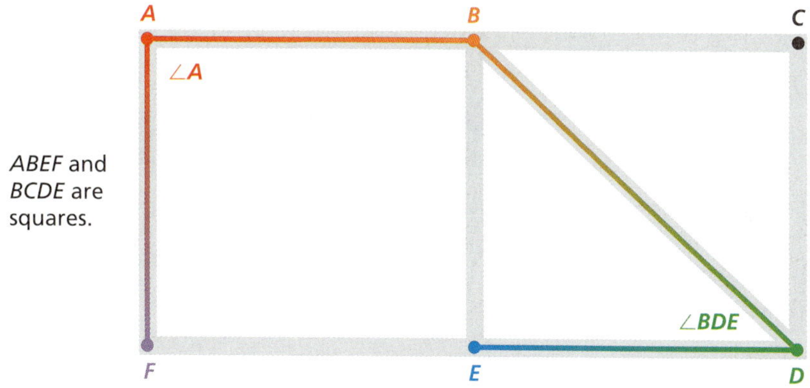

Work with a partner.

a. Name all pairs of complementary angles in the diagram above.

b. Name all pairs of supplementary angles in the diagram above.

What Is Your Answer?

4. **IN YOUR OWN WORDS** How can you classify two angles as complementary or supplementary? Give examples of each type.

5. Find examples of real-life objects that use complementary and supplementary angles. Make a drawing of each object and approximate the degree measure of each angle.

Practice Use what you learned about classifying angles to complete Exercises 3–5 on page 188.

5.1 Lesson

Check It Out
Lesson Tutorials
BigIdeasMath.com

Key Vocabulary

complementary angles, *p. 186*
supplementary angles, *p. 186*
congruent angles, *p. 187*
vertical angles, *p. 187*

Key Ideas

Complementary Angles

Words Two angles are **complementary angles** if the sum of their measures is 90°.

Examples

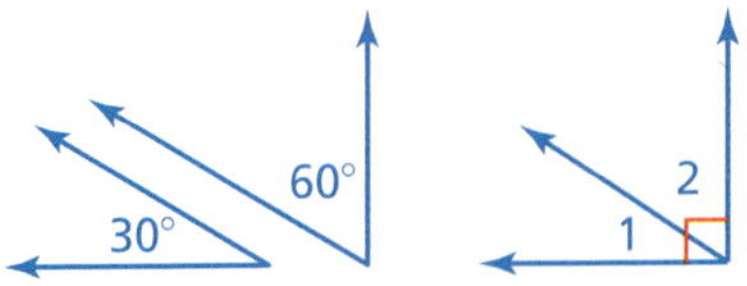

∠1 and ∠2 are complementary angles.

Supplementary Angles

Words Two angles are **supplementary angles** if the sum of their measures is 180°.

Examples

∠3 and ∠4 are supplementary angles.

EXAMPLE 1 Classifying Pairs of Angles

Tell whether the angles are *complementary*, *supplementary*, or *neither*.

a.

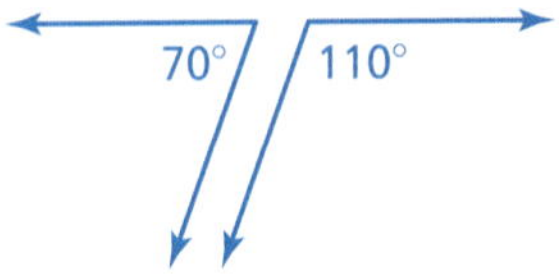

$70° + 110° = 180°$

So, the angles are supplementary.

b.

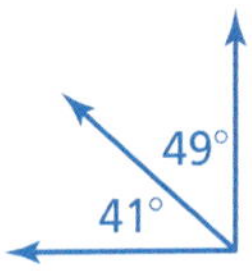

$41° + 49° = 90°$

So, the angles are complementary.

c.

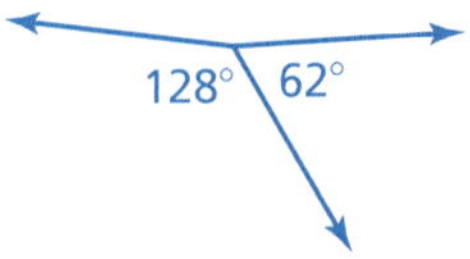

$128° + 62° = 190°$

So, the angles are *neither* complementary nor supplementary.

On Your Own

Now You're Ready
Exercises 6–11

Tell whether the angles are *complementary*, *supplementary*, or *neither*.

1.

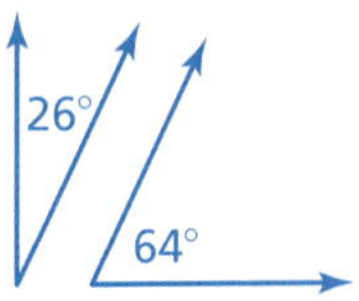

2.

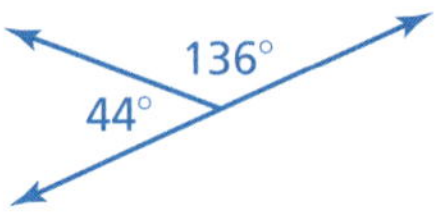

3.

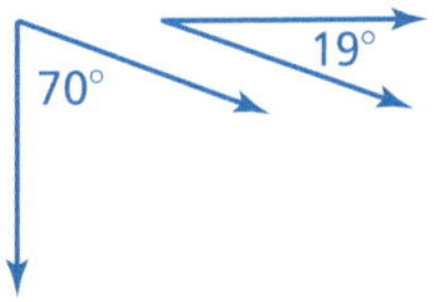

Multi-Language Glossary at BigIdeasMath.com.

Laurie's Notes

Introduction

Connect

- **Yesterday:** Students explored two pairs of angles, complementary and supplementary.
- **Today:** Students will classify several pairs of angles.

Motivate

- Because this chapter will be focusing on geometry, students should know we credit Euclid for the study of geometry. He is often called the Father of Geometry. Euclid was a Greek mathematician best known for his 13 books on geometry known as *The Elements*. This work influenced the development of Western mathematics for more than 2000 years.
- We do not know a lot about Euclid, though one quote is often attributed to him. When a colleague was lamenting about the length of his 13 books, it is reported that Euclid replied, "There is no royal road to Geometry."

Lesson Notes

Key Ideas

- Write the *Key Ideas*. Define and sketch complementary angles and supplementary angles.
- **Common Misconception:** Students sometimes believe that complementary or supplementary angles must be adjacent to (touching) each other because this is the way they are often drawn. In this diagram, they are drawn with an orientation to suggest that the sum is 90° (complementary) or 180° (supplementary), but they do not need to have this orientation. For example, $\angle A$ and $\angle B$ are complementary, however, it is not immediately obvious because of their orientation.

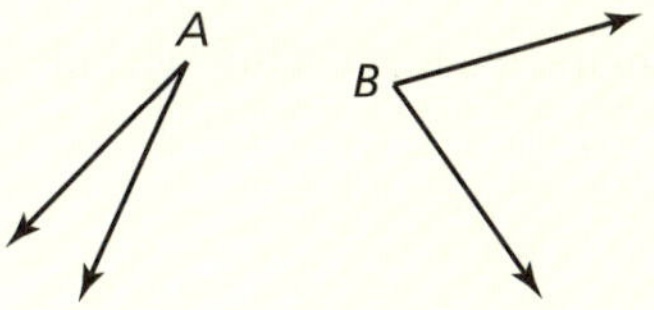

Example 1

- In this example, students sum the angle measures and determine if they add to 90°, 180°, or neither. Make sure students do not rely on their eyesight. They should actually add the angle measures.

On Your Own

- **Think-Pair-Share:** Students should read each question independently and then work with a partner to answer the questions. When they have answered the questions, the pair should compare their answers with another group and discuss any discrepancies.

Goal Today's lesson is classifying angles.

Lesson Materials
Textbook
• scissors

Start Thinking! and Warm Up

Lesson 5.1 Warm Up
For use before Lesson 5.1

Lesson 5.1 Start Thinking!
For use before Lesson 5.1

Complete the statement.

Two angles are _?_ if the sum of their measures is 90°.

Two angles are _?_ if the sum of their measures is 180°.

People often have trouble remembering which is 90° and which is 180°. Make up your own way to help you remember the definitions.

Extra Example 1

Tell whether the angles are *complementary, supplementary,* or *neither*.

a.

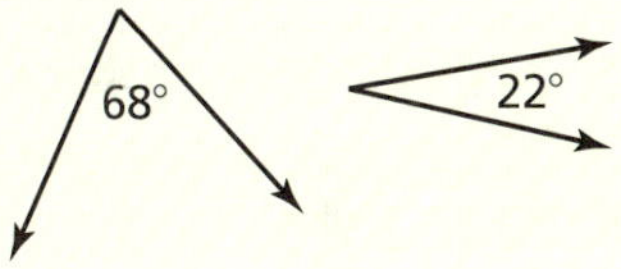

complementary

b.

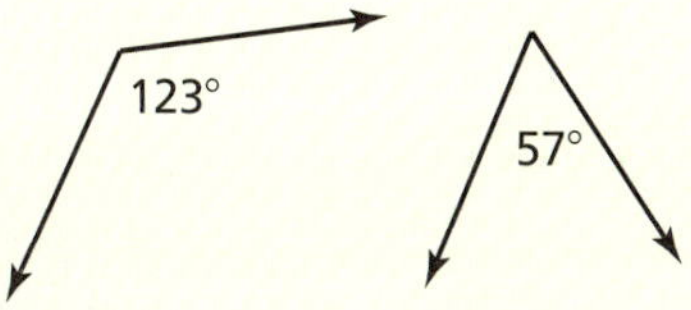

supplementary

c.

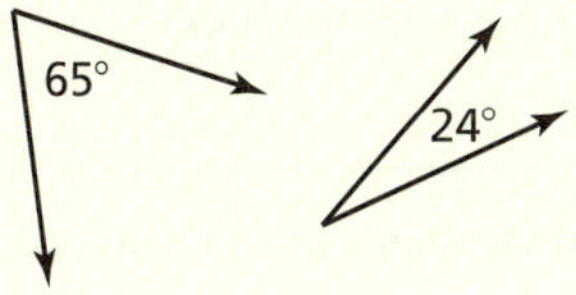

neither

On Your Own

1. complementary
2. supplementary
3. neither

Laurie's Notes

Key Ideas

- Write the Key Ideas. Define and sketch congruent angles and vertical angles.
- **Common Misconception:** The lengths of the rays forming the angle have no bearing on the measure of the angle. The congruent angles shown are congruent because of their angle measures, not because of the length of their rays. Share this example of congruent angles.

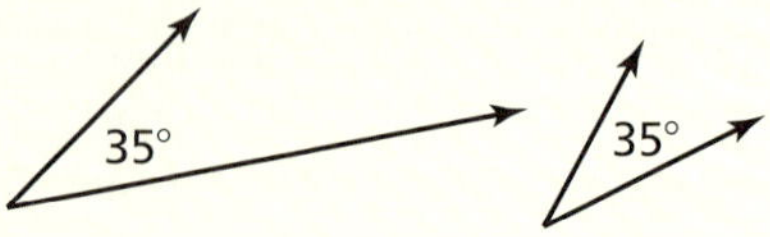

- **Model:** When two lines intersect, two pairs of vertical angles are formed. Vertical angles are congruent. Demonstrate this with a pair of scissors that have straight blades. To make a small cut, you do not open your hands very wide because you want the vertical angle to be small. If you want to make a larger cut, you open your hands wide so that the vertical angle will be greater.

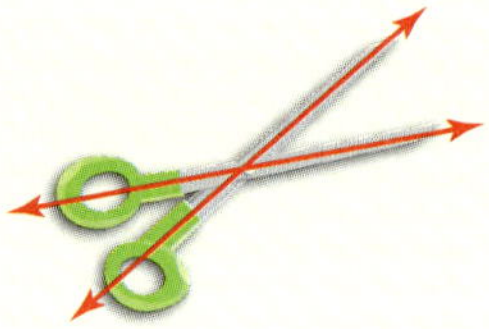

Example 2

- Work through each part as shown.
- **Extension:** In part (a), "What is the measure of the two remaining angles? How do you know?" 110°; The two remaining angles are supplementary with 70°.
- Remind students of the meaning of the symbol used to mark right angles.

Closure

- True or False?
  1. Vertical angles are always acute. false
  2. Supplementary angles could be congruent. true
  3. Complementary angles sum to 180°. false
  4. Vertical angles are congruent. true

Technology For the Teacher

Dynamic Classroom

The Dynamic Planning Tool
Editable Teacher's Resources at *BigIdeasMath.com*

Extra Example 2

Find the value of *x*.

a.

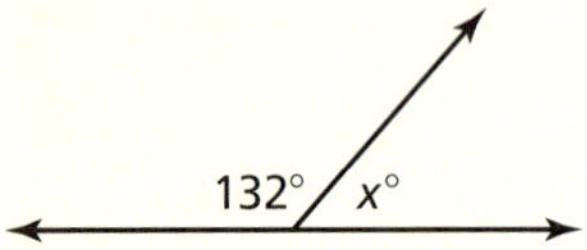

48

b.

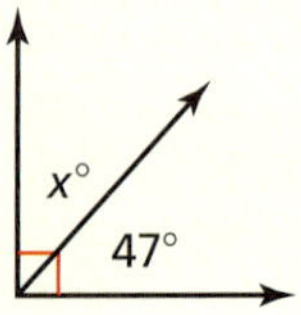

43

On Your Own

4. 95
5. 90
6. 21

Differentiated Instruction

Visual

Help students visualize vertical angles. Draw vertical angles on the board or overhead.

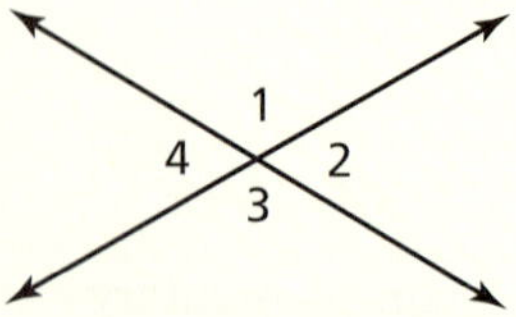

Point out that the lines creating vertical angles form an "X" and that vertical angles do *not* share sides.

Key Ideas

Reading

Arcs are used to indicate congruent angles.

Congruent Angles

Words Two angles are **congruent** if they have the same measure.

Examples

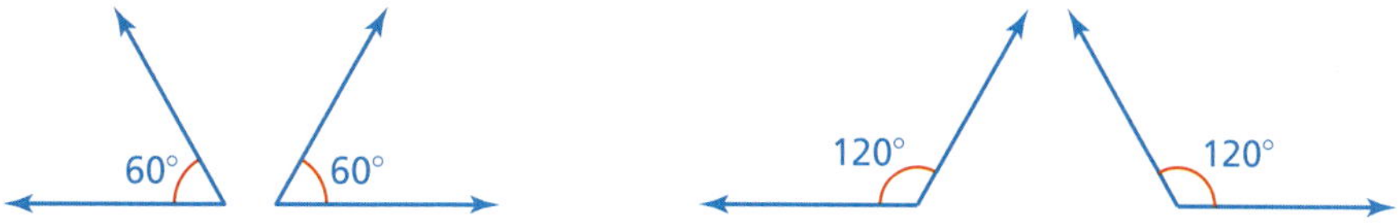

Vertical Angles

Words Two angles are **vertical angles** if they are opposite angles formed by the intersection of two lines. Vertical angles are congruent.

Examples

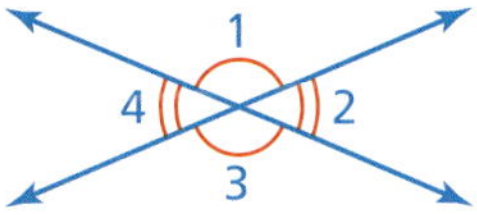

$\angle 1$ and $\angle 3$ are vertical angles.

$\angle 2$ and $\angle 4$ are vertical angles.

EXAMPLE 2 Finding Angle Measures

Find the value of x.

a.

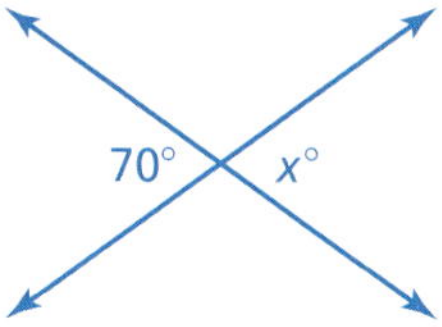

The angles are vertical angles. Because vertical angles are congruent, the angles have the same measure.

So, x is 70.

b.

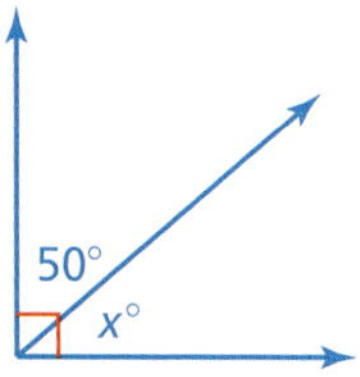

The angles are complementary. So, the sum of their measures is 90°.

$$x + 50 = 90$$

$$x = 40$$

So, x is 40.

On Your Own

Now You're Ready
Exercises 12–14

Find the value of x.

4.

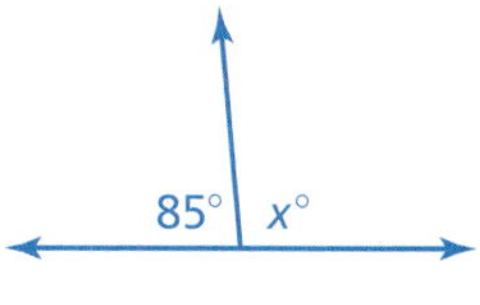

5.

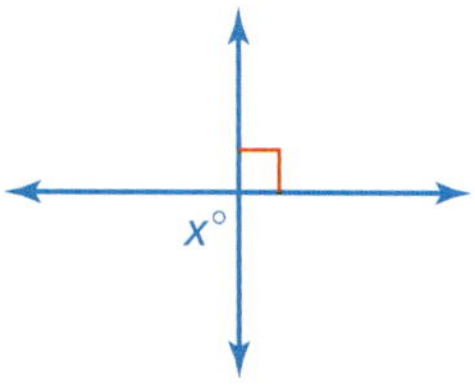

6.

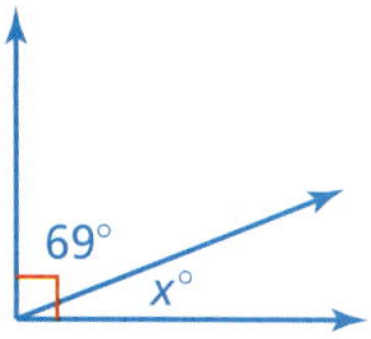

5.1 Exercises

Check It Out
Help with Homework
BigIdeasMath.com

Vocabulary and Concept Check

1. **VOCABULARY** Explain the difference between complementary angles and supplementary angles.

2. **WRITING** When two lines intersect, how many pairs of vertical angles are formed? Explain.

Practice and Problem Solving

Tell whether the statement is *always, sometimes,* or *never* true. Explain.

3. If x and y are supplementary angles, then x is obtuse.

4. If x and y are right angles, then x and y are supplementary angles.

5. If x and y are complementary angles, then y is a right angle.

Tell whether the angles are *complementary, supplementary,* or *neither.*

① 6.
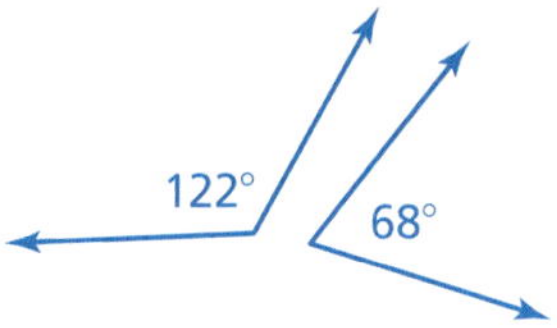

7.
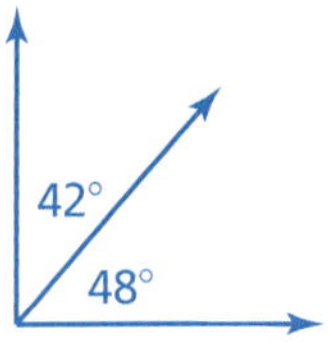

8.
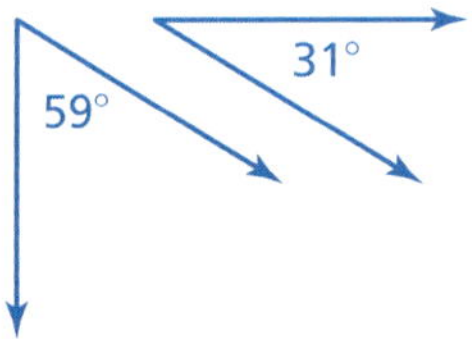

9.
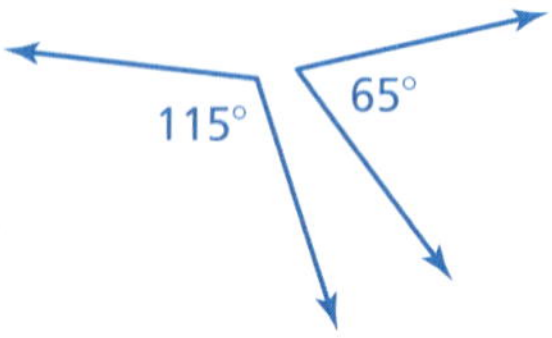

10.
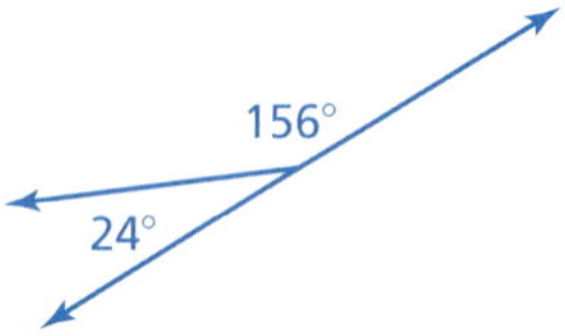

11.
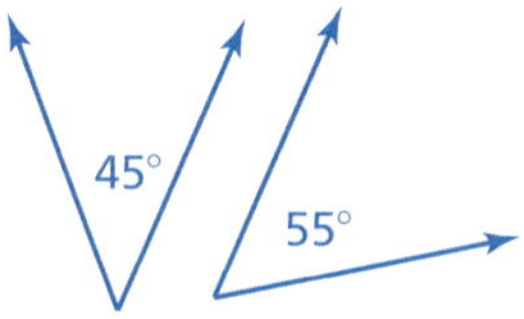

Find the value of x.

② 12.
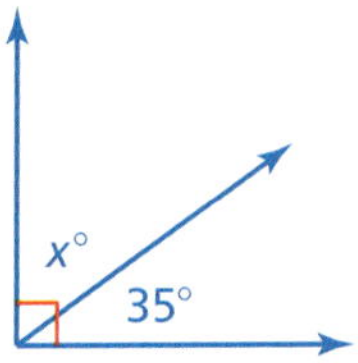

13.
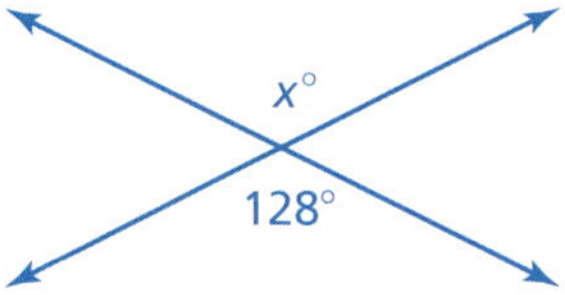

14.
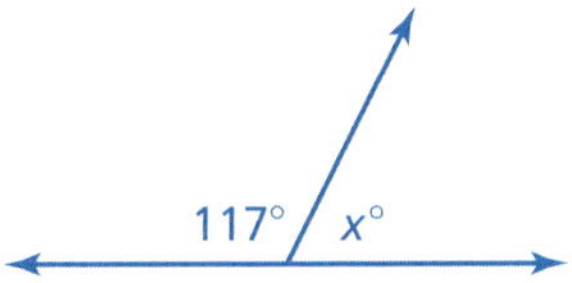

15. **ERROR ANALYSIS** Describe and correct the error in finding the value of x.

✗ The value of x is 55 because vertical angles are complementary.

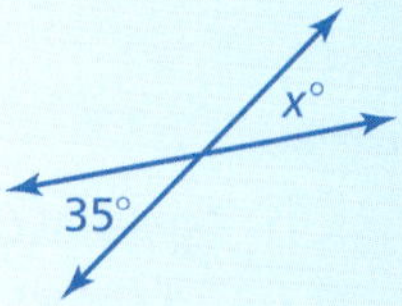

16. **TRIBUTARY** A tributary joins a river at an angle. Find the value of x.

Assignment Guide and Homework Check

Level	Day 1 Activity Assignment	Day 2 Lesson Assignment	Homework Check
Basic	3–5, 26–29	1, 2, 6–16	2, 6, 8, 10, 12, 16
Average	3–5, 26–29	1, 2, 7, 9, 11–15, 17, 19–22	7, 9, 11, 12, 17, 20
Advanced	3–5, 26–29	1, 2, 15, 17–25	18, 20, 23, 24

For Your Information

- **Exercise 21** Students may not understand what a *vanishing point* is. A vanishing point is a point in a perspective drawing to which parallel lines appear to converge.

Common Errors

- **Exercises 6–11** Students may mix up the terms *supplementary* and *complementary*. Remind them of the definitions and use the alliteration that complementary angles are corners and supplementary angles are straight.
- **Exercises 12–14** Students may think that there is not enough information to determine the value of *x*. Remind them of the definitions they have learned in the lesson and ask if any of those could apply to the angles. For example, Exercise 12 is a right angle and there are two angles within. These two angle measures must add to 90°, so a student can use the definition of complementary angles to find *x*.

5.1 Record and Practice Journal

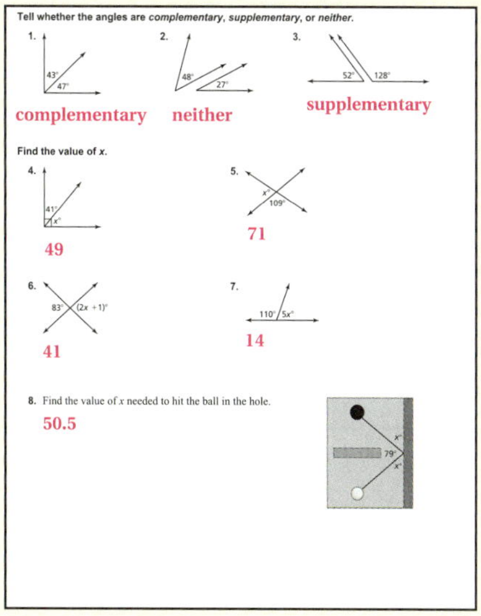
Tell whether the angles are *complementary, supplementary,* or *neither.*

1. complementary
2. neither
3. supplementary

Find the value of *x*.

4. 49
5. 71
6. 41
7. 14
8. Find the value of *x* needed to hit the ball in the hole. 50.5

Technology For the Teacher

Answer Presentation Tool
Quiz*Show*

Vocabulary and Concept Check

1. The sum of the measures of two complementary angles is 90°. The sum of the measures of two supplementary angles is 180°.
2. two pairs; There are two pairs of opposite angles when two lines intersect.

Practice and Problem Solving

3. sometimes; Either *x* or *y* may be obtuse.
4. always; 90° + 90° = 180°
5. never; Because *x* and *y* must both be less than 90° and greater than 0°.
6. neither
7. complementary
8. complementary
9. supplementary
10. supplementary
11. neither
12. 55
13. 128
14. 63
15. Vertical angles are congruent. The value of *x* is 35.
16. 53
17. 37
18. 15
19. 20
20. *Sample answer:* 120°; It is supplementary with a 60° angle, but it is greater than 90°, so it cannot be complementary with another angle.

Practice and Problem Solving

21. **a.** $\angle CBD$ and $\angle DBE$; $\angle ABF$ and $\angle FBE$

b. $\angle ABE$ and $\angle CBE$; $\angle ABD$ and $\angle CBD$; $\angle CBF$ and $\angle ABF$

22. $\angle 1 = 130°$, $\angle 2 = 50°$, $\angle 3 = 130°$

23. $54°$

24. See *Taking Math Deeper.*

25. $7x + y + 90 = 180$
$5x + 2y = 90$
$x = 10; y = 20$

Fair Game Review

26. 75 **27.** 29.3

28. 35 **29.** B

Mini-Assessment

Tell whether the angles are *complementary, supplementary,* or *neither.*

1.

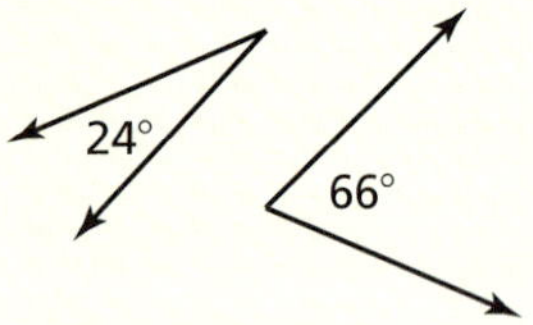

complementary

2.

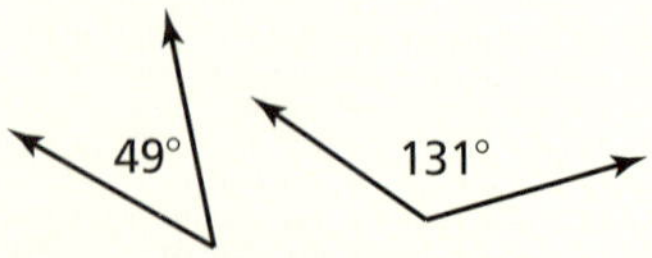

supplementary

3.

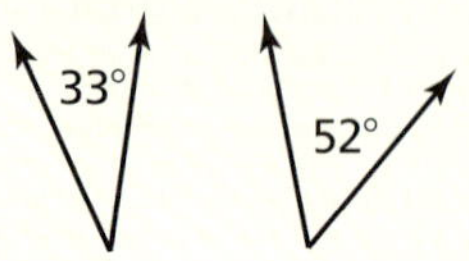

neither

Taking Math Deeper

Exercise 24

This exercise is a good lesson for students. The definition of vertical angles is related to the *position* of the angles. However, the definitions of complementary angles and supplementary angles are only based on the *measures* of the angles and not on the position of the angles.

1 Draw the angles.

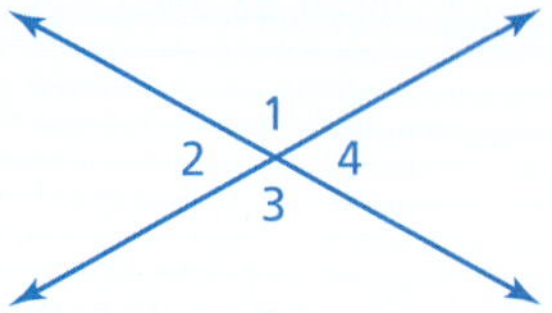

2 $\angle 2$ and $\angle 4$ are complementary.

$\angle 2 = \angle 4$ Vertical angles
$\angle 2 + \angle 4 = 90$ Complementary angles

Solving this system implies that $\angle 2 = 45°$ and $\angle 4 = 45°$.

3 $\angle 2$ and $\angle 4$ are supplementary.

$\angle 2 = \angle 4$ Vertical angles
$\angle 2 + \angle 4 = 180$ Supplementary angles

Solving this system implies that $\angle 2 = 90°$ and $\angle 4 = 90°$.

Project

Look around your classroom, school, home, or anywhere you go. Find examples of complementary and supplementary angles. How do you know they are complementary or supplementary? What is the most common angle you find?

Reteaching and Enrichment Strategies

If students need help. . .	If students got it. . .
Resources by Chapter • Practice A and Practice B • Puzzle Time Record and Practice Journal Practice Differentiating the Lesson Lesson Tutorials Skills Review Handbook	Resources by Chapter • Enrichment and Extension Start the next section

Find the value of *x*.

17.

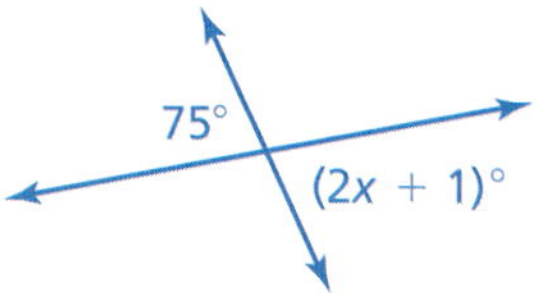

18.

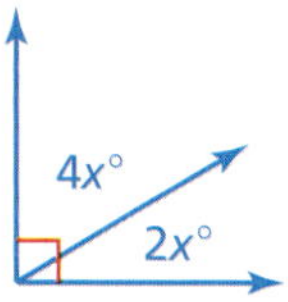

19.

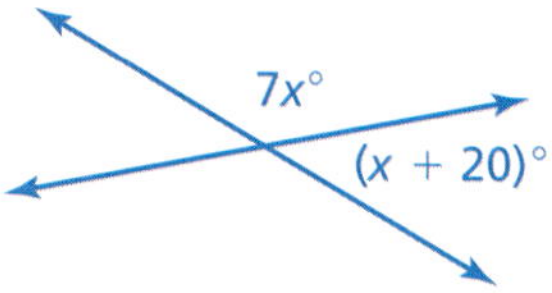

20. **OPEN-ENDED** Give an example of an angle that can be a supplementary angle but cannot be a complementary angle. Explain.

21. **VANISHING POINT** The vanishing point of the picture is represented by point *B*.

a. Name two pairs of complementary angles.

b. Name three pairs of supplementary angles.

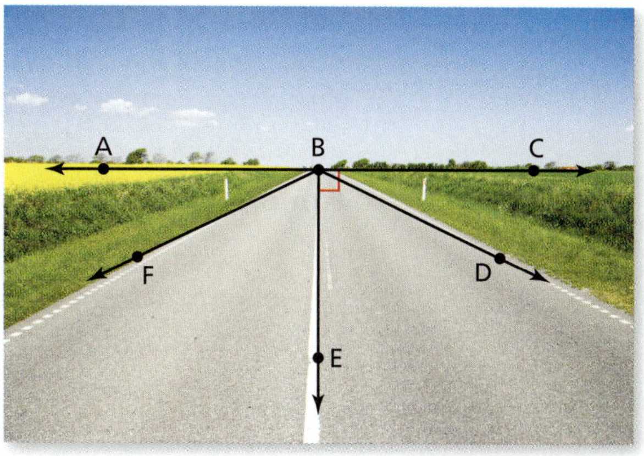

22. **INTERSECTION** What are the measures of the other three angles formed by the intersection?

23. **RATIO** The measures of two complementary angles have a ratio of 3 : 2. What is the measure of the larger angle?

24. **REASONING** Two angles are vertical angles. What are their measures if they are also complementary angles? supplementary angles?

25. **Critical Thinking** Write and solve a system of equations to find the values of *x* and *y*.

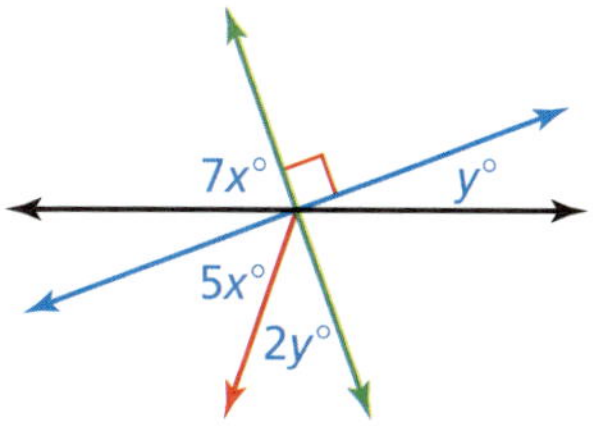

Fair Game Review What you learned in previous grades & lessons

Solve the equation. Check your solution. *(Section 1.1 and Section 1.2)*

26. $x + 60 + 45 = 180$ **27.** $x + 58.5 + 92.2 = 180$ **28.** $x + x + 110 = 180$

29. **MULTIPLE CHOICE** The graph of which equation has a slope of $-\frac{1}{2}$ and passes through the point (6, 4)? *(Section 3.2)*

Ⓐ $y = x + 3$ Ⓑ $y = -\frac{1}{2}x + 7$ Ⓒ $y = -\frac{1}{2}x + 1$ Ⓓ $y = \frac{1}{2}x - 3$

5.2 Angles and Sides of Triangles

Essential Question How can you classify triangles by their angles?

1 ACTIVITY: Exploring the Angles of a Triangle

Work with a partner.

a. Draw a triangle that has an obtuse angle. Label the angles *A*, *B*, and *C*.

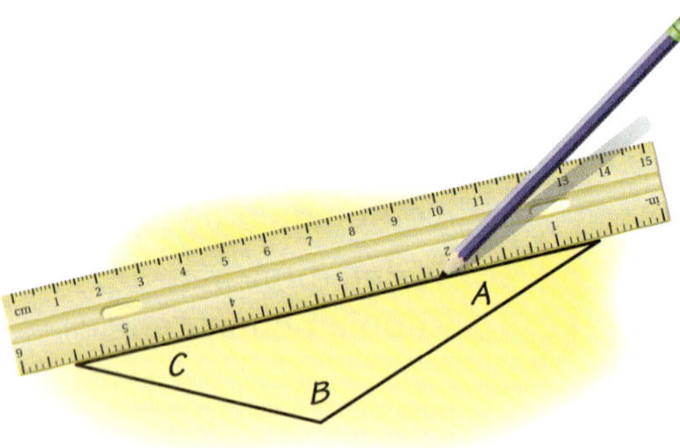

b. Carefully cut out the triangle. Tear off the three corners of the triangle.

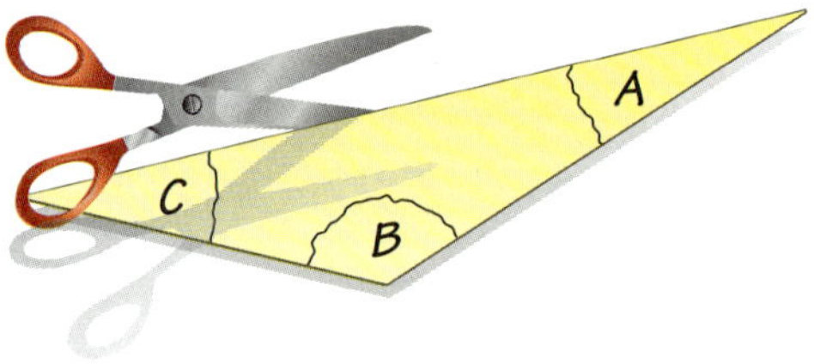

c. Draw a straight line on a piece of paper. Arrange angles *A* and *B* as shown.

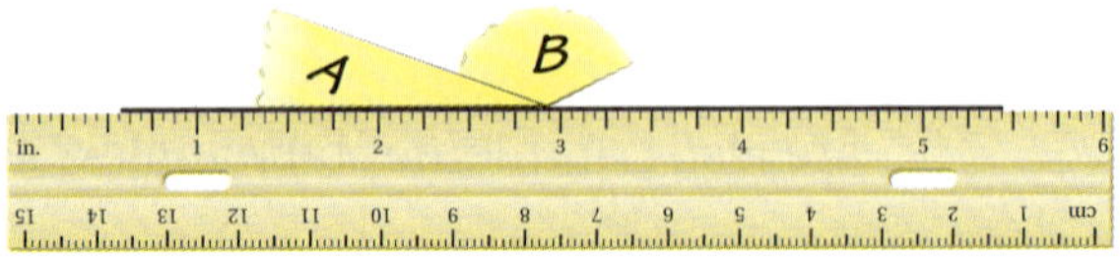

d. Place the third angle as shown. What does this tell you about the sum of the measures of the angles?

e. Draw three other triangles that have different shapes. Repeat parts (b)–(d) for each one. Do you get the same result as in part (d)? Explain.

f. Write a rule about the sum of the measures of the angles of a triangle. Compare your rule with the rule you wrote in Activity 2 in Section 1.1. Did you get the same result? Explain.

Laurie's Notes

Introduction

For the Teacher

- **Goal:** Students will explore the sum of the angle measures of a triangle.
- Many students will have heard of the property they are investigating today. Having heard the property and internalizing it for all triangles are two different levels of knowledge.
- **Management Tip:** There will be torn pieces of scrap paper resulting from this investigation. To help keep the room clean, cluster 4–6 desks together in a circle and tape a recycled paper or plastic bag to the front edge of one of the desks. Students are expected to put scraps of paper in the bag when they are finished with the investigation.

Motivate

- Make teams of three students. Give them 3 minutes to make a list of as many words as they can that begin with the prefix *tri-*.
- Some examples are: triangle, triathlon, tricycle, tri-fold, triangulate, triad, triaxial, trilogy, trimester, trinary, trinity, trio, trilingual, trillium.
- Provide dictionaries if necessary.
- The goal of this activity is to demonstrate that *tri-* is a common prefix.

Activity Notes

Activity 1

- The directions for this activity are direct and easy to follow.
- The sides of the triangle must be straight; otherwise the three angles will not lie adjacent to one another when placed about a point on the line.
- **Teaching Tip:** If you cannot gain access to enough pairs of scissors, you can cut out several triangles in advance using a paper cutter. It is okay to have multiple copies of the same triangle because different pairs of students will get one copy of the triangle.
- The conclusion, or rule, that students should discover is that the angle measures of any triangle will sum to 180°.

Previous Learning

Students should know basic vocabulary associated with angles.

Activity Materials
Textbook
• scissors • straightedge • scrap paper • recycled paper or plastic bags

Start Thinking! and Warm Up

Activity 5.2 **Start Thinking!** For use before Activity 5.2

Activity 5.2 **Warm Up** For use before Activity 5.2

Classify the angle as *acute*, *right*, *obtuse*, or *straight*.

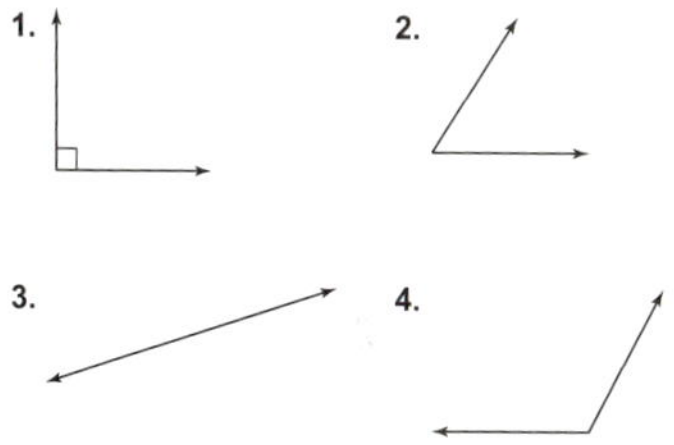

5.2 Record and Practice Journal

Essential Question How can you classify triangles by their angles?

1 ACTIVITY: Exploring the Angles of a Triangle

Work with a partner.

a. Draw a large triangle that has an obtuse angle on a separate piece of paper. Label the angles *A*, *B*, and *C*.

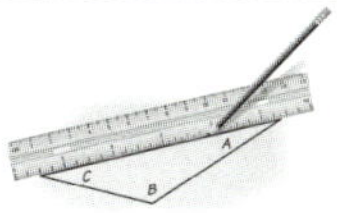

b. Carefully cut out the triangle. Tear off the three corners of the triangle.

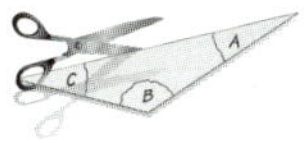

c. Draw a straight line on a piece of paper. Arrange angles *A* and *B* as shown.

d. Place the third angle as shown. What does this tell you about the sum of the measures of the angles?

180°

e. Draw three other triangles that have different shapes. Repeat parts (b)–(d) for each one. Do you get the same results as in part (d)? Explain.

Check students' work.

f. Write a rule about the sum of the measures of the angles of a triangle. Compare your rule with the rule you wrote in Activity 2 in Section 1.1. Did you get the same result? Explain.

Sum of angle measures of a triangle is 180°.

Differentiated Instruction

Kinesthetic

When talking about right, acute, and obtuse angles of a triangle, ask students if it is possible to draw a triangle with 2 right angles. Students should see by drawing the two right angles with a common side that the remaining two sides of the right angles will never meet. So, no triangle can be formed with 2 right angles. Ask students if it is possible for a triangle to have 2 obtuse angles. Students should reach the same conclusion. No triangle can be formed with 2 obtuse angles.

5.2 Record and Practice Journal

2 ACTIVITY: Thinking About Vocabulary

Work with a partner. Talk about the meaning of each name. Use reasoning to define each name. Then match each name with a triangle.

Note: Each triangle has at least one name, but some have more than one name.

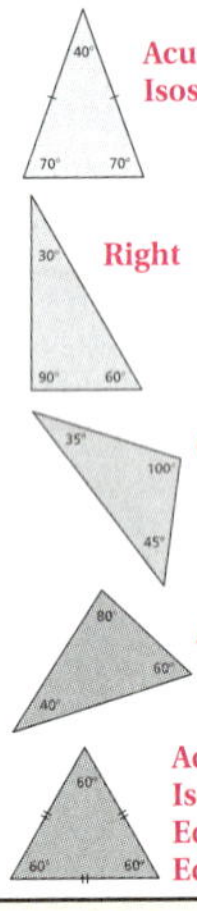

Acute Isosceles

Right

Obtuse

Acute

Acute Isosceles Equilateral Equiangular

a. Right triangle

has one 90° angle

b. Acute triangle

all angles less than 90°

c. Obtuse triangle

has one angle greater than 90°

d. Equiangular triangle

has 3 equal angles

e. Equilateral triangle

has 3 equal side lengths

f. Isosceles triangle

has 2 equal side lengths and 2 equal angles

3 ACTIVITY: Triangles in Art

Work with a partner.

a. Trace four triangles in the painting. Classify each triangle using the names in Activity 2.

Check students' work.

Abstract II by Linda Bahner

b. Design your own abstract art painting. How many different types of triangles did you use in your painting?

Check students' work.

What Is Your Answer?

4. **IN YOUR OWN WORDS** How can you classify triangles by their angles?

Classify them as acute, obtuse, right, or equilangular.

5. Find examples of real-life triangles in architecture. Name each type of triangle that you find.

Check students' work.

Laurie's Notes

Activity 2

- Note that the directions ask students to think about the meaning of each name. Share some historical information about the names.
  - Acute comes from the Latin *acus* which means *needle*. An acute angle is sharp, or pointed.
  - Obtuse comes from the Latin *ob* and *tudere* which means to *beat against*. An object gets blunt or rounded when it is beaten, just like an obtuse angle is blunt compared to an acute angle.
  - The term isosceles derives from the Greek *iso* (same) and *skelos* (leg).
  - The prefix *equi-* means *equal* or *equally*. *Lateral* means *side*.

? "What do the marks on the sides of the blue and orange triangles mean?" The sides have the same length (congruent).

- Remind students that each triangle has at least one name, but some have more than one name. All of the names will be used at least once.

? "Which triangle has the most names? Explain." The orange triangle is acute, equiangular, equilateral, and isosceles. (Students may not call it isosceles. By definition, an isosceles triangle has at least two congruent sides. You can discuss this in the lesson when isosceles is formally defined.)

Activity 3

- Begin with a discussion of abstract art. It appears that artists use random colors and shapes. However, the paintings are often meticulously planned, such as the one shown by Linda Bahner.
- Students should be able to place paper on top of the image and trace four triangles. Because there are more than four triangles, students should reference color and relative position (red, lower left) to remember which triangle they traced.

What Is Your Answer?

- Have students work in pairs.

Closure

- Use a straightedge to draw an example of the following types of triangles.
  1. acute triangle
  2. obtuse triangle
  3. right triangle (Students should use the corner of their paper to make sure it is a right triangle.)

Technology For the Teacher

Dynamic Classroom

The Dynamic Planning Tool
Editable Teacher's Resources at *BigIdeasMath.com*

2 ACTIVITY: Thinking About Vocabulary

Work with a partner. Talk about the meaning of each name. Use reasoning to define each name. Then match each name with a triangle.

Note: Each triangle has at least one name, but some have more than one name.

a. Right triangle

b. Acute triangle

c. Obtuse triangle

d. Equiangular triangle

e. Equilateral triangle

f. Isosceles triangle

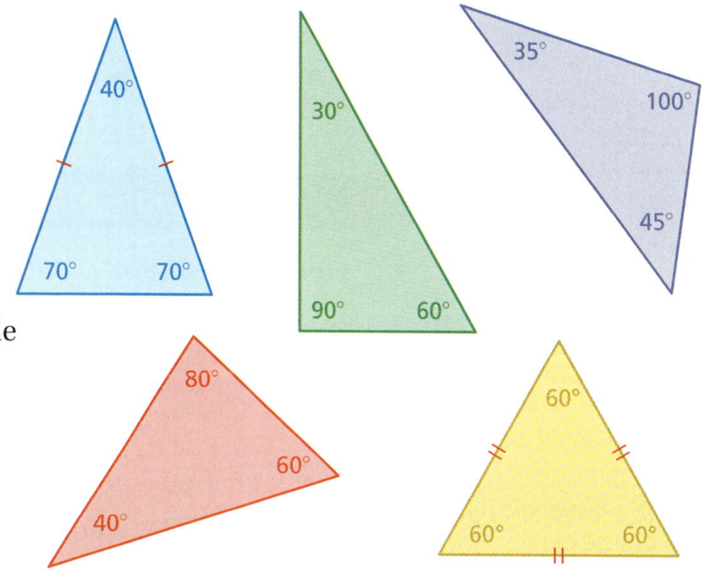

3 ACTIVITY: Triangles in Art

Work with a partner.

a. Trace four triangles in the painting. Classify each triangle using the names in Activity 2.

b. Design your own abstract art painting. How many different types of triangles did you use in your painting?

Abstract II by Linda Bahner

What Is Your Answer?

4. IN YOUR OWN WORDS How can you classify triangles by their angles?

5. Find examples of real-life triangles in architecture. Name each type of triangle that you find.

Practice Use what you learned about angles of triangles to complete Exercises 3–5 on page 194.

5.2 Lesson

Check It Out
Lesson Tutorials
BigIdeasMath.com

Key Vocabulary

isosceles triangle, *p. 192*
congruent sides, *p. 192*
equilateral triangle, *p. 192*
equiangular triangle, *p. 192*

Key Idea

Angle Measures of a Triangle

Words The sum of the angle measures of a triangle is 180°.

Algebra $x + y + z = 180$

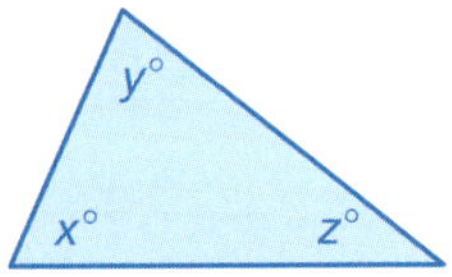

EXAMPLE 1 Finding Angle Measures

Remember

An *acute triangle* has all acute angles.

A *right triangle* has one right angle.

An *obtuse triangle* has one obtuse angle.

Find each value of *x*. Then classify each triangle.

a.

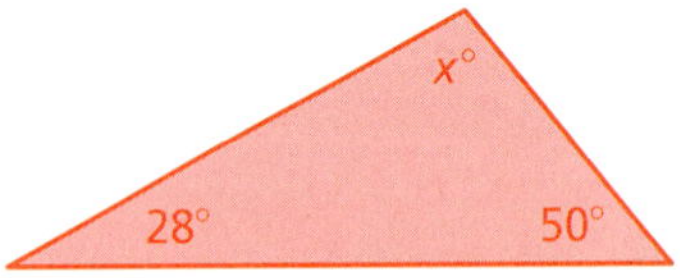

$$x + 28 + 50 = 180$$
$$x + 78 = 180$$
$$x = 102$$

The value of x is 102. The triangle has an obtuse angle. So, it is an obtuse triangle.

b.

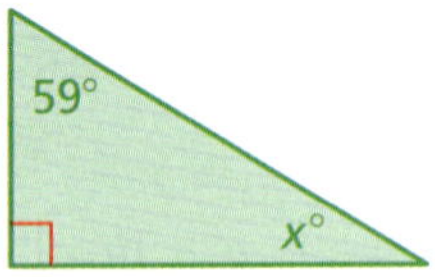

$$x + 59 + 90 = 180$$
$$x + 149 = 180$$
$$x = 31$$

The value of x is 31. The triangle has a right angle. So, it is a right triangle.

On Your Own

Now You're Ready
Exercises 6–8

Find the value of *x*. Then classify the triangle.

1.

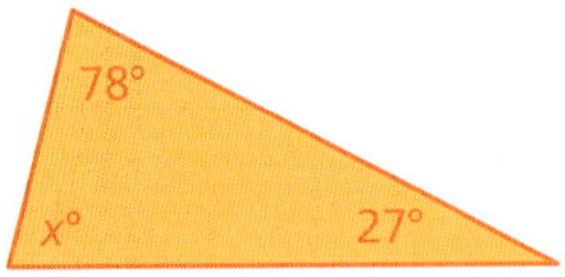

2.

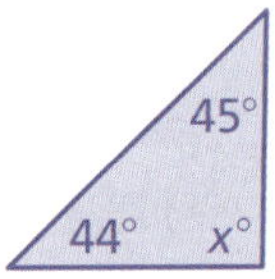

Key Ideas

Reading

Small line segments are used to indicate congruent sides.

Isosceles Triangle

An **isosceles triangle** has at least two sides that are **congruent** (have the same length).

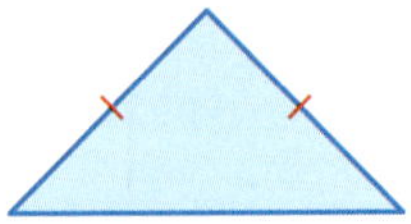

Equilateral Triangle

An **equilateral triangle** has three congruent sides.

An equilateral triangle is also **equiangular** (three congruent angles).

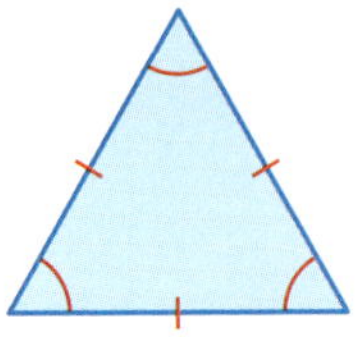

Multi-Language Glossary at BigIdeasMath.com.

Laurie's Notes

Introduction

Connect

- **Yesterday:** Students explored the sum of the angle measures of a triangle and the vocabulary associated with triangles.
- **Today:** Students will find the missing angle measure of a triangle and classify the triangle.

Motivate

? Discuss the Ohio State flag.
The blue triangle represents hills and valleys. The red and white stripes represent roads and waterways. The 13 leftmost stars represent the 13 original colonies. The 4 stars on the right bring the total to 17, representing that Ohio was the 17th state admitted to the Union.

Lesson Notes

Key Idea

- The property is written with variables to suggest that you can solve an equation to find the third angle when you know the other two angles. This is also called the *Triangle Sum Theorem.*

? "What type of angles are the remaining angles of a right triangle? a triangle with an obtuse angle?" Both are acute.

? "Do you think an obtuse triangle could have a right angle? Explain." no; The sum of the angle measures would be greater than 180°.

Example 1

- Some students may argue that all they need to do is add the angle measures and subtract from 180. Remind them that they are practicing a *process*, one that works when the three angle measures are given as algebraic expressions, such as $(x + 10)°$, $(x + 20)°$, and $(x + 30)°$.

Key Ideas

- Define and sketch isosceles and equilateral triangles. Discuss how the sides and angles are marked to show that they are congruent (same length and same measure).
- **Common Misconception:** Draw a series of isosceles and equilateral triangles in different orientations so that students do not think that the triangles must be sitting on a base.
- Make note of the words *at least two sides* in the definition of isosceles. This means that it would be okay for all three sides to be congruent and still call it isosceles. So, all equilateral triangles are isosceles.

? "Are all isosceles triangles equilateral? Explain." No; if the triangle only has two congruent sides, then it only satisfies the definition of isosceles.

Goal Today's lesson is finding angles of a triangle and classifying the triangle.

Start Thinking! and Warm Up

Lesson 5.2 Warm Up
For use before Lesson 5.2

Lesson 5.2 Start Thinking!
For use before Lesson 5.2

In real life, when is it important to know the angle measures of a triangle?

Extra Example 1

Find each value of x. Then classify each triangle.

a.

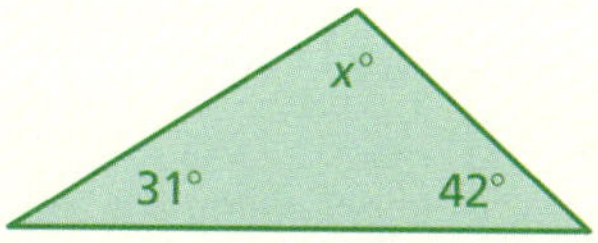

107; obtuse triangle

b.

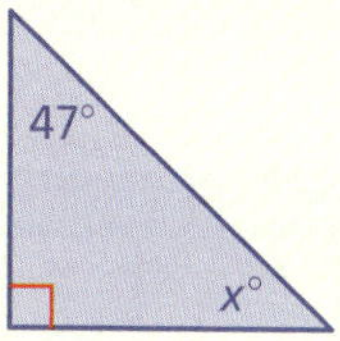

43; right triangle

On Your Own

1. 75; acute triangle
2. 91; obtuse triangle

Laurie's Notes

Extra Example 2

Find the value of x. Then classify the triangle.

25; obtuse triangle

Extra Example 3

A car travels around the park shown below. What is the value of x?

62

On Your Own

3. 30; obtuse isosceles triangle
4. 60; equilateral, equiangular, acute, and isosceles triangle
5. 54.3

English Language Learners

Illustrate

Have students copy the empty table into their notebooks and then complete it with triangles that represent both attributes.

	Acute	Right	Obtuse
Scalene	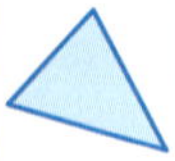	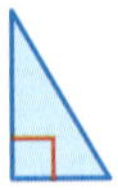	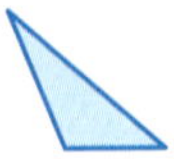
Isosceles		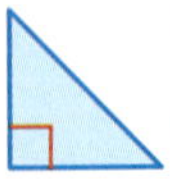	
Equilateral		not possible	not possible

Example 2

- Share the symbolism of each flag.
  - **Jamaica:** The yellow divides the flag into four triangles and represents sunshine and natural resources. Black represents the burdens overcome by the people and the hardships in the future. Green represents the land and hope for the future.
  - **Cuba:** The blue stripes refer to the three old divisions of the island and the two white stripes represent the strength of the independent ideal. The red triangle symbolizes equality, fraternity and freedom, and the blood shed in the struggle for independence. The white star symbolizes the absolute freedom among the Cuban people.
- Set up and solve the equations as shown.

? "How would you classify the green triangle on the Jamaican flag?" obtuse isosceles

- Students may ask if an isosceles triangle also has two congruent angles as in this example. Students will explore this property in Exercise 18.

Example 3

- Add a little interest by sharing information from the Department of the Navy website, *history.navy.mil*. The *Bermuda Triangle* is an imaginary area located off the southeastern Atlantic coast of the U.S. where a supposedly high incidence of unexplained disappearances of ships and aircraft occurs. The vertices of the triangle are Bermuda; Miami, Florida; and San Juan, Puerto Rico.
- Set up the equation and work through the problem as shown.

On Your Own

- **Question 4:** Students may look at the problem and state all of the angles are 60°. Ask what equation they can use to check their answer.

Closure

- **Exit Ticket:** Find the value of x. Then classify the triangle.

a.

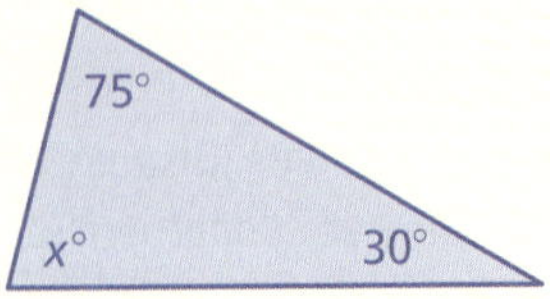

75; acute triangle

b.

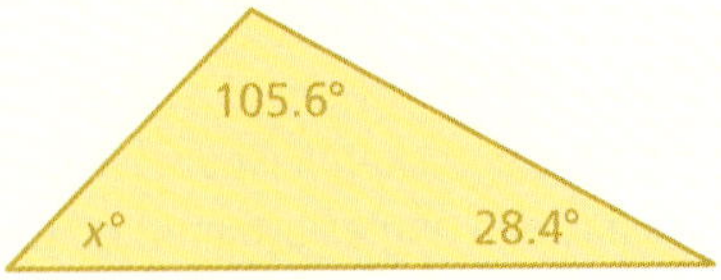

46; obtuse triangle

Technology For the Teacher

Dynamic Classroom

The Dynamic Planning Tool
Editable Teacher's Resources at *BigIdeasMath.com*

EXAMPLE 2 Finding Angle Measures

Find the value of x. Then classify each triangle.

a. Flag of Jamaica

$$x + x + 128 = 180$$
$$2x + 128 = 180$$
$$2x = 52$$
$$x = 26$$

The value of x is 26. Two of the sides are congruent. So, it is an isosceles triangle.

b. Flag of Cuba

$$x + x + 60 = 180$$
$$2x + 60 = 180$$
$$2x = 120$$
$$x = 60$$

The value of x is 60. All three angles are congruent. So, it is an equilateral and equiangular triangle.

EXAMPLE 3 Standardized Test Practice

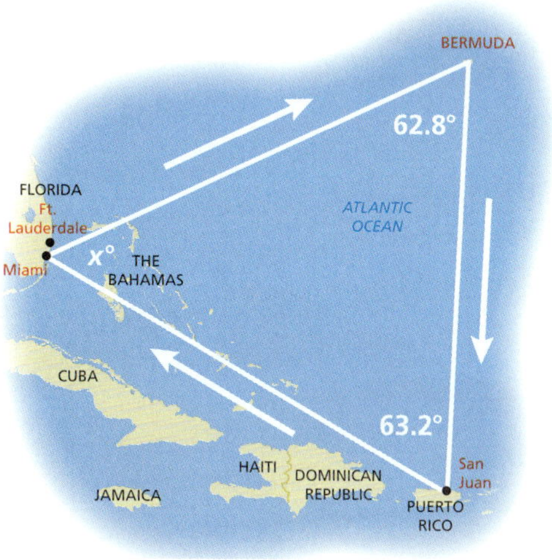

An airplane leaves from Miami and travels around the Bermuda Triangle. What is the value of x?

(A) 26.8 (B) 27.2 (C) 54 (D) 64

Use what you know about the angle measures of a triangle to write an equation.

$x + 62.8 + 63.2 = 180$	Write equation.
$x + 126 = 180$	Add.
$x = 54$	Subtract 126 from each side.

The value of x is 54. The correct answer is (C).

On Your Own

Now You're Ready
Exercises 9–11

Find the value of x. Then classify the triangle in as many ways as possible.

3.

4.

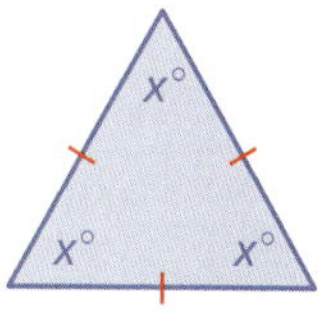

5. In Example 3, the airplane leaves from Fort Lauderdale. The angle measure at Bermuda is 63.9° and the angle measure at San Juan is 61.8°. Find the value of x.

5.2 Exercises

Check It Out
Help with Homework
BigIdeasMath.com

Vocabulary and Concept Check

1. **VOCABULARY** Compare equilateral and isosceles triangles.
2. **REASONING** Describe how to find the missing angle of the triangle.

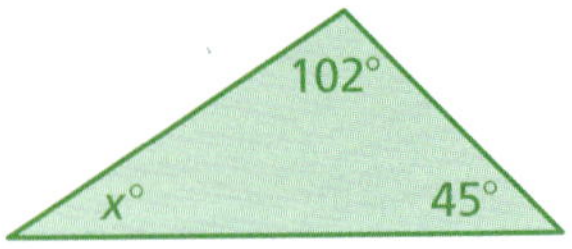

Practice and Problem Solving

Classify the triangle in as many ways as possible.

3. 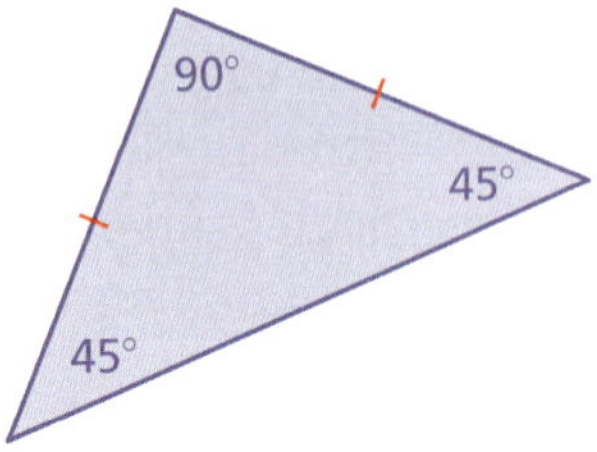

4. 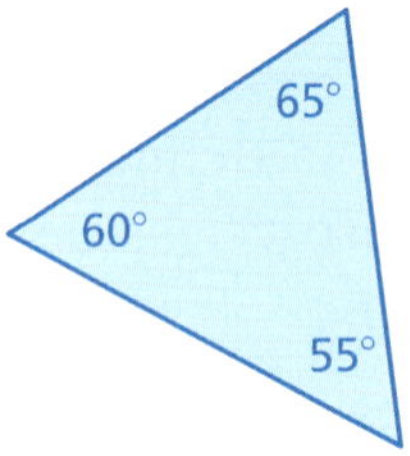

5. 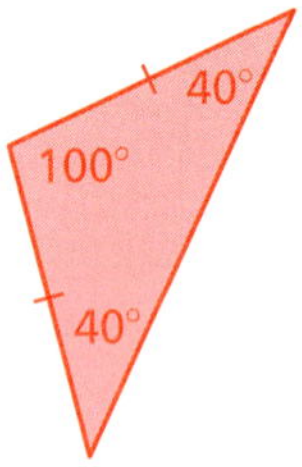

Find the value of x. Then classify the triangle in as many ways as possible.

1 6. 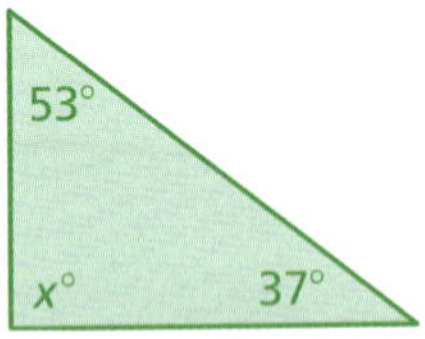

7. 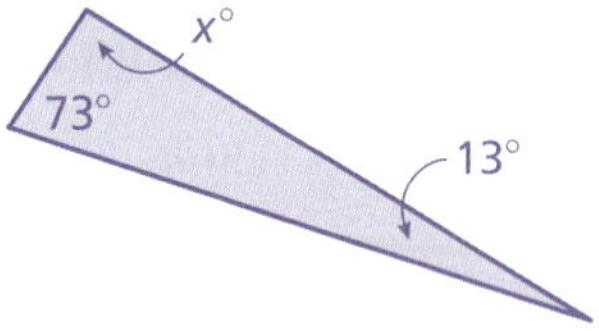

8. 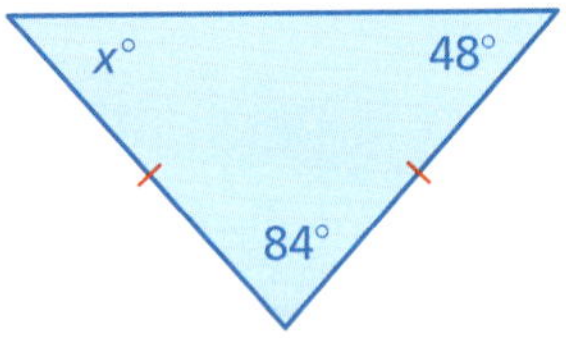

2 9. 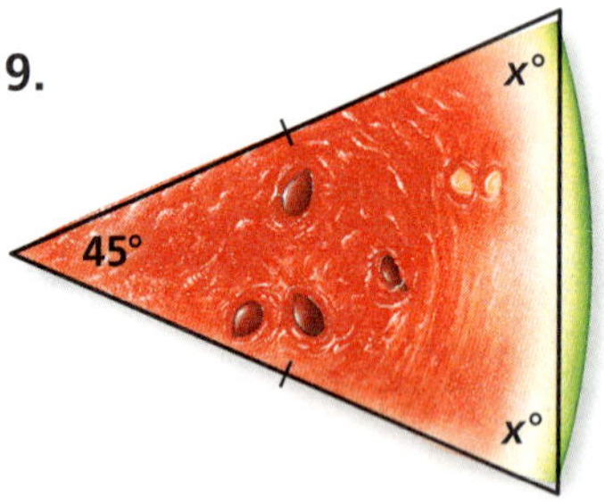

10. 

11. 

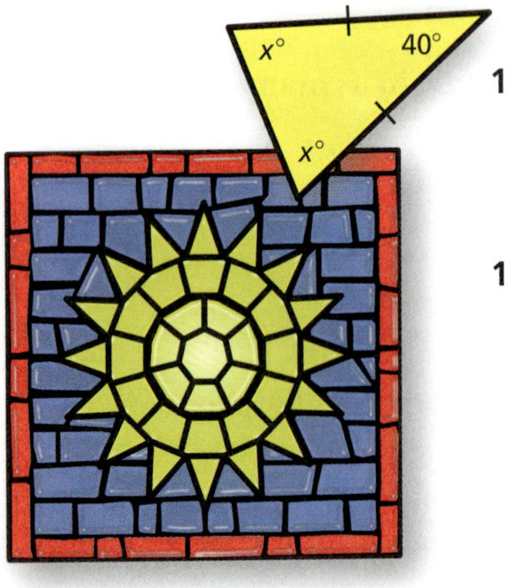

12. **ERROR ANALYSIS** Describe and correct the error in classifying the triangle.

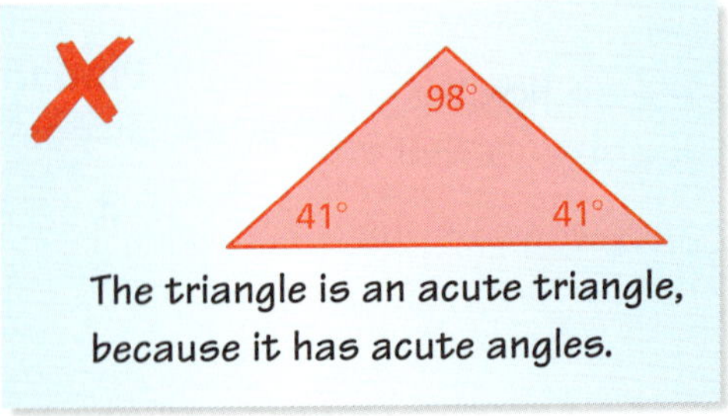

13. **MOSAIC TILE** A mosaic is a pattern or picture made of small pieces of colored material.
    a. Find the value of x.
    b. Classify the triangle used in the mosaic in two ways.

Assignment Guide and Homework Check

Level	Day 1 Activity Assignment	Day 2 Lesson Assignment	Homework Check
Basic	3–5, 21–24	1, 2, 6–13	2, 6, 10, 12
Average	3–5, 21–24	1, 2, 7–17 odd, 12, 18	7, 9, 12, 15, 18
Advanced	3–5, 21–24	1, 2, 12, 14–20	12, 14, 16, 18

Common Errors

- **Exercises 3–11** Students may not classify the triangle using all the possible words. Encourage them to think about the vocabulary.
- **Exercises 6–11** Students may solve for *x*, but forget to classify the triangle. Remind them to read the directions carefully and to answer the question.
- **Exercises 14–17** Students may recognize that the angles do not form a triangle, but may not know how to change the first angle so that all three angles will make a triangle. If the sum of the measure is greater than 180°, encourage them to decrease the first angle measure. If the sum of the measure is less than 180°, encourage them to increase the first angle measure.

5.2 Record and Practice Journal

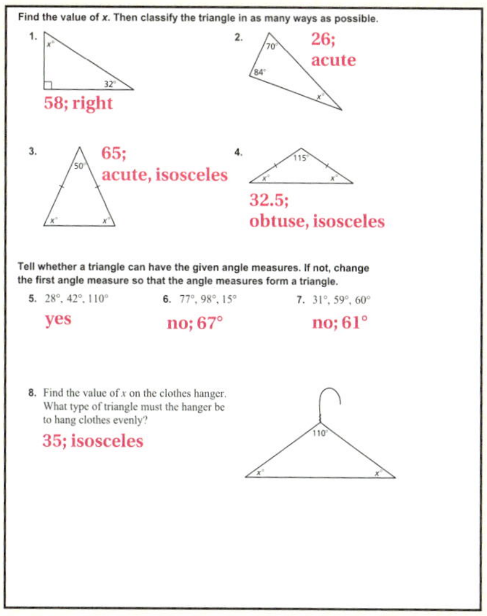
Find the value of x. Then classify the triangle in as many ways as possible.

1. 58; right
2. 26; acute
3. 65; acute, isosceles
4. 32.5; obtuse, isosceles

Tell whether a triangle can have the given angle measures. If not, change the first angle measure so that the angle measures form a triangle.

5. 28°, 42°, 110° yes
6. 77°, 98°, 15° no; 67°
7. 31°, 59°, 60° no; 61°

8. Find the value of x on the clothes hanger. What type of triangle must the hanger be to hang clothes evenly?
35; isosceles

Technology For the Teacher

Answer Presentation Tool
Quiz*Show*

Vocabulary and Concept Check

1. An equilateral triangle has three congruent sides. An isosceles triangle has at least two congruent sides. So, an equilateral triangle is a specific type of isosceles triangle.
2. Subtract 102 and 45 from 180.

Practice and Problem Solving

3. right isosceles triangle
4. acute triangle
5. obtuse isosceles triangle
6. 90; right triangle
7. 94; obtuse triangle
8. 48; acute isosceles triangle
9. 67.5; acute isosceles triangle
10. 60; equilateral, equiangular, acute, and isosceles triangle
11. 24; obtuse isosceles triangle
12. The triangle is not an acute triangle because acute triangles have 3 angles less than 90°. The triangle is an obtuse triangle because it has one angle that is greater than 90°.
13. **a.** 70

    b. acute isosceles triangle
14. yes
15. no; 39.5°
16. no; $28\frac{2}{3}°$
17. yes

Practice and Problem Solving

18. **a.** green: 65; purple: 25; red: 45

b. The angles opposite the congruent sides are congruent.

c. An isosceles triangle has at least two angles that are congruent.

19. See Additional Answers.

20. See *Taking Math Deeper*.

Fair Game Review

21. $x + 2x + 2x + 8 + 5 = 48; 7$

22. $3x + 19 = 28; 3$

23. $4x - 4 + 3\pi = 25.42$ or $2x - 4 = 6; 5$

24. A

Mini-Assessment

Find the value of *x*. Then classify each triangle.

1.

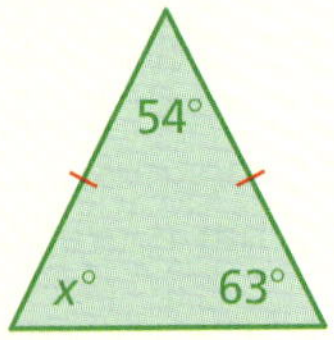

63; acute isosceles triangle

2.

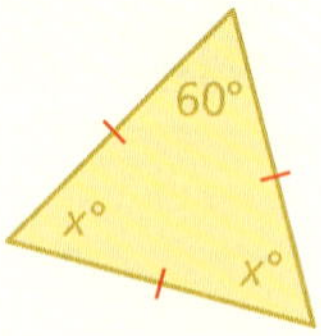

60; equiangular, equilateral, acute, and isosceles triangle

3.

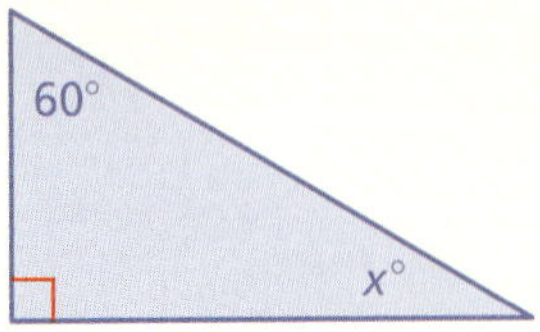

30; right triangle

Taking Math Deeper

Exercise 20

This problem gives students an opportunity to experiment with angle measures. To perform the experiment, students need a flat surface and 15 playing cards. The experiment is easier if the surface is non-slippery, such as a computer mouse pad. Have your students experiment with stacking cards using different angles before describing the limitations for *x*.

1 **a.** Solve for *x*.

$$x + x + 36 = 180$$
$$2x = 144$$
$$x = 72$$

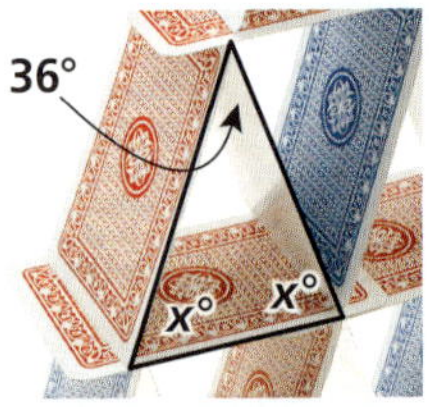

2 Describe the limitations for *x*.

b. If $x = 60$, then the three cards form an equilateral triangle. This is not possible because the two upright cards would have to be exactly on the edges of the base card. So, $x > 60$.

If $x = 90$, then the two upright cards would be vertical, which is not possible. The card structure would not be stable. So, $x < 90$.

3 Test your conclusions.

Use a deck of cards to test your conclusions. In practice, the limits on *x* are probably closer to $70 < x < 80$.

Project

Use a deck of cards to research different ways of building card towers.

Reteaching and Enrichment Strategies

If students need help. . .	If students got it. . .
Resources by Chapter • Practice A and Practice B • Puzzle Time Record and Practice Journal Practice Differentiating the Lesson Lesson Tutorials Skills Review Handbook	Resources by Chapter • Enrichment and Extension Start the next section

Tell whether a triangle can have the given angle measures. If not, change the first angle measure so that the angle measures form a triangle.

14. $76.2^\circ, 81.7^\circ, 22.1^\circ$

15. $115.1^\circ, 47.5^\circ, 93^\circ$

16. $5\frac{2}{3}^\circ, 64\frac{1}{3}^\circ, 87^\circ$

17. $31\frac{3}{4}^\circ, 53\frac{1}{2}^\circ, 94\frac{3}{4}^\circ$

18. **CRITICAL THINKING** Consider the three isosceles triangles.

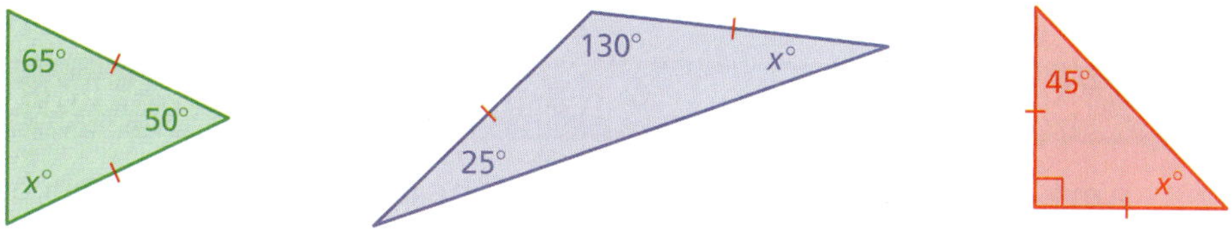

a. Find the value of x for each triangle.

b. What do you notice about the angle measures of each triangle?

c. Write a rule about the angle measures of an isosceles triangle.

19. **REASONING** Explain why all triangles have at least two acute angles.

20. **CARDS** One method of stacking cards is shown.

a. Find the value of x.

b. Critical Thinking Describe how to stack the cards with different angles. Is the value of x limited? If so, what are the limitations? Explain your reasoning.

Fair Game Review What you learned in previous grades & lessons

Write and solve an equation to find x. Use 3.14 for π. *(Skills Review Handbook)*

21. $P = 48$ cm

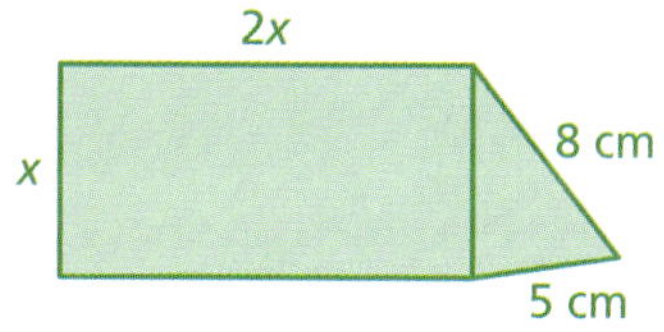

22. $P = 28$ in.

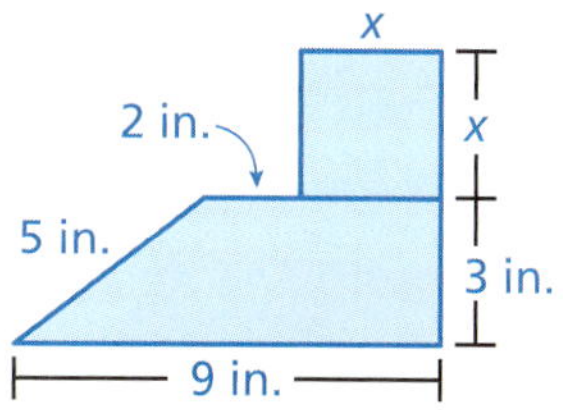

23. $P = 25.42$ m

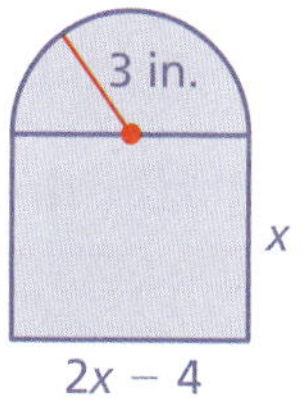

24. **MULTIPLE CHOICE** You have \$10 for text messages. Each message costs \$0.25. Which equation represents the amount of money you have after x messages? *(Section 3.1)*

Ⓐ $y = -0.25x + 10$ Ⓑ $y = 0.25x - 10$

Ⓒ $y = -0.25x - 10$ Ⓓ $y = 0.25x + 10$

5.3 Angles of Polygons

Essential Question How can you find a formula for the sum of the angle measures of any polygon?

1 ACTIVITY: The Sum of the Angle Measures of a Polygon

Work with a partner. Find the sum of the angle measures of each polygon with *n* sides.

a. Sample: Quadrilateral: $n = 4$

Draw a line that divides the quadrilateral into two triangles.

Because the sum of the angle measures of each triangle is 180°, the sum of the angle measures of the quadrilateral is 360°.

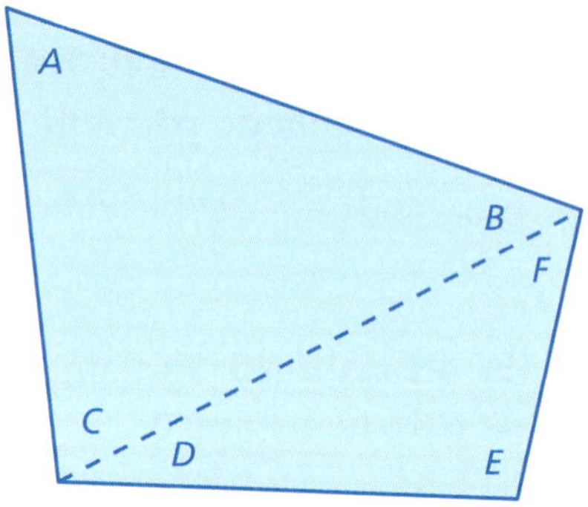

$$(A + B + C) + (D + E + F) = 180° + 180°$$
$$= 360°$$

b. Pentagon: $n = 5$

c. Hexagon: $n = 6$

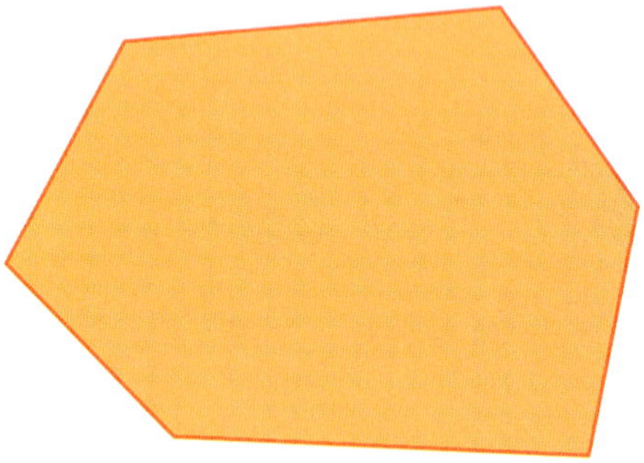

d. Heptagon: $n = 7$

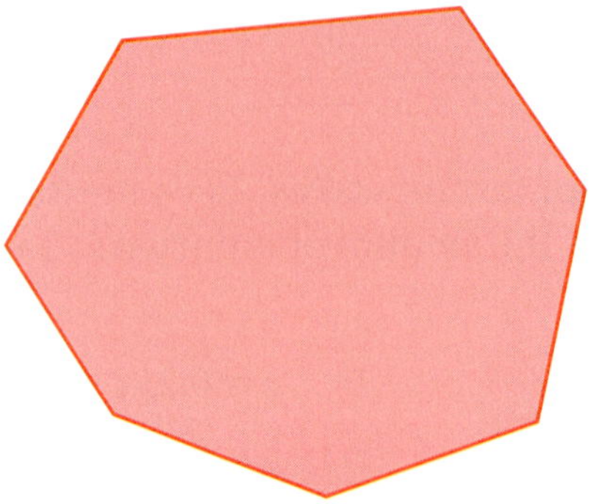

e. Octagon: $n = 8$

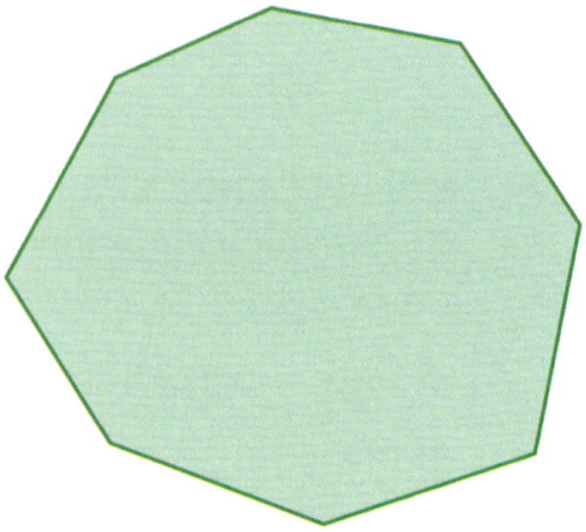

Laurie's Notes

Introduction

For the Teacher

- **Goal:** Students will investigate the sum of the angle measures of several polygons.
- You may need to review the definition of a polygon: A polygon is a closed plane figure made up of 3 or more line segments that intersect only at their endpoints.

Motivate

? "How many of you are looking forward to getting your driver's license?"
- Tell them that they will likely be tested on road signs.
- Draw several shapes and ask students if they know the names of the shapes and what they are used for on highway signs.

Activity Notes

Activity 1

- Work through part (a) with students.
- Draw the quadrilateral on the board.

? Use your non-writing hand to cover up the lower portion of the quadrilateral and ask, "What do you know about the angles of this triangle?" Angle measures sum to 180°.
- Label the angles *A*, *B*, and *C* as shown and write: $A + B + C = 180$.

? Now cover up the top portion of the quadrilateral and ask, "What do you know about the angles of this triangle?" Angle measures sum to 180°.
- Label the angles *D*, *E*, and *F* as shown and write: $D + E + F = 180$.
- Write the two equations together: $(A + B + C) + (D + E + F) = 360$.
- Note that when you combine *C* and *D*, it forms one of the angles of the original quadrilateral. The same is true for *B* and *F*.

? "Would this same technique work for any quadrilateral?" yes

? "What can you conclude about the sum of the angle measures of any quadrilateral?" Sum to 360°.
- Explain that they will use a similar technique to find the sum of the angle measures of polygons with more than 4 sides. It is important that the diagonals drawn divide the polygon into triangles whose angles are part of the original polygon. The simplest method is to draw the diagonals from one vertex.
- Ask volunteers to share their results with the class.
- **Another Way:** Cut polygons out of waxed paper. Tear the angles off, one at each vertex, similar to Activity 1 of Section 5.2. If you place the four angles about a point, they will *fill the space*, meaning 360°. Do this on the overhead. Now do the same thing for the pentagon. Place the 5 angles next to one another, about a point. The 5th angle, and perhaps a portion of the 4th, will overlap the angles below them. The waxed paper should show that you have gone 1.5 revolutions about the point, or 540°.

Previous Learning

Students should know how to solve multi-step equations.

Activity Materials
Introduction
• waxed paper • straightedge

Start Thinking! and Warm Up

Activity 5.3 **Start Thinking!** For use before Activity 5.3

Activity 5.3 **Warm Up** For use before Activity 5.3

Find the value of *y* for the given value of *x*.

1. $y = \frac{1}{2}x - 3;\ x = -2$
2. $y = -x + 2;\ x = 15$
3. $y = 10(x - 2);\ x = 10$
4. $y = 13(x - 3);\ x = 1$
5. $y = \frac{3}{2}x + \frac{5}{2};\ x = -2$
6. $y = 3(x + 4);\ x = -5$

5.3 Record and Practice Journal

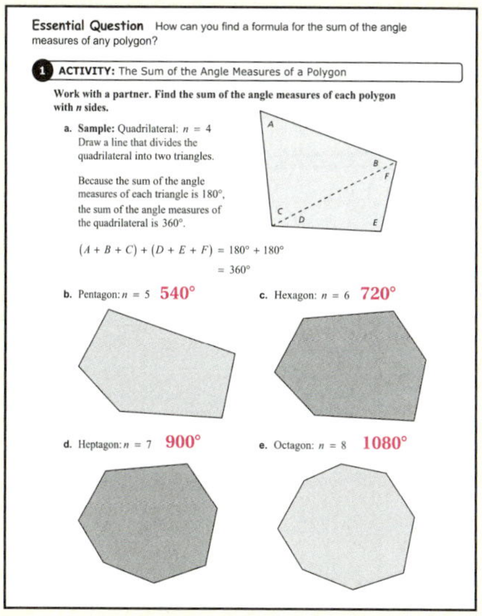

Essential Question How can you find a formula for the sum of the angle measures of any polygon?

1 ACTIVITY: The Sum of the Angle Measures of a Polygon

Work with a partner. Find the sum of the angle measures of each polygon with *n* sides.

a. **Sample:** Quadrilateral: $n = 4$
Draw a line that divides the quadrilateral into two triangles.

Because the sum of the angle measures of each triangle is 180°, the sum of the angle measures of the quadrilateral is 360°.

$$(A + B + C) + (D + E + F) = 180° + 180°$$
$$= 360°$$

b. Pentagon: $n = 5$ 540°

c. Hexagon: $n = 6$ 720°

d. Heptagon: $n = 7$ 900°

e. Octagon: $n = 8$ 1080°

Differentiated Instruction

Kinesthetic

Another way to discover the sum of the angle measures of a polygon is to have the students cut the polygon into triangles. This can be done for convex and concave polygons.

5.3 Record and Practice Journal

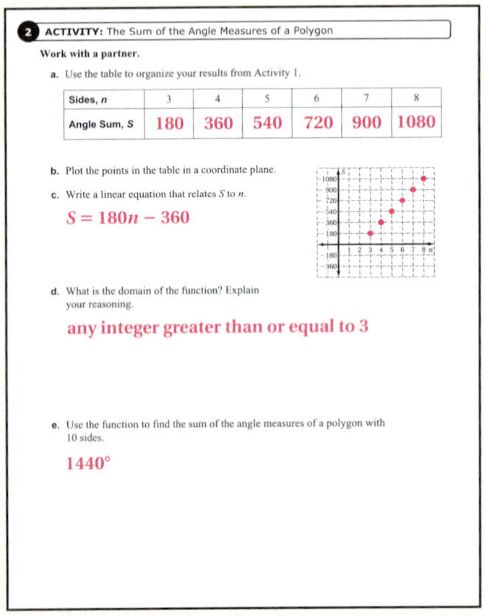

2 ACTIVITY: The Sum of the Angle Measures of a Polygon

Work with a partner.

a. Use the table to organize your results from Activity 1.

Sides, n	3	4	5	6	7	8
Angle Sum, S	180	360	540	720	900	1080

b. Plot the points in the table in a coordinate plane.

c. Write a linear equation that relates S to n.

$S = 180n - 360$

d. What is the domain of the function? Explain your reasoning.

any integer greater than or equal to 3

e. Use the function to find the sum of the angle measures of a polygon with 10 sides.

1440°

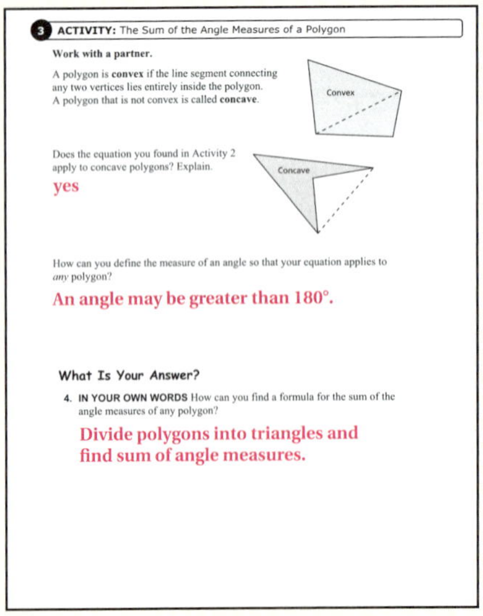

3 ACTIVITY: The Sum of the Angle Measures of a Polygon

Work with a partner.

A polygon is **convex** if the line segment connecting any two vertices lies entirely inside the polygon. A polygon that is not convex is called **concave**.

Does the equation you found in Activity 2 apply to concave polygons? Explain.

yes

How can you define the measure of an angle so that your equation applies to *any* polygon?

An angle may be greater than 180°.

What Is Your Answer?

4. **IN YOUR OWN WORDS** How can you find a formula for the sum of the angle measures of any polygon?

Divide polygons into triangles and find sum of angle measures.

Laurie's Notes

Activity 2

- Students should know that the first value is 180°. They should recognize that the remaining values are multiples of 180. Note that the vertical axis has been scaled by increments of 180 to make plotting easier and more accurate.

? "What is the slope of the line? Explain." slope = 180; When n increases by 1 (run), S increases by 180 (rise).

- Students may have difficulty with part (d). First, the domain must be only whole numbers because you cannot have 3.5 sides. Second, n must be greater than or equal to 3 because a triangle has the least number of sides of any polygon.

? "What do you call a domain that is not continuous?" discrete

- **Connection:** Another way to write the formula $S = 180n - 360$ is $S = 180(n - 2)$. The quantity $(n - 2)$ is the number of triangles that can be made when you draw all the diagonals from one vertex in an n-gon. This number is multiplied by 180 because there are 180° in every one of the triangles.

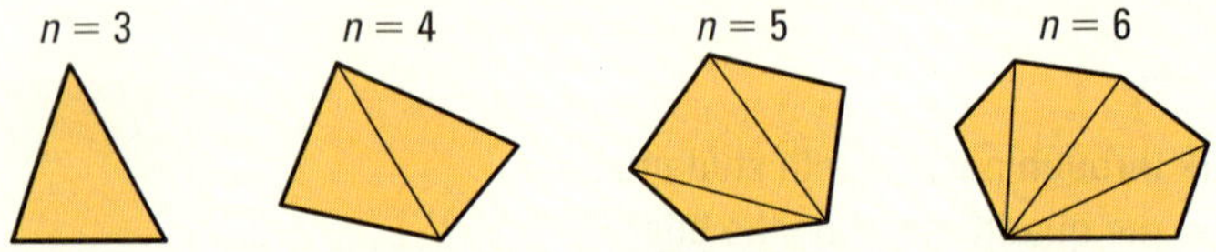

Activity 3

- Students may have heard the words **convex** and **concave** before. These words are often used to describe a lens or mirror. Students will say that a concave polygon *caves in* or is *dented in*.
- **Note:** The rule for the sum of the angle measures of a polygon *does* apply to concave polygons. All you need to do is include angle measures that are greater than 180°.

What Is Your Answer?

- **Neighbor Check** Have students work independently and then have their neighbor check their work. Have students discuss any discrepancies.

Closure

- **Exit Ticket:** Use your equation to find the sum of the angle measures of a polygon with 12 sides. 1800°

Technology For the Teacher

Dynamic Classroom

The Dynamic Planning Tool
Editable Teacher's Resources at *BigIdeasMath.com*

2 ACTIVITY: The Sum of the Angle Measures of a Polygon

Work with a partner.

a. Use the table to organize your results from Activity 1.

Sides, *n*	3	4	5	6	7	8
Angle Sum, *S*						

b. Plot the points in the table in a coordinate plane.

c. Write a linear equation that relates S to n.

d. What is the domain of the function? Explain your reasoning.

e. Use the function to find the sum of the angle measures of a polygon with 10 sides.

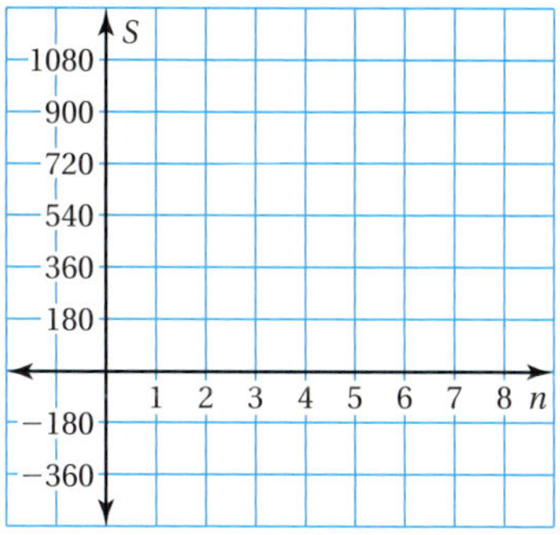

3 ACTIVITY: The Sum of the Angle Measures of a Polygon

Work with a partner.

A polygon is convex if the line segment connecting any two vertices lies entirely inside the polygon. A polygon that is not convex is called concave.

Does the equation you found in Activity 2 apply to concave polygons? Explain.

How can you define the measure of an angle so that your equation applies to *any* polygon?

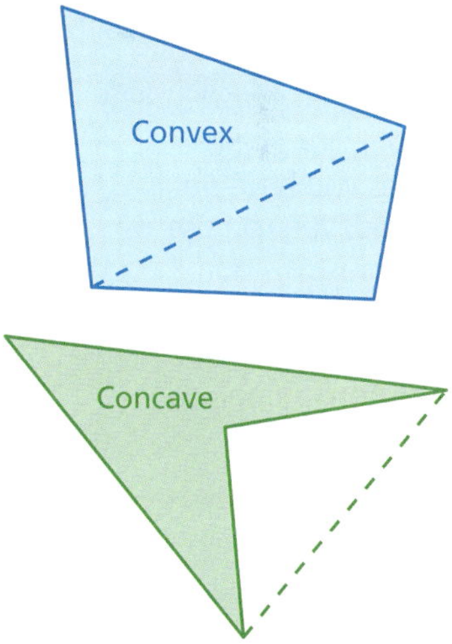

What Is Your Answer?

4. **IN YOUR OWN WORDS** How can you find a formula for the sum of the angle measures of any polygon?

Practice Use what you learned about angles of polygons to complete Exercises 4–6 on page 201.

5.3 Lesson

Check It Out
Lesson Tutorials
BigIdeasMath.com

Key Vocabulary
polygon, *p. 198*
regular polygon, *p. 199*
convex polygon, *p. 200*
concave polygon, *p. 200*

A **polygon** is a closed plane figure made up of three or more line segments that intersect only at their endpoints.

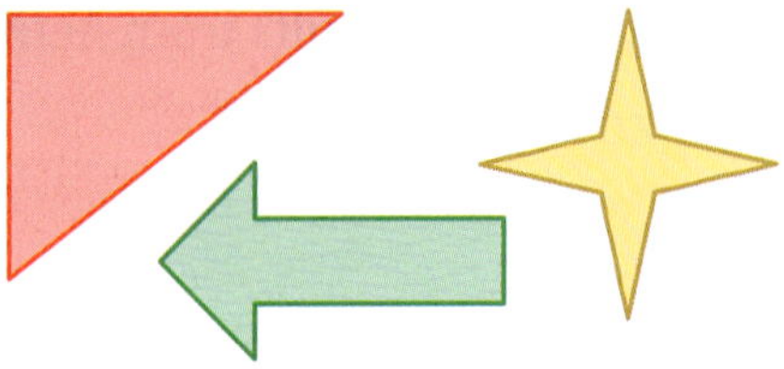

Polygons

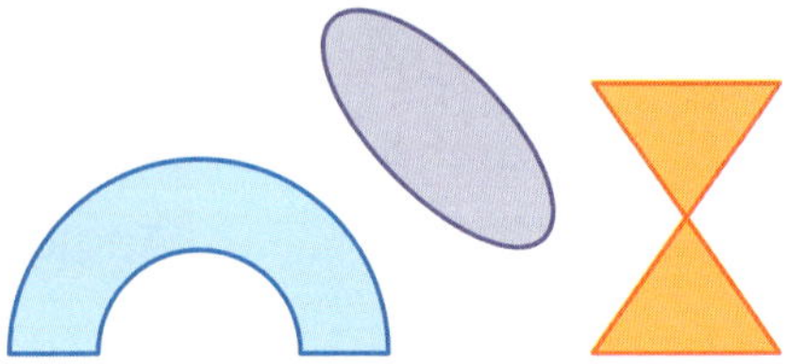

Not polygons

Key Idea

Angle Measures of a Polygon

The sum S of the angle measures of a polygon with n sides is

$$S = (n - 2) \cdot 180^\circ.$$

EXAMPLE 1 Finding the Sum of the Angle Measures of a Polygon

Reading
For polygons whose names you have not learned, you can use the phrase "*n*-gon," where *n* is the number of sides. For example, a 15-gon is a polygon with 15 sides.

Find the sum of the angle measures of the school crossing sign.

The sign is in the shape of a pentagon. It has 5 sides.

$S = (n - 2) \cdot 180^\circ$	Write the formula.
$= (5 - 2) \cdot 180^\circ$	Substitute 5 for *n*.
$= 3 \cdot 180^\circ$	Subtract.
$= 540^\circ$	Multiply.

The sum of the angle measures is 540°.

On Your Own

Now You're Ready
Exercises 7–9

Find the sum of the angle measures of the green polygon.

1.

2.

Laurie's Notes

Introduction

Connect

- **Yesterday:** Students explored finding the sum of the angle measures of a polygon.
- **Today:** Students will use a formula to find the missing angle measure of a polygon.

Motivate

? "Did you ever wonder why bees use a hexagonal structure for their honeycomb? Why not squares? or circles? or octagons?"

- Draw a few cells of the honeycomb.

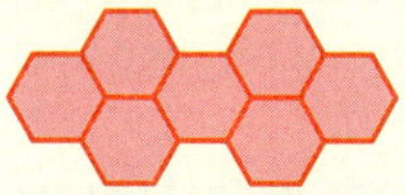

- Mathematicians have concluded that a hexagon is the most appropriate geometric form for the maximum use of a given area. This means that a hexagonal cell requires the minimum amount of wax for construction while it stores the maximum amount of honey.

Lesson Notes

Key Idea

- Write the definition of a polygon. Draw examples of shapes which are and are not polygons. Students should be able to explain why some are not polygons.
- In all of the samples shown, the interior is shaded. The polygon is the figure formed by the line segments. The polygonal region contains the interior of the polygon.
- Write the Key Idea. This is the same equation that students wrote yesterday, but in the more common form. This form highlights the fact that the sum is a multiple of 180°.

Example 1

- Review the names of common polygons: Triangle (3), Quadrilateral (4), Pentagon (5), Hexagon (6), Octagon (8), and Decagon (10). It is also common to say n-gon and replace n with 9 to talk about a 9-sided polygon.
- Read the problem. The polygon is a pentagon.
- Write the equation, substitute 5 for n, and solve.

On Your Own

- **Think-Pair-Share:** Students should read each question independently and then work with a partner to answer the questions. When they have answered the questions, the pair should compare their answers with another group and discuss any discrepancies.

? "What are the names of the polygons in Questions 1 and 2?"
7-gon or heptagon; hexagon

Goal Today's lesson is finding angle measures of a **polygon**.

Start Thinking! and Warm Up

Lesson 5.3 Warm Up
For use before Lesson 5.3

Lesson 5.3 Start Thinking!
For use before Lesson 5.3

Explain how you can draw an octagon on graph paper to make it easy to find the angle measures. Use this method to find each angle measure.

What is the sum of the angle measures? Does this agree with the equation you wrote in Activity 2?

Extra Example 1

Find the sum of the angle measures of the polygon.

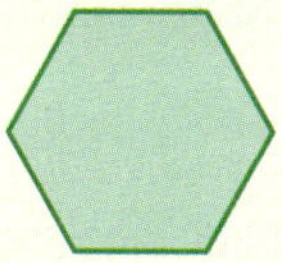

720°

On Your Own

1. 900°
2. 720°

Extra Example 2

Find the value of x.

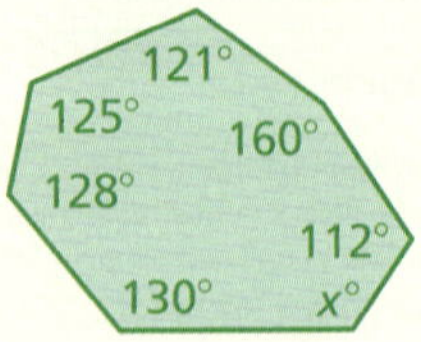

124

On Your Own

3. 105

4. 75

5. 35

Extra Example 3

Find the measure of each angle of a regular 12-gon.

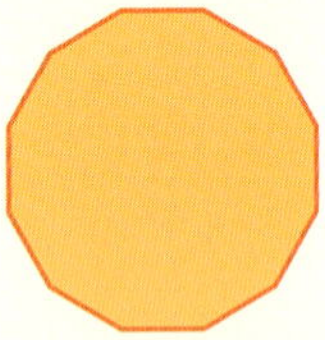

150°

English Language Learners

Vocabulary

Preview the *Key Vocabulary* in this chapter. Understanding geometry depends on understanding the terminology used. Have students write key vocabulary words in their notebooks. Include definitions and examples to help distinguish between words (e.g., convex polygon and concave polygon).

Laurie's Notes

Example 2

- **Connection:** This example integrates equation solving with finding a missing angle.
- ? "How many sides does the polygon have?" 7
- ? "How do you find the sum of the measures of all of the angles of a 7-gon?" Solve $(7 - 2)180 = 900$.
- Once the sum is known, write and solve the equation as shown. Caution students to be careful with their arithmetic.

On Your Own

- Students should check with their neighbor to make sure they are setting up the equation correctly. Each problem has two parts: Determining the sum of all of the angle measures, and then writing the equation to solve for the missing angle.
- In Question 4, remind students that the symbol for a right angle means the angle measures 90°.
- In Question 5, two angles are missing, each with a measure of $2x°$. The sum of the angle measures of this pentagon is 540°, so $2x + 145 + 145 + 2x + 110 = 540$. The steps are to combine like terms, isolate the variable, and solve.
- **Common Error:** Students will solve for the variable correctly, but then forget to substitute this value back into the variable expression to solve for the angle measure. In Question 5, students were only asked to solve for x. If they had been asked to find the measure of the angle, there would be one last step. In this case, $x = 35$ and the two missing angles are each 70°.

Example 3

- Review the definition of a regular polygon. Point out to students that squares and equilateral triangles are examples of regular polygons.
- A regular hexagon has 6 congruent angles. If the angle measures of a hexagon sum to 720° and the 6 angles are congruent, it should make sense to students why they divide 720 by 6.
- Look back to the honeycomb you drew at the beginning of the lesson. There are three 120° angles about one point.
- You can show a video of the cloud system from the website *jpl.nasa.gov*.

EXAMPLE 2 Finding an Angle Measure of a Polygon

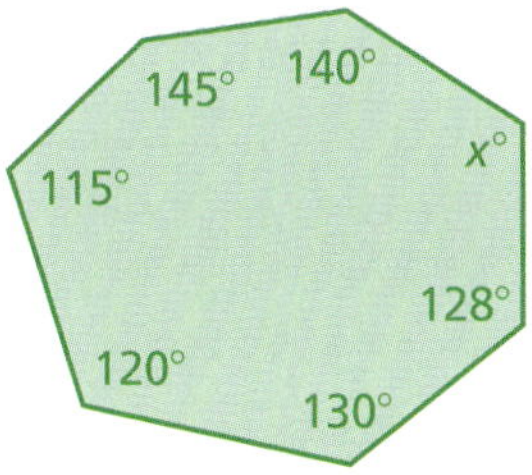

Find the value of x.

Step 1: The polygon has 7 sides. Find the sum of the angle measures.

$S = (n - 2) \cdot 180°$ Write the formula.

$= (7 - 2) \cdot 180°$ Substitute 7 for n.

$= 900°$ Simplify. The sum of the angle measures is 900°.

Step 2: Write and solve an equation.

$$140 + 145 + 115 + 120 + 130 + 128 + x = 900$$

$$778 + x = 900$$

$$x = 122$$

The value of x is 122.

On Your Own

Now You're Ready
Exercises 12–14

Find the value of x.

3.

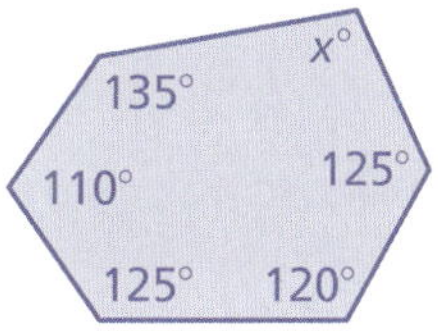

4.

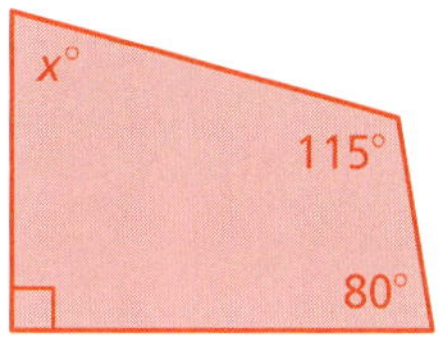

5.

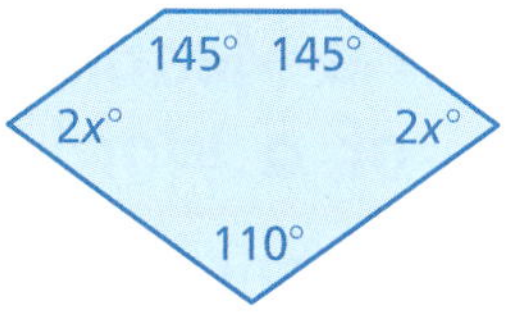

In a **regular polygon**, all of the sides are congruent and all of the angles are congruent.

EXAMPLE 3 Real-Life Application

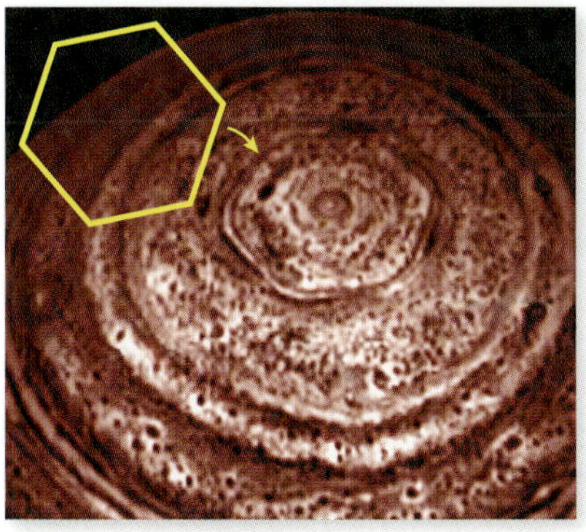

The hexagon is about 15,000 miles across. Approximately four Earths could fit inside it.

A cloud system discovered on Saturn is in the approximate shape of a regular hexagon. Find the measure of each angle of the hexagon.

Step 1: A hexagon has 6 sides. Find the sum of the angle measures.

$S = (n - 2) \cdot 180°$ Write the formula.

$= (6 - 2) \cdot 180°$ Substitute 6 for n.

$= 720°$ Simplify. The sum of the angle measures is 720°.

Step 2: Divide the sum by the number of angles, 6.

$720° \div 6 = 120°$

The measure of each angle is 120°.

On Your Own

Now You're Ready
Exercises 16–18

Find the measure of each angle of the regular polygon.

6. octagon **7.** decagon **8.** 18-gon

Key Idea

Convex and Concave Polygons

A polygon is **convex** if every line segment connecting any two vertices lies entirely inside the polygon.

A polygon is **concave** if at least one line segment connecting any two vertices lies outside the polygon.

EXAMPLE 4 Identifying Convex and Concave Polygons

The Meaning of a Word

Concave

To remember the term con**cave**, think of a polygon that is "**cave**d in."

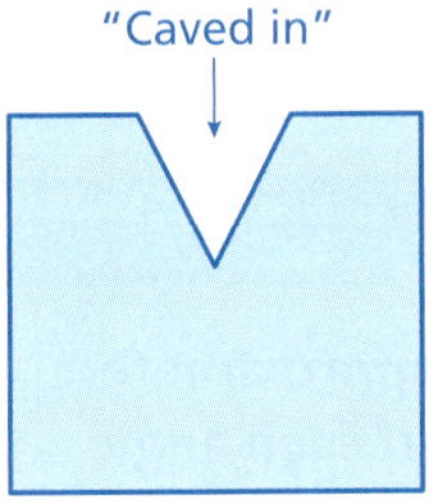

Tell whether the polygon is *convex* or *concave*. Explain.

a.

A line segment connecting two vertices lies outside the polygon. So, the polygon is concave.

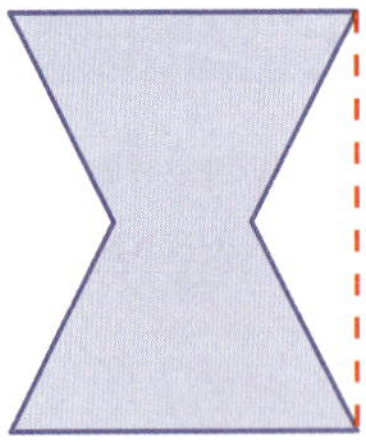

b.

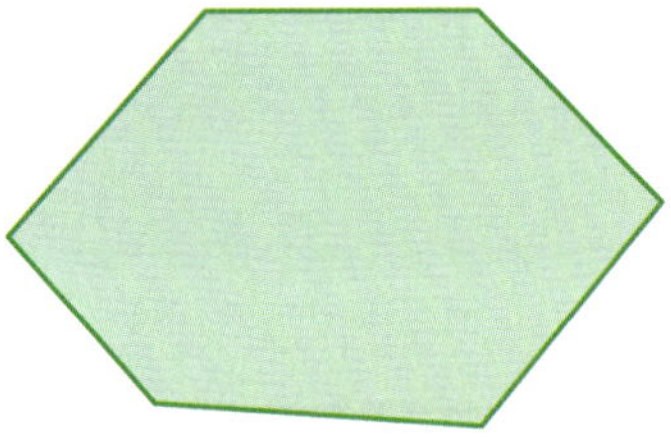

No line segment connecting two vertices lies outside the polygon. So, the polygon is convex.

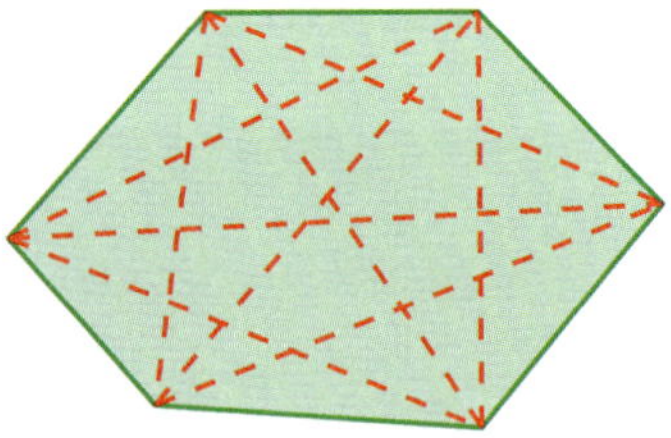

On Your Own

Now You're Ready
Exercises 22–24

Tell whether the polygon is *convex* or *concave*. Explain.

9.

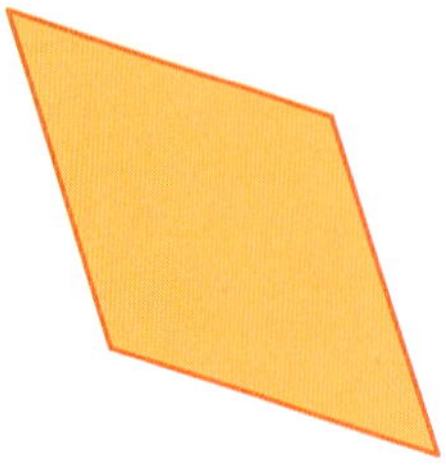

10.

11.

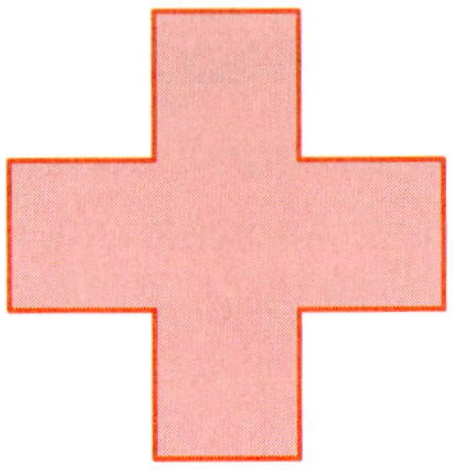

Laurie's Notes

On Your Own

- **Think-Pair-Share:** Students should read each question independently and then work with a partner to answer the questions. When they have answered the questions, the pair should compare their answers with another group and discuss any discrepancies.

Key Idea

- Write the definitions of convex and concave polygons. Draw samples of each. Note the *Meaning of a Word* box.

Example 4

- Students usually do not have trouble making sense of convex and concave polygons.

Closure

- A pentagon has two right angles and the other three angles are all congruent. What is the measure of one of the missing angles? 120°

Technology For the Teacher

Dynamic Classroom

The Dynamic Planning Tool

Editable Teacher's Resources at *BigIdeasMath.com*

On Your Own

6. 135°
7. 144°
8. 160°

Extra Example 4

Tell whether the polygon is *convex* or *concave*. Explain.

a.

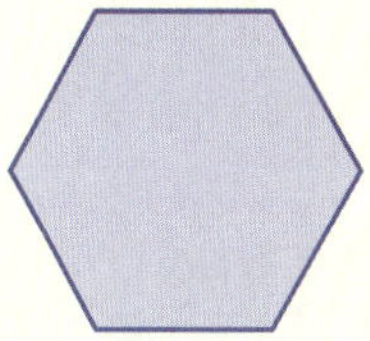

convex; No line segment connecting two vertices lies outside the polygon.

b.

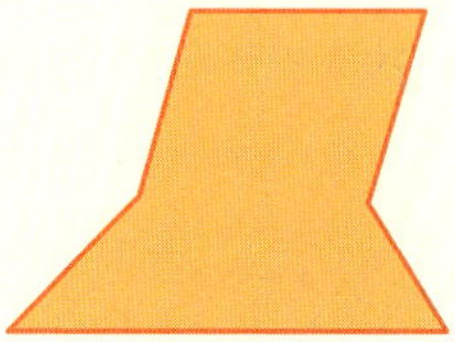

concave; A line segment connecting two vertices lies outside the polygon.

On Your Own

9. convex; No line segment connecting two vertices lies outside the polygon.
10. concave; A line segment connecting two vertices lies outside the polygon.
11. concave; A line segment connecting two vertices lies outside the polygon.

Vocabulary and Concept Check

1. 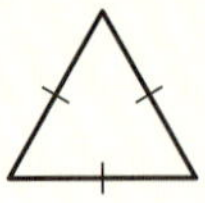

2. The second figure doesn't belong because it is not a polygon.

3. What is the measure of an angle of a regular pentagon?; 108°; 540°

Practice and Problem Solving

4. 360°

5. 1260°

6. 900°

7. 720°

8. 1800°

9. 1080°

10. The right side of the formula is $(n - 2) \cdot 180°$, not $n \cdot 180°$.
$$S = (n - 2) \cdot 180°$$
$$= (13 - 2) \cdot 180°$$
$$= 11 \cdot 180°$$
$$= 1980°$$

11. no; The angle measures given add up to 535°, but the sum of the angle measures of a pentagon is 540°.

12. 43

13. 135

14. 90

15. 140°

16. 60°

17. 140°

18. 150°

Assignment Guide and Homework Check

Level	Day 1 Activity Assignment	Day 2 Lesson Assignment	Homework Check
Basic	4–6, 33–37	1–3, 7, 9–13, 16, 17, 19–23, 27	7, 12, 16, 20, 22
Average	4–6, 33–37	1–3, 7–25 odd, 22, 27–30	7, 13, 17, 22, 30
Advanced	4–6, 33–37	1–3, 11, 14, 15, 18–21, 23–26, 29–32	14, 18, 24, 26, 30

Common Errors

- **Exercises 4–6** Students may struggle dividing the polygon into triangles. Encourage them to trace the polygon in pen in their notebooks, and then draw triangles with a pencil so that they can erase lines if necessary.
- **Exercises 7–9** Students may forget to subtract 2 from the number of sides when using the formula to find the sum of the angle measures. Remind them of the formula and encourage them to write the formula before substituting the number of sides.
- **Exercise 11** Students may say that because the sum of the angle measures is close to the value found when using the formula, a pentagon can have these angle measures. Remind them that the sum of the angle measures must be *exactly* the same as the sum found with the formula for the polygon to be drawn with the given angles.

5.3 Record and Practice Journal

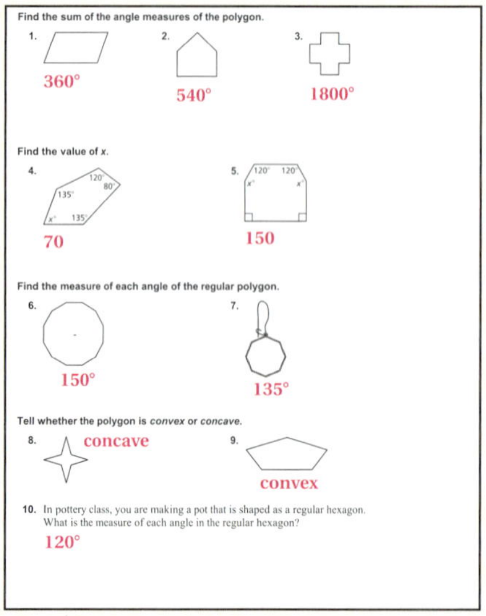

Technology For the Teacher

Answer Presentation Tool

Quiz*Show*

5.3 Exercises

Vocabulary and Concept Check

1. **VOCABULARY** Draw a regular polygon that has three sides.

2. **WHICH ONE DOESN'T BELONG?** Which figure does *not* belong with the other three? Explain your reasoning.

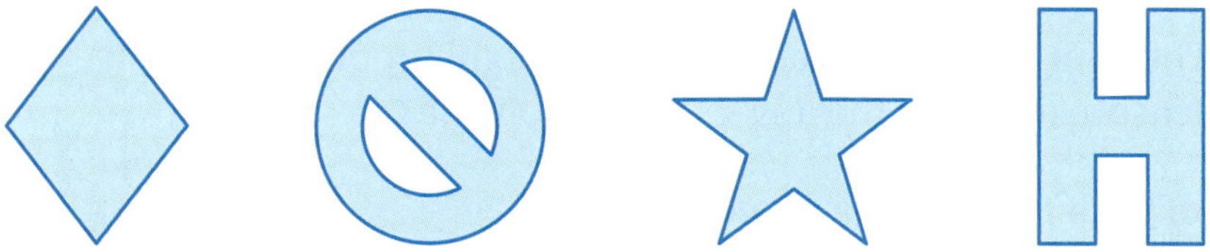

3. **DIFFERENT WORDS, SAME QUESTION** Which is different? Find "both" answers.

What is the measure of an angle of a regular pentagon?	What is the sum of the angle measures of a convex pentagon?
What is the sum of the angle measures of a regular pentagon?	What is the sum of the angle measures of a concave pentagon?

Practice and Problem Solving

Use triangles to find the sum of the angle measures of the polygon.

4. 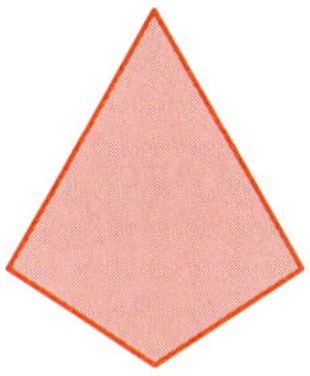

5. 

6. 

Find the sum of the angle measures of the polygon.

1 7. 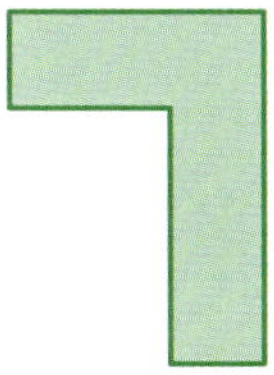

8. 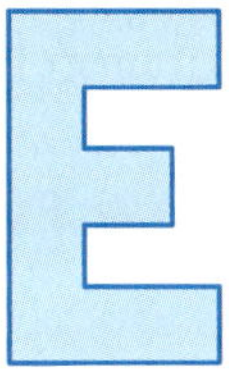

9. 

10. **ERROR ANALYSIS** Describe and correct the error in finding the sum of the angle measures of a 13-gon.

✗
$$S = n \cdot 180^\circ$$
$$= 13 \cdot 180^\circ$$
$$= 2340^\circ$$

11. **NUMBER SENSE** Can a pentagon have angles that measure 120°, 105°, 65°, 150°, and 95°? Explain.

Find the value of x.

② **12.**

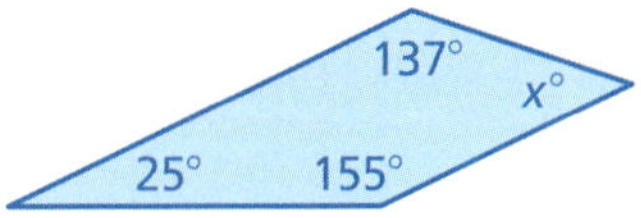

13.

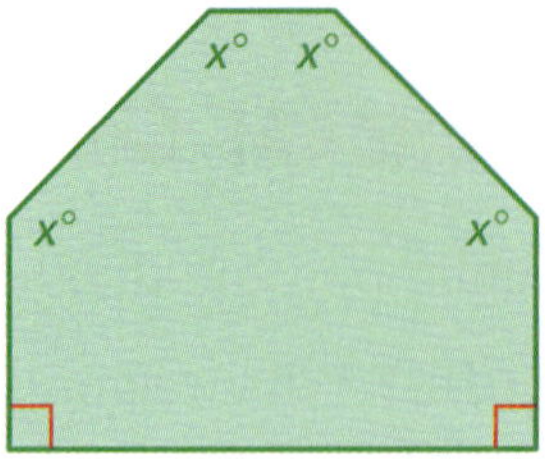

14.

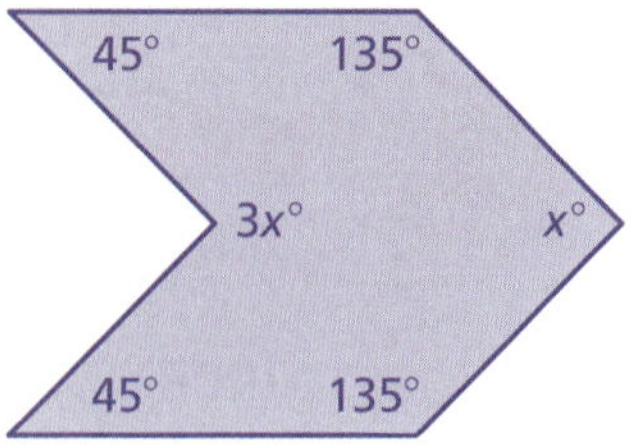

15. REASONING The sum of the angle measures in a regular polygon is 1260°. What is the measure of one of the angles of the polygon?

Find the measure of each angle of the regular polygon.

③ **16.**

17.

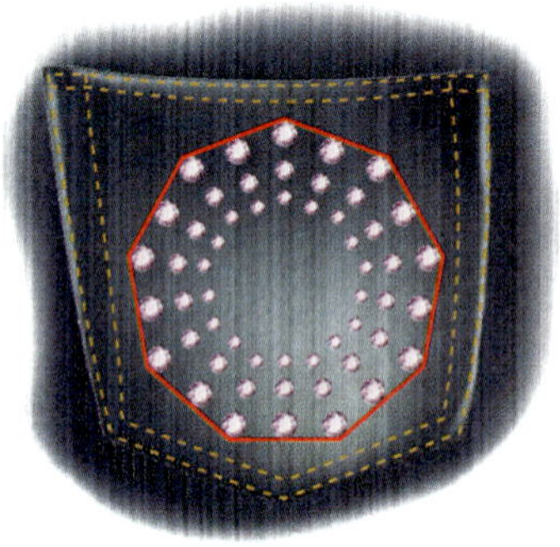

18.

19. ERROR ANALYSIS Describe and correct the error in finding the measure of each angle of a regular 20-gon.

✗
$S = (n - 2) \cdot 180°$
$= (20 - 2) \cdot 180°$
$= 18 \cdot 180°$
$= 3240°$
$3240° \div 18 = 180$
The measure of each angle is 180°.

20. FIRE HYDRANT A fire hydrant bolt is in the shape of a regular pentagon.

a. What is the measure of each angle?

b. Why are fire hydrants made this way?

21. PUZZLE The angles of a regular polygon each measure 165°. How many sides does the polygon have?

Tell whether the polygon is *convex* or *concave*. Explain.

④ **22.**

23.

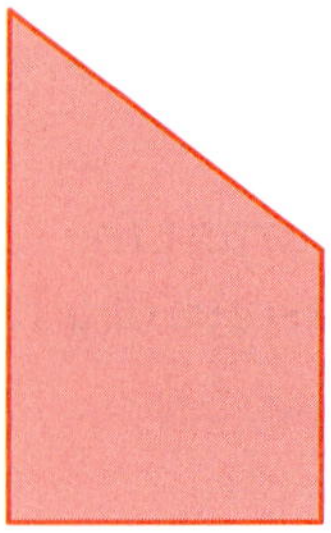

24.

25. CRITICAL THINKING Can a concave polygon be regular? Explain.

26. OPEN-ENDED Draw a polygon that has congruent sides but is not regular.

Common Errors

- **Exercises 12–14** Students may forget to include one or more of the given angles when writing an equation for the missing angles. For example, in Exercise 13, students may write $4x = 720$. Remind them to include all of the angles. Encourage them to write the equation and then count the number of terms to make sure that there are the same number of terms as angles before simplifying.
- **Exercises 16–18** Students may find the sum of the angle measures of the regular polygon, but forget to divide by the number of angles to answer the question. Remind them that they are finding the measure of *one* angle. Because all the angles are congruent (by the definition of a regular polygon), they can divide the sum of the angle measures by the number of angles.
- **Exercises 22–24** Students may not try to connect all the vertices of the polygon and state that the polygon is convex. Remind them to connect all the vertices with all the others. One strategy is to start with one vertex and connect it to all the other vertices, and then rotate clockwise to the next vertex and repeat the process.

English Language Learners

Vocabulary

Many English mathematical names are influenced by Latin and Greek names for numbers.

English	Latin	Greek
one	unus	heis
two	duo	duo
three	tres	tria
four	quattuor	tettara
five	quinque	pente
six	sex	hex
seven	septe	hepta
eight	octo	okto
nine	nove	ennea
ten	dece	deka

Spanish speakers will recognize the similarity of numbers in Spanish to the numbers in Latin.

Practice and Problem Solving

19. The sum of the angle measures should have been divided by the number of angles, 20. $3240° \div 20 = 162°$; The measure of each angle is 162°.

20. **a.** 108°

b. *Sample answer:* So the same wrench can unscrew any fire hydrant bolt.

21. 24 sides

22. concave; A line segment connecting two vertices lies outside the polygon.

23. convex; No line segment connecting two vertices lies outside the polygon.

24. concave; A line segment connecting two vertices lies outside the polygon.

25. no; All of the angles would not be congruent.

26. See Additional Answers.

Practice and Problem Solving

27. 135°

28.

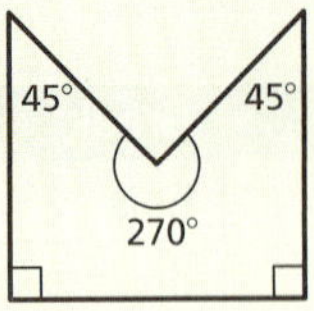

29. 120°

30. **a.** 11 sides

b. 147°

31. See *Taking Math Deeper*.

32. See Additional Answers.

Fair Game Review

33. 9 **34.** 2

35. 3 **36.** 6

37. D

Mini-Assessment

Find the sum of the angle measures of the polygon.

1.

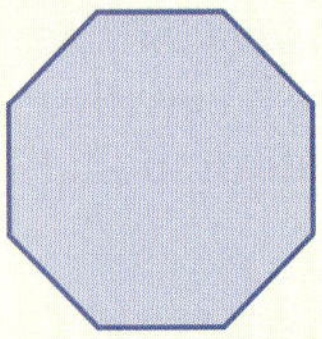

1080°

2.

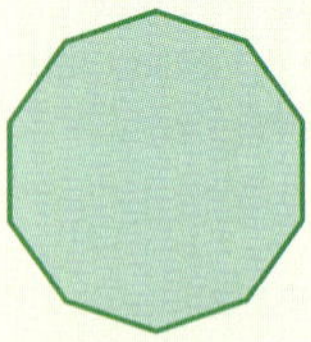

1440°

3. Tell whether the polygon is *convex* or *concave*. Explain.

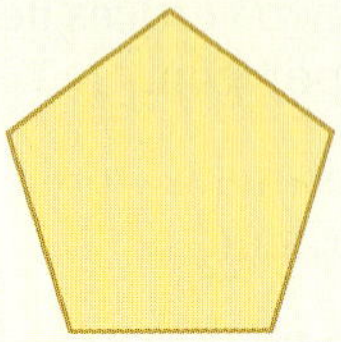

convex; No line segment connecting two vertices lies outside the polygon.

Taking Math Deeper

Exercise 31

This problem is an interesting comparison of linear and nonlinear functions. In Activity 2 on page 197, students discovered that the sum S of the angle measures of a polygon is a linear function of the number n of sides of the polygon.

$$S = (n - 2) \cdot 180$$
$$= 180n - 360$$

However, when you change the formula to represent the angle measure a of each angle in a regular polygon, the function is no longer linear.

$$a = \frac{S}{n} = 180 - \frac{360}{n}$$

1 Complete the table.

Sides of a Regular Polygon, n	3	4	5	6	7	8	9	10
Measure of One Angle (degrees), a	60	90	108	120	128.6	135	140	144

2 **Algebraic Reasoning:** The pattern is not linear because the function

$$a = 180 - \frac{360}{n}$$

cannot be written in slope-intercept form, $a = mn + b$.

3 **Graphical Reasoning:** The function is not linear because the points in the graph do not lie on a line.

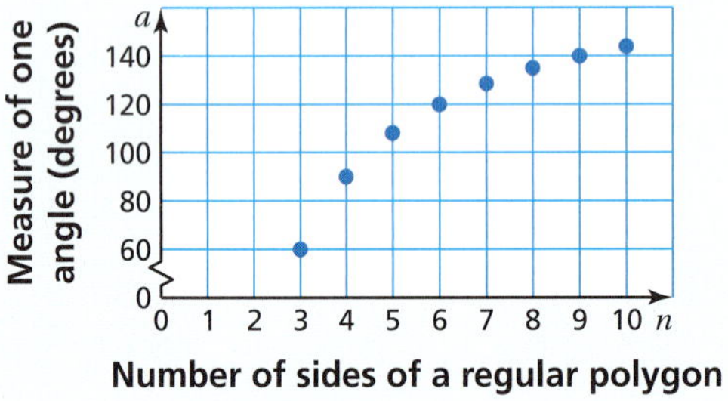

Reteaching and Enrichment Strategies

If students need help. . .	If students got it. . .
Resources by Chapter • Practice A and Practice B • Puzzle Time Record and Practice Journal Practice Differentiating the Lesson Lesson Tutorials Skills Review Handbook	Resources by Chapter • Enrichment and Extension • School-to-Work Start the next section

27. **STAINED GLASS** The center of the stained glass window is in the shape of a regular polygon. What is the measure of each angle of the polygon?

28. **PENTAGON** Draw a pentagon that has two right angles, two 45° angles, and one 270° angle.

29. **GAZEBO** The floor of a gazebo is in the shape of a heptagon. Four of the angles measure 135°. The other angles have equal measures. Find the measure of each of the remaining angles.

30. **MONEY** The border of a Susan B. Anthony dollar is in the shape of a regular polygon.

a. How many sides does the polygon have?

b. What is the measure of each angle of the border? Round your answer to the nearest degree.

31. **REASONING** Copy and complete the table. Does the table represent a linear function? Explain.

Sides of a Regular Polygon, *n*	3	4	5	6	7	8	9	10
Measure of One Angle, *a*								

32. Geometry When tiles can be used to cover a floor with no empty spaces, the collection of tiles is called a *tessellation*.

a. Create a tessellation using equilateral triangles.

b. Find two more regular polygons that form tessellations.

c. Create a tessellation that uses two different regular polygons.

Fair Game Review What you learned in previous grades & lessons

Solve the proportion. *(Skills Review Handbook)*

33. $\frac{x}{12} = \frac{3}{4}$ **34.** $\frac{14}{21} = \frac{x}{3}$ **35.** $\frac{x}{9} = \frac{2}{6}$ **36.** $\frac{4}{10} = \frac{x}{15}$

37. **MULTIPLE CHOICE** The ratio of tulips to daisies is 3 : 5. Which of the following could be the total number of tulips and daisies? *(Skills Review Handbook)*

Ⓐ 6 Ⓑ 10 Ⓒ 15 Ⓓ 16

5 Study Help

Check It Out
Graphic Organizer
BigIdeasMath.com

You can use an **example and non-example chart** to list examples and non-examples of a vocabulary word or item. Here is an example and non-example chart for complementary angles.

Complementary Angles

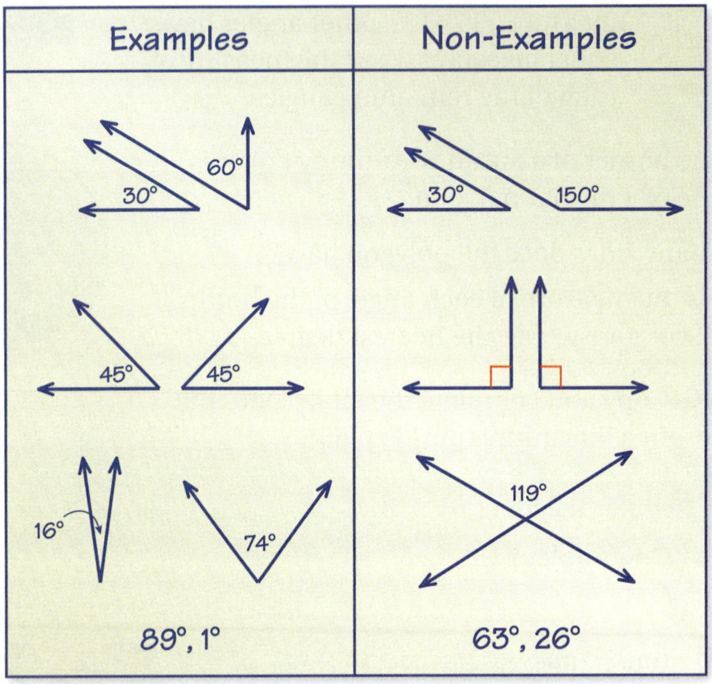

On Your Own

Make an example and non-example chart to help you study these topics.

1. isosceles triangles
2. equilateral triangles
3. regular polygons
4. convex polygons
5. concave polygons

After you complete this chapter, make example and non-example charts for the following topics.

6. similar triangles
7. transversals
8. interior angles

"What do you think of my example & non-example chart for popular cat toys?"

Sample Answers

1. Isosceles Triangles

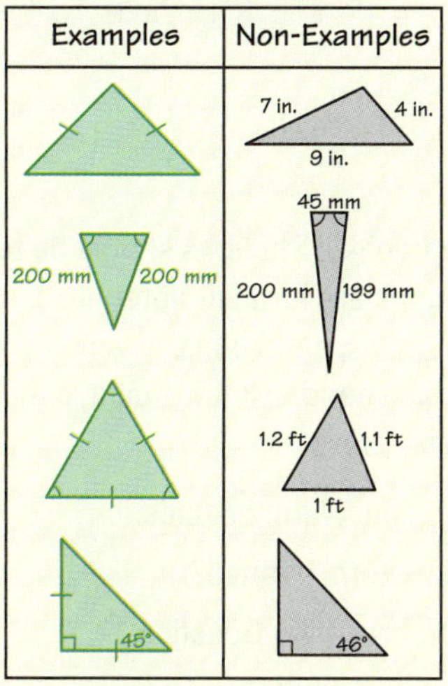

2. Equilateral Triangles

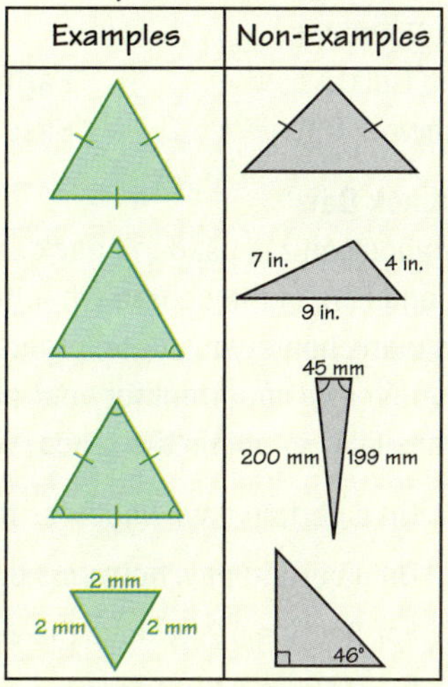

3. Regular Polygons

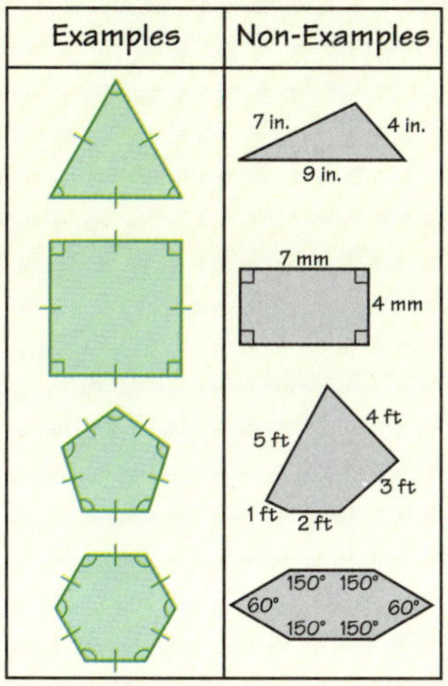

4. Convex Polygons

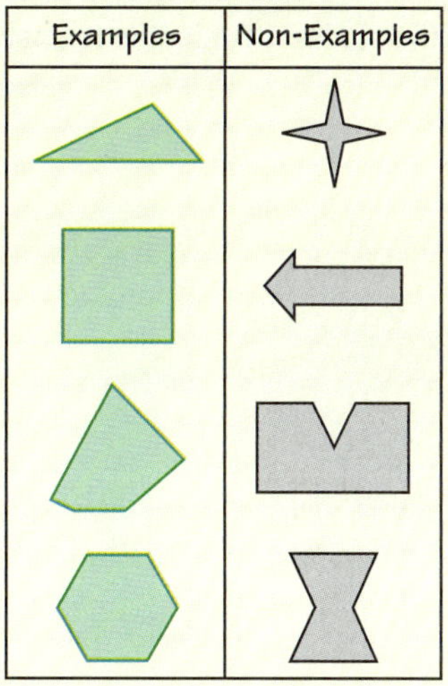

5. Concave Polygons

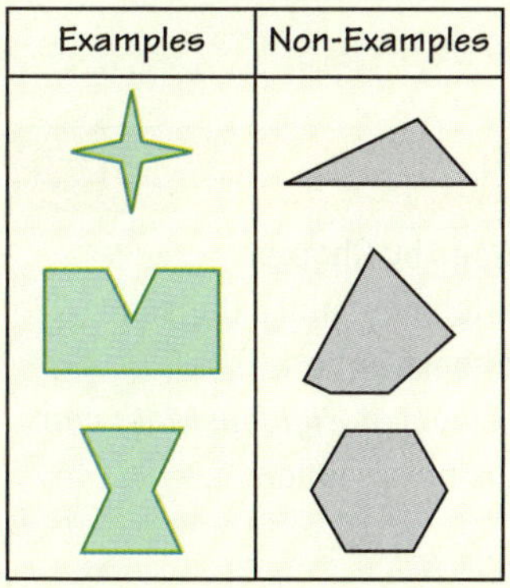

List of Organizers

Available at *BigIdeasMath.com*

Comparison Chart
Concept Circle
Definition (Idea) and Example Chart
Example and Non-Example Chart
Formula Triangle
Four Square
Information Frame
Information Wheel
Notetaking Organizer
Process Diagram
Summary Triangle
Word Magnet
Y Chart

About this Organizer

An **Example and Non-Example Chart** can be used to list examples and non-examples of a vocabulary word or term. Students write examples of the word or term in the left column and non-examples in the right column. This type of organizer serves as a good tool for assessing students' knowledge of pairs of topics that have subtle but important differences, such as complementary and supplementary angles. Blank example and non-example charts can be included on tests or quizzes for this purpose.

Technology For the Teacher

Vocabulary Puzzle Builder

Answers

1. neither
2. complementary
3. supplementary
4. $x = 146$
5. $x = 16$
6. $x = 59$
7. $x = 60$; isosceles, equilaterial, equiangular, acute
8. $x = 115$; obtuse
9. $x = 45$; right, isosceles
10. 1080°
11. concave
12. $x = 58$
13. $x = 126$
14. $x = 70$
15. $\angle 1 = 65°, \angle 2 = 115°, \angle 3 = 65°$
16. 25 sides
17. isosceles, acute

Assessment Book

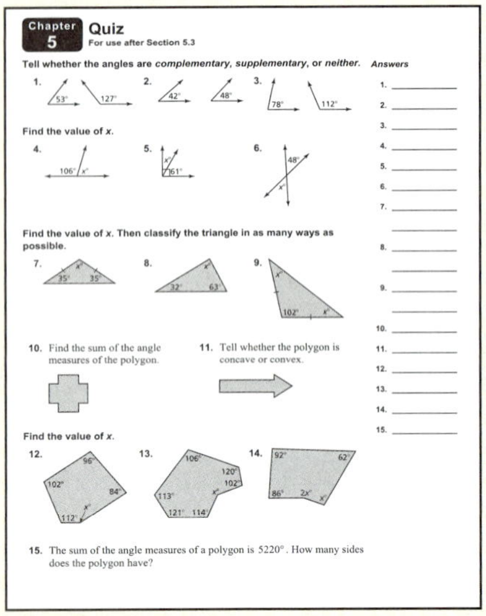

Chapter 5 **Quiz**
For use after Section 5.3

Tell whether the angles are *complementary*, *supplementary*, or *neither*.

1. 53°, 127°
2. 42°, 48°
3. 78°, 112°

Find the value of *x*.

4. 106°, x°
5. x°, 61°
6. 48°, x°

Find the value of *x*. Then classify the triangle in as many ways as possible.

7. x°, 35°, 35°
8. x°, 32°, 63°
9. x°, 102°, x°

10. Find the sum of the angle measures of the polygon.
11. Tell whether the polygon is concave or convex.

Find the value of *x*.

12. 96°, 102°, 84°, 112°, x°
13. 106°, 120°, 102°, x°, 113°, 121°, 114°
14. 92°, 62°, 86°, 2x°, x°

15. The sum of the angle measures of a polygon is 5220°. How many sides does the polygon have?

Answers

1. ______
2. ______
3. ______
4. ______
5. ______
6. ______
7. ______
8. ______
9. ______
10. ______
11. ______
12. ______
13. ______
14. ______
15. ______

Alternative Quiz Ideas

100% Quiz	Math Log
Error Notebook	**Notebook Quiz**
Group Quiz	Partner Quiz
Homework Quiz	Pass the Paper

Notebook Quiz

A notebook quiz is used to check students' notebooks. Students should be told at the beginning of the course what the expectations are for their notebooks: notes, class work, homework, date, problem number, goals, definitions, or anything else that you feel is important for your class. They also need to know that it is their responsibility to obtain the notes when they miss class.

1. On a certain day, what was the answer to the warm up question?
2. On a certain day, how was this vocabulary term defined?
3. For Section 5.2, what is the answer to On Your Own Question 1?
4. For Section 5.3, what is the answer to the Essential Question?
5. On a certain day, what was the homework assignment?

Give the students 5 minutes to answer these questions.

Reteaching and Enrichment Strategies

If students need help. . .	If students got it. . .
Resources by Chapter • Study Help • Practice A and Practice B • Puzzle Time Lesson Tutorials *BigIdeasMath.com* Practice Quiz Practice from the Test Generator	Resources by Chapter • Enrichment and Extension • School-to-Work Game Closet at *BigIdeasMath.com* Start the next section

Technology For the Teacher

Answer Presentation Tool
Big Ideas Test Generator

5.1–5.3 Quiz

Check It Out
Progress Check
BigIdeasMath.com

Tell whether the angles are *complementary*, *supplementary*, or *neither*. *(Section 5.1)*

1.

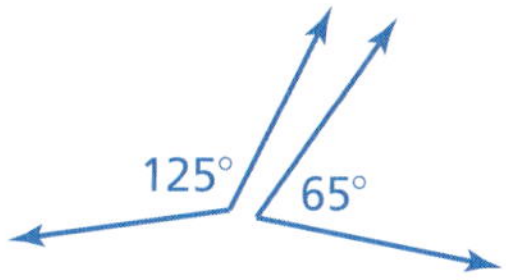

2.

3.

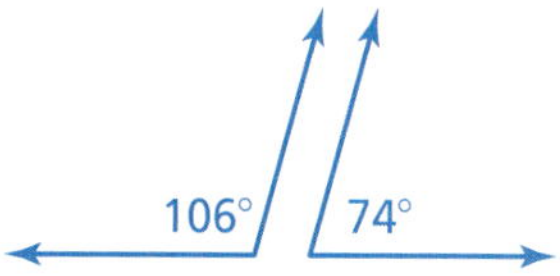

Find the value of *x*. *(Section 5.1)*

4.

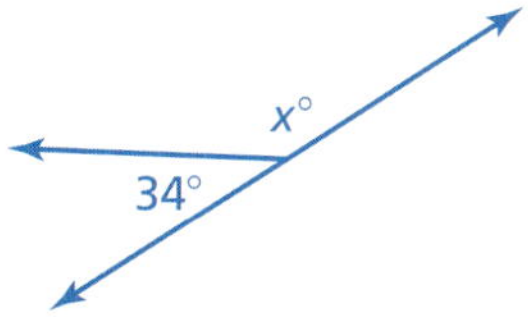

5.

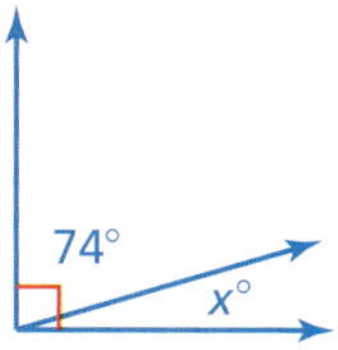

6.

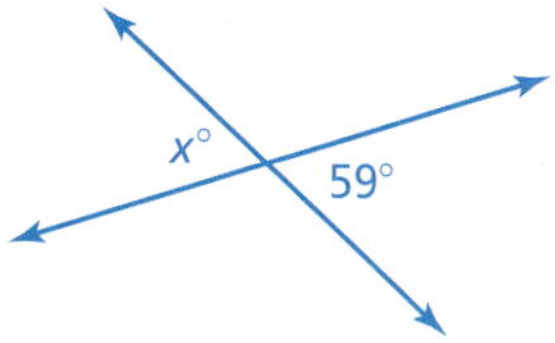

Find the value of *x*. Then classify the triangle in as many ways as possible. *(Section 5.2)*

7.

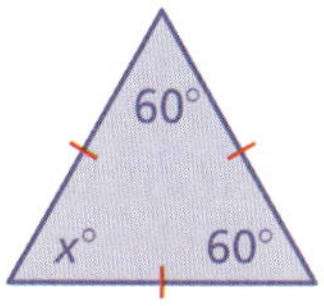

8.

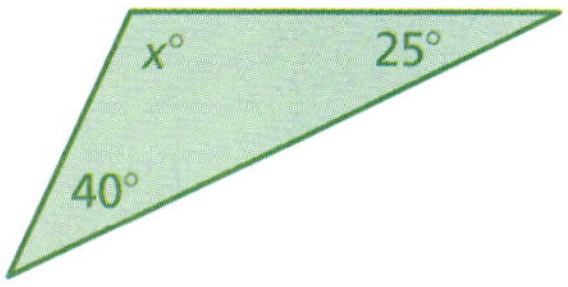

9.

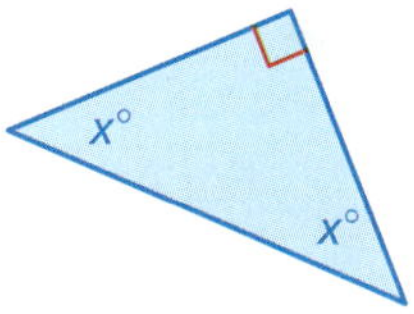

10. Find the sum of the angle measures of the polygon. *(Section 5.3)*

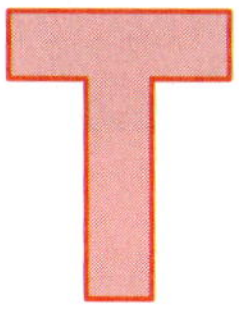

11. Tell whether the polygon is concave or convex. *(Section 5.3)*

Find the value of *x*. *(Section 5.3)*

12.

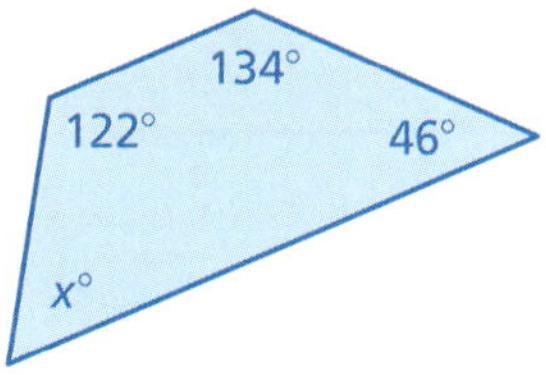

13.

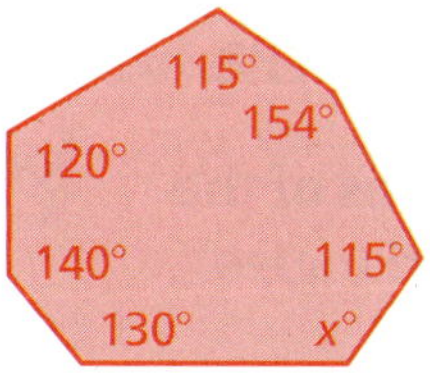

14.

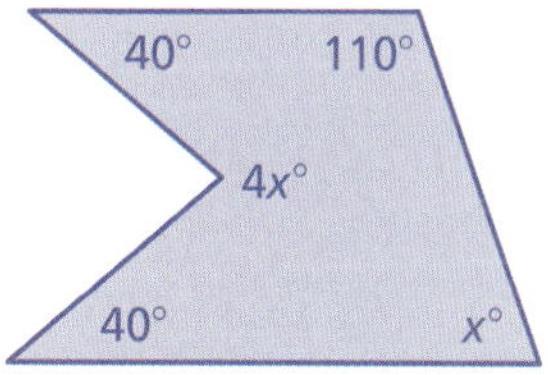

15. RAILROAD CROSSING What are the measures of the other three angles formed by the intersection of the road and the railroad tracks? *(Section 5.1)*

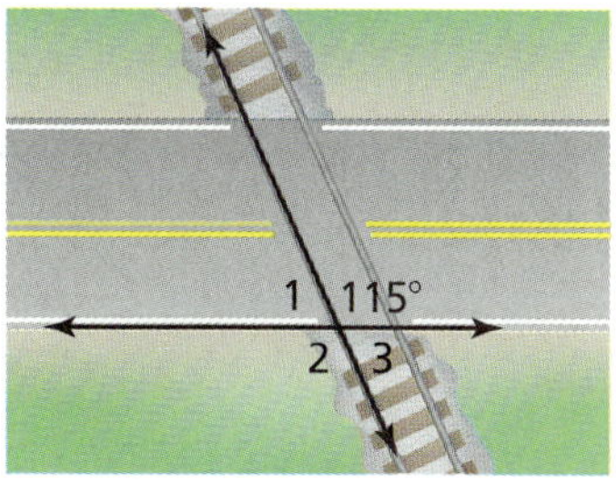

16. REASONING The sum of the angle measures of a polygon is 4140°. How many sides does the polygon have? *(Section 5.3)*

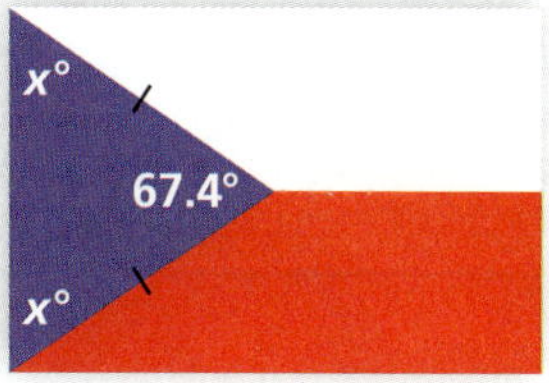

17. FLAG Classify the triangle on the flag of the Czech Republic in as many ways as possible. *(Section 5.2)*

5.4 Using Similar Triangles

Essential Question Which properties of triangles make them special among all other types of polygons?

You already know that two triangles are **similar** if and only if the ratios of their corresponding side lengths are equal.

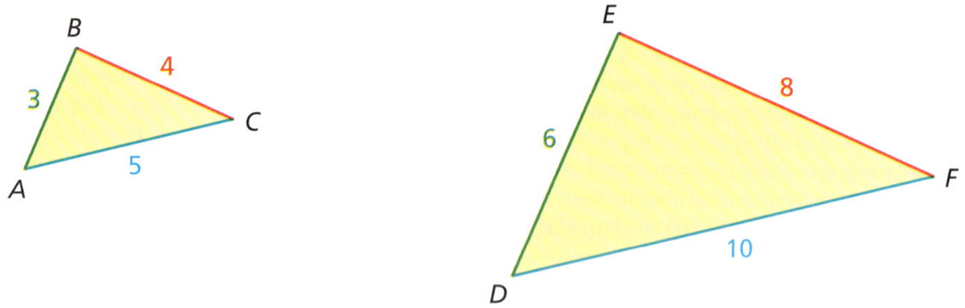

For example, $\triangle ABC$ is similar to $\triangle DEF$ because the ratios of their corresponding side lengths are equal.

$$\frac{6}{3} = \frac{10}{5} = \frac{8}{4}$$

1 ACTIVITY: Angles of Similar Triangles

Work with a partner.

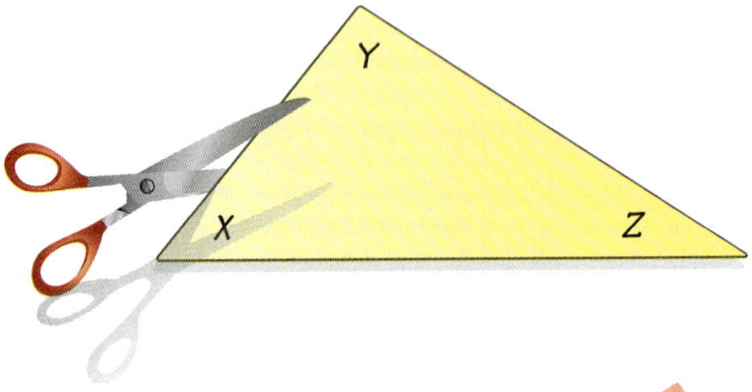

- Discuss how to make a triangle that is larger than $\triangle XYZ$ and has the *same* angle measures as $\triangle XYZ$.
- Measure the lengths of the sides of the two triangles.
- Find the ratios of the corresponding side lengths. Are they all the same? What can you conclude?

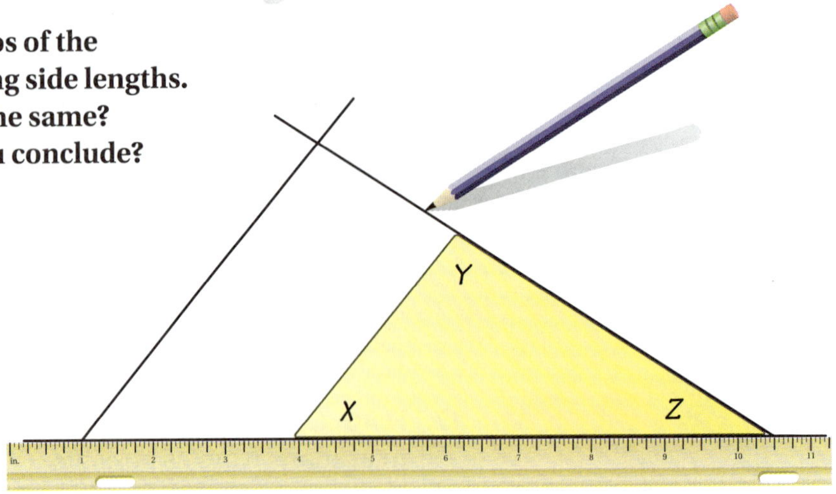

Laurie's Notes

Introduction

For the Teacher

- **Goal:** Students will explore properties of similar triangles.
- **Big Idea:** Triangles are special in several ways because these properties are not true for quadrilaterals.
  - First, when the corresponding sides are proportional, the corresponding angles are congruent. For quadrilaterals, this can be contradicted by a square and a rhombus.

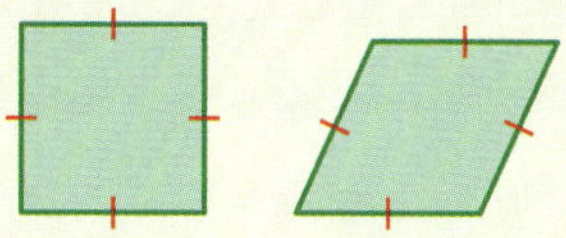

  - Second, when the corresponding angles are congruent, the corresponding sides are proportional. For quadrilaterals, this can be contradicted by a square and a rectangle.

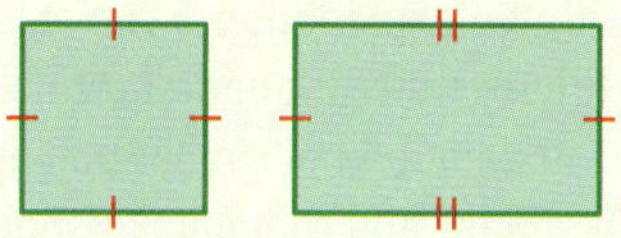

Motivate

- The word *similar* and *simile* sound alike! A simile is a figure of speech in which two essentially unlike things are compared. For example, "He eats like a horse" or "She is as slow as molasses."
- Ask students to give other examples.

Activity Notes

Activity 1

- Explain the definition of similar triangles by drawing the triangles shown. Label the side lengths and write the ratios of corresponding sides.
- The goal of this activity is to make two triangles with the same angle measures whose side lengths are not congruent. To check if the angles are congruent, measure the angles of the second triangle with a protractor.
- **Big Idea:** If two triangles have the same angle measures, then the triangles are similar. This is a property that is only true for triangles.
- **Another Way:** Ask students to make a triangle with angles of 40°, 60°, and 80° using a protractor. Do not specify the side length. Use a metric ruler to measure the 3 sides to the nearest centimeter. Each pair of students will find the ratio of corresponding sides for their two triangles. Using a calculator, compare the three ratios; they should be the same.

Previous Learning

Students should know the definition of similar triangles.

Activity Materials
Textbook
• scissors • ruler • protractor • calculator

Start Thinking! and Warm Up

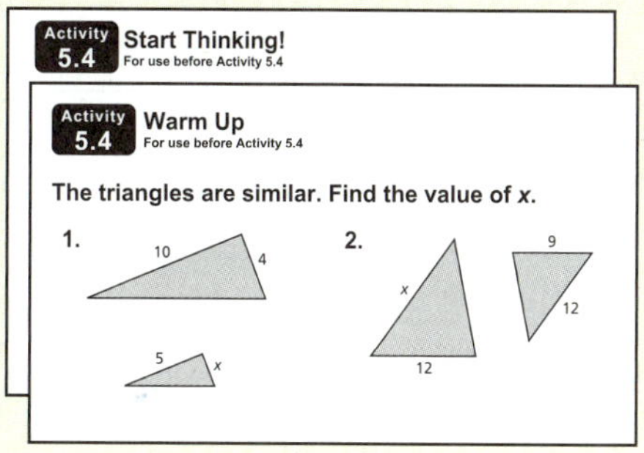

Activity 5.4 Start Thinking!
For use before Activity 5.4

Activity 5.4 Warm Up
For use before Activity 5.4

The triangles are similar. Find the value of x.

1. 10, 4; 5, x

2. x, 12; 9, 12

5.4 Record and Practice Journal

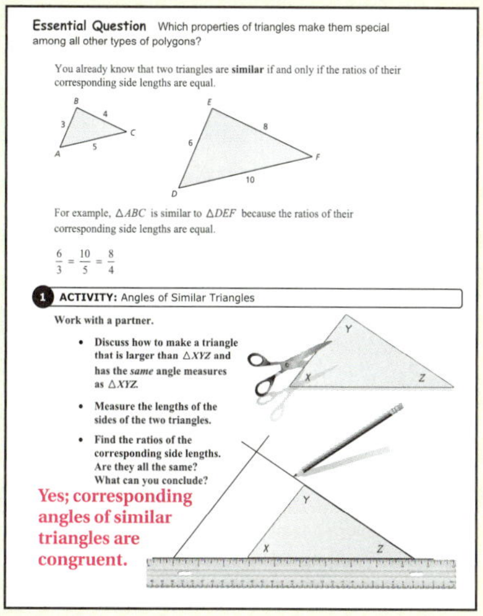

Essential Question Which properties of triangles make them special among all other types of polygons?

You already know that two triangles are **similar** if and only if the ratios of their corresponding side lengths are equal.

For example, $\triangle ABC$ is similar to $\triangle DEF$ because the ratios of their corresponding side lengths are equal.

$\frac{6}{3} = \frac{10}{5} = \frac{8}{4}$

1 ACTIVITY: Angles of Similar Triangles

Work with a partner.

- Discuss how to make a triangle that is larger than $\triangle XYZ$ and has the *same* angle measures as $\triangle XYZ$.
- Measure the lengths of the sides of the two triangles.
- Find the ratios of the corresponding side lengths. Are they all the same? What can you conclude?

Yes; corresponding angles of similar triangles are congruent.

Differentiated Instruction

Visual

Make pairs of similar polygons out of cardboard. Hand out the polygons to your students. Have students find another student in the room with a polygon similar to their own.

5.4 Record and Practice Journal

2 ACTIVITY: Amazing Triangles

Work with a partner. Use what you know about polygons to decide whether each statement is true. In each case, explain your reasoning.

a. If two *triangles* are similar, then the ratios of their corresponding side lengths are equal.

true

If two *quadrilaterals* are similar, then the ratios of their corresponding side lengths are equal.

true

b. If the ratios of the corresponding side lengths of two *triangles* are equal, then the triangles are similar.

true

If the ratios of the corresponding side lengths of two *quadrilaterals* are equal, then the quadrilaterals are similar.

false

c. If two *triangles* are similar, then their corresponding angles are congruent.

true

If two *quadrilaterals* are similar, then their corresponding angles are congruent.

true

d. If the corresponding angles in two triangles are congruent, then the triangles are similar.

true

If the corresponding angles in two quadrilaterals are congruent, then the quadrilaterals are similar.

false

e. If the corresponding sides of two triangles are congruent, then the two triangles have identical shapes.

true

If the corresponding sides of two quadrilaterals are congruent, then the two quadrilaterals have identical shapes.

false

What Is Your Answer?

3. IN YOUR OWN WORDS Which properties of triangles make them special among all other types of polygons? Describe two careers in which the special properties of triangles are used.

If corresponding side lengths of two triangles are proportional, then the triangles are similar. If corresponding angles of two triangles are congruent, then the triangles are similar.

Laurie's Notes

Activity 2

- In this activity, students must read carefully and critically. The questions presented are related to the Big Ideas discussed on the previous page.
- To help students visualize the statements, make simple models out of bendable straws. By making a slit in the straws, one straw can be inserted into another to make polygons. Make samples of different sized equilateral triangles, squares, and rectangles. Because the straws are bendable, the square can become a rhombus, and the rectangle can become a parallelogram.

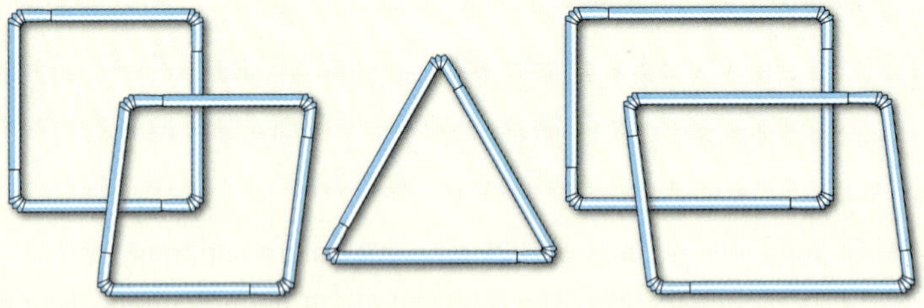

- When students have finished, discuss each statement. Have models available to help with the explanation. Each statement is true for triangles! Only parts (a) and (c) are true for quadrilaterals.
  - Part (a): This is the definition of similar polygons.
  - Part (b): Use the square and rhombus to show it is false.
  - Part (c): This is the definition of similar polygons.
  - Part (d): Use the square and rectangle to show it is false.
  - Part (e): Use the square and rhombus to show it is false.

What Is Your Answer?

- **Think-Pair-Share:** Students should read each question independently and then work with a partner to answer the questions. When they have answered the questions, the pair should compare their answers with another group and discuss any discrepancies.

Closure

- True or False?
  1. All equilateral triangles are similar. true
  2. All squares are similar. true
  3. All rhombuses are similar. false
  4. All rectangles are similar. false

Technology For the Teacher

Dynamic Classroom

The Dynamic Planning Tool
Editable Teacher's Resources at *BigIdeasMath.com*

2 ACTIVITY: Amazing Triangles

Work with a partner. Use what you know about polygons to decide whether each statement is true. In each case, explain your reasoning.

a. 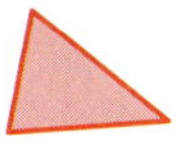 If two triangles are similar, then the ratios of their corresponding side lengths are equal.

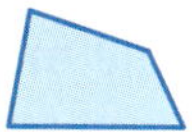 If two quadrilaterals are similar, then the ratios of their corresponding side lengths are equal.

b.  If the ratios of the corresponding sides of two triangles are equal, then the triangles are similar.

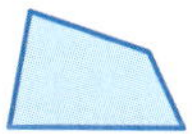 If the ratios of the corresponding sides of two quadrilaterals are equal, then the quadrilaterals are similar.

c.  If two triangles are similar, then their corresponding angles are congruent.

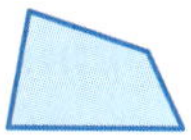 If two quadrilaterals are similar, then their corresponding angles are congruent.

d.  If the corresponding angles in two triangles are congruent, then the triangles are similar.

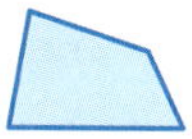 If the corresponding angles in two quadrilaterals are congruent, then the quadrilaterals are similar.

e.  If the corresponding sides of two triangles are congruent, then the two triangles have identical shapes.

 If the corresponding sides of two quadrilaterals are congruent, then the two quadrilaterals have identical shapes.

What Is Your Answer?

3. **IN YOUR OWN WORDS** Which properties of triangles make them special among all other types of polygons? Describe two careers in which the special properties of triangles are used.

Practice Use what you learned about similar triangles to complete Exercises 3 and 4 on page 210.

Check It Out
Lesson Tutorials
BigIdeasMath.com

Key Vocabulary
similar triangles, *p. 208*
indirect measurement, *p. 209*

Study Tip
If two angles in one triangle are congruent to two angles in another triangle, then the third angles are also congruent.

Triangles that have the same shape but not necessarily the same size are **similar triangles**.

Key Idea

Angles of Similar Triangles

Words Two triangles have the same angle measures if and only if they are similar.

Example

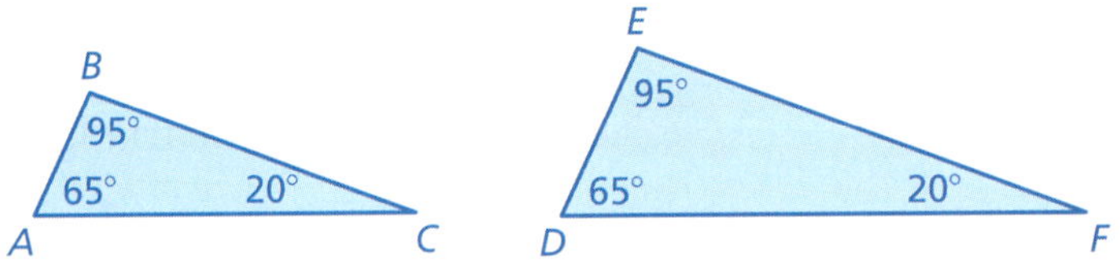

Triangle ABC is similar to triangle DEF: $\triangle ABC \sim \triangle DEF$.

EXAMPLE 1 Identifying Similar Triangles

Tell whether the triangles are similar. Explain.

a.

$$75 + 50 + x = 180 \qquad y + 50 + 55 = 180$$
$$125 + x = 180 \qquad y + 105 = 180$$
$$x = 55 \qquad y = 75$$

The triangles have the same angle measures, 75°, 50°, and 55°. So, they are similar.

b.

$$x + 90 + 42 = 180 \qquad 90 + 38 + y = 180$$
$$x + 132 = 180 \qquad 128 + y = 180$$
$$x = 48 \qquad y = 52$$

The triangles do not have the same angle measures. So, they are not similar.

Multi-Language Glossary at BigIdeasMath.com.

Laurie's Notes

Introduction

Connect

- **Yesterday:** Students explored special properties of similar triangles.
- **Today:** Students will use similar triangles to solve real-life problems.

Motivate

- Have pairs of students put a visual barrier (i.e., a notebook) between them. One student draws a triangle using a straightedge. This student now gives directions to the second student who will draw a triangle based on the information given. The only information the first student may give is angle measure! In fact, after the second angle measure is given, the second student should know the measure of the third angle.
- The triangles should be similar.

? "What do you notice about the triangles?" similar

? "How do you know they are similar?" Listen for same shape, different size; students may also mention yesterday's activities.

Lesson Notes

Key Idea

- Write the informal definition and draw examples of similar triangles.
- Write the Key Idea. This is the first time students have probably seen the phrase *if and only if*. It means two statements have been combined into one statement. The two statements are:
  - If two triangles have the same angle measures, then they are similar.
  - If two triangles are similar, then they have the same angle measures.
- The *if and only if* phrase is an example of a bi-conditional statement.
- Note the Study Tip. Refer to a sample pair of triangles that you drew earlier. Label two of the three angles in each triangle.
- **Another Way:** Work through a few examples to help students see why, when two angles in two triangles are congruent, the third angles are congruent.

? "What do you know about the measure of the third angle in each triangle?" Listen for the numeric answer and a statement that the third angles are congruent.

Example 1

- Draw the two triangles and label the given information. Ask students to solve for the missing angle measure of each triangle.

? "Are the triangles similar? Explain." For the first pair, yes, because the triangles have the same angle measures. For the second pair, no, because the triangles do not have the same angle measures.

- Students are influenced by the orientation of the triangles and how they appear. Students will say that they look similar. Always make students verify or explain their thinking.

Goal Today's lesson is using **similar triangles** to solve problems.

Lesson Materials
Introduction
• protractor • ruler

Start Thinking! and Warm Up

Lesson 5.4 Warm Up
For use before Lesson 5.4

Lesson 5.4 Start Thinking!
For use before Lesson 5.4

Thales was a Greek philosopher and mathematician who lived around 600 B.C. There are several accounts of how he used indirect measurement to find the height of the Great Pyramid in Giza.

According to one account, when his shadow was the same length as his height, he measured the length of the Great Pyramid's shadow.

What does this have to do with similar triangles?

Extra Example 1

Tell whether the triangles are similar. Explain.

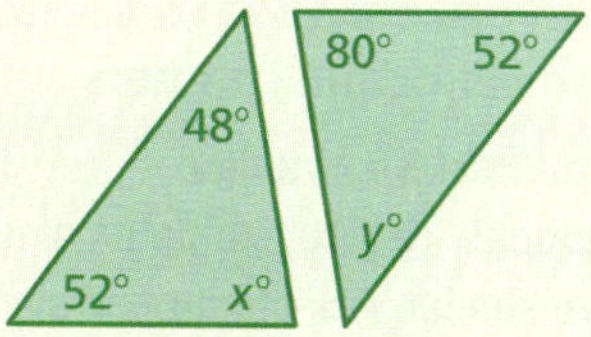

yes; The triangles have the same angle measures, 52°, 48°, and 80°.

Laurie's Notes

On Your Own

1. no; The triangles do not have the same angle measures.
2. yes; The triangles have the same angle measures, 90°, 66°, and 24°.

Extra Example 2

You plan to cross a river and want to know how far it is to the other side. You take measurements on your side of the river and make the drawing shown.

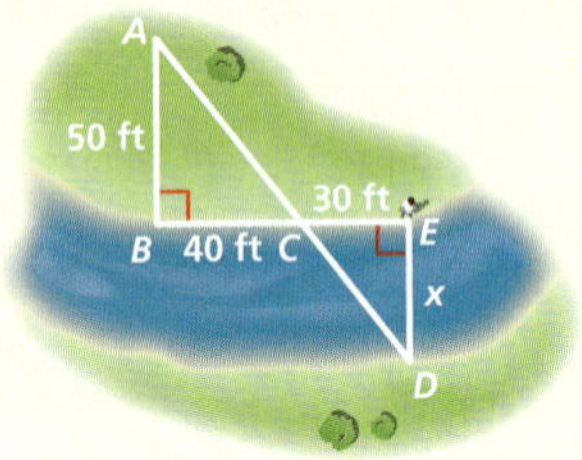

a. Explain why $\triangle ABC$ and $\triangle DEC$ are similar. Because two angles in $\triangle ABC$ are congruent to two angles in $\triangle DEC$, the third angles are also congruent. The triangles have the same angle measures, so they are similar.

b. What is the distance x across the river? 37.5 ft

On Your Own

3. 44 ft

English Language Learners

Build on Past Knowledge

Ask students to give examples of items that are similar. Ask students if similar items are exactly alike. Explain to students that *similar figures* are figures that have the same shape, but not necessarily the same size.

On Your Own

- **Think-Pair-Share:** Students should read each question independently and then work with a partner to answer the questions. When they have answered the questions, the pair should compare their answers with another group and discuss any discrepancies.

Example 2

- Indirect measurement is used when you want to know the measurement of some length (or angle) and you cannot measure the object directly.
- Ask a volunteer to read the problem. Make a rough sketch of the diagram.
- ? "What do you know about the angles in either triangle?" $\angle B$ and $\angle E$ are right angles. The vertical angles are congruent (mark the diagram to show the congruent angles).
- ? "What do you know about the third angle in each triangle?" They are congruent.
- Because the triangles are similar, the corresponding sides will have the same ratio. Setting up the ratios is challenging for students. Talk about the sides in terms of being the shorter leg of the right triangle and the longer leg of the right triangle.
- Use the Multiplication Property of Equality or the Cross Products Property to solve. Check the reasonableness of the answer.

On Your Own

- **Think-Pair-Share:** Students should read the question independently and then work with a partner to answer the question. When they have answered the question, the pair should compare their answer with another group and discuss any discrepancies.

Closure

- **Exit Ticket:** Are the two triangles similar? Explain.

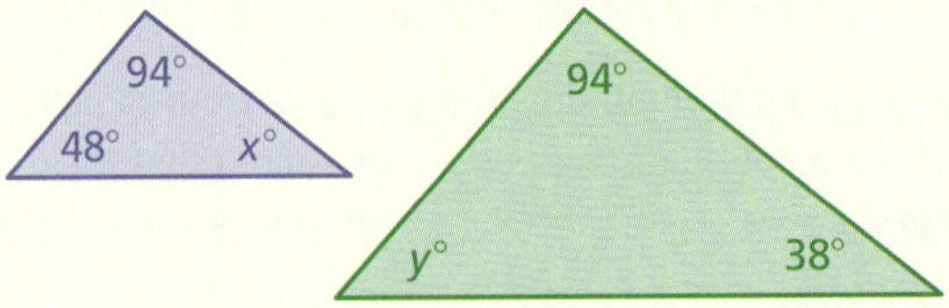

The triangles have the same angle measures, 94°, 48°, and 38°. So, the triangles are similar.

Technology For the Teacher

Dynamic Classroom

The Dynamic Planning Tool
Editable Teacher's Resources at *BigIdeasMath.com*

On Your Own

Now You're Ready
Exercises 5–8

Tell whether the triangles are similar. Explain.

1.

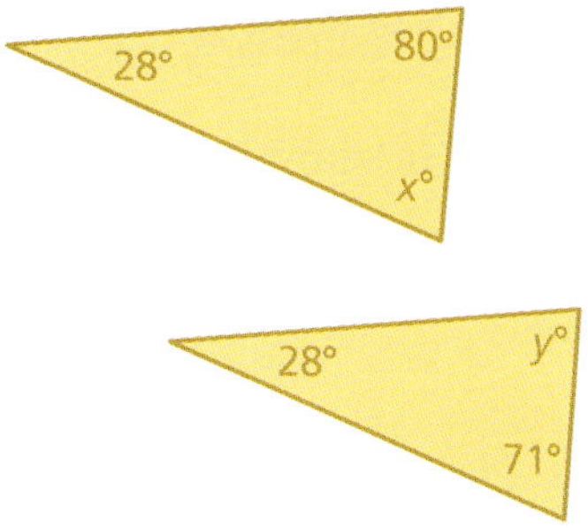

2.

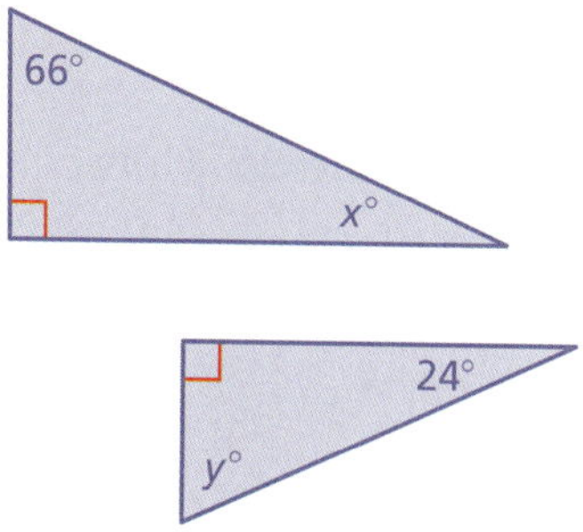

Indirect measurement uses similar figures to find a missing measure when it is difficult to find directly.

EXAMPLE 2 Using Indirect Measurement

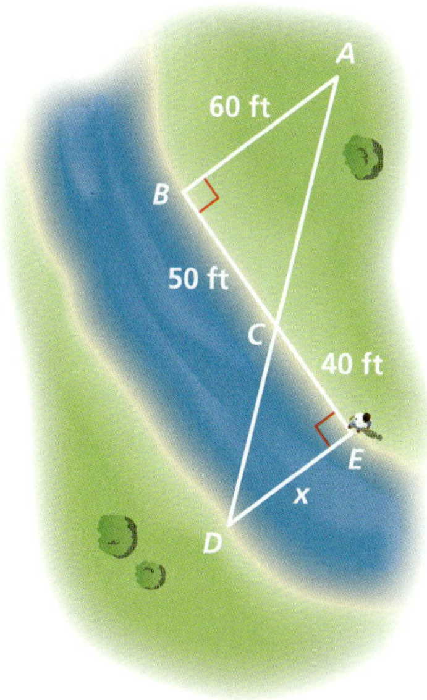

You plan to cross a river and want to know how far it is to the other side. You take measurements on your side of the river and make the drawing shown. (a) Explain why $\triangle ABC$ and $\triangle DEC$ are similar. (b) What is the distance x across the river?

a. $\angle B$ and $\angle E$ are right angles, so they are congruent. $\angle ACB$ and $\angle DCE$ are vertical angles, so they are congruent.

Because two angles in $\triangle ABC$ are congruent to two angles in $\triangle DEC$, the third angles are also congruent. The triangles have the same angle measures, so they are similar.

b. The ratios of the corresponding side lengths in similar triangles are equal. Write and solve a proportion to find x.

$$\frac{x}{60} = \frac{40}{50}$$ Write a proportion.

$$60 \cdot \frac{x}{60} = 60 \cdot \frac{40}{50}$$ Multiply each side by 60.

$$x = 48$$ Simplify.

∴ The distance across the river is 48 feet.

On Your Own

Now You're Ready
Exercises 10 and 11

3. **WHAT IF?** In Example 2, the distance from vertex A to vertex B is 55 feet. What is the distance across the river?

5.4 Exercises

Check It Out
Help with Homework
BigIdeasMath.com

Vocabulary and Concept Check

1. **REASONING** How can you use similar triangles to find a missing measurement?
2. **WHICH ONE DOESN'T BELONG?** Which triangle does *not* belong with the other three? Explain your reasoning.

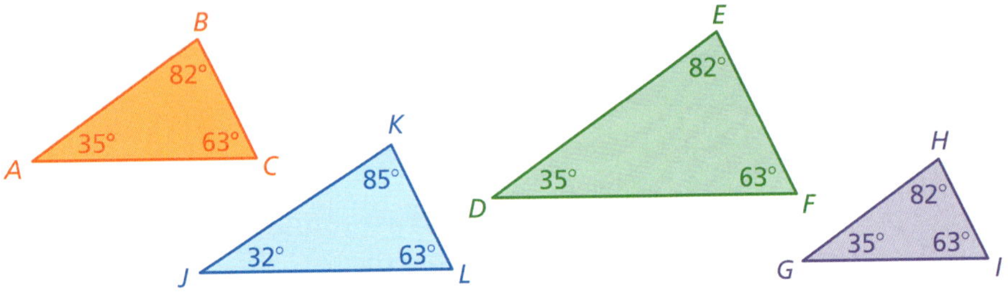

Practice and Problem Solving

Make a triangle that is larger than the one given and has the same angle measures. Find the ratios of the corresponding side lengths.

3. 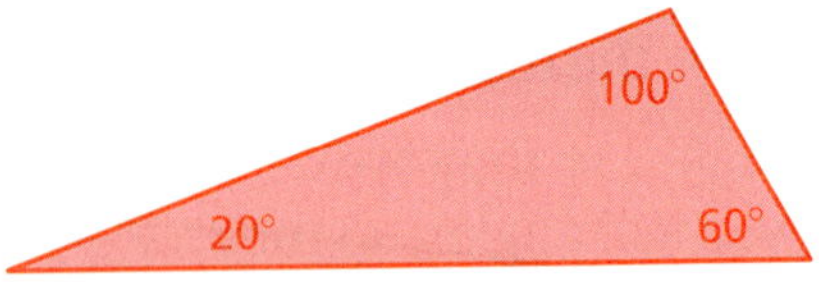

4. 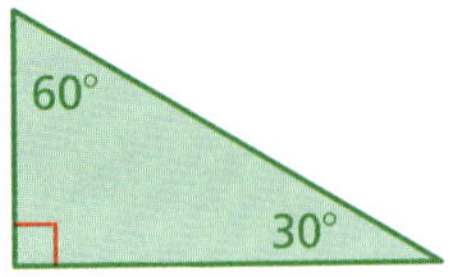

Tell whether the triangles are similar. Explain.

1 5. 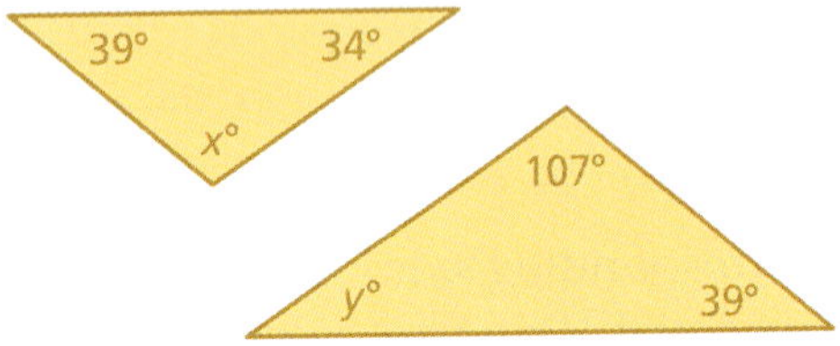

6. 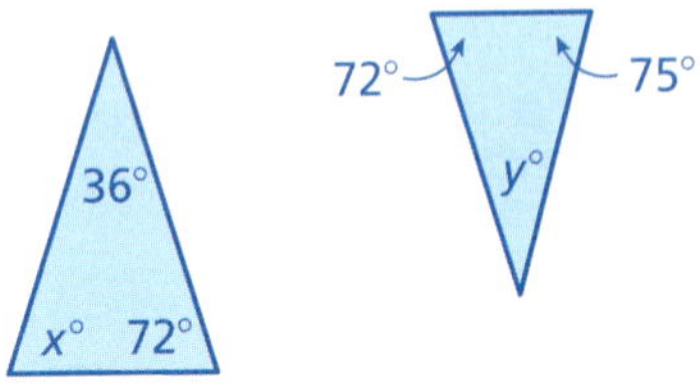

7. 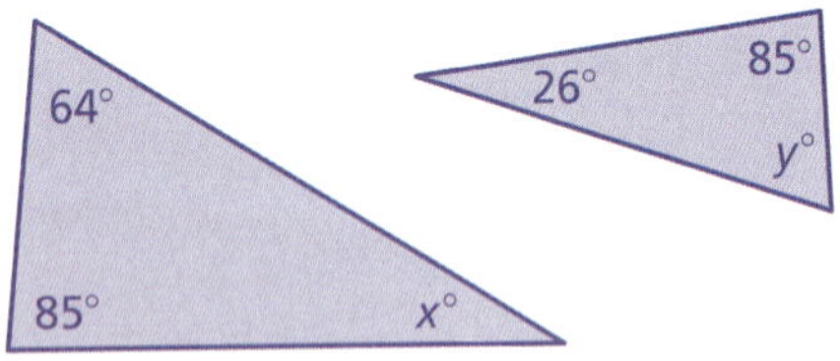

8. 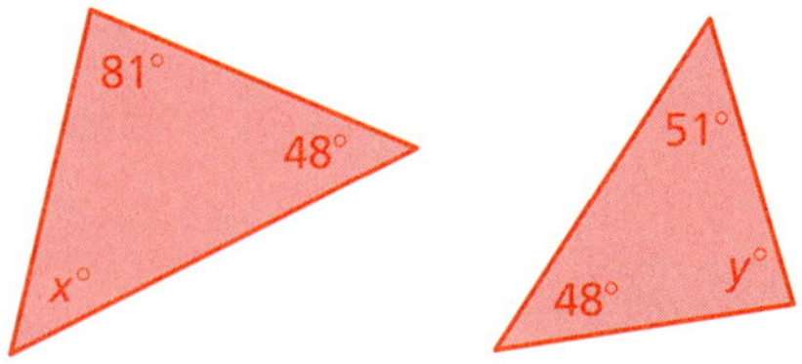

9. **ERROR ANALYSIS** Describe and correct the error in using indirect measurement.

✗

$$\frac{16}{18} = \frac{x}{8}$$

$$18x = 128$$

$$x \approx 7$$

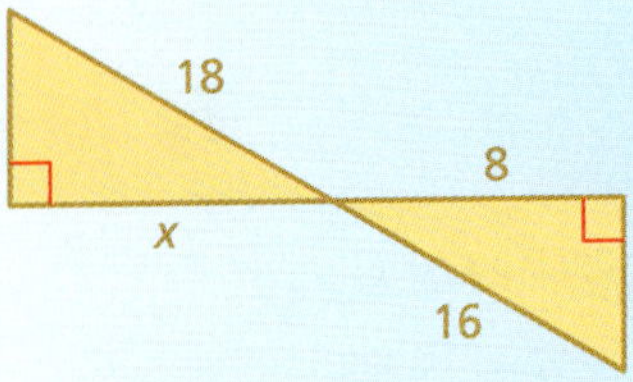

Assignment Guide and Homework Check

Level	Day 1 Activity Assignment	Day 2 Lesson Assignment	Homework Check
Basic	3, 4, 16–20	1, 2, 5–12	2, 6, 10, 12
Average	3, 4, 16–20	1, 2, 5–12, 14	2, 6, 10, 12
Advanced	3, 4, 16–20	1, 2, 8–15	2, 8, 10, 12

Common Errors

- **Exercises 5–8** Students may find the missing angle measure for one of the triangles and then make a decision about the similarity of the triangles. While it is possible to use this method, encourage them to find the missing angles of both triangles to verify that they are correct.
- **Exercises 10 and 11** Students may incorrectly identify congruent angles and find the wrong value for *x*. Encourage them to label the angles of the triangles with *A*, *B*, *C*, and *D*, *E*, *F*, and then write which angles are congruent before solving for *x*.

5.4 Record and Practice Journal

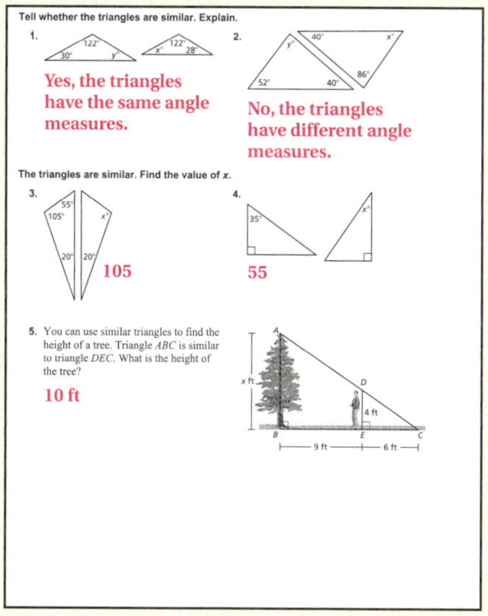
Tell whether the triangles are similar. Explain.

1. Yes, the triangles have the same angle measures.

2. No, the triangles have different angle measures.

The triangles are similar. Find the value of *x*.

3. 105

4. 55

5. You can use similar triangles to find the height of a tree. Triangle *ABC* is similar to triangle *DEC*. What is the height of the tree?

10 ft

Technology For the Teacher

Answer Presentation Tool

Quiz*Show*

Vocabulary and Concept Check

1. Write a proportion that uses the missing measurement because the ratios of corresponding side lengths are equal.
2. $\triangle JKL$ because the other three triangles are similar.

Practice and Problem Solving

3–4. Student should draw a triangle with the same angle measures as the textbook. The ratio of the corresponding side lengths, $\dfrac{\text{student's triangle length}}{\text{book's triangle length}}$, should be greater than 1.

5. yes; The triangles have the same angle measures, 107°, 39°, and 34°.

6. no; The triangles do not have the same angle measures.

7. no; The triangles do not have the same angle measures.

8. yes; The triangles have the same angle measures, 81°, 51°, and 48°.

9. The numerators of the fractions should be from the same triangle.

$$\frac{18}{16} = \frac{x}{8}$$

$$16x = 144$$

$$x = 9$$

10. 50

11. 65

12. 100 steps

Practice and Problem Solving

13. no; Each side increases by 50%, so each side is multiplied by a factor of $\frac{3}{2}$. The area is $\frac{3}{2}\left(\frac{3}{2}\right) = \frac{9}{4}$ or 225% of the original area, which is a 125% increase.

14. See Additional Answers.

15. See *Taking Math Deeper.*

Fair Game Review

16. nonlinear; The equation cannot be rewritten in slope-intercept form.

17. linear; The equation can be rewritten in slope-intercept form.

18. linear; The equation can be rewritten in slope-intercept form.

19. nonlinear; The equation cannot be rewritten in slope-intercept form.

20. C

Mini-Assessment

Tell whether the triangles are similar. Explain.

1.

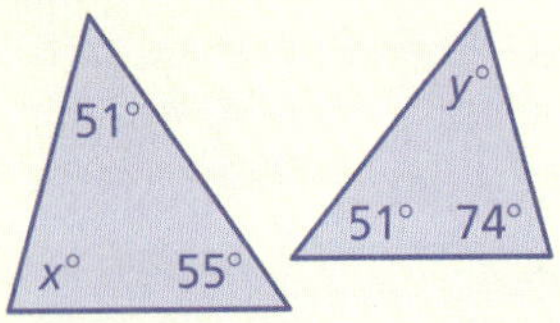

yes; The triangles have the same angle measures, 51°, 55°, and 74°.

2.

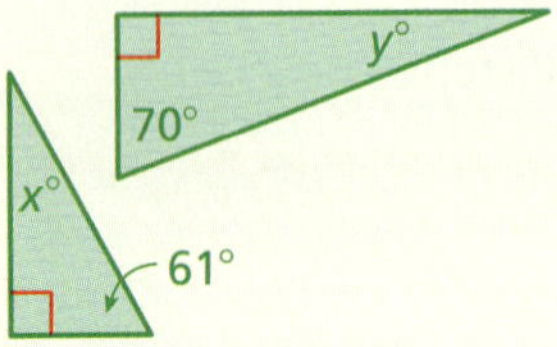

no; The triangles do not have the same angle measures.

Taking Math Deeper

Exercise 15

This problem can be solved using the fact that when two triangles are similar, the ratios of corresponding sides are equal.

1 **a.** Solve for *x*.

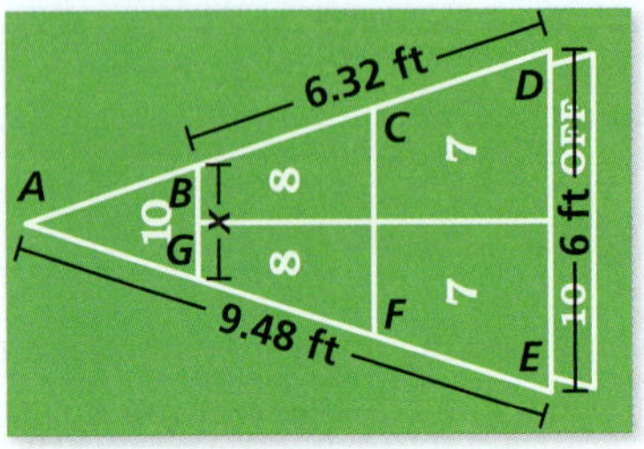

$$\frac{x}{6} = \frac{9.48 - 6.32}{9.48}$$

$$6 \cdot \frac{x}{6} = 6 \bullet \frac{3.16}{9.48}$$

$$x = 2 \text{ ft}$$

2 **b.** Solve for *CF*. Let $y = CF$.

$$\frac{y}{6} = \frac{2(9.48 \div 3)}{9.48}$$

$$6 \cdot \frac{y}{6} = 6 \bullet \frac{6.32}{9.48}$$

$$y = 4 \text{ ft}$$

3 Notice that the bases of the three similar triangles form a linear pattern: 2, 4, and 6.

Project

Research the game of shuffleboard. What are the rules? Are there tournaments? Compare shuffleboard to the Olympic sport of curling.

Reteaching and Enrichment Strategies

If students need help...	If students got it...
Resources by Chapter • Practice A and Practice B • Puzzle Time Record and Practice Journal Practice Differentiating the Lesson Lesson Tutorials Skills Review Handbook	Resources by Chapter • Enrichment and Extension • School-to-Work Start the next section

The triangles are similar. Find the value of x.

10. 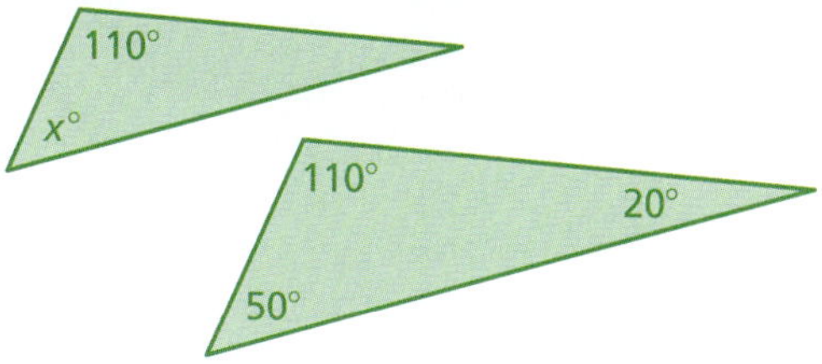

11. 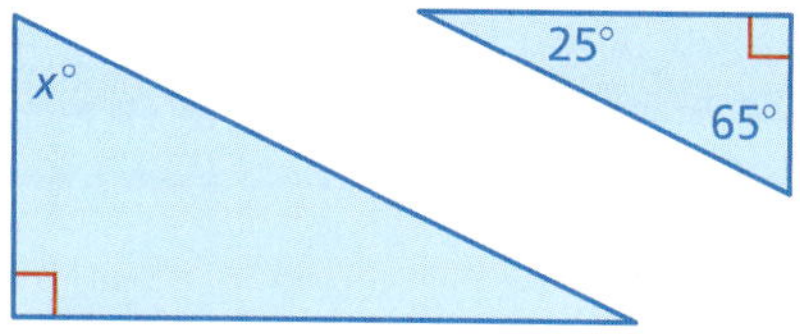

12. **TREASURE** The map shows the number of steps you must take to get to the treasure. However, the map is old and the last dimension is unreadable. How many steps do you take from the pyramids to the treasure?

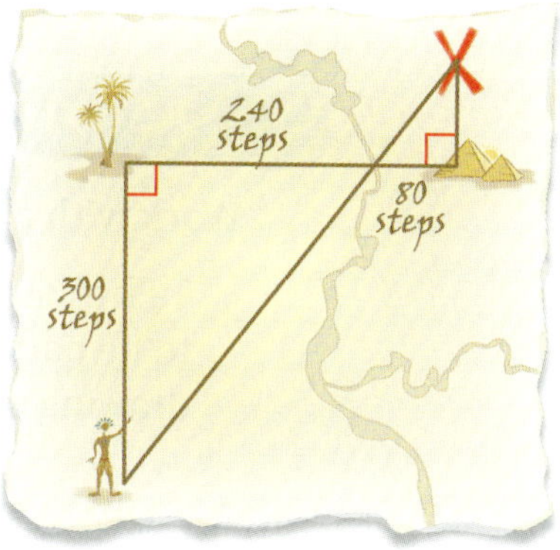

13. **CRITICAL THINKING** The side lengths of a triangle are increased by 50% to make a similar triangle. Does the area increase by 50% as well? Explain.

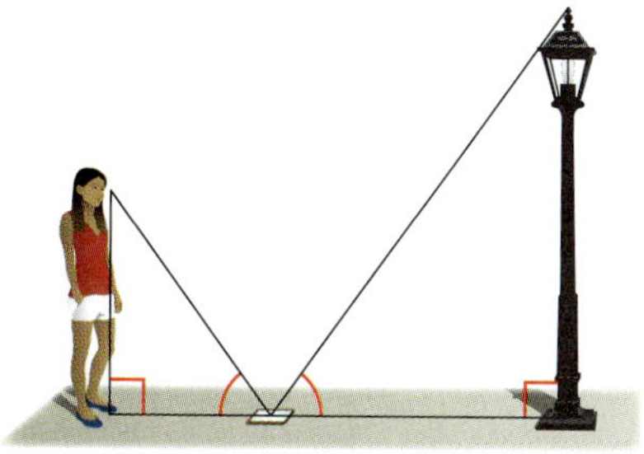

14. **PROJECT** Using a mirror, a tape measure, and indirect measurement, you can find the height of a lamppost. Place the mirror flat on the ground 6 feet from the lamppost. Move away from the mirror and the lamppost until you can see the top of the lamppost in the mirror. Measure the distance between yourself and the mirror. Then use similar triangles to find the height of the lamppost.

15. Geometry The drawing shows the scoring zone of a standard shuffleboard court. $\triangle DAE \sim \triangle BAG \sim \triangle CAF$. The lengths of segments AG, GF, and FE are equal.

a. Find x.

b. Find CF.

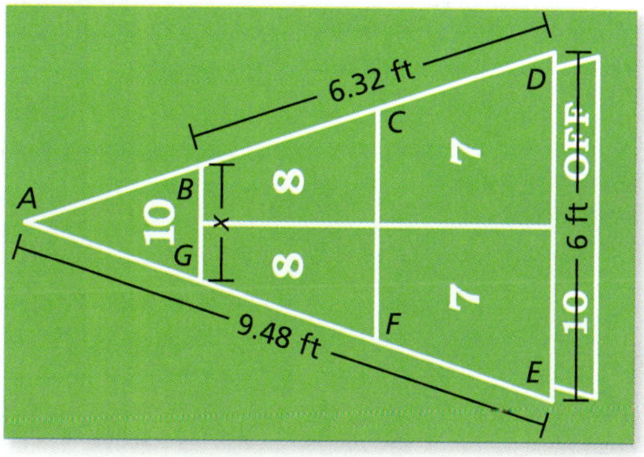

Fair Game Review What you learned in previous grades & lessons

Does the equation represent a *linear* or *nonlinear* function? Explain. *(Section 4.4)*

16. $y = \dfrac{5}{x}$

17. $y = -5.4x + \pi$

18. $y = 2x - 8$

19. $y = 6x^2 + x - 1$

20. **MULTIPLE CHOICE** Which two lines are parallel? *(Section 2.2)*

Ⓐ blue and red

Ⓑ red and green

Ⓒ green and blue

Ⓓ all three are parallel

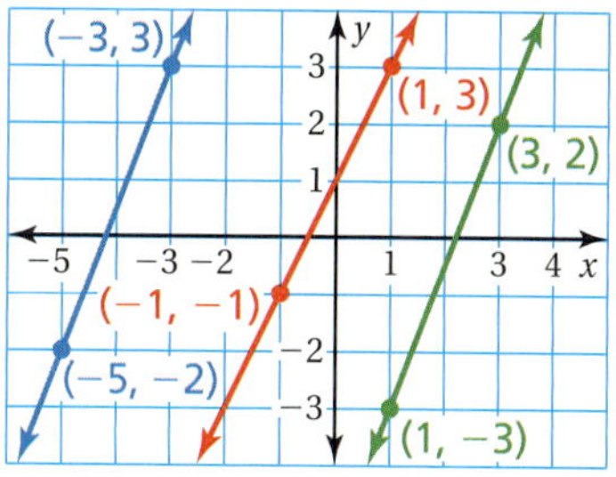

5.5 Parallel Lines and Transversals

Share Your Work at... My.BigIdeasMath.com

Essential Question How can you use properties of parallel lines to solve real-life problems?

1 ACTIVITY: A Property of Parallel Lines

Work with a partner.

- Talk about what it means for two lines to be parallel. Decide on a strategy for drawing two parallel lines.

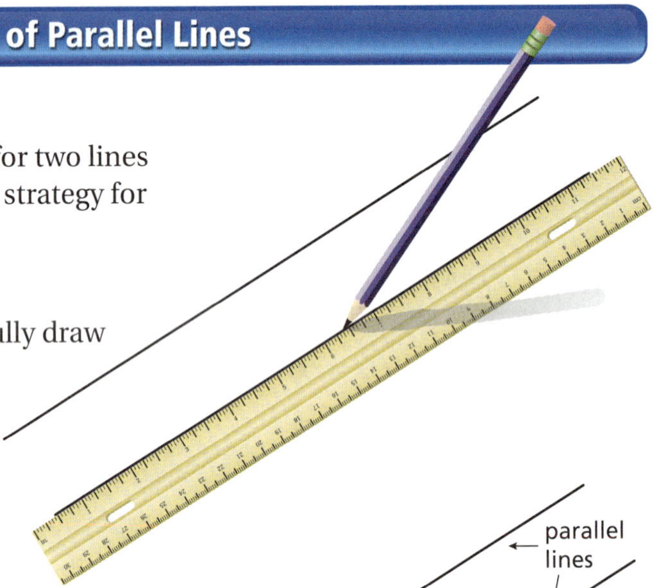

- Use your strategy to carefully draw two lines that are parallel.

- Now, draw a third line that intersects the two parallel lines. This line is called a **transversal**.

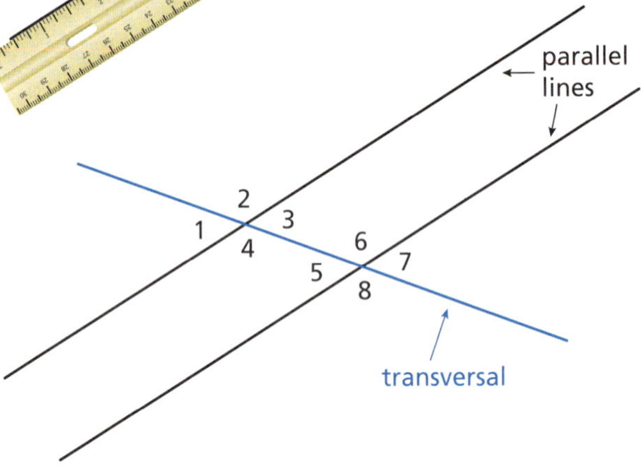

- The two parallel lines and the transversal form eight angles. Which of these angles have equal measures? Explain your reasoning.

2 ACTIVITY: Creating Parallel Lines

Work with a partner.

a. If you were building the house in the photograph, how could you make sure that the studs are parallel to each other?

b. Identify sets of parallel lines and transversals in the photograph.

Laurie's Notes

Introduction

For the Teacher

- **Goal:** Students will explore the properties of angles formed by two parallel lines cut by a transversal.

Motivate

- **Preparation:** Use a transparency that has two parallel lines drawn on it on the overhead projector. Place a grid on top of the parallel lines so that the lines pass through obvious lattice points.
- ? "What appears to be true about the lines?" parallel
- Compute the slope and conclude that the lines are parallel.
- Now remove the grid but keep the transparency with the (parallel) lines.
- ? "Are the lines still parallel? How do you know?" The point of these questions is not for students to give an answer, but for students to consider what it means for two lines to be parallel.

Activity Notes

Activity 1

- The goal of the first part of this activity is for students to devise a method for drawing two parallel lines. A common method is to trace on either side of a ruler. Do not use ruled paper. Challenge students to think of more than one method.
- Spend time discussing different methods.
- Now have students proceed with the second part of Activity 1, where they draw the transversal and investigate the angles formed.
- **FYI:** In Latin, *trans* means *across* and *vers* means *to turn*.
- Students will visually pair up the obtuse angles and then the acute angles. Students will conclude that all of the obtuse angles are congruent and all of the acute angles are congruent. Have tracing paper and/or protractors available for students to check their guess.
- Do not introduce angle vocabulary at this point. Allow students to refer to the angles by number. As shown, the even numbered angles are all obtuse and the odd numbered angles are all acute. Suggest to students that the numbering scheme is arbitrary and determined by the person labeling the diagram.
- **Connection:** Students should be able to identify the four pairs of vertical angles in the diagram. Vertical angles are congruent regardless of whether the lines are parallel or not.

Activity 2

- Students will have different thoughts about how to make sure the studs are parallel to each other. Be prepared for students who do not know much about construction.
- The transversals should be obvious, but remember that the horizontal boards perpendicular to the studs are also transversals.

Previous Learning

Students should know the definition of similar triangles.

Activity Materials	
Introduction	**Textbook**
• transparencies (grid and plain)	• straightedge • protractor • tracing paper

Start Thinking! and Warm Up

Activity 5.5 **Start Thinking!** For use before Activity 5.5

Activity 5.5 **Warm Up** For use before Activity 5.5

Solve the proportion.

1. $\frac{16}{3} = \frac{x}{9}$
2. $\frac{2}{5} = \frac{5}{x}$
3. $\frac{15}{12} = \frac{x}{8}$
4. $\frac{x}{2} = \frac{11}{4}$
5. $\frac{100}{x} = \frac{25}{8}$
6. $\frac{x}{5} = \frac{3}{8}$

5.5 Record and Practice Journal

Essential Question How can you use properties of parallel lines to solve real-life problems?

1 ACTIVITY: A Property of Parallel Lines

Work with a partner.

- Talk about what it means for two lines to be parallel. Decide on a strategy for drawing two parallel lines.
- Use your strategy to carefully draw two lines that are parallel in the space below. **Check students' work.**
- Now, draw a third line that intersects the two parallel lines. This line is called a **transversal**.
- The two parallel lines and the transversal form eight angles. Which of these angles have equal measures? Explain your reasoning.

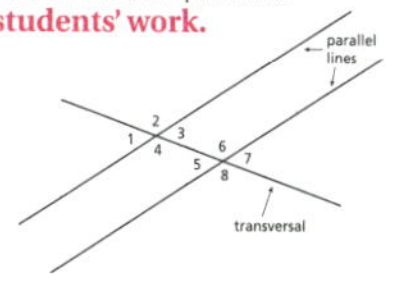

Angles 2, 4, 6, and 8; Angles 1, 3, 5, and 7

Differentiated Instruction

Kinesthetic

When setting up a proportion, have students write each of the three known values and the one unknown value with their units on index cards. On a fifth index card, have the students write an equal sign. Students should then place the cards on their desks to set up the proportion. Discuss the different ways to set up a proportion.

5.5 Record and Practice Journal

2 ACTIVITY: Creating Parallel Lines

Work with a partner.

a. If you were building the house in the photograph, how could you make sure that the studs are parallel to each other?

vertical and corresponding angles should be congruent

b. Identify sets of parallel lines and transversals in the photograph.

Check students' work.

3 ACTIVITY: Indirect Measurement

Work with a partner.

a. Use the fact that two rays from the sun are parallel to explain why $\triangle ABC$ and $\triangle DEF$ are similar.

The rays make the same angle with the ground, so the triangles have 2 congruent corresponding angles.

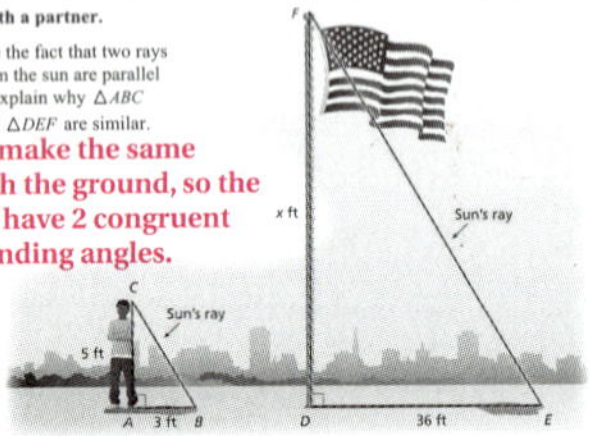

b. Explain how to use similar triangles to find the height of the flagpole.

Use the proportion $\frac{x}{5} = \frac{36}{3}$ and solve for x.

What Is Your Answer?

4. **IN YOUR OWN WORDS** How can you use properties of parallel lines to solve real-life problems? Describe some examples.

can find angle measures and missing dimensions of triangles

5. **INDIRECT MEASUREMENT PROJECT** Work with a partner or in a small group.

a. Explain why the process in Activity 3 is called "indirect" measurement.

You are not measuring the flag pole directly.

b. Use indirect measurement to measure the height of something outside your school (a tree, a building, a flagpole). Before going outside, decide what you need to take with you to do the measurement.

Check students' work.

c. Draw a diagram of the indirect measurement process you used. In the diagram, label the lengths that you actually measured and also the lengths that you calculated.

Check students' work.

Laurie's Notes

Activity 3

- Students need to know that when two triangles have two congruent angles, the triangles are similar. Once they know this concept, they can use it to solve indirect measurement problems.

? "Is your shadow shorter at noon or 5 P.M.? Explain." Noon; The sun is overhead, not at a lower position in the sky.

? "Do adjacent objects of different heights cast the same length shadow? Explain." No; taller objects cast longer shadows.

- The triangles are similar because they both have a right angle and the parallel rays of the sun are at the same angle to the ground.

What Is Your Answer?

- Question 5 takes time. Make sure students have a plan *before* they go outside.

Closure

- Identify five pairs of parallel lines in your classroom. Note that the definition of *line* is modified to include things such as the metal molding on either side of the white board and the side casings on the door frame.

Technology For the Teacher

Dynamic Classroom

The Dynamic Planning Tool
Editable Teacher's Resources at *BigIdeasMath.com*

3 ACTIVITY: Indirect Measurement

Work with a partner.

a. Use the fact that two rays from the Sun are parallel to explain why $\triangle ABC$ and $\triangle DEF$ are similar.

b. Explain how to use similar triangles to find the height of the flagpole.

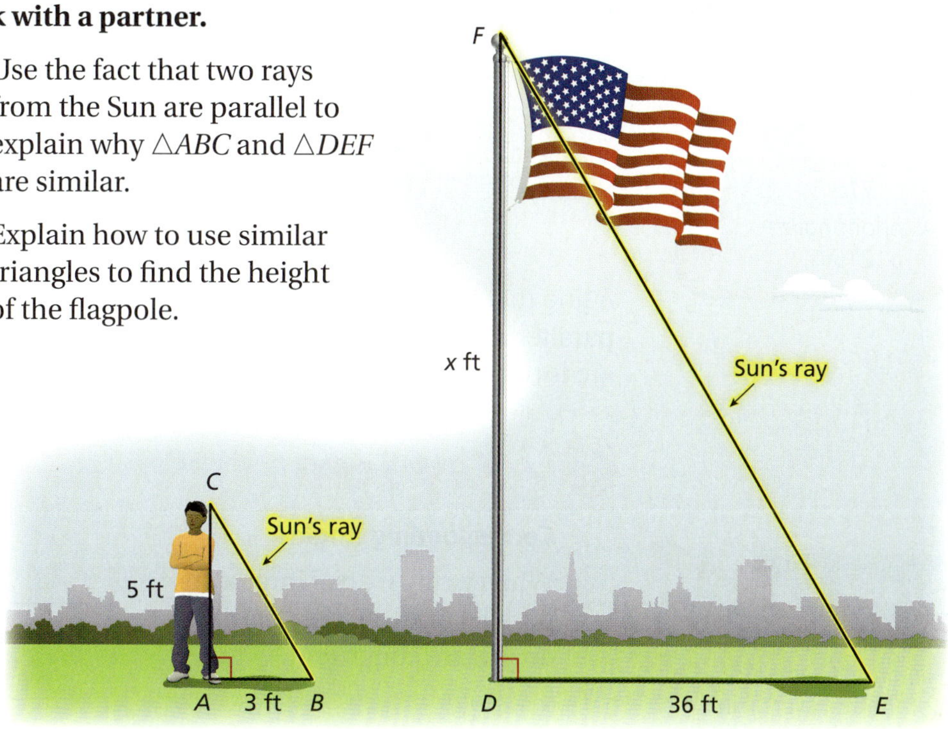

What Is Your Answer?

4. **IN YOUR OWN WORDS** How can you use properties of parallel lines to solve real-life problems? Describe some examples.

5. **INDIRECT MEASUREMENT PROJECT** Work with a partner or in a small group.

   a. Explain why the process in Activity 3 is called "indirect" measurement.

   b. Use indirect measurement to measure the height of something outside your school (a tree, a building, a flagpole). Before going outside, decide what you need to take with you to do the measurement.

   c. Draw a diagram of the indirect measurement process you used. In the diagram, label the lengths that you actually measured and also the lengths that you calculated.

Practice Use what you learned about parallel lines and transversals to complete Exercises 3–6 on page 217.

Key Vocabulary

perpendicular lines, *p. 214*
transversal, *p. 214*
interior angles, *p. 215*
exterior angles, *p. 215*

Lines in the same plane that do not intersect are called parallel lines. Lines that intersect at right angles are called **perpendicular lines**.

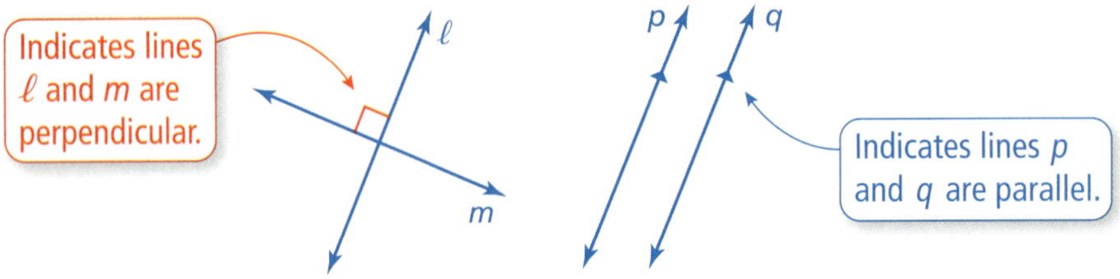

A line that intersects two or more lines is called a **transversal**. When parallel lines are cut by a transversal, several pairs of congruent angles are formed.

Key Idea

Corresponding Angles

When a transversal intersects parallel lines, corresponding angles are congruent.

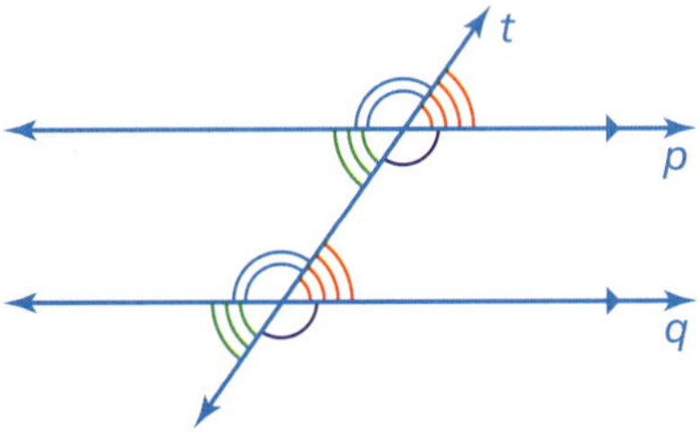

Corresponding angles

Study Tip

Corresponding angles lie on the same side of the transversal in corresponding positions.

EXAMPLE 1 Finding Angle Measures

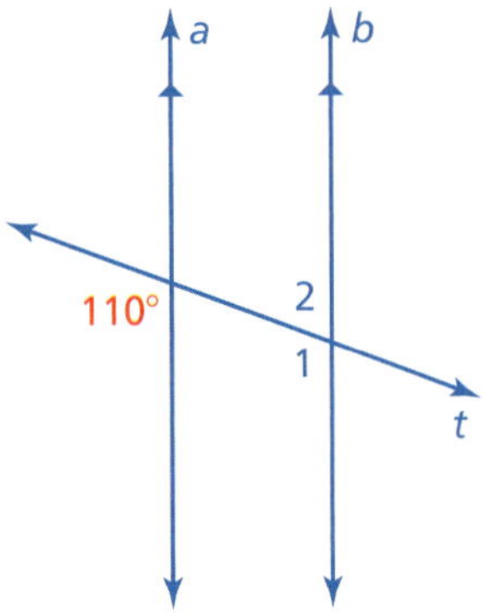

Use the figure to find the measures of (a) ∠1 and (b) ∠2.

a. ∠1 and the 110° angle are corresponding angles. They are congruent.

So, the measure of ∠1 is 110°.

b. ∠1 and ∠2 are supplementary.

$\angle 1 + \angle 2 = 180^\circ$	Definition of supplementary angles
$110^\circ + \angle 2 = 180^\circ$	Substitute 110° for ∠1.
$\angle 2 = 70^\circ$	Subtract 110° from each side.

So, the measure of ∠2 is 70°.

On Your Own

Now You're Ready
Exercises 7–9

Use the figure to find the measure of the angle. Explain your reasoning.

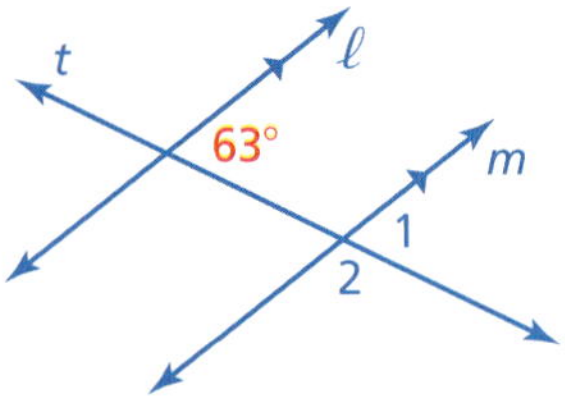

1. ∠1 **2.** ∠2

Laurie's Notes

Introduction

Connect

- **Yesterday:** Students explored angles formed when parallel lines are intersected by a transversal.
- **Today:** Students will find the measures of many types of angles, all formed when parallel lines are cut by a transversal.

Motivate

- **Preparation:** Make a model to help discuss the big ideas of this lesson. Cut 3 strips of card stock; punch holes in the middle of two strips and punch two holes in the third strip. Attach the strips using brass fasteners.

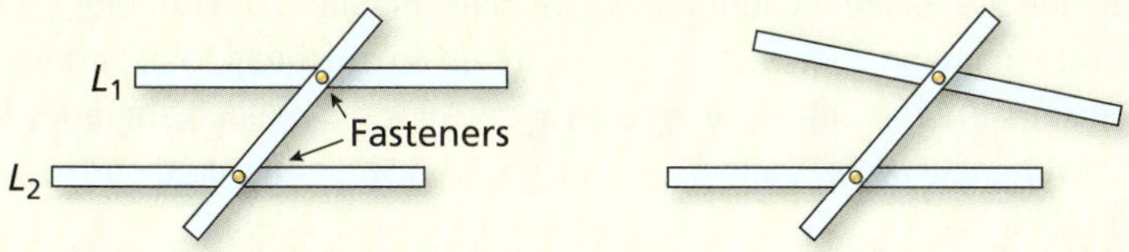

Model A Model B

- Place the model on the overhead. Demonstrate to students that the pieces are moveable, by transforming from Model A to Model B.
- Focus students' attention on the connection between the 4 angles on L_1 and the 4 angles on L_2. Pairs of vertical angles will always be congruent whether or not L_1 and L_2 are parallel.
- Place the model on the overhead and encourage students to point to the angles that they think are congruent. Use models A and B.

Lesson Notes

Key Idea

- Write the informal definitions of parallel lines and perpendicular lines. Draw examples of each and discuss the notation used in the diagram.
- Write the definition of transversal. Explain that a line that intersects two or more lines is called a **transversal** even if the lines are *not* parallel. Only when the lines are parallel are the pairs of angles in this lesson congruent.
- Write the Key Idea. Identify corresponding angles which are color coded in the diagram. Mention the Study Tip.
- Students will ask what is meant by *corresponding position.* The corresponding angles are both above or below the parallel lines (when in horizontal position) and on the same side of the transversal (left or right).

Example 1

? "Are the lines parallel? How do you know?" yes; blue arrow marks

- Students will need to recall the definition of supplementary angles.

Goal Today's lesson is finding measures of angles formed by parallel lines and a **transversal**.

Lesson Materials
Introduction
• brass fasteners • card stock

Start Thinking! and Warm Up

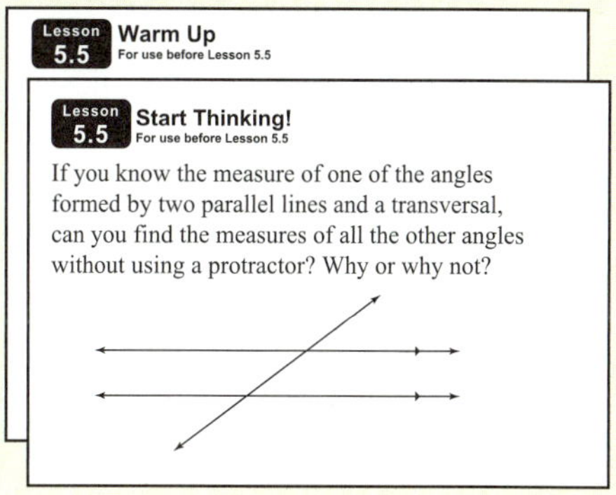
Lesson 5.5 Warm Up For use before Lesson 5.5

Lesson 5.5 Start Thinking! For use before Lesson 5.5

If you know the measure of one of the angles formed by two parallel lines and a transversal, can you find the measures of all the other angles without using a protractor? Why or why not?

Extra Example 1

Use the figure to find the measures of (a) $\angle 1$ and (b) $\angle 2$.

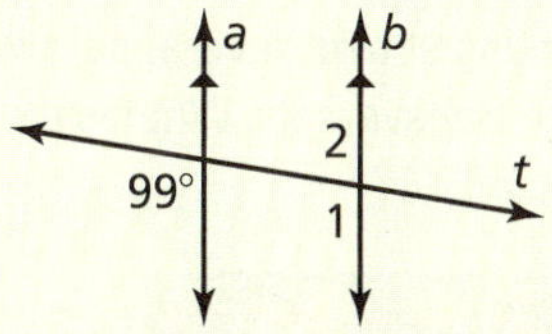

a. 99° **b.** 81°

On Your Own

1. 63°; Corresponding angles are congruent.
2. 117°; $\angle 1$ and $\angle 2$ are supplementary.

Extra Example 2

Use the figure to find the measures of the numbered angles.

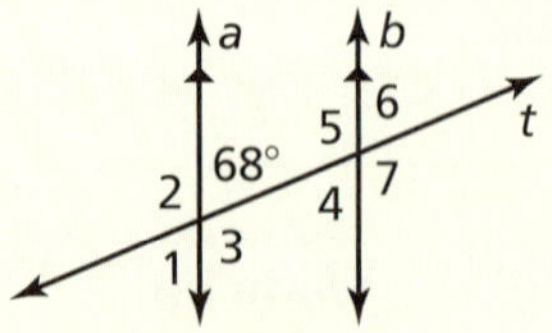

$\angle 1 = 68°$, $\angle 2 = 112°$, $\angle 3 = 112°$, $\angle 4 = 68°$, $\angle 5 = 112°$, $\angle 6 = 68°$, $\angle 7 = 112°$

On Your Own

3. $\angle 1 = 121°$, $\angle 2 = 59°$, $\angle 3 = 121°$, $\angle 4 = 121°$, $\angle 5 = 59°$, $\angle 6 = 121°$, $\angle 7 = 59°$

Extra Example 3

The painting shows several parallel lines and transversals. What is the measure of $\angle 1$?

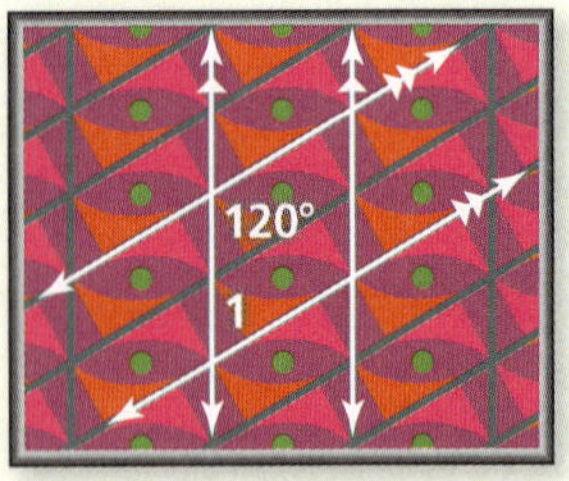

60°

English Language Learners

Vocabulary and Symbols
Make sure that students understand that the arrowhead marks *on* the lines indicate that the lines or line segments are parallel.

Laurie's Notes

Example 2

- Draw the figure on the board or overhead.
- ? "Can you find the measures of all the angles if you only know one angle?" Students may not know the answer at this point, but by the end of this example, they will see that they can.
- **Big Idea:** If you know any angle when a transversal intersects two parallel lines, then you can use vertical, supplementary, and corresponding angles to find all 7 of the other measures. It is not necessary to learn other theorems about alternate exterior or interior angles. **Students should be able to do all of the homework after this example.**
- Once angles 1, 2, and 3 are found, you can use corresponding angles to find the remaining four angles. To help students visualize the corresponding angles, draw the figure on an overhead transparency and cut the transparency in half. Lay the given angle and angles 1, 2, and 3 over angles 4, 5, 6, and 7 to show that they are congruent corresponding angles.

On Your Own

- **Question 3:** Students can say $\angle 3 = 121°$ because of vertical angles *or* because it is the supplement of $\angle 2$.

Discuss

- Use the model from the beginning of class to talk about other pairs of angles. Make the lines parallel even though the angles still have the same definition.
- Identify the four angles that are interior (between the two parallel lines) and the four angles that are exterior (outside the two parallel lines).
- ? "Are there pairs of interior angles that appear congruent?" yes; 3 & 6 and 4 & 5 (in the diagram)
- ? "Are there pairs of exterior angles that appear congruent?" yes; 1 & 8 and 2 & 7 (in the diagram)

Example 3

- ? "Are the two dashed lines parallel? How do you know?" Students may not be sure that they are parallel. Explain that because all the letters are slanted at an 80° angle, the lines are parallel.
- It is helpful to label other angles around $\angle 1$. For example:

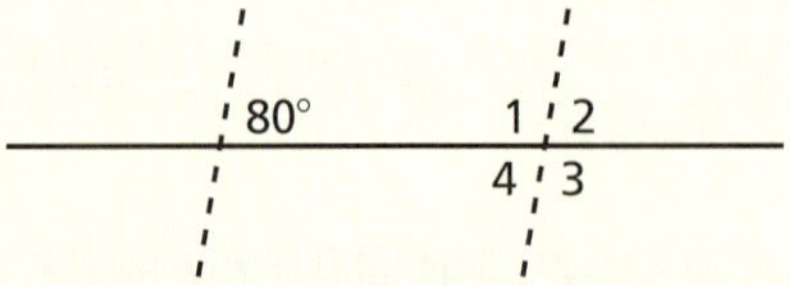

- ? "What angle is congruent to the 80° angle?" $\angle 2$
- ? "How can you find the measure of $\angle 1$?" Because $\angle 2$ and $\angle 1$ are supplementary and $\angle 2$ is congruent to the 80° angle, $\angle 1 = 180° - \angle 2 = 180° - 80° = 100°$.

EXAMPLE 2 Using Corresponding Angles

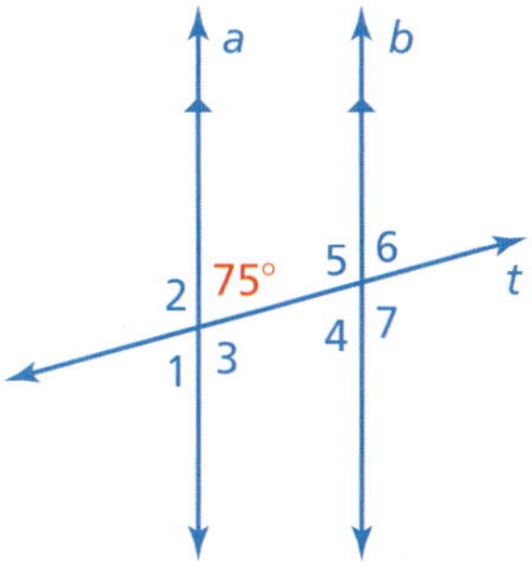

Use the figure to find the measures of the numbered angles.

∠1: ∠1 and the 75° angle are vertical angles. They are congruent.

So, the measure of ∠1 is 75°.

∠2 and ∠3: The 75° angle is supplementary to both ∠2 and ∠3.

$75° + \angle 2 = 180°$	Definition of supplementary angles
$\angle 2 = 105°$	Subtract 75° from each side.

So, the measures of ∠2 and ∠3 are 105°.

∠4, ∠5, ∠6, and ∠7: Using corresponding angles, the measures of ∠4 and ∠6 are 75°, and the measures of ∠5 and ∠7 are 105°.

On Your Own

Now You're Ready
Exercises 15–17

3. Use the figure to find the measures of the numbered angles.

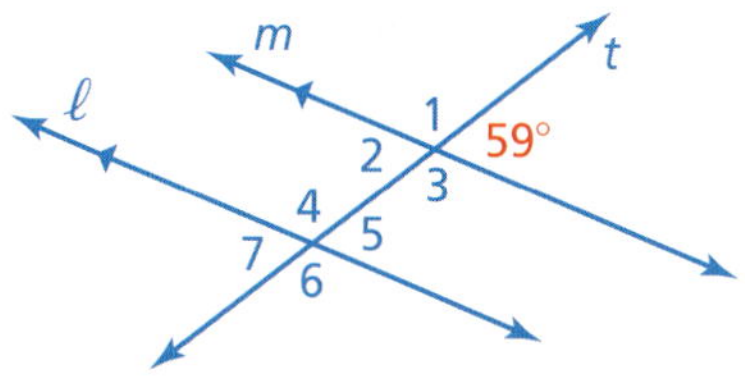

When two parallel lines are cut by a transversal, four **interior angles** are formed on the inside of the parallel lines and four **exterior angles** are formed on the outside of the parallel lines.

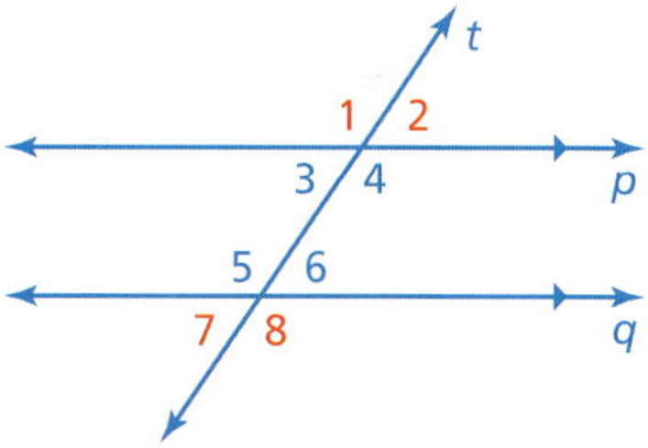

∠3, ∠4, ∠5, and ∠6 are interior angles.
∠1, ∠2, ∠7, and ∠8 are exterior angles.

EXAMPLE 3 Standardized Test Practice

A store owner uses pieces of tape to paint a window advertisement. The letters are slanted at an 80° angle. What is the measure of ∠1?

Ⓐ 80° Ⓑ 100° Ⓒ 110° Ⓓ 120°

Because all of the letters are slanted at an 80° angle, the dashed lines are parallel. The piece of tape is the transversal.

Using the corresponding angles, the 80° angle is congruent to the angle that is supplementary to ∠1, as shown.

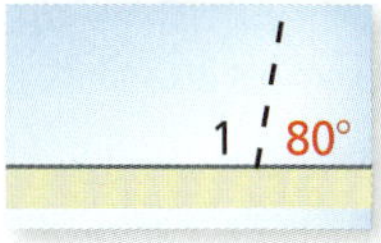

The measure of ∠1 is 180° − 80° = 100°. The correct answer is Ⓑ.

On Your Own

Now You're Ready
Exercises 18 and 19

4. **WHAT IF?** In Example 3, the letters are slanted at a 65° angle. What is the measure of ∠1?

Key Idea

Alternate Interior Angles and Alternate Exterior Angles

When a transversal intersects parallel lines, alternate interior angles are congruent and alternate exterior angles are congruent.

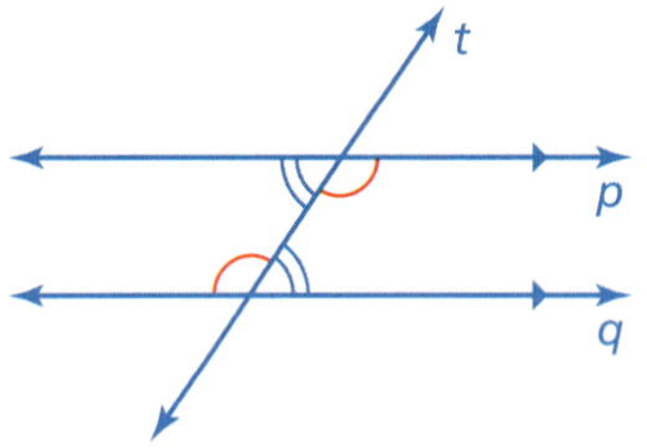

Alternate interior angles

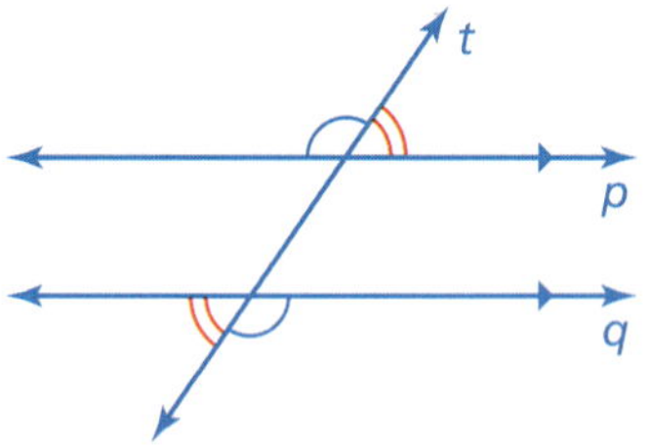

Alternate exterior angles

Study Tip

Alternate interior angles and alternate exterior angles lie on opposite sides of the transversal.

EXAMPLE 4 Identifying Alternate Interior and Alternate Exterior Angles

The photo shows a portion of an airport. Describe the relationship between each pair of angles.

a. ∠3 and ∠6

∠3 and ∠6 are alternate exterior angles.

So, ∠3 is congruent to ∠6.

b. ∠2 and ∠7

∠2 and ∠7 are alternate interior angles.

So, ∠2 is congruent to ∠7.

On Your Own

Now You're Ready
Exercises 20 and 21

In Example 4, the measure of ∠4 is 84°. Find the measure of the angle. Explain your reasoning.

5. ∠3 6. ∠5 7. ∠6

Laurie's Notes

On Your Own

- **Think-Pair-Share:** Students should read the question independently and then work with a partner to answer the question. When they have answered the question, the pair should compare their answer with another group and discuss any discrepancies.

Key Idea

- Write the *Key Idea*. Identify the angles which are marked congruent in the diagram. Mention the *Study Tip*.

Example 4

? "Are the lines parallel? How do you know?" yes; yellow arrow marks

- Work through the explanation as shown. This example helps students identify these new angle pairs.
- **Note:** There is a great deal of vocabulary in this section, so students may need extra practice. It is also important not to draw the parallel lines in the same orientation all of the time, particularly horizontal and vertical.

On Your Own

- Draw the diagram on the board. When students have finished, ask volunteers to come to the board to record their answers and explain their reasoning.
- **Question 7:** Students can say $\angle 6 = 96°$ because of alternate exterior angles *or* because it is the supplement of $\angle 8$.

Closure

- **Exit Ticket:** Find the measure of each angle. Explain your reasoning.

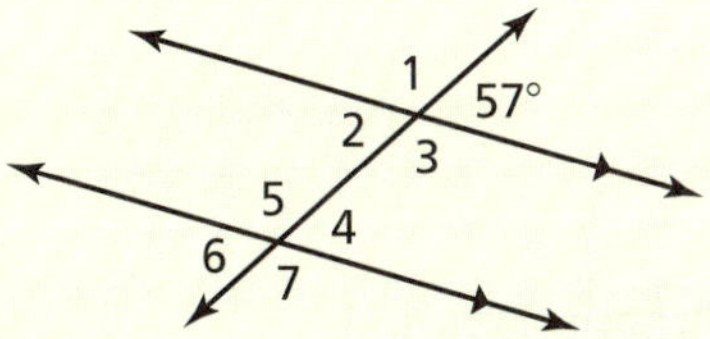

$\angle 1 = 123°$, $\angle 2 = 57°$, $\angle 3 = 123°$, $\angle 4 = 57°$, $\angle 5 = 123°$, $\angle 6 = 57°$, $\angle 7 = 123°$

Technology For the Teacher

Dynamic Classroom

The Dynamic Planning Tool
Editable Teacher's Resources at *BigIdeasMath.com*

On Your Own

4. 115°

Extra Example 4

Describe the relationship between each pair of angles.

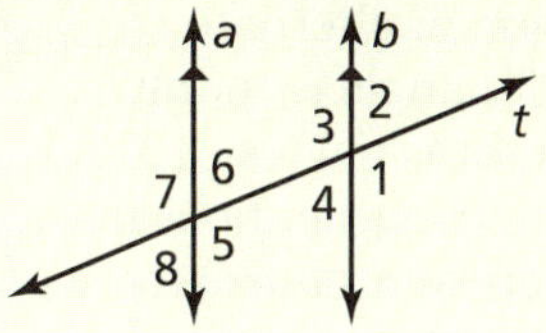

a. $\angle 1$ and $\angle 7$ $\angle 1$ and $\angle 7$ are alternate exterior angles. So, $\angle 1$ is congruent to $\angle 7$.

b. $\angle 3$ and $\angle 5$ $\angle 3$ and $\angle 5$ are alternate interior angles. So, $\angle 3$ is congruent to $\angle 5$.

On Your Own

5. 96°; $\angle 3$ and $\angle 4$ are supplementary.

6. 84°; Alternate interior angles are congruent.

7. 96°; $\angle 5$ and $\angle 6$ are supplementary.

Vocabulary and Concept Check

1. *Sample answer:*

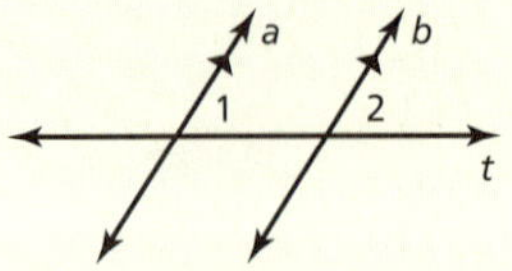

2. "The measure of $\angle 5$" doesn't belong because $\angle 2$, $\angle 6$, and $\angle 8$ are congruent and $\angle 5$ is not a corresponding, alternate interior, or alternate exterior angle with the other three angles. $\angle 2$ and $\angle 8$ are congruent because they are alternate exterior angles. $\angle 6$ and $\angle 8$ are congruent because they are vertical angles.

Practice and Problem Solving

3. m and n
4. t
5. 8
6. $\angle 5$, $\angle 7$, $\angle 1$, and $\angle 3$ are congruent. $\angle 8$, $\angle 6$, $\angle 4$, and $\angle 2$ are congruent.
7. $\angle 1 = 107°$, $\angle 2 = 73°$
8. $\angle 3 = 95°$, $\angle 4 = 85°$
9. $\angle 5 = 49°$, $\angle 6 = 131°$
10. The two lines are not parallel, so $\angle 5 \neq \angle 6$.
11. 60°; Corresponding angles are congruent.
12. *Sample answer:* Railroad tracks are parallel, and the out of bounds lines on a football field are parallel.

Assignment Guide and Homework Check

Level	Day 1 Activity Assignment	Day 2 Lesson Assignment	Homework Check
Basic	3–6, 31–35	1, 2, 7–11, 13–23 odd, 24	8, 10, 15, 19, 24
Average	3–6, 31–35	1, 2, 7–27 odd, 10, 24	7, 10, 15, 19, 24
Advanced	3–6, 31–35	1, 2, 10, 14–24 even, 26–30	10, 14, 16, 22, 26, 28

Common Errors

- **Exercises 7–9** Students may mix up some of the definitions of congruent angles and find incorrect angle measures. Encourage them to look at the Key Ideas and color-code the figure they are given to determine what angles are congruent.
- **Exercise 11** Students may not realize that the line in front of the cars is the transversal. Remind them that lines are infinite and can be extended. Draw a diagram of the parallel parking spaces to help students visualize that $\angle 1$ and $\angle 2$ are corresponding angles.

5.5 Record and Practice Journal

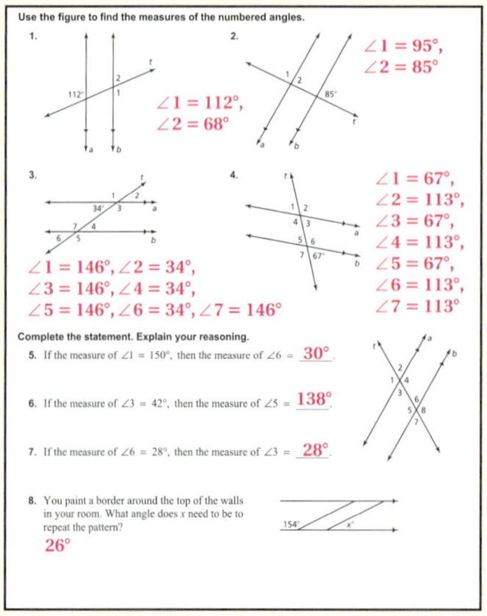

Use the figure to find the measures of the numbered angles.

1. $\angle 1 = 112°$, $\angle 2 = 68°$

2. $\angle 1 = 95°$, $\angle 2 = 85°$

3. $\angle 1 = 146°$, $\angle 2 = 34°$, $\angle 3 = 146°$, $\angle 4 = 34°$, $\angle 5 = 146°$, $\angle 6 = 34°$, $\angle 7 = 146°$

4. $\angle 1 = 67°$, $\angle 2 = 113°$, $\angle 3 = 67°$, $\angle 4 = 113°$, $\angle 5 = 67°$, $\angle 6 = 113°$, $\angle 7 = 113°$

Complete the statement. Explain your reasoning.

5. If the measure of $\angle 1 = 150°$, then the measure of $\angle 6 =$ 30°.
6. If the measure of $\angle 3 = 42°$, then the measure of $\angle 5 =$ 138°.
7. If the measure of $\angle 6 = 28°$, then the measure of $\angle 3 =$ 28°.
8. You paint a border around the top of the walls in your room. What angle does x need to be to repeat the pattern?
   26°

Technology For the Teacher
Answer Presentation Tool
Quiz*Show*

5.5 Exercises

Check It Out
Help with Homework
BigIdeasMath.com

Vocabulary and Concept Check

1. **VOCABULARY** Draw two parallel lines and a transversal. Label a pair of corresponding angles.

2. **WHICH ONE DOES NOT BELONG?** Which statement does *not* belong with the other three? Explain your reasoning. Refer to the figure for Exercises 3–6.

The measure of ∠2	The measure of ∠5
The measure of ∠6	The measure of ∠8

Practice and Problem Solving

In Exercises 3–6, use the figure.

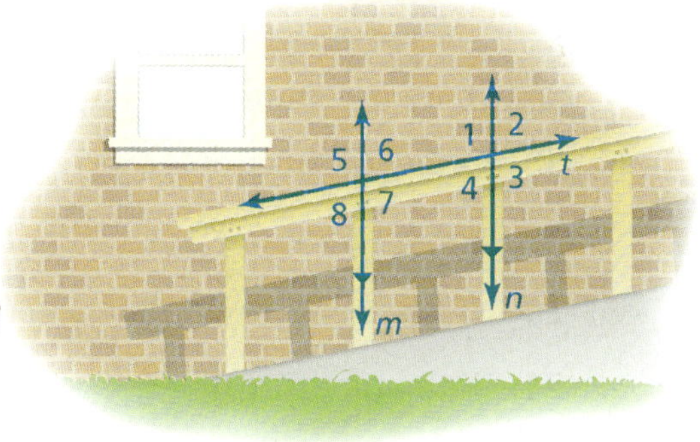

3. Identify the parallel lines.
4. Identify the transversal.
5. How many angles are formed by the transversal?
6. Which of the angles are congruent?

Use the figure to find the measures of the numbered angles.

1 7.

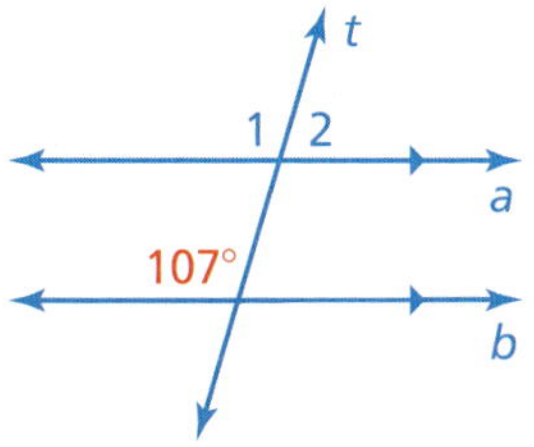

8.

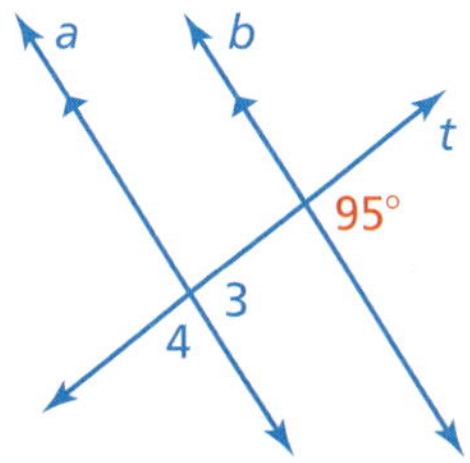

9.

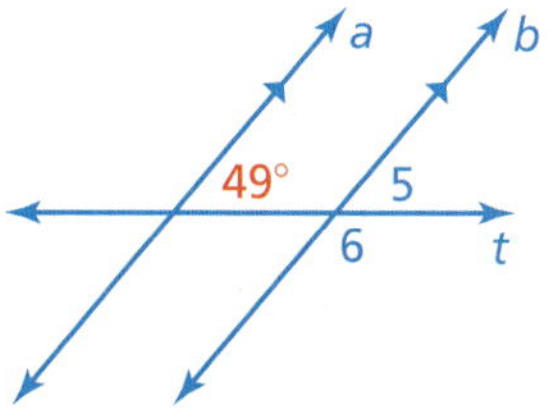

10. **ERROR ANALYSIS** Describe and correct the error in describing the relationship between the angles.

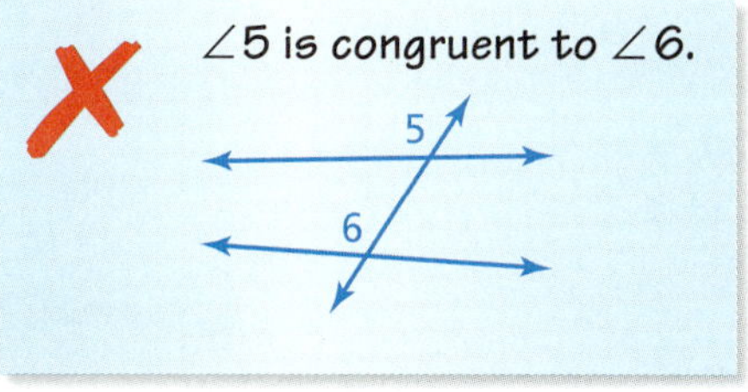

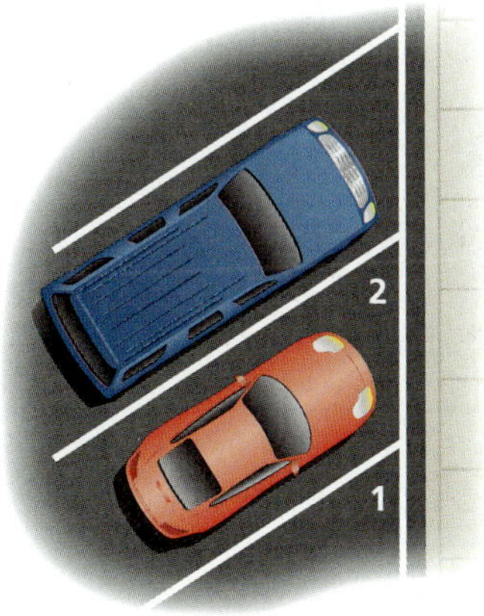

11. **PARKING** The painted lines that separate parking spaces are parallel. The measure of ∠1 is 60°. What is the measure of ∠2? Explain.

12. **OPEN-ENDED** Describe two real-life situations that use parallel lines.

13. **PROJECT** Draw two horizontal lines and a transversal on a piece of notebook paper. Label the angles as shown. Use a pair of scissors to cut out the angles. Compare the angles to determine which angles are congruent.

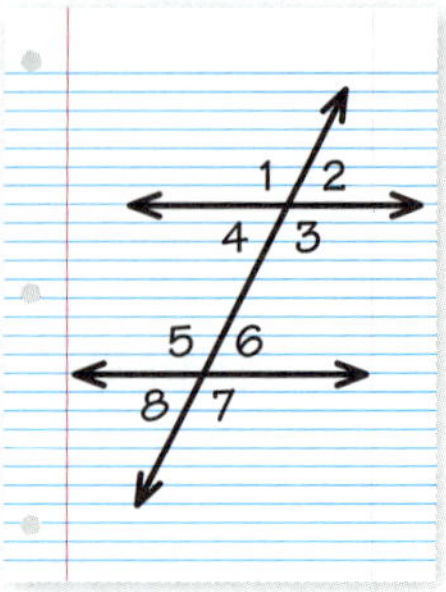

14. **REASONING** Refer to the figure for Exercise 13. What is the least number of angle measures you need to know in order to find the measure of every angle? Explain your reasoning.

Use the figure to find the measures of the numbered angles. Explain your reasoning.

2 15.

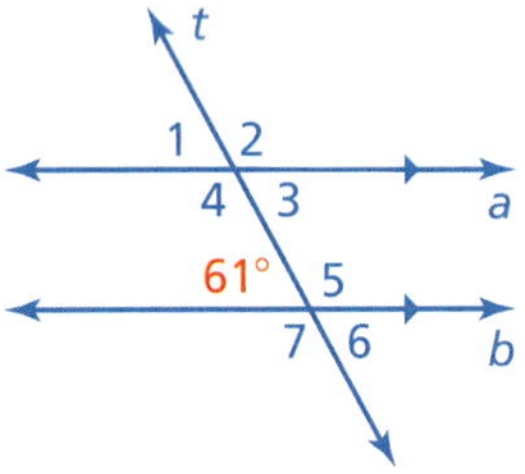

16.

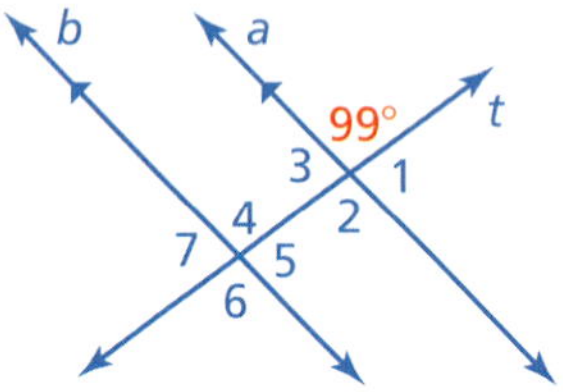

17.

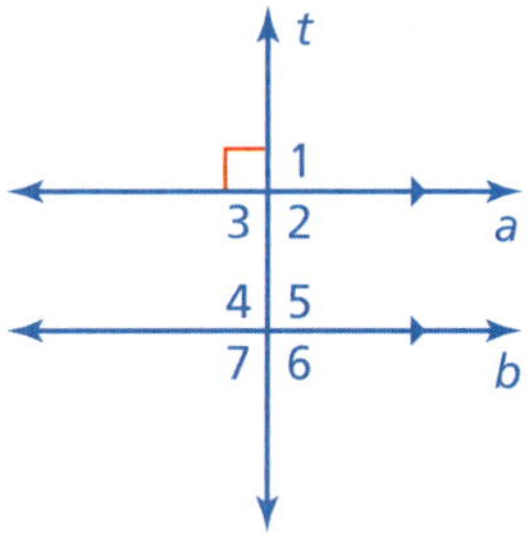

Complete the statement. Explain your reasoning.

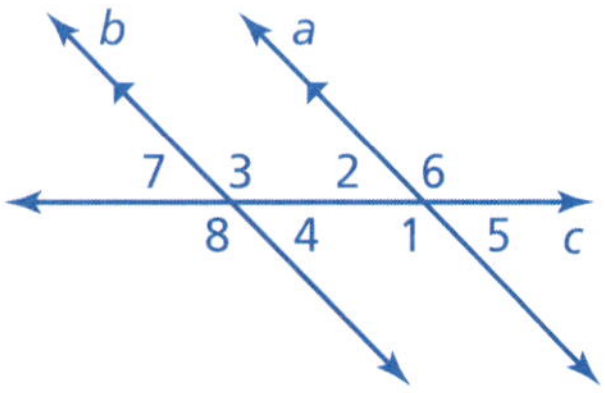

3 18. If the measure of $\angle 1 = 124°$, then the measure of $\angle 4 =$ ___.

19. If the measure of $\angle 2 = 48°$, then the measure of $\angle 3 =$ ___.

4 20. If the measure of $\angle 4 = 55°$, then the measure of $\angle 2 =$ ___.

21. If the measure of $\angle 6 = 120°$, then the measure of $\angle 8 =$ ___.

22. If the measure of $\angle 7 = 50.5°$, then the measure of $\angle 6 =$ ___.

23. If the measure of $\angle 3 = 118.7°$, then the measure of $\angle 2 =$ ___.

24. **RAINBOW** A rainbow is formed when sunlight reflects off raindrops at different angles. For blue light, the measure of $\angle 2$ is 40°. What is the measure of $\angle 1$?

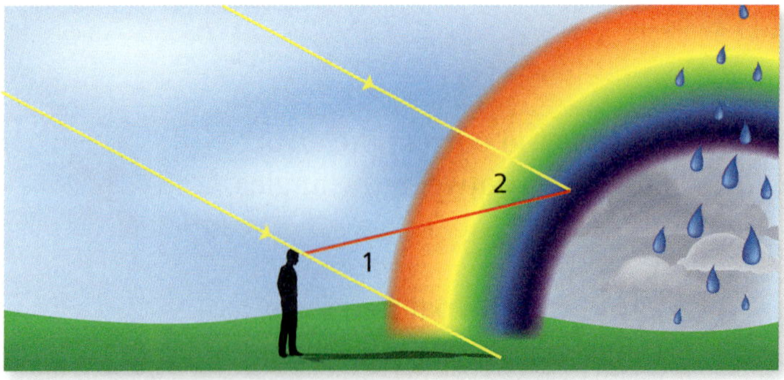

25. **REASONING** If a transversal is perpendicular to two parallel lines, what can you conclude about the angles formed? Explain.

26. **WRITING** Describe two ways you can show that $\angle 1$ is congruent to $\angle 7$.

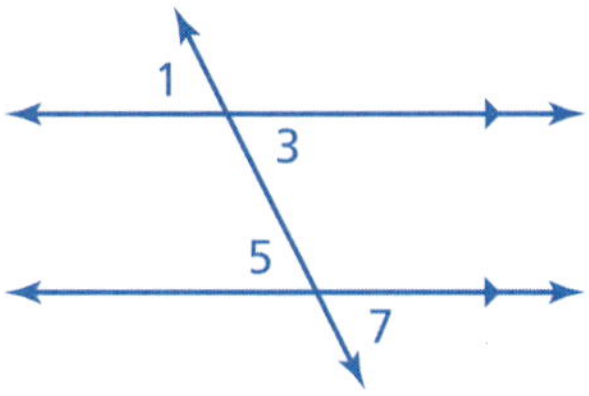

Common Errors

- **Exercises 15–17** Students may not understand alternate interior and exterior angles and say that an exterior angle is congruent to the alternate interior angle. For example, in Exercise 15, a student may say the measure of $\angle 2$ is 61°. Use corresponding angles to show that this is not true.
- **Exercises 18–23** Students may use some of the definitions of congruent angles incorrectly in finding the angle measure of the unknown angle. Review the definitions and give an example with an adequate explanation of how to find the missing angle.
- **Exercises 27 and 28** Students may only see one set of parallel lines and think that they cannot find the measure of the missing angle. Point out the small arrows that denote that two lines are parallel. Encourage them to find the measure of an angle that is near the missing angle and then rotate the figure to help them visualize how to solve for the missing angle.

Differentiated Instruction

Auditory

Students may confuse the measures of *complementary angles* and *supplementary angles*. Show students that *complementary* comes before *supplementary* in the dictionary and that *90* comes before *180* numerically. So, the sum of complementary angles is 90° and the sum of supplementary angles is 180°.

Practice and Problem Solving

13. $\angle 1$, $\angle 3$, $\angle 5$, and $\angle 7$ are congruent. $\angle 2$, $\angle 4$, $\angle 6$, and $\angle 8$ are congruent.

14. You only need one angle because half of the angles are congruent to that angle and you can find the other angles using relationships.

15. $\angle 6 = 61°$; $\angle 6$ and the given angle are vertical angles.
$\angle 5 = 119°$ and $\angle 7 = 119°$; $\angle 5$ and $\angle 7$ are supplementary to the given angle.
$\angle 1 = 61°$; $\angle 1$ and the given angle are corresponding angles.
$\angle 3 = 61°$; $\angle 1$ and $\angle 3$ are vertical angles.
$\angle 2 = 119°$ and $\angle 4 = 119°$; $\angle 2$ and $\angle 4$ are supplementary to $\angle 1$.

16. $\angle 2 = 99°$; $\angle 2$ and the given angle are vertical angles.
$\angle 1 = 81°$ and $\angle 3 = 81°$; $\angle 1$ and $\angle 3$ are supplementary to the given angle.
$\angle 4 = 99°$; $\angle 2$ and $\angle 4$ are alternate interior angles.
$\angle 5 = 81°$ and $\angle 7 = 81°$; $\angle 5$ and $\angle 7$ are supplementary to $\angle 4$.
$\angle 6 = 99°$; $\angle 6$ and the given angle are alternate exterior angles.

17–26. See Additional Answers.

Practice and Problem Solving

27. 130

28. 115

29. a. no; They look like they are spreading apart.

b. Check students' work.

30. See *Taking Math Deeper*.

Fair Game Review

31. 13 **32.** 14

33. 51 **34.** 3

35. B

Mini-Assessment

Use the figure to find the measures of the numbered angles.

1.

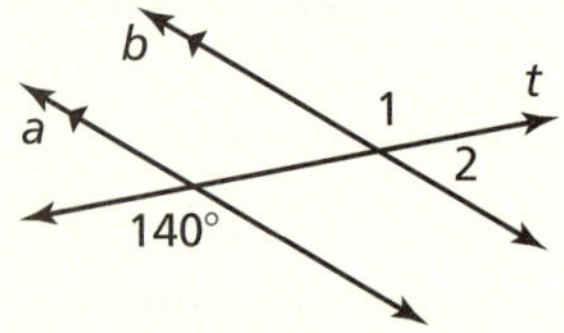

$\angle 1 = 140°$; $\angle 2 = 40°$

2.

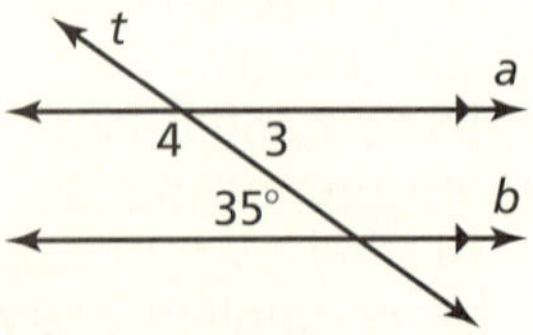

$\angle 3 = 35°$; $\angle 4 = 145°$

3.

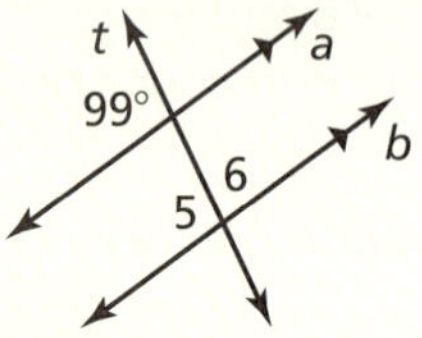

$\angle 5 = 99°$; $\angle 6 = 81°$

Taking Math Deeper

Exercise 30

This problem uses a well-known reflective property in physics. This property applies to mirrors, billiard tables, air hockey tables, and many other objects. The property states that when the hockey puck bounces off the side board, its out-going angle is equal to its in-coming angle.

1 Solve for *m*.

$$m + m + 64 = 180$$
$$2m = 116$$
$$m = 58$$

2 Solve for *x*.

a. Using the property of alternate interior angles, you can determine that $x = 64$.

3 Answer the question.

b. The goal is slightly wider than the hockey puck. So, there is some leeway allowed for the measure of *x*.

By studying the diagram, you can see that *x* cannot be much greater. However, *x* can be a little less and still have the hockey puck go into the goal.

Project

Write a report about at least two other games that use angles as part of their strategy.

Reteaching and Enrichment Strategies

If students need help. . .	If students got it. . .
Resources by Chapter • Practice A and Practice B • Puzzle Time Record and Practice Journal Practice Differentiating the Lesson Lesson Tutorials Skills Review Handbook	Resources by Chapter • Enrichment and Extension • School-to-Work • Financial Literacy Start the next section

CRITICAL THINKING **Find the value of x.**

27.

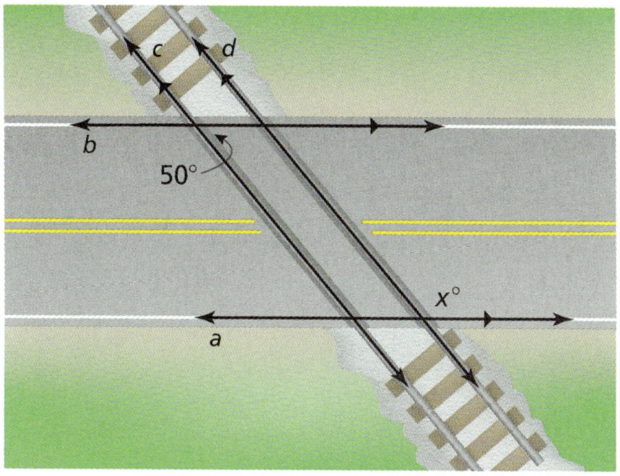

28.

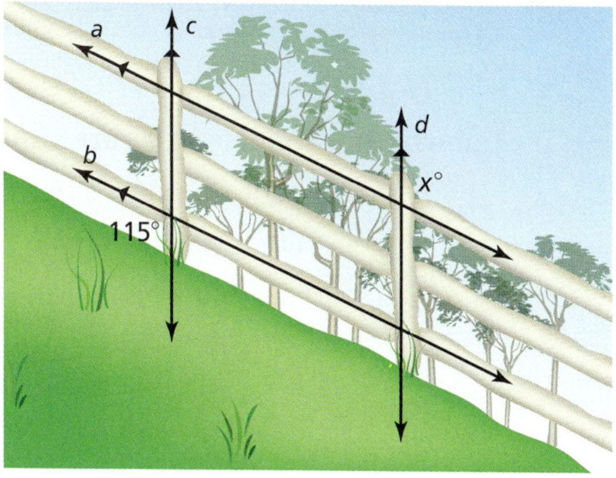

29. **OPTICAL ILLUSION** Refer to the figure.

a. Do the horizontal lines appear to be parallel? Explain.

b. Draw your own optical illusion using parallel lines.

30. Geometry The figure shows the angles used to make a double bank shot in an air hockey game.

a. Find the value of x.

b. Can you still get the red puck in the goal if x is increased by a little? by a lot? Explain.

Fair Game Review What you learned in previous grades & lessons

Evaluate the expression. *(Skills Review Handbook)*

31. $4 + 3^2$ **32.** $5(2)^2 - 6$ **33.** $11 + (-7)^2 - 9$ **34.** $8 \div 2^2 + 1$

35. **MULTIPLE CHOICE** The volume of the cylinder is 20π cubic inches. What is the radius of the base? *(Skills Review Handbook)*

Ⓐ 1 inch Ⓑ 2 inches

Ⓒ 3 inches Ⓓ 4 inches

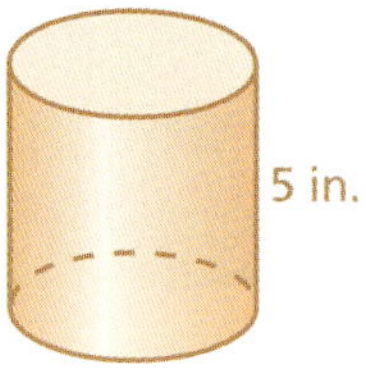

5.4–5.5 Quiz

Check It Out
Progress Check
BigIdeasMath.com

Tell whether the triangles are similar. Explain. *(Section 5.4)*

1.

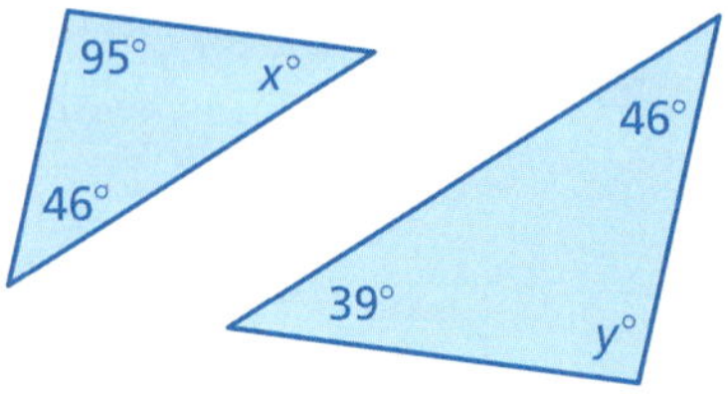

2.

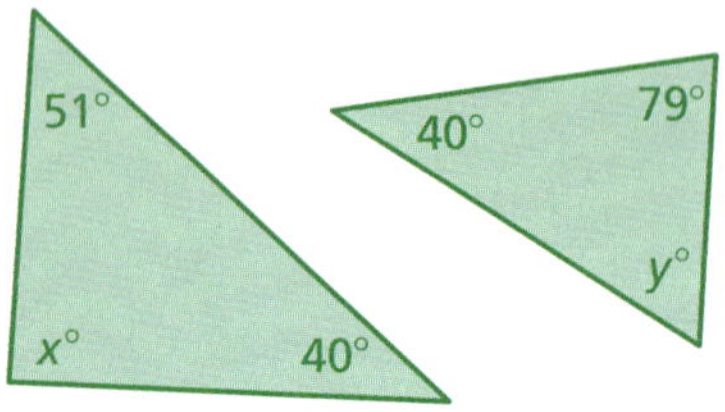

The triangles are similar. Find the value of *x*. *(Section 5.4)*

3.

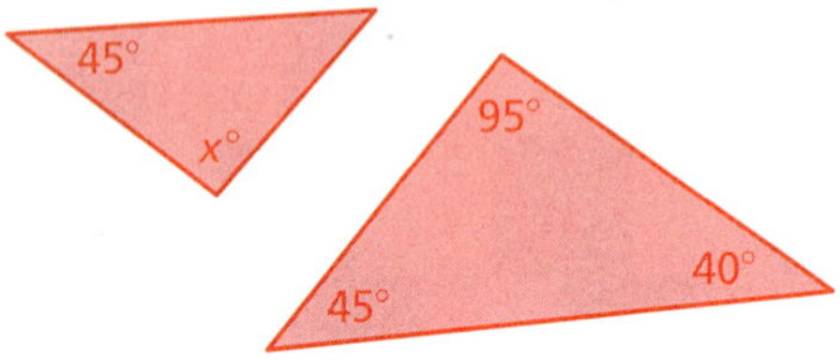

4.

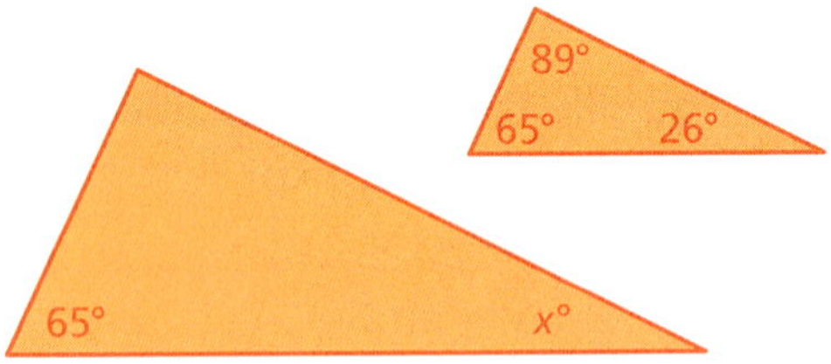

Use the figure to find the measure of the angle. Explain your reasoning. *(Section 5.5)*

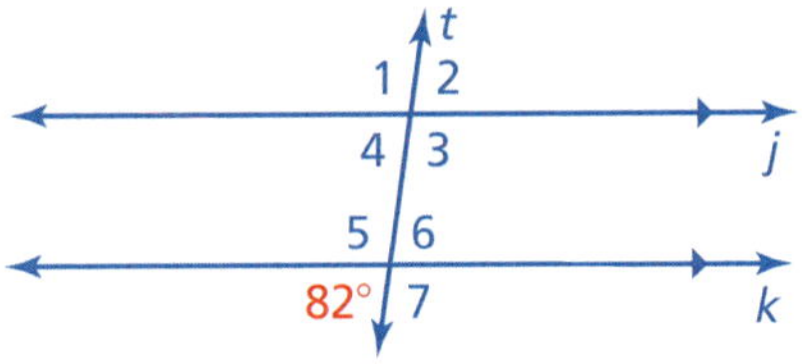

5. ∠2

6. ∠6

7. ∠4

8. ∠1

Complete the statement. Explain your reasoning. *(Section 5.5)*

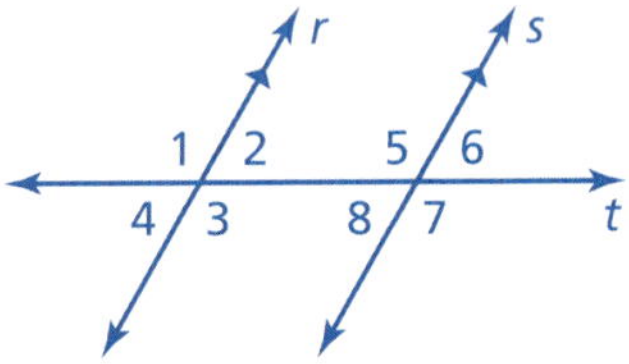

9. If the measure of ∠1 = 123°, then the measure of ∠7 = ____.

10. If the measure of ∠2 = 58°, then the measure of ∠5 = ____.

11. If the measure of ∠5 = 119°, then the measure of ∠3 = ____.

12. If the measure of ∠4 = 60°, then the measure of ∠6 = ____.

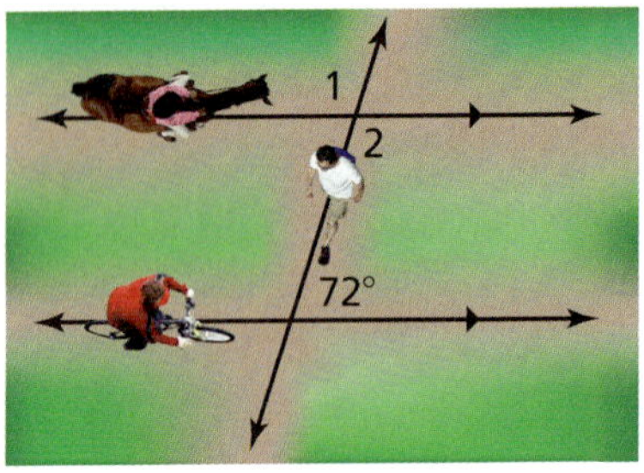

13. PARK In a park, a bike path and a horse riding path are parallel. In one part of the park, a hiking trail intersects the two paths. Find the measures of ∠1 and ∠2. Explain your reasoning. *(Section 5.5)*

14. PERIMETER The side lengths of a right triangle are doubled to make a similar triangle. Does the perimeter double as well? Explain. *(Section 5.4)*

Alternative Assessment Options

Math Chat
Structured Interview
Student Reflective Focus Question
Writing Prompt

Math Chat

- Put students in pairs to complete and discuss the exercises from the quiz. The discussion should include discussing terms such as similar triangles, perpendicular lines, transversals, corresponding angles, interior angles, exterior angles, alternate interior angles, and alternate exterior angles.
- The teacher should walk around the classroom listening to the pairs and ask questions to ensure understanding.

Study Help Sample Answers

Remind students to complete Graphic Organizers for the rest of the chapter.

6. Similar Triangles

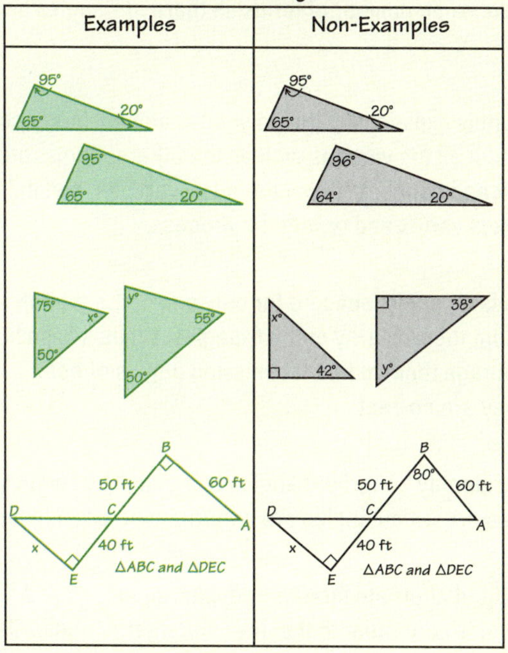

7–8. Available at *BigIdeasMath.com*

Reteaching and Enrichment Strategies

If students need help. . .	If students got it. . .
Resources by Chapter • Study Help • Practice A and Practice B • Puzzle Time Lesson Tutorials *BigIdeasMath.com* Practice Quiz Practice from the Test Generator	Resources by Chapter • Enrichment and Extension • School-to-Work Game Closet at *BigIdeasMath.com* Start the Chapter Review

Answers

1. yes; The triangles have the same angle measures, 95°, 46°, and 39°.
2. no; The triangles do not have the same angle measures.
3. $x = 95$
4. $x = 26$
5. $\angle 2 = 82°$; $\angle 2$ and the given angle are alternate exterior angles.
6. $\angle 6 = 82°$; $\angle 6$ and the given angle are vertical angles.
7. $\angle 4 = 82°$; $\angle 4$ and the given angle are corresponding angles.
8. $\angle 1 = 98°$; $\angle 4$ and $\angle 1$ are supplementary.
9. 123°; $\angle 1$ and $\angle 7$ are alternate exterior angles.
10. 122°; $\angle 2$ and $\angle 8$ are alternate interior angles and $\angle 8$ and $\angle 5$ are supplementary.

11–14. See Additional Answers.

Technology For the Teacher
Answer Presentation Tool

Assessment Book

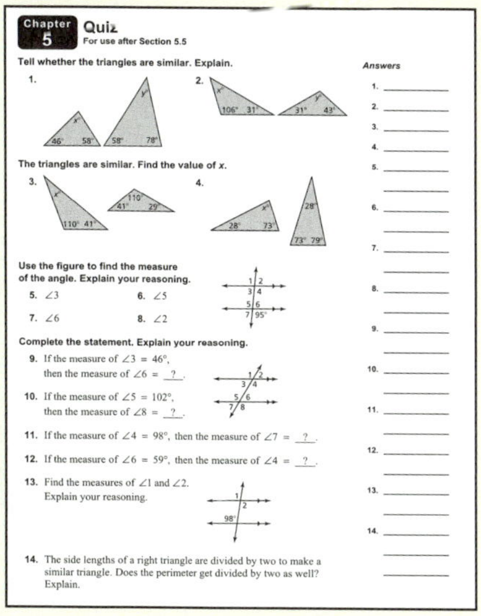
Chapter 5 Quiz
For use after Section 5.5

Tell whether the triangles are similar. Explain.

1. (46°, 58°, x°; 58°, 78°, y°)
2. (x°, 106°, 31°; y°, 31°, 43°)

The triangles are similar. Find the value of *x*.

3. (x°, 110°, 41°; 110°, 41°, 29°)
4. (28°, 73°, x°; 28°, 73°, 79°)

Use the figure to find the measure of the angle. Explain your reasoning.

5. $\angle 3$
6. $\angle 5$
7. $\angle 6$
8. $\angle 2$

Complete the statement. Explain your reasoning.

9. If the measure of $\angle 3 = 46°$, then the measure of $\angle 6 =$? .
10. If the measure of $\angle 5 = 102°$, then the measure of $\angle 8 =$? .
11. If the measure of $\angle 4 = 98°$, then the measure of $\angle 7 =$? .
12. If the measure of $\angle 6 = 59°$, then the measure of $\angle 4 =$? .
13. Find the measures of $\angle 1$ and $\angle 2$. Explain your reasoning.
14. The side lengths of a right triangle are divided by two to make a similar triangle. Does the perimeter get divided by two as well? Explain.

Answers
1. ____ 2. ____ 3. ____ 4. ____ 5. ____ 6. ____ 7. ____ 8. ____ 9. ____ 10. ____ 11. ____ 12. ____ 13. ____ 14. ____

For the Teacher
Additional Review Options

- **Quiz***Show*
- Big Ideas Test Generator
- Game Closet at *BigIdeasMath.com*
- Vocabulary Puzzle Builder
- Resources by Chapter
  - Puzzle Time
  - Study Help

Answers

1. $x = 21$
2. $x = 84$

Review of Common Errors

Exercise 1

- Students may not be able to set up the correct equation. Point out that the two angles make up a right (90°) angle.

Exercise 2

- Students may not realize that the labeled angles are vertical angles and that they are congruent. Review vertical angles with students.

Exercises 3 and 4

- Students may solve for *x*, but forget to classify the triangle. Remind them to read the directions carefully and to answer the question.

Exercises 5–7

- Students may forget to include one or more of the angles when writing an equation to find the value of *x*. Remind students to include all of the angles. Encourage them to write the equation and then count the number of terms to make sure that there is the same number of terms as there are angles before solving.

Exercises 8–10

- Students may not adequately explain why the polygon is convex or concave. Remind students to connect all the vertices with all the others. One strategy is to start with one vertex and connect it to all the other vertices, and then rotate clockwise to the next vertex and repeat the process.

Exercise 11

- Students may find the missing angle measure for only one of the triangles and then make a decision about the similarity of the triangles. While it is possible to use this method, encourage them to find the missing angles of *both* triangles to verify that they are correct.

Exercise 12

- Students may incorrectly identify congruent angles and find the wrong value for *x*. Encourage them to mark which angles are congruent before finding *x*.

Exercises 13–16

- Students may not understand alternate interior and exterior angles and think that an exterior angle is congruent to the alternate interior angle. Use corresponding angles to show that this is not necessarily true.

5 Chapter Review

Check It Out Vocabulary Help BigIdeasMath.com

Review Key Vocabulary

complementary angles, *p. 186*
supplementary angles, *p. 186*
congruent angles, *p. 187*
vertical angles, *p. 187*
isosceles triangle, *p. 192*
congruent sides, *p. 192*
equilateral triangle, *p. 192*
equiangular triangle, *p. 192*
polygon, *p. 198*
regular polygon, *p. 199*
convex polygon, *p. 200*
concave polygon, *p. 200*
similar triangles, *p. 208*
indirect measurement, *p. 209*
perpendicular lines, *p. 214*
transversal, *p. 214*
interior angles, *p. 215*
exterior angles, *p. 215*

Review Examples and Exercises

5.1 Classifying Angles *(pp. 184–189)*

Find the value of *x*.

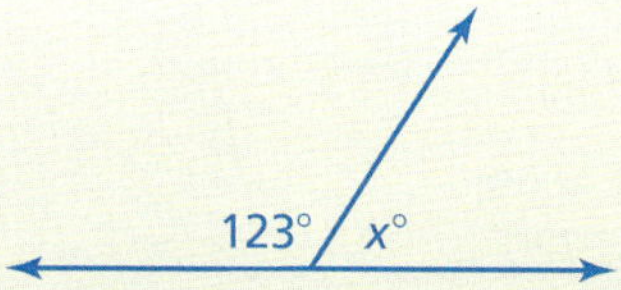

The angles are supplementary angles. So, the sum of their measures is 180°.

$$x + 123 = 180$$

$$x = 57$$

So, x is 57.

Exercises

Find the value of *x*.

1. 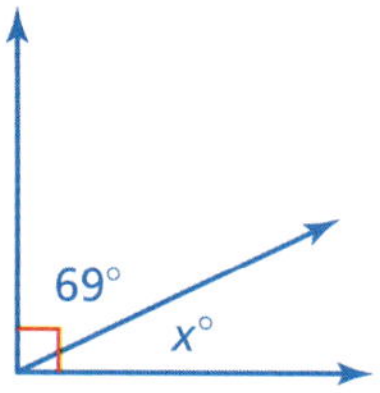

2. 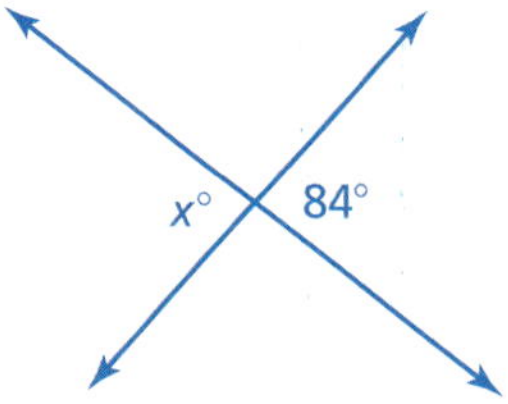

5.2 Angles and Sides of Triangles *(pp. 190–195)*

Find the value of *x*. Then classify the triangle.

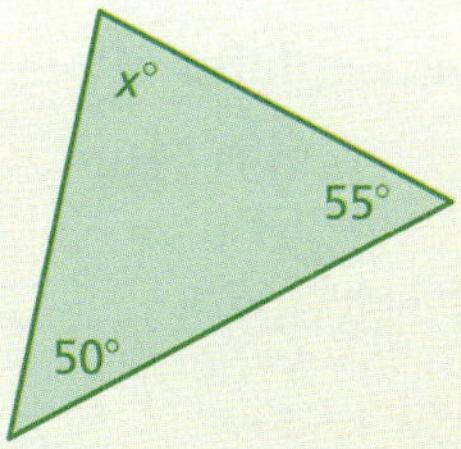

$$x + 50 + 55 = 180$$

$$x + 105 = 180$$

$$x = 75$$

The value of x is 75. The triangle has three acute angle measures, 50°, 55°, and 75°. So, it is an acute triangle.

Exercises

Find the value of x. Then classify the triangle in as many ways as possible.

3.

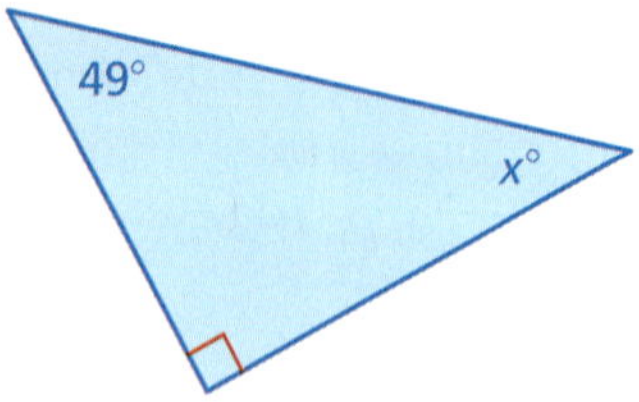

4.

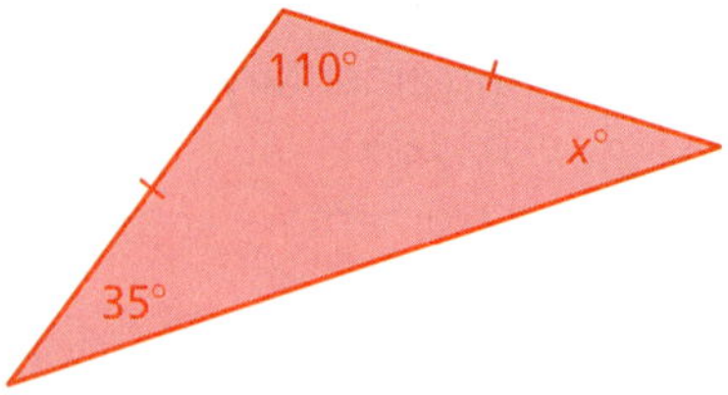

5.3 Angles of Polygons *(pp. 196–203)*

Find the value of x.

Step 1: The polygon has 6 sides. Find the sum of the angle measures.

$S = (n - 2) \cdot 180°$ — Write the formula.

$= (6 - 2) \cdot 180°$ — Substitute 6 for n.

$= 720$ — Simplify. The sum of the angle measures is 720°.

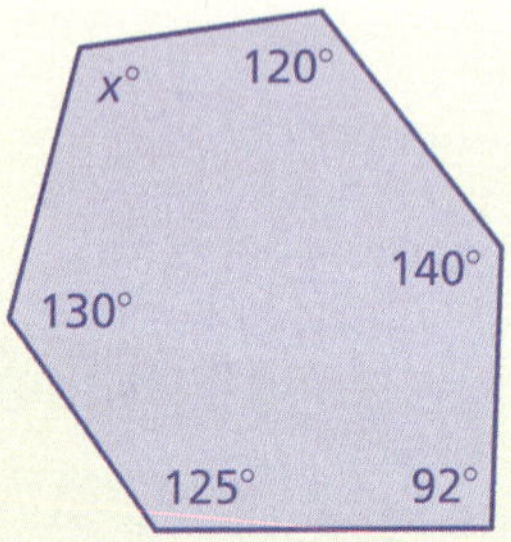

Step 2: Write and solve an equation.

$$130 + 125 + 92 + 140 + 120 + x = 720$$

$$607 + x = 720$$

$$x = 113$$

∴ The value of x is 113.

Exercises

Find the value of x.

5.

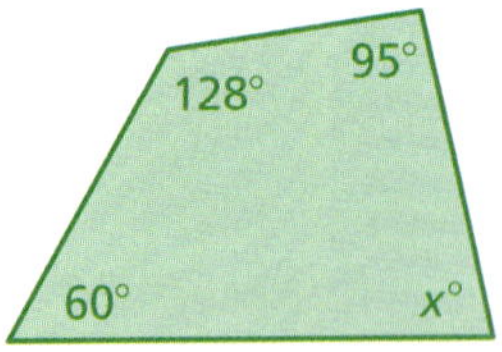

6.

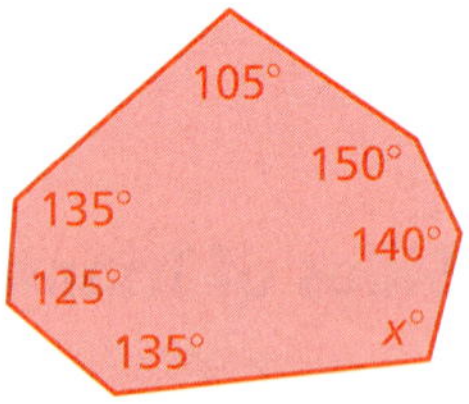

7.

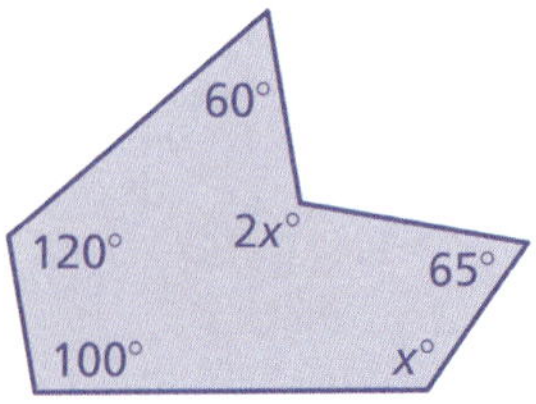

Tell whether the polygon is *convex* or *concave*. Explain.

8.

9.

10.

Review Game

Finding Angle Measures

Big Ideas Game Closet

Materials per Group

- deck of playing cards
- paper
- pencil
- stopwatch

Directions

Divide the class into equally sized groups. A group member lays down two cards next to each other, and below this pair lays down another two cards next to each other. Then they multiply the values of the cards in each pair. (Count kings, queens, jacks, and aces as 10.) These are used to represent the measures of two angles of a triangle. The group member then finds the angle measure of the third angle. Other members time the one working and make sure the computed angle is correct. Each group member takes a turn going through the deck as fast as he or she can. If there is a combination that is impossible to use, they must identify this and move on.

Who wins?

The fastest member in a group after 2 rounds competes against the fastest members in the other groups. The winner is the fastest student.

For the Student Additional Practice

- Lesson Tutorials
- Study Help (textbook)
- Student Website
  - Multi-Language Glossary
  - Practice Assessments

Answers

3. $x = 41$; right

4. $x = 35$; isosceles, obtuse

5. $x = 77$

6. $x = 110$

7. $x = 125$

8. convex; No line segment connecting two vertices lies outside the polygon.

9. concave; A line segment connecting two vertices lies outside the polygon.

10. concave; A line segment connecting two vertices lies outside the polygon.

11. yes; The triangles have the same angle measures, 90°, 68°, and 22°.

12. $x = 50$

13. 140°; $\angle 8$ and the given angle are alternate exterior angles.

14. 140°; $\angle 8$ and $\angle 5$ are vertical angles.

15. 40°; $\angle 8$ and $\angle 7$ are supplementary.

16. 40°; $\angle 2$ and the given angle are supplementary.

My Thoughts on the Chapter

Teacher Tip

Not allowed to write in your teaching edition? Use sticky notes to record your thoughts.

What worked. . .

What did not work. . .

What I would do differently. . .

5.4 Using Similar Triangles (pp. 206–211)

Tell whether the triangles are similar. Explain.

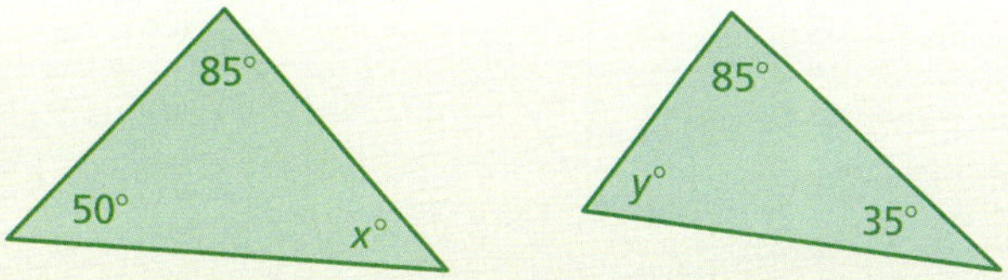

$$50 + 85 + x = 180$$
$$135 + x = 180$$
$$x = 45$$

$$y + 85 + 35° = 180$$
$$y + 120 = 180$$
$$y = 60$$

∴ The triangles do not have the same angle measures. So, they are not similar.

Exercises

11. Tell whether the triangles are similar. Explain.

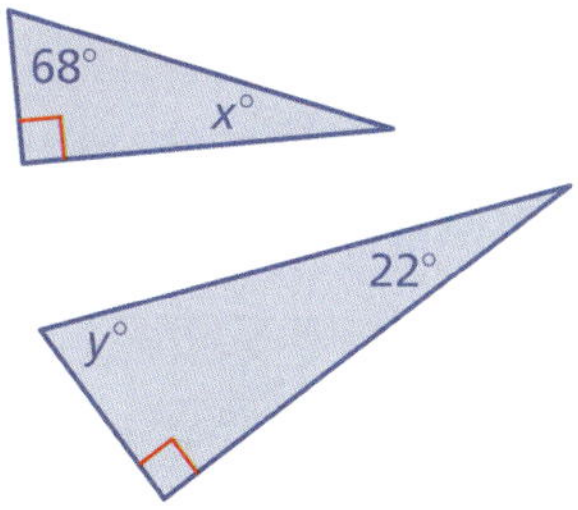

12. The triangles are similar Find the value of x.

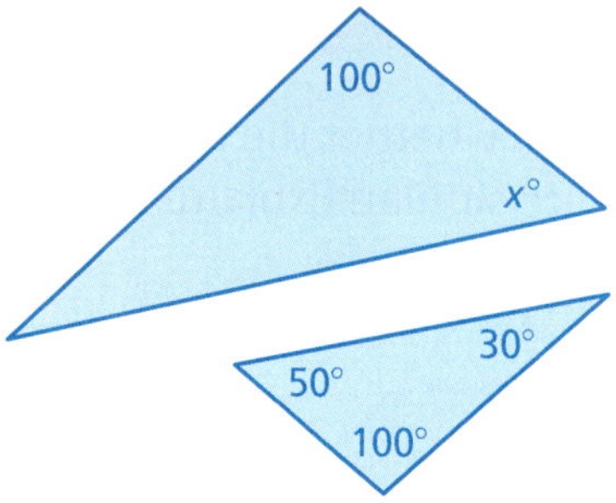

5.5 Parallel Lines and Transversals (pp. 212–219)

Use the figure to find the measure of ∠6.

∠2 and the 55° angle are supplementary.
So, the measure of ∠2 is 180° − 55° = 125°.

∠2 and ∠6 are corresponding angles.
They are congruent.

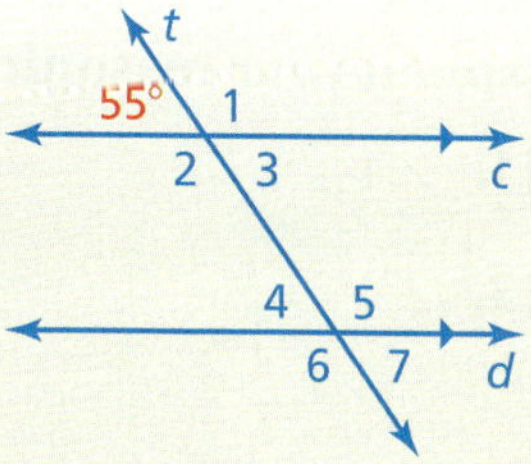

∴ So, the measure of ∠6 is 125°.

Exercises

Use the figure to find the measure of the angle. Explain your reasoning.

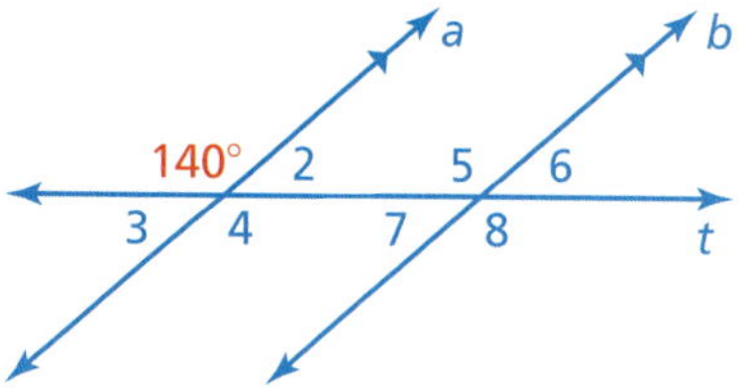

13. ∠8 **14.** ∠5

15. ∠7 **16.** ∠2

5 Chapter Test

Check It Out
Test Practice
BigIdeasMath.com

Find the value of *x*.

1. 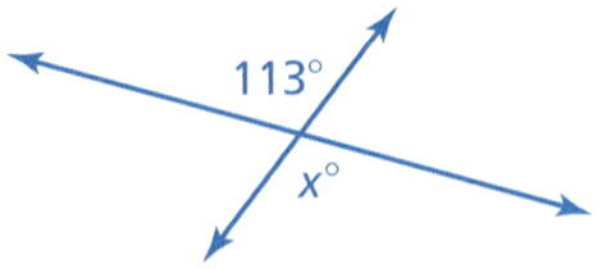

2. 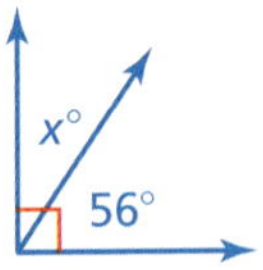

3. 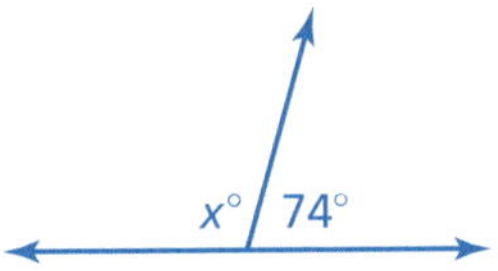

Find the value of *x*. Then classify the triangle in as many ways as possible.

4. 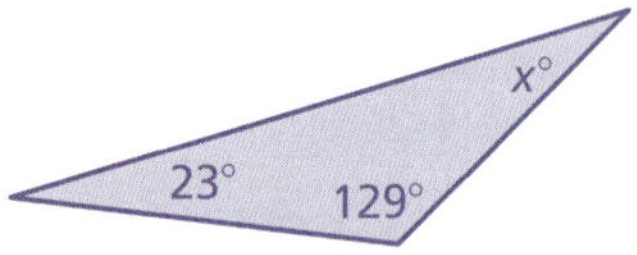

5. 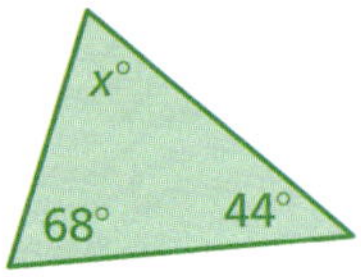

6. 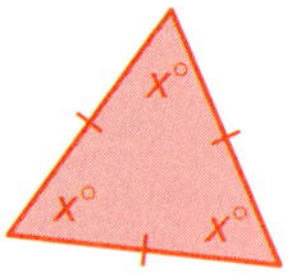

7. Tell whether the polygon is *convex* or *concave*. Explain.

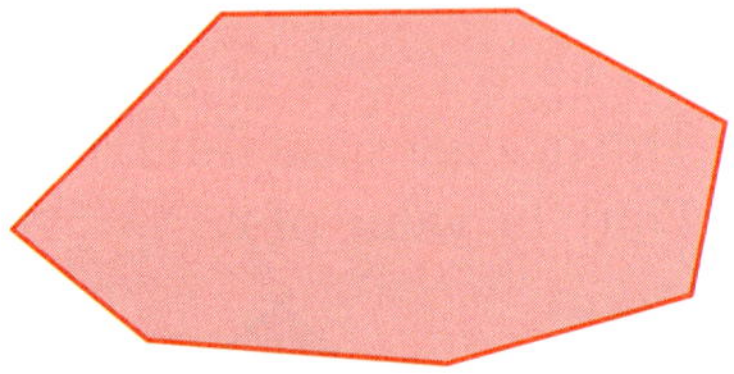

8. Find the value of *x*.

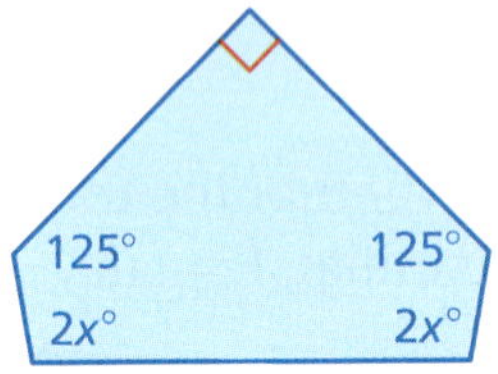

9. Tell whether the triangles are similar. Explain.

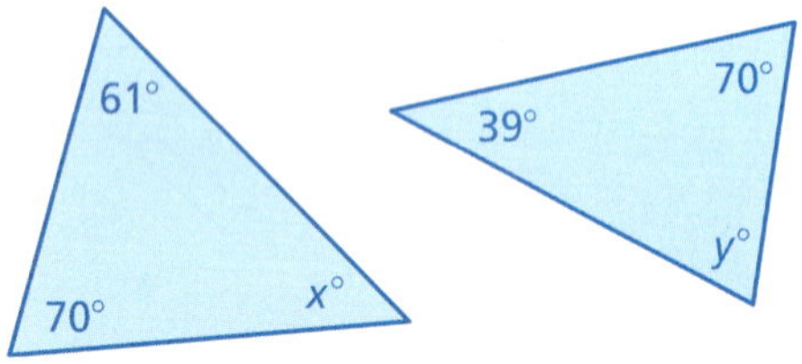

10. The triangles are similar. Find the value of *x*.

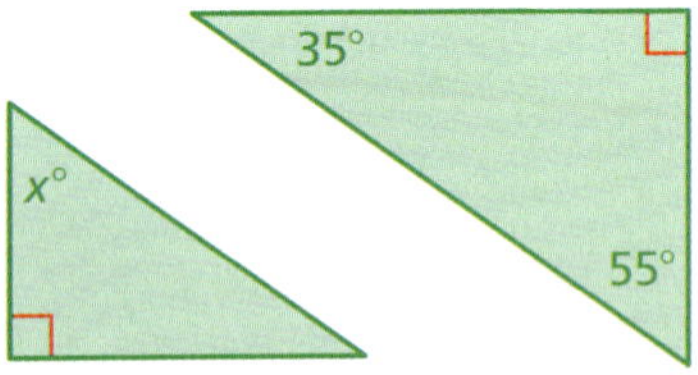

Use the figure to find the measure of the angle. Explain your reasoning.

11. ∠1

12. ∠8

13. ∠4

14. ∠5

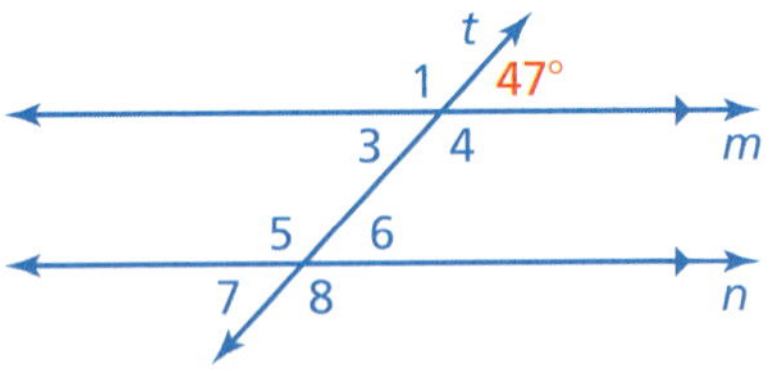

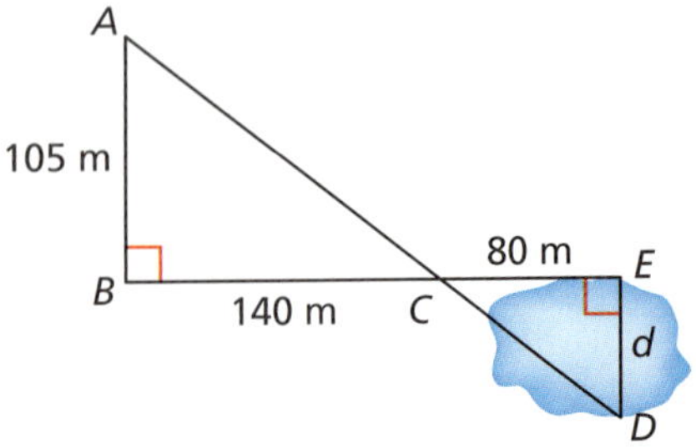

15. POND Use the given measurements to find the distance *d* across the pond.

Test Item References

Chapter Test Questions	Section to Review
1–3	5.1
4–6	5.2
7, 8	5.3
9, 10, 15	5.4
11–14	5.5

Test-Taking Strategies

Remind students to quickly look over the entire test before they start so that they can budget their time. Students should jot down the formula for the sum of interior angles of a polygon on the back of their test before they begin. Students need to use the **Stop** and **Think** strategy before answering questions.

Common Assessment Errors

- **Exercises 2 and 3** Students may not be able to set up the correct equation. Point out that the two angles in Exercise 2 make up a right (90°) angle and that the two angles in Exercise 3 make up a straight (180°) angle.
- **Exercises 4–6** Students may solve for *x*, but forget to classify the triangle. Remind them to read the directions carefully and to answer the question.
- **Exercise 7** Students may not adequately explain why the polygon is convex. Remind students to connect all the vertices with all the others. One strategy is to start with one vertex and connect it to all the other vertices, and then rotate clockwise to the next vertex and repeat the process.
- **Exercise 8** Students may forget to include one or more of the angles when writing an equation to find the value of *x*. Remind students to include all the angles. Encourage them to write the equation and then count the number of terms to make sure that there is the same number of terms as there are angles before solving.
- **Exercise 9** Students may find only one missing angle measure and then make a decision about the similarity of the triangles. While it is possible to use this method, encourage them to find *both* missing angle measures to verify their answer.

Reteaching and Enrichment Strategies

If students need help. . .	If students got it. . .
Resources by Chapter • Practice A and Practice B • Puzzle Time Record and Practice Journal Practice Differentiating the Lesson Lesson Tutorials Practice from the Test Generator Skills Review Handbook	Resources by Chapter • Enrichment and Extension • School-to-Work • Financial Literacy Game Closet at *BigIdeasMath.com* Start Standardized Test Practice

Answers

1. $x = 113$
2. $x = 34$
3. $x = 106$
4. $x = 28$; obtuse
5. $x = 68$; acute, isosceles
6. $x = 60$; acute, isosceles, equilateral, equiangular
7. convex; No line segment connecting two vertices lies outside the polygon.
8. $x = 50$
9. no; The triangles do not have the same angle measures.
10. $x = 55$
11. 133°; $\angle 1$ and the given angle are supplementary.
12. 133°; $\angle 8$ and $\angle 1$ are alternate exterior angles.
13. 133°; $\angle 1$ and $\angle 4$ are vertical angles.
14. 133°; $\angle 4$ and $\angle 5$ are alternate interior angles.
15. 60 m

Assessment Book

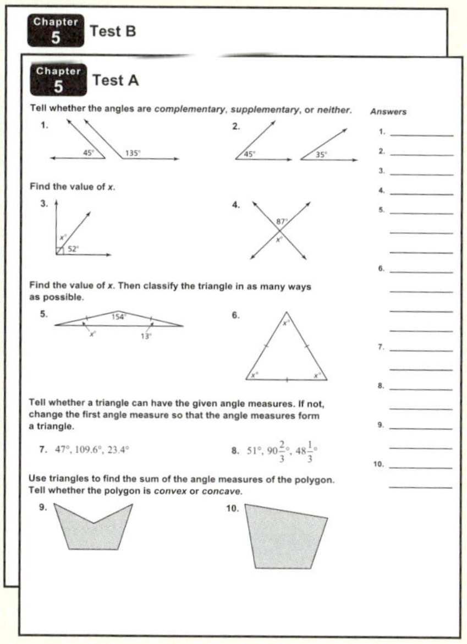

Test-Taking Strategies

Available at *BigIdeasMath.com*

After Answering Easy Questions, Relax
Answer Easy Questions First
Estimate the Answer
Read All Choices before Answering
Read Question before Answering
Solve Directly or Eliminate Choices
Solve Problem before Looking at Choices
Use Intelligent Guessing
Work Backwards

About this Strategy

When taking a multiple choice test, be sure to read each question carefully and thoroughly. Sometimes it is easier to solve the problem and then look for the answer among the choices.

Answers

1. 147°
2. B
3. F
4. 152°
5. D

Technology For the Teacher

Big Ideas Test Generator

Item Analysis

1. **Gridded Response:** Correct answer: 147°

   Common Error: The student might divide 180 by 11.

2. **A.** The student adds 11 and 1.6 together before dividing.

   B. Correct answer

   C. The student divides first and then subtracts.

   D. The student subtracts 1.6 instead of dividing.

3. **F.** Correct answer

   G. The student has the correct slope but the wrong *y*-intercept, perhaps confusing the *x*- and *y*-intercepts.

   H. The student has the correct *y*-intercept but the wrong slope.

   I. The student has the wrong slope and the wrong *y*-intercept, perhaps confusing the *x*- and *y*-intercepts.

4. **Gridded Response:** Correct answer: 152°

   Common Error: The student subtracts 28 from 100, yielding an answer of 72.

5. **A.** The student overlooks the exponent, not realizing that it makes the function nonlinear.

   B. The student overlooks the fact that *x* is in the denominator, not realizing that it makes the function nonlinear.

   C. The student overlooks the fact that *x* and *y* are being multiplied, not realizing that it makes the function nonlinear.

   D. Correct answer

6. **F.** The student misapplies the weights of the laptop and desktop.

   G. The student exchanges quantity and weight in the equations, and misapplies the weights of the laptop and desktop.

   H. Correct answer

   I. The student exchanges quantity and weight in the equations.

7. **A.** Correct answer

   B. The student finds the range.

   C. The student includes both the domain and the range.

   D. The student picks the first ordered pair.

5 Standardized Test Practice

Test-Taking Strategy
Solve Problem Before Looking at Choices

"Solve the problem before looking at the choices. You know 180 − 2(70) = 40. So the answer is C."

1. The border of a Canadian one-dollar coin is shaped like an 11-sided regular polygon. The shape was chosen to help visually-impaired people identify the coin. How many degrees are in each angle along the border? Round your answer to the nearest degree.

2. A public utility charges its residential customers for natural gas based on the number of therms used each month. The formula below shows how the monthly cost C in dollars is related to the number t of therms used.

$$C = 11 + 1.6t$$

Solve this formula for t.

A. $t = \dfrac{C}{12.6}$

B. $t = \dfrac{C - 11}{1.6}$

C. $t = \dfrac{C}{1.6} - 11$

D. $t = C - 12.6$

3. Which equation matches the line shown in the graph?

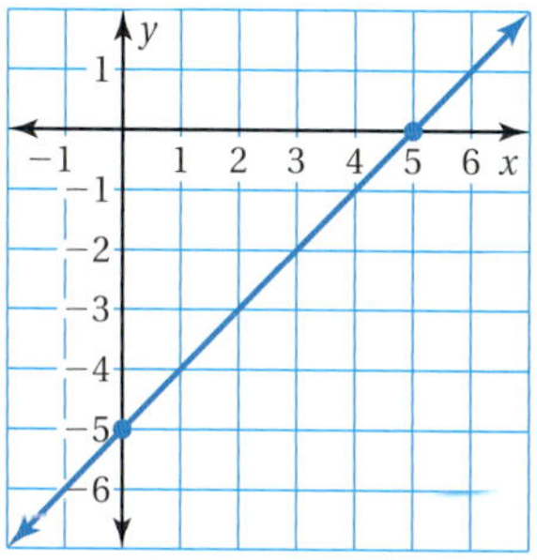

F. $y = x - 5$

G. $y = x + 5$

H. $y = -x - 5$

I. $y = -x + 5$

4. $\angle 1$ and $\angle 2$ form a straight angle. $\angle 1$ has a measure of $28°$. Find the measure of $\angle 2$, in degrees.

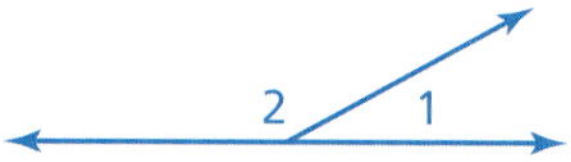

5. Which equation represents a linear function?

A. $y = x^2$

B. $y = \dfrac{2}{x}$

C. $xy = 1$

D. $x + y = 1$

6. A shipment of 2,000 laptop and desktop computers weighs 34,000 pounds. Each laptop computer weighs 8 pounds and each desktop computer weighs 20 pounds. Let ℓ represent the number of laptop computers and d represent the number of desktop computers. Which system of equations could be used to find how many laptop computers are in the shipment?

F. $\ell + d = 2{,}000$
$20\ell + 8d = 34{,}000$

G. $\ell + d = 34{,}000$
$20\ell + 8d = 2{,}000$

H. $\ell + d = 2{,}000$
$8\ell + 20d = 34{,}000$

I. $\ell + d = 34{,}000$
$8\ell + 20d = 2{,}000$

7. What is the domain of the function graphed in the coordinate plane?

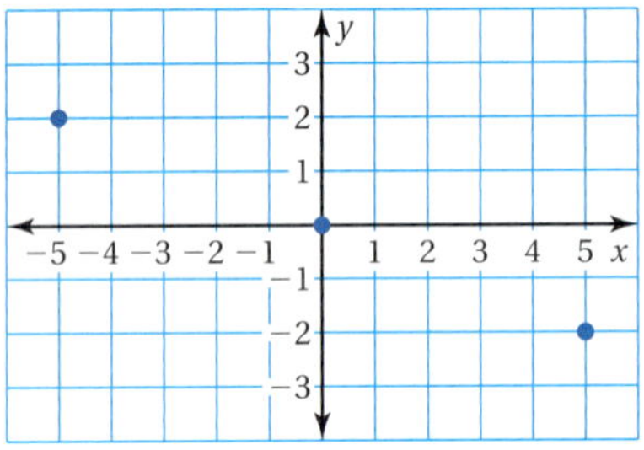

A. −5, 0, 5

B. −2, 0, 2

C. −5, −2, 0, 2, 5

D. −5, 2

8. The sum S of the angle measures of a polygon with n sides can be found using a formula.

Think Solve Explain

Part A Write the formula.

Part B A quadrilateral has angles measuring 100, 90, and 90 degrees. Find the measure of its fourth angle. Show your work and explain your reasoning.

Part C The sum of the measures of the angles of the pentagon shown is 540 degrees. Divide the pentagon into triangles to show why this must be true. Show your work and explain your reasoning.

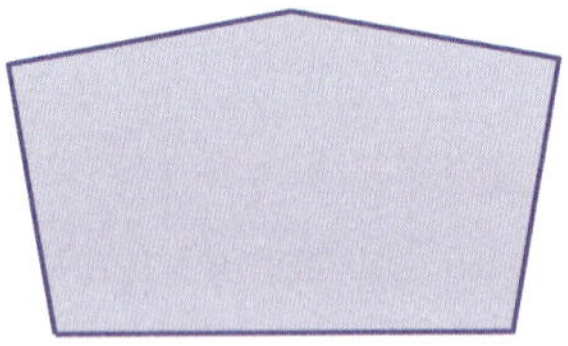

9. The line shown in the graph has a slope of $\frac{2}{5}$.

What is the equation of the line?

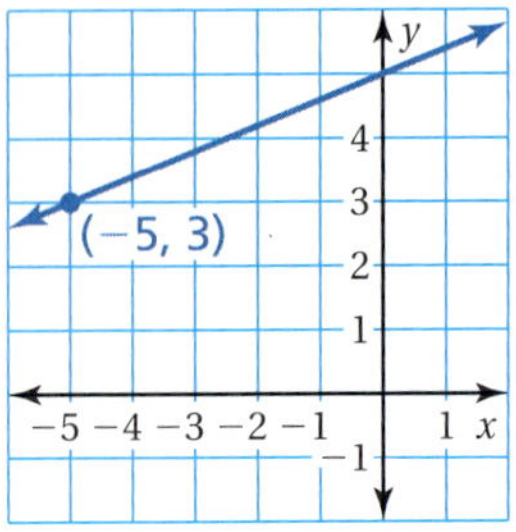

F. $x = \frac{2}{5}y + 5$

G. $y = \frac{2}{5}x + 5$

H. $x = \frac{2}{5}y + 1$

I. $y = \frac{2}{5}x + 1$

Item Analysis (continued)

8. **4 points** The student demonstrates a thorough understanding of writing and applying the angle sum formula for polygons, as well as how it relates to the fact that there are 180 degrees in a triangle. The student's work in Part B shows step-by-step how the fourth angle measures 80 degrees. The student's explanation in part C makes the algebraic-geometric connection clear.

   3 points The student demonstrates an essential but less than thorough understanding. In particular, Parts A and B should be completed fully and clearly, but Part C may lack full explanation of the algebraic-geometric connection.

   2 points The student's work and explanations demonstrate a lack of essential understanding. The formula in Part A should be properly stated, but Part B may show an error in application. Part C lacks any explanation.

   1 point The student demonstrates limited understanding. The student's response is incomplete and exhibits many flaws, including, but not limited to, the inability to state the proper formula in Part A.

   0 points The student provides no response, a completely incorrect or incomprehensible response, or a response that demonstrates insufficient understanding of writing, applying, and understanding the angle sum formula for polygons.

9. **F.** The student interchanges *x* and *y*.
   G. Correct answer
   H. The student interchanges *x* and *y*, and then mishandles slopes, going down 2 and right 5.
   I. The student mishandles slope, going down 2 and right 5.

10. **A.** The student adds 4 degrees to the original temperature.
    B. The student adds 12 degrees to the original temperature.
    C. The student adds 4 degrees to the Heat Index.
    D. Correct answer

11. **F.** The student subtracts 3 instead of adding 3.
    G. Correct answer
    H. The student subtracts 3 instead of adding and then multiplies instead of dividing.
    I. The student multiplies instead of dividing.

12. **A.** The student subtracts 4 from 5 because that is the relationship between sides *BC* and *AB* in triangle *ABC*.
    B. Correct answer
    C. The student finds the length of $\overline{DF}$.
    D. The student assumes that $\overline{DE}$ is congruent to $\overline{AB}$.

Answers

6. H

7. A

8. *Part A* $(n - 2) \cdot 180$

   Part B 80°

   Part C The sum of the angle measures of a triangle is 180 degrees. Because the pentagon can be divided into three triangles, the sum of the angle measures of a pentagon is

   $180 + 180 + 180 = 540$ or
   $(n - 2) \cdot 180 =$
   $(5 - 2) \cdot 180 =$
   $3 \cdot 180 = 540.$

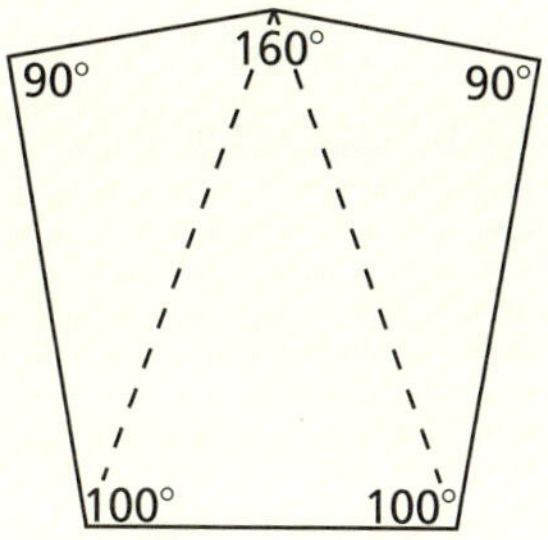

9. G

Answers

10. D

11. G

12. B

13. H

Answer for Extra Example

1. **A.** The student fails to see that these are supplementary, not congruent angles.

   B. The student thinks these are alternate interior (or alternate exterior) angles, and believes they are congruent.

   C. The student thinks these are corresponding angles, and believes they are congruent.

   D. Correct answer

Item Analysis (continued)

13. **F.** The student selects the point where one of the lines crosses the y-axis.

G. The student selects the point where one of the lines crosses the x-axis.

H. Correct answer

I. The student selects the point where one of the lines crosses the x-axis.

Extra Example for Standardized Test Practice

1. In the diagram, lines ℓ and m are parallel. Which angle has the same measure as $\angle 1$?

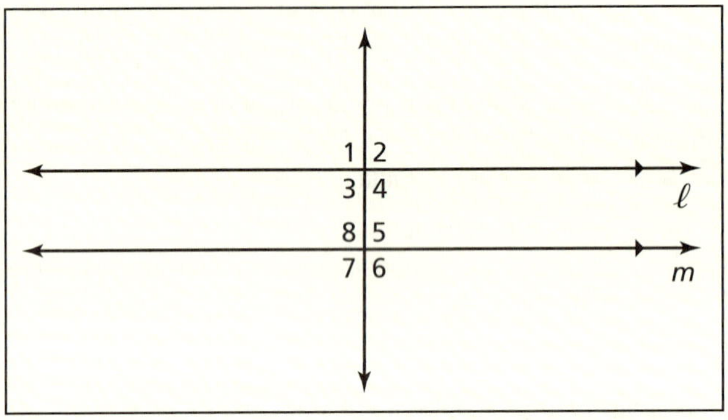

A. $\angle 2$

B. $\angle 5$

C. $\angle 7$

D. $\angle 8$

10. On a hot summer day, the temperature was 95°F, the relative humidity was 75%, and the Heat Index was 122°F. For every degree that the temperature rises, the Heat Index increases by 4 degrees. The temperature rises to 98°F. What is the Heat Index?

A. 99°F
B. 107°F
C. 126°F
D. 134°F

11. Which value of x makes the equation below true?

$$5x - 3 = 11$$

F. 1.6
G. 2.8
H. 40
I. 70

12. In the diagram below, $\triangle ABC \sim \triangle DEF$. What is the value of x?

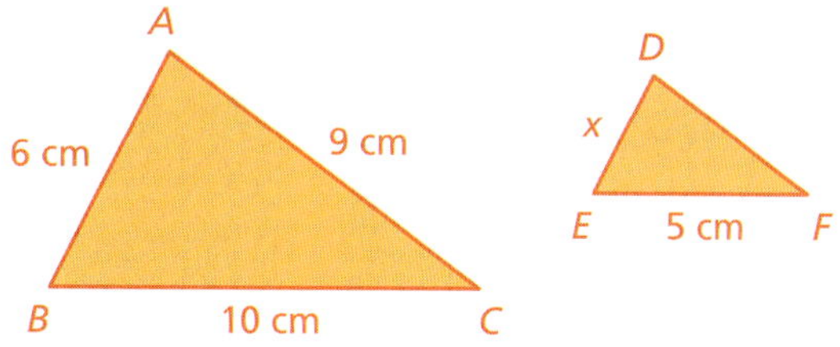

A. 1 cm
B. 3 cm
C. 4.5 cm
D. 6 cm

13. A system of linear equations is shown in the coordinate plane below. What is the solution for this system?

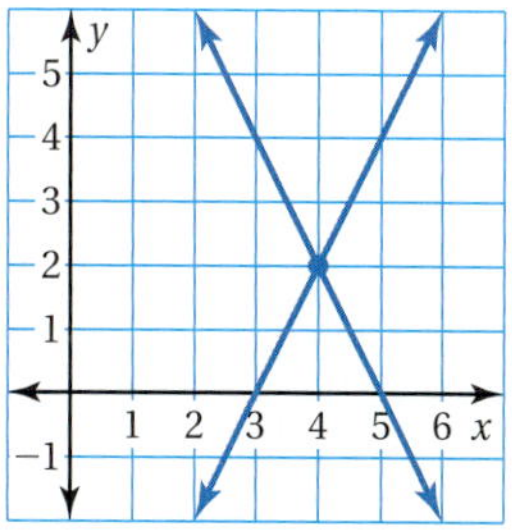

F. (0, 10)
G. (3, 0)
H. (4, 2)
I. (5, 0)

6 Square Roots and the Pythagorean Theorem

"I'm pretty sure that Pythagoras was a Greek."

"I said 'Greek', not 'Geek'."

"Leonardo da Vinci claimed that the human face is made up of golden ratios."

"Let's see if the same is true of a cat's face."

Connections to Previous Learning

- Use order of operations including exponents and parentheses.
- Estimate the results of computations with fractions, decimals, and percents, and verify reasonableness.

- Perform exponential operations with rational bases and whole number exponents.

- Make reasonable approximations of square roots and mathematical expressions that include square roots, and use them to estimate solutions to problems and to compare mathematical expressions involving real numbers and radical expressions.
- Perform operations on real numbers (including radicals, rational numbers, irrational numbers) using multi-step and real world problems.
- Validate and apply the Pythagorean Theorem to find distances in real world situations or between points in the Cartesian plane.

Math in History

Early cultures were aware of square roots. For instance, they were aware that a square with side lengths of 1 unit has a diagonal whose length is $\sqrt{2}$.

★ The following representation of the square root of 2 was found on an old Babylonian tablet.

The symbols are 1, 24, 51, and 10. Because the Babylonians used a base 60 system, this number is $1 + \frac{24}{60} + \frac{51}{60^2} + \frac{10}{60^3} \approx 1.41421$.

★ Another early approximation of the square root of 2 is given in an Indian mathematical text, the Sulbasutras (c. 800–200 B.C.), as follows "Increase the length by its third and this third by its own fourth less the thirty-fourth part of that fourth."

$$1 + \frac{1}{3} + \frac{1}{3 \cdot 4} - \frac{1}{3 \cdot 4 \cdot 34} \approx 1.41422$$

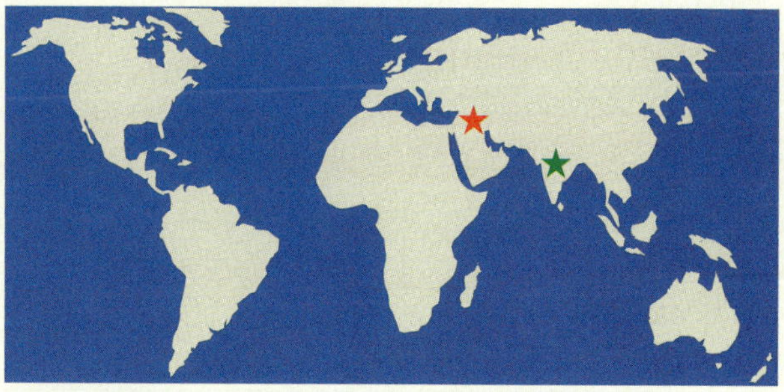

Pacing Guide for Chapter 6

Chapter Opener	1 Day
Section 1 Activity Lesson	 1 Day 1 Day
Section 2 Activity Lesson	 1 Day 1 Day
Study Help / Quiz	1 Day
Section 3 Activity Lesson	 1 Day 2 Days
Section 4 Activity Lesson	 2 Days 1 Day
Section 5 Activity Lesson	 1 Day 1 Day
Quiz / Chapter Review	1 Day
Chapter Test	1 Day
Standardized Test Practice	1 Day
Total Chapter 6	17 Days
Year-to-Date	103 Days

Check Your Resources

- Record and Practice Journal
- Resources by Chapter
- Skills Review Handbook
- Assessment Book
- Worked-Out Solutions

Technology For the Teacher

Dynamic Classroom

The Dynamic Planning Tool
Editable Teacher's Resources at
BigIdeasMath.com

Additional Topics for Review

- Number line
- Converting decimals to fractions
- Order of Operations
- Exponents
- Compare and order decimals and fractions

Try It Yourself

1. = **2.** <

3. <

4. $-0.009, -0.001, 0.01$

5. $-1.75, -1.74, 1.74$

6. $-0.75, 0.74, 0.75$

7. -3 **8.** 181

9. 99

Record and Practice Journal

1. < **2.** >

3. = **4.** >

5–8. Sample answers are given.

5. $-5.2, -5.3, -6.5$

6. $2.56, 2.3, -3.2$

7. $-3.18, -3.1, -2.05$

8. $0.05, 0.3, 1.55$

9. 12.49; 12.495; 12.55; 12.60; 12.63

10. 167 **11.** 3

12. 63 **13.** 116

14. -51 **15.** 1

16. $\frac{24 + 32 + 30 + 28}{2}$; 57

Math Background Notes

Vocabulary Review

- Greater Than
- Less Than
- Order of Operations

Comparing Decimals

- Students should know how to compare decimals.
- **Teaching Tip:** Some students will have difficulty determining which decimal is greater simply by looking. Encourage these students to convert the decimals to fractions with a common denominator and compare the numerators. The fraction with the greater (positive) numerator was produced by the greater decimal.
- **Common Error:** Some students will have difficulty with Example 3 because there is not one "right" answer. Remind them that any number that makes the comparison true is a correct answer. Encourage creativity and remind students their answers will not always match the teacher's answers!

Using Order of Operations

- Students should know how to use the order of operations.
- You may want to review the correct order of operations with students. Many students probably learned the pneumonic device *Please Excuse My Dear Aunt Sally.* Ask a volunteer to explain why this phrase is helpful.
- You may want to review exponents with students. Remind students that the exponent tells you how many times the base is a factor. Exponents express repeated multiplication.

Reteaching and Enrichment Strategies

If students need help. . .	If students got it. . .
Record and Practice Journal • Fair Game Review Skills Review Handbook Lesson Tutorials	Game Closet at *BigIdeasMath.com* Start the next section

What You Learned Before

Comparing Decimals

Complete the number sentence with <, >, or =.

Example 1 1.1 ▭ 1.01

Because $\frac{110}{100}$ is greater than $\frac{101}{100}$, 1.1 is greater than 1.01.

So, $1.1 > 1.01$.

Example 2 −0.3 ▭ −0.003

Because $-\frac{300}{1000}$ is less than $-\frac{3}{1000}$, −0.3 is less than −0.003.

So, $-0.3 < -0.003$.

Example 3 Find three decimals that make the number sentence $-5.12 >$ ▭ true.

Any decimal less than −5.12 will make the sentence true.

Sample answer: −10.1, −9.05, −8.25

Try It Yourself

Complete the number sentence with <, >, or =.

1. 2.10 ▭ 2.1
2. −4.5 ▭ −4.25
3. π ▭ 3.2

Find three decimals that make the number sentence true.

4. $-0.01 \leq$ ▭
5. $1.75 >$ ▭
6. $0.75 \geq$ ▭

Using Order of Operations

Example 4 Evaluate $8^2 \div (32 \div 2) - 2(3 - 5)$.

First:	Parentheses	$8^2 \div (32 \div 2) - 2(3 - 5) = 8^2 \div 16 - 2(-2)$
Second:	Exponents	$= 64 \div 16 - 2(-2)$
Third:	Multiplication and Division (from left to right)	$= 4 + 4$
Fourth:	Addition and Subtraction (from left to right)	$= 8$

Try It Yourself

Evaluate the expression.

7. $15\left(\frac{12}{3}\right) - 7^2 - 2 \cdot 7$
8. $3^2 \cdot 4 \div 18 + 30 \cdot 6 - 1$
9. $-1 + \left(\frac{4}{2}(6 - 1)\right)^2$

6.1 Finding Square Roots

Essential Question How can you find the side length of a square when you are given the area of the square?

When you multiply a number by itself, you square the number.

Symbol for squaring is 2nd power. → $4^2 = 4 \cdot 4$

$= 16$ 4 squared is 16.

To "undo" this, take the **square root** of the number.

Symbol for square root is a radical sign. → $\sqrt{16} = \sqrt{4^2} = 4$ The square root of 16 is 4.

1 ACTIVITY: Finding Square Roots

Work with a partner. Use a square root symbol to write the side length of the square. Then find the square root. Check your answer by multiplying.

a. Sample: $s = \sqrt{121} = 11$ ft

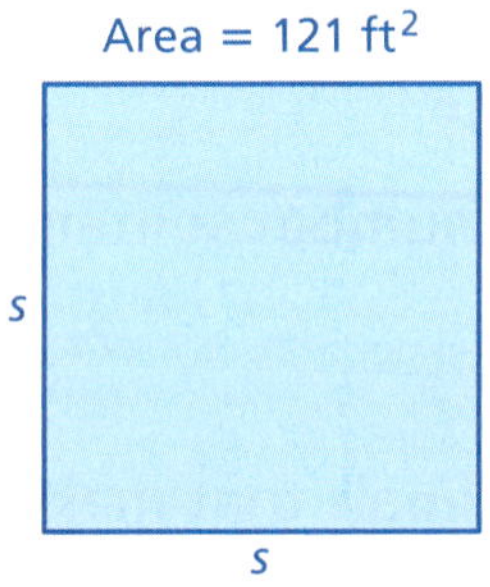

Check

$$\begin{array}{r} 11 \\ \times\ 11 \\ \hline 11 \\ 110 \\ \hline 121 \end{array}$$ ✓

The side length of the square is 11 feet.

b.

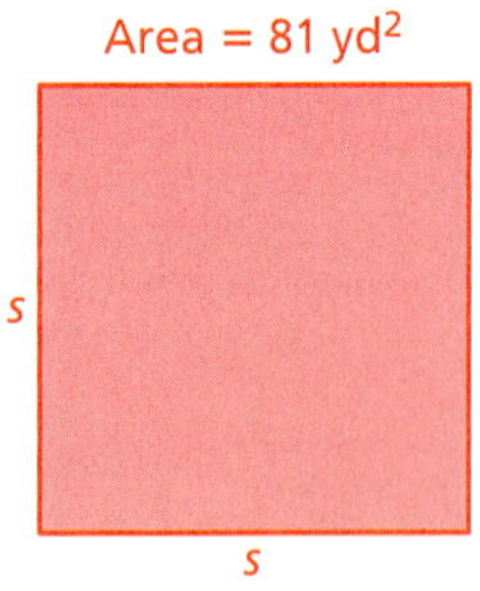

c.

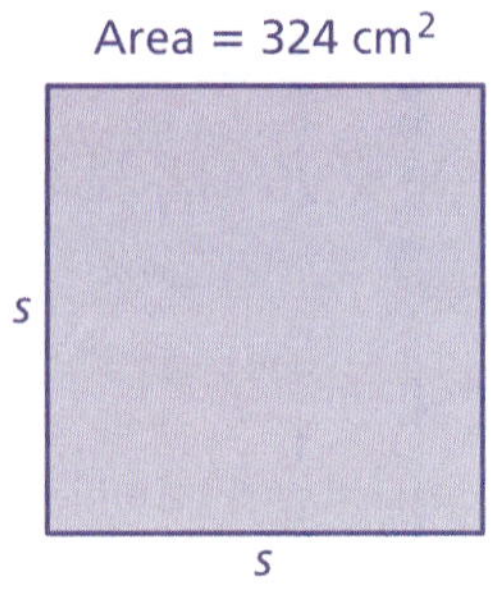

d.

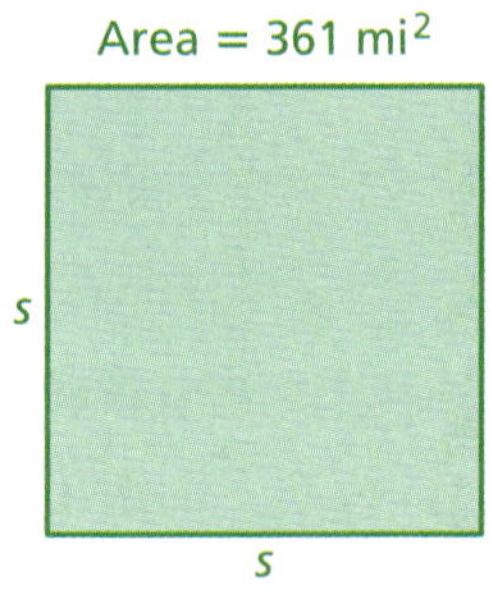

e.

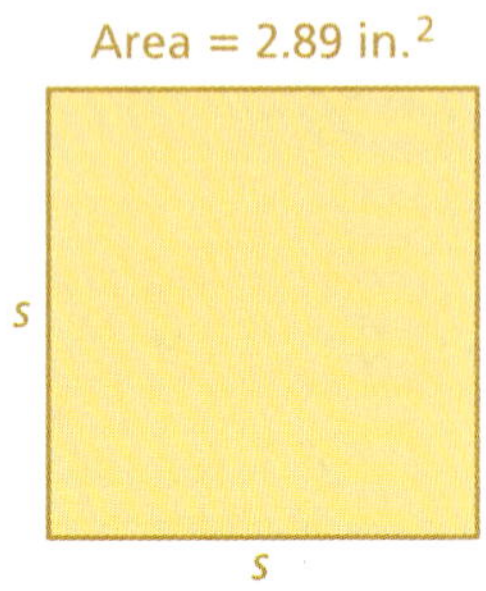

f.

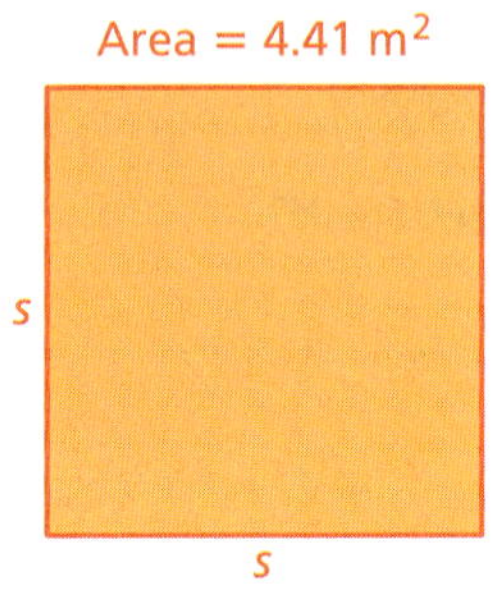

g.

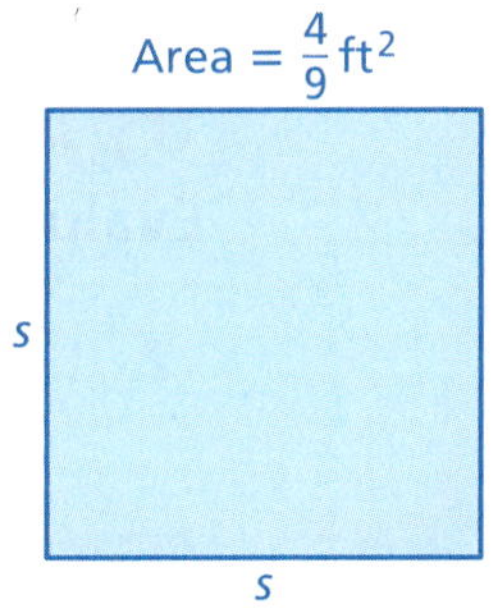

Laurie's Notes

Introduction

For the Teacher

- **Goal:** Students will develop an understanding of perfect squares.
- In this section, square roots of perfect squares are considered. Knowledge of perfect squares will be used in decimal problems. For example, knowing that $12^2 = 144$, you would expect students to know that $1.2^2 = 1.44$.

Motivate

- **Preparation:** Make two (or more) pendulums of different lengths.
- Swing the two pendulums back and forth a few times while telling a story.
- ? "Does it take the same amount of time for the pendulum to go back and forth for each length?" Answers will vary.
- **Extension:** Have a student time you as you swing the pendulum through 10 periods. Divide the total time by 10 to find the time of one period. Repeat for a different length pendulum.

Activity Notes

Activity 1

- **Preparation:** Cut a number of squares from paper. Some should have the area written in the center.
- ? "Could you find the area of any of these squares? Explain." Yes, multiply length by width.
- ? "There are no dimensions marked on the square, so how do I find the area?" Need to measure, then multiply to find area.
- Hold up one of the squares and show students that it has Area = 36 cm^2 (or any number close to actual area) recorded on it.
- ? "If you know the area, how do you find the dimensions?" Ideas will vary.
- **Common Error:** Students will say to divide by 4 (perimeter) or divide by 2. Remind students that the side lengths are the same and they are multiplied together to get 36. So, the side lengths must be 6.
- Introduce the square root symbol and language. Explain that the number 16 is called a square number.

  $4^2 = 16 \rightarrow \text{undo} \rightarrow \sqrt{16} = 4$
- Students should now be ready to begin the activity with a partner.
- As students work the problems, you may need to give a hint to help them remember how to perform operations with decimals and fractions. For example, $11^2 = 121$, so $1.1^2 = 1.21$.
- If students are not using calculators, they will need time for trial and error on these problems. To find $\sqrt{2.89}$, they first need to think *what number multiplied by itself would equal 289.* The answer is 17.
- When students have finished, discuss the answers and the strategies they used.

Previous Learning

Students should know how to multiply fractions and decimals and find the area of a square.

Activity Materials	
Introduction	**Textbook**
• string; weight • stop watch	• different sized squares

Start Thinking! and Warm Up

Activity 6.1 Start Thinking!
For use before Activity 6.1

Activity 6.1 Warm Up
For use before Activity 6.1

Find the product.

1. 12×12 **2.** 9×9 **3.** 18×18

4. 1.6×1.6 **5.** 2.5×2.5 **6.** $\frac{2}{3} \times \frac{2}{3}$

6.1 Record and Practice Journal

Essential Question How can you find the side length of a square when you are given the area of the square?

When you multiply a number by itself, you square the number.

Symbol for squaring is 2nd power. → $4^2 = 4 \bullet 4 = 16$ 4 squared is 16.

To "undo" this, take the **square root** of the number.

Symbol for square root is a radical sign. → $\sqrt{16} = \sqrt{4^2} = 4$ The square root of 16 is 4.

1 ACTIVITY: Finding Square Roots

Work with a partner. Use a square root symbol to write the side length of the square. Then find the square root. Check your answer by multiplying.

a. Sample: $s = \sqrt{121} =$ **Check:**

Area = 121 ft^2

The length of each side of the square is 11 ft.

b. Area = 81 yd^2 **c.** Area = 324 cm^2 **d.** Area = 361 mi^2

$\sqrt{81} = 9$ yd $\sqrt{324} = 18$ cm $\sqrt{361} = 19$ mi

English Language Learners

Build on Past Knowledge

Remind students of inverse operations. Addition and subtraction are inverse operations, as are multiplication and division. Taking the square root of a number is the inverse of squaring a number and squaring a number is the inverse of taking the square root of a number.

6.1 Record and Practice Journal

6.1 Finding Square Roots (continued)

e. Area = 2.89 in.^{2} f. Area = 4.41 m^2 g. Area = $\frac{4}{9}$ ft^2

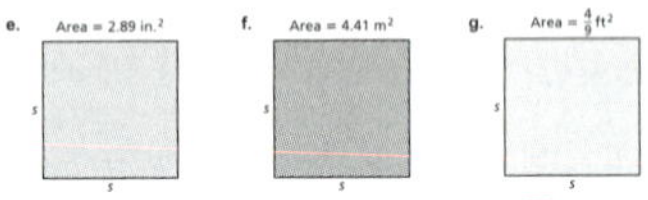

$\sqrt{2.89} = 1.7$ in. $\sqrt{4.41} = 2.1$ m $\sqrt{\frac{4}{9}} = \frac{2}{3}$ ft

2 ACTIVITY: The Period of a Pendulum

Work with a partner.

The period of a pendulum is the time (in seconds) it takes the pendulum to swing back *and* forth.

The period T is represented by $T = 1.1\sqrt{L}$, where L is the length of the pendulum (in feet).

Complete the table. Then graph the function on the next page. Is the function linear?

not linear

L	1.00	1.96	3.24	4.00	4.84	6.25	7.29	7.84	9.00
T	1.1	1.54	1.98	2.2	2.42	2.75	2.97	3.08	3.3

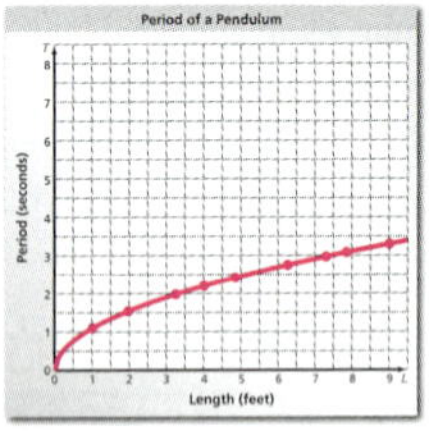

What Is Your Answer?

3. **IN YOUR OWN WORDS** How can you find the side length of a square when you are given the area of a square? Give an example. How can you check your answer?

Find a number whose square is the area.

Laurie's Notes

Activity 2

- Define and model the period of a pendulum.
- Write the formula, $T = 1.1\sqrt{L}$. Explain that if the length is 9 feet, you can find out the time of one period by evaluating the equation for $L = 9$. So, $T = 1.1\sqrt{9} = 1.1(3) = 3.3$ seconds.
- For each of the values in the table, remind students to think about the whole numbers 100, 196, 324, 400, and so on, to help find $\sqrt{1.00}$, $\sqrt{1.96}$, $\sqrt{3.24}$, $\sqrt{4.00}$, and so on.
- When plotting the ordered pairs, students will need to be precise to see the curvature in the graph.
- **Connection:** The equation they are graphing is the function $y = 1.1\sqrt{x}$. This is not a linear function, so the graph is not a line. Find the slope between two different pairs of points on the graph. For (1, 1.1) and (4, 2.2), the slope is $\frac{1.1}{3} = 0.3\overline{6}$. For (4, 2.2) and (9, 3.3), the slope is $\frac{1.1}{5} = 0.22$. The slopes are not the same, so the graph cannot be a line.

What Is Your Answer?

- **Neighbor Check:** Have students work independently and then have their neighbor check their work. Have students discuss any discrepancies.

Closure

- **Matching Activity:** Match each square root with the correct answer.

1. $\sqrt{1600}$ D	**A.** 12
2. $\sqrt{400}$ B	**B.** 20
3. $\sqrt{144}$ A	**C.** 6
4. $\sqrt{36}$ C	**D.** 40

Technology For the Teacher

Dynamic Classroom

The Dynamic Planning Tool
Editable Teacher's Resources at *BigIdeasMath.com*

2 ACTIVITY: The Period of a Pendulum

Work with a partner.

The period of a pendulum is the time (in seconds) it takes the pendulum to swing back *and* forth.

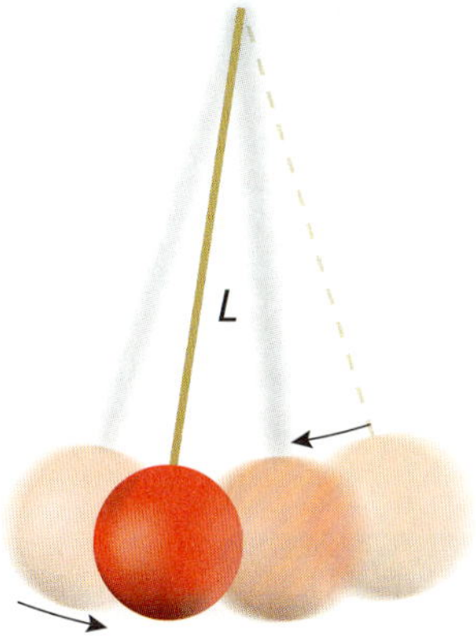

The period T is represented by $T = 1.1\sqrt{L}$, where L is the length of the pendulum (in feet).

Copy and complete the table. Then graph the function. Is the function linear?

L	1.00	1.96	3.24	4.00	4.84	6.25	7.29	7.84	9.00
T									

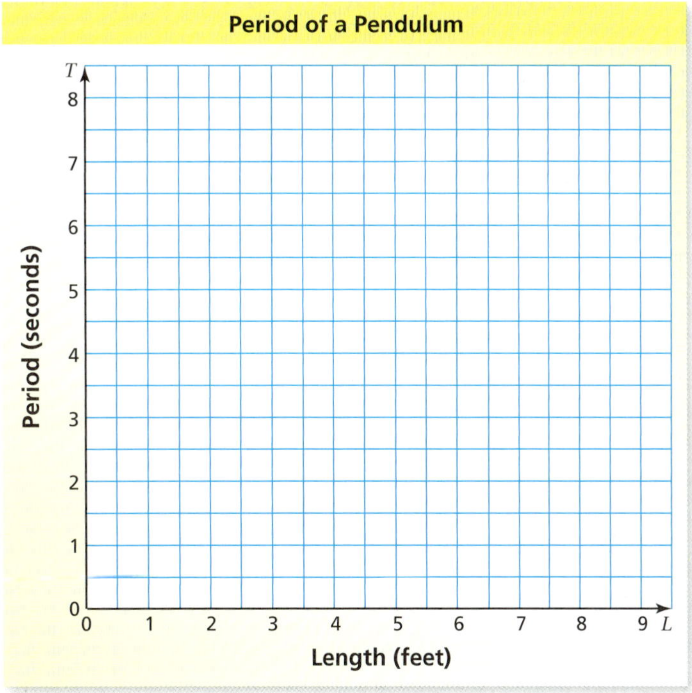

What Is Your Answer?

3. **IN YOUR OWN WORDS** How can you find the side length of a square when you are given the area of the square? Give an example. How can you check your answer?

Practice Use what you learned about finding square roots to complete Exercises 4–6 on page 234.

6.1 Lesson

Check It Out
Lesson Tutorials
BigIdeasMath.com

Key Vocabulary
square root, *p. 232*
perfect square, *p. 232*
radical sign, *p. 232*
radicand, *p. 232*

A **square root** of a number is a number that when multiplied by itself, equals the given number. Every positive number has a positive *and* a negative square root. A **perfect square** is a number with integers as its square roots.

EXAMPLE 1 Finding Square Roots of a Perfect Square

Study Tip
Zero has one square root, which is 0.

Find the two square roots of 49.

$7 \cdot 7 = 49$ and $(-7) \cdot (-7) = 49$

So, the square roots of 49 are 7 and -7.

The symbol $\sqrt{\ }$ is called a **radical sign**. It is used to represent a square root. The number under the radical sign is called the **radicand**.

Positive Square Root $\sqrt{\ }$	Negative Square Root $-\sqrt{\ }$	Both Square Roots $\pm\sqrt{\ }$
$\sqrt{16} = 4$	$-\sqrt{16} = -4$	$\pm\sqrt{16} = \pm 4$

EXAMPLE 2 Finding Square Roots

Find the square root(s).

a. $\sqrt{25}$

$\sqrt{25}$ represents the *positive* square root.

Because $5^2 = 25$, $\sqrt{25} = \sqrt{5^2} = 5$.

b. $-\sqrt{\frac{9}{16}}$

$-\sqrt{\frac{9}{16}}$ represents the *negative* square root.

Because $\left(\frac{3}{4}\right)^2 = \frac{9}{16}$, $-\sqrt{\frac{9}{16}} = -\sqrt{\left(\frac{3}{4}\right)^2} = -\frac{3}{4}$.

c. $\pm\sqrt{2.25}$

$\pm\sqrt{2.25}$ represents both the *positive and negative* square roots.

Because $1.5^2 = 2.25$, $\pm\sqrt{2.25} = \pm\sqrt{1.5^2} = 1.5$ and -1.5.

On Your Own

Now You're Ready
Exercises 7–16

Find the two square roots of the number.

1. 36 **2.** 100 **3.** 121

Find the square root(s).

4. $-\sqrt{1}$ **5.** $\pm\sqrt{\frac{4}{25}}$ **6.** $\sqrt{12.25}$

Multi-Language Glossary at BigIdeasMath.com.

Laurie's Notes

Introduction

Connect

- **Yesterday:** Students explored square roots in the context of finding the side length of a square when the area was known.
- **Today:** Students will find the square roots of a perfect square.

Motivate

- Play the game *Keep it Going!*
- Give the students the first 3 to 4 numbers in a sequence and have them *Keep it Going.* If students are sitting in a row, each person in the row says the next number in the pattern. Keep the pattern going until it becomes too difficult to continue. For example, use the sequence 4, 400, 40,000, . . . (4,000,000, 400,000,000 . . .)

Lesson Notes

Discuss

- Write and discuss the definitions of square root of a number and perfect squares. Mention the *Study Tip.*
- Students are often confused when you say "every positive number has a positive and a negative square root." Use Example 1 to explain.

Example 1

- Note that the direction line is written in words without the square root symbol.

? "What is the product of two positives? two negatives?" Both are positive.

Discuss

- The square root symbol is called a radical sign and the number under the radical sign is the radicand.
- Write and discuss the three examples in the table. Explain to students that the symbol $\pm$ is read as *plus or minus.*
- **Representation:** Students will need to pay attention to how the problem is written, especially if they are asked for more than the positive square root.

Example 2

- Remind students to pay attention to the signs that may precede the radical sign.

? "How do you multiply fractions?" Write the product of the numerators over the product of the denominators.

? "What fraction is multiplied by itself to get $\frac{9}{16}$?" $\frac{3}{4}$

On Your Own

- **Think-Pair-Share:** Students should read each question independently and then work with a partner to answer the questions. When they have answered the questions, the pair should compare their answers with another group and discuss any discrepancies.

Goal Today's lesson is finding the **square root** of a **perfect square**.

Start Thinking! and Warm Up

Lesson 6.1 Warm Up
For use before Lesson 6.1

Lesson 6.1 Start Thinking!
For use before Lesson 6.1

Shelley says that there are two solutions to the equation $x^2 = 400$. Gina says that there is only one solution. Who is correct? Explain.

Extra Example 1

Find the two square roots of 64.
8 and -8

Extra Example 2

Find the square root(s) of $-\sqrt{81}$. -9

On Your Own

1. 6 and -6
2. 10 and -10
3. 11 and -11
4. -1
5. $\pm\frac{2}{5}$
6. 3.5

Extra Example 3

Evaluate $2\sqrt{144} - 10$. 14

Extra Example 4

What is the radius of the circle? Use 3.14 for π. about 4 in.

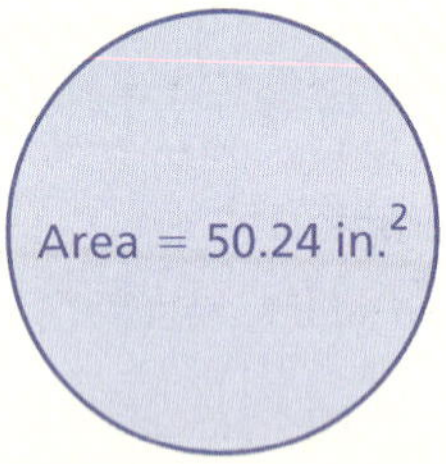

On Your Own

7. -3
8. 4.4
9. -15
10. $3.14r^2 = 2826$; about 30 ft

Differentiated Instruction

Auditory

Ask students to use mental math to answer the following verbal questions.

- "What is the sum of the square root of 9 and 3?" 6
- "What is the difference of 12 and the square root of 144?" 0
- "What is twice the square root of 16?" 8
- "What is one-fourth of the square root of 64?" 2

Laurie's Notes

Example 3

- **Teaching Tip:** In these examples, remind students that square roots are numbers, so you can evaluate numerical expressions that include square roots. Students see a *symbol*, think *variable*, and suddenly they forget things like the order of operations. In part (a), some students think of this as $5x + 7$ and will not know what to do.
- Write the expression in part (a).
- ? "What operations are involved in this problem?" square root, multiplication, and addition
- Work through parts (a) and (b) as shown.
- ? "Can you name inverse operations?" addition and subtraction, multiplication and division, squaring and taking the square root
- Write and discuss that squaring and taking the square root are inverse operations. Use $\sqrt{4^2}$ as an example.

Example 4

- ? "How do you find the area of a circle?" $A = \pi r^2$
- The numbers involved may be overwhelming to students. Reassure them that this is an equation with one variable and that they know how to solve equations!
- Use a calculator or long division.
- When you get to the step $14{,}400 = r^2$, remind students that whatever you do to one side of an equation, you must do to the other side. So, to get r by itself, you need to undo the squaring. This is called taking the square root.
- Discuss with students why a negative square root does not make sense in this context. You cannot have a negative radius, so there will be no negative square root.

On Your Own

- Ask volunteers to share their work at the board for each of the problems.
- Question 8 looks more difficult because of the fraction. You may want to point out that $\frac{28}{7} = 4$.

Closure

- Write 3 numbers of which you know how to take the square root.
- Write 3 numbers of which you do not know how to take the square root.

Technology For the Teacher

Dynamic Classroom

The Dynamic Planning Tool
Editable Teacher's Resources at *BigIdeasMath.com*

EXAMPLE 3 Evaluating Expressions Involving Square Roots

Evaluate the expression.

a. $5\sqrt{36} + 7$

$$5\sqrt{36} + 7 = 5(6) + 7 \quad \text{Evaluate the square root.}$$
$$= 30 + 7 \quad \text{Multiply.}$$
$$= 37 \quad \text{Add.}$$

b. $\frac{1}{4} + \sqrt{\frac{18}{2}}$

$$\frac{1}{4} + \sqrt{\frac{18}{2}} = \frac{1}{4} + \sqrt{9} \quad \text{Simplify.}$$
$$= \frac{1}{4} + 3 \quad \text{Evaluate the square root.}$$
$$= 3\frac{1}{4} \quad \text{Add.}$$

Squaring a positive number and finding a square root are inverse operations. Use this relationship to solve equations involving squares.

EXAMPLE 4 Real-Life Application

The area of a crop circle is 45,216 square feet. What is the radius of the crop circle? Use 3.14 for π.

$$A = \pi r^2 \quad \text{Write the formula for the area of a circle.}$$
$$45{,}216 \approx 3.14r^2 \quad \text{Substitute 45,216 for } A \text{ and 3.14 for } \pi.$$
$$14{,}400 = r^2 \quad \text{Divide each side by 3.14.}$$
$$\sqrt{14{,}400} = \sqrt{r^2} \quad \text{Take positive square root of each side.}$$
$$120 = r \quad \text{Simplify.}$$

∴ The radius of the crop circle is about 120 feet.

On Your Own

Now You're Ready
Exercises 18–23

Evaluate the expression.

7. $12 - 3\sqrt{25}$

8. $\sqrt{\frac{28}{7}} + 2.4$

9. $5(\sqrt{49} - 10)$

10. The area of a circle is 2826 square feet. Write and solve an equation to find the radius of the circle. Use 3.14 for π.

6.1 Exercises

Check It Out
Help with Homework
BigIdeasMath.com

Vocabulary and Concept Check

1. **VOCABULARY** Is 26 a perfect square? Explain.
2. **REASONING** Can the square of an integer be a negative number? Explain.
3. **NUMBER SENSE** Does $\sqrt{256}$ represent the positive square root of 256, the negative square root of 256, or both? Explain.

Practice and Problem Solving

Find the side length of the square. Check your answer by multiplying.

4. Area = 441 cm^2

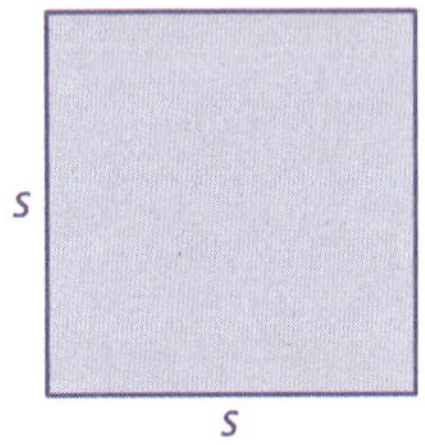

5. Area = 1.69 km^2

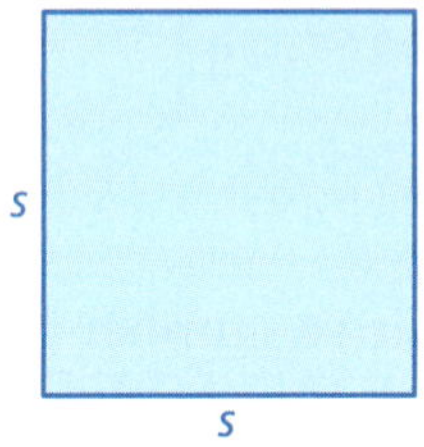

6. Area = $\frac{25}{36}$ yd^2

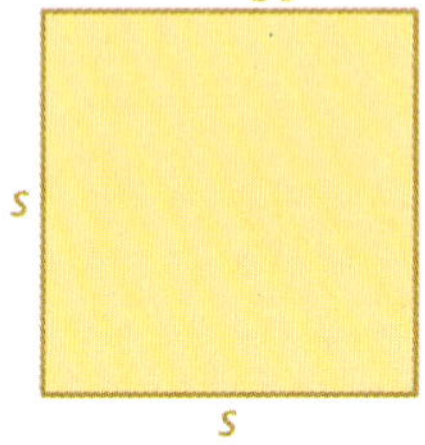

Find the two square roots of the number.

(1) 7. 9   8. 64   9. 4   10. 144

Find the square root(s).

(2) 11. $\sqrt{625}$ 12. $-\sqrt{\frac{9}{100}}$ 13. $\pm\sqrt{\frac{1}{961}}$

14. $\sqrt{7.29}$ 15. $\pm\sqrt{4.84}$ 16. $-\sqrt{361}$

17. **ERROR ANALYSIS** Describe and correct the error in finding the square roots.

✗ $\pm\sqrt{\frac{1}{4}} = \frac{1}{2}$

Evaluate the expression.

(3) 18. $3\sqrt{16} - 5$ 19. $10 - 4\sqrt{\frac{1}{16}}$ 20. $\sqrt{6.76} + 5.4$

21. $8\sqrt{8.41} + 1.8$ 22. $2\left(\sqrt{\frac{80}{5}} - 5\right)$ 23. $4\left(\sqrt{\frac{147}{3}} + 3\right)$

24. **NOTEPAD** The area of the base of a square notepad is 9 square inches. What is the length of one side of the base of the notepad?

25. **CRITICAL THINKING** There are two square roots of 25. Why is there only one answer for the radius of the button?

$A = 25\pi\ \text{mm}^2$

Assignment Guide and Homework Check

Level	Day 1 Activity Assignment	Day 2 Lesson Assignment	Homework Check
Basic	4–6, 35–39	1–3, 7–27 odd, 24	2, 7, 11, 13, 19, 24
Average	4–6, 35–39	1–3, 15–25 odd, 27–32	15, 19, 28, 30, 32
Advanced	4–6, 35–39	1–3, 17, 22, 23, 25, 27–34	22, 28, 30, 32

Common Errors

- **Exercises 7–10** Students may only find the positive square root of the number given. Remind them that a square root can be positive or negative, and the question is asking for both answers.
- **Exercises 11–16** Students may divide the number by two instead of finding a number that, when multiplied by itself, gives the radicand. Remind them that taking the square root of a number is the inverse of squaring a number.
- **Exercises 18–23** Students may not follow the order of operations when evaluating the expression. Remind them of the order of operations. Because taking a square root is the inverse of squaring, it is evaluated before multiplication and division.

6.1 Record and Practice Journal

Find the two square roots of the number.

1. 16 — 4 and −4
2. 100 — 10 and −10
3. 196 — 14 and −14

Find the square root(s).

4. $\sqrt{169}$ — 13
5. $\sqrt{\frac{4}{225}}$ — $\frac{2}{15}$
6. $-\sqrt{12.25}$ — −3.5

Evaluate the expression.

7. $2\sqrt{36} + 9$ — 21
8. $8 - 11\sqrt{\frac{25}{121}}$ — 3
9. $3\left(\sqrt{\frac{125}{5}} - 8\right)$ — −9

10. A trampoline has an area of 49π square feet. What is the diameter of the trampoline?
    14 ft

11. The volume of a cylinder is 75π cubic inches. The cylinder has a height of 3 inches. What is the radius of the base of the cylinder?
    5 in.

Technology For the Teacher

Answer Presentation Tool

Quiz*Show*

Vocabulary and Concept Check

1. no; There is no integer whose square is 26.
2. no; A positive number times a positive number is a positive number, and a negative number times a negative number is a positive number.
3. $\sqrt{256}$ represents the positive square root because there is not a − or a ± in front.

Practice and Problem Solving

4. 21 cm
5. 1.3 km
6. $\frac{5}{6}$ yd
7. 3 and −3
8. 8 and −8
9. 2 and −2
10. 12 and −12
11. 25
12. $-\frac{3}{10}$
13. $\frac{1}{31}$ and $-\frac{1}{31}$
14. 2.7
15. 2.2 and −2.2
16. −19
17. The positive and negative square roots should have been given.
    $\pm\sqrt{\frac{1}{4}} = \frac{1}{2}$ and $-\frac{1}{2}$
18. 7
19. 9
20. 8
21. 25
22. −2
23. 40
24. 3 in.
25. because a negative radius does not make sense

Practice and Problem Solving

26. >

27. =

28. <

29. 9 ft

30. yes; *Sample answer:* Consider the perfect squares, a^2 and b^2. Their product can be written as $a^2b^2 = a \cdot a \cdot b \cdot b = (a \cdot b) \cdot (a \cdot b) = (a \cdot b)^2$.

31. 8 m/sec

32. See *Taking Math Deeper.*

33. 2.5 ft

34. 8 cm

Fair Game Review

35. 25

36. 289

37. 144

38. 49

39. B

Mini-Assessment

Find the square root(s).

1. $\sqrt{169}$ 13

2. $\sqrt{225}$ 15

3. $\pm\sqrt{4.41}$ 2.1 and −2.1

4. $-\sqrt{\frac{16}{25}}$ $-\frac{4}{5}$

5. $\sqrt{\frac{512}{2}}$ 16

Taking Math Deeper

Exercise 32

In this problem, students are given the area of the smaller watch face and are asked to find the radius of the larger watch face.

1 Summarize the given information.

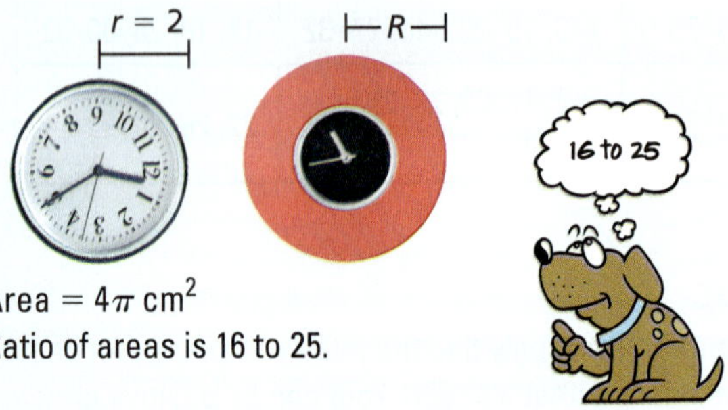

Area = 4π cm^2
Ratio of areas is 16 to 25.

2 Answer the question. The two watch faces are similar, so the ratio of their areas is equal to the square of the ratio of their radii.

$$\frac{\text{Area of small}}{\text{Area of large}} = \left(\frac{\text{radius of small}}{\text{radius of large}}\right)^2$$

$$\frac{16}{25} = \left(\frac{\text{radius of small}}{\text{radius of large}}\right)^2$$

$$\sqrt{\frac{16}{25}} = \frac{\text{radius of small}}{\text{radius of large}}$$

$$\frac{4}{5} = \frac{\text{radius of small}}{\text{radius of large}}$$

a. The ratio of the radius of the smaller watch face to the radius of the larger watch face is $\frac{4}{5}$.

3 **b.** Solve the proportion for R.

$$\frac{4}{5} = \frac{r}{R}$$

$$\frac{4}{5} = \frac{2}{R}$$

$$R = \frac{10}{4}, \text{ or } \frac{5}{2}$$

The radius of the larger watch face is $\frac{5}{2}$ or 2.5 centimeters.

Reteaching and Enrichment Strategies

If students need help. . .	If students got it. . .
Resources by Chapter • Practice A and Practice B • Puzzle Time Record and Practice Journal Practice Differentiating the Lesson Lesson Tutorials Skills Review Handbook	Resources by Chapter • Enrichment and Extension Start the next section

Copy and complete the statement with <, >, or =.

26. $\sqrt{81}$ ☐ 8 **27.** 0.5 ☐ $\sqrt{0.25}$ **28.** $\frac{3}{2}$ ☐ $\sqrt{\frac{25}{4}}$

29. SAILBOAT The area of a sail is $40\frac{1}{2}$ square feet. The base and the height of the sail are equal. What is the height of the sail (in feet)?

30. REASONING Is the product of two perfect squares always a perfect square? Explain your reasoning.

31. ENERGY The kinetic energy K (in joules) of a falling apple is represented by $K = \frac{v^2}{2}$, where v is the speed of the apple (in meters per second). How fast is the apple traveling when the kinetic energy is 32 joules?

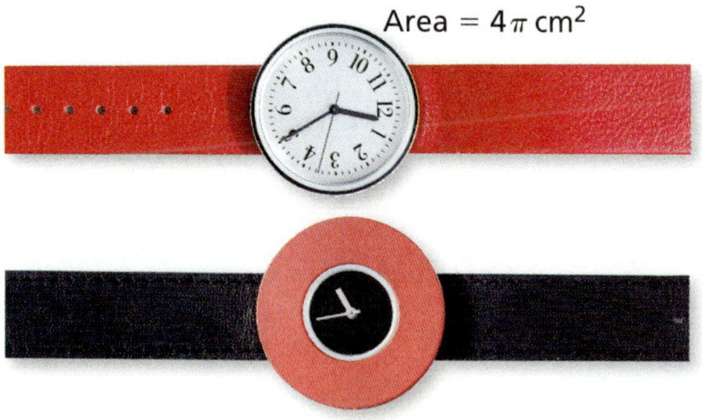

32. WATCHES The areas of the two watch faces have a ratio of 16 : 25.

a. What is the ratio of the radius of the smaller watch face to the radius of the larger watch face?

b. What is the radius of the larger watch face?

33. WINDOW The cost C (in dollars) of making a square window with a side length of n inches is represented by $C = \frac{n^2}{5} + 175$. A window costs \$355. What is the length (in feet) of the window?

34. Geometry The area of the triangle is represented by the formula $A = \sqrt{s(s-21)(s-17)(s-10)}$, where s is equal to half the perimeter. What is the height of the triangle?

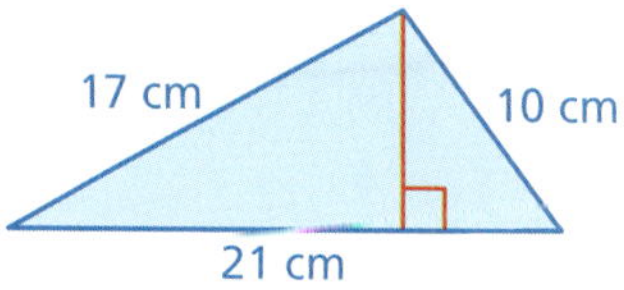

Fair Game Review What you learned in previous grades & lessons

Evaluate the expression. *(Skills Review Handbook)*

35. $3^2 + 4^2$ **36.** $8^2 + 15^2$ **37.** $13^2 - 5^2$ **38.** $25^2 - 24^2$

39. MULTIPLE CHOICE Which of the following describes the triangle? *(Section 5.2)*

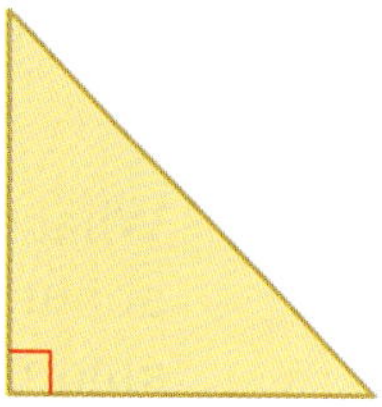

(A) Acute (B) Right

(C) Obtuse (D) Equiangular

6.2 The Pythagorean Theorem

Essential Question How are the lengths of the sides of a right triangle related?

Pythagoras was a Greek mathematician and philosopher who discovered one of the most famous rules in mathematics. In mathematics, a rule is called a **theorem**. So, the rule that Pythagoras discovered is called the Pythagorean Theorem.

Pythagoras
(c. 570 B.C.–c. 490 B.C.)

1 ACTIVITY: Discovering the Pythagorean Theorem

Work with a partner.

a. On grid paper, draw any right triangle. Label the lengths of the two shorter sides (the **legs**) a and b.

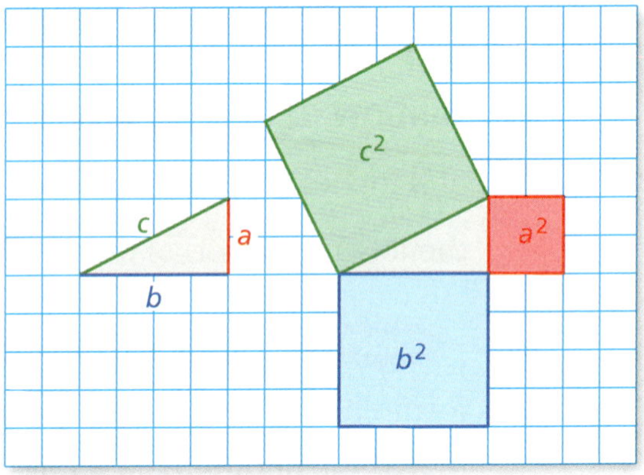

b. Label the length of the longest side (the **hypotenuse**) c.

c. Draw squares along each of the three sides. Label the areas of the three squares a^2, b^2, and c^2.

d. Cut out the three squares. Make eight copies of the right triangle and cut them out. Arrange the figures to form two identical larger squares.

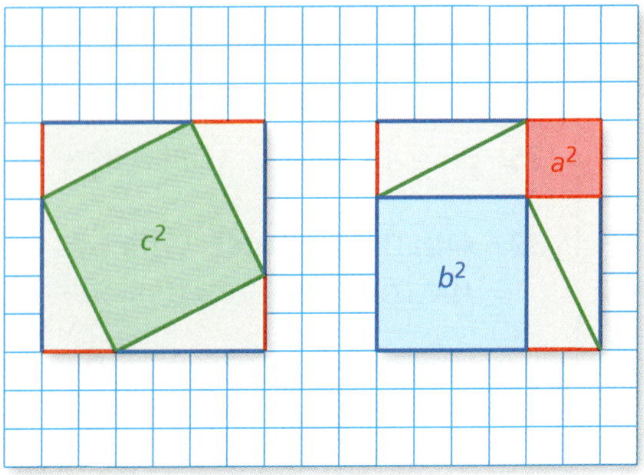

e. What does this tell you about the relationship among a^2, b^2, and c^2?

Laurie's Notes

Introduction

For the Teacher

- **Goal:** Students will explore a geometric proof of the Pythagorean Theorem.
- **Management Tip:** To help keep the room clean, cluster 4 to 6 desks together in a circle and tape a plastic bag to the front edge of one of the desks. Students are expected to put scraps of paper in the bag when they are finished with the investigation.

Motivate

- Share information about Pythagoras, who was born in Greece in 569 B.C.
  - He is known as the *Father of Numbers*.
  - He traveled extensively in Egypt, learning math, astronomy, and music.
  - Pythagoras undertook a reform of the cultural life of Cretona, urging the citizens to follow his religious, political, and philosophical goals.
  - He created a school where his followers, known as Pythagoreans, lived and worked. They observed a rule of silence called *echemythia*, the breaking of which was punishable by death. One had to remain silent for *five years* before he could contribute to the group.

Activity Notes

Activity 1

- **Suggestions:** Use centimeter grid paper for ease of manipulating the cut pieces. Suggest to students that they draw their original triangle in the upper left of the grid paper, and then make a working copy of the triangle towards the middle of the paper. This gives enough room for the squares to be drawn on each side of the triangle.
- Vertices of the triangle need to be on lattice points. You do not want every student in the room to use the same triangle. Suggest other leg lengths (3 and 4, 3 and 6, 2 and 4, 2 and 3, and so on).
- **Model:** Drawing the square on the side of the hypotenuse is the challenging step. Model one technique for accomplishing the task using a triangle with legs 2 units and 5 units.
  - Notice that the hypotenuse has a slope of "right 5 units, up 2 units."
  - Place your pencil on the upper right endpoint and rotate the paper 90° clockwise. Move your pencil right 5 units and up 2 units. Mark a point.
  - Repeat rotating and moving the slope of the hypotenuse until you end at the endpoint of the original hypotenuse.
  - Use a straightedge to connect the four points (two that you marked and two on the endpoints of the hypotenuse) to form the square.
- Before students cut anything, check that they have 3 squares of the correct size.
- **Big Idea:** The two squares formed do have equal area. Referring to areas, if $c^2 + (4 \text{ triangles}) = a^2 + b^2 + (4 \text{ triangles})$, then $c^2 = a^2 + b^2$ by subtracting the 4 triangles from each side of the equation.

Previous Learning

Students should know how to multiply fractions and decimals.

Activity Materials	
Textbook	
• scissors	• plastic bags
• grid paper	• transparency grid

Start Thinking! and Warm Up

Activity 6.2 Start Thinking! For use before Activity 6.2

Activity 6.2 Warm Up For use before Activity 6.2

Find the square root(s).

1. $\sqrt{1.44}$
2. $\pm\sqrt{900}$
3. $\sqrt{\frac{4}{9}}$
4. $-\sqrt{441}$
5. $\pm\sqrt{484}$
6. $-\sqrt{2500}$

6.2 Record and Practice Journal

Essential Question How are the lengths of the sides of a right triangle related?

Pythagoras was a Greek mathematician and philosopher who discovered one of the most famous rules in mathematics. In mathematics, a rule is called a **theorem**. So, the rule that Pythagoras discovered is called the Pythagorean Theorem.

Pythagoras
(c. 570 B.C.–c. 490 B.C.)

1 ACTIVITY: Discovering the Pythagorean Theorem

Work with a partner.

a. On grid paper, draw any right triangle. Label the lengths of the two shorter sides (the **legs**) a and b.

b. Label the length of the longest side (the **hypotenuse**) c.

c. Draw squares along each of the three sides. Label the areas of the three squares a^2, b^2, and c^2.

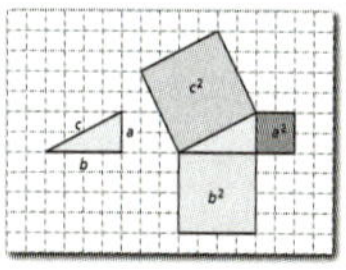

d. Cut out the three squares. Make eight copies of the right triangle and cut them out. Arrange the figures to form two identical larger squares.

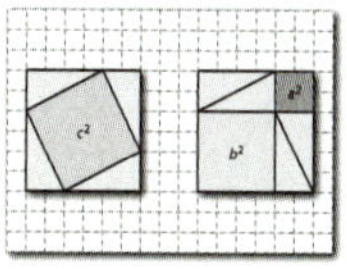

e. What does this tell you about the relationship among a^2, b^2, and c^2?

$a^2 + b^2 = c^2$

English Language Learners

Vocabulary

Help English learners understand the meanings of the words that make up a definition. Provide students with statements containing blanks and a list of the words used to fill in the blanks.

- In any right __, the __ is the side __ the right __.
  Word list: angle, hypotenuse, opposite, triangle
  triangle, hypotenuse, opposite, angle
- In any right __, the __ are the __ sides and the __ is always the __ side.
  Word list: hypotenuse, legs, longest, shorter, triangle
  triangle, legs, shorter, hypotenuse, longest

6.2 Record and Practice Journal

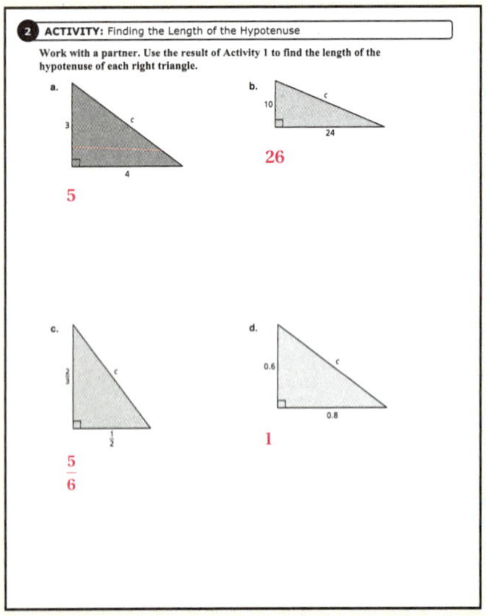
2 ACTIVITY: Finding the Length of the Hypotenuse

Work with a partner. Use the result of Activity 1 to find the length of the hypotenuse of each right triangle.

a. 5

b. 26

c. $\frac{5}{6}$

d. 1

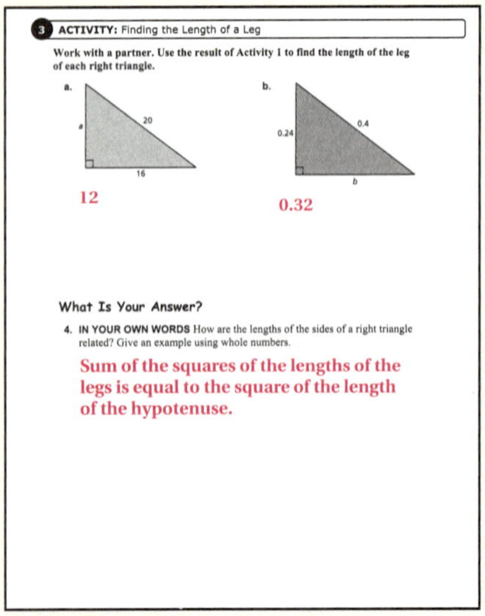
3 ACTIVITY: Finding the Length of a Leg

Work with a partner. Use the result of Activity 1 to find the length of the leg of each right triangle.

a. 12

b. 0.32

What Is Your Answer?

4. IN YOUR OWN WORDS How are the lengths of the sides of a right triangle related? Give an example using whole numbers.

Sum of the squares of the lengths of the legs is equal to the square of the length of the hypotenuse.

Laurie's Notes

Activity 2

- **Part (a):** This triangle is known as a 3-4-5 right triangle. Using the property from the investigation, $3^2 + 4^2 = 9 + 16 = 25$. Students will recognize that $25 = 5^2$, so the length of the hypotenuse is 5.
- **Part (d):** This is related to a 3-4-5 right triangle: $3 \times 0.2 = 0.6$; $4 \times 0.2 = 0.8$; $5 \times 0.2 = 1.0$. Check: $0.6^2 + 0.8^2 = 0.36 + 0.64 = 1.0$ and $\sqrt{1} = 1$.
- Have students share their work for each of these problems.
- **Common Error:** In part (c), when students square a fraction, they sometimes double the numerator and denominator instead of squaring each number. In other words, $\left(\frac{2}{3}\right)^2 \neq \frac{4}{6}$, but $\left(\frac{2}{3}\right)^2 = \frac{4}{9}$.

Activity 3

- The two triangles in this activity have a leg length missing. Building squares on the two legs of the triangle and finding their areas would give a^2 and 16^2 for part (a). The area of the square built on the hypotenuse would be 20^2. The result of Activity 1 says that $a^2 + 16^2 = 20^2$. Students should recognize this as an opportunity to solve an equation.

? "What is the first step in solving the equation $a^2 + 16^2 = 20^2$?" Evaluate 16^2 and 20^2.

? "What is the next step in solving $a^2 + 256 = 400$?" Subtract 256 from each side.

? "Finally, what number squared is 144?" 12

What Is Your Answer?

- **Neighbor Check:** Have students work independently and then have their neighbor check their work. Have students discuss any discrepancies.

Closure

- **Exit Ticket:** If you drew a right triangle with legs of 4 and 6 on grid paper, what would be the area of the square drawn on the hypotenuse of the triangle? 52 square units

Technology For the Teacher

Dynamic Classroom

The Dynamic Planning Tool
Editable Teacher's Resources at *BigIdeasMath.com*

2 ACTIVITY: Finding the Length of the Hypotenuse

Work with a partner. Use the result of Activity 1 to find the length of the hypotenuse of each right triangle.

a.

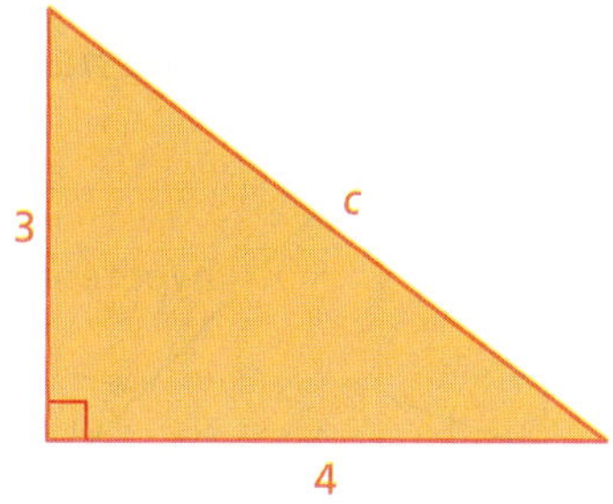

b.

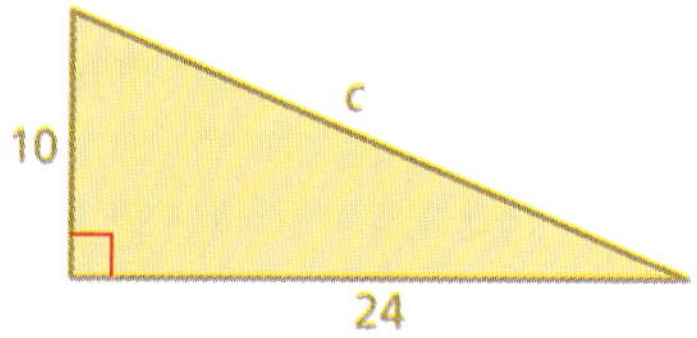

c.

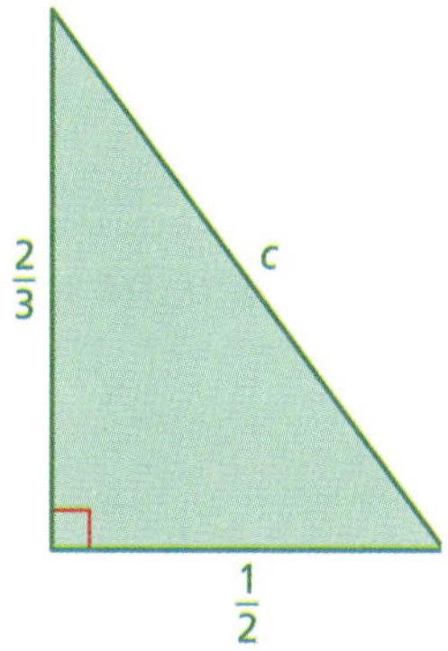

d.

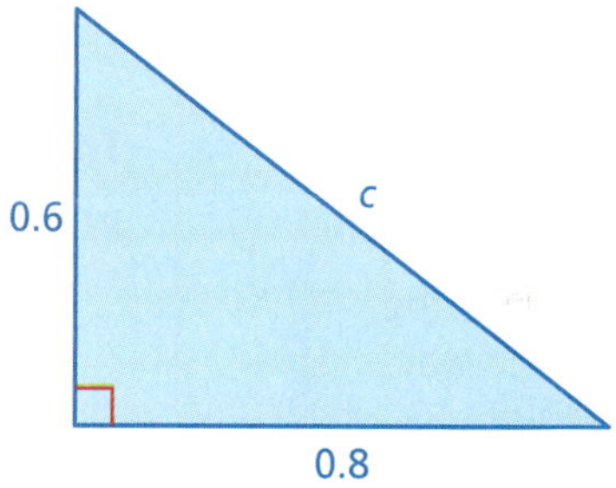

3 ACTIVITY: Finding the Length of a Leg

Work with a partner. Use the result of Activity 1 to find the length of the leg of each right triangle.

a.

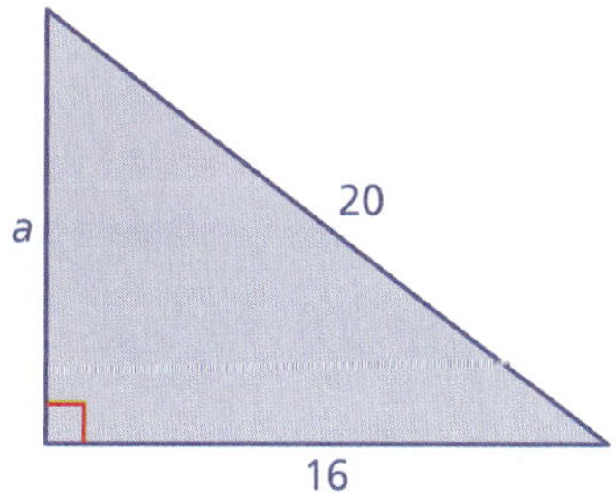

b.

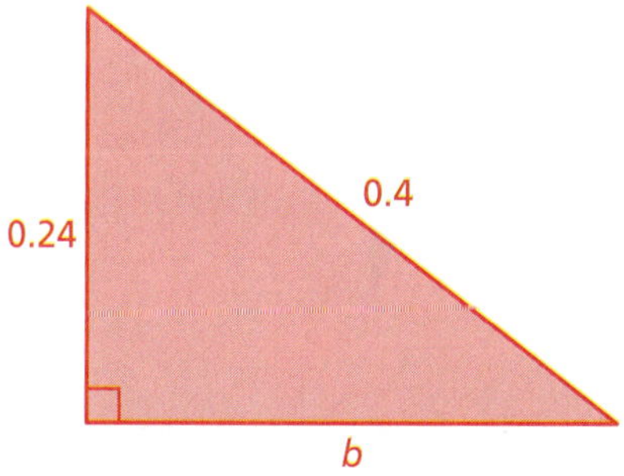

What Is Your Answer?

4. **IN YOUR OWN WORDS** How are the lengths of the sides of a right triangle related? Give an example using whole numbers.

Practice Use what you learned about the Pythagorean Theorem to complete Exercises 3–5 on page 240.

6.2 Lesson

Check It Out
Lesson Tutorials
BigIdeasMath.com

Key Vocabulary
theorem, *p. 236*
legs, *p. 238*
hypotenuse, *p. 238*
Pythagorean Theorem, *p. 238*

Study Tip
In a right triangle, the legs are the shorter sides and the hypotenuse is always the longest side.

Key Ideas

Sides of a Right Triangle

The sides of a right triangle have special names.

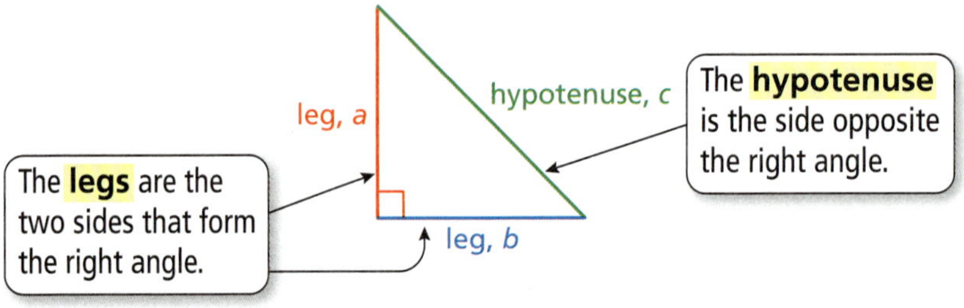

The Pythagorean Theorem

Words In any right triangle, the sum of the squares of the lengths of the legs is equal to the square of the length of the hypotenuse.

Algebra $a^2 + b^2 = c^2$

EXAMPLE 1 Finding the Length of a Hypotenuse

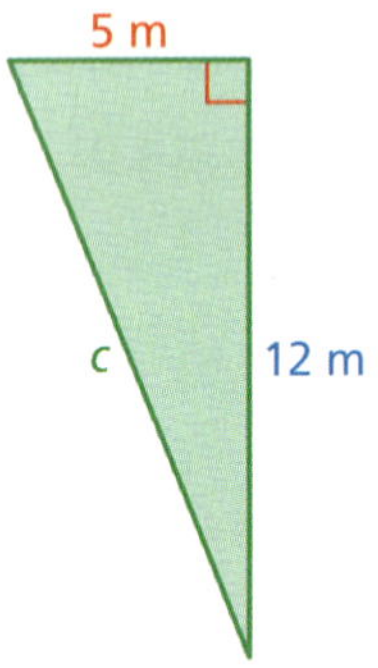

Find the length of the hypotenuse of the triangle.

$a^2 + b^2 = c^2$	Write the Pythagorean Theorem.
$5^2 + 12^2 = c^2$	Substitute 5 for *a* and 12 for *b*.
$25 + 144 = c^2$	Evaluate powers.
$169 = c^2$	Add.
$\sqrt{169} = \sqrt{c^2}$	Take positive square root of each side.
$13 = c$	Simplify.

The length of the hypotenuse is 13 meters.

On Your Own

Find the length of the hypotenuse of the triangle.

1.

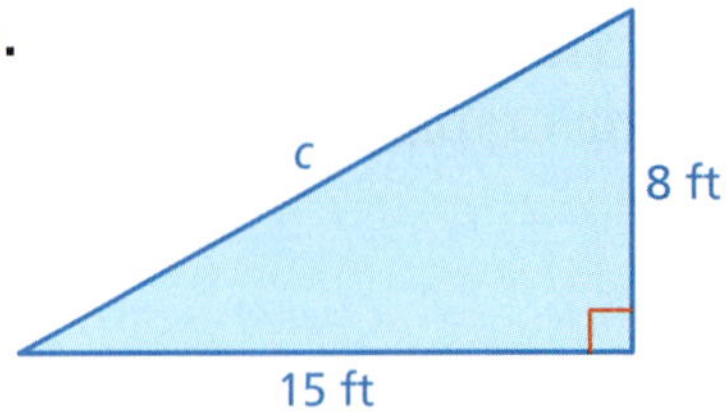

2.

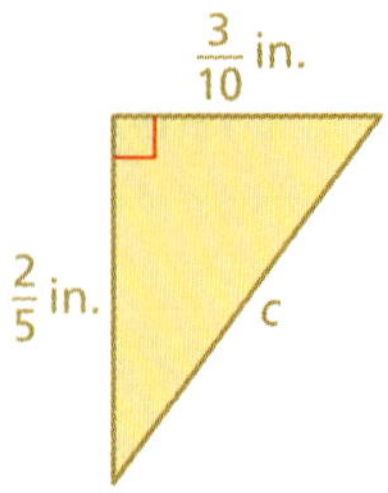

Multi-Language Glossary at BigIdeasMath.com.

Laurie's Notes

Introduction

Connect

- **Yesterday:** Students investigated a visual proof of the Pythagorean Theorem.
- **Today:** Students will use the Pythagorean Theorem to find the missing lengths of a right triangle.

Motivate

- **Preparation:** Cut coffee stirrers (or carefully break spaghetti) so that triangles with the following side lengths can be made: 2-3-4; 3-4-5; 4-5-6.

? "What are consecutive numbers?" numbers in sequential order

- With student aid, use the coffee stirrers to make three triangles: 2-3-4; 3-4-5; and 4-5-6 on the overhead projector. If arranged carefully, all 3 will fit on the screen.
- Ask students to make observations about the 3 triangles. Students may mention that all triangles are scalene; one triangle appears to be acute, one right, one obtuse.
- They should observe that just a small change in the side lengths seems to have made a big change in the angle measures.

Lesson Notes

Key Ideas

- Draw a picture of a right triangle and label the *legs* and the *hypotenuse*. The **hypotenuse** is always opposite the right angle and is the longest side of a right triangle.
- Try not to have all right triangles in the same orientation.
- Write the Pythagorean Theorem.
- **Common Error:** Students often forget that the Pythagorean Theorem is a relationship that is *only* true for right triangles.

Example 1

- Draw and label the triangle. Review the symbol used to show that an angle is a right angle.

? "What information is known for this triangle?" The legs are 5 m and 12 m.

On Your Own

- Give time for students to work the problems. Knowing their perfect squares is helpful.

Goal Today's lesson is using the **Pythagorean Theorem** to solve for lengths of a right triangle.

Lesson Materials
Introduction
• coffee stirrers • scissors

Start Thinking! and Warm Up

Lesson 6.2 Warm Up
For use before Lesson 6.2

Lesson 6.2 Start Thinking!
For use before Lesson 6.2

How can you use the Pythagorean Theorem in sports?

Extra Example 1

Find the length of the hypotenuse of the triangle. 5 in.

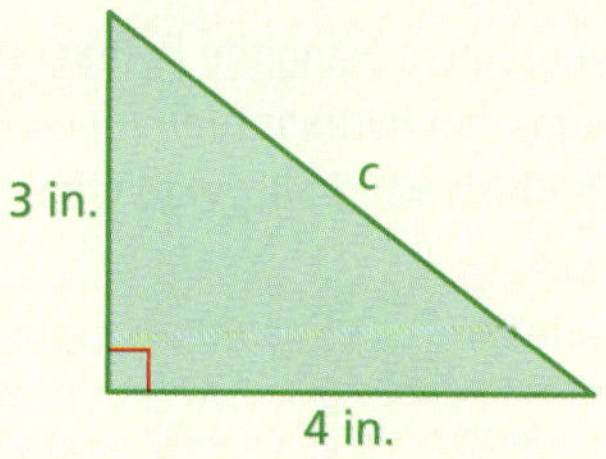

On Your Own

1. 17 ft
2. $\frac{1}{2}$ in.

Extra Example 2

Find the missing length of the triangle. 24 ft

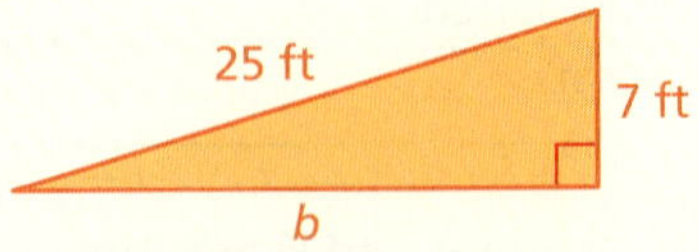

Extra Example 3

In Example 3, Group C leaves the station and hikes 3 kilometers west then 4 kilometers south. How far apart are Group B and Group C? 10 km

On Your Own

3. 30 yd **4.** 4 m

5. 20 km

Differentiated Instruction

Kinesthetic

Have students verify the Pythagorean Theorem by drawing right triangles with legs of a given length, measuring the hypotenuse, and then calculating the hypotenuse using the Pythagorean Theorem. Use Pythagorean triples so that students work only with whole numbers.

Leg Lengths	*Hypotenuse Length*
3, 4	5
6, 8	10
5, 12	13
8, 15	17

Laurie's Notes

Example 2

- ? "What information is known for this triangle?" One leg is 2.1 centimeters and the hypotenuse is 2.9 centimeters.
- Substitute and solve as shown.
- **Common Error:** Students need to be careful with decimal multiplication. It is very common for students to multiply the decimal by 2 instead of multiplying the decimal by itself.
- **FYI:** The side lengths are a scale factor of the triple 20-21-29.

Example 3

- Ask a student to read the example.
- ? "Given the compass directions stated, what is a reasonable way to represent this information?" coordinate plane
- Explain that east and west are in the x-direction and north and south are in the y-direction. Draw the situation in a coordinate plane.
- ? "How can you be sure that the segments representing the hypotenuses of the smaller triangles can be connected to form a straight line?"

  The segments forming the hypotenuses each have a slope of $\frac{4}{3}$.
- Students may claim that the hypotenuses of the smaller triangles can always be connected to form a straight line. To dispute this claim, draw a hiking group that travels 6 miles east and 4 miles north.

Closure

- **Exit Ticket:** Solve for the missing side. $x = 15$ cm, $y = 0.5$ m

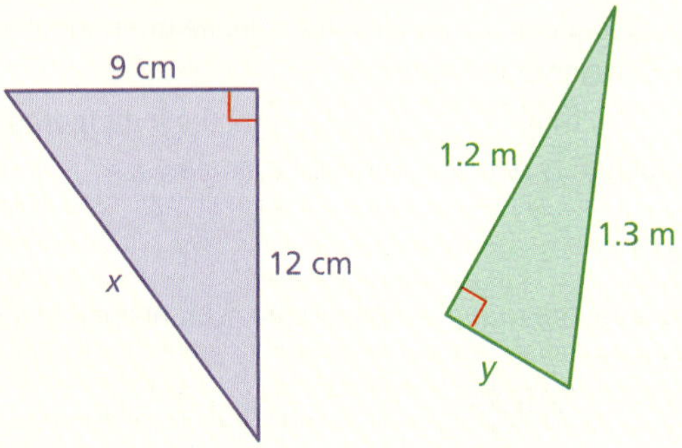

Technology For the Teacher

Dynamic Classroom

The Dynamic Planning Tool
Editable Teacher's Resources at *BigIdeasMath.com*

EXAMPLE 2 Finding the Length of a Leg

Find the missing length of the triangle.

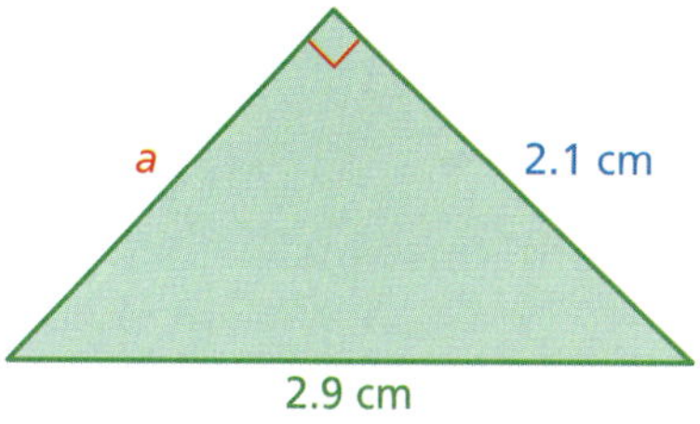

$a^2 + b^2 = c^2$	Write the Pythagorean Theorem.
$a^2 + 2.1^2 = 2.9^2$	Substitute 2.1 for *b* and 2.9 for *c*.
$a^2 + 4.41 = 8.41$	Evaluate powers.
$a^2 = 4$	Subtract 4.41 from each side.
$a = 2$	Take positive square root of each side.

∴ The length of the leg is 2 centimeters.

EXAMPLE 3 Standardized Test Practice

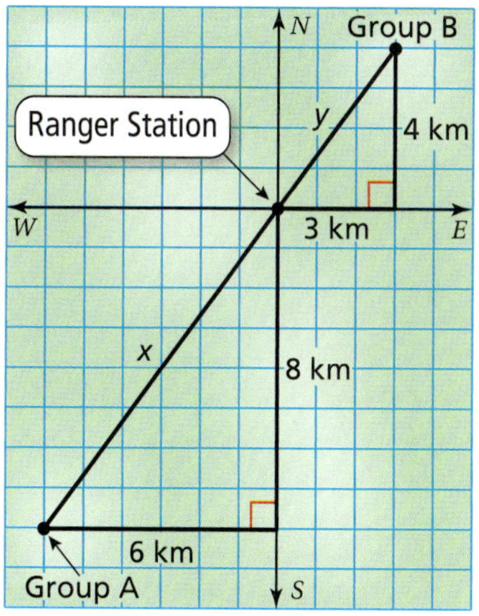

Hiking Group A leaves a ranger station and hikes 8 kilometers south then 6 kilometers west. Group B leaves the station and hikes 3 kilometers east then 4 kilometers north. Using the figure, how far apart are the two groups of hikers?

Ⓐ 5 km Ⓑ 10 km Ⓒ 15 km Ⓓ 21 km

The distance between the groups is the sum of the hypotenuses, *x* and *y*. Use the Pythagorean Theorem to find *x* and *y*.

$a^2 + b^2 = c^2$	Write the Pythagorean Theorem.	$a^2 + b^2 = c^2$
$6^2 + 8^2 = x^2$	Substitute.	$3^2 + 4^2 = y^2$
$36 + 64 = x^2$	Evaluate powers.	$9 + 16 = y^2$
$100 = x^2$	Add.	$25 = y^2$
$10 = x$	Take positive square root of each side.	$5 = y$

∴ The distance between the groups of hikers is $10 + 5 = 15$ kilometers. So, the correct answer is Ⓒ.

On Your Own

Now You're Ready
Exercises 3–8

Find the missing length of the triangle.

3. 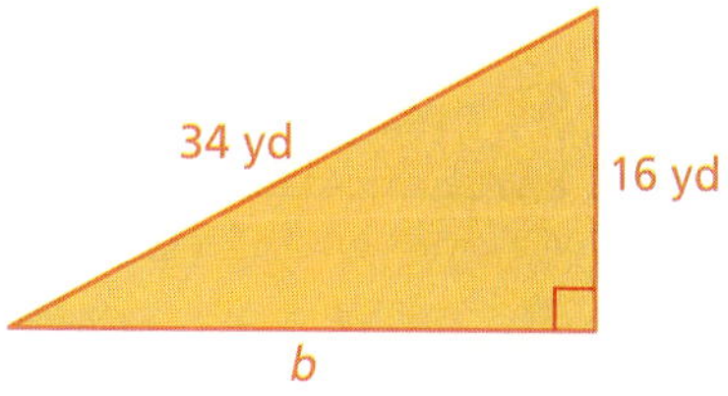

4. 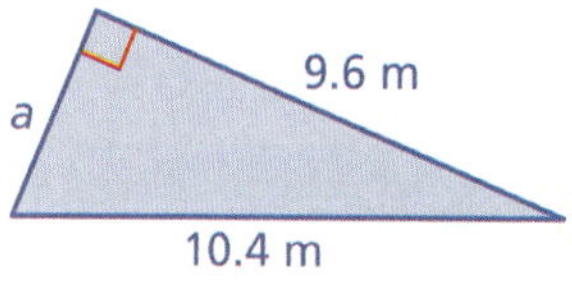

5. **WHAT IF?** In Example 3, Group A hikes 12 kilometers south and 9 kilometers west. How far apart are the hikers?

6.2 Exercises

Check It Out
Help with Homework
BigIdeasMath.com

Vocabulary and Concept Check

1. **VOCABULARY** In a right triangle, how can you tell which sides are the legs and which side is the hypotenuse?

2. **DIFFERENT WORDS, SAME QUESTION** Which is different? Find "both" answers.

Which side is the hypotenuse?

Which side is the longest?

Which side is a leg?

Which side is opposite the right angle?

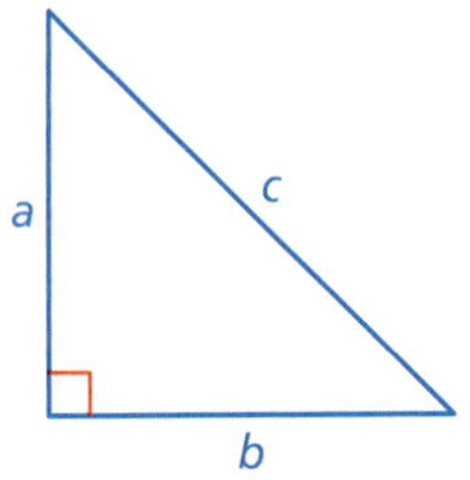

Practice and Problem Solving

Find the missing length of the triangle.

1 2 3.
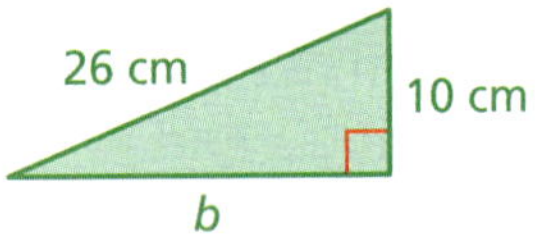

4.
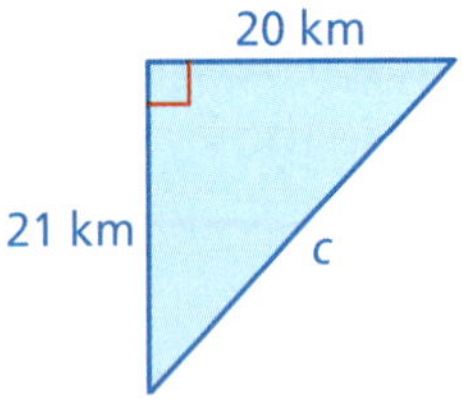

5.
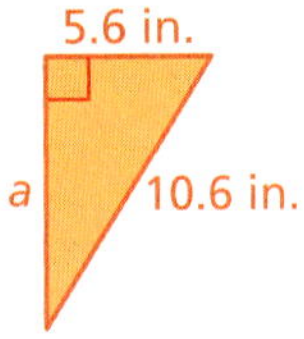

6.
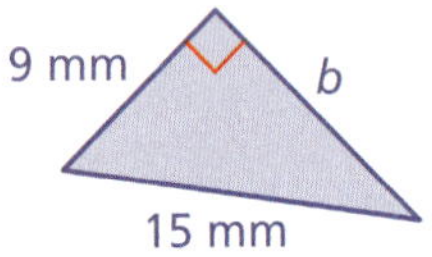

7.
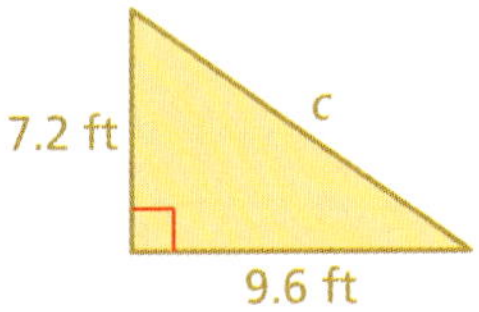

8.
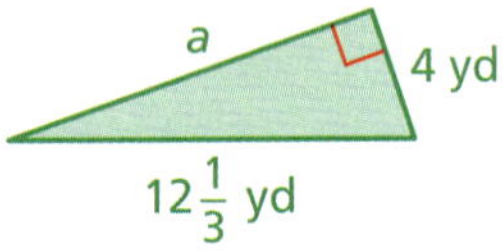

9. **ERROR ANALYSIS** Describe and correct the error in finding the missing length of the triangle.

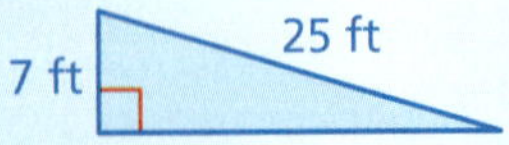

$$a^2 + b^2 = c^2$$
$$7^2 + 25^2 = c^2$$
$$674 = c^2$$
$$\sqrt{674} = c$$

10. **TREE SUPPORT** How long is the wire that supports the tree?

Assignment Guide and Homework Check

Level	Day 1 Activity Assignment	Day 2 Lesson Assignment	Homework Check
Basic	3–5, 19–23	1, 2, 6–11, 14, 15	2, 6, 10, 14
Average	3–5, 19–23	1, 2, 7–11, 14–16	8, 10, 14, 16
Advanced	3–5, 19–23	1, 2, 9, 12–18	12, 14, 16, 17

For Your Information

- **Exercise 17** There is more than one correct drawing for this exercise. Encourage students to start at the origin and move along an axis to begin.

Common Errors

- **Exercises 3–8** Students may substitute the given lengths in the wrong part of the formula. For example, if they are finding one of the legs, they may write $5^2 + 13^2 = c^2$ instead of $5^2 + b^2 = 13^2$. Remind them that the side opposite the right angle is the hypotenuse c.
- **Exercises 3–8** Students may multiply each side length by two instead of squaring the side length. Remind them of the definition of exponents.
- **Exercises 11–13** Students may think that there is not enough information to find the value of x. Tell them that it is possible to find x; however, they may have to make an extra calculation before writing an equation for x.

6.2 Record and Practice Journal

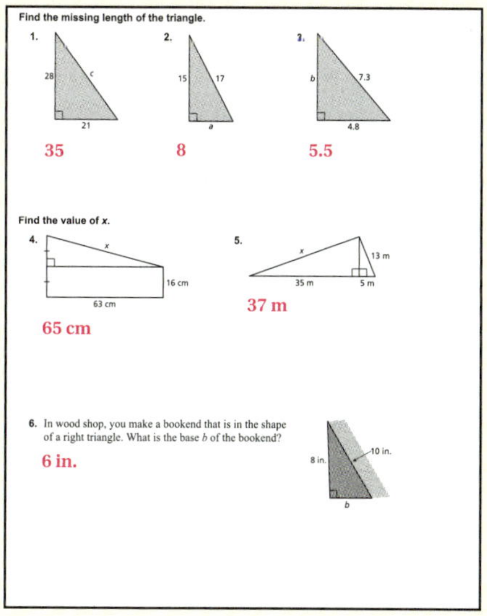
Find the missing length of the triangle.

1. 35
2. 8
3. 5.5

Find the value of *x*.

4. 65 cm
5. 37 m

6. In wood shop, you make a bookend that is in the shape of a right triangle. What is the base *b* of the bookend?
6 in.

Technology For the Teacher

Answer Presentation Tool
Quiz*Show*

Vocabulary and Concept Check

1. The hypotenuse is the longest side and the legs are the other two sides.
2. Which side is a leg?; a or b; c

Practice and Problem Solving

3. 24 cm
4. 29 km
5. 9 in.
6. 12 mm
7. 12 ft
8. $11\frac{2}{3}$ yd
9. The length of the hypotenuse was substituted for the wrong variable.
$a^2 + b^2 = c^2$
$7^2 + b^2 = 25^2$
$49 + b^2 = 625$
$b^2 = 576$
$b = 24$
10. 6.5 ft

Practice and Problem Solving

11. 16 cm

12. 37 mm

13. 10 ft

14. yes; The diagonal of the television is 40 inches.

15. 8.4 cm

16. See *Taking Math Deeper.*

17. a. *Sample answer:*

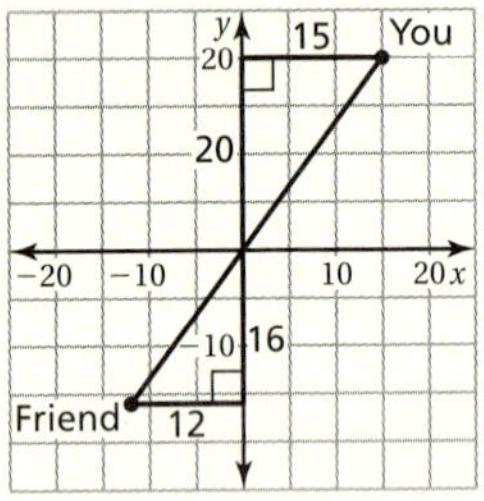

b. 45 ft

18. 7

Fair Game Review

19. 6 and −6 **20.** −11

21. 13 **22.** −15

23. C

Mini-Assessment

Find the missing length of the triangle.

1. 50 ft

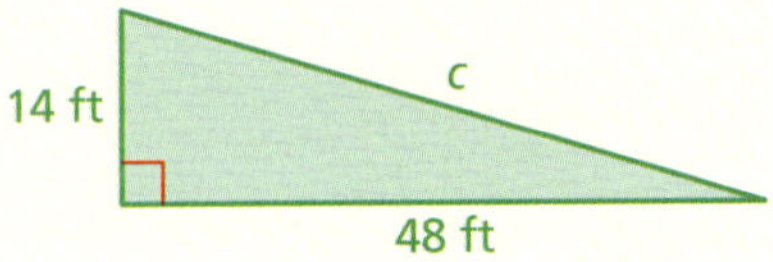

2. 24 mm

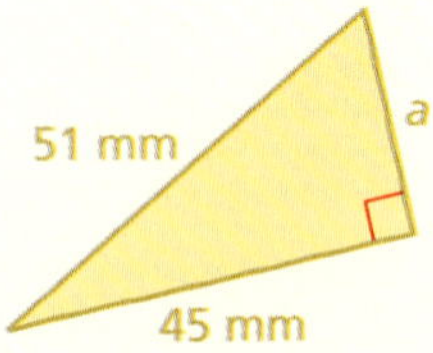

3. 12 in.

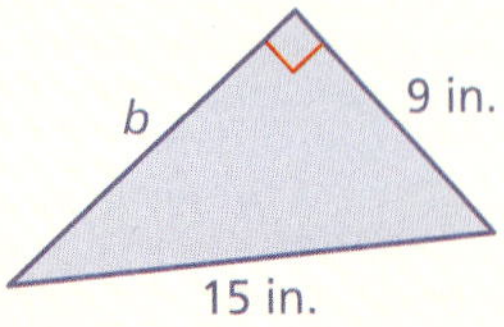

Taking Math Deeper

Exercise 16

The challenging part of this problem is realizing that the hypotenuse of the right triangle is given as 181 yards above the diagram.

1 Find the hypotenuse of the right triangle.

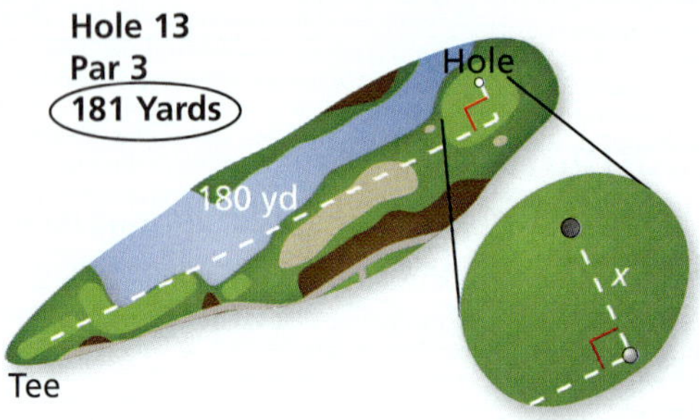

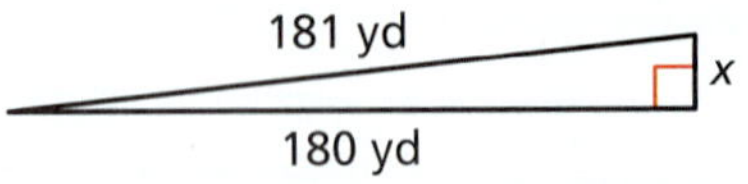

2 Use the Pythagorean Theorem.

$$180^2 + x^2 = 181^2$$
$$32{,}400 + x^2 = 32{,}761$$
$$x^2 = 361$$
$$x = 19$$

3 Answer the question.
The ball is 19 yards from the hole. Using the relationship of 3 feet = 1 yard, 19 yd $\times \frac{3 \text{ ft}}{1 \text{ yd}}$ = 57 ft. So, the ball is 57 feet from the hole.

Reteaching and Enrichment Strategies

If students need help. . .	If students got it. . .
Resources by Chapter • Practice A and Practice B • Puzzle Time Record and Practice Journal Practice Differentiating the Lesson Lesson Tutorials Skills Review Handbook	Resources by Chapter • Enrichment and Extension Start the next section

Find the value of *x*.

11.

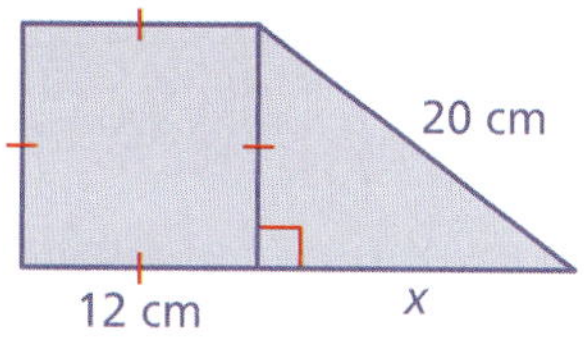

12.

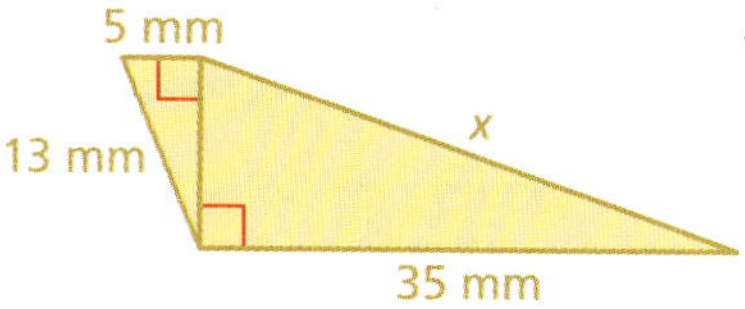

13.

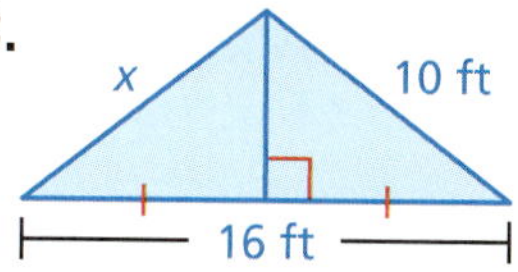

14. FLAT SCREEN Televisions are advertised by the lengths of their diagonals. A store has a sale on televisions 40 inches and larger. Is the television on sale? Explain.

15. BUTTERFLY Approximate the wingspan of the butterfly.

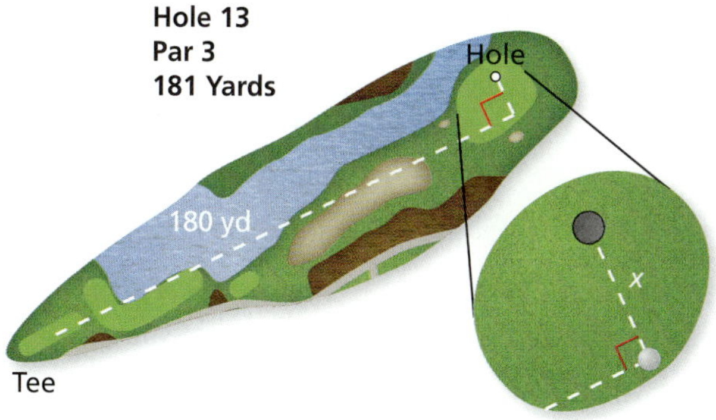

16. GOLF The figure shows the location of a golf ball after a tee shot. How many feet from the hole is the ball?

17. SNOWBALLS You and a friend stand back-to-back. You run 20 feet forward then 15 feet to your right. At the same time, your friend runs 16 feet forward then 12 feet to her right. She stops and hits you with a snowball.

a. Draw the situation in a coordinate plane.

b. How far does your friend throw the snowball?

18. Algebra The legs of a right triangle have lengths of 28 meters and 21 meters. The hypotenuse has a length of $5x$ meters. What is the value of x?

Fair Game Review What you learned in previous grades & lessons

Find the square root(s). *(Section 6.1)*

19. $\pm\sqrt{36}$ **20.** $-\sqrt{121}$ **21.** $\sqrt{169}$ **22.** $-\sqrt{225}$

23. MULTIPLE CHOICE Which type of triangle can have an obtuse angle? *(Section 5.2)*

Ⓐ equiangular Ⓑ right Ⓒ isosceles Ⓓ equilateral

6 Study Help

Check It Out
Graphic Organizer
BigIdeasMath.com

You can use a **summary triangle** to explain a topic. Here is an example of a summary triangle for finding the length of the hypotenuse of a triangle.

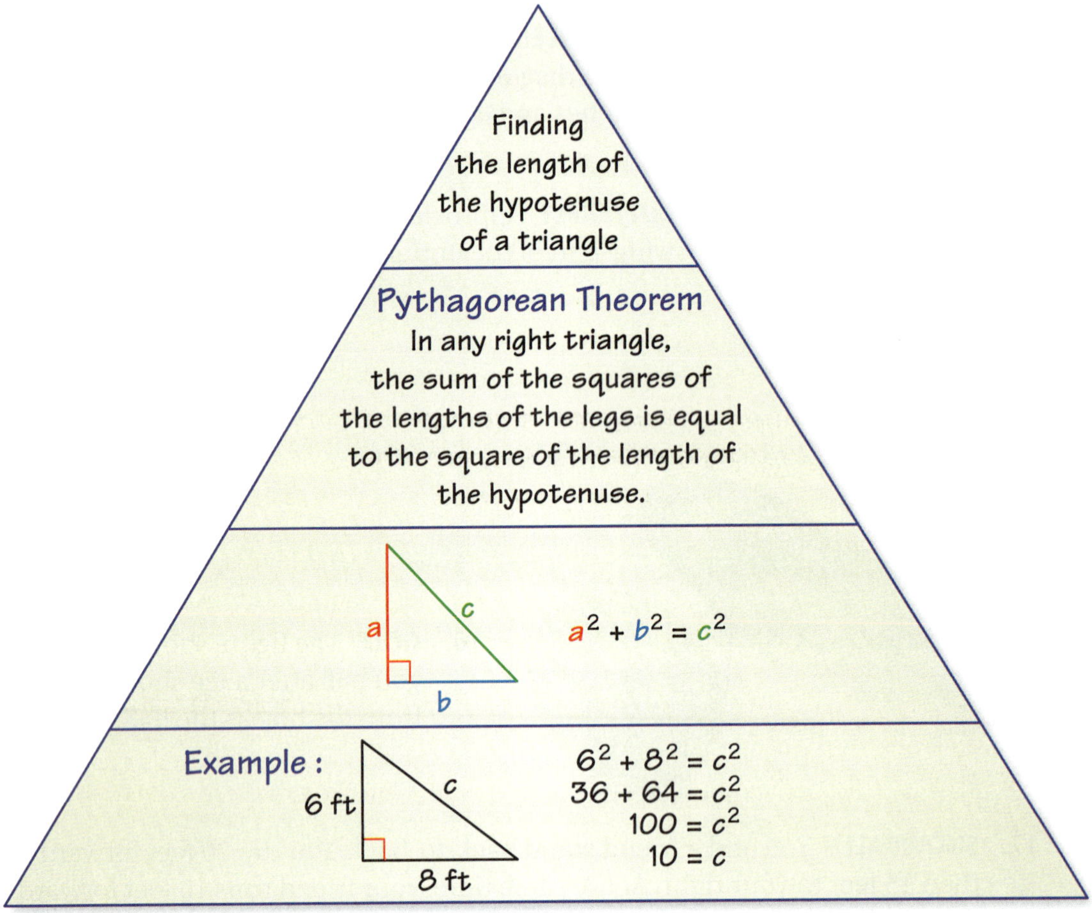

On Your Own

Make a summary triangle to help you study these topics.

1. finding square roots
2. evaluating expressions involving square roots
3. finding the length of a leg of a right triangle

After you complete this chapter, make summary triangles for the following topics.

4. approximating square roots
5. simplifying square roots

"What do you call a cheese summary triangle that isn't yours?"

Sample Answers

1.

Finding square roots

A square root of a number is a number that, when multiplied by itself, equals the given number.

If $a^2 = b$, then a is a square root of b.

Example: The two square roots of 100 are 10 and −10 because $10^2 = 100$ and $(-10)^2 = 100$.

2.

Evaluating expressions involving square roots

Evaluate the square root. Then use the order of operations to evaluate the expression.

Examples:

$3\sqrt{81} - 5 = 3(9) - 5 = 27 - 5 = 22$

$\frac{1}{3} + 2\sqrt{\frac{2}{18}} = \frac{1}{3} + 2\sqrt{\frac{1}{9}} = \frac{1}{3} + 2\left(\frac{1}{3}\right) = 1$

3.

Finding the length of a leg of a right triangle

Pythagorean Theorem

In any right triangle, the sum of the squares of the lengths of the legs is equal to the square of the length of the hypotenuse.

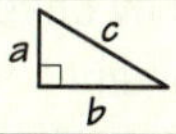

$a^2 + b^2 = c^2$

Example :

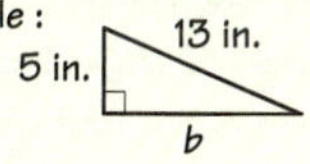

$5^2 + b^2 = 13^2$

$25 + b^2 = 169$

$b^2 = 144$

$b = 12$

List of Organizers

Available at *BigIdeasMath.com*

- Comparison Chart
- Concept Circle
- Definition (Idea) and Example Chart
- Example and Non-Example Chart
- Formula Triangle
- Four Square
- Information Frame
- Information Wheel
- Notetaking Organizer
- Process Diagram
- **Summary Triangle**
- Word Magnet
- Y Chart

About this Organizer

A **Summary Triangle** can be used to explain a concept. Typically, the summary triangle is divided into 3 or 4 parts. In the top part, students write the concept being explained. In the middle part(s), students write any procedure, explanation, description, definition, theorem, and/or formula(s). In the bottom part, students write an example to illustrate the concept. A summary triangle can be used as an assessment tool, in which blanks are left for students to complete. Also, students can place their summary triangles on note cards to use as a quick study reference.

Technology For the Teacher

Vocabulary Puzzle Builder

Answers

1. 14 and −14
2. 7 and −7
3. 20 and −20
4. −2
5. $\frac{4}{5}$
6. 2.5 and −2.5
7. 26
8. −6
9. $5\frac{1}{4}$
10. 41 ft
11. 28 in.
12. 6.3 cm
13. $\frac{1}{2}$ yd
14. $3.14r^2 = 314$; about 20 feet
15. 1000 ft
16. 53 in.

Assessment Book

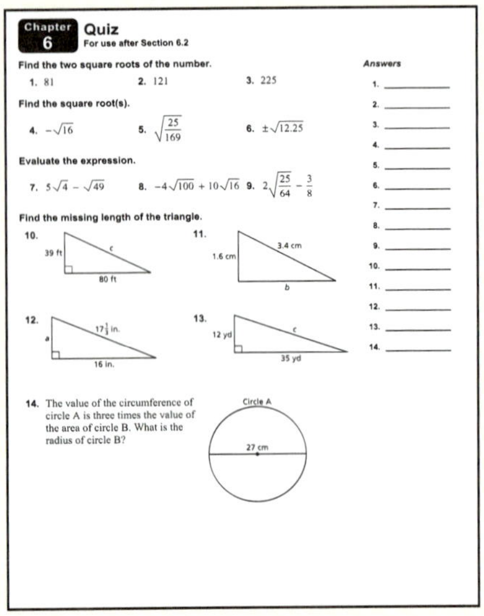
Chapter 6 Quiz
For use after Section 6.2

Find the two square roots of the number.

1. 81 2. 121 3. 225

Find the square root(s).

4. $-\sqrt{16}$ 5. $\sqrt{\frac{25}{169}}$ 6. $\pm\sqrt{12.25}$

Evaluate the expression.

7. $5\sqrt{4} - \sqrt{49}$ 8. $-4\sqrt{100} + 10\sqrt{16}$ 9. $2\sqrt{\frac{25}{64}} - \frac{3}{8}$

Find the missing length of the triangle.

10. 11. 12. 13.

14. The value of the circumference of circle A is three times the value of the area of circle B. What is the radius of circle B?

Answers 1.–14.

Alternative Quiz Ideas

100% Quiz	**Math Log**
Error Notebook	Notebook Quiz
Group Quiz	Partner Quiz
Homework Quiz	Pass the Paper

Math Log

Ask students to keep a math log for the chapter. Have them include diagrams, definitions, and examples. Everything should be clearly labeled. It might be helpful if they put the information in a chart. Students can add to the log as they are introduced to new topics.

Reteaching and Enrichment Strategies

If students need help. . .	If students got it. . .
Resources by Chapter • Study Help • Practice A and Practice B • Puzzle Time Lesson Tutorials *BigIdeasMath.com* Practice Quiz Practice from the Test Generator	Resources by Chapter • Enrichment and Extension • School-to-Work Game Closet at *BigIdeasMath.com* Start the next section

Technology For the Teacher

Answer Presentation Tool
Big Ideas Test Generator

6.1–6.2 Quiz

Check It Out
Progress Check
BigIdeasMath.com

Find the two square roots of the number. *(Section 6.1)*

1. 196 **2.** 49 **3.** 400

Find the square root(s). *(Section 6.1)*

4. $-\sqrt{4}$ **5.** $\sqrt{\frac{16}{25}}$ **6.** $\pm\sqrt{6.25}$

Evaluate the expression. *(Section 6.1)*

7. $3\sqrt{49} + 5$ **8.** $10 - 4\sqrt{16}$ **9.** $\frac{1}{4} + \sqrt{\frac{100}{4}}$

Find the missing length of the triangle. *(Section 6.2)*

10.

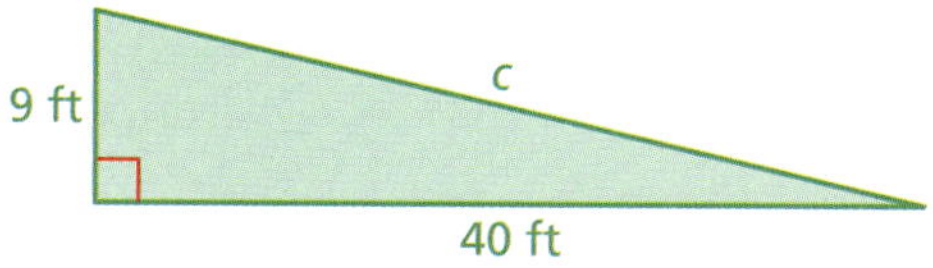

11.

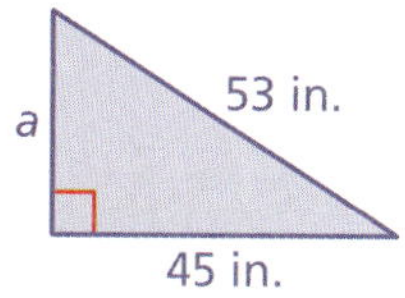

12.

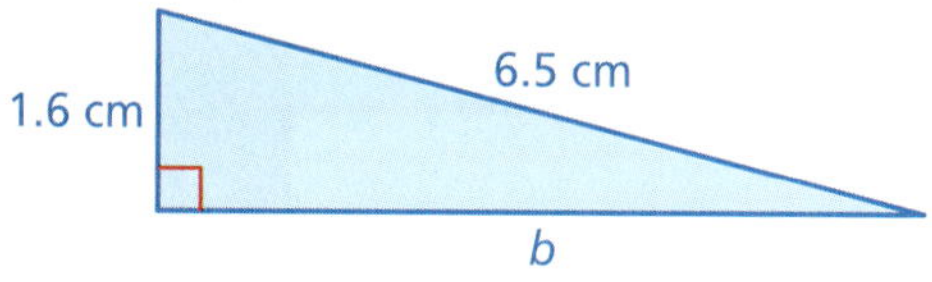

13.

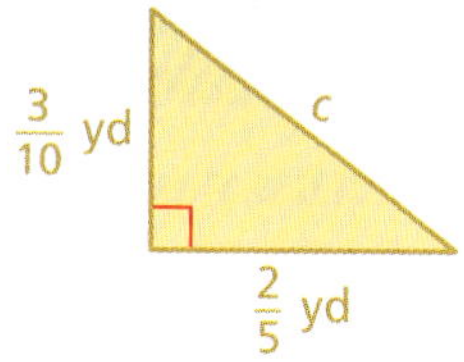

14. POOL The area of a circular pool cover is 314 square feet. Write and solve an equation to find the diameter of the pool cover. Use 3.14 for π. *(Section 6.1)*

15. LAND A square parcel of land has an area of 1 million square feet. What is the length of one side of the parcel? *(Section 6.1)*

16. FABRIC You are cutting a rectangular piece of fabric in half along the diagonal. The fabric measures 28 inches wide and $1\frac{1}{4}$ yards long. What is the length (in inches) of the diagonal? *(Section 6.2)*

6.3 Approximating Square Roots

Essential Question How can you find decimal approximations of square roots that are irrational?

You already know that a rational number is a number that can be written as the ratio of two integers. Numbers that cannot be written as the ratio of two integers are called **irrational**.

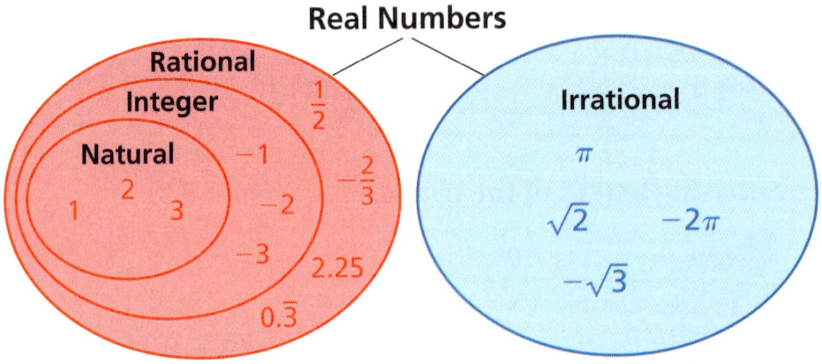

1 ACTIVITY: Approximating Square Roots

Work with a partner.

Archimedes was a Greek mathematician, physicist, engineer, inventor, and astronomer.

Archimedes
(c. 287 B.C.–c. 212 B.C.)

a. Archimedes tried to find a rational number whose square is 3. Here are two that he tried.

$$\frac{265}{153} \text{ and } \frac{1351}{780}$$

Are either of these numbers equal to $\sqrt{3}$? How can you tell?

b. Use a calculator with a square root key to approximate $\sqrt{3}$.

Write the number on a piece of paper. Then enter it into the calculator and square it. Then subtract 3. Do you get 0? Explain.

c. Calculators did not exist in the time of Archimedes. How do you think he might have approximated $\sqrt{3}$?

Laurie's Notes

Introduction

For the Teacher

- **Goal:** Students will investigate the irrational number $\sqrt{3}$ both numerically and geometrically.

Motivate

- Make a large Venn diagram based on student characteristics.
- The diagram can be made on the floor with yarn. Have students write their names on index cards.
- Use the diagram shown for students to place themselves. Sample labels for the groups: A = girls in our class, B = boys in our class, C = wears glasses/contacts, D = brown hair, E = taller than 5' 4", F = wearing a short sleeved T-shirt

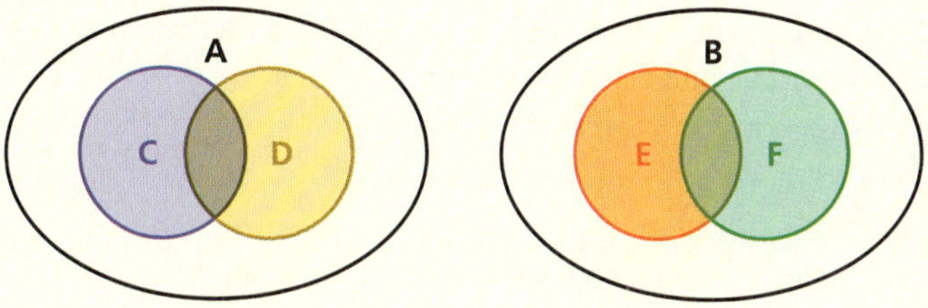

- Discuss what it means to be in certain sets and not in other sets.

Activity Notes

Activity 1

- Draw the Venn diagram shown and write the sets of numbers.
- **FYI:** Archimedes wanted a number x, such that $x^2 = 3$. Students often think $x = 1.5$ or $\frac{3}{2}$ because they are not used to squaring and think of doubling instead. If $x^2 = 3$, then $x = \pm\sqrt{3}$. Archimedes wanted to find a rational number that equaled $\sqrt{3}$.
- Frame this question in the context of what students know: $\sqrt{1} = 1$ and $\sqrt{4} = 2$, so $\sqrt{3}$ will be a decimal between 1 and 2.
- **Part (a):** To square this fraction using a calculator, you can (1) change $\frac{265}{153}$ into a decimal by dividing, then square the decimal, or (2) write the fraction inside parentheses (if available on the calculator), then use the exponent key (if available on the calculator).
- **Part (b):** If students are using their personal calculators, be aware that some calculators have you enter the number and then the square root key, while other calculators have you enter the square root key and then the number.
- **Note:** In part (b), if you get anything other than 0, it means you did not have the exact value for $\sqrt{3}$, because $\sqrt{3}$ is irrational.

Previous Learning

Students should know how to convert between fractions and decimals.

Activity Materials	
Introduction	**Textbook**
• yarn • index cards	• calculators • compasses

Start Thinking! and Warm Up

Activity 6.3 Start Thinking!
For use before Activity 6.3

Activity 6.3 Warm Up
For use before Activity 6.3

Use the Pythagorean Theorem to find the hypotenuse of a right triangle with the given legs.

1. 30, 40	**2.** 10, 24
3. 16, 30	**4.** 9, 40
5. 54, 72	**6.** 2.5, 6

6.3 Record and Practice Journal

Essential Question How can you find decimal approximations of square roots that are irrational?

You already know that a rational number is a number that can be written as the ratio of two integers. Numbers that cannot be written as the ratio of two integers are called **irrational**.

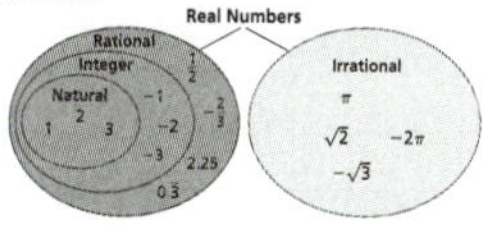

1 ACTIVITY: Approximating Square Roots

Work with a partner.

Archimedes was a Greek mathematician, physicist, engineer, inventor, and astronomer.

a. Archimedes tried to find a rational number whose square is 3. Here are two that he tried.

$\frac{265}{153}$ and $\frac{1351}{780}$

Are either of these numbers equal to $\sqrt{3}$? How can you tell?

no; Each fraction squared is not equal to 3.

b. Use a calculator with a square root key to approximate $\sqrt{3}$. 1.7320508 . . .

Write the number on a piece of paper. Then enter it into the calculator and square it. Then subtract 3. Do you get 0? Explain.

no

Differentiated Instruction

Visual

On the board, create a large Venn diagram similar to the one shown in the activity. Have students place given rational and irrational numbers in the correct spaces on the Venn diagram. Emphasize that a rational number cannot be irrational and an irrational number cannot be rational. You should also reinforce that a number such as 5 is a natural number, as well as an integer and a rational number.

6.3 Record and Practice Journal

c. Calculators did not exist in the time of Archimedes. How do you think he might have approximated $\sqrt{3}$?

Check students' work.

2 ACTIVITY: Approximating Square Roots Geometrically

Work with a partner.

a. Use grid paper and the given scale to draw a horizontal line segment 1 unit in length. Draw your segment near the bottom of the grid. Label this segment AC.

b. Draw a vertical line segment 2 units in length. Draw your segment near the left edge of the grid. Label this segment DC.

c. Set the point of a compass on A. Set the compass to 2 units. Swing the compass to intersect segment DC. Label this intersection as B.

d. Use the Pythagorean Theorem to show that the length of segment BC is $\sqrt{3}$ units.

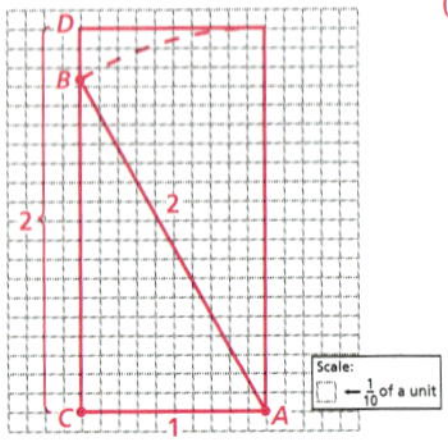

$$(BC)^2 + (AC)^2 = (AB)^2$$
$$(BC)^2 + 1^2 = 2^2$$
$$(BC)^2 + 1 = 4$$
$$(BC)^2 = 3$$
$$BC = \sqrt{3}$$

e. Use the grid paper to approximate $\sqrt{3}$. about 1.7

What Is Your Answer?

3. Repeat Activity 2 for a triangle in which segment CA is 2 units and segment BA is 3 units. Use the Pythagorean Theorem to show that segment BC is $\sqrt{5}$ units. Use the grid paper to approximate $\sqrt{5}$.

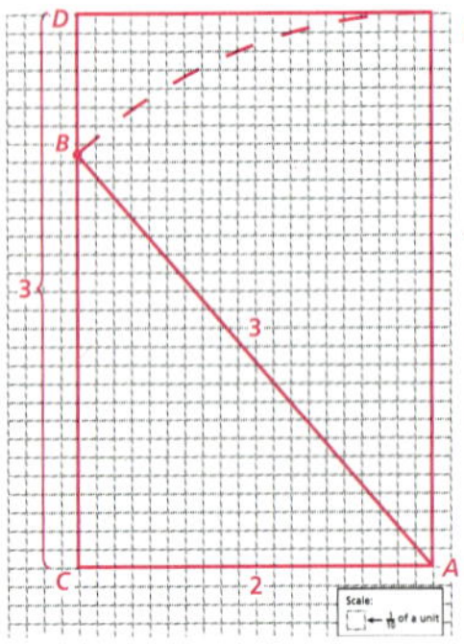

$$(BC)^2 + (AC)^2 = (AB)^2$$
$$(BC)^2 + 2^2 = 3^2$$
$$(BC)^2 + 4 = 9$$
$$(BC)^2 = 5$$
$$BC = \sqrt{5}$$

$\sqrt{5} \approx 2.2$

4. IN YOUR OWN WORDS How can you find decimal approximations of square roots that are irrational?

use geometry

Laurie's Notes

Activity 2

- In Activity 1, students looked at $\sqrt{3}$ as a number. In this activity, students will look at $\sqrt{3}$ geometrically, as a length of a line segment.
- Students should be able to follow the written directions to construct segment BC.
- ? "Why does swinging the compass make AB equal 2 units?" AB is a radius of a circle that you know equals 2.
- ? "What type of triangle is ABC?" right
- ? "What information do you know about the triangle?" The hypotenuse equals 2 and one leg equals 1.
- Ask a volunteer to show how they used the Pythagorean Theorem.
- ? "Is $\sqrt{3}$ greater than or less than 1.5?" greater than

What Is Your Answer?

- **Think-Pair-Share:** Students should read each question independently and then work with a partner to answer the questions. When they have answered the questions, the pair should compare their answers with another group and discuss any discrepancies.

Closure

- **Exit Ticket:** Describe how you would approximate $\sqrt{2}$.
  Listen for students to describe a procedure similar to that used in Activity 2, except with segments 1 unit in length.

Technology For the Teacher

Dynamic Classroom

The Dynamic Planning Tool
Editable Teacher's Resources at *BigIdeasMath.com*

2 ACTIVITY: Approximating Square Roots Geometrically

Work with a partner.

a. Use grid paper and the given scale to draw a horizontal line segment 1 unit in length. Label this segment *AC*.

b. Draw a vertical line segment 2 units in length. Label this segment *DC*.

c. Set the point of a compass on *A*. Set the compass to 2 units. Swing the compass to intersect segment *DC*. Label this intersection as *B*.

d. Use the Pythagorean Theorem to show that the length of segment *BC* is $\sqrt{3}$ units.

e. Use the grid paper to approximate $\sqrt{3}$.

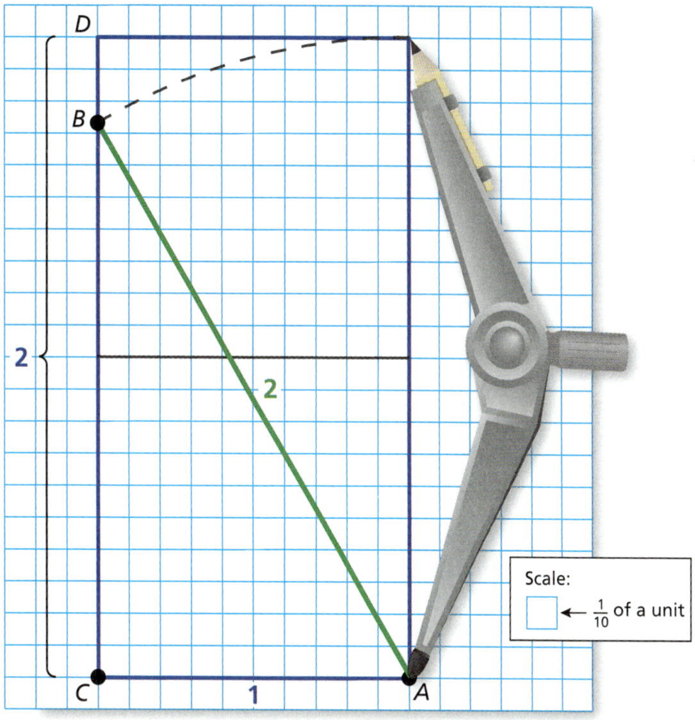

What Is Your Answer?

3. Repeat Activity 2 for a triangle in which segment *CA* is 2 units and segment *BA* is 3 units. Use the Pythagorean Theorem to show that segment *BC* is $\sqrt{5}$ units. Use the grid paper to approximate $\sqrt{5}$.

4. **IN YOUR OWN WORDS** How can you find decimal approximations of square roots that are irrational?

Practice Use what you learned about approximating square roots to complete Exercises 5–8 on page 249.

6.3 Lesson

Check It Out
Lesson Tutorials
BigIdeasMath.com

Key Vocabulary
irrational number, *p. 246*
real numbers, *p. 246*

A rational number is a number that can be written as the ratio of two integers. An **irrational number** cannot be written as the ratio of two integers.

- The square root of any whole number that is not a perfect square is irrational.
- The decimal form of an irrational number neither terminates nor repeats.

Key Idea

Real Numbers

Rational numbers and irrational numbers together form the set of **real numbers**.

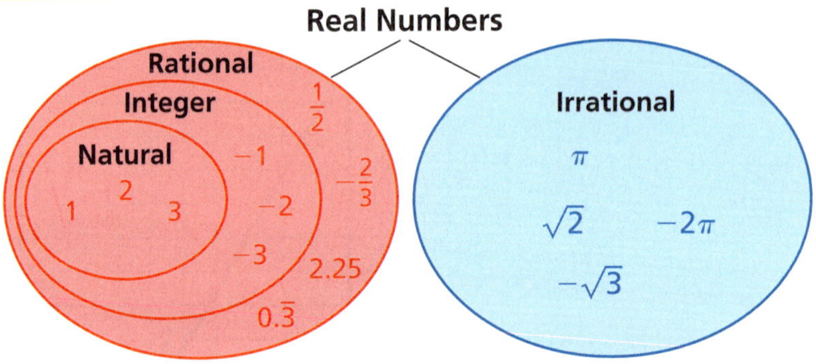

Remember
Decimals that *terminate* or *repeat* are rational.

EXAMPLE 1 Classifying Real Numbers

Tell whether the number is *rational* or *irrational*. Explain.

	Number	Rational or Irrational	Reasoning
a.	$\sqrt{12}$	Irrational	12 is not a perfect square.
b.	$-0.36\overline{4}$	Rational	$-0.36\overline{4}$ is a repeating decimal.
c.	$-1\frac{3}{7}$	Rational	$-1\frac{3}{7}$ can be written as $\frac{-10}{7}$.
d.	0.85	Rational	0.85 can be written as $\frac{17}{20}$.

On Your Own

Now You're Ready
Exercises 9–14

Tell whether the number is *rational* or *irrational*. Explain.

1. 0.121221222. . .
2. $-\sqrt{196}$
3. $\sqrt{2}$

Multi-Language Glossary at BigIdeasMath.com.

Laurie's Notes

Introduction

Connect

- **Yesterday:** Students investigated $\sqrt{3}$ numerically and geometrically.
- **Today:** Students will approximate square roots as being between two integers.

Motivate

- Discuss applications of a periscope and share the following information.
  - A periscope is an optical device for conducting observations from a concealed or protected position.
  - Simple periscopes consist of reflecting mirrors and/or prisms at opposite ends of a tube container. The reflecting surfaces are parallel to each other and at a 45° angle to the axis of the tube.
  - The Navy attributes the invention of the periscope (1902) to Simon Lake and the perfection of the periscope to Sir Howard Grubb.

Lesson Notes

Key Idea

- Explain the definitions of *rational* and *irrational* numbers. Use the Venn diagram to give several examples of each.
- Write the Key Idea. Students saw this Venn diagram yesterday. It is important for students to understand that any real number is either rational or irrational. The two sets do not intersect. Explain that the set of real numbers is composed of rational and irrational numbers.

? "Can you think of a repeating decimal and its fractional equivalent?"

Sample answer: $0.\overline{3} = \frac{1}{3}$

? "Can you think of a terminating decimal and its fractional equivalent?"

Sample answer: $0.5 = \frac{1}{2}$

Example 1

- Students will gain a better understanding of how to identify irrational numbers in this example.

? After working through part (a), ask "Can you think of a number whose square root is rational?" Listen for a perfect square, such as 16 or 25.

On Your Own

- **Think-Pair-Share:** Students should read each question independently and then work with a partner to answer the questions. When they have answered the questions, the pair should compare their answers with another group and discuss any discrepancies.

Goal Today's lesson is approximating square roots.

Lesson Materials
Textbook
• calculators

Start Thinking! and Warm Up

Lesson 6.3 Warm Up
For use before Lesson 6.3

Lesson 6.3 Start Thinking!
For use before Lesson 6.3

How can you find the side length of a square that has the same area as an 8.5-inch by 11-inch piece of paper?

Extra Example 1

Tell whether the number is *rational* or *irrational*. Explain.

a. $\sqrt{15}$ irrational; 15 is not a perfect square.

b. 0.35 rational; 0.35 can be written as $\frac{7}{20}$.

On Your Own

1. irrational; The decimal doesn't terminate or repeat.
2. rational; 196 is a perfect square.
3. irrational; 2 is not a perfect square.

Extra Example 2

Estimate $\sqrt{23}$ to the nearest integer.
Because 23 is closer to 25 than to 16, $\sqrt{23}$ is closer to 5 than to 4. So, $\sqrt{23} \approx 5$.

On Your Own

4. 6 **5.** 9

6. 14 **7.** -3

Extra Example 3

Which is greater, $\sqrt{0.49}$ or 0.71? 0.71

On Your Own

8. $\sqrt{23}$; $\sqrt{23}$ is to the right of $4\frac{1}{5}$.

9. $\sqrt{10}$; $\sqrt{10}$ is positive and $-\sqrt{5}$ is negative.

10. $-\sqrt{2}$; $-\sqrt{2}$ is to the right of -2.

English Language Learners

Vocabulary

Point out to students that the prefix *ir-* means *not*. An *irrational* number is a number that is not rational. Here are other common prefixes that also mean *not*.

dis-	disadvantage, disagree
il-	illiterate, illogical
im-	impolite, improper
in-	independent, indirect
ir-	irrational, irregular
un-	unfair, unfriendly

Laurie's Notes

Example 2

? "What are the first 10 perfect squares?" 1, 4, 9, 16, 25, 36, 49, 64, 81, 100

? "What type of number do you get if you take the square root of any of these perfect squares?" integer

- Write this list on the board as a reference for students. Connect this list to the Venn diagram.
- The integer 52 is not a perfect square. It falls between 49 and 64.
- Draw the number line and work the problem as shown.

On Your Own

- **Question 6:** Students may need help recognizing perfect squares greater than 100.
- **Question 7:** Students should estimate $\sqrt{7}$ first, and then consider the negative sign.

Example 3

- The number line is used as a visual model in each part.
- In part (a), students will ask where to place $\sqrt{5}$. Knowing that $\sqrt{5}$ is between $\sqrt{4}$ and $\sqrt{9}$ does not tell you where to graph it on the number line. Explain that you know it has to be closer to $\sqrt{4}$ than $\sqrt{9}$ because 5 is closer to 4.
- The fraction $2\frac{3}{4}$ is greater than $2\frac{1}{2}$, so it is closer to $\sqrt{9}$.

? "In part (b), what is 0.36 as a fraction?" $\frac{36}{100}$

- Because 36 and 100 are both perfect squares, their square roots are integers. The rational number $\frac{6}{10}$ is equivalent to the decimal 0.6.

On Your Own

- **Question 9:** A positive number is always greater than a negative number.
- **Question 10:** A number line is helpful for this question.

EXAMPLE 2 Approximating Square Roots

Estimate $\sqrt{52}$ to the nearest integer.

Use a number line and the square roots of the perfect squares nearest to the radicand. The nearest perfect square less than 52 is 49. The nearest perfect square greater than 52 is 64.

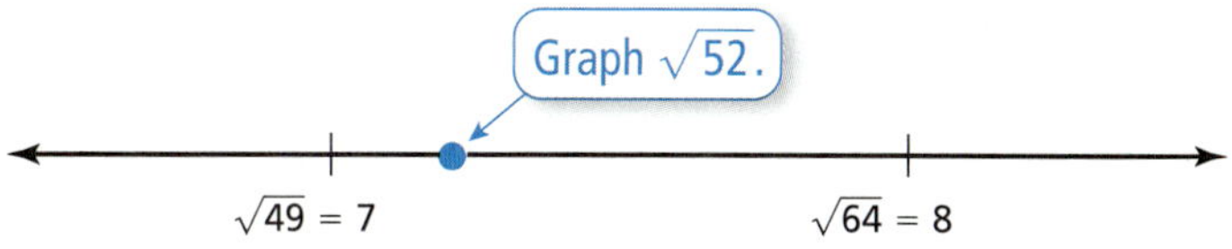

Because 52 is closer to 49 than to 64, $\sqrt{52}$ is closer to 7 than to 8.

So, $\sqrt{52} \approx 7$.

On Your Own

Now You're Ready
Exercises 18–23

Estimate to the nearest integer.

4. $\sqrt{33}$ **5.** $\sqrt{85}$ **6.** $\sqrt{190}$ **7.** $-\sqrt{7}$

EXAMPLE 3 Comparing Real Numbers

a. Which is greater, $\sqrt{5}$ or $2\frac{3}{4}$?

Graph the numbers on a number line.

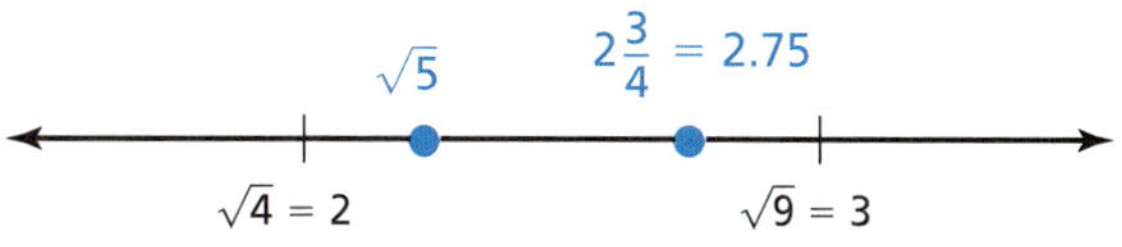

$2\frac{3}{4}$ is to the right of $\sqrt{5}$. So, $2\frac{3}{4}$ is greater.

b. Which is greater, $0.\overline{6}$ or $\sqrt{0.36}$?

Graph the numbers on a number line.

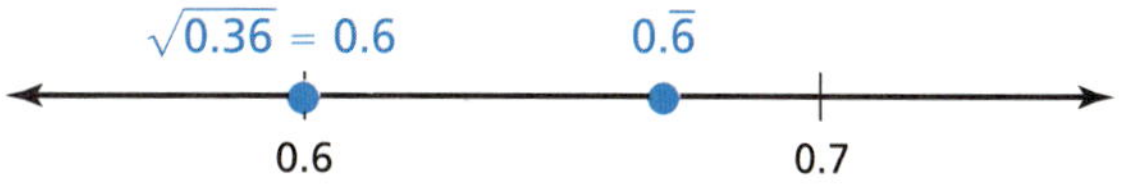

$0.\overline{6}$ is to the right of $\sqrt{0.36}$. So, $0.\overline{6}$ is greater.

On Your Own

Now You're Ready
Exercises 25–30

Which number is greater? Explain.

8. $4\frac{1}{5}, \sqrt{23}$ **9.** $\sqrt{10}, -\sqrt{5}$ **10.** $-\sqrt{2}, -2$

EXAMPLE 4 Approximating an Expression

The radius of a circle with area A is approximately $\sqrt{\frac{A}{3}}$. The area of a circular mouse pad is 51 square inches. Estimate its radius.

$$\sqrt{\frac{A}{3}} = \sqrt{\frac{51}{3}} \qquad \text{Substitute 51 for } A.$$

$$= \sqrt{17} \qquad \text{Divide.}$$

The nearest perfect square less than 17 is 16. The nearest perfect square greater than 17 is 25.

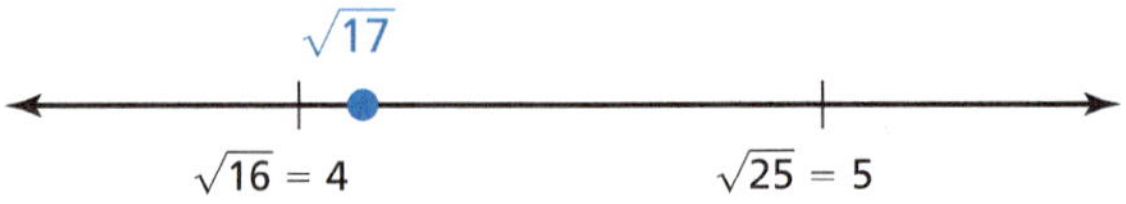

Because 17 is closer to 16 than to 25, $\sqrt{17}$ is closer to 4 than to 5.

∴ The radius is about 4 inches.

On Your Own

11. **WHAT IF?** The area of a circular mouse pad is 64 square inches. Estimate its radius.

EXAMPLE 5 Real-Life Application

The distance (in nautical miles) you can see with a periscope is $1.17\sqrt{h}$, where h is the height of the periscope above the water. Can a periscope that is 6 feet above the water see twice as far as a periscope that is 3 feet above the water? Explain.

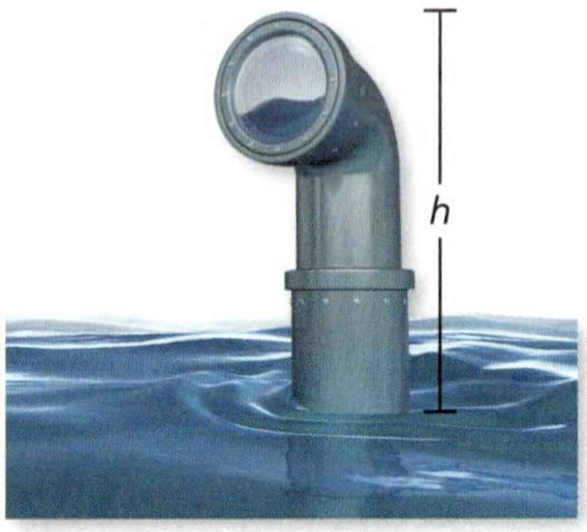

Use a calculator to find the distances.

3 feet above water		*6 feet above water*
$1.17\sqrt{h} = 1.17\sqrt{3}$	Substitute for h.	$1.17\sqrt{h} = 1.17\sqrt{6}$
≈ 2.03	Use a calculator.	≈ 2.87

You can see $\frac{2.87}{2.03} \approx 1.41$ times farther with the periscope that is 6 feet above the water than with the periscope that is 3 feet above the water.

∴ No, the periscope that is 6 feet above the water cannot see twice as far.

On Your Own

12. You use a periscope that is 10 feet above the water. Can you see farther than 4 nautical miles? Explain.

Laurie's Notes

Example 4

- A simple diagram of a circle helps students focus on what is being asked.
- ? "What is the formula for the area of a circle?" $A = \pi r^2$
- Review with students how to solve this formula for r.

$A = \pi r^2$ Write the area formula.

$\frac{A}{\pi} = r^2$ Divide each side by π.

$\sqrt{\frac{A}{\pi}} = r$ Take positive square root of each side.

Because π is close to 3, when approximating the radius, you can replace π with 3. This is the formula presented in this example.

- ? "What information is known in this problem?" You have a circle with an area of 51 square inches.
- ? "What are you trying to find?" an estimate for the radius
- Radicands that are fractions can be intimidating to students.
- Draw the number line and work the problem as shown.

On Your Own

- The quotient $\frac{64}{3}$ is not a whole number, but the question is still completed in the same fashion.

Example 5

- Ask a student to read through the problem.
- You want to compare the distances for a periscope at two different heights above water, so the equation is used twice.
- ? "Do you think you can see twice as far at 6 feet than at 3 feet?" Most will say yes.
- Write the expression $1.17\sqrt{h}$ on the board and evaluate it for each height, as shown.

On Your Own

- Ask volunteers to share their work at the board.

Closure

- Order the numbers from least to greatest: $\sqrt{38}$, $\sqrt{\frac{100}{3}}$, $6.\overline{5}$ $\sqrt{\frac{100}{3}}$, $\sqrt{38}$, $6.\overline{5}$

Technology For the Teacher

Dynamic Classroom

The Dynamic Planning Tool

Editable Teacher's Resources at *BigIdeasMath.com*

Extra Example 4

In Example 4, estimate the radius of a circular mouse pad with an area of 45 square inches. about 4 in.

On Your Own

11. 5 in.

Extra Example 5

In Example 5, a periscope is 8 feet above the water. Can you see farther than 3 nautical miles? Explain. yes; You can see about 3.3 nautical miles.

On Your Own

12. no; You can only see about 3.7 nautical miles.

Vocabulary and Concept Check

1. A rational number can be written as the ratio of two integers. An irrational number cannot be written as the ratio of two integers.
2. 32 is between the perfect squares 25 and 36, but is closer to 36, so $\sqrt{32} \approx 6$.
3. all rational and irrational numbers; *Sample answer:* $-2, \frac{1}{8}, \sqrt{7}$
4. $\sqrt{8}$; $\sqrt{8}$ is irrational and the other three numbers are rational.

Practice and Problem Solving

5. yes
6. no
7. no
8. yes
9. rational; $3.\overline{6}$ is a repeating decimal.
10. irrational; π neither terminates nor repeats.
11. irrational; 7 is not a perfect square.
12. rational; -1.125 is a terminating decimal.
13. rational; $-3\frac{8}{9}$ can be written as the ratio of two integers.
14. irrational; 15 is not a perfect square.
15. 144 is a perfect square. So, $\sqrt{144}$ is rational.
16. no; 52 is not a perfect square.
17. **a.** natural number

    b. irrational number

    c. irrational number

Assignment Guide and Homework Check

Level	Day 1 Activity Assignment	Day 2 Lesson Assignment	Homework Check
Basic	5–8, 41–44	1–4, 9–31 odd, 16, 24, 32	16, 24, 27, 32
Average	5–8, 41–44	1–4, 13–17, 22–24, 25–33 odd, 32, 37	16, 24, 27, 32
Advanced	5–8, 41–44	1–4, 15, 23, 24, 29–40	24, 30, 34, 38

Common Errors

- **Exercises 9–14** Students may think that all the negative numbers are irrational. Remind them of the integers and that negative numbers can be rational or irrational.
- **Exercises 9–14** Students may think that any decimal that does not terminate is irrational. Remind them of fractions that can be written as repeating decimals, such as $\frac{1}{3}$, which are rational.

6.3 Record and Practice Journal

Tell whether the number is *rational* or *irrational*. Explain.

1. $\sqrt{12}$ — Irrational; 12 is not a perfect square.
2. $-\frac{3}{7}$ — rational; It is the ratio of two integers.
3. $0.4\overline{89}$ — rational; It is a repeating decimal.

Estimate to the nearest integer.

4. $\sqrt{8}$ — 3
5. $\sqrt{60}$ — 8
6. $-\sqrt{\frac{172}{25}}$ — -3

Which number is greater? Explain.

7. $\sqrt{88}$, 12 — 12; 12 is to the right of $\sqrt{88}$.
8. $-\sqrt{18}$, -6 — $-\sqrt{18}$; $-\sqrt{18}$ is to the right of -6.
9. 14.5, $\sqrt{220}$ — $\sqrt{220}$; $\sqrt{220}$ is to the right of 14.5.
10. The velocity in meters per second of a ball that is dropped from a window at a height of 10.5 meters is represented by the equation $v = \sqrt{2(9.8)(10.5)}$. Estimate the velocity of the ball. Round your answer to the nearest tenth.

    14.3 m/sec
11. The area of a square table cloth is 30 square feet. Estimate the length of one side of the tablecloth. Round your answer to the nearest tenth.

    5.5 ft

Technology For the Teacher

Answer Presentation Tool

Quiz*Show*

6.3 Exercises

Check It Out
Help with Homework
BigIdeasMath.com

Vocabulary and Concept Check

1. **VOCABULARY** What is the difference between a rational number and an irrational number?
2. **WRITING** Describe a method of approximating $\sqrt{32}$.
3. **VOCABULARY** What are real numbers? Give three examples.
4. **WHICH ONE DOESN'T BELONG?** Which number does *not* belong with the other three? Explain your reasoning.

$-\frac{11}{12}$ 25.075 $\sqrt{8}$ $-3.\overline{3}$

Practice and Problem Solving

Tell whether the rational number is a reasonable approximation of the square root.

5. $\frac{559}{250}$, $\sqrt{5}$
6. $\frac{3021}{250}$, $\sqrt{11}$
7. $\frac{678}{250}$, $\sqrt{28}$
8. $\frac{1677}{250}$, $\sqrt{45}$

Tell whether the number is *rational* or *irrational*. Explain.

1 9. $3.66666\overline{6}$
10. $\frac{\pi}{6}$
11. $-\sqrt{7}$
12. -1.125
13. $-3\frac{8}{9}$
14. $\sqrt{15}$

15. **ERROR ANALYSIS** Describe and correct the error in classifying the number.

✗ $\sqrt{144}$ is irrational.

16. **SCRAPBOOKING** You cut a picture into a right triangle for your scrapbook. The lengths of the legs of the triangle are 4 inches and 6 inches. Is the length of the hypotenuse a rational number? Explain.

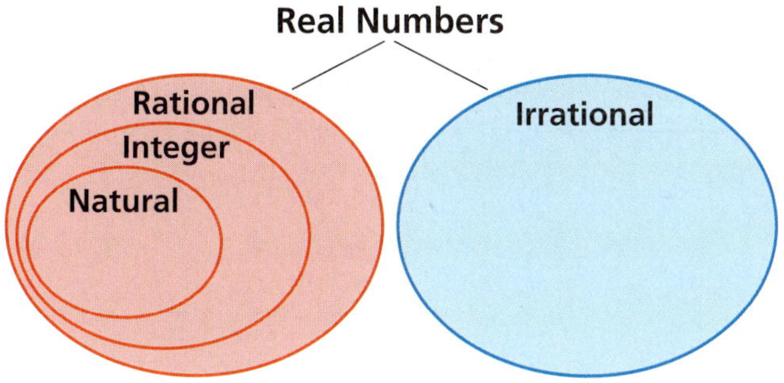

17. **VENN DIAGRAM** Place each number in the correct area of the Venn Diagram.
   a. Your age
   b. The square root of any prime number
   c. The ratio of the circumference of a circle to its diameter

Estimate to the nearest integer.

2 **18.** $\sqrt{24}$ **19.** $\sqrt{685}$ **20.** $-\sqrt{61}$

21. $-\sqrt{105}$ **22.** $\sqrt{\frac{27}{4}}$ **23.** $-\sqrt{\frac{335}{2}}$

24. **CHECKERS** A checkerboard is 8 squares long and 8 squares wide. The area of each square is 14 square centimeters. Estimate the perimeter of the checkerboard.

Which number is greater? Explain.

3 **25.** $\sqrt{20}$, 10 **26.** $\sqrt{15}$, -3.5 **27.** $\sqrt{133}$, $10\frac{3}{4}$

28. $\frac{2}{3}$, $\sqrt{\frac{16}{81}}$ **29.** $-\sqrt{0.25}$, -0.25 **30.** $-\sqrt{182}$, $-\sqrt{192}$

31. **FOUR SQUARE** The area of a four square court is 66 square feet. Estimate the length s of one of the sides of the court.

32. **RADIO SIGNAL** The maximum distance (in nautical miles) that a radio transmitter signal can be sent is represented by the expression $1.23\sqrt{h}$, where h is the height (in feet) above the transmitter.

Estimate the maximum distance x (in nautical miles) between the plane that is receiving the signal and the transmitter. Round your answer to the nearest tenth.

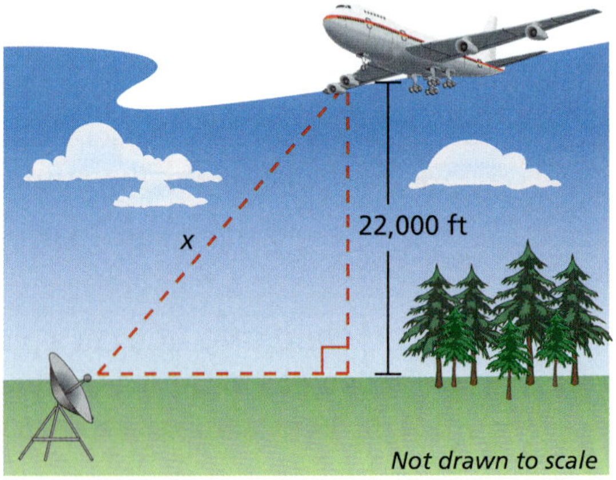

33. **OPEN-ENDED** Find two numbers a and b that satisfy the diagram.

Common Errors

- **Exercises 18–23** Students may struggle with knowing what integer is closest to the given number. To help make comparisons, encourage them to write the first 10 perfect squares. If the number under the radical is greater than 100, then students should use *Guess, Check, and Revise* to find two integers on either side of the number. When determining which integer is closer to the rational number, encourage students to use a number line.
- **Exercises 25–30** Students may guess which is greater just by looking at the numbers. Encourage them to use a number line to compare the numbers. Also remind them to simplify and/or estimate the numbers so that they are easier to compare.
- **Exercises 34–36** Students may struggle estimating the square roots because of the decimals. Remind them of the list they wrote for Exercises 18–23. Remind them to move the decimal point two places to the left. These new perfect squares will help to estimate the square root.

English Language Learners

Illustrate

Draw a number line on the board. Have students locate $\sqrt{1}$, $\sqrt{4}$, and $\sqrt{9}$ on the number line. Now have them approximate the location of $\sqrt{5}$. Visualizing the $\sqrt{5}$ between 2 and 3 helps students form an approximation. Continue doing this with other square roots.

Practice and Problem Solving

18. 5

19. 26

20. -8

21. -10

22. 3

23. -13

24. 128 cm

25. 10; 10 is to the right of $\sqrt{20}$.

26. $\sqrt{15}$; $\sqrt{15}$ is positive and -3.5 is negative.

27. $\sqrt{133}$; $\sqrt{133}$ is to the right of $10\frac{3}{4}$.

28. $\frac{2}{3}$; $\frac{2}{3}$ is to the right of $\sqrt{\frac{16}{81}}$.

29. -0.25; -0.25 is to the right of $-\sqrt{0.25}$.

30. $-\sqrt{182}$; $-\sqrt{182}$ is to the right of $-\sqrt{192}$.

31. 8 ft

32. 182.4 nautical miles

33. *Sample answer:* $a = 82$, $b = 97$

Practice and Problem Solving

34. 0.6

35. 1.1

36. 1.2

37. 30.1 m/sec

38. yes; $\left(\frac{1}{2}\right)^2 = \frac{1}{4}$, so $\sqrt{\frac{1}{4}} = \frac{1}{2}$.

no; $\left(\frac{\sqrt{3}}{4}\right)^2 = \frac{3}{16}$, and $\sqrt{3}$ is irrational.

39. See *Taking Math Deeper.*

40. Sample answers are given.

a. always; The product of two fractions is a fraction. $\frac{2}{3} \cdot \frac{3}{4} = \frac{1}{2}$

b. sometimes; $\pi \cdot 0 = 0$ is rational, but $2 \cdot \sqrt{3}$ is irrational.

c. sometimes; $\sqrt{2} \cdot \pi$ is irrational, but $\pi \cdot \frac{1}{\pi} = 1$ is rational.

Fair Game Review

41. $-3x + 3y$ **42.** $4t - 5\pi$

43. $40k - 9$ **44.** D

Mini-Assessment

Estimate to the nearest integer.

1. $\sqrt{65}$ about 8

2. $\sqrt{99}$ about 10

3. $\sqrt{\frac{15}{2}}$ about 3

Which number is greater?

4. $2\frac{11}{12}, -\sqrt{8}$ $2\frac{11}{12}$

5. $\frac{5}{4}, \sqrt{\frac{49}{64}}$ $\frac{5}{4}$

Taking Math Deeper

Exercise 39

This is a nice science problem. Students learn from this problem that objects do not fall at a linear rate. Their speed increases with each second they are falling.

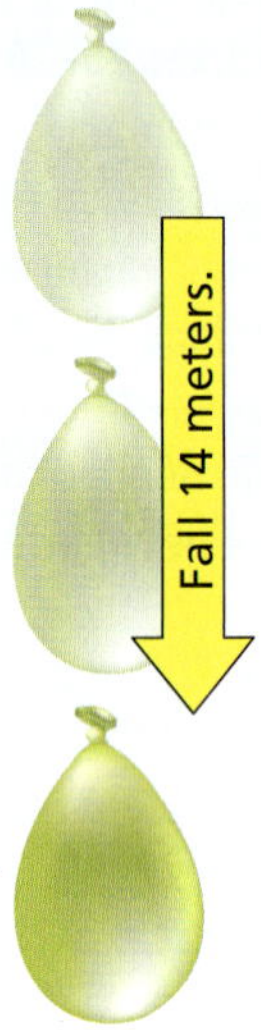

1 Understand the problem.

A water balloon is dropped from a height of 14 meters. How long does it take the balloon to fall to the ground?

2 Use the given formula.

$$t = \sqrt{\frac{d}{4.9}}$$

$$= \sqrt{\frac{14}{4.9}}$$

$$\approx \sqrt{2.86}$$

$$\approx 1.7$$

3 Answer the question.

The water balloon will hit the ground in about 1.7 seconds.

Project

Use a stop watch, a metric tape measure, and several water balloons. Measure the distance from the top of the bleachers at your school. Drop a balloon and record the time it takes to fall to the ground. Use the formula in the problem to calculate the time it should take. Compare.

Reteaching and Enrichment Strategies

If students need help...	If students got it...
Resources by Chapter • Practice A and Practice B • Puzzle Time Record and Practice Journal Practice Differentiating the Lesson Lesson Tutorials Skills Review Handbook	Resources by Chapter • Enrichment and Extension • School-to-Work Start the next section

Estimate to the nearest tenth.

34. $\sqrt{0.39}$

35. $\sqrt{1.19}$

36. $\sqrt{1.52}$

37. ROLLER COASTER The velocity v (in meters per second) of a roller coaster is represented by the equation $v = 3\sqrt{6r}$, where r is the radius of the loop. Estimate the velocity of a car going around the loop. Round your answer to the nearest tenth.

38. Is $\sqrt{\frac{1}{4}}$ a rational number? Is $\sqrt{\frac{3}{16}}$ a rational number? Explain.

39. WATER BALLOON The time t (in seconds) it takes a water balloon to fall d meters is represented by the equation $t = \sqrt{\frac{d}{4.9}}$. Estimate the time it takes the balloon to fall to the ground from a window that is 14 meters above the ground. Round your answer to the nearest tenth.

40. Number Sense Determine if the statement is *sometimes, always,* or *never* true. Explain your reasoning and give an example of each.

a. A rational number multiplied by a rational number is rational.

b. A rational number multiplied by an irrational number is rational.

c. An irrational number multiplied by an irrational number is rational.

Fair Game Review What you learned in previous grades & lessons

Simplify the expression. *(Skills Review Handbook)*

41. $2x + 3y - 5x$

42. $3\pi + 8(t - \pi) - 4t$

43. $17k - 9 + 23k$

44. MULTIPLE CHOICE What is the ratio (red to blue) of the corresponding side lengths of the similar triangles? *(Skills Review Handbook)*

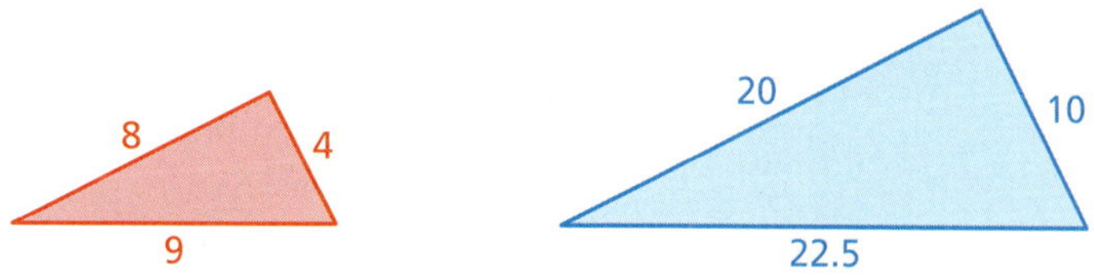

Ⓐ 1:3 Ⓑ 5:2 Ⓒ 3:4 Ⓓ 2:5

6.4 Simplifying Square Roots

Essential Question How can you use a square root to describe the golden ratio?

Two quantities are in the *golden ratio* if the ratio between the sum of the quantities and the greater quantity is the same as the ratio between the greater quantity and the lesser quantity.

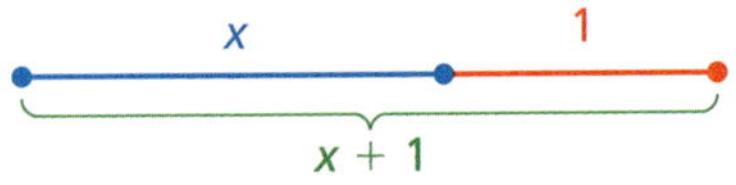

$$\frac{x+1}{x} = \frac{x}{1}$$

In a future algebra course, you will be able to prove that the golden ratio is

$$\frac{1+\sqrt{5}}{2}$$ Golden ratio.

1 ACTIVITY: Constructing a Golden Ratio

Work with a partner.

a. Use grid paper and the given scale to draw a square that is 1 unit by 1 unit (blue).

b. Draw a line from midpoint *C* of one side of the square to the opposite corner *D*, as shown.

c. Use the Pythagorean Theorem to find the length of segment *CD*.

d. Set the point of a compass on *C*. Set the compass radius to the length of segment *CD*. Swing the compass to intersect line *BC* at point *E*.

e. The rectangle *ABEF* is called a *golden rectangle* because the ratio of its side lengths is the golden ratio.

f. Use a calculator to find a decimal approximation of the golden ratio. Round your answer to two decimal places.

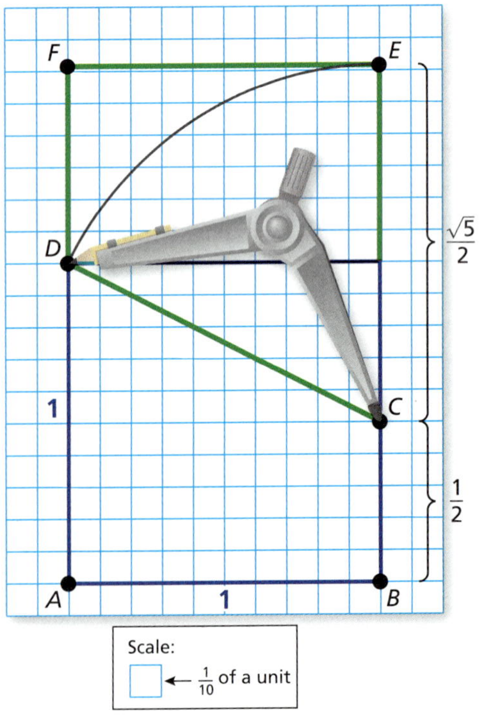

Laurie's Notes

Introduction

For the Teacher

- **Goal:** Students will explore a classic example of a computation that involves a radical, the golden ratio, $\dfrac{1+\sqrt{5}}{2} \approx 1.62$.

Motivate

- Share information with students about the Parthenon in Athens. The Parthenon was perhaps the best example of a mathematical approach to art, with many of its intricate constructions based on the golden ratio.

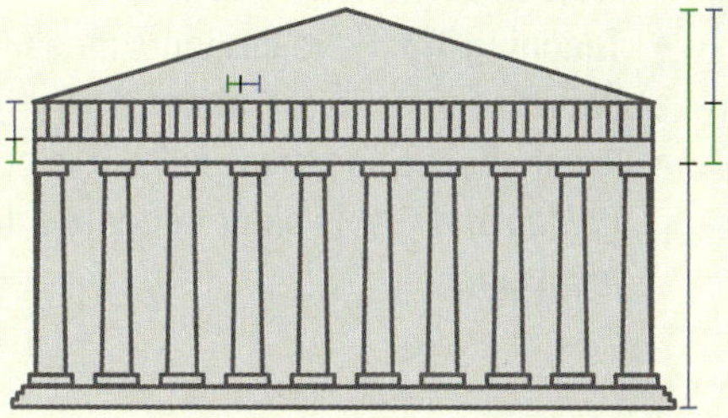

- If possible, show lengths that are in the golden ratio using photographs and diagrams.

Activity Notes

Activity 1

- Note that the midpoint C is not constructed. It is located by knowing the side length is 1 unit. So, the midpoint is 0.5 unit from B.
- ? **Extension:** "What type of quadrilateral is $ABCD$? What is the area of this quadrilateral?" trapezoid; $A = \dfrac{1}{2}\left(1+\dfrac{1}{2}\right)1 = \dfrac{3}{4}$ units^2
- Students may need help with part (c). Set up the Pythagorean Theorem for the right triangle with legs 1 unit and $\frac{1}{2}$ unit. Segment CD is the hypotenuse of the triangle. Let x be the length of segment CD. So, $x^2 = 1^2 + \left(\dfrac{1}{2}\right)^2$ and $x = \sqrt{\dfrac{5}{4}}$.
- The last step involves a property which students will learn formally tomorrow. At this point, you are asking students to accept that $\sqrt{\dfrac{5}{4}} = \dfrac{\sqrt{5}}{\sqrt{4}} = \dfrac{\sqrt{5}}{2}$. In other words, the square root of a quotient is equal to the quotient of the square roots.
- To find the length of the rectangle, find $BC + CE = \dfrac{1}{2} + \dfrac{\sqrt{5}}{2} = \dfrac{1+\sqrt{5}}{2}$. The width of the rectangle is 1 unit, so the ratio of the side lengths is $\dfrac{1+\sqrt{5}}{2}$. This is known as the golden ratio.
- ? "From the grid paper, approximate BE, the golden ratio." about 1.6 units

Activity Materials
Textbook
• calculators • tape measures, yard sticks, or meter sticks

Start Thinking! and Warm Up

Activity 6.4 Start Thinking!
For use before Activity 6.4

Activity 6.4 Warm Up
For use before Activity 6.4

Use a calculator to find a decimal approximation of the expression. Round your answer to the nearest thousandth.

1. $\dfrac{\sqrt{7}}{7}$
2. $\dfrac{\sqrt{3}}{2}$
3. $\dfrac{1+\sqrt{3}}{2}$
4. $\dfrac{\sqrt{3}-1}{3}$
5. $\dfrac{2+\sqrt{2}}{3}$
6. $\dfrac{2-\sqrt{2}}{4}$

6.4 Record and Practice Journal

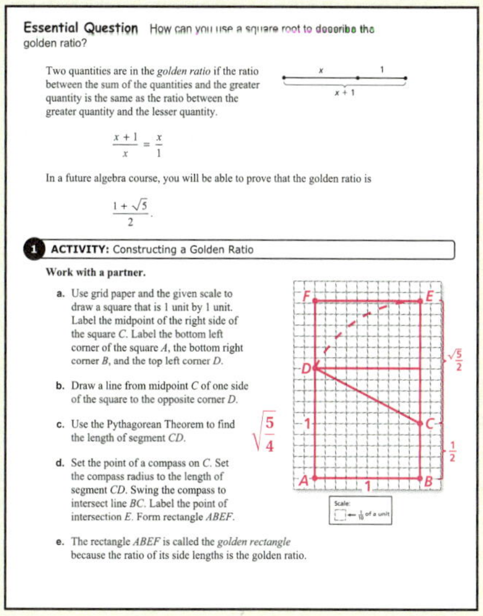

English Language Learners

Vocabulary

For English learners, note the different meanings of the word *radical.* In everyday language, it can be used to describe a considerable departure from the usual or traditional view. In slang, *radical* is something that is excellent or cool. In mathematics, it is a symbol for roots—square roots, cube roots, and so on.

6.4 Record and Practice Journal

f. Use a calculator to find a decimal approximation of the golden ratio. Round your answer to two decimal places.

1.62

2 ACTIVITY: The Golden Ratio and the Human Body

Work with a partner.

Leonardo da Vinci was one of the first to notice that there are several ratios in the human body that approximate the golden ratio.

a. Use a tape measure or two yardsticks to measure the lengths shown in the diagram for both you and your partner. (Take your shoes off before measuring.)

b. Record your results in the first two columns of the tables on the next page.

c. Calculate the ratios shown in the tables.

d. Leonardo da Vinci stated that for many people, the ratios are close to the golden ratio. How close are your ratios?

Check students' work.

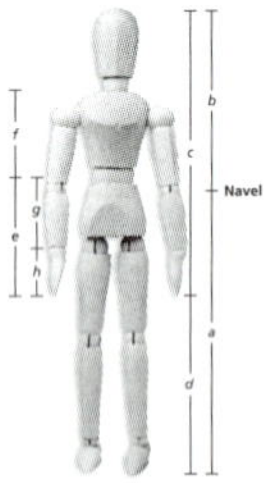

You		
$a =$	$b =$	$\frac{a}{b} =$
$c =$	$d =$	$\frac{c}{d} =$
$e =$	$f =$	$\frac{e}{f} =$
$g =$	$h =$	$\frac{g}{h} =$

Partner		
$a =$	$b =$	$\frac{a}{b} =$
$c =$	$d =$	$\frac{c}{d} =$
$e =$	$f =$	$\frac{e}{f} =$
$g =$	$h =$	$\frac{g}{h} =$

What Is Your Answer?

3. **IN YOUR OWN WORDS** How can you use a square root to describe the golden ratio? Use the Internet or some other reference to find examples of the golden ratio in art and architecture.

Check students' work.

Laurie's Notes

Activity 2

- In this activity, you need to be sensitive to students' feelings about measuring body parts. If you believe there are students in your class who, for whatever reason, will be uncomfortable with having their partner measure certain body parts, you should probably speak to them privately. Refer to the diagram and reassure them that their privacy will not be violated.
- My experience has been that students enjoy this activity and find the results fascinating. Give ample time for students to measure and compute.
- **Teaching Tip:** Have students measure in metric units so that they are using decimals instead of fractions.
- When discussing the results, it is possible that some students are not very *golden* and that is okay. Leonardo da Vinci only implied that it was true for *many* people.

What Is Your Answer?

- **Think-Pair-Share:** Students should read the question independently and then work with a partner to answer the question. When they have answered the question, the pair should compare their answer with another group and discuss any discrepancies.

Closure

- Compute the length to width ratio for each rectangle. Which ratio is closer to the golden ratio? Second rectangle is closer to the golden ratio.

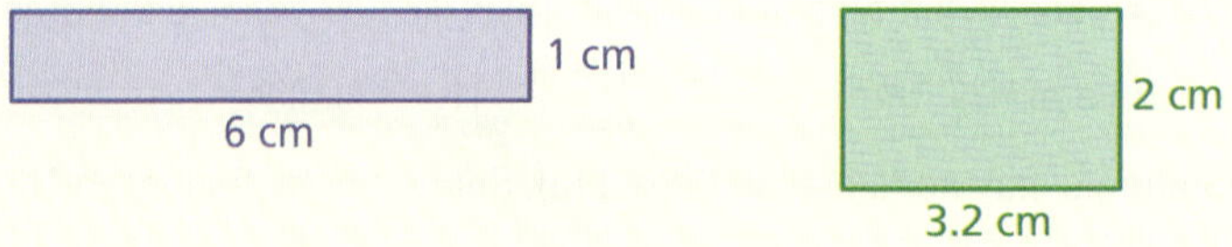

- **FYI:** Most people say that rectangles with dimensions in the ratio close to 1.6 to 1 are more pleasing to look at.

Technology For the Teacher

Dynamic Classroom

The Dynamic Planning Tool

Editable Teacher's Resources at *BigIdeasMath.com*

2 ACTIVITY: The Golden Ratio and the Human Body

Work with a partner.

Leonardo da Vinci was one of the first to notice that there are several ratios in the human body that approximate the golden ratio.

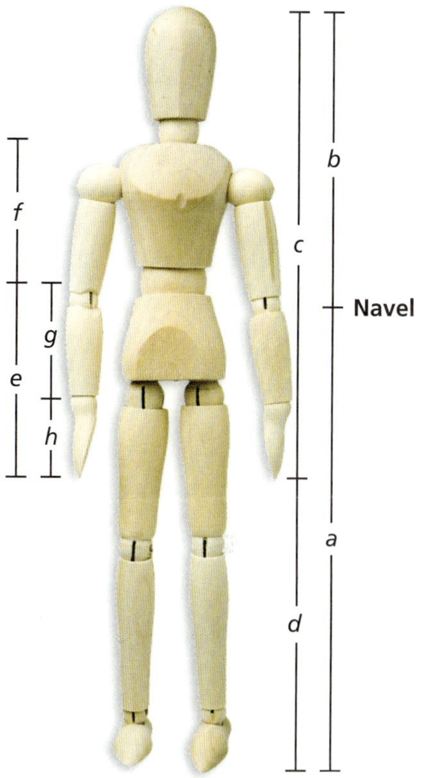

a. Use a tape measure or two yardsticks to measure the lengths shown in the diagram for both you and your partner. (Take your shoes off before measuring.)

b. Copy the tables below. Record your results in the first two columns.

c. Calculate the ratios shown in the tables.

d. Leonardo da Vinci stated that for many people, the ratios are close to the golden ratio. How close are your ratios?

You		
$a =$	$b =$	$\frac{a}{b} =$
$c =$	$d =$	$\frac{c}{d} =$
$e =$	$f =$	$\frac{e}{f} =$
$g =$	$h =$	$\frac{g}{h} =$

Partner		
$a =$	$b =$	$\frac{a}{b} =$
$c =$	$d =$	$\frac{c}{d} =$
$e =$	$f =$	$\frac{e}{f} =$
$g =$	$h =$	$\frac{g}{h} =$

What Is Your Answer?

3. **IN YOUR OWN WORDS** How can you use a square root to describe the golden ratio? Use the Internet or some other reference to find examples of the golden ratio in art and architecture.

Practice Use what you learned about square roots to complete Exercises 3–5 on page 256.

6.4 Lesson

Check It Out
Lesson Tutorials
BigIdeasMath.com

You can add or subtract radical expressions the same way you combine like terms, such as $5x + 4x = 9x$.

EXAMPLE 1 Adding and Subtracting Square Roots

Reading

Do not assume that radicals that have different radicands cannot be simplified.

An expression such as $2\sqrt{4} + \sqrt{1}$ can easily be simplified.

a. Simplify $5\sqrt{2} + 4\sqrt{2}$.

$5\sqrt{2} + 4\sqrt{2} = (5 + 4)\sqrt{2}$ — Use the Distributive Property.

$= 9\sqrt{2}$ — Simplify.

b. Simplify $2\sqrt{3} - 7\sqrt{3}$.

$2\sqrt{3} - 7\sqrt{3} = (2 - 7)\sqrt{3}$ — Use the Distributive Property.

$= -5\sqrt{3}$ — Simplify.

On Your Own

Now You're Ready
Exercises 6–14

Simplify the expression.

1. $\sqrt{5} + \sqrt{5}$
2. $6\sqrt{10} + 4\sqrt{10}$
3. $2\sqrt{7} - \sqrt{7}$

To simplify square roots that are not perfect squares, use the following property.

Key Idea

Product Property of Square Roots

Algebra $\sqrt{xy} = \sqrt{x} \cdot \sqrt{y}$, where $x, y \geq 0$

Numbers $\sqrt{4 \cdot 3} = \sqrt{4} \cdot \sqrt{3} = 2\sqrt{3}$

EXAMPLE 2 Simplifying Square Roots

Study Tip

A square root is simplified when the radicand has no perfect square factors other than 1.

Simplify $\sqrt{50}$.

$\sqrt{50} = \sqrt{25 \cdot 2}$ — Factor using the greatest perfect square factor.

$= \sqrt{25} \cdot \sqrt{2}$ — Use the Product Property of Square Roots.

$= 5\sqrt{2}$ — Simplify.

On Your Own

Now You're Ready
Exercises 16–20

Simplify the expression.

4. $\sqrt{24}$
5. $\sqrt{45}$
6. $\sqrt{98}$

Laurie's Notes

Introduction

Connect

- **Yesterday:** Students investigated the golden ratio.
- **Today:** Students will use properties of square roots to simplify expressions.

Motivate

- To connect today's lesson to variable expressions, have students work with a partner to simplify the following.

1. $12x - 9x$ $3x$ 2. $8m + 6m$ $14m$

3. $2x(3x)$ $6x^2$ 4. $\frac{12m}{4m}$ 3

Lesson Notes

Example 1

? "What are like terms?" terms with the same variables, each raised to the same exponent

- Explain that square root expressions are treated the same as like terms.
- **Teaching Tip:** Write a parallel problem involving x side-by-side with the square root problem.

$5\sqrt{2} + 4\sqrt{2}$	$5x + 4x$
$9\sqrt{2}$	$9x$

- The radicals can be combined because they have the same radicand. You cannot combine $3x + 2y$, so you cannot combine $3\sqrt{5} + 2\sqrt{7}$.

On Your Own

? "How many square roots of 5 are being added together in $\sqrt{5} + \sqrt{5}$?" two; $\sqrt{5} + \sqrt{5} = 2\sqrt{5}$

Key Idea

- **Teaching Tip:** An example that students find helpful is the following.

$$\sqrt{9 \cdot 4} = \sqrt{9} \cdot \sqrt{4}$$
$$\sqrt{36} - 3 \cdot 2$$
$$6 = 6$$

Example 2

? "What are the factors of 50?" 1, 2, 5, 10, 25, 50

- The only square roots that simplify to whole numbers are those which are perfect squares. Write 50 as $25 \cdot 2$.
- Some students might write 50 as $2 \cdot 25$ (Commutative Property). The result will be $\sqrt{50} = \sqrt{2}\sqrt{25} = \sqrt{2} \cdot 5$. Tell students to write the constant first followed by the square root in the same way that you write the coefficient before the variable.
- **Big Idea:** To simplify the square root, find the greatest perfect square factor.

Goal Today's lesson is simplifying expressions involving square roots.

Lesson Materials
Textbook
• calculators

Start Thinking! and Warm Up

Lesson 6.4 Warm Up
For use before Lesson 6.4

Lesson 6.4 Start Thinking!
For use before Lesson 6.4

In previous courses, you have learned how to simplify fractions. When is a fraction simplified?

Square roots can also be simplified.

A square root is simplified when the number under the radical sign has no perfect square factors other than 1.

Which of the following expressions are simplified? Explain why.

$\sqrt{2}$, $\sqrt{4}$, $\sqrt{10}$, $\sqrt{50}$, $3\sqrt{5}$, $3\sqrt{8}$

Extra Example 1

a. Simplify $6\sqrt{7} + 2\sqrt{7}$. $8\sqrt{7}$

b. Simplify $14\sqrt{13} - 17\sqrt{13}$. $-3\sqrt{13}$

On Your Own

1. $2\sqrt{5}$ **2.** $10\sqrt{10}$

3. $\sqrt{7}$

Extra Example 2

Simplify $\sqrt{162}$. $9\sqrt{2}$

On Your Own

4. $2\sqrt{6}$ **5.** $3\sqrt{5}$

6. $7\sqrt{2}$

Extra Example 3

Simplify $\sqrt{\frac{17}{25}}$. $\frac{\sqrt{17}}{5}$

Extra Example 4

Find the volume of the rectangular prism. $12\ m^3$

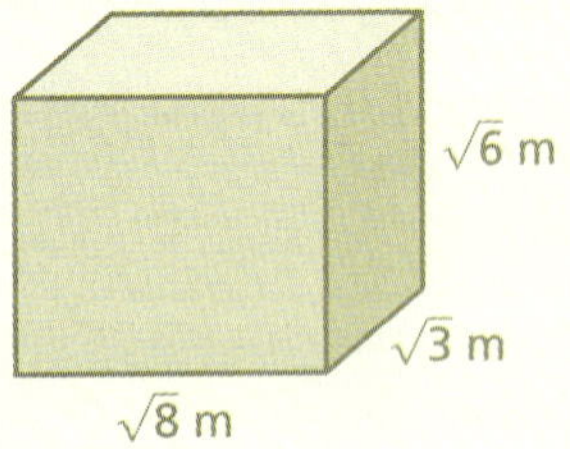

On Your Own

7. $\frac{\sqrt{35}}{6}$ **8.** $\frac{\sqrt{13}}{2}$

9. $\frac{\sqrt{5}}{b}$ **10.** $20\ m^3$

Differentiated Instruction

Visual

Students may have a difficult time understanding that the square root of a number between 0 and 1 is greater than the number. They may think that $\sqrt{\frac{1}{4}} = \frac{1}{16}$, when actually $\sqrt{\frac{1}{16}} = \frac{1}{4}$. Help students to understand this concept by using a 10 by 10 grid to represent the number 1. A square of $\frac{49}{100}$ square units has a side length of $\frac{7}{10}$. Because $\sqrt{\frac{49}{100}} = \frac{7}{10}, \frac{7}{10} > \frac{49}{100}$ (0.7 > 0.49).

Laurie's Notes

Key Idea

- Write the Key Idea.
- **Teaching Tip:** An example that students find helpful is the following.

$$\sqrt{\frac{64}{16}} = \frac{\sqrt{64}}{\sqrt{16}}$$
$$\sqrt{4} = \frac{8}{4}$$
$$2 = 2$$

Example 3

? "How can this problem be rewritten?" listen for the Quotient Property

- Note that only the denominator is a perfect square, so the numerator is left as a square root.

Example 4

? "What is the definition of volume and how do you find the volume of a prism?" Volume is space inside a solid. $V = \ell wh$.

- Write the formula and substitute the variables.
- **Common Misconception:** Students do not always recognize that these expressions can be multiplied. This is the reverse of what was done when they simplified expressions at the beginning of class. Remind students that the Product Property of Square Roots has an equal sign. The left side equals the right side and the right side equals the left side.
- Using the Product Property of Square Roots, multiply the three expressions.
- Finish working through the example as shown.

On Your Own

- For Question 9, remind students that the square root of a squared number is the number. So, the square root of b^2 is b. At this level, you are assuming that $b > 0$. In future math courses, students will learn that $\sqrt{b^2} = |b|$.

Closure

- Simplify the expression.
  $\sqrt{48}$ $4\sqrt{3}$ $\sqrt{10} + 5\sqrt{10}$ $6\sqrt{10}$ $\sqrt{\frac{25}{4}}$ $\frac{5}{2}$

Technology For the Teacher

Dynamic Classroom

The Dynamic Planning Tool
Editable Teacher's Resources at *BigIdeasMath.com*

Key Idea

Quotient Property of Square Roots

Algebra $\sqrt{\dfrac{x}{y}} = \dfrac{\sqrt{x}}{\sqrt{y}}$, where $x \geq 0$ and $y > 0$

Numbers $\sqrt{\dfrac{7}{9}} = \dfrac{\sqrt{7}}{\sqrt{9}} = \dfrac{\sqrt{7}}{3}$

EXAMPLE 3 Simplifying Square Roots

Simplify $\sqrt{\dfrac{11}{16}}$.

$$\sqrt{\frac{11}{16}} = \frac{\sqrt{11}}{\sqrt{16}}$$ Use the Quotient Property of Square Roots.

$$= \frac{\sqrt{11}}{4}$$ Simplify.

EXAMPLE 4 Finding a Volume

Remember

The volume V of a rectangular prism is the product of the area of its base B and its height h.

$V = Bh$

Find the volume of the rectangular prism.

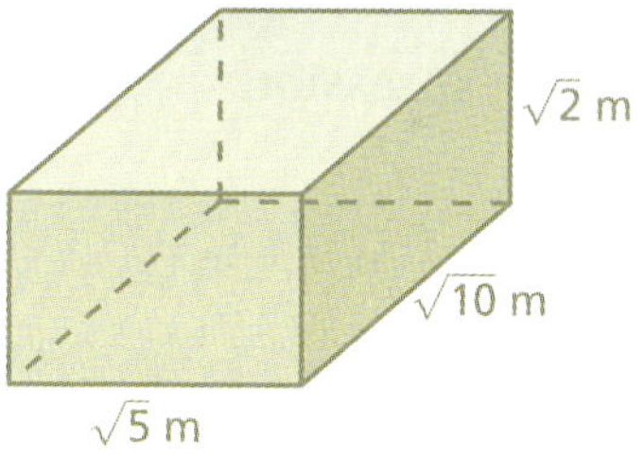

$V = Bh$ Write formula for volume.

$= (\sqrt{5})(\sqrt{10})(\sqrt{2})$ Substitute.

$= \sqrt{5 \cdot 10 \cdot 2}$ Use the Product Property of Square Roots.

$= \sqrt{100}$ Multiply.

$= 10$ Simplify.

∴ The volume is 10 cubic meters.

On Your Own

Now You're Ready
Exercises 21–24

Simplify the expression.

7. $\sqrt{\dfrac{35}{36}}$ **8.** $\sqrt{\dfrac{13}{4}}$ **9.** $\sqrt{\dfrac{5}{b^2}}$

10. WHAT IF? In Example 4, the height of the rectangular prism is $\sqrt{8}$ meters. Find the volume of the prism.

6.4 Exercises

Check It Out
Help with Homework
BigIdeasMath.com

Vocabulary and Concept Check

1. **WRITING** Describe how combining like terms is similar to adding and subtracting square roots.

2. **WRITING** How are the Product Property of Square Roots and the Quotient Property of Square Roots similar?

Practice and Problem Solving

Find the ratio of the side lengths. Is the ratio close to the golden ratio?

3.

4.

5.

Simplify the expression.

(1) 6. $\frac{\sqrt{2}}{9} + \frac{1}{9}$

7. $\frac{\sqrt{7}}{3} + \frac{1}{3}$

8. $\frac{1}{4} + \frac{\sqrt{13}}{4}$

9. $2\sqrt{3} + 4\sqrt{3}$

10. $6\sqrt{7} - 2\sqrt{7}$

11. $\frac{3}{4}\sqrt{5} + \frac{5}{4}\sqrt{5}$

12. $\sqrt{6} - 4\sqrt{6}$

13. $1.5\sqrt{15} - 9.2\sqrt{15}$

14. $\frac{7}{8}\sqrt{11} + \frac{3}{8}\sqrt{11}$

15. **ERROR ANALYSIS** Describe and correct the error in simplifying the expression.

✗ $4\sqrt{5} + 3\sqrt{5} = 7\sqrt{10}$

Simplify the expression.

(2) (3) 16. $\sqrt{18}$

17. $\sqrt{200}$

18. $\sqrt{12}$

19. $\sqrt{48}$

20. $\sqrt{125}$

21. $\sqrt{\frac{23}{64}}$

22. $\sqrt{\frac{65}{121}}$

23. $\sqrt{\frac{17}{49}}$

24. $\sqrt{\frac{22}{c^2}}$

25. **RAIN GUTTER** A rain gutter is made from a single sheet of metal. What is the length of the red cross-section?

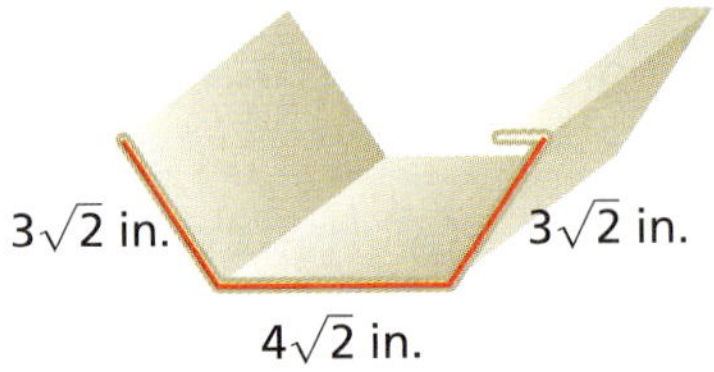

Assignment Guide and Homework Check

Level	Day 1 Activity Assignment	Day 2 Lesson Assignment	Homework Check
Basic	3–5, 34–37	1, 2, 7–29 odd	7, 11, 17, 21, 27
Average	3–5, 34–37	1, 2, 11–31 odd, 30	11, 17, 21, 27, 30
Advanced	3–5, 34–37	1, 2, 13–15, 22–28 even, 29–33	14, 22, 26, 30, 32

Common Errors

- **Exercises 6–14** Students may add the radicands as well as the coefficients, or they may only add the coefficients even if the operation is subtraction. Remind students that they only add the coefficients. Also remind them to perform the correct operation on the numbers.
- **Exercises 16–24** Students may still have difficulty identifying perfect squares to use to simplify the expression. Encourage them to write the first 10 perfect squares at the top of their paper as a reminder and a reference.
- **Exercises 26–28** Students may not know where to start with these problems. Remind them of the order of operations and relate it to the Key Ideas that they learned. They should use the Product Property or the Quotient Property first, and then add or subtract the square roots.

6.4 Record and Practice Journal

Simplify the expression.

1. $\frac{\sqrt{3}}{8} + \frac{1}{8}$ $\frac{1 + \sqrt{3}}{8}$
2. $\frac{2}{9} - \frac{\sqrt{11}}{9}$ $\frac{2 - \sqrt{11}}{9}$
3. $7\sqrt{7} + 3\sqrt{7}$ $10\sqrt{7}$
4. $\frac{3}{2}\sqrt{15} + \frac{1}{2}\sqrt{15}$ $2\sqrt{15}$
5. $12\sqrt{42} - 5\sqrt{42}$ $7\sqrt{42}$
6. $16.4\sqrt{21} - 15.1\sqrt{21}$ $1.3\sqrt{21}$
7. $\sqrt{20}$ $2\sqrt{5}$
8. $\sqrt{32}$ $4\sqrt{2}$
9. $\sqrt{75}$ $5\sqrt{3}$
10. $\sqrt{\frac{29}{81}}$ $\frac{\sqrt{29}}{9}$
11. $\sqrt{\frac{17}{a^2}}$ $\frac{\sqrt{17}}{a}$
12. $\sqrt{40} + 3\sqrt{10}$ $5\sqrt{10}$
13. You build a shed in your backyard.
    a. What is the perimeter of the shed? $28\sqrt{3}$ ft
    b. What is the volume of the shed? $576\sqrt{3}$ ft^3

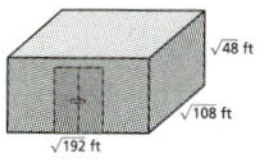

Technology For the Teacher
Answer Presentation Tool
Quiz*Show*

Vocabulary and Concept Check

1–2. Sample answers are given.

1. The square root is like a variable. So, you add or subtract the number in front to simplify.

2. Both allow you to take the square roots of the numbers inside the radical to simplify.

Practice and Problem Solving

3. about 1.62; yes

4. about 1.62; yes

5. about 1.11; no

6. $\frac{\sqrt{2} + 1}{9}$

7. $\frac{\sqrt{7} + 1}{3}$

8. $\frac{1 + \sqrt{13}}{4}$

9. $6\sqrt{3}$

10. $4\sqrt{7}$

11. $2\sqrt{5}$

12. $-3\sqrt{6}$

13. $-7.7\sqrt{15}$

14. $\frac{5}{4}\sqrt{11}$

15. You do not add the radicands.

$4\sqrt{5} + 3\sqrt{5} = 7\sqrt{5}$

16. $3\sqrt{2}$

17. $10\sqrt{2}$

18. $2\sqrt{3}$

19. $4\sqrt{3}$

20. $5\sqrt{5}$

21. $\frac{\sqrt{23}}{8}$

22. $\frac{\sqrt{65}}{11}$

23. $\frac{\sqrt{17}}{7}$

24. $\frac{\sqrt{22}}{c}$

25. $10\sqrt{2}$ in.

Practice and Problem Solving

26. 0

27. $6\sqrt{6}$

28. $\frac{10}{3}\sqrt{7}$

29. 210 ft^3

30. $\sqrt{15}$ or $-\sqrt{15}$

31. **a.** $88\sqrt{2}$ ft

b. 680 ft^2

32. $5\sqrt{15} \approx 19.36$ km

33. See *Taking Math Deeper.*

Fair Game Review

34. 40 m

35. 24 in.

36. 9 cm

37. C

Mini-Assessment

Simplify the expression.

1. $9\sqrt{3} - \sqrt{3}$ $8\sqrt{3}$
2. $\frac{2}{3}\sqrt{5} + \frac{2}{3}\sqrt{5}$ $\frac{4}{3}\sqrt{5}$
3. $\sqrt{18}$ $3\sqrt{2}$
4. $\sqrt{125}$ $5\sqrt{5}$
5. $\sqrt{\frac{21}{196}}$ $\frac{\sqrt{21}}{14}$

Taking Math Deeper

Exercise 33

This problem basically comes down to trying to inscribe a square inside a circle. The square has sides of length s and the circle has a radius of r.

1 Draw a diagram to be sure that students understand the question.

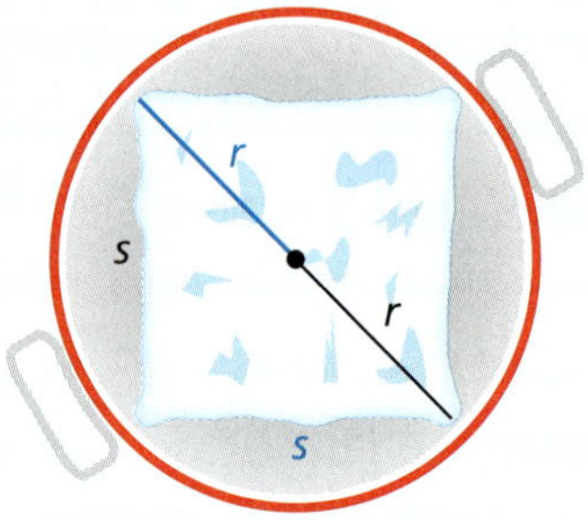

2 Solve the given equation for r.

$$s^2 = 2r^2$$

$$\frac{s^2}{2} = r^2$$

a. $$\frac{s}{\sqrt{2}} = r$$

Note: $s^2 = 2r^2$ came from the Pythagorean Theorem.

$$s^2 + s^2 = (2r)^2$$

$$2s^2 = 4r^2$$

3 Find r when $s = \sqrt{98}$ inches.

$$r = \frac{s}{\sqrt{2}}$$

$$= \frac{\sqrt{98}}{\sqrt{2}}$$

$$= \sqrt{\frac{98}{2}}$$

$$= \sqrt{49}$$

$$= 7$$

b. The cooler must have a radius of 7 inches (or more) to hold the block of ice.

Project

Compare the time it takes for an ice cube to melt inside a cooler with the time it takes for it to melt outside the cooler. Draw a picture to go with your comparison.

Reteaching and Enrichment Strategies

If students need help. . .	If students got it. . .
Resources by Chapter • Practice A and Practice B • Puzzle Time Record and Practice Journal Practice Differentiating the Lesson Lesson Tutorials Skills Review Handbook	Resources by Chapter • Enrichment and Extension • School-to-Work Start the next section

Simplify the expression.

26. $3\sqrt{5} - \sqrt{45}$

27. $\sqrt{24} + 4\sqrt{6}$

28. $\frac{4}{3}\sqrt{7} + \sqrt{28}$

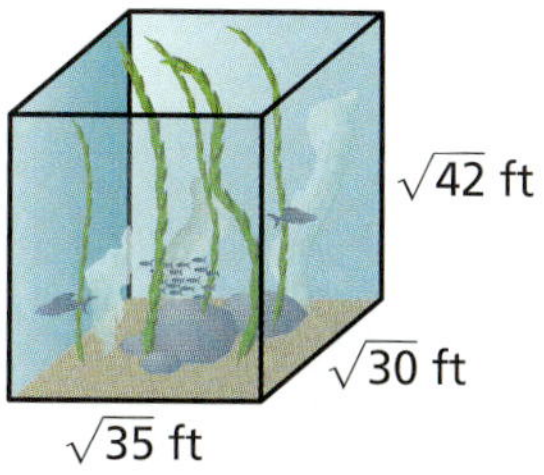

29. VOLUME What is the volume of the aquarium (in cubic feet)?

30. RATIO The ratio $3:x$ is equivalent to the ratio $x:5$. What are the possible values of x?

31. BILLBOARD The billboard has the shape of a rectangle.

a. What is the perimeter of the billboard?

b. What is the area of the billboard?

The height of Mt. Fuji is 3.8 kilometers.

32. MT. FUJI Mt. Fuji is in the shape of a cone with a volume of about 475π cubic kilometers. What is the radius of the base of Mt. Fuji?

33. **Geometry** A block of ice is in the shape of a square prism. You want to put the block of ice in a cylindrical cooler. The equation $s^2 = 2r^2$ represents the minimum radius r needed for the block of ice with side length s to fit in the cooler.

a. Solve the equation for r.

b. Use the equation in part (a) to find the minimum radius needed when the side length of the block of ice is $\sqrt{98}$ inches.

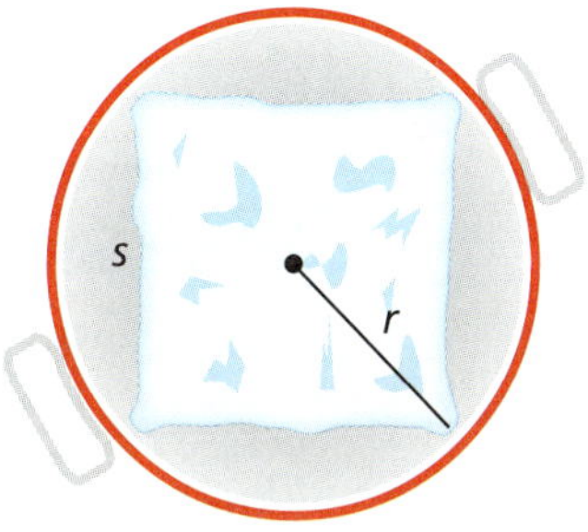

Fair Game Review What you learned in previous grades & lessons

Find the missing length of the triangle. *(Section 6.2)*

34.

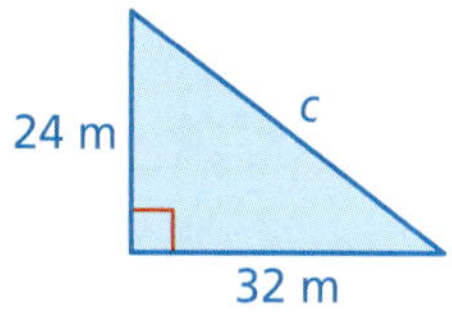

35.

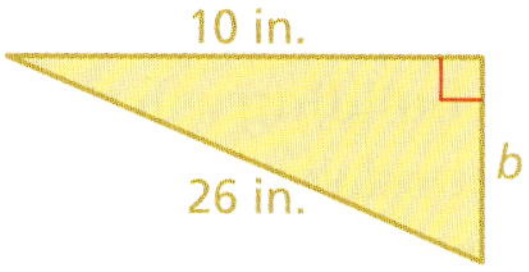

36.

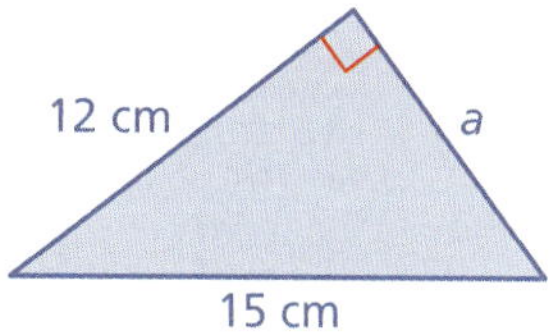

37. MULTIPLE CHOICE Where is $-\sqrt{110}$ on a number line? *(Section 6.3)*

Ⓐ Between -9 and -10

Ⓑ Between 9 and 10

Ⓒ Between -10 and -11

Ⓓ Between 10 and 11

6.5 Using the Pythagorean Theorem

Essential Question How can you use the Pythagorean Theorem to solve real-life problems?

Share Your Work at... My.BigIdeasMath.com

1 ACTIVITY: Using the Pythagorean Theorem

Work with a partner.

a. A baseball player throws a ball from second base to home plate. How far does the player throw the ball? Include a diagram showing how you got your answer. Decide how many decimal points of accuracy are reasonable. Explain your reasoning.

b. The distance from the pitcher's mound to home plate is 60.5 feet. Does this form a right triangle with first base? Explain your reasoning.

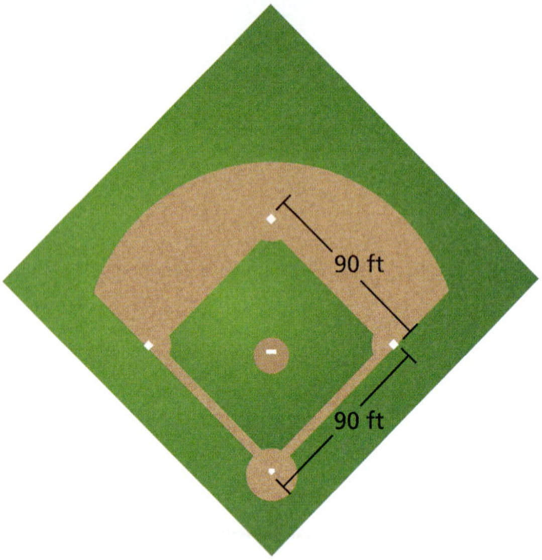

2 ACTIVITY: Firefighting and Ladders

Work with a partner.

The recommended angle for a firefighting ladder is 75°.

When a 110-foot ladder is put up against a building at this angle, the base of the ladder is about 28 feet from the building.

The base of the ladder is 8 feet above the ground.

How high on the building will the ladder reach? Round your answer to the nearest tenth.

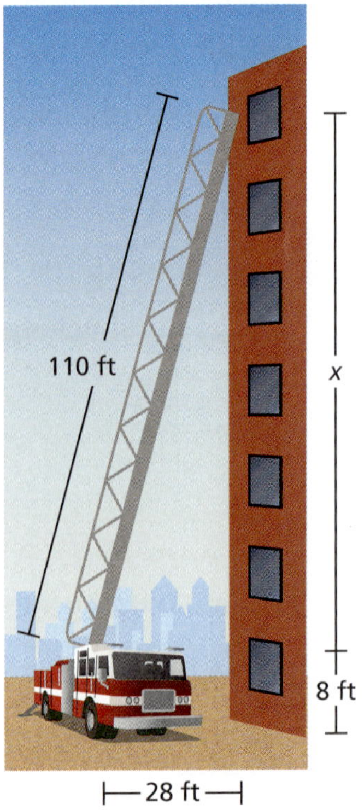

Laurie's Notes

Introduction

For the Teacher

- **Goal:** Students will explore solving real-life problems that can be modeled by a right triangle.

Motivate

- ? "How many of you have ridden on a fire truck?" anticipate a small number
- ? "Did you know that you can hire a fire truck for a party, reunion, or other special events?" unlikely

Activity Notes

Activity 1

- ? "In baseball, which is farther, home to first base or home to second base? How do you know?" Home to second is the hypotenuse of a right triangle and home to first is a leg.
- Not all students will realize there is a right triangle because the third side of the triangle is not drawn.
- ? "What are the dimensions of a major league *diamond*?" 90 feet between the bases
- The baseball *diamond* is a square. Draw and label the right triangle. Students should be ready to work the problem with their partner.
- ? "What is the exact distance from second to home?" $\sqrt{16{,}200}$ ft
- In discussing the approximate distance, round to the nearest whole number, 127 feet.
- **Part (b):** Have students share their reasoning. If students use the Pythagorean Theorem to find the distance from the pitcher's mound to first base, remind them that you need to know it is a right triangle first. You cannot use the theorem to find the length, and then use the length to prove it is a right triangle. That is circular reasoning.

Activity 2

- In this activity, the right angle is not drawn. Ask students to make a sketch of the problem. Check to see that they represent the base of the triangle 8 feet off the ground.
- ? "What are the measures of the three angles of this triangle?" 90°, 75°, 15°
- ? "What is the relationship between the two acute angles?" complementary
- Students should use the Pythagorean Theorem to solve for x, a leg of the right triangle.

Activity Materials
Textbook
• calculators

Start Thinking! and Warm Up

Activity 6.5 Start Thinking! For use before Activity 6.5

Activity 6.5 Warm Up For use before Activity 6.5

Find the missing length of the triangle. Round your answer to the nearest tenth.

1.

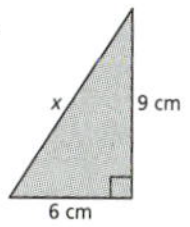

2.

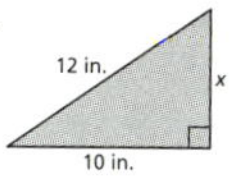

3.

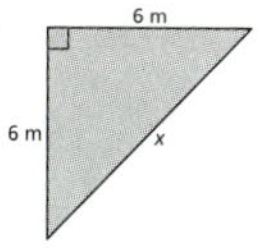

4.

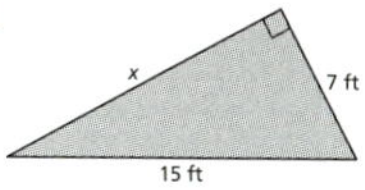

6.5 Record and Practice Journal

Essential Question How can you use the Pythagorean Theorem to solve real-life problems?

1 ACTIVITY: Using the Pythagorean Theorem

Work with a partner.

a. A baseball player throws a ball from second base to home plate. How far does the player throw the ball? Include a diagram showing how you got your answer. Decide how many decimal points of accuracy are reasonable. Explain your reasoning.

about 127 ft

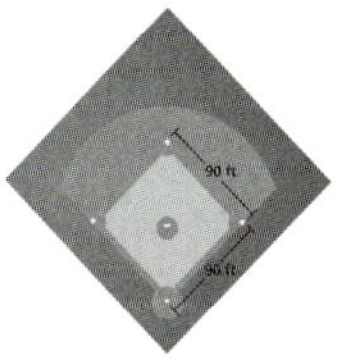

b. The distance from the pitcher's mound to home plate is 60.5 feet. Does this form a right triangle with first base? Explain your reasoning.

no; To form a right triangle, it would need to be half the distance from home plate to second base, which is about 63.5 feet.

English Language Learners

Visual Aid

Create a poster to display the connections between a right triangle and the Pythagorean Theorem. Use color to indicate corresponding parts of the triangle and the equation. Give examples of solving for the hypotenuse and solving for a leg of the triangle.

6.5 Record and Practice Journal

2 ACTIVITY: Firefighting and Ladders

Work with a partner.

The recommended angle for a firefighting ladder is 75°.

When a 110-foot ladder is put up against a building at this angle, the base of the ladder is about 28 feet from the building.

The base of the ladder is 8 feet above the ground.

How high on the building will the ladder reach? Round your answer to the nearest tenth.

114.4 ft

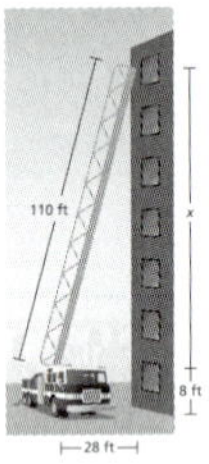

3 ACTIVITY: Finding Perimeters

Work with a partner.

Find the perimeter of each figure. Round your answer to the nearest tenth. Did you use the Pythagorean Theorem? If so, explain.

a. Right triangle b. Trapezoid c. Parallelogram

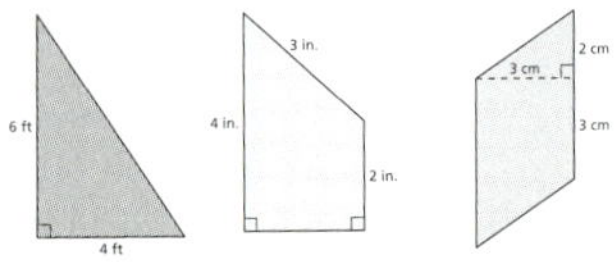

a. 17.2 ft b. 11.2 in. c. 17.2 cm

4 ACTIVITY: Writing a Formula

Work with a partner.

a. Write a formula for the area of an equilateral triangle with side length s.

$\frac{\sqrt{3}}{4}s^2$

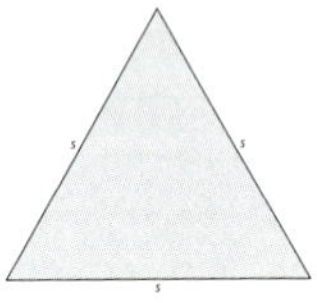

b. Use your formula to find the area of an equilateral triangle with a side length of 10 inches.

about 43.3 in.^{2}

What Is Your Answer?

5. **IN YOUR OWN WORDS** How can you use the Pythagorean Theorem to solve real-life problems?

Sample answer: height of a building, length of a wheelchair ramp

6. Describe a situation in which you could use the Pythagorean Theorem to help make decisions. Give an example of a real-life problem.

Sample answer: measurement of a TV

Laurie's Notes

Activity 3

- ? "How do you find the perimeter of a polygon?" Sum the lengths of the sides.
- ? "Do you know, or can you find, the lengths of all sides of the polygons? Explain." Yes; Use the Pythagorean Theorem to solve for the missing hypotenuse in parts (a) and (c) and for the missing leg in part (b).
- This is the first time students have to solve more than a trivial problem in order to find a missing side of a polygon.

Activity 4

- ? "How do you find the area of a triangle?" $A = \frac{1}{2}bh$
- ? "Can you sketch in a height for this triangle? Do you know anything about the height and where it intersects the base?" Answers vary.
- Students need to recognize that for an equilateral triangle, the height is perpendicular to the base and intersects it at its midpoint. The height will make the two right triangles shown. You may need to work through the mathematics of this problem. The use of variables in the general case is very challenging for students.

$$\left(\frac{s}{2}\right)^2 + h^2 = s^2$$

$$h^2 = s^2 - \left(\frac{s}{2}\right)^2$$

$$h^2 = s^2 - \frac{s^2}{4}$$

$$h^2 = \frac{3}{4}s^2$$

$$h = \sqrt{\frac{3}{4}s^2}$$

$$h = \frac{\sqrt{3}}{2}s$$

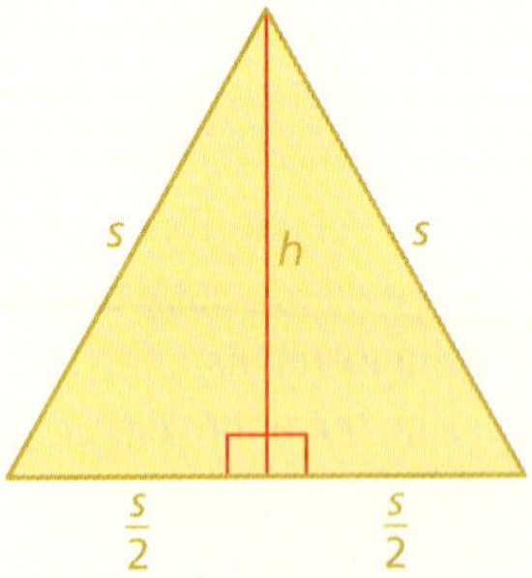

- Once students know the height of an equalateral triangle in terms of its side length s, they can write the formula for the area.
- Give students time to work part (b) with their partner.

Closure

- A little league baseball diamond is 60 feet between the bases. What is the distance from home to second base? about 85 feet

Technology For the Teacher

Dynamic Classroom

The Dynamic Planning Tool

Editable Teacher's Resources at *BigIdeasMath.com*

3 ACTIVITY: Finding Perimeters

Work with a partner.

Find the perimeter of each figure. Round your answer to the nearest tenth. Did you use the Pythagorean Theorem? If so, explain.

a. Right triangle

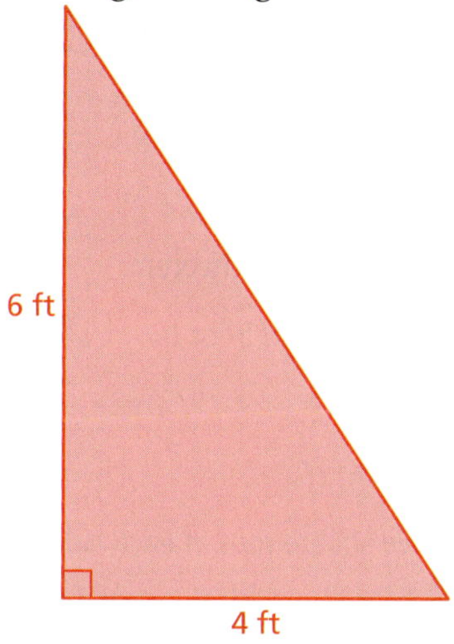

b. Trapezoid

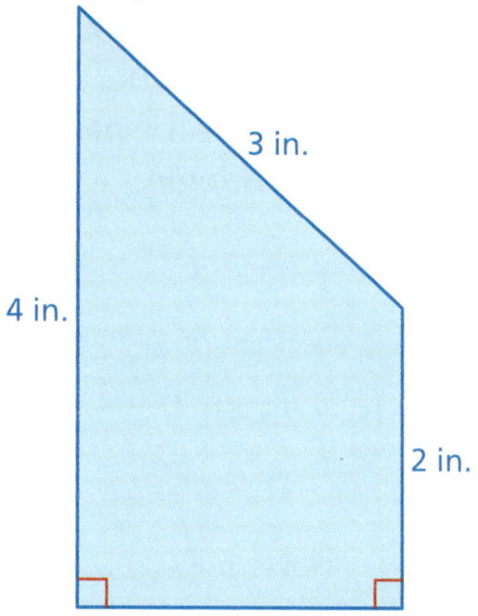

c. Parallelogram

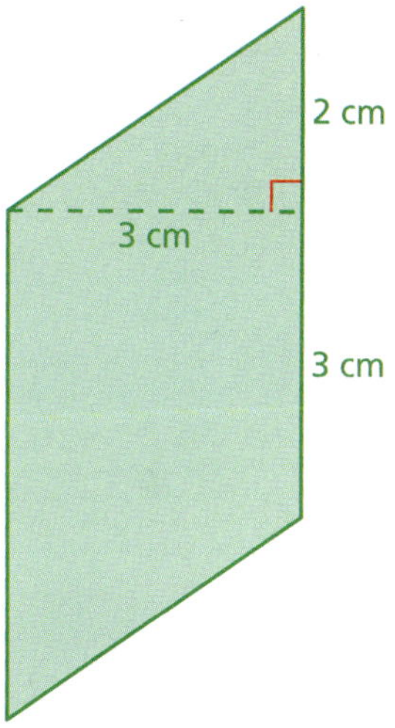

4 ACTIVITY: Writing a Formula

Work with a partner.

a. Write a formula for the area of an equilateral triangle with side length s.

b. Use your formula to find the area of an equilateral triangle with a side length of 10 inches.

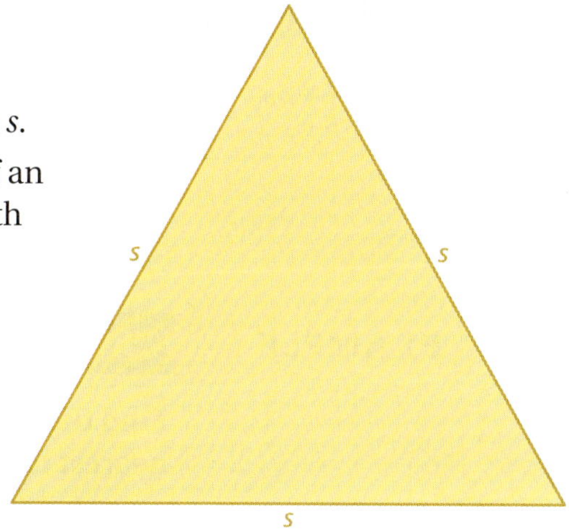

What Is Your Answer?

5. **IN YOUR OWN WORDS** How can you use the Pythagorean Theorem to solve real-life problems?

6. Describe a situation in which you could use the Pythagorean Theorem to help make decisions. Give an example of a real-life problem.

Practice Use what you learned about using the Pythagorean Theorem to complete Exercises 3–5 on page 262.

6.5 Lesson

Check It Out
Lesson Tutorials
BigIdeasMath.com

EXAMPLE 1 Finding a Distance in a Coordinate Plane

Key Vocabulary
Pythagorean triple, p. 261

The park is 5 miles east of your home. The library is 4 miles north of the park. How far is your home from the library? Round your answer to the nearest tenth.

Plot a point for your home at the origin in a coordinate plane. Then plot points for the locations of the park and the library to form a right triangle.

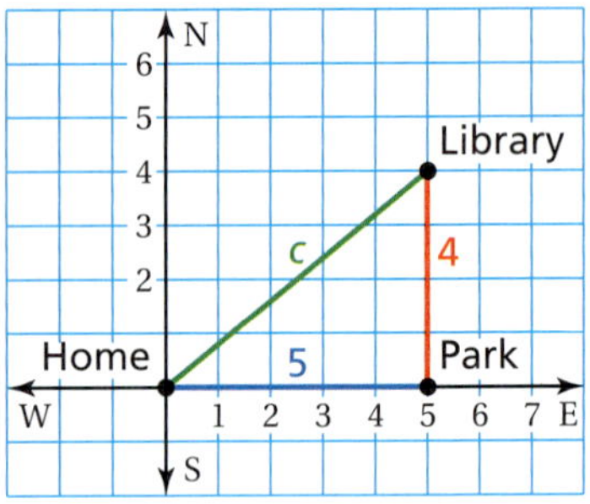

$a^2 + b^2 = c^2$	Write the Pythagorean Theorem.
$4^2 + 5^2 = c^2$	Substitute 4 for a and 5 for b.
$16 + 25 = c^2$	Evaluate powers.
$41 = c^2$	Add.
$\sqrt{41} = \sqrt{c^2}$	Take positive square root of each side.
$6.4 \approx c$	Use a calculator.

Your home is about 6.4 miles from the library.

On Your Own

Now You're Ready
Exercises 6–8

1. The post office is 3 miles west of your home. Your school is 2 miles north of the post office. How far is your home from your school? Round your answer to the nearest tenth.

EXAMPLE 2 Real-Life Application

Find the height of the firework. Round your answer to the nearest tenth.

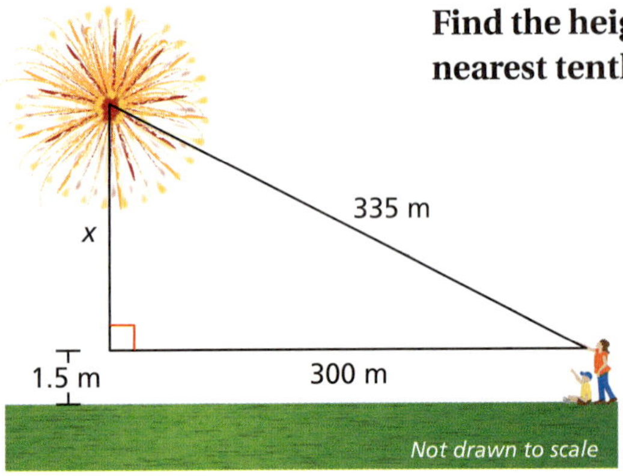

$a^2 + b^2 = c^2$	Write the Pythagorean Theorem.
$x^2 + 300^2 = 335^2$	Substitute.
$x^2 + 90{,}000 = 112{,}225$	Evaluate powers.
$x^2 = 22{,}225$	Subtract 90,000 from each side.
$\sqrt{x^2} = \sqrt{22{,}225}$	Take positive square root of each side.
$x \approx 149.1$	Use a calculator.

The height of the firework is about $149.1 + 1.5 = 150.6$ meters.

Multi-Language Glossary at BigIdeasMath.com.

Laurie's Notes

Introduction

Connect

- **Yesterday:** Students investigated real-life problems that required the Pythagorean Theorem to solve.
- **Today:** Students will solve right triangle problems and determine if three side lengths form a right triangle.

Motivate

- Explain the converse of an if-then statement. The converse of *if p then q* is *if q then p*. Example:

  Original—If all sides of the triangle are congruent, then the triangle is equilateral.

  Converse—If the triangle is equilateral, then all three sides are congruent.
- Both the statement and its converse are true. Putting the two statements together gives the definition of an equilateral triangle.

? "If an if-then statement is true, do you think the converse is always true?" Answers will vary.

- The converse of a true statement is not always true.
- Ask students to think of an example (not related to math) where the converse is true and an example where the converse is false. Something related to your school would be helpful. Converse true: If the doors are closed, then you must open them to get in. Converse not true: If it is snowing, then it is cold outside.

Lesson Notes

Example 1

- Ask a volunteer to read the problem.
- Draw a coordinate plane and sketch the example.

? "What are you solving for, a leg or the hypotenuse?" hypotenuse

- Set up the problem and solve.
- Before using a calculator, ask students to estimate $\sqrt{41}$ using a number line.

? **Extension:** "If your house had been located at $(-2, 3)$, what would be the coordinates of the park and the library?" $(3, 3)$ and $(3, 7)$ "What would be the distance to the library?" the same, 6.4 miles

On Your Own

? "Does it matter where you locate your house?" no

Example 2

- Information to solve the problem is given in the illustration.

? "What are you solving for, a leg or the hypotenuse?" leg

- Set up the problem and solve.

? "Why do you add 1.5 to the answer?" to account for height of person

Goal Today's lesson is using the Pythagorean Theorem to find distances.

Lesson Materials
Textbook
• calculators

Start Thinking! and Warm Up

Lesson 6.5 Warm Up
For use before Lesson 6.5

Lesson 6.5 Start Thinking!
For use before Lesson 6.5

Write a word problem that can be solved using the Pythagorean Theorem. Be sure to include a sketch of the situation.

Extra Example 1

In Example 1, your friend's house is 6 miles east of the library. A grocery store is 5 miles north of your friend's house. How far is the grocery store from the library? Round your answer to the nearest tenth. about 7.8 mi

On Your Own

1. 3.6 mi

Extra Example 2

In Example 2, the distance between you and the firework is 370 meters. Estimate the height of the firework. Round your answer to the nearest tenth.
about 218.1 m

On Your Own

2. 181.8 m

Extra Example 3

Tell whether the triangle is a right triangle. *not* a right triangle

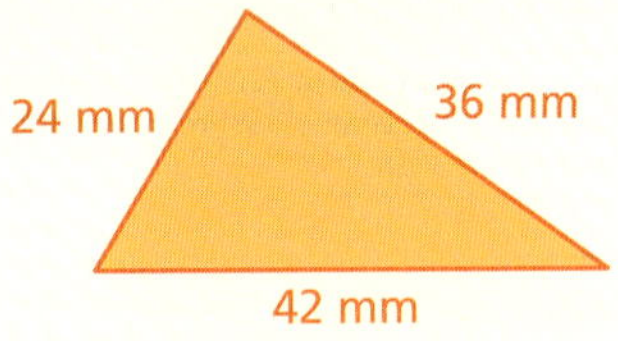

On Your Own

3. yes **4.** no

5. no **6.** yes

Differentiated Instruction

Kinesthetic

Give students copies of the diagram. Have them cut out and physically move the pieces of the two smaller squares to create the larger square. A hands-on approach will help kinesthetic learners remember the relationship.

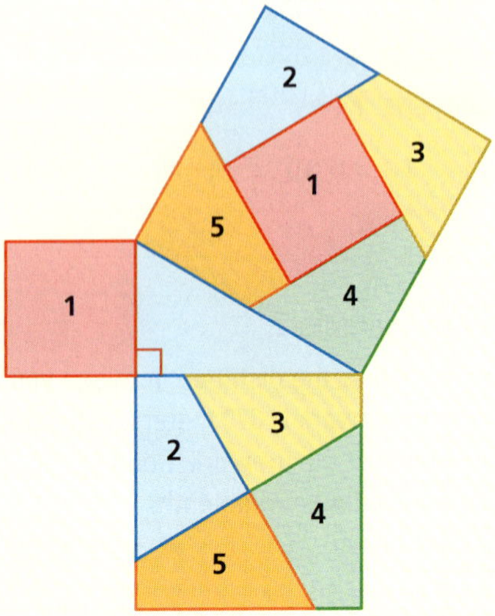

Laurie's Notes

On Your Own

- **Neighbor Check:** Have students work independently and then have their neighbor check their work. Have students discuss any discrepancies.

Key Idea

- **Pythagorean Triple:** A Pythagorean triple is three positive integers, *a*, *b*, and *c*, where $a^2 + b^2 = c^2$. Remembering the basic triples, and knowing that multiplying each number by the same scalar produces another triple, can be very helpful.
  Example: $3^2 + 4^2 = 5^2 \rightarrow$ multiply by 2 $\rightarrow 6^2 + 8^2 = 10^2$
- The Converse of the Pythagorean Theorem says that when the relationship $a^2 + b^2 = c^2$ is true, you have a right triangle.
- Previously, you were given a right triangle and concluded that $a^2 + b^2 = c^2$ was true. Now you are given side lengths of a triangle. If $a^2 + b^2 = c^2$ is true, then the triangle is a right triangle.

Example 3

- If the three numbers satisfy the theorem, then it is a right triangle. Explain that you are not using eyesight to decide if it is a right triangle.
- Substitute the side lengths for each triangle. Remind students that the longest side is substituted for *c*.
- Because 9, 40, and 41 satisfy the theorem, they are Pythagorean triples and the triangle is a right triangle.
- Because 12, 18, and 24 do not satisfy the theorem, it is not a right triangle.
- **Extension:** The 12, 18, 24 triangle is similar to a triangle with side lengths 2, 3, and 4 (scale factor is 6). The 2, 3, 4 triangle is not a right triangle either.

On Your Own

- **Common Error:** Students may not substitute the longest side for *c*. This is particularly true for Question 6 because the measures are listed longest to shortest.

Closure

- **Writing Prompt:** To determine if three lengths are the sides of a right triangle, …

Technology For the Teacher

Dynamic Classroom

The Dynamic Planning Tool
Editable Teacher's Resources at *BigIdeasMath.com*

On Your Own

Now You're Ready
Exercises 9–11

2. **WHAT IF?** In Example 2, the distance between you and the firework is 350 meters. Find the height of the firework. Round your answer to the nearest tenth.

A **Pythagorean triple** is a set of three positive integers a, b, and c where $a^2 + b^2 = c^2$.

Key Idea

Converse of the Pythagorean Theorem

If the equation $a^2 + b^2 = c^2$ is true for the side lengths of a triangle, then the triangle is a right triangle.

When using the converse of the Pythagorean Theorem, always substitute the length of the longest side for c.

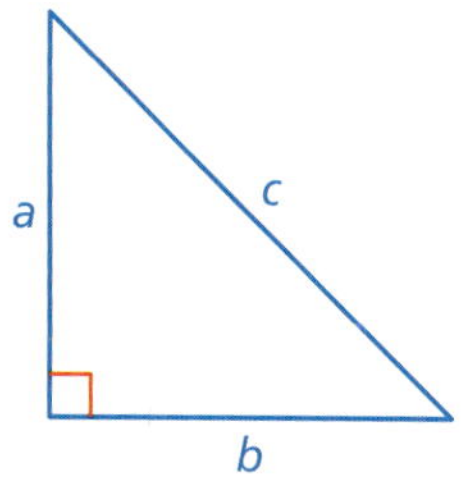

EXAMPLE 3 Identifying a Right Triangle

Tell whether the given triangle is a right triangle.

a.

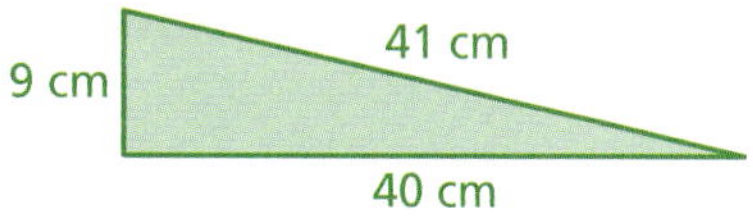

$$a^2 + b^2 = c^2$$
$$9^2 + 40^2 \stackrel{?}{=} 41^2$$
$$81 + 1600 \stackrel{?}{=} 1681$$
$$1681 = 1681 \checkmark$$

It *is* a right triangle.

b.

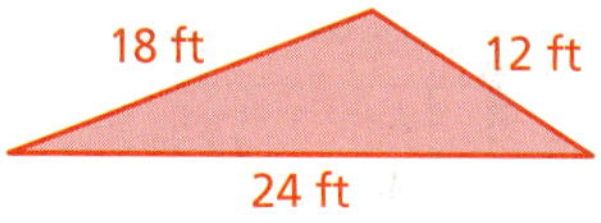

$$a^2 + b^2 = c^2$$
$$12^2 + 18^2 \stackrel{?}{=} 24^2$$
$$144 + 324 \stackrel{?}{=} 576$$
$$468 \neq 576 \text{ ✗}$$

It is *not* a right triangle.

On Your Own

Now You're Ready
Exercises 13–18

Tell whether the triangle with the given side lengths is a right triangle.

3.

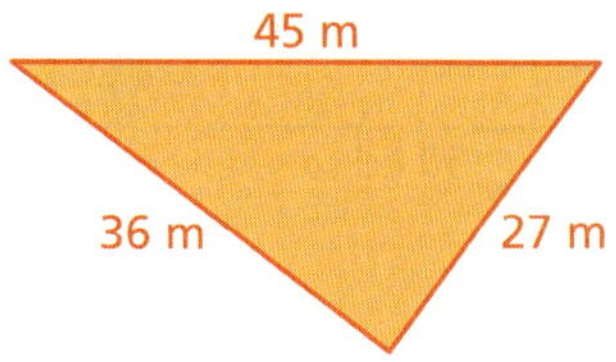

4.

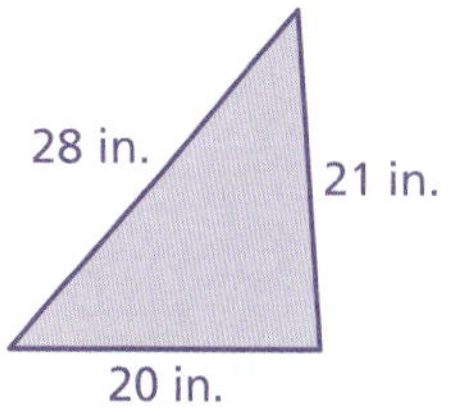

5. $1\frac{1}{2}$ yd, $2\frac{1}{2}$ yd, $3\frac{1}{2}$ yd

6. 1.25 mm, 1 mm, 0.75 mm

6.5 Exercises

Check It Out
Help with Homework
BigIdeasMath.com

Vocabulary and Concept Check

1. **WRITING** How can the Pythagorean Theorem be used to find distances in a coordinate plane?

2. **WHICH ONE DOESN'T BELONG?** Which set of numbers does *not* belong with the other three? Explain your reasoning.

3, 6, 8	6, 8, 10	5, 12, 13	7, 24, 25

Practice and Problem Solving

Find the perimeter of the figure. Round your answer to the nearest tenth.

3. Right triangle

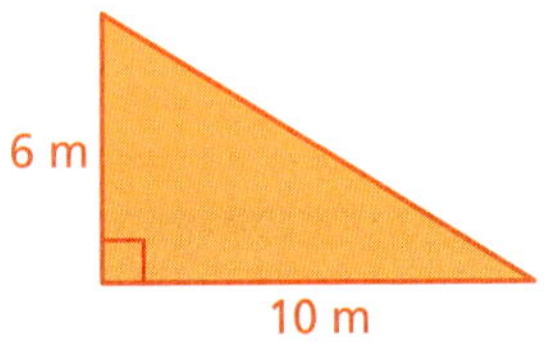

4. Parallelogram

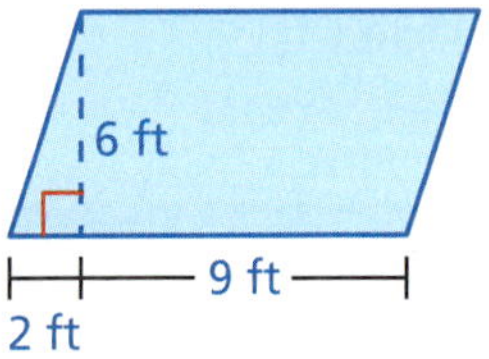

5. Square

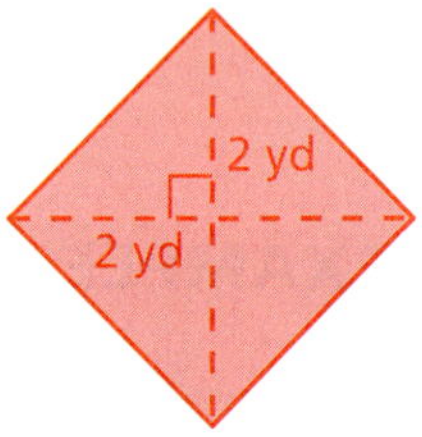

Find the distance d. Round your answer to the nearest tenth.

1 6.

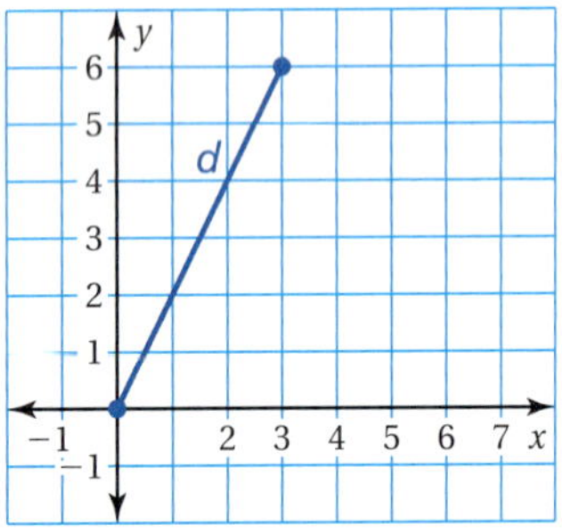

7.

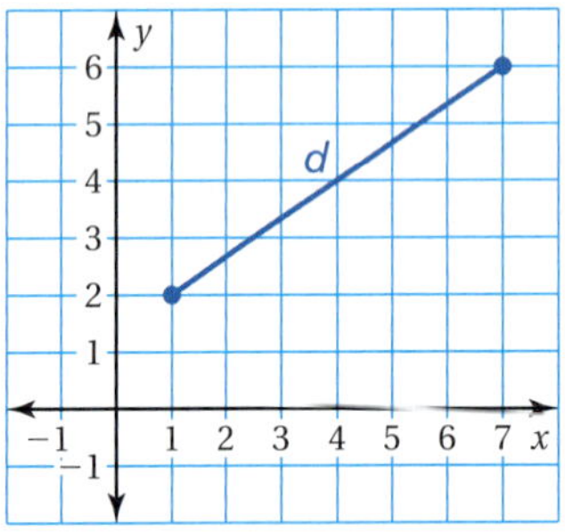

8.

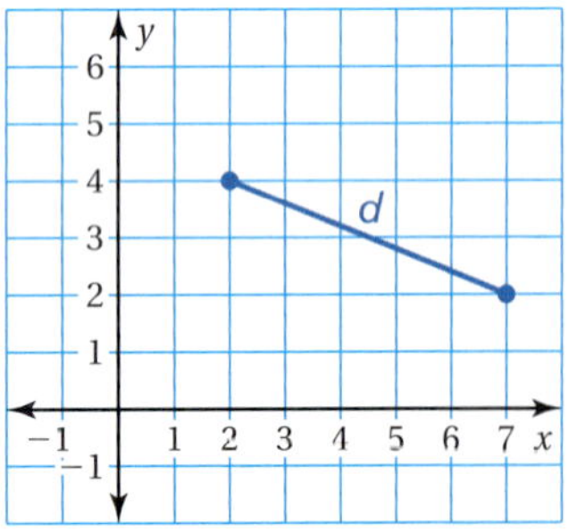

Find the height x. Round your answer to the nearest tenth.

2 9.

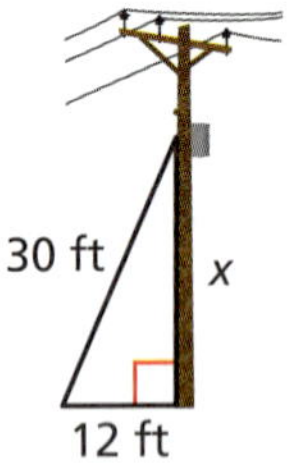

10.

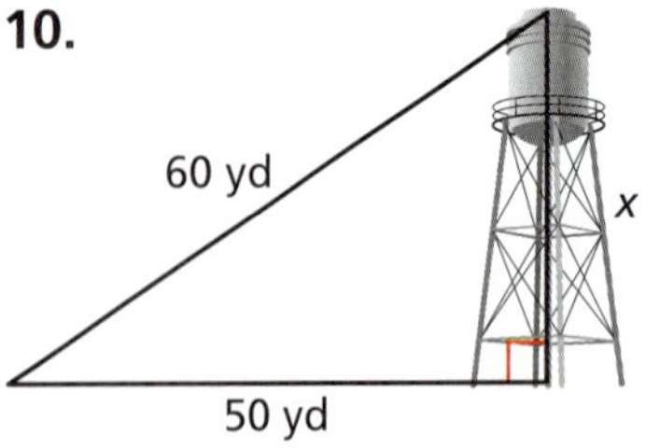

11.

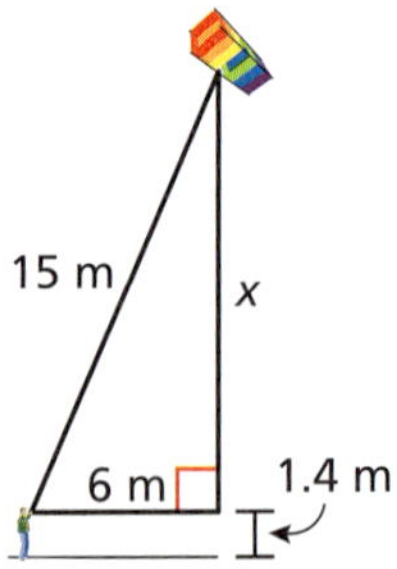

12. **BICYCLE** You ride your bicycle along the outer edge of a park. Then you take a shortcut back to where you started. Find the length of the shortcut. Round your answer to the nearest tenth.

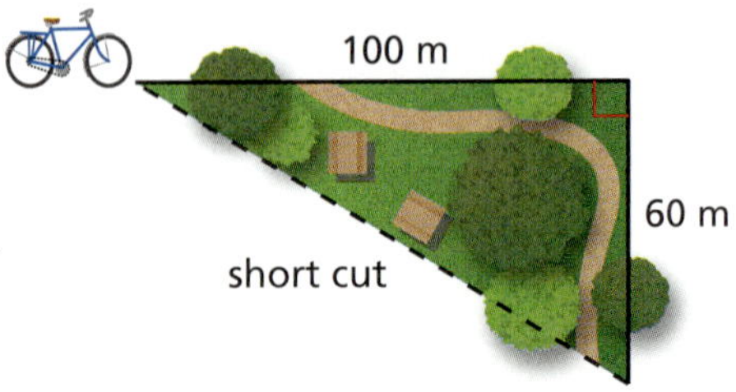

Assignment Guide and Homework Check

Level	Day 1 Activity Assignment	Day 2 Lesson Assignment	Homework Check
Basic	3–5, 23–26	1, 2, 6–13, 15, 16	6, 10, 12, 16
Average	3–5, 23–26	1, 2, 6–11, 13–19 odd, 20	6, 10, 17, 20
Advanced	3–5, 23–26	1, 2, 8–18 even, 19–22	8, 10, 14, 20

Common Errors

- **Exercises 6–8** Students may not know how to find the distance because there are no triangles drawn. Remind them of the arrows that they drew for change in x and change in y when finding the slope of a line. Encourage them to use a similar technique to find the distance of the line. Instead of writing a ratio, they will use the Pythagorean Theorem.
- **Exercise 11** Students may only find the value of x and forget to add the additional height labeled in the picture. Remind them that they are not solving for x, but they are finding the height of the object. So, they need to include any additional heights after using the Pythagorean Theorem.
- **Exercises 13–15** Students may substitute the wrong value for c in the Pythagorean Theorem. Remind them that c will be the longest side, so they should substitute the greatest value for c.

6.5 Record and Practice Journal

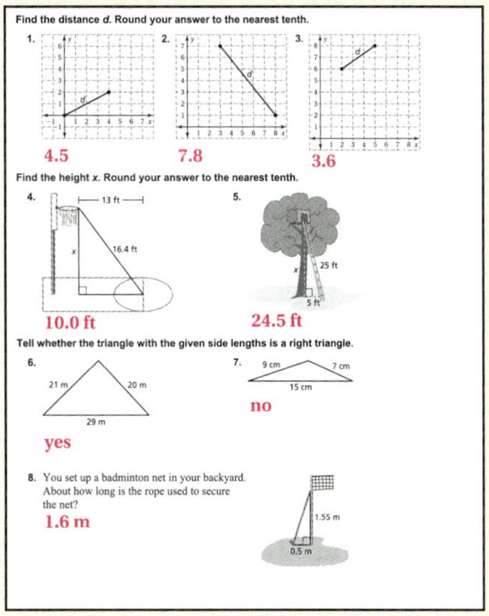

Find the distance d. Round your answer to the nearest tenth.

1. **4.5**
2. **7.8**
3. **3.6**

Find the height x. Round your answer to the nearest tenth.

4. **10.0 ft**
5. **24.5 ft**

Tell whether the triangle with the given side lengths is a right triangle.

6. **yes**
7. **no**

8. You set up a badminton net in your backyard. About how long is the rope used to secure the net? **1.6 m**

Technology For the Teacher

Answer Presentation Tool

Quiz*Show*

Vocabulary and Concept Check

1. *Sample answer:* You can plot a point at the origin and then draw lengths that represent the legs. Then, you can use the Pythagorean Theorem to find the hypotenuse of the triangle.
2. 3, 6, 8; It is the only set that is not a Pythagorean triple.

Practice and Problem Solving

3. 27.7 m
4. 34.6 ft
5. 11.3 yd
6. 6.7 units
7. 7.2 units
8. 5.4 units
9. 27.5 ft
10. 33.2 yd
11. 15.1 m
12. 116.6 m
13. yes
14. yes
15. no
16. no
17. yes
18. yes

Practice and Problem Solving

19. 12.8 ft

20. See *Taking Math Deeper.*

21. **a.** *Sample answer:* 5 in., 7 in., 3 in.

b. *Sample answer:* $BC \approx 8.6$ in.; $AB \approx 9.1$ in.

c. Check students' work.

22. See Additional Answers.

Fair Game Review

23. mean: 13; median: 12.5; mode: 12

24. mean: 21; median: 21; no mode

25. mean: 58; median: 59; mode: 59

26. B

Mini-Assessment

Estimate the height. Round your answer to the nearest tenth.

1. about 39.8 ft

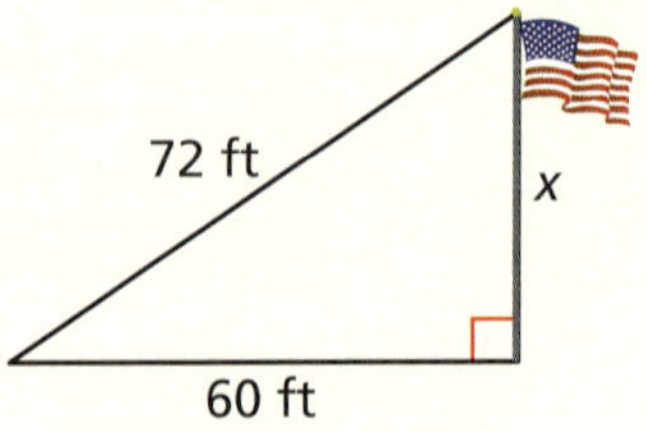

2. about 20.5 ft

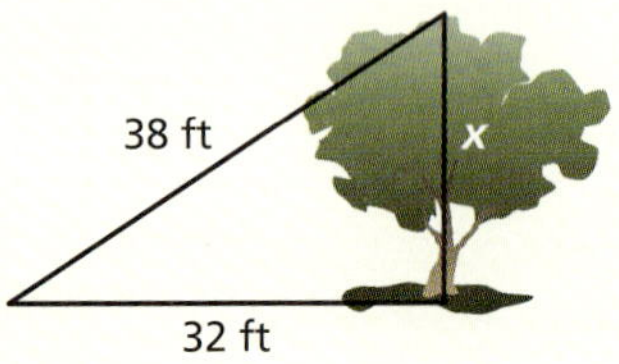

Taking Math Deeper

Exercise 20

At first this seems like a simple question. Plane A is 5 kilometers from the tower and Plane B is 7 kilometers from the tower. So, Plane A seems closer. However, on second glance, you see that Plane A is much higher than Plane B. So, to see which is closer, you need to compute the diagonal distance of each.

1 Find the distance for Plane A.

$$x^2 = 5^2 + 6.1^2$$
$$x^2 = 62.21$$
$$x \approx 7.89 \text{ km}$$

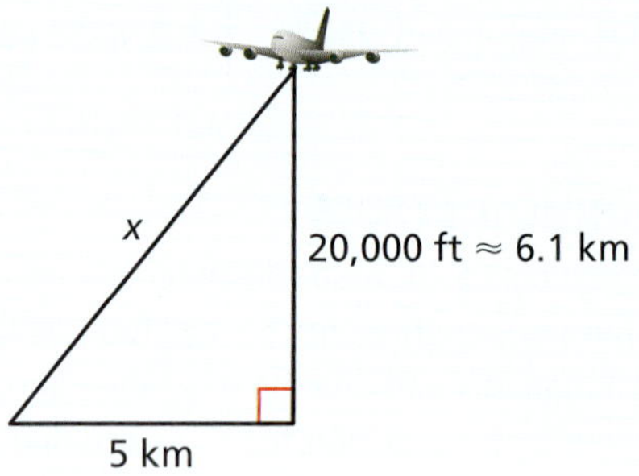

2 Find the distance for Plane B.

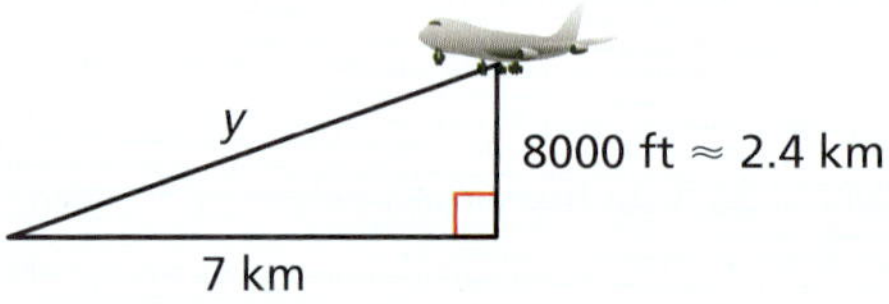

$$y^2 = 7^2 + 2.4^2$$
$$y^2 = 54.76$$
$$y = 7.4 \text{ km}$$

3 Answer the question.

Plane B is slightly closer to the tower.

Reteaching and Enrichment Strategies

If students need help...	If students got it...
Resources by Chapter • Practice A and Practice B • Puzzle Time Record and Practice Journal Practice Differentiating the Lesson Lesson Tutorials Skills Review Handbook	Resources by Chapter • Enrichment and Extension • School-to-Work • Financial Literacy Start the next section

Tell whether the triangle with the given side lengths is a right triangle.

13.

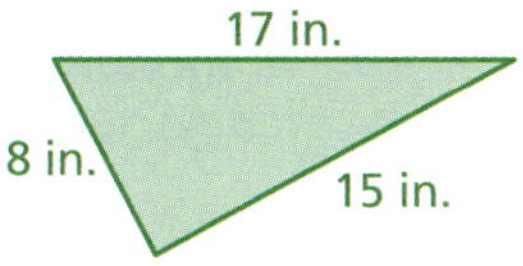

14.

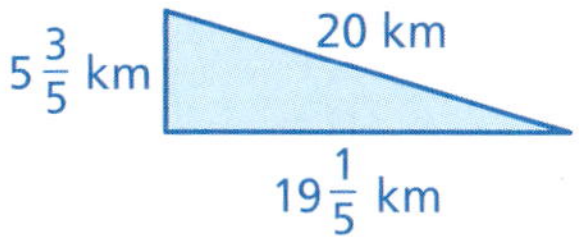

15.

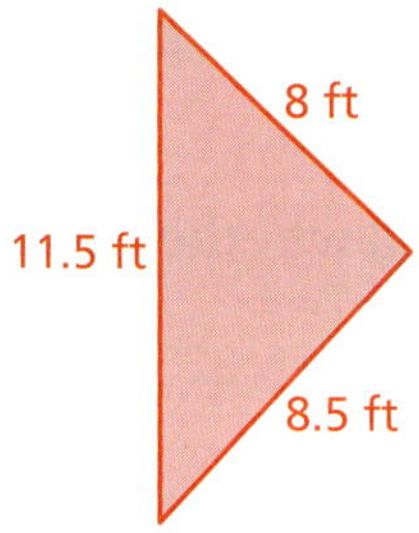

16. 14 mm, 19 mm, 23 mm

17. $\frac{9}{10}$ mi, $1\frac{1}{5}$ mi, $1\frac{1}{2}$ mi

18. 1.4 m, 4.8 m, 5 m

19. STAIRS There are 12 steps in the staircase. Find the distance from point *A* to point *B* (in feet). Round your answer to the nearest tenth.

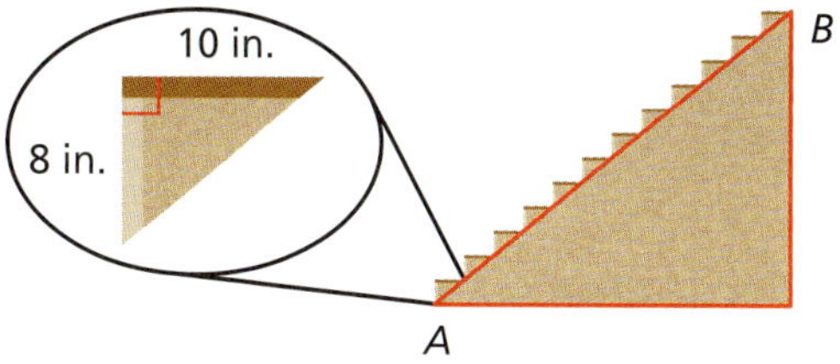

20. AIRPORT Which plane is closer to the tower? Explain.

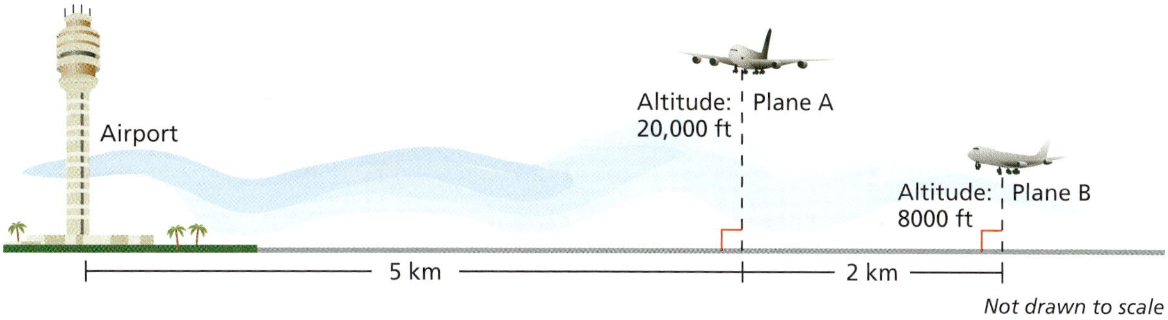

21. PROJECT Find a shoebox or some other small box.

a. Measure the dimensions of the box.

b. Without measuring, find length *BC* and length *AB*.

c. Use a piece of string and a ruler to check the lengths you found in part (b).

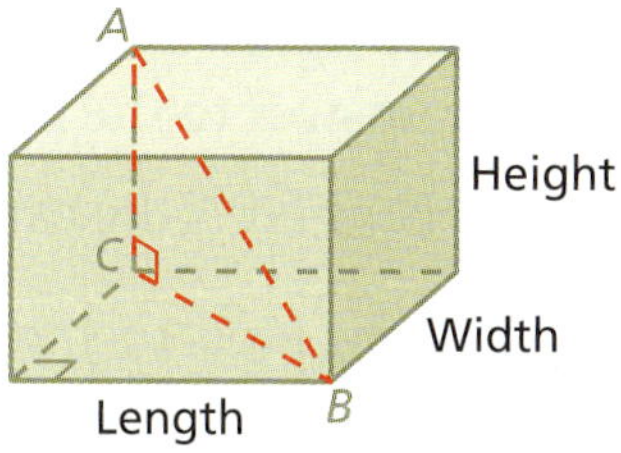

22. Critical Thinking Plot the points $(-1, -2)$, $(2, 1)$, and $(-3, 6)$ in a coordinate plane. Are the points the vertices of a right triangle? Explain.

Fair Game Review What you learned in previous grades & lessons

Find the mean, median, and mode of the data. *(Skills Review Handbook)*

23. 12, 9, 17, 15, 12, 13

24. 21, 32, 16, 27, 22, 19, 10

25. 67, 59, 34, 71, 59

26. MULTIPLE CHOICE What is the sum of the angle measures of an octagon? *(Section 5.3)*

Ⓐ 720° Ⓑ 1080° Ⓒ 1440° Ⓓ 1800°

6.3–6.5 Quiz

Check It Out
Progress Check
BigIdeasMath.com

Tell whether the number is *rational* or *irrational*. Explain. *(Section 6.3)*

1. $-\sqrt{225}$ **2.** $-1\frac{1}{9}$ **3.** $\sqrt{41}$

Estimate to the nearest integer. *(Section 6.3)*

4. $\sqrt{38}$ **5.** $-\sqrt{99}$ **6.** $\sqrt{172}$

Which number is greater? Explain. *(Section 6.3)*

7. $\sqrt{11}, 3\frac{3}{5}$ **8.** $\sqrt{1.44}, 1.1\overline{8}$

Simplify the expression. *(Section 6.4)*

9. $\sqrt{2} + 2\sqrt{2}$ **10.** $3\sqrt{15} - 7\sqrt{15}$ **11.** $\sqrt{\frac{6}{25}}$

Find the volume of the rectangular prism. *(Section 6.4)*

12.

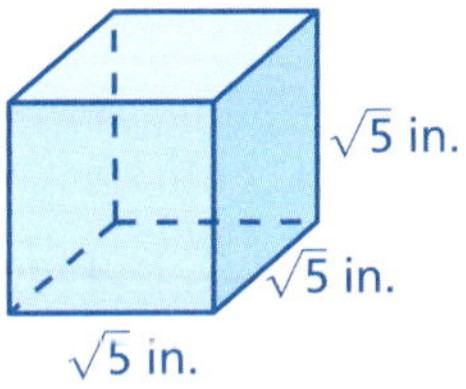

13.

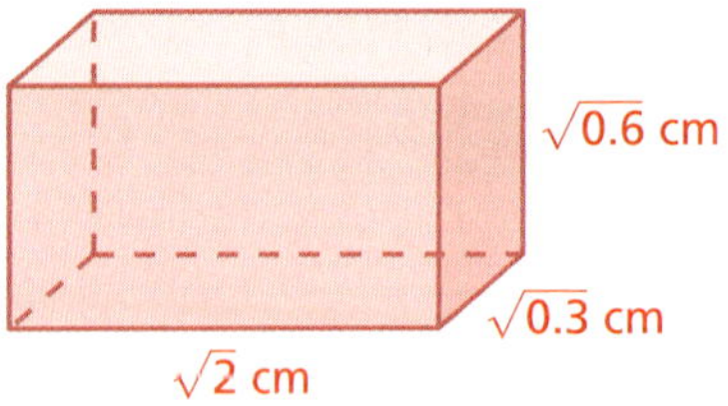

Use the figure to answer Exercises 14–17. Round your answer to the nearest tenth. *(Section 6.5)*

14. How far is the cabin from the peak?

15. How far is the fire tower from the lake?

16. How far is the lake from the peak?

17. You are standing at $(-5, -6)$. How far are you from the lake?

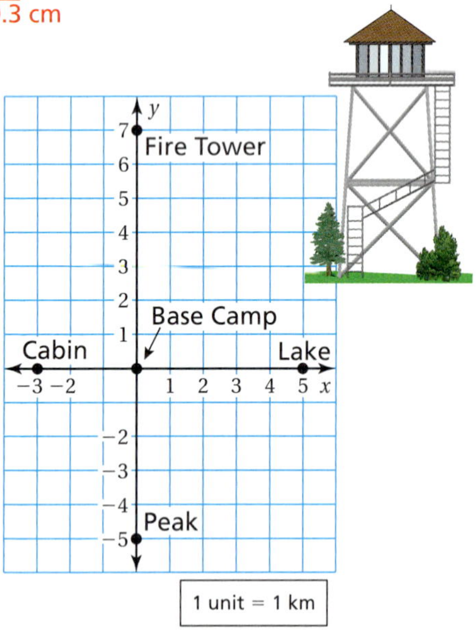

Tell whether the triangle with the given side lengths is a right triangle. *(Section 6.5)*

18.

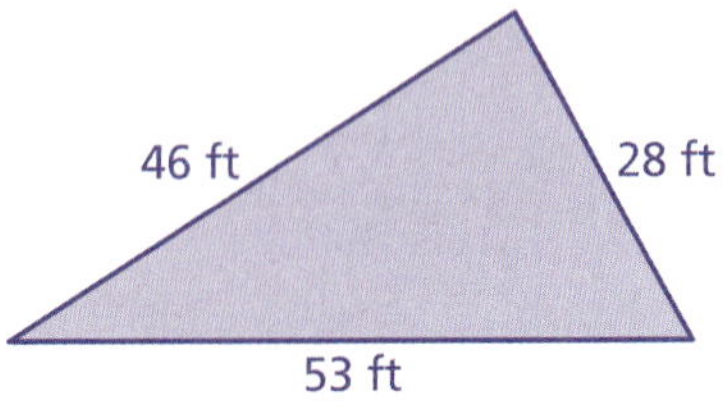

19.

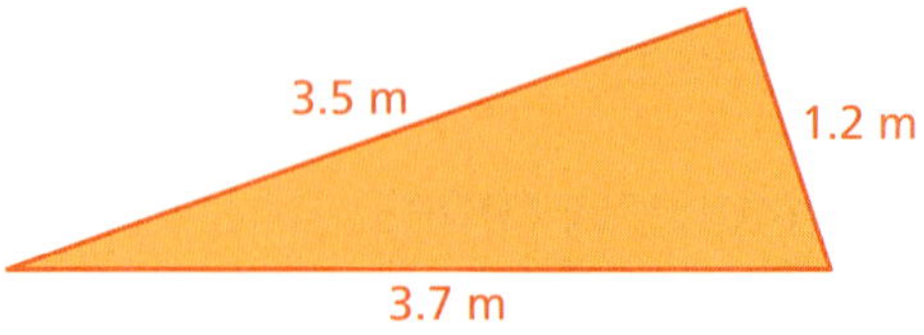

Alternative Assessment Options

Math Chat	Student Reflective Focus Question
Structured Interview	Writing Prompt

Math Chat

- Have students work in pairs. Assign Quiz Exercises 14–17 to each pair. Each student works through all four problems. After the students have worked through the problems, they take turns talking through the processes that they used to get each answer. Students analyze and evaluate the mathematical thinking and strategies used.
- The teacher should walk around the classroom listening to the pairs and ask questions to ensure understanding.

Study Help Sample Answers

Remind students to complete Graphic Organizers for the rest of the chapter.

4.

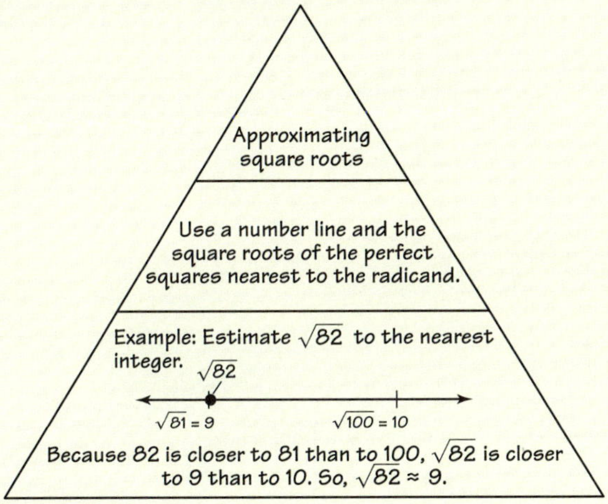

5. Available at *BigIdeasMath.com.*

Reteaching and Enrichment Strategies

If students need help. . .	If students got it. . .
Resources by Chapter • Study Help • Practice A and Practice B • Puzzle Time Lesson Tutorials *BigIdeasMath.com* Practice Quiz Practice from the Test Generator	Resources by Chapter • Enrichment and Extension • School-to-Work Game Closet at *BigIdeasMath.com* Start the Chapter Review

Answers

1. rational; 225 is a perfect square.
2. rational; $-1\frac{1}{9}$ can be written as the ratio of two integers.
3. irrational; 41 is not a perfect square.
4. 6
5. -10
6. 13
7. $3\frac{3}{5}$; $3\frac{3}{5}$ is to the right of $\sqrt{11}$.
8. $\sqrt{1.44}$; $\sqrt{1.44}$ is to the right of $1.1\overline{8}$.
9. $3\sqrt{2}$
10. $-4\sqrt{15}$
11. $\frac{\sqrt{6}}{5}$
12. $5\sqrt{5}$ in.^{3}
13. 0.6 cm^3
14. 5.8 km
15. 8.6 km
16. 7.1 km
17. 11.7 km
18. no
19. yes

Technology For the Teacher

Answer Presentation Tool

Assessment Book

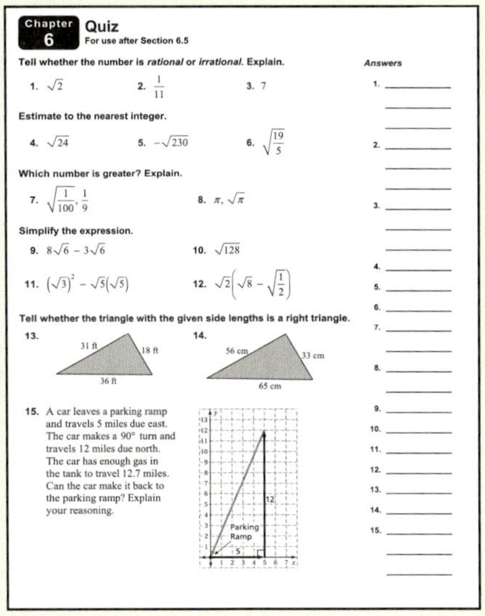

Chapter 6 Quiz
For use after Section 6.5

Tell whether the number is *rational* or *irrational*. Explain.

1. $\sqrt{2}$
2. $\frac{1}{11}$
3. 7

Estimate to the nearest integer.

4. $\sqrt{24}$
5. $-\sqrt{230}$
6. $\sqrt{\frac{19}{5}}$

Which number is greater? Explain.

7. $\sqrt{\frac{1}{100}}, \frac{1}{9}$
8. $\pi, \sqrt{\pi}$

Simplify the expression.

9. $8\sqrt{6} - 3\sqrt{6}$
10. $\sqrt{128}$
11. $(\sqrt{3})^2 - \sqrt{5}(\sqrt{5})$
12. $\sqrt{2}\left(\sqrt{8} - \sqrt{\frac{1}{2}}\right)$

Tell whether the triangle with the given side lengths is a right triangle.

13. 31 ft, 18 ft, 36 ft
14. 56 cm, 33 cm, 65 cm

15. A car leaves a parking ramp and travels 5 miles due east. The car makes a 90° turn and travels 12 miles due north. The car has enough gas in the tank to travel 12.7 miles. Can the car make it back to the parking ramp? Explain your reasoning.

Answers

1. ______
2. ______
3. ______
4. ______
5. ______
6. ______
7. ______
8. ______
9. ______
10. ______
11. ______
12. ______
13. ______
14. ______
15. ______

For the Teacher
Additional Review Options

- **Quiz***Show*
- Big Ideas Test Generator
- Game Closet at *BigIdeasMath.com*
- Vocabulary Puzzle Builder
- Resources by Chapter
  - Puzzle Time
  - Study Help

Answers

1. 4 and -4
2. 30 and -30
3. 50 and -50
4. 1
5. $-\frac{3}{5}$
6. 1.4 and -1.4
7. -1
8. $3\frac{2}{3}$
9. -30

Review of Common Errors

Exercises 7–9

- Remind students of the order of operations. Because taking a square root is the inverse of squaring, it is evaluated before multiplication and division.

Exercises 10 and 11

- Students may substitute the given lengths in the wrong part of the formula. Remind them that the side opposite the right angle is the hypotenuse *c*.

Exercises 12–17

- Encourage students to write the first 10 perfect squares at the top of their papers as a reminder and a reference.

6 Chapter Review

Check It Out
Vocabulary Help
BigIdeasMath.com

Review Key Vocabulary

square root, *p. 232*
perfect square, *p. 232*
radical sign, *p. 232*
radicand, *p. 232*
theorem, *p. 236*
legs, *p. 238*
hypotenuse, *p. 238*
Pythagorean Theorem, *p. 238*
irrational number, *p. 246*
real numbers, *p. 246*
Pythagorean triple, *p. 261*

Review Examples and Exercises

6.1 Finding Square Roots *(pp. 230–235)*

Find the square root(s).

a. $-\sqrt{36}$

$-\sqrt{36}$ represents the *negative* square root.

Because $6^2 = 36$, $-\sqrt{36} = -\sqrt{6^2} = -6$.

b. $\sqrt{1.96}$

$\sqrt{1.96}$ represents the *positive* square root.

Because $1.4^2 = 1.96$, $\sqrt{1.96} = \sqrt{1.4^2} = 1.4$.

c. $\pm\sqrt{\frac{16}{81}}$

$\pm\sqrt{\frac{16}{81}}$ represents both the *positive and negative* square roots.

Because $\left(\frac{4}{9}\right)^2 = \frac{16}{81}$, $\pm\sqrt{\frac{16}{81}} = \pm\sqrt{\left(\frac{4}{9}\right)^2} = \frac{4}{9}$ and $-\frac{4}{9}$.

Exercises

Find the two square roots of the number.

1. 16

2. 900

3. 2500

Find the square root(s).

4. $\sqrt{1}$

5. $-\sqrt{\frac{9}{25}}$

6. $\pm\sqrt{1.96}$

Evaluate the expression.

7. $15 - 4\sqrt{16}$

8. $\sqrt{\frac{54}{6}} + \frac{2}{3}$

9. $10\left(\sqrt{81} - 12\right)$

6.2 The Pythagorean Theorem (pp. 236–241)

Find the length of the hypotenuse of the triangle.

$a^2 + b^2 = c^2$	Write the Pythagorean Theorem.
$7^2 + 24^2 = c^2$	Substitute.
$49 + 576 = c^2$	Evaluate powers.
$625 = c^2$	Add.
$\sqrt{625} = \sqrt{c^2}$	Take positive square root of each side.
$25 = c$	Simplify.

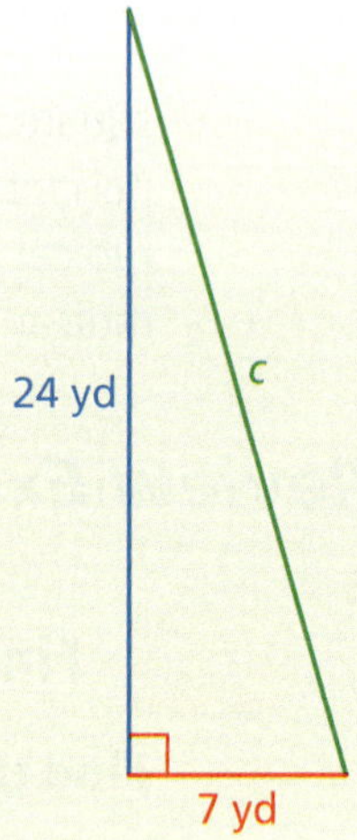

The length of the hypotenuse is 25 yards.

Exercises

Find the missing length of the triangle.

10.

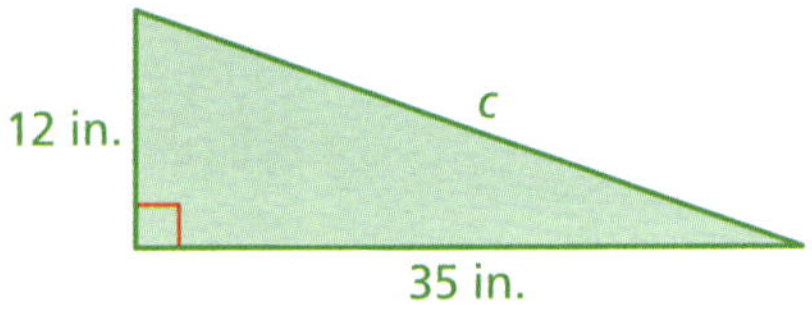

11.

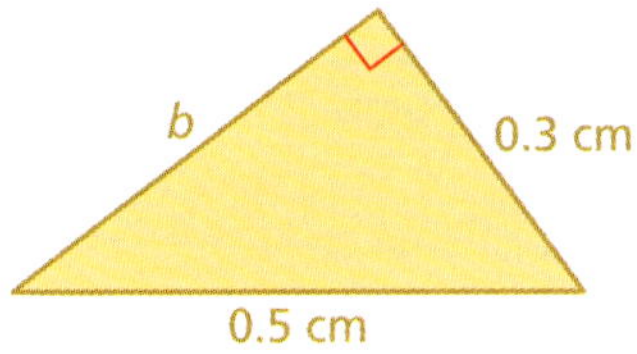

6.3 Approximating Square Roots (pp. 244–251)

Estimate $\sqrt{34}$ to the nearest integer.

Use a number line and the square roots of the perfect squares nearest to the radicand. The nearest perfect square less than 34 is 25. The nearest perfect square greater than 34 is 36.

Because 34 is closer to 36 than to 25, $\sqrt{34}$ is closer to 6 than to 5.

So, $\sqrt{34} \approx 6$.

Exercises

Estimate to the nearest integer.

12. $\sqrt{14}$ **13.** $\sqrt{90}$ **14.** $\sqrt{175}$

Review Game

Significant Square Roots

Big Ideas Game Closet

Materials per group

- piece of paper
- pencil

Directions

Divide the class into groups of 3 or 4.

Each group is to come up with 5 significant numbers and compute the exact square root of each number.

Examples of significant numbers:

School address: 1764 Knowledge Road; $\sqrt{1764} = 42$

Year of presidential election: 1936—Franklin D. Roosevelt elected to his second term; $\sqrt{1936} = 44$

Age for driver's license: 16 years old; $\sqrt{16} = 4$

Who wins?

The first group to come up with five significant numbers and correct square roots wins.

For the Student Additional Practice

- Lesson Tutorials
- Study Help (textbook)
- Student Website
  - Multi-Language Glossary
  - Practice Assessments

Answers

10. 37 in.

11. 0.4 cm

12. 4

13. 9

14. 13

15. $\frac{3\sqrt{11}}{10}$

16. $4\sqrt{6}$

17. $5\sqrt{3}$

18. 32.2 ft

19. 36 ft

My Thoughts on the Chapter

> **Teacher Tip**
>
> Not allowed to write in your teaching edition? Use sticky notes to record your thoughts.

What worked. . .

What did not work. . .

What I would do differently. . .

6.4 Simplifying Square Roots (pp. 252–257)

Simplify $\sqrt{28}$.

$\sqrt{28} = \sqrt{4 \cdot 7}$ — Factor using the greatest perfect square factor.

$= \sqrt{4} \cdot \sqrt{7}$ — Use the Product Property of Square Roots.

$= 2\sqrt{7}$ — Simplify.

Simplify $\sqrt{\frac{13}{64}}$.

$\sqrt{\frac{13}{64}} = \frac{\sqrt{13}}{\sqrt{64}}$ — Use the Quotient Property of Square Roots.

$= \frac{\sqrt{13}}{8}$ — Simplify.

Exercises

Simplify the expression.

15. $\sqrt{\frac{99}{100}}$ **16.** $\sqrt{96}$ **17.** $\sqrt{75}$

6.5 Using the Pythagorean Theorem (pp. 258–263)

Find the height of the stilt walker. Round your answer to the nearest tenth.

$a^2 + b^2 = c^2$ — Write the Pythagorean Theorem.

$6^2 + x^2 = 13^2$ — Substitute.

$36 + x^2 = 169$ — Evaluate powers.

$x^2 = 133$ — Subtract 36 from each side.

$\sqrt{x^2} = \sqrt{133}$ — Take positive square root of each side.

$x \approx 11.5$ — Use a calculator.

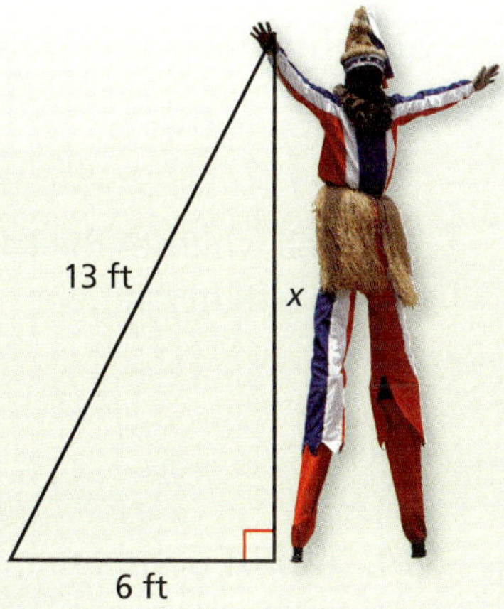

The height of the stilt walker is about 11.5 feet.

Exercises

Find the height x. Round your answer to the nearest tenth, if necessary.

18.

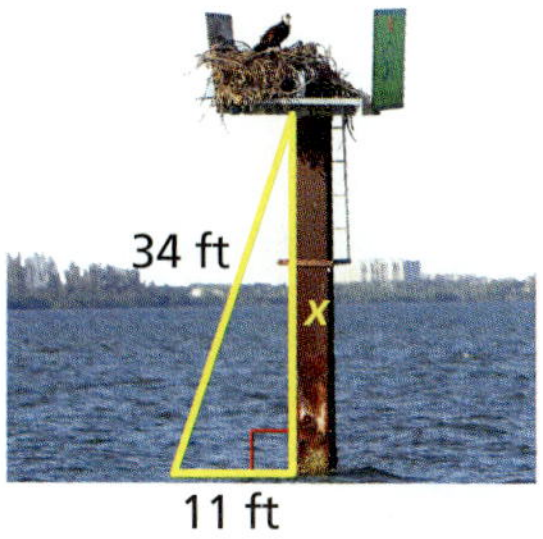

19.

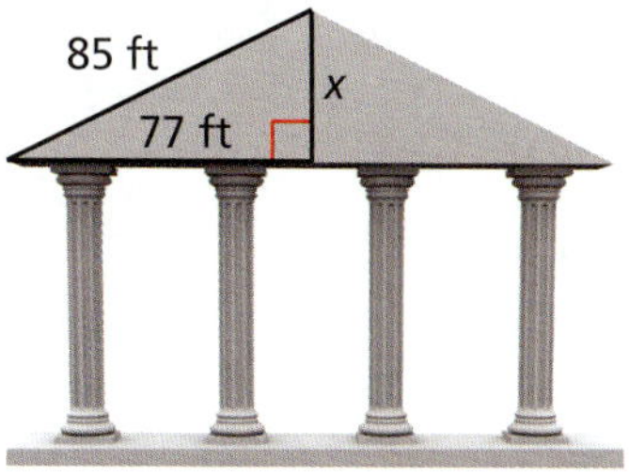

6 Chapter Test

Check It Out
Test Practice
BigIdeasMath.com

Find the square root(s).

1. $-\sqrt{1600}$
2. $\sqrt{\frac{25}{49}}$
3. $\pm\sqrt{\frac{100}{9}}$

Evaluate the expression.

4. $12 + 8\sqrt{16}$
5. $\frac{1}{2} + \sqrt{\frac{72}{2}}$
6. Find the missing length of the triangle.

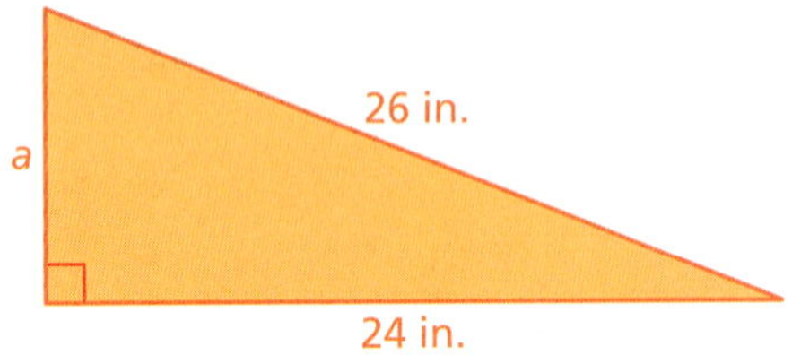

Tell whether the number is *rational* or *irrational*. Explain.

7. 16π
8. $-\sqrt{49}$

Which number is greater? Explain.

9. $\sqrt{0.16}, \frac{1}{2}$
10. $\sqrt{45}, 6.\overline{3}$

Simplify the expression.

11. $6\sqrt{5} + 5\sqrt{5}$
12. $\sqrt{250}$
13. Tell whether the triangle is a right triangle.

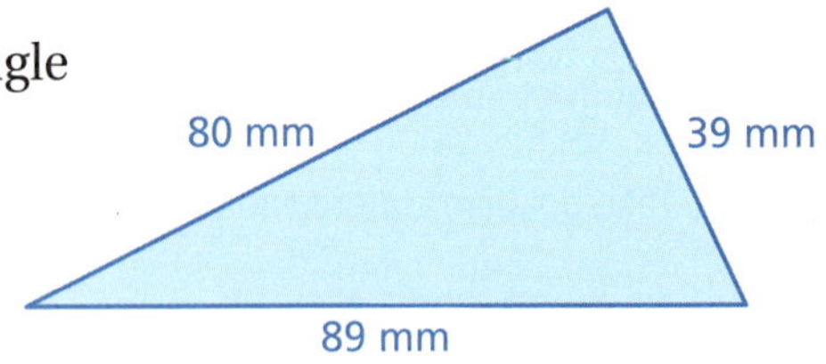

14. **ROBOT** Find the height of the dinosaur robot.

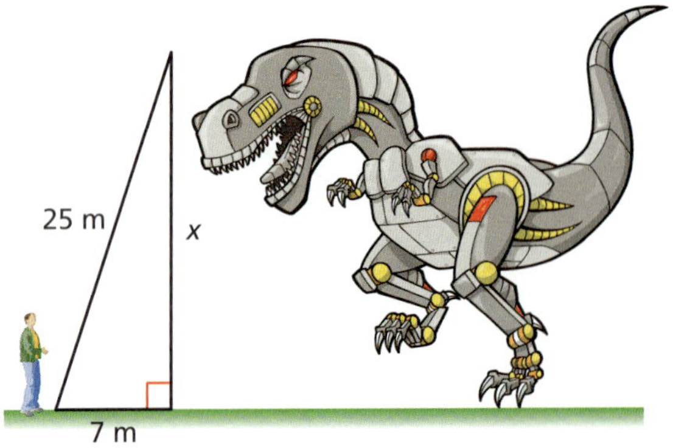

15. **SUPERHERO** Find the altitude of the superhero balloon.

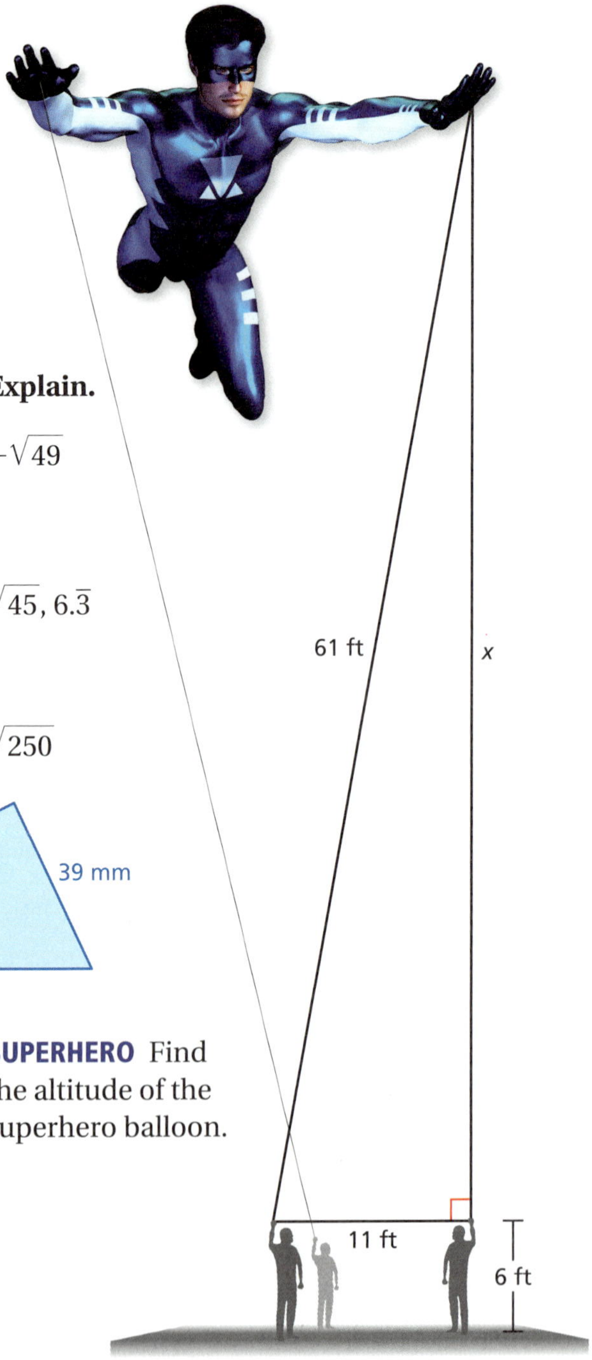

Test Item References

Chapter Test Questions	Section to Review
1–5	6.1
6	6.2
7–10	6.3
11, 12	6.4
13–15	6.5

Test-Taking Strategies

Remind students to quickly look over the entire test before they start so that they can budget their time. Students should estimate and check their answers for reasonableness as they work through the test. Teach the students to use the Stop and Think strategy before answering. **Stop** and carefully read the question, and **Think** about what the answer should look like.

Common Assessment Errors

- **Exercises 1–3** Remind students that a square root can be positive or negative.
- **Exercises 4 and 5** Remind students of the order of operations. Because taking a square root is the inverse of squaring, it is evaluated before multiplication and division.
- **Exercises 6, 13–15** Students may substitute the given lengths in the wrong part of the formula. Remind them that the side opposite the right angle is the hypotenuse *c*.
- **Exercise 11** Students may add the radicands as well as the coefficients. Remind them to only add the coefficients because the square roots are like terms.
- **Exercise 12** Students may still have difficulty identifying perfect squares to use to simplify the expression. Encourage them to write the first 10 perfect squares at the top of their papers as a reminder and a reference.

Reteaching and Enrichment Strategies

If students need help. . .	If students got it. . .
Resources by Chapter • Practice A and Practice B • Puzzle Time Record and Practice Journal Practice Differentiating the Lesson Lesson Tutorials Practice from the Test Generator Skills Review Handbook	Resources by Chapter • Enrichment and Extension • School-to-Work • Financial Literacy Game Closet at *BigIdeasMath.com* Start Standardized Test Practice

Answers

1. -40
2. $\frac{5}{7}$
3. $\frac{10}{3}$ and $-\frac{10}{3}$
4. 44
5. $6\frac{1}{2}$
6. 10 in.
7. irrational; π is irrational.
8. rational; 49 is a perfect square.
9. $\frac{1}{2}$; $\frac{1}{2}$ is to the right of $\sqrt{0.16}$.
10. $\sqrt{45}$; $\sqrt{45}$ is to the right of $6.\overline{3}$.
11. $11\sqrt{5}$
12. $5\sqrt{10}$
13. yes
14. 24 m
15. 66 ft

Assessment Book

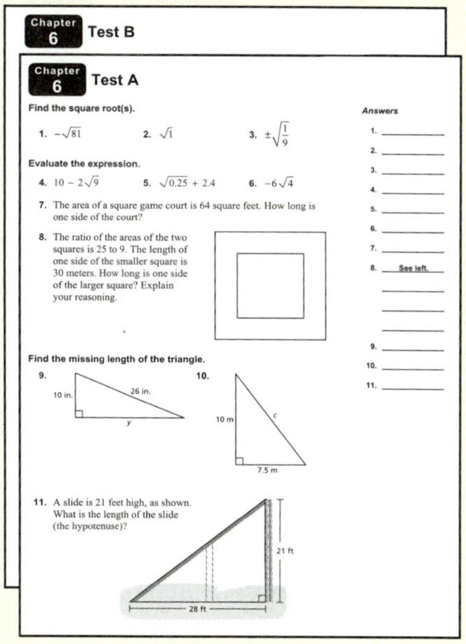
Chapter 6 Test B

Chapter 6 Test A

Find the square root(s).

1. $-\sqrt{81}$ 2. $\sqrt{1}$ 3. $\pm\sqrt{\frac{1}{9}}$

Evaluate the expression.

4. $10 - 2\sqrt{9}$ 5. $\sqrt{0.25} + 2.4$ 6. $-6\sqrt{4}$

7. The area of a square game court is 64 square feet. How long is one side of the court?

8. The ratio of the areas of the two squares is 25 to 9. The length of one side of the smaller square is 30 meters. How long is one side of the larger square? Explain your reasoning.

Find the missing length of the triangle.

9. 10.

11. A slide is 21 feet high, as shown. What is the length of the slide (the hypotenuse)?

Test-Taking Strategies

Available at *BigIdeasMath.com*

After Answering Easy Questions, Relax
Answer Easy Questions First
Estimate the Answer
Read All Choices before Answering
Read Question before Answering
Solve Directly or Eliminate Choices
Solve Problem before Looking at Choices
Use Intelligent Guessing
Work Backwards

About this Strategy

When taking a multiple choice test, be sure to read each question carefully and thoroughly. When taking a timed test, it is often best to skim the test and answer the easy questions first. Be careful that you record your answer in the correct position on the answer sheet.

Answers

1. D

2. H

3. A

4. G

Technology For the Teacher

Big Ideas Test Generator

Item Analysis

1. **A.** The student adds 1.1 and 4.

B. The student multiplies 1.1 by 4.

C. The student adds 1.1 and 2.

D. Correct answer

2. **F.** The student misunderstands what the question is asking.

G. The student misunderstands what the question is asking and incorrectly identifies the slope formula.

H. Correct answer

I. The student misunderstands the slope-intercept form of an equation.

3. **A.** Correct answer

B. The student adds 60 and 8 together.

C. The student interchanges the roles of 60 and 8 in the equation.

D. The student adds 8 and y together instead of multiplying them.

4. **F.** The student misunderstands the meaning of continuous and discrete.

G. Correct answer

H. The student makes an error by assuming the function is linear from the information given.

I. The student assumes that range will be the same as domain.

5. **2 points** The student demonstrates a thorough understanding of how to apply the Pythagorean Theorem to the problem, explains the work fully, and calculates the distance accurately. The distance between opposite corners is $\sqrt{16{,}000} = 40\sqrt{10} \approx 126.5$ yards.

1 point The student's work and explanations demonstrate a lack of essential understanding. The Pythagorean Theorem is misstated or, if stated correctly, is applied incorrectly to the problem.

0 points The student provides no response, a completely incorrect or incomprehensible response, or a response that demonstrates insufficient understanding of the Pythagorean Theorem.

6. **Gridded Response:** Correct answer: 15 hours

Common Error: The student adds 50 to the right hand side of the equation instead of subtracting, yielding an answer of 17.5.

7. **A.** The student interchanges the roles of n and S.

B. Correct answer

C. The student subtracts 2 from S instead of adding it.

D. The student adds 2 first, then divides through by 180.

6 Standardized Test Practice

1. The period T of a pendulum is the time, in seconds, it takes the pendulum to swing back and forth. The period can be found using the formula $T = 1.1\sqrt{L}$, where L is the length, in feet, of the pendulum. A pendulum has a length of 4 feet. Find its period.

   A. 5.1 sec
   B. 4.4 sec
   C. 3.1 sec
   D. 2.2 sec

Test-Taking Strategy
Answer Easy Questions First

"Scan the test and answer the easy questions first. You know the square root of 4 is 2."

2. The steps Pat took to write the equation in slope-intercept form are shown below. What should Pat change in order to correctly rewrite the equation in slope-intercept form?

$$3x - 6y = 1$$
$$3x = 6y + 1$$
$$x = 2y + \frac{1}{3}$$

   F. Use the formula $m = \frac{\text{rise}}{\text{run}}$.
   G. Use the formula $m = \frac{\text{run}}{\text{rise}}$.
   H. Subtract $3x$ from both sides of the equation and divide every term by -6.
   I. Subtract 1 from both sides of the equation and divide every term by 3.

3. A housing community started with 60 homes. In each of the following years, 8 more homes were built. Let y represent the number of years that have passed since the first year and let n represent the number of homes. Which equation describes the relationship between n and y?

   A. $n = 8y + 60$
   B. $n = 68y$
   C. $n = 60y + 8$
   D. $n = 60 + 8 + y$

4. The domain of a function is 0, 1, 2, 3, 4, 5. What can you conclude?

   F. The domain is continuous.
   G. The domain is discrete.
   H. The function is linear.
   I. The range is 0, 1, 2, 3, 4, 5.

5. A football field is 40 yards wide and 120 yards long. Find the distance between opposite corners of the football field. Show your work and explain your reasoning.

Think Solve Explain

6. A computer consultant charges \$50 plus \$40 for each hour she works. The consultant charged \$650 for one job. This can be represented by the equation below, where h represents the number of hours worked.

$$40h + 50 = 650$$

How many hours did the consultant work?

7. The formula below can be used to find the number S of degrees in a polygon with n sides. Solve the formula for n.

$$S = 180(n - 2)$$

A. $n = 180(S - 2)$

B. $n = \dfrac{S}{180} + 2$

C. $n = \dfrac{S}{180} - 2$

D. $n = \dfrac{S}{180} + \dfrac{1}{90}$

8. The table below shows a linear pattern. Which linear function relates y to x?

x	1	2	3	4	5
y	4	2	0	−2	−4

F. $y = 2x + 2$

G. $y = 4x$

H. $y = -2x + 2$

I. $y = -2x + 6$

9. What is the value of x in the right triangle shown?

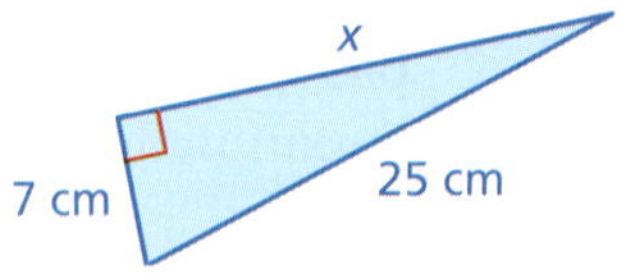

A. 16 cm

B. 18 cm

C. 24 cm

D. $\sqrt{674}$ cm

10. Find the height of the tree in the diagram.

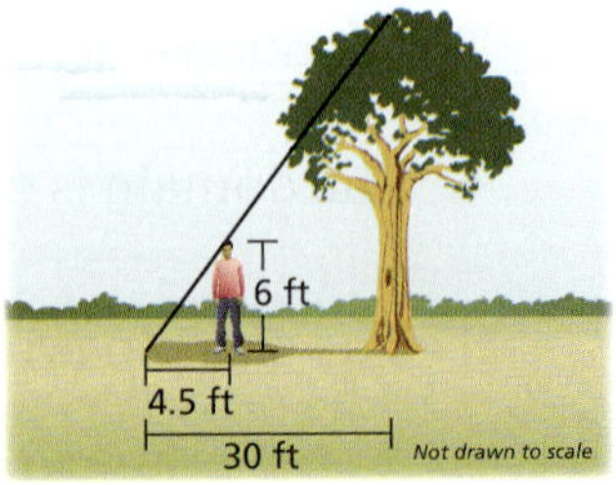

F. 22.5 ft

G. 31.5 ft

H. 35 ft

I. 40 ft

Item Analysis (continued)

8. **F.** The student has the wrong sign for the slope and makes a mistake finding the y-intercept.
   G. The student bases the equation on the first column only.
   H. The student finds the correct slope, but makes a mistake finding the y-intercept.
   I. Correct answer

9. **A.** The student takes the average of the two sides.
   B. The student subtracts the shorter from the longer side.
   C. Correct answer
   D. The student treats the missing leg as the hypotenuse.

10. **F.** The student treats shadow length as tree height (i.e., reverses the ratio).
    G. The student adds 1.5 to the shadow length because the man is 1.5 feet taller than his shadow.
    H. The student guesses, and picks an answer slightly greater than 30.
    I. Correct answer

11. **A.** The student simplifies $\sqrt{24}$ correctly, but evaluates $\sqrt{4}$ as 4 instead of as 2.
    B. The student reverses the numbers.
    C. Correct answer
    D. The student adds 12 and 24 under the radical.

12. **F.** Correct answer
    G. The student mistakes supplement for complement.
    H. The student reverses the order of the subtraction.
    I. The student mistakes supplement for complement and reverses the order of the subtraction.

13. **A.** The student divides the right side of the equation by 2.5, but makes a decimal error by dividing the left side of the equation by 25.
    B. Correct answer
    C. The student sets g equal to 0.
    D. The student adds 3 to 18 instead of subtracting.

14. **Gridded Response:** Correct answer: 65 mi

    Common Error: The student adds the two distances to get 89 miles as an answer.

Answers

5. about 126.5 yards
6. 15 hours
7. B
8. I
9. C
10. I

Answers

11. C

12. F

13. B

14. 65 mi

15. F

Answer for Extra Example

1. **A.** Correct answer

   B. The student misreads "greatest speed a car is allowed legally" to mean strict inequality.

   C. The student mistakes the inequality symbols.

   D. The student mistakes the inequality symbols and assumes strict inequality.

Item Analysis (continued)

15. **F.** Correct answer

G. The student picks slope of 2 instead of -2.

H. The student picks y-intercept of 2 instead of -2.

I. The student picks y-intercept of 2 instead of -2 and slope of 2 instead of -2.

Extra Example for Standardized Test Practice

1. The sign below shows the speed limit on a highway.

The speed limit indicates the **greatest** speed a car is legally allowed. If s represents speed, how can this speed limit be written algebraically?

A. $s \leq 65$

B. $s < 65$

C. $s \geq 65$

D. $s > 65$

11. Which expression is equivalent to $12\sqrt{24}$?

A. $48\sqrt{6}$

B. $24\sqrt{12}$

C. $24\sqrt{6}$

D. 6

12. The measure of an angle is x degrees. What is the measure of its complement?

F. $(90 - x)°$

G. $(180 - x)°$

H. $(x - 90)°$

I. $(x - 180)°$

13. You fill up the gas tank of your car and begin driving on the interstate. You drive at an average speed of 60 miles per hour. The amount g, in gallons, of gas left in your car can be estimated. Use the formula shown below, where h is the number of hours you have been driving.

$$g = 18 - 2.5h$$

You will fill up when you have 3 gallons of gas left in the gas tank. How long after you start driving will you fill up again?

A. about 36 min

B. about 6.0 h

C. about 7.2 h

D. about 8.4 h

14. An airplane flies 56 miles due north and then 33 miles due east. How many miles is the plane from its starting point?

15. Which graph represents the linear equation $y = -2x - 2$?

F.
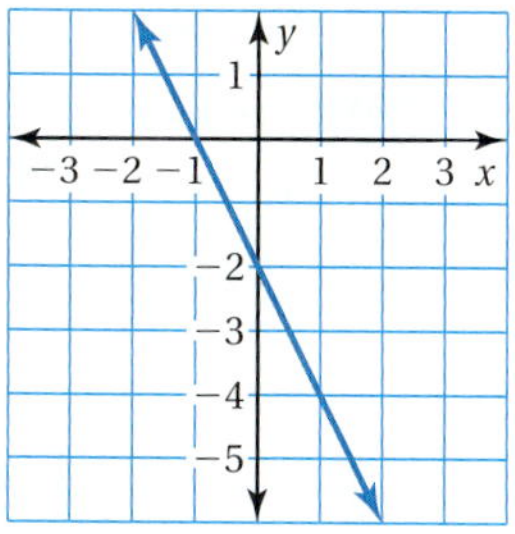

H.
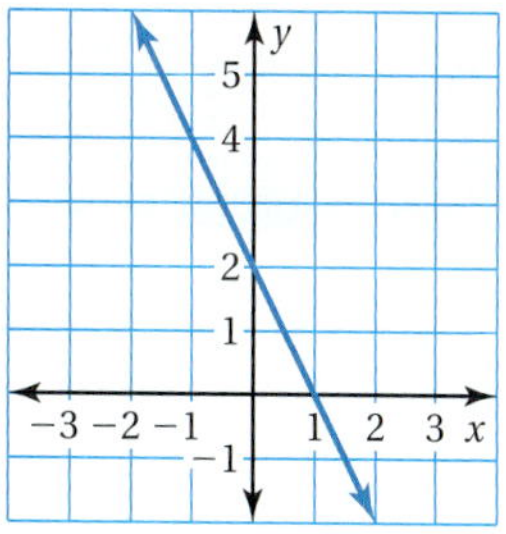

G.
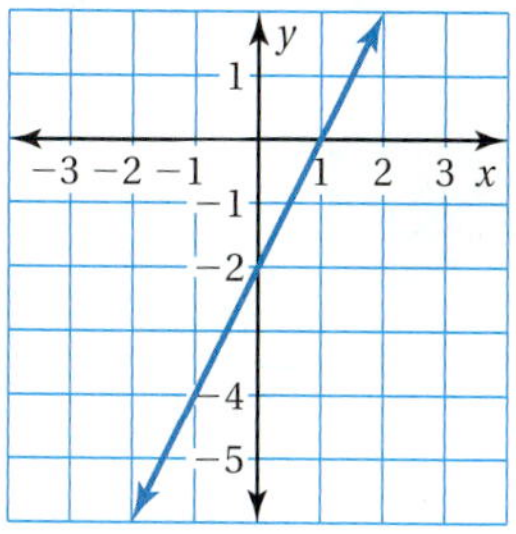

I.
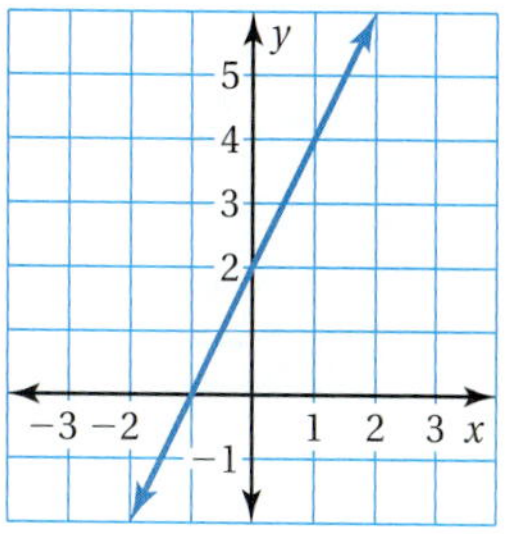

7 Data Analysis and Displays

"Wow. The number of minutes I can dog paddle is growing like crazy!"

"Please hold still. I am trying to find the mean of 6, 8, and 10 by dividing their sum into three equal piles."

Connections to Previous Learning

- Construct and analyze line graphs and double bar graphs.
- Differentiate between continuous and discrete data, and choose ways to represent it.
- Determine measures of central tendency including mean, median, mode, and range (variability).
- Select appropriate measures of central tendency to describe a data set.

- Evaluate a sample to generalize a population.
- Construct histograms, stem-and-leaf plots, and circle graphs.

- Determine and describe how changes in data values impact measures of central tendency.
- Select, organize, and construct appropriate data displays, including box and whisker plots, scatter plots, and lines of best fit to convey information and make conjectures about possible relationships.

Math in History

Many ancient cultures used a counting frame or abacus to perform calculations. Most of these cultures used a base ten system like the Roman or Chinese system. The abacus shown below is representing the number 2786.

★ Examples of the use of an abacus in Rome date back to around the first century A.D.

★ Examples of the use of an abacus in China date back to around the 14th century A.D.

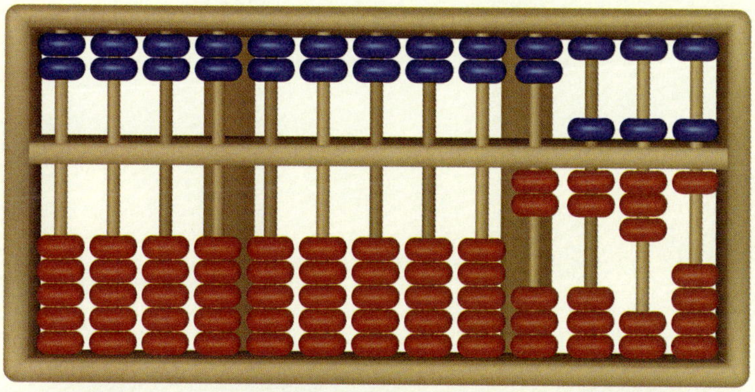

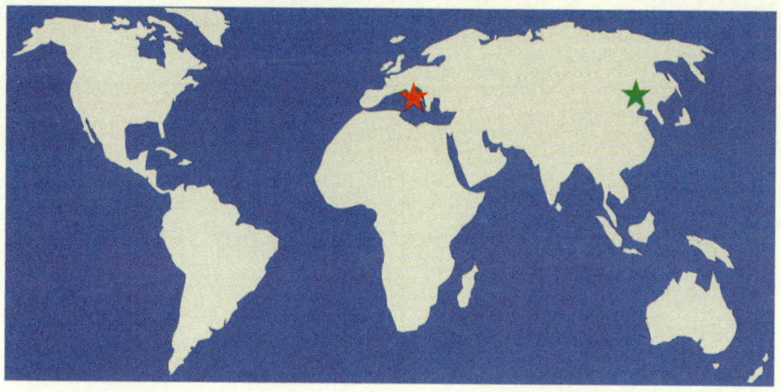

Pacing Guide for Chapter 7

Chapter Opener	1 Day
Section 1 Activity Lesson	 1 Day 1 Day
Section 2 Activity Lesson	 1 Day 1 Day
Study Help / Quiz	1 Day
Section 3 Activity Lesson	 1 Day 2 Days
Section 4 Activity Lesson	 1 Day 2 Days
Quiz / Chapter Review	1 Day
Chapter Test	1 Day
Standardized Test Practice	1 Day
Total Chapter 7	15 Days
Year-to-Date	118 Days

Check Your Resources

- Record and Practice Journal
- Resources by Chapter
- Skills Review Handbook
- Assessment Book
- Worked-Out Solutions

Technology For the Teacher

Dynamic Classroom

The Dynamic Planning Tool
Editable Teacher's Resources at
BigIdeasMath.com

Additional Topics for Review

- Multiplying fractions
- Circle graphs
- Histograms
- Frequency tables
- Writing the equations of lines

Try It Yourself

1. Answers will vary.
2. Answers will vary.

Record and Practice Journal

1.

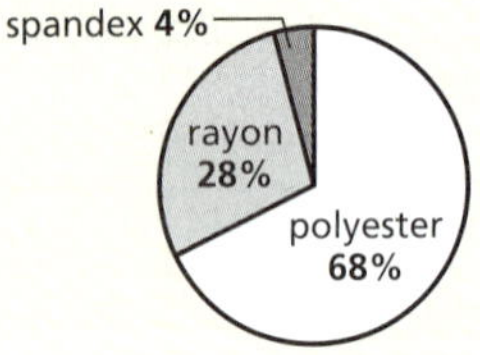

2.

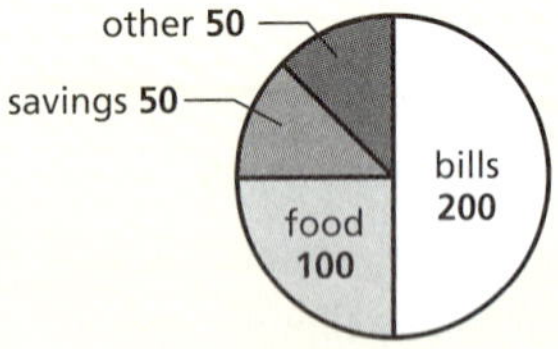

3.

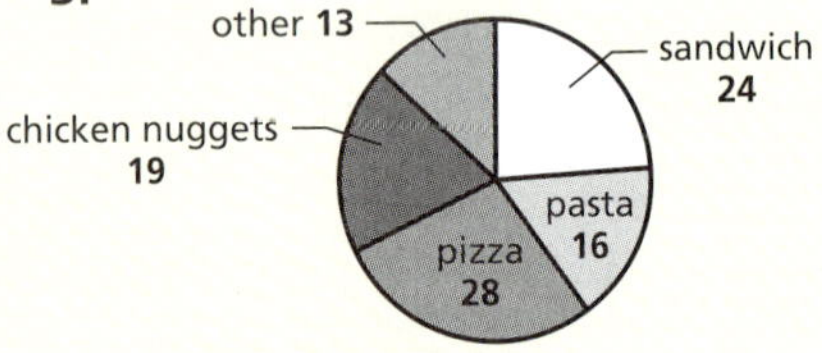

4.

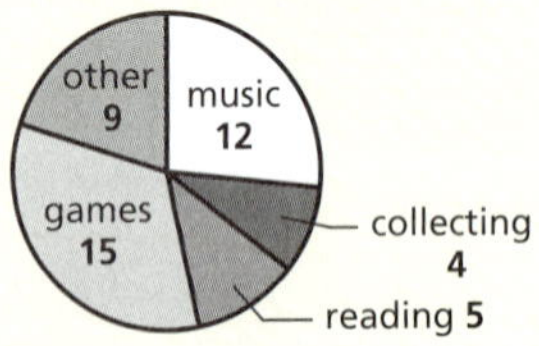

5.

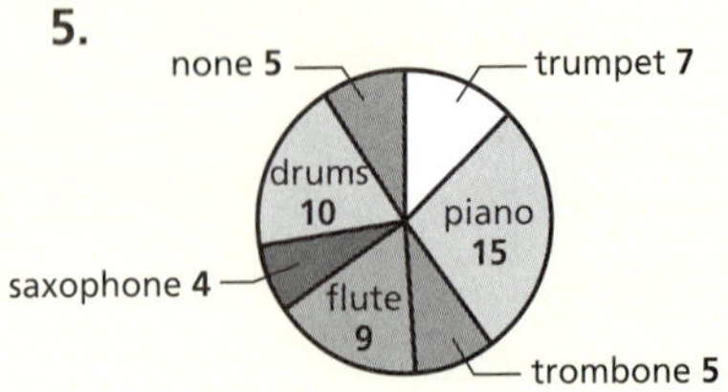

6–10. See Additional Answers.

Math Background Notes

Vocabulary Review

- Frequency Table
- Histogram
- Survey

Displaying Data

- Students have collected, analyzed, and displayed data.
- **Teaching Tip:** Example 1 provides a great opportunity for review. Remind students that a circle contains 360°. This is also a good time to review using a protractor.
- Remind students that it is helpful to know the total number of people surveyed before constructing the circle graph. This number will serve as the whole (denominator of the fraction).
- **Teaching Tip:** Example 2 provides an excellent opportunity to explore students' prerequisite knowledge. Students should be familiar with bar graphs, double bar graphs, and histograms. Consider using a Venn diagram to compare and contrast the three types of displays. This will create a nice visual representation to show which characteristics go with which display.
- **Common Error:** Students might forget that a histogram uses intervals rather than individual data values. Remind students that the horizontal axis of the graph will be labeled with intervals and the vertical axis of the graph will be labeled with frequency.
- **Common Error:** Even if an interval has a frequency of zero, it must appear on the histogram.
- **Teaching Tip:** The exercises in this set require students to take a survey to collect data and then display the data using a circle graph and a histogram. This provides a good opportunity to revisit topics such as population and sample size. What characteristics make for a fair survey? How will students ensure that the data they collect is a fair representation of the population?
- You can adapt the context of the survey to personalize it to your class.

Reteaching and Enrichment Strategies

If students need help. . .	If students got it. . .
Record and Practice Journal • Fair Game Review Skills Review Handbook Lesson Tutorials	Game Closet at *BigIdeasMath.com* Start the next section

What You Learned Before

"Okay, I have the box. But, I need your help to complete my box-and-whisker plot."

Displaying Data

Example 1 **The table shows the results of a survey. Display the data in a circle graph.**

Class Trip Location	Water park	Museum	Zoo	Other
Students	25	11	5	4

A total of 45 students took the survey.

Water park:

$\frac{25}{45} \cdot 360^\circ = 200^\circ$

Museum:

$\frac{11}{45} \cdot 360^\circ = 88^\circ$

Zoo:

$\frac{5}{45} \cdot 360^\circ = 40^\circ$

Other:

$\frac{4}{45} \cdot 360^\circ = 32^\circ$

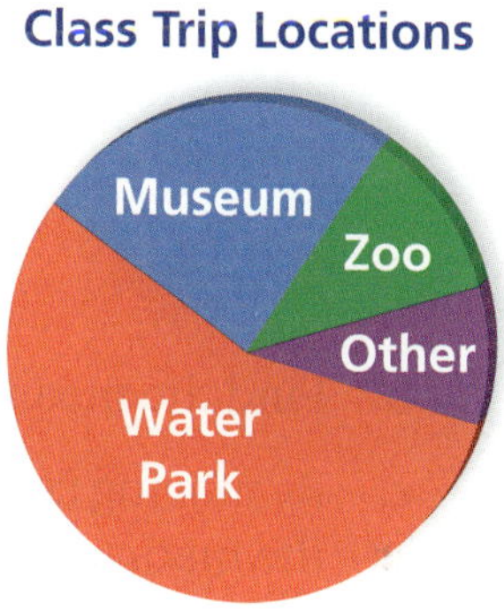

Example 2 **The frequency table shows the number of books that 12 people read last month. Display the data in a histogram.**

Books Read Last Month	Frequency
0–1	6
2–3	4
4–5	0
6–7	2

Try It Yourself

1. Conduct a survey to determine the after-school activities of students in your class. Display the results in a circle graph.
2. Conduct a survey to determine the number of pets owned by students in your class. Display the results in a histogram.

7.1 Measures of Central Tendency

Essential Question How can you use measures of central tendency to distribute an amount evenly among a group of people?

1 ACTIVITY: Exploring Mean, Median, and Mode

Work with a partner. Forty-five coins are arranged in nine stacks.

a. Record the number of coins in each stack in a table.

Stack	1	2	3	4	5	6	7	8	9
Coins									

b. Find the mean, median, and mode of the number of coins in each stack.

c. By moving coins from one stack to another, can you change the mean? the median? the mode? Explain.

d. Is it possible to arrange the coins in stacks so that the median is 6? 8? Explain.

2 EXAMPLE: Drawing a Line Plot

Work with a partner.

a. Draw a number line. Label the tick marks from 1 to 10.

b. Place each stack of coins in Activity 1 above the number of coins in the stack.

c. Draw an × to represent each stack. This graph is called a *line plot*.

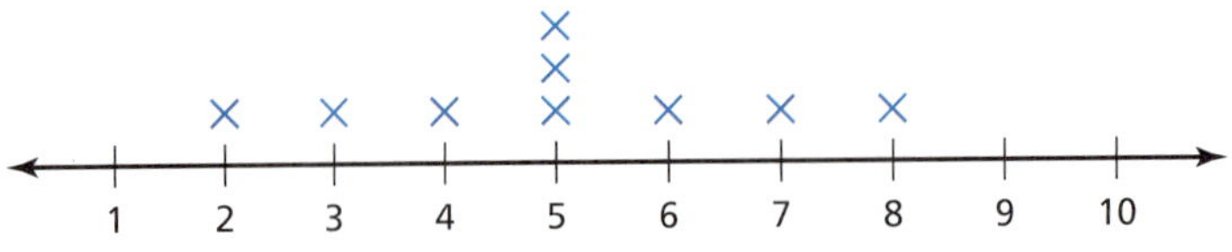

Laurie's Notes

Introduction

For the Teacher

- **Goal:** Students will study measures of central tendency as well as fair and unfair distributions.
- **Preparation:** Set up zipped baggies with 45 circular disks in each bag.

Motivate

- Explain the work of the U.S. Census Bureau. Share the following U.S. information from the Census Bureau with students. In each case, ask students to interpret what the statistic means.
  - The mean travel time to work for workers age 16 and older is 25.1 minutes.
  - The median household income in 2007 was $50,007.
  - The *average* family size is 3.19 people.

Activity Notes

Activity 1

- For this first activity, students benefit by having some sort of manipulatives, such as coins, circular disks, square tiles, or cubes. If the manipulatives are stackable, the activity will be easier for students to follow.
- This is a nice activity to get students thinking about the distribution of the data, and that there could be outliers and/or gaps in the data.
- Review *mean*, *median*, and *mode*.
- Give sufficient time for students to work through each part of this activity.
- Discuss students' explanations for parts (c) and (d).
- **Big Idea:** As the median increases, the distribution becomes more skewed. In order to have a median of 8 coins, five of the stacks would each have 8 coins and the remaining four stacks would share 5 coins.

Example 2

- This example reviews line plots.
- Reading the line plot shown, there are 3 stacks that have 5 coins in them, no stacks that have 1, 9, or 10 coins in them, and all of the other possibilities have one stack each.
- This plot is reviewed so that it can be used to explore fair distributions in the next activity.

Previous Learning

Students should know how to find the mean, median, and mode.

Activity Materials
Textbook
• circular disks or other stackable manipulative

Start Thinking! and Warm Up

Activity 7.1 Start Thinking!
For use before Activity 7.1

Activity 7.1 Warm Up
For use before Activity 7.1

Find the mean of the data set.

1. 1, 4, 6, 7, 8, 8, 8, 12
2. 1, 1, 3, 3, 5, 5, 7, 7, 9, 9
3. 17, 18, 19, 19, 26, 27, 28
4. 5, 5, 5, 5, 5, 5, 5, 34
5. 11, 38, 39, 40, 44, 44
6. 1, 5, 7, 3, 8, 5, 3, 2, 0, 9, 1

7.1 Record and Practice Journal

Essential Question How can you use measures of central tendency to distribute an amount evenly among a group of people?

1 ACTIVITY: Exploring Mean, Median, and Mode

Work with a partner. Forty-five coins are arranged in nine stacks.

a. Record the number of coins in each stack in a table.

Stack	1	2	3	4	5	6	7	8	9
Coins	5	4	3	6	2	5	8	7	5

b. Find the mean, median, and mode of the number of coins in each stack.

mean: 5; median: 5; mode: 5

c. By moving coins from one stack to another, can you change the mean? the median? the mode? Explain.

mean cannot change; median and mode can change

d. Is it possible to arrange the coins in stacks so that the median is 6? 8? Explain. yes; yes;

Sample answer:
median 6: 1, 2, 3, 4, 6, 6, 7, 8, 8
median 8: 1, 1, 1, 2, 8, 8, 8, 8, 8

English Language Learners

Graphic Organizer

Provide small groups with different data sets and a copy of the organizer. Have each group find the mean, median, mode, and range of their data set and present their work to the class.

Data set:	
Mean	
Median	
Mode	
Range	

7.1 Record and Practice Journal

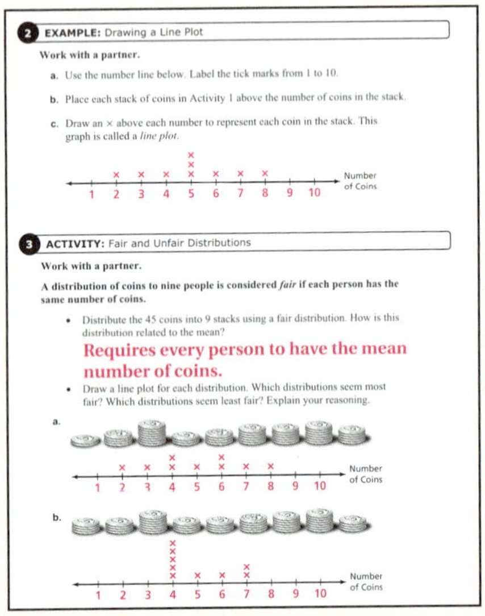

2 **EXAMPLE:** Drawing a Line Plot

Work with a partner.

a. Use the number line below. Label the tick marks from 1 to 10.

b. Place each stack of coins in Activity 1 above the number of coins in the stack.

c. Draw an × above each number to represent each coin in the stack. This graph is called a *line plot*.

3 **ACTIVITY:** Fair and Unfair Distributions

Work with a partner.

A distribution of coins to nine people is considered *fair* if each person has the same number of coins.

- Distribute the 45 coins into 9 stacks using a fair distribution. How is this distribution related to the mean?

  Requires every person to have the mean number of coins.

- Draw a line plot for each distribution. Which distributions seem most fair? Which distributions seem least fair? Explain your reasoning.

a.

b.

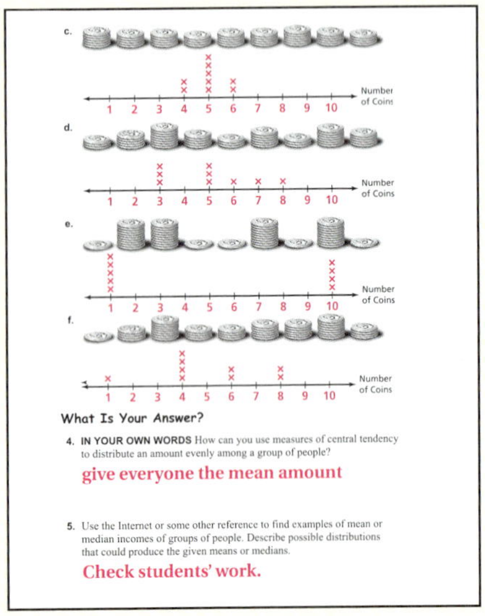

c.

d.

e.

f.

What Is Your Answer?

4. **IN YOUR OWN WORDS** How can you use measures of central tendency to distribute an amount evenly among a group of people?

   give everyone the mean amount

5. Use the Internet or some other reference to find examples of mean or median incomes of groups of people. Describe possible distributions that could produce the given means or medians.

   Check students' work.

Laurie's Notes

Activity 3

- Read the definition of a fair distribution and discuss what a fair distribution has to do with the mean.
- When students construct the line plot for each part, they should get the sense that it is similar to constructing a bar graph. The heights of the bars in a bar graph are similar to the heights of the Xs in a line plot.
- Discuss student responses to which distributions seem most and least fair.
- **Extension:** Given the requirement that each stack has to have at least one coin, ask students to make a line plot for the most fair distribution possible and the least fair distribution possible. The most fair would be 5 coins in each stack, so the line plot would have nine Xs above the 5 and no Xs elsewhere. The least fair would likely be eight Xs above the 1 and one X above the 37, meaning the line plot would need to be extended and there would be a big gap.
- **Big Idea:** The concept of the mean is that it is a sharing process. You are trying to level out the stacks. If the average amounts of money 3 students have is $10, it is possible that they all have $10 (most fair distribution) or one has $30 and the others have no money (least fair distribution). If all 3 students are known to have some money, pooling it and spreading it out into 3 stacks would level out at $10 per stack—the mean.

What Is Your Answer?

- For Question 4, the mean is the only measure of central tendency that can be used to distribute an amount evenly.

Closure

- Distribute 50 coins in 10 stacks. Make a line plot of the distribution.

Technology For the Teacher

Dynamic Classroom

The Dynamic Planning Tool

Editable Teacher's Resources at *BigIdeasMath.com*

3 ACTIVITY: Fair and Unfair Distributions

Work with a partner.

A distribution of coins to nine people is considered *fair* if each person has the same number of coins.

- Distribute the 45 coins into 9 stacks using a fair distribution. How is this distribution related to the mean?
- Draw a line plot for each distribution. Which distributions seem most fair? Which distributions seem least fair? Explain your reasoning.

a.

b.

c.

d.

e.

f.

What Is Your Answer?

4. **IN YOUR OWN WORDS** How can you use measures of central tendency to distribute an amount evenly among a group of people?
5. Use the Internet or some other reference to find examples of mean or median incomes of groups of people. Describe possible distributions that could produce the given means or medians.

Practice Use what you learned about measures of central tendency to complete Exercise 4 on page 278.

7.1 Lesson

Check It Out
Lesson Tutorials
BigIdeasMath.com

Key Vocabulary
measure of central tendency, *p. 276*

A **measure of central tendency** is a measure that represents the center of a data set. The *mean*, *median*, and *mode* are measures of central tendency.

Key Ideas

Mean

The *mean* of a data set is the sum of the data divided by the number of data values.

Median

Order the data. For a set with an odd number of values, the *median* is the middle value. For a set with an even number of values, the *median* is the mean of the two middle values.

Mode

The *mode* of a data set is the value or values that occur most often.

Remember

Data can have one mode, more than one mode, or no mode. When each value occurs only once, there is no mode.

EXAMPLE 1 Finding the Mean, Median, and Mode

Students' Hourly Wages	
$3.87	$7.25
$8.75	$8.45
$8.25	$7.25
$6.99	$7.99

An amusement park hires students for the summer. The students' hourly wages are given in the table. Find the mean, median, and mode of the hourly wages.

Mean: $\frac{58.8}{8} = 7.35$ (sum of the data / number of values)

Median: 3.87, 6.99, 7.25, 7.25, 7.99, 8.25, 8.45, 8.75 — Order the data.

$\frac{15.24}{2} = 7.62$ — Mean of two middle values

Mode: 3.87, 6.99, 7.25, 7.25, 7.99, 8.25, 8.45, 8.75

The value 7.25 occurs most often.

∴ The mean is $7.35, the median is $7.62, and the mode is $7.25.

On Your Own

Now You're Ready
Exercises 5–8

1. **WHAT IF?** In Example 1, the park hires another student at an hourly wage of $6.99. How does this additional value affect the mean, median, and mode? Explain.

Laurie's Notes

Introduction

Connect

- **Yesterday:** Students explored the connection between the mean and fair distributions. They used a line plot to display results.
- **Today:** Students will explore how an outlier affects the three measures of central tendency.

Motivate

- Time to play M & M's! No, it's not the candy; it's mean, median, mode time.
- To help students think about the three measures, ask a series of questions. Give the results and have students come up with the data.
  Examples:
  - Name 3 different numbers whose mean is 10.
  - Name 3 different numbers whose mean is 10 and whose median is 12.
  - Name 5 different numbers whose mean is 10 and whose median is 10.
  - Name 5 different numbers whose mean and median are 10, and whose mode is 8.
- Continue to ask questions. Knowing the number of values and the mean tells you the sum of all the data. The median tells you the middle value.

Lesson Notes

Key Ideas

- Write the definition of measure of central tendency, noting that the mean, median, and mode are all measures of central tendency.

Example 1

? "Have any of you had a summer or part-time job?" Answers will vary.

- Have a general discussion of different compensation methods: hourly, hourly plus tips, salaried, and commission.

? "Are there any observations about the wages listed?" Students may mention $3.87 as an outlier. Most wages are around $7 or $8 per hour.

? "What do you need to do to compute the mean?" Sum the data and divide by 8.

- You may wish to have students use calculators in this lesson.
- **Common Error:** When finding the median, students forget to sort the data first. In this example, there are an even number of data values, so you need to sort and then find the mean of the middle two values.

? "Why do you think the mean might be less than the median?" Listen for the affect of the outlier, although students may not have a sense of this yet.

On Your Own

- Give time for students to actually compute the three measures.
- Discuss the results.

Goal Today's lesson is determining the effect an outlier has on the **measures of central tendency** for a data set.

Lesson Materials
Textbook
• calculator

Start Thinking! and Warm Up

Lesson 7.1 **Warm Up**
For use before Lesson 7.1

Lesson 7.1 **Start Thinking!**
For use before Lesson 7.1

Record the number of siblings of each person in your class.

Find the mean, median, and mode of the data.

If you had instead recorded the number of children in each person's family, how would the mean, median, and mode have changed? Explain.

Extra Example 1

An amusement park hires students for the summer. The students' hourly wages are given in the table. Find the mean, median, and mode of the hourly wages.

Students' Hourly Wages	
$3.74	$7.75
$7.30	$8.43
$7.90	$7.83
$8.15	$8.50

mean: $7.45; median: $7.87; no mode

On Your Own

1. mean: $7.31, decreases; median: $7.25, decreases; Because the hourly wage of the student is less than the mean and median, both mean and median decrease. modes: $6.99 and $7.25; The data now has two modes instead of one mode.

Extra Example 2

Identify the outlier in Extra Example 1. How does the outlier affect the mean, median, and mode? By removing the outlier, \$3.74, the mean increases \$7.98 − \$7.45 = \$0.53, the median increases \$7.90 − \$7.87 = \$0.03, and there still is no mode.

Extra Example 3

In Extra Example 1, the park increases each hourly wage \$0.30. How does this increase affect the mean, median, and mode? The mean and median both increase by \$0.30, and there still is no mode.

On Your Own

2. $4\frac{1}{5}$ mi; By removing the outlier, the mean decreases $1.85 - 1.38 = 0.47$ mile; the median decreases $1.45 - 1.4 = 0.05$ mile; and there is still no mode.
3. The mean and median both increase by $1\frac{1}{2}$ miles. There is still no mode.

Differentiated Instruction

Vocabulary

Have students add a glossary to their math notebook. Key vocabulary words should be added as they are introduced. Vocabulary words for this lesson are *measure of central tendency, outlier, mean, median,* and *mode.* Mean, median, and mode are often confusing to students. Drawing illustrations next to the vocabulary words will help in understanding and reinforcing their meanings.

Laurie's Notes

Example 2

- Remind students of what an outlier is.
- Before the three measures are computed, ask students to predict what they think will happen.
- Work through the computations in the example.
- Remind students to use the correct number of values when computing the mean after removing an outlier.

? "Why did the mean increase in this example?" The data value eliminated was much less than the mean. So, the sum was not affected much and you divide by a lesser number.

? "The median increased in this example. Could it have decreased or stayed the same? Explain." Yes. Depending on the middle of the data set, the median can increase or stay the same if the outlier is a low value, *or* the median can decrease or stay the same if the outlier is a high value.

- Discuss students' predictions and the actual results.

Example 3

? "What impact will there be if everyone receives a \$0.40 raise?" Answers will vary. Encourage students to reason about the problem.

- Work through the example.
- Discuss students' predictions and the actual results.
- **Big Idea:** When the same amount is added to each data value, the three measures increase by that same amount.
- **Extension:** If time permits, explore what happens if everyone receives a 10% increase.

On Your Own

- These questions integrate a review of fraction operations.
- For Question 3, check to see if students simply add $1\frac{1}{2}$ to each of their previous answers or if they compute the mean, median, and mode again.

Closure

- Explain the affect of an outlier on each of the three measures of central tendency.

Technology For the Teacher

Dynamic Classroom

The Dynamic Planning Tool
Editable Teacher's Resources at *BigIdeasMath.com*

EXAMPLE 2 Removing an Outlier

Remember

An *outlier* is a data value that is much greater or much less than the other values.

Identify the outlier in Example 1. How does the outlier affect the mean, median, and mode?

The value \$3.87 is low compared to the other wages. It is the outlier.

Find the mean, median, and mode without the outlier.

Mean: $\frac{54.93}{7} \approx 7.85$

Median: 6.99, 7.25, 7.25, 7.99, 8.25, 8.45, 8.75 — The middle value, 7.99, is the median.

Mode: 6.99, 7.25, 7.25, 7.99, 8.25, 8.45, 8.75 — The mode is 7.25.

By removing the outlier, the mean increases \$7.85 − \$7.35 = \$0.50, the median increases \$7.99 − \$7.62 = \$0.37, and the mode is the same.

EXAMPLE 3 Changing the Values of a Data Set

In Example 1, each hourly wage increases \$0.40. How does this increase affect the mean, median, and mode?

Make a new table by adding \$0.40 to each hourly wage.

Students' Hourly Wages	
\$4.27	\$7.65
\$9.15	\$8.85
\$8.65	\$7.65
\$7.39	\$8.39

Mean: $\frac{62}{8} = 7.75$

Median: 4.27, 7.39, 7.65, 7.65, 8.39, 8.65, 8.85, 9.15 — Order the data.

$\frac{16.04}{2} = 8.02$ — Mean of two middle values

Mode: 4.27, 7.39, 7.65, 7.65, 8.39, 8.65, 8.85, 9.15 — The mode is 7.65.

By increasing each hourly wage \$0.40, the mean, median, and mode all increase \$0.40.

On Your Own

Now You're Ready
Exercises 16 and 17

The figure shows the altitudes of several airplanes.

2. Identify the outlier. How does the outlier affect the mean, median, and mode? Explain.

3. Each airplane increases its altitude $1\frac{1}{2}$ miles. How does this affect the mean, median and mode? Explain.

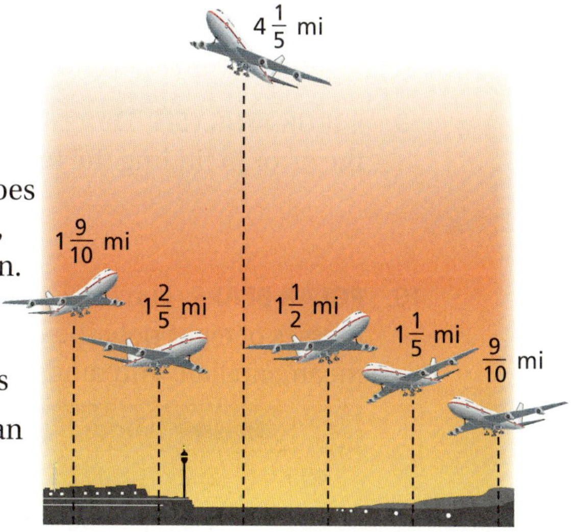

7.1 Exercises

Check It Out
Help with Homework
BigIdeasMath.com

Vocabulary and Concept Check

1. **VOCABULARY** Can a data value be an outlier *and* a measure of central tendency of the same data set? Explain.
2. **OPEN-ENDED** Create a data set that has more than one mode.
3. **WRITING** Describe how removing an outlier from a data set affects the mean of the data set.

Practice and Problem Solving

4. Draw a line plot of the data. Then find the mean, median, and mode of the data.

Bag	1	2	3	4	5	6	7	8	9
Strawberries	10	13	11	15	8	14	7	11	12

Find the mean, median, and mode of the data.

1 5.

Golf Scores		
3	−2	1
6	4	−1
−3	−1	2

6.

Changes in Stock Value (dollars)			
1.05	2.03	−1.78	−2.41
−2.64	0.67	4.02	1.39
0.66	−0.38	−3.01	2.20

7.

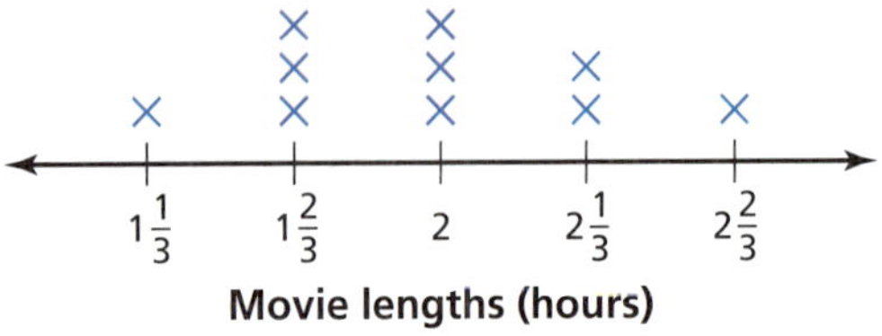

8. **Available Memory**

Stem	Leaf
6	5
7	0 5 5
8	0 4 5
9	4

Key: 7 | 5 = 75 megabytes

9. **ERROR ANALYSIS** Describe and correct the error in finding the median.

✗ Test scores: 98, 90, 80, 80, 90, 90

The median is $\frac{528}{6} = 88$.

10. **POLAR BEARS** The table shows the masses of eight polar bears. Find the mean, median, and mode of the masses.

Masses (kilograms)			
455	262	471	358
364	553	352	467

Assignment Guide and Homework Check

Level	Day 1 Activity Assignment	Day 2 Lesson Assignment	Homework Check
Basic	4, 19–21	1–3, 5–10, 15, 17	2, 6, 8, 15
Average	4, 19–21	1–3, 5–9, 11–17 odd	2, 6, 8, 11, 15
Advanced	4, 19–21	1–3, 6, 8, 9, 12, 14–18	6, 8, 12, 16

Common Errors

- **Exercises 5–8** When finding the mean, students may forget to divide by the total number of data values and instead divide by the maximum value. Remind them that the definition of mean is an "average," so they must take into account the total number of items or numbers to get an average. Explain to students that it is as if they are dividing the total evenly among the number of groups.
- **Exercises 5–8** Students may try to identify the median without ordering the data first. Remind them that it is essential to order the data first and then find the median. This also makes finding the mode easier.
- **Exercises 11–14** Students may not know how to find the missing value. Remind them of the definition of mean and median. Encourage students to use these definitions to write an equation to find the value of *x*.

7.1 Record and Practice Journal

Find the mean, median, and mode of the data.

1.

Song Lengths (minutes)		
2.6	3.25	4.15
2.52	3.67	3.1
3.78	4.9	3.8

mean: 3.53; median: 3.67; mode: none

2.

mean: 4.25; median: 4; mode: 4

Find the value of *x*.

3. Mean is 20; 6, 22, x, 7, 36

   $x = 29$

4. Median is 28; 16, 24, x, 48

   $x = 32$

5. A statistician records the winning scores of five basketball teams.

   105, 98, 92, 108, 70

   a. Identify the outlier.

   70

   b. Which measure of central tendency will be most affected by removing the outlier?

   mean

   c. Calculate the mean and median with and without the outlier.

   mean with outlier: 94.6; mean without outlier: 100.75; median with outlier: 98; median without outlier: 101.5

Technology For the Teacher

Answer Presentation Tool

Quiz*Show*

Vocabulary and Concept Check

1. no; The definition of an outlier implies that it is not in the center of the data.
2. *Sample answer:* 1, 2, 2, 3, 6, 8, 8, 9, 12
3. If the outlier is greater than the mean, removing it will decrease the mean. If the outlier is less than the mean, removing it will increase the mean.

Practice and Problem Solving

4. 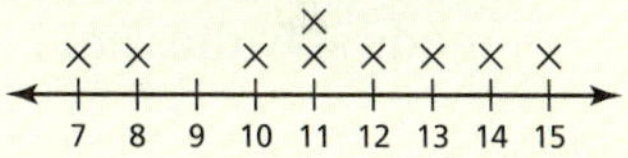

   mean: $11.\overline{2}$
   median: 11
   mode: 11

5. mean: 1
   median: 1
   mode: -1

6. mean: \$0.15
   median: \$0.665
   mode: none

7. mean: $1\frac{29}{30}$ h
   median: 2 h
   modes: $1\frac{2}{3}$ h and 2 h

8. mean: 78.5 MB
   median: 77.5 MB
   mode: 75 MB

9. They calculated the mean, not the median.
   Test scores:
   80, 80, 90, 90, 90, 98

   $$\text{Median} = \frac{90 + 90}{2} = \frac{180}{2} = 90$$

10. mean: 410.25 kg
    median: 409.5 kg
    mode: none

Practice and Problem Solving

11. 4 **12.** −6.5

13. 16 **14.** 57

15. **a.** 105°F

b. mean

16. See *Taking Math Deeper*.

17. The mean and median both decrease by $0.05. There is still no mode.

18. **a.** mean: 19.37 yr
median: 19 yr
mode: 18 yr

b. 37 yr; The mean decreases about $19.37 - 19.19 = 0.18$ year. The median and mode stay the same.

Fair Game Review

19. −8, −5, −3, 1, 4, 7

20. $-4.7, -2.8, -\frac{2}{3}, 1.2, \frac{3}{2}, 5.4$

21. B

Mini-Assessment

Find the mean, median, and mode of the data.

1. 10, −4, 3, −1, 12 4, 3, no mode
2. 1.25, 3.80, −0.65, −2.40 0.5, 0.3, no mode
3. 5, 15, 8, 13, 10, 8, 6, 4, 12 9, 8, 8

Find the value of *x*.

4. Mean is 2; −4, −2, 3, *x*, 9 4
5. Median is 16.5; 8, 11, *x*, 18, 24, 26 15

Taking Math Deeper

Exercise 16

The Appalachian Trail is the longest marked trail in the U.S. at 2178 miles. The 11 shelters in this problem are in Massachusetts.

1 Order the data.

0.1, 1.8, 3.3, 5.3, 6.3, 8.8, 8.8, 14, 14.3, 16.7

2 Find the mean, median, and mode.

a. $\text{Mean} = \frac{79.4}{10} = 7.94$ miles

$\text{Median} = \frac{(6.3 + 8.8)}{2} = 7.55$ miles

The mode is 8.8 miles.

3 A hiker begins the trail at Shelter 2 and therefore skips the 0.1-mile distance of the trail. The mean, median, and mode of the remaining distances are as follows.

1.8, 3.3, 5.3, 6.3, 8.8, 8.8, 14, 14.3, 16.7

Find the mean, median, and mode.

b. $\text{Mean} = \frac{79.3}{9} \approx 8.8$ miles

Median = 8.8 miles

The mode is 8.8 miles.

The mean increases by about 0.86 mile.
The median increases by 1.25 miles.
The mode does not change.

Project

Plan a hiking trip. List the things you need to take with you. Include the approximate amount of time you think it will take.

Reteaching and Enrichment Strategies

If students need help. . .	If students got it. . .
Resources by Chapter • Practice A and Practice B • Puzzle Time Record and Practice Journal Practice Differentiating the Lesson Lesson Tutorials Skills Review Handbook	Resources by Chapter • Enrichment and Extension Start the next section

Find the value of x.

11. Mean is 6; 2, 8, 9, 7, 6, x

12. Mean is 0; 11.5, 12.5, -10, -7.5, x

13. Median is 14; 9, 10, 12, x, 20, 25

14. Median is 51; 30, 45, x, 100

② **15. TEMPERATURES** An environmentalist records the average temperatures of five regions.

a. Identify the outlier.

b. Which measure of central tendency will be most affected by removing the outlier?

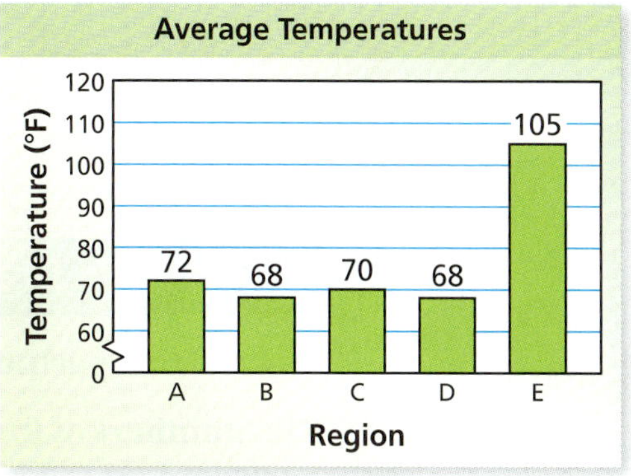

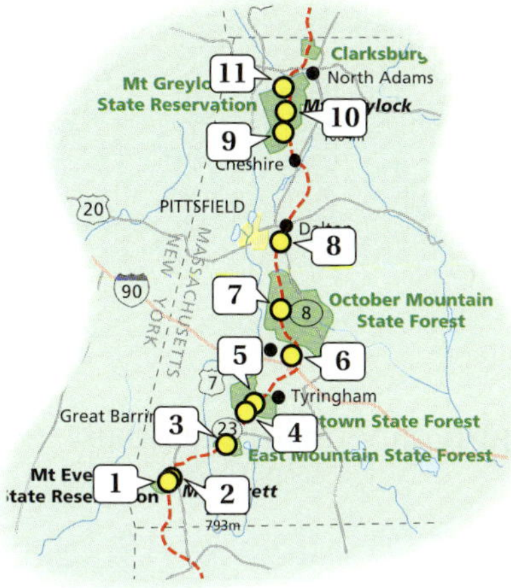

16. TRAIL The map shows the locations of 11 shelters along the Appalachian Trail. The distances (in miles) between these shelters are 0.1, 14.3, 5.3, 1.8, 14, 8.8, 8.8, 16.7, 6.3, and 3.3.

a. Find the mean, median, and mode of the distances.

b. A hiker starts at Shelter 2 and hikes to Shelter 11. How does this affect the mean, median, and mode? Explain.

③ **17. REASONING** The value of each stock in Exercise 6 decreases \$0.05. How does this affect the mean, median, and mode? Explain.

18. Critical Thinking The circle graph shows the ages of 200 students in a college psychology class.

a. Find the mean, median, and mode of the students' ages.

b. Identify the outliers. How do the outliers affect the mean, median, and mode?

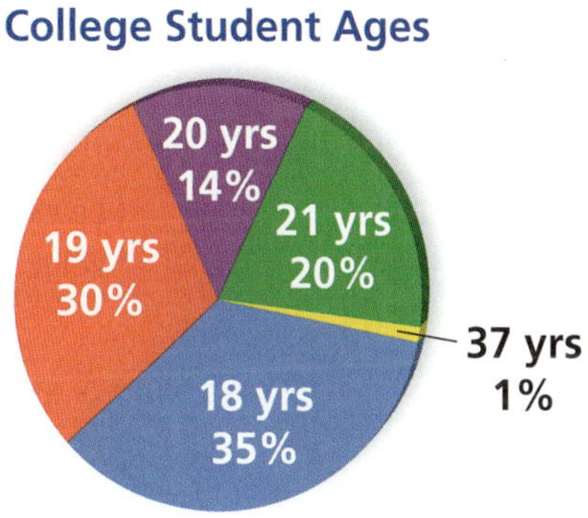

Fair Game Review What you learned in previous grades & lessons

Order the values from least to greatest. *(Skills Review Handbook)*

19. 1, -3, -8, 4, 7, -5

20. 1.2, -2.8, $\frac{3}{2}$, 5.4, -4.7, $-\frac{2}{3}$

21. MULTIPLE CHOICE Which equation represents a linear function? *(Section 4.4)*

Ⓐ $y = x^2$ Ⓑ $y = 2x$ Ⓒ $y = \frac{2}{x}$ Ⓓ $xy = 2$

7.2 Box-and-Whisker Plots

Essential Question How can you use a box-and-whisker plot to describe a population?

1 ACTIVITY: Drawing a Box-and-Whisker Plot

Work with a partner.

The numbers of first cousins of each student in an eighth-grade class are shown.

A box-and-whisker plot uses a number line to represent the data visually.

Numbers of First Cousins

3	10	18	8
9	3	0	32
23	19	13	8
6	3	3	10
12	45	1	5
13	24	16	14

a. Order the data set and write it on a strip of grid paper with 24 equally spaced boxes.

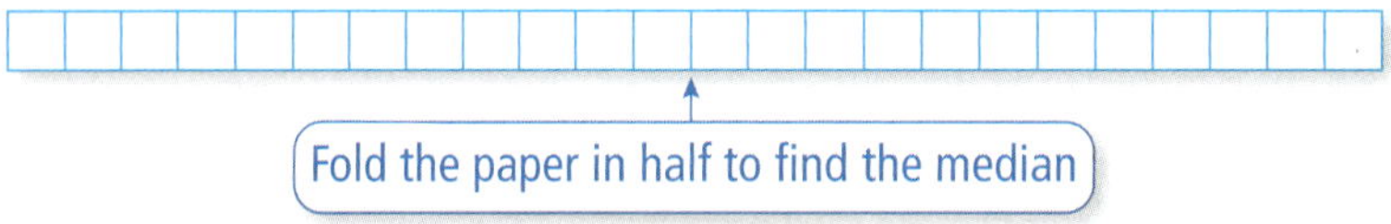

b. Fold the paper in half again to divide the data into four groups. Because there are 24 numbers in the data set, each group should have six numbers.

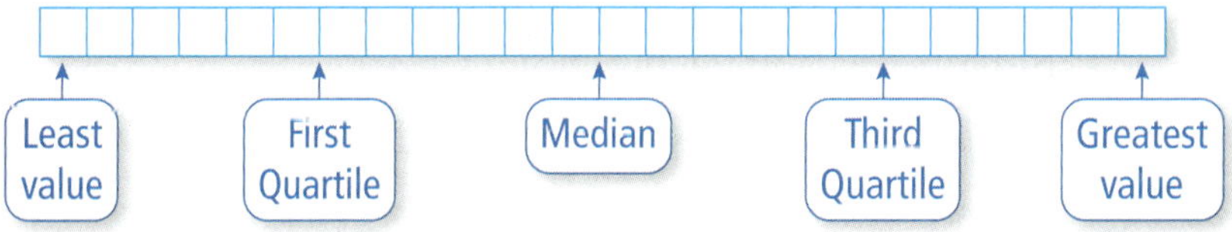

c. Draw a number line that includes the least value and the greatest value in the data set. Graph the five numbers that you found in part (b).

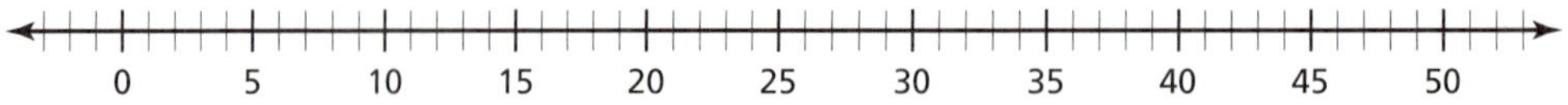

d. Explain how the box-and-whisker plot shown below represents the data set.

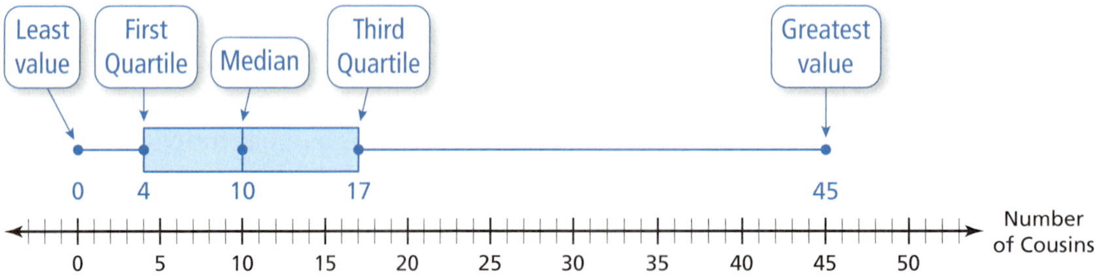

Laurie's Notes

Introduction

For the Teacher

- **Goal:** Students will gain a general understanding of how a box-and-whisker plot is constructed.
- The box-and-whisker plot is used to display information about very large data sets. However, in order to make learning the technique manageable for students, the actual data sets used today are small.
- Students quickly recognize that box-and-whisker plots involve finding five key values. Constructing the plot is not difficult. Analyzing the plot is the challenging part for students.
- **Connection:** When discussing the meaning of percentiles and quartiles with my students, I had one boy who said, "It's like at the doctor's office when they tell me what percentile I'm in for height and weight." The connection was immediate for the class and *they got it*!

Motivate

- Share a story about your commute to school today—perhaps the traffic, something you saw, or a stop you made for coffee. Conclude with how many minutes it took for your commute.
- Collect class data about the numbers of minutes it took your students to commute to school this morning, from the time they left their front doors until they walked into the school. If this is awkward data to collect, change the question. Data can be collected on slips of paper.
- Record the data on the board and leave it for later. You may want to take time to have students make comments about the data set.

Activity Notes

Activity 1

? "How many pieces of data are there?" 24

- Explain that today they are going to construct a box-and-whisker plot, a data display that is generally used for very large data sets. Data are *not* graphed, but characteristics of the data set are still conveyed. For instance, the results of a state test for all 8th graders could be displayed using a box-and-whisker plot.
- The box-and-whisker plot uses a number line to visually represent the data.

? "How many numbers did you graph in making the box-and-whisker plot?" 5

- **Big Idea:** The 5 numbers graphed summarize the entire data set. The least and greatest values are the boundaries. The median separates the data into two parts. The first (or lower) quartile is the median of the lower half. The third (or upper) quartile is the median of the upper half. The box encloses the middle 50% of the data.

? "What percent of the data is represented by the upper whisker?" 25%
"How many data values are in the upper whisker?" 6

? "What percent of the data is represented by the lower whisker?" 25%
"How many data values are in the lower whisker?" 6

Previous Learning

Students should know how to find the median of a data set.

Start Thinking! and Warm Up

Activity 7.2 Start Thinking!
For use before Activity 7.2

Activity 7.2 Warm Up
For use before Activity 7.2

Find the median, least value, and greatest value in the data set.

1. 1, 3, 4, 5, 5, 6, 8, 10, 12
2. 0.5, 0.6, 0.8, 1.1, 1.2, 1.7
3. 45, 55, 55, 57, 59, 89
4. 5, 14, 54, 56, 78, 79, 100, 120, 222

7.2 Record and Practice Journal

Essential Question How can you use a box-and-whisker plot to describe a population?

1 ACTIVITY: Drawing a Box-and-Whisker Plot

Work with a partner.

The numbers of first cousins of each student in an eighth-grade class are shown.

A box-and-whisker plot uses a number line to represent the data visually.

Numbers of First Cousins

3	10	18	8
9	3	0	32
23	19	13	8
6	3	3	10
12	45	1	5
13	24	16	14

a. Order the data set and write it on a strip of grid paper with 24 equally spaced boxes.

0 1 3 3 3 3 5 6 8 8 9 10 10 12 13 13 14 16 18 19 23 24 32 45

Fold the paper in half to find the median

b. Fold the paper in half again to divide the data into four groups. Because there are 24 numbers in the data set, each group should have six numbers.

0 1 3 3 3 3 5 6 8 8 9 10 10 12 13 13 14 16 18 19 23 24 32 45

Least value — First Quartile — Median — Third Quartile — Greatest value

c. Use the number line. Graph the five numbers that you found in part (b).

0 5 10 15 20 25 30 35 40 45 50

English Language Learners

Visual

Use a diagram of a generic box-and-whisker plot on an overhead as a visual aid for English learners. Have students identify the parts of the box-and-whisker plot: *median, first quartile, third quartile, least value,* and *greatest value.* Make sure students understand that they can interpret a box-and-whisker plot that does not have a scale.

7.2 Record and Practice Journal

d. Explain how the box-and-whisker plot shown below represents the data set.

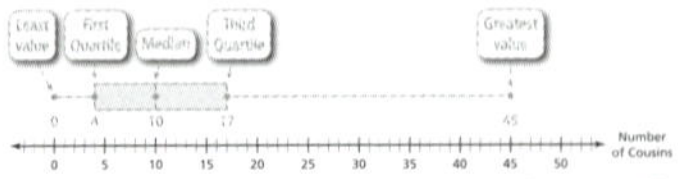

represents median, range, and distribution of the data set

2 ACTIVITY: Conducting a Survey

Conduct a survey in your class. Ask each student to write the number of his or her first cousins on a piece of paper.

Two people are first cousins if they share at least one grandparent, but do not share a parent.

Collect the pieces of paper and write the data on the chalkboard.

Now, work with a partner to draw a box-and-whisker plot of the data.

Check students' work.

3 ACTIVITY: Reading a Box-and-Whisker Plot

Work with a partner. The box-and-whisker plots show the test score distributions of two eighth-grade standardized tests. The tests were taken by the same group of students. One test was taken in the fall and the other was taken in the spring.

a. Compare and contrast the test results.

b. Decide which box-and-whisker plot represents the results of which test. How did you make your decision?

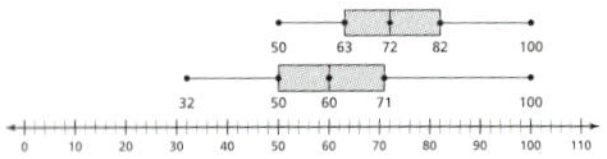

(a) Upper plot shows higher scores.
(b) Spring; Fall

What Is Your Answer?

4. **IN YOUR OWN WORDS** How can you use a box-and-whisker plot to describe test scores?

show distribution, median, and range

5. Describe who might be interested in test score distributions like those shown in Activity 3. Explain why it is important for such people to know test score distributions.

Sample answer: Teachers; if they know when scores are lower, they can determine the reason and address it.

Laurie's Notes

Activity 2

- Explain that you want to practice making a box-and-whisker plot with data collected from students. You will gather and record information about the number of first cousins for each student in your class. If this is awkward, ask a different question.
- Students should follow the steps from Activity 1 to construct this plot. If there is an odd number of students in the class, the median is the middle value of the sorted data. To find the first quartile, exclude the median and find the median of the lower half of the data. To find the third quartile, exclude the median and find the median of the upper half of the data.
- As students are making the plot, make the same plot at the overhead projector for discussion purposes.
- Ask questions about the plot: median, range, number of data values considered versus number of data values graphed.

Activity 3

- One advantage of box-and-whisker plots is that multiple plots can be displayed and analyzed using the same number line. For instance, state test scores for 5 different schools could be displayed on the same number line.
- Read the information given and analyze the two plots.

? "Which test is represented by which plot? Explain." Listen for students discussing the location of the median and the third quartile for each plot. The spring test is the top plot.

? "True or false: 50% of the top plot is greater than 75% of the bottom plot." true

What Is Your Answer?

- **Think-Pair-Share:** Students should read each question independently and then work with a partner to answer the questions. When they have answered the questions, the pair should compare their answers with another group and discuss any discrepancies.

Closure

- Make a box-and-whisker plot of the data collected at the beginning of class. Write one or two observations about the plot.

Technology For the Teacher

Dynamic Classroom

The Dynamic Planning Tool
Editable Teacher's Resources at *BigIdeasMath.com*

2 ACTIVITY: Conducting a Survey

Conduct a survey in your class. Ask each student to write the number of his or her first cousins on a piece of paper. Collect the pieces of paper and write the data on the chalkboard.

Now, work with a partner to draw a box-and-whisker plot of the data.

Two people are first cousins if they share at least one grandparent, but do not share a parent.

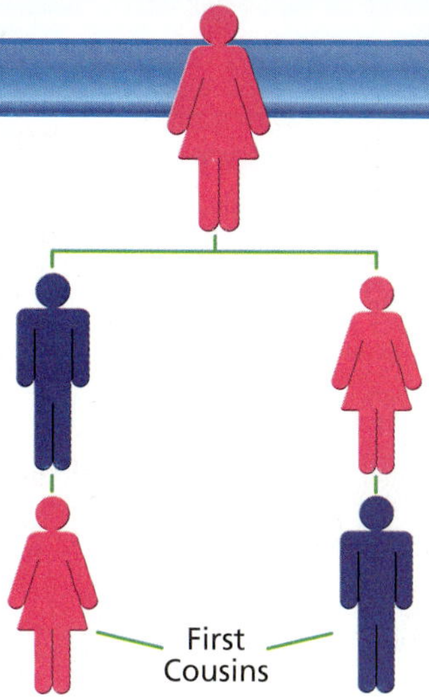

3 ACTIVITY: Reading a Box-and-Whisker Plot

Work with a partner. The box-and-whisker plots show the test score distributions of two eighth-grade standardized tests. The tests were taken by the same group of students. One test was taken in the fall and the other was taken in the spring.

a. Compare and contrast the test results.

b. Decide which box-and-whisker plot represents the results of which test. How did you make your decision?

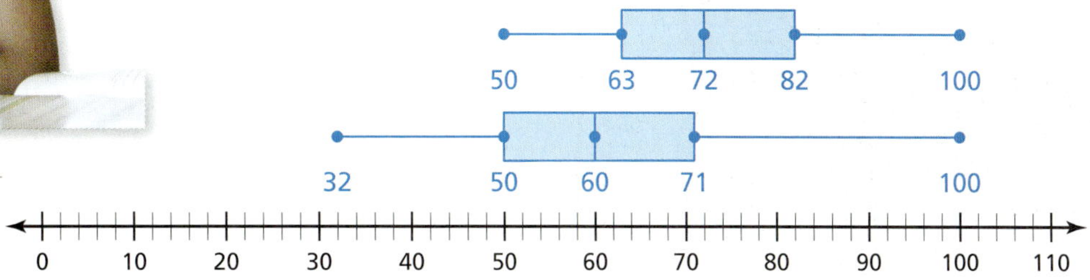

What Is Your Answer?

4. **IN YOUR OWN WORDS** How can you use a box-and-whisker plot to describe test scores?

5. Describe who might be interested in test score distributions like those shown in Activity 3. Explain why it is important for such people to know test score distributions.

Practice Use what you learned about box-and-whisker plots to complete Exercise 4 on page 284.

7.2 Lesson

Check It Out
Lesson Tutorials
BigIdeasMath.com

Key Vocabulary
box-and-whisker plot, *p. 282*
quartiles, *p. 282*

Study Tip
A box-and-whisker plot shows the *variability* of a data set.

Key Idea

Box-and-Whisker Plot

A **box-and-whisker plot** displays a data set along a number line using medians. **Quartiles** divide the data set into four equal parts. The median (second quartile) divides the data set into two halves. The median of the lower half is the first quartile. The median of the upper half is the third quartile.

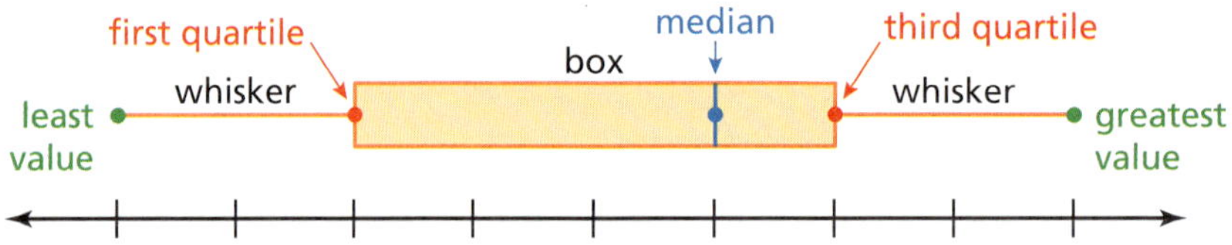

EXAMPLE 1 Making a Box-and-Whisker Plot

Make a box-and-whisker plot for the ages of the members of the 2008 U.S. women's wheelchair basketball team.

24, 30, 30, 22, 25, 22, 18, 25, 28, 30, 25, 27

Step 1: Order the data. Find the median and the quartiles.

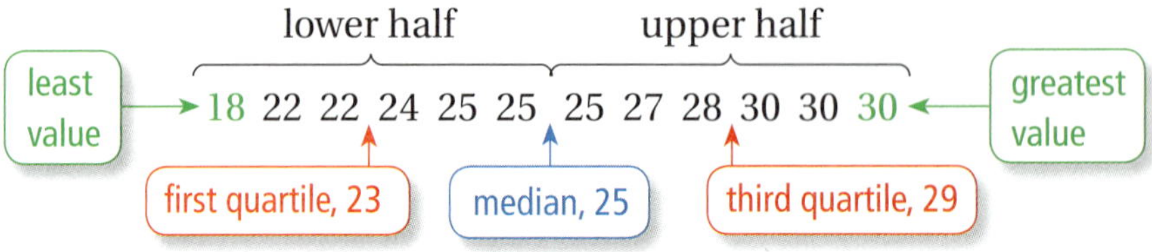

Step 2: Draw a number line that includes the least and greatest values. Graph points above the number line for the least value, greatest value, median, first quartile, and third quartile.

Step 3: Draw a box using the quartiles. Draw a line through the median. Draw whiskers from the box to the least and greatest values.

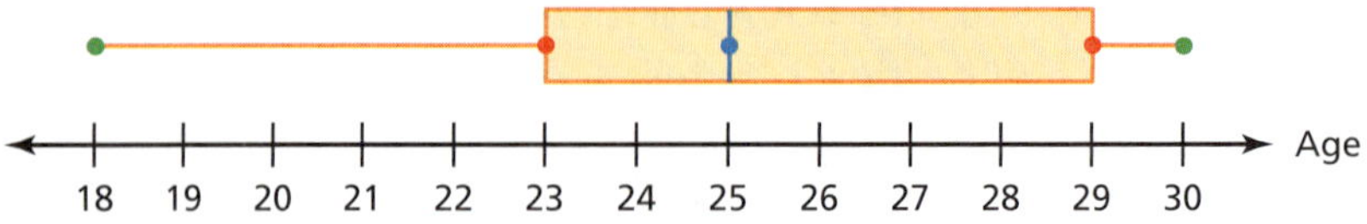

On Your Own

Now You're Ready
Exercises 5–7

1. A basketball player scores 14, 16, 20, 5, 22, 30, 16, and 28 points during a tournament. Make a box-and-whisker plot for the points scored by the player.

Laurie's Notes

Introduction

Connect

- **Yesterday:** Students gained a general understanding of how a box-and-whisker plot is constructed.
- **Today:** Students will construct and analyze box-and-whisker plots.

Motivate

- The physical involvement of making a human box-and-whisker plot makes a lasting impression on students.
- **Preparation:** Give each student an index card with a number written on it. Decide on an interesting feature of the plot, such as an outlier or two at one end.
- ? "What is the first step in making a box-and-whisker plot?" sort the data
- Students should stand up and sort themselves. Have the median, the first and third quartiles, and the least and greatest data values take one step forward. If there is an even number of data values, the middle two students must figure out how to represent the mean and take a step forward.
- Make a number line on the floor or on the board. Position the 5 key values. If the plot is done on the floor, use string to form the whiskers. Students will have to visualize the box.
- Discuss features of the plot. If the plot includes an outlier, the length of the string becomes an instant topic of conversation. Students recognize that the same number of data values (25% of the class) is being represented by each whisker, yet the lengths of string are very different.

Lesson Notes

Key Idea

- Define the box-and-whisker plot constructed by graphing the three quartiles of the data. Draw the sample plot and discuss the process and vocabulary.
- **Discuss:** The box-and-whisker plot shows the *variability* of the data. Refer to this idea in each example done today.

Example 1

- Remind students that the data must always be sorted before finding the median of a data set.
- Notice that the first quartile (23) and the third quartile (29) are not data values from the set. This is fine. The five values are simply giving a marker for how the sorted data is spread out into four groups.
- ? "How many players are represented in each quartile?" 3
- ? "What percent of the players were older than 23?" 75%
- ? "What is the range of ages for the team?" 12

On Your Own

- This is a very small data set to make it manageable for students.

Goal Today's lesson is constructing and analyzing a **box-and-whisker plot**.

Lesson Materials
Introduction
• index cards • string or yarn

Start Thinking! and Warm Up

Lesson 7.2 Warm Up
For use before Lesson 7.2

Lesson 7.2 Start Thinking!
For use before Lesson 7.2

Do you think it costs more to see a professional baseball game or basketball game?

The box-and-whisker plots show the average ticket price for each team in Major League Baseball (MLB) and the National Basketball Association (NBA).

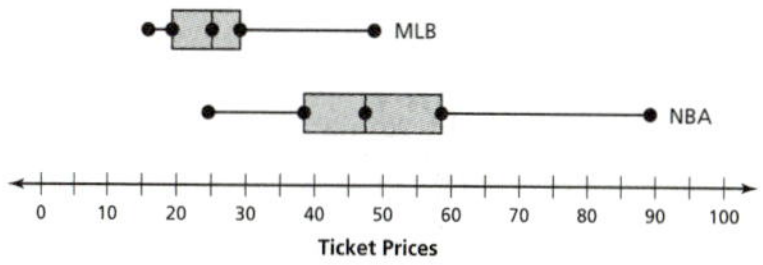

Analyze the box-and-whisker plots to compare the prices in MLB and the NBA.

Extra Example 1

Make a box-and-whisker plot for the ages of the members of a women's basketball team.

25, 22, 18, 23, 27, 20, 18, 25, 28, 17, 23, 18

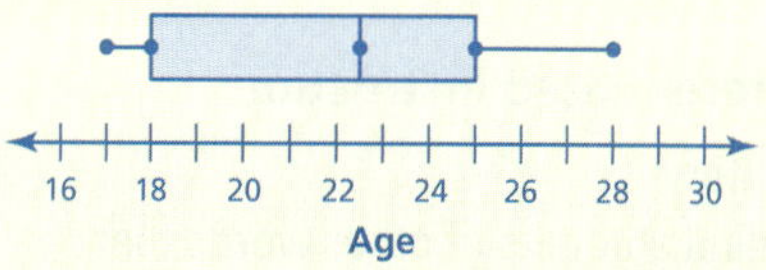

On Your Own

1. See Additional Answers.

Extra Example 2

In Extra Example 1, what does the box-and-whisker plot tell you about the data? *Sample answer:* The right whisker is longer than the left whisker. So, the data are more spread out above the third quartile than below the first quartile. The range of the data is 28 − 17 = 11 years.

Extra Example 3

Compare the in-line skate prices of Shop A and Shop B. What are three conclusions you can make from the double box-and-whisker plot?

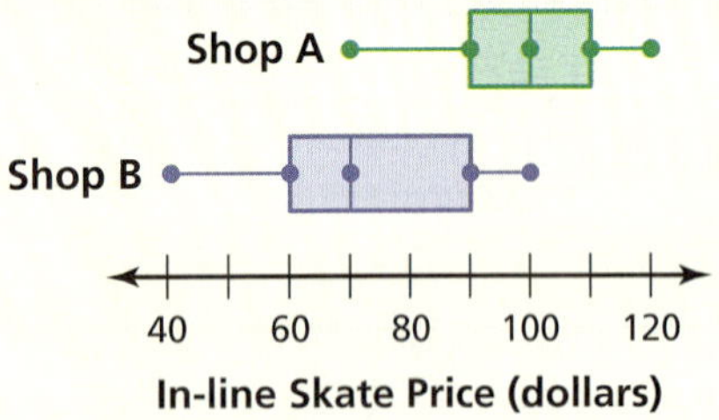

Sample answer: The range of the prices at Shop A is less than the range at Shop B. The Shop B prices are less than or equal to half of the Shop A prices. The median price at Shop A is greater than the median price at Shop B.

On Your Own

2. *Sample answer:* Shop A has more expensive prices on average. Shop B sells bargain surfboards. The prices for Shop A are more spread out.

Differentiated Instruction

Auditory

Remind students of other words that have the same root as *quartile:* quarter and quart, for example. Define the words as *four* or *fourths.* Mention that one-fourth is the same as 25%.

Laurie's Notes

Example 2

- This example does not state how many pieces of data the plot represents. This can be true of box-and-whisker plots found in print materials.
- ? "Can someone read the 5 key values from the plot and what they mean in the context of the problem?" least value = 50 in.; first quartile = 58 in.; median = 64 in.; third quartile = 70 in.; greatest value = 72 in.
- ? "What is the range and how did you find it?" 22 in.; The range of a data set is the difference between the greatest data value and the least data value.
- Explain what the lengths of the whiskers and the box tell you about the range of the data. The data below the first quartile (0% to 25%) has the greatest range, while the data above the third quartile (75% to 100%) has the least range.
- Stress that each of the four parts of the plot represent 25% or $\frac{1}{4}$ of the data.

Example 3

- Read each of the 4 statements and assess if they are true or not.
- ? Ask additional questions about the plots.
  - "50% of your class did better than what percent of your friend's class?" 75%
  - "Were there more students who took the test in your friend's class than in your class?" cannot tell
  - "Is it true that your class did better than your friend's class on the test?" generally true

On Your Own

- **Whole Class Discussion:** Have students analyze the plots shown and make two valid statements about the plots. Share the statements as a class.

Closure

- **Exit Ticket:**
  - What are the 5 key values that are graphed in a box-and-whisker plot? least value, first quartile, median, third quartile, greatest value
  - How does an outlier affect a box-and-whisker plot? increases the length of one of the whiskers
  - Explain why two data sets of different sizes can be graphed on the same number line. Box-and-whisker plots show the distribution of the data, not individual data points.

Technology For the Teacher

Dynamic Classroom

The Dynamic Planning Tool
Editable Teacher's Resources at *BigIdeasMath.com*

EXAMPLE 2 Interpreting a Box-and-Whisker Plot

Study Tip

A long whisker or box indicates data is more spread out.

What does the box-and-whisker plot tell you about the data?

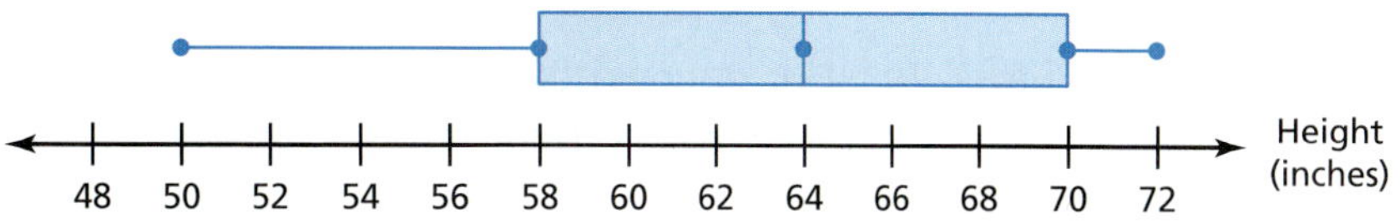

- The left whisker is longer than the right whisker. So, the data are more spread out below the first quartile than above the third quartile.
- The range of the data is $72 - 50 = 22$ inches.

EXAMPLE 3 Standardized Test Practice

Which statement is true about the double box-and-whisker plot?

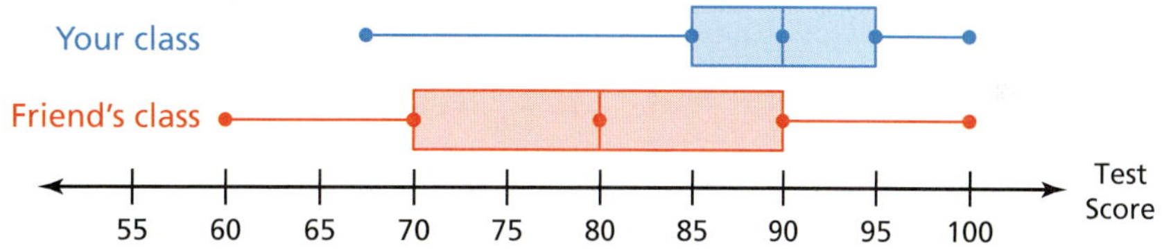

Ⓐ Half of the test scores in your class are between 85 and 100.

Ⓑ 25% of the test scores in your friend's class are 80 or above.

Ⓒ The medians are the same for both classes.

Ⓓ The test scores in your friend's class are more spread out than the test scores in your class.

The range of the test scores in your class is less than the range in your friend's class. Also, the box for your friend's class is longer than the box for your class. So, the test scores in your friend's class are more spread out than the test scores in your class.

∴ The correct answer is Ⓓ.

On Your Own

Now You're Ready
Exercise 10

2. Compare the surfboard prices of Shop A and Shop B. What are three conclusions you can make from the double box-and-whisker plot?

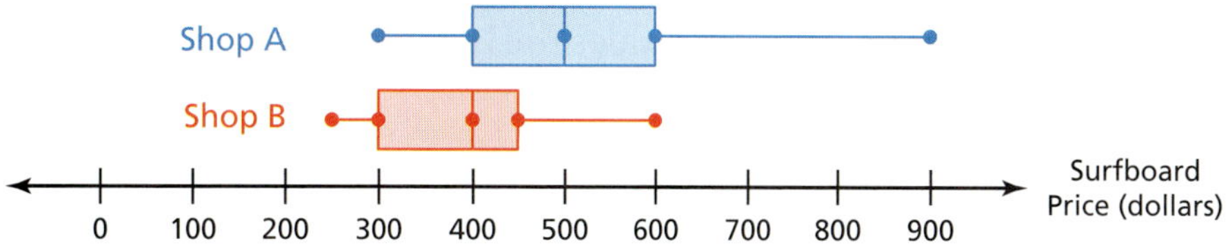

7.2 Exercises

Check It Out
Help with Homework
BigIdeasMath.com

Vocabulary and Concept Check

1. **VOCABULARY** In a box-and-whisker plot, what percent of the data is represented by each whisker? the box?

2. **WRITING** Describe how to find the first quartile of a data set.

3. **NUMBER SENSE** What does the length of the box-and-whisker plot tell you about the data?

Practice and Problem Solving

4. The box-and-whisker plots show the monthly car sales for a year for two sales representatives. Compare and contrast the sales of the two representatives.

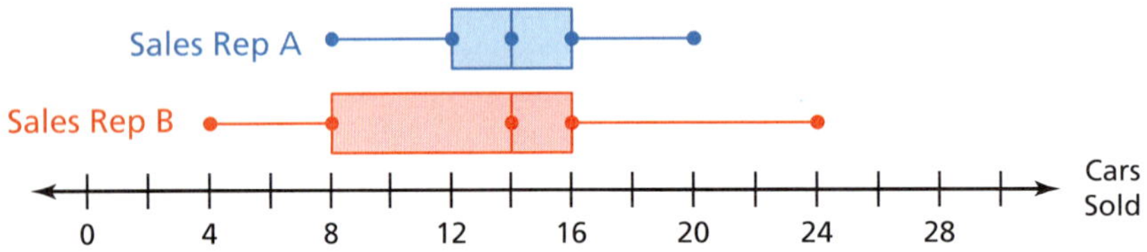

Make a box-and-whisker plot for the data.

(1) 5. Hours of television watched: 0, 3, 4, 5, 3, 4, 6, 5

6. Lengths (in inches) of cats: 16, 18, 20, 25, 17, 22, 23, 21

7. Elevations (in feet): −2, 0, 5, −4, 1, −3, 2, 0, 2, −3, 6, −1

8. **ERROR ANALYSIS** Describe and correct the error in making a box-and-whisker plot for the data.

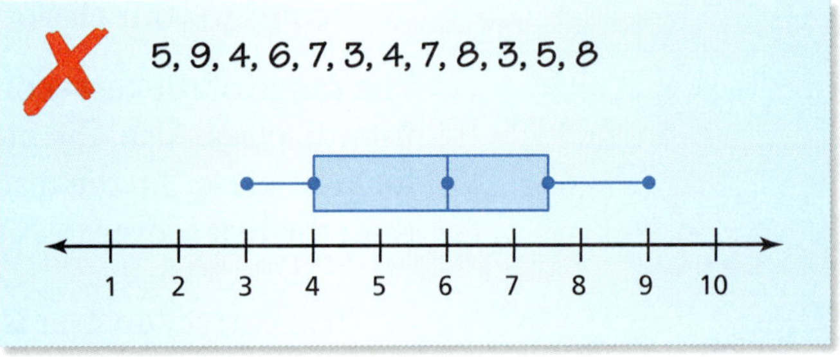

9. **FISH** The lengths (in inches) of the fish caught on a fishing trip are 9, 10, 12, 8, 13, 10, 12, 14, 7, 14, 8, and 14. Make a box-and-whisker plot for the data. What is the range of the data?

(2) 10. **INCHWORM** The table shows the lengths of 12 inchworms. Make a box-and-whisker plot for the data. What does the box-and-whisker plot tell you about the data?

Length (cm)	2.5	2.4	2.3	2.5	2.7	2.1	2.8	2.6	2.1	2.6	2.9	2.0

Assignment Guide and Homework Check

Level	Day 1 Activity Assignment	Day 2 Lesson Assignment	Homework Check
Basic	4, 17–21	1–3, 5–10	2, 6, 8, 10
Average	4, 17–21	1–3, 7–12	2, 8, 10, 12
Advanced	4, 17–21	1–3, 8–10, 12–16	2, 8, 10, 12

Common Errors

- **Exercises 5–7** Students may forget to order the data before beginning their box-and-whisker plot. Remind them that they need to order the data to find the median and each quartile.
- **Exercises 5–7** Students may find an incorrect median or quartile when it is between two numbers. Remind them that when the median is between two numbers, they need to find the *mean* of those two values.
- **Exercise 11** Students may make a calculation mistake when they are making the box-and-whisker plot without the outlier. Encourage them to rewrite the data before finding the new median and quartiles.

Vocabulary and Concept Check

1. 25%; 50%
2. Order the data. Take the first half of the data, and find the median of that data.
3. The length gives the range of the data set. This tells how much the data vary.

Practice and Problem Solving

4. *Sample answer:* Both Sales Reps have the same median number of cars sold, but Sales Rep B's numbers are more spread out than Sales Rep A's numbers.

5–10. See Additional Answers.

7.2 Record and Practice Journal

Make a box-and-whisker plot for the data.

1. Hours of reading: 1, 6, 7, 5, 5, 8, 4, 8

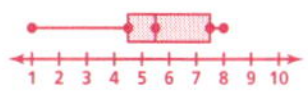

2. Golf scores: −5, −12, 0, 2, −4, 3, −3, −7, −1, −3, −5, 0

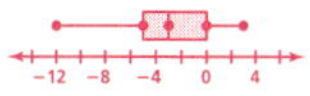

3. The table shows quiz scores of 10 students. Make a box-and-whisker plot for the data. What does the box-and-whisker plot tell you about the data?

Quiz Scores (points)	
20	19
17	18
16	18
22	20
24	25

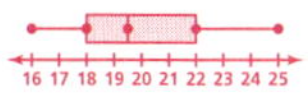

Sample answer: **The data are more spread out above the third quartile than below the first quartile.**

4. The box-and-whisker plot shows the number of pages in a stack of books. What is the range of the lower 25% of the data?

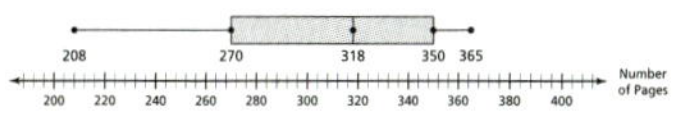

62 pages

Technology For the Teacher

Answer Presentation Tool

Quiz*Show*

Practice and Problem Solving

11. See Additional Answers.

12. See *Taking Math Deeper.*

13. *Sample answer:* 0, 5, 10, 10, 10, 15, 20

14. *Sample answer:* 0, 2, 2, 3, 4, 4, 6

15. *Sample answer:* 1, 7, 9, 10, 11, 11, 12

16. *Sample answer:* 1, 2, 4, 5, 10, 10, 10

Fair Game Review

17. $y = 3x + 2$

18. $y = -\frac{4}{3}x - 1$

19. $y = -\frac{1}{4}x$

20. $y = \frac{1}{2}x + 4$

21. B

Mini-Assessment

Make a box-and-whisker plot for the data.

1. Hours online: 2, 4, 8, 6, 2, 10, 5, 12, 8, 5, 0, 1

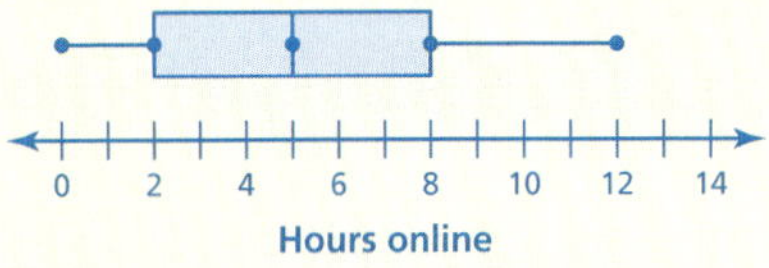

2. Height (in inches) of students: 58, 60, 72, 70, 65, 68, 62, 73, 69, 66

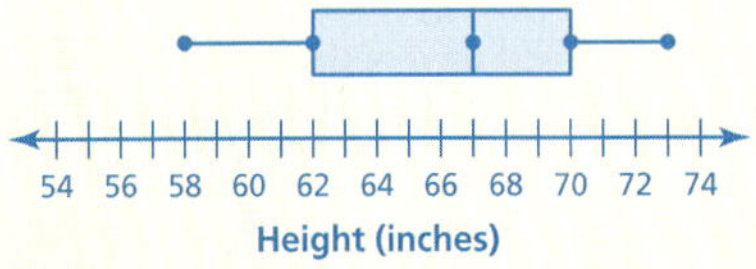

Taking Math Deeper

Exercise 12

This problem can help students see the benefits and limitations of box-and-whisker plots.

1 Copy the box-and-whisker plots.

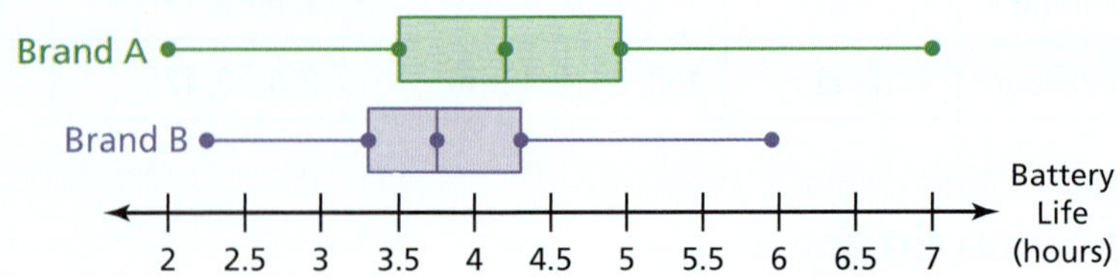

2 **a.** Find the range of the top 75% of the data.

Brand A Range: $7 - 3.5 = 3.5$ hours

Brand B Range: $6 - 3.3 = 2.7$ hours

3 Which brand is better?

This question is open-ended. For Brand A (with 12 batteries), the best and worst case scenarios are as follows.

Mean ≈ 4.1 hours

Brand A: 2, 2, 3.5, 3.5, 3.5, 4.2, 4.2, 4.2, 5, 5, 5, 7
(worst) 2 3.5 4.2 5 7

Mean ≈ 4.5 hours

Brand A: 2, 3.5, 3.5, 3.5, 4.2, 4.2, 4.2, 5, 5, 5, 7, 7
(best) 2 3.5 4.2 5 7

For Brand B, the best and worst case scenarios are as follows.

Mean ≈ 3.7 hours

Brand B: 2.2, 2.2, 3.3, 3.3, 3.3, 3.75, 3.75, 3.75, 4.3, 4.3, 4.3, 6
(worst) 2.2 3.3 3.75 4.3 6

Mean ≈ 4.0 hours

Brand B: 2.2, 3.3, 3.3, 3.3, 3.75, 3.75, 3.75, 4.3, 4.3, 4.3, 6, 6
(best) 2.2 3.3 3.75 4.3 6

b. It is probably safe to say that Brand A is better. But, by unlucky sampling, its mean might be only slightly better than the mean for Brand B.

Reteaching and Enrichment Strategies

If students need help. . .	If students got it. . .
Resource by Chapter • Practice A and Practice B • Puzzle Time Record and Practice Journal Practice Differentiating the Lesson Lesson Tutorials Skills Review Handbook	Resources by Chapter • Enrichment and Extension Start the next section

11. **CALORIES** The table shows the number of calories burned per hour for nine activities.

a. Make a box-and-whisker plot for the data.

b. Identify the outlier.

c. Make another box-and-whisker plot without the outlier.

d. **WRITING** Describe how the outlier affects the whiskers, the box, and the quartiles of the box-and-whisker plot.

Calories Burned per Hour	
Fishing	207
Mowing the lawn	325
Canoeing	236
Bowling	177
Hunting	295
Fencing	354
Bike racing	944
Horseback riding	236
Dancing	266

12. **CELL PHONES** The double box-and-whisker plot compares the battery life (in hours) of two brands of cell phones.

a. What is the range of the upper 75% of each brand?

b. Which battery has a longer battery life? Explain.

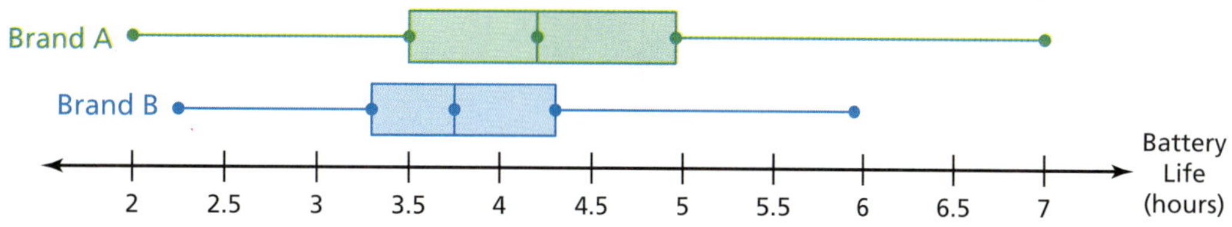

Critical Thinking **Create a set of data values whose box-and-whisker plot has the given characteristic(s).**

13. The least value, greatest value, quartiles, and median are all equally spaced.

14. Both whiskers are the same length as the box.

15. The box between the median and the first quartile is three times as long as the box between the median and the third quartile.

16. There is no right whisker.

Fair Game Review What you learned in previous grades & lessons

Write an equation of the line that passes through the points. *(Section 3.3)*

17. (−4, −10), (2, 8)

18. (−3, 3), (0, −1)

19. (−4, 1), (4, −1)

20. (6, 7), (8, 8)

21. **MULTIPLE CHOICE** You run 10 feet per second. What is this rate in miles per hour? *(Section 1.5)*

Ⓐ 0.11 mi/h Ⓑ 6.82 mi/h Ⓒ 10.23 mi/h Ⓓ 14.67 mi/h

7 Study Help

Check It Out
Graphic Organizer
BigIdeasMath.com

You can use a **word magnet** to organize information associated with a vocabulary word. Here is an example of a word magnet for measures of central tendency.

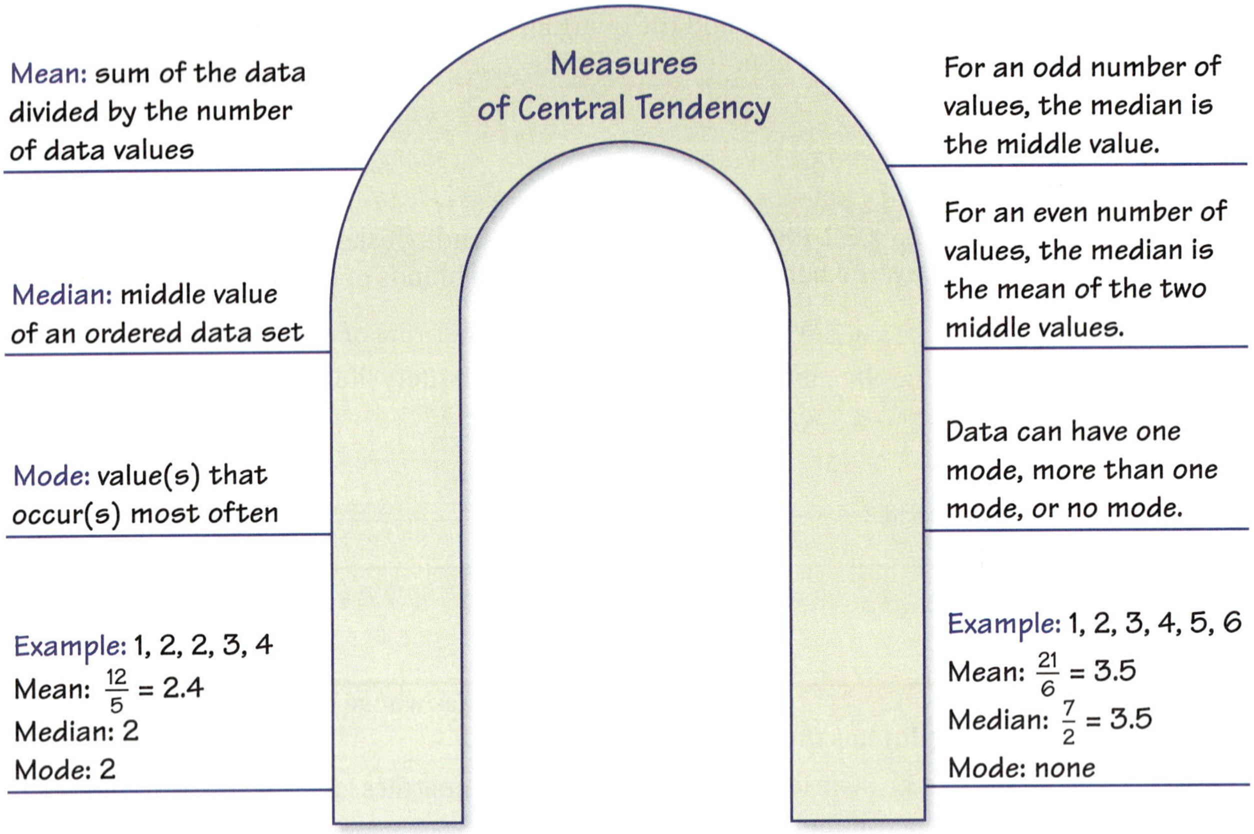

On Your Own

Make a word magnet to help you study these topics.

1. outliers
2. box-and-whisker plots

After you complete this chapter, make word magnets for the following topics.

3. scatter plots
4. lines of best fit
5. data displays

"How do you like the word magnet I made for 'Beagle'?"

Sample Answers

1.

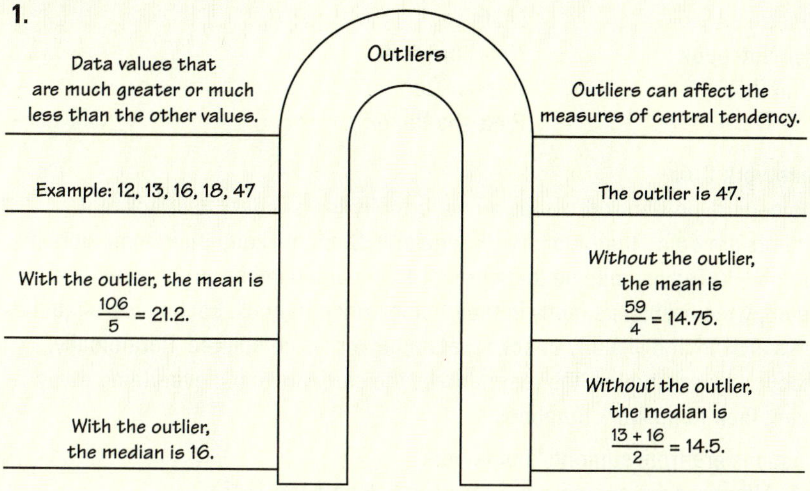

2.

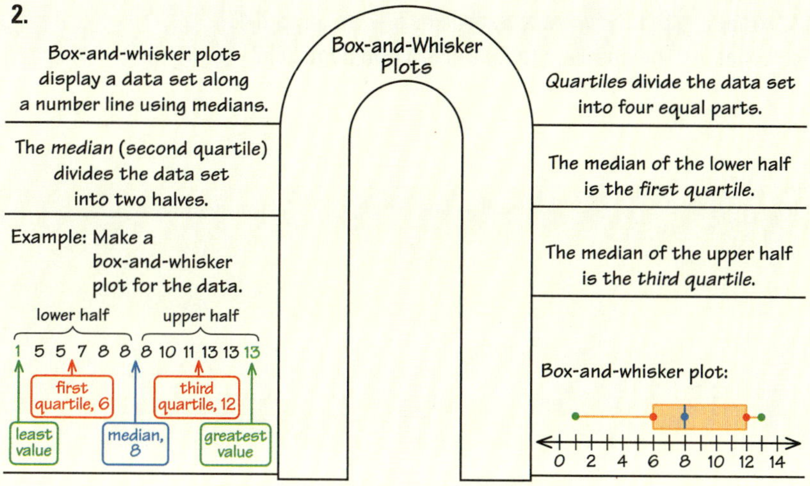

List of Organizers

Available at *BigIdeasMath.com*

- Comparison Chart
- Concept Circle
- Definition (Idea) and Example Chart
- Example and Non-Example Chart
- Formula Triangle
- Four Square
- Information Frame
- Information Wheel
- Notetaking Organizer
- Process Diagram
- Summary Triangle
- **Word Magnet**
- Y Chart

About this Organizer

A **Word Magnet** can be used to organize information associated with a vocabulary word or term. As shown, students write the word or term inside the magnet. Students write associated information on the blank lines that "radiate" from the magnet. Associated information can include, but is not limited to: other vocabulary words or terms, definitions, formulas, procedures, examples, and visuals. This type of organizer serves as a good summary tool because any information related to a topic can be included.

Technology For the Teacher

Vocabulary Puzzle Builder

Answers

1. mean: 6.8
   median: 6.5
   mode: 6

2. mean: 20
   median: 30
   mode: 40

3. mean: 3

   median: $3\frac{1}{4}$

   mode: $3\frac{1}{2}$

4. mean: 106
   median: 104
   mode: 113

5–7. See Additional Answers.

8. **a.** mean: 401.25
   median: 407.5
   mode: none

   b. The mean increases $425 - 401.25 = \$23.75$, the median increases $495 - 407.5 = \$87.50$, and the mode is now \$615.

9. See Additional Answers.

10. **a.** 40

    b. The mean because the median and mode are unaffected.

Assessment Book

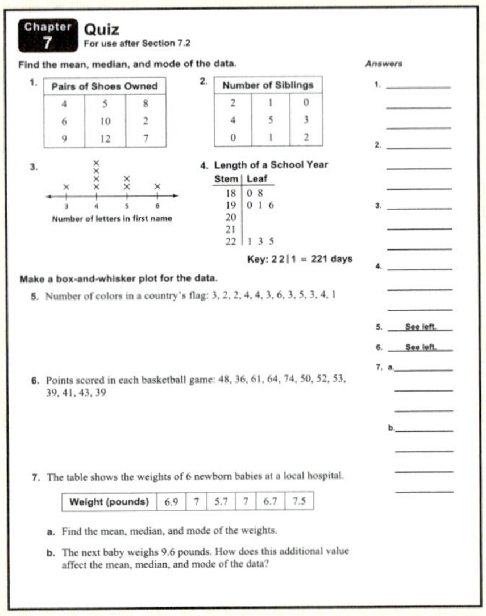

Chapter 7 Quiz
For use after Section 7.2

Find the mean, median, and mode of the data.

1.

Pairs of Shoes Owned		
4	5	8
6	10	2
9	12	7

2.

Number of Siblings		
2	1	0
4	5	3
0	1	2

3.

4. Length of a School Year

Stem	Leaf
18	0 8
19	0 1 6
20	
21	
22	1 3 5

Key: 22 | 1 = 221 days

Make a box-and-whisker plot for the data.

5. Number of colors in a country's flag: 3, 2, 2, 4, 4, 3, 6, 3, 5, 3, 4, 1

6. Points scored in each basketball game: 48, 36, 61, 64, 74, 50, 52, 53, 39, 41, 43, 39

7. The table shows the weights of 6 newborn babies at a local hospital.

Weight (pounds)	6.9	7	5.7	7	6.7	7.5

   a. Find the mean, median, and mode of the weights.

   b. The next baby weighs 9.6 pounds. How does this additional value affect the mean, median, and mode of the data?

Answers

1. ______
2. ______
3. ______
4. ______
5. See left.
6. See left.
7. a. ______
   b. ______

Alternative Quiz Ideas

100% Quiz
Error Notebook
Group Quiz
Homework Quiz
Math Log
Notebook Quiz
Partner Quiz
Pass the Paper

Homework Quiz

A homework notebook provides an opportunity for teachers to check that students are doing their homework regularly. Students keep their homework in a notebook. They should be told to record the page number, problem number, and copy the problem exactly in their homework notebook. Each day the teacher walks around and visually checks that homework is completed. Periodically, without advance notice, the teacher tells the students to put everything away except their homework notebook.

Questions are from students' homework.

1. What are the answers to Exercises 5–8 on page 278?
2. What are the answers to Exercises 11–14 on page 279?
3. What are the answers to Exercises 5–7 on page 284?
4. What are the answers to Exercise 11 on page 285?

Reteaching and Enrichment Strategies

If students need help...	If students got it...
Resources by Chapter • Study Help • Practice A and Practice B • Puzzle Time Lesson Tutorials *BigIdeasMath.com* Practice Quiz Practice from the Test Generator	Resources by Chapter • Enrichment and Extension • School-to-Work Game Closet at *BigIdeasMath.com* Start the next section

Technology For the Teacher

Answer Presentation Tool
Big Ideas Test Generator

7.1–7.2 Quiz

Check It Out
Progress Check
BigIdeasMath.com

Find the mean, median, and mode of the data. *(Section 7.1)*

1.

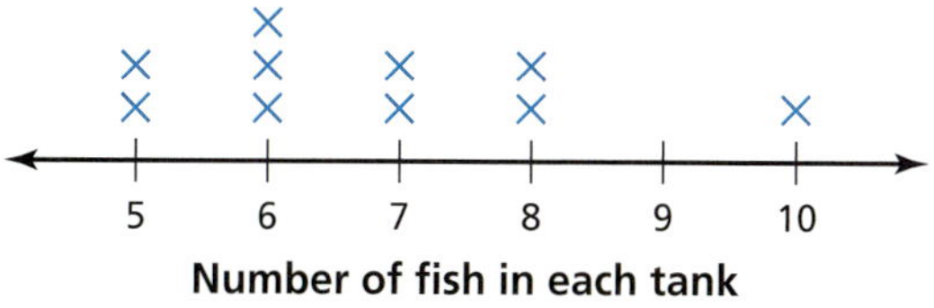

2.

Checkbook Balances (dollars)		
40	10	−20
0	−10	40
30	40	50

3.

Hours Spent on Project		
$3\frac{1}{2}$	5	$2\frac{1}{2}$
3	$3\frac{1}{2}$	$\frac{1}{2}$

4. Students in a Grade

Stem	Leaf
9	4 9
10	1 2 6
11	3 3
12	0

Key: 10 | 6 = 106 students

Make a box-and-whisker plot for the data. *(Section 7.2)*

5. Hours spent on each babysitting job: 2, 4, 7, 5, 4, 1, 7, 4

6. Minutes of violin practice: 20, 50, 60, 40, 40, 30, 60, 40, 50, 20, 20, 35

7. Players' scores at end of first round: 200, −100, 100, 350, −50, 0, −50, 300

8. The table shows the prices of eight acoustic guitars at a music store. *(Section 7.1)*

Prices of Acoustic Guitars (dollars)			
650	225	320	615
595	495	200	110

a. Find the mean, median, and mode of the prices.

b. The store gets a ninth guitar in stock. The guitar costs $615. How does this additional value affect the mean, median, and mode of the data?

9. ANOLES The table shows the lengths of 12 green anoles. Make a box-and-whisker plot for the data. What does the box-and-whisker plot tell you about the data? *(Section 7.2)*

Length (cm)	17.5	17.3	16.5	16.8	17.0	16.5	17.0	16.7	16.5	17.0	17.4	17.1

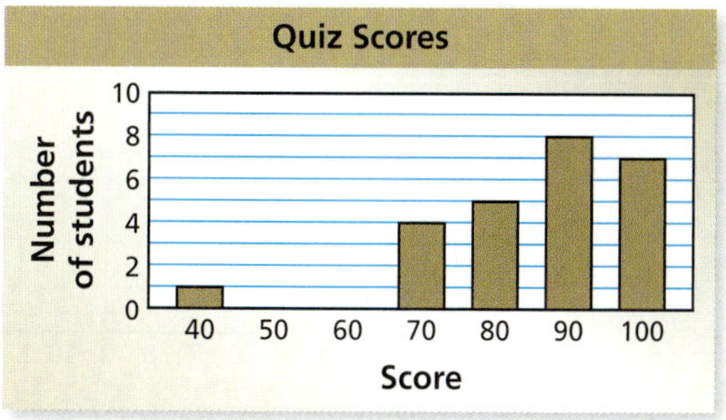

10. QUIZ SCORES The graph shows the quiz scores of students in a class. *(Section 7.1)*

a. Identify the outlier.

b. Which measure of central tendency will be most affected by removing the outlier? Explain.

7.3 Scatter Plots and Lines of Best Fit

Essential Question How can you use data to predict an event?

Share Your Work at... My.BigIdeasMath.com

1 ACTIVITY: Representing Data by a Linear Equation

Work with a partner. You have been working on a science project for 8 months. Each month, you have measured the length of a baby alligator.

The table shows your measurements.

September (month 0) — April (month 7)

Month, x	0	1	2	3	4	5	6	7
Length (in.), y	22.0	22.5	23.5	25.0	26.0	27.5	28.5	29.5

Use the following steps to predict the baby alligator's length next September.

a. Graph the data in the table.

b. Draw the straight line that you think best approximates the points.

c. Write an equation of the line you drew.

d. Use the equation to predict the baby alligator's length next September.

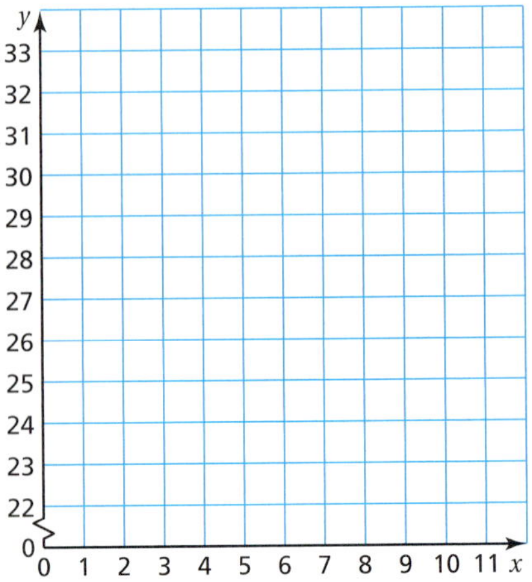

Laurie's Notes

Introduction

For the Teacher

- **Goal:** Students will gain an intuitive understanding of how to construct scatter plots and how to write an equation of the line of best fit.
- **Big Idea:** A scatter plot differs from previous data displays in that it is bi-variate (paired data). Each data point is associated with two numbers, *x* and *y*, unlike uni-variate data where each is a single value.
- My experience has been that students are comfortable with plotting ordered pairs. Summarizing the data using the *line of best fit* is challenging.

Motivate

- Solicit information about what students know about alligators. Share alligator facts with students as a warm-up. (See the next page.)
- That should be enough information to set the context for this first activity!

Activity Notes

Activity 1

- ? "Look at the table of values. What do the ordered pairs represent?" (month, length of alligator)
- ? "Does the data represent the first 7 months of growth of a baby alligator? Explain." No, it does not suggest that this is from birth to age 7 months.
- ? "Are there any observations about the data in the table?" Months are increasing by 1. Lengths are increasing by about one-half to an inch each month.
- Students will ask what drawing a line "that best approximates the points" means. You should explain that it is a line that passes as closely as possible to all the points. Use a straightedge to lightly draw the line.
- ? "What does the jagged symbol at the bottom of the *y*-axis mean?" broken axis
- ? "Do you think everyone in class drew the exact same line? Explain." no; They will be close, but they do not have to be exactly the same.
- ? "How did you write the equation for the line?" Listen for an approximation of the slope (rise over run) and the *y*-intercept (close to 22). Write the equation in slope-intercept form.
- ? "Does everyone have the same slope?" No, but they should be relatively close and should match the observations made about the data when looking at the table.
- Have students interpret the slope and *y*-intercept in the context of the problem.
- ? "How does the equation help you answer part (d)?" Substitute 12 for *x*, and find *y*.
- ? "Without the equation, can you predict the length of the alligator next September?" Yes, but you need to extend the graph and use eyesight to approximate the ordered pair.

Previous Learning

Students should know how to plot ordered pairs and write equations in slope-intercept form.

Activity Materials
Textbook
• straightedge

Start Thinking! and Warm Up

Activity 7.3 Start Thinking!
For use before Activity 7.3

Activity 7.3 Warm Up
For use before Activity 7.3

Write an equation of the line that passes through the two points.

1. (0, 4) and (5, 3)
2. (0, 6) and (2, 0)
3. (8, 3) and (2, 6)
4. (1, 2) and (5, 6)
5. (9, 3) and (3, 1)
6. (4, 16) and (2, 12)

7.3 Record and Practice Journal

Essential Question How can you use data to predict an event?

1 ACTIVITY: Representing Data by a Linear Equation

Work with a partner. You have been working on a science project for 8 months. Each month, you have measured the length of a baby alligator.

The table shows your measurements.

September (month 0), April (month 7)

Month, *x*	0	1	2	3	4	5	6	7
Length (in.), *y*	22.0	22.5	23.5	25.0	26.0	27.5	28.5	29.5

Use the following steps to predict the baby alligator's length next September.

a. Graph the data in the table.

b. Draw the straight line that you think best approximates the points.

c. Write an equation of the line you drew.

Sample answer: $y = 1.1x + 22$

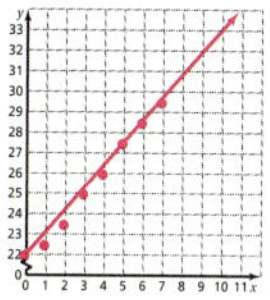

d. Use the equation to predict the baby alligator's length next September.

Sample answer: 35.2 in.

English Language Learners

Class Activity

Provide English learners with an opportunity to interact while learning the concept. Draw a coordinate plane on poster board. Label the horizontal axis *shoe size* and the vertical axis *height*. Have students place a sticker on the ordered pair that represents their shoe size and height. Then have the class fit a line to the data and write an equation of the line.

7.3 Record and Practice Journal

2 ACTIVITY: Representing Data by a Linear Equation

Work with a partner. You are a biologist and are studying bat populations.

You are asked to predict the number of bats that will be living in an abandoned mine in 3 years.

To start, you find the number of bats that have been living in the mine during the past 8 years.

The table shows the results of your research.

7 years ago ↓ (Year 0) this year ↓ (Year 7)

Year, x	0	1	2	3	4	5	6	7
Bats (thousands), y	327	306	299	270	254	232	215	197

Use the following steps to predict the number of bats that will be living in the mine after 3 years.

a. Graph the data in the table.

b. Draw the straight line that you think best approximates the points.

c. Write an equation of the line you drew.

Sample answer: $y = -19.5x + 327$

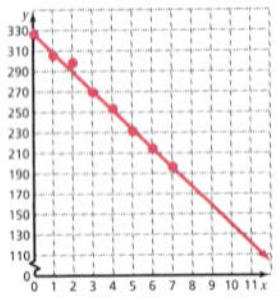

d. Use the equation to predict the number of bats in 3 years.

Sample answer: 132,000 bats

What Is Your Answer?

3. **IN YOUR OWN WORDS** How can you use data to predict an event?

Plot data and line of best fit.
Use equation of line to predict.

4. Use the Internet or some other reference to find data that appear to have a linear pattern. List the data in a table and graph the data. Use an equation that is based on the data to predict a future event.

Check students' work.

Laurie's Notes

Activity 2

- In this activity, the data has not been collected from an experiment. The data has been collected from a documented source and recorded in the table.
- Read the introduction. The purpose of making a scatter plot is stated. You want to make a prediction about the future by examining known data.

? "Are there any observations about the data in the table?" Students may recognize that as the years increase, the number of bats is decreasing by about 15–20 (in thousands) per year.

- Discuss equations written by students. Record students' results on the board. There will likely be a bit more variation of results than in the first activity.
- Have students interpret what the slope and y-intercept mean in the context of the problem.
- Discuss how the equation allows us to make predictions about the future.

What Is Your Answer?

- Question 4 can become a project due at the conclusion of the chapter.

Closure

- **Exit Ticket:** Describe the difference in the source of data for Activity 1 versus Activity 2. The data in Activity 1 are the result of gathering actual data from an experiment. The data in Activity 2 has been collected from a documented source and recorded in a table.

More about Alligators

- The American alligator (Alligator mississippiensis) is the largest reptile in North America. The first reptiles appeared 300 million years ago. Ancestors of the American alligator appeared 200 million years ago.
- The name alligator comes from early Spanish explorers who called them "El legarto" or "big lizard" when they first saw these giant reptiles.
- Louisiana and Florida have the most alligators. There are over one million wild alligators in each state with over a quarter million more on alligator farms.
- Alligators are about 10–12 inches in length when they are hatched from eggs. Growth rates vary from 2 inches per year to 12 inches per year, depending on the habitat, sex, size, and age of the alligator.
- Females can grow to about 9 feet in length and over 200 pounds. Males can grow to about 13 feet in length and over 500 pounds.
- The largest alligator was taken in Louisiana and measured 19 feet 2 inches.
- Alligators live about as long as humans, an average of 70 years.

Technology For the Teacher

Dynamic Classroom

The Dynamic Planning Tool
Editable Teacher's Resources at *BigIdeasMath.com*

2 ACTIVITY: Representing Data by a Linear Equation

Work with a partner. You are a biologist and are studying bat populations.

You are asked to predict the number of bats that will be living in an abandoned mine in 3 years.

To start, you find the number of bats that have been living in the mine during the past 8 years.

The table shows the results of your research.

7 years ago ↓ (year 0) this year ↓ (year 7)

Year, x	0	1	2	3	4	5	6	7
Bats (thousands), y	327	306	299	270	254	232	215	197

Use the following steps to predict the number of bats that will be living in the mine after 3 years.

a. Graph the data in the table.

b. Draw the straight line that you think best approximates the points.

c. Write an equation of the line you drew.

d. Use the equation to predict the number of bats in 3 years.

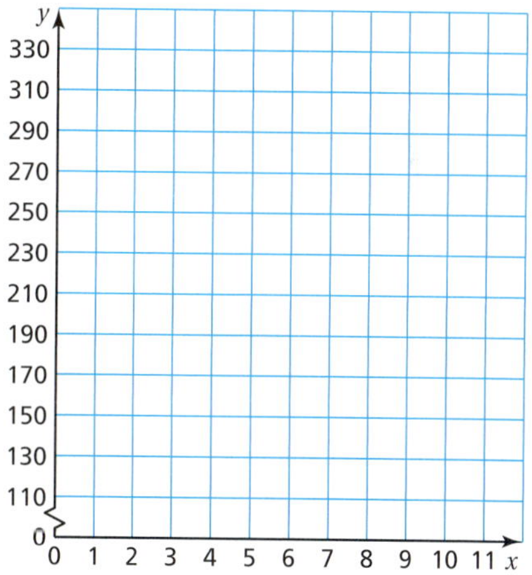

What Is Your Answer?

3. **IN YOUR OWN WORDS** How can you use data to predict an event?

4. Use the Internet or some other reference to find data that appear to have a linear pattern. List the data in a table and graph the data. Use an equation that is based on the data to predict a future event.

Practice Use what you learned about scatter plots and lines of best fit to complete Exercise 3 on page 293.

7.3 Lesson

Check It Out
Lesson Tutorials
BigIdeasMath.com

Key Vocabulary
scatter plot, *p. 290*
line of best fit, *p. 292*

Key Idea

Scatter Plot

A **scatter plot** is a graph that shows the relationship between two data sets. The two sets of data are graphed as ordered pairs in a coordinate plane.

EXAMPLE 1 Interpreting a Scatter Plot

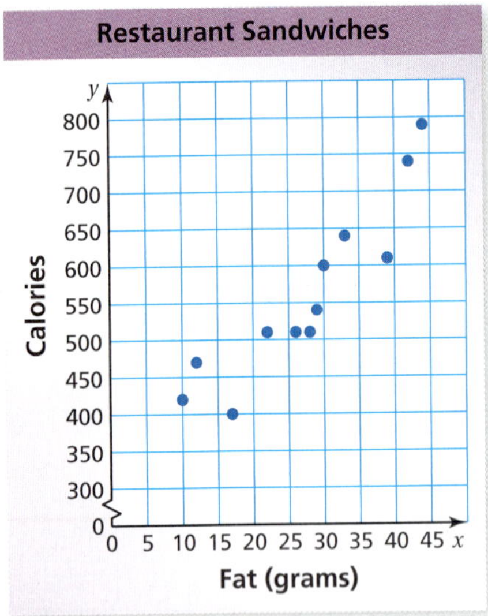

The scatter plot at the left shows the total fat (in grams) and the total calories in 12 restaurant sandwiches.

a. How many calories are in the sandwich that contains 17 grams of fat?

Draw a horizontal line from the point that has an x-value of 17. It crosses the y-axis at 400.

So, the sandwich has 400 calories.

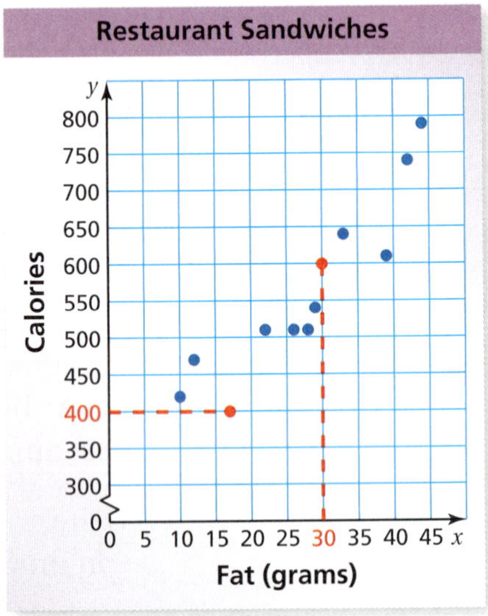

b. How many grams of fat are in the sandwich that contains 600 calories?

Draw a vertical line from the point that has a y-value of 600. It crosses the x-axis at 30.

So, the sandwich has 30 grams of fat.

c. What tends to happen to the number of calories as the number of grams of fat increases?

Looking at the graph, the plotted points go up from left to right.

So, as the number of grams of fat increases, the number of calories increases.

On Your Own

Now You're Ready
Exercises 4 and 5

1. **WHAT IF?** A sandwich has 650 calories. Based on the scatter plot in Example 1, how many grams of fat would you expect the sandwich to have? Explain your reasoning.

Multi-Language Glossary at BigIdeasMath.com.

Laurie's Notes

Introduction

Connect

- **Yesterday:** Students gained an intuitive understanding of how to construct scatter plots and write an equation of the line of best fit.
- **Today:** Students will construct scatter plots, draw the line of best fit, and analyze the equation.

Motivate

- **Preparation:** Stop by any fast food restaurant to pick up a pamphlet, or go online to find nutritional information about the menu items.
- ? "Do you think there is a relationship between the grams of fat and number of calories in the sandwich?" yes
- Share the information about a few of the sandwiches from your pamphlet or printout to confirm students' opinions.

Lesson Notes

Key Idea

- Explain that the plot they are going to make today displays the relationship, if any, between two variables, such as grams of fat and calories.
- Define scatter plot.
- Discuss the two scatter plots made yesterday. In Activity 1, the two sets of data were months and alligator length. In Activity 2, the two sets of data were years and number of bats.

Example 1

- This example helps students understand how a scatter plot is read and interpreted. Discuss the labels on the axes and what an ordered pair represents: (grams of fat, number of calories). There are 12 different sandwiches that are represented.
- To read information from the plot, move horizontally to the x-value, find the ordered pair, and then move to the y-axis to read the y-value. It is helpful to use your hands to demonstrate the motion.
- A scatter plot allows you to see trends in the data. You read a scatter plot from left to right. As the x-coordinate increases, is the y-coordinate increasing, decreasing, staying the same, *or* is there no pattern?

On Your Own

- This question implies that because you can see a particular trend in the data, you are able to make estimates about points which are not part of the data set but would fall within the trend in the data. Although it is possible that the 650 calorie sandwich has 10 grams of fat, you would not predict it based upon this scatter plot.

Goal Today's lesson is making a **scatter plot** and identifying a **line of best fit**.

Lesson Materials
Textbook
• transparency grid

Start Thinking! and Warm Up

Lesson 7.3 **Warm Up** For use before Lesson 7.3

Lesson 7.3 **Start Thinking!** For use before Lesson 7.3

In which graph is the line shown most representative of the data? Explain.

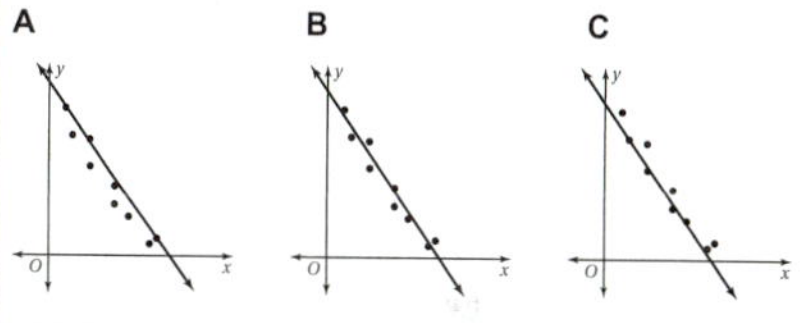

Extra Example 1

Use the scatter plot in Example 1.

a. How many grams of fat are in a sandwich that contains 740 calories? about 42 g

b. How many calories are in a sandwich that contains 33 grams of fat? about 640 calories

On Your Own

1. about 35 g; The point just below $y = 650$ has an x-value just below $x = 35$.

Extra Example 2

Tell whether the data show a *positive*, a *negative*, or *no* relationship.

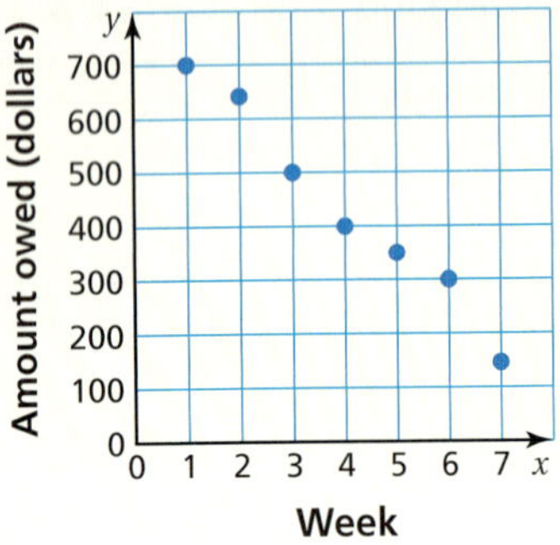

negative relationship

On Your Own

2. See Additional Answers.

3. 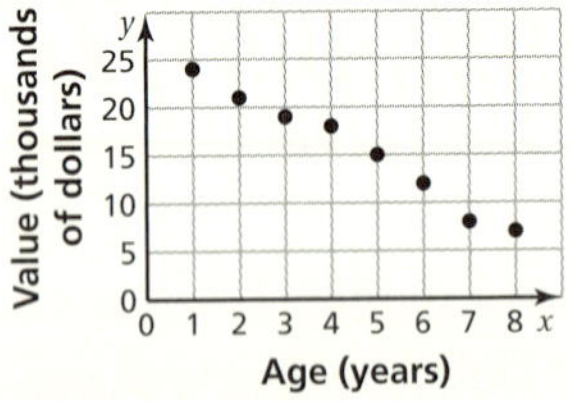

negative relationship

Differentiated Instruction

Visual

Some students may find it easier to draw a line of best fit before determining if the data have a positive relationship, a negative relationship, or no relationship. A line with a *positive* slope means the data have a *positive* relationship. A line with a *negative* slope means the data have a *negative* relationship. If a line cannot be drawn, the data have *no* relationship.

Laurie's Notes

Discuss

- There are three general cases that describe the relationship between two data sets. Draw a quick example of each case.
- The alligator data was an example of a positive relationship and the bat data was an example of a negative relationship.

Example 2

- Have students review the two scatter plots shown.
- Ask students to complete this sentence. As the size of the television increases, the price increases. This is an example of a positive relationship.
- **Connection:** By this point, some students have made the connection between the slope of a line and the relationship between the two data sets. A positive relationship is related to a positive slope.
- ? "Should there be a relationship between a person's age and the number of pets they own?" no
- Part (b) makes sense to students. There should be no trend in the data.

On Your Own

- Give time for students to complete these two scatter plots.
- A common difficulty for students is deciding how to scale the axes. Students should look at the range of numbers that need to be displayed, and then decide if it is necessary to start their axes at 0 or if another starting point (broken axes) makes sense.
- Have transparency grids available so that results can be shared quickly as a class.

A scatter plot can show that a relationship exists between two data sets.

Positive Relationship

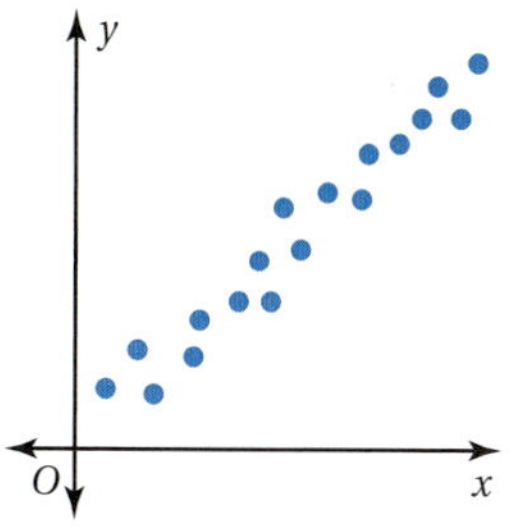

As x increases, y increases.

Negative Relationship

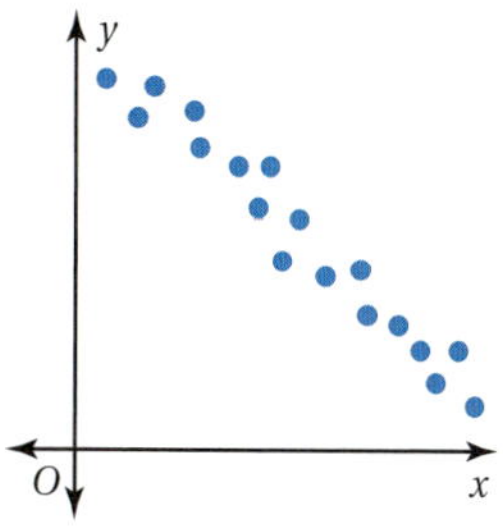

As x increases, y decreases.

No Relationship

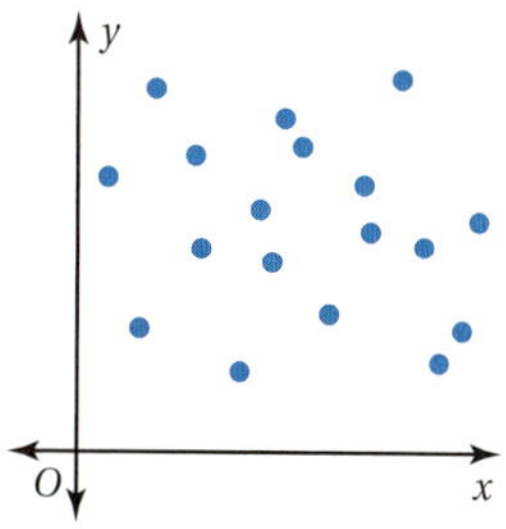

The points show no pattern.

EXAMPLE 2 Identifying a Relationship

Tell whether the data show a *positive*, a *negative*, or *no* relationship.

a. Television size and price

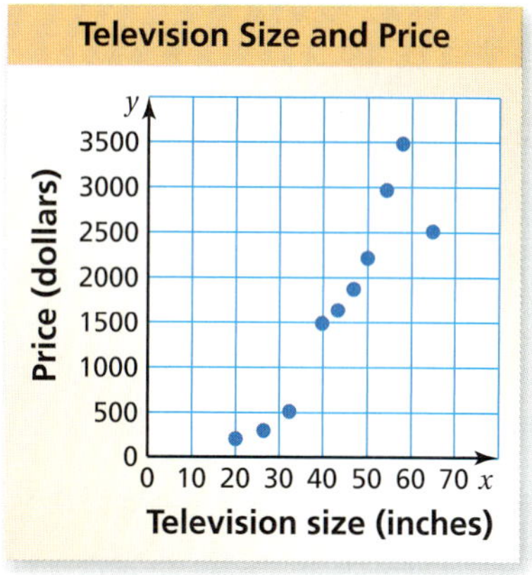

As the size of the television increases, the price increases.

So, the scatter plot shows a positive relationship.

b. Age and number of pets owned

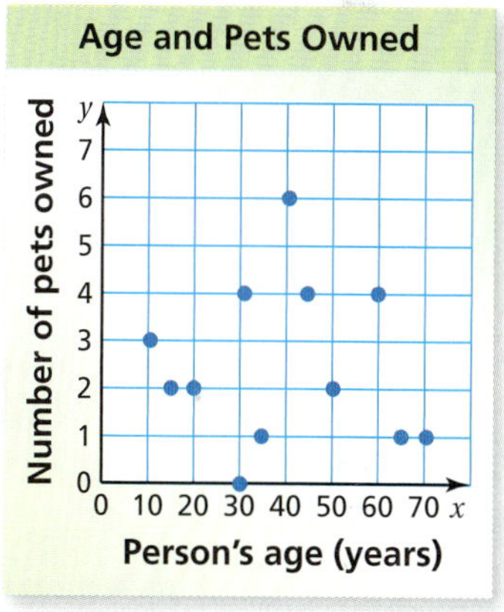

The number of pets owned does not depend on a person's age.

So, the scatter plot shows no relationship.

On Your Own

Now You're Ready
Exercises 6–8

Make a scatter plot of the data. Tell whether the data show a *positive*, a *negative*, or *no* relationship.

2.

Study Time (min), x	30	20	60	90	45	10	30	75	120	80
Test Score, y	87	74	92	97	85	62	83	90	95	91

3.

Age of a Car (years), x	1	2	3	4	5	6	7	8
Value (thousands), y	\$24	\$21	\$19	\$18	\$15	\$12	\$8	\$7

A **line of best fit** is a line drawn on a scatter plot that is close to most of the data points. It can be used to estimate data on a graph.

EXAMPLE 3 Finding a Line of Best Fit

Week, x	Sales (millions), y
1	$19
2	$15
3	$13
4	$11
5	$10
6	$8
7	$7
8	$5

The table shows the weekly sales of a DVD and the number of weeks since its release. (a) Make a scatter plot of the data. (b) Draw a line of best fit. (c) Write an equation of the line of best fit. (d) Predict the sales in week 9.

a. Plot the points in a coordinate plane. The scatter plot shows a negative relationship.

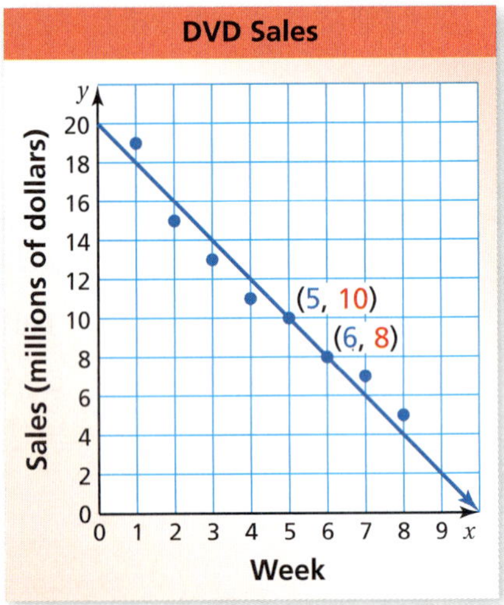

b. Draw a line that is close to the data points. Try to have as many points above the line as below it.

c. The line passes through (5, 10) and (6, 8).

$$\text{slope} = \frac{\text{rise}}{\text{run}} = \frac{-2}{1} = -2$$

Because the line crosses the y-axis at (0, 20), the y-intercept is 20.

So, the equation of the line of best fit is $y = -2x + 20$.

d. To predict the sales for week 9, substitute 9 for x in the equation of the line of best fit.

$y = -2x + 20$	Line of best fit
$= -2(9) + 20$	Substitute 9 for x.
$= 2$	Evaluate.

The sales in week 9 should be about $2 million.

Study Tip

A line of best fit does not need to pass through any of the data points.

On Your Own

Now You're Ready
Exercise 11

4. The table shows the number of people who have attended a neighborhood festival over an 8-year period.

Year, x	1	2	3	4	5	6	7	8
Attendance, y	420	500	650	900	1100	1500	1750	2400

a. Make a scatter plot of the data.

b. Draw a line of best fit.

c. Write an equation of the line of best fit.

d. Predict the number of people who will attend the festival in year 10.

Laurie's Notes

Discuss

- **Model:** Define and discuss a line of best fit. I find it helpful to model this with a piece of spaghetti. Use a scatter plot from the previous page that was completed on the transparency. Model how the spaghetti can approximate the trend of the data.
- Move the spaghetti so that it does *not* represent the data, and then move the spaghetti so that it does. You will use your eyesight when judging where to draw the line.

Example 3

? "What observations can you make about the sales of the DVD as the weeks go on?" Sales are decreasing.

- Carefully plot the ordered pairs on a transparency grid.
- When drawing the line of best fit, try to put as many points above the line as below it.
- Students may draw different lines of best fit and still get a reasonable answer.
- In this example, the line passes through two actual data points, (5, 10) and (6, 8). As noted in the *Study Tip*, a line of best fit does not need to pass through any of the data points.
- Finish working the problem as shown.
- **Big Idea:** The purpose of writing the equation of the line of best fit is to make predictions. The equation becomes a model for the data, describing its behavior.

On Your Own

- This is a nice summary problem. Students should quickly observe the positive relationship just from the table of values.
- Share results of this problem as a whole class.

Closure

- **Exit Ticket:** In Example 3, interpret the slope and y-intercept in the context of the problem. Why does the slope make sense for this problem? The slope, -2, represents the decrease in sales in millions of dollars per week. The y-intercept, 20, represents the sales in millions of dollars for week 0. The slope makes sense for this problem because interest in buying a DVD is greatest when the DVD is first released, and then sales fall. The y-intercept does not make sense in this problem, because there would not have been any sales before the first week.

Technology For the Teacher

Dynamic Classroom

The Dynamic Planning Tool
Editable Teacher's Resources at *BigIdeasMath.com*

Extra Example 3

The table shows the weekly sales of a DVD and the number of weeks since its release.

Week, x	Sales (millions), y
1	\$15
2	\$13
3	\$12
4	\$9
5	\$6
6	\$4
7	\$3

a. Make a scatter plot of the data.
b. Draw a line of best fit.

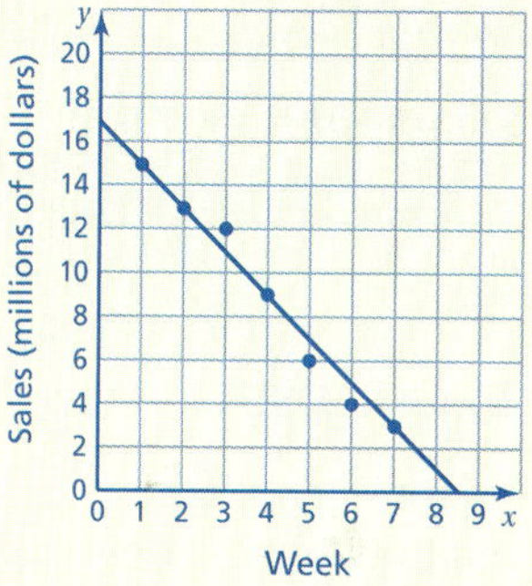

c. Write an equation of the line of best fit. $y = -2x + 17$

d. Predict the sales in week 8. about \$1 million

On Your Own

4. a–b.

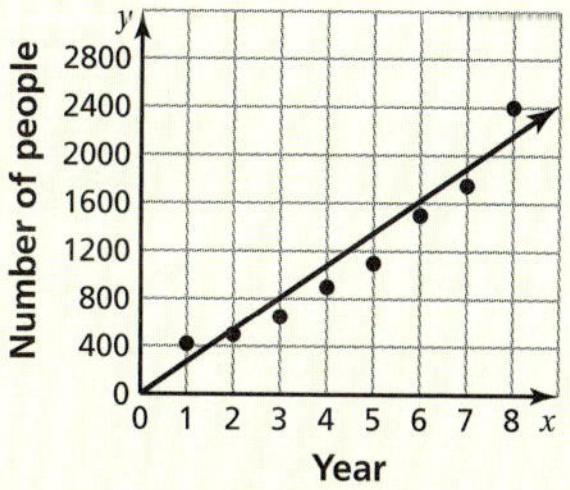

c. *Sample answer:* $y = 270x$

d. *Sample answer:* about 2700 people

Vocabulary and Concept Check

1. They must be ordered pairs so there are equal amounts of x- and y-values.
2. You can estimate and predict values.

Practice and Problem Solving

3. a–b.

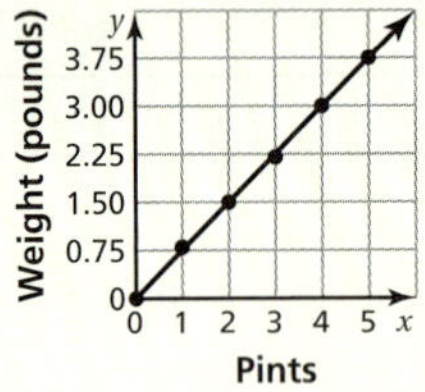

c. *Sample answer:* $y = 0.75x$

d. *Sample answer:* 7.5 lb

e. *Sample answer:* \$16.88

4. a. 2007

b. about 875 SUVs

c. There is a negative relationship between year and number of SUVs sold.

5. a. 3.5 h

b. \$85

c. There is a positive relationship between hours worked and earnings.

Assignment Guide and Homework Check

Level	Day 1 Activity Assignment	Day 2 Lesson Assignment	Homework Check
Basic	3, 17–20	1, 2, 4–9, 11–13	2, 4, 6, 11, 12
Average	3, 17–20	1, 2, 4–9, 11–13, 15	2, 4, 6, 11, 12
Advanced	3, 17–20	1, 2–16 even, 11, 15	2, 4, 6, 12, 14

Common Errors

- **Exercise 3** Students may use inconsistent increments or forget to label their graphs. Students should use consistent increments to represent the data. Remind them to label the axes so that information can be read from the graph.
- **Exercises 4 and 5** When finding values from the graph, students may accidentally shift over or up too far and get an answer that is off by an increment. Encourage them to start at the given value and trace the graph to where the point or line of best fit is, and then trace down or left to the other axis for the answer.

7.3 Record and Practice Journal

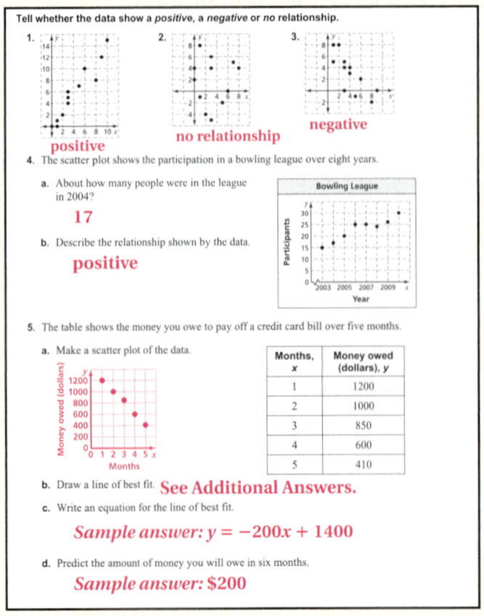

Tell whether the data show a *positive*, a *negative* or *no* relationship.

1. positive

2. no relationship

3. negative

4. The scatter plot shows the participation in a bowling league over eight years.

a. About how many people were in the league in 2004?

17

b. Describe the relationship shown by the data.

positive

Bowling League

5. The table shows the money you owe to pay off a credit card bill over five months.

a. Make a scatter plot of the data.

Months, x	Money owed (dollars), y
1	1200
2	1000
3	850
4	600
5	410

b. Draw a line of best fit. See Additional Answers.

c. Write an equation for the line of best fit.

Sample answer: $y = -200x + 1400$

d. Predict the amount of money you will owe in six months.

Sample answer: \$200

Technology For the Teacher

Answer Presentation Tool

Quiz*Show*

7.3 Exercises

Check It Out
Help with Homework
BigIdeasMath.com

Vocabulary and Concept Check

1. **VOCABULARY** What type of data are needed to make a scatter plot? Explain.
2. **WRITING** Explain why a line of best fit is helpful when analyzing data.

Practice and Problem Solving

3. **BLUEBERRIES** The table shows the weights y of x pints of blueberries.

Number of Pints, x	0	1	2	3	4	5
Weight (pounds), y	0	0.8	1.50	2.20	3.0	3.75

a. Graph the data in the table.
b. Draw the straight line that you think best approximates the points.
c. Write an equation of the line you drew.
d. Use the equation to predict the weight of 10 pints of blueberries.
e. Blueberries cost \$2.25 per pound. How much do 10 pints of blueberries cost?

1 4. **SUVS** The scatter plot shows the number of sport utility vehicles sold in a city from 2005 to 2010.

a. In what year were 1000 SUVs sold?
b. About how many SUVs were sold in 2009?
c. Describe the relationship shown by the data.

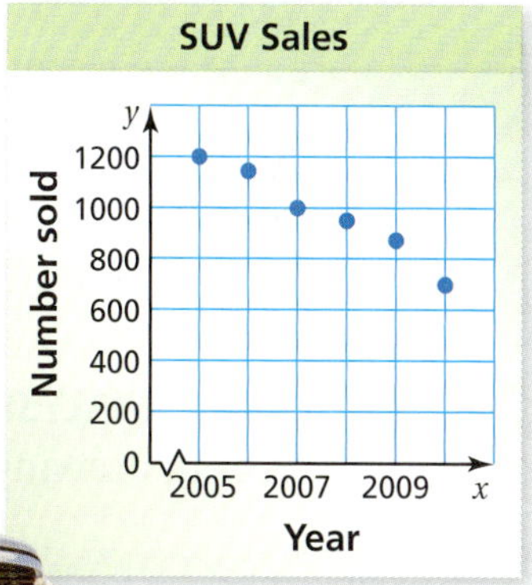

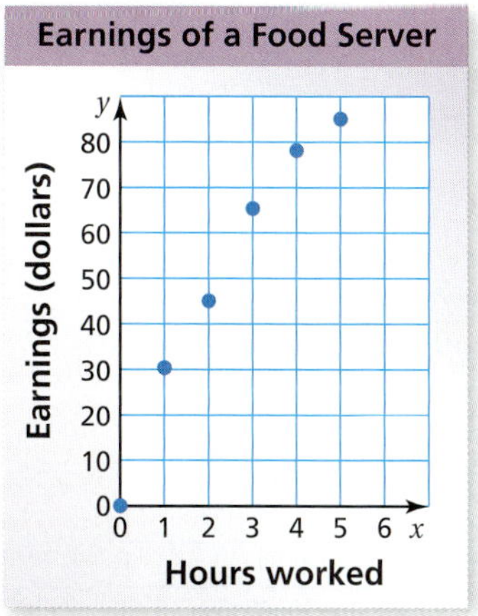

5. **EARNINGS** The scatter plot shows the total earnings (wages and tips) of a food server during 1 day.

a. About how many hours must the server work to earn \$70?
b. About how much did the server earn for 5 hours of work?
c. Describe the relationship shown by the data.

Tell whether the data show a *positive*, a *negative*, or *no* relationship.

2 **6.**

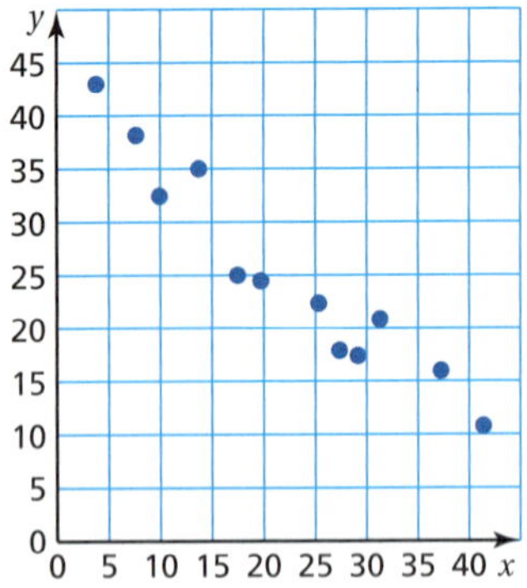

7.

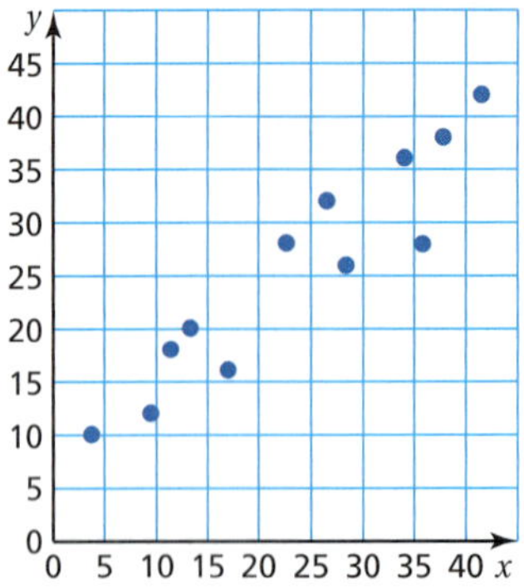

8.

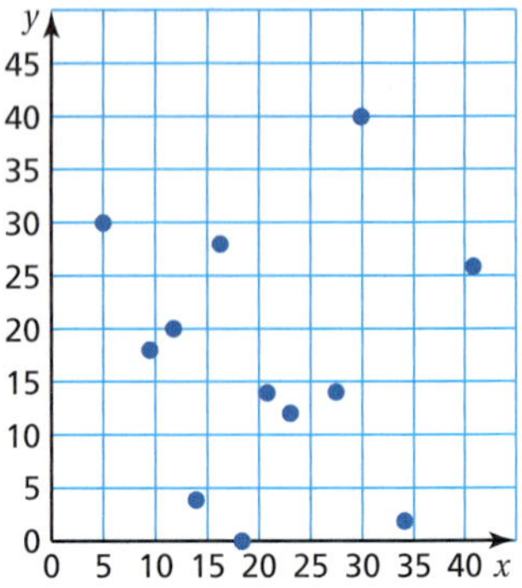

9. **HONEYBEES** The table shows the number of honeybee colonies in the United States from 2003 to 2006. What type of relationship do the data show?

Year, x	2003	2004	2005	2006
Honeybee Colonies (millions), y	2.599	2.556	2.413	2.392

10. **OPEN-ENDED** Describe a set of real-life data that has a positive relationship.

3 **11.** **VACATION** The table shows the distance you travel over a 6-hour period.

a. Make a scatter plot of the data.

b. Draw a line of best fit.

c. Write an equation of the line of best fit.

d. Predict the distance you will travel in 7 hours.

Hours, x	Distance (miles), y
1	62
2	123
3	188
4	228
5	280
6	344

12. **ERROR ANALYSIS** Describe and correct the error in drawing the line of best fit.

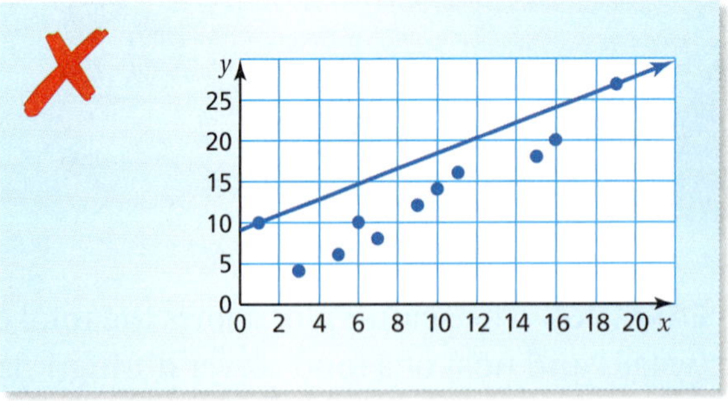

13. **TEST SCORES** The scatter plot shows the relationship between the number of minutes spent studying and the test scores for a science class.

a. What type of relationship does the data show?

b. Interpret the relationship.

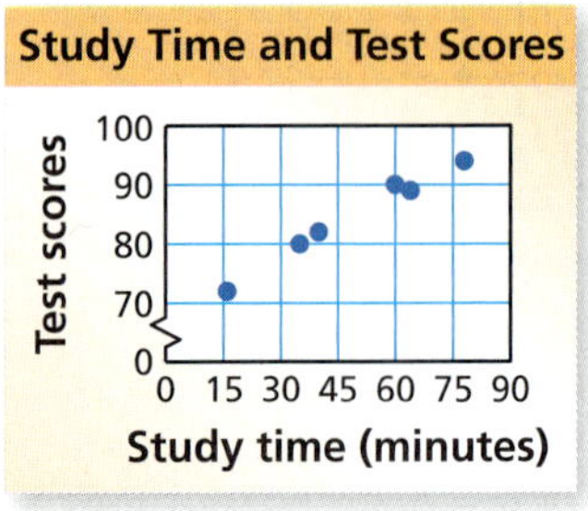

Common Errors

- **Exercises 6–8** Students may mix up positive and negative relationships. Remind them about slope. The slope is positive when the line rises from left to right and negative when it falls from left to right. The same is true for relationships in a scatter plot. If the data rises from left to right, it is a positive relationship. If it falls from left to right, it is a negative relationship.
- **Exercise 11** Students may draw a line of best fit that does not accurately reflect the data trend. Remind them that the line does not have to go through any of the data points. Also remind them that the line should go through the middle of the data so that about half of the data points are above the line and half are below. One strategy is to draw an oval around the data and then draw a line through the middle of the oval. For example:

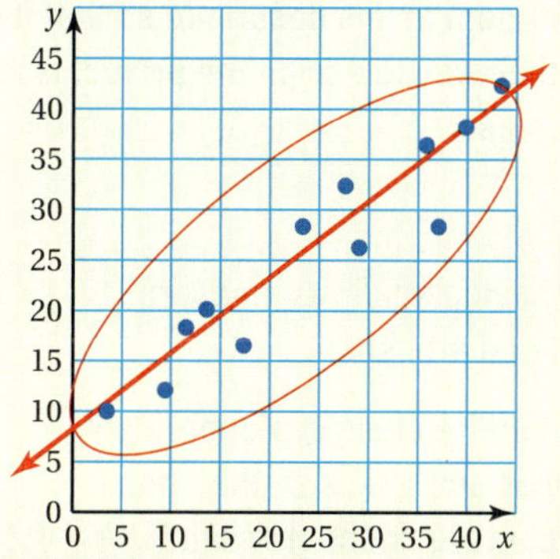

- **Exercise 11** Students may struggle writing the equation for the line of best fit. When drawing the line, encourage them to try to make the line go through a lattice point. Also, students can use lattice points that are very close to the line to help them find the slope.

Differentiated Instruction

Kinesthetic

Form groups of 8 to 10 students who will create life-size models of a positive relationship, a negative relationship, and no relationship. Give two pairs of students 10-foot lengths of string and have them form the x- and y-axes of a coordinate plane. The remaining students will be the data points. Have these students represent a *positive relationship* in the coordinate plane. After students have had a few minutes, check their positions. Continue by having students represent a *negative relationship* and *no relationship.* Extend the activity by having two students hold a third string to show the line of best fit.

Practice and Problem Solving

6. negative relationship

7. positive relationship

8. no relationship

9. negative relationship

10. *Sample answer:* age and height of a person

11. a–b.

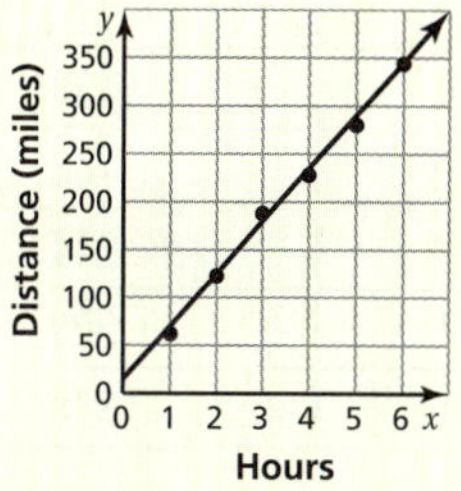

c. *Sample answer:* $y = 55x + 15$

d. *Sample answer:* 400 mi

12. The line does not fit the data.

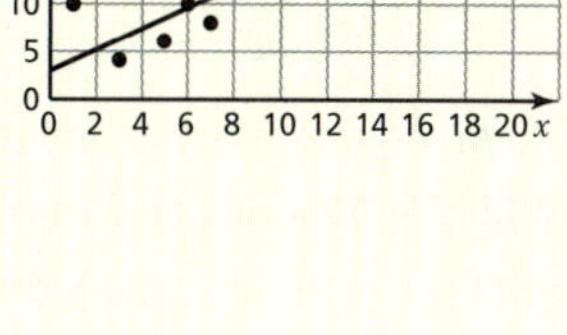

Practice and Problem Solving

13. **a.** positive relationship

b. The more time spent studying, the better the test score.

14. no; There is no line that lies close to most of the points.

15. See *Taking Math Deeper.*

16. See Additional Answers.

Fair Game Review

17. 2	**18.** 8
19. −4	**20.** B

Mini-Assessment

The table shows the distance you travel over a 6-hour period.

Hours, x	Distance (miles), y
1	60
2	130
3	186
4	244
5	300
6	378

a. Make a scatter plot of the data.

b. Draw a line of best fit.

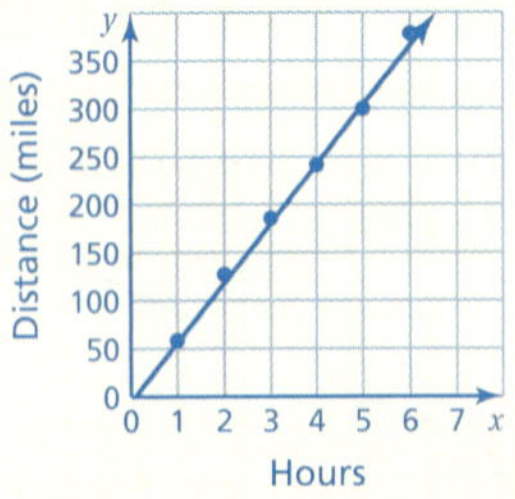

c. Write an equation of the line of best fit. *Sample answer:* $y = 60x$

d. Predict the distance traveled after 7 hours. *Sample answer:* about 420 mi

Taking Math Deeper

Exercise 15

The drawing for Exercise 15 is a stylized version of Leonardo da Vinci's famous drawing called "The Vitruvian Man." Leonardo created the drawing in 1487. It depicts a male figure in two superimposed positions with his arms and legs apart, and inscribed in a circle and square. The drawing and text are sometimes called the Proportions of Man. It is stored in the Gallerie dell'Accademia in Venice, Italy, but is only displayed on special occasions. Leonardo's drawing shows that the height and arm span of a typical human are equal.

1. The project in the student text is described so that it can be assigned as homework. Another way to assign the project is to ask students to do the project in class.

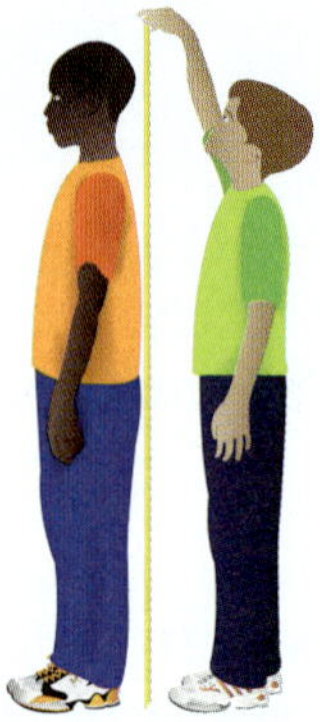

2. Gather the data by having students measure each other's height and arm span.

   Plot the data for the entire class in a coordinate plane. Mark the x-axis and y-axis so that the heights of your students (in inches) will fit.

3. 

   Ask your students to describe the relationship between height and arm span. If your class' results are similar to those described by Leonardo da Vinci, the slope of the line of best fit should be approximately equal to 1.

Project

Research Leonardo da Vinci's drawing of the *Vitruvian Man*. Explain the concept behind the drawing.

Reteaching and Enrichment Strategies

If students need help. . .	If students got it. . .
Resources by Chapter • Practice A and Practice B • Puzzle Time Record and Practice Journal Practice Differentiating the Lesson Lesson Tutorials Skills Review Handbook	Resources by Chapter • Enrichment and Extension • School-to-Work Start the next section

14. **REASONING** A data set has no relationship. Is it possible to find the line of best fit for the data? Explain.

15. **PROJECT** Use a ruler or a yardstick to find the height and arm span of three people.

    a. Make a scatter plot using the data you collected. Then draw the line of best fit for the data.

    b. Use your height and the line of best fit to predict your arm span.

    c. Measure your arm span. Compare the result with your prediction in part (b).

    d. Is there a relationship between a person's height x and arm span y? Explain.

16. Critical Thinking The table shows the price of admission to a local theater and the yearly attendance for several years.

Price of Admission (dollars), x	Yearly Attendance, y
19.50	50,000
21.95	48,000
23.95	47,500
24.00	40,000
24.50	45,000
25.00	43,500

a. Identify the outlier.

b. How does the outlier affect the line of best fit? Explain.

c. Make a scatter plot of the data and draw the line of best fit.

d. Use the line of best fit to predict the attendance when the admission cost is $27.

Fair Game Review What you learned in previous grades & lessons

Use a graph to solve the equation. Check your solution. *(Section 2.7)*

17. $5x = 2x + 6$

18. $7x + 3 = 9x - 13$

19. $\frac{2}{3}x = -\frac{1}{3}x - 4$

20. **MULTIPLE CHOICE** The circle graph shows the super powers chosen by a class. What percent of the students want strength as their super power? *(Skills Review Handbook)*

    Ⓐ 10.5% Ⓑ 12.5%

    Ⓒ 15% Ⓓ 25%

Super Powers

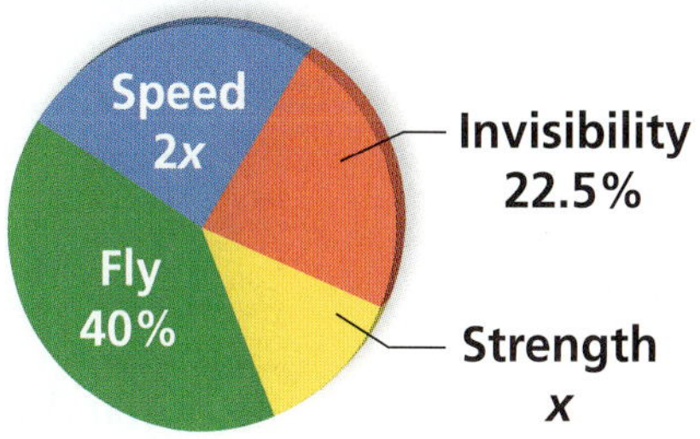

7.4 Choosing a Data Display

Essential Question How can you display data in a way that helps you make decisions?

Share Your Work at... My.BigIdeasMath.com

1 ACTIVITY: Displaying Data

Work with a partner. Analyze and display each data set in a way that best describes the data. Explain your choice of display.

a. **ROAD KILL** A group of schools in New England participated in a 2-month study and reported 3962 dead animals.

Birds 307	Mammals 2746
Amphibians 145	Reptiles 75
Unknown 689	

b. **BLACK BEAR ROAD KILL** The data below show the number of black bears killed on Florida roads from 1987 to 2006.

1987	30	1994	47	2001	99
1988	37	1995	49	2002	129
1989	46	1996	61	2003	111
1990	33	1997	74	2004	127
1991	43	1998	88	2005	141
1992	35	1999	82	2006	135
1993	43	2000	109		

c. **RACCOON ROAD KILL** A 1-week study along a 4-mile section of road found the following weights (in pounds) of raccoons that had been killed by vehicles.

13.4	14.8	17.0	12.9
21.3	21.5	16.8	14.8
15.2	18.7	18.6	17.2
18.5	9.4	19.4	15.7
14.5	9.5	25.4	21.5
17.3	19.1	11.0	12.4
20.4	13.6	17.5	18.5
21.5	14.0	13.9	19.0

d. What do you think can be done to minimize the number of animals killed by vehicles?

Laurie's Notes

Introduction

For the Teacher

- **Goal:** Students will review data displays.
- An alternative to making the visual displays in Activity 1 is for students to describe their choices for each part without actually creating the display, or to have different groups of students working on one of the three parts simultaneously. It is an instructional decision based upon length of class time, and if you want time for students to investigate Activity 2.

Motivate

- The theme for the first activity is road kill. While students may giggle at the thought, automobile accidents involving large animals can be serious. I had my first and only accident with a deer 5 years ago. I was 2 miles from home and I was traveling 40 miles per hour. The deer was killed, my daughter and I were not injured, and repairs to my car were about $1400.
- See the next page for some statistics on animal-vehicle accidents.
- Allow time for students to share personal stories.

Activity Notes

Discuss

- Discuss the data displays with which students are familiar: pictograph, bar graph, line graph, circle graph, stem-and-leaf plot, histogram, line plot, box-and-whisker plot, and scatter plot. Have students describe the feature(s) of each display.
- Discuss the different numerical tools they have for describing data: mean, median, mode, range, and outlier.

Activity 1

- The data given in parts (a) and (b) are actual data. The data for part (c) is fictional; however, it is based on the average weight of raccoons.
- Students need to decide what display makes sense for the type of data that they have. There may be more than one appropriate answer.
- Discuss students' choices and their explanations.
- Possible data displays:
  - Part (a): a circle graph (what part of the whole set is each animal) or a bar graph (compare the different categories, although there is a large difference in bar heights: 75 to 2746)
  - Part (b): a scatter plot and line of best fit (pair data, show trend over time, and make predictions for the future)
  - Part (c): a stem-and-leaf plot (spread of data), along with calculating the mean (about 16.7) and median (17.1)
  - Part (d): As a class, discuss students' ideas for minimizing the number of animals killed by vehicles.

Previous Learning

Students should know how to construct a variety of data displays from this year and past years.

Start Thinking! and Warm Up

Activity 7.4 Start Thinking! For use before Activity 7.4

Activity 7.4 Warm Up For use before Activity 7.4

Make a circle graph of the data.

1. A middle school has 390 sixth graders, 310 seventh graders, and 300 eighth graders.
2. The results of a survey in which 20 people were asked to name their favorite color are shown in the table.

Color	blue	red	green	pink	other
People	7	4	3	2	4

7.4 Record and Practice Journal

Essential Question How can you display data in a way that helps you make decisions?

1 ACTIVITY: Displaying Data

Work with a partner. Analyze and display each data set in a way that best describes the data. Explain your choice of display.

a. **ROAD KILL** A group of schools in New England participated in a 2-month study and reported 3962 dead animals.

Birds 307
Mammals 2746
Amphibians 145
Reptiles 75
Unknown 689

Sample answer:

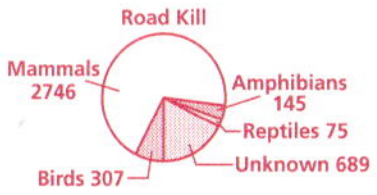

shows kinds of animals as parts of a whole

b. **BLACK BEAR ROAD KILL** The data below show the number of black bears killed on Florida roads from 1987 to 2006.

1987	30	1997	74
1988	37	1998	88
1989	46	1999	82
1990	33	2000	109
1991	43	2001	99
1992	35	2002	129
1993	43	2003	111
1994	47	2004	127
1995	49	2005	141
1996	61	2006	135

Sample answer:

shows relationship over time

English Language Learners

Vocabulary

English learners may need help understanding the word *scale*. There are several meanings in the English language. Some of the common meanings are:

a series of musical notes,
the covering of a reptile,
a device for weighing,
a ratio,
to climb.

In bar graphs, the scale is a series of markings used for measuring. Most scales start at 0 and go to (at least) the greatest value of the data.

7.4 Record and Practice Journal

c. RACCOON ROAD KILL A 1-week study along a 4-mile section of road found the following weights (in pounds) of raccoons that had been killed by vehicles.

13.4	14.8	17.0	12.9	21.3	21.5	16.8	14.8
15.2	18.7	18.6	17.2	18.5	9.4	19.4	15.7
14.5	9.5	25.4	21.5	17.3	19.1	11.0	12.4
20.4	13.6	17.5	18.5	21.5	14.0	13.9	19.0

See Additional Answers.

d. What do you think can be done to minimize the number of animals killed by vehicles?

Check students' work.

2 ACTIVITY: Statistics Project

ENDANGERED SPECIES PROJECT Use the Internet or some other reference to write a report about an animal species that is (or has been) endangered. Include graphical displays of the data you have gathered.

Sample: Florida Key Deer
In 1939, Florida banned the hunting of Key deer. The numbers of Key deer fell to about 100 in the 1940s.

About half of Key deer deaths are due to vehicles.

In 1947, public sentiment was stirred by 11-year-old Glenn Allen from Miami. Allen organized Boy Scouts and others in a letter-writing campaign that led to the establishment of the National Key Deer Refuge in 1957. The approximately 8600-acre refuge includes 2280 acres of designated wilderness.

One of two Key deer wildlife underpasses on Big Pine Key.

Key Deer Refuge has increased the population of Key deer. A recent study estimated the total Key deer population to be between 700 and 800.

Check students' work.

What Is Your Answer?

3. **IN YOUR OWN WORDS** How can you display data in a way that helps you make decisions? Use the Internet or some other reference to find examples of the following types of data displays.
   - Bar graph
   - Circle graph
   - Scatter plot
   - Stem-and-leaf plot
   - Box-and-whisker plot

Data displays make it easy to interpret data and make conclusions.

Laurie's Notes

Activity 2

- Ask a volunteer to read the information presented about Key deer. Discuss how the actions of one person can often make a big difference.
- It would be ideal if the library or computer room is available. If not, you or your students could bring in newspapers and magazines that contain graphical displays.
- If you assign this project, students will need several days.

What Is Your Answer?

- I find that many students can make the displays, if they are told which display to use. It is equally important that students be able to select the display based upon the data and the question you hope to answer from making the display.
- The information gathered by students can be made into classroom posters.

Closure

- **Class Discussion:** Have students present their answers to Question 3. Then have students discuss features of each display, and what types of data lend itself to each data display.

Statistics on Animal-Vehicle Accidents

- Making the roads safer for humans and animals is a goal of many highway transportation agencies.
- You may wish to share additional road kill data.

4 million	miles of roads in the U.S.
226 million	number of vehicles registered in the U.S.
23 trillion	vehicle miles traveled in the U.S. in 2002
6.3 million	number of automobile accidents annually in the U.S.
253,000	number of animal-vehicle accidents annually
50	estimated percent of large animal-vehicle collisions that go unreported
90	percent of animal-vehicle collisions that involve deer
2000	average minimum cost (in dollars) for repairing a vehicle after a collision with a deer
200	number of human deaths annually resulting from vehicle-wildlife collisions
40	percent by which deer-vehicle collisions were reduced after installation of a deer crosswalk system in northeast Utah

Technology For the Teacher

Dynamic Classroom

The Dynamic Planning Tool
Editable Teacher's Resources at *BigIdeasMath.com*

2 ACTIVITY: Statistics Project

ENDANGERED SPECIES PROJECT **Use the Internet or some other reference to write a report about an animal species that is (or has been) endangered. Include graphical displays of the data you have gathered.**

Sample: Florida Key Deer

In 1939, Florida banned the hunting of Key deer. The numbers of Key deer fell to about 100 in the 1940s.

In 1947, public sentiment was stirred by 11-year-old Glenn Allen from Miami. Allen organized Boy Scouts and others in a letter-writing campaign that led to the establishment of the National Key Deer Refuge in 1957. The approximately 8600-acre refuge includes 2280 acres of designated wilderness.

Key Deer Refuge has increased the population of Key deer. A recent study estimated the total Key deer population to be between 700 and 800.

About half of Key deer deaths are due to vehicles.

One of two Key deer wildlife underpasses on Big Pine Key

What Is Your Answer?

3. **IN YOUR OWN WORDS** How can you display data in a way that helps you make decisions? Use the Internet or some other reference to find examples of the following types of data displays.

- Bar graph
- Circle graph
- Scatter plot
- Stem-and-leaf plot
- Box-and-whisker plot

Practice Use what you learned about choosing data displays to complete Exercise 3 on page 300.

Check It Out
Lesson Tutorials
BigIdeasMath.com

Key Idea

Data Display — **What does it do?**

Pictograph — shows data using pictures

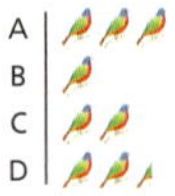

Bar Graph — shows data in specific categories

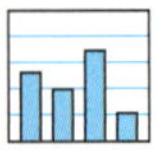

Circle Graph — shows data as parts of a whole

Line Graph — shows how data change over time

Histogram — shows frequencies of data values in intervals of the same size

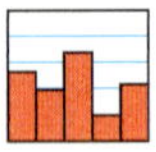

Stem-and-Leaf Plot — orders numerical data and shows how they are distributed

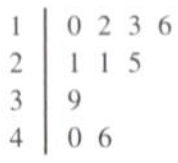

Box-and-Whisker Plot — shows the variability of a data set using quartiles

Line Plot — shows the number of times each value occurs in a data set

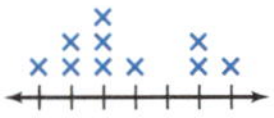

Scatter Plot — shows the relationship between two data sets using ordered pairs in a coordinate plane

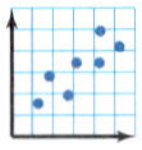

EXAMPLE 1 Choosing an Appropriate Data Display

Choose an appropriate data display for the situation. Explain your reasoning.

a. the number of students in a marching band each year

A line graph shows change over time. So, a line graph is an appropriate data display.

b. comparison of people's shoe sizes and their heights

You want to compare two different data sets. So, a scatter plot is an appropriate data display.

On Your Own

Now You're Ready
Exercises 4–7

Choose an appropriate data display for the situation. Explain your reasoning.

1. the population of the United States divided into age groups
2. the percents of students in your school who speak Spanish, French, or Haitian Creole

Laurie's Notes

Introduction

Connect

- **Yesterday:** Students reviewed data displays.
- **Today:** Students will choose and construct an appropriate data display.

Motivate

- Make a quick sketch of the two bar graphs shown and ask students to comment on each.

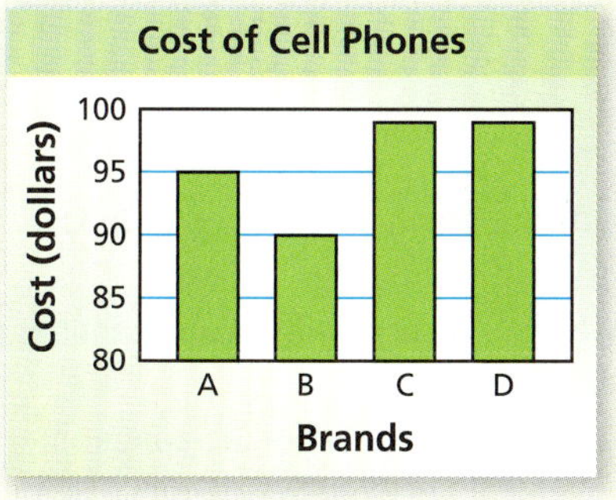

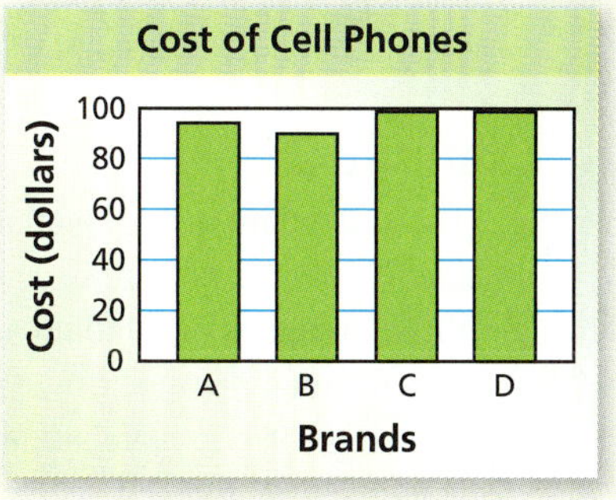

? "Who do you think printed the graph on the left and why?" company selling brand B; Appears to be a lot less expensive than the other brands.

Lesson Notes

Key Idea

- Write the Key Idea. This is a terrific summary of data displays that students have learned to make.
- Emphasize that *choosing an appropriate display* is more poetry than science. On the other hand, it is clearly possible to use any of the graphs in misleading ways. This is science.
- There may be examples of each of these displays around your room.

Example 1

- Read each problem. Students should not have difficulty determining the appropriate data display for each problem.

On Your Own

- **Think-Pair-Share:** Students should read each question independently and then work with a partner to answer the questions. When they have answered the questions, the pair should compare their answers with another group and discuss any discrepancies.

Goal Today's lesson is choosing and constructing an appropriate data display.

Start Thinking! and Warm Up

Lesson 7.4 **Warm Up**
For use before Lesson 7.4

Lesson 7.4 **Start Thinking!**
For use before Lesson 7.4

How are a bar graph and a histogram similar? How are they different?

How are a line graph and a line plot similar? How are they different?

How are a stem-and-leaf plot and box-and-whisker plot similar? How are they different?

Differentiated Instruction

Auditory

Ask students what data display would best represent the given data.

- the number of baseball cards each boy in the class has
  box-and-whisker plot
- the number of hours studying for a test and the test scores of students in a class scatter plot

Extra Example 1

You conduct a survey at your school about insects that students fear the most. Choose an appropriate data display. Explain your reasoning.
Sample answers: Circle graph: shows data as parts of a whole; Bar graph: shows data in specific categories; Pictograph: shows data using pictures.

On Your Own

1. *Sample answer:* histogram; Shows frequencies of ages (data values) in intervals of the same size.
2. *Sample answer:* bar graph; Shows data in specific categories.

Extra Example 2

Which line graph is misleading? Explain.

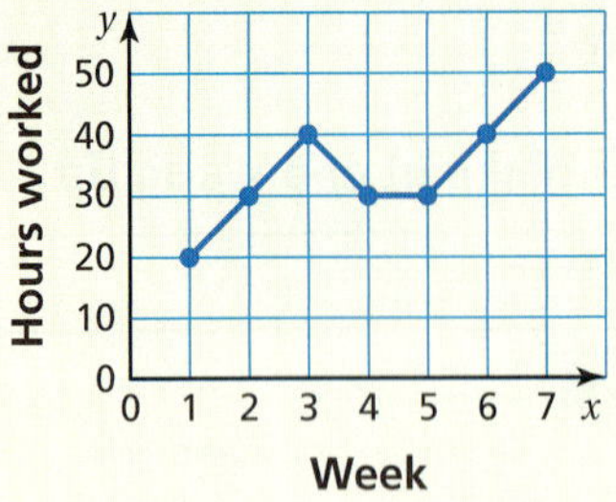

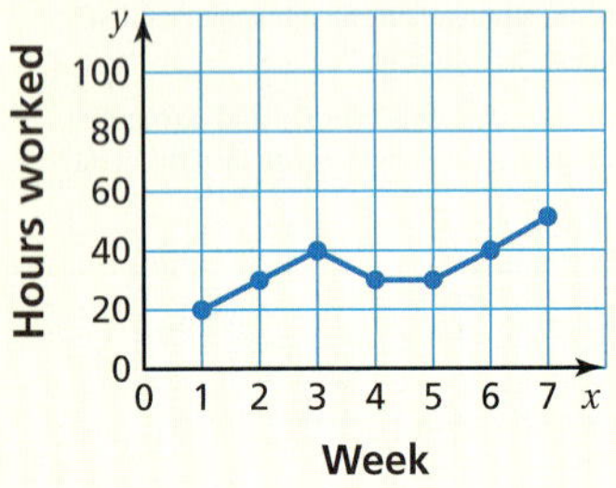

the second graph; The y-scale makes the change from week to week appear smaller.

Extra Example 3

Explain why the data display is misleading.

Favorite Pets

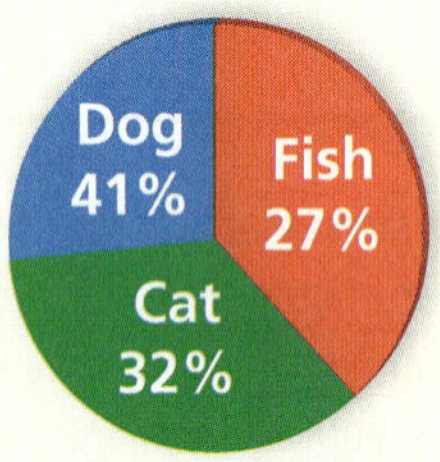

The size of each part of the circle is not proportional to the percent each choice represents.

On Your Own

3. The tickets vary in width and the break in the vertical axis makes the difference in ticket prices appear to be greater.

4. The bars become wider as the years progress, making the increase in profit appear greater.

Laurie's Notes

Discuss

- I have a collection of misleading data displays. When you find a data display in the newspaper or magazine that is misleading, cut it out and save it for later use. Ask colleagues in your school to do the same.
- Often what makes a graph misleading is the scale selected for one, or both, of the axes. By spreading out the scale, or condensing it, the graph becomes misleading.
- As I always tell my students, the person who makes the data display influences how we will view it. They control the extent to which we can see, or not see, features of the data.

Example 2

? "The same data is displayed in each line graph. How do the graphs differ?" The vertical scale is different.

? "Which graph is misleading and why?" first graph; It makes it appear that there has been a rapid growth in box office gross.

- **Extension:** I like to have students pretend that both graphs appear in the newspaper with an article, and I ask them what they would use for a headline for each article. What story does the author want readers to see when they look at each graph?

Example 3

- Have students "read" the pictograph and ask them to summarize what information it describes. Many students will conclude that the amount of cans and the amount of boxes is about the same due to the horizontal distance each set of icons takes up.

? "Approximately how many cans of food and boxes of food have been donated?" 11 cans $\times$ 20 = 220 cans; 6 boxes $\times$ 20 = 120 boxes

- Almost twice as many cans of food have been donated as boxes, so this is misleading. The box icon is too large. It should be the same width as the can.

On Your Own

- **Think-Pair-Share:** Students should read each question independently and then work with a partner to answer the questions. When they have answered the questions, the pair should compare their answers with another group and discuss any discrepancies.

Closure

- **Exit Ticket:** Make a pictograph for the data in Example 3 that would not be misleading.

Technology For the Teacher

Dynamic Classroom

The Dynamic Planning Tool

Editable Teacher's Resources at *BigIdeasMath.com*

EXAMPLE 2 Identifying a Misleading Data Display

Which line graph is misleading? Explain.

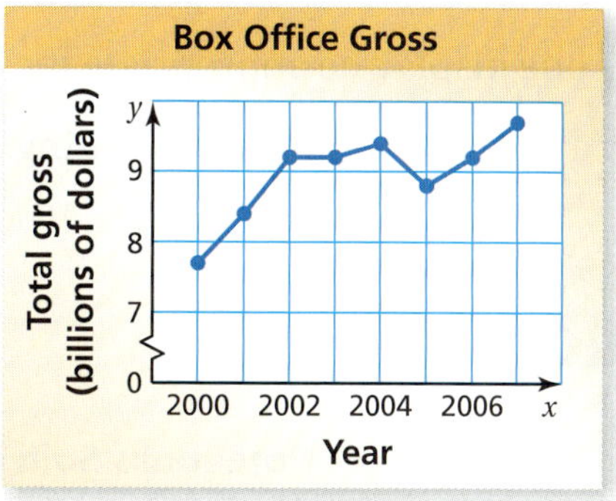

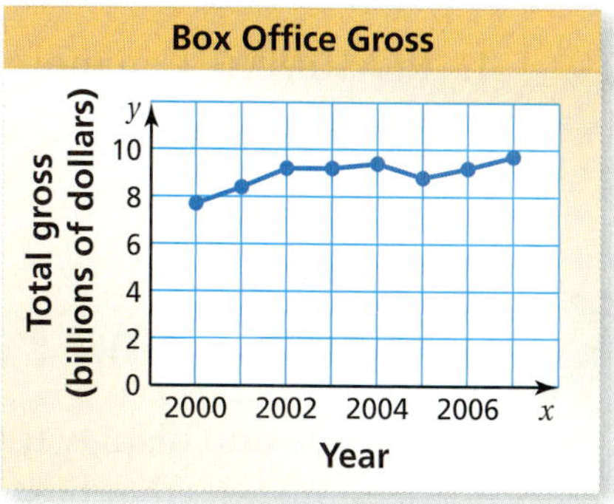

The vertical axis of the line graph on the left has a break (≶) and begins at 7. This graph makes it appear that the total gross increased rapidly from 2000 to 2004. The graph on the right has an unbroken axis. It is more honest and shows that the total gross increased slowly.

∴ So, the graph on the left is misleading.

EXAMPLE 3 Analyzing a Misleading Data Display

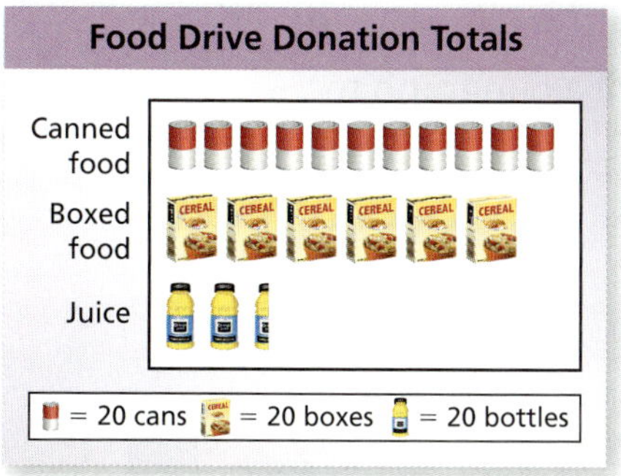

A volunteer concludes that the number of cans of food and boxes of food donated were about the same. Is this conclusion accurate? Explain.

Each icon represents the same number of items. Because the box icon is larger than the can icon, it looks like the number of boxes is about the same as the number of cans, but the number of boxes is actually about half of the number of cans.

∴ So, the conclusion is not accurate.

On Your Own

Now You're Ready
Exercises 9–12

Explain why the data display is misleading.

3.

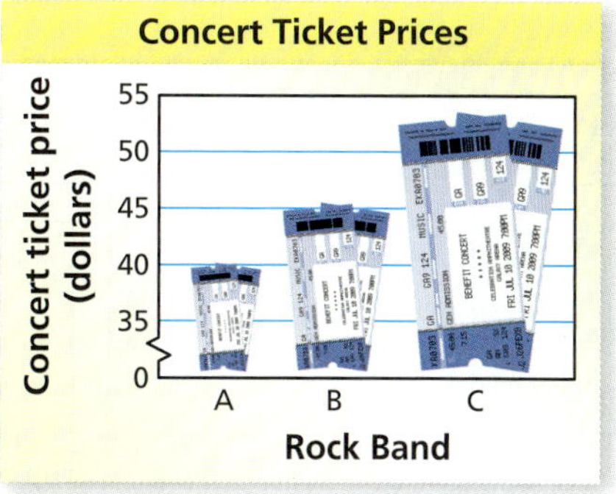

4.

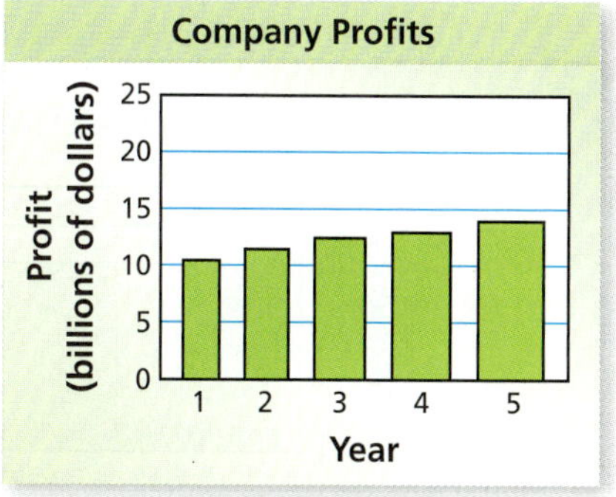

7.4 Exercises

Check It Out
Help with Homework
BigIdeasMath.com

Vocabulary and Concept Check

1. **REASONING** Can more than one display be appropriate for a data set? Explain.
2. **OPEN-ENDED** Describe how a histogram can be misleading.

Practice and Problem Solving

3. Analyze and display the data in a way that best describes the data. Explain your choice of display.

Notebooks Sold in One Week				
192 red	170 green	203 black	183 pink	230 blue
165 yellow	210 purple	250 orange	179 white	218 other

Choose an appropriate data display for the situation. Explain your reasoning.

1 4. a student's test scores and how the scores are spread out

5. the distance a person drives each month

6. the outcome of rolling a number cube

7. homework problems assigned each day

8. **WRITING** When would you choose a histogram instead of a bar graph to display data?

Explain why the data display is misleading.

2 3 9.

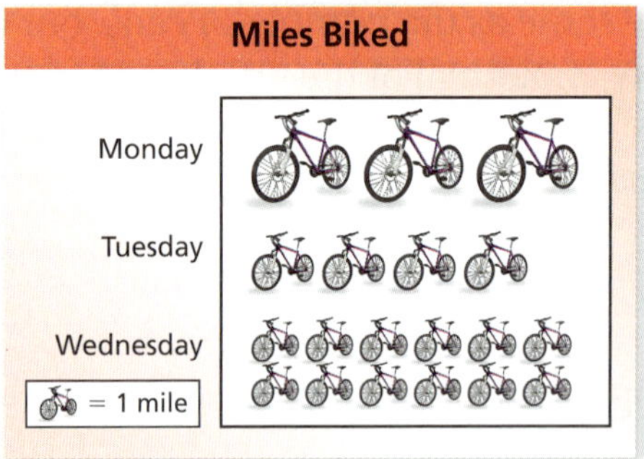

10.

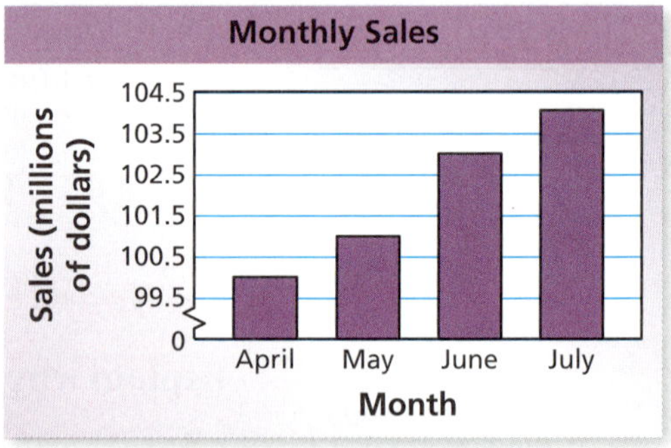

11.

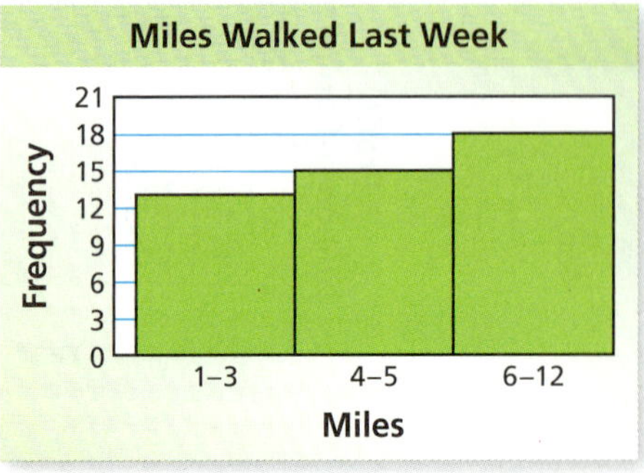

12.

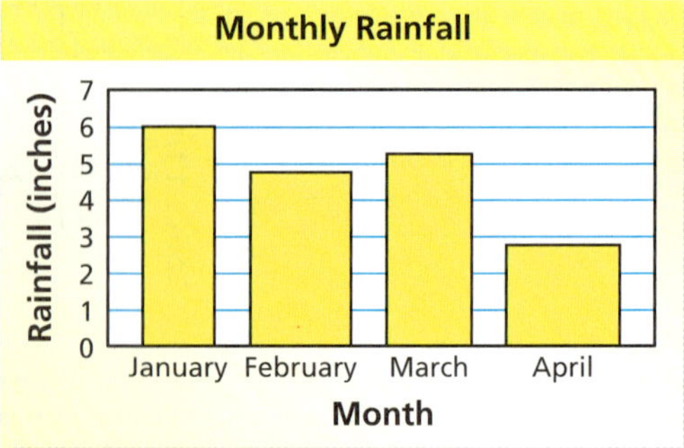

Assignment Guide and Homework Check

Level	Day 1 Activity Assignment	Day 2 Lesson Assignment	Homework Check
Basic	3, 18–20	1, 2, 4–14	4, 8, 10, 14
Average	3, 18–20	1, 2, 4–14	4, 8, 10, 14
Advanced	3, 18–20	1, 2–12 even, 13–17	8, 10, 14, 16

Common Errors

- **Exercises 4–7** Students may get some of the data displays mixed up. For example, students may say that a line plot should be used when they mean a line graph. Remind students to write an explanation for why they chose that type of data display based on the description given. Have them check the title of the type of data display next to the description.
- **Exercises 9–12** Students may not be able to recognize why the data display is misleading. As a class, make a list of things to examine when analyzing a data display. For example, check the increments or intervals for the axes.
- **Exercise 15** Students may say that the best data display for showing the mode is a stem-and-leaf plot because the leaves that have more repeated data will be higher. However, this display could have other data in the leaf. Remind them that a line plot isolates each data value and shows the frequency of each individual number, so this is the best data display.

7.4 Record and Practice Journal

Choose an appropriate data display for the situation. Explain your reasoning.

1. the number of people that donated blood over the last 5 years

   Sample answer: **line graph; shows how data changes over time**

2. percent of class participating in school clubs

   Sample answer: **circle graph; shows the data as part of a whole**

Explain why the data display is misleading.

3. 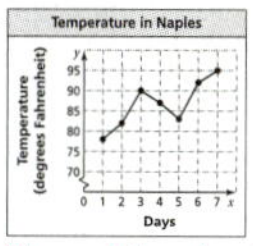

   The *y*-axis has a break, making the data look as if it changed rapidly and by large amounts.

4. 

   See Additional Answers.

5. A team statistician wants to use a data display to show the points scored per game during the season. Choose an appropriate data display for the situation. Explain your reasoning.

   Sample answer: **stem-and-leaf plot; can easily read each data value**

Technology For the Teacher

Answer Presentation Tool

Quiz*Show*

Vocabulary and Concept Check

1. yes; Different displays may show different aspects of the data.
2. *Sample answer:* The pictures of the objects could be made larger so there appears to be more of them, or the vertical axis could be too small or too large.

Practice and Problem Solving

3. *Sample answer:*

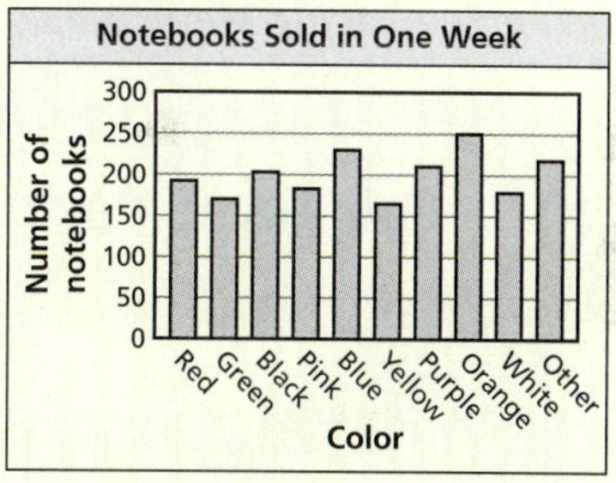

   A bar graph shows the data in different color categories.

4. *Sample answer:* Stem-and-leaf plot: shows how data is distributed.
5. *Sample answer:* Line graph: shows changes over time.
6. *Sample answer:* Line plot: shows the number of times each outcome occurs.
7. *Sample answer:* Line graph: shows changes over time.
8. when the data is in terms of intervals of one category, as opposed to multiple categories
9. The pictures of the bikes are larger on Monday, which makes it seem like the distance is the same each day.

10–12. See Additional Answers.

Practice and Problem Solving

13. *Sample answer:* bar graph; Each bar can represent a different vegetable.

14. yes; The vertical axis has a scale that increases by powers of 10, which makes the data appear to have a linear relationship.

15. *Sample answer:* line plot

16. a. The percents do not sum to 100%.

    b. *Sample answer:* bar graph; It would show the frequency of each sport.

17. See *Taking Math Deeper.*

Fair Game Review

18 $x + 3 = 5$

19. $8x = 24$

20. A

Mini-Assessment

Choose an appropriate data display for the situation. Explain your reasoning.

1. the outcome of flipping a coin
   Sample answers: Pictograph: shows number of times heads or tails appears using picture of coins; Bar graph: shows number of times you get heads or tails; Line plot: show number of times you get a heads or tails.
2. comparison of student's test scores and how long students studied
   Sample answer: Scatter plot: you want to compare two data sets.
3. the number of students participating in after-school sports each year
   Sample answer: Line graph: shows how data change over time.

Taking Math Deeper

Exercise 17

This exercise introduces students to an amazing property of the number pi. Pi is an irrational number and therefore its decimal representation is not repeating. Even so, the ten digits from 0 to 9 each occur about 10 percent of the time, when one considers thousands of digits.

1 Display the data in a bar graph.

a.

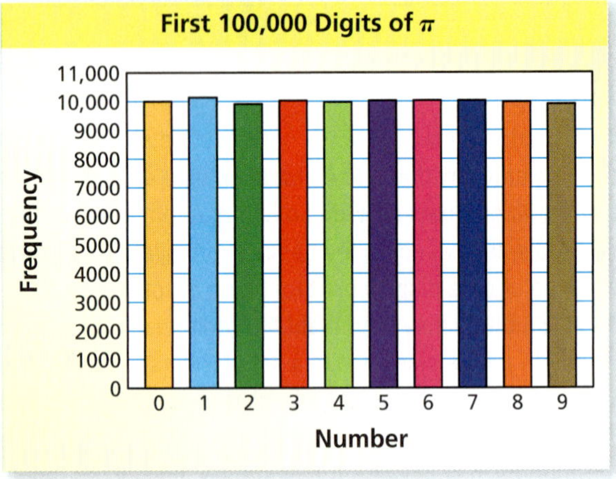

2 Display the data in a circle graph.

b.

3 **c. and d.** Compare the two displays.

Both graphs show that each digit occurs about 10% of the time. The bar graph has a slight advantage because it shows that some digits occur slightly more than others.

Reteaching and Enrichment Strategies

If students need help...	If students got it...
Resources by Chapter • Practice A and Practice B • Puzzle Time Record and Practice Journal Practice Differentiating the Lesson Lesson Tutorials Skills Review Handbook	Resources by Chapter • Enrichment and Extension • School-to-Work Start the next section

13. VEGETABLES A nutritionist wants to use a data display to show the favorite vegetables of the students at a school. Choose an appropriate data display for the situation. Explain your reasoning.

14. CHEMICALS A scientist gathers data about a decaying chemical compound. The results are shown in the scatter plot. Is the data display misleading? Explain.

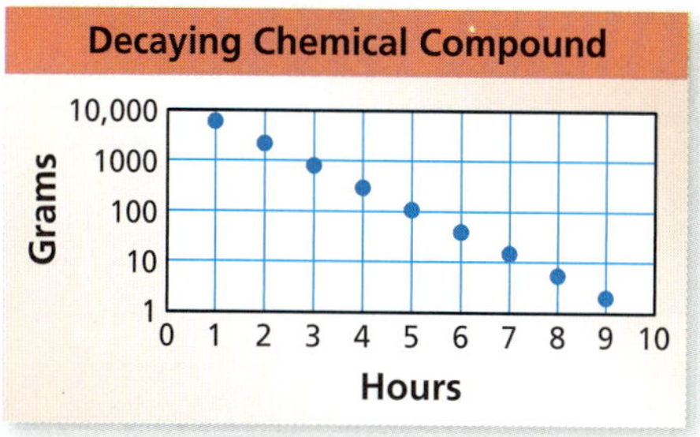

15. REASONING What type of data display is appropriate for showing the mode of a data set?

16. SPORTS A survey asked 100 students to choose their favorite sports. The results are shown in the circle graph.

Favorite Sports

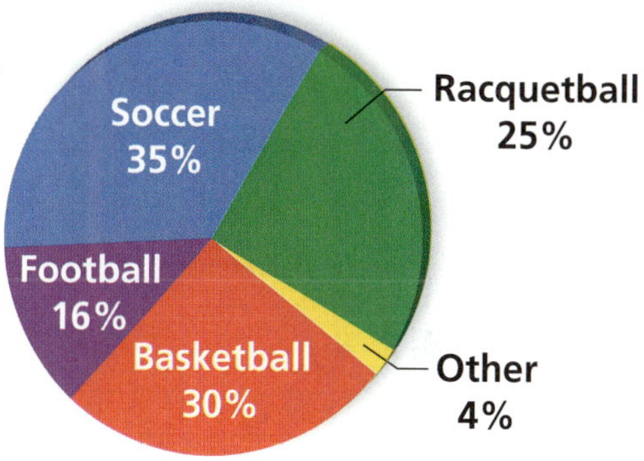

a. Explain why the graph is misleading.

b. What type of data display would be more appropriate for the data? Explain.

17. Critical Thinking With the help of computers, mathematicians have computed and analyzed billions of digits of the irrational number π. One of the things they analyze is the frequency of each of the numbers 0 through 9. The table shows the frequency of each number in the first 100,000 digits of π.

a. Display the data in a bar graph.

b. Display the data in a circle graph.

c. Which data display is more appropriate? Explain.

d. Describe the distribution.

Number	0	1	2	3	4	5	6	7	8	9
Frequency	9999	10,137	9908	10,025	9971	10,026	10,029	10,025	9978	9902

Fair Game Review What you learned in previous grades & lessons

Write the verbal statement as an equation. *(Skills Review Handbook)*

18. A number plus 3 is 5.

19. 8 times a number is 24.

20. MULTIPLE CHOICE What is 20% of 25% of 400? *(Skills Review Handbook)*

Ⓐ 20 Ⓑ 200 Ⓒ 240 Ⓓ 380

7.3–7.4 Quiz

Check It Out
Progress Check
BigIdeasMath.com

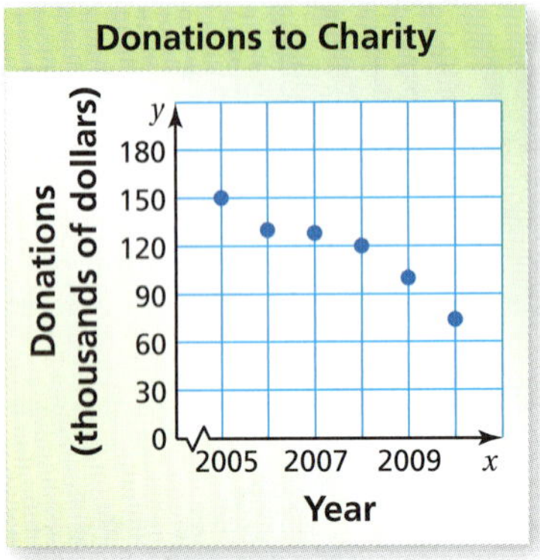

1. The scatter plot shows the number of donations a charity received from 2005 to 2010. *(Section 7.3)*

a. In what year did the charity receive \$150,000?

b. How much did the charity receive in 2008?

c. Describe the relationship shown by the data.

Tell whether the data show a *positive*, a *negative*, or *no* relationship. *(Section 7.3)*

2.

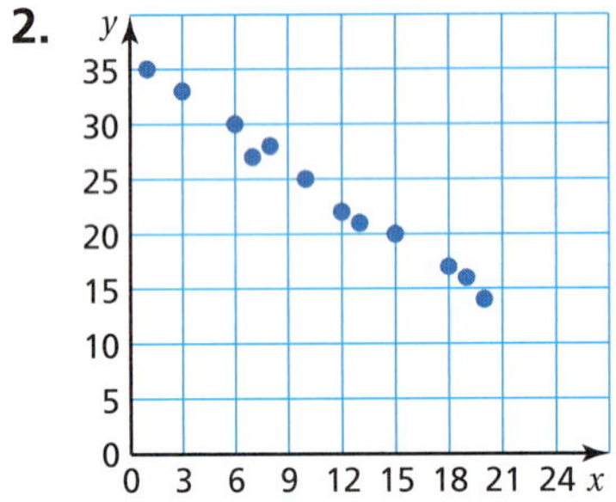

3.

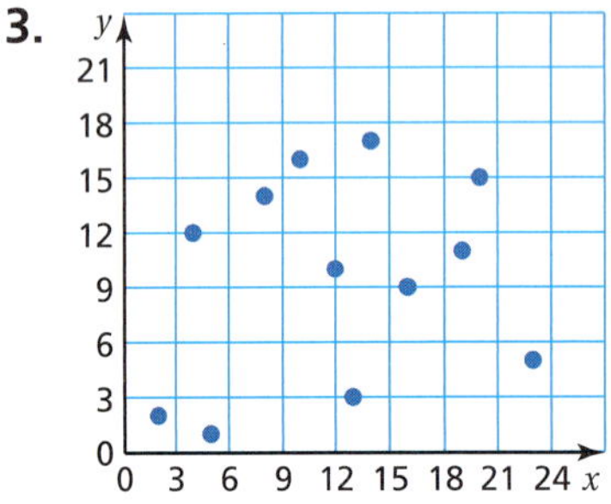

4.

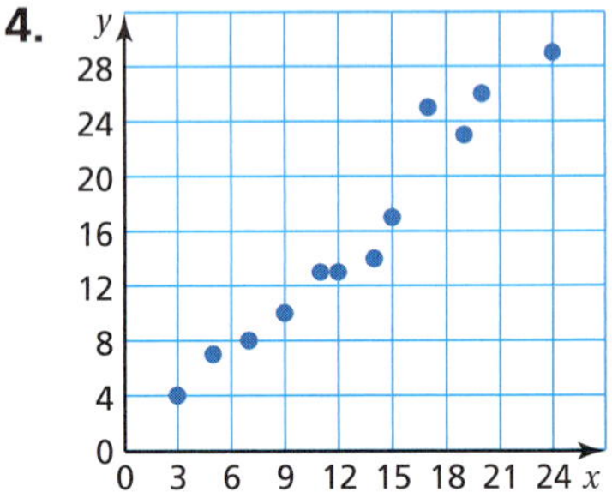

Choose an appropriate data display for the situation. Explain your reasoning. *(Section 7.4)*

5. percent of band students in each section

6. company's profit for each week

7. **FUNDRAISER** The graph shows the amount of money that the eighth-grade students at a school raised each month to pay for the class trip. Is the data display misleading? Explain. *(Section 7.4)*

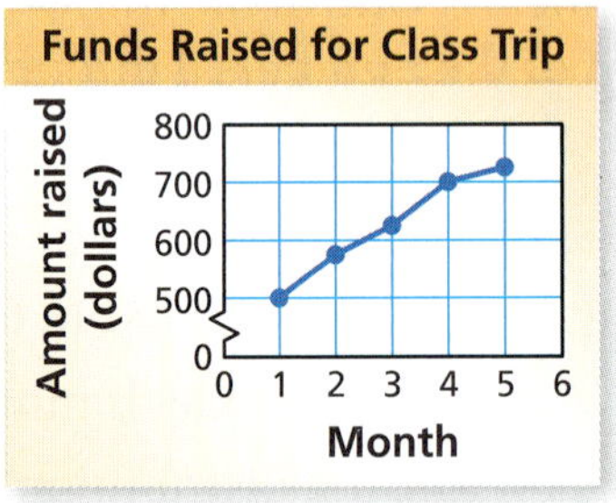

8. **CATS** The table shows the number of cats adopted from an animal shelter each month. *(Section 7.3)*

a. Make a scatter plot of the data.

b. Draw a line of best fit.

c. Write an equation of the line of best fit.

d. Predict how many cats will be adopted in month 10.

Month	1	2	3	4	5	6	7	8	9
Cats	3	6	7	11	13	14	15	18	19

Alternative Assessment Options

Math Chat | Student Reflective Focus Question
Structured Interview | Writing Prompt

Math Chat

Ask students to use their own words to summarize how they would choose an appropriate data display. Be sure that they include examples. Select students at random to present to the class.

Study Help Sample Answers

Remind students to complete Graphic Organizers for the rest of the chapter.

3.

Scatter Plots

- Graphs that show the relationship between two data sets
- Scatter plots can show that a relationship exists between two data sets.
- The two sets of data are plotted as ordered pairs in a coordinate plane.
- *Positive relationship:* As x increases, y increases.
- *Negative relationship:* As x increases, y decreases.
- *No relationship:* The points show no pattern.

4.

Lines of Best Fit

- A *line of best fit* is a line drawn on a scatter plot that is close to most of the data points.
- A line of best fit does not need to pass through any of the data points.
- A line of best fit can be used to estimate data on a graph.

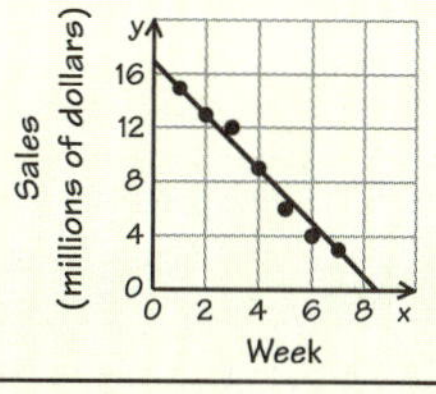

5. Available at *BigIdeasMath.com.*

Reteaching and Enrichment Strategies

If students need help. . .	If students got it. . .
Resources by Chapter • Study Help • Practice A and Practice B • Puzzle Time Lesson Tutorials *BigIdeasMath.com* Practice Quiz Practice from the Test Generator	Resources by Chapter • Enrichment and Extension • School-to-Work Game Closet at *BigIdeasMath.com* Start the Chapter Review

Answers

1. **a.** 2005 **b.** \$120,000
   c. There is a negative relationship between year and number of donations.
2. negative relationship
3. no relationship
4. positive relationship
5. *Sample answer:* circle graph; shows data as parts of a whole
6. *Sample answer:* line graph; shows changes over time
7. yes; The break in the vertical axis makes it appear that the amount of money raised increased very rapidly from month to month.
8. See Additional Answers.

Technology For the Teacher

Answer Presentation Tool

Assessment Book

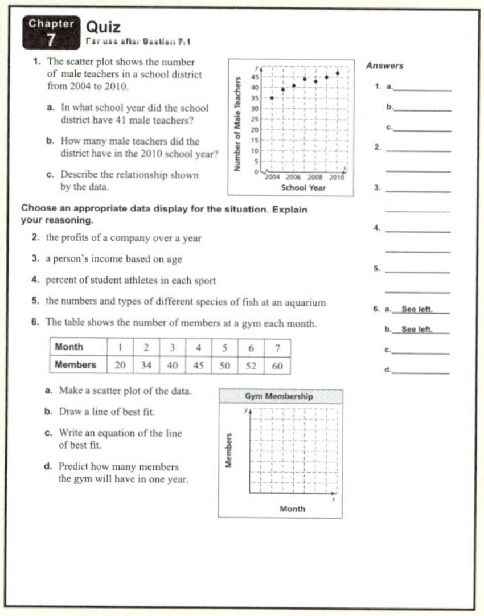
Chapter 7 Quiz
For use after Section 7.1

1. The scatter plot shows the number of male teachers in a school district from 2004 to 2010.
   a. In what school year did the school district have 41 male teachers?
   b. How many male teachers did the district have in the 2010 school year?
   c. Describe the relationship shown by the data.

Choose an appropriate data display for the situation. Explain your reasoning.

2. the profits of a company over a year
3. a person's income based on age
4. percent of student athletes in each sport
5. the numbers and types of different species of fish at an aquarium
6. The table shows the number of members at a gym each month.

Month	1	2	3	4	5	6	7
Members	20	34	40	45	50	52	60

   a. Make a scatter plot of the data.
   b. Draw a line of best fit.
   c. Write an equation of the line of best fit.
   d. Predict how many members the gym will have in one year.

Answers
1. a. ____ b. ____ c. ____
2. ____
3. ____
4. ____
5. ____
6. a. See left. b. See left. c. ____ d. ____

For the Teacher
Additional Review Options

- **Quiz***Show*
- Big Ideas Test Generator
- Game Closet at *BigIdeasMath.com*
- Vocabulary Puzzle Builder
- Resources by Chapter
  - Puzzle Time
  - Study Help

Answers

1. The mean stays the same, the median decreases $4.2 - 4.1 = 0.1$ kilometer, and the mode changes to 4.0 and 4.3.
2. mean: 1.7
   median: 1
   mode: 1
3. mean: 4
   median: 3
   mode: 10

Review of Common Errors

Exercises 1–3

- When finding the mean, students may forget to divide by the total number of data values and instead divide by the maximum value. Remind them that the definition of mean is an "average," so they must take into account the total number of items or numbers to get an average. Explain to students that it is as if they are dividing the total evenly among the number of groups that there are, similar to the activities.
- Students may try to identify the median without ordering the data first. Remind them that it is essential to order the data first and then find the median. This also makes finding the mode easier.

Exercises 4 and 5

- Students may forget to order the data before making the box-and-whisker plot. Remind them that they need to order the data first because finding each quartile is similar to finding a median.
- Students may find an incorrect median or quartile when it is between two numbers. Remind them that when the median is between two numbers, they need to find the mean of those two values.

Exercise 6

- When finding values from the graph, students may accidentally shift over or up too far and get an answer that is off by an increment. Encourage them to start at the given value and trace the graph to where the point or line of best fit is, and then trace down or left to the other axis for the answer.

Exercises 9–12

- Students may mix up the data displays. Remind students to write an explanation for why they chose that type of data display based on the description given.

7 Chapter Review

Check It Out
Vocabulary Help
BigIdeasMath.com

Review Key Vocabulary

measure of central tendency, *p. 276*
box-and-whisker plot, *p. 282*
quartiles, *p. 282*
scatter plot, *p. 290*
line of best fit, *p. 292*

Review Examples and Exercises

7.1 Measures of Central Tendency *(pp. 274–279)*

The table shows the number of kilometers you ran each day for the past 10 days. Find the mean, median, and mode of the distances.

Kilometers Run	
3.5	4.1
4.0	4.3
4.4	4.5
3.9	2.0
4.3	5.0

Mean: $\frac{40}{10} = 4$ (40 = sum of the data; 10 = number of values)

Median: 2.0, 3.5, 3.9, 4.0, 4.1, 4.3, 4.3, 4.4, 4.5, 5.0 — Order the data.

$\frac{8.4}{2} = 4.2$ — Mean of two middle values

Mode: 2.0, 3.5, 3.9, 4.0, 4.1, 4.3, 4.3, 4.4, 4.5, 5.0

The value 4.3 occurs most often.

∴ The mean is 4 kilometers, the median is 4.2 kilometers, and the mode is 4.3 kilometers.

Exercises

1. Use the data in the example above. You run 4.0 miles on day 11. How does this additional value affect the mean, median, and mode? Explain.

Find the mean, median, and mode of the data.

2.
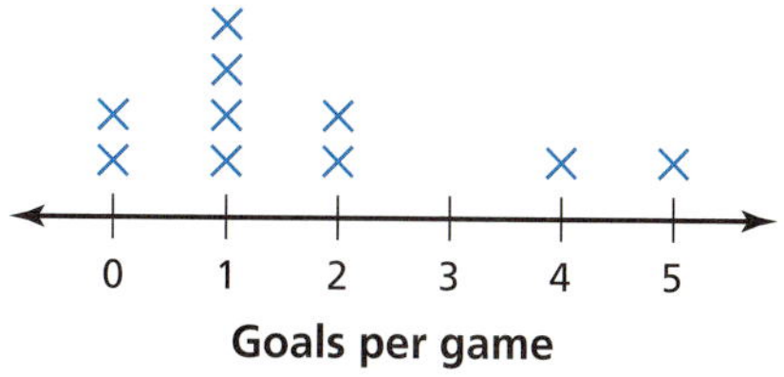

3.

Ski Resort Temperatures (°F)		
11	3	3
0	−9	−2
10	10	10

7.2 Box-and-Whisker Plots (pp. 280–285)

Make a box-and-whisker plot for the weights (in pounds) of pumpkins sold at a market.

16, 20, 14, 15, 12, 8, 8, 19, 14, 10, 8, 16

Step 1: Order the data. Find the median and the quartiles.

lower half | upper half

least value → 8 8 8 10 12 14 | 14 15 16 16 19 20 ← greatest value

first quartile, 9 — median, 14 — third quartile, 16

Step 2: Draw a number line that includes the least and greatest values. Graph points above the number line for the least value, greatest value, median, first quartile, and third quartile.

Step 3: Draw a box using the quartiles. Draw a line through the median. Draw whiskers from the box to the least and greatest values.

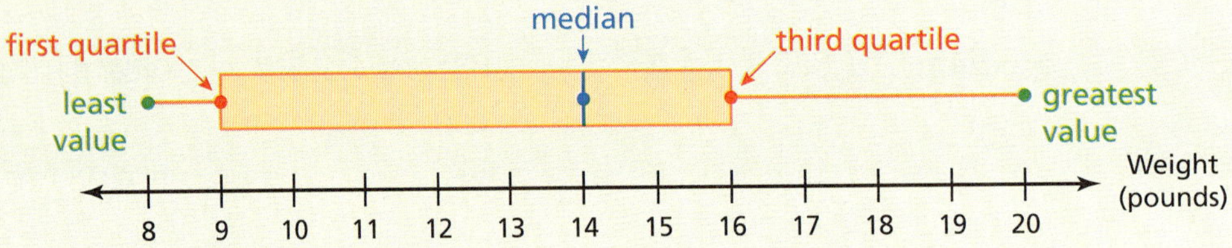

Exercises

Make a box-and-whisker plot for the data.

4. Ages of volunteers at a hospital: 14, 17, 20, 16, 17, 14, 21, 18

5. Masses (in kilograms) of lions: 120, 200, 180, 150, 200, 200, 230, 160

7.3 Scatter Plots and Lines of Best Fit (pp. 288–295)

Your school is ordering custom T-shirts. The scatter plot shows the number of T-shirts ordered and the cost per shirt. What tends to happen to the cost per shirt as the number of T-shirts ordered increases?

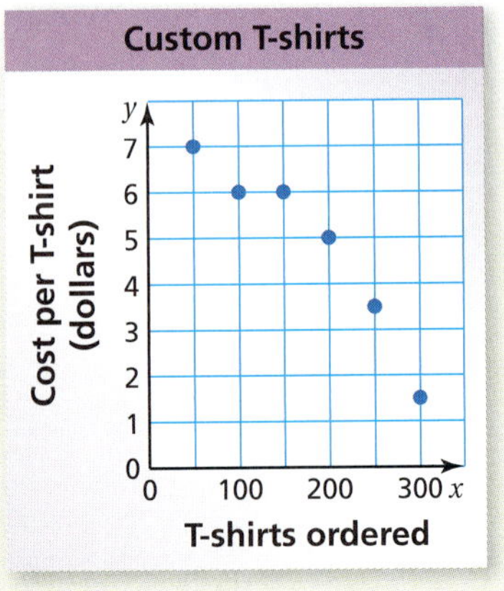

Looking at the graph, the plotted points go down from left to right.

So, as the number of T-shirts ordered increases, the cost per shirt decreases.

Review Game

Rolling for Data

Big Ideas Game Closet

Materials per Pair

- two number cubes
- paper
- pencil

Directions

Students should work in pairs. Students in each pair take turns rolling the two number cubes one at a time. The first number they roll represents the tens digit and the second number represents the ones digit of a whole number. For example, If a 1 is rolled and then a 6, the whole number is 16. Students record the whole numbers in a stem-and-leaf plot and keep rolling until they have 10 leaves for any one stem. Once a pair acquires the 10 leaves, they race to find the mean, median, and mode of all their whole numbers and make a box-and-whisker plot to display the data.

Who Wins?

The first pair to finish all tasks wins 10 points, the second 9 points, the third 8 points, and so on. The game can be repeated as many times as desired. The pair with the most points after a predetermined number of rounds or amount of time wins.

For the Student Additional Practice

- Lesson Tutorials
- Study Help (textbook)
- Student Website
  - Multi-Language Glossary
  - Practice Assessments

Answers

4–5. See Additional Answers.

6. **a.** 2009

b. 225 geese

c. positive relationship

7. negative relationship

8. no relationship

9. *Sample answer:* line graph; shows changes over time

10. *Sample answer:* circle graph; shows data as parts of a whole

11. *Sample answer:* pictograph; shows data using pictures

12. *Sample answer:* scatter plot; shows the relationship between two data sets

My Thoughts on the Chapter

What worked. . .

> **Teacher Tip**
>
> Not allowed to write in your teaching edition? Use sticky notes to record your thoughts.

What did not work. . .

What I would do differently. . .

Exercises

6. The scatter plot shows the number of geese that migrated to a park each season.

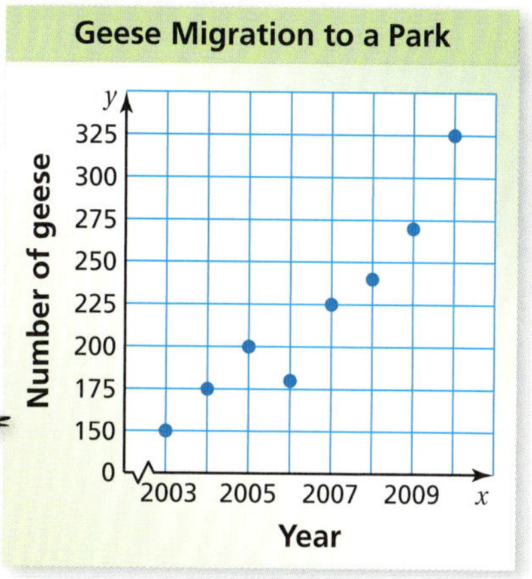

a. In what year did 270 geese migrate?

b. How many geese migrated in 2007?

c. Describe the relationship shown by the data.

Tell whether the data show a *positive*, a *negative*, or *no* relationship.

7. 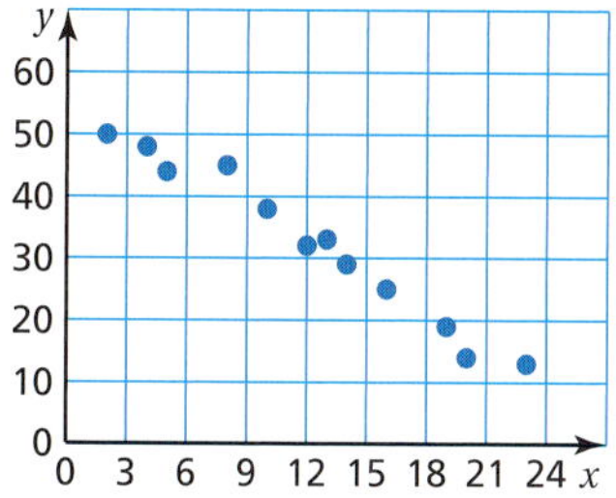

8. 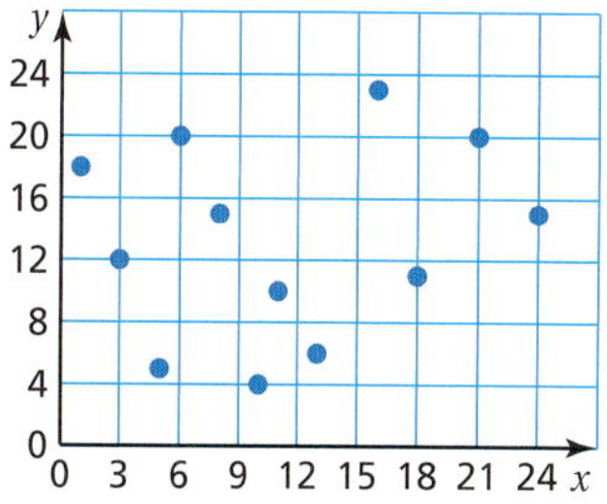

7.4 Choosing a Data Display (pp. 296–301)

Choose an appropriate data display for the situation. Explain your reasoning.

a. the percent of votes that each candidate received in an election

A circle graph shows data as parts of a whole. So, a circle graph is an appropriate data display.

b. the distribution of the ages of U.S. presidents

A stem-and-leaf plot orders numerical data and shows how they are distributed. So, a stem-and-leaf plot is an appropriate data display.

Exercises

Choose an appropriate data display for the situation. Explain your reasoning.

9. the number of pairs of shoes sold by a store each week

10. the outcomes of spinning a spinner with 3 equal sections numbered 1, 2, and 3

11. comparison of the number of cans of food donated by each eighth-grade class

12. comparison of the heights of brothers and sisters

7 Chapter Test

Check It Out
Test Practice
BigIdeasMath.com

Find the mean, median, and mode of the data.

1. **Distances (feet) Above or Below Water Level in Pool**

−3	0	−3
3	10	0
11	−6	−3

2. **Cooking Time (minutes)**

Stem	Leaf
3	5 8
4	0 1 8
5	0 4 4 4 5 9
6	0

Key: 4 | 1 = 41 minutes

Make a box-and-whisker plot for the data.

3. Ages (in years) of dogs at a vet's office: 1, 3, 5, 11, 5, 7, 5, 9

4. Lengths (in inches) of fish in a pond: 12, 13, 7, 8, 14, 6, 13, 10

5. Hours practiced each week: 7, 6, 5, 4.5, 3.5, 7, 7.5, 2, 8, 7, 7.5, 6.5

6. **POPULATION** The graph shows the population (in millions) of the United States from 1960 to 2000.

   a. In what year was the population of the United States about 180 million?

   b. What was the approximate population of the United States in 1990?

   c. Describe the relationship shown by the data.

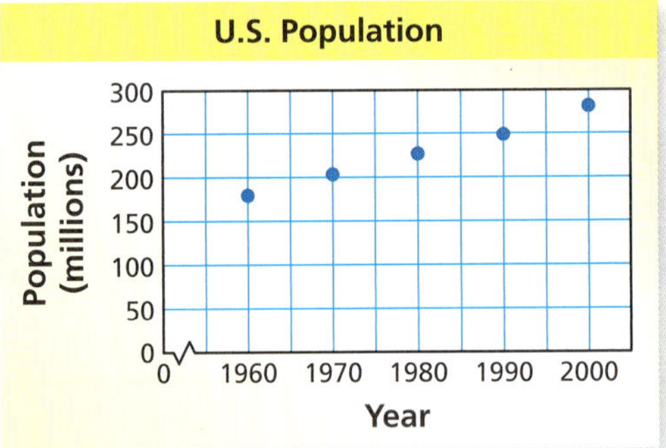

Choose an appropriate data display for the situation. Explain your reasoning.

7. magazine sales grouped by price

8. distance a person hikes each week

9. **ALLIGATORS** The table shows the lengths of 12 alligators. Make a box-and-whisker plot of the data. What does the box-and-whisker plot tell you about the data?

Length (meters)	2.0	1.9	2.2	2.8	3.0	2.0	2.2	3.0	2.5	1.8	2.1	3.0

10. **REASONING** Name two types of data displays that are appropriate for showing the median of a data set.

11. **NEWBORNS** The table shows the lengths and weights of several newborn babies.

    a. Make a scatter plot of the data.

    b. Draw the line of best fit.

    c. Write an equation of the line of best fit.

    d. Use the equation to predict the weight of a newborn that is 19.75 inches long.

Length (inches)	Weight (pounds)
19	6
19.5	7
20	7.75
20.25	8.5
20.5	8.5
22.5	11

Test Item References

Chapter Test Questions	Section to Review
1, 2, 10	7.1
3–5, 9	7.2
6, 11	7.3
7, 8, 10	7.4

Test-Taking Strategies

Remind students to quickly look over the entire test before they start so that they can budget their time. When they receive their test, students should list the different types of data displays. Students should also be encouraged to always order the given data before they start the problem. Teach the students to use the Stop and Think strategy before answering. **Stop** and carefully read the question, and **Think** about what the answer should look like.

Common Assessment Errors

- **Exercises 1–5, 9** Students may try to identify the median without ordering the data first. Remind them to order the data first and then find the median.
- **Exercise 6** When finding values from the graph, students may accidentally shift over or up too far and get an answer that is off by an increment. Encourage them to start at the given value and trace the graph to where the point or line of best fit is, and then trace down or left to the other axis for the answer.
- **Exercises 7, 8, 10** Students may get the data displays mixed up. Remind students to write an explanation for why they chose that type of data display based on the description given.
- **Exercise 11** Students may use inconsistent increments on their graphs or forget to label their graphs. Students should use consistent increments to represent the data. Remind them to label the axes so that information can be read from the graph.

Reteaching and Enrichment Strategies

If students need help. . .	If students got it. . .
Resources by Chapter • Practice A and Practice B • Puzzle Time Record and Practice Journal Practice Differentiating the Lesson Lesson Tutorials Practice from the Test Generator Skills Review Handbook	Resources by Chapter • Enrichment and Extension • School-to-Work Game Closet at *BigIdeasMath.com* Start Standardized Test Practice

Answers

1. mean: 1; median: 0; mode: -3
2. mean: 49; median: 52; mode: 54

3–5. See Additional Answers.

6. a. 1960
   b. about 250 million
   c. There is a positive relationship between year and population.
7. *Sample answer:* histogram; shows frequencies of data values in intervals of the same size
8. *Sample answer:* line graph; shows how data change over time
9. See Additional Answers.
10. box-and-whisker plot; stem-and-leaf plot
11. See Additional Answers.

Assessment Book

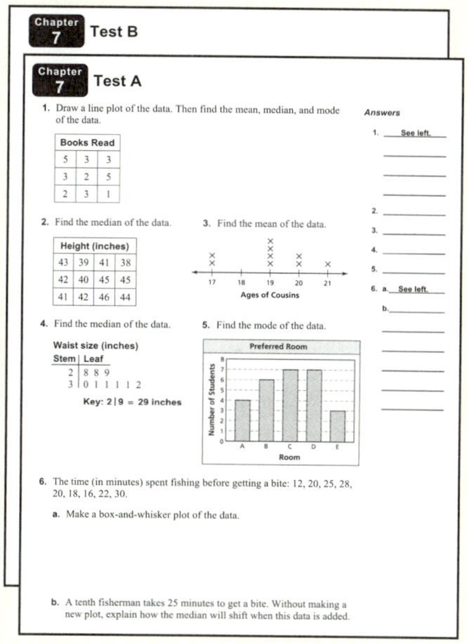
Chapter 7 Test B

Chapter 7 Test A

1. Draw a line plot of the data. Then find the mean, median, and mode of the data.

Books Read		
5	3	3
3	2	5
2	3	1

2. Find the median of the data.

Height (inches)			
43	39	41	38
42	40	45	45
41	42	46	44

3. Find the mean of the data.

Ages of Cousins (17, 18, 19, 20, 21)

4. Find the median of the data.

Waist size (inches)

Stem	Leaf
2	8 8 9
3	0 1 1 1 1 2

Key: 2 | 9 = 29 inches

5. Find the mode of the data.

Preferred Room (Number of Students; Room A, B, C, D, E)

6. The time (in minutes) spent fishing before getting a bite: 12, 20, 25, 28, 20, 18, 16, 22, 30.

a. Make a box-and-whisker plot of the data.

b. A tenth fisherman takes 25 minutes to get a bite. Without making a new plot, explain how the median will shift when this data is added.

Answers
1. See left.
6. a. See left.

Test-Taking Strategies

Available at *BigIdeasMath.com*

After Answering Easy Questions, Relax
Answer Easy Questions First
Estimate the Answer
Read All Choices before Answering
Read Question before Answering
Solve Directly or Eliminate Choices
Solve Problem before Looking at Choices
Use Intelligent Guessing
Work Backwards

About this Strategy

When taking a multiple choice test, be sure to read each question carefully and thoroughly. Look closely for words that change the meaning of the question, such as *not, never, all, every,* and *always.*

Answers

1. D
2. F
3. B

Technology For the Teacher

Big Ideas Test Generator

Item Analysis

1. **A.** The student misreads the plot, thinking that the median is the number at its right end.

   B. The student misreads the plot, thinking that 8 is the median.

   C. The student misreads the plot, thinking that the lower whisker represents 50% of the data.

   D. Correct answer

2. **F.** Correct answer

   G. The student finds the square root of $\frac{40}{4}$.

   H. The student thinks the square root of 40 is equal to 20.

   I. The student finds the square root of 40, and ignores the denominator.

3. **A.** The student finds the hourly amount.

   B. Correct answer

   C. The student finds the average hourly amount from the 2-hour job.

   D. The student chooses the amount for a 1-hour job.

4. **F.** Correct answer

   G. The student picks the angle on the same side of the transversal as angle 6, but not the corresponding angle.

   H. The student confuses alternate interior angles with corresponding angles.

   I. The student confuses vertical angles with corresponding angles.

7 Standardized Test Practice

Test-Taking Strategy
Read Question Before Answering

"Be sure to read the question before choosing your answer. You may find a word that changes the meaning."

1. Research scientists are measuring the number of days lettuce seeds take to germinate. In a study, 500 seeds were planted. Of these, 473 seeds germinated. The box-and-whisker plot summarizes the number of days it took the seeds to germinate. What can you conclude from the box-and-whisker plot?

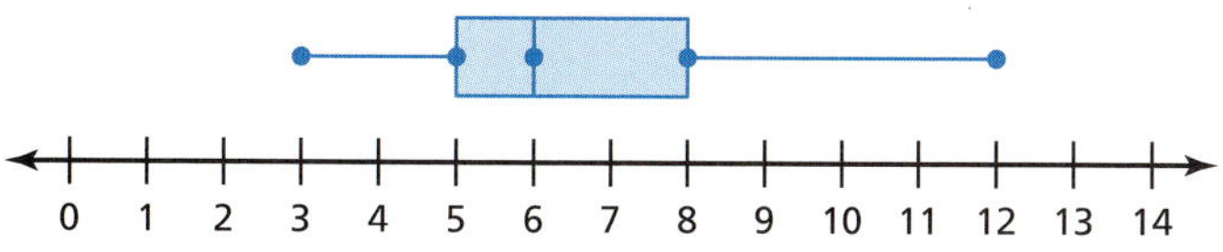

A. The median number of days for the seeds to germinate is 12.

B. 50% of the seeds took more than 8 days to germinate.

C. 50% of the seeds took less than 5 days to germinate.

D. The median number of days for the seeds to germinate was 6.

2. An object dropped from a height will fall under the force of gravity. The time t, in seconds, it takes to fall a distance d, in feet, can be found using the formula below.

$$t = \frac{\sqrt{d}}{4}$$

A ball is dropped from the top of a building that is 40 feet tall. Approximately how many seconds will it take for the ball to reach the ground?

F. 1.6 sec

G. 3.2 sec

H. 5 sec

I. 6.3 sec

3. A plumber charges a fixed amount for a house call plus an amount based on the number of hours worked. A job lasting 1 hour costs \$95 and a job lasting 2 hours costs \$140. What is the fixed amount charged by the plumber?

A. \$45

B. \$50

C. \$70

D. \$95

4. The diagram below shows parallel lines cut by a transversal. Which angle is the corresponding angle for $\angle 6$?

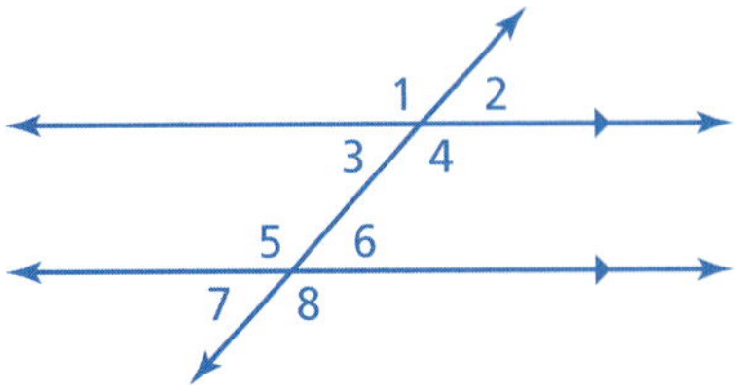

F. $\angle 2$

G. $\angle 3$

H. $\angle 4$

I. $\angle 8$

5. Which value of x makes the equation below true?

$$3x - 9 = 2(x + 4)$$

6. As part of a probability experiment, students were asked to roll two number cubes and find the sum of the numbers obtained. One group of students did this 600 times and obtained the results shown in the table. Which sum was the mode for this group's 600 rolls?

Sum	2	3	4	5	6	7	8	9	10	11	12
Number of Rolls	16	30	51	66	83	98	93	64	47	35	17

A. 4

B. 7

C. 12

D. 98

7. Which expression is equivalent to $\frac{\sqrt{32}}{\sqrt{18}}$?

F. $\sqrt{14}$

G. $\frac{4}{3}$

H. $\frac{8}{3}$

I. $\frac{16}{9}$

8. Which point lies on the graph of the line given by $y = -\frac{1}{2}x + 7$?

A. $(5, 4)$

B. $(-4, 5)$

C. $(20, 3)$

D. $(40, -13)$

Item Analysis (continued)

5. **Gridded Response:** Correct answer: 17

   Common Error: The student may fail to distribute 2 to the expression $x + 4$, yielding an answer of 13.

6. **A.** The student orders the data by number of rolls and picks the median sum in that ordered set.

   B. Correct answer

   C. The student chooses the greatest sum, not the sum with the greatest number of rolls.

   D. The student chooses the greatest number of rolls, not the sum with the greatest number of rolls.

7. **F.** The student subtracts 18 from 32.

   G. Correct answer

   H. The student makes a mistake dividing out $\frac{\sqrt{2}}{\sqrt{2}}$, yielding 2 instead of 1.

   I. The student factors correctly, but leaves $\frac{16}{9}$ as the answer instead of taking its square root.

8. **A.** The student reverses the roles of x and y.

   B. The student makes a mistake operating with signed integers.

   C. The student makes a mistake operating with signed integers.

   D. Correct answer

9. **F.** The student confuses positive and negative relationships.

   G. The student confuses no relationship with a negative relationship.

   H. The student confuses a constant function with a negative relationship.

   I. Correct answer

10. **Gridded Response:** Correct answer: 17 cm

    Common Error: The student adds the legs for an answer of 23 centimeters.

Answers

4. F

5. 17

6. B

7. G

8. D

Answers

9. I

10. 17 cm

11. *Part A* 22, 24, 25, 28, 28, 30, 31, 37, 37, 39, 40, 40, 44, 51, 58, 62

Part B 37

Part C

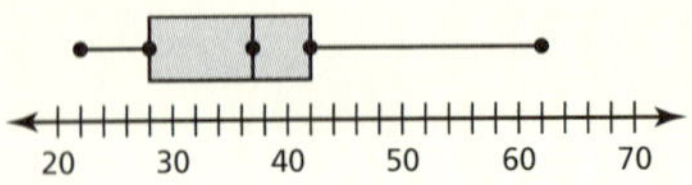

Answers for Extra Examples

1. **A.** The student thinks mean is mode.

   B. The student finds the mean of the four scores (91) and adds 1 to 92 because 91 is 1 less than 92.

   C. Correct answer

   D. The student makes an arithmetic error finding the sum.

2. **F.** The student omits (overlooks) the point on the *y*-axis.

   G. Correct answer

   H. The student reports the domain.

   I. The student combines the domain and range.

Item Analysis (continued)

11. **4 points** The student demonstrates a thorough understanding of how to make a box-and-whisker plot. The data is ordered correctly. The median of 37, first quartile of 28, third quartile of 42, least value of 22, and greatest value of 62 are identified correctly and graphed accurately.

3 points The student demonstrates an essential but less than thorough understanding of how to make a box-and-whisker plot. The data is ordered correctly and the median is found correctly. But, there may be a small mistake made subsequently, e.g., when finding the third quartile or when trying to accurately match the plot to the number line given.

2 points The student demonstrates a partial understanding of how to make a box-and-whisker plot. The student's work and explanations demonstrate a lack of essential understanding. For example, the data is ordered correctly, but the quartiles are misidentified.

1 point The student demonstrates very limited understanding of making a box-and-whisker plot. E.g., the only key value that the student can accurately find is the median, and the data may not be correctly ordered.

0 points The student provides no response, a completely incorrect or incomprehensible response, or a response that demonstrates insufficient understanding of making a box-and-whisker plot.

Extra Examples for Standardized Test Practice

1. A student took five tests this marking period and had a mean score of 92. Her scores on the first four tests were 90, 96, 86, and 92. What was her score on the fifth test?

   A. 92 **C.** 96

   B. 93 **D.** 98

2. What is the range of the function represented by the graph?

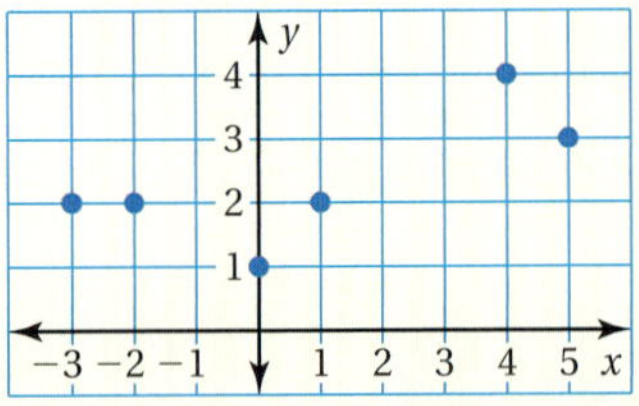

   F. 2, 3, 4 **H.** −3, −2, 0, 1, 4, 5

   G. 1, 2, 3, 4 **I.** −3, −2, 0, 1, 2, 3, 4, 5

9. Which scatter plot shows a negative relationship between x and y?

F.

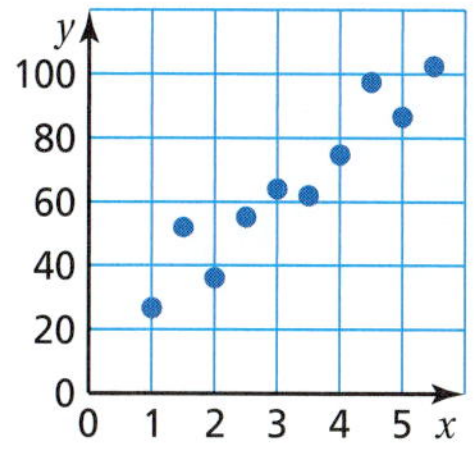

G.

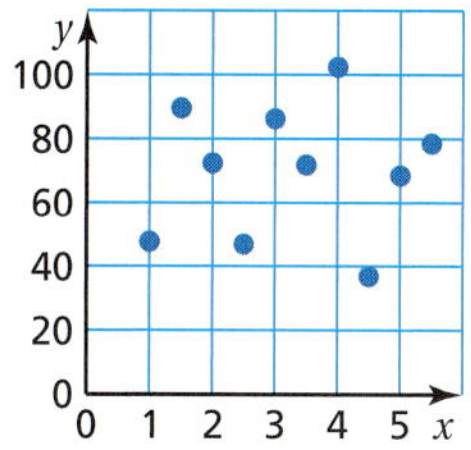

H.

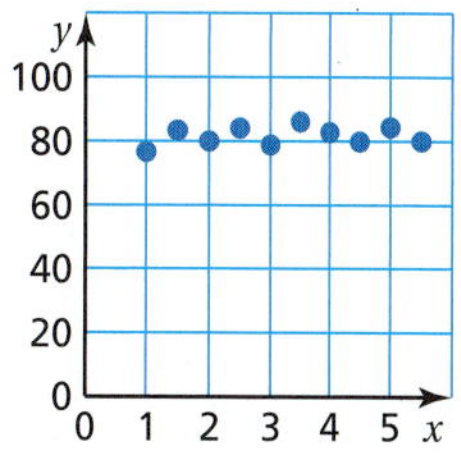

I.

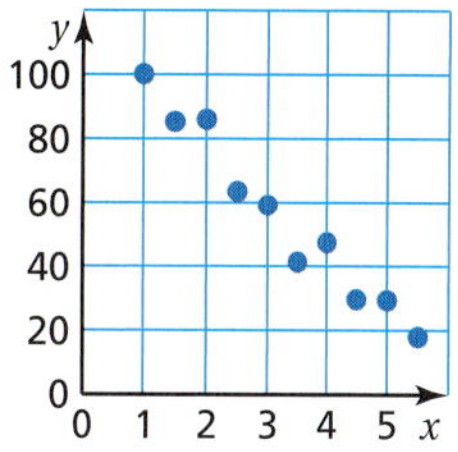

10. The legs of a right triangle have the lengths of 8 centimeters and 15 centimeters. What is the length of the hypotenuse, in centimeters?

11. The 16 members of a camera club have the ages listed below.

40, 22, 24, 58, 30, 31, 37, 25, 62, 40, 39, 37, 28, 28, 51, 44

Part A Order the ages from least to greatest.

Part B Find the median of the ages.

Part C Make a box-and-whisker plot for the ages of the camera club members.

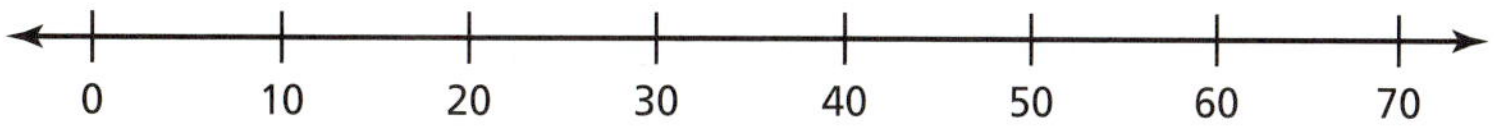

8 Linear Inequalities

8.1 Writing and Graphing Inequalities

8.2 Solving Inequalities Using Addition or Subtraction

8.3 Solving Inequalities Using Multiplication or Division

8.4 Solving Multi-Step Inequalities

Connections to Previous Learning

• Write, solve, and graph one-step and two-step linear inequalities.
• Formulate and use different strategies to solve one-step and two-step linear equations, including equations with rational coefficients.
• Solve and graph one-step and two-step inequalities in one variable.

Math in History

Long distance communication occurred in two basic ways in ancient cultures: by sight and by sound. Because the different sights (smoke signals) and sounds (drum beats) were limited, people had to restrict the communication to important topics, such as safety or danger.

★ The smoke signal messages that Native American tribes sent were simple, but important. Here are three of the signals used by the Apache Indian tribe.

- One puff: Something unusual is going on, but there's no cause for alarm or imminent danger.
- Two puffs: All is well. Camp is established and safe.
- Three puffs: This was an alarm signal, just as it is with the Boy Scouts today. A continuous column of smoke indicated great danger and a call for help.

★ In Africa, New Guinea, and South America, people used drum telegraphy to communicate with each other from great distances. For instance, when European expeditions came into the jungles to explore, they were often surprised to find that the message of their coming and their intention was carried through the woods in advance of their arrival.

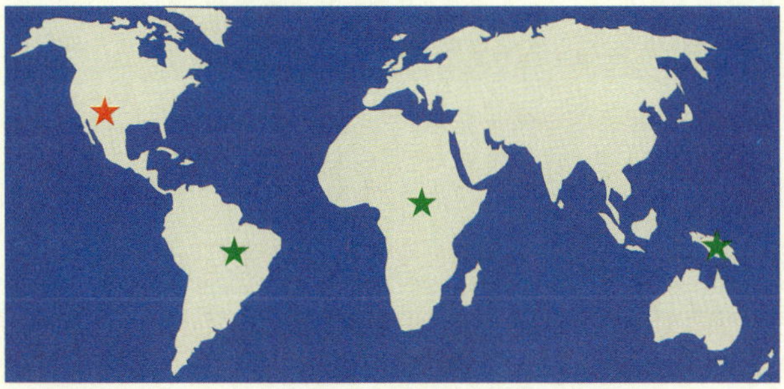

Pacing Guide for Chapter 8

Chapter Opener	1 Day
Section 1 Activity Lesson	 1 Day 1 Day
Section 2 Activity Lesson	 1 Day 1 Day
Study Help / Quiz	1 Day
Section 3 Activity Lesson	 1 Day 2 Days
Section 4 Activity Lesson	 2 Days 1 Day
Quiz / Chapter Review	1 Day
Chapter Test	1 Day
Standardized Test Practice	1 Day
Total Chapter 8	15 Days
Year-to-Date	133 Days

Check Your Resources

- Record and Practice Journal
- Resources by Chapter
- Skills Review Handbook
- Assessment Book
- Worked-Out Solutions

Technology For the Teacher

Dynamic Classroom

The Dynamic Planning Tool
Editable Teacher's Resources at
BigIdeasMath.com

Additional Topics for Review

- Approximating square roots
- Converting between fractions and decimals
- Inequalities

Try It Yourself

1. $=$ **2.** $<$

3. $<$

4. 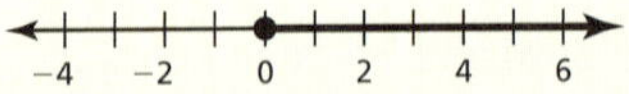

5. 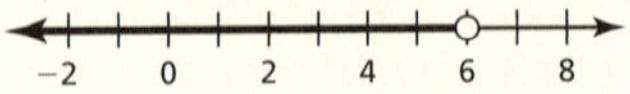

6. 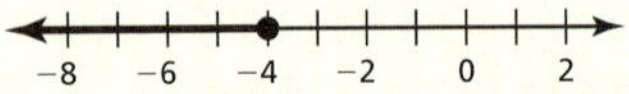

7. 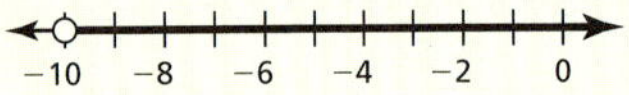

Record and Practice Journal

1. $>$ **2.** $=$

3. $>$ **4.** $<$

5. $<$ **6.** $<$

7. your friend; 5.6 ft is about 5 ft and 7 in.

8–14. See Additional Answers.

Math Background Notes

Vocabulary Review

- Inequality
- Rational Number
- Solution Set
- Irrational Number

Comparing Real Numbers

- Students should be able to compare integers. Students have studied real numbers. Comparing irrational and rational numbers is a relatively new skill.
- Encourage students to convert the real numbers into a similar form before comparing them. In Example 1, it is helpful to convert the given numbers into fractions. In Example 2, it is helpful to convert to decimals.
- **Teaching Tip:** Encourage students to use reasoning before reaching for a calculator. In Example 2, students may see $\sqrt{6}$ and immediately reach for a calculator. Remind students that the square root of a whole number greater than 1 is always less than the radicand. Remembering this fact will allow students to complete the example without a calculator and save them time.

Graphing Inequalities

- Students should be able to graph inequalities on a number line.
- Remind students that an equation produces a finite number of solutions, but an inequality produces an entire set of solutions. That is why an inequality requires you to shade the number line to describe the solutions.
- Remind students that inequalities containing $\leq$ or $\geq$ will require a closed circle. Inequalities containing $<$ or $>$ will require an open circle.
- **Teaching Tip:** Some students have difficulty deciding which side of the number line to shade. Encourage students to pick a test value on each side of the circle. Substitute each test value for *x*. Only one of the resulting inequalities will be true. Shade the number line on the side of the circle from which the valid test value was selected.

Reteaching and Enrichment Strategies

If students need help. . .	If students got it. . .
Record and Practice Journal • Fair Game Review Skills Review Handbook Lesson Tutorials	Game Closet at *BigIdeasMath.com* Start the next section

What You Learned Before

"Some people remember which is bigger by thinking that < is the mouth of a hungry alligator who is trying to eat the LARGER number."

Comparing Real Numbers

Complete the number sentence with <, >, or =.

Example 1 $\frac{1}{3}$ ▭ 0.3

$\frac{1}{3} = \frac{10}{30}$, $0.3 = \frac{3}{10} = \frac{9}{30}$

Because $\frac{10}{30}$ is greater than $\frac{9}{30}$,

$\frac{1}{3}$ is greater than 0.3.

∴ So, $\frac{1}{3} > 0.3$.

Example 2 $\sqrt{6}$ ▭ 6

Use a calculator to estimate $\sqrt{6}$.

$\sqrt{6} \approx 2.45$

Because 2.45 is less than 6, $\sqrt{6}$ is less than 6.

∴ So, $\sqrt{6} < 6$.

Try It Yourself

Complete the number sentence with <, >, or =.

1. $\frac{1}{4}$ ▭ 0.25 **2.** 0.1 ▭ $\frac{1}{9}$ **3.** π ▭ $\sqrt{10}$

Graphing Inequalities

Example 3 **Graph $x \geq 3$.**

Use a closed circle because 3 is a solution.

Shade the number line on the side where you found the solution.

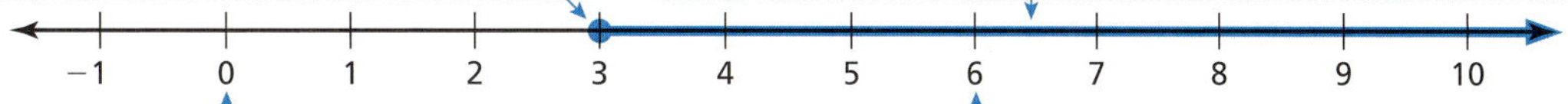

Test a number to the left of 3. $x = 0$ is *not* a solution.

Test a number to the right of 3. $x = 6$ is a solution.

Example 4 **Graph $x < -2$.**

Shade the number line on the side where you found the solution.

Use an open circle because −2 is *not* a solution.

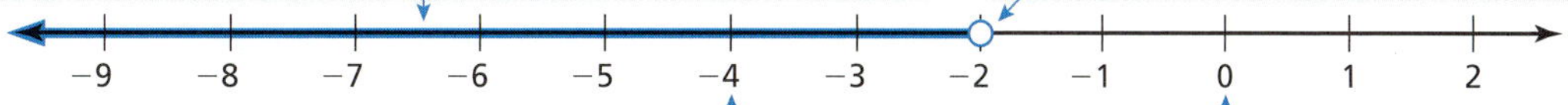

Test a number to the left of −2. $x = -4$ is a solution.

Test a number to the right of −2. $x = 0$ is *not* a solution.

Try It Yourself

Graph the inequality.

4. $x \geq 0$ **5.** $x < 6$ **6.** $x \leq -4$ **7.** $x > -10$

8.1 Writing and Graphing Inequalities

Essential Question How can you use an inequality to describe a real-life statement?

1 ACTIVITY: Writing and Graphing Inequalities

Work with a partner. Write an inequality for the statement. Then sketch the graph of all the numbers that make the inequality true.

a. Statement: The temperature t in Minot, North Dakota has never been below -36°F.

Inequality:

Graph:

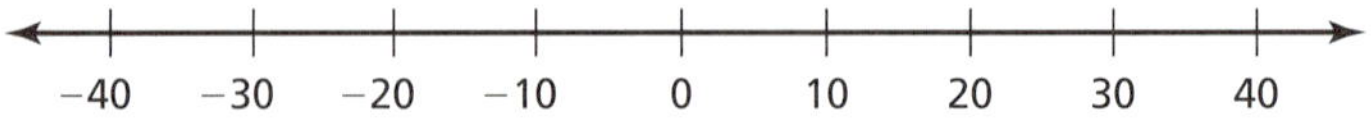

b. Statement: The elevation e in Wisconsin is at most 1951.5 feet above sea level.

Inequality:

Graph:

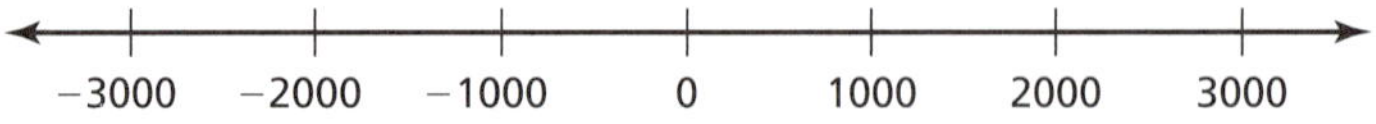

2 ACTIVITY: Writing and Graphing Inequalities

Work with a partner. Write an inequality for the graph. Then, in words, describe all the values of x that make the inequality true.

a.

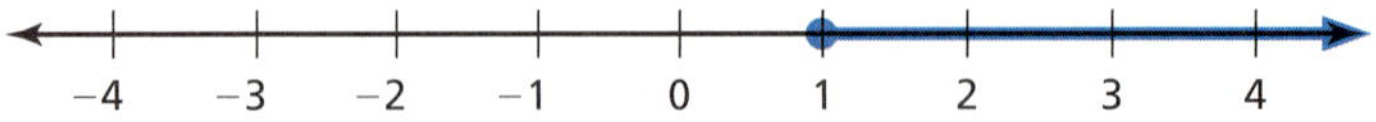

b.

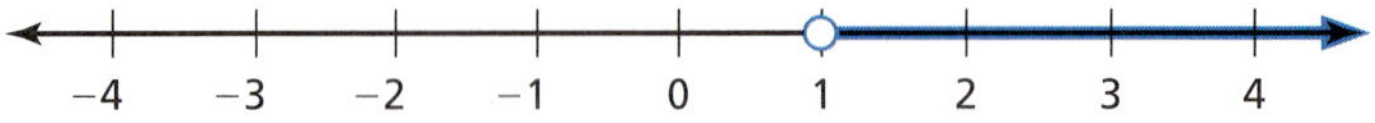

c.

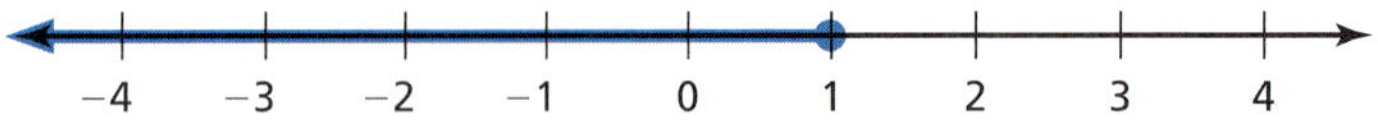

d.

Laurie's Notes

Introduction

For the Teacher

- **Goal:** Students will review how to graph and write an inequality.
- This lesson has an algebraic theme with geometric concepts integrated within.
- Students will be working with inequalities involving integers.

Motivate

- **Preparation:** Write 8 inequalities on index cards. Draw the matching graphs on 8 strips of paper large enough to be seen by students across your room. Tape the 8 graphs in different locations around your room.
- Examples of inequalities to explore: $x > 4$; $x \leq -4$; $x > -4$; $x \leq 4$; $x < -2.5$; $x \leq -2.5$; $x > 3.5$; $x \geq 3.5$
- Explain to students that they are starting a new chapter today. Express your confidence in them, knowing that they will have little difficulty with graphing inequalities.
- Select 8 students at random and hand each an index card. Ask students to find their graphs and to go stand next to the graphs.
- ? After students have matched their cards to the graphs, ask each student to explain how they know their match is correct. What features of the graph did they look for? Listen for: open circle versus closed circle, shading the correct side of the number line.
- After all of the students have made their explanations, collect their cards. Next, ask 8 different students to go to one of the graphs and say aloud the inequality that is shown by the graph.

Activity Notes

Discuss

- Remind students of the symbols used to express inequalities and the open circle/closed circle notation used when graphing an inequality.

Activity 1

- Students should work with their partner on this activity. Caution students to read carefully.
- In part (a), did students graph the temperatures that Minot experienced or did *not* experience?

Activity 2

- Students should be familiar with the direction of the inequality and with the open/closed notation.
- This activity assesses a student's ability to distinguish between $x > 1$ and $x \geq 1$, and between $x \leq 1$ and $x < 1$.

Previous Learning

Students should know how to graph numbers on a number line, solve single variable equations, and solve single variable inequalities using whole numbers.

Activity Materials	
Introduction	**Textbook**
• index cards • paper strips	• spaghetti • metric ruler

Start Thinking! and Warm Up

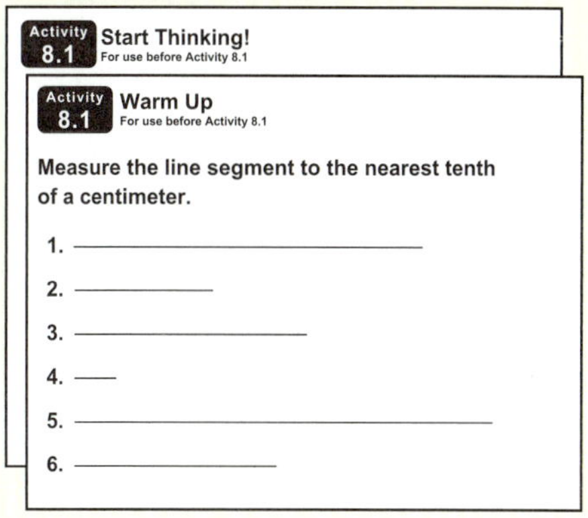

Activity 8.1 Start Thinking! For use before Activity 8.1

Activity 8.1 Warm Up For use before Activity 8.1

Measure the line segment to the nearest tenth of a centimeter.

1.
2.
3.
4.
5.
6.

8.1 Record and Practice Journal

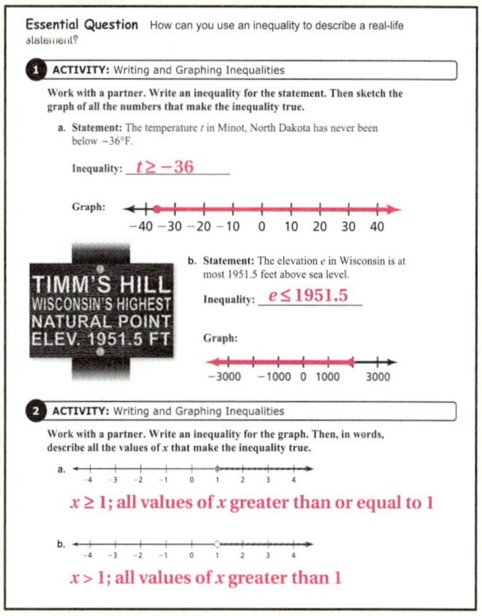

Essential Question How can you use an inequality to describe a real-life statement?

1 ACTIVITY: Writing and Graphing Inequalities

Work with a partner. Write an inequality for the statement. Then sketch the graph of all the numbers that make the inequality true.

a. Statement: The temperature t in Minot, North Dakota has never been below $-36°$F.

Inequality: $t \geq -36$

Graph: −40 −30 −20 −10 0 10 20 30 40

b. Statement: The elevation e in Wisconsin is at most 1951.5 feet above sea level.

Inequality: $e \leq 1951.5$

Graph: −3000 −1000 0 1000 3000

2 ACTIVITY: Writing and Graphing Inequalities

Work with a partner. Write an inequality for the graph. Then, in words, describe all the values of x that make the inequality true.

a. −4 −3 −2 −1 0 1 2 3 4

$x \geq 1$; all values of x greater than or equal to 1

b. −4 −3 −2 −1 0 1 2 3 4

$x > 1$; all values of x greater than 1

English Language Learners

Vocabulary and Symbols

Students should review the vocabulary and symbols for inequalities. Have students add a table of symbols and what the symbols mean to their notebooks. Students should add to the table as new phrases are used in the chapter.

Symbol	Phrase
$=$	is equal to
$\neq$	is not equal to
$<$	is less than
$\leq$	is less than or equal to
$>$	is greater than
$\geq$	is greater than or equal to

8.1 Record and Practice Journal

c. 

$x \leq 1$; all values of x less than or equal to 1

d. 

$x < 1$; all values of x less than 1

3 ACTIVITY: Triangle Inequality

Work with a partner. Use 8 to 10 pieces of spaghetti.

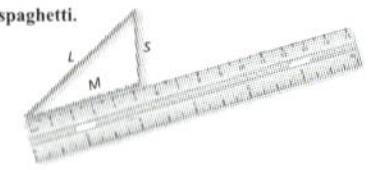

- Break one piece of spaghetti into three parts that can be used to form a triangle.
- Form a triangle and use a centimeter ruler to measure each side. Round the side lengths to the nearest tenth.
- Record the side lengths in the table.
- Repeat the process with two other pieces of spaghetti.

Check students' work.

Side Lengths That Form a Triangle			
Small	Medium	Large	$S + M$

- Repeat the experiment by breaking pieces of spaghetti into three pieces that *do not* form a triangle. Record the lengths in a table.

Check students' work.

Side Lengths That Do Not Form a Triangle			
Small	Medium	Large	$S + M$

- **INDUCTIVE REASONING** Write a rule that uses an inequality to compare the lengths of three sides of a triangle.

$S + M > L$

- Use your rule to decide whether the following triangles are possible. Explain.

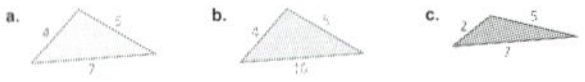

a. yes $4 + 5 > 7$
b. no $4 + 5 < 10$
c. no $2 + 5 = 7$

What Is Your Answer?

4. **IN YOUR OWN WORDS** How can you use an inequality to describe a real-life statement? Give two examples of real-life statements that can be represented by inequalities.

when a value has a limit, but also has many possible values

Laurie's Notes

Activity 3

- **Management Tip:** In this activity, students will explore another property of triangles. This investigation uses spaghetti. Tell students your expectation is that the floor will remain spaghetti free.
- Distribute metric rulers and pieces of spaghetti to each pair of students.
- Circulate as students are working on the activity. Check to see that students are measuring to the nearest tenth of a centimeter.
- **Whole Class:** Discuss results with the class. Some students may not have observed a pattern for when the three lengths form a triangle.

? "Is there a group that would like to share their observation about when the lengths form a triangle and when they don't? Explain." Sum of the two shorter sides has to be greater than the longest side.

- **FYI:** Even though a triangle is shown for the last three parts of Activity 3, they may not be drawn to scale. In fact, parts (b) and (c) are *not* triangles.

What Is Your Answer?

- **Big Idea:** This is known as the *Triangle Inequality Theorem*. When the sum of the two shorter sides is less than the length of the longest side, a triangle is not formed, that is, the ends do *not* meet.

Closure

- **Exit Ticket:** Write a word description with a real-life context for each inequality. Then graph the inequality.

$x > 8$ *Sample answer:* You need to work more than 8 hours a day.

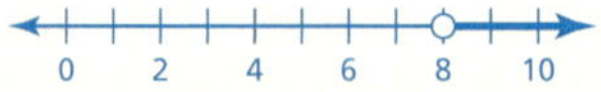

$x \leq -10$ *Sample answer:* The diver will stay at least 10 feet below sea level.

Technology For the Teacher

Dynamic Classroom

The Dynamic Planning Tool

Editable Teacher's Resources at *BigIdeasMath.com*

3 ACTIVITY: Triangle Inequality

Work with a partner. Use 8 to 10 pieces of spaghetti.

- Break one piece of spaghetti into three parts that can be used to form a triangle.
- Form a triangle and use a centimeter ruler to measure each side. Round the side lengths to the nearest tenth.
- Record the side lengths in a table.
- Repeat the process with two other pieces of spaghetti.

Side Lengths That Form a Triangle			
Small	Medium	Large	$S + M$

- Repeat the experiment by breaking pieces of spaghetti into three pieces that *do not* form a triangle. Record the lengths in a table.

Side Lengths That Do Not Form a Triangle			
Small	Medium	Large	$S + M$

- **INDUCTIVE REASONING** Write a rule that uses an inequality to compare the lengths of three sides of a triangle.
- Use your rule to decide whether the following triangles are possible. Explain.

a.

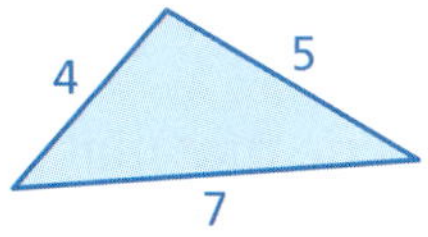

b.

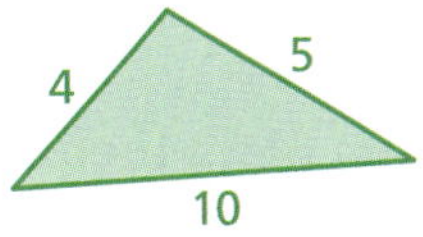

c.

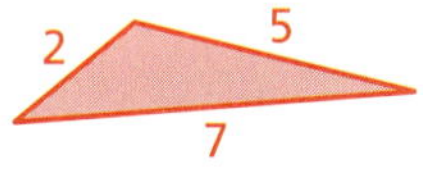

What Is Your Answer?

4. **IN YOUR OWN WORDS** How can you use an inequality to describe a real-life statement? Give two examples of real-life statements that can be represented by inequalities.

Practice Use what you learned about writing and graphing inequalities to complete Exercises 4 and 5 on page 316.

8.1 Lesson

Check It Out
Lesson Tutorials
BigIdeasMath.com

Key Vocabulary
inequality, *p. 314*
solution of an inequality, *p. 314*
solution set, *p. 314*
graph of an inequality, *p. 315*

An **inequality** is a mathematical sentence that compares expressions. It contains the symbols $<$, $>$, $\leq$, or $\geq$. To write an inequality, look for the following phrases to determine where to place the inequality symbol.

Inequality Symbols				
Symbol	$<$	$>$	$\leq$	$\geq$
Key Phrases	• is less than • is fewer than	• is greater than • is more than	• is less than or equal to • is at most • is no more than	• is greater than or equal to • is at least • is no less than

EXAMPLE 1 Writing an Inequality

A number w minus 3.5 is less than or equal to -2. Write this sentence as an inequality.

A number w minus 3.5 is less than or equal to -2.

$w - 3.5$ $\leq$ -2

An inequality is $w - 3.5 \leq -2$.

On Your Own

Now You're Ready
Exercises 6–9

Write the word sentence as an inequality.

1. A number b is fewer than 30.4.
2. Twice a number k is at least $-\frac{7}{10}$.

A **solution of an inequality** is a value that makes the inequality true. An inequality can have more than one solution. The set of all solutions of an inequality is called the **solution set**.

Value of x	$x + 5 \geq -2$	Is the inequality true?
-6	$-6 + 5 \overset{?}{\geq} -2$ $-1 \geq -2$ ✓	yes
-7	$-7 + 5 \overset{?}{\geq} -2$ $-2 \geq -2$ ✓	yes
-8	$-8 + 5 \overset{?}{\geq} -2$ $-3 \not\geq -2$ ✗	no

Reading
The symbol $\not\geq$ means "is not greater than or equal to."

Multi-Language Glossary at BigIdeasMath.com.

Laurie's Notes

Introduction

Connect

- **Yesterday:** Students reviewed how to graph and write an inequality.
- **Today:** Students will translate inequalities from words to symbols and check to see if a value is a solution of the inequality.

Motivate

- **Story Time:** You are planning to visit several theme parks and notice in doing your research that some of the rides have height restrictions.

Attraction	Restriction	Inequality
Dinosaur	Minimum is now 40 inches	$h \geq 40$
Primeval Whirl	Must be at least 48 inches	$h \geq 48$
Bay Slide	Must be under 60 inches	$h < 60$

- Ask students to write each as an inequality, where *h* is the rider's height.
- In today's lesson, they will be translating words to symbols.

Lesson Notes

Discuss

- Write the definition of an inequality.
- Review the four inequality symbols and key phrases or words that suggest each inequality.

Example 1

? "Would $3.5 - w \leq -2$ be equivalent to $w - 3.5 \leq -2$? Explain." no; Subtraction is *not* commutative.

? "Is there another way to say $w - 3.5$? Explain." yes; the difference of *w* and 3.5

On Your Own

- **Think-Pair-Share:** Students should read each question independently and then work with a partner to answer the questions. When they have answered the questions, the pair should compare their answers with another group and discuss any discrepancies.

Discuss

- Discuss what is meant by a solution of an inequality. Inequalities can, and generally do, have more than one solution. All of the solutions are collectively referred to as the **solution set**.
- It is helpful to write the inequality and substitute the value you are checking, as shown in the table.
- **Common Error:** Students will often make the mistake of thinking $-1 \leq -2$, forgetting that relationships are reversed on the negative side of 0; $-1 \geq -2$.

Goal Today's lesson is writing and graphing **inequalities**.

Start Thinking! and Warm Up

Lesson 8.1 Warm Up
For use before Lesson 8.1

Lesson 8.1 Start Thinking!
For use before Lesson 8.1

Write a sentence involving a real-life situation that can be modeled using an inequality.

Which inequality symbol applies: $<$, $\leq$, $>$, or $\geq$?

Extra Example 1

A number *b* plus 2.7 is greater than or equal to 3. Write this sentence as an inequality. $b + 2.7 \geq 3$

On Your Own

1. $b < 30.4$
2. $2k \geq -\frac{7}{10}$

Extra Example 2

Tell whether -2 is a solution of the inequality.

a. $x - 4 < -10$ no

b. $2.3x > -5$ yes

On Your Own

3. yes

4. no

5. yes

Extra Example 3

Graph $y \geq -5$.

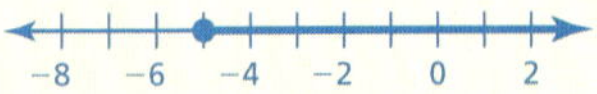

On Your Own

6.
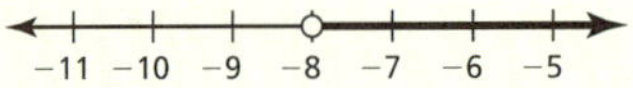

7.
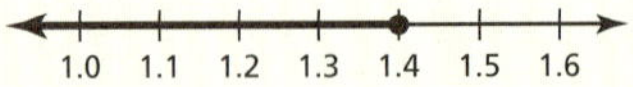

8.
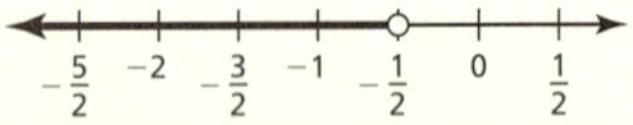

9.
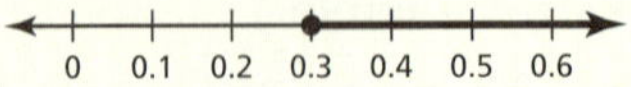

Differentiated Instruction

Auditory

Stress to students the importance of reading a statement and translating it into an expression, equation, or inequality. The word "is" plays an important role in the meaning of the statement. For instance, *six less than a number* translates to $x - 6$, while *six is less than a number* translates to $6 < x$.

Laurie's Notes

Example 2

? "How do you determine if -4 is a solution of an inequality?" Substitute -4 for the variable, simplify, and decide if the inequality is true.

- Work through each example as shown. In part (b), students must recall that the product of two negatives is a positive.

On Your Own

- **Common Error:** In Question 4, when students substitute for *m*, the result is $5 - (-6)$ which is 11.
- Ask volunteers to share their work at the board.

Discuss

- Discuss what is meant by the graph of an inequality. Remind students of the difference between the open and closed circles.

Example 3

- A number is tested on each side of the boundary point. This is a technique that demonstrates what it means to have a boundary point. On one side of the boundary point are all of the values which satisfy the inequality, and on the other side are all of the values which do *not* satisfy the inequality.

On Your Own

- In Question 8, check to see that students locate $-\frac{1}{2}$ correctly.
- In Question 9, students must first evaluate $\sqrt{0.09}$.
- Ask students to share their graphs at the board.

Closure

- **Writing Prompt:** To decide if a number is a solution of the inequality, you . . .

Technology For the Teacher

Dynamic Classroom

The Dynamic Planning Tool

Editable Teacher's Resources at *BigIdeasMath.com*

EXAMPLE 2 Checking Solutions

Tell whether −4 is a solution of the inequality.

a. $x + 8 < -3$

$x + 8 < -3$	Write the inequality.	
$-4 + 8 \overset{?}{<} -3$	Substitute −4 for *x*.	
$4 \not< -3$ ✗	Simplify.	

4 is *not* less than −3.

∴ So, −4 is *not* a solution of the inequality.

b. $-4.5x > -21$

$-4.5x > -21$

$-4.5(-4) \overset{?}{>} -21$

$18 > -21$ ✓

18 is greater than −21.

∴ So, −4 is a solution of the inequality.

On Your Own

Now You're Ready
Exercises 11–16

Tell whether −6 is a solution of the inequality.

3. $c + 4 < -1$ **4.** $5 - m \le 10$ **5.** $21 \div x \ge -3.5$

The **graph of an inequality** shows all of the solutions of the inequality on a number line. An open circle ○ is used when a number is *not* a solution. A closed circle ● is used when a number is a solution. An arrow to the left or right shows that the graph continues in that direction.

EXAMPLE 3 Graphing an Inequality

Graph $y \le -3$.

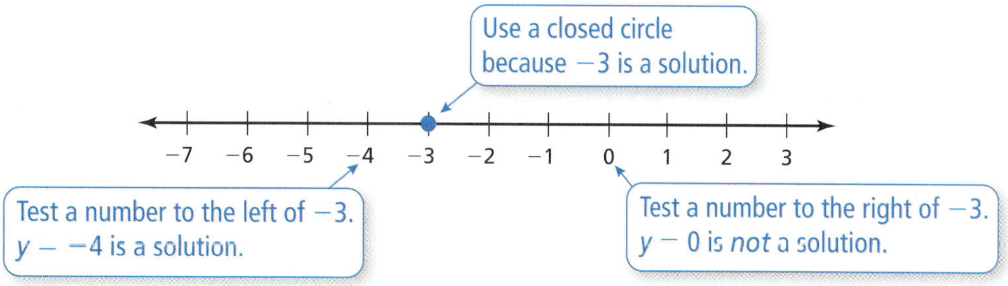

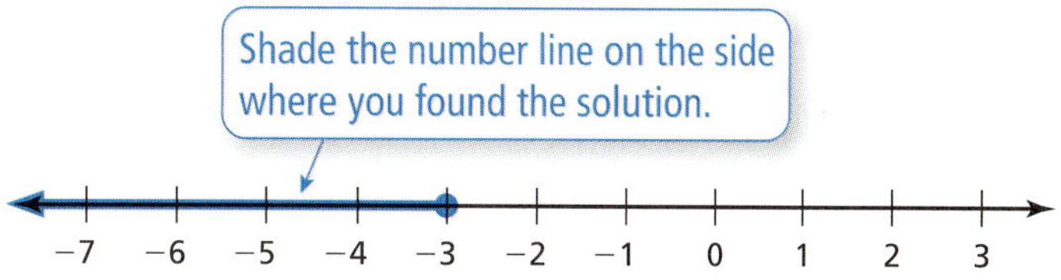

On Your Own

Now You're Ready
Exercises 17–20

Graph the inequality on a number line.

6. $b > -8$ **7.** $g \le 1.4$ **8.** $r < -\frac{1}{2}$ **9.** $v \ge \sqrt{0.09}$

8.1 Exercises

Check It Out
Help with Homework
BigIdeasMath.com

Vocabulary and Concept Check

1. **VOCABULARY** Would an open circle or a closed circle be used in the graph of the inequality $k < 250$? Explain.

2. **DIFFERENT WORDS, SAME QUESTION** Which is different? Write "both" inequalities.

   - w is greater than or equal to -7.
   - w is no less than -7.
   - w is no more than -7.
   - w is at least -7.

3. **REASONING** Do $x \geq -9$ and $-9 \geq x$ represent the same inequality? Explain.

Practice and Problem Solving

Write an inequality for the graph. Then, in words, describe all the values of x that make the inequality true.

4. 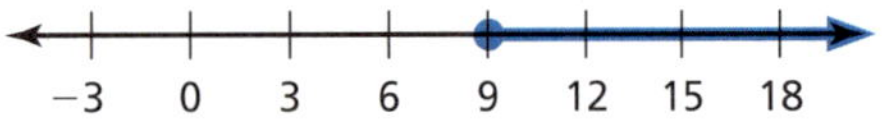

5. 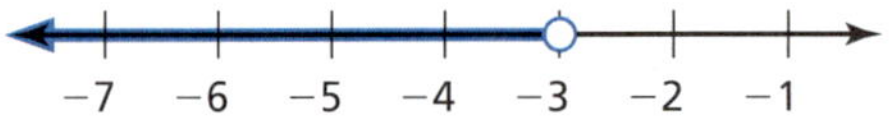

Write the word sentence as an inequality.

(1) 6. A number x is no less than -4.

7. A number y added to 5.2 is less than 23.

8. A number b multiplied by -5 is at most $-\frac{3}{4}$.

9. A number k minus 8.3 is greater than 48.

10. **ERROR ANALYSIS** Describe and correct the error in writing the word sentence as an inequality.

    ✗ Twice a number c is at least $-\frac{4}{9}$.

    $2c \leq -\frac{4}{9}$

Tell whether the given value is a solution of the inequality.

(2) 11. $s + 6 \leq 12;\ s = 4$

12. $15n > -3;\ n = -2$

13. $a - 2.5 \leq 1.6;\ a = 4.1$

14. $-3.3q > -13;\ q = 4.6$

15. $\frac{4}{5}h \geq -4;\ h = -15$

16. $\frac{1}{12} - p < \frac{1}{3};\ p = \frac{1}{6}$

Graph the inequality on a number line.

(3) 17. $g \geq -6$

18. $q > 1.25$

19. $z < 11\frac{1}{4}$

20. $w \leq -\sqrt{289}$

21. **DRIVING** When you are driving with a learner's license, a licensed driver who is 21 years of age or older must be with you. Write an inequality that represents this situation.

Assignment Guide and Homework Check

Level	Day 1 Activity Assignment	Day 2 Lesson Assignment	Homework Check
Basic	4, 5, 28–31	1–3, 6–10, 11–21 odd	8, 10, 13, 17
Average	4, 5, 28–31	1–3, 7–19 odd, 10, 18, 23–25	9, 10, 13, 17, 24
Advanced	4, 5, 28–31	1–3, 10–20 even, 22–27	10, 12, 20, 22, 24

Common Errors

- **Exercises 6–9** Students may struggle with knowing which inequality symbol to use. Encourage them to put the word sentence into a real-life context and to use the table in the lesson that explains what symbol matches each phrase.
- **Exercises 11–16** Students may try to solve for the variable instead of substituting the given value into the inequality and determining if that value is a solution of the inequality. Remind them that they are not solving inequalities yet, just checking a number to see if it is a solution.
- **Exercises 17–20** Students may use a closed circle instead of an open circle and vice versa. They may also shade the wrong side of the number line. Review how to graph inequalities and encourage students to test a value on each side of the circle.

8.1 Record and Practice Journal

Write the word sentence as an inequality.

1. A number p is no greater than -6.
   $p \leq -6$

2. A number n divided by -2 is no less than $\frac{1}{2}$.
   $\frac{n}{-2} \geq \frac{1}{2}$

Tell whether the given value is a solution of the inequality.

3. $q + 7 \geq 8;\ q = 10$
   solution

4. $-12r < -6;\ r = -2$
   not a solution

5. $-2.4k \geq -4;\ k = 0.5$
   solution

6. $\frac{x}{4} < x - 9;\ x = 8$
   not a solution

Graph the inequality on a number line.

7. $p \leq 4\frac{1}{2}$

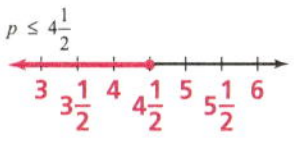

8. $z > -8.3$

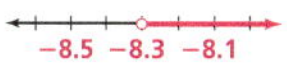

9. For your birthday, you want to invite some friends to join you at the movies. Movie tickets cost \$8. You can spend no more than \$35. Write an inequality to represent this situation. Then solve the inequality to find the greatest number of people you can invite.
   $8x \leq 35$; 4

Technology For the Teacher

Answer Presentation Tool

Quiz*Show*

Vocabulary and Concept Check

1. An open circle would be used because 250 is not a solution.
2. w is no more than -7.; $w \leq -7$; $w \geq -7$
3. no; $x \geq -9$ is all values of x greater than or equal to -9. $-9 \geq x$ is all values of x less than or equal to -9.

Practice and Problem Solving

4. $x \geq 9$; all values of x greater than or equal to 9
5. $x < -3$; all values of x less than -3
6. $x \geq -4$
7. $y + 5.2 < 23$
8. $-5b \leq -\frac{3}{4}$
9. $k - 8.3 > 48$
10. The inequality symbol is reversed. $2c \geq -\frac{4}{9}$
11. yes
12. no
13. yes
14. no
15. no
16. yes
17.

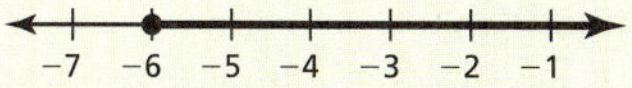

18.

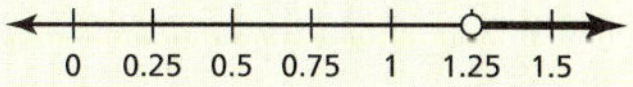

19.

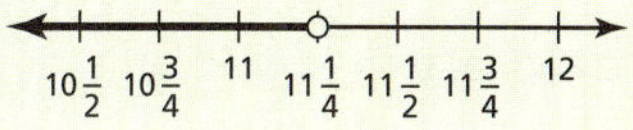

20.

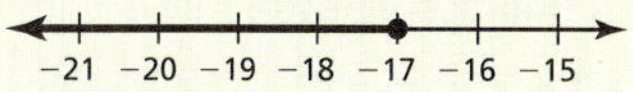

21. $x \geq 21$

Practice and Problem Solving

22. yes **23.** yes

24. maybe; If your friend is 10, 11, or 12, then your friend can play "E 10+" games, but is not old enough for "T" games. If your friend is 13 or older, then your friend can play "T" games.

25. See Additional Answers.

26. See *Taking Math Deeper.*

27. a. $m < n$; $n \le p$

b. $m < p$

c. no; Because n is no more than p and m is less than n, m cannot be equal to p.

Fair Game Review

28. 15 **29.** -1.7

30. 10π **31.** D

Mini-Assessment

Write the word sentence as an inequality.

1. A number m multiplied by -4.9 is at most 5. $-4.9m \le 5$
2. A number p minus 1.1 is greater than or equal to $-\frac{2}{3}$. $p - 1.1 \ge -\frac{2}{3}$
3. A number h divided by 4 is less than -7.5. $\frac{h}{4} < -7.5$

Graph the inequality on a number line.

4. $x > -2.9$

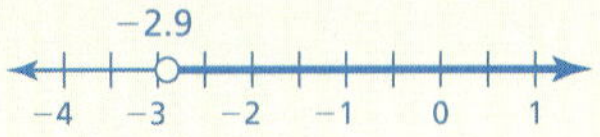

5. $a \le -5.25$

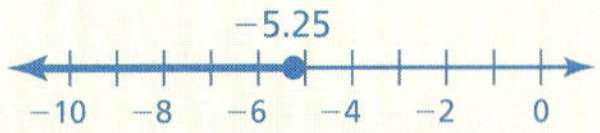

Taking Math Deeper

Exercise 26

This is a practical problem for anyone who is planning to fly. This size restriction applies only to carry-on luggage, not to luggage that is checked. For students who have not thought of the differences, it might be interesting for them to think about the advantages of carry-on luggage.

- No chance of luggage not arriving
- No waiting for luggage at destination
- No extra fees for luggage

1 Draw a diagram showing the length, width, and height of a carry-on bag.

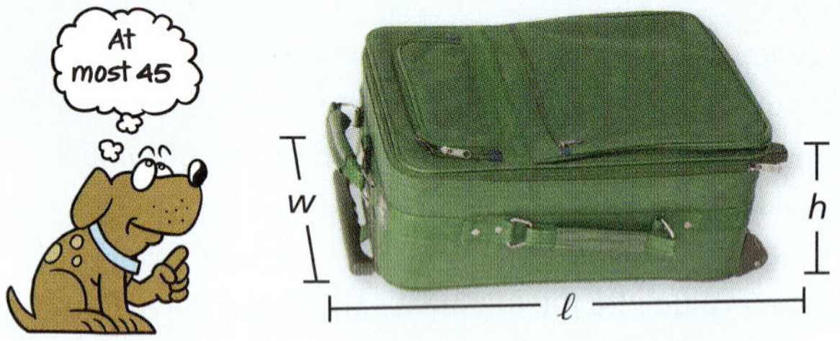

2 Find some possible combinations for which $\ell + w + h \le 45$.

Bag	ℓ	w	h
A (Standard size)	22 in.	14 in.	9 in.
B	20 in.	14 in.	11 in.
C	18 in.	14 in.	13 in.

3 Students might find it interesting to discover which of their three choices has the greatest volume. In the three examples in the table, the volumes are 2272, 3080, and 3276 cubic inches.

In general, the more cube-like the luggage, the greater the volume. So, the maximum volume would be with luggage that is 15 inches by 15 inches by 15 inches, which has a volume of 3375 cubic inches.

Reteaching and Enrichment Strategies

If students need help...	If students got it...
Resources by Chapter • Practice A and Practice B • Puzzle Time Record and Practice Journal Practice Differentiating the Lesson Lesson Tutorials Skills Review Handbook	Resources by Chapter • Enrichment and Extension Start the next section

Tell whether the given value is a solution of the inequality.

22. $3p > 5 + p;\ p = 4$

23. $\frac{y}{2} \geq y - 11;\ y = 18$

24. VIDEO GAME RATINGS Each rating is matched with the inequality that represents the recommended ages of players. Your friend is old enough to play "E 10+" games. Is your friend old enough to play "T" games? Explain.

$x \geq 3$ | $x \geq 6$ | $x \geq 10$ | $x \geq 13$ | $x \geq 17$

The ESRB rating icons are registered trademarks of the Entertainment Software Association.

25. SCUBA DIVING Three requirements for a scuba diving training course are shown.

a. Write and graph three inequalities that represent the requirements.

b. You can swim 10 lengths of a 25-yard pool. Do you satisfy the swimming requirement of the course? Explain.

26. LUGGAGE On an airplane, the maximum sum of the length, width, and height of a carry-on bag is 45 inches. Find three different sets of dimensions that are reasonable for a carry-on bag.

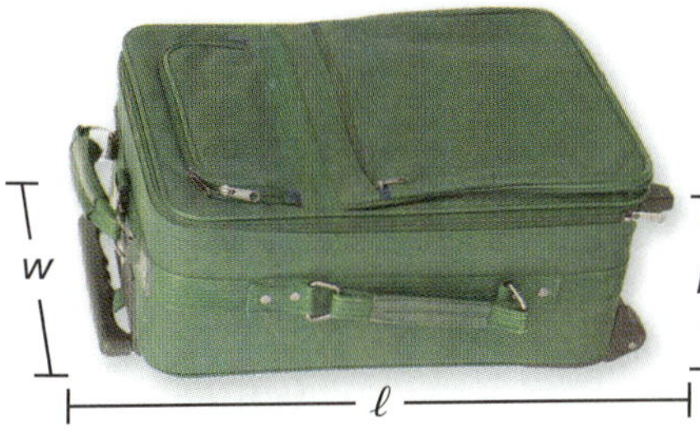

27. Critical Thinking A number m is less than another number n. The number n is less than or equal to a third number p.

a. Write two inequalities representing these relationships.

b. Describe the relationship between m and p.

c. Can m be equal to p? Explain.

Fair Game Review What you learned in previous grades & lessons

Solve the equation. Check your solution. *(Section 1.1)*

28. $r - 12 = 3$

29. $4.2 + p = 2.5$

30. $n - 3\pi = 7\pi$

31. MULTIPLE CHOICE Which linear function relates y to x? *(Section 4.3)*

Ⓐ $y = -0.5x - 3$

Ⓑ $y = 2x + 3$

Ⓒ $y = 0.5x - 3$

Ⓓ $y = 2x - 3$

x	-1	0	1	2
y	-5	-3	-1	1

8.2 Solving Inequalities Using Addition or Subtraction

Essential Question How can you use addition or subtraction to solve an inequality?

1 ACTIVITY: Quarterback Passing Efficiency

Work with a partner. The National Collegiate Athletic Association (NCAA) uses the following formula to rank the passing efficiency *P* of quarterbacks.

$$P = \frac{8.4Y + 100C + 330T - 200N}{A}$$

Y = total length of all completed passes (in Yards)

C = Completed passes

T = passes resulting in a Touchdown

N = iNtercepted passes

A = Attempted passes

M = incoMplete passes

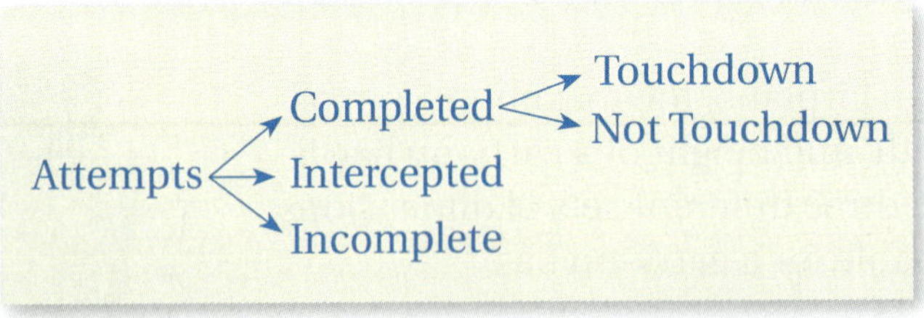

Which of the following equations or inequalities are true relationships among the variables? Explain your reasoning.

a. $C + N < A$ **b.** $C + N \leq A$ **c.** $T < C$ **d.** $T \leq C$

e. $N < A$ **f.** $A > T$ **g.** $A - C \geq M$ **h.** $A = C + N + M$

2 ACTIVITY: Quarterback Passing Efficiency

Work with a partner. Which of the following quarterbacks has a passing efficiency rating that satisfies the inequality $P > 100$? Show your work.

Player	Attempts	Completions	Yards	Touchdowns	Interceptions
A	149	88	1065	7	9
B	400	205	2000	10	3
C	426	244	3105	30	9
D	188	89	1167	6	15

Laurie's Notes

Introduction

For the Teacher

- **Goal:** Students will explore inequalities and solve simple inequalities using mental math.
- Students have solved whole number inequalities and equations. This lesson is a natural extension of familiar content.
- Wear a football related piece of clothing today, if you own one.

Motivate and Discuss

- Set the tone by tossing a few passes in class with a small foam football. Ask a statistician to record your efforts in a table at the board. Use 3 columns: **C**ompleted, I**N**tercepted, and Inco**M**plete.
- I recommend **A**ttempting 10 short passes to students nearby. You may need to give permission to have a pass intercepted.

? Ask the following questions.

- "How many passes did I attempt?" 10 Record this next to the table.
- "Can I complete more passes than I attempt?" no
- "Are *completed passes* + *incomplete passes* always *less than or equal to attempted passes*?" yes
- "Are *completed passes* + *incomplete passes* always *less than attempted passes*?" No, they could be equal.

Activity Notes

Activity 1

- The tree diagram should be a helpful aid to students.
- Discuss students' answers and their reasoning when they have finished.
- For parts (c) and (d), point out the need to pay attention to the inequality symbol. It is possible, though unlikely, that $T = C$. In that case, the inequality $T < C$ may *not* be true, while the inequality $T \leq C$ is true.
- There will be some heated discussion about the inequalities, but remember to ask, is it *possible* versus is it *probable*.

Activity 2

- You may want to allow calculators to increase speed and accuracy, or you may want to use this as an opportunity to review computation skills.
- **Common Error:** Students may forget order of operations. The computation in the numerator must be completed before dividing by the denominator. On a calculator, this can be done by using parentheses, or simply by pressing the *Enter* key before dividing by the denominator.
- Suggest to students that they write the formula, and then rewrite it substituting the values for the variables.

? Ask the following questions.

- "Which player(s) were above average, meaning $P > 100$?" A and C
- "Which player(s) were average, meaning $P = 100$?" B
- "So, was player D below average, meaning $P < 100$?" yes

Previous Learning

Students should know how to solve equations using addition and subtraction. Students should be able to evaluate expressions.

Activity Materials	
Introduction	**Textbook**
• foam football	• calculator

Start Thinking! and Warm Up

Activity 8.2 Start Thinking!
For use before Activity 8.2

Activity 8.2 Warm Up
For use before Activity 8.2

Evaluate the expression when $x = 2$, $y = -6$, and $z = 8$.

1. $xy + 3z$ **2.** $1.5x - y + 5z$

3. $5x + 3y - 8z$ **4.** $x(8y - z)$

5. $\dfrac{-x + y + 2z}{z}$ **6.** $\dfrac{4x + 2y - z}{2x}$

8.2 Record and Practice Journal

Essential Question How can you use addition or subtraction to solve an inequality?

1 ACTIVITY: Quarterback Passing Efficiency

Work with a partner. The National Collegiate Athletic Association (NCAA) uses the following formula to rank the passing efficiency P of quarterbacks.

$$P = \frac{8.4Y + 100C + 330T - 200N}{A}$$

Y = total length of all completed passes (in **Y**ards)
C = **C**ompleted passes
T = passes resulting in a **T**ouchdown
N = i**N**tercepted passes
A = **A**ttempted passes
M = inco**M**plete passes

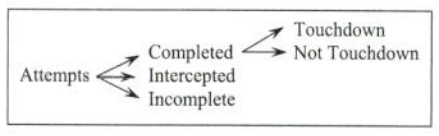

Which of the following equations or inequalities are true relationships among the variables? Explain your reasoning.

a. $C + N < A$ **b.** $C + N \leq A$ **c.** $T < C$ **d.** $T \leq C$

b and d are true

e. $N < A$ **f.** $A > T$ **g.** $A - C \geq M$ **h.** $A = C + N + M$

g and h are true

English Language Learners

Vocabulary

It is important that English learners understand the difference between *is less than* and *is less than or equal to,* as well as *is greater than* and *is greater than or equal to.* Give each student a card with one of the numbers $-10, -9, -8, -7, -6, \ldots, 10$. (Include more numbers if your class is larger.) Tell students to stand up if their number *is less than* (say a number), and then *is less than or equal to* that same number. The class should discuss the difference between the two. This can be repeated with *is greater than* and *is greater than or equal to.*

8.2 Record and Practice Journal

2 ACTIVITY: Quarterback Passing Efficiency

Work with a partner. Which of the following quarterbacks has a passing efficiency rating that satisfies the inequality $P > 100$? Show your work.

Player	Attempts	Completions	Yards	Touchdowns	Interceptions
A	149	88	1065	7	9
B	400	205	2000	10	3
C	426	244	3105	30	9
D	188	89	1167	6	15

A and C

3 ACTIVITY: Finding Solutions of Inequalities

Work with a partner. Use the passing efficiency formula to create a passing record that makes the inequality true. Then describe the values of P that make the inequality true. Sample answers are given.

a. $P < 0$

Attempts	Completions	Yards	Touchdowns	Interceptions
10	2	10	0	2

all values of P less than 0

b. $P + 100 \geq 250$

Attempts	Completions	Yards	Touchdowns	Interceptions
390	244	3105	30	9

all values of P greater than or equal to 150

c. $180 < P - 50$

Attempts	Completions	Yards	Touchdowns	Interceptions
390	350	4700	50	2

all values of P greater than 230

d. $P + 30 \geq 120$

Attempts	Completions	Yards	Touchdowns	Interceptions
188	89	1167	6	15

all values of P greater than or equal to 90

e. $P - 250 > -80$

Attempts	Completions	Yards	Touchdowns	Interceptions
350	275	3105	50	2

all values of P greater than 170

What Is Your Answer?

4. Write a rule that describes how to solve inequalities like those in Activity 3. Then use your rule to solve each of the inequalities in Activity 3.

   Add or subtract the same number from each side to get the variable alone.
   a. $P < 0$, b. $P \geq 150$, c. $P > 230$, d. $P \geq 90$, e. $P > 170$

5. **IN YOUR OWN WORDS** How can you use addition or subtraction to solve an inequality?

   Add or subtract the same number from each side.

6. How is solving the inequality $x + 3 < 4$ similar to solving the equation $x + 3 = 4$? How is it different?

   Subtract 3 from each side in both but the solutions are different.

Laurie's Notes

Activity 3

- Now that students have had the opportunity to work with the formula as stated, it is time to put a twist on the problem.
- Notice that four of the inequalities in this activity force students to think about solving the inequality before they begin. In the second problem, if $P + 100 \geq 250$, then it means $P \geq 150$. This should make sense to students.
- Remind students that yards do not count unless the pass is completed.
- Answers will vary for this activity, but suggest to students that they keep the numbers as simple as possible. For instance, one possible answer for the first question is (1, 0, 1, 0, 0), where only 1 pass is attempted and it is intercepted. This is the $P = -200$ result, and so $P < 0$.
- Students will need to do a little trial and error with these problems. They should start to ask themselves, "*what happens when I increase this variable, but decrease another variable.*"

What Is Your Answer?

- **Think-Pair-Share:** Students should read each question independently and then work with a partner to answer the questions. When they have answered the questions, the pair should compare their answers with another group and discuss any discrepancies.

Closure

- **Exit Ticket:**
  If $a < b$ is true, is $a \leq b$ also true? Explain. yes; Because in both cases a is less than b.
  If $a \leq b$ is true, is $a < b$ also true? Explain. no; Because if a equals b, then a cannot be less than b.

Technology For the Teacher

Dynamic Classroom

The Dynamic Planning Tool
Editable Teacher's Resources at *BigIdeasMath.com*

3 ACTIVITY: Finding Solutions of Inequalities

Work with a partner. Use the passing efficiency formula to create a passing record that makes the inequality true. Then describe the values of *P* that make the inequality true.

a. $P < 0$

Attempts	Completions	Yards	Touchdowns	Interceptions

b. $P + 100 \geq 250$

Attempts	Completions	Yards	Touchdowns	Interceptions

c. $180 < P - 50$

Attempts	Completions	Yards	Touchdowns	Interceptions

d. $P + 30 \geq 120$

Attempts	Completions	Yards	Touchdowns	Interceptions

e. $P - 250 > -80$

Attempts	Completions	Yards	Touchdowns	Interceptions

What Is Your Answer?

4. Write a rule that describes how to solve inequalities like those in Activity 3. Then use your rule to solve each of the inequalities in Activity 3.

5. **IN YOUR OWN WORDS** How can you use addition or subtraction to solve an inequality?

6. How is solving the inequality $x + 3 < 4$ similar to solving the equation $x + 3 = 4$? How is it different?

Practice Use what you learned about solving inequalities using addition or subtraction to complete Exercises 3–5 on page 322.

8.2 Lesson

Check It Out
Lesson Tutorials
BigIdeasMath.com

Key Ideas

Addition Property of Inequality

Words If you add the same number to each side of an inequality, the inequality remains true.

Numbers

$$\begin{array}{rcr} -3 & < & 2 \\ \underline{+4} & & \underline{+4} \\ 1 & < & 6 \end{array}$$

Algebra

$$\begin{array}{rcr} x - 3 & > & -10 \\ \underline{+3} & & \underline{+3} \\ x & > & -7 \end{array}$$

Subtraction Property of Inequality

Words If you subtract the same number from each side of an inequality, the inequality remains true.

Numbers

$$\begin{array}{rcr} -3 & < & 1 \\ \underline{-5} & & \underline{-5} \\ -8 & < & -4 \end{array}$$

Algebra

$$\begin{array}{rcr} x + 7 & > & -20 \\ \underline{-7} & & \underline{-7} \\ x & > & -27 \end{array}$$

These properties are also true for $\leq$ and $\geq$.

Study Tip

You can solve inequalities the same way you solve equations. Use inverse operations to get the variable by itself.

EXAMPLE 1 Solving an Inequality Using Addition

Solve $x - 6 \geq -10$. Graph the solution.

$$\begin{array}{rcrl} x - 6 & \geq & -10 & \text{Write the inequality.} \\ \underline{+6} & & \underline{+6} & \text{Add 6 to each side.} \\ x & \geq & -4 & \text{Simplify.} \end{array}$$

Undo the subtraction. → $+6$

The solution is $x \geq -4$.

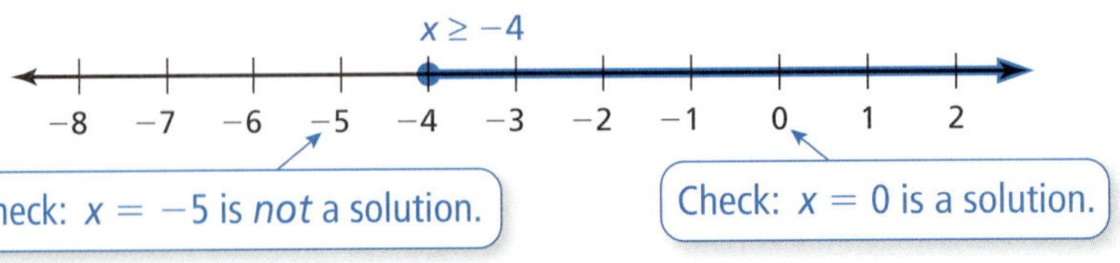

Study Tip

To check a solution, you check some numbers that are solutions and some that are not.

On Your Own

Solve the inequality. Graph the solution.

1. $b - 2 > -9$
2. $m - 3.8 \leq 5$
3. $\frac{1}{4} > y - \frac{1}{4}$

Laurie's Notes

Introduction

Connect

- **Yesterday:** Students explored inequalities and solved simple inequalities using mental math.
- **Today:** Students will use the Addition and Subtraction Properties of Inequality to solve inequalities.

Motivate

- Airlines have guidelines for the maximum weight of luggage, depending upon whether it is carry-on or checked luggage.

? "If there is a maximum weight restriction of 50 pounds for a checked bag, what inequality does this suggest?" $w \leq 50$

- Suggest different scenarios. If my bag weighs 40.5 pounds, how much more can I add? If my bag weighs 56.4 pounds, how much must I remove?
- Today's lesson involves solving inequalities of this type.

Lesson Notes

Key Ideas

- These properties should look very familiar, as they are similar to the Addition and Subtraction Properties of Equality used in solving equations.
- **Teaching Tip:** I summarize these two properties in the following way:
  I am older than you. In two years, I will still be older than you.

Laurie's age > Student's age	if $a > b$
Laurie's age + 2 > Student's age + 2	then $a + c > b + c$

Two years ago, I was older than you.

Laurie's age > Student's age	if $a > b$
Laurie's age − 2 > Student's age − 2	then $a - c > b - c$

Example 1

? "How do you isolate the variable, meaning get *x* by itself?" Add 6 to each side of the inequality.

- Adding 6 is the inverse operation of subtracting 6.
- Solve, graph, and check. Note the *Study Tip*.

On Your Own

- **Think-Pair-Share:** Students should read each question independently and then work with a partner to answer the questions. When they have answered the questions, the pair should compare their answers with another group and discuss any discrepancies.
- I often have students who like to rewrite inequalities so that the variable is on the left. I try to discourage this. If they do, they must be extremely careful. The inequality symbol must be reversed. Note if $4 < x$, then $x > 4$.
- These problems integrate a review of fraction and decimal operations.

Goal Today's lesson is solving inequalities using addition or subtraction.

Start Thinking! and Warm Up

Lesson 8.2 Warm Up
For use before Lesson 8.2

Lesson 8.2 Start Thinking!
For use before Lesson 8.2

Students are selling magazine subscriptions for a fundraiser. The student who sells the most magazines wins a prize. You have sold 26 subscriptions. Your friend is currently in first place, having sold 41. How many subscriptions do you have to sell in order to move into first place?

What does this have to do with solving inequalities using addition or subtraction?

Extra Example 1

Solve $f - 4 \leq 1$. Graph the solution.
$f \leq 5$

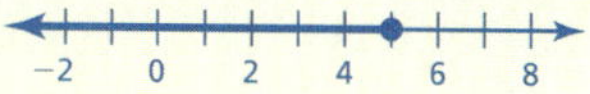

On Your Own

1. $b > -7$;

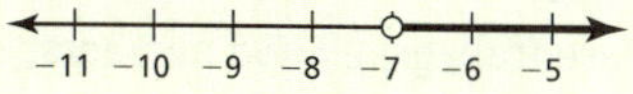

2. $m \leq 8.8$;

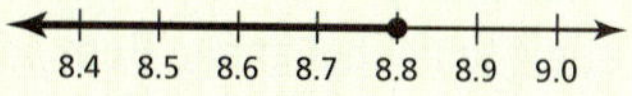

3. $\frac{1}{2} > y$;

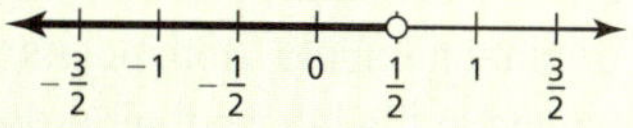

Laurie's Notes

Extra Example 2

Solve $-7 \geq 2.3 + x$. Graph the solution.
$x \leq -9.3$

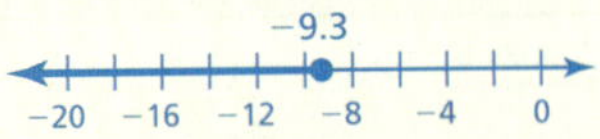

On Your Own

4. $k \leq -8$;

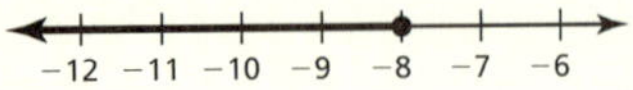

5–6. See Additional Answers.

Extra Example 3

You have raised \$225 for a charity. Your goal is to raise at least \$600. Write and solve an inequality that represents the amount of money you need to raise to reach your goal. $225 + x \geq 600$; $x \geq \$375$

On Your Own

7. $32.5 + w \leq 50$;
$w \leq 17.5$ lb

Differentiated Instruction

Visual
To show students that the Addition and Subtraction Properties of Inequalities are true, graph two numbers, −2 and 5, on the number line. Write the ordered relationship, $-2 < 5$, on the board. Now add 4 to each number. Both points move the same distance to the right, so their ordered positions remain the same, $2 < 9$. Now subtract 6 from each of the original numbers. Both points move 6 units to the left, so their ordered positions remain the same, $-8 < -1$.

Example 2

- ? "What operation is being performed on the right side of the inequality?" addition
- ? "How do you undo an addition problem?" subtract
- In a problem such as this, point out that the inequality can be rewritten as $-8 > x + 1.4$. This is possible because of the Commutative Property of Addition. Some students feel more comfortable with the inequality written in this form.
- Solve, graph, and check.

On Your Own

- **Think-Pair-Share:** Students should read each question independently and then work with a partner to answer the questions. When they have answered the questions, the pair should compare their answers with another group and discuss any discrepancies.
- These problems integrate a review of fraction and decimal operations.

Example 3

- Ask a volunteer to read the problem.
- Take time to write each stage of the solution: words, variables, and inequality. Notice the use of color coding.
- Set up the inequality and solve as shown. Note that the constant terms are aligned vertically.

On Your Own

- **Neighbor Check:** Have students work independently and then have their neighbor check their work. Have students discuss any discrepancies.

Closure

- **Exit Ticket:** Solve and graph.

$x + 3.8 \leq -9$ $\quad x \leq -12.8$

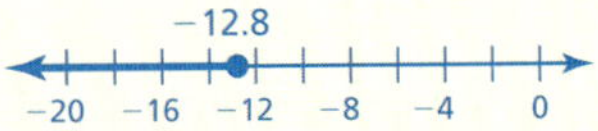

$\frac{2}{5} > x - \frac{3}{4}$ $\quad x < 1\frac{3}{20}$

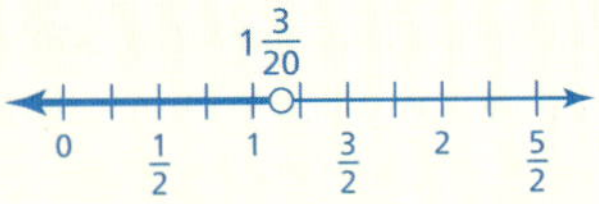

Technology For the Teacher

Dynamic Classroom

The Dynamic Planning Tool
Editable Teacher's Resources at *BigIdeasMath.com*

EXAMPLE 2 Solving an Inequality Using Subtraction

Solve $-8 > 1.4 + x$. Graph the solution.

	$-8 > 1.4 + x$	Write the inequality.
Undo the addition.	$-1.4 \quad -1.4$	Subtract 1.4 from each side.
	$-9.4 > x$	Simplify.

Reading

The inequality $-9.4 > x$ is the same as $x < -9.4$.

The solution is $x < -9.4$.

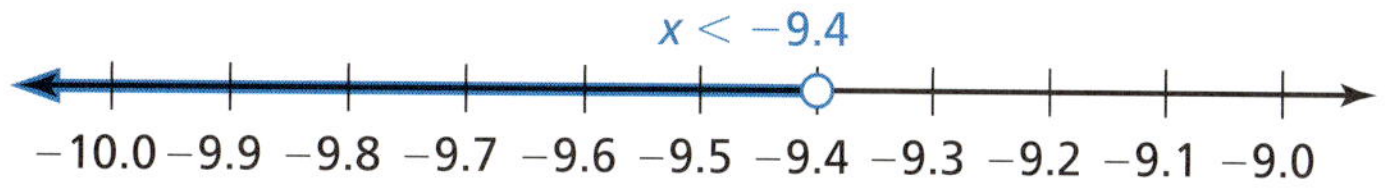

On Your Own

Now You're Ready
Exercises 6–17

Solve the inequality. Graph the solution.

4. $k + 5 \le -3$

5. $\frac{5}{6} \le z + \frac{2}{3}$

6. $p + 0.7 > -2.3$

EXAMPLE 3 Real-Life Application

On a train, carry-on bags can weigh no more than 50 pounds. Your bag weighs 24.8 pounds. Write and solve an inequality that represents the amount of weight you can add to your bag.

Words	Weight of your bag	plus	amount of weight you can add	is no more than	the weight limit.
Variable	Let w be the possible weight you can add.				
Inequality	24.8	+	w	$\le$	50

$24.8 + w \le 50$	Write the inequality.
$-24.8 \quad -24.8$	Subtract 24.8 from each side.
$w \le 25.2$	Simplify.

You can add no more than 25.2 pounds to your bag.

On Your Own

7. **WHAT IF?** Your carry-on bag weighs 32.5 pounds. Write and solve an inequality that represents the possible weight you can add to your bag.

8.2 Exercises

Check It Out
Help with Homework
BigIdeasMath.com

Vocabulary and Concept Check

1. **REASONING** Is the inequality $r - 5 \le 8$ the same as $8 \le r - 5$? Explain.

2. **WHICH ONE DOESN'T BELONG?** Which inequality does *not* belong with the other three? Explain your reasoning.

$c + \frac{7}{2} \le \frac{3}{2}$	$c + \frac{7}{2} \ge \frac{3}{2}$	$\frac{3}{2} \ge c + \frac{7}{2}$	$c - \frac{3}{2} \le -\frac{7}{2}$

Practice and Problem Solving

Use the formula in Activity 1 to create a passing record that makes the inequality true.

3. $P \ge 180$
4. $P + 40 < 110$
5. $280 \le P - 20$

Solve the inequality. Graph the solution.

6. $y - 3 \ge 7$
7. $t - 8 > -4$
8. $n + 11 \le 20$
9. $a + 7 > -1$
10. $5 < v - \frac{1}{2}$
11. $\frac{1}{5} > d + \frac{4}{5}$
12. $-\frac{2}{3} \le g - \frac{1}{3}$
13. $m + \frac{7}{4} \le \frac{11}{4}$
14. $11.2 \le k + 9.8$
15. $h - 1.7 < -3.2$
16. $0 > s + \pi$
17. $5 \ge u - 4.5$

18. **ERROR ANALYSIS** Describe and correct the error in graphing the solution of the inequality.

✗ $5 \ge x - 5$
$10 \ge x$

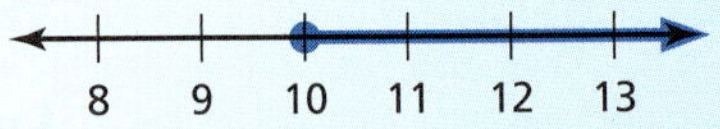

19. **PELICAN** The maximum volume of a great white pelican's bill is about 700 cubic inches.

   a. A pelican scoops up 100 cubic inches of water. Write and solve an inequality that represents the additional volume the bill can contain.

   b. A pelican's stomach can contain about one-third the maximum amount that its bill can contain. Write an inequality that represents the volume of the pelican's stomach.

Assignment Guide and Homework Check

Level	Day 1 Activity Assignment	Day 2 Lesson Assignment	Homework Check
Basic	3–5, 28–32	1, 2, 7–21 odd, 18	7, 11, 18, 21
Average	3–5, 28–32	1, 2, 9–17 odd, 18, 21, 24, 25	11, 15, 21, 24
Advanced	3–5, 28–32	1, 2, 12–18 even, 21–24, 26, 27	12, 22, 24, 26

Common Errors

- **Exercises 6–17** When solving the inequality, students may use the same operation instead of the opposite operation. Remind them that solving inequalities is similar to solving equations, so they should use the opposite operation to solve the inequality.
- **Exercises 6–17** Students may reverse the direction of the inequality symbol when adding or subtracting. Remind them that the inequality symbol does not change direction when adding to or subtracting from both sides. Review the *Numbers* part of the *Key Idea.*
- **Exercises 20–22** Students may write the wrong formula before solving the inequality. Encourage them to write a formula with variables and then substitute the values given in the figure to solve.

8.2 Record and Practice Journal

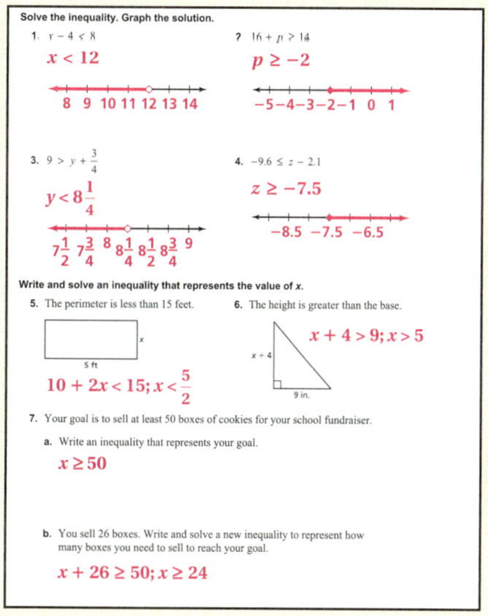
Solve the inequality. Graph the solution.

1. $x - 4 < 8$
$x < 12$

2. $16 + p \geq 14$
$p \geq -2$

3. $9 > y + \frac{3}{4}$
$y < 8\frac{1}{4}$

4. $-9.6 \leq z - 2.1$
$z \geq -7.5$

Write and solve an inequality that represents the value of x.

5. The perimeter is less than 15 feet.
$10 + 2x < 15; x < \frac{5}{2}$

6. The height is greater than the base.
$x + 4 > 9; x > 5$

7. Your goal is to sell at least 50 boxes of cookies for your school fundraiser.

a. Write an inequality that represents your goal.
$x \geq 50$

b. You sell 26 boxes. Write and solve a new inequality to represent how many boxes you need to sell to reach your goal.
$x + 26 \geq 50; x \geq 24$

Technology For the Teacher

Answer Presentation Tool

Quiz*Show*

Vocabulary and Concept Check

1. no; The solution of $r - 5 \leq 8$ is $r \leq 13$ and the solution of $8 \leq r - 5$ is $r \geq 13$.
2. $c + \frac{7}{2} \geq \frac{3}{2}$; It is the only one whose solution is $c \geq -2$. The solution of the other three inequalities is $c \leq -2$.

Practice and Problem Solving

3. *Sample answer:* $A = 350$, $C = 275$, $Y = 3105$, $T = 50$, $N = 2$
4. *Sample answer:* $A = 500$, $C = 205$, $Y = 1700$, $T = 10$, $N = 17$
5. *Sample answer:* $A = 400$, $C = 380$, $Y = 6510$, $T = 83$, $N = 0$
6. $y \geq 10$;

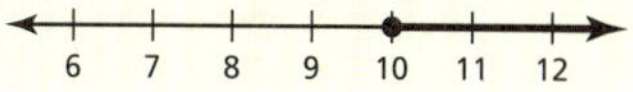

7. $t > 4$;

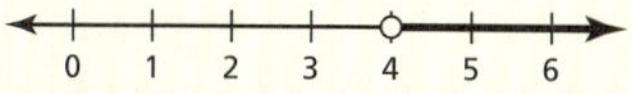

8. $n \leq 9$;

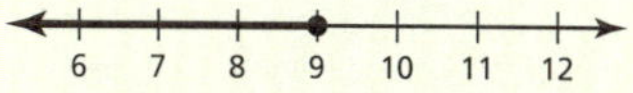

9. $a > -8$;

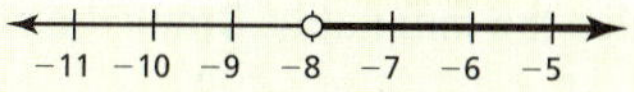

10–17. See Additional Answers.

18. The wrong side of the number line is shaded.

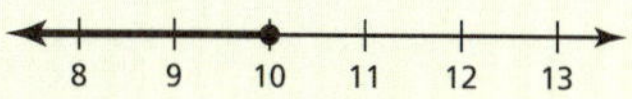

19. a. $100 + V \leq 700$; $V \leq 600 \text{ in.}^3$

b. $V \leq \frac{700}{3} \text{ in.}^3$

Practice and Problem Solving

20. $4 + 4 + x < 16; x < 8$ ft

21. $x + 2 > 10; x > 8$ m

22. $10 + 10 + 12 + 12 + x \le 60;$ $x \le 16$ in.

23. 5

24. $x - 2 \ge 4; x \ge 6$ ft

25. See Additional Answers.

26. See *Taking Math Deeper.*

27. $2\pi h + 2\pi \le 15\pi; h \le 6.5$ mm

Fair Game Review

28. 2

29. 10

30. $\frac{15}{4}$

31. 12

32. B

Mini-Assessment

Solve the inequality. Graph the solution.

1. $-6 \le u - 4$ $u \ge -2$

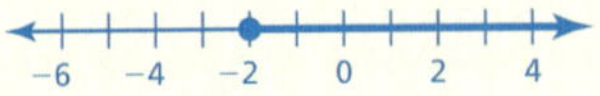

2. $q - 2.5 \ge 6.3$ $q \ge 8.8$

3. $s + 9 < 33$ $s < 24$

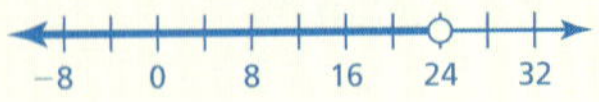

4. $-\frac{2}{3} \ge f + \frac{1}{3}$ $f \le -1$

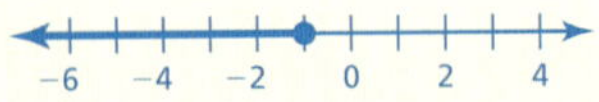

Taking Math Deeper

Exercise 26

Some students may not have had experiences with an electrical circuit that overloads and triggers the circuit breaker. This problem is a nice opportunity to familiarize students with the fact that different appliances use different amounts of electricity.

1. Write an inequality.

   Let x = amount of additional electricity used.

   1100 = amount used by the microwave oven.

   1800 = amount of electricity that overloads the circuit.

   $1100 + x < 1800$

2. Solve the inequality.

   a. $1100 + x < 1800$

   $x < 700$ watts

3. Answer the question.

   You cannot plug in the hot plate or the toaster because each one uses more than 700 watts.

   b. You *can* plug in the clock radio and the blender without overloading the circuit.

Appliance	Watts
Clock radio	50
Blender	300
Hot plate	1200
Toaster	800

Project

It might be interesting for students to research different types of appliances and how much electricity each one uses. In this example, you can see that appliances that generate heat use a lot of electricity. It would follow that a crockpot uses quite a bit less electricity than an oven or a range.

Reteaching and Enrichment Strategies

If students need help. . .	If students got it. . .
Resources by Chapter • Practice A and Practice B • Puzzle Time Record and Practice Journal Practice Differentiating the Lesson Lesson Tutorials Skills Review Handbook	Resources by Chapter • Enrichment and Extension Start the next section

Write and solve an inequality that represents the value of x.

20. The perimeter is less than 16 feet.

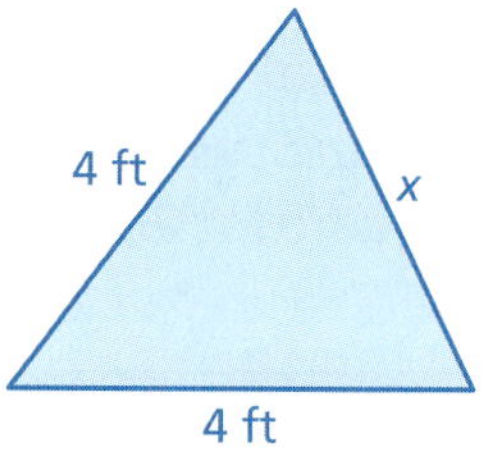

21. The base is greater than the height.

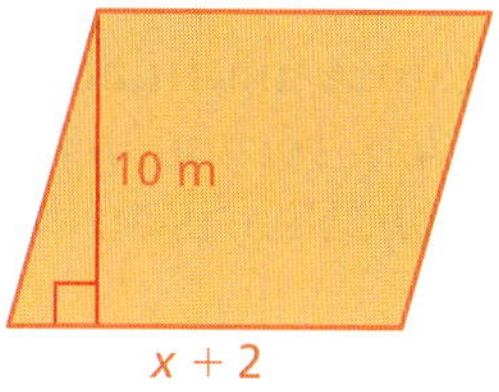

22. The perimeter is less than or equal to 5 feet.

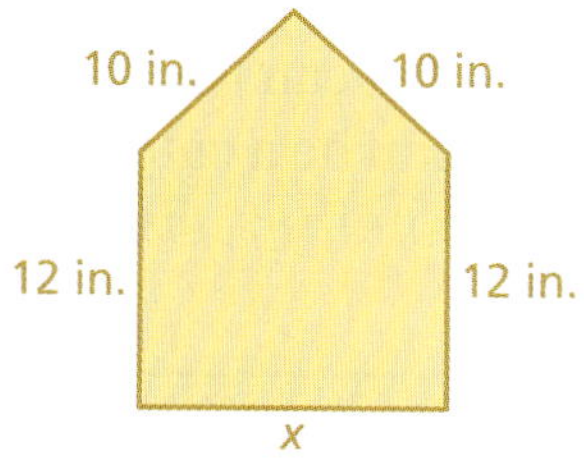

23. REASONING The solution of $w + c \leq 8$ is $w \leq 3$. What is the value of c?

24. FENCE The hole for a fence post is 2 feet deep. The top of the fence post needs to be at least 4 feet above the ground. Write and solve an inequality that represents the required length of the fence post.

25. VIDEO GAME You need at least 12,000 points to advance to the next level of a video game.

a. Write and solve an inequality that represents the number of points you need to advance.

b. You find a treasure chest that increases your score by 60%. How does this change the inequality?

26. POWER A circuit overloads at 1800 watts of electricity. A microwave that uses 1100 watts of electricity is plugged into the circuit.

a. Write and solve an inequality that represents the additional number of watts you can plug in without overloading the circuit.

b. In addition to the microwave, what two appliances in the table can you plug in without overloading the circuit?

Appliance	Watts
Clock radio	50
Blender	300
Hot plate	1200
Toaster	800

27. Critical Thinking The maximum surface area of the solid is 15π square millimeters. Write and solve an inequality that represents the height of the cylinder.

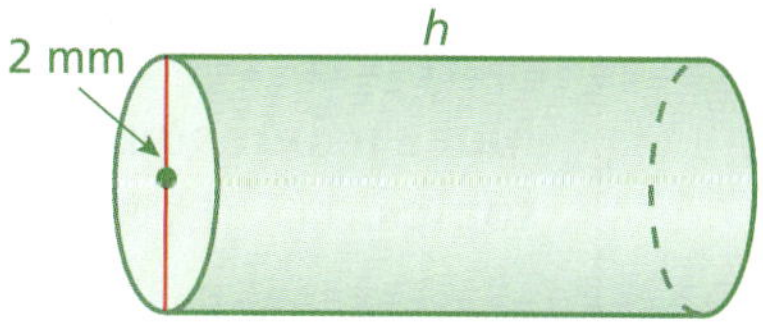

Fair Game Review What you learned in previous grades & lessons

Solve the equation. *(Section 1.1)*

28. $6 = 3x$ **29.** $\frac{r}{5} = 2$ **30.** $4c = 15$ **31.** $8 = \frac{2}{3}b$

32. MULTIPLE CHOICE Which fraction is equivalent to 3.8? *(Skills Review Handbook)*

Ⓐ $\frac{5}{19}$ Ⓑ $\frac{19}{5}$ Ⓒ $\frac{12}{15}$ Ⓓ $\frac{12}{5}$

8 Study Help

Check It Out
Graphic Organizer
BigIdeasMath.com

You can use a **four square** to organize information about a topic. Each of the four squares can be a category, such as *definition, vocabulary, example, non-example, words, algebra, table, numbers, visual, graph,* or *equation*. Here is an example of a four square for an inequality.

INEQUALITY (<)

Definition:

A mathematical sentence that uses an inequality symbol (<).

Vocabulary:

Key phrases that can represent an inequality:
- is less than
- is fewer than

Example:

Words: A number x is less than 2.
Symbols: $x < 2$

Graph:

To graph $x < 2$, draw an open circle at $x = 2$. Then draw an arrow pointing to the left.

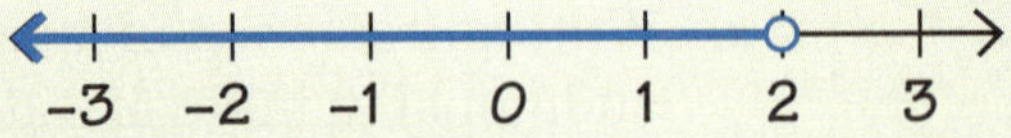

On Your Own

Make a four square to help you study these topics.

1. inequality (>)
2. inequality (≤)
3. inequality (≥)
4. solving an inequality using addition
5. solving an inequality using subtraction

After you complete this chapter, make four squares for the following topics.

6. solving an inequality using multiplication
7. solving an inequality using division

"Sorry, but I have limited space in my four square. I needed pet names with only three letters."

Sample Answers

1.
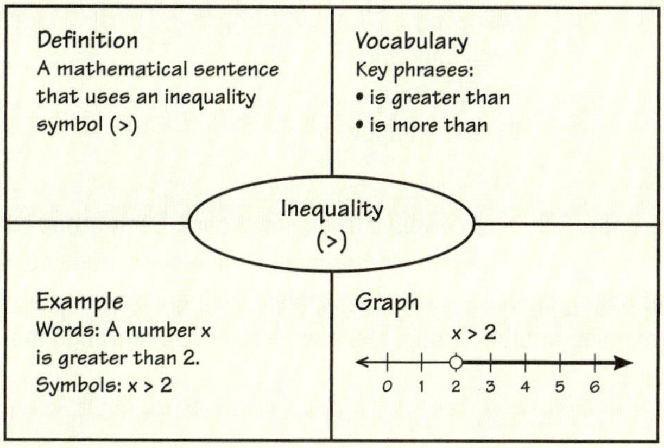

2.
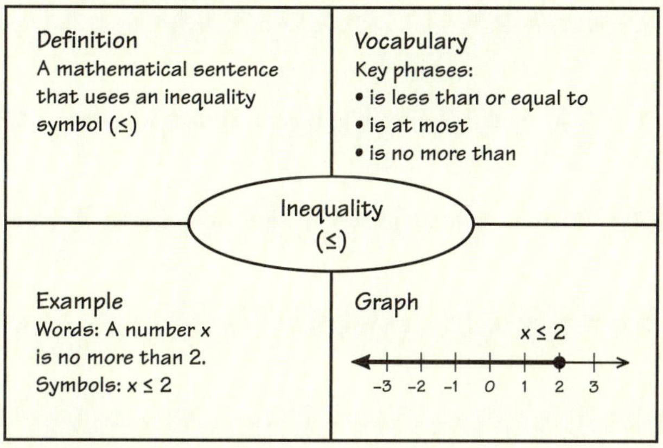

3.
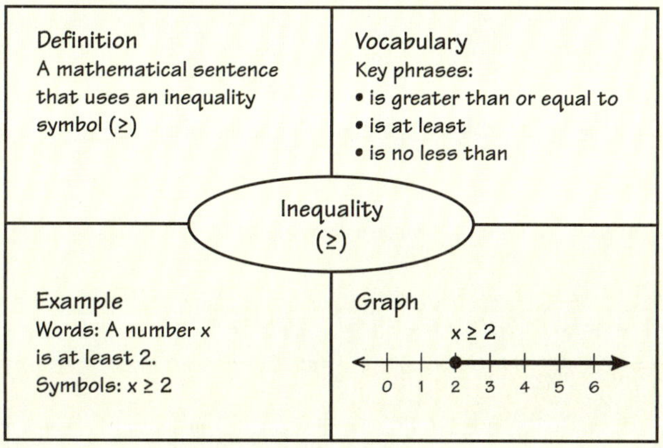

4–5. Available at *BigIdeasMath.com*.

List of Organizers

Available at *BigIdeasMath.com*

Comparison Chart
Concept Circle
Definition (Idea) and Example Chart
Example and Non-Example Chart
Formula Triangle
Four Square
Information Frame
Information Wheel
Notetaking Organizer
Process Diagram
Summary Triangle
Word Magnet
Y Chart

About this Organizer

A **Four Square** can be used to organize information about a topic. Students write the topic in the "bubble" in the middle of the four square. Then students write concepts related to the topic in the four squares surrounding the bubble. Any concept related to the topic can be used. Encourage students to include concepts that will help them learn the topic. Students can place their four squares on note cards to use as a quick study reference.

Technology For the Teacher

Vocabulary Puzzle Builder

Answers

1. $x + 1 < -13$
2. $t - 1.6 \leq 9$
3. yes
4. no
5. 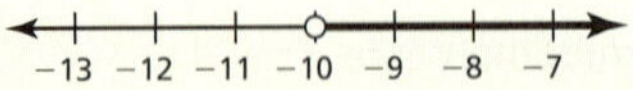

6. 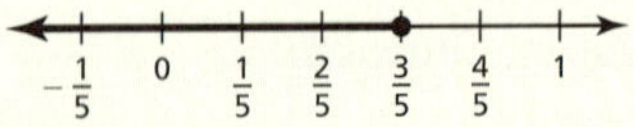

7. 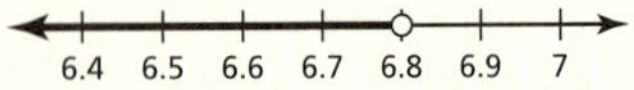

8. $x < 6$
9. $g \geq 16$
10. $h \leq -8$
11. $s > -10$
12. $v < \frac{3}{4}$
13. $1 < p$
14. $x \geq 14$
15. 7
16. $3.5 + x \leq 8$; $x \leq 4.5$ gigabytes
17. See Additional Answers.

Assessment Book

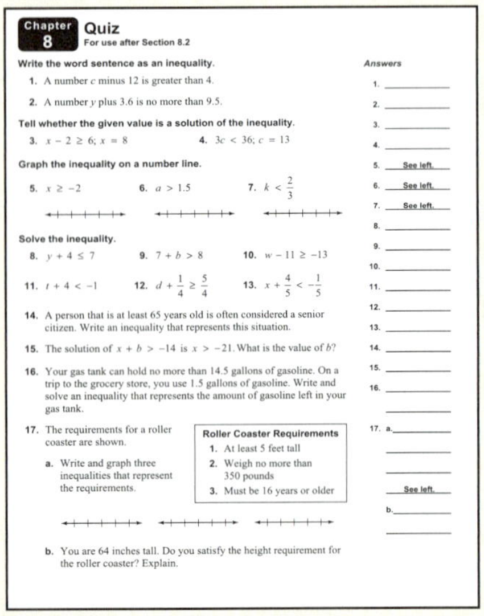

Chapter 8 Quiz
For use after Section 8.2

Write the word sentence as an inequality.

1. A number c minus 12 is greater than 4.
2. A number y plus 3.6 is no more than 9.5.

Tell whether the given value is a solution of the inequality.

3. $x - 2 \geq 6$; $x = 8$
4. $3c < 36$; $c = 13$

Graph the inequality on a number line.

5. $x \geq -2$
6. $a > 1.5$
7. $k < \frac{2}{3}$

Solve the inequality.

8. $y + 4 \leq 7$
9. $7 + b > 8$
10. $w - 11 \geq -13$
11. $t + 4 < -1$
12. $d + \frac{1}{4} \geq \frac{5}{4}$
13. $x + \frac{4}{5} < -\frac{1}{5}$
14. A person that is at least 65 years old is often considered a senior citizen. Write an inequality that represents this situation.
15. The solution of $x + b > -14$ is $x > -21$. What is the value of b?
16. Your gas tank can hold no more than 14.5 gallons of gasoline. On a trip to the grocery store, you use 1.5 gallons of gasoline. Write and solve an inequality that represents the amount of gasoline left in your gas tank.
17. The requirements for a roller coaster are shown.
    a. Write and graph three inequalities that represent the requirements.
    b. You are 64 inches tall. Do you satisfy the height requirement for the roller coaster? Explain.

Roller Coaster Requirements
1. At least 5 feet tall
2. Weigh no more than 350 pounds
3. Must be 16 years or older

Answers
1. ____
2. ____
3. ____
4. ____
5. See left.
6. See left.
7. See left.
8. ____
9. ____
10. ____
11. ____
12. ____
13. ____
14. ____
15. ____
16. ____
17. a. See left.
b. ____

Alternative Quiz Ideas

100% Quiz	Math Log
Error Notebook	Notebook Quiz
Group Quiz	Partner Quiz
Homework Quiz	Pass the Paper

Group Quiz

Students work in groups. Give each group a large index card. Each group writes five questions that they feel evaluate the material they have been studying. On a separate piece of paper, students solve the problems. When they are finished, they exchange cards with another group. The new groups work through the questions on the card.

Reteaching and Enrichment Strategies

If students need help. . .	If students got it. . .
Resources by Chapter • Study Help • Practice A and Practice B • Puzzle Time Lesson Tutorials *BigIdeasMath.com* Practice Quiz Practice from the Test Generator	Resources by Chapter • Enrichment and Extension • School-to-Work Game Closet at *BigIdeasMath.com* Start the next section

Technology For the Teacher

Answer Presentation Tool
Big Ideas Test Generator

8.1–8.2 Quiz

Check It Out
Progress Check
BigIdeasMath.com

Write the word sentence as an inequality. *(Section 8.1)*

1. A number x plus 1 is less than -13.
2. A number t minus 1.6 is at most 9.

Tell whether the given value is a solution of the inequality. *(Section 8.1)*

3. $12n < -2;\ n = -1$
4. $y + 4 < -3;\ y = -7$

Graph the inequality on a number line. *(Section 8.1)*

5. $x > -10$
6. $y \le \frac{3}{5}$
7. $w < 6.8$

Solve the inequality. *(Section 8.2)*

8. $x - 2 < 4$
9. $g + 14 \ge 30$
10. $h - 1 \le -9$
11. $s + 3 > -7$
12. $v - \frac{3}{4} < 0$
13. $\frac{3}{2} < p + \frac{1}{2}$

14. **WATERCRAFT** In many states, you must be at least 14 years old to operate a personal watercraft. Write an inequality that represents this situation. *(Section 8.1)*

15. **REASONING** The solution of $x - a > 4$ is $x > 11$. What is the value of a? *(Section 8.2)*

16. **MP3 PLAYER** Your MP3 player can store up to 8 gigabytes of media. You transfer 3.5 gigabytes of media to the MP3 player. Write and solve an inequality that represents the amount of memory available on the MP3 player. *(Section 8.2)*

17. **LIFEGUARD** Three requirements for a lifeguard training course are shown. *(Section 8.1)*

    a. Write and graph three inequalities that represent the requirements.

    b. You can swim 350 feet. Do you satisfy the swimming requirement of the course? Explain.

8.3 Solving Inequalities Using Multiplication or Division

Essential Question How can you use multiplication or division to solve an inequality?

1 ACTIVITY: Using a Table to Solve an Inequality

Work with a partner.

- **Copy and complete the table.**
- **Decide which graph represents the solution of the inequality.**
- **Write the solution of the inequality.**

a. $3x \le 6$

x	-1	0	1	2	3	4	5
$3x$							
$3x \overset{?}{\le} 6$							

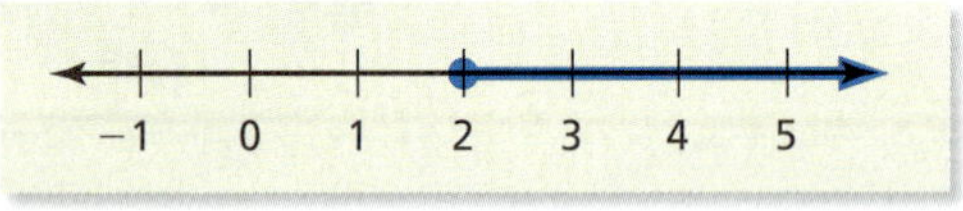

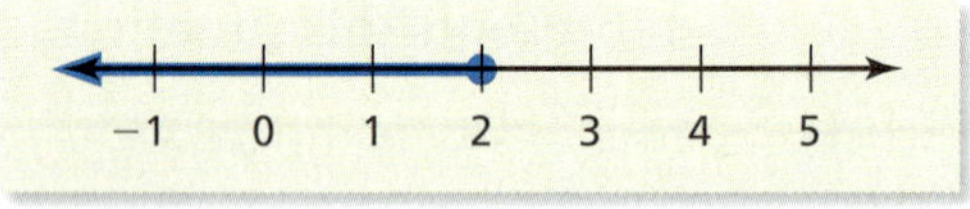

b. $-2x > 4$

x	-5	-4	-3	-2	-1	0	1
$-2x$							
$-2x \overset{?}{>} 4$							

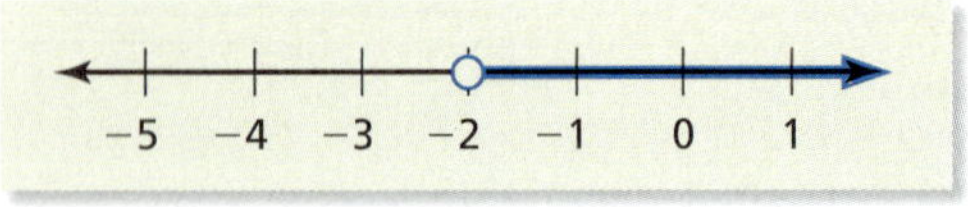

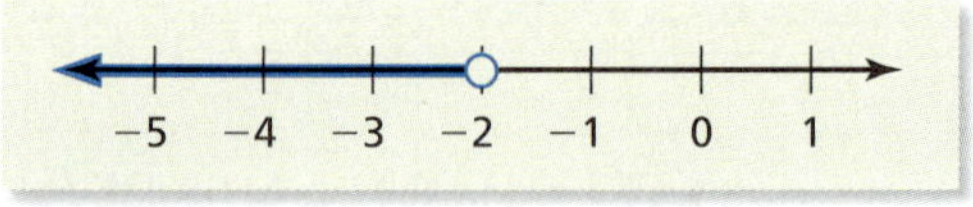

2 ACTIVITY: Writing a Rule

Work with a partner. Use a table to solve each inequality.

a. $3x > 3$ **b.** $4x \le 4$ **c.** $-2x \ge 6$ **d.** $-5x < 10$

Write a rule that describes how to solve inequalities like those in Activity 1. Then use your rule to solve each of the four inequalities above.

Laurie's Notes

Introduction

For the Teacher

- **Goal:** Students will gain an intuitive understanding of solving inequalities involving multiplication and division.
- The approach in this investigation is to use a table of values to see what numbers satisfy the inequality. The problems involving positive coefficients behave as expected. It is the problems involving negative coefficients that seem not to work as expected—from the student perspective.

Motivate

- Ask a series of questions and record the students' solutions.
  - ? "What integers are solutions of $x > 4$?" 5, 6, 7, ...
  - ? "What integers are solutions of $-x > 4$, meaning what numbers have an opposite that is greater than 4?" $-5, -6, -7, \ldots$
  - ? "What integers are solutions of $x < -4$?" $-5, -6, -7, \ldots$
- Leave these 3 problems on the board and refer to them at the end of class.

Activity Notes

Activity 1

- Explain to students that for each inequality, they are to evaluate one side of the inequality and then decide if the inequality is satisfied. This means, *is the value of x a solution of the inequality?* Students will write *yes* or *no* in the third row of the table to indicate if the value is a solution or not.
- Using the information in the table, students decide which graph represents the solution. Finally, they write the solution.

? Discuss the results with your students.
- "What did you find as the solution for part (a)?" $x \le 2$
- "Is this what you would have expected?" Likely, they will say yes.
- "What did you find as the solution for part (b)?" $x < -2$
- "Is this what you would have expected?" Likely, they will say no.
- Do not tell students a rule at this point. Simply say that perhaps they need to try a few more problems to help figure out what is going on.

Activity 2

? "Did any of the inequalities have solutions that you expected?" Yes, the first two inequalities; Listen for students to say that for the last two inequalities, the number part of the solution was expected but the inequality symbol was switched.

? "Did you notice a difference in the first two inequalities versus the last two?" They noticed the positive coefficient in parts (a) and (b), versus the negative coefficient in parts (c) and (d).

- Students might not be quite ready to write a rule, but they sense the issue has to do with the negative coefficient. Let the uncertainty be unresolved. You don't need to solve every problem immediately.

Previous Learning

Students should know how to solve equations using multiplication and division. Students should be able to evaluate expressions and decide if a number is a solution of an inequality.

Start Thinking! and Warm Up

Activity 8.3 Start Thinking! For use before Activity 8.3

Activity 8.3 Warm Up For use before Activity 8.3

Complete the statement with < or >.

1. -7 __?__ -5
2. 2 __?__ -2
3. 7 __?__ -10
4. -13 __?__ -11
5. -1 __?__ -2
6. -8 __?__ 8

8.3 Record and Practice Journal

Essential Question How can you use multiplication or division to solve an inequality?

1 ACTIVITY: Using a Table to Solve an Inequality

Work with a partner.

- **Complete the table.**
- **Decide which graph represents the solution of the inequality.**
- **Write the solution of the inequality.**

a. $3x \le 6$

x	-1	0	1	2	3	4	5
$3x$	-3	0	3	6	9	12	15
$3x \overset{?}{\le} 6$	T	T	T	T	F	F	F

$x \le 2$

b. $-2x > 4$

x	-5	-4	-3	-2	-1	0	1
$-2x$	10	8	6	4	2	0	-2
$-2x \overset{?}{>} 4$	T	T	T	F	F	F	F

$x < -2$

English Language Learners

Pair Activity

Create index cards with problems similar to those in Activities 2 and 4. Pair English learners with English speakers and give 5 cards to each pair. Have students work together to solve the inequalities. When students have completed the problems, check their work and give them another set of cards.

8.3 Record and Practice Journal

2 ACTIVITY: Writing a Rule

Work with a partner. Use a table to solve each inequality. *Sample answer:*

a. $3x > 3$ $x > 1$ b. $4x \le 4$ $x \le 1$ c. $-2x \ge 6$ $x \le -3$ d. $-5x < 10$ $x > -2$

x	−5	−4	−3	−2	−1	0	1	2	3
$3x$	−15	−12	−9	−6	−3	0	3	6	9
$4x$	−20	−16	−12	−8	−4	0	4	8	12
$-2x$	10	8	6	4	2	0	−2	−4	−6
$-5x$	25	20	15	10	5	0	−5	−10	−15

Write a rule that describes how to solve inequalities like those in Activity 1. Then use your rule to solve each of the four inequalities above.

Check students' work.

3 ACTIVITY: Using a Table to Solve an Inequality

Work with a partner.

- **Complete the table.**
- **Decide which graph represents the solution of the inequality.**
- **Write the solution of the inequality.**

a. $\frac{x}{2} \ge 1$

x	−1	0	1	2	3	4	5
$\frac{x}{2}$	$-\frac{1}{2}$	0	$\frac{1}{2}$	1	$\frac{3}{2}$	2	$\frac{5}{2}$
$\frac{x}{2} \overset{?}{\ge} 1$	F	F	F	T	T	T	T

$x \ge 2$

b. $\frac{x}{-3} < \frac{2}{3}$

x	−5	−4	−3	−2	−1	0	1
$\frac{x}{-3}$	$\frac{5}{3}$	$\frac{4}{3}$	1	$\frac{2}{3}$	$\frac{1}{3}$	0	$-\frac{1}{3}$
$\frac{x}{-3} \overset{?}{<} \frac{2}{3}$	F	F	F	F	T	T	T

$x > -2$

4 ACTIVITY: Writing a Rule

Work with a partner. Use a table to solve each inequality. *Sample answer:*

a. $\frac{x}{4} \ge 1$ $x \ge 4$ b. $\frac{x}{2} < \frac{3}{2}$ $x < 3$ c. $\frac{x}{-2} > 2$ $x < -4$ d. $\frac{x}{-5} \le \frac{1}{5}$ $x \ge -1$

x	−5	−4	−3	−2	−1	0	1	2	3	4	5
$\frac{x}{4}$	$-\frac{5}{4}$	−1	$-\frac{3}{4}$	$-\frac{1}{2}$	$-\frac{1}{4}$	0	$\frac{1}{4}$	$\frac{1}{2}$	$\frac{3}{4}$	1	$\frac{5}{4}$
$\frac{x}{2}$	$-\frac{5}{2}$	−2	$-\frac{3}{2}$	−1	$-\frac{1}{2}$	0	$\frac{1}{2}$	1	$\frac{3}{2}$	2	$\frac{5}{2}$
$\frac{x}{-2}$	$\frac{5}{2}$	2	$\frac{3}{2}$	1	$\frac{1}{2}$	0	$-\frac{1}{2}$	−1	$-\frac{3}{2}$	−2	$-\frac{5}{2}$
$\frac{x}{-5}$	1	$\frac{4}{5}$	$\frac{3}{5}$	$\frac{2}{5}$	$\frac{1}{5}$	0	$-\frac{1}{5}$	$-\frac{2}{5}$	$-\frac{3}{5}$	$-\frac{4}{5}$	−1

Write a rule that describes how to solve inequalities like those in Activity 3. Then use your rule to solve each of the four inequalities above.

Check students' work.

What Is Your Answer?

5. **IN YOUR OWN WORDS** How can you use multiplication or division to solve an inequality?

Check students' work.

Laurie's Notes

Activity 3

- Explain that the next two activities are similar to the first two except they involve using multiplication to solve the inequality instead of division.
- Give time for students to work through the two problems.

? Discuss the results with your students.

  - "What did you find as the solution for part (a)?" $x \ge 2$
  - "Is this what you would have expected?" Likely, they will say yes.
  - "What did you find as the solution for part (b)?" $x > -2$
  - "Is this what you would have expected?" Likely, they will say no.

- Again, resist the temptation to simply tell students a rule. Suggest that trying additional problems might help them.

Activity 4

- Give time for students to work through the four problems with their partner.

? "Did any of the inequalities have solutions that you expected?" Again, listen for the same comments from students as before. The first two problems have expected solutions. The last two problems had the number part of the solution expected, but the inequality symbol switched.

- After working through all four problems, students should have a sense that solving these inequalities is the same as solving equations except when the coefficient is negative.

What Is Your Answer?

- **Neighbor Check** Have students work independently and then have their neighbor check their work. Have students discuss any discrepancies.

Closure

- Refer to the three inequalities written at the beginning of class.

$$x > 4 \qquad -x > 4 \qquad x < -4$$

? "Which inequalities have the same solution?" $-x > 4$ and $x < -4$

? "Is this consistent with what you discovered in the activities? Explain." yes; Listen for comments about the negative coefficient of x and the switching of the inequality symbol.

Technology For the Teacher

Dynamic Classroom

The Dynamic Planning Tool

Editable Teacher's Resources at *BigIdeasMath.com*

3 ACTIVITY: Using a Table to Solve an Inequality

Work with a partner.

- **Copy and complete the table.**
- **Decide which graph represents the solution of the inequality.**
- **Write the solution of the inequality.**

a. $\frac{x}{2} \geq 1$

x	-1	0	1	2	3	4	5
$\frac{x}{2}$							
$\frac{x}{2} \overset{?}{\geq} 1$							

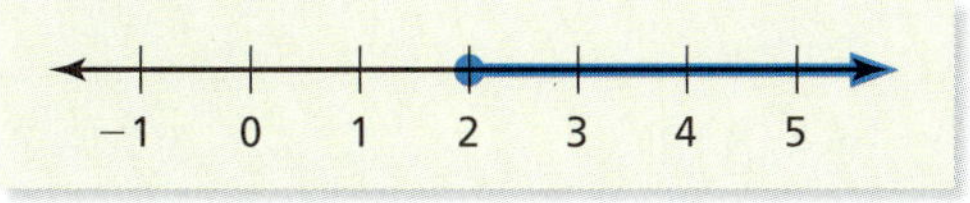

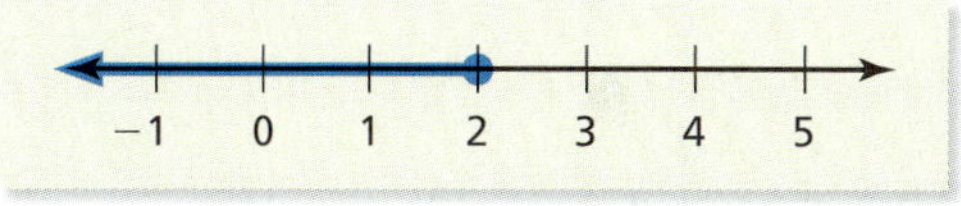

b. $\frac{x}{-3} < \frac{2}{3}$

x	-5	-4	-3	-2	-1	0	1
$\frac{x}{-3}$							
$\frac{x}{-3} \overset{?}{<} \frac{2}{3}$							

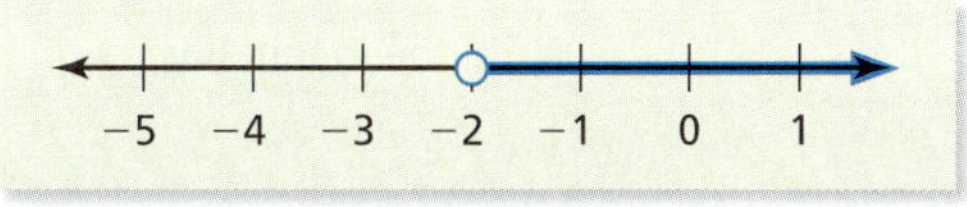

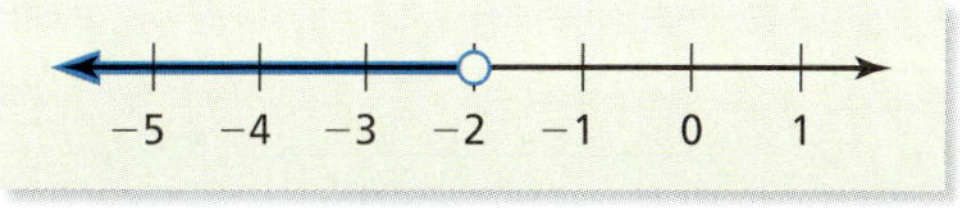

4 ACTIVITY: Writing a Rule

Work with a partner. Use a table to solve each inequality.

a. $\frac{x}{4} \geq 1$ **b.** $\frac{x}{2} < \frac{3}{2}$ **c.** $\frac{x}{-2} > 2$ **d.** $\frac{x}{-5} \leq \frac{1}{5}$

Write a rule that describes how to solve inequalities like those in Activity 3. Then use your rule to solve each of the four inequalities above.

What Is Your Answer?

5. **IN YOUR OWN WORDS** How can you use multiplication or division to solve an inequality?

Practice Use what you learned about solving inequalities using multiplication or division to complete Exercises 4–9 on page 331.

8.3 Lesson

Check It Out
Lesson Tutorials
BigIdeasMath.com

Remember

Multiplication and division are inverse operations.

Key Idea

Multiplication and Division Properties of Inequality (Case 1)

Words If you multiply or divide each side of an inequality by the same *positive* number, the inequality remains true.

Numbers

$-6 < 8$	$6 > -8$
$2 \cdot (-6) < 2 \cdot 8$	$\frac{6}{2} > \frac{-8}{2}$
$-12 < 16$	$3 > -4$

Algebra

$\frac{x}{2} < -9$	$4x > -12$
$2 \cdot \frac{x}{2} < 2 \cdot (-9)$	$\frac{4x}{4} > \frac{-12}{4}$
$x < -18$	$x > -3$

These properties are also true for $\leq$ and $\geq$.

EXAMPLE 1 Solving an Inequality Using Multiplication

Solve $\frac{x}{8} > -5$. Graph the solution.

$\frac{x}{8} > -5$ Write the inequality.

Undo the division. → $8 \cdot \frac{x}{8} > 8 \cdot (-5)$ Multiply each side by 8.

$x > -40$ Simplify.

∴ The solution is $x > -40$.

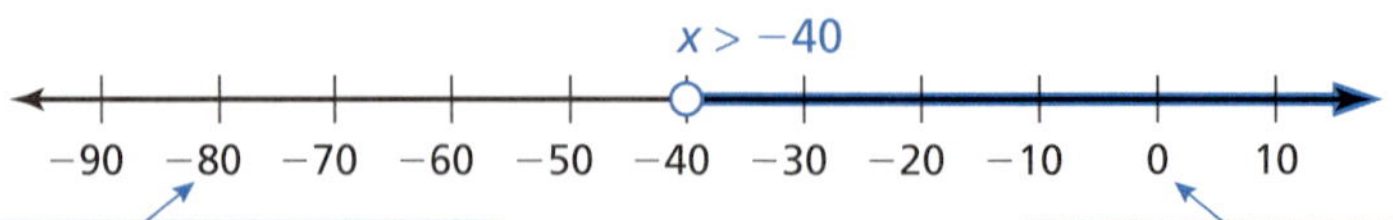

Check: $x = -80$ is *not* a solution.

Check: $x = 0$ is a solution.

On Your Own

Solve the inequality. Graph the solution.

1. $a \div 2 < 4$ **2.** $\frac{n}{7} \geq -1$ **3.** $-6.4 \geq \frac{w}{5}$

Laurie's Notes

Introduction

Connect

- **Yesterday:** Students gained an intuitive understanding of solving inequalities involving multiplication and division.
- **Today:** Students will use the Multiplication and Division Properties of Inequality to solve inequalities.

Motivate

? "Have you heard of Ultimate?" Answers will vary.

- It is a sport played with a flying disc at colleges, high schools, and some middle schools. There are 10 simple rules, one of which is there aren't any officials! Pretty cool.
- The popularity of the sport has skyrocketed, but there is *at most* one-fifth the numbers of students playing Ultimate as there are playing lacrosse. If there are 26 students playing Ultimate, what is the minimum number playing lacrosse? $\frac{1}{5}x \geq 26$; $x \geq 130$ players

Lesson Notes

Key Idea

- These properties should look familiar, as they are similar to the Multiplication and Division Properties of Equality used in solving equations.
- Note that the properties are restricted to multiplying and dividing by a *positive* number. This is very important.

Example 1

? "How do you isolate the variable, meaning get x by itself?" Multiply by 8 on each side of the inequality.

- Multiplying by 8 is the inverse operation of dividing by 8.
- **Representation:** Multiplication is represented by the dot notation, and -5 is enclosed in parentheses for clarity only. Otherwise, students might become confused and think 5 is being subtracted.

On Your Own

- **Think-Pair-Share:** Students should read each question independently and then work with a partner to answer the questions. When they have answered the questions, the pair should compare their answers with another group and discuss any discrepancies.
- Division is represented in different ways in Questions 1 and 2. The second representation is more common in algebra (higher mathematics).
- After solving the inequality in Question 3, the result will be $-32 \geq w$. Students can also rewrite this as $w \leq -32$. The direction of the inequality symbol is reversed *only* because the solution is being rewritten with the variable on the left side of the inequality statement.

Goal Today's lesson is solving inequalities using multiplication or division.

Start Thinking! and Warm Up

Lesson 8.3 Warm Up
For use before Lesson 8.3

Lesson 8.3 Start Thinking!
For use before Lesson 8.3

In football, a team has four attempts to advance the ball at least 10 yards, which earns them a first down.

Write an inequality to describe the average yards per down for a team who has earned a first down in a set of 4 downs.

Extra Example 1

Solve $\frac{d}{5} < -7$. Graph the solution.

$d < -35$

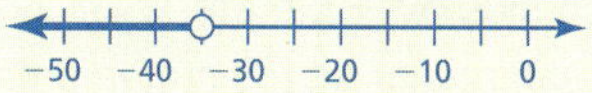

On Your Own

1. $a < 8$;

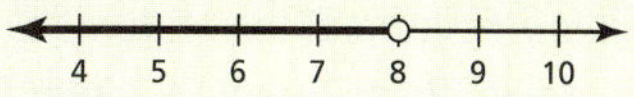

2. $n \geq -7$;

3. $w \leq -32$;

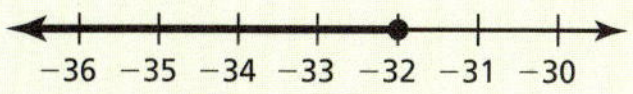

Laurie's Notes

Extra Example 2

Solve $2x \geq 12$. Graph the solution.

$x \geq 6$

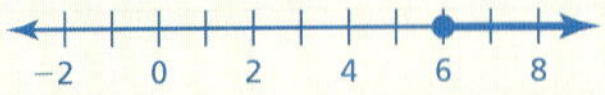

On Your Own

4. $b \geq 9$;

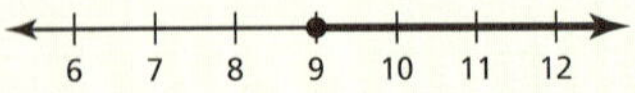

5. $k > -5$;

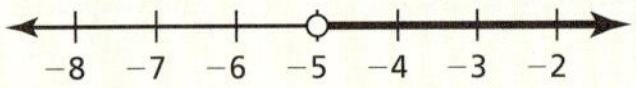

6. $q < -12$;

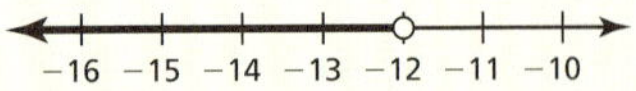

Differentiated Instruction

Visual

Use the inequality $6 < 9$ to show students why it is necessary to reverse the inequality symbol when multiplying or dividing by a negative number.

Add -3 to each side. The result is $3 < 6$, a true statement.

Subtract -3 from each side. The result is $9 < 12$, a true statement.

Multiply each side by -3. If the inequality is *not* reversed, the statement $-18 < -27$ is false. By reversing the inequality, the statement $-18 > -27$ is true.

Divide each side by -3. If the inequality is *not* reversed, the statement $-2 < -3$ is false. By reversing the inequality, the statement $-2 > -3$ is true.

Example 2

? "What operation is being performed on the left side of the inequality?" multiplication

? "How do you undo a multiplication problem?" divide

- Solve, graph, and check.

On Your Own

- **Think-Pair-Share:** Students should read each question independently and then work with a partner to answer the questions. When they have answered the questions, the pair should compare their answers with another group and discuss any discrepancies.
- Notice that although all of the coefficients are positive, sometimes the constant is negative. This is important in helping students understand when the direction of the inequality symbol is going to be reversed. The focus is on the sign of the coefficient, not the sign of the constant.
- For Question 6, remind students that after solving this inequality, the result will be $-12 > q$. Students can also rewrite this as $q < -12$. The direction of the inequality symbol is reversed *only* because the solution is being rewritten with the variable on the left side of the inequality statement.
- These problems integrate review of decimal operations.

Key Idea

- These properties look identical to what they have been using in the lesson, *except* now the direction of the inequality symbol must be reversed for the inequality to remain true because they are multiplying or dividing by a *negative* quantity!
- The short version of the property: When you multiply or divide by a negative quantity, reverse the direction of the inequality symbol.
- **Common Error:** When students solve $2x < -4$, they sometimes reverse the inequality symbol because there's a negative number in the problem. The inequality symbol is reversed *only* when the coefficient is negative, not when the constant is negative.

EXAMPLE 2 Solving an Inequality Using Division

Solve $3x \le -24$. Graph the solution.

	$3x \le -24$	Write the inequality.
Undo the multiplication. →	$\frac{3x}{3} \le \frac{-24}{3}$	Divide each side by 3.
	$x \le -8$	Simplify.

The solution is $x \le -8$.

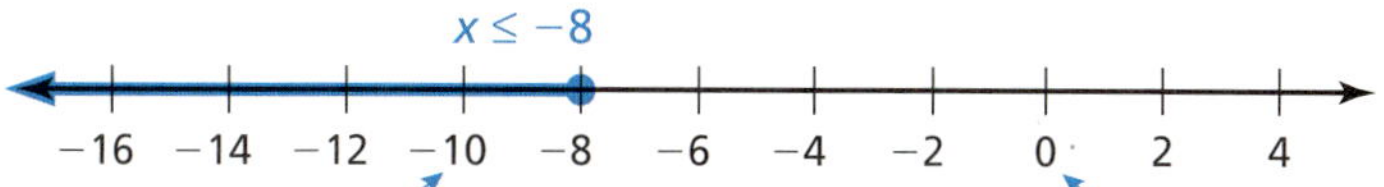

Check: $x = -10$ is a solution.

Check: $x = 0$ is *not* a solution.

On Your Own

Now You're Ready
Exercises 10–18

Solve the inequality. Graph the solution.

4. $4b \ge 36$ **5.** $2k > -10$ **6.** $-18 > 1.5q$

Key Idea

Multiplication and Division Properties of Inequality (Case 2)

Words If you multiply or divide each side of an inequality by the same *negative* number, the direction of the inequality symbol must be reversed for the inequality to remain true.

Numbers

$$-6 < 8 \qquad\qquad 6 > -8$$

$$(-2) \cdot (-6) > (-2) \cdot 8 \qquad\qquad \frac{6}{-2} < \frac{-8}{-2}$$

$$12 > -16 \qquad\qquad -3 < 4$$

Algebra

$$\frac{x}{-6} < 3 \qquad\qquad -5x > 30$$

$$-6 \cdot \frac{x}{-6} > -6 \cdot 3 \qquad\qquad \frac{-5x}{-5} < \frac{30}{-5}$$

$$x > -18 \qquad\qquad x < -6$$

These properties are also true for ≤ and ≥.

Common Error

A negative sign in an inequality does not necessarily mean you must reverse the inequality symbol.

Only reverse the inequality symbol when you multiply or divide both sides by a negative number.

EXAMPLE 3 Solving an Inequality Using Multiplication

Solve $\frac{y}{-3} > 2$. Graph the solution.

$$\frac{y}{-3} > 2$$ Write the inequality.

Undo the division. → $$-3 \cdot \frac{y}{-3} < -3 \cdot 2$$ Multiply each side by −3. Reverse the inequality symbol.

$$y < -6$$ Simplify.

∴ The solution is $y < -6$.

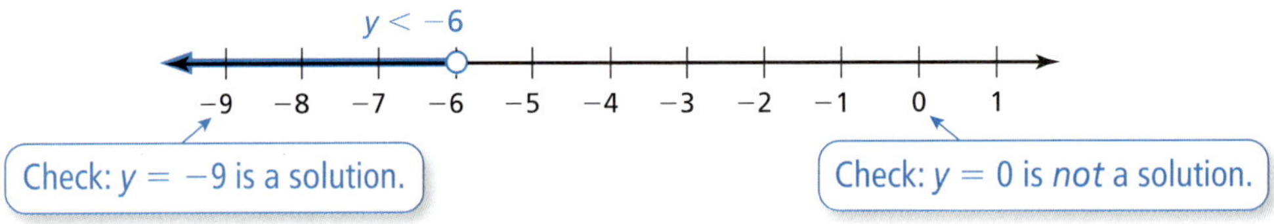

EXAMPLE 4 Solving an Inequality Using Division

Solve $-7y \le -35$. Graph the solution.

$$-7y \le -35$$ Write the inequality.

Undo the multiplication. → $$\frac{-7y}{-7} \ge \frac{-35}{-7}$$ Divide each side by −7. Reverse the inequality symbol.

$$y \ge 5$$ Simplify.

∴ The solution is $y \ge 5$.

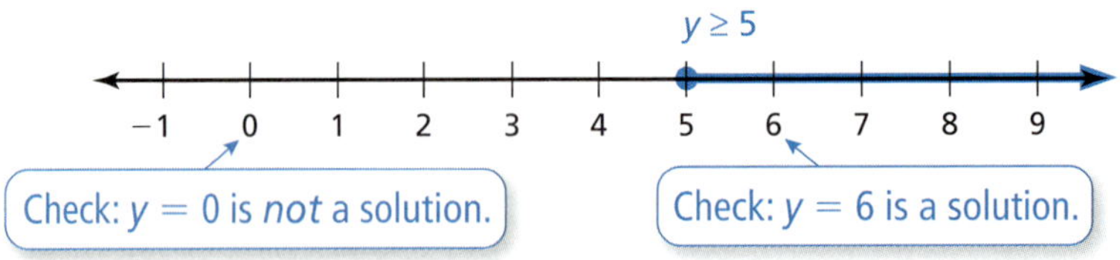

On Your Own

Now You're Ready
Exercises 27–35

Solve the inequality. Graph the solution.

7. $\frac{p}{-4} < 7$

8. $\frac{x}{-5} \le -5$

9. $1 \ge -\frac{1}{10}z$

10. $-9m > 63$

11. $-2r \ge -22$

12. $-0.4y \ge -12$

Laurie's Notes

Lesson Notes

Example 3

- Write the problem.
- ? "What operation is being performed?" division by -3
- ? "How do you undo dividing by -3?" multiply by -3
- Solve as usual, but remember to reverse the direction of the inequality symbol.
- When graphing, remember to use an open circle because the inequality is strictly *less than.*

Example 4

- Write the example.
- ? "What operation is being performed?" multiplication by -7
- ? "How do you undo multiplying by -7?" divide by -7
- Solve as usual, but remember to reverse the direction of the inequality symbol. Remember, the quotient of two negatives is positive.
- ? "Should you use an open or closed circle?" Use a closed circle because the inequality is greater than or equal to.

On Your Own

- **Neighbor Check:** Have students work independently and then have their neighbor check their work. Have students discuss any discrepancies.
- Have students share their work at the board.

Closure

- **Exit Ticket:** Solve and graph.

$\frac{x}{-3} \le -9$ $x \ge 27$

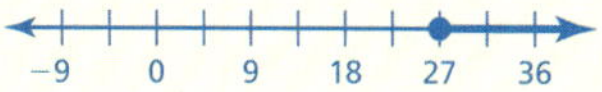

$-8 > 4x$ $x < -2$

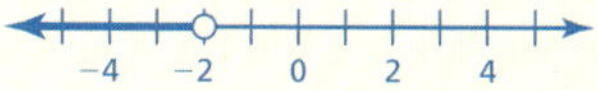

Technology For the Teacher

Dynamic Classroom

The Dynamic Planning Tool

Editable Teacher's Resources at *BigIdeasMath.com*

Extra Example 3

Solve $\frac{c}{-4} \le 3$. Graph the solution.

$c \ge -12$

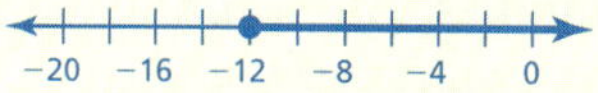

Extra Example 4

Solve $-3j > -9$. Graph the solution.

$j < 3$

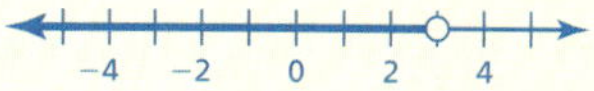

On Your Own

7. $p > -28$;

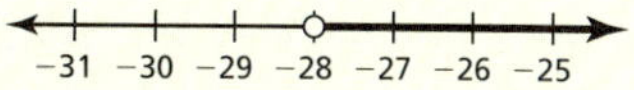

8. $x \ge 25$;

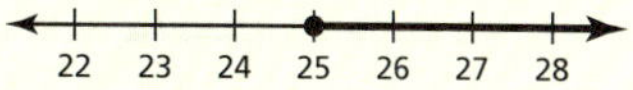

9. $z \ge -10$;

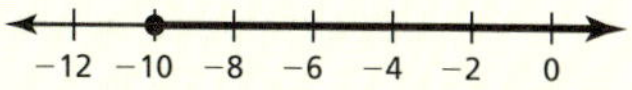

10. $m < -7$;

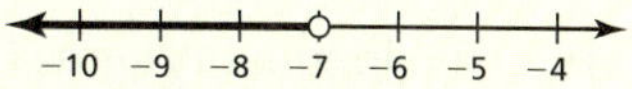

11. $r \le 11$;

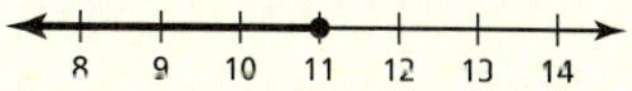

12. $y \le 30$;

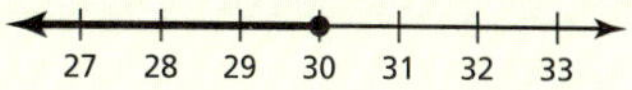

Vocabulary and Concept Check

1. Multiply each side of the inequality by 6.
2. The first inequality is divided by a positive number. The second inequality is divided by a negative number. Because this inequality is divided by a negative number, the direction of the inequality symbol must be reversed.
3. *Sample answer:* $-3x < 6$

Practice and Problem Solving

4. $x < 1$
5. $x \geq -1$
6. $x < -3$
7. $x \leq -3$
8. $x < -5$
9. $x \leq \frac{3}{2}$
10. $n > 6$;

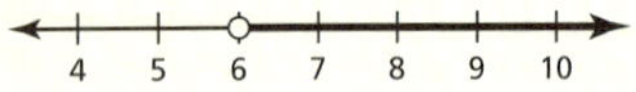

11. $c \leq -36$;

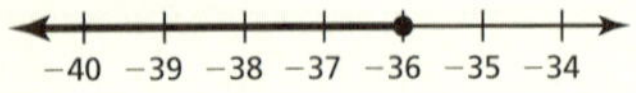

12–18. See Additional Answers.

19. The inequality sign should not have been reversed.

$$\frac{x}{2} < -5$$
$$2 \cdot \frac{x}{2} < 2 \cdot (-5)$$
$$x < -10$$

20. $\frac{x}{3} \leq 4$; $x \leq 12$
21. $\frac{x}{8} < -2$; $x < -16$
22. $4x \geq -12$; $x \geq -3$
23. $5x > 20$; $x > 4$
24. $9.5x \geq 247$; $x \geq 26$ h

Assignment Guide and Homework Check

Level	Day 1 Activity Assignment	Day 2 Lesson Assignment	Homework Check
Basic	4–9, 48–52	1–3, 11–35 odd, 24, 36, 38	15, 21, 29, 36, 38
Average	4–9, 48–52	1–3, 13–41 odd, 36, 38	15, 21, 29, 36, 39
Advanced	4–9, 48–52	1–3, 16–22 even, 19, 25, 32–42 even, 37, 44–47	16, 20, 34, 36, 42

Common Errors

- **Exercises 10–18** Students may perform the same operation on both sides instead of the opposite operation when solving the inequality. Remind them that solving inequalities is similar to solving equations.
- **Exercises 10–18** When there is a negative in the inequality, students may reverse the direction of the inequality symbol. Remind them that they only reverse the direction when they are multiplying or dividing by a negative number. All of these exercises keep the same inequality symbol.

8.3 Record and Practice Journal

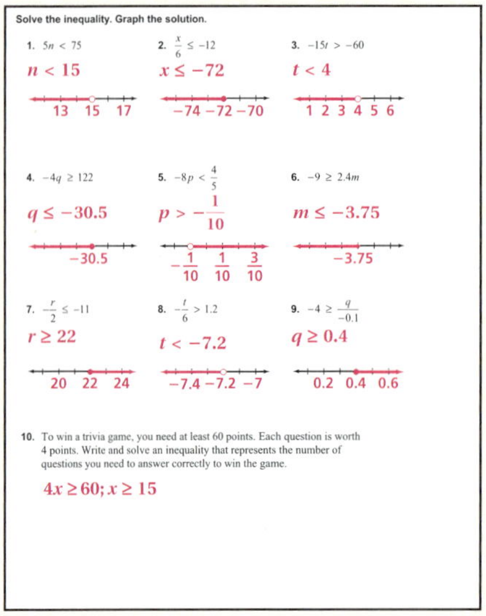

Solve the inequality. Graph the solution.

1. $5n < 75$ — $n < 15$ (13 15 17)
2. $\frac{x}{6} \leq -12$ — $x \leq -72$ (−74 −72 −70)
3. $-15t > -60$ — $t < 4$ (1 2 3 4 5 6)
4. $-4q \geq 122$ — $q \leq -30.5$ (−30.5)
5. $-8p < \frac{4}{5}$ — $p > -\frac{1}{10}$ ($-\frac{1}{10}$ $\frac{1}{10}$ $\frac{3}{10}$)
6. $-9 \geq 2.4m$ — $m \leq -3.75$ (−3.75)
7. $-\frac{r}{2} \leq -11$ — $r \geq 22$ (20 22 24)
8. $-\frac{t}{6} > 1.2$ — $t < -7.2$ (−7.4 −7.2 −7)
9. $-4 \geq \frac{q}{-0.1}$ — $q \geq 0.4$ (0.2 0.4 0.6)
10. To win a trivia game, you need at least 60 points. Each question is worth 4 points. Write and solve an inequality that represents the number of questions you need to answer correctly to win the game.

$4x \geq 60$; $x \geq 15$

Technology For the Teacher
Answer Presentation Tool
Quiz*Show*

8.3 Exercises

Check It Out
Help with Homework
BigIdeasMath.com

Vocabulary and Concept Check

1. **VOCABULARY** Explain how to solve $\frac{x}{6} < -5$.
2. **WRITING** Explain how solving $2x < -8$ is different from solving $-2x < 8$.
3. **OPEN-ENDED** Write an inequality that is solved using the Division Property of Inequality where the inequality symbol needs to be reversed.

Practice and Problem Solving

Use a table to solve the inequality.

4. $4x < 4$
5. $-2x \le 2$
6. $-5x > 15$
7. $\frac{x}{-3} \ge 1$
8. $\frac{x}{-2} > \frac{5}{2}$
9. $\frac{x}{4} \le \frac{3}{8}$

Solve the inequality. Graph the solution.

(1) (2) 10. $3n > 18$

11. $\frac{c}{4} \le -9$
12. $1.2m < 12$
13. $-14 > x \div 2$
14. $\frac{w}{5} \ge -2.6$
15. $5 < 2.5k$
16. $4x \le -\frac{3}{2}$
17. $2.6y \le -10.4$
18. $10.2 > \frac{b}{3.4}$
19. **ERROR ANALYSIS** Describe and correct the error in solving the inequality.

✗
$$\frac{x}{2} < -5$$
$$2 \cdot \frac{x}{2} > 2 \cdot (-5)$$
$$x > -10$$

Write the word sentence as an inequality. Then solve the inequality.

20. The quotient of a number and 3 is at most 4.
21. A number divided by 8 is less than -2.
22. Four times a number is at least -12.
23. The product of 5 and a number is greater than 20.

24. **CAMERA** You earn $9.50 per hour at your summer job. Write and solve an inequality that represents the number of hours you need to work in order to buy a digital camera that costs $247.

25. **COPIES** You have \$3.65 to make copies. Write and solve an inequality that represents the number of copies you can make.

26. **SPEED LIMIT** The maximum speed limit for a school bus is 55 miles per hour. Write and solve an inequality that represents the number of hours it takes to travel 165 miles in a school bus.

Solve the inequality. Graph the solution.

③ ④ 27. $-2n \le 10$

28. $-5w > 30$

29. $\frac{h}{-6} \ge 7$

30. $-8 < -\frac{1}{3}x$

31. $-2y < -11$

32. $-7d \ge 56$

33. $2.4 > -\frac{m}{5}$

34. $\frac{k}{-0.5} \le 18$

35. $-2.5 > \frac{b}{-1.6}$

36. **ERROR ANALYSIS** Describe and correct the error in solving the inequality.

✗

$-4m \ge 16$

$\frac{-4m}{-4} \ge \frac{16}{-4}$

$m \ge -4$

37. **CRITICAL THINKING** Are all numbers greater than zero solutions of $-x > 0$? Explain.

38. **TRUCKING** In many states, the maximum height (including freight) of a vehicle is 13.5 feet.

a. Write and solve an inequality that represents the number of crates that can be stacked vertically on the bed of the truck.

b. Five crates are stacked vertically on the bed of the truck. Is this legal? Explain.

Write and solve an inequality that represents the value of *x*.

39. Area $\ge$ 102 cm^2

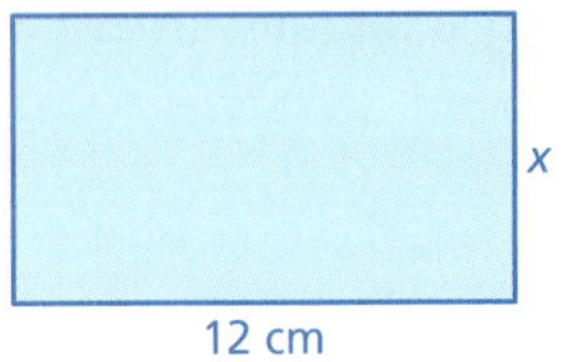

40. Area < 30 ft^2

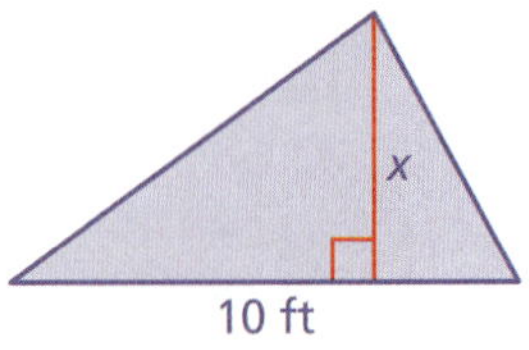

Common Errors

- **Exercise 25** Students may forget to change the cost per copy to a decimal. In the photo, the cost is given as a whole number of cents. Some students will write $25c < 3.65$. Point out to them that they need to write the cost per copy in dollars (0.25). Unit analysis can and should be used when working with rates.
- **Exercises 27–35** Students may forget to reverse the inequality symbol when multiplying or dividing by a negative number. Remind them of this rule. Encourage students to substitute values into the original inequality to check that the solution is correct.
- **Exercise 41** Students may write an incorrect inequality before solving. They may write $\frac{c}{3} < 80$ because there are three friends. However, the student is included in the trip as well, so there are 4 people going on the trip. The inequality should be $\frac{c}{4} < 80$.

Differentiated Instruction

Visual

Students may question why they are asked to graph the solution of an inequality, but not asked to graph the solution of an equation. The graph of an equation is just a point on the number line. The graph of an inequality provides more information, because of the infinite number of possible solutions.

Practice and Problem Solving

25. $0.25x \leq 3.65$; $x \leq 14.6$; You can make at most 14 copies.

26. $55x \geq 165$; $x \geq 3$ h

27. $n \geq -5$;

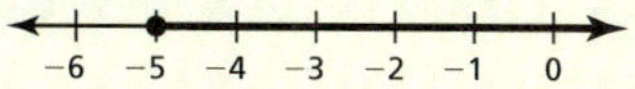

28. $w < -6$;

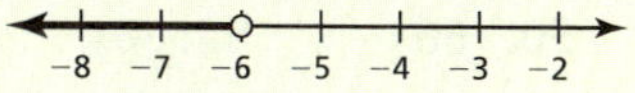

29–35. See Additional Answers.

36. They forgot to reverse the inequality symbol.

$$-4m \geq 16$$

$$\frac{-4m}{-4} \leq \frac{16}{-4}$$

$$m \leq -4$$

37. no; You need to solve the inequality for x. The solution is $x < 0$. Therefore, numbers greater than 0 are not solutions.

38. **a.** $\frac{28}{12}x \leq 10$; $x \leq \frac{30}{7}$, or $4\frac{2}{7}$ At most 4 crates can be stacked.

b. no; The maximum height allowed is 162 inches, and 5 crates on the truck has a height of 182 inches.

39. $12x \geq 102$; $x \geq 8.5$ cm

40. $5x < 30$; $x < 6$ ft

Practice and Problem Solving

41. $\frac{x}{4} < 80$; $x < \$320$

42–43. See Additional Answers.

44. See *Taking Math Deeper*.

45. $n \geq -6$ and $n \leq -4$;

46. $x \geq 2$;

47. $m < 20$;

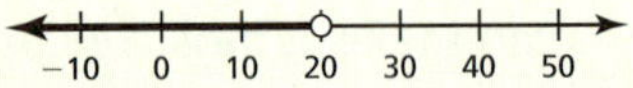

Fair Game Review

48. 4 **49.** $8\frac{1}{4}$

50. 66 **51.** 84

52. C

Mini-Assessment

Solve the inequality. Graph the solution.

1. $2z < 4$ $z < 2$

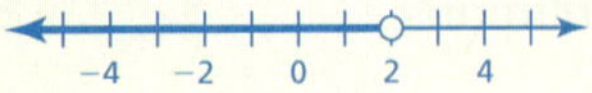

2. $15 \geq 5k$ $k \leq 3$

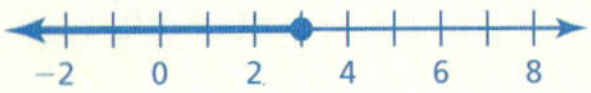

3. $\frac{m}{4} \geq -3$ $m \geq -12$

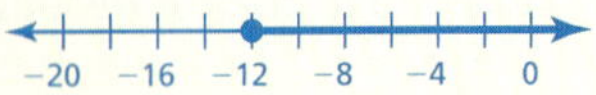

4. $3 < \frac{\ell}{-6}$ $\ell < -18$

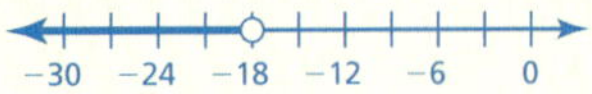

5. $-6p < -36$ $p > 6$

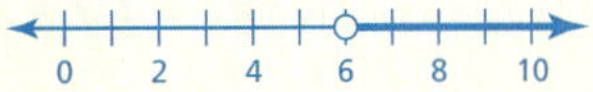

Taking Math Deeper

Exercise 44

Double (or compound) inequalities, like those in Exercises 44–47, can often be written using a single inequality statement.

1. Begin by graphing each inequality.

$3m > -12$ or $m > -4$

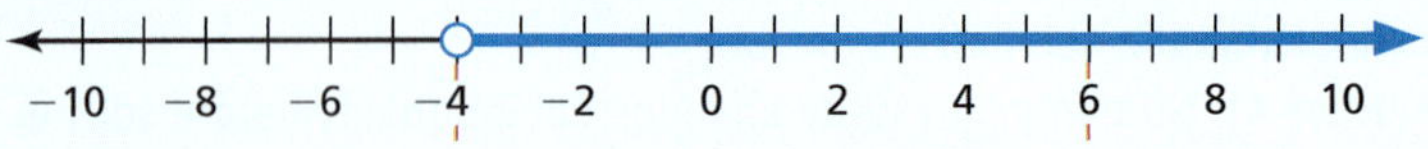

$2m < 12$ or $m < 6$

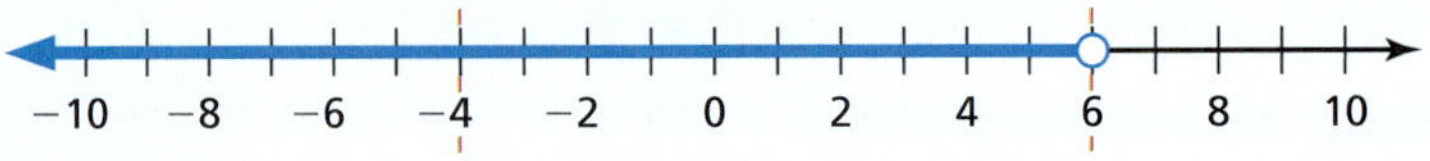

2. Combine the two graphs.

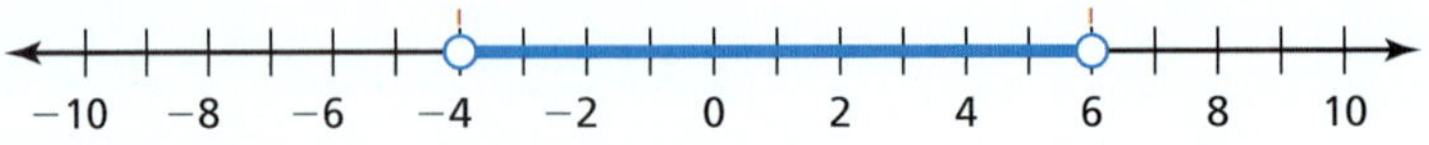

3. The numbers that satisfy both inequalities are all numbers greater than −4 and less than 6. If you rewrite $m > -4$ as $-4 < m$, you can write the statement as a single inequality, $-4 < m < 6$.

Project

Use the newspaper, Internet, TV, radio, or any other source to record and graph at least 10 different uses of inequalities.

Reteaching and Enrichment Strategies

If students need help. . .	If students got it. . .
Resources by Chapter • Practice A and Practice B • Puzzle Time Record and Practice Journal Practice Differentiating the Lesson Lesson Tutorials Skills Review Handbook	Resources by Chapter • Enrichment and Extension • School-to-Work Start the next section

41. **TRIP** You and three friends are planning a trip. You want to keep the cost below \$80 per person. Write and solve an inequality that represents the total cost of the trip.

42. **REASONING** Explain why the direction of the inequality symbol must be reversed when multiplying or dividing by the same negative number.

43. **PROJECT** Choose two musical artists to research.

a. Use the Internet or a magazine to complete the table.

b. Find the average number of copies sold per month for each CD.

c. Use the release date to write and solve an inequality that represents the minimum average number of copies sold per month for each CD.

d. In how many months do you expect the number of copies of the second top selling CD to surpass the current number of copies of the top selling CD?

Artist	Name of CD	Release Date	Current Number of Copies Sold
1.			
2.			

Number Sense **Describe all numbers that satisfy *both* inequalities. Include a graph with your description.**

44. $3m > -12$ and $2m < 12$

45. $\frac{n}{2} \geq -3$ and $\frac{n}{-4} \geq 1$

46. $2x \geq -4$ and $2x \geq 4$

47. $\frac{m}{-4} > -5$ and $\frac{m}{4} < 10$

Fair Game Review What you learned in previous grades & lessons

Solve the equation. *(Section 1.2)*

48. $-4w + 5 = -11$

49. $4(x - 3) = 21$

50. $\frac{v}{6} - 7 = 4$

51. $\frac{m + 300}{4} = 96$

52. **MULTIPLE CHOICE** Which measure can have more than one value for a given data set? *(Section 7.1)*

Ⓐ mean   Ⓑ median   Ⓒ mode   Ⓓ range

8.4 Solving Multi-Step Inequalities

Essential Question How can you use an inequality to describe the area and perimeter of a composite figure?

1 ACTIVITY: Areas and Perimeters of Composite Figures

Work with a partner.

a. For what values of x will the area of the blue region be greater than 12 square units?

b. For what values of x will the sum of the inner and outer perimeters of the blue region be greater than 20 units?

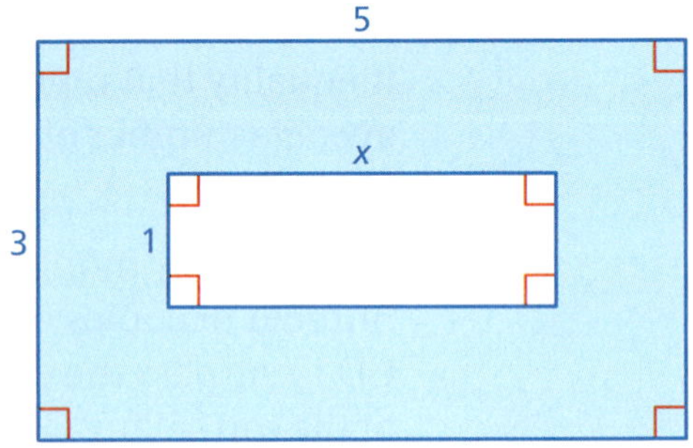

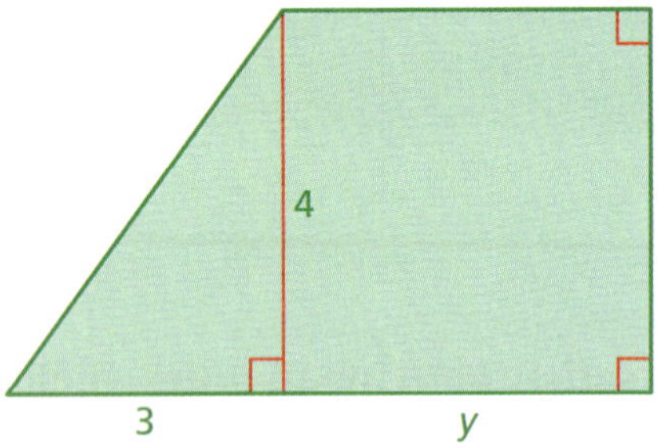

c. For what values of y will the area of the trapezoid be less than or equal to 10 square units?

d. For what values of y will the perimeter of the trapezoid be less than or equal to 16 units?

e. For what values of w will the area of the red region be greater than or equal to 36 square units?

f. For what values of w will the sum of the inner and outer perimeters of the red region be greater than 47 units?

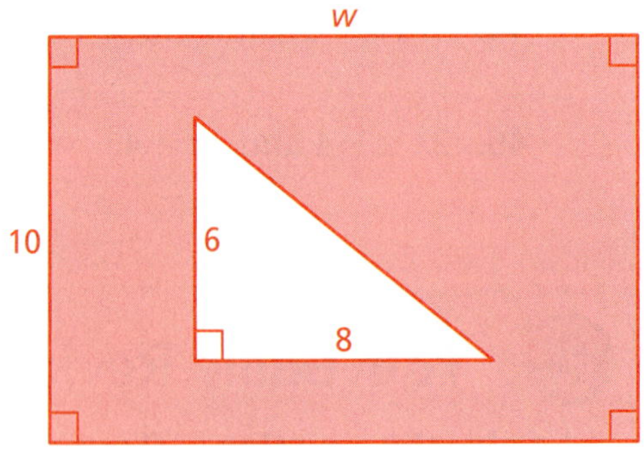

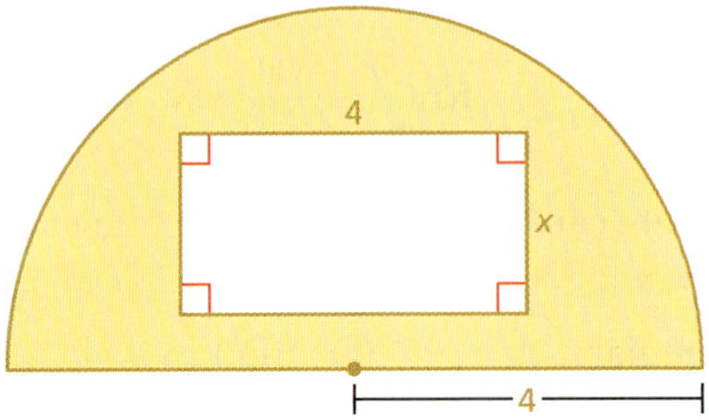

g. For what values of x will the area of the yellow region be less than 4π square units?

h. For what values of x will the sum of the inner and outer perimeters of the yellow region be less than $4\pi + 20$ units?

Laurie's Notes

Introduction

For the Teacher

- **Goal:** Students will review geometric formulas and gain an intuitive understanding of solving multi-step inequalities.
- In this activity, students are expected to recall common formulas for area and perimeter. This is a nice integrated review of many skills.

Motivate

- A regulation Ultimate field is 64 meters long and 37 meters wide, with two end zones of 18 meters each. Sometimes the length (64 meters) varies due to using existing fields. Let the length be x. Draw a sketch and find the area of the field in terms of x.

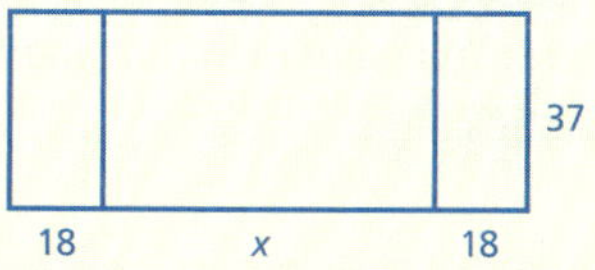

Area $= 37(x + 36) = (37x + 1332)$ m^2

Activity Notes

Activity 1

- This activity uses composite figures. Sometimes instead of adding figures together, one figure is removed from another.
- For each problem, students will need to write a variable expression for the composite area or perimeter. This expression will then become part of an inequality to be solved.
- Model the first problem.
- ? "How do you find the area of just the blue region?" Subtract the area of the inner rectangle from the area of the larger rectangle.
- ? "How do you find the total perimeter of the blue region?" Add the perimeter of the inner rectangle to the perimeter of the larger rectangle.
- Discuss with students that $x > 0$. At $x = 0$, the inner rectangle no longer exists. This is true for any geometric figure.
- These are neither trivial, nor simple problems. They require that students be familiar with the formulas and solve inequalities. My students always say that it feels like they're solving an equation and at the end they go back to make sure that they've paid attention to the inequality symbol.
- In parts (e) and (f), students should notice that the width of the red rectangle must be greater than 8, the base of the inner triangle.
- In part (g), the length of a segment from the midpoint of the base of the rectangle to an upper corner of the rectangle must be less than 4, the radius of the circle. So, $x^2 + 2^2 < 4^2$, or $x < 2\sqrt{3}$.
- As students are solving the problems, you may need to remind them about collecting like terms and the Distributive Property.

Previous Learning

Students should know how to solve multi-step equations and one-step inequalities.

Start Thinking! and Warm Up

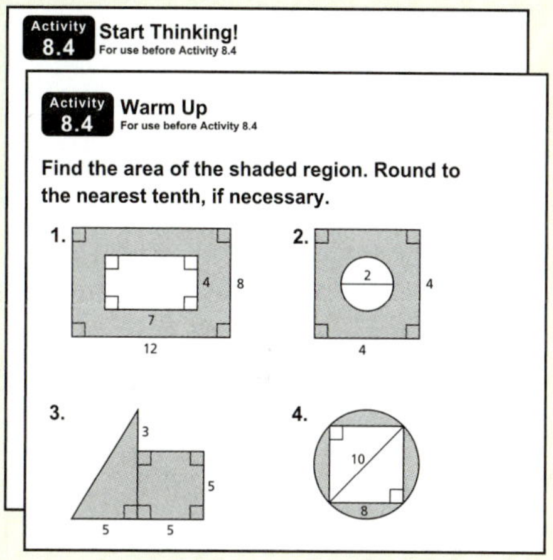

Activity 8.4 Start Thinking! For use before Activity 8.4

Activity 8.4 Warm Up For use before Activity 8.4

Find the area of the shaded region. Round to the nearest tenth, if necessary.

1. 2. 3. 4.

8.4 Record and Practice Journal

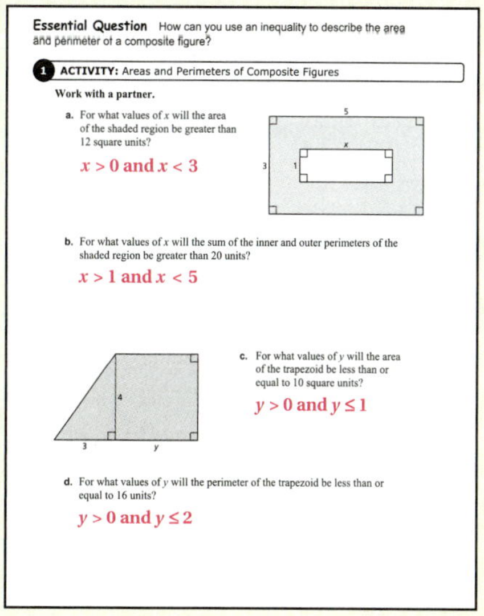

Essential Question How can you use an inequality to describe the area and perimeter of a composite figure?

1 ACTIVITY: Areas and Perimeters of Composite Figures

Work with a partner.

a. For what values of x will the area of the shaded region be greater than 12 square units?
$x > 0$ and $x < 3$

b. For what values of x will the sum of the inner and outer perimeters of the shaded region be greater than 20 units?
$x > 1$ and $x < 5$

c. For what values of y will the area of the trapezoid be less than or equal to 10 square units?
$y > 0$ and $y \leq 1$

d. For what values of y will the perimeter of the trapezoid be less than or equal to 16 units?
$y > 0$ and $y \leq 2$

Differentiated Instruction

Visual

Discuss with students the composite figure in Activity 1(a). Ask students if the number −1 makes sense as a solution. Ask students if the number 6 makes sense as a solution. Point out that in application problems, students need to determine whether the solution of an inequality makes sense as an answer.

8.4 Record and Practice Journal

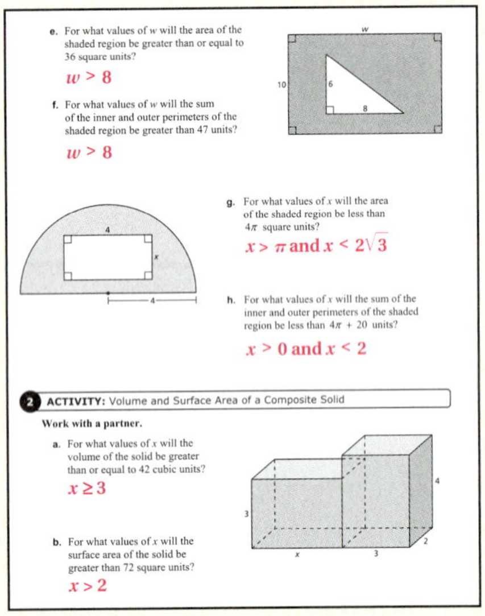

e. For what values of w will the area of the shaded region be greater than or equal to 36 square units?

$w > 8$

f. For what values of w will the sum of the inner and outer perimeters of the shaded region be greater than 47 units?

$w > 8$

g. For what values of x will the area of the shaded region be less than 4π square units?

$x > \pi$ and $x < 2\sqrt{3}$

h. For what values of x will the sum of the inner and outer perimeters of the shaded region be less than $4\pi + 20$ units?

$x > 0$ and $x < 2$

2 ACTIVITY: Volume and Surface Area of a Composite Solid

Work with a partner.

a. For what values of x will the volume of the solid be greater than or equal to 42 cubic units?

$x \geq 3$

b. For what values of x will the surface area of the solid be greater than 72 square units?

$x > 2$

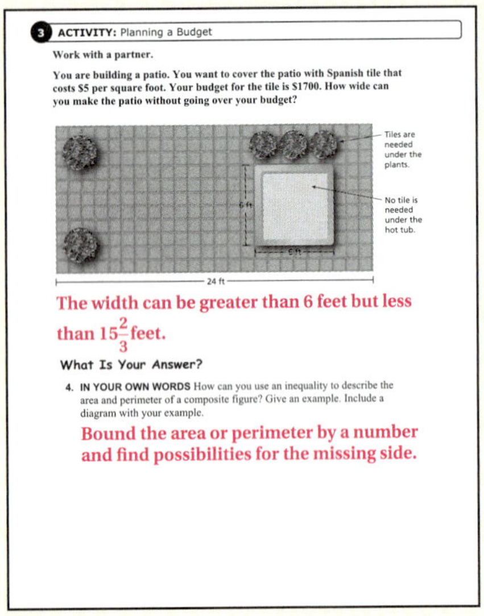

3 ACTIVITY: Planning a Budget

Work with a partner.

You are building a patio. You want to cover the patio with Spanish tile that costs $5 per square foot. Your budget for the tile is $1700. How wide can you make the patio without going over your budget?

The width can be greater than 6 feet but less than $15\frac{2}{3}$ feet.

What Is Your Answer?

4. **IN YOUR OWN WORDS** How can you use an inequality to describe the area and perimeter of a composite figure? Give an example. Include a diagram with your example.

Bound the area or perimeter by a number and find possibilities for the missing side.

Laurie's Notes

Activity 2

- While eating in a restaurant, I noticed a cardboard model that had been folded from a net. It was a great example of a composite solid, a rectangular prism and a trapezoidal prism that looked like a fire truck! Of course, I asked the waiter if I could take it home.
- In part (a), students find the volume of each prism separately and add, $V = 6x + 24$. Then, they solve $6x + 24 \geq 42$.
- In part (b), students find the surface area of each prism and subtract the rectangular region that has been counted twice.

  Left prism: $S = 2(2x) + 2(6) + 2(3x) = 10x + 12$

  Right prism: $S = 2(12) + 2(6) + 2(8) = 52$

  Inside surface: $A = 6 \times 2 = 12$ (to be subtracted)

  Summary: Surface Area $= 10x + 12 + 52 - 12 = 10x + 52$

  Solve: $10x + 52 > 72$
- Ask two volunteers who have their work displayed in a neat and organized fashion to share their solutions at the board. You want to model good problem solving, as well as good mathematics.

Activity 3

- Ask a student to read the problem.
- ? "Do you know the finished size of the patio?" no
- The width x is going to vary depending upon the budget. You want the patio as wide as possible.
- ? "How do you find the area of the patio if it is going to vary in size?" Use a variable to express the width.
- Area $= 24x - 36$; Cost: $5(24x - 36) < 1700$

$$120x - 180 < 1700$$
$$120x < 1880$$
$$x < 15\frac{2}{3}$$

Closure

- Refer to the Ultimate playing field from the beginning of class. How long can the field be so that the perimeter, including end zones, is less than 268 meters? less than 61 m

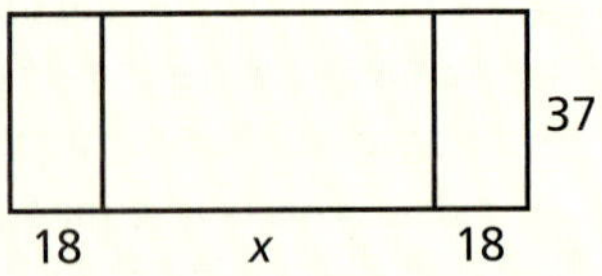

Technology For the Teacher

Dynamic Classroom

The Dynamic Planning Tool

Editable Teacher's Resources at *BigIdeasMath.com*

2 ACTIVITY: Volume and Surface Area of a Composite Solid

Work with a partner.

a. For what values of x will the volume of the solid be greater than or equal to 42 cubic units?

b. For what values of x will the surface area of the solid be greater than 72 square units?

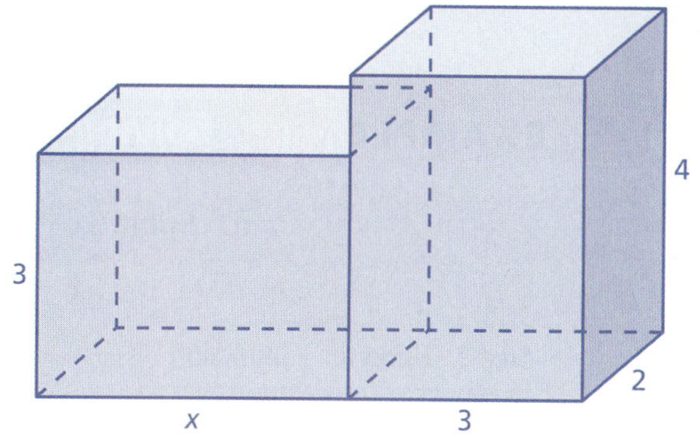

3 ACTIVITY: Planning a Budget

Work with a partner.

You are building a patio. You want to cover the patio with Spanish tile that costs \$5 per square foot. Your budget for the tile is \$1700. How wide can you make the patio without going over your budget?

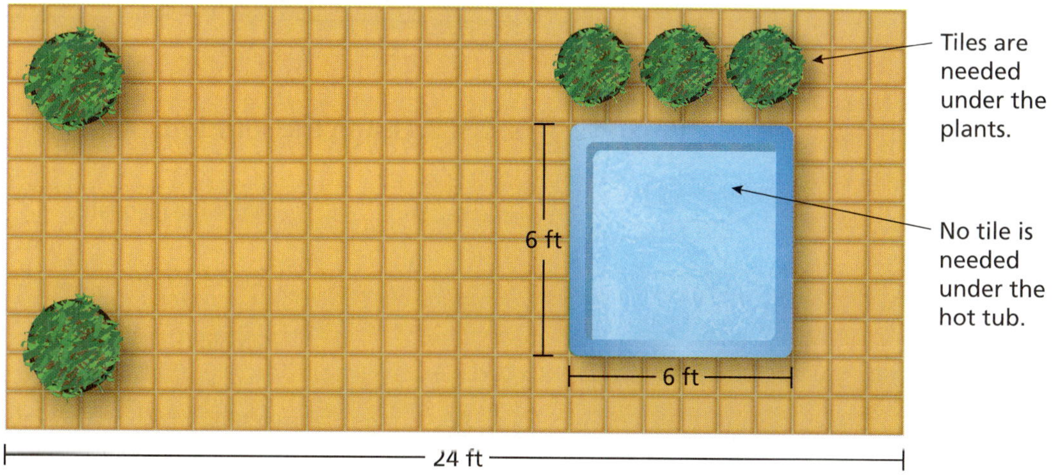

What Is Your Answer?

4. **IN YOUR OWN WORDS** How can you use an inequality to describe the area and perimeter of a composite figure? Give an example. Include a diagram with your example.

Practice Use what you learned about solving multi-step inequalities to complete Exercises 3 and 4 on page 338.

You can solve multi-step inequalities the same way you solve multi-step equations.

EXAMPLE 1 Solving Two-Step Inequalities

a. Solve $5x - 4 \geq 11$. Graph the solution.

	$5x - 4 \geq 11$	Write the inequality.
Step 1: Undo the subtraction. →	$+4 \quad +4$	Add 4 to each side.
	$5x \geq 15$	Simplify.
Step 2: Undo the multiplication. →	$\dfrac{5x}{5} \geq \dfrac{15}{5}$	Divide each side by 5.
	$x \geq 3$	Simplify.

∴ The solution is $x \geq 3$.

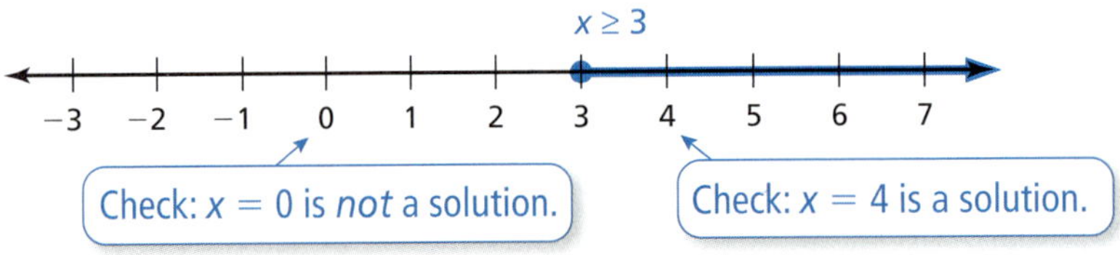

b. Solve $\dfrac{y}{-6} + 7 < 9$. Graph the solution.

$\dfrac{y}{-6} + 7 < 9$	Write the inequality.
$-7 \quad -7$	Subtract 7 from each side.
$\dfrac{y}{-6} < 2$	Simplify.
$-6 \cdot \dfrac{y}{-6} > -6 \cdot 2$	Multiply each side by −6. Reverse the inequality symbol.
$y > -12$	Simplify.

∴ The solution is $y > -12$.

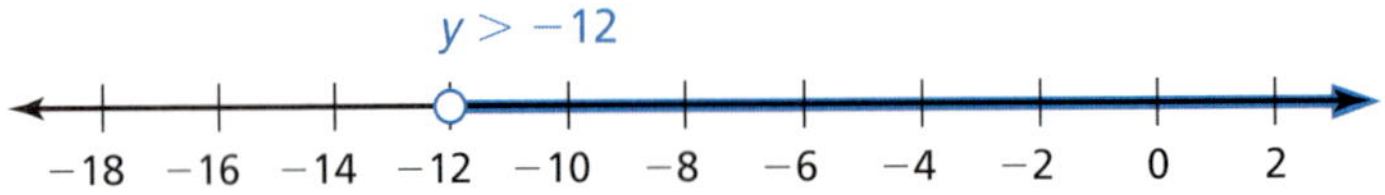

On Your Own

Now You're Ready
Exercises 5–10

Solve the inequality. Graph the solution.

1. $4b - 1 < 7$ **2.** $8 + 9c \geq -28$ **3.** $\dfrac{n}{-2} + 11 > 12$

Laurie's Notes

Introduction

Connect

- **Yesterday:** Students reviewed geometric formulas and gained an intuitive understanding of solving multi-step inequalities.
- **Today:** Students will solve and graph multi-step inequalities.

Motivate

? "How many of you have played the game *Trivial Pursuit*® or a variation of it?" Answers will vary.

- This popular trivia game was created in 1979 by two friends who had lost some pieces to the game Scrabble®. They decided to create a new game. The rest is history. *Trivial Pursuit*® has sold more than 88 million copies in 26 countries, and in 17 languages. There are versions of the game that focus on sports, pop culture, and regional geography.
- In Example 3, students use the context of a trivia game to solve a multi-step inequality.

Lesson Notes

Discuss

- You solve multi-step inequalities the same way you solve multi-step equations. You only need to remember to change the direction of the inequality symbol if you multiply or divide by a negative quantity.
- Recall that solving an equation undoes the evaluating in reverse order. The goal is to isolate the variable.

Example 1

? "What operations are being performed on the left side of the inequality? multiplication and subtraction

? "What is the first step in isolating the variable, meaning getting the x-term by itself?" Add 4 to each side of the inequality.

- Notice that subtracting 4 would have been the last step if evaluating the left side, so its inverse operation is the first step in solving the inequality.

? "To solve for x, what is the last step?" Divide both sides by 5.

- Because you are dividing by a positive quantity, the inequality symbol does not change. The solution is $x \geq 3$. Graph and check.
- Part (b) is solved in a similar fashion.

? "To solve $\frac{y}{-6} < 2$, what do you need to do?" Multiply both sides by -6 and change the direction of the inequality symbol.

- Graph and check. Remember to use an open circle because the variable cannot equal -12.

On Your Own

- These are all straightforward problems. Students should not be confused by them.
- Ask volunteers to share their solutions at the board.

Goal Today's lesson is solving multi-step inequalities.

Start Thinking! and Warm Up

Lesson 8.4 **Warm Up**
For use before Lesson 8.4

Lesson 8.4 **Start Thinking!**
For use before Lesson 8.4

What are your math quiz grades so far this grading period?

What must you earn on the next quiz in order to have at least a C quiz average? a B? an A?

Extra Example 1

Solve $17 \leq 3y - 4$. Graph the solution.
$y \geq 7$

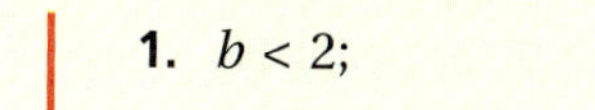

On Your Own

1. $b < 2$;

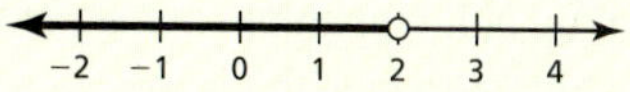

2. $c \geq -4$;

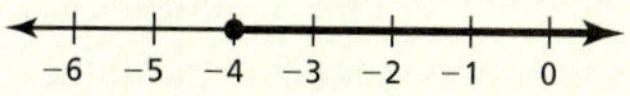

3. $n < -2$;

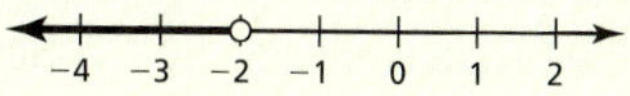

Laurie's Notes

Extra Example 2

Solve $12 > -2(y - 4)$. Graph the solution. $y > -2$

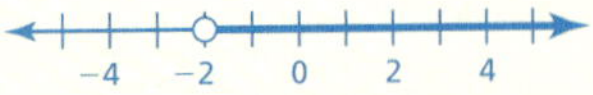

Extra Example 3

In Example 3, suppose you need a mean score of at least 85 to advance to the next round of the trivia game. What score do you need on the fifth game to advance? You need at least 73 points to advance to the next round.

On Your Own

4. $k < 8$;

5. $n > 2$;

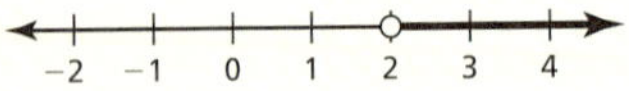

6. $y \geq -14$;

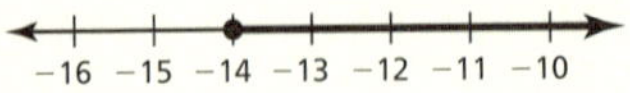

7. at least 88 points

English Language Learners

Vocabulary

Give English learners the opportunity to use precise language to solve an inequality. Write the inequality $-2x + 4 > 8$ on the board. Have one student come to the board. For each step of the solution, call on another student to give the instruction for solving. The instructions should be given in complete sentences. The instructions for the inequality are:

(1) Subtract 4 from each side.

(2) Simplify.

(3) Divide each side by −2. Reverse the inequality symbol.

(4) Simplify.

Example 2

- **Another Way:** The inequality has two factors on the left side: −7 and $(x + 3)$. Instead of distributing, divide both sides by −7. Dividing by a negative number changes the direction of the inequality symbol.

$-7(x + 3) \leq 28$	Write the inequality.
$\dfrac{-7(x + 3)}{-7} \geq \dfrac{28}{-7}$	Divide each side by −7. Reverse the inequality symbol.
$x + 3 \geq -4$	Simplify.
$\underline{-3} \quad \underline{-3}$	Subtract 3 from each side.
$x \geq -7$	Simplify.

- Discuss each method with students.

Example 3

- This is a classic problem. Students always want to know what they have to score on a test in order to have a (mean) average of ____. This is the same type of problem.

? "How do you compute a mean?" Sum the data and divide by the number of data values.

- Set up the problem to compute the mean. Because you want your score to be a minimum of 90, you need to set the mean greater than or equal to 90.
- You need at least a 98 to advance to the next level. Hopefully this score is attainable. Often with my students, the score they need to achieve isn't possible on one test!

On Your Own

- **Common Error:** If students solve Question 5 by distributing the −4, it is very possible they will write $-4n - 40$ instead of $-4n + 40$. For the factor $n - 10$, they need to remember to *add the opposite* so that the initial equation could be written as $-4[n + (-10)] < 32$. Then distribute the −4.
- Students may need guidance on Question 6. Distributing 0.5 results in $-3 \leq 4 + 0.5y$.

Closure

- **Writing:** How are these problems alike? How are they different?

  $3n - 4 = -25$ $\qquad$ $3n - 4 > -25$ $\qquad$ $3n - 4 \leq -25$

  Sample answer: They are alike because they each use the expressions $(3n - 4)$ and (-25). They are different because of the way the expressions are related: equal to, greater than, and less than or equal to.

Technology For the Teacher

Dynamic Classroom

The Dynamic Planning Tool

Editable Teacher's Resources at *BigIdeasMath.com*

EXAMPLE 2 Standardized Test Practice

Which graph represents the solution of $-7(x + 3) \le 28$?

Ⓐ
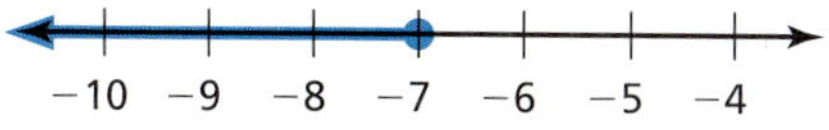

Ⓑ
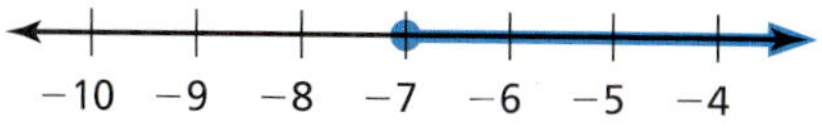

Ⓒ
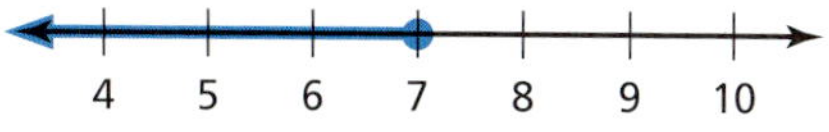

Ⓓ
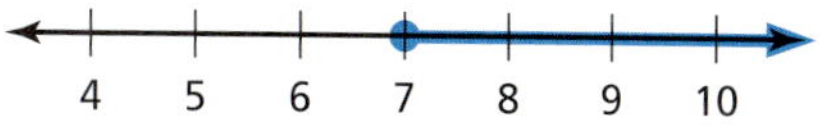

$-7(x + 3) \le 28$ Write the inequality.

$-7x - 21 \le 28$ Use Distributive Property.

$+21 \quad +21$ Add 21 to each side.

$-7x \le 49$ Simplify.

$\frac{-7x}{-7} \ge \frac{49}{-7}$ Divide each side by −7. Reverse the inequality symbol.

$x \ge -7$ Simplify.

∴ The correct answer is Ⓑ.

EXAMPLE 3 Real-Life Application

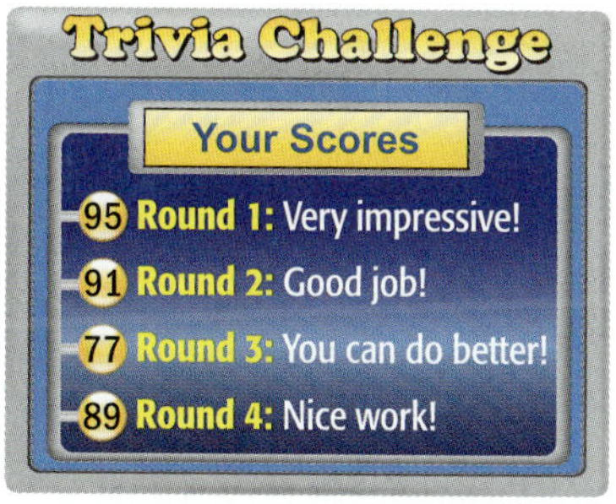

You need a mean score of at least 90 to advance to the next round of the trivia game. What score do you need on the fifth game to advance?

Use the definition of mean to write and solve an inequality. Let x be the score on the fifth game.

$$\frac{95 + 91 + 77 + 89 + x}{5} \ge 90$$

The phrase "at least" means greater than or equal to.

$\frac{352 + x}{5} \ge 90$ Simplify.

$5 \cdot \frac{352 + x}{5} \ge 5 \cdot 90$ Multiply each side by 5.

$352 + x \ge 450$ Simplify.

$-352 \quad -352$ Subtract 352 from each side.

$x \ge 98$ Simplify.

∴ You need at least 98 points to advance to the next round.

Remember

The mean in Example 3 is equal to the sum of the game scores divided by the number of games.

On Your Own

Now You're Ready
Exercises 12–17

Solve the inequality. Graph the solution.

4. $2(k - 5) < 6$ **5.** $-4(n - 10) < 32$ **6.** $-3 \le 0.5(8 + y)$

7. **WHAT IF?** In Example 3, you need a mean score of at least 88 to advance to the next round of the trivia game. What score do you need on the fifth game to advance?

8.4 Exercises

Check It Out
Help with Homework
BigIdeasMath.com

Vocabulary and Concept Check

1. **WRITING** Compare and contrast solving multi-step inequalities and solving multi-step equations.

2. **OPEN-ENDED** Describe how to solve the inequality $3(a + 5) < 9$.

Practice and Problem Solving

3. For what values of k will the perimeter of the octagon be less than or equal to 64 units?

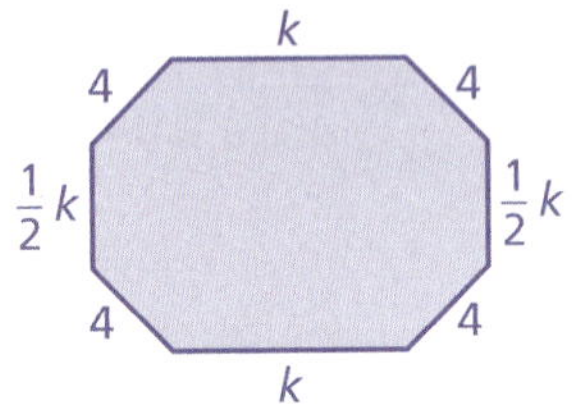

4. For what values of h will the surface area of the solid be greater than 46 square units?

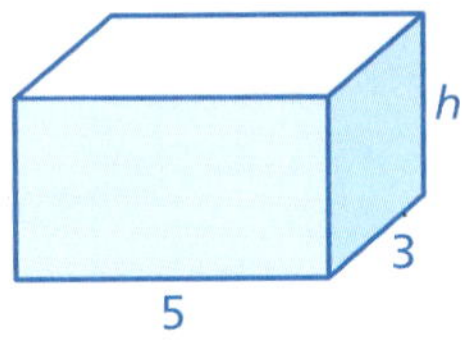

Solve the inequality. Graph the solution.

(1) 5. $7b + 4 \geq 11$

6. $2v - 4 < 8$

7. $1 - \frac{m}{3} \leq 6$

8. $\frac{4}{5} < 3w - \frac{11}{5}$

9. $1.8 < 0.5 - 1.3p$

10. $-2.4r + 9.6 \geq 4.8$

11. **ERROR ANALYSIS** Describe and correct the error in solving the inequality.

✗
$\frac{x}{4} + 6 \geq 3$
$x + 6 \geq 12$
$x \geq 6$

Solve the inequality. Graph the solution.

(2) 12. $6(g + 2) \leq 18$

13. $2(y - 5) \leq 16$

14. $-10 \geq \frac{5}{3}(h - 3)$

15. $-\frac{1}{3}(u + 2) > 5$

16. $2.7 > 0.9(n - 1.7)$

17. $10 > -2.5(z - 3.1)$

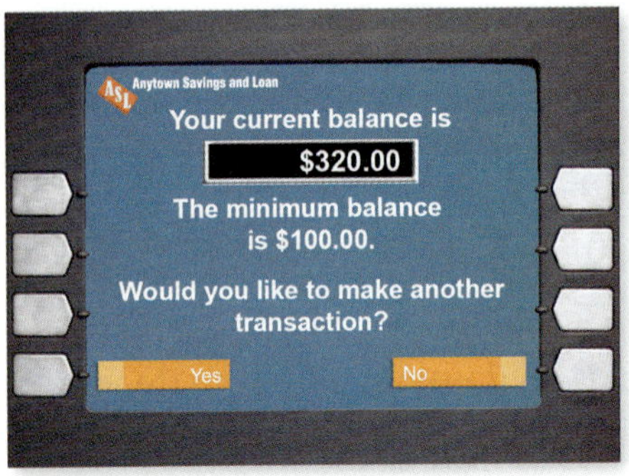

18. **ATM** Write and solve an inequality that represents the number of $20 bills you can withdraw from the account without going below the minimum balance.

Assignment Guide and Homework Check

Level	Day 1 Activity Assignment	Day 2 Lesson Assignment	Homework Check
Basic	3, 4, 26–29	1, 2, 5–21 odd, 18	7, 15, 18, 19
Average	3, 4, 26–29	1, 2, 7–21 odd, 18, 22	7, 15, 18, 19
Advanced	3, 4, 26–29	1, 2, 11, 14, 16, 19–25	14, 20, 22, 23

Common Errors

- **Exercises 5–10** Students may incorrectly multiply or divide before adding to or subtracting from both sides. Remind them that they should work backward through the order of operations, or that they should start away from the variable and move toward it.
- **Exercises 5–10, 12–17** Students may forget to reverse the inequality symbol when multiplying or dividing by a negative number. Encourage them to write the inequality symbol that they should have in the solution before solving.
- **Exercises 12–17** If students distribute before solving, they may forget to distribute the number to the second term. Remind them that they need to distribute to everything within the parentheses. Encourage students to draw arrows to represent the multiplication.

8.4 Record and Practice Journal

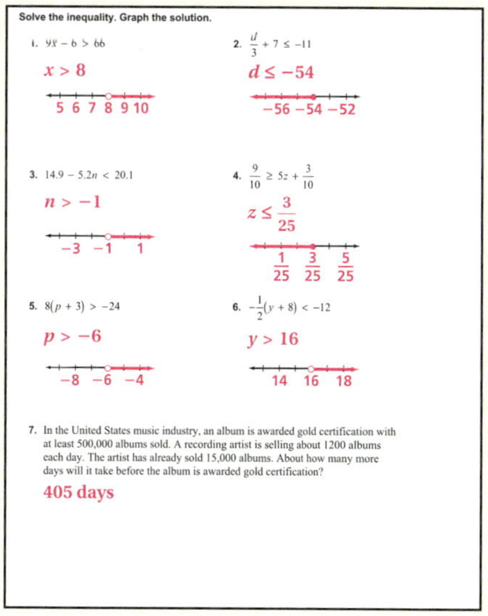
Solve the inequality. Graph the solution.

1. $9x - 6 > 66$
$x > 8$

2. $\frac{d}{3} + 7 \le -11$
$d \le -54$

3. $14.9 - 5.2n < 20.1$
$n > -1$

4. $\frac{9}{10} \ge 5z + \frac{3}{10}$
$z \le \frac{3}{25}$

5. $8(p + 3) > -24$
$p > -6$

6. $-\frac{1}{2}(y + 8) < -12$
$y > 16$

7. In the United States music industry, an album is awarded gold certification with at least 500,000 albums sold. A recording artist is selling about 1200 albums each day. The artist has already sold 15,000 albums. About how many more days will it take before the album is awarded gold certification?
405 days

Technology For the Teacher

Answer Presentation Tool

Quiz*Show*

Vocabulary and Concept Check

1. *Sample answer:* They use the same techniques, but when solving an inequality, you must be careful to reverse the inequality symbol when you multiply or divide by a negative number.
2. *Sample answer:* Divide both sides by 3 and then subtract 5 from both sides.

Practice and Problem Solving

3. $k > 0$ and $k \le 16$ units
4. $h > 1$ unit
5. $b \ge 1$;

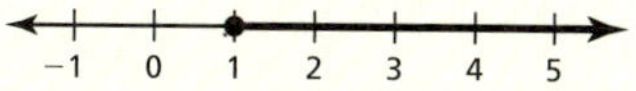

6. $v < 6$;

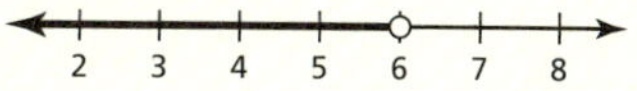

7. $m \ge -15$;

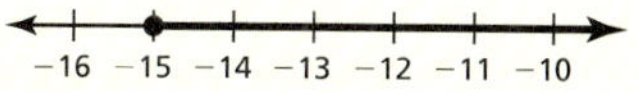

8. $w > 1$;

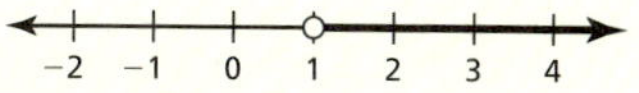

9. $p < -1$;

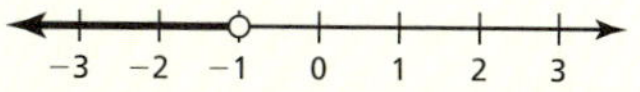

10. $r \le 2$;

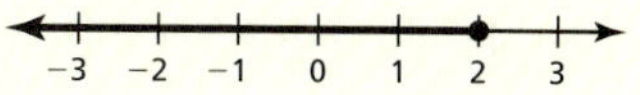

11. They did not perform the operations in proper order.

$$\frac{x}{4} + 6 \ge 3$$
$$\frac{x}{4} \ge -3$$
$$x \ge -12$$

12–18. See Additional Answers.

Practice and Problem Solving

19. $x \leq 6$;

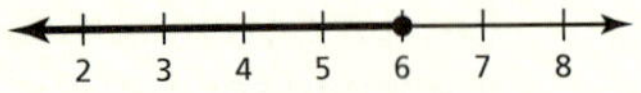

20. $b < 3$;

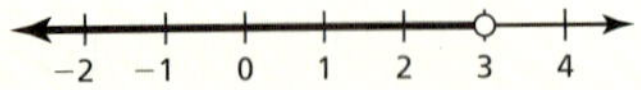

21. $\frac{3}{16}x + 2 \leq 11$;
$x > 0$ and $x \leq 48$ lines

22. $500 - 20x \geq 100$;
$x > \$0$ and $x \leq \$20$ per hour

23. See *Taking Math Deeper.*

24–25. See Additional Answers.

Fair Game Review

26. 100π mm^2

27. 625π in.^{2}

28. 1089π m^2

29. A

Mini-Assessment

Solve the inequality. Graph the solution.

1. $2x + 4 < 10$ $x < 3$

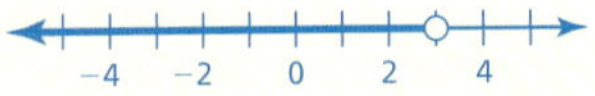

2. $3 \leq \frac{y}{-5} + 7$ $y \leq 20$

3. $-4.2 - 1.1b \leq 2.4$ $b \geq -6$

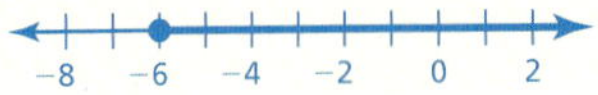

4. $\frac{2}{3}m + \frac{2}{3} \geq -\frac{1}{3}$ $m \geq -\frac{3}{2}$

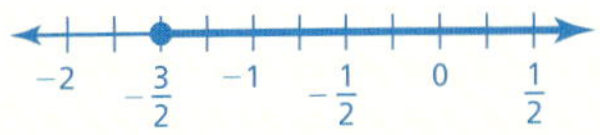

Taking Math Deeper

Exercise 23

Many inequality problems are easier to solve using equations. This is an example of such a problem.

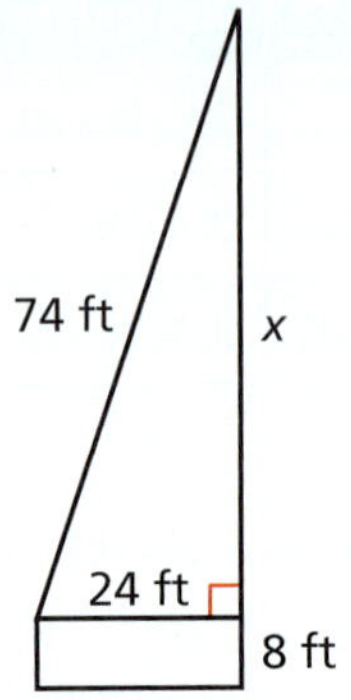

1. Draw a diagram and label the dimensions.

2. Use the Pythagorean Theorem to solve for x.

$$x^2 + 24^2 = 74^2$$
$$x^2 + 576 = 5476$$
$$x^2 = 4900$$
$$x = 70 \text{ ft}$$

3. Answer the question.

When the fire truck is exactly 24 feet from the building, the ladder can reach a total height of 78 feet. If the fire truck is farther away from the building, the ladder will not reach as high.

Let S represent the number of stories. An inequality for S is given by:

$$S \leq \frac{\text{number of feet ladder can reach}}{\text{10 feet per story}}$$

$$S \leq \frac{78}{10}$$

$$S \leq 7.8$$

Project

Many buildings are much taller than the ladder on a fire truck can reach. Use the internet to research what fire codes exist for these buildings. What special safety measures are required for skyscrapers?

Reteaching and Enrichment Strategies

If students need help. . .	If students got it. . .
Resources by Chapter • Practice A and Practice B • Puzzle Time Record and Practice Journal Practice Differentiating the Lesson Lesson Tutorials Skills Review Handbook	Resources by Chapter • Enrichment and Extension • School-to-Work Start the next section

Solve the inequality. Graph the solution.

19. $5x - 2x + 7 \le 15 + 10$

20. $7b - 12b + 1.4 > 8.4 - 22$

21. TYPING One line of text on a page uses about $\frac{3}{16}$ of an inch. There are 1-inch margins at the top and bottom of a page. Write and solve an inequality to find the number of lines that can be typed on a page that is 11 inches long.

22. WOODWORKING A woodworker builds a cabinet in 20 hours. The cabinet is sold at a store for \$500. Write and solve an inequality that represents the hourly wage the store can pay the woodworker and still make a profit of at least \$100.

23. FIRE TRUCK The height of one story of a building is about 10 feet. The bottom of the ladder on the fire truck must be at least 24 feet away from the building. Write and solve an inequality to find the number of stories the ladder can reach.

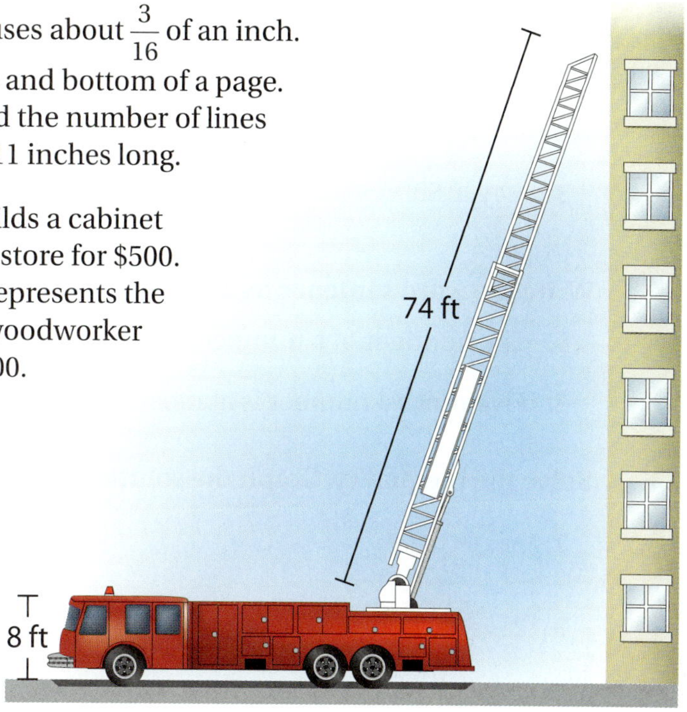

24. DRIVE-IN A drive-in movie theater charges \$3.50 per car. The drive-in has already admitted 100 cars. Write and solve an inequality to find the number of cars the drive-in needs to admit to make at least \$500.

25. Challenge For what values of r will the area of the shaded region be greater than or equal to $9(\pi - 2)$?

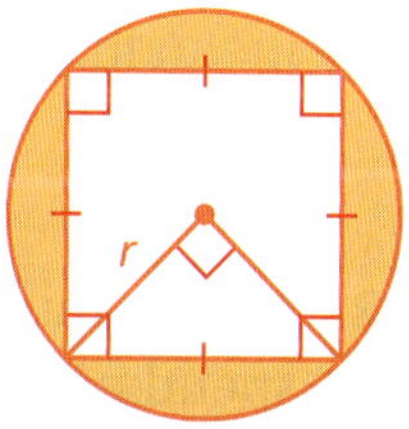

Fair Game Review What you learned in previous grades & lessons

Find the area of the circle. *(Skills Review Handbook)*

26.

27.

28.

29. MULTIPLE CHOICE What is the volume of the cube? *(Skills Review Handbook)*

Ⓐ 8 ft^3 Ⓑ 16 ft^3

Ⓒ 24 ft^3 Ⓓ 32 ft^3

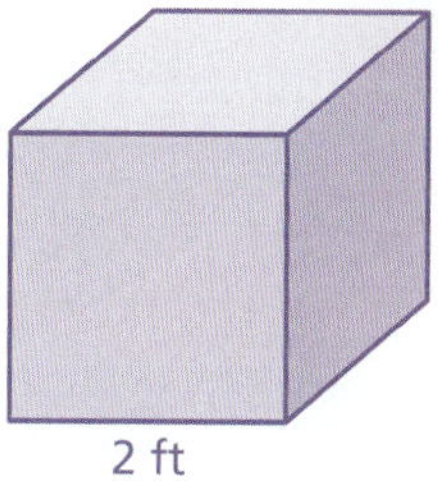

8.3–8.4 Quiz

Check It Out
Progress Check
BigIdeasMath.com

Solve the inequality. Graph the solution. *(Section 8.3)*

1. $x \div 4 > 12$

2. $\frac{n}{-6} \geq -2$

3. $-4y \geq 60$

4. $-2.3 \geq \frac{p}{5}$

Write the word sentence as an inequality. Then solve the inequality. *(Section 8.3)*

5. The quotient of a number and 6 is more than 9.

6. Five times a number is at most -10.

Solve the inequality. Graph the solution. *(Section 8.4)*

7. $2m + 1 \geq 7$

8. $\frac{n}{6} - 8 \leq 2$

9. $2 - \frac{j}{5} > 7$

10. $\frac{5}{4} > -3w - \frac{7}{4}$

11. FLOWERS A soccer team needs to raise \$200 for new uniforms. The team earns \$0.50 for each flower sold. Write and solve an inequality to find the number of flowers it must sell to meet or exceed its fundraising goal. *(Section 8.3)*

12. PARTY You buy lunch for guests at a party. You can spend no more than \$100. You will spend \$20 on beverages and \$10 per guest on sandwiches. Write and solve an inequality to find the number of guests you can invite to the party. *(Section 8.4)*

13. BOOKS You have a gift card worth \$50. You want to buy several paperback books that cost \$6 each. Write and solve an inequality to find the number of books you can buy and still have at least \$20 on the gift card. *(Section 8.4)*

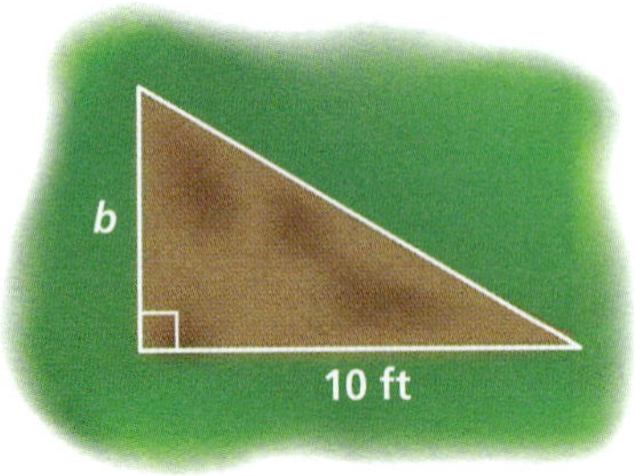

14. GARDEN The area of the triangular garden must be less than 35 square feet. Write and solve an inequality that represents the value of b. *(Section 8.3)*

Alternative Assessment Options

Math Chat | **Student Reflective Focus Question**
Structured Interview | Writing Prompt

Student Reflective Focus Question

Ask students to summarize the similarities and differences between multiplying or dividing each side of an inequality by the same positive or negative number. Be sure that they include examples. Select students at random to present to the class.

Study Help Sample Answers

Remind students to complete Graphic Organizers for the rest of the chapter.

6.

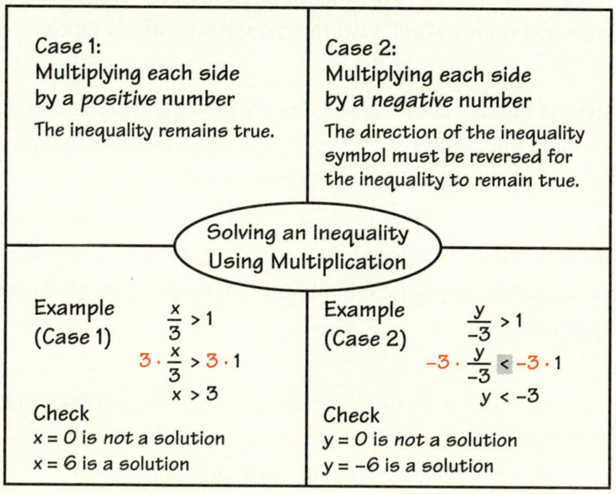

7. Available at *BigIdeasMath.com*.

Answers

1. $x > 48$;

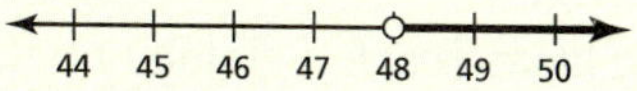

2. $n \le 12$;

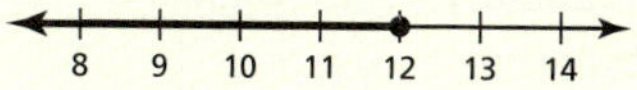

3. $y \le -15$;

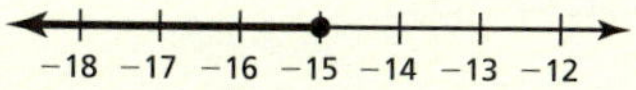

4. $p \le -11.5$;

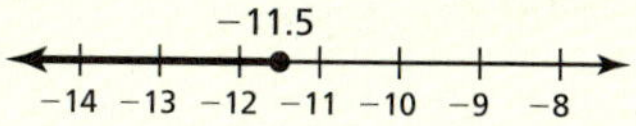

5. $\frac{x}{6} > 9$; $x > 54$

6. $5x \le -10$; $x \le -2$

7. $m \ge 3$;

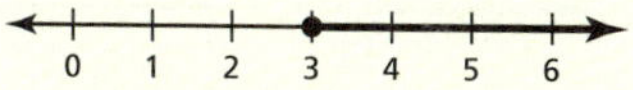

8. $n \le 60$;

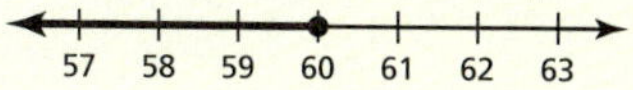

9–12. See Additional Answers.

13. $6x + 20 \le 50$; $x \le 5$ books

14. $5b < 35$; $b < 7$ ft

Technology For the Teacher

Answer Presentation Tool

Assessment Book

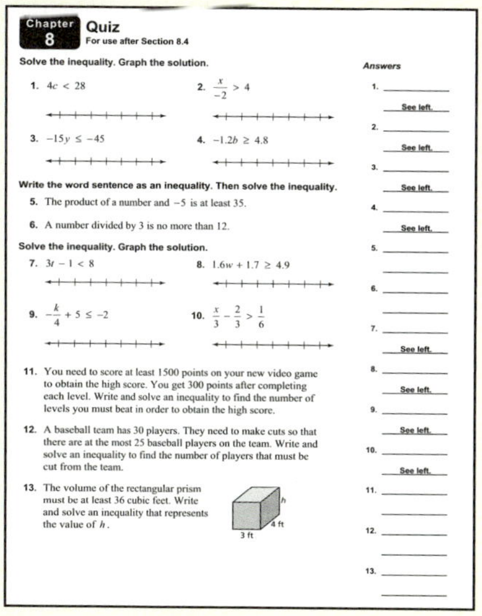

Chapter 8 Quiz
For use after Section 8.4

Solve the inequality. Graph the solution.

1. $4c < 28$
2. $\frac{x}{-2} > 4$
3. $-15y \le -45$
4. $-1.2b \ge 4.8$

Write the word sentence as an inequality. Then solve the inequality.

5. The product of a number and -5 is at least 35.
6. A number divided by 3 is no more than 12.

Solve the inequality. Graph the solution.

7. $3t - 1 < 8$
8. $1.6w + 1.7 \ge 4.9$
9. $-\frac{k}{4} + 5 \le -2$
10. $\frac{x}{3} - \frac{2}{3} > \frac{1}{6}$

11. You need to score at least 1500 points on your new video game to obtain the high score. You get 300 points after completing each level. Write and solve an inequality to find the number of levels you must beat in order to obtain the high score.

12. A baseball team has 30 players. They need to make cuts so that there are at the most 25 baseball players on the team. Write and solve an inequality to find the number of players that must be cut from the team.

13. The volume of the rectangular prism must be at least 36 cubic feet. Write and solve an inequality that represents the value of h.

Answers

1. See left.
2. See left.
3. See left.
4. See left.
5.
6.
7. See left.
8. See left.
9. See left.
10. See left.
11.
12.
13.

Reteaching and Enrichment Strategies

If students need help. . .	If students got it. . .
Resources by Chapter • Study Help • Practice A and Practice B • Puzzle Time Lesson Tutorials *BigIdeasMath.com* Practice Quiz Practice from the Test Generator	Resources by Chapter • Enrichment and Extension • School-to-Work Game Closet at *BigIdeasMath.com* Start the Chapter Review

For the Teacher
Additional Review Options

- **Quiz***Show*
- Big Ideas Test Generator
- Game Closet at *BigIdeasMath.com*
- Vocabulary Puzzle Builder
- Resources by Chapter
  - Puzzle Time
  - Study Help

Answers

1. $v < -2$
2. $x - \frac{1}{4} \le -\frac{3}{4}$
3. no
4. yes
5. 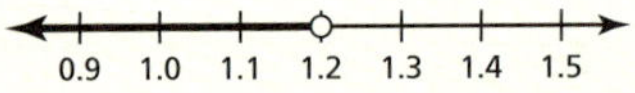

6. 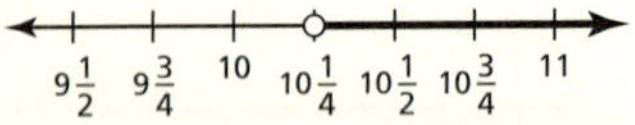

Review of Common Errors

Exercises 1 and 2

- Students may struggle knowing which inequality symbol to use. Encourage them to put the word sentence into a real-life context and to use the table in the lesson that explains what symbol matches each phrase.

Exercises 3 and 4

- Students may try to solve for the variable instead of substituting the given value into the inequality and determining if that value is a solution of the inequality.

Exercises 5 and 6

- Students may use a closed circle instead of an open circle and vice versa. They may shade the wrong side of the number line. Review how to graph inequalities. Encourage them to test a value on each side of the circle.

Exercises 7–9

- Students may reverse the direction of the inequality symbol when adding or subtracting. Remind them that the inequality symbol does not change direction when adding to or subtracting from both sides.

Exercises 7–12

- Students may perform the same operation instead of the inverse operation when solving the inequality symbol. Remind them that solving inequalities is similar to solving equations.

Exercises 10–18

- When there is a negative in the inequality, students may reverse the direction of the inequality symbol. Remind them that they only reverse the direction when they are multiplying or dividing by a negative number.
- Students may forget to reverse the inequality symbol when multiplying or dividing by a negative number. Encourage them to write the inequality symbol that they should have in the solution before solving.

Exercises 13–15

- Students may incorrectly multiply or divide before adding to or subtracting from both sides. Remind them that they should work backwards through the order of operations.

Exercises 16–18

- If students distribute before solving, they may forget to distribute the coefficient to the second term. Remind them that they need to distribute to everything within the parentheses. Encourage students to draw arrows to represent the multiplication.

Check It Out
Vocabulary Help
BigIdeasMath.com

Review Key Vocabulary

inequality, *p. 314*
solution of an inequality, *p. 314*
solution set, *p. 314*
graph of an inequality, *p. 315*

Review Examples and Exercises

8.1 Writing and Graphing Inequalities *(pp. 312–317)*

a. Four plus a number w is at least $-\frac{1}{2}$. Write this sentence as an inequality.

Four plus a number w | is at least | $-\frac{1}{2}$.

$4 + w$ | $\geq$ | $-\frac{1}{2}$

An inequality is $4 + w \geq -\frac{1}{2}$.

b. Graph $m > 4$.

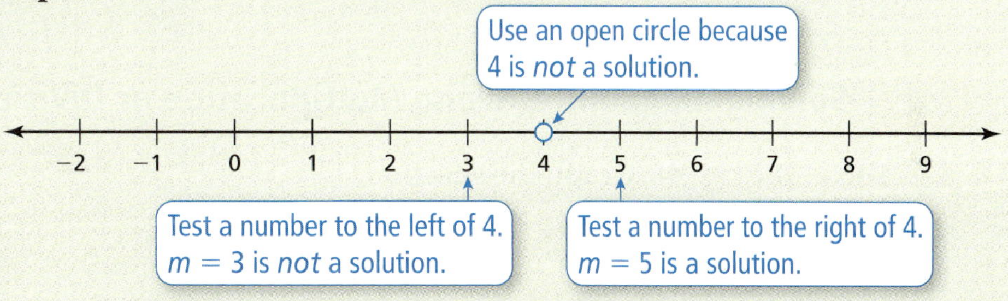

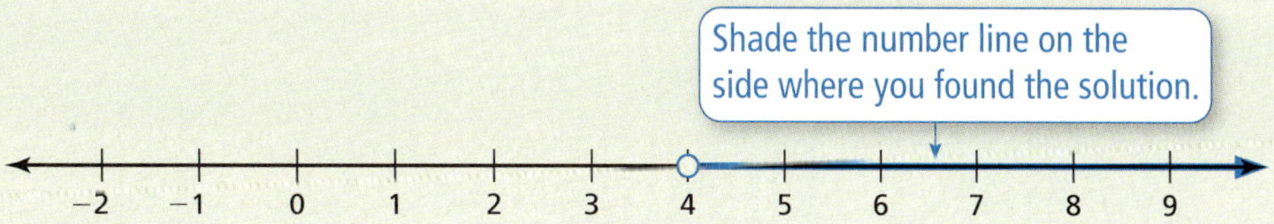

Exercises

Write the word sentence as an inequality.

1. A number v is less than -2.
2. A number x minus $\frac{1}{4}$ is no more than $-\frac{3}{4}$.

Tell whether the given value is a solution of the inequality.

3. $10 - q < 3;\ q = 6$
4. $12 \div m \geq -4;\ m = -3$

Graph the inequality on a number line.

5. $p < 1.2$
6. $n > 10\frac{1}{4}$

8.2 Solving Inequalities Using Addition or Subtraction (pp. 318–323)

Solve $-4 < n - 3$. Graph the solution.

	$-4 < n - 3$	Write the inequality.
Undo the subtraction. →	$\underline{+3} \quad \underline{+3}$	Add 3 to each side.
	$-1 < n$	Simplify.

The solution is $n > -1$.

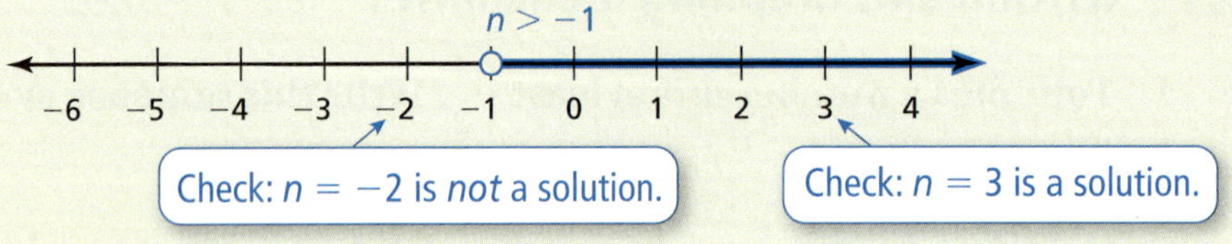

Exercises

Solve the inequality. Graph the solution.

7. $b + 13 < 18$ **8.** $x - 3 \leq 10$ **9.** $y + 1 \geq -2$

8.3 Solving Inequalities Using Multiplication or Division (pp. 326–333)

Solve $-8a \geq -48$. Graph the solution.

	$-8a \geq -48$	Write the inequality.
Undo the multiplication. →	$\frac{-8a}{-8} \leq \frac{-48}{-8}$	Divide each side by -8. Reverse the inequality symbol.
	$a \leq 6$	Simplify.

The solution is $a \leq 6$.

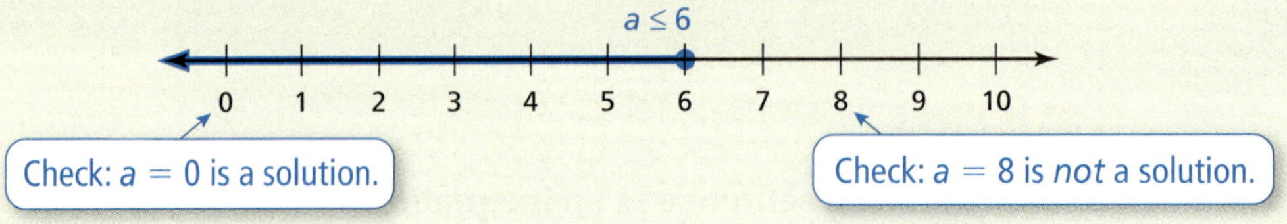

Exercises

Solve the inequality. Graph the solution.

10. $\frac{x}{2} \geq 4$ **11.** $4z < -44$ **12.** $-2q \geq -18$

Review Game

Musical Toss

Big Ideas Game Closet

Materials

- soft object that can be tossed around
- a device to play music
- old homework, quiz, and test questions

Directions

Divide the class into pairs (groups of two).

Designate one pair of students to play the music and write the problems on the board. Pairs of students should be switched periodically.

The remaining members of the class will stand in a circle with each pair clearly identifiable.

When the music starts, the soft object is tossed to a pair of students and the problem is written on the board. That pair has to solve the problem and toss the object to another pair before the music stops.

Who wins?

The group holding the object when the music stops is eliminated. This will continue until there is one group remaining, the winner.

For the Student Additional Practice

- Lesson Tutorials
- Study Help (textbook)
- Student Website
  - Multi-Language Glossary
  - Practice Assessments

Answers

7. $b < 5$;

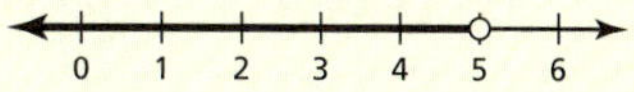

8. $x \leq 13$;

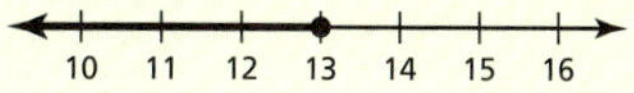

9. $y \geq -3$;

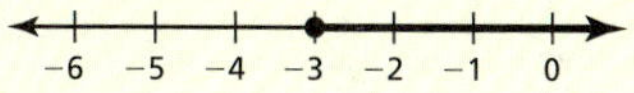

10. $x \geq 8$;

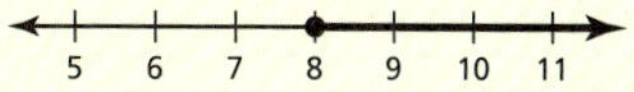

11. $z < -11$;

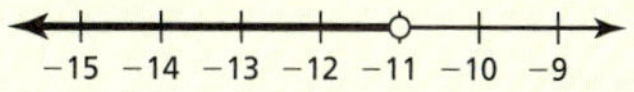

12. $q \leq 9$;

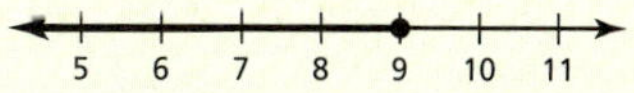

13. $x < 2$;

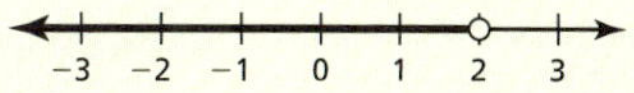

14. $z \geq -16$;

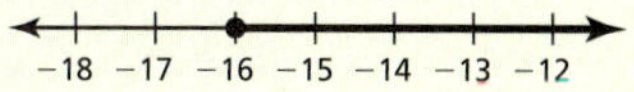

15. $w < -4$;

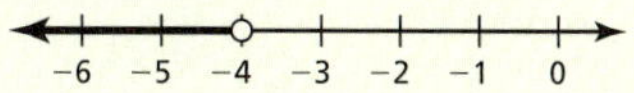

16–18. See Additional Answers.

My Thoughts on the Chapter

Teacher Tip

Not allowed to write in your teaching edition? Use sticky notes to record your thoughts.

What worked. . .

What did not work. . .

What I would do differently. . .

8.4 Solving Multi-Step Inequalities (pp. 334–339)

a. Solve $2x - 3 \le -9$. Graph the solution.

	$2x - 3 \le -9$	Write the inequality.
Step 1: Undo the subtraction. →	$+3 \quad +3$	Add 3 to each side.
	$2x \le -6$	Simplify.
Step 2: Undo the multiplication. →	$\frac{2x}{2} \le \frac{-6}{2}$	Divide each side by 2.
	$x \le -3$	Simplify.

The solution is $x \le -3$.

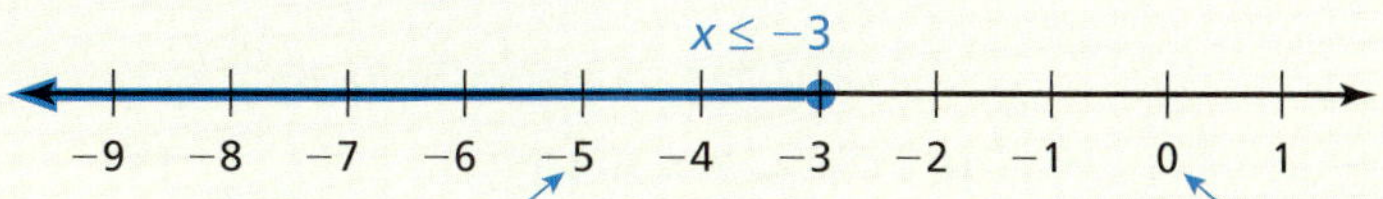

Check: $x = -5$ is a solution.

Check: $x = 0$ is *not* a solution.

b. Solve $\frac{t}{-3} + 4 > 7$. Graph the solution.

	$\frac{t}{-3} + 4 > 7$	Write the inequality.
Step 1: Undo the addition. →	$-4 \quad -4$	Subtract 4 from each side.
	$\frac{t}{-3} > 3$	Simplify.
Step 2: Undo the division. →	$-3 \cdot \frac{t}{-3} < -3 \cdot 3$	Multiply each side by -3. Reverse the inequality symbol.
	$t < -9$	Simplify.

The solution is $t < -9$.

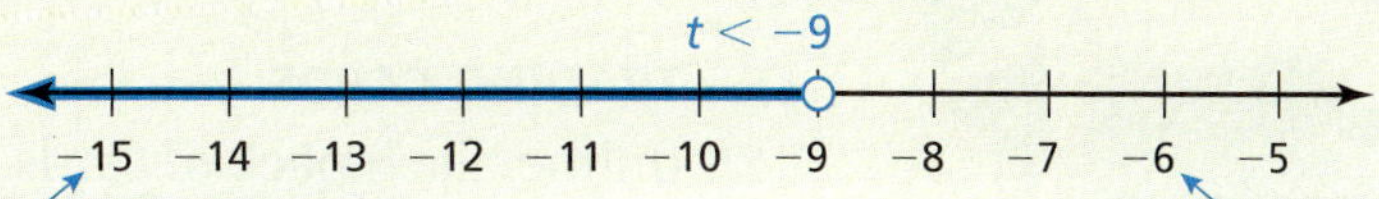

Check: $t = -15$ is a solution.

Check: $t = -6$ is *not* a solution.

Exercises

Solve the inequality. Graph the solution.

13. $4x + 3 < 11$

14. $\frac{z}{-4} - 3 \le 1$

15. $-3w - 4 > 8$

16. $8(q + 2) < 40$

17. $-\frac{1}{2}(p + 4) \le 18$

18. $1.5(k + 3.2) \ge 6.9$

8 Chapter Test

Check It Out
Test Practice
BigIdeasMath.com

Write the word sentence as an inequality.

1. A number j plus 20.5 is greater than or equal to 50.
2. A number r multiplied by $\frac{1}{7}$ is less than -14.

Tell whether the given value is a solution of the inequality.

3. $v - 2 \le 7;\ v = 9$
4. $\frac{3}{10}p < 0;\ p = 10$
5. $-3n \ge 6;\ n = -3$

Solve the inequality. Graph the solution.

6. $n - 3 > -3$
7. $x - \frac{7}{8} \le \frac{9}{8}$
8. $-6b \ge -30$
9. $\frac{y}{-4} \ge 13$
10. $3v - 7 \ge -13.3$
11. $-5(t + 11) < -60$

12. **VOTING** U.S. citizens must be at least 18 years of age on Election Day to vote. Write an inequality that represents this situation.

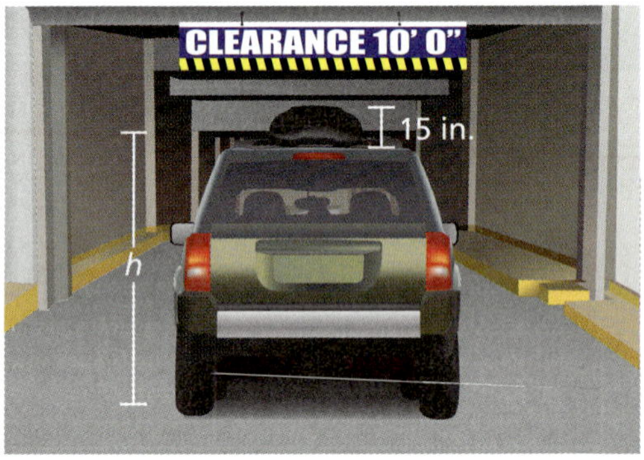

13. **GARAGE** The vertical clearance for a hotel parking garage is 10 feet. Write and solve an inequality that represents the height (in feet) of the vehicle.

14. **LUNCH BILL** A lunch bill, including tax, is divided equally among you and five friends. Everyone pays less than \$8.75. Write and solve an inequality that describes the total amount of the bill.

15. **TRADING CARDS** You have \$25 to buy trading cards online. Each pack of cards costs \$4.50. Shipping costs \$2.95. Write and solve an inequality to find the number of packs of trading cards you can buy.

16. **SCIENCE QUIZZES** The table shows your scores on four science quizzes. What score do you need on the fifth quiz to have a mean score of at least 80?

Test	1	2	3	4	5
Score (%)	76	87	73	72	?

Test Item References

Chapter Test Questions	Section to Review
1–5, 12	8.1
6, 7, 13	8.2
8, 9, 14	8.3
10, 11, 15, 16	8.4

Test-Taking Strategies

Remind students to quickly look over the entire test before they start so that they can budget their time. When writing word phrases as inequalities, students can get confused by the subtle differences in wording, such as "is no more than" and "is no less than." Encourage students to think very carefully about which inequality symbol is implied by the wording. Teach the students to use the Stop and Think strategy before answering. **Stop** and carefully read the question, and **Think** about what the answer should look like.

Common Assessment Errors

- **Exercises 1 and 2** Students may not use the correct inequality symbol. Remind them to put the word sentence into a real-life context.
- **Exercises 6–11** Remind them that they only reverse the direction of the inequality symbol when they are multiplying or dividing by a negative number.
- **Exercises 6–11** Students may perform the same operation instead of the inverse operation when solving the inequality. Remind them that solving inequalities is similar to solving equations.
- **Exercises 6–11** Students may use the wrong circle and/or shade the wrong side of the number line. Remind them to test a value of each side of the circle.

Answers

1. $j + 20.5 \geq 50$
2. $\frac{1}{7}r < -14$
3. yes
4. no
5. yes
6. $n > 0$;

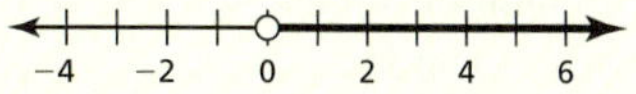

7. $x \leq 2$;

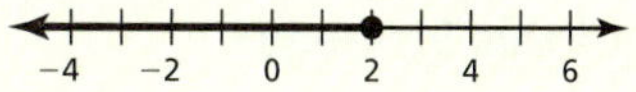

8. $b \leq 5$;

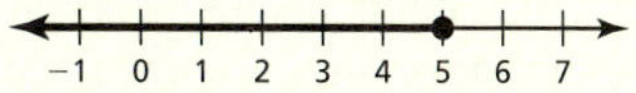

9. $y \leq -52$;

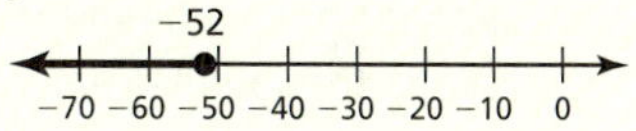

10–11. See Additional Answers.

12. $x \geq 18$
13. $h + 1.25 < 10$; $h < 8.75$ feet
14. $\frac{x}{6} < 8.75$; $x < \$52.50$
15. $4.5x + 2.95 \leq 25$; $x \leq 4.9$; at most 4 packs of trading cards
16. at least 92%

Assessment Book

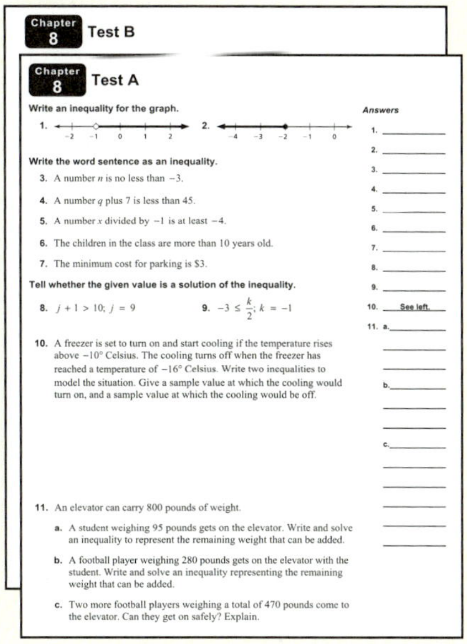

Chapter 8 Test B

Chapter 8 Test A

Write an inequality for the graph.

1. (number line: −2, −1, 0, 1, 2) 2. (number line: −4, −3, −2, −1, 0)

Write the word sentence as an inequality.

3. A number n is no less than −3.
4. A number q plus 7 is less than 45.
5. A number x divided by −1 is at least −4.
6. The children in the class are more than 10 years old.
7. The minimum cost for parking is $3.

Tell whether the given value is a solution of the inequality.

8. $j + 1 > 10$; $j = 9$ 9. $-3 \leq \frac{k}{2}$; $k = -1$

10. A freezer is set to turn on and start cooling if the temperature rises above −10° Celsius. The cooling turns off when the freezer has reached a temperature of −16° Celsius. Write two inequalities to model the situation. Give a sample value at which the cooling would turn on, and a sample value at which the cooling would be off.

11. An elevator can carry 800 pounds of weight.
 a. A student weighing 95 pounds gets on the elevator. Write and solve an inequality to represent the remaining weight that can be added.
 b. A football player weighing 280 pounds gets on the elevator with the student. Write and solve an inequality representing the remaining weight that can be added.
 c. Two more football players weighing a total of 470 pounds come to the elevator. Can they get on safely? Explain.

Answers

1. ____ 2. ____ 3. ____ 4. ____ 5. ____ 6. ____ 7. ____ 8. ____ 9. ____ 10. See left. 11. a. ____ b. ____ c. ____

Reteaching and Enrichment Strategies

If students need help. . .	If students got it. . .
Resources by Chapter • Practice A and Practice B • Puzzle Time Record and Practice Journal Practice Differentiating the Lesson Lesson Tutorials Practice from the Test Generator Skills Review Handbook	Resources by Chapter • Enrichment and Extension • School-to-Work Game Closet at *BigIdeasMath.com* Start Standardized Test Practice

Test-Taking Strategies
Available at *BigIdeasMath.com*

After Answering Easy Questions, Relax

Answer Easy Questions First

Estimate the Answer

Read All Choices before Answering

Read Question before Answering

Solve Directly or Eliminate Choices

Solve Problem before Looking at Choices

Use Intelligent Guessing

Work Backwards

About this Strategy
When taking a multiple choice test, be sure to read each question carefully and thoroughly. After skimming the test and answering the easy questions, stop for a few seconds, take a deep breath, and relax. Work through the remaining questions carefully, using your knowledge and test-taking strategies. Remember, you already completed many of the questions on the test!

Answers
1. D
2. G
3. B
4. H

Technology For the Teacher

Big Ideas Test Generator

Item Analysis

1. **A.** The student confuses the area and perimeter formulas (or the concepts), and writes the less than symbol instead of the greater than symbol.

   B. The student writes the less than symbol instead of the greater than symbol.

   C. The student confuses the area and perimeter formulas (or the concepts).

   D. Correct answer

2. **F.** The student does not realize that the underlying idea here is to show change over time, something of which a circle graph is not capable.

   G. Correct answer

   H. The student does not realize that the underlying idea here is to show change over time, something of which a box-and-whisker plot is not capable.

   I. The student does not realize that the underlying idea here is to show change over time, something of which a stem-and-leaf plot is not capable.

3. **A.** The student calculates correctly but fails to take a square root.

   B. Correct answer

   C. The student adds the differences between the two sets of coordinates together (instead of the squared differences).

   D. The student adds the differences between the two sets of coordinates together (instead of the squared differences), and then takes the square root.

4. **F.** The student confuses a system with no solution and a system with an infinite number of solutions.

   G. The student thinks the similarity of the equations implies an infinite number of solutions.

   H. Correct answer

   I. The student makes a mistake converting the first equation to slope-intercept form.

5. **A.** Correct answer

   B. The student adds 4 and 25.

   C. The student multiplies 4 and 25 (ignoring the square root symbol).

   D. The student multiplies 4 and 25 (keeping the square root symbol).

8 Standardized Test Practice

Test-Taking Strategy
After Answering Easy Questions, Relax

"After answering the easy questions, relax and try the harder ones. For this, x = $946. So, it's D."

1. The perimeter of the triangle shown below is greater than 50 centimeters. Which inequality represents this algebraically?

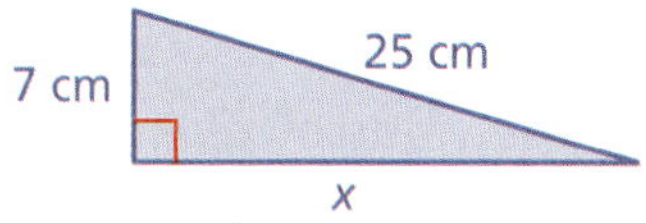

A. $\frac{1}{2}(7x) < 50$ **C.** $\frac{1}{2}(7x) > 50$

B. $x + 32 < 50$ **D.** $x + 32 > 50$

2. A store has recorded total dollar sales each month for the past three years. Which type of graph would best show how sales have increased over this time period?

F. circle graph **H.** box-and-whisker plot

G. line graph **I.** stem-and-leaf plot

3. What is the length d in the coordinate plane?

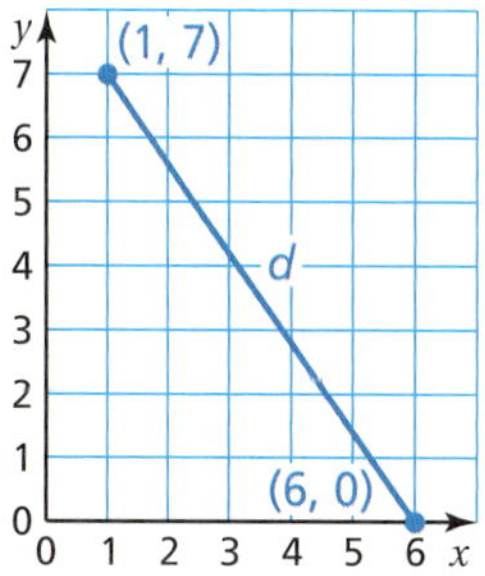

A. 74 **C.** 12

B. $\sqrt{74}$ **D.** $\sqrt{12}$

4. Which system of equations has infinitely many solutions?

F. $x + y = 1$
$x + y = 2$

H. $4x - 2y = 9$
$y = 2x - 4.5$

G. $y = x$
$y = -x$

I. $3x + 4y = 12$
$y = \frac{3}{4}x + 3$

5. Which is equivalent to $4\sqrt{25}$?

A. 20

B. $\sqrt{29}$

C. 100

D. $\sqrt{100}$

6. The triangles shown below are similar. What is the value of x?

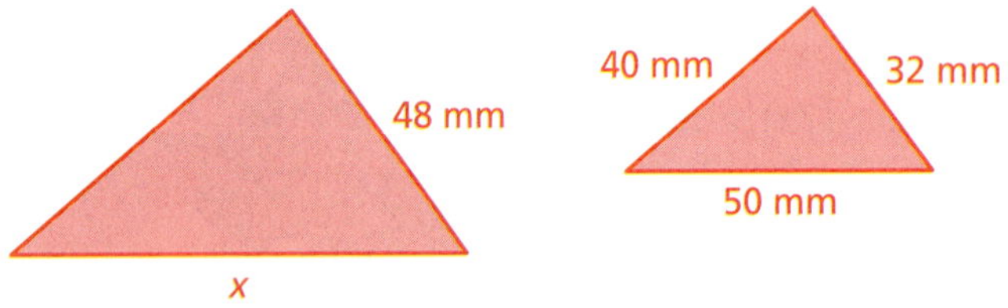

7. The table lists the mean, median, and mode salaries at a company. Suppose a new worker is hired at a salary of $70,000. Which statement is true?

Mean	Median	Mode
$62,000	$58,000	$54,000

F. The mean annual salary must increase.

G. The median annual salary must increase.

H. The median annual salary must remain the same.

I. The mode annual salary must increase.

8. Does squaring a number always make it greater? Is the inequality shown below true for all numbers?

Think
Solve
Explain

$$x^2 > x$$

Show your work and explain your reasoning.

Item Analysis (continued)

6. **Gridded Response:** Correct answer: 75 mm

   Common Error: The student adds 16 and 50 to get 66 as an answer, because the pair of corresponding sides given in the diagram differ by 16.

7. **F.** Correct answer

   G. The student fails to realize that several workers could earn the median salary.

   H. The student doesn't realize that the addition of one higher salary could alter the median.

   I. The student misunderstands mode.

8. **2 points** The student demonstrates a thorough understanding of how to read and interpret an inequality, and then finds a value of x for which the inequality is not true (for example, $x = 0.5$). A clear and complete explanation is provided.

   1 points The student demonstrates a partial understanding of how to read and interpret an inequality. He or she is able to see what the inequality means, but fails to go far enough to provide a value of x for which it is false.

   0 points The student provides no response, a completely incorrect or incomprehensible response, or a response that demonstrates insufficient understanding of inequalities.

9. **A.** The student makes an error subtracting or dividing integers.

   B. The student makes an error subtracting or dividing integers, and then fails to reverse the direction of the inequality symbol when dividing by a negative coefficient.

   C. Correct answer

   D. The student fails to reverse the direction of the inequality symbol when dividing by a negative coefficient.

Answers

5. A
6. 75 mm
7. F
8. No. For example, the inequality is not true when $x = 0.5$.

Answers

9. C

10. G

11. 45 in.

Answer for Extra Example

1. A. The student confuses one plot for the other.

   B. The student reads the first quartile as the median.

   C. The student applies a property of one grade to both grades.

   D. Correct answer

Item Analysis (continued)

10. F. The student mistakes a discrete variable as continuous.

    G. Correct answer

    H. The student identifies range as domain and calls it continuous when it is discrete.

    I. The student identifies range as domain.

11. **Gridded Response:** Correct answer: 45 in.

    Common Error: The student adds 27 and 36 to get 63 inches as an answer.

Extra Example for Standardized Test Practice

1. The box-and-whisker plots summarize how many minutes it takes 8th and 9th graders in Fulton Township to get to school each morning. What can you conclude from the box-and-whisker plots?

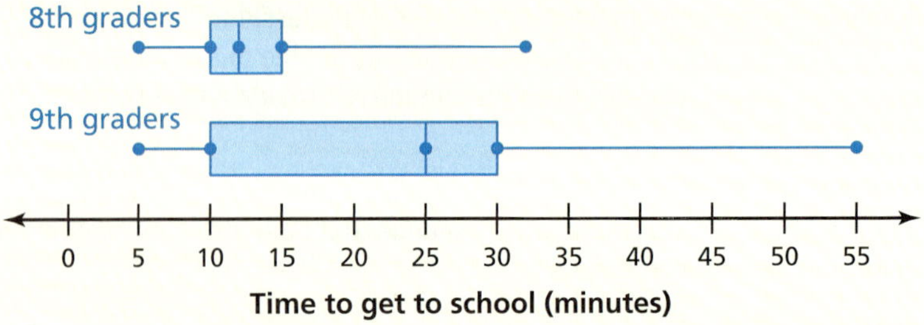

   A. Eighth graders take longer than ninth graders, on average, to get to school.

   B. The median time is the same for both grades.

   C. About $\frac{1}{4}$ of the students in both grades take more than 30 minutes to get to school.

   D. About $\frac{3}{4}$ of the students in both grades take more than 10 minutes to get to school.

9. Which graph represents the inequality below?

$$-2x + 3 < 1$$

A. 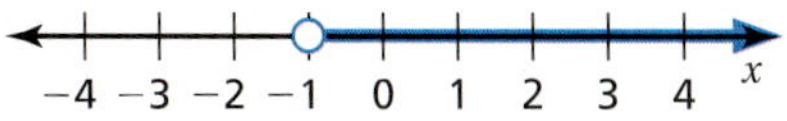

B. 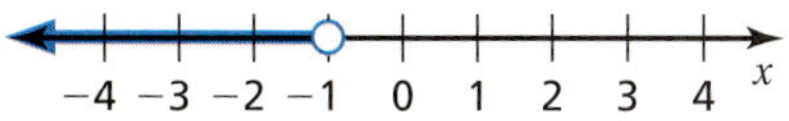

C. 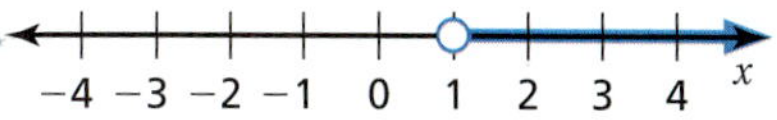

D. 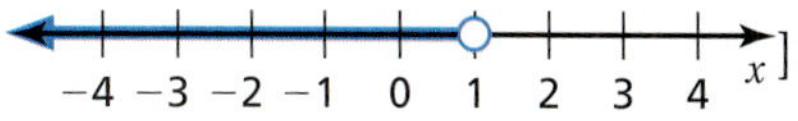

10. The function $y = 29.95x$ represents the total cost y of purchasing x day passes to a water park. Which statement is true?

F. The domain represents day passes and it is continuous.

G. The domain represents day passes and it is discrete.

H. The domain represents total cost and it is continuous.

I. The domain represents total cost and it is discrete.

11. A television screen measures 36 inches across and 27 inches high. What is the length, in inches, of the television screen's diagonal?

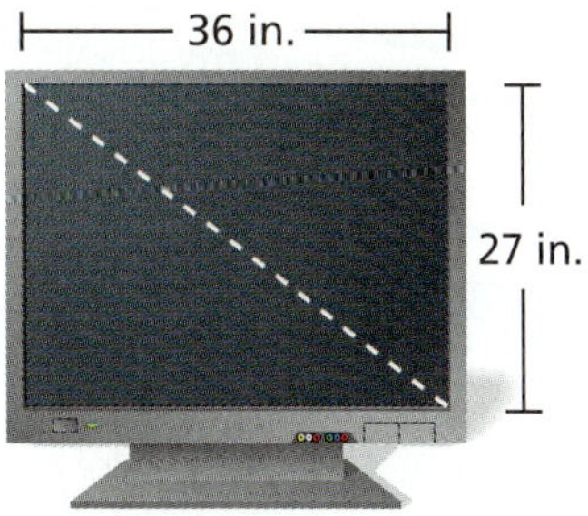

9 Exponents and Scientific Notation

9.1 Exponents

9.2 Product of Powers Property

9.3 Quotient of Powers Property

9.4 Zero and Negative Exponents

9.5 Reading Scientific Notation

9.6 Writing Scientific Notation

Connections to Previous Learning

- Use order of operations including exponents and parentheses.
- Explain and justify procedures for multiplication and division.
- Multiply and divide decimals efficiently, including real-world problems.

- Perform exponential operations with rational bases and whole number exponents.

- Use exponents and scientific notation to write large and small numbers and vice versa, and to solve problems.
- Simplify real number expressions using the laws of exponents.
- Perform operations on real numbers (including integer exponents, scientific notation, rational numbers, irrational numbers) using multi-step and real world problems.

Math in History

Before the use of calculators, it was common for textbooks to have a table of square roots. These were calculated by various methods.

★ Isaac Newton (1643–1727) was a famous English mathematician who invented calculus. Here is his method for calculating a square root.

1. You want to find the square root of x.
2. Let G be your best guess.
3. To find the next G, call it G'. $G' = \dfrac{x + G^2}{2G}$
4. This G' becomes your next G. Go back to Step 2 and continue until you have reached your desired accuracy.

★ During the 17th century, many people invented calculating machines. Here is one by Frenchman Blaise Pascal (1623–1662).

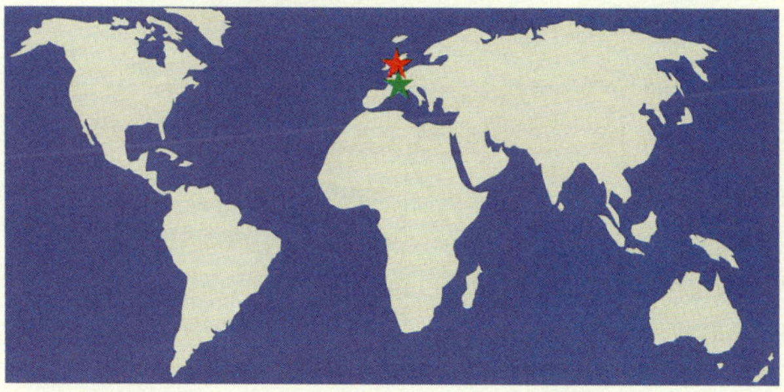

Pacing Guide for Chapter 9

Chapter Opener	1 Day
Section 1	
Activity	1 Day
Lesson	1 Day
Section 2	
Activity	1 Day
Lesson	1 Day
Section 3	
Activity	1 Day
Lesson	1 Day
Study Help / Quiz	1 Day
Section 4	
Activity	1 Day
Lesson	1 Day
Section 5	
Activity	1 Day
Lesson	1 Day
Section 6	
Activity	1 Day
Lesson	1 Day
Quiz / Chapter Review	1 Day
Chapter Test	1 Day
Standardized Test Practice	1 Day
Total Chapter 9	17 Days
Year-to-Date	150 Days

Check Your Resources

- Record and Practice Journal
- Resources by Chapter
- Skills Review Handbook
- Assessment Book
- Worked-Out Solutions

Technology For the Teacher

Dynamic Classroom

The Dynamic Planning Tool
Editable Teacher's Resources at
BigIdeasMath.com

Additional Topics for Review

- The division algorithm
- Place value
- Adding, subtracting, multiplying, and dividing rational numbers
- Perfect squares
- Exponents

Try It Yourself

1. 3.737	**2.** 2.13
3. 1.099	**4.** 1.109
5. 5.022	**6.** 4.203
7. 1.433	**8.** 1.103
9. 0.35	**10.** 0.84
11. 30.229	**12.** 0.1788
13. 60	**14.** 13.9
15. 24	**16.** 1800

Record and Practice Journal

1. 12.21	**2.** 20.658
3. 5.565	**4.** 8.42
5. 0.85	**6.** 3.814
7. a. \$12.38	**b.** \$7.20
8. 6.45 yards	
9. 2.352	**10.** 0.1014
11. 6.0048	**12.** 9
13. 1.5	**14.** 2700
15. \$6.93	**16.** \$0.27

Math Background Notes

Vocabulary Review

- Divisor
- Dividend

Adding and Subtracting Decimals

- Students should be able to add and subtract decimals.
- Remind students that the important rule for adding and subtracting decimals is alignment. Only numbers sharing the same place value position can be combined. Line up the decimal points and then add or subtract each vertical column.
- **Teaching Tip:** Some students have a difficult time adding and subtracting decimals. Encourage these students to do their problems using rainbow writing. Have students write the problem so that each place value is a different color. Only numbers that are the same color can combine.
- Encourage students to "fill in" any empty spaces with zeros. This will make subtraction problems requiring borrowing and renaming easier to complete.

Multiplying and Dividing Decimals

- Students should be able to multiply and divide decimals.
- Encourage students to ignore the decimal point and use the normal multiplication algorithm when multiplying decimals. After the multiplication is complete, count the total number of digits in both factors that appear to the right of the decimal point. Then put that many digits to the right of the decimal point in the answer.
- **Common Error:** In the horizontal division problem, some students have difficulty determining which number is the dividend and which is the divisor. Encourage students to read the problem aloud as they rewrite it. Each time a student says the words "divided by," he or she should be trapping a number inside the division box.
- **Common Error:** Some students may try to clear the decimal point from the dividend instead of the divisor. Remind students that having a decimal inside the division box as part of the dividend is fine. Also remind students that the divisor and dividend must both be multiplied by the same power of ten so as not to change the value of the problem.

Reteaching and Enrichment Strategies

If students need help. . .	If students got it. . .
Record and Practice Journal • Fair Game Review Skills Review Handbook Lesson Tutorials	Game Closet at *BigIdeasMath.com* Start the next section

What You Learned Before

"It's called the Power of Negative One, Descartes!"

Adding and Subtracting Decimals

Example 1 Find 2.65 + 5.012.

$$\begin{array}{r} 2.650 \\ +\ 5.012 \\ \hline 7.662 \end{array}$$

Example 2 Find 3.7 − 0.48.

$$\begin{array}{r} \overset{6\ 10}{3.\not{7}\ \not{0}} \\ -\ 0.4\ 8 \\ \hline 3.2\ 2 \end{array}$$

Try It Yourself

Find the sum or difference.

1. 2.73 + 1.007 **2.** 3.4 − 1.27 **3.** 0.35 + 0.749 **4.** 1.019 + 0.09

5. 6.03 − 1.008 **6.** 4.21 − 0.007 **7.** 0.228 + 1.205 **8.** 3.003 − 1.9

Multiplying and Dividing Decimals

Example 3 Find 2.1 • 0.35.

$$\begin{array}{rl} 2.1 & \leftarrow \text{1 decimal place} \\ \times\ 0.3\,5 & \leftarrow \text{+ 2 decimal places} \\ \hline 1\,0\,5 & \\ 6\,3\ \ & \\ \hline 0.7\,3\,5 & \leftarrow \text{3 decimal places} \end{array}$$

Example 4 Find 1.08 ÷ 0.9.

$0.9\overline{)1.08}$ Multiply each number by 10.

$$\begin{array}{r} 1.2 \\ 9\overline{)10.8} \\ -\ 9\ \ \\ \hline 1\,8 \\ -\ 1\,8 \\ \hline 0 \end{array}$$

Place the decimal point above the decimal point in the dividend 10.8.

Try It Yourself

Find the product or quotient.

9. 1.75 • 0.2 **10.** 1.4 • 0.6 **11.** $\begin{array}{r} 7.03 \\ \times\ 4.3 \\ \hline \end{array}$ **12.** $\begin{array}{r} 0.894 \\ \times\ 0.2 \\ \hline \end{array}$

13. 5.40 ÷ 0.09 **14.** 4.17 ÷ 0.3 **15.** $0.15\overline{)3.6}$ **16.** $0.004\overline{)7.2}$

9.1 Exponents

Essential Question How can you use exponents to write numbers?

The expression 3^5 is called a **power**. The **base** is 3. The **exponent** is 5.

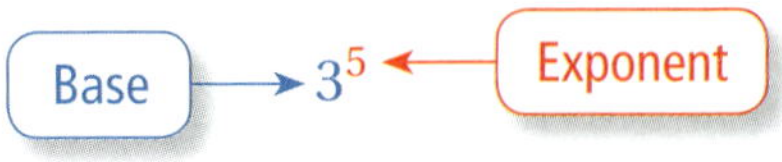

1 ACTIVITY: Using Exponent Notation

Work with a partner.

a. Copy and complete the table.

Power	Repeated Multiplication Form	Value
$(-3)^1$	-3	-3
$(-3)^2$	$(-3) \cdot (-3)$	9
$(-3)^3$		
$(-3)^4$		
$(-3)^5$		
$(-3)^6$		
$(-3)^7$		

b. Describe what is meant by the expression $(-3)^n$. How can you find the value of $(-3)^n$?

2 ACTIVITY: Using Exponent Notation

Work with a partner.

a. The cube at the right has \$3 in each of its small cubes. Write a single power that represents the total amount of money in the large cube.

b. Evaluate the power to find the total amount of money in the large cube.

Laurie's Notes

Introduction

For the Teacher

- **Goal:** Students will explore writing numbers with exponents.

Motivate

? "How big is a cubic millimeter?" A grain of salt is a reasonable estimate.
- Use a metric ruler and your fingers to show what a millimeter is. A cubic millimeter means it is 1 mm × 1 mm × 1 mm.

? "How big is a cubic meter?" about the size of a baby's play pen
- Use 3 meter sticks to demonstrate what 1 m × 1 m × 1 m looks like.

? "How many cubic millimeters are there in a cubic meter?" Students may or may not have an answer.
- Give students time to think. Someone might ask how many millimeters there are in a meter. The prefix milli- means 1000.
- The volume of a cubic meter in terms of cubic millimeters is $1000 \times 1000 \times 1000 = 1{,}000{,}000{,}000 = 1$ billion mm^3.

? "Can 1 billion be expressed using exponents?" 1000^3 or 10^9

Activity Notes

Activity 1

- Review the vocabulary associated with exponents.
- Have students work with their partner to complete the table. Students should recognize that a calculator is not necessary. They only need to multiply their previous product by −3.
- When students have finished, discuss the problem.

? "What did you notice about the values in the third column?" Values alternate—negative, positive, negative…
- Students might also mention: all values are odd and divisible by 3; last digits repeat in a cluster of 4: 3, 9, 7, 1, 3, 9, 7, …; there are two 1-digit numbers, two 2-digit numbers, two 3-digit numbers, and predict two 4-digit numbers
- **Part (b):** Listen for students to describe the exponent n in $(-3)^n$ as how many times −3 is multiplied by itself. Try to have students say that the exponent tells the number of times the base is used as a factor.

? "How can you find the value of $(-3)^n$?" Multiply −3 by itself n times.

Activity 2

- Although not all of the cubes are visible, students generally know that the cube contains 3^3 or 27 smaller cubes. So, at \$3 per small cube, $3 \times 3^3 = 3^4$.
- The expression is 3^4 and the answer is \$81.

Previous Learning

Students should know how to raise a number to a power.

Activity Materials
Introduction
• 3 meter sticks • metric ruler

Start Thinking! and Warm Up

Activity 9.1 Start Thinking!
For use before Activity 9.1

Activity 9.1 Warm Up
For use before Activity 9.1

Find the product.

1. $5 \times 5 \times 5$
2. $10 \times 10 \times 10$
3. $(-3) \times (-3) \times (-3)$
4. $10 \times 10 \times 10 \times 10 \times 10$
5. $4 \times 4 \times 4 \times 4$
6. $(-2) \times (-2) \times (-2) \times (-2)$

9.1 Record and Practice Journal

Essential Question How can you use exponents to write numbers?

The expression 3^5 is called a **power.** The **base** is 3. The **exponent** is 5.

Base → 3^5 ← Exponent

1 ACTIVITY: Using Exponent Notation

Work with a partner.

a. Complete the table.

Power	Repeated Multiplication Form	Value
$(-3)^1$	-3	-3
$(-3)^2$	$(-3) \cdot (-3)$	9
$(-3)^3$	$(-3) \cdot (-3) \cdot (-3)$	-27
$(-3)^4$	$(-3) \cdot (-3) \cdot (-3) \cdot (-3)$	81
$(-3)^5$	$(-3) \cdot (-3) \cdot (-3) \cdot (-3) \cdot (-3)$	-243
$(-3)^6$	$(-3) \cdot (-3) \cdot (-3) \cdot (-3) \cdot (-3) \cdot (-3)$	729
$(-3)^7$	$(-3) \cdot (-3) \cdot (-3) \cdot (-3) \cdot (-3) \cdot (-3) \cdot (-3)$	-2187

b. Describe what is meant by the expression $(-3)^n$. How can you find the value of $(-3)^n$?

Use (-3) as a factor n times and multiply.

Differentiated Instruction

Visual

Have students create pyramids of factors -10, -5, -2, 2, 5, and 10. Ask them to write the exponential form and evaluate each row. Ask "What is the product of 3 factors of -2?" and "What is the product of 4 factors of 5?" -8; 625

$$-2$$
$$(-2) \times (-2)$$
$$(-2) \times (-2) \times (-2)$$

$$5$$
$$5 \times 5$$
$$5 \times 5 \times 5$$
$$5 \times 5 \times 5 \times 5$$

9.1 Record and Practice Journal

2 ACTIVITY: Using Exponent Notation

Work with a partner.

a. The cube at the right has \$3 in each of its small cubes. Write a single power that represents the total amount of money in the large cube.

3^4

b. Evaluate the power to find the total amount of money in the large cube.

\$81

3 ACTIVITY: Writing Powers as Whole Numbers

Work with a partner. Write each distance as a whole number. Which numbers do you know how to write in words? For instance, in words, 10^3 is equal to *one thousand*.

a. 10^{26} meters: Diameter of observable universe

b. 10^{21} meters: Diameter of Milky Way Galaxy

a. 100,000,000,000,000,000,000,000,000 m

b. 1,000,000,000,000,000,000,000 m

c. 10^{16} meters: Diameter of Solar System

10,000,000,000,000,000 m

d. 10^7 meters: Diameter of Earth

10,000,000 m

e. 10^6 meters: Length of Lake Erie Shoreline

1,000,000 m

f. 10^5 meters: Width of Lake Erie

100,000 m

4 ACTIVITY: Writing a Power

Work with a partner. Write the number of kits, cats, sacks, and wives as a power.

As I was going to St. Ives
I met a man with seven wives
and every wife had seven sacks
and every sack had seven cats
and every cat had seven kits
Kits, cats, sacks, wives
How many were going to St. Ives?

Nursery Rhyme, 1730

wives: 7^1, sacks: 7^2, cats: 7^3, kits: 7^4

What Is Your Answer?

5. **IN YOUR OWN WORDS** How can you use exponents to write numbers? Give some examples of how exponents are used in real life.

represents repeated multiplication of the same factor

Laurie's Notes

Activity 3

- The distances in this activity represent magnitudes, powers of 10, not exact distances.
- **FYI:** There is a classic video made by two designers in the late 1970's called *Powers of Ten,* which is easily available on the Internet.

? "Which numbers do you know the names for in parts (a)–(f)?" part (d): ten million; part (e): one million; part (f): one hundred thousand

- Share other vocabulary that might be of interest:

million $= 10^6$	billion $= 10^9$	trillion $= 10^{12}$
quadrillion $= 10^{15}$	quintillion $= 10^{18}$	hexillion $= 10^{21}$
heptillion $= 10^{24}$	octillion $= 10^{27}$	nonillion $= 10^{30}$
decillion $= 10^{33}$	unodecillion $= 10^{36}$	duodecillion $= 10^{39}$

- The prefixes should look familiar. They are the same prefixes used in naming polygons.

Activity 4

- This is a classic rhyme. In the original rhyme, the answer is one because the man and his wives were coming *from* St. Ives.
- For this problem, suggest to students that they draw a picture or diagram to help them solve the problem.
- **Summary:** wives $= 7^1$, sacks $= 7^2$, cats $= 7^3$, kits $= 7^4$

What Is Your Answer?

- **Neighbor Check:** Have students work independently and then have their neighbor check their work. Have students discuss any discrepancies.

Closure

- Compare the following powers using >, <, or =.

1. 2^{10} _____ 10^2 > **2.** 10^3 _____ 3^{10} <

Technology For the Teacher

Dynamic Classroom

The Dynamic Planning Tool

Editable Teacher's Resources at *BigIdeasMath.com*

3 ACTIVITY: Writing Powers as Whole Numbers

Work with a partner. Write each distance as a whole number. Which numbers do you know how to write in words? For instance, in words, 10^3 is equal to *one thousand*.

a. 10^{26} meters: Diameter of observable universe

b. 10^{21} meters: Diameter of Milky Way Galaxy

c. 10^{16} meters: Diameter of Solar System

d. 10^{7} meters: Diameter of Earth

e. 10^{6} meters: Length of Lake Erie Shoreline

f. 10^{5} meters: Width of Lake Erie

4 ACTIVITY: Writing a Power

Work with a partner. Write the number of kits, cats, sacks, and wives as a power.

As I was going to St. Ives
I met a man with seven wives
And every wife had seven sacks
And every sack had seven cats
And every cat had seven kits
Kits, cats, sacks, wives
How many were going to St. Ives?

Nursery Rhyme, 1730

What Is Your Answer?

5. **IN YOUR OWN WORDS** How can you use exponents to write numbers? Give some examples of how exponents are used in real life.

Practice Use what you learned about exponents to complete Exercises 3–5 on page 354.

9.1 Lesson

Check It Out
Lesson Tutorials
BigIdeasMath.com

Key Vocabulary
power, *p. 352*
base, *p. 352*
exponent, *p. 352*

A **power** is a product of repeated factors. The **base** of a power is the common factor. The **exponent** of a power indicates the number of times the base is used as a factor.

base → $\frac{1}{2}$; exponent → 5

$$\left(\frac{1}{2}\right)^5 = \frac{1}{2} \cdot \frac{1}{2} \cdot \frac{1}{2} \cdot \frac{1}{2} \cdot \frac{1}{2}$$

power; $\frac{1}{2}$ is used as a factor 5 times.

EXAMPLE 1 Writing Expressions Using Exponents

Study Tip
Use parentheses to write powers with negative bases.

Write each product using exponents.

a. $(-7) \cdot (-7) \cdot (-7)$

Because -7 is used as a factor 3 times, its exponent is 3.

So, $(-7) \cdot (-7) \cdot (-7) = (-7)^3$.

b. $\pi \cdot \pi \cdot r \cdot r \cdot r$

Because π is used as a factor 2 times, its exponent is 2. Because r is used as a factor 3 times, its exponent is 3.

So, $\pi \cdot \pi \cdot r \cdot r \cdot r = \pi^2 r^3$.

On Your Own

Now You're Ready
Exercises 3–10

Write the product using exponents.

1. $\frac{1}{4} \cdot \frac{1}{4} \cdot \frac{1}{4} \cdot \frac{1}{4} \cdot \frac{1}{4}$

2. $0.3 \cdot 0.3 \cdot 0.3 \cdot 0.3 \cdot x \cdot x$

EXAMPLE 2 Evaluating Expressions

Evaluate the expression.

a. $(-2)^4$

$(-2)^4 = (-2) \cdot (-2) \cdot (-2) \cdot (-2)$ — Write as repeated multiplication.

$= 16$ — Simplify.

The factor is -2.

b. -2^4

$-2^4 = -(2 \cdot 2 \cdot 2 \cdot 2)$ — Write as repeated multiplication.

$= -16$ — Simplify.

The factor is 2.

Multi-Language Glossary at BigIdeasMath.com.

Laurie's Notes

Introduction

Connect

- **Yesterday:** Students explored writing numbers with exponents.
- **Today:** Students will write expressions involving exponents and evaluate powers.

Motivate

- Because U.S. currency has coins and bills which are powers of 10, find out what your students know about the people on the coins and bills. You can create a matching activity or have a few questions and answers ready for the class.

Lesson Notes

Example 1

- Write the definitions of power, base, and exponent. Note the use of *factor* instead of *multiplying the base by itself.*
- Write the example shown. When this power is evaluated, the answer is $\frac{1}{32}$. You say, "$\frac{1}{32}$ is the 5th power of $\frac{1}{2}$."
- Exponents are used to rewrite an expression involving repeated factors.

? "Is it necessary to write the multiplication dot between the factors?" no

- **Common Error:** Parentheses *must* be used when you write a power with a negative base. This is a common error that students will make. For example: $(-2)^2 = (-2)(-2) = 4$
  $-2^2 = -(2)(2) = -4$
  Without the parentheses, the number being squared is 2, and then you multiply the product by -1. With the parentheses, the number being squared is -2, and the product is 4. The underlying property is the order of operations. Exponents are performed before multiplication.
- **Part (b):** Variables and constants are expressed using exponents in a similar fashion.

On Your Own

- In Question 1, $\frac{1}{4}$ needs to be written with parentheses so that the denominator is raised to a power. Without parentheses, it is read as $\frac{1^5}{4}$.

Example 2

? "What is the base in each problem? What will be used as the factor in each problem?" Bases are -2 and 2. Factors are -2 and 2.

- This example addresses the need to write the base within parentheses when the base is negative.

Goal Today's lesson is writing and evaluating **powers**.

Start Thinking! and Warm Up

Lesson 9.1 Warm Up
For use before Lesson 9.1

Lesson 9.1 Start Thinking!
For use before Lesson 9.1

Use the nursery rhyme on page 351 of your textbook to guide you in writing your own "power poem."

Extra Example 1

Write each product using exponents.

a. $4 \cdot 4 \cdot 4 \cdot 4$ 4^4

b. $(-2) \cdot (-2) \cdot x \cdot x \cdot x$ $(-2)^2x^3$

On Your Own

1. $\left(\frac{1}{4}\right)^5$
2. $(0.3)^4x^2$

Extra Example 2

Evaluate the expression.

a. $(-3)^3$ -27

b. -3^3 -27

Extra Example 3

Evaluate the expression.

a. $2 - 4 \cdot 5^2$ -98

b. $5 + 6^2 \div 4$ 14

On Your Own

3. -625 **4.** $-\frac{1}{216}$

5. 1 **6.** -7

Extra Example 4

In Example 4, the diameter of the inner sphere is 2.2 meters. What is the volume of the inflated space? about 8.56 m^3

On Your Own

7. about 11.08 m^3

English Language Learners

Labels

Have students label and practice reading statements of powers.

base, exponent

$2^3 = 2 \times 2 \times 2 = 8$

power, factor

"Two to the third power equals two times two times two."

"The base is 2, the exponent is 3, and 2 is written as a factor 3 times."

Laurie's Notes

Example 3

- You may wish to review the order of operations or wait to see how students evaluate the expressions.
- **Common Error:** Students may evaluate the problem left to right, performing the addition first. They need to be reminded of the order of operations.

? "How do you start to evaluate this expression?" powers first

- Continue to evaluate the problem as shown.
- In part (b), there are two powers to evaluate. After that is done, division is performed before subtraction.

On Your Own

- Encourage students to write out the steps in their solutions. Discourage them from performing multiple steps in their head.

Example 4

- **FYI:** Ask if any of your students have heard of sphering. You can find information about this sport online at *zorb.com.*
- You want to find the volume of the inflated space. Reassure students that they are not expected to memorize this formula.
- Write the volume of a sphere: $V = \frac{4}{3}\pi r^3$. From the photo, the diameter of each sphere is known. The radius is half the diameter. Substitute the radius of each sphere and simplify.
- **Extension:** The inner sphere has a volume of a little more than 4 cubic meters. Find a region of your classroom that is approximately 4 cubic meters.

Closure

- **Exit Ticket:** Evaluate.

1. $4^2 - 8(2) + 3^3$ 27 **2.** $\left(-\frac{2}{3}\right)^3 + \left|5^2 - 2 \cdot 15\right|$ $\frac{127}{27} \approx 4.7$

Technology For the Teacher

Dynamic Classroom

The Dynamic Planning Tool

Editable Teacher's Resources at *BigIdeasMath.com*

EXAMPLE 3 Using Order of Operations

Evaluate the expression.

a. $3 + 2 \cdot 3^4$

$$3 + 2 \cdot 3^4 = 3 + 2 \cdot 81 \quad \text{Evaluate the power.}$$
$$= 3 + 162 \quad \text{Multiply.}$$
$$= 165 \quad \text{Add.}$$

b. $3^3 - 8^2 \div 2$

$$3^3 - 8^2 \div 2 = 27 - 64 \div 2 \quad \text{Evaluate the powers.}$$
$$= 27 - 32 \quad \text{Divide.}$$
$$= -5 \quad \text{Subtract.}$$

On Your Own

Now You're Ready
Exercises 11–16 and 21–26

Evaluate the expression.

3. -5^4 **4.** $\left(-\frac{1}{6}\right)^3$ **5.** $\left|-3^3 \div 27\right|$ **6.** $9 - 2^5 \cdot 0.5$

EXAMPLE 4 Real-Life Application

In sphering, a person is secured inside a small, hollow sphere that is surrounded by a larger sphere. The space between the spheres is inflated with air. What is the volume of the inflated space? (The volume V of a sphere is $V = \frac{4}{3}\pi r^3$. Use 3.14 for π.)

Outer sphere		*Inner sphere*
$V = \frac{4}{3}\pi r^3$	Write formula.	$V = \frac{4}{3}\pi r^3$
$= \frac{4}{3}\pi(1.5)^3$	Substitute.	$= \frac{4}{3}\pi(1)^3$
$= \frac{4}{3}\pi(3.375)$	Evaluate the power.	$= \frac{4}{3}\pi(1)$
≈ 14.13	Multiply.	≈ 4.19

So, the volume of the inflated space is about $14.13 - 4.19$, or 9.94 cubic meters.

On Your Own

7. WHAT IF? In Example 4, the diameter of the inner sphere is 1.8 meters. What is the volume of the inflated space?

9.1 Exercises

Check It Out
Help with Homework
BigIdeasMath.com

Vocabulary and Concept Check

1. **VOCABULARY** Describe the difference between an exponent and a power. Can the two words be used interchangeably?

2. **WHICH ONE DOESN'T BELONG?** Which one does *not* belong with the other three? Explain your reasoning.

5^3 The exponent is 3.	5^3 The power is 5.	5^3 The base is 5.	5^3 Five is used as a factor 3 times.

Practice and Problem Solving

Write the product using exponents.

1 3. $3 \cdot 3 \cdot 3 \cdot 3$

4. $(-6) \cdot (-6)$

5. $\left(-\frac{1}{2}\right) \cdot \left(-\frac{1}{2}\right) \cdot \left(-\frac{1}{2}\right)$

6. $\frac{1}{3} \cdot \frac{1}{3} \cdot \frac{1}{3}$

7. $\pi \cdot \pi \cdot \pi \cdot x \cdot x \cdot x \cdot x$

8. $(-4) \cdot (-4) \cdot (-4) \cdot y \cdot y$

9. $8 \cdot 8 \cdot 8 \cdot 8 \cdot b \cdot b \cdot b$

10. $(-t) \cdot (-t) \cdot (-t) \cdot (-t) \cdot (-t)$

Evaluate the expression.

2 11. 5^2

12. -11^3

13. $(-1)^6$

14. $\left(\frac{1}{2}\right)^6$

15. $\left(-\frac{1}{12}\right)^2$

16. $-\left(\frac{1}{9}\right)^3$

17. **ERROR ANALYSIS** Describe and correct the error in evaluating the expression.

✗ $6^3 = 6 \cdot 3 = 18$

18. **PRIME FACTORIZATION** Write the prime factorization of 675 using exponents.

19. **NUMBER SENSE** Write $-\left(\frac{1}{4} \cdot \frac{1}{4} \cdot \frac{1}{4} \cdot \frac{1}{4}\right)$ using exponents.

20. **RUSSIAN DOLLS** The largest doll is 12 inches tall. The height of each of the other dolls is $\frac{7}{10}$ the height of the next larger doll. Write an expression for the height of the smallest doll. What is the height of the smallest doll?

Assignment Guide and Homework Check

Level	Day 1 Activity Assignment	Day 2 Lesson Assignment	Homework Check
Basic	3–5, 30–33	1, 2, 7–13, 17–27 odd, 20	7, 8, 12, 20, 21
Average	3–5, 30–33	1, 2, 8–16 even, 17–19, 21–27 odd, 28	8, 12, 18, 21, 28
Advanced	3–5, 30–33	1, 2, 8, 10, 15–19, 24–29	8, 15, 18, 26, 28

Common Errors

- **Exercises 3–10** Students may count the wrong number of factors for the variable or number. Remind them to be careful when counting the number of factors, and that the number of factors is the exponent.
- **Exercises 11–16** Students may have the wrong sign in their answer. Remind them that when the sign is inside the parentheses, it is part of the base and should be evaluated that way. If the sign is outside the parentheses, it is the same as multiplying the expression by −1. Because of the order of operations, students should multiply by −1 after evaluating the exponent.
- **Exercises 21–26** Students may not remember the definition of absolute value or the correct order of operations. Remind them that absolute value is the distance from zero and to follow the order of operations.

9.1 Record and Practice Journal

Write the product using exponents.

1. $4 \cdot 4 \cdot 4 \cdot 4 \cdot 4$

   4^5

2. $\left(-\frac{1}{8}\right) \cdot \left(-\frac{1}{8}\right) \cdot \left(-\frac{1}{8}\right)$

   $\left(-\frac{1}{8}\right)^3$

3. $5 \cdot 5 \cdot (-x) \cdot (-x) \cdot (-x) \cdot (-x)$

   $5^2(-x)^4$

4. $9 \cdot 9 \cdot y \cdot y \cdot y \cdot y \cdot y \cdot y$

   9^2y^6

Evaluate the expression.

5. 10^3

   1000

6. $(-7)^4$

   2401

7. $-\left(\frac{1}{6}\right)^5$

   $-\frac{1}{7776}$

8. $3 + 6 \cdot (-5)^2$

   153

9. $\left|-\frac{1}{3}(1^{10} + 9 - 2^3)\right|$

   $\frac{2}{3}$

10. A foam toy is 2 inches wide. It doubles in size for every minute it is in water. Write an expression for the width of the toy after 5 minutes. What is the width after 5 minutes?

    2^6; 64 inches

Technology For the Teacher

Answer Presentation Tool

Quiz*Show*

Vocabulary and Concept Check

1. An exponent describes the number of times the base is used as a factor. A power is the entire expression (base and exponent). A power tells you the value of the factor and the number of factors. No, the two cannot be used interchangeably.
2. 5^3, The power is 5; The power is 5^3. Five is the base.

Practice and Problem Solving

3. 3^4
4. $(-6)^2$
5. $\left(-\frac{1}{2}\right)^3$
6. $\left(\frac{1}{3}\right)^3$
7. π^3x^4
8. $(-4)^3y^2$
9. 8^4b^3
10. $(-t)^5$
11. 25
12. −1331
13. 1
14. $\frac{1}{64}$
15. $\frac{1}{144}$
16. $-\frac{1}{729}$
17. The exponent 3 describes how many times the base 6 should be used as a factor. Three should not appear as a factor in the product. $6^3 = 6 \cdot 6 \cdot 6 = 216$
18. $3^3 \cdot 5^2$
19. $-\left(\frac{1}{4}\right)^4$
20. $12 \cdot \left(\frac{7}{10}\right)^3$; 4.116 in.

Practice and Problem Solving

21. 29

22. 65

23. 5

24. 5

25. 66

26. 2

27. See Additional Answers.

28. **a.** about 99.95 g

b. 99.95%

29. See *Taking Math Deeper*.

Fair Game Review

30. Commutative Property of Multiplication

31. Associative Property of Multiplication

32. Identity Property of Multiplication

33. B

Mini-Assessment

Evaluate the expression.

1. 7^2 49
2. -3^4 -81
3. $(-3)^4$ 81
4. $4 + 6 \cdot (-2)^3$ -44
5. $\frac{3}{4}\left(2^5 - 6 \div \left(\frac{1}{2}\right)^2\right)$ 6

Taking Math Deeper

Exercise 29

This exercise is based on the 12-note chromatic scale used in western music. Asian music uses a 9-note scale. Other cultures use 8-note and 10-note scales.

1 Use a calculator or spreadsheet to calculate the frequency of each note.

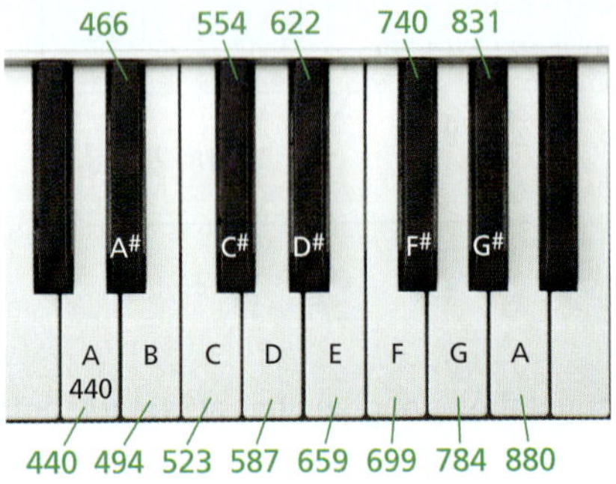

In the formula, the number 1.0595 is the 12th root of 2. A more accurate representation of this number is 1.0594630944.

2 **a.** It takes 12 notes to travel from A-440 to A-880. That is why the western scale is called a 12-note scale. The term *octave*, which is based on the number 8, refers only to the white keys on the piano.

3 **b.** The frequency of the A above A-440 is about 880 vibrations per second. Piano tuners use A-440 as the basic key to tune all the other notes. The A below A-440 has a frequency of 220 vibrations per second.

c. Because $(1.0594630944)^{12} \approx 2$, it follows that the A above A-440 has twice the frequency of A.

Note: The scale given by this formula is called the *equal temperament scale*. Notes whose frequencies have many common divisors "harmonize" more.

Project

Use the school library or the Internet to research instruments used in other countries, such as China and Japan. Many have different scales. Compare the scales of the instruments you find to that of the piano.

Reteaching and Enrichment Strategies

If students need help. . .	If students got it. . .
Resources by Chapter • Practice A and Practice B • Puzzle Time Record and Practice Journal Practice Differentiating the Lesson Lesson Tutorials Skills Review Handbook	Resources by Chapter • Enrichment and Extension Start the next section

Evaluate the expression.

3 **21.** $5 + 3 \cdot 2^3$

22. $2 + 7 \cdot (-3)^2$

23. $(13^2 - 12^2) \div 5$

24. $\frac{1}{2}(4^3 - 6 \cdot 3^2)$

25. $\left| \frac{1}{2}(7 + 5^3) \right|$

26. $\left| \left(-\frac{1}{2}\right)^3 \div \left(\frac{1}{4}\right)^2 \right|$

27. MONEY You have a part-time job. One day your boss offers to pay you either $2^h - 1$ or 2^{h-1} dollars for each hour h you work that day. Copy and complete the table. Which option should you choose? Explain.

h	1	2	3	4	5
$2^h - 1$					
2^{h-1}					

28. CARBON-14 DATING Carbon-14 dating is used by scientists to determine the age of a sample.

a. The amount C (in grams) of a 100-gram sample of carbon-14 remaining after t years is represented by the equation $C = 100(0.99988)^t$. Use a calculator to find the amount of carbon-14 remaining after 4 years.

b. What percent of the carbon-14 remains after 4 years?

29. Critical Thinking The frequency (in vibrations per second) of a note on a piano is represented by the equation $F = 440(1.0595)^n$, where n is the number of notes above A-440. Each black or white key represents one note.

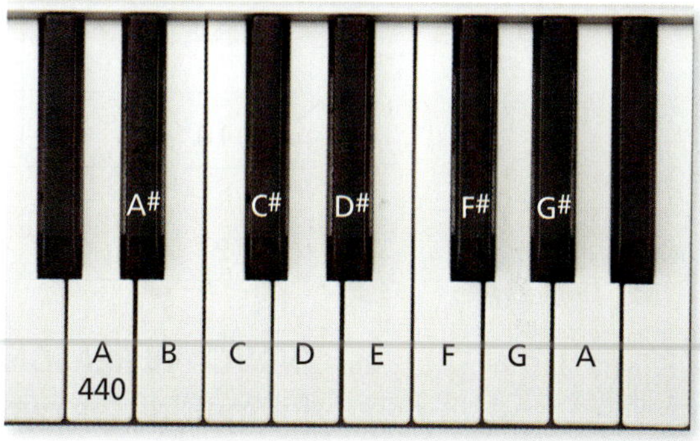

a. How many notes do you take to travel from A-440 to A?

b. What is the frequency of A?

c. Describe the relationship between the number of notes between A-440 and A and the frequency of the notes.

Fair Game Review What you learned in previous grades & lessons

Tell which property is illustrated by the statement. *(Skills Review Handbook)*

30. $8 \cdot x = x \cdot 8$

31. $(2 \cdot 10)x = 2(10 \cdot x)$

32. $3(x \cdot 1) = 3x$

33. MULTIPLE CHOICE A cone of yarn has a surface area of 16π square inches. What is the slant height of the cone of yarn? *(Skills Review Handbook)*

Ⓐ 4 in. Ⓑ 6 in.

Ⓒ 8 in. Ⓓ 10 in.

9.2 Product of Powers Property

Essential Question How can you multiply two powers that have the same base?

1 ACTIVITY: Finding Products of Powers

Work with a partner.

a. Copy and complete the table.

Product	Repeated Multiplication Form	Power
$2^2 \cdot 2^4$	$2 \cdot 2 \cdot 2 \cdot 2 \cdot 2 \cdot 2$	2^6
$(-3)^2 \cdot (-3)^4$	$(-3) \cdot (-3) \cdot (-3) \cdot (-3) \cdot (-3) \cdot (-3)$	$(-3)^6$
$7^3 \cdot 7^2$		
$5.1^1 \cdot 5.1^6$		
$(-4)^2 \cdot (-4)^2$		
$10^3 \cdot 10^5$		
$\left(\frac{1}{2}\right)^5 \cdot \left(\frac{1}{2}\right)^5$		

b. **INDUCTIVE REASONING** Describe the pattern in the table. Then write a rule for multiplying two powers that have the same base.

$$a^m \cdot a^n = a^{\square}$$

c. Use your rule to simplify the products in the first column of the table above. Does your rule give the results in the third column?

2 ACTIVITY: Using a Calculator

Work with a partner.

Some calculators have *exponent keys* that are used to evaluate powers.

Use a calculator with an exponent key to evaluate the products in Activity 1.

Laurie's Notes

Introduction

For the Teacher

- **Goal:** Students will explore how to multiply two powers with the same base.

Motivate

- **Story Time:** Tell students that the superintendent has agreed to put you on a special salary schedule for one month. On day 1 you will receive 1¢, on day 2 you will receive 2¢, day 3 is 4¢, and so on, with your salary doubling every school day for the month. There are 23 school days this month. Should you take the new salary?
- Give time for students to start the tabulation. Let them use a calculator for speed. The table below shows the daily pay.

1	$2 = 2^1$	$4 = 2^2$	$8 = 2^3$
$16 = 2^4$	$32 = 2^5$	$64 = 2^6$	$128 = 2^7$
$256 = 2^8$	$512 = 2^9$	$1024 = 2^{10}$	$2048 = 2^{11}$
$4096 = 2^{12}$	$8192 = 2^{13}$	$16{,}384 = 2^{14}$	$32{,}768 = 2^{15}$
$65{,}536 = 2^{16}$	$131{,}072 = 2^{17}$	$262{,}144 = 2^{18}$	$524{,}288 = 2^{19}$
$1{,}048{,}576 = 2^{20}$	$2{,}097{,}152 = 2^{21}$	$4{,}194{,}304 = 2^{22}$	

- If the superintendent is looking for additional math teachers, they will be lined up at the door.
- In this penny doubling problem, each day you are paid a power of 2. Your salary is actually the *sum* of all of these amounts.

Activity Notes

Activity 1

- Have students work with their partner to complete the table. A calculator is not necessary for this activity. The first problem has been done as a sample to follow. Notice the color coding.
- ? "What did you notice about the number of factors in the middle column and the exponent used to write the power?" same number
- **Part (b):** Students will recognize that the exponents are added together, but it may not be obvious to them how to write this fact using variables.
- **Big Idea:** Write the summary statement: $a^m \cdot a^n = a^{m+n}$. Stress that the bases must be the same. That is why a is the base for both powers. This rule tells us nothing about how to simplify a problem such as $3^3 \cdot 4^2$.

Activity 2

- If students are going to use calculators, they need to know how to use them correctly. Different calculators have different ways in which the powers are entered.

Previous Learning

Students should know how to raise a number to a power.

Activity Materials	
Introduction	**Textbook**
• calculator	• calculator

Start Thinking! and Warm Up

Activity 9.2 Start Thinking! For use before Activity 9.2

Activity 9.2 Warm Up For use before Activity 9.2

Evaluate the power.

1. 3^5 2. 5^4 3. 10^6
4. $(-4)^3$ 5. $(-3)^2$ 6. $(-2)^5$

9.2 Record and Practice Journal

Essential Question How can you multiply two powers that have the same base?

1 ACTIVITY: Finding Products of Powers

Work with a partner.

a. Complete the table.

Product	Repeated Multiplication Form	Power
$2^2 \cdot 2^4$	$2 \cdot 2 \cdot 2 \cdot 2 \cdot 2 \cdot 2$	2^6
$(-3)^2 \cdot (-3)^4$	$(-3) \cdot (-3) \cdot (-3) \cdot (-3) \cdot (-3) \cdot (-3)$	$(-3)^6$
$7^3 \cdot 7^2$	$7 \cdot 7 \cdot 7 \cdot 7 \cdot 7$	7^5
$5.1^1 \cdot 5.1^6$	$(5.1) \cdot (5.1) \cdot (5.1) \cdot (5.1) \cdot (5.1) \cdot (5.1) \cdot (5.1)$	$(5.1)^7$
$(-4)^2 \cdot (-4)^2$	$(-4) \cdot (-4) \cdot (-4) \cdot (-4)$	$(-4)^4$
$10^3 \cdot 10^5$	$10 \cdot 10 \cdot 10 \cdot 10 \cdot 10 \cdot 10 \cdot 10 \cdot 10$	10^8
$\left(\frac{1}{2}\right)^5 \cdot \left(\frac{1}{2}\right)^5$	$\left(\frac{1}{2}\right) \cdot \left(\frac{1}{2}\right) \cdot \left(\frac{1}{2}\right) \cdot \left(\frac{1}{2}\right) \cdot \left(\frac{1}{2}\right) \cdot \left(\frac{1}{2}\right) \cdot \left(\frac{1}{2}\right) \cdot \left(\frac{1}{2}\right) \cdot \left(\frac{1}{2}\right) \cdot \left(\frac{1}{2}\right)$	$\left(\frac{1}{2}\right)^{10}$

b. **INDUCTIVE REASONING** Describe the pattern in the table. Then write a rule for multiplying two powers that have the same base.

Number of factors is equal to the sum of the exponents.

$a^m \cdot a^n = a^{m+n}$

c. Use your rule to simplify the products in the first column of the table above. Does your rule give the results in the third column?

yes

English Language Learners

Pair Activity

Have students work in pairs to simplify exponential expressions. Each student simplifies a different expression. When both students are done, they take turns explaining the solution while the other person follows along.

9.2 Record and Practice Journal

2 ACTIVITY: Using a Calculator

Work with a partner.

Some calculators have *exponent keys* that are used to evaluate powers.

Use a calculator with an exponent key to evaluate the products in Activity 1. $2^6 = 64$, $(-3)^6 = 729$, $7^5 = 16{,}807$, $5.1^7 \approx 89{,}741.1$, $(-4)^4 = 256$, $10^8 = 100{,}000{,}000$, $\left(\frac{1}{2}\right)^{10} = \frac{1}{1024}$

3 ACTIVITY: The Penny Puzzle

Work with a partner.

- The rows y and columns x of a chess board are numbered as shown.
- Each position on the chess board has a stack of pennies. (Only the first row is shown.)
- The number of pennies in each stack is $2^x \cdot 2^y$.

a. How many pennies are in the stack in location (3, 5)?

256

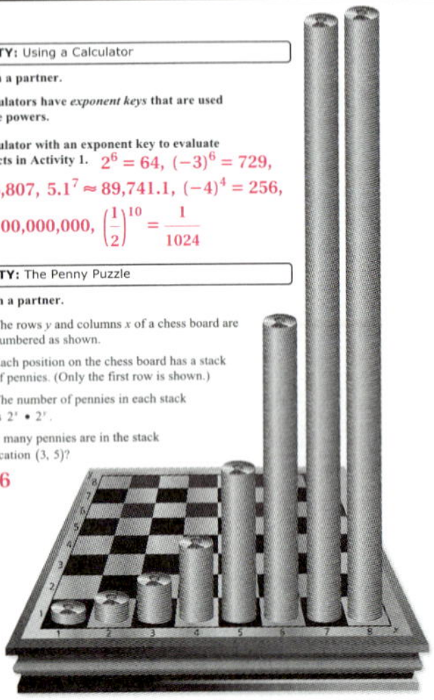

b. Which locations have 32 pennies in their stacks?

(1, 4), (2, 3), (3, 2), (4, 1)

c. How much money (in dollars) is in the location with the tallest stack?

\$655.36

d. A penny is about 0.06 inch thick. About how tall (in inches) is the tallest stack?

3932.16 in.

What Is Your Answer?

4. IN YOUR OWN WORDS How can you multiply two powers that have the same base? Give two examples of your rule.

Add their exponents then evaluate.

Laurie's Notes

Activity 3

- Take time to discuss the notation. In position (1, 1), the amount of pennies is $2^1 \cdot 2^1 = 4$. In position (2, 1), the amount of pennies is $2^2 \cdot 2^1 = 8$. Answer any questions about notation or how to find the number of pennies on any square.
- There are many patterns and interesting extensions to this problem that may surface as they explore the questions presented.
- **Part (a):** There are $2^3 \cdot 2^5 = 2^8 = 256$ pennies in location (3, 5).
- **Part (b):** Because $32 = 2^5$, the exponents need to sum to 5. The locations include (1, 4), (4, 1), (2, 3), and (3, 2).
- **Part (c):** The most money will be in the location where x and y have the greatest sum. This will occur at (8, 8), where the value is $2^8 \cdot 2^8 = 2^{16} = 65{,}536 = \$655.36$.
- **Part (d):** Multiply the number of pennies by the thickness, $65{,}536 \times 0.06 = 3932.16$ inches.

What Is Your Answer?

- **Neighbor Check:** Have students work independently and then have their neighbor check their work. Have students discuss any discrepancies.

Closure

- Refer back to the penny doubling problem from the beginning of the lesson. On what day was your salary more than \$1000? day 18

Technology For the Teacher

Dynamic Classroom

The Dynamic Planning Tool

Editable Teacher's Resources at *BigIdeasMath.com*

3 ACTIVITY: The Penny Puzzle

Work with a partner.

- The rows y and columns x of a chess board are numbered as shown.
- Each position on the chess board has a stack of pennies. (Only the first row is shown.)
- The number of pennies in each stack is

  $2^x \cdot 2^y$.

a. How many pennies are in the stack in location (3, 5)?

b. Which locations have 32 pennies in their stacks?

c. How much money (in dollars) is in the location with the tallest stack?

d. A penny is about 0.06 inch thick. About how tall (in inches) is the tallest stack?

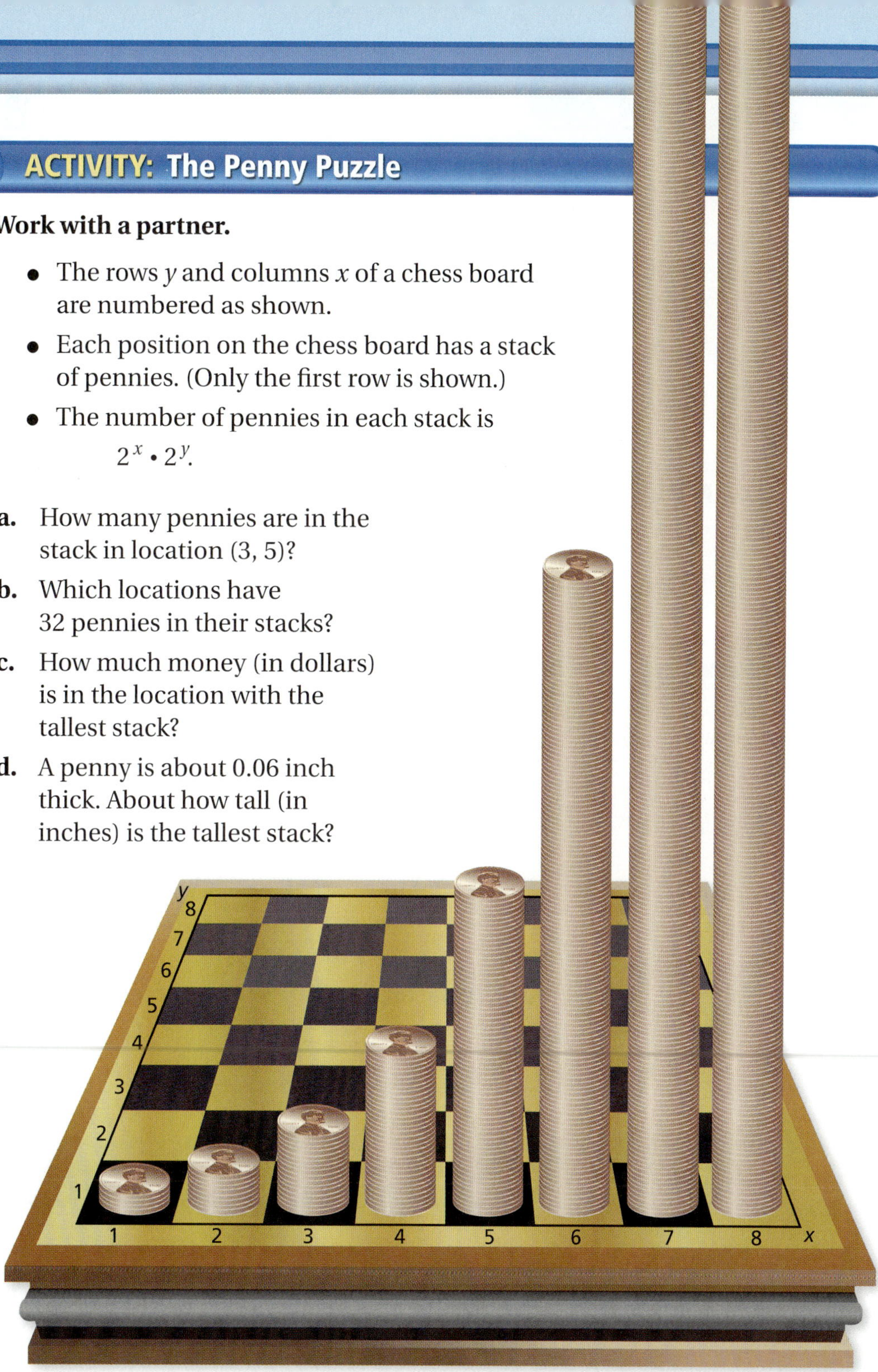

What Is Your Answer?

4. **IN YOUR OWN WORDS** How can you multiply two powers that have the same base? Give two examples of your rule.

Practice Use what you learned about the Product of Powers Property to complete Exercises 3–5 on page 360.

Check It Out
Lesson Tutorials
BigIdeasMath.com

Key Idea

Product of Powers Property

Words To multiply powers with the same base, add their exponents.

Numbers $4^2 \cdot 4^3 = 4^{2+3} = 4^5$ **Algebra** $a^m \cdot a^n = a^{m+n}$

EXAMPLE 1 Multiplying Powers with the Same Base

a. $2^4 \cdot 2^5 = 2^{4+5}$ The base is 2. Add the exponents.

$= 2^9$ Simplify.

b. $-5 \cdot (-5)^6 = (-5)^1 \cdot (-5)^6$ Rewrite -5 as $(-5)^1$.

$= (-5)^{1+6}$ The base is -5. Add the exponents.

$= (-5)^7$ Simplify.

c. $x^3 \cdot x^7 = x^{3+7}$ The base is x. Add the exponents.

$= x^{10}$ Simplify.

Study Tip

When a number is written without an exponent, its exponent is 1.

On Your Own

Simplify the expression. Write your answer as a power.

1. $6^2 \cdot 6^4$
2. $\left(-\frac{1}{2}\right)^3 \cdot \left(-\frac{1}{2}\right)^6$
3. $z \cdot z^{12}$

EXAMPLE 2 Raising a Power to a Power

a. $(3^4)^3 = 3^4 \cdot 3^4 \cdot 3^4$ Write as repeated multiplication.

$= 3^{4+4+4}$ The base is 3. Add the exponents.

$= 3^{12}$ Simplify.

b. $(w^5)^4 = w^5 \cdot w^5 \cdot w^5 \cdot w^5$ Write as repeated multiplication.

$= w^{5+5+5+5}$ The base is w. Add the exponents.

$= w^{20}$ Simplify.

On Your Own

Now You're Ready
Exercises 3–14

Simplify the expression. Write your answer as a power.

4. $(4^4)^3$
5. $(y^2)^4$
6. $(\pi^3)^3$
7. $\left((-4)^3\right)^2$

Laurie's Notes

Introduction

Connect

- **Yesterday:** Students explored exponents.
- **Today:** Students will use the Product of Powers Property to simplify expressions.

Motivate

- More money talk! The $10,000 bill, which is no longer in circulation, would be much easier to carry than the same amount in pennies.

? Ask a few questions about money.

  - "How many pennies equal $10,000?" $100 \times 10{,}000 = 1{,}000{,}000$ or 10^6
  - "How many dimes equal $10,000?" $10 \times 10{,}000 = 100{,}000$ or 10^5
  - "How many $10 bills equal $10,000?" $\frac{1}{10} \times 10{,}000 = 1000$ or 10^3
  - "How many $100 bills equal $10,000?" $\frac{1}{100} \times 10{,}000 = 100$ or 10^2

Lesson Notes

Key Idea

- Write the Key Idea. Explain the Words, Numbers, and Algebra.

Example 1

- **Part (a)**: Write and simplify the expression. The base is 2 for each power, so add the exponents.

? "In part (b), what is the base for each power? To what power is each base raised?" base = −5; exponents = 1 and 6

- **Common Error:** When the exponent is 1, it is not written. When it is not written, students will sometimes forget to add 1 in their answer.
- **Part (c):** The property applies to variables as well as numbers.

On Your Own

- **Think-Pair-Share:** Students should read each question independently and then work with a partner to answer the questions. When they have answered the questions, the pair should compare their answers with another group and discuss any discrepancies.

Example 2

? "In the expression $(3^4)^3$, what does the exponent of 3 tell us to do?" Use 3^4 as a factor three times.

- Expanding the expression, use the Product of Powers Property and add the exponents.

On Your Own

- Students struggle with Questions 5 and 6. Assure them that the questions are evaluated the same for variables (y) and irrational numbers (π).

Goal Today's lesson is using the Product of Powers Property to simplify expressions.

Start Thinking! and Warm Up

Lesson 9.2 Warm Up
For use before Lesson 9.2

Lesson 9.2 Start Thinking!
For use before Lesson 9.2

Think of an example in geometry where you would need to use the Product of Powers Property.

Extra Example 1

Simplify the expression. Write your answer as a power.

a. $6^2 \cdot 6^7$ 6^9

b. $-2 \cdot (-2)^3$ $(-2)^4$

c. $x^2 \cdot x^5$ x^7

On Your Own

1. 6^6
2. $\left(-\frac{1}{2}\right)^9$
3. z^{13}

Extra Example 2

Simplify the expression. Write your answer as a power.

a. $(5^2)^3$ 5^6

b. $(y^4)^6$ y^{24}

On Your Own

4. 4^{12}
5. y^8
6. π^9
7. $(-4)^6$

Extra Example 3

Simplify the expression.

a. $(4x)^2$ $16x^2$

b. $(wz)^3$ w^3z^3

On Your Own

8. $625y^4$ **9.** $0.25n^2$

10. a^5b^5

Extra Example 4

In Example 4, the total storage space of a computer is 32 gigabytes. How many bytes of total storage space does the computer have? 2^{35} bytes

On Your Own

11. 2^{34} bytes

Differentiated Instruction

Visual

Remind students that the Product of Powers Property can only be applied to powers having the same base. Have students highlight each unique base with a different color. Then add the exponents.

$$\begin{aligned}2^4 \cdot 2^5 - (3^2)^2 &= 2^4 \cdot 2^5 - 3^2 \cdot 3^2\\ &= 2^{4+5} - 3^{2+2}\\ &= 2^9 - 3^4\\ &= 512 - 81\\ &= 431\end{aligned}$$

Laurie's Notes

Example 3

- **Common Misconception:** Students sometimes think of this as an application of the Distributive Property and distribute the exponent. Be careful and deliberate with language when simplifying these problems.
- ? "In part (a), what does the exponent of 3 tell us to do in the expression $(2x)^3$?" Use $2x$ as a factor three times.
- Write the factor $2x$ three times. Properties of Multiplication (Associative and Commutative) allow you to reorder the terms. Notice that you can identify six factors: three 2's and three x's. Use exponents to write the factors. Finally, 2^3 is rewritten as 8 and the final answer is $8x^3$.
- **Part (b):** Follow the same procedure of writing xy as a factor twice. It is very common for students to write $x \cdot x = 2x$. Do not assume that students will see this error.

On Your Own

- Encourage students to write out the steps in their solution. Show each product as a factor the appropriate number of times.
- **Common Error:** $(0.5)^2 \neq 1$; $(0.5)^2 = 0.25$
- Have volunteers write their solutions on the board.

Example 4

- Writing the verbal model is necessary in this problem because the terms gigabytes and bytes may not be familiar to all. The first sentence is a conversion fact: 1 GB = 2^{30} bytes. There are 64 GB of total storage. Students may naturally think 64×2^{30} to solve the problem.
- ? "What is wrong with writing 64×2^{30} to solve the problem?" The answer choices have a base of 2, so try to write each factor in the product with a base of 2.
- Rewrite 64 as a power with a base of 2, $64 = 2^6$. Now you can solve the example.

On Your Own

- Students may respond with $\frac{1}{4}$ as much storage space and mean $\frac{1}{4}$ the total storage space. In fact, $\frac{1}{4}$ of $2^{36} = 2^{34}$. However, this is not an obvious step. Students should model the problem after Example 4.

Closure

- **Exit Ticket:** Simplify. $5^3 \cdot 5^4$ 5^7 $(-3x)^3$ $-27x^3$

Technology For the Teacher

Dynamic Classroom

The Dynamic Planning Tool
Editable Teacher's Resources at *BigIdeasMath.com*

EXAMPLE 3 Raising a Product to a Power

a. $(2x)^3 = 2x \cdot 2x \cdot 2x$ — Write as repeated multiplication.

$= (2 \cdot 2 \cdot 2) \cdot (x \cdot x \cdot x)$ — Group like bases using properties of multiplication.

$= 2^{1+1+1} \cdot x^{1+1+1}$ — The bases are 2 and x. Add the exponents.

$= 2^3 \cdot x^3 = 8x^3$ — Simplify.

b. $(xy)^2 = xy \cdot xy$ — Write as repeated multiplication.

$= (x \cdot x) \cdot (y \cdot y)$ — Group like bases using properties of multiplication.

$= x^{1+1} \cdot y^{1+1}$ — The bases are x and y. Add the exponents.

$= x^2y^2$ — Simplify.

On Your Own

Now You're Ready
Exercises 17–22

Simplify the expression.

8. $(5y)^4$ **9.** $(0.5n)^2$ **10.** $(ab)^5$

EXAMPLE 4 Standardized Test Practice

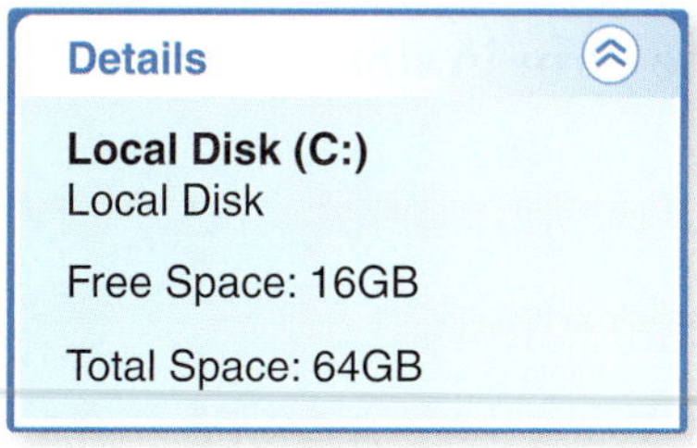

A gigabyte (GB) of computer storage space is 2^{30} bytes. The details of a computer are shown. How many bytes of total storage space does the computer have?

Ⓐ 2^{34} Ⓑ 2^{36} Ⓒ 2^{180} Ⓓ 128^{30}

The computer has 64 gigabytes of total storage space. Notice that 64 can be written as a power, 2^6. Use a model to solve the problem.

Total number of bytes = Number of bytes in a gigabyte • Number of gigabytes

$= 2^{30} \cdot 2^6$ — Substitute.

$= 2^{30+6}$ — Add exponents.

$= 2^{36}$ — Simplify.

The computer has 2^{36} bytes of total storage space. The correct answer is Ⓑ.

On Your Own

11. How many bytes of free storage space does the computer have?

9.2 Exercises

Check It Out
Help with Homework
BigIdeasMath.com

Vocabulary and Concept Check

1. **REASONING** When should you use the Product of Powers Property?

2. **CRITICAL THINKING** Can you use the Product of Powers Property to multiply powers with different bases? Explain.

Practice and Problem Solving

Simplify the expression. Write your answer as a power.

(1) (2) 3. $3^2 \cdot 3^2$

4. $8^{10} \cdot 8^4$

5. $(-4)^5 \cdot (-4)^7$

6. $a^3 \cdot a^3$

7. $h^6 \cdot h$

8. $\left(\frac{2}{3}\right)^2 \cdot \left(\frac{2}{3}\right)^6$

9. $\left(-\frac{5}{7}\right)^8 \cdot \left(-\frac{5}{7}\right)^9$

10. $(-2.9) \cdot (-2.9)^7$

11. $(5^4)^3$

12. $(b^{12})^3$

13. $(3.8^3)^4$

14. $\left(\left(-\frac{3}{4}\right)^5\right)^2$

ERROR ANALYSIS **Describe and correct the error in simplifying the expression.**

15. ✗
$$5^2 \cdot 5^9 = (5 \cdot 5)^{2+9}$$
$$= 25^{11}$$

16. ✗
$$(r^6)^4 = r^{6+4}$$
$$= r^{10}$$

Simplify the expression.

(3) 17. $(6g)^3$

18. $(-3v)^5$

19. $\left(\frac{1}{5}k\right)^2$

20. $(1.2m)^4$

21. $(rt)^{12}$

22. $\left(-\frac{3}{4}p\right)^3$

23. **CRITICAL THINKING** Is $3^2 + 3^3$ equal to 3^5? Explain.

24. **ARTIFACT** A display case for the artifact is in the shape of a cube. Each side of the display case is three times longer than the width of the artifact.

   a. Write an expression for the volume of the case. Write your answer as a power.

   b. Simplify the expression.

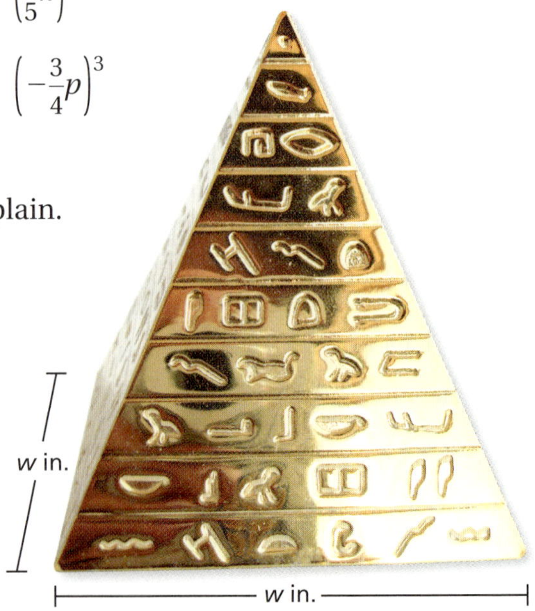

Assignment Guide and Homework Check

Level	Day 1 Activity Assignment	Day 2 Lesson Assignment	Homework Check
Basic	3–5, 33–37	1, 2, 6–13, 15–20, 23, 24	10, 12, 16, 18, 23, 24
Average	3–5, 33–37	1, 2, 6–16 even, 17–31 odd, 24, 28	10, 12, 16, 19, 25, 28
Advanced	3–5, 33–37	1, 2, 13–16, 20–23, 26–32	14, 22, 26, 30

Common Errors

- **Exercises 3–14** Students may multiply the bases as well as add the exponents. Remind them that the base stays the same and only the exponent changes when simplifying using the Product of Powers Property.
- **Exercises 11–14** Students may add the exponents instead of multiplying. Encourage them to write the expression as repeated multiplication and then add the exponents as shown in Example 2.
- **Exercises 17–22** Students may forget to raise the coefficient to the power. Remind them that everything within the parentheses will be raised to the power separately. Encourage them to write the expression as repeated multiplication.

9.2 Record and Practice Journal

Simplify the expression. Write your answer as a power.

1. $(-6)^5 \cdot (-6)^4$
   $(-6)^9$
2. $x^1 \cdot x^9$
   x^{10}
3. $\left(\frac{4}{5}\right)^3 \cdot \left(\frac{4}{5}\right)^{12}$
   $\left(\frac{4}{5}\right)^{15}$
4. $(-1.5)^{11} \cdot (-1.5)^{11}$
   $(-1.5)^{22}$
5. $(y^{10})^{20}$
   y^{200}
6. $\left(\left(-\frac{2}{9}\right)^8\right)^7$
   $\left(-\frac{2}{9}\right)^{56}$

Simplify the expression.

7. $(2a)^6$
   $64a^6$
8. $(-4b)^4$
   $256b^4$
9. $\left(-\frac{9}{10}p\right)^2$
   $\frac{81}{100}p^2$
10. $(xy)^{15}$
   $x^{15}y^{15}$
11. $10^5 \cdot 10^3 - (10^1)^8$
   0
12. $7^2(7^4 \cdot 7^4)$
   282,475,249
13. The surface area of the sun is about $4 \times 3.141 \times (7 \times 10^5)^2$ square kilometers. Simplify the expression.
   6,156,360,000,000 square kilometers

Technology For the Teacher

Answer Presentation Tool

Quiz*Show*

Vocabulary and Concept Check

1. When multiplying powers with the same base
2. no; The bases must be the same.

Practice and Problem Solving

3. 3^4
4. 8^{14}
5. $(-4)^{12}$
6. a^6
7. h^7
8. $\left(\frac{2}{3}\right)^8$
9. $\left(-\frac{5}{7}\right)^{17}$
10. $(-2.9)^8$
11. 5^{12}
12. b^{36}
13. 3.8^{12}
14. $\left(-\frac{3}{4}\right)^{10}$
15. The bases should not be multiplied.
   $$5^2 \cdot 5^9 = 5^{2+9} = 5^{11}$$
16. The exponents should not be added. Write the expression as repeated multiplication.
   $$(r^6)^4 = r^6 \cdot r^6 \cdot r^6 \cdot r^6 = r^{6+6+6+6} = r^{24}$$
17. $216g^3$
18. $-243v^5$
19. $\frac{1}{25}k^2$
20. $2.0736m^4$
21. $r^{12}t^{12}$
22. $-\frac{27}{64}p^3$
23. no; $3^2 + 3^3 = 9 + 27 = 36$ and $3^5 = 243$

Practice and Problem Solving

24. **a.** $(3w)^3$

b. $27w^3$

25. 496 **26.** x^4

27. 78,125 **28.** 3^9 ft

29. **a.** $16\pi \approx 50.24$ in.^{3}

b. $192\pi \approx 602.88$ in.^{3} Squaring each of the dimensions causes the volume to be 12 times larger.

30. $V = \frac{3}{4}b^2h$

31. See *Taking Math Deeper.*

32. **a.** 3

b. 4

Fair Game Review

33. 4 **34.** 25

35. 3 **36.** 6

37. B

Mini-Assessment

Simplify the expression. Write your answer as a power.

1. $b^2 \cdot b^6$ b^8
2. $(-2)^3 \cdot (-2)^2$ $(-2)^5$
3. $(c^8)^3$ c^{24}

Simplify the expression.

4. $(-5w)^4$ $625w^4$
5. $(st)^{11}$ $s^{11}t^{11}$

Taking Math Deeper

Exercise 31

This exercise gives students some practice in representing large numbers as powers.

1. Summarize the given information.

Mail delivered each second: $2^6 \cdot 5^3 = 8000$
Seconds in 6 days: $2^8 \cdot 3^4 \cdot 5^2 = 518{,}400$
How many pieces of mail in 6 days?

2. Multiply to find the number of pieces of mail delivered in 6 days.

$$(2^6 \cdot 5^3)(2^8 \cdot 3^4 \cdot 5^2) = 2^6 \cdot 5^3 \cdot 2^8 \cdot 3^4 \cdot 5^2$$
$$= 2^{14} \cdot 3^4 \cdot 5^5$$

3. Write the number in normal decimal form.

If you expand this number, you find that the U.S. postal service delivers about 4 billion pieces of mail each week (6 days not counting Sunday). This is an average of 13 pieces of mail per week for each person in the United States!

Project

Research the price of a postage stamp. How many times has it changed? How often has it changed? What has been the range in the cost over the last one hundred years?

Reteaching and Enrichment Strategies

If students need help. . .	If students got it. . .
Resources by Chapter • Practice A and Practice B • Puzzle Time Record and Practice Journal Practice Differentiating the Lesson Lesson Tutorials Skills Review Handbook	Resources by Chapter • Enrichment and Extension Start the next section

Simplify the expression.

25. $2^4 \cdot 2^5 - (2^2)^2$

26. $16\left(\frac{1}{2}x\right)^4$

27. $5^2(5^3 \cdot 5^2)$

28. **CLOUDS** The lowest altitude of an altocumulus cloud is about 3^8 feet. The highest altitude of an altocumulus cloud is about 3 times the lowest altitude. What is the highest altitude of an altocumulus cloud? Write your answer as a power.

29. **PYTHON EGG** The volume V of a python egg is given by the formula $V = \frac{4}{3}\pi abc$. For the python egg shown, $a = 2$ inches, $b = 2$ inches, and $c = 3$ inches.

a. Find the volume of the python egg.

b. Square the dimensions of the python egg. Then evaluate the formula. How does this volume compare to your answer in part (a)?

30. **PYRAMID** The volume of a square pyramid is $V = \frac{1}{3}b^2h$, where b is the length of one side of the base and h is the height of the pyramid. The length of each side of the base increases by 50%. Write a formula for the volume of the new pyramid.

31. **MAIL** The United States Postal Service delivers about $2^6 \cdot 5^3$ pieces of mail each second. There are $2^8 \cdot 3^4 \cdot 5^2$ seconds in 6 days. How many pieces of mail does the United States Postal Service deliver in 6 days? Write your answer as a power.

32. **Critical Thinking** Find the value of x in the equation without evaluating the power.

a. $2^5 \cdot 2^x = 256$

b. $\left(\frac{1}{3}\right)^2 \cdot \left(\frac{1}{3}\right)^x = \frac{1}{729}$

Fair Game Review What you learned in previous grades & lessons

Simplify. *(Skills Review Handbook)*

33. $\frac{4 \cdot 4}{4}$

34. $\frac{5 \cdot 5 \cdot 5}{5}$

35. $\frac{2 \cdot 3}{2}$

36. $\frac{8 \cdot 6 \cdot 6}{6 \cdot 8}$

37. **MULTIPLE CHOICE** What is the measure of each angle of the regular polygon? *(Section 5.3)*

(A) 45°

(B) 135°

(C) 1080°

(D) 1440°

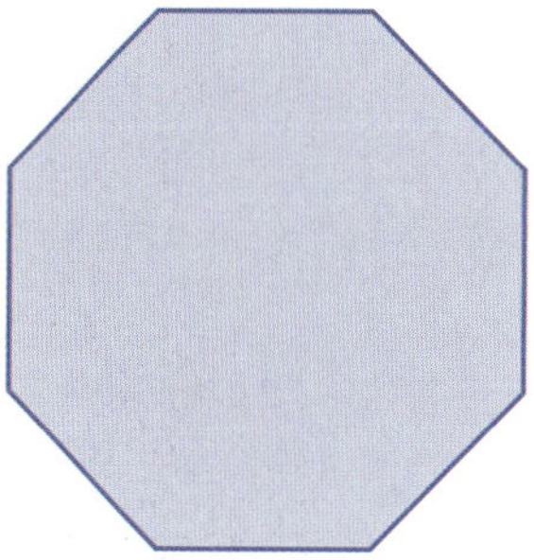

9.3 Quotient of Powers Property

Essential Question How can you divide two powers that have the same base?

1 ACTIVITY: Finding Quotients of Powers

Work with a partner.

a. Copy and complete the table.

Quotient	Repeated Multiplication Form	Power
$\frac{2^4}{2^2}$	$\frac{\cancel{2} \cdot \cancel{2} \cdot 2 \cdot 2}{\cancel{2} \cdot \cancel{2}}$	2^2
$\frac{(-4)^5}{(-4)^2}$	$\frac{\cancel{(-4)} \cdot \cancel{(-4)} \cdot (-4) \cdot (-4) \cdot (-4)}{\cancel{(-4)} \cdot \cancel{(-4)}}$	$(-4)^3$
$\frac{7^7}{7^3}$		
$\frac{8.5^9}{8.5^6}$		
$\frac{10^8}{10^5}$		
$\frac{3^{12}}{3^4}$		
$\frac{(-5)^7}{(-5)^5}$		
$\frac{11^4}{11^1}$		

b. **INDUCTIVE REASONING** Describe the pattern in the table. Then write a rule for dividing two powers that have the same base.

$$\frac{a^m}{a^n} = a^{\square}$$

c. Use your rule to simplify the quotients in the first column of the table above. Does your rule give the results in the third column?

Laurie's Notes

Introduction

For the Teacher

- **Goal:** Students will explore how to divide two powers with the same base.
- Remember to use correct vocabulary in this lesson. The numbers are not *canceling*. There is no mathematical definition of *cancel*. It is the factors that are common in the numerator and the denominator that are being divided out, similar to simplifying fractions. The fraction $\frac{2}{4} = \frac{1}{2}$ because there is a common factor of 2 in both the numerator and denominator that divide out. This same concept of dividing out common factors is why the Quotient of Powers Property works.

Motivate

- Tell students that you spent last evening working on a very long problem and you want them to give it a try. Write the problem on the board.

$$\frac{1}{2} \cdot \frac{2}{3} \cdot \frac{3}{4} \cdot \frac{4}{5} \cdot \frac{5}{6} \cdot \frac{6}{7} \cdot \frac{7}{8} \cdot \frac{8}{9} \cdot \frac{9}{10}$$

- It is likely that at least one of your students will recognize the answer immediately after you finish writing the problem. Act surprised and ask for their strategy…because you spent a long time on the problem.
- You want all students to recognize that the common factors in the numerator divide out with common factors in the denominator, leaving only $\frac{1}{10}$ as the final answer.

Activity Notes

Activity 1

- Have students work with their partner to complete the table. The first two problems have been done as a sample to follow. Notice the color coding. Also notice that integers and decimals are used as bases.

? "What do you notice about the number of factors in the numerator and denominator of the middle column, and the exponent used to write the power?" When you subtract the number of factors in the denominator from the number of factors in the numerator, it equals the exponent in the power.

- Students may need help in writing the summary statement: $\frac{a^m}{a^n} = a^{m-n}$. Stress that the bases must be the same in order to use this property.

Previous Learning

Students should know how to simplify fractions by dividing out common factors.

Activity Materials	
Introduction	**Textbook**
• calculator	• plastic cubes

Start Thinking! and Warm Up

Activity 9.3 **Start Thinking!** For use before Activity 9.3

Activity 9.3 **Warm Up** For use before Activity 9.3

Write the power as repeated multiplication.

1. 5^4 **2.** 7^3 **3.** 6^7

4. $(-4)^3$ **5.** $(-3)^5$ **6.** $(-1)^3$

9.3 Record and Practice Journal

Essential Question How can you divide two powers that have the same base?

1 ACTIVITY: Finding Quotients of Powers

Work with a partner.

a. Complete the table.

Quotient	Repeated Multiplication Form	Power
$\frac{2^4}{2^2}$	Check students' work.	2^2
$\frac{(-4)^5}{(-4)^2}$		$(-4)^3$
$\frac{7^7}{7^3}$		7^4
$\frac{8.5^9}{8.5^6}$		8.5^3
$\frac{10^8}{10^5}$		10^3
$\frac{3^{12}}{3^4}$		3^8
$\frac{(-5)^7}{(-5)^5}$		$(-5)^2$
$\frac{11^4}{11^1}$		11^3

b. INDUCTIVE REASONING Describe the pattern in the table. Then write a rule for dividing two powers that have the same base.

$\frac{a^m}{a^n} = a^{m-n}$ the difference of the exponents (numerator to denominator) of quotient column

Differentiated Instruction

Kinesthetic

Use algebra tiles or slips of paper to help students understand the Quotient of Powers Property. Have students model the quotient $\frac{x^4}{x^2}$.

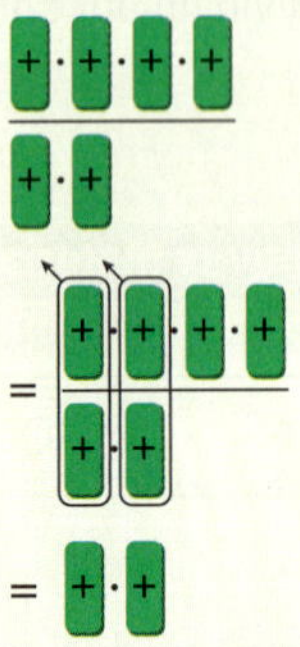

So, $\frac{x^4}{x^2} = x^2$.

9.3 Record and Practice Journal

c. Use your rule to simplify the quotients in the first column of the table on the previous page. Does your rule give the results in the third column?

yes

2 ACTIVITY: Comparing Volumes

Work with a partner.

How many of the smaller cubes will fit inside the larger cube? Record your results in the table on the next page. Describe the pattern in the table.

a. Sample: b. c. d.

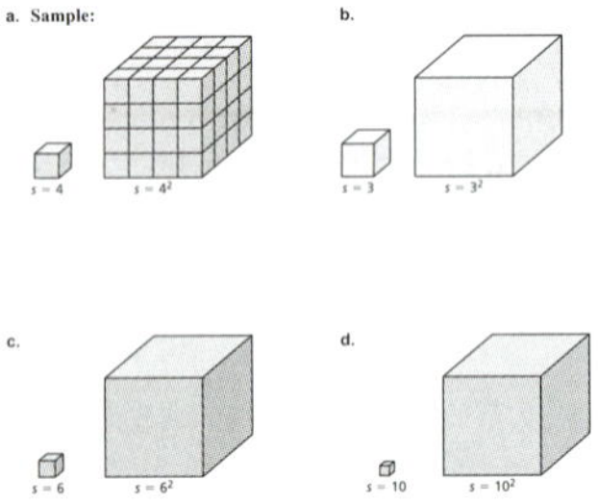

	Volume of Smaller Cube	Volume of Larger Cube	$\frac{\text{Larger Volume}}{\text{Smaller Volume}}$	Answer
a.	4^3	$(4^2)^3 = 4^6$	$\frac{4^6}{4^3}$	4^3
b.	3^3	$(3^2)^3 = 3^6$	$\frac{3^6}{3^3}$	3^3
c.	6^3	$(6^2)^3 = 6^6$	$\frac{6^6}{6^3}$	6^3
d.	10^3	$(10^2)^3 = 10^6$	$\frac{10^6}{10^3}$	10^3

What Is Your Answer?

3. **IN YOUR OWN WORDS** How can you divide two powers that have the same base? Give two examples of your rule.

Subtract their exponents then evaluate.

Laurie's Notes

Activity 2

- If you have small wooden or plastic cubes available, model one of these problems or a similar problem to start.
- Point out to students that $s = 4$ means the edge (or side) length is 4. In part (a), the side length of the larger cube is 4^2 or 16, which is 4 times as long as the small red cube. You can see it is four times longer by looking at the additional lines that have been drawn on the larger cube. Those same markings do not appear on the remaining cubes.
- When completing the table, it is necessary for students to simplify a power raised to a power as shown in the sample. Recall, $(4^2)^3 = 4^2 \cdot 4^2 \cdot 4^2 = 4^{2+2+2} = 4^6$.
- Students will work with their partner to complete the table.

? "How do you find the volume of the small cube each time?" Cube the side length; s^3.

? "How do you find the volume of the larger cube each time?" Cube the side length; s^3; The side length for the larger cube, however, is a power.

- When finding the ratio of the volumes, students will need to divide out the common factors.

? "What do you notice about the volume of the small cube and the answer?" The answer is always the same as the volume of the small cube.

What Is Your Answer?

- **Think-Pair-Share:** Students should read the question independently and then work with a partner to answer the question. When they have answered the question, the pair should compare their answer with another group and discuss any discrepancies.

Closure

- Simplify.

1. $\frac{2^2}{2} \cdot \frac{2^3}{2^2} \cdot \frac{2^4}{2^3}$ 2^3 **2.** $\frac{(-3)^7}{(-3)^4}$ $(-3)^3$

Technology For the Teacher

Dynamic Classroom

The Dynamic Planning Tool

Editable Teacher's Resources at *BigIdeasMath.com*

2 ACTIVITY: Comparing Volumes

Work with a partner.

How many of the smaller cubes will fit inside the larger cube? Record your results in the table. Describe the pattern in the table.

a. Sample:

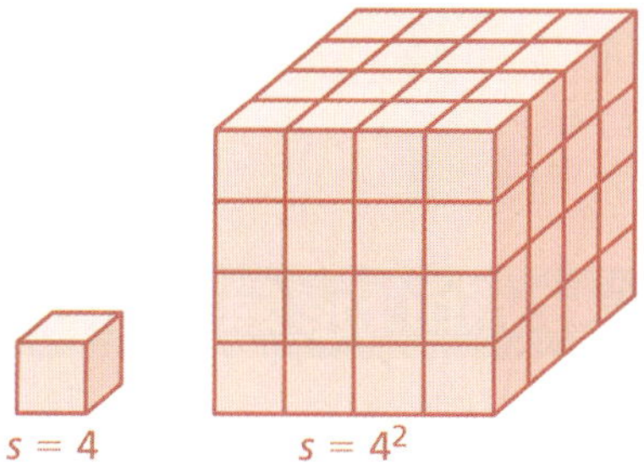

b.

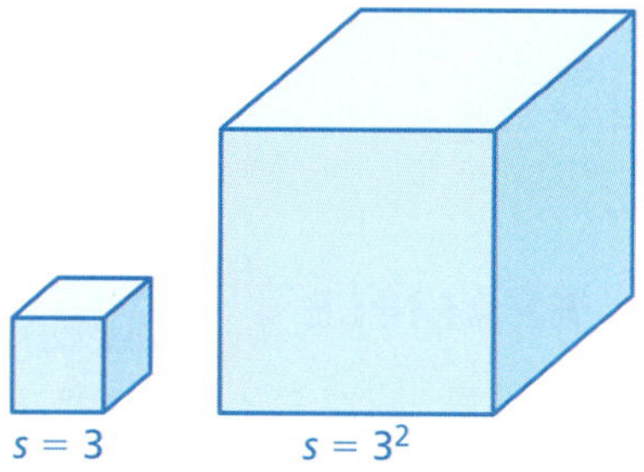

c.

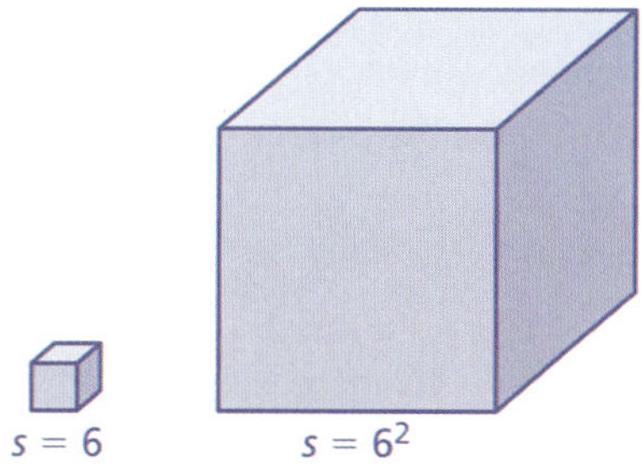

d.

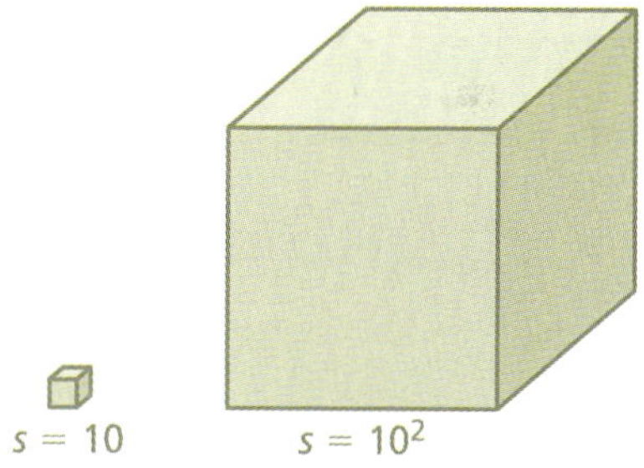

	Volume of Smaller Cube	Volume of Larger Cube	$\frac{\text{Larger Volume}}{\text{Smaller Volume}}$	Answer
a.	4^3	$(4^2)^3 = 4^6$	$\frac{4^6}{4^3}$	4^3
b.				
c.				
d.				

What Is Your Answer?

3. **IN YOUR OWN WORDS** How can you divide two powers that have the same base? Give two examples of your rule.

Practice Use what you learned about the Quotient of Powers Property to complete Exercises 3–6 on page 366.

9.3 Lesson

Check It Out
Lesson Tutorials
BigIdeasMath.com

Key Idea

Quotient of Powers Property

Words To divide powers with the same base, subtract their exponents.

Numbers $\dfrac{4^5}{4^2} = 4^{5-2} = 4^3$ **Algebra** $\dfrac{a^m}{a^n} = a^{m-n}$, where $a \neq 0$

EXAMPLE 1 Dividing Powers with the Same Base

a. $\dfrac{2^6}{2^4} = 2^{6-4}$ The base is 2. Subtract the exponents.

$= 2^2$ Simplify.

b. $\dfrac{(-7)^9}{(-7)^3} = (-7)^{9-3}$ The base is −7. Subtract the exponents.

$= (-7)^6$ Simplify.

c. $\dfrac{h^7}{h^6} = h^{7-6}$ The base is h. Subtract the exponents.

$= h^1 = h$ Simplify.

Common Error

When dividing powers, do not divide the bases.

$\dfrac{2^6}{2^4} = 2^2$, not 1^2.

On Your Own

Now You're Ready
Exercises 7–14

Simplify the expression. Write your answer as a power.

1. $\dfrac{9^7}{9^4}$
2. $\dfrac{4.2^6}{4.2^5}$
3. $\dfrac{(-8)^8}{(-8)^4}$
4. $\dfrac{x^8}{x^3}$

EXAMPLE 2 Simplifying an Expression

Simplify $\dfrac{3^4 \cdot 3^2}{3^3}$. Write your answer as a power.

The numerator is a product of powers.

$\dfrac{3^4 \cdot 3^2}{3^3} = \dfrac{3^{4+2}}{3^3}$ Add the exponents in the numerator.

$= \dfrac{3^6}{3^3}$ Simplify.

$= 3^{6-3}$ The base is 3. Subtract the exponents.

$= 3^3$ Simplify.

Laurie's Notes

Introduction

Connect

- **Yesterday:** Students explored exponents.
- **Today:** Students will use the Quotient of Powers Property to simplify expressions.

Motivate

- **Preparation:** Find the area of your classroom in square feet. Select two smaller regions of your room that make logical sense given the shape of your room. My classroom is shown. I found the area of the entire room (A + B); the area of B; and the area of C.

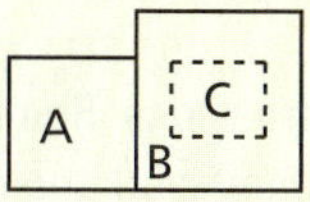

- Have students stand around the room so that they are an arm's length away from everyone else.
- Ask them to stand only in region B (which includes C).
- Ask them to move into region C. It should be very tight.
- Ask them to describe the three regions and how they felt about personal space. Then discuss population density and compute it for each region.

Lesson Notes

Key Idea

- Write the Key Idea. Discuss the Words, Numbers, and Algebra.

Example 1

? **Part (a):** Write and simplify the expression. The base is 2 for each power. Ask the following questions to help develop correct vocabulary.
  - "How many factors of 2 are in the numerator?" 6
  - "How many factors of 2 are in the denominator?" 4
  - "How many factors of 2 are common in *both* the numerator and denominator?" 4
  - "How many factors of 2 remain after you divide out the common factors?" 2
- Repeat similar questions for parts (b) and (c).

Example 2

? This example combines two properties. Ask the following questions.
  - "How many factors of 3 are in the numerator?" $4 + 2 = 6$
  - "How many factors of 3 are in the denominator?" 3
  - "How many factors of 3 are common in *both* the numerator and denominator?" 3
  - "How many factors of 3 remain after you divide out the common factors?" 3

Goal Today's lesson is using the Quotient of Powers Property to simplify expressions.

Start Thinking! and Warm Up

Lesson 9.3 **Warm Up**
For use before Lesson 9.3

Lesson 9.3 **Start Thinking!**
For use before Lesson 9.3

Scott learned about the Quotient of Powers Property in math class, but he is not convinced that it is helpful. For example, he thinks that it is just as easy to simplify $\frac{2^5}{2^2}$ by calculating $2^5 = 32$ and dividing by $2^2 = 4$ to get 8. Do you agree or disagree with Scott? Give reasons to support your answer.

Extra Example 1

Simplify the expression. Write your answer as a power.

a. $\frac{4^5}{4^2}$ 4^3

b. $\frac{(-2)^{10}}{(-2)^3}$ $(-2)^7$

c. $\frac{p^7}{p^6}$ p

On Your Own

1. 9^3 **2.** 4.2

3. $(-8)^4$ **4.** x^5

Extra Example 2

Simplify $\frac{5^6 \cdot 5^2}{5^4}$. Write your answer as a power. 5^4

Extra Example 3

Simplify $\frac{z^6}{z^2} \cdot \frac{z^8}{z^5}$. Write your answer as a power. z^7

On Your Own

5. 2^7 6. d^5

Extra Example 4

The projected population of Hawaii in 2020 is about $5.48 \cdot 2^{18}$. The land area of Hawaii is about 2^{14} square kilometers. Predict the average number of people per square kilometer in 2020. about 88 people per km^2

On Your Own

7. 36 people per km^2

English Language Learners

Organization

Have students organize the *Key Ideas* of this chapter in their notebooks. This will provide them with easy access to the material and concepts of the chapter.

Key Idea	Product of Powers Property	Quotient of Powers Property
Example	$(-3)^2(-3)^4$	$\frac{5^3}{5^2}$
Answer	$(-3)^6 = 729$	5
Method	Add exponents: $2 + 4$	Subtract exponents: $3 - 2$

Laurie's Notes

Example 3

- This example also combines two properties.
- Work through the problem as shown.
- Discuss the approach with students. Each quotient was simplified first and then the product of the two expressions was found.

? "Will the answer be the same if the product of the two expressions is found and then the quotient is simplified? Explain." Yes; It is the same as multiplying two fractions and then simplifying the answer.

- Simplify the expression using the alternate approach in the Study Tip.

$$\frac{a^{10}}{a^6} \cdot \frac{a^7}{a^4} = \frac{a^{10} \cdot a^7}{a^6 \cdot a^4} = \frac{a^{10+7}}{a^{6+4}} = \frac{a^{17}}{a^{10}} = a^{17-10} = a^7$$

On Your Own

- There is more than one way to simplify these expressions. Remind students to think about the number of factors as they work the problems.
- **Question 6:** Students may forget that $d = d^1$.
- Have volunteers write their solutions on the board.

Example 4

- This problem is about population density, the number of people per square unit. In this case, it is the projected number of people in Tennessee per square mile in 2030.
- When working through this problem, notice that the factor 5 in the numerator is not the same base as the other two factors.

? "Why can you move 5 out of the numerator and write it as a whole number times the quotient of $(5.9)^8$ and $(5.9)^6$?" definition of multiplying fractions

- Simplify the quotient and multiply by 5.
- Use local landmarks to help students visualize the size of a square mile.

Closure

- Explain how the Quotient of Powers Property is related to simplifying fractions. You divide out the common factors.

Technology For the Teacher

Dynamic Classroom

The Dynamic Planning Tool
Editable Teacher's Resources at *BigIdeasMath.com*

EXAMPLE 3 Simplifying an Expression

Study Tip

You can also simplify the expression in Example 3 as follows.

$$\frac{a^{10}}{a^6} \cdot \frac{a^7}{a^4} = \frac{a^{10} \cdot a^7}{a^6 \cdot a^4}$$
$$= \frac{a^{17}}{a^{10}}$$
$$= a^{17-10}$$
$$= a^7$$

Simplify $\frac{a^{10}}{a^6} \cdot \frac{a^7}{a^4}$. Write your answer as a power.

$\frac{a^{10}}{a^6} \cdot \frac{a^7}{a^4} = a^{10-6} \cdot a^{7-4}$ — Subtract the exponents.

$= a^4 \cdot a^3$ — Simplify.

$= a^{4+3}$ — Add the exponents.

$= a^7$ — Simplify.

On Your Own

Now You're Ready
Exercises 16–21

Simplify the expression. Write your answer as a power.

5. $\frac{2^{15}}{2^3 \cdot 2^5}$

6. $\frac{d^5}{d} \cdot \frac{d^9}{d^8}$

EXAMPLE 4 Real-Life Application

The projected population of Tennessee in 2030 is about $5 \cdot 5.9^8$. Predict the average number of people per square mile in 2030.

Use a model to solve the problem.

$$\text{People per square mile} = \frac{\text{Population in 2030}}{\text{Land area}}$$

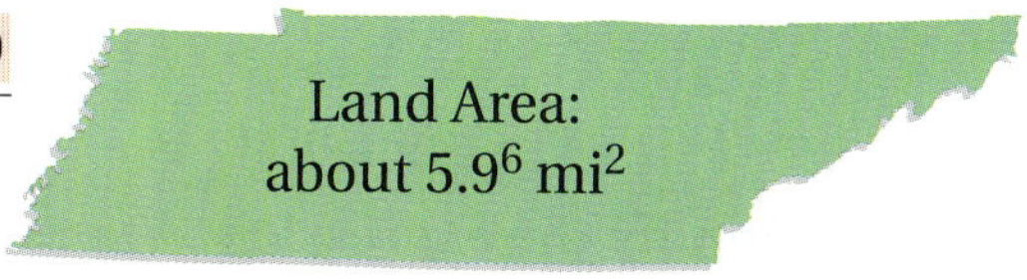

$= \frac{5 \cdot 5.9^8}{5.9^6}$ — Substitute.

$= 5 \cdot \frac{5.9^8}{5.9^6}$ — Rewrite.

$= 5 \cdot 5.9^2$ — Subtract the exponents.

$= 174.05$ — Evaluate.

There will be about 174 people per square mile in Tennessee in 2030.

On Your Own

Now You're Ready
Exercises 23–28

7. The projected population of Alabama in 2020 is about $2.25 \cdot 2^{21}$. The land area of Alabama is about 2^{17} square kilometers. Predict the average number of people per square kilometer in 2020.

9.3 Exercises

Check It Out
Help with Homework
BigIdeasMath.com

Vocabulary and Concept Check

1. **WRITING** Explain in your own words what it means to divide powers.

2. **WHICH ONE DOESN'T BELONG?** Which quotient does *not* belong with the other three? Explain your reasoning.

$\frac{(-10)^7}{(-10)^2}$ $\frac{6^3}{6^2}$ $\frac{(-4)^8}{(-3)^4}$ $\frac{5^6}{5^3}$

Practice and Problem Solving

Simplify the expression. Write your answer as a power.

3. $\frac{6^{10}}{6^4}$
4. $\frac{8^9}{8^7}$
5. $\frac{(-3)^4}{(-3)^1}$
6. $\frac{4.5^5}{4.5^3}$

① 7. $\frac{5^9}{5^3}$
8. $\frac{64^4}{64^3}$
9. $\frac{(-17)^5}{(-17)^2}$
10. $\frac{(-7.9)^{10}}{(-7.9)^4}$

11. $\frac{(-6.4)^8}{(-6.4)^6}$
12. $\frac{\pi^{11}}{\pi^7}$
13. $\frac{b^{24}}{b^{11}}$
14. $\frac{n^{18}}{n^7}$

15. **ERROR ANALYSIS** Describe and correct the error in simplifying the quotient.

✗ $\frac{6^{15}}{6^5} = 6^{\frac{15}{5}}$
$= 6^3$

Simplify the expression. Write your answer as a power.

② ③ 16. $\frac{7^5 \cdot 7^3}{7^2}$
17. $\frac{2^{19} \cdot 2^5}{2^{12} \cdot 2^3}$
18. $\frac{(-8.3)^8}{(-8.3)^7} \cdot \frac{(-8.3)^4}{(-8.3)^3}$

19. $\frac{\pi^{30}}{\pi^{18} \cdot \pi^4}$
20. $\frac{c^{22}}{c^8 \cdot c^9}$
21. $\frac{k^{13}}{k^5} \cdot \frac{k^{17}}{k^{11}}$

22. **SOUND INTENSITY** The sound intensity of a normal conversation is 10^6 times greater than the quietest noise a person can hear. The sound intensity of a jet at takeoff is 10^{14} times greater than the quietest noise a person can hear. How many times more intense is the sound of a jet at takeoff than the sound of a normal conversation?

Assignment Guide and Homework Check

Level	Day 1 Activity Assignment	Day 2 Lesson Assignment	Homework Check
Basic	3–6, 33–37	1, 2, 7–21 odd, 22–25	9, 17, 21, 22, 25
Average	3–6, 33–37	1, 2, 7–25 odd, 29, 30	9, 17, 21, 25, 30
Advanced	3–6, 33–37	1, 2, 13–15, 20, 21, 26–32	14, 20, 26, 30, 31

Common Errors

- **Exercises 3–14** Students may divide the exponents instead of subtracting the exponents. Remind them that the Quotient of Powers Property states that the exponents are subtracted.
- **Exercises 16–21** Students may multiply and/or divide the bases when simplifying the expression. Remind them that the base does not change when they use the Quotient of Powers Property to simplify an expression.
- **Exercises 23–28** Students may try to combine unlike terms when simplifying. Remind them of the rules of combining like terms, and that the Quotient of Powers and Product of Powers Properties can only be used with like terms.

9.3 Record and Practice Journal

Simplify the expression. Write your answer as a power.

1. $\frac{7^6}{7^5}$ — **7**
2. $\frac{(-21)^{15}}{(-21)^9}$ — $(-21)^6$
3. $\frac{8.6^{11}}{8.6^4}$ — $(8.6)^7$
4. $\frac{(3.9)^{20}}{(3.9)^{10}}$ — $(3.9)^{10}$
5. $\frac{t^7}{t^3}$ — t^4
6. $\frac{d^{32}}{d^{16}}$ — d^{16}
7. $\frac{8^7 \bullet 8^4}{8^9}$ — 8^2
8. $\frac{(-1.1)^{13} \bullet (-1.1)^{12}}{(-1.1)^{10} \bullet (-1.1)^1}$ — $(-1.1)^{14}$
9. $\frac{m^{50}}{m^{22}} \bullet \frac{m^{17}}{m^{15}}$ — m^{30}

Simplify the expression.

10. $\frac{k \bullet 3^9}{3^5}$ — $81k$
11. $\frac{x^4 \bullet y^{10} \bullet 2^{11}}{y^8 \bullet 2^7}$ — $16x^4y^2$
12. $\frac{a^{15}b^{19}}{a^6b^{12}}$ — a^9b^7

13. The radius of a basketball is about 3.6 times greater than the radius of a tennis ball. How many times greater is the volume of a basketball than the volume of a tennis ball? $\left(\text{Note: The volume of a sphere is } V = \frac{4}{3}\pi r^3.\right)$

46.656

Technology For the Teacher

Answer Presentation Tool

Quiz*Show*

Vocabulary and Concept Check

1. To divide powers means to divide out the common factors of the numerator and denominator. To divide powers with the same base, write the power with the common base and an exponent found by subtracting the exponent in the denominator from the exponent in the numerator.
2. $\frac{(-4)^8}{(-3)^4}$; The other quotients have powers with the same base.

Practice and Problem Solving

3. 6^6
4. 8^2
5. $(-3)^3$
6. 4.5^2
7. 5^6
8. 64
9. $(-17)^3$
10. $(-7.9)^6$
11. $(-6.4)^2$
12. π^4
13. b^{13}
14. n^{11}
15. You should subtract the exponents instead of dividing them.

$$\frac{6^{15}}{6^5} = 6^{15-5}$$
$$= 6^{10}$$

16. 7^6
17. 2^9
18. $(-8.3)^2$
19. π^8
20. c^5
21. k^{14}
22. 10^8 times

Practice and Problem Solving

23. $64x$ **24.** $6w$

25. $125a^3b^2$ **26.** $125cd^2$

27. x^7y^6 **28.** m^9n

29. See *Taking Math Deeper.*

30. a. *Sample answer:* $m = 5$, $n = 3$

b. yes; Any two numbers that satisfy the equation $m - n = 2$.

31. 10^{13} galaxies

32. 10; The difference in the exponents needs to be 9. To find x, solve the equation $3x - (2x + 1) = 9$.

Fair Game Review

33. -9 **34.** -8

35. 61 **36.** -4

37. B

Mini-Assessment

Simplify the expression. Write your answer as a power.

1. $\frac{(-4)^3}{(-4)^1}$ $(-4)^2$
2. $\frac{9.7^7}{9.7^3}$ 9.7^4
3. $\frac{5^4 \cdot 5^2}{5^3}$ 5^3
4. $\frac{m^{10}}{m^5 \cdot m^2}$ m^3
5. $\frac{y^{17}}{y^{10}} \cdot \frac{y^6}{y^3}$ y^{10}

Taking Math Deeper

Exercise 29

This is an interesting problem. The memory in the different styles of MP3 Players increases exponentially, but the price increases linearly.

1 Compare Player D with Player B.

a. $\frac{2^4}{2^2} = 2^2 = 4$ times more memory

2 Compare the memory with the price.

If you plot the five points representing the memory and the prices, you get the following graph.

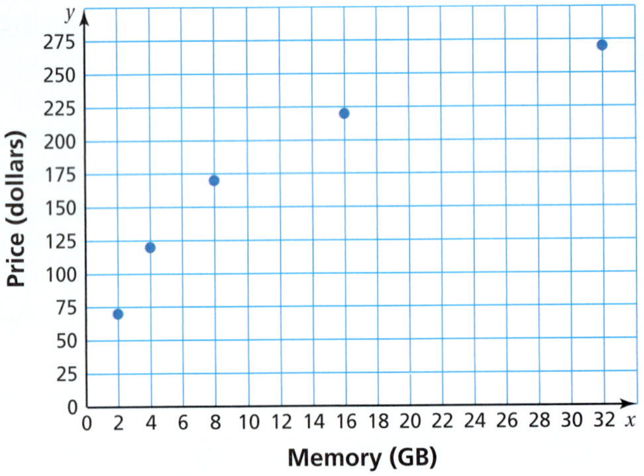

3 Answer the question.

b. This graph does not show a constant rate of change. In other words, the relationship between price and memory is not linear. However, the differences in price between consecutive sizes reflect a constant rate of change.

Project

What changes in technology have occurred over the past 50 years? What do you predict will change over the next 50 years?

Reteaching and Enrichment Strategies

If students need help. . .	If students got it. . .
Resources by Chapter • Practice A and Practice B • Puzzle Time Record and Practice Journal Practice Differentiating the Lesson Lesson Tutorials Skills Review Handbook	Resources by Chapter • Enrichment and Extension • School-to-Work Start the next section

Simplify the expression.

4 23. $\frac{x \cdot 4^8}{4^5}$

24. $\frac{6^3 \cdot w}{6^2}$

25. $\frac{a^3 \cdot b^4 \cdot 5^4}{b^2 \cdot 5}$

26. $\frac{5^{12} \cdot c^{10} \cdot d^2}{5^9 \cdot c^9}$

27. $\frac{x^{15}y^9}{x^8y^3}$

28. $\frac{m^{10}n^7}{m^1n^6}$

MP3 Player	Memory (GB)	Price
A	2^1	\$70
B	2^2	\$120
C	2^3	\$170
D	2^4	\$220
E	2^5	\$270

29. **MEMORY** The memory capacities and prices of five MP3 players are shown in the table.

a. How many times more memory does MP3 Player D have than MP3 Player B?

b. Do the differences in price between consecutive sizes reflect a constant rate of change?

30. **CRITICAL THINKING** Consider the equation $\frac{9^m}{9^n} = 9^2$.

a. Find two numbers m and n that satisfy the equation.

b. Are there any other pairs of numbers that satisfy the equation? Explain.

Milky Way Galaxy
$10 \cdot 10^{10}$ stars

31. **STARS** There are about 10^{24} stars in the Universe. Each galaxy has approximately the same number of stars as the Milky Way Galaxy. About how many galaxies are in the Universe?

32. Number Sense Find the value of x that makes $\frac{8^{3x}}{8^{2x+1}} = 8^9$ true. Explain how you found your answer.

Fair Game Review What you learned in previous grades & lessons

Subtract. *(Skills Review Handbook)*

33. $-4 - 5$

34. $-23 - (-15)$

35. $33 - (-28)$

36. $18 - 22$

37. **MULTIPLE CHOICE** What is the value of x? *(Section 5.1)*

Ⓐ 20

Ⓑ 30

Ⓒ 45

Ⓓ 60

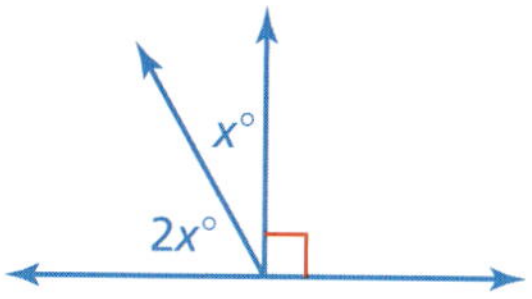

9 Study Help

Check It Out
Graphic Organizer
BigIdeasMath.com

You can use an **information wheel** to organize information about a topic. Here is an example of an information wheel for exponents.

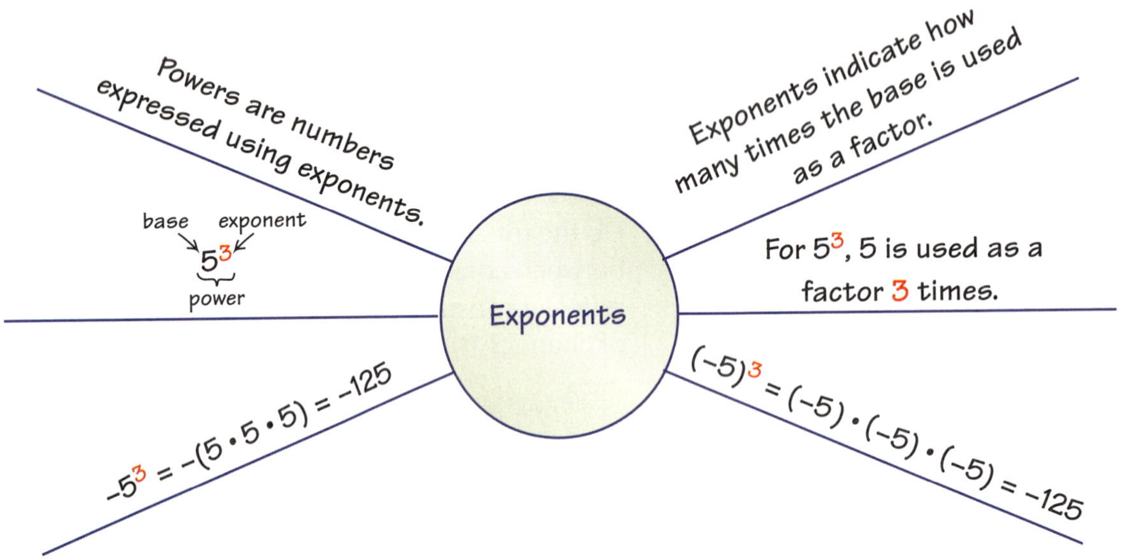

On Your Own

Make an information wheel to help you study these topics.

1. order of operations
2. Product of Powers Property
3. Quotient of Powers Property

After you complete this chapter, make information wheels for the following topics.

4. zero exponents
5. negative exponents
6. writing numbers in scientific notation
7. writing numbers in standard form
8. Choose three other topics you studied earlier in this course. Make an information wheel for each topic to summarize what you know about them.

"My information wheel for Fluffy has matching adjectives and nouns."

Sample Answers

1.

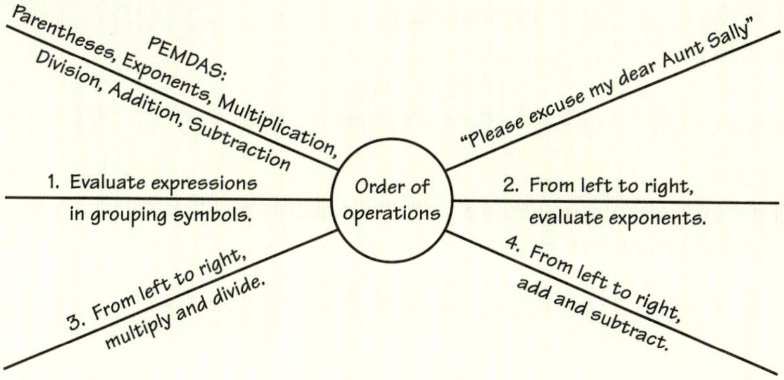

2.

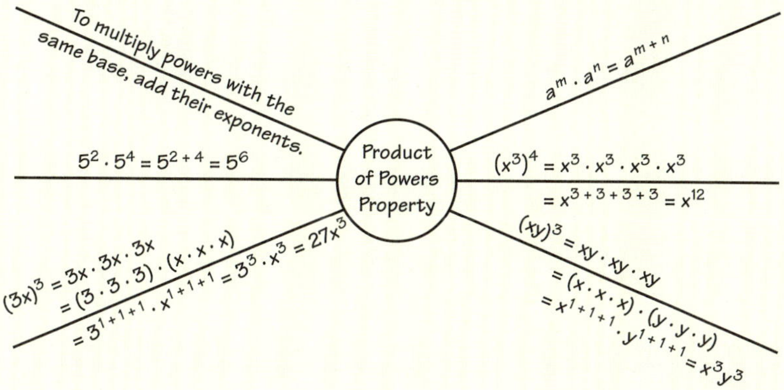

3.

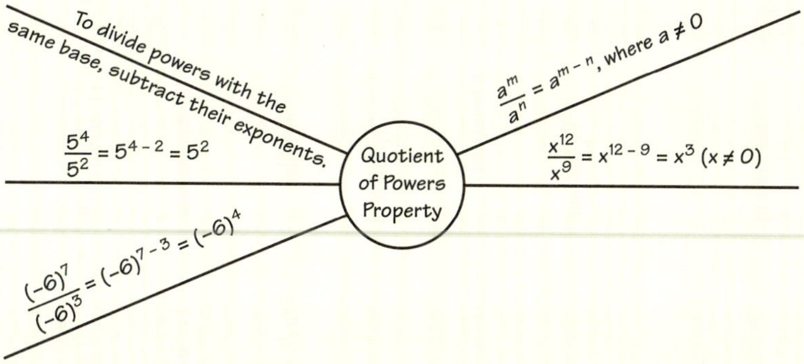

List of Organizers

Available at *BigIdeasMath.com*

Comparison Chart
Concept Circle
Definition (Idea) and Example Chart
Example and Non-Example Chart
Formula Triangle
Four Square
Information Frame
Information Wheel
Notetaking Organizer
Process Diagram
Summary Triangle
Word Magnet
Y Chart

About this Organizer

An **Information Wheel** can be used to organize information about a concept. Students write the concept in the middle of the "wheel." Then students write information related to the concept on the "spokes" of the wheel. Related information can include, but is not limited to: vocabulary words or terms, definitions, formulas, procedures, examples, and visuals. This type of organizer serves as a good summary tool because any information related to a concept can be included.

Technology For the Teacher

Vocabulary Puzzle Builder

Answers

1. $(-5)^4$
2. $\left(\frac{1}{6}\right)^5$
3. $(-x)^6$
4. 7^2m^3
5. 625
6. 64
7. 3^9
8. a^{15}
9. $81c^4$
10. $\frac{4}{49}p^2$
11. 8^3
12. 6^8
13. π^3
14. t^{10}
15. 9; 99; 999; 9999
16. no; $(ab)^2 = (ab) \cdot (ab) = a \cdot a \cdot b \cdot b = a^2b^2$
17. 10^5 times

Assessment Book

Chapter 9 Quiz
For use after Section 9.3

Write the product using exponents.

1. $\frac{1}{5} \cdot \frac{1}{5} \cdot \frac{1}{5} \cdot \frac{1}{5} \cdot \frac{1}{5}$
2. $(-2) \cdot (-2) \cdot (-2)$
3. $y \cdot y \cdot y \cdot y \cdot y \cdot y$
4. $4 \cdot 4 \cdot 4 \cdot c \cdot c$

Evaluate the expression.

5. 2^4
6. $(-3)^3$

Simplify the expression. Write your answer as a power.

7. $(-1)^3 \cdot (-1)^2$
8. $(b^4)^2$

Simplify the expression.

9. $(4f)^3$
10. $\left(-\frac{3}{8}t^2\right)^2$

Simplify the expression. Write your answer as a power.

11. $\frac{(-2)^{10}}{(-2)^5}$
12. $\frac{5^4 \cdot 5^9}{5^6}$
13. $\frac{x^{14}}{x^4 \cdot x^2}$
14. $\frac{y^3}{y^{11}} \cdot \frac{y^{21}}{y^9}$
15. The nth term of a sequence can be found by evaluating $2^n + 5$. Complete the table to find the first four terms of the sequence.

n	$2^n + 5$
1	
2	
3	
4	

16. Is x^3x^5 equivalent to x^{15}? Explain. If not, what expression is equivalent to x^3x^5?
17. Find the value of x that makes $\frac{(-2)^{4x+2}}{(-2)^{2x}} = (-2)^{12}$ true.

Answers

1. ______
2. ______
3. ______
4. ______
5. ______
6. ______
7. ______
8. ______
9. ______
10. ______
11. ______
12. ______
13. ______
14. ______
15. See left.
16. ______
17. ______

Alternative Quiz Ideas

100% Quiz
Error Notebook
Group Quiz
Homework Quiz
Math Log
Notebook Quiz
Partner Quiz
Pass the Paper

100% Quiz

This is a quiz where students are given the answers and then they have to explain and justify each answer.

Reteaching and Enrichment Strategies

If students need help. . .	If students got it. . .
Resources by Chapter • Study Help • Practice A and Practice B • Puzzle Time Lesson Tutorials *BigIdeasMath.com* Practice Quiz Practice from the Test Generator	Resources by Chapter • Enrichment and Extension • School-to-Work Game Closet at *BigIdeasMath.com* Start the next section

Technology For the Teacher

Answer Presentation Tool
Big Ideas Test Generator

9.1–9.3 Quiz

Check It Out
Progress Check
BigIdeasMath.com

Write the product using exponents. *(Section 9.1)*

1. $(-5) \cdot (-5) \cdot (-5) \cdot (-5)$

2. $\frac{1}{6} \cdot \frac{1}{6} \cdot \frac{1}{6} \cdot \frac{1}{6} \cdot \frac{1}{6}$

3. $(-x) \cdot (-x) \cdot (-x) \cdot (-x) \cdot (-x) \cdot (-x)$

4. $7 \cdot 7 \cdot m \cdot m \cdot m$

Evaluate the expression. *(Section 9.1)*

5. 5^4

6. $(-2)^6$

Simplify the expression. Write your answer as a power. *(Section 9.2)*

7. $3^8 \cdot 3$

8. $(a^5)^3$

Simplify the expression. *(Section 9.2)*

9. $(3c)^4$

10. $\left(-\frac{2}{7}p\right)^2$

Simplify the expression. Write your answer as a power. *(Section 9.3)*

11. $\frac{8^7}{8^4}$

12. $\frac{6^3 \cdot 6^7}{6^2}$

13. $\frac{\pi^{15}}{\pi^3 \cdot \pi^9}$

14. $\frac{t^{13}}{t^5} \cdot \frac{t^8}{t^6}$

15. **SEQUENCE** The nth term of a sequence can be found by evaluating $10^n - 1$. Copy and complete the table to find the first four terms of the sequence. *(Section 9.1)*

n	$10^n - 1$
1	
2	
3	
4	

16. **CRITICAL THINKING** Is $(ab)^2$ equivalent to ab^2? Explain. *(Section 9.2)*

17. **EARTHQUAKES** An earthquake of magnitude 3.0 is 10^2 times stronger than an earthquake of magnitude 1.0. An earthquake of magnitude 8.0 is 10^7 times stronger than an earthquake of magnitude 1.0. How many times stronger is an earthquake of magnitude 8.0 than an earthquake of magnitude 3.0? *(Section 9.3)*

9.4 Zero and Negative Exponents

Essential Question How can you define zero and negative exponents?

1 ACTIVITY: Finding Patterns and Writing Definitions

Work with a partner.

a. Talk about the following notation.

Thousands, Hundreds, Tens, Ones

$$4327 = 4 \cdot 10^3 + 3 \cdot 10^2 + 2 \cdot 10^1 + 7 \cdot 10^{\square}$$

What patterns do you see in the first three exponents?
Continue the pattern to find the fourth exponent.
How would you define 10^0? Explain.

b. Copy and complete the table.

n	5	4	3	2	1	0
2^n						

What patterns do you see in the first six values of 2^n?
How would you define 2^0? Explain.

c. Use the Quotient of Powers Property to complete the table.

$\frac{3^5}{3^2} = 3^{5-2} = 3^3 = 27$
$\frac{3^4}{3^2} = 3^{4-2} = \square = \square$
$\frac{3^3}{3^2} = 3^{3-2} = \square = \square$
$\frac{3^2}{3^2} = 3^{2-2} = \square = \square$

What patterns do you see in the first four rows of the table?
How would you define 3^0? Explain.

Laurie's Notes

Introduction

For the Teacher

- **Goal:** Students will explore negative powers and define a^0 to be equal to 1.
- The examples presented do not *prove* that $a^0 = 1$.

Motivate

- Use U.S. currency to introduce zero and negative exponents.
- Write the chart on the board. The middle row is the ratio of the currency to \$1. Another way of saying this is, how many one dollar bills are there in the currency *c*? For denominations greater than \$1, this is a whole number. For denominations less than \$1, this is a fraction.

Currency, *c*	\$1000	\$100	\$10	\$1	\$.10	\$.01
Ratio of *c*: one dollar	1000	100	10	1	$\frac{1}{10}$	$\frac{1}{100}$
Ratio as power of 10	10^3	10^2	10^1	$10^?$	$10^?$	$10^?$

- When you get to the last three cells, ask students what they think the exponents should be. It is okay if students do not have specific suggestions.

Activity Notes

Activity 1

- Have students work through all three parts of this activity before there is a class discussion. Each presents a different method for why it seems natural to define a^0 as equal to 1. There are many patterns that can be explored.

? "In part (a), when numbers are written in expanded notation, does it seem reasonable that the ones place should also be a power of 10?" Answers vary; most students recognize that all of the other place values are powers of 10, and the ones should be also.

- **Connection:** You can use the same model from part (a) to help students understand negative exponents as well. The tenths place value is 10^{-1}, hundredths place value is 10^{-2}, and so on.

? "How would you define 10^0? Why?" Listen for decreasing exponents in the powers of 10; others may say it must be 1, otherwise when you multiply by the 7 in the ones places, you get something other than 7.

- In part (b), the powers of 2 are decreasing. Students should describe a pattern where each successive power of 2 is half the previous power.

? "How would you define 2^0? Why?" Listen for, "each power is half the previous," and "$\frac{1}{2}$ of 2 is 1."

- Notice how this pattern will naturally lead into negative exponents. Half of 1 is $\frac{1}{2}$, which equals 2^{-1}.
- In part (c), the Quotient of Powers Property is used to help students reach the same conclusion: define 3^0 as equal to 1.

Previous Learning

Students should know how fractions are multiplied, simplified, changed to a mixed number, and converted to a decimal.

Start Thinking! and Warm Up

Activity 9.4 Start Thinking!
For use before Activity 9.4

Activity 9.4 Warm Up
For use before Activity 9.4

Simplify the expression. Write your answer as a power.

1. $\frac{5^4 \bullet 5^2}{5^3}$
2. $\frac{2^{11} \bullet 2^5}{2^{13}}$
3. $\frac{4^5 \bullet 4^3}{4^2}$
4. $\frac{a^{13} \bullet a^{11}}{a^{12}}$
5. $\frac{c^9 \bullet c^5}{c^{10}}$
6. $\frac{n^7 \bullet n^{14}}{n^{11}}$

9.4 Record and Practice Journal

Essential Question How can you define zero and negative exponents?

1 ACTIVITY: Finding Patterns and Writing Definitions

Work with a partner.

a. Talk about the following notation.

Thousands, Hundreds, Tens, Ones

$4327 = 4 \bullet 10^3 + 3 \bullet 10^2 + 2 \bullet 10^1 + 7 \bullet 10^{\square}$ **0**

What patterns do you see in the first three exponents?

decrease by one each time

Continue the pattern to find the fourth exponent.

How would you define 10^0? Explain.

$10^0 = 1$

b. Complete the table.

n	5	4	3	2	1	0
2^n	32	16	8	4	2	1

c. Use the Quotient of Powers Property to complete the table.

$\frac{3^5}{3^2} = 3^{5-2} =$	3^3	$= 27$
$\frac{3^4}{3^2} = 3^{4-2} =$	3^2	$= 9$
$\frac{3^3}{3^2} = 3^{3-2} =$	3^1	$= 3$
$\frac{3^2}{3^2} = 3^{2-2} =$	3^0	$= 1$

Differentiated Instruction

Visual

Help students to understand zero and negative exponents using methods already known to them.

Evaluating and then simplifying:

$$\frac{3^2}{3^3} = \frac{9}{27} = \frac{9 \div 9}{27 \div 9} = \frac{1}{3}$$

Dividing out common factors:

$$\frac{3^2}{3^3} = \frac{3^1 \cdot 3^1}{3^1 \cdot 3^1 \cdot 3} = \frac{1}{3}$$

Quotient of Powers Property:

$$\frac{3^2}{3^3} = 3^{2-3} = 3^{-1} = \frac{1}{3}$$

9.4 Record and Practice Journal

What patterns do you see in the first four rows of the table on the previous page? How would you define 3^0? Explain.

Each time exponent in numerator decreases by 1, the answer is $\frac{1}{3}$ of the previous row; $3^0 = 1$.

2 ACTIVITY: Comparing Volumes

Work with a partner.

The quotients show three ratios of the volumes of the solids. Identify each ratio, find its value, and describe what it means.

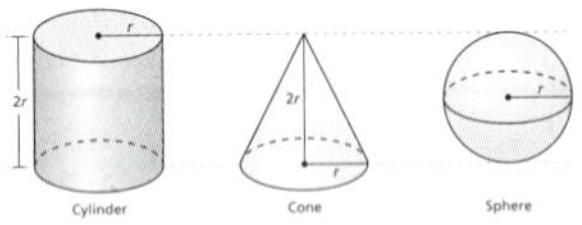

a. $2\pi r^3 \div \frac{2}{3}\pi r^3 =$ *volume of cylinder divided by volume of cone; 3; Volume of the cylinder is 3 times the volume of the cone.*

b. $\frac{4}{3}\pi r^3 \div \frac{2}{3}\pi r^3 =$ *volume of sphere divided by volume of cone; 2; Volume of the sphere is 2 times the volume of the cone.*

c. $2\pi r^3 \div \frac{4}{3}\pi r^3 =$ *volume of cylinder divided by volume of sphere; $\frac{3}{2}$; Volume of the cylinder is $1\frac{1}{2}$ times the volume of the sphere.*

3 ACTIVITY: Writing a Definition

Work with a partner.

Compare the two methods used to simplify $\frac{3^2}{3^5}$. Then describe how you can rewrite a power with a negative exponent as a fraction.

Method 1

$$\frac{3^2}{3^5} = \frac{3^1 \cdot 3^1}{{}_1 3 \cdot {}_1 3 \cdot 3 \cdot 3 \cdot 3} = \frac{1}{3^3}$$

Method 2

$$\frac{3^2}{3^5} = 3^{2-5} = 3^{-3}$$

Method 1 lists factors and reduces the fraction. Method 2 uses Quotient of Powers Property.; rewrite as a fraction with 1 in the numerator and base to the absolute value of the exponent in the denominator.

What Is Your Answer?

4. **IN YOUR OWN WORDS** How can you define zero and negative exponents? Give two examples of each.

A base to zero power equals one and negative exponents result when exponent in denominator is greater than exponent in numerator.

Laurie's Notes

Activity 2

- Students know the formulas for the volume of a cylinder and a cone.
- Note that the radius of all 3 solids is the same, r, and the height (or diameter) of all 3 solids is the same, $2r$.
- The question looks at the ratio of the volumes of two solids. The volume of the cone is less than either of the other two solids. It is also true that the volume of the cylinder exceeds the volume of the sphere. Discuss these observations before students examine the ratios.
- **Teaching Tip:** The numerator and denominator each contain three types of factors: the numerical constant, a π term, and an r^3 term. Divide the common factors and you are left with the numerical constants.
- Remind students how fractions are divided.

Part (a): $2 \div \frac{2}{3} = 2 \times \frac{3}{2} = 3$

Part (b): $\frac{4}{3} \div \frac{2}{3} = \frac{4}{3} \times \frac{3}{2} = \frac{4}{2} = 2$

Part (c): $2 \div \frac{4}{3} = 2 \times \frac{3}{4} = \frac{3}{2}$

- Note that the definition of $r^0 = 1$ makes sense in this problem. There are the same number of factors of r in the numerator as in the denominator. Just as $\frac{5}{5} = 1$, $\frac{r^3}{r^3} = r^0 = 1$.

Activity 3

- This activity provides a brief look at negative exponents.
- While students can follow both methods presented, the symbolism of how negative exponents are defined does not occur to them.
- Be satisfied if students can follow each method and give an explanation of the mathematics that is going on.

Closure

- Refer to the chart made at the beginning of class. Complete the last three cells. Discuss why these powers make sense.

Currency, c	\$1000	\$100	\$10	\$1	\$.10	\$.01
Ratio of c: one dollar	1000	100	10	1	$\frac{1}{10}$	$\frac{1}{100}$
Ratio as power of 10	10^3	10^2	10^1	10^0	10^{-1}	10^{-2}

Technology For the Teacher

Dynamic Classroom

The Dynamic Planning Tool

Editable Teacher's Resources at *BigIdeasMath.com*

2 ACTIVITY: Comparing Volumes

Work with a partner.

The quotients show three ratios of the volumes of the solids. Identify each ratio, find its value, and describe what it means.

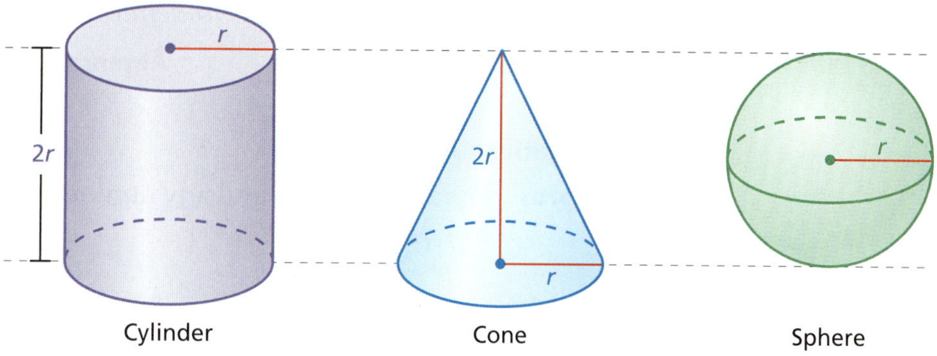

a. $2\pi r^3 \div \frac{2}{3}\pi r^3 = \square$

b. $\frac{4}{3}\pi r^3 \div \frac{2}{3}\pi r^3 = \square$

c. $2\pi r^3 \div \frac{4}{3}\pi r^3 = \square$

3 ACTIVITY: Writing a Definition

Work with a partner.

Compare the two methods used to simplify $\frac{3^2}{3^5}$. Then describe how you can rewrite a power with a negative exponent as a fraction.

Method 1

$$\frac{3^2}{3^5} = \frac{\overset{1}{\cancel{3}} \cdot \overset{1}{\cancel{3}}}{\underset{1}{\cancel{3}} \cdot \underset{1}{\cancel{3}} \cdot 3 \cdot 3 \cdot 3}$$

$$= \frac{1}{3^3}$$

Method 2

$$\frac{3^2}{3^5} = 3^{2-5}$$

$$= 3^{-3}$$

What Is Your Answer?

4. **IN YOUR OWN WORDS** How can you define zero and negative exponents? Give two examples of each.

Practice Use what you learned about zero and negative exponents to complete Exercises 5–8 on page 374.

9.4 Lesson

Check It Out
Lesson Tutorials
BigIdeasMath.com

Key Ideas

Zero Exponents

Words Any nonzero number to the zero power is equal to 1. Zero to the zero power, 0^0, is *undefined*.

Numbers $4^0 = 1$ **Algebra** $a^0 = 1$, where $a \neq 0$

Negative Exponents

Words For any integer n and any number a not equal to 0, a^{-n} is equal to 1 divided by a^n.

Numbers $4^{-2} = \frac{1}{4^2}$ **Algebra** $a^{-n} = \frac{1}{a^n}$, where $a \neq 0$

EXAMPLE 1 Evaluating Expressions

a. $3^{-4} = \frac{1}{3^4}$ Definition of negative exponent

$= \frac{1}{81}$ Evaluate power.

b. $(-8.5)^{-4} \cdot (-8.5)^4 = (-8.5)^{-4+4}$ Add the exponents.

$= (-8.5)^0$ Simplify.

$= 1$ Definition of zero exponent

c. $\frac{2^6}{2^8} = 2^{6-8}$ Subtract the exponents.

$= 2^{-2}$ Simplify.

$= \frac{1}{2^2}$ Definition of negative exponent

$= \frac{1}{4}$ Evaluate power.

On Your Own

Now You're Ready
Exercises 9–16

Evaluate the expression.

1. 4^{-2}
2. $(-2)^{-5}$
3. $6^{-8} \cdot 6^8$
4. $\frac{(-3)^5}{(-3)^6}$
5. $\frac{1}{5^7} \cdot \frac{1}{5^{-4}}$
6. $\frac{4^5 \cdot 4^{-3}}{4^2}$

Laurie's Notes

Introduction

Connect

- **Yesterday:** Students explored zero and negative exponents.
- **Today:** Students will use the definitions of zero and negative exponents to evaluate and simplify expressions.

Motivate

? "Did you know that a faucet leaking 30 drops per minute wastes 54 gallons a month?"

? "Can you visualize 54 gallons? Is that more than a standard kitchen sink? Is it more than a bath tub? What does it compare to?" Most students have difficulty estimating capacity. If there is a 50-gallon waste barrel in your school's cafeteria, you could compare it to that.

Lesson Notes

Key Ideas

- The definition of zero exponents is easily understood by students. Writing a negative exponent as a unit fraction with a positive exponent in the denominator takes time for students to understand. They need to see multiple examples using the Quotient of Powers Property, where the exponent in the denominator is greater than the exponent in the numerator.
- Example: $\frac{5^2}{5^3} = 5^{2-3} = 5^{-1}$ $\qquad \frac{5^2}{5^3} = \frac{\cancel{5} \cdot \cancel{5}}{\cancel{5} \cdot \cancel{5} \cdot 5} = \frac{1}{5}$

  The Quotient of Powers Property is used to simplify the expression on the left. Dividing out common factors is used on the right. Because both problems begin with the same expression, the results must be equivalent. In this example, you see that $5^{-1} = \frac{1}{5^1}$.

Example 1

- **Part (a):** This is a direct application of the definition of negative exponents.
- **Part (b):** The bases are the same, so the exponents are added. The sum of the exponents is 0, so the expression equals 1.
- In part (b), the two powers have a product of 1, so $(-8.5)^{-4}$ and $(-8.5)^4$ are reciprocals.
- **Part (c):** The Quotient of Powers Property is used first, resulting in a negative exponent. Use the definition of negative exponent and simplify.

On Your Own

- Question 5 can be done by thinking about simple fractions and how they are multiplied. The product of these two fractions is $\frac{1}{5^7 \cdot 5^{-4}} = \frac{1}{5^3}$.

Goal Today's lesson is evaluating and simplifying expressions with negative and zero exponents.

Start Thinking! and Warm Up

Lesson 9.4 Warm Up
For use before Lesson 9.4

Lesson 9.4 Start Thinking!
For use before Lesson 9.4

Can a number raised to a negative power ever be greater than 1? If so, give an example. If not, explain why not.

Can a number raised to a negative power ever be less than 0? If so, give an example. If not, explain why not.

Extra Example 1

Evaluate the expression.

a. 4^{-3} $\frac{1}{64}$

b. $(-3.7)^{-2} \cdot (-3.7)^2$ 1

c. $\frac{3^6}{3^9}$ $\frac{1}{27}$

On Your Own

1. $\frac{1}{16}$ **2.** $-\frac{1}{32}$

3. 1 **4.** $-\frac{1}{3}$

5. $\frac{1}{125}$ **6.** 1

Extra Example 2

Simplify. Write the expression using only positive exponents.

a. $-2x^0$ -2

b. $\frac{4b^{-4}}{b^7}$ $\frac{4}{b^{11}}$

On Your Own

7. $\frac{8}{x^2}$

8. $\frac{1}{b^{10}}$

9. $\frac{1}{15z^3}$

Extra Example 3

In Example 3, the faucet leaks water at a rate of 4^{-6} liter per second. How many liters of water leak from the faucet in 1 hour? about 0.88 L

On Your Own

10. 1.152 L

English Language Learners

Vocabulary

Remind English learners that when they see a negative exponent, they should think *reciprocal.* Review the meaning of the word reciprocal. Students often think that because the exponent is negative, the expression is negative. Remind them that a number of the form x^a cannot be negative unless the base is negative.

Laurie's Notes

Example 2

- **Common Error:** In part (a), students see the zero exponent and immediately think the answer is 1. Remind students that only the variable is being raised to the 0 power, the -5 is not.
- **Common Error:** In part (b), the constant 9 is not being raised to a power, only the variables are. Students need to distinguish this. In the last step, when the expression has been simplified to $9y^{-8}$ ask, "What is being raised to the -8?" In other words, what is the base for the exponent?
- Work through the steps slowly. It takes time for students to make sense of all that is going on in each problem. Because there is often more than one approach to simplifying the expression, it can confuse students. Instead of seeing it as a way to show that the properties are all connected, students see it as a way of trying to confuse them.

On Your Own

- Have students share their work at the board, *and* explain aloud what they did. Students need to hear the words and see the work. It also helps students become better communicators when they have the opportunity to practice their skills.

Example 3

- Ask a student to read the problem. The information known in this problem is the rate of leaking (in seconds) and the amount it leaks in liters (from the illustration).

? "What are you solving for in this problem?" amount faucet leaks in 1 hour

- First, use unit analysis to convert 1 hour to 3600 seconds. Then, because the faucet leaks 50^{-2} liter every second and there are 3600 seconds in an hour, multiply to find how many liters leaked in the hour.

On Your Own

- Before students begin, ask them which faucet was leaking more, the faucet that leaked 50^{-2} liter per second or 5^{-5} liter per second.

Closure

- **Exit Ticket:** Simplify.

1. $4x^{-3}$ $\frac{4}{x^3}$

2. $\frac{6^3}{6^5}$ $\frac{1}{36}$

3. $\frac{4n^0}{n^2}$ $\frac{4}{n^2}$

Technology For the Teacher

Dynamic Classroom

The Dynamic Planning Tool

Editable Teacher's Resources at *BigIdeasMath.com*

EXAMPLE 2 Simplifying Expressions

a. $-5x^0 = -5(1)$ Definition of zero exponent

$= -5$ Multiply.

b. $\dfrac{9y^{-3}}{y^5} = 9y^{-3-5}$ Subtract the exponents.

$= 9y^{-8}$ Simplify.

$= \dfrac{9}{y^8}$ Definition of negative exponent

On Your Own

Now You're Ready
Exercises 20–27

Simplify. Write the expression using only positive exponents.

7. $8x^{-2}$ **8.** $b^0 \cdot b^{-10}$ **9.** $\dfrac{z^6}{15z^9}$

EXAMPLE 3 Real-Life Application

A drop of water leaks from a faucet every second. How many liters of water leak from the faucet in 1 hour?

Drop of water: 50^{-2} L

Convert 1 hour to seconds.

$$1\ \cancel{h} \times \frac{60\ \cancel{min}}{1\ \cancel{h}} \times \frac{60\ sec}{1\ \cancel{min}} = 3600\ sec$$

Water leaks from the faucet at a rate of 50^{-2} liter per second. Multiply the time by the rate.

$3600 \cdot 50^{-2} = 3600 \cdot \dfrac{1}{50^2}$ Definition of negative exponent

$= 3600 \cdot \dfrac{1}{2500}$ Evaluate power.

$= \dfrac{3600}{2500}$ Multiply.

$= 1\dfrac{11}{25} = 1.44$ Simplify.

So, 1.44 liters of water leak from the faucet in 1 hour.

On Your Own

10. **WHAT IF?** In Example 4, the faucet leaks water at a rate of 5^{-5} liter per second. How many liters of water leak from the faucet in 1 hour?

9.4 Exercises

Check It Out
Help with Homework
BigIdeasMath.com

Vocabulary and Concept Check

1. **VOCABULARY** If a is a nonzero number, does the value of a^0 depend on the value of a? Explain.
2. **WRITING** Explain how to evaluate 10^{-3}.
3. **NUMBER SENSE** Without evaluating, order 5^0, 5^4, and 5^{-5} from least to greatest.
4. **DIFFERENT WORDS, SAME QUESTION** Which is different? Find "both" answers.

Rewrite $\frac{1}{3 \cdot 3 \cdot 3}$ using a negative exponent.	Write 3 to the negative third power.
Write $\frac{1}{3}$ cubed as a power.	Write $(-3) \cdot (-3) \cdot (-3)$ as a power.

Practice and Problem Solving

5. Use the Quotient of Powers Property to copy and complete the table.

n	4	3	2	1
$\frac{5^n}{5^2}$				

6. What patterns do you see?
7. How would you define 5^0? Why?
8. How can you rewrite 5^{-1} as a fraction?

Evaluate the expression.

1 9. 6^{-2}

10. 158^0

11. $\frac{4^3}{4^5}$

12. $\frac{-3}{(-3)^2}$

13. $(-2)^{-8} \cdot (-2)^8$

14. $3^{-3} \cdot 3^{-2}$

15. $\frac{1}{5^{-3}} \cdot \frac{1}{5^6}$

16. $\frac{(1.5)^2}{(1.5)^{-2} \cdot (1.5)^4}$

17. **ERROR ANALYSIS** Describe and correct the error in evaluating the expression.

$$(4)^{-3} = (-4)(-4)(-4) = -64$$

18. **SAND** The mass of a grain of sand is about 10^{-3} gram. About how many grains of sand are in the bag of sand?

19. **CRITICAL THINKING** How can you write the number 1 as 2 to a power? 10 to a power?

Assignment Guide and Homework Check

Level	Day 1 Activity Assignment	Day 2 Lesson Assignment	Homework Check
Basic	5–8, 36–39	1–4, 9–31 odd, 18	9, 15, 18, 25, 27
Average	5–8, 36–39	1–4, 13–17, 19, 24–27, 31–33	14, 25, 27, 32
Advanced	5–8, 36–39	1–4, 15–17, 19, 24–27, 31–35	16, 27, 32, 34

Common Errors

- **Exercises 9–16** Students may forget that any nonzero number to the zero power is 1 and say that the number to the zero power is zero. Remind them of the definition of zero exponents and refer them to the activities to remind them.
- **Exercises 9–16** Students may simplify the expression, but forget to evaluate the exponents. Remind them of the directions and Example 1 for how to evaluate these expressions. There should be no exponents, other than 1, in their answers.
- **Exercises 20–27** Students may say that the coefficient has the same exponent as the variable, but this is not the case unless there are parentheses around both. Remind them that the coefficient is multiplied by the variable and that they can have different exponents.

9.4 Record and Practice Journal

Evaluate the expression.

1. 29^0 — **1**
2. 12^{-1} — $\frac{1}{12}$
3. $(-15)^{-2} \bullet (-15)^2$ — **1**
4. $10^{-4} \bullet 10^{-6}$ — $\frac{1}{10{,}000{,}000{,}000}$
5. $\frac{1}{3^{-3}} \bullet \frac{1}{3^5}$ — $\frac{1}{9}$
6. $\frac{(4.1)^8}{(4.1)^5 \bullet (4.1)^7}$ — $\frac{1}{282.5761}$

Simplify. Write the expression using only positive exponents.

7. $19x^{-6}$ — $\frac{19}{x^6}$
8. $\frac{14a^{-5}}{a^{-8}}$ — $14a^3$
9. $\frac{16y^4}{4y^{10}}$ — $\frac{4}{y^6}$
10. $3t^6 \bullet 8t^{-6}$ — **24**
11. $7k^{-2} \bullet 5m^0 \bullet k^9$ — $35k^7$
12. $\frac{12s^{-1} \bullet 4^{-2} \bullet r^3}{s^2 \bullet r^5}$ — $\frac{3}{4r^2s^3}$
13. The density of a proton is about $\frac{1.64 \times 10^{-24}}{3.7 \times 10^{-38}}$ grams per cubic centimeter. Simplify the expression.
about 44,300,000,000,000 grams per cubic centimeter

Technology For the Teacher
Answer Presentation Tool
Quiz*Show*

Vocabulary and Concept Check

1. no; Any nonzero base raised to the zero power is always 1.
2. Use the definition of negative exponents to rewrite it as $\frac{1}{10^3}$. Then evaluate the power to get $\frac{1}{1000}$.
3. $5^{-5}, 5^0, 5^4$
4. Write $(-3) \bullet (-3) \bullet (-3)$ as a power.; $(-3)^3$; 3^{-3}

Practice and Problem Solving

5. $5^2 = 25$; $5^1 = 5$; $5^0 = 1$; $5^{-1} = \frac{1}{5}$
6. As n decreases by one, the expression is $\frac{1}{5}$ of the value of the previous expression.
7. One-fifth of 5^1; $5^0 = \frac{1}{5}(5^1) = 1$
8. Write as $\frac{1}{5}$ of 5^0.; $5^{-1} = \frac{1}{5}(5^0) = \frac{1}{5}(1) = \frac{1}{5}$
9. $\frac{1}{36}$
10. 1
11. $\frac{1}{16}$
12. $-\frac{1}{3}$
13. 1
14. $\frac{1}{243}$
15. $\frac{1}{125}$
16. 1
17. The negative sign goes with the exponent, not the base. $(4)^{-3} = \frac{1}{4^3} = \frac{1}{64}$
18. 10,000,000 grains of sand
19. 2^0; 10^0

Practice and Problem Solving

20. $\frac{6}{y^4}$
21. $\frac{a^7}{64}$
22. $9c^7$
23. $5b$
24. $\frac{4}{x^6}$
25. 12
26. $\frac{n^3}{m^2}$
27. $\frac{w^6}{9}$
28. 100 mm
29. 10,000 micrometers
30. 1,000,000 nanometers
31. 1,000,000 micrometers
32. **a.** 10^{-9} m

    b. equal to
33. See *Taking Math Deeper.*
34. *Sample answer:* 2^{-4}; 4^{-2}
35. If $a = 0$, then $0^n = 0$. Because you can not divide by 0, the expression $\frac{1}{0}$ is undefined.

Fair Game Review

36. 10^9
37. 10^3
38. 10^4
39. D

Mini-Assessment

Evaluate the expression.

1. 5^{-3} $\frac{1}{125}$
2. 9^0 1
3. $\frac{3^6}{3^{10}}$ $\frac{1}{81}$
4. $\frac{1}{2^{-2}} \cdot \frac{1}{2^5}$ $\frac{1}{8}$
5. $\frac{(2.3)^4}{(2.3)^{-3} \cdot (2.3)^8}$ $\frac{10}{23}$

Taking Math Deeper

Exercise 33

To solve this problem, students need to notice that the blood sample shown has a volume of 500 mL or 500 milliliters. This is a good problem to help students with *unit analysis.*

1 **a.** How many white blood cells are in the donation?

$$500\,\cancel{\text{mL}} \cdot \frac{1\,\cancel{\text{mm}^3}}{10^{-3}\,\cancel{\text{mL}}} \cdot \frac{10^4 \text{ white blood cells}}{1\,\cancel{\text{mm}^3}}$$

$$= 500 \cdot \frac{10^4 \text{ white blood cells}}{10^{-3}}$$

$$= 500 \cdot 10^7 \text{ white blood cells}$$

$$= 5{,}000{,}000{,}000 \text{ white blood cells}$$

$$= 5 \text{ billion white blood cells}$$

2 **b.** How many red blood cells are in the donation?

$$500\,\cancel{\text{mL}} \cdot \frac{1\,\cancel{\text{mm}^3}}{10^{-3}\,\cancel{\text{mL}}} \cdot \frac{5 \cdot 10^6 \text{ red blood cells}}{1\,\cancel{\text{mm}^3}}$$

$$= 2500 \cdot \frac{10^6 \text{ red blood cells}}{10^{-3}}$$

$$= 2500 \cdot 10^9 \text{ red blood cells}$$

$$= 2{,}500{,}000{,}000{,}000 \text{ red blood cells}$$

$$= 2.5 \text{ trillion red blood cells}$$

3 **c.** The ratio of red blood cells to white blood cells is

$$\frac{2{,}500{,}000{,}000{,}000 \text{ red blood cells}}{5{,}000{,}000{,}000 \text{ white blood cells}} = \frac{500}{1}.$$

Red blood cells are responsible for picking up carbon dioxide from our blood and for transporting oxygen. White blood cells are responsible for fighting foreign organisms that enter the body.

Reteaching and Enrichment Strategies

If students need help. . .	If students got it. . .
Resources by Chapter • Practice A and Practice B • Puzzle Time Record and Practice Journal Practice Differentiating the Lesson Lesson Tutorials Skills Review Handbook	Resources by Chapter • Enrichment and Extension • School-to-Work Start the next section

Simplify. Write the expression using only positive exponents.

2 **20.** $6y^{-4}$ **21.** $8^{-2} \cdot a^{7}$ **22.** $\dfrac{9c^{3}}{c^{-4}}$ **23.** $\dfrac{5b^{-2}}{b^{-3}}$

24. $\dfrac{8x^{3}}{2x^{9}}$ **25.** $3d^{-4} \cdot 4d^{4}$ **26.** $m^{-2} \cdot n^{3}$ **27.** $\dfrac{3^{-2} \cdot k^{0} \cdot w^{0}}{w^{-6}}$

METRIC UNITS **In Exercises 28–31, use the table.**

28. How many millimeters are in a decimeter?

29. How many micrometers are in a centimeter?

30. How many nanometers are in a millimeter?

31. How many micrometers are in a meter?

Unit of Length	Length
decimeter	10^{-1} m
centimeter	10^{-2} m
millimeter	10^{-3} m
micrometer	10^{-6} m
nanometer	10^{-9} m

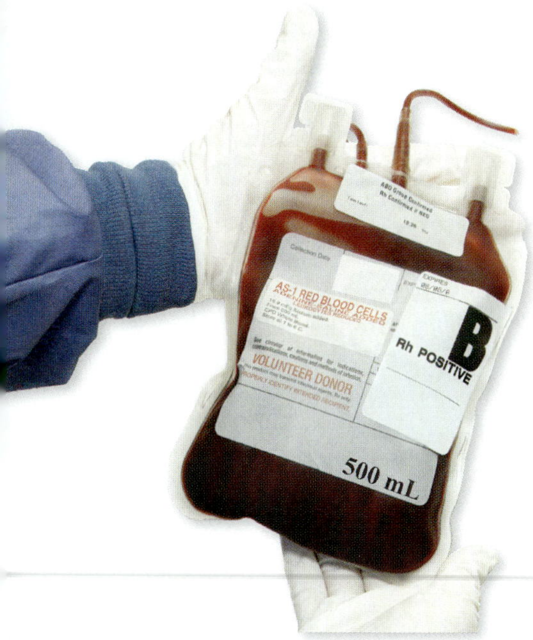

32. **MICROBES** A species of bacteria is 10 micrometers long. A virus is 10,000 times smaller than the bacteria.

a. Using the table above, find the length of the virus in meters.

b. Is the answer to part (a) *less than*, *greater than*, or *equal to* one nanometer?

33. **BLOOD DONATION** Every 2 seconds, someone in the United States needs blood. A sample blood donation is shown. ($1 \text{ mm}^3 = 10^{-3}$ mL)

a. One cubic millimeter of blood contains about 10^4 white blood cells. How many white blood cells are in the donation? Write your answer in words.

b. One cubic millimeter of blood contains about 5×10^6 red blood cells. How many red blood cells are in the donation? Write your answer in words.

c. Compare your answers for parts (a) and (b).

34. **OPEN-ENDED** Write two different powers with negative exponents that have the same value.

35. **Reasoning** The rule for negative exponents states that $a^{-n} = \dfrac{1}{a^n}$. Explain why this rule does not apply when $a = 0$.

Fair Game Review What you learned in previous grades & lessons

Simplify the expression. *(Section 9.2 and Section 9.3)*

36. $10^{3} \cdot 10^{6}$ **37.** $10^{2} \cdot 10$ **38.** $\dfrac{10^{8}}{10^{4}}$

39. **MULTIPLE CHOICE** Which data display best shows the variability of a data set? *(Section 7.2)*

Ⓐ bar graph Ⓑ circle graph

Ⓒ scatter plot Ⓓ box-and-whisker plot

9.5 Reading Scientific Notation

Essential Question How can you read numbers that are written in scientific notation?

1 ACTIVITY: Very Large Numbers

Work with a partner.

- Use a calculator. Experiment with multiplying large numbers until your calculator gives an answer that is *not* in standard form.
- When the calculator at the right was used to multiply 2 billion by 3 billion, it listed the result as

  6.0E+18.

- Multiply 2 billion by 3 billion by hand. Use the result to explain what 6.0E+18 means.
- Check your explanation using products of other large numbers.
- Why didn't the calculator show the answer in standard form?
- Experiment to find the maximum number of digits your calculator displays. For instance, if you multiply 1000 by 1000 and your calculator shows 1,000,000, then it can display 7 digits.

2 ACTIVITY: Very Small Numbers

Work with a partner.

- Use a calculator. Experiment with multiplying very small numbers until your calculator gives an answer that is *not* in standard form.
- When the calculator at the right was used to multiply 2 billionths by 3 billionths, it listed the result as

  6.0E–18.

- Multiply 2 billionths by 3 billionths by hand. Use the result to explain what 6.0E–18 means.
- Check your explanation using products of other very small numbers.

Laurie's Notes

Introduction

For the Teacher

- **Goal:** Students will explore scientific notation using a calculator.
- The lesson today requires that students work with a calculator. A four-function calculator is fine, but a scientific calculator is better.

Motivate

? "Have you had your millionth heartbeat?" Answers will vary.

- Assume that your heart beats once per second, and it has since you were born. Convert 1 million seconds to days.

$$10^6 \text{ sec} \cdot \frac{1 \text{ min}}{60 \text{ sec}} \cdot \frac{1 \text{ h}}{60 \text{ min}} \cdot \frac{1 \text{ day}}{24 \text{ h}} \approx 11.57 \text{ days}$$

- Clearly, all of your students have had their millionth heart beat. But have they had their billionth? Because there are 1000 million in 1 billion,

$$11.57 \text{ days} \times 1000 = 11{,}570 \text{ days} \div 365 \approx 31.7 \text{ years.}$$

Activity Notes

Activity 1

? "What does standard form mean?" Numbers are written using digits; example: 123.

? "What does expanded notation mean?" Numbers are written showing the value of each digit; example: $123 = 1 \times 100 + 2 \times 10 + 3 \times 1$.

- This activity gives students time to explore how scientific notation is displayed on their calculator.
- Students should be able to explain how they determined the number of digits the calculator displays.

? "If the display on my calculator reads 4.5 E+8, what does this mean?" The decimal point is located 8 places to the right from where it currently is.

- Write on the board: 4.5 E+8 = 450,000,000.

Activity 2

- This activity is the same as Activity 1, except with very small numbers.
- Review the place values less than 1.
- **Summary:** Check to see that everyone was successful in getting both large and small numbers to display. Has everyone figured out what the E+4 and E−6 notations mean on the calculator? Does everyone know how many digits their calculator displays?

? "If the display on my calculator reads 6.2 E−6, what does this mean?" The decimal point is located 6 places to the left from where it currently is.

- Write on the board: 6.2 E−6 = 0.0000062.

Previous Learning

Students should know the base 10 place value system.

Activity Materials	
Introduction	**Textbook**
• calculator	• calculator

Start Thinking! and Warm Up

Activity 9.5 Start Thinking!
For use before Activity 9.5

Activity 9.5 Warm Up
For use before Activity 9.5

Evaluate the power.

1. 10^3 2. 10^{-4} 3. 10^5
4. 10^{-2} 5. 10^{10} 6. 10^{-5}

9.5 Record and Practice Journal

Essential Question How can you read numbers that are written in scientific notation?

1 ACTIVITY: Very Large Numbers

Work with a partner.

- Use a calculator. Experiment with multiplying large numbers until your calculator gives an answer that is *not* in standard form.
- When the calculator on the right was used to multiply 2 billion by 3 billion, it listed the result as

  6.0E+18.

- Multiply 2 billion by 3 billion by hand. Use the result to explain what 6.0E+18. means.

  6 followed by 18 zeros

- Check your explanation using products of other large numbers.
- Why didn't the calculator show the answer in standard form?

  not enough display places for all the zeros

- Experiment to find the maximum number of digits your calculator displays. For instance, if you multiply 1000 by 1000 and your calculator shows 1,000,000, then it can display 7 digits.

  Check students' work.

Differentiated Instruction

Kinesthetic

Have students use grid paper when converting numbers from scientific notation to standard form and from standard form to scientific notation. Write the number with one digit in each square. Place the decimal point on the line between the squares. Students may find it easier to count the number of squares than the number of digits.

9.5 Record and Practice Journal

2 ACTIVITY: Very Small Numbers

Work with a partner.

- Use a calculator. Experiment with multiplying very small numbers until your calculator gives an answer that is *not* in standard form.
- When the calculator at the right was used to multiply 2 billionths by 3 billionths, it listed the result as

  6.0E−18.

- Multiply 2 billionths by 3 billionths by hand. Use the result to explain what 6.0E−18. means.

  17 zeros followed by a 6, all to the right of the decimal point
- Check your explanation using products of other very small numbers.

3 ACTIVITY: Reading Scientific Notation

Work with a partner.

Each description gives an example of a number written in scientific notation. Answer the question in the description. Write your answer in standard form.

a. Nearly 1.0×10^5 dust mites can live in 1 square yard of carpet.

How many dust mites can live in 100 square yards of carpet?

10,000,000 dust mites

b. A micron is about 4.0×10^{-5} inch. The length of a dust mite is 250 microns.

How long is a dust mite in inches?

0.01 inch

c. About 1.0×10^{15} bacteria live in a human body.

How many bacteria are living in the humans in your classroom?

Check students' work.

d. A micron is about 4.0×10^{-5} inch. The length of a bacterium is about 0.5 micron.

How many bacteria could lie end-to-end on your finger?

Check students' work.

e. Earth has only about 1.5×10^8 kilograms of gold. Earth has a mass of 6.0×10^{24} kilograms.

What percent of Earth's mass is gold?

0.0000000000000025%

f. A gram is about 0.035 ounce. An atom of gold weighs about 3.3×10^{-22} gram.

How many atoms are in an ounce of gold?

about 86,580,000,000,000,000,000,000 atoms per ounce

What Is Your Answer?

4. **IN YOUR OWN WORDS** How can you read numbers that are written in scientific notation? Why do you think this type of notation is called "scientific notation?" Why is scientific notation important?

   It is used frequently in scientific fields.

Laurie's Notes

Activity 3

- Tell students to use the samples on the board from Activities 1 and 2 if needed. They should not need their calculators if they use the properties of exponents that they have studied.
- Each statement involves a number written in scientific notation. Students need to do something with that number. There are also many different units used in this lesson. Make sure students have the correct units.
  - Part (a): $\dfrac{1.0 \times 10^5 \text{ mites}}{1 \text{ yd}^2} \cdot 100 \text{ yd}^2 = 1.0 \times 10^7 \text{ mites}$
  - Part (b): $250 \text{ microns} \cdot \dfrac{4.0 \times 10^{-5} \text{ in.}}{1 \text{ micron}} = 1.0 \times 10^{-2} \text{ in.}$
  - Part (c): $\dfrac{1.0 \times 10^{15} \text{ bacteria}}{1 \text{ human}} \cdot k \text{ humans} = k(1.0 \times 10^{15}) \text{ bacteria}$
  - Part (d): Students should estimate that their finger is approximately 2 inches, so $\dfrac{1 \text{ bacteria}}{0.5 \text{ micron}} \cdot \dfrac{1 \text{ micron}}{4.0 \times 10^{-5} \text{ in.}} \cdot 2 \text{ in.} = 100{,}000 \text{ bacteria}$
  - Part (e): $\dfrac{1.5 \times 10^8 \text{ kg of gold}}{6.0 \times 10^{24} \text{ kg Earth}} = 2.5 \times 10^{-17} = 2.5 \times 10^{-15}\%$
  - Part (f): $1 \text{ oz} \cdot \dfrac{1 \text{ g}}{0.035 \text{ oz}} \cdot \dfrac{1 \text{ atom of gold}}{3.3 \times 10^{-22} \text{ g}} \approx 8.66 \times 10^{22} \text{ atoms of gold}$

What Is Your Answer?

- **Think-Pair-Share:** Students should read the question independently and then work with a partner to answer the question. When they have answered the question, the pair should compare their answer with another group and discuss any discrepancies.

Closure

- What examples have you read or heard about that involve very large or very small numbers?

Technology For the Teacher

Dynamic Classroom

The Dynamic Planning Tool

Editable Teacher's Resources at *BigIdeasMath.com*

3 ACTIVITY: Reading Scientific Notation

Work with a partner.

Each description gives an example of a number written in scientific notation. Answer the question in the description. Write your answer in standard form.

a. Nearly 1.0×10^5 dust mites can live in 1 square yard of carpet.

How many dust mites can live in 100 square yards of carpet?

b. A micron is about 4.0×10^{-5} inch. The length of a dust mite is 250 microns.

How long is a dust mite in inches?

c. About 1.0×10^{15} bacteria live in a human body.

How many bacteria are living in the humans in your classroom?

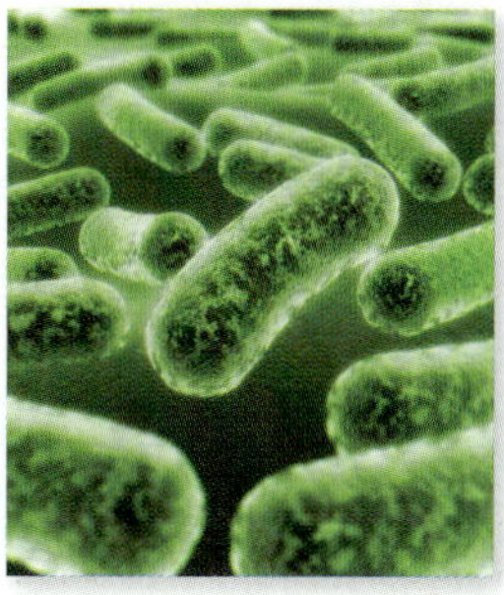

d. A micron is about 4.0×10^{-5} inch. The length of a bacterium is about 0.5 micron.

How many bacteria could lie end-to-end on your finger?

e. Earth has only about 1.5×10^8 kilograms of gold. Earth has a mass of 6.0×10^{24} kilograms.

What percent of Earth's mass is gold?

f. A gram is about 0.035 ounce. An atom of gold weighs about 3.3×10^{-22} gram.

How many atoms are in an ounce of gold?

What Is Your Answer?

4. **IN YOUR OWN WORDS** How can you read numbers that are written in scientific notation? Why do you think this type of notation is called "scientific notation?" Why is scientific notation important?

Practice Use what you learned about reading scientific notation to complete Exercises 3–5 on page 380.

9.5 Lesson

Check It Out
Lesson Tutorials
BigIdeasMath.com

Key Vocabulary
scientific notation, *p. 378*

Study Tip
Scientific notation is used to write very small and very large numbers.

Key Idea

Scientific Notation

A number is written in **scientific notation** when it is represented as the product of a factor and a power of 10. The factor must be at least 1 and less than 10.

The factor is at least 1 and less than 10. → 8.3×10^{-7} ← The power of 10 has an integer exponent.

EXAMPLE 1 Identifying Numbers Written in Scientific Notation

Tell whether the number is written in scientific notation. Explain.

a. 5.9×10^{-6}

The factor is at least 1 and less than 10. The power of 10 has an integer exponent. So, the number is written in scientific notation.

b. 0.9×10^{8}

The factor is less than 1. So, the number is not written in scientific notation.

Key Idea

Writing Numbers in Standard Form

When writing a number from scientific notation to standard form, the absolute value of the exponent tells you how many places to move the decimal point.

- If the exponent is negative, move the decimal point to the left.
- If the exponent is positive, move the decimal point to the right.

EXAMPLE 2 Writing Numbers in Standard Form

a. Write 3.22×10^{-4} in standard form.

$3.22 \times 10^{-4} = 0.000322$ (4) Move decimal point $|-4| = 4$ places to the left.

b. Write 7.9×10^{5} in standard form.

$7.9 \times 10^{5} = 790{,}000$ (5) Move decimal point $|5| = 5$ places to the right.

Multi-Language Glossary at BigIdeasMath.com.

Laurie's Notes

Introduction

Connect

- **Yesterday:** Students explored very large and very small numbers written in scientific notation.
- **Today:** Students will read numbers in scientific notation and write them in standard form.

Motivate

- Share some information about the Florida Keys.
- The Florida Keys are made up of approximately 1700 islands, most of which are uninhabited. The islands, or Keys, that are populated are connected by 42 bridges and stretch some 126 miles from the mainland to the last Key, Key West.

? "Do you know how big a square foot is?" Model with their hands

- The Florida Keys are approximately 3.83×10^9 square feet.

? "Is the area of the Keys more than or less than a billion square feet?" Students must be able to write 3.83×10^9 square feet in standard form to answer.

Lesson Notes

Key Idea

- Write the Key Idea. There are two parts to the definition; the factor is a number n, with $1 \le n < 10$, and it is multiplied by a power of 10 with an integer exponent.

Example 1

- Work through each example as described.

Key Idea

- Write the Key Idea. From the activities yesterday, students should find it reasonable that the exponent of 10 is connected to place value. If the exponent is positive, the number will be larger, so the decimal point moves to the right. Conversely, if the exponent is negative, the number will be smaller, so the decimal point moves to the left.

? Have students fill in the blank with less than or greater than.

- "A power of 10 with a positive exponent is _____ 1." greater than
- "A power of 10 with a negative exponent is _____ 1." less than

Example 2

- In part (a), 3.22 is the factor and −4 is the exponent. The number in standard form will be less than 3.22, so the decimal point moves to the left 4 places.
- In part (b), 7.9 is the factor and 5 is the exponent. The number in standard form will be greater than 7.9, so the decimal point moves to the right 5 places.

Goal Today's lesson is reading numbers in **scientific notation** and writing them in standard form.

Start Thinking! and Warm Up

Lesson 9.5 Warm Up
For use before Lesson 9.5

Lesson 9.5 Start Thinking!
For use before Lesson 9.5

Use the Internet to find the mass of an electron and the mass of Earth.

Before you begin your research, do you expect that the masses will be given in scientific notation? Why or why not?

Were you correct?

Extra Example 1

Tell whether the number is written in scientific notation. Explain.

a. 2.5×10^{-9} yes; The factor is at least 1 and less than 10. The power of 10 has an integer exponent.

b. 0.5×10^6 no; The factor is less than 1.

Extra Example 2

a. Write 2.75×10^{-3} in standard form. 0.00275

b. Write 6.38×10^7 in standard form. 63,800,000

Laurie's Notes

On Your Own

? **Extension:** "How could you write the number in Question 1 in scientific notation?" 1.2×10^5

- **Think-Pair-Share:** Students should read each question independently and then work with a partner to answer the questions. When they have answered the questions, the pair should compare their answers with another group and discuss any discrepancies.

Example 3

- Ask a student to read the problem. Discuss the density of a substance if they have encountered it in science class, otherwise give an explanation of what density means. Density equals mass divided by volume.
- Work through the example as shown.
- **Another Way:** Show students that you can compare numbers without writing them in standard form. Rewrite the numbers so that they all have the same exponent, then compare the first factor of each number. For example, 1.0×10^3, 1.84×10^3, and 0.641×10^3.

Example 4

- Before students get too concerned about (vampire) fleas, remind them that 10^{-5} is a small number!
- Students should be comfortable multiplying by powers of 10.

? "Can you multiply 10^{-5} by 100 first, and then multiply by 1.4? Explain." yes; Multiplication is commutative.

Closure

- **Exit Ticket:** Write the number in standard form.
  1. 1.56×10^7 15,600,000 **2.** 6.3×10^{-5} 0.000063

Technology For the Teacher

Dynamic Classroom

The Dynamic Planning Tool
Editable Teacher's Resources at *BigIdeasMath.com*

On Your Own

1. no; The factor is greater than 10.
2. 60,000,000
3. 0.000099 **4.** 12,850

Extra Example 3

In Example 3, the density of an ear of corn is 7.21×10^2 kilograms per cubic meter. What happens when an ear of corn is placed in water? The ear of corn is less dense than water, so it will float.

Extra Example 4

In Example 4, a dog has 50 female fleas. How many milliliters of blood do the fleas consume per day? 0.7 milliliter of blood per day

On Your Own

5. It will sink.
6. 1.05 mL

English Language Learners

Word Problems

Have students work in groups of 3 or 4, including both English learners and English speakers. Provide them with poster board and markers. Assign each group a problem-solving exercise with scientific notation. Each group is to solve their problem showing all of the steps and using scientific notation. English learners will benefit by having the opportunity to restate the problem and gain a deeper understanding of the concept.

On Your Own

Now You're Ready
Exercises 6–23

1. Is 12×10^4 written in scientific notation? Explain.

Write the number in standard form.

2. 6×10^7
3. 9.9×10^{-5}
4. 1.285×10^4

EXAMPLE 3 Comparing Numbers in Scientific Notation

An object with a lesser density than water will float. An object with a greater density than water will sink. Use each given density (in kilograms per cubic meter) to explain what happens when you place a brick and an apple in water.

Water: 1.0×10^3 **Brick:** 1.84×10^3 **Apple:** 6.41×10^2

Write each density in standard form.

Water	Brick	Apple
$1.0 \times 10^3 = 1000$	$1.84 \times 10^3 = 1840$	$6.41 \times 10^2 = 641$

∴ The apple is less dense than water, so it will float. The brick is denser than water, so it will sink.

EXAMPLE 4 Real-Life Application

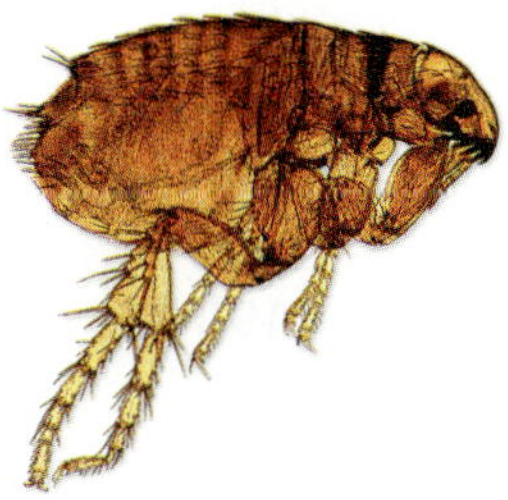

A female flea consumes about 1.4×10^{-5} liter of blood per day.

A dog has 100 female fleas. How many milliliters of blood do the fleas consume per day?

$$1.4 \times 10^{-5} \cdot 100 = 0.000014 \cdot 100 \quad \text{Write in standard form.}$$
$$= 0.0014 \quad \text{Multiply.}$$

∴ The fleas consume about 0.0014 liter, or 1.4 milliliters of blood per day.

On Your Own

Now You're Ready
Exercise 27

5. **WHAT IF?** In Example 3, the density of lead is 1.14×10^4 kilograms per cubic meter. What happens when lead is placed in water?
6. **WHAT IF?** In Example 4, a dog has 75 female fleas. How many milliliters of blood do the fleas consume per day?

9.5 Exercises

Check It Out
Help with Homework
BigIdeasMath.com

Vocabulary and Concept Check

1. **WRITING** Describe the difference between scientific notation and standard form.

2. **WHICH ONE DOESN'T BELONG?** Which number does *not* belong with the other three? Explain.

2.8×10^{15} 4.3×10^{-30} 1.05×10^{28} 10×9.2^{-13}

Practice and Problem Solving

Write your answer in standard form.

3. A micrometer is 1.0×10^{-6} meter. How long is 150 micrometers in meters?

4. An acre is about 4.05×10^{7} square centimeters. How many square centimeters are in 4 acres?

5. A cubic millimeter is about 6.1×10^{-5} cubic inches. How many cubic millimeters are in 1.22 cubic inches?

Tell whether the number is written in scientific notation. Explain.

1
6. 1.8×10^{9}
7. 3.45×10^{14}
8. 0.26×10^{-25}
9. 10.5×10^{12}
10. 46×10^{-17}
11. 5×10^{-19}
12. 7.814×10^{-36}
13. 0.999×10^{42}
14. 6.022×10^{23}

Write the number in standard form.

2
15. 7×10^{7}
16. 8×10^{-3}
17. 5×10^{2}
18. 2.7×10^{-4}
19. 4.4×10^{-5}
20. 2.1×10^{3}
21. 1.66×10^{9}
22. 3.85×10^{-8}
23. 9.725×10^{6}

24. **ERROR ANALYSIS** Describe and correct the error in writing the number in standard form.

✗ $4.1 \times 10^{-6} = 4100000$

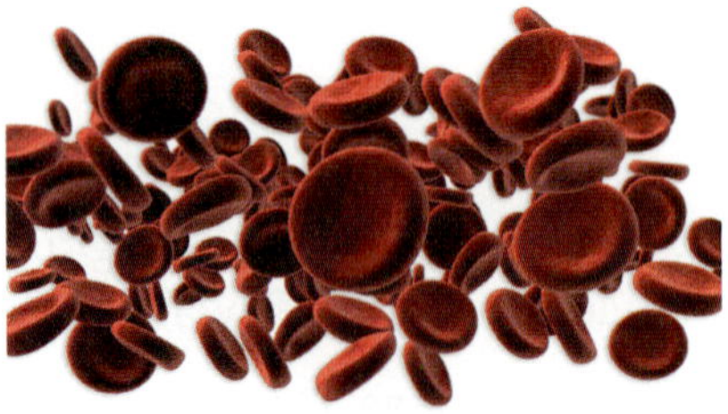

2.7×10^{8} platelets per milliliter

25. **PLATELETS** Platelets are cell-like particles in the blood that help form blood clots.

a. How many platelets are in 3 milliliters of blood? Write your answer in standard form.

b. An adult body contains about 5 liters of blood. How many platelets are in an adult body?

Assignment Guide and Homework Check

Level	Day 1 Activity Assignment	Day 2 Lesson Assignment	Homework Check
Basic	3–5, 32–36	1, 2, 7–27 odd, 24	9, 19, 24, 27
Average	3–5, 32–36	1, 2, 12–14, 21–29	12, 22, 24, 27
Advanced	3–5, 32–36	1, 2, 12–14, 22–27, 29–31	12, 22, 27, 30

Common Errors

- **Exercises 6–14** Students may think that all of the numbers are in scientific notation because all of the exponents are integers. Remind them that the factor at the beginning of the number must be at least 1 and less than 10.
- **Exercises 15–23** Students may move the decimal point in the wrong direction. Remind them that when the exponent is negative they move to the left, and when it is positive they move to the right.
- **Exercise 27** Students may order the surface temperatures by the factor at the beginning of the number without taking into account the power of 10. Encourage them to write the numbers in standard form before comparing the numbers.

9.5 Record and Practice Journal

Tell whether the number is written in scientific notation. Explain.

1. 14×10^8 No; the factor is greater than 10.
2. 2.6×10^{12} Yes; the factor is greater than 1 and less than 10.
3. 4.79×10^{-8} Yes; the factor is greater than 1 and less than 10.
4. 3.99×10^{16} Yes; the factor is greater than 1 and less than 10.
5. 0.15×10^{22} No; the factor is less than 1.
6. 6×10^3 Yes; the factor is greater than 1 and less than 10.

Write the number in standard form.

7. 4×10^9 4,000,000,000
8. 2×10^{-5} 0.00002
9. 3.7×10^6 3,700,000
10. 4.12×10^{-3} 0.00412
11. 7.62×10^{10} 76,200,000,000
12. 9.908×10^{-12} 0.000000000009908
13. Light travels at 3×10^8 meters per second.
    a. Write the speed of light in standard form.
    300,000,000
    b. How far has light traveled after 5 seconds?
    1,500,000,000 meters

Technology For the Teacher
Answer Presentation Tool
Quiz*Show*

Vocabulary and Concept Check

1. Scientific notation uses a factor of at least one but less than 10 multiplied by a power of 10. A number in standard form is written out with all the zeros and place values included.
2. 10×9.2^{-13}; All of the other numbers are written in scientific notation.

Practice and Problem Solving

3. 0.00015 m
4. 162,000,000 cm^2
5. 20,000 mm^3
6. yes; The factor is at least 1 and less than 10. The power of 10 has an integer exponent.
7. yes; The factor is at least 1 and less than 10. The power of 10 has an integer exponent.
8. no; The factor is less than 1.
9. no; The factor is greater than 10.
10. no; The factor is greater than 10.
11. yes; The factor is at least 1 and less than 10. The power of 10 has an integer exponent.
12. yes; The factor is at least 1 and less than 10. The power of 10 has an integer exponent.
13. no; The factor is less than 1.
14. yes; The factor is at least 1 and less than 10. The power of 10 has an integer exponent.
15. 70,000,000
16. 0.008

Practice and Problem Solving

17. 500 **18.** 0.00027

19. 0.000044 **20.** 2100

21. 1,660,000,000

22. 0.0000000385

23. 9,725,000

24. The negative exponent means the decimal point will move left, not right, when the number is written in standard form.
$4.1 \times 10^{-6} = 0.0000041$

25. a. 810,000,000 platelets

b. 1,350,000,000,000 platelets

26. 100 zeros

27. a. Bellatrix

b. Betelgeuse

28. $1555.2\ \text{km}^2$

29. $5 \times 10^{12}\ \text{km}^2$

30. $35{,}000{,}000\ \text{km}^3$

31. See *Taking Math Deeper*.

Fair Game Review

32. 6^8 **33.** 10^7

34. $\frac{1}{8^{12}}$ **35.** $\frac{1}{10^{16}}$

36. B

Mini-Assessment

Write the number in standard form.

1. 5×10^{-4} 0.0005
2. 2.5×10^{-3} 0.0025
3. 1.66×10^{3} 1660
4. 3.89×10^{-5} 0.0000389
5. 4.576×10^{8} 457,600,000

Taking Math Deeper

Exercise 31

This is an interesting problem in physics. It is about the speed of light in different mediums. The problem gives students practice in unit analysis, and it also points out that to compare speeds, you need to compare apples to apples, not apples to oranges.

1 Make a table and convert units.

Medium	Speed	Speed (m per sec)
Air	$\frac{6.7 \times 10^8\ \text{mi}}{\text{h}}$	$\frac{3.0 \times 10^8\ \text{m}}{\text{sec}}$
Glass	$\frac{6.6 \times 10^8\ \text{ft}}{\text{sec}}$	$\frac{2.0 \times 10^8\ \text{m}}{\text{sec}}$
Ice	$\frac{2.3 \times 10^5\ \text{km}}{\text{sec}}$	$\frac{2.3 \times 10^8\ \text{m}}{\text{sec}}$
Vacuum	$\frac{3.0 \times 10^8\ \text{m}}{\text{sec}}$	$\frac{3.0 \times 10^8\ \text{m}}{\text{sec}}$
Water	$\frac{2.3 \times 10^{10}\ \text{cm}}{\text{sec}}$	$\frac{2.3 \times 10^8\ \text{m}}{\text{sec}}$

2 **a.** For the significant digits given, the speed of light is the same in air or in a vacuum. Light is fastest in these two mediums.

b. Of the five mediums listed, light travels the slowest in glass.

3 If students take a course in physics, they will learn that light is slowed down in transparent media such as air, water, ice, and glass. The ratio by which it is slowed is called the *refractive index* of the medium and is always greater than one. This was discovered by Jean Foucault in 1850. The refractive index of air is 1.0003, which means that light travels slightly slower in air than in a vacuum.

When people talk about "the speed of light" in a general context, they usually mean "the speed of light in a vacuum." This quantity is also refered to as c. It is famous from Einstein's equation $E = mc^2$.

Reteaching and Enrichment Strategies

If students need help. . .	If students got it. . .
Resources by Chapter • Practice A and Practice B • Puzzle Time Record and Practice Journal Practice Differentiating the Lesson Lesson Tutorials Skills Review Handbook	Resources by Chapter • Enrichment and Extension • School-to-Work • Financial Literacy Start the next section

26. **REASONING** A googol is 1.0×10^{100}. How many zeros are in a googol?

3 27. **STARS** The table shows the surface temperatures of five stars.

a. Which star has the highest surface temperature?

b. Which star has the lowest surface temperature?

Star	Betelgeuse	Bellatrix	Sun	Aldebaran	Rigel
Surface Temperature (°F)	6.2×10^3	3.8×10^4	1.1×10^4	7.2×10^3	2.2×10^4

28. **CORAL REEF** The area of the Florida Keys National Marine Sanctuary is about 9.6×10^3 square kilometers. The area of the Florida Reef Tract is about 16.2% of the area of the sanctuary. What is the area of the Florida Reef Tract in square kilometers?

29. **REASONING** A gigameter is 1.0×10^6 kilometers. How many square kilometers are in 5 square gigameters?

30. **WATER** There are about 1.4×10^9 cubic kilometers of water on Earth. About 2.5% of the water is fresh water. How much fresh water is on Earth?

31. **Critical Thinking** The table shows the speed of light through five media.

a. In which medium does light travel the fastest?

b. In which medium does light travel the slowest?

Medium	Speed
Air	6.7×10^8 mi/h
Glass	6.6×10^8 ft/sec
Ice	2.3×10^5 km/sec
Vacuum	3.0×10^8 m/sec
Water	2.3×10^{10} cm/sec

Fair Game Review What you learned in previous grades & lessons

Simplify. Write the expression using only positive exponents. *(Section 9.2 and Section 9.4)*

32. $6^3 \cdot 6^5$　　33. $10^2 \cdot 10^5$　　34. $8^{-1} \cdot 8^{-11}$　　35. $10^{-6} \cdot 10^{-10}$

36. **MULTIPLE CHOICE** What is the length of the hypotenuse of the right triangle? *(Section 6.5)*

Ⓐ $\sqrt{18}$ in.　　Ⓑ $\sqrt{41}$ in.

Ⓒ 18 in.　　Ⓓ 41 in.

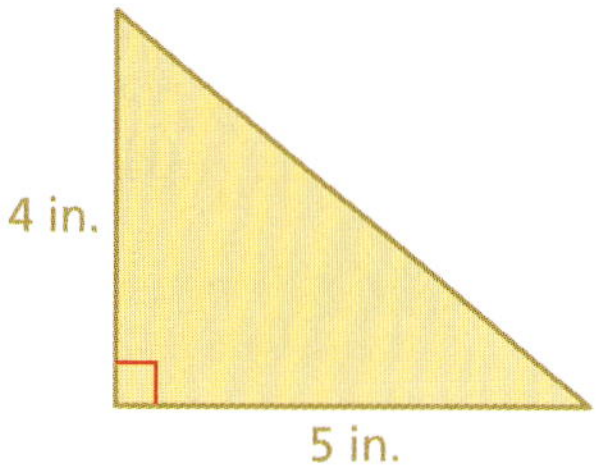

9.6 Writing Scientific Notation

Essential Question How can you write a number in scientific notation?

1 ACTIVITY: Finding pH Levels

Work with a partner. In chemistry, pH is a measure of the activity of dissolved hydrogen ions (H^+). Liquids with low pH values are called acids. Liquids with high pH values are called bases.

Find the pH of each liquid. Is the liquid a base, neutral, or an acid?

a. Lime juice:
$[H^+] = 0.01$

b. Egg:
$[H^+] = 0.00000001$

c. Distilled water:
$[H^+] = 0.0000001$

d. Ammonia water:
$[H^+] = 0.00000000001$

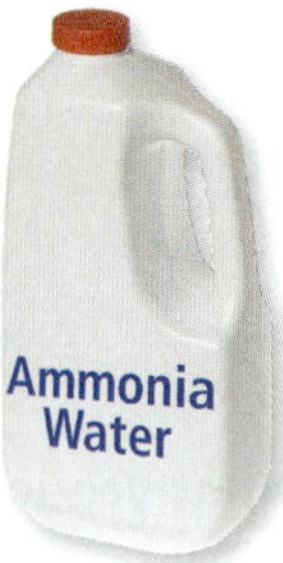

e. Tomato juice:
$[H^+] = 0.0001$

f. Hydrochloric acid:
$[H^+] = 1$

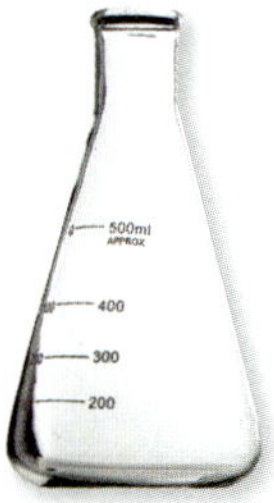

pH	$[H^+]$	
14	1×10^{-14}	Bases
13	1×10^{-13}	
12	1×10^{-12}	
11	1×10^{-11}	
10	1×10^{-10}	
9	1×10^{-9}	
8	1×10^{-8}	
7	1×10^{-7}	Neutral
6	1×10^{-6}	Acids
5	1×10^{-5}	
4	1×10^{-4}	
3	1×10^{-3}	
2	1×10^{-2}	
1	1×10^{-1}	
0	1×10^{0}	

Laurie's Notes

Introduction

For the Teacher

- **Goal:** Students will encounter numbers in standard form and write the numbers in scientific notation.

Motivate

- **Preparation:** If possible, borrow a pH meter or a few strips of litmus paper from the science department. If they are not available, move on to Activity 1.
- Without explanation, have 3 containers (coffee cups or paper cups) with liquids at the front of the room.
- Use the litmus paper or pH meter to test the liquids. Students should guess that you are doing a pH test, which leads into Activity 1.

Words of Wisdom

- **Safety:** For reasons of safety you should not consume, nor allow the students to consume, any of the liquids.

Activity Notes

Activity 1

? "How many of you have heard of pH or studied pH in science?"

? "What does pH level refer to? Can anyone explain?" Listen for the measure of activity of dissolved hydrogen ions; liquids with low pH are called acids; liquids with high pH are called bases.

- Have students refer to the pH chart in the book. The pH level is a number from 0 to 14, the opposite of the exponent (0 to −14) measuring the activity of the dissolved hydrogen ions. At the middle of the chart, a pH of 7 is called neutral.
- **FYI:** Pure water is neutral with a pH value of 7. Low on the scale are acids, which have a sour taste like lemons. High on the scale are bases, which have a bitter taste like soap.
- Notice that all of the pH values are given in standard form. Students will need to think about how these numbers would be written in scientific notation.

? "How did you compare the numbers in standard form with the scale in scientific notation?" Listen for idea of the number of place values away from 1.

Previous Learning

Students should know how to write powers of 10.

Activity Materials	
Introduction	**Textbook**
• pH digital meter or litmus paper • sample liquids	• calculator • strips of paper

Start Thinking! and Warm Up

Activity 9.6 **Start Thinking!** For use before Activity 9.6

Activity 9.6 **Warm Up** For use before Activity 9.6

Write the number in standard form.

1. 6×10^3
2. 4×10^{-4}
3. 2×10^5
4. 2.6×10^{-2}
5. 5.25×10^{10}
6. 8.52×10^{-5}

9.6 Record and Practice Journal

Essential Question How can you write a number in scientific notation?

1 ACTIVITY: Finding pH Levels

Work with a partner. In chemistry, pH is a measure of the activity of dissolved hydrogen ions (H^+). Liquids with low pH values are called acids. Liquids with high pH values are called bases.

Find the pH of each liquid. Is the liquid a base, neutral, or an acid?

a. Lime juice: $[H^+] = 0.01$
2; acid

b. Egg: $[H^+] = 0.00000001$
8; base

c. Distilled water: $[H^+] = 0.0000001$
7; neutral

d. Ammonia water: $[H^+] = 0.00000000001$
11; base

e. Tomato juice: $[H^+] = 0.0001$
4; acid

f. Hydrochloric acid: $[H^+] = 1$
0; acid

pH	$[H^+]$	
14	1×10^{-14}	Bases
13	1×10^{-13}	
12	1×10^{-12}	
11	1×10^{-11}	
10	1×10^{-10}	
9	1×10^{-9}	
8	1×10^{-8}	
7	1×10^{-7}	Neutral
6	1×10^{-6}	Acids
5	1×10^{-5}	
4	1×10^{-4}	
3	1×10^{-3}	
2	1×10^{-2}	
1	1×10^{-1}	
0	1×10^{0}	

English Language Learners

Culture

Have students use the Internet or library resources to find the land area of their native country, the state where they currently live, and two other states in the U.S. that have approximately the same land area as their native country. Students can organize the information in a table giving the land area, the land area rounded to the nearest thousand square miles, and the land area in scientific notation. Create a classroom display of the tables.

9.6 Record and Practice Journal

2 ACTIVITY: Writing Scientific Notation

Work with a partner. Match each planet with its description. Then write each of the following in scientific notation.

- **Distance from the Sun (in miles)**
- **Distance from the Sun (in feet)**
- **Mass (in kilograms)**

a. **Distance:** 1,800,000,000 miles — 1.8×10^9 mi; 9.504×10^{12} ft
Mass: 87,000,000,000,000,000,000,000,000 kg — 8.7×10^{25} kg; Uranus

b. **Distance:** 67,000,000 miles — 6.7×10^7 mi; 3.5376×10^{11} ft
Mass: 4,900,000,000,000,000,000,000,000 kg — 4.9×10^{24} kg; Venus

c. **Distance:** 890,000,000 miles — 8.9×10^8 mi; 4.6992×10^{12} ft
Mass: 570,000,000,000,000,000,000,000,000 kg — 5.7×10^{26} kg; Saturn

d. **Distance:** 93,000,000 miles — 9.3×10^7 mi; 4.9104×10^{11} ft
Mass: 6,000,000,000,000,000,000,000,000 kg — 6×10^{24} kg; Earth

e. **Distance:** 140,000,000 miles — 1.4×10^8 mi; 7.392×10^{11} ft
Mass: 640,000,000,000,000,000,000,000 kg — 6.4×10^{23} kg; Mars

f. **Distance:** 2,800,000,000 miles — 2.8×10^9 mi; 1.4784×10^{13} ft
Mass: 100,000,000,000,000,000,000,000,000 kg — 1×10^{26} kg; Neptune

g. **Distance:** 480,000,000 miles — 4.8×10^8 mi; 2.5344×10^{12} ft
Mass: 1,900,000,000,000,000,000,000,000,000 kg — 1.9×10^{27} kg; Jupiter

h. **Distance:** 36,000,000 miles — 3.6×10^7 mi; 1.9008×10^{11} ft
Mass: 330,000,000,000,000,000,000,000 kg — 3.3×10^{23} kg; Mercury

3 ACTIVITY: Making a Scale Drawing

Work with a partner. The illustration in Activity 2 is not drawn to scale. Make a scale drawing of the distances in our solar system.

- **Cut a sheet of paper into three strips of equal width. Tape the strips together.**
- **Draw a long number line. Label the number line in hundreds of millions of miles.**
- **Locate each planet's position on the number line.**

Check students' work.

What Is Your Answer?

4. **IN YOUR OWN WORDS** How can you write a number in scientific notation?

See Additional Answers.

Laurie's Notes

Activity 2

- This activity demonstrates the need for scientific notation. The distance of each planet from the sun is a very large number (in magnitude). The mass of each planet is a number that most of us do not know how to read.
- Students should be able to get started right away without much explanation.

? "How do you match the planets with the descriptions?" Listen for ordering distances least to greatest and looking at the illustration.

? "How do you write the distances (in miles) from the sun in scientific notation?" Move the decimal point to the left until you have a number that is at least 1 and less than 10. Count the number of places you moved the decimal point. This number becomes the exponent of the power of 10.

? "How do you write the distances (in feet) from the sun in scientific notation?" Use the conversion factor $\frac{5280 \text{ ft}}{1 \text{ mi}} = \frac{5.28 \times 10^3 \text{ ft}}{1 \text{ mi}}$.

- Ask students about the process for writing the mass in scientific notation.
- Because of the context, and the visual illustration, students should not have difficulty with this activity. They understand how to write the numbers in scientific notation.

Activity 3

- To help facilitate this activity, you may want to prepare strips of paper in advance.
- Even though the scale is given (hundreds of millions of miles), just writing multiples of this scale is a challenge for students. You may want to make a model on the front board.
- If students have studied the solar system, they should have some sense about how far away Neptune is.

What Is Your Answer?

- **Neighbor Check:** Have students work independently and then have their neighbor check their work. Have students discuss any discrepancies.

Closure

- **Writing Prompt:** Why is it useful to write very large or very small numbers in scientific notation?

Technology For the Teacher

Dynamic Classroom

The Dynamic Planning Tool
Editable Teacher's Resources at *BigIdeasMath.com*

2 ACTIVITY: Writing Scientific Notation

Work with a partner. Match each planet with its description. Then write each of the following in scientific notation.

- **Distance from the Sun (in miles)**
- **Distance from the Sun (in feet)**
- **Mass (in kilograms)**

a. Distance: 1,800,000,000 miles
Mass: 87,000,000,000,000,000,000,000,000 kg

b. Distance: 67,000,000 miles
Mass: 4,900,000,000,000,000,000,000,000 kg

c. Distance: 890,000,000 miles
Mass: 570,000,000,000,000,000,000,000,000 kg

d. Distance: 93,000,000 miles
Mass: 6,000,000,000,000,000,000,000,000 kg

e. Distance: 140,000,000 miles
Mass: 640,000,000,000,000,000,000,000 kg

f. Distance: 2,800,000,000 miles
Mass: 100,000,000,000,000,000,000,000,000 kg

g. Distance: 480,000,000 miles
Mass: 1,900,000,000,000,000,000,000,000,000 kg

h. Distance: 36,000,000 miles
Mass: 330,000,000,000,000,000,000,000 kg

3 ACTIVITY: Making a Scale Drawing

Work with a partner. The illustration in Activity 2 is not drawn to scale. Make a scale drawing of the distances in our solar system.

- **Cut a sheet of paper into three strips of equal width. Tape the strips together.**
- **Draw a long number line. Label the number line in hundreds of millions of miles.**
- **Locate each planet's position on the number line.**

What Is Your Answer?

4. **IN YOUR OWN WORDS** How can you write a number in scientific notation?

Practice Use what you learned about writing scientific notation to complete Exercises 3–5 on page 386.

9.6 Lesson

Check It Out
Lesson Tutorials
BigIdeasMath.com

Key Idea

Writing Numbers in Scientific Notation

Step 1: Move the decimal point to the right of the first nonzero digit.

Step 2: Count the number of places you moved the decimal point. This determines the exponent of the power of 10.

Number greater than or equal to 10

Use a positive exponent when you move the decimal point to the left.

$$8600 = 8.6 \times 10^3$$

(decimal point moved 3 places)

Number between 0 and 1

Use a negative exponent when you move the decimal point to the right.

$$0.0024 = 2.4 \times 10^{-3}$$

(decimal point moved 3 places)

EXAMPLE 1 Writing Large Numbers in Scientific Notation

Google purchased YouTube for $1,650,000,000. Write this number in scientific notation.

The number is greater than 1. So, move the decimal point 9 places to the left.

$$1{,}650{,}000{,}000 = 1.65 \times 10^9$$

(9 places)

The exponent is positive.

EXAMPLE 2 Writing Small Numbers in Scientific Notation

The 2004 Indonesian earthquake slowed the rotation of Earth, making the length of a day 0.00000268 second shorter. Write this number in scientific notation.

The number is between 0 and 1. So, move the decimal point 6 places to the right.

$$0.00000268 = 2.68 \times 10^{-6}$$

(6 places)

The exponent is negative.

On Your Own

Now You're Ready
Exercises 3–11

Write the number in scientific notation.

1. 50,000 **2.** 25,000,000 **3.** 683

4. 0.005 **5.** 0.00000033 **6.** 0.000506

Laurie's Notes

Introduction

Connect

- **Yesterday:** Students explored very large and very small numbers written in standard form.
- **Today:** Students will convert numbers in standard form to scientific notation.

Motivate

- Share with students some information about the U.S. economy.
- Numbers reported during the fall of 2008 and into the winter months of 2009 were so large and so common, many became numb to their magnitude. The place values; millions, billions, and trillions, differ by only a letter or two, yet their magnitudes are significantly different.
- The reason to bring these numbers up is that after you move beyond a million, many calculators will not display the number in standard form.

Lesson Notes

Key Idea

- Review with students the definition of scientific notation.
- Write the Key Idea. Relate the steps to the activities from the investigation.
- ? "Why is the decimal point moved to the right of the first nonzero digit?" Need the factor n to be in the interval $1 \leq n < 10$.

Example 1

- **Teaching Tip:** Have students underline the first two nonzero digits. In scientific notation, the decimal point is placed between these two digits.
- ? "How do you read the number?" one billion six hundred fifty million dollars
- **FYI:** Drawing the movement of the decimal point under the numbers helps students keep track of their counting.
- ? "How do you know if the exponent for the power of 10 will be positive or negative?" If the standard form of the number is greater than 1, positive exponent; if the standard form of the number is less than 1, negative exponent.

Example 2

- Discuss the context of this number.
- Note that reading the number takes time, and you have to count place values. The number is 268 hundred millionths.
- ? "Why is the exponent negative?" Original number is less than 1.

On Your Own

- **Think-Pair-Share:** Students should read each question independently and then work with a partner to answer the questions. When they have answered the questions, the pair should compare their answers with another group and discuss any discrepancies.

Goal Today's lesson is writing numbers in scientific notation.

Start Thinking! and Warm Up

Lesson 9.6 **Warm Up**
For use before Lesson 9.6

Lesson 9.6 **Start Thinking!**
For use before Lesson 9.6

Estimate the population of the world.

Go to *www.census.gov* to find the actual world population.

Write the population in scientific notation. How did you choose to round the number and why?

Extra Example 1

Write 2,450,000 in scientific notation. 2.45×10^6

On Your Own

1. 5×10^4
2. 2.5×10^7
3. 6.83×10^2

Extra Example 2

Write 0.0000045 in scientific notation. 4.5×10^{-6}

On Your Own

4. 5×10^{-3}
5. 3.3×10^{-7}
6. 5.06×10^{-4}

Extra Example 3

In Example 3, an album has sold 750,000 copies. How many more copies does it need to sell to receive the award? Write your answer in scientific notation. 9.25×10^6 copies

On Your Own

7. 9.045×10^6

Extra Example 4

Find $(2 \times 10^{-4}) \times (6 \times 10^{-3})$. Write your answer in scientific notation.
1.2×10^{-6}

On Your Own

8. 5×10^{11}
9. 2×10^{-8}
10. 2.7×10^{-12}
11. 2.1×10^{8}

Differentiated Instruction

Kinesthetic

Students may incorrectly count the number of zeros in a number and use that as the exponent in scientific notation. Have students write the number and place an arrow where the decimal point will be placed in the factor. Then have students count the number of places the decimal point moves.

$54{,}000 = 5.4 \times 10^4$ (↑ arrow after 5)

$0.00000675 = 6.75 \times 10^{-6}$ (↑ arrow after 6)

Laurie's Notes

Example 3

- Encourage students to read the numbers when they are in standard form: 10 million; 8 million 780 thousand.
- The subtraction is performed on the numbers in standard form, with the result written in scientific notation.
- **Extension:** Ask students if it is possible to subtract numbers in scientific notation. The answer is no, unless the powers are the same. They need to be like terms.

On Your Own

- **Think-Pair-Share:** Students should read each question independently and then work with a partner to answer the questions. When they have answered the questions, the pair should compare their answers with another group and discuss any discrepancies.

Example 4

- Even though you cannot add or subtract numbers in scientific notation (unless they have the same power), you can multiply them.
- Write the problem and ask students how they would multiply the numbers. Most students will immediately suggest multiplying the factors, and then multiplying the powers of 10. This process makes sense to them. You want to make sure that students realize that the Commutative and Associative Properties allow this to happen, and then the Product of Powers Property is used to multiply the powers of 10.

On Your Own

- These are nice problems to review mental math and properties of exponents.

Closure

- The land area of Rhode Island is about 1000 square miles. The land area of Alaska is about 570,000 square miles. The United States land area is about 3,500,000 square miles. Write each of these in scientific notation.
1×10^3, 5.7×10^5, 3.5×10^6

Technology For the Teacher

Dynamic Classroom

The Dynamic Planning Tool
Editable Teacher's Resources at *BigIdeasMath.com*

EXAMPLE 3 Standardized Test Practice

An album receives an award when it sells 10,000,000 copies.

An album has sold 8,780,000 copies. How many more copies does it need to sell to receive the award?

Ⓐ 1.22×10^{-7} Ⓑ 1.22×10^{-6}

Ⓒ 1.22×10^{6} Ⓓ 1.22×10^{7}

Use a model to solve the problem.

Remaining sales needed for award = Sales required for award − Current sales total

$= 10{,}000{,}000 - 8{,}780{,}000$

$= 1{,}220{,}000$

$= 1.22 \times 10^{6}$

The album must sell 1.22×10^{6} more copies to receive the award. The correct answer is Ⓒ.

On Your Own

7. An album has sold 955,000 copies. How many more copies does it need to sell to receive the award? Write your answer in scientific notation.

EXAMPLE 4 Multiplying Numbers in Scientific Notation

Find $(3 \times 10^{-5}) \times (5 \times 10^{-2})$. Write your answer in scientific notation.

$(3 \times 10^{-5}) \times (5 \times 10^{-2})$	
$= 3 \times 5 \times 10^{-5} \times 10^{-2}$	Commutative Property of Multiplication
$= (3 \times 5) \times (10^{-5} \times 10^{-2})$	Associative Property of Multiplication
$= 15 \times 10^{-7}$	Simplify.
$= 1.5 \times 10^{1} \times 10^{-7}$	Write factor in scientific notation.
$= 1.5 \times 10^{-6}$	Simplify.

Study Tip

You can check your answer using standard form.

$(3 \times 10^{-5}) \times (5 \times 10^{-2})$
$= 0.00003 \times 0.05$
$= 0.0000015$
$= 1.5 \times 10^{-6}$

On Your Own

Now You're Ready Exercises 14–19

Multiply. Write your answer in scientific notation.

8. $(2.5 \times 10^{8}) \times (2 \times 10^{3})$

9. $(2 \times 10^{-4}) \times (1 \times 10^{-4})$

10. $(5 \times 10^{-4}) \times (5.4 \times 10^{-9})$

11. $(7 \times 10^{2}) \times (3 \times 10^{5})$

9.6 Exercises

Check It Out
Help with Homework
BigIdeasMath.com

Vocabulary and Concept Check

1. **REASONING** How do you know whether a number written in standard form will have a positive or negative exponent when written in scientific notation?

2. **WRITING** Describe how to write a number in scientific notation.

Practice and Problem Solving

Write the number in scientific notation.

① ② 3. 0.0021 4. 5,430,000 5. 321,000,000

6. 0.00000625 7. 0.00004 8. 10,700,000

9. 45,600,000,000 10. 0.000000000009256 11. 840,000

ERROR ANALYSIS Describe and correct the error in writing the number in scientific notation.

12. ✗ 0.000036 (5) 3.6×10^5

13. ✗ 72,500,000 (6) 72.5×10^6

Multiply. Write your answer in scientific notation.

④ 14. $(4 \times 10^4) \times (2 \times 10^6)$

15. $(3 \times 10^{-8}) \times (3 \times 10^{-2})$

16. $(5 \times 10^{-7}) \times (3 \times 10^6)$

17. $(8 \times 10^3) \times (2 \times 10^4)$

18. $(6 \times 10^8) \times (1.4 \times 10^{-5})$

19. $(7.2 \times 10^{-1}) \times (4 \times 10^{-7})$

20. **HAIR** What is the diameter of a human hair in scientific notation?

Diameter: 0.000099 meter

21. **EARTH** What is the circumference of Earth in scientific notation?

Circumference at the equator: about 40,100,000 meters

22. **WATERFALLS** During high flow, more than 44,380,000 gallons of water go over Niagara Falls every minute. Write this number in scientific notation.

Assignment Guide and Homework Check

Level	Day 1 Activity Assignment	Day 2 Lesson Assignment	Homework Check
Basic	3–5, 30–32	1, 2, 7–23 odd, 12, 20, 22	7, 12, 15, 20, 23
Average	3–5, 30–32	1, 2, 10–13, 18–21, 23–26	10, 18, 20, 24, 25
Advanced	3–5, 30–32	1, 2, 10–12, 18–20, 23–26, 28, 29	10, 18, 24, 28

For Your Information

- **Exercise 27** Remind students that population density is the average number of people for some amount of land area. To find the population density, you divide the population by the land area.

Common Errors

- **Exercises 3–11** Students may write an exponent with the opposite sign of what is correct. Remind them that large numbers have a positive exponent in scientific notation and that small numbers have a negative exponent in scientific notation.
- **Exercises 14–19** Students may multiply the factors and leave the number greater than 10. Remind them that the factor in scientific notation must be at least one and less than 10. If the number is less than 10 and greater than or equal to 1, the exponent will be 0.

9.6 Record and Practice Journal

Write the number in scientific notation.

1. 4,200,000 — 4.2×10^6
2. 0.038 — 3.8×10^{-2}
3. 600,000 — 6×10^5
4. 0.0000808 — 8.08×10^{-5}
5. 0.0007 — 7×10^{-4}
6. 29,010,000,000 — 2.901×10^{10}

Multiply. Write your answer in scientific notation.

7. $(6 \times 10^8) \times (4 \times 10^6)$ — 2.4×10^{15}
8. $(9 \times 10^{-3}) \times (9 \times 10^{-3})$ — 8.1×10^{-5}
9. $(7 \times 10^{-7}) \times (5 \times 10^{10})$ — 3.5×10^4
10. $(1.4 \times 10^{-2}) \times (2 \times 10^{-15})$ — 2.8×10^{-17}
11. A patient has 0.0000075 gram of iron in 1 liter of blood. The normal level is between 6×10^{-7} gram and 1.6×10^{-5} gram. Is the patient's iron level normal? Write the patient's amount of iron in scientific notation.

    yes; 7.5×10^{-6}

Technology For the Teacher

Answer Presentation Tool

Quiz*Show*

Vocabulary and Concept Check

1. If the number is greater than or equal to 10, the exponent will be positive. If the number is less than 1 and greater than 0, the exponent will be negative.
2. Move the decimal point to the right of the first nonzero digit. Then count the number of places you moved the decimal point. This determines the exponent of the power of 10. Use a positive exponent when you move the decimal point to the left. Use a negative exponent when you move the decimal point to the right.

Practice and Problem Solving

3. 2.1×10^{-3}
4. 5.43×10^6
5. 3.21×10^8
6. 6.25×10^{-6}
7. 4×10^{-5}
8. 1.07×10^7
9. 4.56×10^{10}
10. 9.256×10^{-12}
11. 8.4×10^5
12. The decimal point moved 5 places to the right, so the exponent should be negative. 3.6×10^{-5}
13. 72.5 is not less than 10. The decimal point needs to move one more place to the left. 7.25×10^7
14. 8×10^{10}
15. 9×10^{-10}
16. 1.5×10^0
17. 1.6×10^8
18. 8.4×10^3
19. 2.88×10^{-7}
20. 9.9×10^{-5} m
21. 4.01×10^7 m
22. 4.438×10^7 gal

Practice and Problem Solving

23. 5.612×10^{14} cm^2

24. 9×10^{-7} ft^2

25. 9.75×10^9 newton-meters per second

26. *Sample answer:* 670,000,000; $6 \times 10^8 + 7 \times 10^7$; 6.7×10^8

27. *Answer should include, but is not limited to:* Make sure calculations using scientific notation are done correctly.

28. See *Taking Math Deeper*.

29. **a.** 2.65×10^8

b. 2.2×10^{-4}

Fair Game Review

30. 9 **31.** 200

32. A

Mini-Assessment

Write the number in scientific notation.

1. 0.00035 3.5×10^{-4}
2. 0.0000000000567 5.67×10^{-11}
3. 25,500,000 2.55×10^7

Multiply. Write your answer in scientific notation.

4. $(3 \times 10^5) \times (2.5 \times 10^3)$ 7.5×10^8
5. $(1.7 \times 10^{-2}) \times (4.3 \times 10^{-4})$ 7.31×10^{-6}

Taking Math Deeper

Exercise 28

We are all familiar with the look of a DVD (digital versatile discs), but how many of us know what the surface of a DVD looks like? This problem gives students some idea of what this digital storage device actually looks like.

1. Summarize the given information.

Width of each ridge = 0.000032 cm
Width of each valley = 0.000074 cm
Diameter of center portion = 4.26 cm

2. Find the diameter of the DVD.

$$\begin{aligned} \text{ridges} + \text{valleys} &= 73{,}000(0.000032) + 73{,}000(0.000074) \\ &= 73{,}000(0.000032 + 0.000074) \\ &= 73{,}000(0.000106) \\ &= 7.738 \text{ cm} \end{aligned}$$

OR using scientific notation

$$\begin{aligned} \text{ridges} + \text{valleys} &= (7.3 \times 10^4)(3.2 \times 10^{-5}) + (7.3 \times 10^4)(7.4 \times 10^{-5}) \\ &= (7.3 \times 10^4)(3.2 \times 10^{-5} + 7.4 \times 10^{-5}) \\ &= (7.3 \times 10^4)(10.6 \times 10^{-5}) \\ &= 77.38 \times 10^{-1} \\ &= 7.738 \text{ cm} \end{aligned}$$

$$\begin{aligned} \text{diameter} &= \text{ridges} + \text{valleys} + \text{center portion} \\ &= 7.738 + 4.26 \\ &= 11.998 \text{ cm} \\ &\approx 12 \text{ cm} \end{aligned}$$

3. Here's a fun fact.

The microscopic dimensions of the bumps make the spiral track on a DVD extremely long. If you could lift the data track off a single layer of a DVD, and stretch it out into a straight line, it would be almost 7.5 miles long!

Project

Write a report on the invention of the DVD and the DVD player.

Reteaching and Enrichment Strategies

If students need help. . .	If students got it. . .
Resources by Chapter • Practice A and Practice B • Puzzle Time Record and Practice Journal Practice Differentiating the Lesson Lesson Tutorials Skills Review Handbook	Resources by Chapter • Enrichment and Extension • School-to-Work • Financial Literacy • Technology Connection Start the Chapter Review

Find the area of the figure. Write your answer in scientific notation.

23.

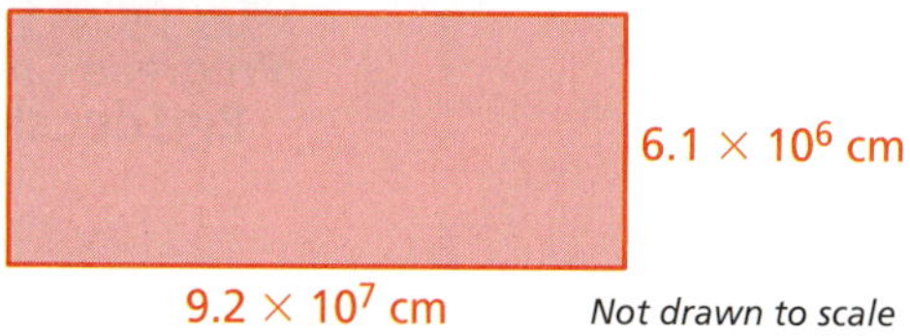

24.

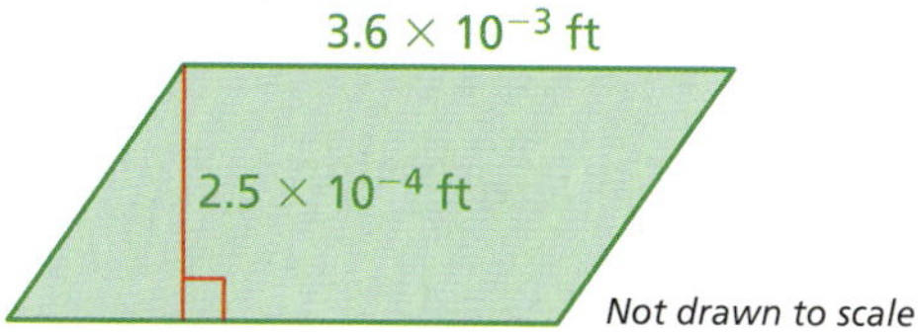

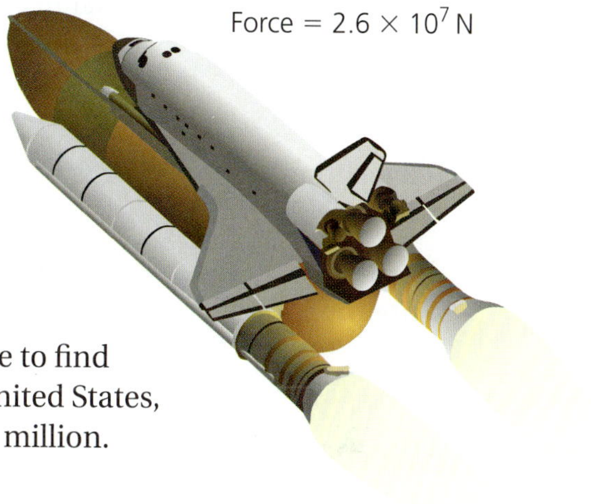

25. SPACE SHUTTLE The power of a space shuttle during launch is the force of the solid rocket boosters multiplied by the velocity. The velocity is 3.75×10^2 meters per second. What is the power (in newton-meters per second) of the shuttle shown during launch?

26. NUMBER SENSE Write 670 million in three ways.

27. PROJECT Use the Internet or some other reference to find the populations of India, China, Argentina, the United States, and Egypt. Round each population to the nearest million.

a. Write each population in scientific notation.

b. Use the Internet or some other reference to find the population density for each country.

c. Use the results of parts (a) and (b) to find the area of each country.

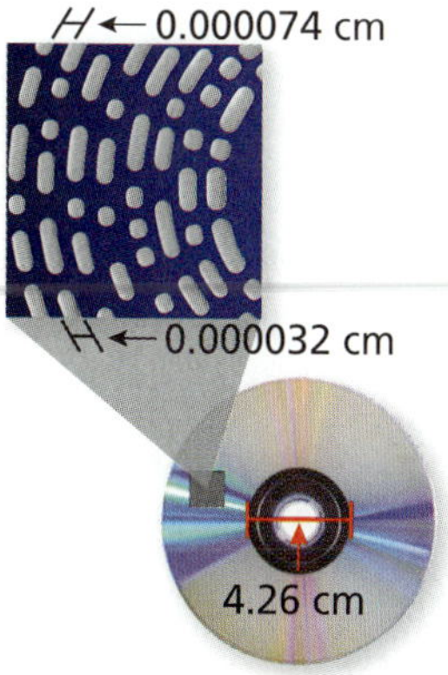

28. DVDS On a DVD, information is stored on bumps that spiral around the disk. There are 73,000 ridges (with bumps) and 73,000 valleys (without bumps) across the diameter of the DVD. What is the diameter of the DVD in centimeters?

29. Number Sense Simplify. Write your answer in scientific notation.

a. $\dfrac{(53{,}000{,}000)(0.002)}{(0.0004)}$ **b.** $\dfrac{(0.33)(60{,}000)}{(90{,}000{,}000)}$

Fair Game Review What you learned in previous grades & lessons

Write and solve an equation to answer the question. *(Skills Review Handbook)*

30. 15% of 60 is what number?

31. 85% of what number is 170?

32. MULTIPLE CHOICE What is the domain of the function represented by the table? *(Section 4.1)*

x	−2	−1	0	1	2
y	−6	−2	2	6	10

Ⓐ −2, −1, 0, 1, 2 Ⓑ −6, −2, 2, 6, 10

Ⓒ all integers Ⓓ all whole numbers

9.4–9.6 Quiz

Check It Out
Progress Check
BigIdeasMath.com

Evaluate the expression. *(Section 9.4)*

1. $(-4.8)^{-9} \cdot (-4.8)^{9}$

2. $\dfrac{5^4}{5^7}$

Simplify. Write the expression using only positive exponents. *(Section 9.4)*

3. $8d^{-6}$

4. $\dfrac{12x^5}{4x^7}$

Tell whether the number is written in scientific notation. Explain. *(Section 9.5)*

5. 23×10^{9}

6. 0.6×10^{-7}

Write the number in standard form. *(Section 9.5)*

7. 8×10^{6}

8. 1.6×10^{-2}

Write the number in scientific notation. *(Section 9.6)*

9. 0.00524

10. 892,000,000

Multiply. Write your answer in scientific notation. *(Section 9.6)*

11. $(9 \times 10^{3}) \times (4 \times 10^{4})$

12. $(2 \times 10^{-5}) \times (3.1 \times 10^{-2})$

13. PLANETS The table shows the equatorial radii of the eight planets in our solar system. *(Section 9.5)*

a. Which planet has the second smallest equatorial radius?

b. Which planet has the second greatest equatorial radius?

Planet	Equatorial Radius (km)
Mercury	2.44×10^{3}
Venus	6.05×10^{3}
Earth	6.38×10^{3}
Mars	3.4×10^{3}
Jupiter	7.15×10^{4}
Saturn	6.03×10^{4}
Uranus	2.56×10^{4}
Neptune	2.48×10^{4}

14. OORT CLOUD The Oort cloud is a spherical cloud that surrounds our solar system. It is about 2×10^{5} astronomical units from the Sun. An astronomical unit is about 1.5×10^{8} kilometers. How far is the Oort cloud from the Sun in kilometers? *(Section 9.6)*

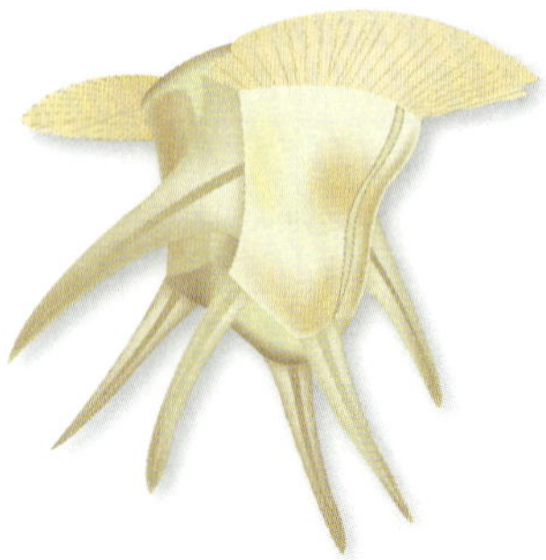

15. ORGANISM A one-celled, aquatic organism called a dinoflagellate is 1000 micrometers long. *(Section 9.4)*

a. One micrometer is 10^{-6} meter. What is the length of the dinoflagellate in meters?

b. Is the length of the dinoflagellate equal to 1 millimeter or 1 kilometer? Explain.

Alternative Assessment Options

Math Chat	Student Reflective Focus Question
Structured Interview	Writing Prompt

Math Chat

- Have individual students work problems from the quiz on the board. The student explains the process used and justifies each step. Students in the class ask questions of the student presenting.
- The teacher probes the thought process of the student presenting, but does not teach or ask leading questions.

Study Help Sample Answers

Remind students to complete Graphic Organizers for the rest of the chapter.

4.

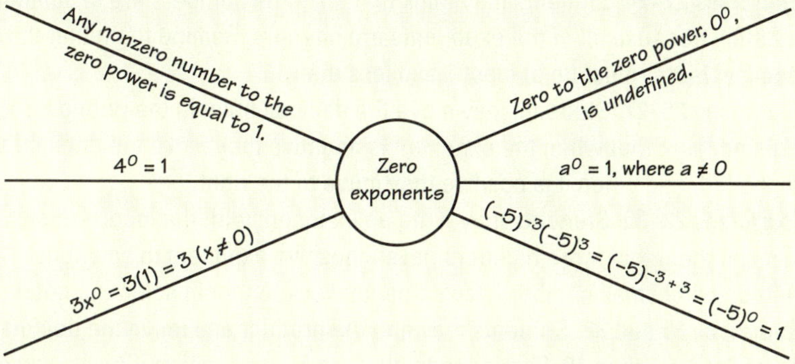

5.

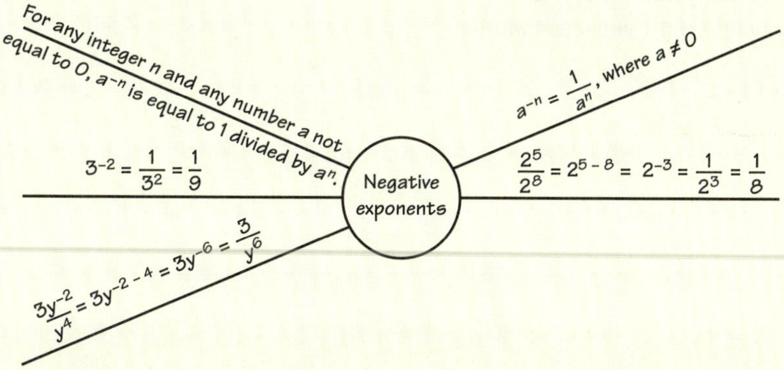

6–8. Available at *BigIdeasMath.com*

Reteaching and Enrichment Strategies

If students need help. . .	If students got it. . .
Resources by Chapter • Study Help • Practice A and Practice B • Puzzle Time Lesson Tutorials *BigIdeasMath.com* Practice Quiz Practice from the Test Generator	Resources by Chapter • Enrichment and Extension • School-to-Work Game Closet at *BigIdeasMath.com* Start the Chapter Review

Answers

1. 1
2. $\frac{1}{125}$
3. $\frac{8}{d^6}$
4. $\frac{3}{x^2}$
5. no; The factor is greater than 10.
6. no; The factor is less than 1.
7. 8,000,000
8. 0.016
9. 5.24×10^{-3}
10. 8.92×10^8
11. 3.6×10^8
12. 6.2×10^{-7}
13. **a.** Mars
    b. Saturn
14. 3×10^{13} km
15. **a.** 1×10^{-3} m
    b. 1 millimeter; The length is less than 1 meter and a millimeter is smaller than a meter.

Technology For the Teacher

Answer Presentation Tool

Assessment Book

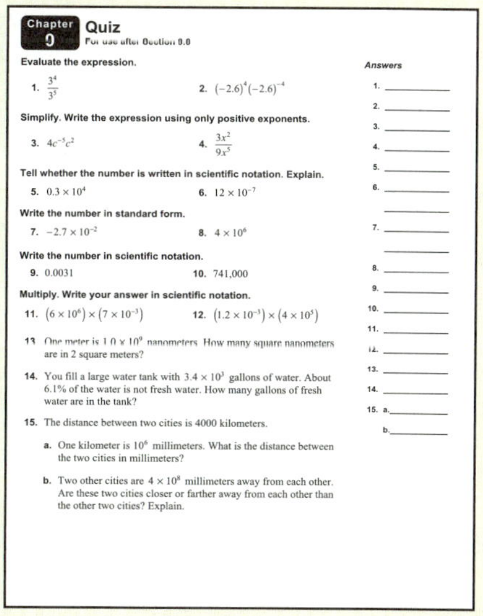
Chapter 9 **Quiz**

For use after Section [illegible]

Evaluate the expression.

1. $\frac{3^4}{3^5}$
2. $(-2.6)^4(-2.6)^{-4}$

Simplify. Write the expression using only positive exponents.

3. $4c^{-5}c^2$
4. $\frac{3x^2}{9x^5}$

Tell whether the number is written in scientific notation. Explain.

5. 0.3×10^4
6. 12×10^{-7}

Write the number in standard form.

7. -2.7×10^{-2}
8. 4×10^6

Write the number in scientific notation.

9. 0.0031
10. 741,000

Multiply. Write your answer in scientific notation.

11. $(6 \times 10^6) \times (7 \times 10^{-3})$
12. $(1.2 \times 10^{-3}) \times (4 \times 10^5)$
13. One meter is 1.0×10^9 nanometers. How many square nanometers are in 2 square meters?
14. You fill a large water tank with 3.4×10^3 gallons of water. About 6.1% of the water is not fresh water. How many gallons of fresh water are in the tank?
15. The distance between two cities is 4000 kilometers.
    a. One kilometer is 10^6 millimeters. What is the distance between the two cities in millimeters?
    b. Two other cities are 4×10^8 millimeters away from each other. Are these two cities closer or farther away from each other than the other two cities? Explain.

Answers

1. ______
2. ______
3. ______
4. ______
5. ______
6. ______
7. ______
8. ______
9. ______
10. ______
11. ______
12. ______
13. ______
14. ______
15. a. ______
    b. ______

For the Teacher
Additional Review Options

- **Quiz***Show*
- Big Ideas Test Generator
- Game Closet at *BigIdeasMath.com*
- Vocabulary Puzzle Builder
- Resources by Chapter
  - Puzzle Time
  - Study Help

Review of Common Errors

- **Exercises 1 and 2** Students may count the wrong number of factors.
- **Exercises 3 and 4** Students may have the wrong sign in the answer. Remind them that when the sign is inside the parentheses, it is part of the base.
- **Exercise 5** Students may not use the correct order of operations when they evaluate the expression. Review the order of operations with students.
- **Exercises 6–9** Students may multiply the bases as well as add the exponents. Remind them that the base stays the same and only the exponent changes when simplifying using the Product of Powers Property.
- **Exercises 10–15** Students may divide the exponents instead of subtracting them. Remind them of the Quotient of Powers Property.
- **Exercises 16–21** Students may forget the rules for zero and negative exponents. Reviewing these rules may be helpful.
- **Exercises 22–24** Students may think that all of the numbers are in scientific notation because all of the exponents are integers. Remind them that the decimal factor must be at least 1 and less than 10.
- **Exercises 25–27** Students may move the decimal point in the wrong direction. Remind them that when the exponent is negative they move the decimal point to the left, and when it is positive they move to the right.
- **Exercises 28–30** Students may write an exponent with the incorrect sign. Remind them that large numbers have a positive exponent in scientific notation and small numbers have a negative exponent in scientific notation.
- **Exercises 31 and 32** Students may find the product and leave the decimal factor greater than 10. Remind them that the factor in scientific notation must be at least 1 and less than 10.

Answers

1. $(-9)^5$
2. 2^3n^2
3. 216
4. $-\frac{1}{16}$
5. 100
6. p^7
7. n^{22}
8. $125y^3$
9. $16k^4$

9 Chapter Review

Check It Out
Vocabulary Help
BigIdeasMath.com

Review Key Vocabulary

power, *p. 352*
base, *p. 352*
exponent, *p. 352*
scientific notation, *p. 378*

Review Examples and Exercises

9.1 Exponents *(pp. 350–355)*

Write $(-4) \cdot (-4) \cdot (-4) \cdot y \cdot y$ using exponents.

Because -4 is used as a factor 3 times, its exponent is 3. Because y is used as a factor 2 times, its exponent is 2.

So, $(-4) \cdot (-4) \cdot (-4) \cdot y \cdot y = (-4)^3y^2$.

Exercises

Write the product using exponents.

1. $(-9) \cdot (-9) \cdot (-9) \cdot (-9) \cdot (-9)$
2. $2 \cdot 2 \cdot 2 \cdot n \cdot n$

Evaluate the expression.

3. 6^3
4. $-\left(\frac{1}{2}\right)^4$
5. $\left|\frac{1}{2}(16 - 6^3)\right|$

9.2 Product of Powers Property *(pp. 356–361)*

a. $\left(-\frac{1}{8}\right)^7 \cdot \left(-\frac{1}{8}\right)^4 = \left(-\frac{1}{8}\right)^{7+4}$ The base is $-\frac{1}{8}$. Add the exponents.

$= \left(-\frac{1}{8}\right)^{11}$ Simplify.

b. $(3m)^2 = 3m \cdot 3m$ Write as repeated multiplication.

$= (3 \cdot 3) \cdot (m \cdot m)$ Use properties of multiplication.

$= 3^{1+1} \cdot m^{1+1}$ The bases are 3 and m. Add the exponents.

$= 3^2 \cdot m^2 = 9m^2$ Simplify.

Exercises

Simplify the expression.

6. $p^5 \cdot p^2$
7. $(n^{11})^2$
8. $(5y)^3$
9. $(-2k)^4$

9.3 Quotient of Powers Property (pp. 362–367)

a. $\frac{(-4)^9}{(-4)^6} = (-4)^{9-6}$ The base is -4. Subtract the exponents.

$= (-4)^3$ Simplify.

b. $\frac{x^4}{x^3} = x^{4-3}$ The base is x. Subtract the exponents.

$= x^1$

$= x$ Simplify.

Exercises

Simplify the expression. Write your answer as a power.

10. $\frac{8^8}{8^3}$

11. $\frac{5^2 \cdot 5^9}{5}$

12. $\frac{w^8}{w^7} \cdot \frac{w^5}{w^2}$

Simplify the expression.

13. $\frac{2^2 \cdot 2^5}{2^3}$

14. $\frac{(6c)^3}{c}$

15. $\frac{m^8}{m^6} \cdot \frac{m^{10}}{m^9}$

9.4 Zero and Negative Exponents (pp. 370–375)

a. $10^{-3} = \frac{1}{10^3}$ Definition of negative exponent

$= \frac{1}{1000}$ Evaluate power.

b. $(-0.5)^{-5} \cdot (-0.5)^5 = (-0.5)^{-5+5}$ Add the exponents.

$= (-0.5)^0$ Simplify.

$= 1$ Definition of zero exponent

Exercises

Evaluate the expression.

16. 2^{-4}

17. 95^0

18. $\frac{8^2}{8^4}$

19. $(-12)^{-7} \cdot (-12)^7$

20. $\frac{1}{7^9} \cdot \frac{1}{7^{-6}}$

21. $\frac{9^4 \cdot 9^{-2}}{9^2}$

Review Game

Comparing Values in Scientific Notation

Big Ideas Game Closet

Materials per Group
- pencil
- paper
- computer with Internet access

Directions
- The game can be completed in one to two class periods, but you may want to give the students one or two days to complete the necessary research.
- Each student comes up with three different values written in scientific notation. (Example: the length of an ant in meters, the weight of a person in ounces, the volume of a car's gas tank in cups)
- One value should be length, one value should be weight, and one value should be volume.
- Divide the class into an even number of groups.
- Randomly call on two groups and have them complete the following:
  - Each group writes the length (in scientific notation) of one of their items on the board.
  - The members of each group work together to write each number in standard form and determine which value is the least.
  - They write their answer on a piece of paper and submit it to the teacher.
  - One point is awarded to each group that answers correctly.
  - Note: Be sure students are aware that when they are comparing these values, the units of the values must be considered.
- Repeat this process for the remaining length, weight, and volume values.

Who wins?
After all groups have compared their values, the group(s) with the most points wins.

For the Student Additional Practice
- Lesson Tutorials
- Study Help (textbook)
- Student Website
  Multi-Language Glossary
  Practice Assessments

Answers

10. 8^5

11. 5^{10}

12. w^4

13. 16

14. $216c^2$

15. m^3

16. $\frac{1}{16}$

17. 1

18. $\frac{1}{64}$

19. 1

20. $\frac{1}{343}$

21. 1

22. no; The factor is less than 1.

23. yes; The factor is at least 1 and less than 10. The power of 10 has an integer exponent.

24. no; The factor is greater than 10.

25. 20,000,000

26. 0.0048

27. 625,000

28. 3.6×10^{-4}

29. 8×10^5

30. 7.92×10^7

31. 8×10^5

32. 1.2×10^{-11}

My Thoughts on the Chapter

What worked. . .

Teacher Tip

Not allowed to write in your teaching edition? Use sticky notes to record your thoughts.

What did not work. . .

What I would do differently. . .

9.5 Reading Scientific Notation (pp. 376–381)

a. Write 5.9×10^4 in standard form.

$5.9 \times 10^4 = 59{,}000$ (4) — Move decimal point 4 places to the right.

b. Write 7.31×10^{-6} in standard form.

$7.31 \times 10^{-6} = 0.00000731$ (6) — Move decimal point 6 places to the left.

Exercises

Tell whether the number is written in scientific notation. Explain.

22. 0.9×10^9 **23.** 3.04×10^{-11} **24.** 15×10^{26}

Write the number in standard form.

25. 2×10^7 **26.** 4.8×10^{-3} **27.** 6.25×10^5

9.6 Writing Scientific Notation (pp. 382–387)

a. In 2010, the population of the United States was about 309,000,000. Write this number in scientific notation.

The number is greater than 1. So, move the decimal point 8 places to the left.

$309{,}000{,}000 = 3.09 \times 10^8$ (8)

The exponent is positive.

b. The cornea of an eye is 0.00056 meter thick. Write this number in scientific notation.

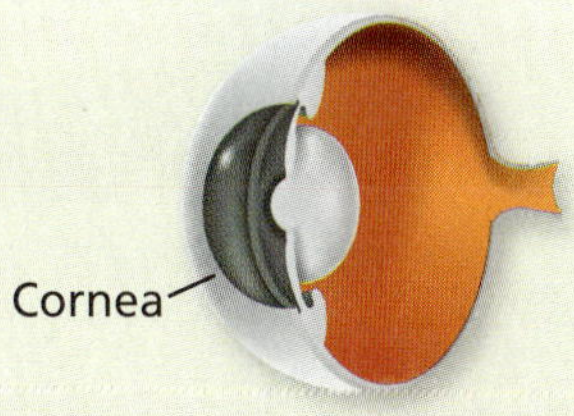

The number is between 0 and 1. So, move the decimal point 4 places to the right.

$0.00056 = 5.6 \times 10^{-4}$ (4)

The exponent is negative.

Exercises

Write the number in scientific notation.

28. 0.00036 **29.** 800,000 **30.** 79,200,000

Multiply. Write your answer in scientific notation.

31. $(4 \times 10^3) \times (2 \times 10^2)$ **32.** $(1.5 \times 10^{-9}) \times (8 \times 10^{-3})$

9 Chapter Test

Check It Out
Test Practice
BigIdeasMath.com

Write the product using exponents.

1. $(-15) \cdot (-15) \cdot (-15)$

2. $\left(\frac{1}{12}\right) \cdot \left(\frac{1}{12}\right) \cdot \left(\frac{1}{12}\right) \cdot \left(\frac{1}{12}\right) \cdot \left(\frac{1}{12}\right)$

Evaluate the expression.

3. -2^3

4. $10 + 3^3 \div 9$

Simplify the expression. Write your answer as a power.

5. $9^{10} \cdot 9$

6. $\frac{(-3.5)^{13}}{(-3.5)^9}$

Evaluate the expression.

7. $5^{-2} \cdot 5^2$

8. $\frac{-8}{(-8)^3}$

Write the number in standard form.

9. 3×10^7

10. 9.05×10^{-3}

Multiply. Write your answer in scientific notation.

11. $(7 \times 10^3) \times (5 \times 10^2)$

12. $(3 \times 10^{-5}) \times (2 \times 10^{-3})$

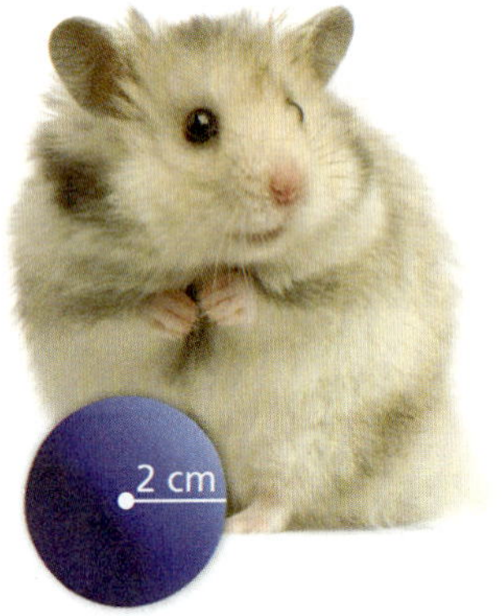

13. **HAMSTER** A hamster toy is in the shape of a sphere. The volume V of a sphere is represented by $V = \frac{4}{3}\pi r^3$, where r is the radius of the sphere. What is the volume of the toy? Round your answer to the nearest cubic centimeter. Use 3.14 for π.

14. **CRITICAL THINKING** Is $(xy^2)^3$ the same as $(xy^3)^2$? Explain.

15. **RICE** A grain of rice weighs about 3^3 milligrams. About how many grains of rice are in one scoop?

16. **TASTE BUDS** There are about 10,000 taste buds on a human tongue. Write this number in scientific notation.

17. **LEAD** From 1978 to 2008, the amount of lead allowed in the air in the United States was 1.5×10^{-6} gram per cubic meter. In 2008, the amount allowed was reduced by 90%. What is the new amount of lead allowed in the air?

One scoop of rice weighs about 3^9 milligrams.

Test Item References

Chapter Test Questions	Section to Review
1–4, 13	9.1
5, 14	9.2
6, 15	9.3
7, 8	9.4
9, 10, 17	9.5
11, 12, 16	9.6

Test-Taking Strategies

Remind students to quickly look over the entire test before they start so that they can budget their time. Have students use the **Stop** and **Think** strategy before they answer each question.

Common Assessment Errors

- **Exercises 1 and 2** Students may count the wrong number of factors. Remind them to check their work.
- **Exercises 3, 6, and 8** Students may have the wrong sign in their answer. Remind them that when the negative sign is inside parentheses, it is part of the base. Point out that in Exercise 3, the negative sign is not part of the base.
- **Exercise 4** Students may not use the correct order of operations when they evaluate the expression. Review the order of operations with students.
- **Exercises 5 and 7** Students may multiply the bases as well as add the exponents. Remind them that the base stays the same and only the exponent changes when using the Product of Powers Property.
- **Exercises 6 and 8** Students may divide the exponents instead of subtracting them. Remind them of the Quotient of Powers Property.
- **Exercises 9 and 10** Students may move the decimal point in the wrong direction. Remind them that when the exponent is negative they move the decimal point to the left, and when the exponent is positive they move the decimal point to the right.
- **Exercise 11** Students may find the product and leave the decimal factor greater than 10. Remind them that the factor in scientific notation must be at least 1 and less than 10.

Reteaching and Enrichment Strategies

If students need help. . .	If students got it. . .
Resources by Chapter • Practice A and Practice B • Puzzle Time Record and Practice Journal Practice Differentiating the Lesson Lesson Tutorials Practice from the Test Generator Skills Review Handbook	Resources by Chapter • Enrichment and Extension • School-to-Work • Financial Literacy Game Closet at *BigIdeasMath.com* Start Standardized Test Practice

Answers

1. $(-15)^3$
2. $\left(\frac{1}{12}\right)^5$
3. -8
4. 13
5. 9^{11}
6. $(-3.5)^4$
7. 1
8. $\frac{1}{64}$
9. $30{,}000{,}000$
10. 0.00905
11. 3.5×10^6
12. 6×10^{-8}
13. 33 cm^3
14. no; $(xy^2)^3 = (xy^2) \cdot (xy^2) \cdot (xy^2) = x \cdot x \cdot x \cdot y^2 \cdot y^2 \cdot y^2 = x^3y^6$
    $(xy^3)^2 = (xy^3) \cdot (xy^3) = x \cdot x \cdot y^3 \cdot y^3 = x^2y^6$
15. 3^6 or 729 grains
16. 1×10^4
17. 1.5×10^{-7} gram per cubic meter

Assessment Book

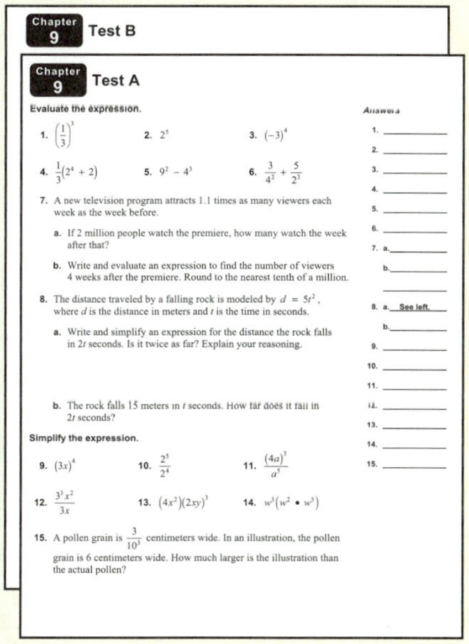
Chapter 9 Test B

Chapter 9 Test A

Evaluate the expression.

1. $\left(\frac{1}{3}\right)^3$ 2. 2^5 3. $(-3)^4$

4. $\frac{1}{3}(2^4 + 2)$ 5. $9^2 - 4^3$ 6. $\frac{3}{4^2} + \frac{5}{2^3}$

7. A new television program attracts 1.1 times as many viewers each week as the week before.
   a. If 2 million people watch the premiere, how many watch the week after that?
   b. Write and evaluate an expression to find the number of viewers 4 weeks after the premiere. Round to the nearest tenth of a million.

8. The distance traveled by a falling rock is modeled by $d = 5t^2$, where d is the distance in meters and t is the time in seconds.
   a. Write and simplify an expression for the distance the rock falls in $2t$ seconds. Is it twice as far? Explain your reasoning.
   b. The rock falls 15 meters in t seconds. How far does it fall in $2t$ seconds?

Simplify the expression.

9. $(3x)^4$ 10. $\frac{2^5}{2^4}$ 11. $\frac{(4a)^3}{a^5}$

12. $\frac{3^3x^2}{3x}$ 13. $(4x^2)(2xy)^3$ 14. $w^3(w^2 \bullet w^5)$

15. A pollen grain is $\frac{3}{10^3}$ centimeters wide. In an illustration, the pollen grain is 6 centimeters wide. How much larger is the illustration than the actual pollen?

Answers
1. ____ 2. ____ 3. ____ 4. ____ 5. ____ 6. ____ 7. a. ____ b. ____ 8. a. See left. b. ____ 9. ____ 10. ____ 11. ____ 12. ____ 13. ____ 14. ____ 15. ____

Test-Taking Strategies

Available at *BigIdeasMath.com*

After Answering Easy Questions, Relax
Answer Easy Questions First
Estimate the Answer
Read All Choices before Answering
Read Question before Answering
Solve Directly or Eliminate Choices
Solve Problem before Looking at Choices
Use Intelligent Guessing
Work Backwards

About this Strategy

When taking a multiple choice test, be sure to read each question carefully and thoroughly. Sometimes you don't know the answer. So...guess intelligently! Look at the choices and choose the ones that are possible answers.

Answers

1. C
2. I
3. D
4. 68
5. G

Technology For the Teacher

Big Ideas Test Generator

Item Analysis

1. **A.** The student assumes there should be 7 zeroes.
   B. The student miscounts when adding zeroes.
   C. Correct answer
   D. The student miscounts when adding zeroes.

2. **F.** The student is finding the sum of the angle measures of a quadrilateral.
   G. The student is finding the sum of the measures of two acute angles of a right triangle.
   H. The student finds the measure of the wrong angle.
   I. Correct answer

3. **A.** The student multiplies exponents instead of adding them.
   B. The student multiplies the bases and adds the exponents.
   C. The student multiplies all of the numbers in the expression.
   D. Correct answer

4. **Gridded Response:** Correct answer: 68

   Common Error: The student finds that the average of the first three rounds is 72. Because this is 1 greater than the desired mean, he or she picks 70, the score that is 1 less than the desired mean, for the fourth score.

5. **F.** The student misinterprets "never been above" to mean "always been below."
   G. Correct answer
   H. The student confuses less than and greater than symbols, and uses strict inequality inappropriately.
   I. The student confuses less than and greater than symbols.

6. **A.** The student adds $5000 to the account every ten years.
   B. The student doubles the amount every ten years, but loses track of the number of doublings needed.
   C. Correct answer
   D. The student doubles the amount every ten years, but loses track of the number of doublings needed, counting 1940 as one of the times to double.

9 Standarized Test Practice

Test-Taking Strategy

Use Intelligent Guessing

"It can't be 40 or 5000 because they aren't divisible by 3. So, you can intelligently guess between 3000 and 3072."

1. Mercury's distance to the Sun is approximately 5.79×10^7 kilometers. Write this distance in standard form.

A. 5,790,000,000 km

B. 579,000,000 km

C. 57,900,000 km

D. 5,790,000 km

2. The steps Jim took to answer the question are shown below. What should Jim change to correctly answer the question?

How many degrees are in the largest angle in the triangle below?

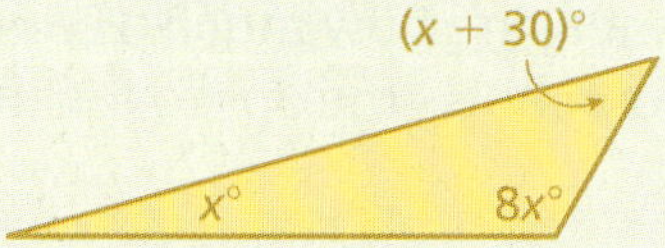

$$x + 8x + x + 30 = 180$$

$$10x = 150$$

$$x = 15$$

F. The left side of the equation should equal 360° instead of 180°.

G. The sum of the acute angles should equal 90°.

H. Evaluate the smallest angle when $x = 15$.

I. Evaluate the largest angle when $x = 15$.

3. Which expression is equivalent to the expression below?

$$2^4 2^3$$

A. 2^{12}

B. 4^7

C. 48

D. 128

4. Your mean score for four rounds of golf was 71. Your scores on the first three rounds were 76, 70, and 70. What was your score on the fourth round?

5. The temperature in Frostbite Falls has never been above 38 degrees Fahrenheit. Let t represent the temperature, in degrees Fahrenheit. Write this as an inequality.

F. $t < 38$

G. $t \leq 38$

H. $t > 38$

I. $t \geq 38$

6. A bank account pays interest so that the amount in the account doubles every 10 years. The account started with \$5,000 in 1940. How much would be in the account in the year 2010?

A. \$40,000

B. \$320,000

C. \$640,000

D. \$1,280,000

7. Which expression is equivalent to $5\sqrt{5} + 2\sqrt{5}$?

F. $7\sqrt{5}$

G. $10\sqrt{5}$

H. $7\sqrt{10}$

I. $10\sqrt{10}$

8. The gross domestic product (GDP) is a way to measure how much a country produces economically in a year. The table below shows the approximate population and GDP for the United States.

Think Solve Explain

United States 2008	
Population	300 million (300,000,000)
GDP	14.4 trillion dollars (\$14,400,000,000,000)

Part A Find the GDP per person for the United States. Show your work and explain your reasoning.

Part B Write the population and GDP using scientific notation.

Part C Find the GDP per person for the United States using your answers from Part B. Write your answer in scientific notation. Show your work and explain your reasoning.

9. What is the equation of the line shown in the graph?

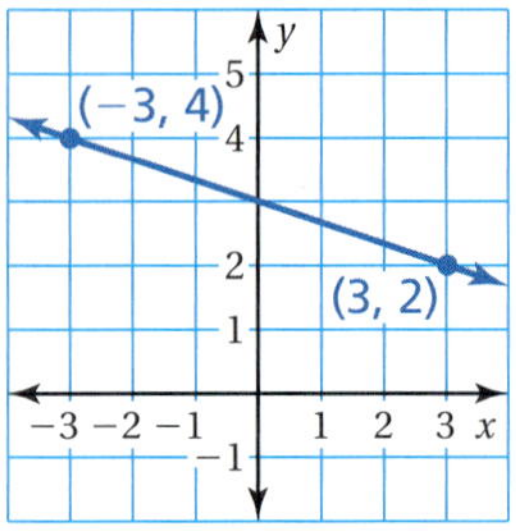

A. $y = -\frac{1}{3}x + 3$

B. $y = \frac{1}{3}x + 1$

C. $y = -3x + 3$

D. $y = 3x - \frac{1}{3}$

Item Analysis (continued)

7. **F.** Correct answer

   G. The student multiplies coefficients.

   H. The student adds the radicands.

   I. The student multiplies the coefficients and adds the radicands.

8. **4 points** The student demonstrates a thorough understanding of how to work arithmetically with large numbers both in standard form and in scientific notation. In Part A, an answer of $48,000 is obtained. In Part B, the data are written as 3×10^8 and 1.44×10^{13}. In Part C, the student works out the quotient $\frac{1.44 \times 10^{13}}{3 \times 10^8}$ step-by-step.

   3 points The student demonstrates an essential but less than thorough understanding. In particular, Parts A and B should be completed correctly, but the steps taken in Part C may show gaps or be incomplete.

   2 points The student demonstrates an understanding of how to write the data using scientific notation, but is otherwise limited in understanding how to approach the problem arithmetically.

   1 point The student demonstrates limited understanding of working with large numbers arithmetically. The student's response is incomplete and exhibits many flaws.

   0 points The student provides no response, a completely incorrect or incomprehensible response, or a response that demonstrates insufficient understanding of how to work with large numbers.

9. **A.** Correct answer

   B. The student miscalculates slope as positive, and then chooses the equation that has (3, 2) as a solution.

   C. The student reverses the roles of *x* and *y* in finding slope, and then uses the correct intercept, interpolated from the graph.

   D. The student reverses slope and *y*-intercept in the equation.

Answers

6. C
7. F
8. *Part A* $48,000

   Part B 3×10^8; 1.44×10^{13}

   Part C $\frac{1.44 \times 10^{13}}{3 \times 10^8} = 4.8 \times 10^4$
9. A

Answers

10. I

11. 0.16 or $\frac{4}{25}$

12. B

13. H

Answer for Extra Example

1. **A.** The student rewrites $\frac{4^8}{4^4}$ as 2 (by dividing out the bases of 4).

   B. Correct answer

   C. The student multiplies and divides exponents, instead of adding and subtracting.

   D. The student multiplies bases in the numerator (and adds exponents).

Item Analysis (continued)

10. **F.** The student makes an arithmetic mistake when adding 1.5 to both sides, and chooses strict inequality incorrectly.

G. The student chooses strict inequality incorrectly.

H. The student makes an arithmetic mistake when adding 1.5 to both sides.

I. Correct answer

11. **Gridded Response:** Correct answer: 0.16 or $\frac{4}{25}$

Common Error: The student writes the answer as a negative number.

12. **A.** The student fails to realize that a box-and-whisker plot is inappropriate. The display must allow a reader to compare parts of a whole.

B. Correct answer

C. The student fails to realize that a line graph is inappropriate. The display must allow a reader to compare parts of a whole.

D. The student fails to realize that a scatter plot is inappropriate. The display must allow a reader to compare parts of a whole.

13. **F.** The student uses addition where multiplication is appropriate.

G. The student incorrectly interprets "more than" as "greater than or equal to" and uses addition where multiplication is called for.

H. Correct answer

I. The student incorrectly interprets "more than" as "greater than or equal to."

Extra Example for Standardized Test Practice

1. Which of the following is equivalent to $\frac{4^8 4^5}{4^4}$?

   A. $2 \cdot 4^5$

   B. 4^9

   C. 4^{10}

   D. $\frac{16^{13}}{4^4}$

10. Which graph represents the inequality shown below?

$$x - 1.5 \leq -1$$

F. 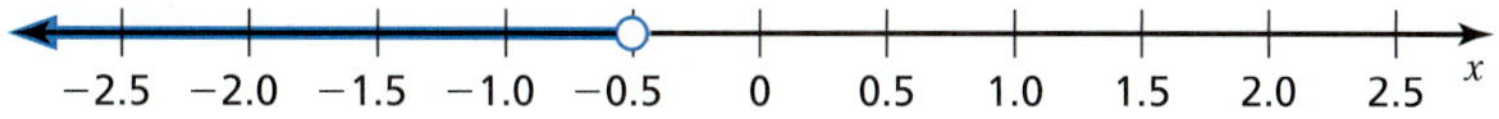

G. 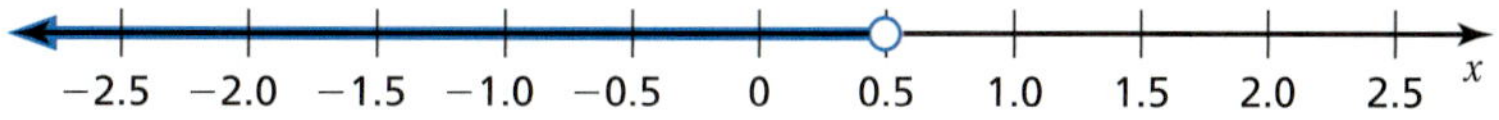

H. 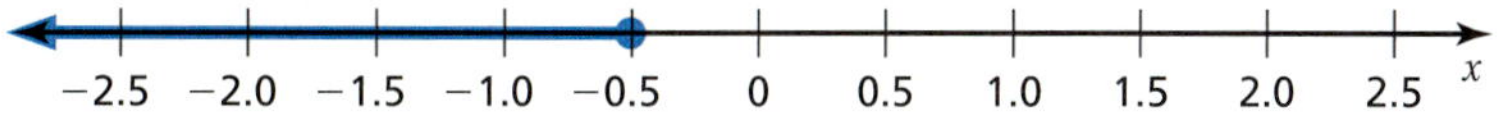

I. 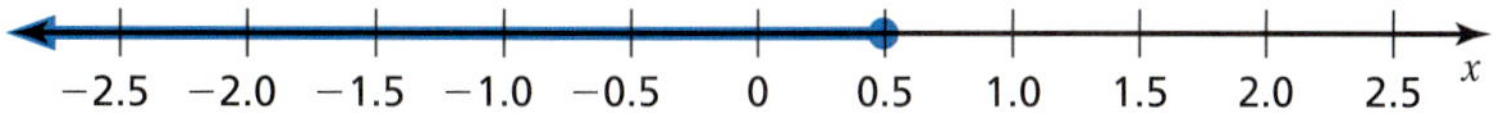

11. Find $(-2.5)^{-2}$.

12. The director of a research lab wants to present data to donors, showing how a great deal of donated money is used for research and how only a small amount of money is used for other expenses. Which type of display is best suited for showing this data?

A. box-and-whisker plot

B. circle graph

C. line graph

D. scatter plot

13. You earn \$14.75 per hour at your job. Your goal is to earn more than \$2000 next month. If you work h hours next month, which inequality represents this situation algebraically?

F. $14.75 + h > 2000$

G. $14.75 + h \geq 2000$

H. $14.75h > 2000$

I. $14.75h \geq 2000$

Appendix A
My Big Ideas Projects

"I'm nervous. The IRS dog is coming today to see if I have buried any bones that I didn't report."

"It's really simple, Descartes."

"If you simply earn interest on simple principal, that's simple interest."

"But if the simple interest earns more simple interest, that's compound interest."

About the Appendix

- The interdisciplinary projects can be used anytime throughout the year.
- The projects offer students an opportunity to build on prior knowledge, to take mathematics to a deeper level, and to develop organizational skills.
- Students will use the Essential Questions to help them form "need to knows" to focus their research.

Essential Question

- **Literature Project**
  How does the knowledge of mathematics provide you and your family with survival skills?
- **History Project**
  How have tools and knowledge from the past influenced modern day mathematics?
- **Art Project**
  How do polyhedra influence the design of games and architecture?
- **Science Project**
  How do the characteristics of a planet influence whether or not it can sustain life?

Additional Resources

BigIdeasMath.com

Pacing Guide for Appendix B

Section 1	
Activity	1 Day
Lesson	1 Day
Section 2	
Activity	1 Day
Lesson	1 Day
Section 3	
Activity	1 Day
Lesson	1 Day
Section 4	
Activity	1 Day
Lesson	1 Day
Study Help / Quiz	1 Day
Section 5	
Activity	1 Day
Lesson	1 Day
Section 6	
Activity	1 Day
Lesson	1 Day
Quiz / Chapter Review	1 Day
Chapter Test	1 Day
Total Appendix B	15 Days
Year-to-Date	165 Days

Check Your Resources

- Record and Practice Journal
- Resources by Chapter
- Skills Review Handbook
- Assessment Book
- Worked-Out Solutions

About the Appendix

- This Appendix reviews and extends the mathematics taught throughout the year. It is offered as a full chapter and can be used after your state's assessment.
- This Appendix includes the following features:
  - Study Help
  - Quizzes
  - Review
  - Test
  - Ancillary support

Appendix B
Financial Literacy

A.1 Literature Project

Share Your Work at... MyBigIdeasMath.com

Swiss Family Robinson

1 Getting Started

Swiss Family Robinson is a novel about a Swiss family who was shipwrecked in the East Indies. The story was written by Johann David Wyss, and was first published in 1812.

Essential Question How does the knowledge of mathematics provide you and your family with survival tools?

Read *Swiss Family Robinson*. As you read the exciting adventures, think about the mathematics the family knew and used to survive.

Sample: The tree house built by the family was accessed by a long rope ladder. The ladder was about 30 feet long with a rung every 10 inches. To make the ladder, the family had to plan how many rungs were needed. They decided the number was $1 + 12(30) \div 10$. Why?

Project Notes

Introduction

For the Teacher

- **Goal:** Students will read *Swiss Family Robinson* by Johann David Wyss and write a report about the mathematics they find in the story. Samples of things that could be included in students' reports are discussed below.
- **Management Tip:** You may want to have students work together in groups.

Essential Question

- How does the knowledge of mathematics provide you and your family with survival tools?

Things to Think About

Summary of *Swiss Family Robinson*

- A husband, wife, and four young sons, were shipwrecked as a result of a turbulent storm. Abandoned by the crew, they were left to their own ingenuity. The ship was smashed on a rock, but within sight of land. They built a raft out of barrels to transport themselves and material from the ship to the shore. After repeated trips, they removed as much as possible from the ship, which was full of livestock, equipment, and food. There were even pieces of a sailboat onboard, which they assembled and sailed.
- The family established temporary quarters in a tent made with sailcloth. They hunted, planted, and explored. They built a treehouse to use as a safe place to live. They named the tree house *Falconhurst.* It was very comfortable until the rainy season.
- The family built several other structures to serve as shelters on their various expeditions. Each structure was well furnished using either the things from the ship or furniture that they had made.
- The father was extremely innovative, able to make anything they needed from the materials at hand. He seemed to know how to use every resource for either the appropriate or improvised purpose.
- The island was lush with both flowers and animals. Vegetables and fruits from potatoes to pineapples were abundant. The family planted additional crops using the seeds obtained from the ship. Animals, from apes to kangaroos, were captured and tamed or used for food. The family established farms with animals that they captured or rescued from the ship.
- The family considered themselves very fortunate, knowing that things could have been much worse. On the one-year anniversary of the shipwreck, they observed a day of thanksgiving.

References

Go to *BigIdeasMath.com* to access links related to this project.

Cross-Curricular Instruction

Meet with a reading or language arts teacher and review curriculum maps to identify whether students have or have yet to read *Swiss Family Robinson*. If the book has been read, you may want to discuss the work students have completed and review the book with them. If the book has not been read, perhaps you can both work simultaneously and share notes. Or, you may want to explore activities that the reading or language arts teacher has done in the past to support student learning in this particular area.

Project Notes

- Fritz, the eldest son, became restless after 10 years. So, he ventured out to explore, and discovered an albatross with a message tied to its leg. He found and rescued Jenny (Miss Montrose), an English woman who had also been shipwrecked. She lived alone on an island until Fritz noticed her signal fire. A rescue ship came shortly after Fritz found Jenny. Some of the occupants decided to stay at New Switzerland (as they liked to call their dominion). Others, including Jenny, Fritz, and Franz (the youngest son) decided to board the ship and go to England.

Mathematics Used in the Story

Some examples of mathematics used are illustrated in the following excerpts from the story.

- "This convinced me that we must not be far from the equator, for twilight results from the refraction of the sun's rays; the more obliquely these rays fall, the further does the partial light extend, while the more perpendicularly they strike the earth the longer do they continue their undiminished force, ..."
- "Jack showed me where he thought the bridge should be, and I certainly saw no better place, as the banks were at that point tolerably close to each other, steep, and of about equal height. 'How shall we find out if our planks are long enough to reach across? ... A surveyor's table would be useful now.' 'What do you say to a ball of string, father? ... Tie one end to a stone, throw it across, then draw it back, and measure the line!' ... we speedily ascertained the distance across to be eighteen feet. Then allowing three feet more at each side, I calculated twenty-four feet as the necessary length of the boards."
- In constructing the raft, they used empty water-casks, "arranging twelve of them side by side in rows of three" and placing planks on top for the floor.
- When making a canoe out of birch bark, they "cut the bark through in a circle" in two places, "took a narrow perpendicular slip of bark entirely out," and then loosened and separated the bark from the tree.
- While getting next to a whale they estimated, "the length being from sixty to sixty-five feet, and the girth between thirty and forty, while the weight could not have been less than 50,000 lbs. ...the enormous head about one-third the length of the entire hulk..."

Closure

- **Rubric** An editable rubric for this project is available at *BigIdeasMath.com.*
- Students may present their reports to the class or school as a television report or public information broadcast.

2 Things to Include

- Suppose you lived in the 18th century. Plan a trip from Switzerland to Australia. Describe your route. Estimate the length of the route and the number of miles you will travel each day. About how many days will the entire trip take?
- Suppose that your family is shipwrecked on an island that has no other people. What do you need to do to survive? What types of tools do you hope to salvage from the ship? Describe how mathematics could help you survive.
- Suppose that you are the oldest of four children in a shipwrecked family. Your parents have made you responsible for the education of your younger siblings. What type of mathematics would you teach them? Explain your reasoning.

3 Things to Remember

- You can download each part of the book at *BigIdeasMath.com.*
- Add your own illustrations to your project.
- Organize your math stories in a folder, and think of a title for your report.

A.2 History Project

Share Your Work at… My.BigIdeasMath.com

Mathematics in Ancient China

1 Getting Started

Mathematics was developed in China independently of the mathematics that was developed in Europe and the Middle East. For example, the Pythagorean Theorem and the computation of pi were used in China prior to the time when China and Europe began communicating with each other.

Essential Question How have tools and knowledge from the past influenced modern day mathematics?

Sample: Here are the names and symbols that were used in ancient China to represent the digits from 1 through 10.

1	yi	一
2	er	二
3	san	三
4	si	四
5	wu	五
6	liu	六
7	qi	七
8	ba	八
9	jiu	九
10	shi	十

Life-size Terra-cotta Warriors

A Chinese Abacus

Project Notes

Introduction

For the Teacher

- **Goal:** Students will discover how mathematics was used in ancient China.
- **Management Tip:** Students can work in groups to research the required topics and generate a report.

Essential Question

- How have tools and knowledge from the past influenced modern day mathematics?

Things to Think About

? What is the ancient Chinese book *The Nine Chapters on the Mathematical Art* (c. 100 B.C.)?

- The book is one of the earliest surviving mathematics texts of ancient China. It is a collection of scholarly math writings that were written over a span of more than a thousand years. It assisted the ancient Chinese in solving problems dealing with trade, taxation, surveying, and engineering. It consists of 246 problems, a solution to each problem, and an explanation of how to get each solution.

? What types of mathematics are contained in *The Nine Chapters on the Mathematical Art*?

- **Chapter One:** Areas of triangles, rectangles, circles, and trapeziums; operations on fractions; finding greatest common divisor
- **Chapter Two:** Rates of exchange; proportions; percentages
- **Chapter Three:** Direct, inverse, and compound proportions
- **Chapter Four:** Keeping the area of a rectangle the same while increasing its width; square roots and cube roots
- **Chapter Five:** Volumes of prisms, pyramids, cylinders, and cones
- **Chapter Six:** Ratios and proportions
- **Chapter Seven:** Linear equations
- **Chapter Eight:** Systems of linear equations
- **Chapter Nine:** Right triangles; similar triangles; Pythagoean Theorem (called the Gougu rule)

? What is an abacus?

- The Chinese abacus consists of a wooden frame with 13 vertical rods and a horizontal wooden divider. There are seven beads on each rod. Five of the seven beads on each rod are located below the wooden divider, which the Chinese called the earth. The other two beads on each rod are located above the wooden divider, called the heaven. Earth beads are worth one point each, while the heaven beads are worth five points each.

? How is a number represented on the abacus?

- To start, all of the beads in both sections are pushed away from the divider. To represent numbers on the abacus, the beads are pushed toward the divider. The rod on the far right is the ones rod, then the one to its left is the tens rod, then the hundreds rod, then the thousands rod, etc.

References

Go to *BigIdeasMath.com* to access links related to this project.

Cross-Curricular Instruction

Meet with a history teacher and review curriculum maps to identify whether students have covered or have yet to discuss ancient China. If the topic has been covered, you may want to discuss the work students have completed and review prior knowledge with them. If the history teacher has not discussed these concepts, perhaps you can both work simultaneously on these concepts and share notes. Or, you may want to explore activities that the history teacher has done in the past to support student learning in this particular area.

Project Notes

- Suppose you want to represent the number 827 on the abacus. On the ones rod, move one of the heaven beads to the divider (5 points) and two of the earth beads to the divider (2 points). On the tens rod, move none of the heaven beads (0 points) and two of the earth beads (2 points). On the hundreds rod, move one of the heaven beads (5 points) and three of the earth beads (3 points).

? How is an abacus used to add or subtract numbers?

- Suppose you want to add 827 and 122. With 827 already on the abacus, add 122, digit by digit, starting with the ones rod. On the ones rod, move two more earth beads to the divider, making that rod represent a nine. On the tens rod, move two more earth beads to the divider, making that rod represent a four. On the hundreds rod, move one more earth bead to the divider, making that rod represent a nine. So, 827 + 122 = 949.
- To subtract two numbers, follow the same format, except in reverse, moving beads away from the divider instead of toward it.

? How did the ancient Chinese write numbers that are greater than 10?

- The ancient Chinese had symbols to represent each of the numbers from zero through nine.

零	一	二	三	四	五	六	七	八	九
0	**1**	**2**	**3**	**4**	**5**	**6**	**7**	**8**	**9**

- They also had symbols to represent 10, 100, 1000, and 10,000.

十	百	千	萬
10	**100**	**1000**	**10,000**

- To write a number such as 467, they had to write the number as $4 \cdot 100 + 6 \cdot 10 + 7$.

467 = 四百六十七

Closure

- **Rubric** An editable rubric for this project is available at *BigIdeasMath.com.*
- You may hold a class debate where students can compare, defend, and discuss their findings with another student or group of students.

2 Things to Include

- Describe the ancient Chinese book *The Nine Chapters on the Mathematical Art* (c. 100 B.C.). What types of mathematics are contained in this book?
- How did the ancient Chinese use the abacus to add and subtract numbers? How is the abacus related to base 10?
- How did the ancient Chinese use mathematics to build large structures, such as the Great Wall and the Forbidden City?
- How did the ancient Chinese write numbers that are greater than 10?
- Describe how the ancient Chinese used mathematics. How does this compare with the ways in which mathematics is used today?

Ancient Chinese Teapot

The Great Wall of China

3 Things to Remember

- Add your own illustrations to your project.
- Organize your math stories in a folder, and think of a title for your report.

Chinese Guardian Fu Lions

A.3 Art Project

Share Your Work at... My.BigIdeasMath.com

Polyhedra in Art

1 Getting Started

Polyhedra is the plural of *polyhedron*. Polyhedra have been used in art for many centuries, in cultures all over the world.

Essential Question Do polyhedra influence the design of games and architecture?

Some of the most famous polyhedra are the five Platonic solids. They have faces that are congruent, regular, convex polygons.

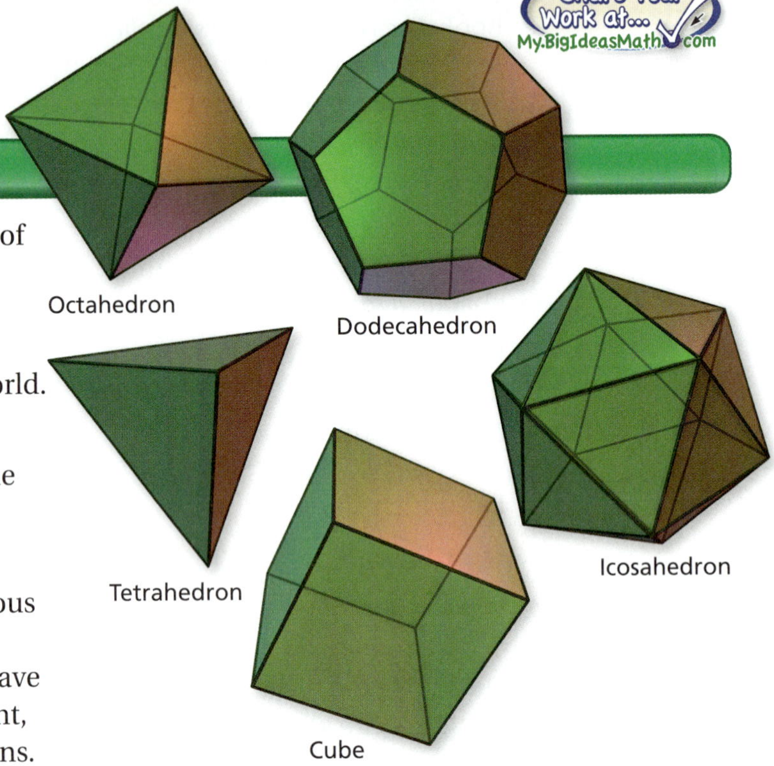

Sample: Three of the Platonic solids have faces that are equilateral triangles. One of these is the octahedron. Here is a net for an octahedron.

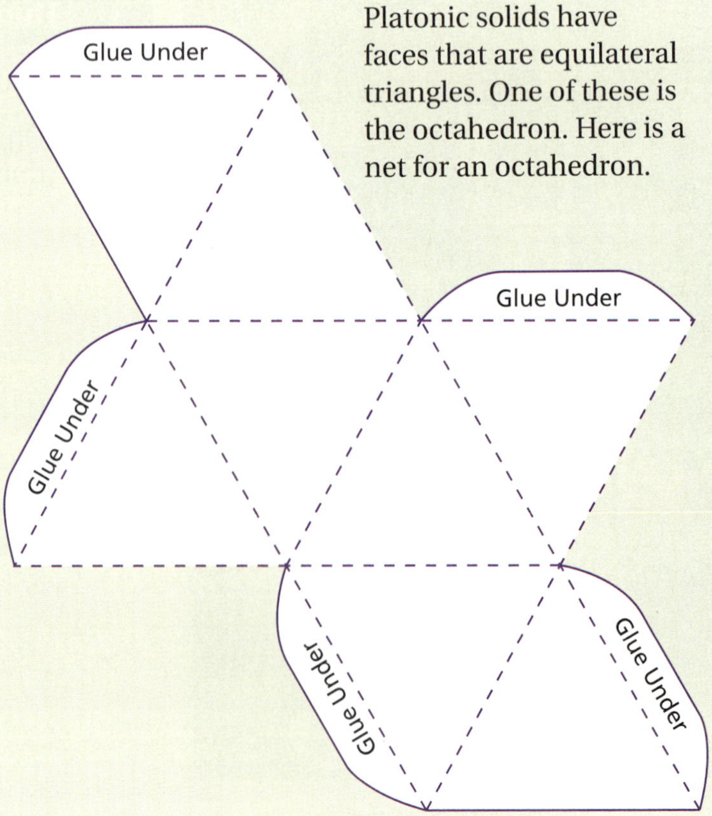

Mosaic by Paolo Uccello, 1430 A.D.

Project Notes

Introduction

For the Teacher

- **Goal:** Students will research and discover how polyhedra are used in games and architecture. They will also create their own polyhedra.
- **Management Tip:** Students may wish to search online or visit *BigIdeasMath.com* for links to websites containing instructions on how to make their own icosahedron or dodecahedron.

Essential Question

- Do polyhedra influence the design of games and architecture?

Things to Think About

? What are polyhedra?

- Polyhedra are three-dimensional shapes that have faces (sides) that are polygons. The polygons meet along straight line segments (edges) and the edges meet at vertices.

? What did Plato associate with the five Platonic solids?

- The Platonic solids, named for Plato, are a prominent feature of his philosophy. In his dialogue *Timaeus c. 360 B.C.*, Plato associated each of the four classical elements with a regular solid.

  "There was intuitive justification for these associations: the heat of fire feels sharp and stabbing (like little tetrahedra). Air is made of the octahedron; its minuscule components are so smooth that one can barely feel it. Water the icosahedron, flows out of one's hand when picked up, as if it is made of tiny little balls. By contrast, a highly un-spherical solid, the hexahedron (cube) represents earth. These clumsy little solids cause dirt to crumble and break when picked up, in stark difference to the smooth flow of water."

- Because there were only the four elements, there was need for only the four polyhedra. However, a fifth polyhedra, the dodecahedron, was used for the universe. Plato remarked, "*...the god used for arranging the constellations on the whole heaven.*"

? Why are polyhedra used in architecture?

- From the Renaissance to present time, many artists saw the use of polyhedra as a challenging way to demonstrate the mastery of perspective. Others, such as Plato, thought that polyhedra symbolized philosophical truths or religious beliefs. Other artists used polyhedra simply because of their symmetrical beauty. From the Great Pyramids in Egypt to the Spaceship Earth geosphere at Epcot, polyhedra have been a part of architecture since prehistoric times.

Materials
• List of materials needed to construct a polyhedron is available at *BigIdeasMath.com*

References

Go to *BigIdeasMath.com* to access links related to this project.

Cross-Curricular Instruction

Meet with an art teacher and review curriculum maps to identify whether students have covered or have yet to discuss polyhedra. If the topic has been covered, you may want to discuss the work students have completed and review prior knowledge with them. If the art teacher has not discussed these concepts, perhaps you can both work simultaneously on these concepts and share notes. Or, you may want to explore activities that the art teacher has done in the past to support student learning in this particular area.

Project Notes

? What is one of the most popular uses of polyhedra in games?

- Polyhedra dice games are a popular way to explore mathematics. The following are examples of polyhedra dice.

? What are the Archimedean Solids?

- The Archimedean Solids are symmetric, semi-regular convex polyhedra. They are composed of two or more types of regular polygons meeting at identical vertices. Three are listed below. Visit *BigIdeasMath.com* to view all 13 Archimedean Solids.

Name	rhombicuboctahedron	truncated cuboctahedron	rhombicosidodecahedron
Solid			
Net			
Faces	26	26	62
Faces (by type)	8 triangles; 18 squares	12 squares; 8 hexagons; 6 octagons	20 triangles; 30 squares; 12 pentagons
Edges	48	72	120

Closure

- **Rubric** An editable rubric for this project is available at *BigIdeasMath.com.*
- Students may present their reports to a parent panel or community members.

2 Things to Include

- Explain why the platonic solids are sometimes referred to as the cosmic figures.
- Draw a net for an icosahedron or a dodecahedron. Cut out the net and fold it to form the polyhedron.
- Describe the 13 polyhedra that are called Archimedean solids. What is the definition of this category of polyhedra? Draw a net for one of them. Then cut out the net and fold it to form the polyhedron.
- Find examples of polyhedra in games and architecture.

Faceted Cut Gem

Origami Polyhedron

3 Things to Remember

- Add your own illustrations or paper creations to your project.
- Organize your report in a folder, and think of a title for your report.

Concrete Tetrahedrons by Ocean

Bulatov Sculpture

A.4 Science Project

Share Your Work at... My.BigIdeasMath.com

Our Solar System

1 Getting Started

Our solar system consists of four inner planets, four outer planets, dwarf planets such as Pluto, several moons, and many asteroids and comets.

Essential Question How do the characteristics of a planet influence whether or not it can sustain life?

Sample: The average temperatures of the eight planets in our solar system are shown in the graph.

The average temperature tends to drop as the distance between the Sun and the planet increases.

An exception to this rule is Venus. It has a higher average temperature than Mercury, even though Mercury is closer to the Sun.

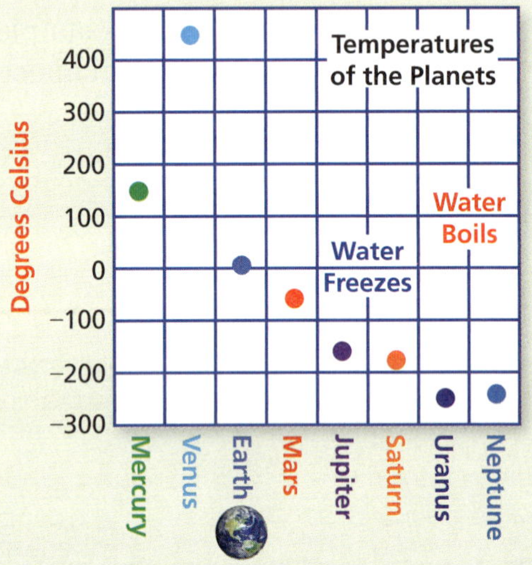

Project Notes

Introduction

For the Teacher

- **Goal:** Students will discover facts about objects in our solar system.
- **Management Tip:** Students can work in groups to create a report on our solar system.

Essential Question

- How do the characteristics of a planet influence whether or not it can sustain life?

Things to Think About

Inner System

	Mercury	Venus	Earth	Mars
Distance from the Sun	5.8×10^7 km	1.08×10^8 km	1.5×10^8 km	2.3×10^8 km
Diameter	4.8×10^3 km	1.2×10^4 km	1.27×10^4 km	6.7×10^3 km
Mass	3.3×10^{23} kg	4.9×10^{24} kg	6.0×10^{24} kg	6.4×10^{23} kg
Gravitational pull	38% of Earth	91% of Earth	100% of Earth	38% of Earth
Length of day	1407 hours, 30 minutes	5832 hours,	1 Earth Day	24 hours, 37 minutes
Length of year	88 days	225 days	365.25 days	687 days
Range or average temperature	−279°F to 801°F	864°F	45°F	−125°F to 23°F
Support human life?	No	No	Yes	No

? Have the planets been visited by humans?

- **Mercury** has been or will be visited by two spacecrafts; *Mariner 10* in 1974 and 1975 and *Messenger* which was launched by NASA in 2004. *Messenger* did several "flybys" in 2008 and is set to orbit Mercury beginning in 2011.

References

Go to *BigIdeasMath.com* to access links related to this project.

Cross-Curricular Instruction

Meet with a science teacher and review curriculum maps to identify whether students have covered or have yet to discuss the planets in our solar system. If the topic has been covered, you may want to discuss the work students have completed and review prior knowledge with them. If the science teacher has not discussed these concepts, perhaps you can both work simultaneously on these concepts and share notes. Or, you may want to explore activities that the science teacher has done in the past to support student learning in this particular area.

Project Notes

- **Venus** has been visited by at least 20 spacecrafts. *Mariner 2* was the first in 1962. Others have included *Pioneer Venus*, the Soviet's *Venera 7* and *Venera 9*, and ESA's *Venus Express*.
- **Mars** has been visited by several spacecrafts, the first being *Mariner 4* in 1965. Spacecrafts that have landed on Mars include: *Mars 2*, two *Viking* landers, *Mars Pathfinder*, and Mars Expedition Rovers *Spirit* and *Opportunity*.
- **Jupiter** has been visited by the spacecrafts *Pioneer 10*, *Pioneer 11*, *Voyager 1*, *Voyage 2*, *Ulysses*, and *Galileo*.
- **Saturn** has been visited by *Pioneer 11* in 1979, *Voyager 1*, *Voyager 2*, and *Cassini*.
- **Uranus** *Voyager 2* is the only spacecraft to have visited Uranus.
- **Neptune** *Voyager 2* is the only spacecraft to have visited Neptune.

Outer System

	Jupiter	Saturn	Uranus	Neptune
Distance from the Sun	7.7×10^8 km	1.4×10^9 km	2.8×10^9 km	4.5×10^9 km
Diameter	1.4×10^5 km	1.2×10^5 km	5.1×10^4 km	4.9×10^4 km
Mass	2.0×10^{27} kg	5.7×10^{26} kg	8.7×10^{25} kg	1.0×10^{26} kg
Gravitational pull	260% of Earth	117% of Earth	85% of Earth	120% of Earth
Length of day	9 hours, 56 minutes	10 hours, 39 minutes	17 hours, 5 minutes	16 hours, 7 minutes
Length of year	4331 days	10,759 days	30,687 days	60,190 days
Range or average temperature	-234°F	-288°F	-357°F	-353°F
Support human life?	No	No	No	No

Closure

- **Rubric** An editable rubric for this project is available at *BigIdeasMath.com.*
- Students may present their reports to the class or compare their report with other students' reports.

2 Things to Include

- Compare the masses of the planets.
- Compare the gravitational forces of the planets.
- How long is a “day” on each planet? Why?
- How long is a “year” on each planet? Why?
- Which planets or moons have humans explored?
- Which planets or moons could support human life? Explain your reasoning.

Mars Rover

3 Things to Remember

- Add your own drawings or photographs to your report. You can download photographs of the solar system and space travel at *NASA.gov*.
- Organize your report in a folder, and think of a title for your report.

Hubble Image of Space

Hubble Spacecraft

B.1 Simple and Compound Interest

Essential Question How can you find the balance in an account that earns simple interest or compound interest?

1 ACTIVITY: Comparing Simple and Compound Interest

Work with a partner. Interest that is calculated only on principal is simple interest. Interest that is calculated on principal *and* previously earned interest is compound interest.

You deposit $1000 in a savings account that earns 6% interest per year.

a. Copy and complete the first table that shows the balance after 10 years with simple interest.

b. Copy and complete the second table that shows the balance after 10 years with interest that is compounded annually.

c. Which type of interest gives the greater balance?

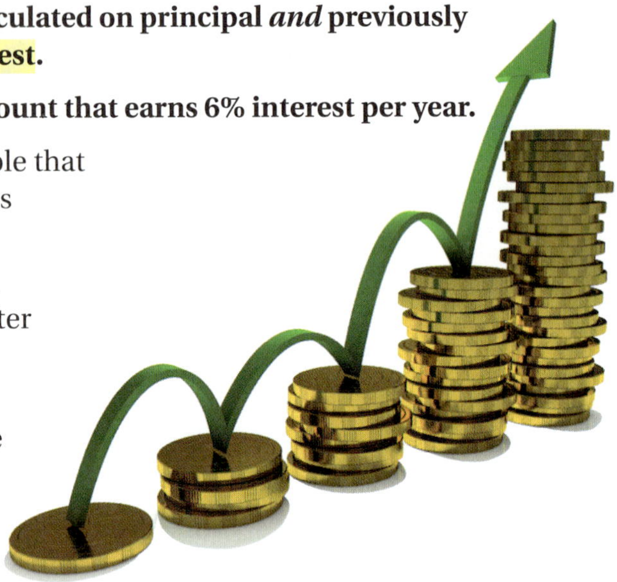

$I = Prt$
$= 1000(0.06)(1)$

Simple Interest			
t	Principal	Annual Interest	Balance at End of Year
1	$1000.00	$60.00	$1060.00
2	$1000.00	$60.00	$1120.00
3			
4			
5			
6			
7			
8			
9			
10			

Compound Interest			
t	Principal and Interest	Annual Interest	Balance at End of Year
1	$1000.00	$60.00	$1060.00
2	$1060.00	$63.60	$1123.60
3			
4			
5			
6			
7			
8			
9			
10			

Laurie's Notes

Introduction

For the Teacher

- **Goal:** Students will explore simple and compound interest on a deposit of $1000.
- I recommend both activities be done using a spreadsheet. If this is not possible, a calculator is needed. You do not want students to do paper and pencil multiplication in this context. They will lose interest and motivation, and will not care about the results of the investigation.
- Please note that no formulas are introduced in this activity, however, you can still use a formula to fill the cells in a spreadsheet.
- **Compound interest** is interest earned on the principal *and* on the previously earned interest.

Motivate

? "What does the word *bank* mean?"

- This question should generate more questions than answers! There are many different meanings for the word *bank*. Give time for students to work with their partner to generate a list.
- Here are a few examples to help you with your own list.

Noun	Financial institution
Noun	Ground breaking waters (cliff, ledge, edge)
Noun	Row or tier of objects (array, group, series)
Noun	Group of items gathered haphazardly (mound, mass)
Verb	Collect money or advantage (amass, hoard, invest)
Verb	Lean or tilt (incline, slant, slope)
Verb	To place (money) in a bank (deposit, lay away)
Phrasal verb	To place trust or confidence in (count on, rely)

Activity Notes

Activity 1

- Discuss each table. The first row is the same in both. The interest earned on $1000 at 6% interest for 1 year is $60.

? "Where is the extra $3.60 in interest coming from in row 2 of the second table?" interest calculated on interest earned in year 1: $I = (60)(0.06)(1) = \$3.60$

- **Big Idea:** The interest earned in the second year is different. The compound interest is computed on the principal and the $60 interest from the first year.
- Remind students that they should round answers to the nearest penny as they work through the table.
- Students should compare answers frequently with their partner to avoid an early error that impacts all subsequent calculations.

? "After 10 years, how much of a difference is there between the two methods of computing interest?" $1790.85 − $1600.00 = $190.85

? "Is it worth shopping around for better interest rates?" absolutely

Previous Learning

Students should know how to write percents as decimals. Students should know how to compute simple interest.

Activity Materials
Textbook
• spreadsheet or calculator

Start Thinking! and Warm Up

Activity B.1 Start Thinking!
For use before Activity B.1

Activity B.1 Warm Up
For use before Activity B.1

Evaluate the expression *Prt* for the given values of *P*, *r*, and *t*.

1. $P = 500,\ r = 0.05,\ t = 2$
2. $P = 2500,\ r = 0.02,\ t = 1$
3. $P = 100,\ r = 0.1,\ t = 10$
4. $P = 8000,\ r = 0.06,\ t = 4$
5. $P = 16{,}000,\ r = 0.065,\ t = 5$
6. $P = 50{,}000,\ r = 0.04,\ t = 30$

B.1 Record and Practice Journal

Essential Question How can you find the balance in an account that earns simple interest or compound interest?

1 ACTIVITY: Comparing Simple Interest and Compound Interest

Work with a partner. Interest that is calculated only on principal is simple interest. Interest that is calculated on principal *and* previously earned interest is compound interest.

You deposit $1000 in a savings account that earns 6% interest per year.

a. Complete the first table that shows the balance after 10 years with simple interest.

b. Complete the second table that shows the balance after 10 years with interest that is compounded annually.

Simple Interest			
t	Principal	Annual Interest	Balance at End of Year
1	$1000.00	$60.00	$1060.00
2	$1000.00	$60.00	$1120.00
3	$1000.00	$60.00	$1180.00
4	$1000.00	$60.00	$1240.00
5	$1000.00	$60.00	$1300.00
6	$1000.00	$60.00	$1360.00
7	$1000.00	$60.00	$1420.00
8	$1000.00	$60.00	$1480.00
9	$1000.00	$60.00	$1540.00
10	$1000.00	$60.00	$1600.00

Compound Interest			
t	Principal and Interest	Annual Interest	Balance at End of Year
1	$1000.00	$60.00	$1060.00
2	$1060.00	$63.60	$1123.60
3	$1123.60	$67.42	$1191.02
4	$1191.02	$71.46	$1262.48
5	$1262.48	$75.75	$1338.23
6	$1338.23	$80.29	$1418.52
7	$1418.52	$85.11	$1503.63
8	$1503.63	$90.22	$1593.85
9	$1593.85	$95.63	$1689.48
10	$1689.48	$101.37	$1790.85

Differentiated Instruction

Auditory

Discuss the meaning of the word *interest*. An interest rate is often expressed as an annual percent of the principal.

Note: Students should assume that deposits are made at the beginning of the interest period in all banking problems, unless otherwise stated.

B.1 Record and Practice Journal

c. Which type of interest gives the greater balance?

compound interest

2 ACTIVITY: Comparing Simple and Compound Interest

Work with a partner.

a. Graph the end-of-year balances for each type of interest in Activity 1.

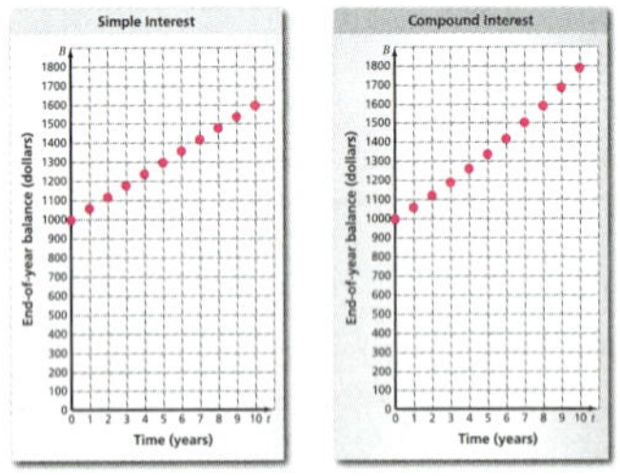

b. Which graph is linear? Explain your reasoning.

simple interest

c. For the linear graph on the previous page, write a linear function that represents the balance after t years.

$B = 60t + 1000$

What Is Your Answer?

3. **IN YOUR OWN WORDS** How can you find the balance in an account that earns simple interest or compound interest?

See Additional Answers.

4. Use what you learned in Activity 2. About how many years will it take for the balance to double with simple interest? with compound interest?

about 17 years; about 12 years

Laurie's Notes

Activity 2

- This activity uses the data from Activity 1. Check with students to ensure that their tables of data are correct.
- The graphs are very quick and easy to create if you had your students use a spreadsheet for Activity 1.
- Students need to be very accurate if they create their graphs by hand.
- The scales on both axes are the same, and the graphs will look similar, but not exact.

? "How can you tell that the simple interest data is linear?" The ratio of the change in B (rise) to the change in t (run) is contant. The balance increases \$60 each year.

- The linear function for simple interest is $B = 1000 + 60t$. The B-intercept makes sense. At time t equal to 0, the balance is \$1000. The slope makes sense. The balance is increasing \$60 each year.
- If students have plotted accurately, they should see a slight curvature in the graph for compound interest. Even if their graph looks linear, they should be able to explain why they know it is not. The table of values shows that the change in the balance is not the same each year.

What Is Your Answer?

- **Think-Pair-Share:** Students should read each question independently and then work with a partner to answer the questions. When they have answered the questions, the pair should compare their answers with another group and discuss any discrepancies.

Closure

- **Exit Ticket:** What is the balance at the end of 2 years if you deposit \$1000 in an account that earns 4% simple interest per year? 4% compound interest? \$1080; \$1081.60

Technology For the Teacher

Dynamic Classroom

The Dynamic Planning Tool

Editable Teacher's Resources at *BigIdeasMath.com*

2 ACTIVITY: Comparing Simple and Compound Interest

Work with a partner.

a. Graph the end-of-year balances for each type of interest in Activity 1.

b. Which graph is linear? Explain your reasoning.

c. For the linear graph, write a linear function that represents the balance after t years.

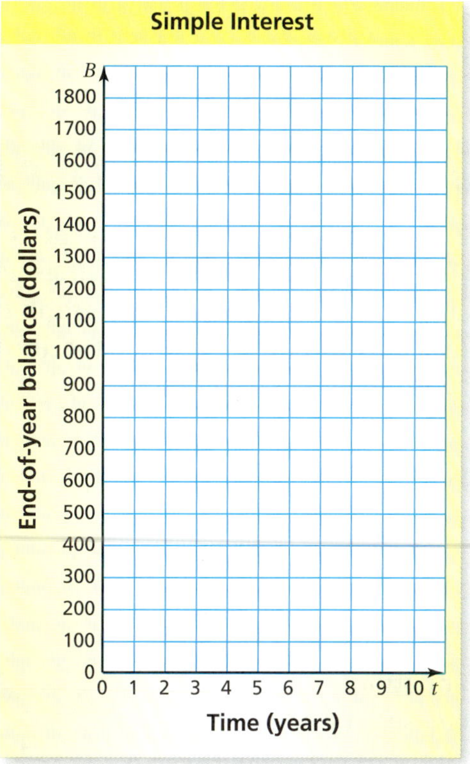

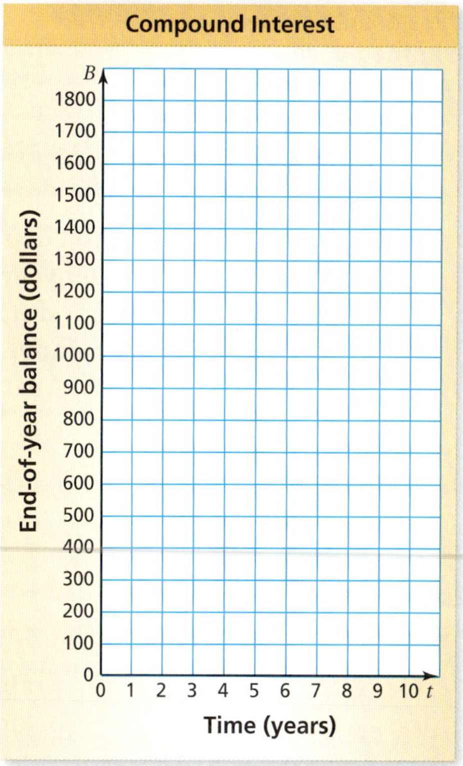

What Is Your Answer?

3. **IN YOUR OWN WORDS** How can you find the balance in an account that earns simple interest or compound interest?

4. Use what you learned in Activity 2. About how many years will it take for the balance to double with simple interest? with compound interest?

Practice Use what you learned about simple and compound interest to complete Exercise 3 on page A14.

Check It Out
Lesson Tutorials
BigIdeasMath.com

Key Vocabulary
compound interest, *p. A13*

Remember
Interest is money paid or earned for the use of money. The *principal* is the amount of money borrowed or deposited.

Key Idea

Balance in a Simple Interest Account

The balance B of an account that earns simple interest is

$$B = P(1 + rt).$$

- Balance: B
- Principal: P
- Annual interest rate (in decimal form): r
- Time (in years): t

EXAMPLE 1 Comparing Simple Interest Accounts

You deposit \$275 in a savings account that earns 4% simple interest per year. Your friend deposits \$300 in a savings account that earns 2% simple interest per year. (a) Write and graph two equations for the balance B in each account after t years. Describe the equations. (b) Are the account balances ever equal? Explain.

a. ***You*** / ***Your Friend***

You		Your Friend
$B = P(1 + rt)$	Write balance formula.	$B = P(1 + rt)$
$= 275(1 + 0.04t)$	Substitute values.	$= 300(1 + 0.02t)$
$= 275 + 11t$	Simplify.	$= 300 + 6t$

Reading
Notice that both equations are of the form
$y = mx + b$.

Graph the equations. Your account balance increases at a constant rate of \$11 per year. Your friend's account balance increases at a constant rate of \$6 per year.

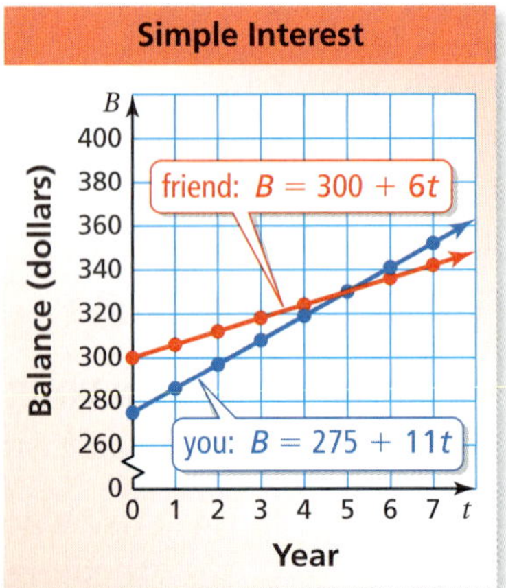

b. Yes, the graphs appear to intersect when $t = 5$. So, after 5 years, the accounts are equal.

On Your Own

Now You're Ready
Exercise 4

1. **WHAT IF?** In Example 1, your friend deposits \$600 in the account. Are the account balances ever equal? Explain.

Multi-Language Glossary at BigIdeasMath.com.

Laurie's Notes

Introduction

Connect

- **Yesterday:** Students explored simple and compound interest on a deposit of $1000.
- **Today:** Students will calculate account balances using simple and compound interest.

Motivate

? "What does the word *interest* mean?"

- Again, this question should generate more questions than answers! There are many different meanings for the word *interest*. Give students time to work with their partner to generate a list.
- Here are a few examples to help you with your own list!

Noun	Attraction, curiosity, absorption
Noun	Advantage (benefit, gain, welfare)
Noun	Share, investment (accrual, earnings, gain)
Noun	Mental acquisitiveness (curiousness, inquisitiveness)
Noun	Curiosity about or attention to someone or something
Verb	Hold the attention of (affect, arouse, attract, grab, hook)
Verb	To arouse the interest and attention of (attract, intrigue)
Adjective	Absorbed, immersed, intent, engrossed

Lesson Notes

Key Idea

- The balance of a simple interest account is the sum of the principal and the money earned on the principal. The principal is the amount initially invested for a period of time. The annual interest rate, written as a decimal, is the percent earned on the principal.

? "What operation is implied when variables are written adjacent to one another?" multiplication

Example 1

- The big picture in the problem is that two people invest different amounts of money at different rates. Will the balances ever be the same?
- Read carefully. Like the problem in yesterday's activity, the equation is the balance in the account.
- Write the formula for the balance in a simple interest account and substitute the information known for each investment.
- **Common Error:** 4% is 0.04, not 0.4. Likewise, 2% is 0.02, not 0.2.

? "What does a slope of 11 mean in the context of the problem?" You earn $11 interest every year.

? "What does a *B*-intercept of 275 mean in the context of the problem?" You start with a balance of $275 in your account.

Goal Today's lesson is computing simple and **compound interest**.

Lesson Materials
Textbook
• calculator

Start Thinking! and Warm Up

Lesson B.1 Warm Up
For use before Lesson B.1

Lesson B.1 Start Thinking!
For use before Lesson B.1

Would you rather open a savings account with simple interest or compound interest? Why?

Extra Example 1

In Example 1, your friend deposits $700 in the account. Are the account balances ever equal? Explain. no; Your friend deposits more money and earns more interest each year. So, the balances are never equal.

On Your Own

1. no; Your friend deposits more money and earns more interest each year. So, the balances are never equal.

Extra Example 2

In Example 2, what is the account balance after 6 years? $491.71

Extra Example 3

In Example 3, how much more does Account B earn than Account A after 3 more years? $202.55

On Your Own

2. $475.08
3. $133.22

English Language Learners

Vocabulary

Review with English learners the mathematical meanings of *principal, interest,* and *balance* because these words have multiple meanings in the English language. They should understand that interest is paid to customers when they deposit money into an account. When a person borrows money, the person pays interest to the bank on the money the person borrows.

Laurie's Notes

Discuss

- Discuss the difference between *compound interest* and simple interest.

Example 2

? "How do you write 3.5% as a decimal?" 0.035

- Work through the computations. There are no short-cuts in this process yet. The interest earned for the first year is added to the principal *before* interest is calculated for the second year. Note the color coding, which should draw attention to this key concept.

? "How much more interest was earned in the second year than in the first year?" $0.49

- The $0.49 is compound interest. While some of my students try to suggest that $0.49 is not a big deal, I always reply with two comments: $0.49 will always look better in my pocket than in yours and the numbers are small to facilitate computation. The amounts consumers deposit or borrow are much greater, for a longer period of time, and at higher interest rates.

Example 3

- Read and discuss the problem. The calculations are done. Discuss the reason for the differences in the two tables.
- Students are able to see the better deal. The big question is "Do they understand how Account B ended up with the greater balance?"

On Your Own

- For Question 2, encourage students to set up a table to organize their work.

Closure

- What is the balance of an account in which you invest $4000 at 4.5% simple interest per year after 2 years? $4360
- What is the balance of an account in which you invest $4000 at 4.5% interest compounded annually after 2 years? $4368.10

Technology For the Teacher

Dynamic Classroom

The Dynamic Planning Tool
Editable Teacher's Resources at *BigIdeasMath.com*

Compound interest is interest earned on the principal *and* on the previously earned interest.

EXAMPLE 2 Calculating Compound Interest

You deposit $400 in an account that earns 3.5% interest compounded annually. What is the account balance after 2 years?

> **Study Tip**
> The simple interest formula $I = Prt$ can be used to find the balance in an account that earns compound interest.

Interest for 1st year: $I = Prt$ — Write simple interest formula.

$= 400(0.035)(1)$ — Substitute.

$= 14$ — Simplify.

The account balance after 1 year is $400 + $14 = $414.

Interest for 2nd year: $I = Prt$ — Write simple interest formula.

$= 414(0.035)(1)$ — Substitute. Use 414 for P.

$= 14.49$ — Simplify.

∴ The account balance after 2 years is $414 + $14.49 = $428.49.

EXAMPLE 3 Comparing Simple and Compound Interest

You want to invest $8000 for 3 years. Which account should you choose?

- **Account A earns 4% simple interest per year.**
- **Account B earns 4% interest compounded annually.**

Make two tables that show the account balances for 3 years.

Account A			
Year	Principal	4% Simple Interest	Balance at End of Year
1	$8000	$320	$8320
2	$8000	$320	$8640
3	$8000	$320	$8960

Account B			
Year	Principal and Interest	4% Compound Interest	Balance at End of Year
1	$8000.00	$320.00	$8320.00
2	$8320.00	$332.80	$8652.80
3	$8652.80	$346.11	$8998.91

∴ Account B earns $8998.91 − $8960 = $38.91 more than Account A after 3 years. So, you should choose Account B.

On Your Own

Now You're Ready
Exercises 5, 6, and 8

2. In Example 2, what is the account balance after 5 years?
3. In Example 3, how much more does Account B earn than Account A after 2 more years?

B.1 Exercises

Check It Out
Help with Homework
BigIdeasMath.com

Vocabulary and Concept Check

1. **VOCABULARY** What type of interest is money earned only on the principal?
2. **WRITING** How are simple interest and compound interest similar? How are they different?

Practice and Problem Solving

3. You deposit $500 in a savings account that earns 3% interest per year.

   a. Copy and complete the tables that show the balances after 5 years with simple interest and compound interest.

   b. Which type of interest gives the greater balance?

Simple Interest			
t	Principal	Annual Interest	Balance at End of Year
1	$500	$15	$515
2	$500	$15	$530
3			
4			
5			

Compound Interest			
t	Principal and Interest	Annual Interest	Balance at End of Year
1	$500	$15.00	$515.00
2	$515	$15.45	$530.45
3			
4			
5			

1 4. You deposit $600 in a savings account that earns 4% simple interest per year. Your friend deposits $400 in a savings account that earns 5% simple interest per year.

   a. Write and graph two equations for the balance B in each account after t years.

   b. Are the account balances ever equal? Explain.

2 5. You deposit $1200 in a savings account that earns 5.4% interest compounded annually. What is the account balance after 3 years?

6. You deposit $300 in a savings account that earns 7.2% interest compounded annually. What is the account balance after 2 years?

7. **ERROR ANALYSIS** Describe and correct the error in finding the balance of the simple interest account after two years.

✗ Principal: $700 Rate: 3% per year

t	Principal	Annual Interest	Balance at End of Year
1	$700.00	$21.00	$721.00
2	$721.00	$21.63	$742.63

Assignment Guide and Homework Check

Level	Day 1 Activity Assignment	Day 2 Lesson Assignment	Homework Check
Basic	3, 12–16	1, 2, 4–8	2, 4, 6, 8
Average	3, 12–16	1, 2, 4–9	2, 4, 6, 8
Advanced	3, 12–16	1, 2–10 even, 7, 11	2, 4, 6, 8

Common Errors

- **Exercise 4** Students may forget to convert the percent to a decimal before writing the formula. Remind them that the rate is a decimal and that they need to divide the percent by 100 to find the correct decimal.
- **Exercises 5 and 6** Students may forget to include previously earned interest when calculating annual interest. Remind them of the definition of compound interest.
- **Exercise 8** Students may use the wrong formula for one of the accounts, or perform all the arithmetic on their calculator and enter the wrong numbers. Encourage them to write the formulas for each account and to make a table of the interest earned and the balance at the end of each year.

B.1 Record and Practice Journal

1. You deposit \$250 in a savings account that earns 3% simple interest per year. Your friend deposits \$200 in a savings account that earns 3.5% simple interest per year.

   a. Write and graph two equations for the balance B in each account after t years. **You: $B = 7.5t + 250$; Friend: $B = 7t + 200$**

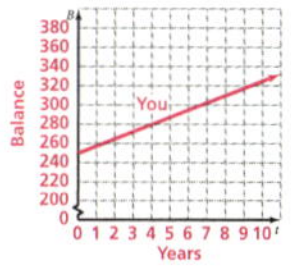

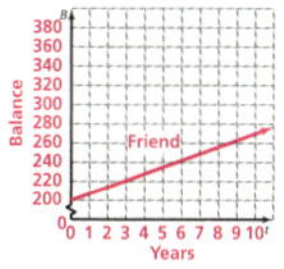

   b. Are the account balances ever equal? Explain.

   no; You deposit more money than your friend and you earn more simple interest each year.

2. You deposit \$500 in an account that earns 4.5% interest compounded annually. What is the account balance after 2 years?

   \$546.01

3. You want to invest \$10,000 for 4 years. Which account should you choose? Explain.

   - Account A earns 4% interest compounded annually.
   - Account B earns 4% simple interest per year.

   Account A; Account A earns \$11,698.59 − \$11,600 = \$98.59 more than Account B

Technology For the Teacher

Answer Presentation Tool

Quiz*Show*

Vocabulary and Concept Check

1. simple interest
2. They are similar because they both earn interest. They are different because simple interest is interest earned only on the principal, and compound interest is interest earned on the principal and on the previously earned interest.

Practice and Problem Solving

3. See Additional Answers.
4. a. You: $B = 600 + 24t$
   Friend: $B = 400 + 20t$

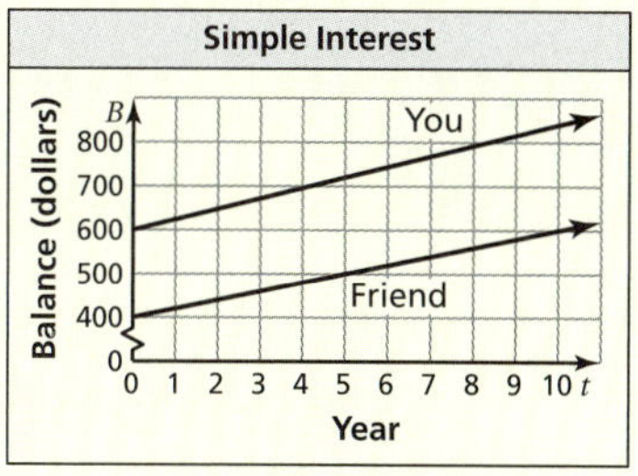

   b. no; You deposit more money and earn more interest each year. So, the balances are never equal.
5. \$1405.09
6. \$344.76
7. The table shows compound interest.

t	Principal	Annual Interest	Balance at End of Year
1	\$700.00	\$21	\$721.00
2	\$700.00	\$21	\$742.00

Practice and Problem Solving

8. **a.** Account B; It earns more interest because it earns interest on the interest.

b. \$186.08

9. **a.** \$300

b. \$13

10. \$228.38

11. See *Taking Math Deeper.*

Fair Game Review

12. 12^7 **13.** $3.375x^6$

14. $7.2p^{12}$ **15.** $29.16d^{18}$

16. A

Mini-Assessment

1. You deposit \$850 in a savings account that earns 5% simple interest per year. What is the account balance after 3 years? \$977.50
2. You deposit \$1300 in a savings account that earns 4.5% interest compounded annually. What is the account balance after 4 years? \$1550.27
3. You deposit \$400 in a savings account that earns 4.5% simple interest per year. Your friend deposits \$500 in a savings account that earns 5% interest compounded annually. Are the account balances ever equal? Explain. no; Your friend deposits more money and earns more interest each year. So, the balances are never equal.

Taking Math Deeper

Exercise 11

This exercise demonstrates the classic difference between simple and compound interest. In the Activity and the examples for this lesson, students learned that with simple interest, a balance grows with a linear pattern. However, with compound interest, a balance grows exponentially.

1. Make a table showing the balances.

Balance at End of Year	Account A	Account B
0	\$600.00	\$600.00
1	\$621.00	\$619.80
2	\$642.00	\$640.25
3	\$663.00	\$661.38
4	\$684.00	\$683.21
5	\$705.00	\$705.76
6	\$726.00	\$729.05
7	\$747.00	\$753.11

2. Compare 3 years and 6 years.

a. Account A earns more in 3 years.
Account B earns more in 6 years.

3. Compare the long-term benefits.

b. After 5 years, Account B has overtaken Account A.

c. After 7 years, Account B's balance has exceeded Account A's balance by more than \$5.

To see more dramatic differences in the account balances, students need to look at the very long-term differences. For instance, after 25 years, Account B will exceed Account A by more than \$225.

Project

Research banks in your area. What type of savings accounts do they offer? Include interest rates and fees. Which savings account would be best? Explain.

Reteaching and Enrichment Strategies

If students need help...	If students got it...
Resources by Chapter • Practice A and Practice B • Puzzle Time Record and Practice Journal Practice Differentiating the Lesson Lesson Tutorials Skills Review Handbook	Resources by Chapter • Enrichment and Extension Start the next section

3 8. **INVESTMENT** The owners of a company want to invest \$12,000 for 4 years.

a. Which account should they choose? Explain.
- Account A earns 5% simple interest per year.
- Account B earns 5% interest compounded annually.

b. How much more do they earn by choosing the better account?

9. **GAMES** After 7 years at 3% simple interest per year, your savings account earns \$63.

a. What is the principal?

b. How much money do you have left after buying the video game system?

10. **DIRT BIKE** Your friend borrows \$1050 from you to buy a new dirt bike. Your friend pays you back the principal plus 7.25% simple interest per year in 3 years. How much money do you earn?

11. **Critical Thinking** You want to deposit \$600 in a savings account. Account A earns 3.5% simple interest per year. Account B earns 3.3% interest compounded annually.

a. Which account should you choose if you invest your money for 3 years? 6 years? Explain.

b. After how many years is the balance in Account B greater than the balance in Account A?

c. After how many years is the difference between the account balances greater than \$5?

Fair Game Review What you learned in previous grades & lessons

Simplify the expression. *(Section 9.2)*

12. $12^2 \cdot 12^5$ 13. $(1.5x^2)^3$ 14. $3p^4 \cdot 2.4p^8$ 15. $(5.4d^9)^2$

16. **MULTIPLE CHOICE** The ratio of the number of students with Attention Deficit Hyperactivity Disorder to the number of students in an entire school is 1 : 20. What percent of students in the school have Attention Deficit Hyperactivity Disorder? *(Skills Review Handbook)*

Ⓐ 5% Ⓑ 8%

Ⓒ 20% Ⓓ 25%

B.2 Formula for Compound Interest

Essential Question How can you use a formula to find the balance in an account that earns compound interest?

1 ACTIVITY: Finding a Formula for Compound Interest

Work with a partner. You deposit \$1000 in a savings account that earns 6% interest compounded annually. Explain why parts (a)–(d) in the table are true.

a. Sample: Balance at End of Year = Principal + Annual Interest

$$= \$1000 + \$1000(0.06)$$

$$= \$1000(1 + 0.06)$$

$$= \$1000(1.06)$$

t	Principal and Interest	Annual Interest	Balance at End of Year
1	\$1000.00	\$1000(0.06)	**a.** $B = \$1000(1.06)$
2	\$1000(1.06)	**b.** \$1000(1.06)(0.06)	**c.** $B = \$1000(1.06)^2$
3	$\$1000(1.06)^2$	$\$1000(1.06)^2(0.06)$	$B = \$1000(1.06)^3$
4	$\$1000(1.06)^3$	$\$1000(1.06)^3(0.06)$	$B = \$1000(1.06)^4$
5	$\$1000(1.06)^4$	$\$1000(1.06)^4(0.06)$	**d.** $B = \$1000(1.06)^5$
6	$\$1000(1.06)^5$	$\$1000(1.06)^5(0.06)$	$B = \$1000(1.06)^6$
7	$\$1000(1.06)^6$	$\$1000(1.06)^6(0.06)$	$B = \$1000(1.06)^7$
8	$\$1000(1.06)^7$	$\$1000(1.06)^7(0.06)$	$B = \$1000(1.06)^8$
9	$\$1000(1.06)^8$	$\$1000(1.06)^8(0.06)$	$B = \$1000(1.06)^9$
10	$\$1000(1.06)^9$	$\$1000(1.06)^9(0.06)$	$B = \$1000(1.06)^{10}$

e. Use the pattern in the table to find the balance after 20 years.

Laurie's Notes

Introduction

For the Teacher

- **Goal:** Students will gain an understanding of how to compute compound interest.
- Students have difficulty understanding how a rate of 4% relates to 1.04 in the formula.

Motivate

- **Preparation:** Write percents of 5, 6, 10, 20, 25, and 50 on index cards.
- Hand the cards out to 6 students. Tell them they are store owners purchasing an item from you (the wholesaler). They need to determine the retail price based on the percent on their card.
- On the board, write a table. Use the headings *percent markup*, *wholesale price*, *amount markup*, *retail price*, and *relationship*.
- Tell one student a wholesale price (\$40–\$200). The student announces the amount of markup and the retail price (wholesale price + amount of markup). Another student completes the table.
- In the last column, the class writes the relationship between the retail and wholesale costs. Examples: \$50 is 125% of \$40; \$176 is 110% of \$160
- **Connection:** Ask students how percents greater than 100% are written. For instance, 125% = 1.25 and 106% = 1.06.

Activity Notes

Activity 1

- Students should study the table and look for patterns.
- In part (a), the expression is simplified (not evaluated) using the Distributive Property. The common factor is \$1000.
- In the second year, \$1000(1.06) and the amount of interest for the second year \$1000(1.06)(0.06) are added to get the balance at the end of the second year.

Balance $= \$1000(1.06) + \$1000(1.06)(0.06)$	Common factors: \$1000(1.06)
$= \$1000(1.06)(1 + 0.06)$	Distributive Property
$= \$1000(1.06)(1.06)$	Add.
$= \$1000(1.06)^2$	Simplify.

- Several patterns can be seen.
  - The amount that earns interest each year is the same as the end-of-year balance of the previous year.
  - The common factors of the expression for the balance are the factors of the previous end-of-year balance.
  - Each time you simplify the expression using the Distributive Property, you get another factor of 1.06.
  - Each year, your balance is 106% of what you began the year with. So, at the end of 5 years, you have earned 6% five times.

Previous Learning

Students should know how to write percents as decimals. Students should know how to compute simple interest.

Activity Materials	
Introduction	**Textbook**
• index cards	• calculator

Start Thinking! and Warm Up

Activity B.2 Start Thinking!
For use before Activity B.2

Activity B.2 Warm Up
For use before Activity B.2

Evaluate the expression.

1. $50(1.5)^2(5)$
2. $1500(1.05)^3(3)$
3. $1000(1.1)^4(10)$
4. $100(1.06)^2(4)$
5. $500(1.2)^2(2)$
6. $2000(1.02)^5(6)$

B.2 Record and Practice Journal

Essential Question How can you use a formula to find the balance in an account that earns compound interest?

1 ACTIVITY: Finding a Formula for Compound Interest

Work with a partner. You deposit \$1000 in a savings account that earns 6% interest compounded annually. Explain why parts (a)–(d) in the table are true.

a. Sample: Balance at End of Year = Principal + Annual Interest

$= \$1000 + \$1000(0.06)$

$= \$1000(1 + 0.06)$

$= \$1000(1.06)$

t	Principal and Interest	Annual Interest	Balance at End of Year
1	\$1000.00	$\$1000(0.06)$	**a.** $B = \$1000(1.06)$
2	$\$1000(1.06)$	**b.** $\$1000(1.06)(0.06)$	**c.** $B = \$1000(1.06)^2$
3	$\$1000(1.06)^2$	$\$1000(1.06)^2(0.06)$	$B = \$1000(1.06)^3$
4	$\$1000(1.06)^3$	$\$1000(1.06)^3(0.06)$	$B = \$1000(1.06)^4$
5	$\$1000(1.06)^4$	$\$1000(1.06)^4(0.06)$	**d.** $B = \$1000(1.06)^5$
6	$\$1000(1.06)^5$	$\$1000(1.06)^5(0.06)$	$B = \$1000(1.06)^6$
7	$\$1000(1.06)^6$	$\$1000(1.06)^6(0.06)$	$B = \$1000(1.06)^7$
8	$\$1000(1.06)^7$	$\$1000(1.06)^7(0.06)$	$B = \$1000(1.06)^8$
9	$\$1000(1.06)^8$	$\$1000(1.06)^8(0.06)$	$B = \$1000(1.06)^9$
10	$\$1000(1.06)^9$	$\$1000(1.06)^9(0.06)$	$B = \$1000(1.06)^{10}$

See Additional Answers.

Differentiated Instruction

Visual

Have students copy the simple interest formula and compound interest formula into their notebooks. They should identify the variables that both formulas have in common.

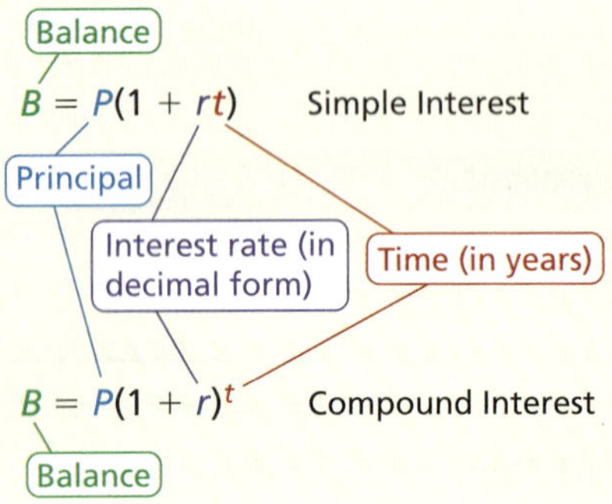

B.2 Record and Practice Journal

e. Use the pattern in the table on the previous page to find the balance after 20 years.

$3207.14

2 ACTIVITY: Writing a Formula

Work with a partner. Use the pattern in Activity 1 to write a formula for the balance in an account that earns interest compounded annually after t years.

P = Principal (initial deposit)

r = Annual interest rate (in decimal form)

t = Time (in years)

B = Balance after t years

$B = P(1 + r)^t$

3 ACTIVITY: A Penny Saved

Work with a partner. In his will, Benjamin Franklin gave $4000 to the state of Pennsylvania. He instructed that the money be invested for 200 years. After 200 years, the money should be used to do good. Franklin died in 1790. In 1990, his gift had grown to over $2 million.

Use your formula from Activity 2 to complete the table on the next page. Then approximate the annual interest rate that Benjamin Franklin's gift earned.

"A penny saved is a penny earned." Benjamin Franklin

Annual Interest Rate	Balance After 200 Years
3.0%	$1,477,423.26
3.1%	$1,793,878.60
3.2%	$2,177,707.01
3.3%	$2,643,165.12
3.4%	$3,207,507.96

about 3.2%

What Is Your Answer?

4. **IN YOUR OWN WORDS** How can you use a formula to find the balance in an account that earns compound interest? How much would Benjamin Franklin's account have been worth if it had earned 6% interest each year?

Substitute values for P, r, and t in the formula $B = P(1 + r)^t$; $460,503,615.51

Laurie's Notes

Activity 2

- Give students time to think about this before you jump in to rescue them.
- Students should see the pattern in Activity 1 and come up with the compound interest formula, $B = P(1 + r)^t$.

? "What does the 1 in the formula represent?" 100%

Activity 3

- Have a student read the information about Benjamin Franklin.
- Students need to use the formula from Activity 2, with $P = \$4000$ and $t = 200$. Try other interest rates.

? "How do you write 3.2% as a decimal?" 0.032

What Is Your Answer?

- **Think-Pair-Share:** Students should read the question independently and then work with a partner to answer the question. When they have answered the question, the pair should compare their answer with another group and discuss any discrepancies.

Closure

- **Exit Ticket:** You deposit $1000 in an account that earns 10% interest compounded annually for 5 years. Your friend deposits $1000 in an account that earns 5% interest compounded annually for 10 years. Compare the two account balances. Your balance is $1610.51, your friend's balance is $1628.89; It takes the second account almost twice as long to have the same balance.

Technology For the Teacher

Dynamic Classroom

The Dynamic Planning Tool

Editable Teacher's Resources at *BigIdeasMath.com*

2 ACTIVITY: Writing a Formula

Work with a partner. Use the pattern in Activity 1 to write a formula for the balance in an account that earns interest compounded annually after *t* years.

P = Principal (initial deposit)

r = Annual interest rate (in decimal form)

t = Time (in years)

B = Balance after t years

$B =$

3 ACTIVITY: A Penny Saved

Work with a partner. In his will, Benjamin Franklin gave $4000 to the state of Pennsylvania. He instructed that the money be invested for 200 years. After 200 years, the money should be used to do good. Franklin died in 1790. In 1990, his gift had grown to over $2 million.

Use your formula from Activity 2 to copy and complete the table. Then approximate the annual interest rate that Benjamin Franklin's gift earned.

"A penny saved is a penny earned." Benjamin Franklin

Annual Interest Rate	Balance After 200 years
3.0%	
3.1%	
3.2%	
3.3%	
3.4%	

What Is Your Answer?

4. **IN YOUR OWN WORDS** How can you use a formula to find the balance in an account that earns compound interest? How much would Benjamin Franklin's account have been worth if it had earned 6% interest each year?

Practice Use what you learned about the formula for compound interest to complete Exercises 3 and 4 on page A20.

B.2 Lesson

Check It Out
Lesson Tutorials
BigIdeasMath.com

Key Idea

Interest Compounded Annually

$$B = P(1 + r)^t$$

Balance (B)
Principal (P)
Annual interest rate (in decimal form) (r)
Time (in years) (t)

EXAMPLE 1 Finding an Account Balance

You deposit \$1000 in a savings account. The account earns 3% interest compounded annually. (a) What is the balance after 2 years? (b) What is the interest earned?

a.

$B = P(1 + r)^t$	Write compound interest formula.
$= 1000(1 + 0.03)^2$	Substitute.
$= 1000(1.03)^2$	Add.
$= 1060.9$	Simplify.

The balance is \$1060.90 after 2 years.

b. The interest earned is the difference of the balance and the principal.

So, the interest earned is \$1060.90 − \$1000.00 = \$60.90.

Study Tip

For small values of t, you can use the simple interest formula to estimate the interest.

$I = Prt$
$= 1000(0.03)(2)$
$= 60$

EXAMPLE 2 Finding a Principal

An account opened on January 1, 2010 earns 5% interest compounded annually. The only change in the account is interest earned. What is the principal?

Descartes the Cat
9 Lives Lane
Meow Town, USA

Account No.
3-141592654

Date	Description	Deposits	Balance
01/01/2012	Previous Balance		\$8820.00
12/31/2012	Interest Earned	\$441.00	\$9261.00

The balance is \$9261.00 after 3 years.
Use the formula to find the principal.

$B = P(1 + r)^t$	Write formula.
$9261.00 = P(1 + 0.05)^3$	Substitute.
$9261.00 = P(1.05)^3$	Add.
$8000 = P$	Solve for P.

The principal is \$8000.

Laurie's Notes

Introduction

Connect

- **Yesterday:** Students gained an understanding of how to compute with compound interest.
- **Today:** Students will use the compound interest formula.

Motivate

- **Financial Literacy:** Yesterday, students learned an interesting fact about an amazing man. Here are a few quotes from Benjamin Franklin that connect to financial literacy.
  - "An investment in knowledge always pays the best interest."
  - "Early to bed and early to rise makes a man healthy, wealthy, and wise."
  - "Genius without education is like silver in the mine."
  - "Contentment makes poor men rich; discontentment makes rich men poor."

Lesson Notes

Key Idea

- The principal is the amount initially invested. The annual interest rate, written as a decimal, is the percent earned on the principal.
- The interest rate must be written as a decimal. Remember that 1 is the decimal form of 100%.

Example 1

- Using a calculator, find the balance. Check to make sure students know how to enter the exponent on their calculator. On many calculators, the expression can be entered as written, without having to find the intermediate values.
- ? "What do you call the difference between the balance and the principal?" interest earned

Example 2

- Read the information given and discuss the account statement shown.
- ? "In the compound interest formula, how many variables are there?" four
- Within the scope of this class, two variables can be solved for, *B* and *P*. In this example, students are asked to solve for *P*, the principal.
- **Common Misconception:** When the formula is simplified to $9261.00 = P(1.05)^3$, students believe there is nothing else they can do. They do not recognize that $(1.05)^3$ is the coefficient of *P*. To solve for *P*, divide both sides by the coefficient.
- The last step is to divide both sides by 1.05^3 or 1.157625.
- Point out the need to know how equations are solved. Properties of Equality are still very important.

Goal Today's lesson is using the compound interest formula to find an account balance and to find the principal invested.

Lesson Materials
Textbook
• calculator

Start Thinking! and Warm Up

Lesson B.2 **Warm Up** For use before Lesson B.2

Lesson B.2 **Start Thinking!** For use before Lesson B.2

Have you ever heard of a mnemonic device? It is a way of helping a person to memorize information. Often, mnemonic devices are words, phrases, rhymes, or short poems.

Come up with a mnemonic device to help you memorize the formula you wrote in Activity 2 on page A17 in your textbook.

Extra Example 1

You deposit $800 in a savings account. The account earns 3.5% interest compounded annually.

a. What is the balance after 3 years? $886.97

b. What is the interest earned? $86.97

Extra Example 2

An account opened on January 1, 2011 earns 4% interest compounded annually. The only change in the account is interest earned. The balance is $1637.80 after 4 years. What is the principal? $1400

On Your Own

1. a. $2653.02

   b. $153.02

2. $10,000

Extra Example 3

In Example 3, which account should the business owner choose if the money is being invested for 3 years? Explain. Account A: $11,500; Account B: $11,576.25; The business owner should choose Account B because its balance is greater than the balance of Account A after 3 years.

On Your Own

3. Account B because its balance is greater than the balance of Account A after 5 years.

English Language Learners

Money

Have English learners and English speakers work together to find how much money can be saved for college using a compound interest account. Set up groups of 2 to 4 students. Assign each group an initial amount and the number of years they have to save. Students should organize their results in a table showing how different rates affect the amount that can be saved.

Laurie's Notes

On Your Own

- These two questions are modeled after Example 1 and Example 2.
- When students have finished, it would be helpful to have several students share their solutions at the board.

Example 3

- Read and discuss the problem.
- Both accounts have a deposit of $10,000, an interest rate of 5%, and a length of 10 years.
- For Account A, the information is substituted into the simple interest balance formula.
- For Account B, the information is substituted into the compound interest formula and the calculations performed.

On Your Own

- **Big Idea:** Given the same principal, rate, and time, compound interest will always be greater than or equal to simple interest.

Closure

- Use the compound interest formula. You invest $1000 at 8% interest compounded annually. How many years will it take before your balance is $2000? Use trial and error. about 9 years

Technology For the Teacher

Dynamic Classroom

The Dynamic Planning Tool
Editable Teacher's Resources at *BigIdeasMath.com*

On Your Own

Now You're Ready
Exercises 5 and 6

1. You deposit \$2500 in a savings account. The account earns 2% interest compounded annually. (a) What is the balance after 3 years? (b) What is the interest earned?

2. An account earns 1% interest compounded annually. The balance after 2 years is \$10,201. What is the principal?

EXAMPLE 3 Choosing an Account

A business owner wants to invest \$10,000 for 10 years. Which account should the business owner choose? Explain.

Account A

Account B

Account A

$B = P(1 + rt)$	Write simple interest balance formula.
$= 10{,}000(1 + 0.05(10))$	Substitute.
$= 15{,}000$	Simplify.

The balance after 10 years is \$15,000.

Account B

$B = P(1 + r)^t$	Write compound interest formula.
$= 10{,}000(1 + 0.05)^{10}$	Substitute.
$= 10{,}000(1.05)^{10}$	Add.
$\approx 16{,}288.95$	Use a calculator.

The balance after 10 years is \$16,288.95.

The business owner should choose Account B because its balance is greater than the balance of Account A after 10 years.

On Your Own

Now You're Ready
Exercises 8 and 9

3. **WHAT IF?** In Example 3, which account should the business owner choose if the money is being invested for 5 years? Explain.

B.2 Exercises

Check It Out
Help with Homework
BigIdeasMath.com

Vocabulary and Concept Check

1. **WRITING** For a given interest rate, how does compound interest differ from simple interest?

2. **DIFFERENT WORDS, SAME QUESTION** You deposit \$1000 in an account that earns 4% interest compounded annually. Which question is different? Find "both" answers.

   What is the balance after 1 year?

   What is the interest earned after 1 year?

   By how much money did your account increase after 1 year?

   What is the difference of the balance and the principal after 1 year?

Practice and Problem Solving

Copy and complete the table for an account that earns interest compounded annually.

3. Principal: \$500

Annual Interest Rate	Balance After 5 Years
3.0%	
3.5%	
4.0%	
4.5%	

4. Principal: \$1000

Annual Interest Rate	Balance After 10 Years
3.5%	
4.0%	
4.5%	
5.0%	

(1) 5. You deposit \$600 in a savings account. The account earns 3.5% interest compounded annually.

   a. What is the balance after 4 years?

   b. What is the interest earned?

(2) 6. An account earns 1.5% interest compounded annually. The balance after 2 years is \$8241.80. What is the principal?

7. **ERROR ANALYSIS** Describe and correct the error in finding the balance of a savings account with interest compounded annually.

   ✗

   Principal: \$2000
   Annual Interest Rate: 4%
   Time: 2 years

   $B = P(1 + r)^t$
   $= 2000(1 + 4)^2$
   $= 2000(5)^2$
   $= 50{,}000$

Assignment Guide and Homework Check

Level	Day 1 Activity Assignment	Day 2 Lesson Assignment	Homework Check
Basic	3, 4, 13–17	1, 2, 5–10	5, 6, 8, 10
Average	3, 4, 13–17	1, 2, 5–10	5, 6, 8, 10
Advanced	3, 4, 13–17	1, 2, 7–12	8, 9, 10, 11

Common Errors

- **Exercise 5** Students may use the simple interest formula to find the interest earned on the account. Remind them that this account earns compound interest, so the simple interest formula can only be used to estimate the interest earned.
- **Exercise 6** Students may evaluate part of the equation incorrectly when solving. Remind them of the order of operations and that they should simplify the equation before solving for the principal.
- **Exercises 8 and 9** Students may state that because the simple interest rate is higher, the account with simple interest will earn more over the time period. This is not necessarily the case. Remind them that they should calculate the new balances for both accounts to confirm their answer.

B.2 Record and Practice Journal

1. You deposit $800 in a savings account. The account earns 4.5% interest compounded annually.

   a. What is the balance after 3 years?

   $912.93

   b. What is the interest earned?

   $112.93

2. An account earns 2.4% interest compounded annually. The balance after 3 years is $9533.12. What is the principal?

   $8878.41

3. A business owner wants to invest $8000 for 5 years. Which account should the business owner choose?

Account A

Account B

Account A; Account A earns $9600 − $9274.19 = $325.81 more than Account B.

Technology For the Teacher

Answer Presentation Tool

Quiz*Show*

Vocabulary and Concept Check

1. Compound interest earns more than simple interest for a given interest rate because it is interest earned on the principal and on the previously earned interest.
2. What is the balance after 1 year?; $1040; $40

Practice and Problem Solving

3. $579.64; $593.84; $608.33; $623.09
4. $1410.60; $1480.24; $1552.97; $1628.89
5. **a.** $688.51

   b. $88.51
6. $8000
7. The interest rate should be in decimal form.

$$\begin{aligned} B &= P(1 + r)^t \\ &= 2000(1 + 0.04)^2 \\ &= 2000(1.04)^2 \\ &= \$2163.20 \end{aligned}$$

Practice and Problem Solving

8. Account A; Its balance is greater after 8 years.

9. Account B; Its balance is greater after 8 years.

10. \$10,695.13

11. See *Taking Math Deeper.*

12. The account that earns interest compounded monthly earns more interest because the interest is compounded more often.

Fair Game Review

13. 27,915 **14.** 0.003

15. 0.4 **16.** 19,344.96

17. D

Mini-Assessment

1. **You deposit \$850 in a savings account. The account earns 4% interest compounded annually.**
   - **a.** What is the balance after 4 years? \$994.38
   - **b.** What is the interest earned? \$144.38
2. **You deposit \$1700 in a savings account. The account earns 5.5% interest compounded annually.**
   - **a.** What is the balance after 6 years? \$2344.03
   - **b.** What is the interest earned? \$644.03

Taking Math Deeper

Exercise 11

This exercise looks forward to interest compounded n times per year (instead of just once per year). Part (b) looks at what happens when n increases. In advanced courses, students will learn that there is a limit to the advantage of frequent compounding. This limit is called *continuous compounding* and its formula involves the irrational number $e \approx 2.718281828$.

1 Summarize the values of the variables.

$P = \$1000$ (deposit)
$r = 0.03$ (3%)
$n = 4$ (times per year)
$t = 2$ (years)

2 Use the formula.

$$B = P\left(1 + \frac{r}{n}\right)^{nt}$$

$$= 1000\left(1 + \frac{0.03}{4}\right)^{(4)(2)}$$

a. $\approx \$1061.60$

3 Look at the balances for greater values of n.

$n = 4$ (quarterly) \$1061.60
$n = 12$ (monthly) \$1061.76
$n = 52$ (weekly) \$1061.82
$n = 365$ (daily) \$1061.83

The greatest balance would be given by continuous compounding (comparable to every second). The formula for this is

$$B = Pe^{rt} = 1000e^{(0.03)(2)} = \$1061.84.$$

b. So, as n increases, the balance increases slightly.

Project

Compare investing in an interest bearing savings account at the bank and other types of investments, such as money market accounts and the stock market.

Reteaching and Enrichment Strategies

If students need help. . .	If students got it. . .
Resources by Chapter • Practice A and Practice B • Puzzle Time Record and Practice Journal Practice Differentiating the Lesson Lesson Tutorials Skills Review Handbook	Resources by Chapter • Enrichment and Extension Start the next section

A business owner wants to invest \$7000 for 8 years. Which account should the business owner choose? Explain.

8. Account A

Account B

9. Account A

Account B

10. **COLLEGE** You want to save \$13,000 in 4 years for college tuition. About how much money should you deposit in a savings account that earns 5% interest compounded annually?

11. **CRITICAL THINKING** Interest can be compounded more than once per year. The formula for interest compounded more than once per year is $B = P\left(1 + \frac{r}{n}\right)^{nt}$, where n is the number of times the balance is compounded each year.

a. You deposit \$1000 in a savings account. The account earns 3% interest compounded 4 times per year. What is the balance after 2 years?

b. What happens to the balance after 2 years as n increases?

12. Critical Thinking Two accounts have the same interest rate, but one account earns interest compounded annually and the other earns interest compounded monthly. Which account earns more interest? Explain.

Fair Game Review What you learned in previous grades & lessons

Simplify the expression. *(Skills Review Handbook)*

13. $465.25 \cdot 60$ **14.** $\frac{1}{12} \cdot 0.036$ **15.** $\frac{4.8}{12}$ **16.** $268.68 \cdot 72$

17. **MULTIPLE CHOICE** Which graph represents the solution of the inequality $-\frac{c}{4} \geq -5$? *(Section 8.3)*

Ⓐ 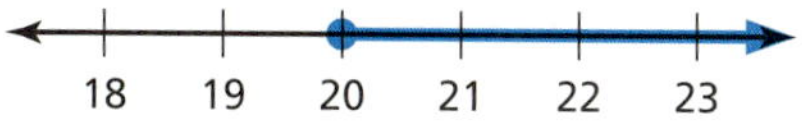

Ⓑ 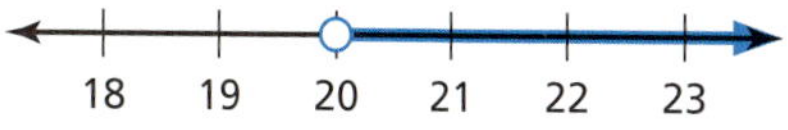

Ⓒ 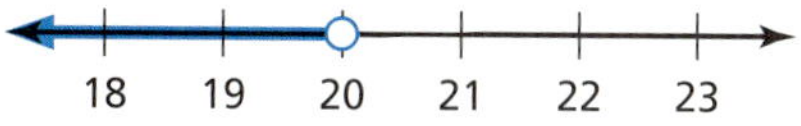

Ⓓ 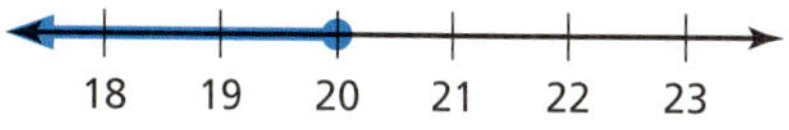

B.3 Installment Loans

Essential Question How can you find the total amount of interest paid on an installment loan?

1 ACTIVITY: Monthly Loan Payments

Work with a partner.

You borrow $1000 for 6 months. Your annual interest rate is 12%, which means that your monthly interest rate is

$$\text{Monthly Interest Rate} = \frac{\text{Annual Interest Rate}}{\text{Months in a Year}} = \frac{0.12}{12} = 0.01.$$

LOAN STATEMENT	
Amount of Loan	$1000.00
Monthly Payment	$172.55
Annual Interest Rate	12.00%
Monthly Interest Rate	1.00%
Payment Due Date	1st of Month
Term	6 Months

a. Copy and complete the table.

t	Balance Before Payment	Monthly Payment	Monthly Interest	Balance After Payment
1	$1000.00	$172.55	$1000.00(0.01) = $10.00	$1000.00 − 172.55 + 10.00 = $837.45
2	$837.45	$172.55	$837.45(0.01) = $8.37	$837.45 − 172.55 + 8.37 = $673.27
3		$172.55		
4		$172.55		
5		$172.55		
6		$172.53		

b. How much interest do you pay during the 6 months?

c. How is your answer to part (b) related to the expression 6(172.55) − 1000?

Laurie's Notes

Introduction

For the Teacher

- **Goal:** Students will explore the cost of loans.
- Students need to know the other side of interest, when consumers *pay* interest for the convenience of being able to borrow money from the bank. Just as with investments, the variables involved are the amount borrowed, interest rate, and a period of time.

Motivate

- Discuss with students the payments that most consumers are responsible for each month: mortgage, car payment, utilities, insurance, food, and credit cards. In addition to these, you have daily living expenses such as gas, clothing, and entertainment.
- Discuss budgets and the need to plan.

Activity Notes

Activity 1

- Discuss taking out a loan, information about credit rating, and shopping around for rates. Once the loan is taken, the borrower must repay the loan in monthly payments that stay fixed for the duration of the loan. Each monthly payment contains a portion that pays off the principal and a portion, called interest, that is the fee to the bank.
- Read the *Loan Statement* information and discuss each item.
- The word *annual* means for the year. The monthly interest rate is $\frac{1}{12}$ the annual rate.
- Discuss the calculations in the first two rows of the table.
  - The borrower is paying monthly interest on the remaining balance.
  - When the payment is made, the fixed payment is subtracted from the balance and the monthly interest is added.
  - The balance at the end of the month is brought forward for the start of the next month.
  - The interest calculated for the next month is based on the lesser balance, and is still 1%.
- **Neighbor Check:** Advise students to check with one another to ensure that they are not going off track.
- Notice that the final monthly payment is slightly less because of rounding during the life of the loan.

? "How much did you pay in monthly payments?" $1035.28

Previous Learning

Students should know how to write percents as decimals. Students should know how to compute simple and compound interest.

Activity Materials
Textbook
• calculator

Start Thinking! and Warm Up

Activity B.3 Start Thinking!
For use before Activity B.3

Activity B.3 Warm Up
For use before Activity B.3

Write the percent as a decimal.

1. 25% **2.** 37.5%

3. 5% **4.** 12%

5. 0.8% **6.** 1.6%

B.3 Record and Practice Journal

Essential Question How can you find the total amount of interest paid on an installment loan?

1 ACTIVITY: Monthly Loan Payments

Work with a partner.

You borrow $1000 for 6 months. Your annual interest rate is 12%, which means that your monthly interest rate is

$$\text{Monthly Interest Rate} = \frac{\text{Annual Interest Rate}}{\text{Months in a Year}} = \frac{0.12}{12} = 0.01.$$

LOAN STATEMENT	
Amount of Loan	$1000.00
Monthly Payment	$172.55
Annual Interest Rate	12.00%
Monthly Interest Rate	1.00%
Payment Due Date	1st of Month
Term	6 Months

a. Complete the table.

t	Balance Before Payment	Monthly Payment	Monthly Interest	Balance After Payment
1	$1000.00	$172.55	$1000(0.01) = $10.00	$1000 − 172.55 + 10.00 = $837.45
2	$837.45	$172.55	$837.45(0.01) = $8.37	$837.45 − 172.55 + 8.37 = $673.27
3	$673.27	$172.55	$6.73	$507.45
4	$507.45	$172.55	$5.07	$339.97
5	$339.97	$172.55	$3.40	$170.82
6	$170.82	$172.53	$1.71	$0.00

English Language Learners

Visual Aid

Give this demonstration of writing a percent as a decimal.

Locate the decimal point in the 7%.

7.%

Draw two arrows to show that the decimal point moves two places *left*.

7.%

Write zeros to the left of the number if needed.

007.%

Rewrite as a decimal with the decimal point two places to the left and without the percent sign.

0.07

B.3 Record and Practice Journal

b. How much interest do you pay during the 6 months?

$35.28

c. How is your answer to part (b) related to the expression 6(172.55) − 1000?

about the same

2 ACTIVITY: Finding the Total Interest Paid

Work with a partner. Use the result of Activity 1.

a. **CAR LOAN** You borrow $18,000 for a car. Make a table that shows about how much interest you pay for each annual interest rate and term.

Monthly Payments					
Term \ Annual Interest Rate	6%	8%	10%	12%	14%
36 months	$547.59	$564.05	$580.81	$597.86	$615.20
48 months	$422.73	$439.43	$456.53	$474.01	$491.88
60 months	$347.99	$364.98	$382.45	$400.40	$418.83
72 months	$298.31	$315.60	$333.47	$351.90	$370.90

See Additional Answers.

b. **HOME MORTGAGE** You borrow $180,000 to buy a house. Make a table that shows about how much interest you pay for each annual interest rate and term.

Monthly Payments					
Term \ Annual Interest Rate	4%	6%	8%	10%	12%
20 years	$1090.76	$1289.58	$1505.59	$1737.04	$1981.96
30 years	$859.35	$1079.19	$1320.78	$1579.63	$1851.50
40 years	$752.29	$990.38	$1251.56	$1528.46	$1815.30
50 years	$694.27	$947.53	$1222.69	$1510.39	$1804.61

See Additional Answers.

What Is Your Answer?

3. **IN YOUR OWN WORDS** How can you find the total amount of interest paid on an installment loan? Why is it important to search for low interest rates and short terms when you are applying for an installment loan?

Multiply the monthly payment by the total number of payments, then subtract the loan amount; to minimize the total amount of interest paid.

Laurie's Notes

Activity 2

- The two tables show the monthly payment for loans of varying length and various interest rates. Looking at the car loan table, ask the following questions.
- ? "For any length of a loan, how does the monthly payment change as the interest rate changes?" As the interest rate increases, the monthly payment increases.
- ? "For any interest rate of a loan, how does the monthly payment change as the length of the loan changes?" As the length of the loan increases, the monthly payment decreases.
- ? **Extension:** "Do you always want to shop for the lowest interest rate and longest loan period?" Students probably don't have a sense about this yet. They need to find the cost of each loan.
- Give students time to calculate the cost of each loan. Students should set up a table as in Activity 1.
- In each case (car and house), the least cost of the loan occurs when you have the lowest interest rate and shortest loan.
- **Common Misconception:** Students want to know why you would not always go for the shortest loan. The issue is being able to afford the higher monthly payment. If you can afford the higher monthly payment, take it.
- This table shows students why 50-year mortgages are rarely used. The monthly payment is not much lower, but the interest the consumer pays back is *much* higher. Even a 40-year mortgage is a bad choice for consumers.

What Is Your Answer?

- **Neighbor Check:** Have students work independently and then have their neighbor check their work. Have students discuss any discrepancies.

Closure

- **Writing Prompt:** To determine the cost of a loan, you …

Technology For the Teacher

Dynamic Classroom

The Dynamic Planning Tool

Editable Teacher's Resources at *BigIdeasMath.com*

2 ACTIVITY: Finding the Total Interest Paid

Work with a partner. Use the result of Activity 1.

a. **CAR LOAN** You borrow $18,000 for a car. Make a table that shows about how much interest you pay for each annual interest rate and term.

Monthly Payments					
Annual Interest Rate / Term	6%	8%	10%	12%	14%
36 months	$547.59	$564.05	$580.81	$597.86	$615.20
48 months	$422.73	$439.43	$456.53	$474.01	$491.88
60 months	$347.99	$364.98	$382.45	$400.40	$418.83
72 months	$298.31	$315.60	$333.47	$351.90	$370.90

b. **HOME MORTGAGE** You borrow $180,000 to buy a house. Make a table that shows about how much interest you pay for each annual interest rate and term.

Monthly Payments					
Annual Interest Rate / Term	4%	6%	8%	10%	12%
20 years	$1090.76	$1289.58	$1505.59	$1737.04	$1981.96
30 years	$859.35	$1079.19	$1320.78	$1579.63	$1851.50
40 years	$752.29	$990.38	$1251.56	$1528.46	$1815.30
50 years	$694.27	$947.53	$1222.69	$1510.39	$1804.61

What Is Your Answer?

3. **IN YOUR OWN WORDS** How can you find the total amount of interest paid on an installment loan? Why is it important to search for low interest rates and short terms when you are applying for an installment loan?

Practice Use what you learned about installment loans to complete Exercise 3 on page A26.

B.3 Lesson

Check It Out
Lesson Tutorials
BigIdeasMath.com

Key Vocabulary
loan, *p. A24*
term, *p. A24*
installment loan, *p. A24*
mortgage, *p. A25*

A **loan** is money borrowed for a specified amount of time, called the **term** of the loan. The lender may add interest to the principal, increasing the total amount due. When the total amount due is paid back in equal increments over the term of the loan, it is called an **installment loan**.

EXAMPLE 1 Paying Interest on a Loan

A student borrows money for college. The loan summary is shown below. How much interest is paid on the loan?

LOAN AMOUNT	\$1450.00
MONTHLY PAYMENT AMOUNT	\$250.20
ANNUAL INTEREST RATE	12% annual interest
TERM	6 months

Study Tip
Due to rounding, the last payment of an installment loan may be a few cents more or less than the regular monthly payment.

The annual interest rate is 12%. So, the monthly interest rate is $\frac{0.12}{12} = 0.01$.

Use a table to track the payment and the balance each month.

Balance Before Payment	Monthly Payment	Monthly Interest	Balance After Payment
\$1450.00	\$250.20	\$1450.00(0.01) = \$14.50	\$1450.00 − 250.20 + 14.50 = \$1214.30
\$1214.30	\$250.20	\$1214.30(0.01) ≈ \$12.14	\$1214.30 − 250.20 + 12.14 = \$976.24
\$976.24	\$250.20	\$976.24(0.01) ≈ \$9.76	\$976.24 − 250.20 + 9.76 = \$735.80
\$735.80	\$250.20	\$735.80(0.01) ≈ \$7.36	\$735.80 − 250.20 + 7.36 = \$492.96
\$492.96	\$250.20	\$492.96(0.01) ≈ \$4.93	\$492.96 − 250.20 + 4.93 = \$247.69
\$247.69	\$250.17	\$247.69(0.01) ≈ \$2.48	\$247.69 − 250.17 + 2.48 = \$0.00

Find the sum of the monthly interest totals in the table.

$$\$14.50 + \$12.14 + \$9.76 + \$7.36 + \$4.93 + \$2.48 = \$51.17$$

So, \$51.17 in interest is paid on the loan.

On Your Own

Now You're Ready
Exercises 4–7

1. In Example 1, the loan amount is \$1750 and the monthly payment amount is \$301.96. How much interest is paid on the loan?

Laurie's Notes

Introduction

Connect

- **Yesterday:** Students explored the cost of loans.
- **Today:** Students will find the cost of installment loans.

Motivate

- Students love to do the what-ifs of a problem. To get students thinking about the costs of loans today, let them first find out how much the monthly payment would be for a purchase. I usually have a student who would immediately want to know the payments for a Lamborghini!
- Do an online search for "monthly payment calculator" and use it during class to answer students' what-ifs. If online access is not possible from your classroom, you could do the search in advance and record 5 to 6 scenarios in which you know your students would be interested. Try to keep them reasonable, so that not all of them are expensive cars!

Lesson Notes

Discuss

- A **loan** is money borrowed for a specified amount of time, called the **term** of the loan. The borrower pays back the amount borrowed, plus interest, in equal increments or payments. This is an **installment loan**.

Example 1

- **Discuss:** Read the information given for the installment loan. Identify the amount of the loan, rate, time, and monthly payment. Discuss the *Study Tip*.
- The table shows the payment schedule information for the life of the loan, which is 6 months.

? "The monthly payment is a fixed amount. Is the amount of interest paid each month a fixed amount? Explain." No; Interest decreases each month as the amount paid towards the outstanding balance increases.

? "What is the total amount paid in interest?" $51.17

- There are two ways in which the interest paid can be determined. Because the table of monthly payments has been calculated, the monthly interest can be summed. If the table of information was not available, calculate the difference between the total paid in monthly payments and the amount borrowed.

Goal Today's lesson is finding the cost of a **loan**.

Lesson Materials
Textbook
• calculator

Start Thinking! and Warm Up

Lesson B.3 Warm Up
For use before Lesson B.3

Lesson B.3 Start Thinking!
For use before Lesson B.3

Suppose that your friend wants to buy something from you, like your bicycle, MP3 player, or cell phone, and offers to pay you in installments.

What would be the conditions of your installment loan? Decide on the loan amount, annual interest rate, and term of the loan. Then calculate the monthly payment. Do you think your friend would pay that?

Extra Example 1

In Example 1, the loan amount is $2080 and the monthly payment amount is $358.90. How much interest is paid on the loan? $73.40

On Your Own

1. $61.76

Laurie's Notes

Extra Example 2

The principal on a home mortgage is \$150,000, the monthly payment is \$805.23, and the term is 30 years. How much interest is paid on the loan?
\$139,882.80

On Your Own

2. \$144,798.40

Differentiated Instruction

Visual
When dividing decimals, students may forget which number to make the whole number (the *divisor*), or they may forget to multiply the dividend by the same power of 10. Students may be careless in placing the decimal point in the quotient. The decimal point should be placed above the new decimal point in the dividend. Create a bulletin board display as a visual reminder.

whole number $0.05\overline{)2.655}$

Example 2

- One of the largest loans a consumer takes out is for the purchase of a home. This loan is called a **mortgage**. The amount borrowed is quite significant, and therefore the terms are generally longer than they would be for a car.
- **Discuss:** Give time for students to study the table. It usually does not take long before a student makes the observation you are paying a great deal in interest, and in one case you are paying more in interest than what you borrowed! That is not uncommon with mortgages.

? "What are the advantages and disadvantages of the 15-year mortgage?" Primary advantage is that cost of the loan is the least of all the options. Primary disadvantage is the monthly payment is the highest of all the options.

- Discuss with students the need to balance the monthly cost of the loan (can I afford the payment?) and the long-range cost of the loan.

On Your Own

- **Think-Pair-Share:** Students should read the question independently and then work with a partner to answer the question. When they have answered the question, the pair should compare their answer with another group and discuss any discrepancies.

Closure

- **Exit Ticket:** You borrow \$24,000 for a car. The monthly payment is \$608.70 for 4 years with an interest rate of 10%. What does the car loan cost you? $608.70(12)(4) - 24{,}000 = \$5217.60$

Technology For the Teacher

Dynamic Classroom

The Dynamic Planning Tool
Editable Teacher's Resources at *BigIdeasMath.com*

A **mortgage** is an installment loan used to buy a house. The term of a mortgage is usually greater than the terms of other types of loans.

EXAMPLE 2 Paying a Mortgage

The table shows loan options for a $100,000 mortgage. (a) How much interest is paid on each option? (b) Discuss an advantage and a disadvantage of the 15-year mortgage.

Monthly Payments		
Term \ Annual Interest Rate	5%	7%
15 years	$790.79	$898.83
20 years	$659.96	$775.30
30 years	$536.82	$665.30

a. Find the number of monthly payments for each term.

15 years	***20 years***	***30 years***
$15 \cdot 12 = 180$	$20 \cdot 12 = 240$	$30 \cdot 12 = 360$

Use a table to find the interest I paid on each option. The total amount paid for the mortgage is the product of the monthly payment and the number of payments. The interest paid is the difference of this product and the principal.

Total Interest Paid		
Term \ Annual Interest Rate	5%	7%
15 years	$I = 180(790.79) - 100{,}000$ $= \$42{,}342.20$	$I = 180(898.83) - 100{,}000$ $= \$61{,}789.40$
20 years	$I = 240(659.96) - 100{,}000$ $= \$58{,}390.40$	$I = 240(775.30) - 100{,}000$ $= \$86{,}072.00$
30 years	$I = 360(536.82) - 100{,}000$ $= \$93{,}255.20$	$I = 360(665.30) - 100{,}000$ $= \$139{,}508.00$

b. You pay less interest with the 15-year mortgage than with the 20-year and 30-year mortgages, but the monthly payment for the 15-year mortgage is higher than the monthly payments for the 20-year and 30-year mortgages.

On Your Own

Now You're Ready
Exercises 9 and 10

2. The principal on a home mortgage is $125,000, the monthly payment is $749.44, and the term is 30 years. How much interest is paid on the loan?

B.3 Exercises

Check It Out
Help with Homework
BigIdeasMath.com

Vocabulary and Concept Check

1. **VOCABULARY** Compare the principal of a loan and the total amount due.
2. **REASONING** As the term of a loan increases, what happens to the monthly payment? What happens to the total amount of interest paid?

Practice and Problem Solving

3. Copy and complete the table. How much interest do you pay during the 5 months? The annual interest rate is 12%

Balance Before Payment	Monthly Payment	Monthly Interest	Balance After Payment
$1135.00	$233.86		
	$233.86		
	$233.86		
	$233.86		
	$233.84		

A loan summary is shown. How much interest is paid on the loan?

1 4.

LOAN AMOUNT	$1020.00
MONTHLY PAYMENT	$176.00
ANNUAL INTEREST RATE	12% annual interest
TERM	6 months

5.

LOAN AMOUNT	$1635.00
MONTHLY PAYMENT	$277.29
ANNUAL INTEREST RATE	6% annual interest
TERM	6 months

6.

LOAN AMOUNT	$1500.00
MONTHLY PAYMENT	$129.10
ANNUAL INTEREST RATE	6% annual interest
TERM	12 months

7.

LOAN AMOUNT	$920.00
MONTHLY PAYMENT	$81.74
ANNUAL INTEREST RATE	12% annual interest
TERM	12 months

8. **ERROR ANALYSIS** Describe and correct the error in finding the balance after payment.

✗
Balance Before Payment: $1000
Monthly Payment: $172.55
Monthly Interest: $10
Balance After Payment:
$1000 − $172.55 − $10 = $817.45

Assignment Guide and Homework Check

Level	Day 1 Activity Assignment	Day 2 Lesson Assignment	Homework Check
Basic	3, 14–16	1, 2, 5, 7–11	2, 5, 8, 10
Average	3, 14–16	1, 2, 5, 7–11	2, 5, 8, 10
Advanced	3, 14–16	1, 2, 6–12 even, 11, 13	6, 8, 10, 12

Common Errors

- **Exercises 4–7** Students may find the interest using the monthly payment instead of the principal. Remind them how to calculate monthly interest. Encourage them to make a table of information to keep track of their calculations.
- **Exercises 9 and 10** Students may find the total amount paid, but forget to find the interest paid on the loan. Remind them that to find the interest, they need to find the total amount paid for the loan and subtract the original amount of the loan.

B.3 Record and Practice Journal

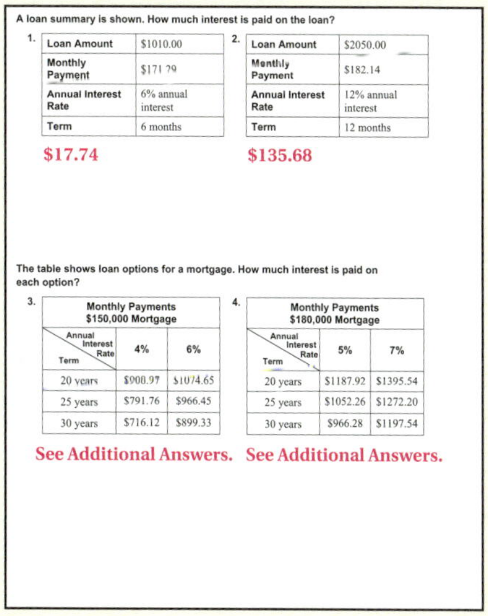
A loan summary is shown. How much interest is paid on the loan?

1.

Loan Amount	$1010.00
Monthly Payment	$171.29
Annual Interest Rate	6% annual interest
Term	6 months

$17.74

2.

Loan Amount	$2050.00
Monthly Payment	$182.14
Annual Interest Rate	12% annual interest
Term	12 months

$135.68

The table shows loan options for a mortgage. How much interest is paid on each option?

3. Monthly Payments $150,000 Mortgage

Term \ Annual Interest Rate	4%	6%
20 years	$908.97	$1074.65
25 years	$791.76	$966.45
30 years	$716.12	$899.33

See Additional Answers.

4. Monthly Payments $180,000 Mortgage

Term \ Annual Interest Rate	5%	7%
20 years	$1187.92	$1395.54
25 years	$1052.26	$1272.20
30 years	$966.28	$1197.54

See Additional Answers.

Technology For the Teacher

Answer Presentation Tool

Quiz*Show*

Vocabulary and Concept Check

1. The principal of a loan is less than the total amount due because interest is included in the amount due.
2. The monthly payment decreases. The total amount of interest paid increases.

Practice and Problem Solving

3.

Balance Before Payment	Monthly Payment	Monthly Interest	Balance After Payment
$1135.00	$233.86	$11.35	$912.49
$912.49	$233.86	$9.12	$687.75
$687.75	$233.86	$6.88	$460.77
$460.77	$233.86	$4.61	$231.52
$231.52	$233.84	$2.32	$0.00

$34.28

4. $36.00
5. $28.74
6. $49.20
7. $60.88
8. The monthly interest should be added to balance after payment. The balance after payment should be $\$1000 - \$172.55 + \$10 = \$837.45$.

Practice and Problem Solving

9.

Term \ Annual Interest Rate	Interest Paid	
	5%	7%
20	$70,068.00	$103,286.40
25	$90,453.00	$134,442.00
30	$111,908.40	$167,409.60

10.

Term \ Annual Interest Rate	Interest Paid	
	4%	6%
20	$61,336.80	$97,123.20
25	$78,774.00	$125,943.00
30	$97,023.60	$156,380.40

11. about 24.9%

12. about 46.3%

13. See *Taking Math Deeper.*

Fair Game Review

14. 1151.56

15. −28.06

16. D

Mini-Assessment

A mortgage loan summary is shown. How much interest is paid on the loan?

1.

LOAN AMOUNT	$80,000.00
MONTHLY PAYMENT	$505.65
ANNUAL INTEREST RATE	6.5% annual interest
TERM	30 years

$102,034.00

2.

LOAN AMOUNT	$100,000.00
MONTHLY PAYMENT	$687.89
ANNUAL INTEREST RATE	5.5% annual interest
TERM	20 years

$65,093.60

Taking Math Deeper

Exercise 13

This problem involves a classic budget decision for people. The mathematics of the problem can only partially answer the question. The question of "what should you do?" involves a person's personality more than it involves the differences in the costs.

1. Compute the expenses for Loan A.

- Keep your savings of $5000. At the end of 3 years, your balance will be $5543.59.
- Borrow the full amount of $16,000 for 3 years. Your payments will total 36(494.03) = $17,785.08.
- Your total expense is

  $17,785.08 − $5543.59 = $12,241.49.

2. Compute the expenses for Loan B.

- Use your savings of $5000 as a downpayment.
- Borrow the remaining $11,000 for 2 years. Your payments will total 24(492.50) = $11,820.00.

3. Compare the two plans.

- Loan A costs you $12,241.49.
- Loan B costs you $11,820.00.

You can save $421.49 by following Plan B. Some people might still be willing to pay the extra $421.49 for the security of having $5000 in their savings account.

Project

Use newspapers or the Internet to find the price of your "dream car." Using the current rates at the bank and the savings you have, how long will it take before you have enough money for the car? What can you do to shorten the time it takes to save for the car?

Reteaching and Enrichment Strategies

If students need help. . .	If students got it. . .
Resources by Chapter • Practice A and Practice B • Puzzle Time Record and Practice Journal Practice Differentiating the Lesson Lesson Tutorials Skills Review Handbook	Resources by Chapter • Enrichment and Extension • School-to-Work Start the next section

The table shows loan options for a mortgage. How much interest is paid on each option?

2 **9.**

Monthly Payments $120,000 Mortgage		
Term \ Annual Interest Rate	5%	7%
20 years	$791.95	$930.36
25 years	$701.51	$848.14
30 years	$644.19	$798.36

10.

Monthly Payments $135,000 Mortgage		
Term \ Annual Interest Rate	4%	6%
20 years	$818.07	$967.18
25 years	$712.58	$869.81
30 years	$644.51	$809.39

11. TUITION You pay off a $15,000 college tuition loan in 10 years. The monthly payment is $166.53. What percent of the total amount paid is the interest on the loan?

12. CONDOMINIUM A luxury condominium costs $13 million. The monthly payment for a 30-year mortgage on the condominium is $77,941.57. What percent of the total amount paid is the principal of the loan?

13. Critical Thinking You have $5000 in a savings account that earns 3.5% interest compounded annually. You need $16,000 to buy a car. What should you do? Explain.

- Let your savings earn interest for 3 years and get Loan A.
- Use your savings to get Loan B.

Loan A

APPLY NOW!
$16,000 for 36 months
7% annual interest rate
$494.03 monthly payment

Loan B

New Loan Special
$11,000 for 24 months
7% annual interest rate
$492.50 monthly payment

Fair Game Review What you learned in previous grades & lessons

Find the sum or difference. *(Skills Review Handbook)*

14. 666.35 + 485.21

15. 872.38 − 900.44

16. MULTIPLE CHOICE Which data display is most appropriate for the information given in thc table? *(Section 7.4)*

Ⓐ box-and-whisker plot
Ⓑ stem-and-leaf plot
Ⓒ circle graph
Ⓓ line graph

Year	Enrollment
2006	5042 students
2007	4921 students
2008	5198 students
2009	5003 students
2010	5117 students

B.4 Checking Accounts

Essential Question How do checking accounts work?

Share Your Work at... My.BigIdeasMath.com

1 ACTIVITY: Writing and Entering a Check

Work with a partner. You receive a bill for three pairs of jeans.

a. Copy and complete the check for the bill.

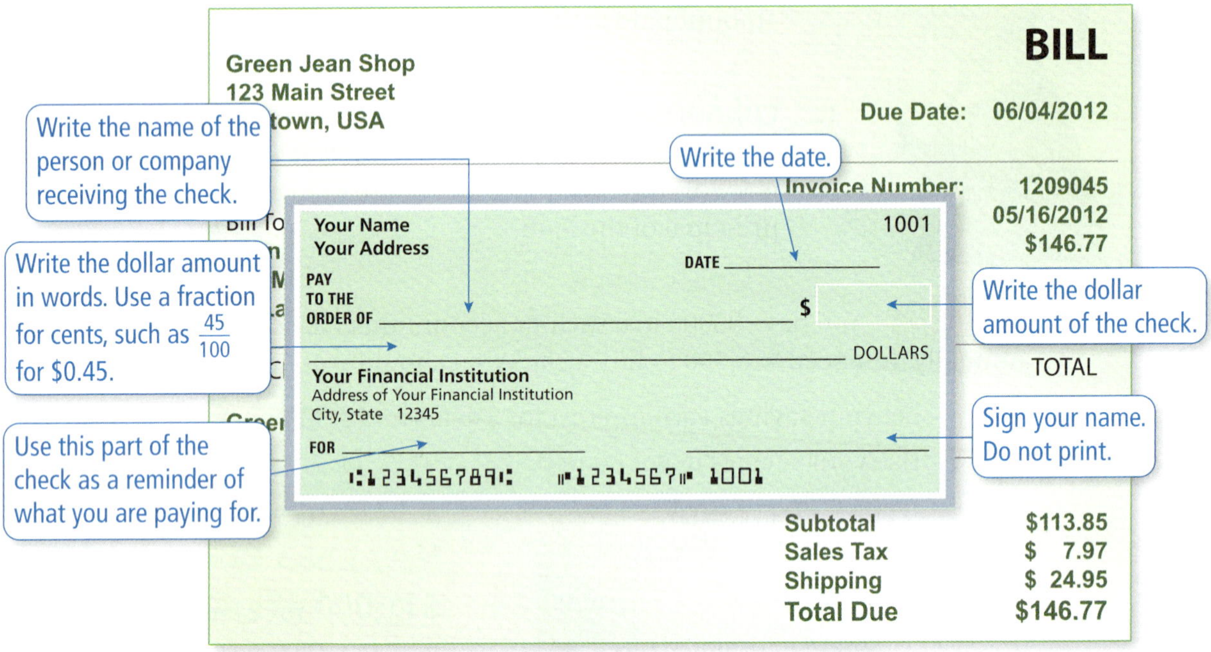

b. Enter the check in your checkbook registry below.

c. Copy and complete your checkbook registry.

Date	Check #	Transaction	Credit	Debit	Balance
		Balance Forward			100.00
5/11/12	996	Grocery Store		29.55	
5/11/12		Deposit Paycheck	482.75		
5/16/12		ATM Withdrawal		100.00	
5/17/12	997	Gym Membership		22.35	
5/17/12	998	Cell Phone Company		58.00	
5/17/12	999	Car Payment		82.66	
5/18/12		Deposit Paycheck	501.50		
5/18/12	1000	Birthday Gift		41.28	

Laurie's Notes

Introduction

For the Teacher

- **Goal:** Students will explore the features of managing a checkbook.
- A checking account is something that most adults and an increasing number of teenagers have. Checks provide a convenient way to pay bills and purchase goods without having to carry cash. Many people now make payments electronically, with the money being directly withdrawn from the checking account.
- **Misconception:** I have had students in the past who believed that checkbooks were "free money." They did not understand that the consumer first puts money into the checking account and then records the balance in the checkbook registry. Once this is done, checks may be written against that balance.
- **Discuss:** You should not write checks for amounts which exceed the balance. (It is illegal!) When you do, your account becomes *overdrawn*. In most cases, the bank charges an additional fee (about $20–$30), which is taken directly out of your account. The business you wrote the check to will also charge you a fee, *and* you still owe the business the amount of the check.
- Local banks in my area provide sample checks and checkbook registries for use by educators. Check with banks in your area.

Motivate

- Ask how many students have a checking account, debit card, or ATM (Automated Teller Machine) card.
- Ask them not to divulge any personal information, but have a discussion about the reasons why people use these consumer services.

Activity Notes

Activity 1

- This activity involves two skills, writing a check and recording the check in the checkbook registry.
- Discuss how to write out a check.
  - What you need: legible handwriting, pen.
  - What you need to know: date, payee, amount of the check in numbers and words.
  - What you must do: sign the check.
- It is a good practice to write the invoice number or the account number on the memo line.
- Each time a check is written, it should be recorded immediately in the checkbook registry. It is very easy to forget details about the check if you do not do it immediately.
- Discuss entries in the checkbook registry. A *credit* is a deposit of money into the account, while a *debit* is an amount that is subtracted.
- Students will practice each skill, writing a check and making an entry in the checkbook registry.

Previous Learning

Students should know how to write dollar amounts in words.

Start Thinking! and Warm Up

Activity B.4 Start Thinking!
For use before Activity B.4

Activity B.4 Warm Up
For use before Activity B.4

Find the sum.

1. $150.00 + (-29.65) + (-18.44)$
2. $69.10 + (-33.06) + 25.00$
3. $34.08 + (-33.98) + 78.67$
4. $118.05 + (-12.50) + (-54.00) + 30.00$
5. $115.00 + (-12.99) + (-16.97) + (-15.66)$
6. $50.95 + (-13.05) + (-27.57) + (-12.13)$

B.4 Record and Practice Journal

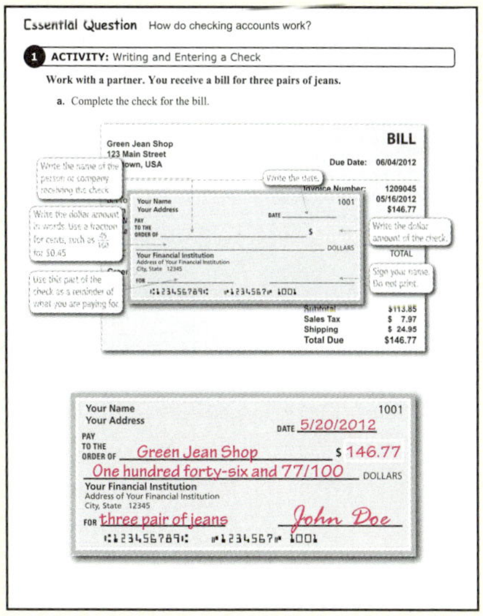

English Language Learners

Vocabulary

English learners may not be familiar with terms used in banking, such as *checking account, checkbook registry, debit card, bad check,* and *ATM*. Take time to explain these terms.

B.4 Record and Practice Journal

b. Enter the check in your checkbook registry below.

c. Complete your checkbook registry. See Additional Answers.

Date	Check #	Transaction	Credit	Debit	Balance
		Balance Forward			100.00
5/11/12	996	Grocery Store		29.55	
5/11/12		Deposit Paycheck	482.75		
5/16/12		ATM Withdrawal		100.00	
5/17/12	997	Gym Membership		22.35	
5/17/12	998	Cell Phone Company		58.00	
5/17/12	999	Car Payment		82.66	
5/18/12		Deposit Paycheck	501.50		
5/18/12	1000	Birthday Gift		41.28	

2 ACTIVITY: Reconciling Your Account

Work with a partner. Your monthly bank statement shows that the balance of your account is $602.96. This balance does not agree with your balance. You notice that you did not enter the two transactions shown below in your checkbook registry. See Additional Answers.

- 5/16/12 ATM Fee $2.50 (debit)
- 5/24/12 Interest $1.82 (credit)

Enter these transactions and the check you wrote in Activity 1 in your checkbook registry. Does your balance now agree with the bank statement?

Date	Check #	Transaction	Credit	Debit	Balance
5/17/12	998	Cell Phone Company		58.00	372.85
5/17/12	999	Car Payment		82.66	
5/18/12		Deposit Paycheck	501.50		
5/18/12	1000	Birthday Gift		41.28	

3 ACTIVITY: Choosing a Checkbook Account

RESEARCH PROJECT Research the checking accounts offered by three different banks. (Do this online or by visiting banks and collecting brochures.)

Write a report that compares the different features of the checking accounts. Then decide which account is best for you.

- **Does the account pay interest?**
- **Is there a minimum initial deposit?**
- **Are the checks free?**
- **What is the overdraft fee?**
- **Is the ATM card free to use?**
- **Is online banking free?**

Check students' work.

What Is Your Answer?

4. IN YOUR OWN WORDS How do checking accounts work?

allows you to deposit money in an account and then write personal checks using money from the account

Laurie's Notes

Activity 2

- Discuss the need and process of reconciling the account. Explain that reconciling means you check your balance against what the bank says is your balance. You should reconcile your account each month when your bank statement arrives. This information is also available electronically.
- In this activity, students will enter additional information in the registry and then reconcile the account.

? "What is an ATM and why would there be a $2.50 fee?" ATM stands for Automatic Teller Machine, where you can perform certain banking transactions at a location away from your home bank. For the convenience of being able to do this, some banks charge a service fee, $2.50 in this case.

? "Why would interest be credited to a checking account?" Some checking accounts pay interest on the balance you have in your account.

Activity 3

- This project could be assigned to groups of students to be completed over a period of time.

What Is Your Answer?

- **Neighbor Check:** Have students work independently and then have their neighbor check their work. Have students discuss any discrepancies.

Closure

- Your checking account balance is $484.20. You write a check for a new shirt that costs $24.95. You deposit $40 from a birthday gift. What is your new balance? $499.25

Technology For the Teacher

Dynamic Classroom

The Dynamic Planning Tool

Editable Teacher's Resources at *BigIdeasMath.com*

2 ACTIVITY: Reconciling Your Account

Work with a partner. Your monthly bank statement shows that the balance of your account is $602.96. This balance does not agree with your balance. You notice that you did not enter the two transactions shown below in your checkbook registry.

- 5/16/12 ATM Fee $2.50 (debit)
- 5/24/12 Interest $1.82 (credit)

Enter these transactions and the check you wrote in Activity 1 in your checkbook registry. Does your balance now agree with the bank statement?

Date	Check #	Transaction	Credit	Debit	Balance
5/17/12	998	Cell Phone Company		58.00	372.85
5/17/12	999	Car Payment		82.66	
5/18/12		Deposit Paycheck	501.50		
5/18/12	1000	Birthday Gift		41.28	

3 ACTIVITY: Choosing a Checking Account

RESEARCH PROJECT Research the checking accounts offered by three different banks. (Do this online or by visiting banks and collecting brochures.)

Write a report that compares the different features of the checking accounts. Then decide which account is best for you.

- Does the account pay interest?
- Is there a minimum initial deposit?
- Are the checks free?
- What is the overdraft fee?
- Is the ATM card free to use?
- Is online banking free?

What Is Your Answer?

4. **IN YOUR OWN WORDS** How do checking accounts work?

Practice Use what you learned about checking accounts to complete Exercise 3 on page A32.

B.4 Lesson

Check It Out
Lesson Tutorials
BigIdeasMath.com

Key Vocabulary
checking account, *p. A30*
checkbook registry, *p. A30*
debit card, *p. A30*
bad check, *p. A31*

A **checking account** is a service provided by a financial institution (such as a bank or credit union) that allows you to deposit money in an account and then write personal checks using money from the account. A **checkbook registry** is used to record credits to an account and debits from an account.

EXAMPLE 1 Completing a Checkbook Registry

You balance your checkbook at the end of the month.

a. Complete the balance column. What is the ending balance?

Date	Check #	Transaction	Credit	Debit	Balance
		Balance Forward			275.58
9/2/12	1308	Grocery Store		70.89	204.69
9/5/12		Deposit Paycheck	402.98		607.67
9/5/12	1309	Cable Company		35.60	572.07
9/7/12	1310	Cell Phone Company		65.40	506.67
9/9/12	1311	Car Payment		239.17	267.50
9/12/12	1312	Electric Company		72.54	194.96
9/12/12		Deposit Paycheck	402.95		597.91
9/16/12	1313	Birthday Gift		25.00	572.91
9/19/12		Deposit Paycheck	402.57		975.48
9/23/12	1314	Dinner		42.00	933.48
9/25/12	1315	Gas Company		45.75	887.73
9/26/12		Deposit Paycheck	403.08		1290.81
9/29/12	1316	Rent		450.00	840.81

Calculations
275.58 − 70.89
204.69 + 402.98
607.67 − 35.60
572.07 − 65.40
506.67 − 239.17
267.50 − 72.54
194.96 + 402.95
597.91 − 25.00
572.91 + 402.57
975.48 − 42.00
933.48 − 45.75
887.73 + 403.08
1290.81 − 450.00

The ending balance is $840.81.

b. Your monthly bank statement shows a balance of $1290.81. Explain why your balance does not agree with the bank statement.

The balance on the bank statement is equal to your balance on 9/26/12. Because Check #1316 is close to the end of the month, it likely was not processed in time to appear on your statement.

On Your Own

Now You're Ready
Exercise 4

1. You write a check on 9/30/12 for $74.95 to the auto shop, and another on 10/1/12 for $120.85 to the plumber. Complete two more rows in the registry for these checks. What is your new balance?

Reading
ATM stands for Automated Teller Machine.

Most banks include a **debit card**, or check card, with a checking account. This card allows you to withdraw cash from ATMs *and* pay for purchases in stores. The money is taken directly from your account.

Laurie's Notes

Introduction

Connect

- **Yesterday:** Students explored features of managing a checkbook.
- **Today:** Students will learn how to keep a checkbook registry.

Motivate

- Starting in 1939, Luther Simjian began registering patents for an automatic teller machine (ATM). Testing began in 1961, and by 1968 the ATM was introduced by the company now known as Citicorp. After 6 months, there was little demand and its use was discontinued.
- In the 1960s, John Sheppard-Barron introduced a cash dispensing machine in London that used a personal identification number (PIN). While in Scotland, James Goodfellow invented a cash dispensing machine that would read a card encrypted with information.
- In the 1970s, ATMs went from being standalone machines to online systems, able to access the user's bank account information.
- In the 1980s, ATMs with 24/7 access became increasingly popular with bank customers.
- Now shoppers use ATM cards to pay for their store purchases.

Lesson Notes

Discuss

- Review the terms: *checking account* and *checkbook registry.*

Example 1

- Read the entries made in the checkbook registry. Review the difference between a credit and a debit.
- Explain that the numbers to the right show the calculation that is done in order to *balance* or *reconcile* your account. This is typically done with a calculator. Some people also keep their checkbook registry electronically by using software that tracks all of the credits and debits.
- In part (b), share with students that this is a fairly common occurrence when trying to reconcile your account. You have written a check, but the person you wrote the check to has not cashed it yet. In reconciling, you need to account for transactions you have made that the bank has not yet recorded.

On Your Own

- Students should check their balance with others.

Discuss

- Discuss what a **debit card** is and how it is similar and different from a credit card. A debit card is linked to your checking and/or savings account. Your debit card does not let you spend more money than what is in your personal account(s). The credit card does! Because you cannot spend more money than what is in your account, it is important to keep track of your debits.

Goal Today's lesson is keeping a **checkbook registry**.

Lesson Materials
Textbook
• calculator

Start Thinking! and Warm Up

Lesson B.4 Warm Up
For use before Lesson B.4

Lesson B.4 Start Thinking!
For use before Lesson B.4

What factors would influence what bank you would use to open a checking account?

Extra Example 1

In Example 1, your balance is $840.81. You write another check on 9/29/12 for $123.45 to the grocery store, and another check on 10/1/12 for $74.55 to your friend. What is your new balance? $642.81

On Your Own

1. See Additional Answers.

Laurie's Notes

Extra Example 2

In Example 2, there is a $25 monthly fee for *Basic Checking*. How many ATM withdrawals must you make each month for the cost of *Regular Checking* to be greater than the cost of *Basic Checking*? 13 or more withdrawals

Extra Example 3

In Example 3, write an inequality that describes the possible amounts for Check #417 that may result in a bad check. Explain. $x > \$190.60$; The account balance prior to Check #417 is $190.60. So, a check greater than $190.60 may result in a bad check.

On Your Own

2. at least 8 withdrawals each month
3. $x > \$362.93$; The account balance prior to check #420 is $362.93. So, a check greater than $362.93 may result in a bad check.

Differentiated Instruction

Auditory

Start with a discussion about reading a real-life problem and how to determine what question is being asked. Ask students to identify key words, numbers, and units that are needed to solve the problem. Ask why it is important to identify these words. Ask students how unit analysis helps when writing an expression for a real-life problem.

Example 2

- Read the information given for each account. Be sure students understand the language of the ad. Ask them to explain the difference.
- Write the verbal model. Note the color coding to help students translate from words to symbols.
- Continue to solve the problem as shown.

Discuss

- Discuss what a *bad check* or bounced check is.
- Remind students that you may only write checks for an amount *less than or equal to* your balance.
- Some reasons a bad check may occur:
  - lack of organizational skills
  - Slow clearing of a deposit, which allows the check writer to think they have more money than they actually have available for checks.
  - A bank error where a bank employee misreads the amount of a hand-written check.

Example 3

- Students should read through the registry to find the bad check.

? "If the $100 deposit had been 4 days earlier, would the check have been bad?" no

On Your Own

- **Think-Pair-Share:** Students should read each question independently and then work with a partner to answer the questions. When they have answered the questions, the pair should compare their answers with another group and discuss any discrepancies.
- Discuss these questions as a whole class.

Closure

- You have a balance of $185.42 and write a check of $75 for a class you are going to take. You also want to make a payment toward your credit card balance. What is the maximum payment you can make? $110.42

EXAMPLE 2 Choosing a Checking Account

New Customers

Basic Checking
- $20 monthly fee
- free ATM usage

Regular Checking
- no monthly fee
- $2 charge for each ATM withdrawal

How many ATM withdrawals must you make each month for the cost of *Regular Checking* to be greater than the cost of *Basic Checking*?

Use a model to write and solve an inequality.

Words	Cost per withdrawal	times	the number of withdrawals	is greater than	the monthly fee.
Variable	Let w be the number of withdrawals.				
Inequality	2	$\cdot$	w	$>$	20

$2w > 20$ Write the inequality.

$w > 10$ Solve for w.

You must make more than 10 withdrawals each month.

A **bad check**, or "bounced" check, occurs when the amount written on the check is greater than the balance available in the checking account. Many banks charge a fee for a bad check.

EXAMPLE 3 Finding a Bad Check

Which check is the bad check?

Date	Check #	Transaction	Credit	Debit	Balance
		Balance Forward			190.60
6/2/12	417	Clothing Store		69.95	120.65
6/3/12		Deposit Paycheck	370.54		491.19
6/3/12	418	Gas Company		52.37	438.82
6/5/12	419	Cell Phone Company		75.89	362.93
6/7/12	420	Car Payment		275.10	87.83
6/9/12	421	Water Company		93.75	–5.92
6/13/12		Deposit	100.00		94.08
6/14/12	422	Birthday Gift		25.00	69.08

The amount of Check #421 is greater than the available balance.

So, Check #421 is the bad check.

On Your Own

Now You're Ready
Exercises 5 and 7

2. **WHAT IF?** In Example 2, there is a $15 monthly fee for *Basic Checking*. How many ATM withdrawals must you make each month for the cost of *Regular Checking* to be greater than the cost of *Basic Checking*?

3. Write an inequality that describes the possible car payments for Check #420 in Example 3 that may result in a bad check. Explain.

B.4 Exercises

Check It Out
Help with Homework
BigIdeasMath.com

Vocabulary and Concept Check

1. **VOCABULARY** What information do you need to know to enter a transaction in your checkbook registry?
2. **VOCABULARY** How do you know if a check is a bad check?

Practice and Problem Solving

3. Your checkbook registry is shown.

   a. Copy and complete your checkbook registry.

Date	Check #	Transaction	Credit	Debit	Balance
		Balance Forward			300.00
3/4/12	506	Dinner		45.32	
3/7/12		ATM Withdrawal		80.00	
3/8/12	507	Electric Company		39.76	
3/14/12		Deposit	150.00		
3/18/12	508	Credit Card Payment		129.50	

   b. Your monthly bank statement shows that your account balance is $81.12. You notice that you did not enter the two transactions shown below in your checkbook registry. After entering these transactions in your checkbook registry, does your balance now agree with the bank statement?

   - 3/7/12 ATM Fee $2.00
   - 3/10/12 Department Store $72.30

1 4. You balance your checkbook at the end of the month.

   a. Complete the balance column. What is the ending balance?

Date	Check #	Transaction	Credit	Debit	Balance
		Balance Forward			756.90
1/2/12	1563	Cell Phone Company		78.83	
1/5/12	1564	Grocery Store		132.70	
1/8/12		Deposit Paycheck	1567.00		
1/11/12	1565	Mortgage Payment		876.00	
1/13/12	1566	Credit Card Payment		651.25	
1/18/12		ATM Withdrawal		40.00	
1/20/12	1567	Grocery Store		143.78	
1/24/12		Deposit Paycheck	1567.00		
1/29/12	1568	Gas Company		68.23	

   b. Your monthly bank statement shows a balance of $1968.34. Explain why your balance does not agree with the bank statement.

Assignment Guide and Homework Check

Level	Day 1 Activity Assignment	Day 2 Lesson Assignment	Homework Check
Basic	3, 9–13	1, 2, 4–7	2, 4, 6, 7
Average	3, 9–13	1, 2, 4–8	2, 4, 6, 7
Advanced	3, 9–13	1, 2, 4–8	2, 4, 6, 7

For Your Information

- **Exercises 7 and 8** Here are four reasons to balance your checkbook.
  1. Balancing your checkbook is a method of verifying that your records match the bank's records.
  2. Banks often make mistakes. You typically have only 60 days in which to inform the bank of the error.
  3. If you overdraw your account, your bank will charge you a fee for each bounced check.
  4. If you discover a problem, it is easier to find it in one month's records than in several months' records.

Common Errors

- **Exercises 3 and 4** Students may add or subtract incorrectly. Remind them to be careful when finding the balance. When there is a debit, they should subtract and when there is a credit, they should add.
- **Exercise 5** Students may write an incorrect inequality to answer the question. Encourage them to write the inequality as a word sentence before substituting numbers and variables.
- **Exercise 7** Students may make an error adding or subtracting when finding the balance and may not be able to identify the bad check. Encourage them to check their answer for each calculation and to estimate what the balance will be for each debit or credit.

Technology For the Teacher

Answer Presentation Tool

Quiz*Show*

B.4 Record and Practice Journal

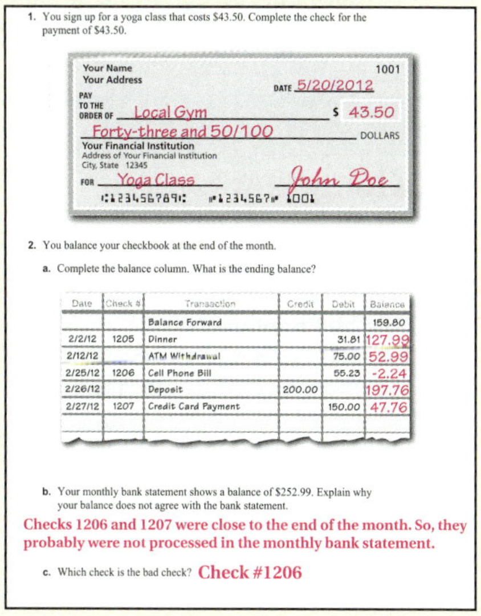

1. You sign up for a yoga class that costs $43.50. Complete the check for the payment of $43.50.

Your Name
Your Address
1001
DATE 5/20/2012
PAY TO THE ORDER OF Local Gym $ 43.50
Forty-three and 50/100 DOLLARS
Your Financial Institution
Address of Your Financial Institution
City, State 12345
FOR Yoga Class John Doe
⑆123456789⑆ ⑈1234567⑈ 1001

2. You balance your checkbook at the end of the month.

a. Complete the balance column. What is the ending balance?

Date	Check #	Transaction	Credit	Debit	Balance
		Balance Forward			159.80
2/2/12	1205	Dinner		31.81	127.99
2/12/12		ATM Withdrawal		75.00	52.99
2/25/12	1206	Cell Phone Bill		55.23	-2.24
2/26/12		Deposit	200.00		197.76
2/27/12	1207	Credit Card Payment		150.00	47.76

b. Your monthly bank statement shows a balance of $252.99. Explain why your balance does not agree with the bank statement.

Checks 1206 and 1207 were close to the end of the month. So, they probably were not processed in the monthly bank statement.

c. Which check is the bad check? **Check #1206**

Vocabulary and Concept Check

1. date, check number (if one is written), transaction description, credit or debit amount
2. The check is greater than the available balance.

Practice and Problem Solving

3. a. 254.68; 174.68; 134.92; 284.92; 155.42

   b. yes

4. a. 678.07; 545.37; 2112.37; 1236.37; 585.12; 545.12; 401.34; 1968.34; 1900.11
   The ending balance is $1900.11.

   b. Because Check #1568 is close to the end of the month, it likely was not processed in time to appear on the monthly statement.

Practice and Problem Solving

5. more than 20 withdrawals each month
6. The deposit should be entered in the "Credit" column and added to the balance. The correct balance is $880.

7–8. See *Taking Math Deeper*.

Fair Game Review

9. 7.9×10^{-8}
10. 1.5×10^{6}
11. 3.6×10^{11}
12. 2.4×10^{-5}
13. C

Mini-Assessment

In the check registry from Exercise 4 on page A32, your monthly bank statement shows that your account balance is $1760.33. You notice that you did not enter the two transactions shown below in your checkbook registry. After entering these transactions in your checkbook registry, does your balance now agree with the bank statement?

- 1/15/12 Dinner $47.68
- 1/25/12 Grocery Store $92.10

yes

Taking Math Deeper

Exercises 7 and 8

This problem is straightforward, but it involves an important lesson for students. Balance your checkbook with every transaction!

1 Balance the check registry.

7.

Date	Check #	Transaction	Credit	Debit	Balance
		Balance Forward			230.50
11/3/12	1021	Electric Company		37.29	193.21
11/8/12	1022	Shoe Store		49.21	144.00
11/9/12	1023	Credit Card Payment		173.80	-29.80
11/12/12		Deposit Paycheck	275.00		245.20
11/17/12	1024	Car Payment		133.61	111.59
11/26/12		Deposit Paycheck	275.00		386.59
11/29/12	1025	Rent		375.00	11.59

Check #1023 (credit card payment) is the bad check.

2 **8a.** In Exercise 8, notice the problems that the bad check caused. The bank charges a $30 penalty fee for the check, which means that your rent payment also bounced, and you then incurred a second penalty fee of $30.

Date	Check #	Transaction	Credit	Debit	Balance
		Balance Forward			230.50
11/3/12	1021	Electric Company		37.29	193.21
11/8/12	1022	Shoe Store		49.21	144.00
11/9/12	1023	Credit Card Payment		173.80	-29.80
11/12/12		Overdraft fee		30.00	-59.80
11/17/12	1024	Deposit Paycheck	275.00		215.20
11/26/12		Car Payment		133.61	81.59
11/29/12	1025	Deposit Paycheck	275.00		356.59
11/26/12		Rent		375.00	-18.41
11/29/12	1025	Overdraft fee		30.00	-48.41

3 **8b.** Not keeping track of ATM withdrawals and debit purchases makes it appear that you have more money in your account than you really do. You can avoid writing a bad check by balancing your checkbook after each transaction.

Reteaching and Enrichment Strategies

If students need help...	If students got it...
Resources by Chapter • Practice A and Practice B • Puzzle Time Record and Practice Journal Practice Differentiating the Lesson Lesson Tutorials Skills Review Handbook	Resources by Chapter • Enrichment and Extension • School-to-Work Start the next section

5. How many ATM withdrawals must you make each month for the cost of *Basic Checking* to be greater than the cost of *Advanced Checking*?

6. **ERROR ANALYSIS** Describe and correct the error in the checkbook registry.

Date	Check #	Transaction	Credit	Debit	Balance
		Balance Forward			760.00
4/5/12	1462	Gas Company		34.00	726.00
4/6/12		Deposit		154.00	572.00

7. Complete the balance column. Which check is the bad check?

Date	Check #	Transaction	Credit	Debit	Balance
		Balance Forward			230.50
11/3/12	1021	Electric Company		37.29	
11/8/12	1022	Shoe Store		49.21	
11/9/12	1023	Credit Card Payment		173.80	
11/12/12		Deposit Paycheck	275.00		
11/17/12	1024	Car Payment		133.61	
11/26/12		Deposit Paycheck	275.00		
11/29/12	1025	Rent		375.00	

8. **Critical Thinking** When you write a bad check, a bank charges a fee of $30.

a. How will the bad check fee change the checkbook registry in Exercise 7?

b. What mistakes can lead to writing a bad check? How can you avoid writing a bad check?

Fair Game Review What you learned in previous grades & lessons

Write the number in scientific notation. *(Section 9.6)*

9. 0.000000079

10. 1,500,000

11. 360,000,000,000

12. 0.000024

13. **MULTIPLE CHOICE** An account earns 4% interest compounded annually. The balance after 3 years is $618.68. What is the principal? *(Section B.2)*

Ⓐ $500 Ⓑ $525 Ⓒ $550 Ⓓ $600

B Study Help

Check It Out
Graphic Organizer
BigIdeasMath.com

You can use an **information frame** to help you organize and remember topics. Here is an example of an information frame for simple interest.

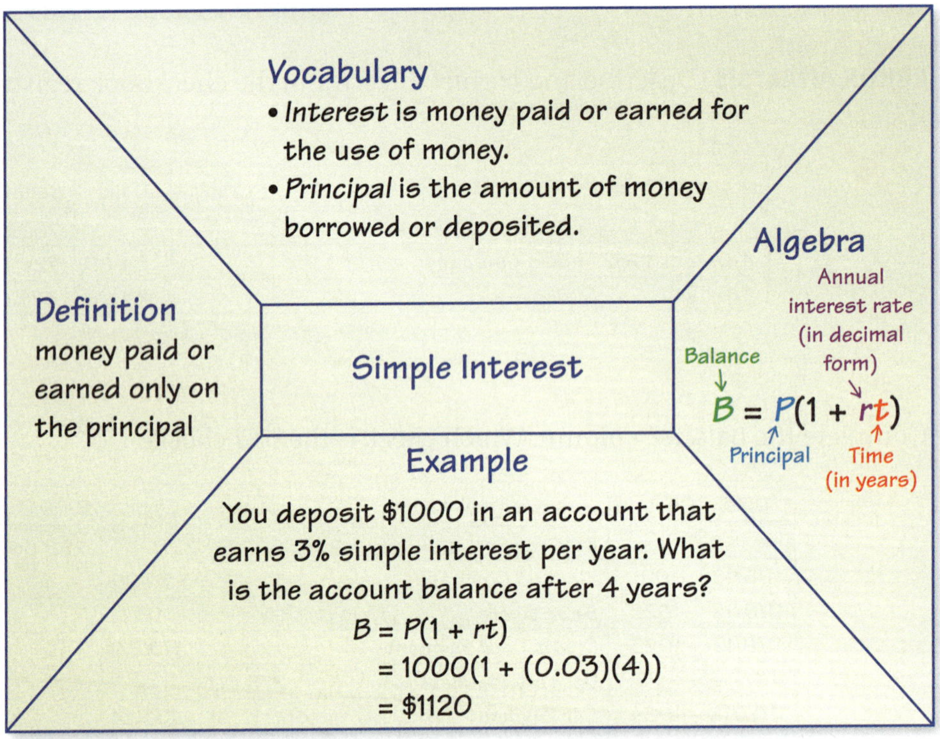

On Your Own

Make an information frame to help you study these topics.

1. compound interest
2. loan
3. mortgage
4. checking account

After you complete this chapter, make information frames for the following topics.

5. credit card
6. payroll statement

"Dear Teacher, I am emailing my information frame showing the characteristics of circles."

Sample Answers

1.

Vocabulary

- *Interest* is money paid or earned for the use of money.
- *Principal* is the amount of money borrowed or deposited.

Definition

Interest earned on the principal *and* on the previously earned interest.

Compound Interest

Algebra

Interest Compounded Annually

$B = P(1 + r)^t$

Balance (B); Principal (P); Annual interest rate (in decimal form) (r); Time (in years) (t)

Example

You deposit $1000 in an account that earns 3% interest compounded annually. What is the account balance after 4 years?

$B = P(1 + r)^t = 1000(1 + 0.03)^4 \approx \$1125.51$

2.

Vocabulary

An *installment loan* is paid back in equal increments over the term of the loan.

Definition

Money borrowed for a specified amout of time, called the *term* of the loan.

Loan

Vocabulary Help

- The lender may add interest to the principal of the loan, increasing the total amount due.
- Due to rounding, the last payment of an installment loan may be a few cents more or less than the monthly payment.

Example: A $1000 loan had a term of 3 months, an annual interest rate of 4%, and a monthly payment of $335.56. The amount of interest paid on the loan is $3.33 + $2.23 + $1.11 = $6.67.

Balance Before Payment	Monthly Payment	Monthly Interest	Balance After Payment
$1000	$335.56	$\$1000\left(\frac{0.04}{12}\right) \approx \$3.33$	$1000 − 335.56 + 3.33 = $667.77
$667.77	$335.56	$\$667.77\left(\frac{0.04}{12}\right) \approx \$2.23$	$667.77 − 335.56 + 2.23 = $334.44
$334.44	$335.55	$\$334.44\left(\frac{0.04}{12}\right) \approx \$1.11$	$334.44 − 335.55 + 1.11 = $0.00

3.

Vocabulary

- A *loan* is money borrowed for a specific amount of time, called the *term* of the loan.
- An *installment loan* is a loan that is paid back in equal increments over the term of the loan.

Definition

An installment loan that is used to buy a house.

Mortgage

Vocabulary Help

The term of a mortgage is usually greater than the terms of other types of loans.

Example

A $200,000 mortgage has a 20-year term, an annual interest rate of 5%, and a monthly payment of $1319.91. How much interest is paid on the mortgage?

$I = 240(1319.91) - 200{,}000 = \$116{,}778.40$

4. Available at *BigIdeasMath.com*.

List of Organizers

Available at *BigIdeasMath.com*

- Comparison Chart
- Concept Circle
- Definition (Idea) and Example Chart
- Example and Non-Example Chart
- Formula Triangle
- Four Square
- **Information Frame**
- Information Wheel
- Notetaking Organizer
- Process Diagram
- Summary Triangle
- Word Magnet
- Y Chart

About this Organizer

An **Information Frame** can be used to help students organize and remember concepts. Students write the topic in the middle rectangle. Then students write related concepts in the spaces around the rectangle. Related concepts can include *Words, Numbers, Algebra, Example, Non-Example, Formula, Definition, Visual, Procedure, Details, Solution,* and *Vocabulary.* Students can place their information frames on note cards to use as a quick study reference.

Technology For the Teacher

Vocabulary Puzzle Builder

Answers

1. See Additional Answers.
2. a. $1297.92

   b. $97.92
3. $5000
4. $75.34
5. a. 58.12; −6.41; 617.09; 467.09; 424.30; 99.30; The ending balance is $99.30.

   b. Because Check #315 is close to the end of the month, it likely was not processed in time to appear on the monthly statement.

   c. check #312
6. $209,262
7. $60.75

Assessment Book

Appendix B Quiz
For use after Section B.4

1. You deposit $1600 in a savings account that earns 3.5% simple interest per year. Your friend deposits $1200 in a savings account that earns 6% simple interest per year.
   a. Write two equations for the balance B in each account after t years.
   b. Are the account balances ever equal? Explain.
2. You deposit $900 in a savings account that earns 5.5% interest compounded annually.
   a. What is the balance after 3 years?
   b. What is the interest earned?
3. An account earns 3% interest compounded annually. The balance after 4 years is $1125.51. What is the principal?
4. A loan summary is shown. How much interest is paid on the loan?

Loan Amount	$1660.00
Monthly Payment	$286.43
Annual Interest Rate	12%
Term	6 months

5. You balance your checkbook at the end of the month.
   a. Complete the balance column. What is the ending balance? (Assume there is no fee for a bad check.)

Date	Check #	Transaction	Credit	Debit	Balance
		Balance Forward			101.75
10/6/12	101	Gas Station		41.25	
10/15/12		Deposit Paycheck	740.50		
10/21/12	102	Grocery Store		53.26	
10/26/12	103	Electric Company		48.31	

   b. Your monthly bank statement shows a balance of $747.74. Explain why your balance does not agree with the bank statement.
6. The principal on a home mortgage is $200,000, the monthly payment is $1025.75, and the term is 30 years. How much interest is paid on the loan?
7. Your brother borrows $220 from you for a new bicycle. Your brother pays you back the principal plus 10% simple interest per year in 2 years. How much money do you earn?

Answers

1. a. ______
   ______
   b. ______
   ______
2. a. ______
   b. ______
3. ______
4. ______
5. a. ______
   b. ______
   ______
   ______
6. ______
7. ______

Alternative Quiz Ideas

100% Quiz
Error Notebook
Group Quiz
Homework Quiz
Math Log
Notebook Quiz
Partner Quiz
Pass the Paper

100% Quiz
This is a quiz where students are given the answers and then they have to explain and justify each answer.

Reteaching and Enrichment Strategies

If students need help. . .	If students got it. . .
Resources by Chapter • Study Help • Practice A and Practice B • Puzzle Time Lesson Tutorials *BigIdeasMath.com* Practice Quiz Practice from the Test Generator	Resources by Chapter • Enrichment and Extension • School-to-Work Game Closet at *BigIdeasMath.com* Start the next section

Technology For the Teacher
Answer Presentation Tool
Big Ideas Test Generator

B.1–B.4 Quiz

Check It Out
Progress Check
BigIdeasMath.com

1. You deposit $1200 in a savings account that earns 2.5% simple interest per year. Your friend deposits $1000 in a savings account that earns 5% simple interest per year. *(Section B.1)*

   a. Write and graph two equations for the balance B in each account after t years.

   b. Are the account balances ever equal? Explain.

2. You deposit $1200 in a savings account that earns 4% interest compounded annually. *(Section B.2)*

   a. What is the balance after 2 years?

   b. What is the interest earned?

3. An account earns 2% interest compounded annually. The balance after 3 years is $5306.04. What is the principal? *(Section B.2)*

4. A loan summary is shown. How much interest is paid on the loan? *(Section B.3)*

LOAN AMOUNT	$2135.00
MONTHLY PAYMENT	$368.39
ANNUAL INTEREST RATE	12%
TERM	6 months

5. You balance your checkbook at the end of the month. *(Section B.4)*

   a. Complete the balance column. What is the ending balance? (Assume there is no fee for a bad check.)

Date	Check #	Transaction	Credit	Debit	Balance
		Balance Forward			98.12
11/1/12		ATM Withdrawal		40.00	
11/5/12	312	Grocery Store		64.53	
11/15/12		Deposit Paycheck	623.50		
11/16/12	313	Credit Card Payment		150.00	
11/21/12	314	Electric Company		42.79	
11/28/12	315	Rent		325.00	

   b. Your monthly bank statement shows a balance of $424.30. Explain why your balance does not agree with the bank statement.

   c. Which check is the bad check?

6. **MORTGAGE** The principal on a home mortgage is $150,000, the monthly payment is $997.95, and the term is 30 years. How much interest is paid on the loan? *(Section B.3)*

7. **KEYBOARD** Your friend borrows $450 from you for a new electronic keyboard. Your friend pays you back the principal plus 9% simple interest per year in 1.5 years. How much money do you earn? *(Section B.1)*

B.5 Credit Cards

Essential Question How do credit cards work?

1 ACTIVITY: Running a Business

Work with a partner. You own a shop that sells greeting cards. A customer uses a credit card to pay the total shown on the receipt.

a. How much will the credit card company deposit in your bank account?

b. Of the amount in part (a), how much do you have to send to the state and county?

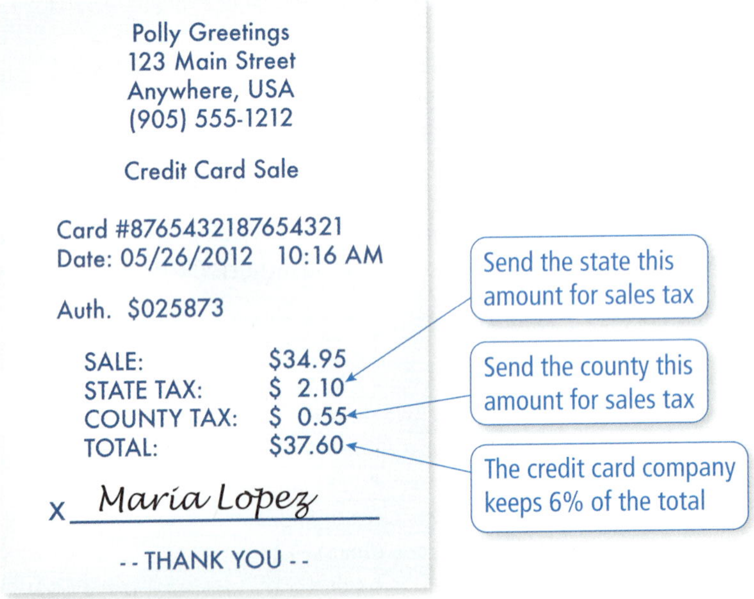

Polly Greetings
123 Main Street
Anywhere, USA
(905) 555-1212

Credit Card Sale

Card #8765432187654321
Date: 05/26/2012 10:16 AM

Auth. $025873

SALE:	$34.95
STATE TAX:	$ 2.10
COUNTY TAX:	$ 0.55
TOTAL:	$37.60

X *Maria Lopez*

-- THANK YOU --

2 ACTIVITY: Credit Rating

RESEARCH PROJECT **Your credit score is a number from 300 to 850 that rates your credit history. Write a report on the types of things that increase your credit score and also the types of things that decrease it.**

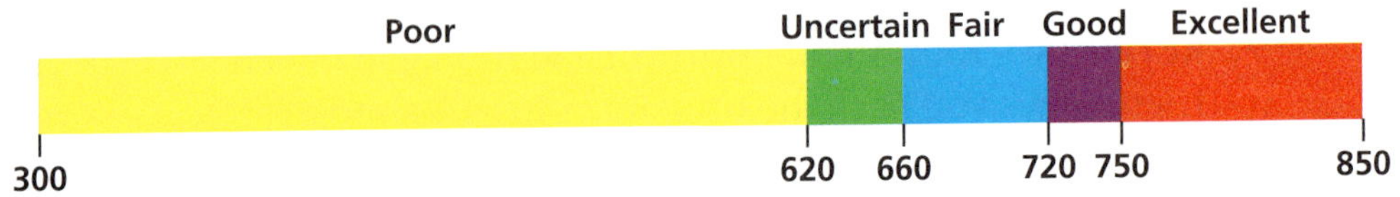

Laurie's Notes

Introduction

For the Teacher

- **Goal:** Students will explore the use of a credit card.

Motivate

- Ask students what they know about credit cards and how they work.
- Discuss with students the wise use of credit cards. You want to share information with students so that they may make informed decisions.
- You do *not* want students sharing personal information about their families.
- Credit cards provide a convenient way to pay bills and purchase goods.
- **Caution:** The real danger is spending money you don't have. It becomes a loan at a much higher interest rate than a conventional bank loan.

? "How does a credit card company make money?" Most students will say from the interest paid by the consumer.

? "If every consumer paid their balance each month, would the credit card company still make money?" yes; Credit card companies charge businesses a percent of the total sales to process the transaction.

Activity Notes

Activity 1

? "When Maria receives her monthly credit card statement, what amount will show on her statement for this transaction? Is this fair? Explain." \$37.60; yes; It is the amount she would have paid in cash.

? "If Maria pays the full amount of her monthly credit card statement, what does the transaction cost her?" \$37.60

? "How much does the business receive from the transaction when the credit card company pays them? 94% of \$37.60 = \$35.34.

- Your business needs to send the state and county the money collected for sales tax. The credit card company kept 6% of the sales tax. So, the business received less than the 94% of Maria's bill.

$$\$35.34 - (\$2.10 + \$0.55) = \$32.69$$

? "What percent of \$34.95 is \$32.69?" ≈ 93.5%

? "Why would businesses allow customers to use credit cards if they don't receive 100% of the purchase price?" *Sample answers:* cost of doing business; consumer might not make the purchase if it couldn't be charged; consumer might buy more if they don't have to pay in cash

Activity 2

- Your credit history tracks credit card accounts, loans, and other financial accounts.
- Creditors and lenders update your credit history based on your financial activity, such as late payments.
- Discuss the advantages of a good credit rating and the disadvantages of a poor credit rating.

Previous Learning

Students should know how to solve percent problems.

Activity Materials
Textbook
• calculator

Start Thinking! and Warm Up

Activity B.5 Start Thinking!
For use before Activity B.5

Activity B.5 Warm Up
For use before Activity B.5

Find the mean of the numbers.

1. 5, 5, 7, 3, 8, 1, 6
2. 10, 18, 16, 18, 29, 15, 14, 15
3. 1, 1, 1, 1, 4, 5, 5, 6, 7, 8
4. 350, 350, 350, 350, 340, 300
5. 12, 15, 17, 15, 16, 19, 21, 20
6. 1.5, 3.5, 2.5, 9.6, 1.2, 4.4, 6.5, 3.4

B.5 Record and Practice Journal

Essential Question How do credit cards work?

1 ACTIVITY: Running a Business

Work with a partner. You own a shop that sells greeting cards. A customer uses a credit card to pay the total shown on the receipt.

a. How much will the credit card company deposit in your bank account?

\$35.34

Polly Greetings
123 Main Street
Anywhere, USA
(905) 555-1212

Credit Card Sale

Card #8765432187654321
Date: 05/26/2012 10:16 AM

Auth. S025873

SALE: \$34.95
STATE TAX: \$ 2.10
COUNTY TAX: \$ 0.55
TOTAL: \$37.60

x Maria Lopez

-- THANK YOU --

Send the state this amount for sales tax

Send the county this amount for sales tax

The credit card company keeps 6% of the total

b. Of the amount in part (a), how much do you have to send to the state and county?

\$2.65

2 ACTIVITY: Credit Rating

RESEARCH PROJECT Your credit score is a number from 300 to 850 that rates your credit history. Write a report on the types of things that increase your credit score and also the types of things that decrease it.

Check students' work.

English Language Learners

Vocabulary

Display on the board or overhead the following forms of division to reinforce the key words in division. Have students make a copy of these in their notebooks.

$$\text{dividend} \div \text{divisor} = \text{quotient}$$

$$\text{divisor}\overline{)\text{dividend}}\ \ \text{(quotient above)}$$

$$\frac{\text{dividend}}{\text{divisor}} = \text{quotient}$$

B.5 Record and Practice Journal

3 ACTIVITY: Credit Card Bills

Work with a partner. Maria paid $140.00 of her previous month's credit card balance of $490.00. Her credit card company charged her interest on the average daily balance for the current month.

The average daily balance for the first week is

$$\frac{490.00 + 490.00 + 490.00 + 350.00 + 350.00 + 350.00 + 350.00}{7} = 410.00.$$

a. Complete the table for the entire month of May to find the average daily balance for the month. **about $370.83**

Date	Activity Since Last Statement	Amount	Balance
05/01/2012			490.00
05/02/2012			490.00
05/03/2012			490.00
05/04/2012	Payment	–140.00	350.00
05/05/2012			350.00
⋮	⋮	⋮	⋮
05/25/2012			350.00
05/26/2012	Polly Greetings	37.60	387.60
05/27/2012			387.60
05/28/2012			387.60
05/29/2012			387.60
05/30/2012			387.60
05/31/2012			387.60

b. Maria's credit card company charges her $5.56 interest on her average daily balance for the month. What is her monthly interest rate?

about 1.5%

c. Find the average daily balance for the month without the purchase at Polly Greetings. Use the interest rate you found in part (b) to calculate the interest charge.

$363.55; $5.45

d. What does the difference between the interest charges in parts (b) and (c) represent?

interest paid for purchase at Polly Greetings

What Is Your Answer?

4. **IN YOUR OWN WORDS** How do credit cards work?

Credit card company keeps a percentage of total and sends remaining amount to the seller. Seller pays taxes due on purchase.

Laurie's Notes

Activity 3

- You are being charged interest on your credit card balance. If you only make one purchase, it seems fair that it should cost you less making the purchase at the end of the month than at the beginning of the month. This is not a simple idea. Take time to work through the math.
- Work through the calculation with your students. For 3 days the balance is $490, for the next 22 days the balance is $350, and for the last 6 days the balance is $387.60.

$$\text{Average Daily Balance} = \frac{(3)490 + (22)350 + (6)387.60}{31} \approx \$370.83$$

? "What percent of $370.83 is $5.56?" ≈ 0.015 = 1.5% monthly (which is 18% annually)

- In part (c), without the purchase there are 28 days at $350.

$$\text{Average Daily Balance} = \frac{(3)490 + (28)350}{31} \approx \$363.55$$

$$\text{Interest charge} = 1.5\% \text{ of } \$363.55 = 0.015 \times \$363.55 \approx \$5.45$$

- The difference between the two interest charges is $0.11, which represents 6 days of interest on the purchase at Polly Greetings.

What Is Your Answer?

- **Think-Pair-Share:** Students should read the question independently and then work with a partner to answer the question. When they have answered the question, the pair should compare their answer with another group and discuss any discrepancies.

Closure

- **Exit Ticket:** How do you avoid interest payments on your credit card? Pay the balance in full and on time each month.

3 ACTIVITY: Credit Card Bills

Work with a partner. Maria paid $140.00 of her previous month's credit card balance of $490.00. Her credit card company charged her interest on the average daily balance for the current month.

The average daily balance for the first week is

$$\frac{490.00 + 490.00 + 490.00 + 350.00 + 350.00 + 350.00 + 350.00}{7} = 410.00.$$

a. Copy and complete the table for the entire month of May to find the average daily balance for the month.

Date	Activity Since Last Statement	Amount	Balance
05/01/2012			490.00
05/02/2012			490.00
05/03/2012			490.00
05/04/2012	Payment	−140.00	350.00
05/05/2012			350.00
⋮	⋮	⋮	⋮
05/25/2012			350.00
05/26/2012	Polly Greetings	37.60	
05/27/2012			
05/28/2012			
05/29/2012			
05/30/2012			
05/31/2012			

b. Maria's credit card company charges her $5.56 interest on her average daily balance for the month. What is her monthly interest rate?

c. Find the average daily balance for the month without the purchase at Polly Greetings. Use the interest rate you found in part (b) to calculate the interest charge.

d. What does the difference between the interest charges in parts (b) and (c) represent?

What Is Your Answer?

4. IN YOUR OWN WORDS How do credit cards work?

Practice Use what you learned about credit cards to complete Exercises 3–5 on page A40.

B.5 Lesson

Check It Out
Lesson Tutorials
BigIdeasMath.com

Key Vocabulary
credit card, *p. A38*

A **credit card** is used to postpone payment on a purchase. To allow this, the credit card company keeps a percentage of the total and sends the remaining amount to the seller. The seller pays the taxes due on the purchase.

EXAMPLE 1 Analyzing a Credit Card Transaction

Louis uses his credit card to buy CDs and guitar strings at a music store. The credit card company keeps 6% of the total.

Caja de Música

Louis A. Cardholder
Credit card # xxxx xxxx xxxx 8080
5:19 P.M. 08/10/12

ITEM	AMOUNT
2 CDs	22.98
Pkg Guitar Strings	12.05
SUBTOTAL	35.03
7% State Tax	2.45
TOTAL	**37.48**

-THANK YOU-
Customer Copy

a. How much money does the credit card company keep from the transaction?

The credit card company keeps 6% of the total. Find 6% of \$37.48.

$$6\% \text{ of } \$37.48 = 0.06 \cdot 37.48$$
$$\approx 2.25$$

∴ The credit card company keeps \$2.25.

b. How much money does the music store receive from the credit card company?

The music store receives the remaining amount.

∴ So, the music store receives \$37.48 − \$2.25 = \$35.23.

c. How much money does the music store keep from the transaction?

The music store sends the state sales tax to the state.

∴ So, the music store keeps \$35.23 − \$2.45 = \$32.78.

On Your Own

Now You're Ready
Exercises 6–9

1. **WHAT IF?** In Example 1, Louis charges \$90.00, including sales tax, to his credit card. How much money does the music store receive from the credit card company?

Laurie's Notes

Introduction

Connect

- **Yesterday:** Students explored the use of a credit card.
- **Today:** Students will analyze the cost of having a credit card.

Motivate

- Share additional information about credit cards with your students.
  - Many credit cards have an annual fee associated with them. A typical fee might be $50 per year.
  - The interest rates charged on unpaid balances does vary by company. I have received different offers in the mail from credit card companies where the annual interest rate ranges from 14% to 24%.
  - When you do not pay the monthly balance in full, the interest rate is applied toward the average daily balance and will appear on the statement the following month.

Lesson Notes

Example 1

- This example is similar to Activity 1. It helps to explain how credit cards work.
- Discuss information shown on the receipt. Work through the problem as shown.

On Your Own

- Remind students that the credit card company keeps 6%.

Goal Today's lesson is understanding how a **credit card** works.

Lesson Materials
Textbook
• calculator

Start Thinking! and Warm Up

Lesson B.5 **Warm Up**
For use before Lesson B.5

Lesson B.5 **Start Thinking!**
For use before Lesson B.5

Who gets paid when you pay with a credit card?

Extra Example 1

In Example 1, Louis charges $120.00, including sales tax, to his credit card. How much money does the credit card company keep from the transaction? $7.20

On Your Own

1. $84.60

Laurie's Notes

Discuss

- Discuss payments made on credit cards.
- Students need to understand that if you make your payment in full, and on time, the credit card does not cost you any additional money to use beyond the annual fee.

Example 2

- Read through the credit card statement, explaining information contained on the statement.
- Continue to solve the problem as shown.

? "How do you write 1.5% as a decimal?" 0.015

? "If the monthly interest rate is 1.5%, what is the annual interest rate?" 18%

On Your Own

- **Think-Pair-Share:** Students should read the question independently and then work with a partner to answer the question. When they have answered the question, the pair should compare their answer with another group and discuss any discrepancies.

Closure

- You have an average daily balance of $1580.24 from the previous month's statement. What will be the interest charge at a monthly rate of 1.5%? $23.70

Extra Example 2

In Example 2, if Louis had paid only $300.00 on last month's bill, the average daily balance would be $735.37. Find the interest charge and the new balance. $11.03; $811.03

On Your Own

2. $13.69; $1063.69

Differentiated Instruction

Visual

Help students that have difficulty with rounding by breaking the process down into steps.

Round 3.548 to the nearest tenth.

3.548	**Step 1** Underscore the place value to which you are rounding.
3.548	**Step 2** Circle the number to the right.
3.5	**Step 3** If the value of the circled number is 0 to 4, keep the underscored value as is. If the value of the circled number is 5 to 9, increase the value of the underscored number by 1.

Technology For the Teacher

Dynamic Classroom

The Dynamic Planning Tool

Editable Teacher's Resources at *BigIdeasMath.com*

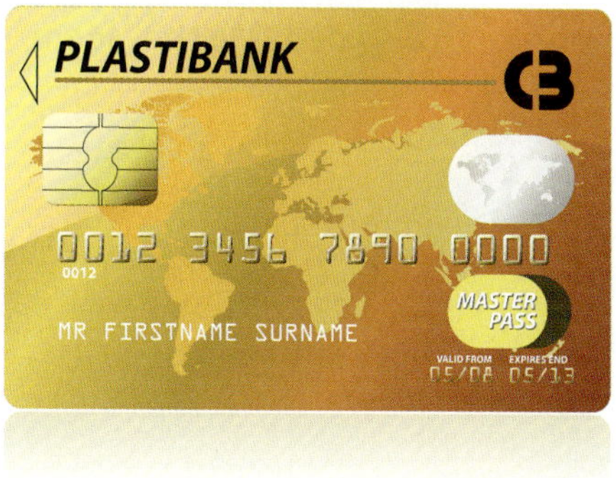

Credit card owners are sent a monthly bill. The new balance shown on the bill may consist of unpaid balances, new purchases, interest, or other fees.

Many credit card companies charge interest based on the average daily balance on the account during a statement period. The interest charge is the product of the monthly interest rate and the average daily balance.

EXAMPLE 2 Paying a Credit Card Bill

Louis receives a credit card bill in the mail. His monthly interest rate is 1.5%. Find the interest charge and the new balance.

Credit Card Statement

Account Number XXXX XXXX XXXX 8080

Date	Activity Since Last Statement	Amount
08/10/12	Online Payment	-600.00
	---- **Purchases** ----	
08/02/12	Gas Station	59.61
08/06/12	Cell Phone Company	58.86
08/07/12	Restaurant	28.65
08/10/12	Caja de Música	37.48
08/15/12	Electric Bill	50.47
08/20/12	Car Repair Shop	164.93

Average Daily Balance	$522.47
Statement Closing Date	09/01/12
Previous Balance	$700.00
Payments	-600.00
Unpaid Balance	$100.00
New Purchases	$400.00
Interest Charge	
New Balance	

Find the product of the monthly interest rate and the average daily balance.

$$0.015 \cdot 522.47 \approx \$7.84$$

The previous balance was not fully paid. So, the new balance is the sum of the unpaid balance, the new purchases, and the interest charge.

$$\$100.00 + \$400.00 + \$7.84 = \$507.84$$

So, the interest charge is \$7.84 and the new balance is \$507.84.

On Your Own

Now You're Ready
Exercise 10

2. **WHAT IF?** In Example 2, if Louis had paid only \$50.00 on last month's bill, the average daily balance would be \$912.79. Find the interest charge and the new balance.

B.5 Exercises

Check It Out
Help with Homework
BigIdeasMath.com

Vocabulary and Concept Check

1. **VOCABULARY** How is a credit card similar to a debit card? How is a credit card similar to a loan? Explain.
2. **REASONING** Why is it important to know the annual interest rate of a credit card?

Practice and Problem Solving

You own a bookstore. A customer uses a credit card to pay the total shown on the receipt.

3. How much money is sent to the state?
4. How much money is sent to the county?
5. The credit card company keeps 5% of the total. How much money does the credit card company keep?

SALE:	$45.00
STATE TAX:	$ 2.70
COUNTY TAX:	$ 0.72
TOTAL:	$48.42

In Exercises 6–9, a credit card company keeps the given percent of the total. Find the amount of money (a) the credit card company keeps from the transaction, (b) the store receives from the credit card company, and (c) the store keeps from the transaction.

1 6. 4%

Office Supply Store
Maria Lopez
Credit card # xxxx xxxx xxxx 4321
11:24 A.M. 05/13/11

SALE:	$62.00
SALES TAX:	$ 4.03
TOTAL:	**$66.03**

-THANK YOU-

7. 3%

ROY'S HARDWARE Store
Michael Jones
Credit card # xxxx xxxx xxxx 7914
9:45 A.M. 06/02/12

SALE:	$105.80
SALES TAX:	$ 5.29
TOTAL:	**$111.09**

THANK YOU

8. 3.5%

Pretzel Shop
Gustav Tanner
Credit card # xxxx xxxx xxxx 5059
1:24 P.M. 08/24/11

SALE:	$8.25
SALES TAX:	$0.50
TOTAL:	**$8.75**

THANK YOU

9. 5%

Arts and Crafts Outlet
Gloria Moore
Credit card # xxxx xxxx xxxx 6338
3:12 P.M. 11/30/12

SALE:	$75.00
SALES TAX:	$ 3.75
TOTAL:	**$78.75**

-THANK YOU-

Assignment Guide and Homework Check

Level	Day 1 Activity Assignment	Day 2 Lesson Assignment	Homework Check
Basic	3–5, 14–18	1, 2, 7, 9–12	2, 7, 10, 11
Average	3–5, 14–18	1, 2, 7, 9–12	2, 7, 10, 11
Advanced	3–5, 14–18	1, 2, 8, 10–13	2, 8, 10, 11

For Your Information

- **Exercise 13** This exercise assumes that the bill is being paid on the closing date so that the average daily balance each month is a simple calculation.

Common Errors

- **Exercises 6–9** Students may forget to convert the percent to a decimal when finding how much money the credit card company keeps from the transaction. Remind them that to convert a percent to a decimal, they divide by 100.
- **Exercise 10** Students may use the total amount due to find the interest instead of the average daily balance. Remind them that the interest charged by the credit card company is calculated from the average daily balance.

Vocabulary and Concept Check

1. *Sample answer:* You can use both cards to make a purchase, instead of using cash or a check.
   Sample answer: A credit card is similar to a loan in that payment is postponed and interest is charged.
2. So you can calculate the monthly interest rate that is applied to your average daily balance. This product is the interest charge.

Practice and Problem Solving

3. $2.70
4. $0.72
5. $2.42
6. **a.** $2.64
   b. $63.39
   c. $59.36
7. **a.** $3.33
   b. $107.76
   c. $102.47
8. **a.** $0.31
   b. $8.44
   c. $7.94
9. **a.** $3.94
   b. $74.81
   c. $71.06

B.5 Record and Practice Journal

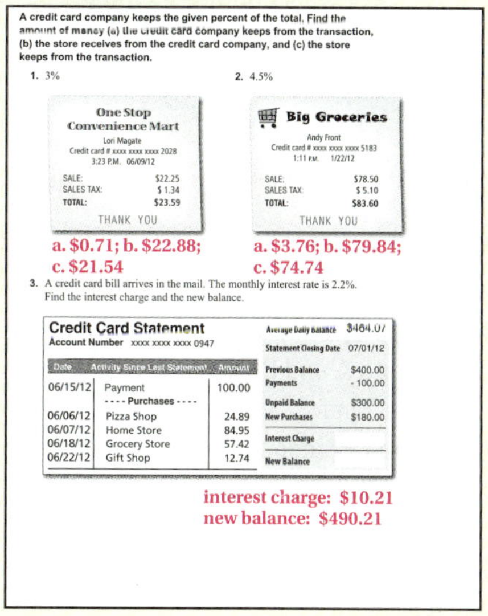

A credit card company keeps the given percent of the total. Find the amount of money (a) the credit card company keeps from the transaction, (b) the store receives from the credit card company, and (c) the store keeps from the transaction.

1. 3%

One Stop Convenience Mart
Lori Magate
Credit card # xxxx xxxx xxxx 2028
3:23 P.M. 06/09/12
SALE: $22.25
SALES TAX: $ 1.34
TOTAL: $23.59
THANK YOU

a. $0.71; b. $22.88; c. $21.54

2. 4.5%

Big Groceries
Andy Front
Credit card # xxxx xxxx xxxx 5183
1:11 P.M. 1/22/12
SALE: $78.50
SALES TAX: $ 5.10
TOTAL: $83.60
THANK YOU

a. $3.76; b. $79.84; c. $74.74

3. A credit card bill arrives in the mail. The monthly interest rate is 2.2%. Find the interest charge and the new balance.

Credit Card Statement
Account Number xxxx xxxx xxxx 0947

Date	Activity Since Last Statement	Amount
06/15/12	Payment	100.00
	---- Purchases ----	
06/06/12	Pizza Shop	24.89
06/07/12	Home Store	84.95
06/18/12	Grocery Store	57.42
06/22/12	Gift Shop	12.74

Average Daily Balance	$464.07
Statement Closing Date	07/01/12
Previous Balance	$400.00
Payments	- 100.00
Unpaid Balance	$300.00
New Purchases	$180.00
Interest Charge	
New Balance	

interest charge: $10.21
new balance: $490.21

Technology For the Teacher
Answer Presentation Tool
Quiz*Show*

Practice and Problem Solving

10. \$6.91; \$556.91

11. See *Taking Math Deeper.*

12. You don't have to pay interest.

13. \$740.98; \$240.00; \$59.02; The amount you paid is more than four times the decrease in the amount due. You end up decreasing your balance by about \$5 each month because of the interest. You should pay more than the minimum so that more money goes toward decreasing the balance, rather than paying mostly interest.

Fair Game Review

14. $4\sqrt{3}$ **15.** 12

16. $\frac{\sqrt{10}}{7}$ **17.** $6\sqrt{19}$

18. C

Mini-Assessment

A credit card company keeps 4.5% of the total. Find the amount of money (a) the credit card company keeps from the transaction, (b) the store receives from the credit card company, and (c) the store keeps from the transaction.

1. Sale: \$49.50

   Sales Tax: \$2.97

   Total: \$52.47

   (a) \$2.36; (b) \$50.11; (c) \$47.14

2. Sale: \$14.99

   Sales Tax: \$0.97

   Total: \$15.96

   (a) \$0.72; (b) \$15.24; (c) \$14.27

Taking Math Deeper

Exercise 11

One thing that students might not realize is that credit card companies make money in two basic ways. (1) They charge fees and interest to the credit card holder. (2) They also charge a fee to the vendor. This fee varies from 2% up to 6% of the purchase price.

1 How much does the store keep?

	Charge	\$100.00
	Sales Tax	\$ 6.00
	Total	\$106.00
	Fee	−\$ 2.12
	Vendor Gets	\$103.88
	Sales Tax	−\$ 6.00
a.	Vendor Keeps	\$ 97.88

2 What percent of the \$106.00 does the store keep?

b. The store keeps about 92.3% of the transaction.

3 Can the store keep \$100 from the sale by diverting the 2% credit card fee to the buyer?

Charge	\$100.00
CC Fee	\$ 2.00
Sales Tax	\$ 6.12
Total	\$108.12
Fee	−\$ 2.16
Vendor Gets	\$105.96
Sales Tax	−\$ 6.12
Vendor Keeps	\$ 99.84

c. Almost, but not quite! With this plan, the store keeps \$99.84.

Project

Research the differences in credit card companies. Compare the rates of interest companies charge. What things should be considered before using the credit card to pay for merchandise? What are some things that you think should never be charged? Why?

Reteaching and Enrichment Strategies

If students need help. . .	If students got it. . .
Resources by Chapter • Practice A and Practice B • Puzzle Time Record and Practice Journal Practice Differentiating the Lesson Lesson Tutorials Skills Review Handbook	Resources by Chapter • Enrichment and Extension • School-to-Work • Financial Literacy Start the next section

2 **10.** A credit card bill arrives in the mail. The monthly interest rate is 1.25%. Find the interest charge and the new balance.

Credit Card Statement

Account Number XXXX XXXX XXXX 0947

Date	Activity Since Last Statement	Amount
02/15/11	Payment	250.00
	- - - - Purchases - - - -	
02/03/11	Town Mall	75.08
02/07/11	Grocery Store	92.65
02/13/11	Costume Shop	18.33
02/20/11	Camping Store	48.50
02/25/11	Shoe Outlet	65.44

Average Daily Balance	$552.93
Statement Closing Date	03/01/11
Previous Balance	$500.00
Payments	-250.00
Unpaid Balance	$250.00
New Purchases	$300.00
Interest Charge	
New Balance	

11. **CLOTHING** You use a credit card to buy $100 worth of merchandise at a clothing store. The sales tax is 6%. The credit card company keeps 2% of the total.

a. How much money does the store keep from the transaction?

b. About what percent of the transaction does the store keep?

c. The clothing store decides to charge a 2% fee for using a credit card. The fee is added to the purchase price before taxes are calculated. Does the store keep $100 from the sale now? Explain.

12. **LAYAWAY** *Layaway* is a payment method used to purchase an item without paying the entire cost at one time. A customer pays a deposit to put an item "on layaway" for a given period of time. The store reserves the item until it is paid for in full. What is an advantage of putting an item "on layaway" compared to using a credit card?

13. Number Sense The statement closing date on your credit card is the first of each month. Your new balance on January 1 is $800. You stop making purchases with the card and pay the minimum payment of $20 on the first of each month, starting January 1. The monthly interest rate is 2%. What is the balance one year later? How much have you paid after one year? How much did the amount due decrease? Explain why you should pay more than the minimum payment.

Fair Game Review What you learned in previous grades & lessons

Simplify the expression. *(Section 6.4)*

14. $6\sqrt{3} - 2\sqrt{3}$ **15.** $\sqrt{8} \cdot \sqrt{18}$ **16.** $\sqrt{\frac{10}{49}}$ **17.** $5\sqrt{19} + \sqrt{19}$

18. **MULTIPLE CHOICE** What is $\frac{30}{800}$ as a percent? *(Skills Review Handbook)*

Ⓐ 0.0375% Ⓑ 0.375% Ⓒ 3.75% Ⓓ 37.5%

B.6 Payroll Deductions

Essential Question How can you read a payroll earnings statement?

1 ACTIVITY: Reading a Payroll Earnings Statement

Work with a partner. Talk about each entry on the payroll statement. Do you understand each entry?

a. How much does John get paid per hour?

b. How often does John get paid?

c. How many paychecks has John received so far in 2012?

d. How is the gross pay related to the net pay?

e. How much does John pay for:

- federal income tax?
- Social Security tax?
- Medicare tax?

NAME
John Doe

EARNINGS STATEMENT

HOURS	RATE	AMOUNT
80	17.50	1400.00

YTD GROSS PAY	GROSS PAY
14,000.00	1400.00

PAY ENDING	PAY DATE	DEPARTMENT	NET PAY
05/18/12	05/18/12	312	983.87

PAYROLL TAX WITHHELD

FEDERAL	SOC. SEC.	HEALTH INS.	MEDICARE	401K
192.85	83.70	50.00	19.58	70.00

YEAR TO DATE

FEDERAL	SOC. SEC.	HEALTH INS.	MEDICARE	401K
1928.50	837.00	500.00	195.80	700.00

PAYROLL CHECK

EMPLOYEE ID NO.
876 54 3210

DEPT.	CHECK NO.	PAY DATE
312	12789	05/18/12

PAY EXACTLY ***Nine Hundred Eighty Three and 87/100*** DOLLARS

BANK NAME
CITY, STATE, ZIP CODE

AMOUNT
***983.87

PAY TO THE ORDER OF
John Doe
Street Address
Your City, USA

Susan Florencia

Laurie's Notes

Introduction

For the Teacher

- **Goal:** Students will explore a payroll statement and federal income tax.
- One of the biggest realizations for students when they take their first job is that there are payroll deductions. They don't get to keep all of the money that they earn. In this activity, students will explore various payroll deductions.

Motivate

- Ask students what they know about payroll deductions.
- One deduction that appears is FICA—Federal Insurance Contributions Act, better known as Social Security.
- The Social Security Act was made into law in 1935 following the Great Depression and under the Franklin D. Roosevelt Administration. Social security pays retirement benefits, disability benefits, and survivor's benefits.
- The Social Security program is funded through payroll deductions and is under review annually by politicians.

Activity Notes

Activity 1

- Give students time to read and discuss the payroll statement with their partner before having a class discussion.
- An hourly wage of $17.50 sounds like a lot of money to an 8th grade student.
- The standard work week is 40 hours (5 days × 8 hours per day). Most employees are paid weekly, bi-weekly (every two weeks), or monthly.
- You can assume that this is a 2-week pay statement.
- ? "How do you know the number of paychecks John has received so far?" *Sample answers:* Divide the *YTD* (year to date) *Gross Pay* by the weekly *Gross Pay;* Use the *Pay Ending* date to determine the number of pay periods.
- You may need to define gross pay (before deductions) and net pay (after deductions—your take home pay).
- ? "Who decides how much John pays for health insurance?" The $50 is likely a percent of the total cost of the insurance, with John's employer paying the rest. Insurance may be a negotiated benefit.
- ? "What does 401(k) refer to and who decides how much John pays?" 401(k) is a retirement plan that John elects to pay into.
- ? "What percent of his gross pay does John net?" $983.87 ÷ $1400 ≈ 70.3%
- ? "Do you think it is common to have deductions be 30% of your gross pay?" Answers will vary. Note to students that it could be higher than 30%!

Previous Learning

Students should know how to solve percent problems and how to make a scatter plot.

Activity Materials
Textbook
• calculator

Start Thinking! and Warm Up

Activity B.6 Start Thinking! For use before Activity B.6

Activity B.6 Warm Up For use before Activity B.6

Make a scatter plot of the data.

1.

x	0	1	2	3
y	0	2	4	6

2.

x	0	1	2	3
y	2	5	8	11

3.

x	0	1	2	3
y	5	3	1	–1

4.

x	0	1	2	3
y	–11	–7	–3	1

B.6 Record and Practice Journal

Essential Question How can you read a payroll earnings statement?

1 ACTIVITY: Reading a Payroll Earnings Statement

Work with a partner. Talk about each entry on the payroll statement. Do you understand each entry?

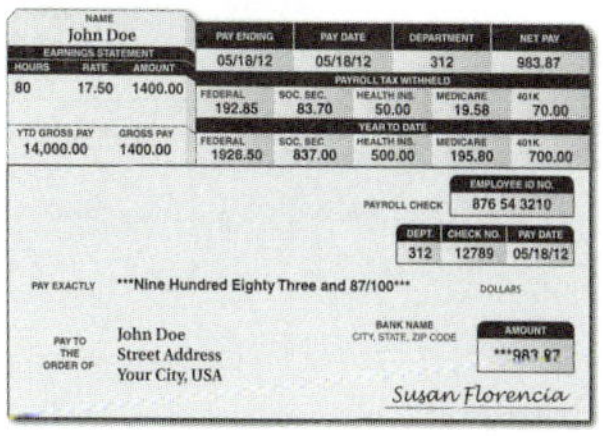

a. How much does John get paid per hour? $17.50

b. How often does John get paid? every 2 weeks

c. How many paychecks has John received so far in 2012? 10

d. How is the gross pay related to the net pay?
See Additional Answers.

English Language Learners

Vocabulary

Students should be warned not to drop the *trailing zeros* before placing the decimal point. Students need to include the zeros in the number of decimal places to be counted. Trailing zeros may be dropped after the decimal point has been placed.

$$\begin{array}{r} 3.25 \\ \times \quad 0.8 \\ \hline 2.600 \end{array}$$

$3.25 \times 0.8 = 2.6$

B.6 Record and Practice Journal

e. How much does John pay for:

- federal income tax? $192.85
- Social Security tax? $83.70
- Medicare tax? $19.58

2 ACTIVITY: Federal Income Tax Rates

Work with a partner. Use the table to find the federal income tax for 2008 for the given taxable income.

a. $7982.00 — $798.20

b. $26,500.00 — $3573.75

c. $54,780.00 — $10,038.75

d. $125,348.40 — $29,075.80

e. $235,893.70 — $63,595.67

f. $356,789.00 — $103,491.12

Filing Status: Single		
If taxable income is between...	**Tax bracket is:**	**Federal income tax is:**
$0 and $8,024	10%	10% of the income over $0
$8,025 and $32,549	15%	$802.50 plus 15% of the income over $8,025
$32,550 and $78,849	25%	$4,481.25 plus 25% of the income over $32,550
$78,850 and $164,549	28%	$16,056.25 plus 28% of the income over $78,850
$164,550 and $357,699	33%	$40,052.25 plus 33% of the income over $164,550
$357,700 and Above	35%	$103,791.75 plus 35% of the income over $357,700

3 ACTIVITY: Federal Income Tax Rates

Work with a partner.

a. Make a scatter plot of the data in Activity 2.

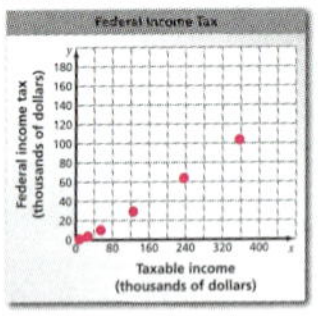

b. Federal income tax is called a *graduated tax*. What does this mean?

tax rate increases as taxable income increases

c. Graduated income tax graphs are not linear. Does the graph show a linear pattern?

no

What Is Your Answer?

4. **IN YOUR OWN WORDS** How can you read a payroll earnings statement?

it shows the gross pay, deductions, and net pay for a pay period and year to date

Laurie's Notes

Activity 2

- Explain the term *taxable income* before students begin. Roughly speaking, taxable income refers to the amount of income that is subject to income taxes. There are certain deductions that are exempt from tax.
- Your tax bracket is based on your yearly taxable income. The more you earn, the higher the percent of tax you pay.

? "How do you calculate a percent of a number?" Write the percent as a decimal and multiply by the number.

- Give students time to compute the percents.
- **Extension:** Compare the taxable income and federal income tax for a person with a taxable income of $358,780.00 and a person with a taxable income of $356,789.00 [part (f)].

	Taxable Income	Federal Income Tax
	$358,780.00	$104,169.75
f.	$356,789.00	$103,491.12
Difference	$ 1,991.00	$ 678.63

Activity 3

- Students will use the data computed in Activity 2 in order to make the scatter plot in this activity.
- Students may come up with an explanation for the meaning of *graduated tax*, if the tax bracket model is making sense to them. Essentially it means that all people who earn an amount of money within a certain range are all assessed the same percent of tax; then they *graduate* to the next tax bracket where the percent increases.
- If students have been careful with the graph, it won't be linear, although it appears to be close.

What Is Your Answer?

- **Neighbor Check:** Have students work independently and then have their neighbor check their work. Have students discuss any discrepancies.

Closure

- **Exit Ticket:** In Activity 1, assuming all of John Doe's income is taxable, what would his federal income tax be? taxable income: 26($1400) = $36,400;tax bracket: 25%;
federal income tax: $4481.25 + (0.25)(36,400 − 32,550) = $5443.75

Technology For the Teacher

Dynamic Classroom

The Dynamic Planning Tool
Editable Teacher's Resources at *BigIdeasMath.com*

2 ACTIVITY: Federal Income Tax Rates

Work with a partner. Use the table to find the federal income tax for 2008 for the given taxable income.

a. \$7982.00 **b.** \$26,500.00 **c.** \$54,780.00

d. \$125,348.40 **e.** \$235,893.70 **f.** \$356,789.00

Filing Status: Single		
If taxable income is between. . .	**Tax bracket is:**	**Federal income tax is:**
\$0 and \$8,024	10%	10% of the income over \$0
\$8,025 and \$32,549	15%	\$802.50 plus 15% of the income over \$8,025
\$32,550 and \$78,849	25%	\$4,481.25 plus 25% of the income over \$32,550
\$78,850 and \$164,549	28%	\$16,056.25 plus 28% of the income over \$78,850
\$164,550 and \$357,699	33%	\$40,052.25 plus 33% of the income over \$164,550
\$357,700 and Above	35%	\$103,791.75 plus 35% of the income over \$357,700

3 ACTIVITY: Federal Income Tax Rates

Work with a partner.

a. Make a scatter plot of the data in Activity 2.

b. Federal income tax is called a *graduated tax.* What does this mean?

c. Graduated income tax graphs are not linear. Does the graph show a linear pattern?

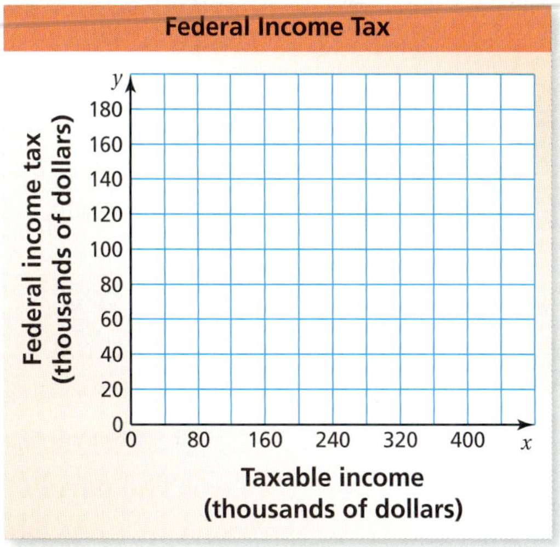

What Is Your Answer?

4. **IN YOUR OWN WORDS** How can you read a payroll earnings statement?

Practice Use what you learned about payroll deductions to complete Exercises 3–5 on page A46.

Check It Out
Lesson Tutorials
BigIdeasMath.com

Key Vocabulary
gross pay, *p. A44*
net pay, *p. A44*
payroll statement, *p. A44*

Study Tip
Other deductions, such as health insurance, may appear on a payroll statement.

The amount of money earned by an employee during a pay period is called **gross pay**. Several deductions are made from an employee's gross pay. Here are some of the deductions that are required by U.S. law.

- ***Federal income tax*** A tax based on an employee's earned income that is paid directly to the U.S. federal government.
- ***Social Security tax*** A tax used to fund the Social Security system, which provides benefits mainly for retired workers.
- ***Medicare tax*** A tax used to fund Medicare, which is a medical plan administered by the federal government mainly for people age 65 and older.

The amount of money that an employee receives after all the deductions have been made is called **net pay**. A **payroll statement** shows the gross pay, the deductions, and the net pay.

EXAMPLE 1 Reading a Payroll Statement

According to the payroll statement, how often does Cara get paid?

NAME
Cara Jimenez

EARNINGS STATEMENT

HOURS	RATE	AMOUNT
168	14.50	2436.00

YTD GROSS PAY	GROSS PAY
14,616.00	2436.00

PAY ENDING	PAY DATE	DEPARTMENT	NET PAY
06/30/12	06/30/12	M	1845.39

PAYROLL TAX WITHHELD

FEDERAL	SOC. SEC.	HEALTH INS.	MEDICARE	401K
170.94	143.55	120.75	33.57	121.80

YEAR TO DATE

FEDERAL	SOC. SEC.	HEALTH INS.	MEDICARE	401K
1025.64	861.30	724.50	201.42	730.80

PAYROLL CHECK

EMPLOYEE ID NO.
987 65 4320

DEPT.	CHECK NO.	PAY DATE
M	10259	06/30/12

PAY EXACTLY ***ONE THOUSAND EIGHT HUNDRED FORTY FIVE and 39/100*** DOLLARS

PAY TO THE ORDER OF
Cara Jimenez
314 Pi Circle
Mathville, USA

BANK NAME
CITY, STATE, ZIP CODE

AMOUNT
***1845.39

Matthew Pascal

The paycheck is for 168 hours and is dated for the end of June.

So, you can conclude that Cara is paid once per month.

On Your Own

Now You're Ready
Exercises 6–11

1. What is Cara's hourly wage?
2. What is the difference between Cara's gross pay and net pay?
3. How much does Cara pay annually for health insurance?

Multi-Language Glossary at BigIdeasMath.com.

Laurie's Notes

Introduction

Connect

- **Yesterday:** Students explored a payroll statement and federal income tax.
- **Today:** Students will analyze a monthly payroll statement.

Motivate

- Discuss with students the different ways in which a person's salary might be stated. salaried (like a teacher), commission (sales); job (contractor); royalty (entertainers and authors)

Lesson Notes

Discuss

- Discuss any vocabulary or concepts not previously discussed. *Gross pay* and *net pay* are certainly two concepts you want students to understand.

Example 1

- This example is similar to Activity 1. The difference is that it is a monthly statement.
- If time permits, you might ask students if they would want to have a monthly or a weekly pay check. Budgeting certainly becomes more critical for the monthly pay check!
- Discuss information shown on the payroll statement.
- Work through the problem as shown.

On Your Own

- Ask students to share their thinking when all have finished.

Goal Today's lesson is understanding payroll deductions.

Lesson Materials
Textbook
• calculator

Start Thinking! and Warm Up

Lesson B.6 Warm Up
For use before Lesson B.6

Lesson B.6 Start Thinking!
For use before Lesson B.6

What do you know about the following terms?

- Federal income tax
- Social Security
- Medicare
- 401(k)

Extra Example 1

In Example 1, how much does Cara put in her 401(k) annually? $1461.60

On Your Own

1. $14.50
2. $590.61
3. $1449.00

Laurie's Notes

Discuss

- The 401(k) is an optional retirement savings plan. It has a special purpose, to save money for your retirement.
- One deduction that does not appear is for automatic savings. I try not to preach, but I do discuss with students that they should set money aside on a regular basis. They need to know that it could be a regular deduction, which makes it easier to save.

Example 2

- Calculate the percent of gross pay put into the 401(k).

Discuss

- There are a great deal of vocabulary words and several big concepts (Social Security and Medicare) involved with payroll deductions.
- Assure students that they will have many different types of payroll statements over the course of their lifetime.
- They should feel comfortable asking questions of their employer or the business manager who makes out the statements. Errors can be made, so they need to read their payroll statements carefully and ask when they have a question—just like in math class.

Example 3

- Students should be familiar with finding what percent one number is of another number.
- **FYI:** In addition to what workers pay into Social Security, employers match this amount. Self-employed workers make the entire payment, roughly 12.4%.

On Your Own

- Ask students to share their work at the board while they share the process aloud.

Closure

- **Exit Ticket:** If Cara puts 8% of her gross pay in a 401(k), what amount will she have contributed after one year? $2338.56

Technology For the Teacher

Dynamic Classroom

The Dynamic Planning Tool
Editable Teacher's Resources at *BigIdeasMath.com*

Extra Example 2

In Example 2, Cara changes her 401(k) contribution to 7% of her gross pay. How much will Cara put in her 401(k) each month? $170.52

Extra Example 3

Your payroll statement shows the following amounts.

Gross Pay: $340.00
Federal Income Tax: $25.00
Social Security: $21.08
Health Insurance: $15.00
Medicare: $4.76
401(k): $17.00

What percent do you pay for federal income tax? about 8.1%

On Your Own

4. $194.88
5. 1.45%

Differentiated Instruction

Inclusion
Remind students that one difference between multiplying and dividing decimals is the placement of the decimal point. In multiplication, the decimal point is placed after the decimals are multiplied. In division, the placement of the decimal point is determined before dividing.

A 401(k) plan is an optional retirement savings plan offered by some companies to their employees. The amount an employee can contribute toward a 401(k) plan is limited by U.S. law. The amount of money in a 401(k) is not taxed until it is withdrawn.

EXAMPLE 2 Analyzing a Payroll Statement

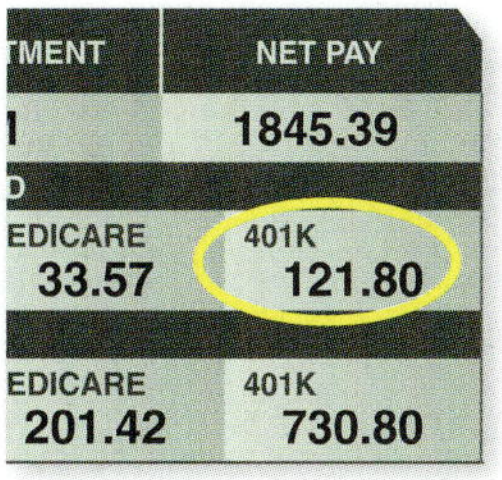

In Example 1, what percent of Cara's gross pay is put in her 401(k)?

Divide the amount put in her 401(k) by her gross pay.

$$\frac{\text{amount put in 401(k)}}{\text{gross pay}} = \frac{121.80}{2436.00} = 0.05 = 5\%$$

∴ 5% of her gross pay is put in her 401(k).

Social Security and Medicare taxes are calculated after the health insurance amount is subtracted from the gross pay. Federal income taxes are calculated after the health insurance amount *and* the 401(k) contribution are subtracted from the gross pay.

EXAMPLE 3 Finding Tax Percents

a. What percent does Cara pay for Social Security tax?

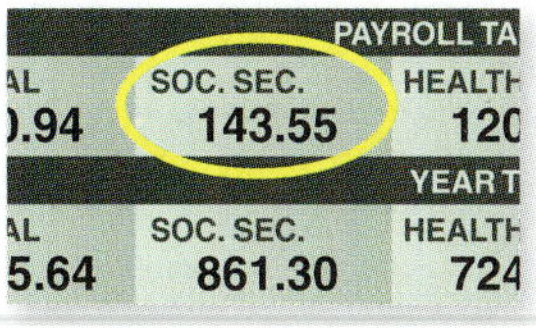

Divide the amount of Social Security tax by the difference of her gross pay and health insurance amount.

$$\frac{\text{Social Security tax}}{\text{gross pay} - \text{health insurance}} = \frac{143.55}{2436.00 - 120.75} = \frac{143.55}{2315.25}$$

$$\approx 0.062 = 6.2\%$$

∴ Cara pays 6.2% for Social Security tax.

b. What percent does Cara pay for federal income tax?

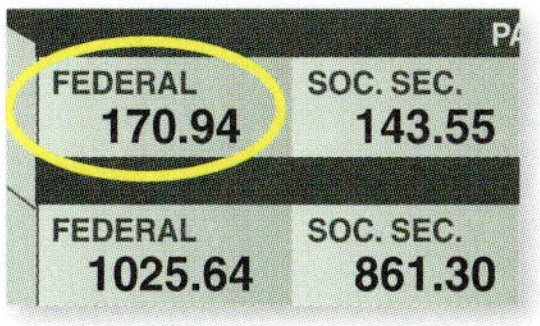

Divide the amount of federal income tax by the difference of her gross pay and the sum of the health insurance amount and 401(k) contribution.

$$\frac{\text{federal income tax}}{\text{gross pay} - [\text{health insurance} + 401(k)]} = \frac{170.94}{2436.00 - 242.55}$$

$$\approx 0.078 = 7.8\%$$

∴ Cara pays 7.8% for federal income tax.

On Your Own

Now You're Ready
Exercises 13–15

4. Cara changes her 401(k) contribution to 8% of her gross pay. How much will Cara put in her 401(k) each month?
5. What percent does Cara pay for Medicare tax?

B.6 Exercises

Check It Out
Help with Homework
BigIdeasMath.com

Vocabulary and Concept Check

1. **REASONING** Is the net pay *sometimes, always,* or *never* greater than the gross pay? Explain.
2. **REASONING** Why do you think the amount of money an employee can contribute toward a 401(k) plan is limited by U.S. law?

Practice and Problem Solving

In Exercises 3–5, use the table in Activity 2 to find the federal income tax for a person who had the given amount of taxable income.

3. \$10,500.00
4. \$58,435.00
5. \$6120.75

In Exercises 6–12, use the payroll statement.

NAME			PAY ENDING	PAY DATE	DEPARTMENT	NET PAY	
Al Smith			04/20/12	04/20/12	216	671.85	
EARNINGS STATEMENT			PAYROLL TAX WITHHELD				
HOURS	RATE	AMOUNT	FEDERAL	SOC. SEC.	HEALTH INS.	MEDICARE	401K
72	12.50	900.00	97.62	53.11	43.40	12.42	21.60
			YEAR TO DATE				
YTD GROSS PAY		GROSS PAY	FEDERAL	SOC. SEC.	HEALTH INS.	MEDICARE	401K
7200.00		900.00	780.96	424.88	347.20	99.36	172.80

PAYROLL CHECK

EMPLOYEE ID NO.
812 84 9628

DEPT.	CHECK NO.	PAY DATE
216	14320	04/20/12

PAY EXACTLY ***SIX HUNDRED SEVENTY ONE AND 85/100*** DOLLARS

PAY TO THE ORDER OF
Al Smith
983 Any Street
My Town, USA

BANK NAME
CITY, STATE, ZIP CODE

AMOUNT
***671.85

Louisa Fermat

1 6. What is Al's hourly wage?
7. How often does Al get paid?
8. What is Al's gross pay for April?
9. What is Al's annual net pay?
10. How many paychecks has Al received so far in 2012?
11. What is the difference between Al's gross pay and net pay?
12. **ERROR ANALYSIS** Describe and correct the error in finding Al's annual gross pay.

✗
Gross Pay: \$900.00
\$900.00 • 12 = \$10,800
His annual gross pay is \$10,800.

Assignment Guide and Homework Check

Level	Day 1 Activity Assignment	Day 2 Lesson Assignment	Homework Check
Basic	3–5, 19–23	1, 2, 6–15	2, 8, 10, 12, 14
Average	3–5, 19–23	1, 2, 6–15, 17	2, 8, 10, 12, 14
Advanced	3–5, 19–23	1, 2, 7–15, 17, 18	2, 8, 10, 12, 14

Common Errors

- **Exercises 6–11** Students may use the wrong numbers or make calculation errors when answering these questions. Remind them to read the payroll statement carefully. Encourage them to write down the information they need to know to answer the question. For example, in Exercise 9, students need to know net pay and pay period.
- **Exercises 13–15** Students may use an incorrect formula for finding the percent of taxes paid. Remind them that Social Security and Medicare are determined after the health insurance amount is subtracted from gross pay. Federal tax is determined after the health insurance amount and the 401(k) contribution are subtracted from gross pay.

B.6 Record and Practice Journal

In Exercises 1–6, use the payroll statement.

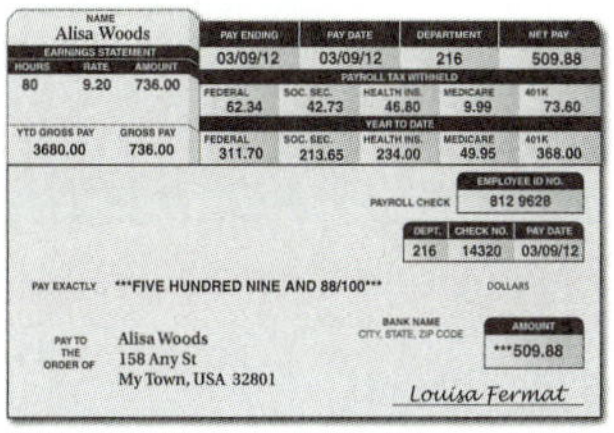

1. What is Alisa's hourly wage?
   $9.20
2. How often does Alisa get paid?
   every 2 weeks
3. What is Alisa's gross pay?
   $736.00
4. How many paychecks has Alisa received so far in 2012?
   5 paychecks
5. What percent of Alisa's gross pay is put in her 401(k)?
   10%
6. What percent does Alisa pay for Social Security tax?
   6.2%

Technology For the Teacher

Answer Presentation Tool

Quiz*Show*

Vocabulary and Concept Check

1. never; At most, net pay can equal gross pay if there are no deductions, but it will never be greater than gross pay.
2. to limit the amount of earnings that an individual can defer taxes on

Practice and Problem Solving

3. $1173.75
4. $10,952.50
5. $612.08
6. $12.50
7. every 2 weeks
8. $1800.00
9. $17,468.10
10. 8 paychecks
11. $228.15
12. Al gets paid every 2 weeks, so there are 26 pay periods in a year. The gross pay should be multiplied by 26, not 12. His annual gross pay is $\$900.00 \cdot 26 = \$23,400.00$.

Practice and Problem Solving

13–15. See *Taking Math Deeper.*

16. a. *Answer should include, but is not limited to:* Make sure the federal income tax brackets are for the head of household filing status.

b. *Answer should include, but is not limited to:* Make sure the federal income taxes are found using the formula and not by multiplying the taxable income by the percent. A taxable income of \$30,000 in 2008 is in the 15% tax bracket and the federal income tax is \$3927.50. A taxable income of \$380,000 in 2008 is in the 35% tax bracket and the federal income tax is \$108,409.00

c. *Answer should include, but is not limited to:* The federal income tax should have increased for each amount.

17. See Additional Answers.

18. \$4276.80

Fair Game Review

19. 3 **20.** 125

21. 343 **22.** $\frac{1}{36}$

23. D

Mini-Assessment

In Exercises 1–3, use the payroll statement on page A46.

1. What percent of Al's gross pay is put in his 401(k)? 2.4%
2. How much does Al pay annually for health insurance? \$1128.40
3. What is Al's YTD gross pay? \$7200

Taking Math Deeper

Exercises 13–15

This lesson on payroll deductions may surprise students. That is, when a person (like Al in the Exercise) earns \$900, he doesn't get to keep a significant part of it. In this example, Al's take-home pay is \$671.85, which is roughly 75% of his gross earnings. As Al's salary increases, the percent he will retain drops. If Al were to earn \$50 an hour (instead of \$12.50), he would retain less than 50% of his gross pay. (This depends on the state and county he lives in.)

① **13.** Find the percent paid for Social Security tax. $\frac{53.11}{900 - 43.40} \approx 6.2\%$

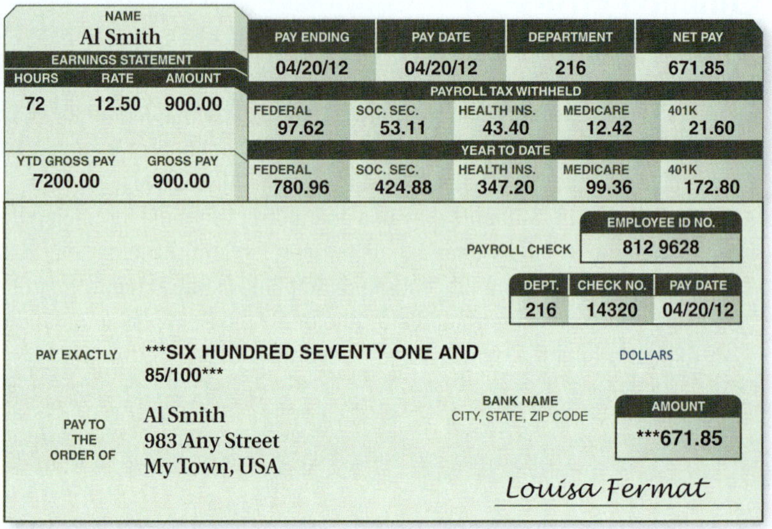

NAME: Al Smith

EARNINGS STATEMENT

HOURS	RATE	AMOUNT
72	12.50	900.00

YTD GROSS PAY	GROSS PAY
7200.00	900.00

PAY ENDING	PAY DATE	DEPARTMENT	NET PAY
04/20/12	04/20/12	216	671.85

	FEDERAL	SOC. SEC.	HEALTH INS.	MEDICARE	401K
PAYROLL TAX WITHHELD	97.62	53.11	43.40	12.42	21.60
YEAR TO DATE	780.96	424.88	347.20	99.36	172.80

PAYROLL CHECK — EMPLOYEE ID NO. 812 9628

DEPT.	CHECK NO.	PAY DATE
216	14320	04/20/12

PAY EXACTLY ***SIX HUNDRED SEVENTY ONE AND 85/100*** DOLLARS

PAY TO THE ORDER OF Al Smith, 983 Any Street, My Town, USA

BANK NAME, CITY, STATE, ZIP CODE

AMOUNT ***671.85

Louisa Fermat

② **14.** Find the percent paid for Medicare tax. $\frac{12.42}{900 - 43.40} \approx 1.45\%$

③ Find the percent paid for federal income tax. Federal income tax is not charged on amounts that are deposited into a 401(k) plan or into a health insurance plan.

15. $\frac{97.62}{900 - (43.40 + 21.60)} \approx 11.7\%$

Reteaching and Enrichment Strategies

If students need help. . .	If students got it. . .
Resources by Chapter • Practice A and Practice B • Puzzle Time Record and Practice Journal Practice Differentiating the Lesson Lesson Tutorials Skills Review Handbook	Resources by Chapter • Enrichment and Extension • School-to-Work • Financial Literacy • Technology Connection Start the next section

In Exercises 13–15, use the payroll statement on page A46.

2 3 **13.** What percent does Al pay for Social Security tax?

14. What percent does Al pay for Medicare tax?

15. What percent does Al pay for federal income tax?

16. **PROJECT** The table shows the federal income tax brackets for 2008.

Filing Status: Head of Household		
If taxable income is between. . .	**Tax bracket is:**	**Federal income tax is:**
\$0 and \$11,449	10%	10% of the income over \$0
\$11,450 and \$43,649	15%	\$1,145 plus 15% of the income over \$11,450
\$43,650 and \$112,649	25%	\$5,975 plus 25% of the income over \$43,650
\$112,650 and \$182,399	28%	\$23,225 plus 28% of the income over \$112,650
\$182,400 and \$357,699	33%	\$42,755 plus 33% of the income over \$182,400
\$357,700 and Above	35%	\$100,604 plus 35% of the income over \$357,700

a. Use the Internet to research the federal income tax brackets for the current year. Make a table similar to the one above.

b. Describe the change in the federal income tax from 2008 to the current year for each tax bracket.

c. Compare the federal income tax for 2008 and the federal income tax for the current year for a person whose taxable income was \$30,000.

17. **STATE TAXES** In addition to federal income tax, you may have to pay state income tax. Use the Internet to research the income taxes for Colorado, Florida, and Massachusetts. Which state has the lowest income tax? the highest income tax?

18. **Critical Thinking** Your annual gross pay is \$32,400. Five percent of your gross pay is put in your 401(k) and 7% pays for health insurance. You pay 15% for federal income tax. What amount of federal income tax do you pay?

Fair Game Review What you learned in previous grades & lessons

Evaluate the expression. *(Section 9.4)*

19. $3^{-1} \cdot 3^2$ **20.** $5^0 \cdot 5^3$ **21.** $\frac{7^2 \cdot 7^{-4}}{7^{-5}}$ **22.** $\frac{6^{-1} \cdot 6^2}{6^3}$

23. **MULTIPLE CHOICE** Which inequality represents "20 is less than a number x"? *(Section 8.1)*

Ⓐ $x < 20$ Ⓑ $20 \geq x$ Ⓒ $20 \leq x$ Ⓓ $20 < x$

B.5–B.6 Quiz

Check It Out
Progress Check
BigIdeasMath.com

1. A credit card bill arrives in the mail. The annual interest rate is 15%. Find the interest charge and the new balance. *(Section B.5)*

Credit Card Statement

Account Number xxxx xxxx xxxx 6003

Date	Activity Since Last Statement	Amount
07/11/12	Payment	50.00
	- - - - Purchases - - - -	
07/02/12	Hair Salon	26.18
07/12/12	Farmer's Market	35.05
07/24/12	Office Supply Store	63.42
07/30/12	Gas Station	25.35

Average Daily Balance	$132.08
Statement Closing Date	08/01/12
Previous Balance	$100.00
Payments	-50.00
Unpaid Balance	$50.00
New Purchases	$150.00
Interest Charge	
New Balance	

In Exercises 2–5, use the payroll statement. *(Section B.6)*

NAME: Sally Wong

EARNINGS STATEMENT

HOURS	RATE	AMOUNT
72	11.25	810.00

YTD GROSS PAY	GROSS PAY
7290.00	810.00

PAY ENDING	PAY DATE	DEPARTMENT	NET PAY
05/04/12	05/04/12	103	623.58

	FEDERAL	SOC. SEC.	HEALTH INS.	MEDICARE	401K
PAYROLL TAX WITHHELD	75.90	48.21	32.40	11.28	18.63
YEAR TO DATE	683.10	433.89	291.60	101.52	167.67

2. What is Sally's hourly wage?

3. What would you expect Sally's net pay to be for the entire year?

4. How much does Sally put in her 401(k) annually?

5. What percent does Sally pay for Medicare tax?

In Exercises 6 and 7, a credit card company keeps the given percent of the total. Find the percent of the transaction the store keeps. *(Section B.5)*

6. 3%

Ice Cream Shoppe

Pablo Rodriguez
Credit card # xxxx xxxx xxxx 5631
8:10 P.M. 07/22/12

SALE:	$18.75
SALES TAX:	$ 0.94
TOTAL:	**$19.69**

-THANK YOU-

7. 4%

Appliance OUTLET — #1 IN CUSTOMER SATISFACTION

Celia Danielson
Credit card # xxxx xxxx xxxx 9103
2:08 P.M. 09/30/12

SALE:	$845.15
SALES TAX:	$ 50.71
TOTAL:	**$895.86**

THANK YOU

Alternative Assessment Options

Math Chat	**Student Reflective Focus Question**
Structured Interview	Writing Prompt

Student Reflective Focus Question

Ask students to summarize the advantages and disadvantages of using cash versus using credit cards. Be sure that they include examples. Select students at random to present to the class.

Study Help Sample Answers

Remind students to complete Graphic Organizers for the rest of the chapter.

5.

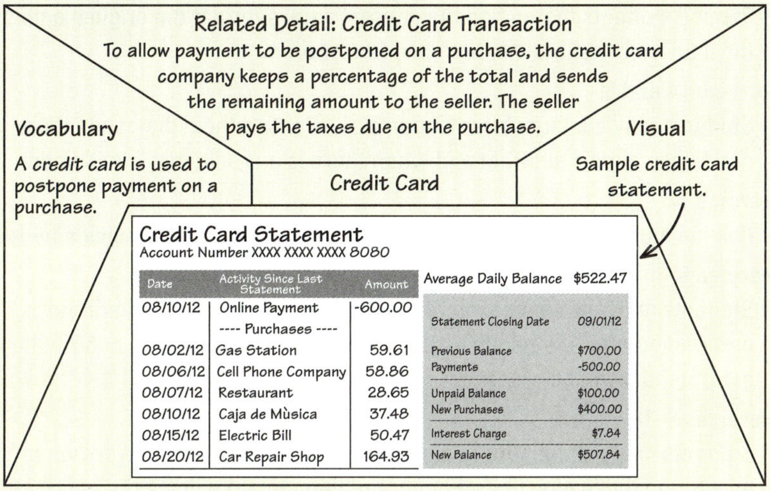

6. Available at *BigIdeasMath.com*.

Reteaching and Enrichment Strategies

If students need help. . .	If students got it. . .
Resources by Chapter • Study Help • Practice A and Practice B • Puzzle Time Lesson Tutorials *BigIdeasMath.com* Practice Quiz Practice from the Test Generator	Resources by Chapter • Enrichment and Extension • School-to-Work Game Closet at *BigIdeasMath.com* Start the Chapter Review

Answers

1. $1.65; $201.65
2. $11.25
3. $16,213.08
4. $484.38
5. about 1.45%
6. about 92.2%
7. about 90.3%

Technology For the Teacher

Answer Presentation Tool

Assessment Book

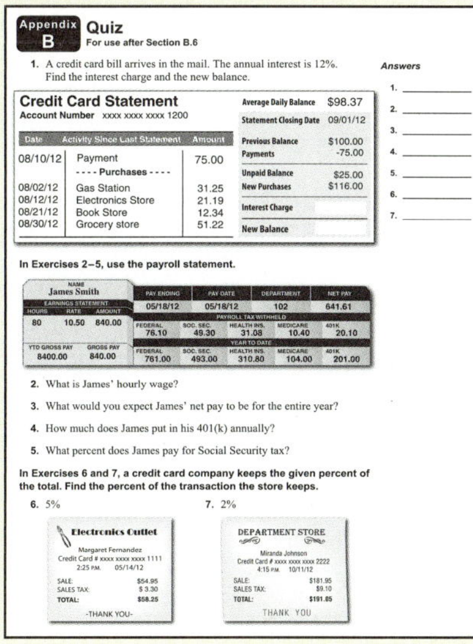

Appendix B **Quiz**
For use after Section B.6

1. A credit card bill arrives in the mail. The annual interest is 12%. Find the interest charge and the new balance.

Credit Card Statement
Account Number xxxx xxxx xxxx 1200

Date	Activity Since Last Statement	Amount
08/10/12	Payment	75.00
	---- Purchases ----	
08/02/12	Gas Station	31.25
08/12/12	Electronics Store	21.19
08/21/12	Book Store	12.34
08/30/12	Grocery store	51.22

Average Daily Balance	$98.37
Statement Closing Date	09/01/12
Previous Balance	$100.00
Payments	-75.00
Unpaid Balance	$25.00
New Purchases	$116.00
Interest Charge	
New Balance	

Answers

1. ______
2. ______
3. ______
4. ______
5. ______
6. ______
7. ______

In Exercises 2–5, use the payroll statement.

NAME James Smith

EARNINGS STATEMENT

HOURS	RATE	AMOUNT
80	10.50	840.00

YTD GROSS PAY	GROSS PAY
8400.00	840.00

PAY ENDING	PAY DATE	DEPARTMENT	NET PAY
05/18/12	05/18/12	102	641.61

	FEDERAL	SOC. SEC.	HEALTH INS.	MEDICARE	401K
PAYROLL TAX WITHHELD	76.10	49.30	31.08	10.40	20.10
YEAR TO DATE	761.00	493.00	310.80	104.00	201.00

2. What is James' hourly wage?
3. What would you expect James' net pay to be for the entire year?
4. How much does James put in his 401(k) annually?
5. What percent does James pay for Social Security tax?

In Exercises 6 and 7, a credit card company keeps the given percent of the total. Find the percent of the transaction the store keeps.

6. 5%

Electronics Outlet
Margaret Fernandez
Credit Card # xxxx xxxx xxxx 1111
2:25 P.M. 05/14/12
SALE: $54.95
SALES TAX: $ 3.30
TOTAL: $58.25
-THANK YOU-

7. 2%

DEPARTMENT STORE
Miranda Johnson
Credit Card # xxxx xxxx xxxx 2222
4:15 P.M. 10/11/12
SALE: $181.95
SALES TAX: $9.10
TOTAL: $191.05
THANK YOU

For the Teacher
Additional Review Options

- **Quiz***Show*
- Big Ideas Test Generator
- Game Closet at *BigIdeasMath.com*
- Vocabulary Puzzle Builder
- Resources by Chapter
  - Puzzle Time
  - Study Help

Answers

1. $540.80

Review of Common Errors

Exercise 1

- Students may forget to include previously earned interest when calculating annual interest. Remind them of the definition of compound interest.

Exercise 2

- Students may use the simple interest formula to find the interest earned on the account instead of the compound interest formula.
- Students may evaluate part of the equation incorrectly when solving. Remind them of the order of operations and that they should simplify the equation before solving for the principal.

Exercise 3

- Remind students that to find the interest paid, subtract the original amount of the loan from the total amount paid for the loan.

Exercises 4 and 5

- Students may add or subtract incorrectly. Remind them that when there is a debit, they should subtract and when there is a credit, they should add.

Exercise 7

- Remind students that to convert a percent to a decimal, they divide by 100.

Exercise 8

- Students may use an incorrect formula. Remind them that federal tax is determined after the health insurance amount and the 401(k) contribution are subtracted from the gross pay.

Exercises 8–10

- Students may use the wrong numbers or make calculation errors when answering these questions. Encourage them to write down the information they need to know to answer the question.

B Appendix Review

Check It Out
Vocabulary Help
BigIdeasMath.com

Review Key Vocabulary

compound interest, *p. A13*
loan, *p. A24*
term, *p. A24*
installment loan, *p. A24*
mortgage, *p. A25*
checking account, *p. A30*
checkbook registry, *p. A30*
debit card, *p. A30*
bad check, *p. A31*
credit card, *p. A38*
gross pay, *p. A44*
net pay, *p. A44*
payroll statement, *p. A44*

Review Examples and Exercises

B.1 Simple and Compound Interest *(pp. A10–A15)*

You want to invest \$400 for 2 years. Which account should you choose?

- **Account A earns 4% simple interest per year.**
- **Account B earns 3% interest compounded annually.**

Make two tables that show the account balances for 2 years.

Account A			
Year	Principal	4% Simple Interest	Balance at End of Year
1	\$400	\$16	\$416
2	\$400	\$16	\$432

Account B			
Year	Principal and Interest	3% Compound Interest	Balance at End of Year
1	\$400.00	\$12.00	\$412.00
2	\$412.00	\$12.36	\$424.36

Account A earns \$432 − \$424.36 = \$7.64 more than Account B after 2 years. So, you should choose Account A.

Exercises

1. You deposit \$500 in an account that earns 4% interest compounded annually. What is the account balance after 2 years?

B.2 Formula for Compound Interest *(pp. A16–A21)*

You deposit \$1800 in a savings account. The account earns 3% interest compounded annually. What is the balance after 7 years?

$B = P(1 + r)^t$ — Write compound interest formula.

$= 1800(1 + 0.03)^7$ — Substitute.

$= 1800(1.03)^7$ — Add.

≈ 2213.77 — Simplify.

The balance is \$2213.77 after 7 years.

Exercises

2. An account earns 4% interest compounded annually. The balance after 5 years is $7421.58. What is the principal?

B.3 Installment Loans (pp. A22–A27)

The principal on a home mortgage is $145,000, the monthly payment is $1081.08, and the term is 20 years. How much interest is paid on the loan?

The total amount paid for the mortgage is the product of the monthly payment and the number of payments. The interest paid on the loan is the difference of the total amount paid for the mortgage and the principal.

$1081.08 \cdot 240 = 259{,}459.20$ — Multiply the monthly payment by number of payments.

$259{,}459.20 - 145{,}000 = 114{,}459.20$ — Subtract the principal from the total amount paid.

∴ $114,459.20 in interest is paid on the loan.

Exercises

3. The principal on a home mortgage is $250,000, the montly payment is $1580.17, and the term is 30 years. How much interest is paid on the loan?

B.4 Checking Accounts (pp. A28–A33)

Which check is the bad check?

Date	Check #	Transaction	Credit	Debit	Balance
		Balance Forward			89.37
11/1/12	528	Cell Phone Company		65.11	24.26
11/5/12	529	Grocery Store		52.08	-27.82
11/15/12		Deposit Paycheck	418.94		391.12
11/16/12	530	Credit Card Payment		200.00	191.12
11/30/12		Charity Donation		30.00	161.12
11/30/12		Magazine Subscription		20.00	141.12

The amount of Check #529 is greater than the available balance.

∴ So, Check #529 is the bad check.

Exercises

4. Your monthly bank statement shows a balance of $191.12. Explain why your balance does not agree with the bank statement.

5. Your next two transactions are shown below. What is your new balance?

- 12/1/12 Deposit $300.00
- 12/2/12 ATM Withdrawal $60.00

Review Game

Make a Million

Big Ideas Game Closet

Materials per group

- game board
- number cube or spinner
- game token for each player (counter tile, algebra tile, button, etc.)

Directions

- Each person starts with an account balance of $10,000.
- To determine who plays first, each person rolls the number cube or spins the spinner once. The person with the greatest number plays first, then the person to their left plays second and so on. If there is a tie, repeat this process.
- Each person rolls the number cube or spins the spinner when it is his or her turn. The person moves that number of times on the game board.
- Each space on the game board will either add or subtract money from the person's balance. Examples of game spaces:
  - Earn 3% simple interest for 4 years
  - Earn 5% compound interest for 6 years
  - Withdraw $2000
  - Deposit $500
  - Lose 2% of your balance
- Continue playing until someone's balance reaches one million dollars or time runs out.

Who wins?

The first person to earn at least 1 million dollars or, if time is limited, the person with the highest account balance.

The game board for the "Make a Million" game is available at *BigIdeasMath.com*.

For the Student
Additional Practice

- Lesson Tutorials
- Study Help (textbook)
- Student Website
  - Multi-Language Glossary
  - Practice Assessments

Answers

2. $6100.00

3. $318,861.20

4. The last two debit purchases are close to the end of the month and likely were not processed in time for your monthly bank statement.

5. $381.12

6. $607.10

7. $5.47; $405.47

8. about 12.3%

9. $1404

10. $28,080.00

My Thoughts on the Chapter

Teacher Tip

Not allowed to write in your teaching edition? Use sticky notes to record your thoughts.

What worked. . .

What did not work. . .

What I would do differently. . .

B.5 Credit Cards (pp. A36–A41)

Rocco receives a credit card bill in the mail. The monthly interest rate is 1.1%. Find the interest charge.

Find the product of the monthly interest rate and the average daily balance.

$$0.011 \cdot 645.79 \approx \$7.10$$

∴ The interest charge is $7.10.

Average Daily Balance	$645.79
Statement Closing Date	06/01/12
Previous Balance	$800.00
Payments	-400.00
Unpaid Balance	$400.00
New Purchases	$200.00
Interest Charge	
New Balance	

Exercises

6. What is the new balance?

7. If Rocco had paid $600 on last month's bill, the average daily balance would be $497.41. Find the interest charge and the new balance.

B.6 Payroll Deductions (pp. A42–A47)

What percent does Raj pay for health insurance?

NAME		
Raj Singh		
EARNINGS STATEMENT		
HOURS	RATE	AMOUNT
80	13.50	1080.00
YTD GROSS PAY		GROSS PAY
4320.00		1080.00

PAY ENDING	PAY DATE	DEPARTMENT		NET PAY
02/24/12	02/24/12	P		744.21
PAYROLL TAX WITHHELD				
FEDERAL	SOC. SEC.	HEALTH INS.	MEDICARE	401K
115.37	61.33	90.75	14.34	54.00
YEAR TO DATE				
FEDERAL	SOC. SEC.	HEALTH INS.	MEDICARE	401K
461.48	245.32	363.00	57.36	216.00

$$\frac{\text{amount for health insurance}}{\text{gross pay}} = \frac{90.75}{1080.00}$$

$$\approx 0.084 = 8.4\%$$

∴ Raj pays 8.4% for health insurance.

Exercises

8. What percent does Raj pay for federal income tax?

9. How much will Raj put in his 401(k) for the entire year?

10. What is Raj's annual gross pay?

B Appendix Test

Check It Out
Test Practice
BigIdeasMath.com

1. You deposit \$800 in a savings account that earns 4% simple interest per year. Your friend deposits \$900 in a savings account that earns 3% simple interest per year.

a. Write and graph two equations for the balance B in each account after t years.

b. Are the account balances ever equal? Explain.

2. An account earns 3% interest compounded annually. The balance after 2 years is \$4774.05. What is the principal?

3. A bank offers two types of checking accounts. How many transactions must you make each month for the cost of *Basic Checking* to be greater than the cost of *Advanced Checking*?

Basic Checking
- no monthly fee
- \$0.15 per transaction

ASL Anytown Savings and Loan
Anytown, USA

Advanced Checking
- \$12 monthly fee
- no fee per transaction

ASL Anytown Savings and Loan
Anytown, USA

4. You use a credit card to pay for a \$186.90 purchase at a sporting goods store. The credit card company keeps 3% of the total. Find the amount of money the store receives from the credit card company for the transaction.

5. Use the payroll statement.

NAME	PAY ENDING	PAY DATE	DEPARTMENT	NET PAY
Karen Walker	06/30/12	06/30/12	J	1990.98

EARNINGS STATEMENT

HOURS	RATE	AMOUNT
168	15.60	2620.80

YTD GROSS PAY	GROSS PAY
15,724.80	2620.80

	FEDERAL	SOC. SEC.	HEALTH INS.	MEDICARE	401K
PAYROLL TAX WITHHELD	189.37	155.18	117.94	36.29	131.04
YEAR TO DATE	1136.22	931.08	707.64	217.74	786.24

a. How often does Karen get paid?

b. What percent does Karen pay for federal income tax?

c. How much will Karen put in her 401(k) for the entire year?

6. USED CAR The average sale price of a used car at a car dealership is \$6580. A savings account earns 4% interest compounded annually. About how much should you deposit in the account to have enough to buy a car in 3 years?

7. MORTGAGE The principal on a home mortgage is \$128,000, the monthly payment is \$1097.50, and the term is 15 years. How much interest is paid on the loan?

Test Item References

Chapter Test Questions	Section to Review
1	B.1
2, 6	B.2
7	B.3
3	B.4
4	B.5
5	B.6

Test-Taking Strategies

Remind students to quickly look over the entire test before they start so that they can budget their time. When working through the test, students should mentally read the numbers to ensure correct decimal point placement. Teach the students to use the Stop and Think strategy before answering. **Stop** and carefully read the question, and **Think** about what the answer should look like.

Common Assessment Errors

- **Exercise 1** Students may forget to convert the percent to a decimal before writing the formula. Remind them that the rate should be in decimal form. They divide the percent by 100 to put the rate in decimal form.
- **Exercises 2 and 6** Students may evaluate part of the equation incorrectly when solving. Remind them of the order of operations and that they should simplify the equation before solving for the principal.
- **Exercise 3** Students may write an incorrect inequality to answer the question. Encourage them to write the inequality as a word sentence before substituting numbers and variables.
- **Exercise 4** Students may forget to convert the percent to a decimal when finding how much money the credit card company keeps from the transaction.
- **Exercise 5** Students may use the wrong numbers or make calculation errors when answering these questions. Remind them to read the payroll statement carefully. Encourage them to write down the information they need to know to answer the question.
- **Exercise 7** Students may find the total amount paid, but forget to find the interest paid on the loan. Remind them to subtract the original amount of the loan from the total amount paid for the loan.

Reteaching and Enrichment Strategies

If students need help. . .	If students got it. . .
Resources by Chapter • Practice A and Practice B • Puzzle Time Record and Practice Journal Practice Differentiating the Lesson Lesson Tutorials Practice from the Test Generator Skills Review Handbook	Resources by Chapter • Enrichment and Extension • School-to-Work • Financial Literacy Game Closet at *BigIdeasMath.com* Start Standardized Test Practice

Answers

1. **a.** You: $B = 800 + 32t$
   Friend: $B = 900 + 27t$

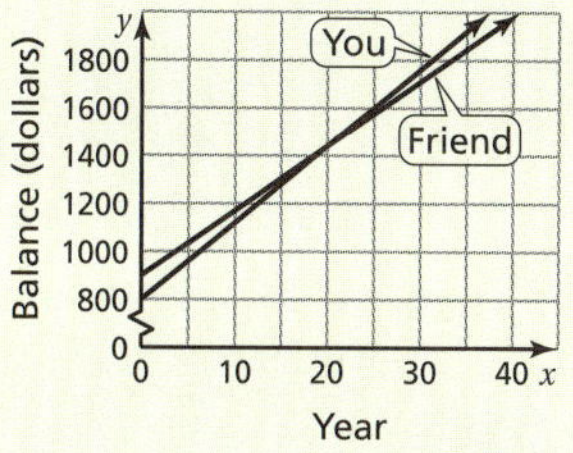

   b. Yes, the graphs appear to intersect when $t = 20$. So, after 20 years, the accounts are equal.

2. $4500.00
3. more than 80 transactions
4. $181.29
5. **a.** once per month

   b. about 8%

   c. $1572.48
6. $5849.60
7. $69,550.00

Assessment Book

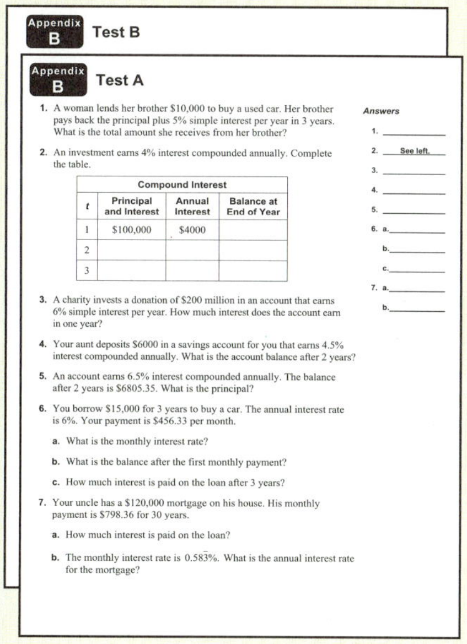

Appendix B Test B

Appendix B Test A

1. A woman lends her brother $10,000 to buy a used car. Her brother pays back the principal plus 5% simple interest per year in 3 years. What is the total amount she receives from her brother?
2. An investment earns 4% interest compounded annually. Complete the table.

Compound Interest			
t	Principal and Interest	Annual Interest	Balance at End of Year
1	$100,000	$4000	
2			
3			

3. A charity invests a donation of $200 million in an account that earns 6% simple interest per year. How much interest does the account earn in one year?
4. Your aunt deposits $6000 in a savings account for you that earns 4.5% interest compounded annually. What is the account balance after 2 years?
5. An account earns 6.5% interest compounded annually. The balance after 2 years is $6805.35. What is the principal?
6. You borrow $15,000 for 3 years to buy a car. The annual interest rate is 6%. Your payment is $456.33 per month.
   a. What is the monthly interest rate?
   b. What is the balance after the first monthly payment?
   c. How much interest is paid on the loan after 3 years?
7. Your uncle has a $120,000 mortgage on his house. His monthly payment is $798.36 for 30 years.
   a. How much interest is paid on the loan?
   b. The monthly interest rate is $0.58\overline{3}\%$. What is the annual interest rate for the mortgage?

Answers

1. ____
2. See left.
3. ____
4. ____
5. ____
6. a. ____ b. ____ c. ____
7. a. ____ b. ____

Photo Credits

inside front cover ©iStockphoto.com/sweetym; **iv** Big Ideas Learning, LLC; **vi** *top* ©iStockphoto.com/Lisa Thornberg, ©iStockphoto.com/Ann Marie Kurtz; *bottom* Evok20; **vii** *top* ©iStockphoto.com/Jonathan Larsen; *bottom* Apollofoto; **viii** *top* ©iStockphoto.com/Angel Rodriguez, ©iStockphoto.com/Ann Marie Kurtz; *bottom* ©iStockphoto.com/daaronj; **ix** ©iStockphoto.com/Ron Sumners, ©iStockphoto.com/Ann Marie Kurtz; *bottom* Apollofoto; **x** *top* ©iStockphoto.com/Stefan Klein; *bottom* ©iStockphoto.com/ronen; **xi** *top* ©iStockphoto.com/Michael Flippo, ©iStockphoto.com/Ann Marie Kurtz; *bottom* Heather Prosch-Jensen; **xii** *top* ©iStockphoto.com/Alistair Cotton; *bottom* Peter Close; **xiii** *top* ©iStockphoto.com/Angel Rodriguez, ©iStockphoto.com/Ann Marie Kurtz; *bottom* ©iStockphoto.com/ranplett; **xiv** *top* ©iStockphoto.com/Varina and Jay Patel, ©iStockphoto.com/Ann Marie Kurtz; *bottom* Jane Norton; **xv** *top* ©iStockphoto.com/Björn Kindler; *bottom* ©iStockphoto.com/Thomas Perkins; **xvii** Gelpi; **xxxvii** ©iStockphoto.com/Andresr; **xl, xli** Big Ideas Learning, LLC; **xliv** ©iStockphoto.com/Ekaterina Monakhova; **1** ©iStockphoto.com/Lisa Thornberg, ©iStockphoto.com/Ann Marie Kurtz; **6** ©iStockphoto.com/David Freund; **7** ©iStockphoto.com/shapecharge; **8** NASA; **9** ©iStockphoto.com/Judson Lane; **12** ©iStockphoto.com/harley_mccabe; **13** ©iStockphoto.com/Jacom Stephens; **14** ©iStockphoto.com/Harry Hu; **15** ©Paul Slaughter; **20** ©iStockphoto.com/Andrey Krasnov; **29** *top right* ©iStockphoto.com/Alan Crawford; *center left* ©iStockphoto.com/Julio Yeste; *center right* ©iStockphoto.com/Mark Stay; **31** ©iStockphoto.com/Michel de Nijs; **33** Steve MacAulay Photography; **34** *top* Benedict S. Gibson; *bottom* Frantz Petion; **35** Dehk; **36** *bottom right* ©iStockphoto.com/Paul Tessier; *center left* Tom Uhlenberg; *top right* ©iStockphoto.com/polarica; **37** ©iStockphoto.com/Winston Davidian; **38** ©iStockphoto.com/Steve Mcsweeny; **41** ©iStockphoto.com/machinim; **46** ©iStockphoto.com/Jonathan Larsen; **51** NASA; **52** ©iStockphoto.com/David Morgan; **53** *center left* ©iStockphoto.com/Jakub Semeniuk; *top right* NASA; **60** ©iStockphoto.com/Amanda Rohde; **61** Julián Rovagnati; **66** ©iStockphoto.com/whitechild; **67** *top right* Jerry Horbert; *bottom left* ©iStockphoto.com/Chris Schmidt; **69** ©iStockphoto.com/Peter Finnie; **72** ©iStockphoto.com/Stephen Pothier; **73** *bottom* Dewayne Flowers; *top* Gina Smith; **75** Philip Lange; **76** Howard Sandler, ©iStockphoto.com/Dori O'Connell; **79** Yuri Arcurs; **80** ©iStockphoto.com/Kathy Hicks; **81** ©iStockphoto.com/webphotographeer; **82** ©iStockphoto.com/walik; **86** ©iStockphoto.com/Corina Estepa; **87** ©iStockphoto.com/Tomislav Forgo; **91** Kateryna Potrokhova; **92** ©iStockphoto.com/Jason Stitt; **93** ©iStockphoto.com/Gabor Izso, ©iStockphoto.com/Dawn Jagroop; **94** ©iStockphoto.com/ZanyZeus; **98** ©iStockphoto.com/Justin Horrocks; **104** ©iStockphoto.com/Angel Rodriguez, ©iStockphoto.com/Ann Marie Kurtz; **109** Printed courtesy of the Florida Department of Transportation; **110** ©iStockphoto.com/Adam Mattel; **T-111** ©iStockphoto.com/Pawel Liprec, ©iStockphoto.com/beetle8; **111** *center right* ©iStockphoto.com/Pawel Liprec, ©iStockphoto.com/beetle8; *top left* ©iStockphoto.com/Gene Chutka; **113** ©iStockphoto.com/Kirsty Pargeter, **115** ©iStockphoto.com/Connie Maher; **116** ©iStockphoto.com/Jacom Stephens; **117** *center left* ©iStockphoto.com/Andrea Krause; *top right* ©iStockphoto.com/Petr Podzemny; **121** ©iStockphoto.com/Michael Chen; **123** *bottom* ©iStockphoto.com/adrian beesley; *top* ©iStockphoto.com/Brian McEntire; **125** *center right* ©iStockphoto.com/John Kounadeas; *bottom right* ©iStockphoto.com/Rich Yasick; **128** ©iStockphoto.com/Ryan Putnam; **129** ©iStockphoto.com/ iLexx; **130** ©iStockphoto.com/Jeremy Edwards; **131** ©iStockphoto.com/Robert Kohlhuber; **136** *center left* ©iStockphoto.com/George Peters; *center right* ©iStockphoto.com/Olga Shelego; *bottom left* ©iStockphoto.com/Valerie Loiseleux; **137** ©iStockphoto.com/Duncan Walker; **140** ©iStockphoto.com/ Marcio Silva; **142** ©iStockphoto.com/Carmen Martínez Banús; **146** ©iStockphoto.com/Ron Sumners, ©iStockphoto.com/Ann Marie Kurtz; **149** *right* ©2008 Zappos.com, Inc.; *left* ©iStockphoto.com/Andrew Johnson; **151** ©iStockphoto.com/alohaspirit; **152** ©iStockphoto.com/Timur Kulgarin; **153** *center right* Primo Ponies Photography; *center left* ©iStockphoto.com/Wayne Johnson; **155** Digital Vision Royalty Free Photograph/Getty Images; **158** ©iStockphoto.com/Hannu Liivaar; **T-159** ©iStockphoto.com/Justin Horrocks, ©iStockphoto.com/Ana Abejon, ©iStockphoto.com/Huchen Lu; **159** *center left* ©iStockphoto.com/LoopAll; *center right* ©iStockphoto.com/Justin Horrocks, ©iStockphoto.com/Ana Abejon, ©iStockphoto.com/Huchen Lu; **161** ©iStockphoto.com/Lisa F. Young; **165** ©iStockphoto.com/technotr; **167** *top* ©iStockphoto.com/Alexander Hafemann; *bottom* ©iStockphoto.com/medobear; **169** *left* ©iStockphoto.com/PeskyMonkey; *right* ©iStockphoto.com/shapecharge; **173** *top right* ©iStockphoto.com/Dean Turner; *bottom* ©iStockphoto.com/Mladen Mladenov; *center* ©iStockphoto.com/Tom Buttle; **176** Junial Enterprises; **178** ©iStockphoto.com/Louis Aguinaldo; **182** ©iStockphoto.com/Stefan Klein; **189** ©iStockphoto.com/Jorgen Jacobsen; **194** *exercise 9* ©iStockphoto.com/Chih-Feng Chen; *exercise 10* ©iStockphoto.com/Andreas Gradin; *exercise 11* ©iStockphoto.com/Jim Lopes; **T-195** ©iStockphoto.com/Zoran Kolundzija; **195** ©iStockphoto.com/Zoran Kolundzija; **198** *center* Booker Middle School; *bottom left* ©iStockphoto.com/Black Jack 3D; *bottom right* ©iStockphoto.com/Vadym Volodin; **199** NASA; **202** ©iStockphoto.com/Evelyn Peyton; **203** *top left* ©iStockphoto.com/Lora Clark; *top right* ©iStockphoto.com/Terraxplorer; **212** Estate Craft Homes, Inc.; **228** ©iStockphoto.com/Michael Flippo, ©iStockphoto.com/Ann Marie Kurtz; **233** Perfectblue97; **234** ©iStockphoto.com/Benjamin Lazare; **T-235** ©iStockphoto.com/Sheldon Kralstein; **235** *center* ©iStockphoto.com/Jill Chen; *top right* ©iStockphoto.com/Sheldon Kralstein; **236** ©ImageState **240** ©iStockphoto.com/Melissa Carroll; **241** *center right* ©iStockphoto.com/Cathy Keifer; *center left* ©iStockphoto.com/Alex Slobodkin, ©iStockphoto.com/Sebastian Duda; **243** *bottom right* ©iStockphoto.com/MACIEJ NOSKOWSKI; *center left* ©iStockphoto.com/Yvan Dubé **244** ©iStockphoto.com/Kais Tolmats; **248** *top left* ©iStockphoto.com/Don Bayley; *bottom right* ©iStockphoto.com/iLexx; **251** ©iStockphoto.com/Marcio Silva; **253** Luminis; **256** *exercise 3* Joshua Haviv; *exercise 4* ©iStockphoto.com/William D Fergus McNeill; *exercise 5* ©iStockphoto.com/Klaas Jan Schraa; **257** *left* ©iStockphoto.com/Parema; *right* ©iStockphoto.com/Nikontiger; **267** *center right* Hasan Shaheed; *bottom left* ©iStockphoto.com/JenDen2005; *bottom right* Orla; **268** *center right* CD Lenzen; *bottom left* red06; **272** ©iStockphoto.com/Alistair Cotton **278** ©iStockphoto.com/Jan Will; **279** Elena Elisseeva; **281** Laurence Gough; **282** Frederic J. Brown/Getty Images; **284** ©iStockphoto.com/Mike Elliott; **285** Monkey Business Images; **287** ©iStockphoto.com/David15; **288** Gina Brockett; **289** ©iStockphoto.com/Craig RJD; **293** ©iStockphoto.com/Jill Fromer; **294** ©iStockphoto.com/Janis Litavnieks; **295** DLD; **296** *top right* Florida Park Service; *bottom right* ©iStockphoto.com/Eric Isselée; *center left* ©iStockphoto.com/Tony Campbell; **297** *top right* Larry Korhnak; *center right* Photo by Andy Newman; **301** *center left* ©iStockphoto.com/Jane norton; *center right* ©iStockphoto.com/Krzysztof Zmij; **302** Dwight Smith; **304** Nikola Bilic; **305** WizData, inc.; **310** ©iStockphoto.com/Angel Rodriguez, ©iStockphoto.com/Ann Marie Kurtz; **313** ©iStockphoto.com/Floortje; **T-317** ©iStockphoto.com/Richard Goerg; **317** *top* The ESRB rating icons are registered trademarks of the Entertainment Software Association; *center right* ©iStockphoto.com/Richard Goerg; **318** ©iStockphoto.com/George Peters; **319** ©iStockphoto.com/George Peters; **322** pandapaw; **323** ©iStockphoto.com/Daniel Van Beek; **325** ©iStockphoto.com/Eric Simard; **331** Alexander Kalina; **332** *top right* ©iStockphoto.com/Martin Firus; *bottom* Robert Pernell; **333** ©iStockphoto.com/Trevor Fisher; **338** ©iStockphoto.com/fotoVoyager; **339** *exercise 26* ©iStockphoto.com/Jill Chen; *exercise 27* ©iStockphoto.com/7unit; *exercise 28* ©iStockphoto.com/George Clerk; **340** Tatiana Popova; **344** ©iStockphoto.com/Heather Shimmin, ©iStockphoto.com/itographer; **348** ©iStockphoto.com/Varina and Jay Patel, ©iStockphoto.com/Ann Marie Kurtz; **350** ©iStockphoto.com/Franck Boston; **351** *activity 3a* NASA; *activity 3b* NASA/JPL-Caltech/R. Hurt (SSC); *activity 3c* NASA; *activity 3d* NASA; *center right;* Stevyn Colgan; **353** ©iStockphoto.com/Philippa Banks; **354** ©iStockphoto.com/Clotilde Hulin; **T-355** ©iStockphoto.com/Boris Yankov; **355** ©iStockphoto.com/Boris Yankov; **356** ©iStockphoto.com/John Tomaselli; **360** ©iStockphoto.com/Viktoriia Kulish; **361** *top right* ©iStockphoto.com/Paul Tessier; *center left* ©iStockphoto.com/Marie-france Bélanger, ©iStockphoto.com/Valerie Loiseleux, ©iStockphoto.com/Linda Steward; **366** ©iStockphoto.com/Andrey Volodin; **367** *top right* Dash; **367** *center left* NASA/JPL-Caltech/L. Cieza (UT Austin); **369** ©iStockphoto.com/Dan Moore; **373** ©iStockphoto.com/Aliaksandr Autayeu; **374** EugeneF; **375** ©iStockphoto.com/Nancy Louie; **376** ©iStockphoto.com/Kais Tolmats; **377** *center* ©iStockphoto.com/Henrik Jonsson; *top* ©iStockphoto.com/Sebastian Kaulitzki; *bottom* 7artscreensavers.com; **379** *top center* ©iStockphoto.com/Frank Wright; *top left* ©iStockphoto.com/Mark Stay; *bottom left* ©iStockphoto.com/Oliver Sun Kim; *top right* ©iStockphoto.com/Evgeniy Ivanov; **380** ©iStockphoto.com/ChristianAnthony; **381** Microgen; **382** ©iStockphoto.com/camilla wisbauer, ©iStockphoto.com/David Freund, ©iStockphoto.com/Joe Belanger, ©iStockphoto.com/thumb, ©iStockphoto.com/Marie-france Bélanger, ©iStockphoto.com/Susan Trigg; **383** NASA; **384** *center left* Google logo ©Google Inc., 2008. Reprinted with Permission.; *center right* YouTube logo ©Google Inc., 2008. Reprinted with Permission.; **385** Elaine Barker; **386** *bottom left* ©iStockphoto.com/angelhell; *bottom right* ©iStockphoto.com/Jan Rysavy; **387** BORTEL Pavel; **392** *bottom right* ©iStockphoto.com/Eric Holsinger; *center left* Eric Isselée; **A1** ©iStockphoto.com/Björn Kindler; **A4** *top right* ©iStockphoto.com/Hsing-WenHsu; *center right* ©iStockphoto.com/blackred; *bottom left* ©iStockphoto.com/Thomas Kuest; *bottom right* ©iStockphoto.com/Lim ChewHow; **A5** *top right* ©iStockphoto.com/Richard Cano; *bottom left* ©iStockphoto.com/best-photo; *bottom right* ©iStockphoto.com/Mika Makkonen; **A6** Emmer, Michele, ed., The Visual Mind: Art and Mathematics, Plate 2, ©1993 Massachusetts Institute of Technology, by permission of The MIT Press.; **A7** *top right* Elena Borodynkina; *center right* ©iStockphoto.com/Andrew Cribb; *bottom left* ©iStockphoto.com/Matthew Okimi; *bottom right* Vladimir Bulatov; **A8** NASA; **A9** NASA; **A10** Koksharov Dmitry; **A11** Valva Vitaly; **A15** *top right* Dmitry Terentjev; *center left* ©iStockphoto.com/James Ferrie; *bottom right* ©iStockphoto.com/Nicholas Monu; **A17** patrimonio; **A19** ©iStockphoto.com/Roel Smart; **A21** ©iStockphoto.com/Jacom Stephens; **A23** *top right* ©iStockphoto.com/Brian sullivan; *center left* ©iStockphoto.com/Justin Horrocks; **A25** Denis Miraniuk; **A27** ©iStockphoto.com/Rich Vintage; **A29** Andresr; **A36** ©iStockphoto.com/Roberto A Sanchez, ©iStockphoto.com/Frank Leung; **A38** Elnur; **A39** Laurent Renault; **A41** Yuri Arcurs; **A42** ©iStockphoto.com/Feng Yu

Cartoon illustrations Tyler Stout
Cover image Lechner & Benson Design

Selected Answers

Section 1.1 Solving Simple Equations
(pages 7–9)

1. $+$ and $-$ are inverses. $\times$ and $\div$ are inverses.

3. $x - 3 = 6$; It is the only equation that does not have $x = 6$ as a solution.

5. $x = 57$

7. $x = -5$

9. $p = 21$

11. $x = 9\pi$

13. $d = \frac{1}{2}$

15. $n = -4.9$

17. **a.** $105 = x + 14$; $x = 91$

    b. no; Because $82 + 9 = 91$, you did not knock down the last pin with the second ball of the frame.

19. $n = -5$

21. $m = 7.3\pi$

23. $k = 1\frac{2}{3}$

25. $p = -2\frac{1}{3}$

27. They should have added 1.5 to each side.

    $-1.5 + k = 8.2$

    $k = 8.2 + 1.5$

    $k = 9.7$

29. $6.5x = 42.25$; \$6.50 per hour

31. $420 = \frac{7}{6}b$, $b = 360$; \$60

33. $h = -7$

35. $q = 3.2$

37. $x = -1\frac{4}{9}$

39. greater than; Because a negative number divided by a negative number is a positive number.

41. 3 mg

43. 8 in.

45. $7x - 4$

47. $\frac{25}{4}g - \frac{2}{3}$

Section 1.2 Solving Multi-Step Equations
(pages 14 and 15)

1. $2 + 3x = 17$; $x = 5$

3. $k = 45$; $45°, 45°, 90°$

5. $b = 90$; $90°, 135°, 90°, 90°, 135°$

7. $c = 0.5$

9. $h = -9$

11. $x = -\frac{2}{9}$

13. 20 watches

15. $4(b + 3) = 24$; 3 in.

17. $\frac{2580 + 2920 + x}{3} = 3000$; 3500 people

19. $<$

21. $>$

Section 1.3 Solving Equations with Variables on Both Sides
(pages 20 and 21)

1. no; When 3 is substituted for x, the left side simplifies to 4 and the right side simplifies to 3.

3. $x = 13.2$ in.

5. $x = 7.5$ in.

7. $k = -0.75$

9. $p = -48$

11. $n = -3.5$

13. $x = -4$

15. The 4 should have been added to the right side.

$$3x - 4 = 2x + 1$$
$$3x - 2x - 4 = 2x + 1 - 2x$$
$$x - 4 = 1$$
$$x - 4 + 4 = 1 + 4$$
$$x = 5$$

17. $15 + 0.5m = 25 + 0.25m$; 40 mi

19. 7.5 units

21. Remember that the box is with priority mail and the envelope is with express mail.

23. 10 mL

25. square: 12 units; triangle: 10 units, 19 units, 19 units

27. 24 in.^{3}

29. C

Section 1.4 Rewriting Equations and Formulas
(pages 28 and 29)

1. no; The equation only contains one variable.

3. a. $A = \frac{1}{2}bh$ b. $b = \frac{2A}{h}$ c. $b = 12$ mm

5. $y = 4 - \frac{1}{3}x$

7. $y = \frac{2}{3} - \frac{4}{9}x$

9. $y = 3x - 1.5$

11. The y should have a negative sign in front of it.

$$2x - y = 5$$
$$-y = -2x + 5$$
$$y = 2x - 5$$

13. a. $t = \frac{I}{Pr}$

b. $t = 3$ yr

15. $m = \frac{e}{c^2}$

17. $\ell = \dfrac{A - \frac{1}{2}\pi w^2}{2w}$

19. $w = 6g - 40$

21. a. $F = 32 + \frac{9}{5}(K - 273.15)$

b. 32°F

c. liquid nitrogen

23. $r^3 = \frac{3V}{4\pi}$; $r = 4.5$ in.

25. $6\frac{2}{5}$

27. $1\frac{1}{4}$

Section 1.5 Converting Units of Measure
(pages 35–37)

1. yes; Because 1 centimeter is equal to 10 millimeters, the conversion factor equals 1.

3. 6.25 ft; The other three represent the same length.

5. 11 yd, 33 ft

7. 12.63

9. 1.22

11. 0.19

13. 37.78

15. 14.4

17. a. about 60.67 m

b. about 8.04 km

19. 1320

21. 112.5

23. 0.001

25. about 0.99 mL/sec

27. 80

29. a. spine-tailed swift; mallard

b. yes, It is faster than all of the other birds in the table. Its dive speed is about 201.25 miles per hour.

31. 34,848

33. 3,000,000,000

35. 0.00042

Section 1.5 Converting Units of Measure *(continued)*

(pages 35–37)

37. **a.** 120 in.^{3} **b.** 138 tissues

39. 113,000 mm^3

41–43.

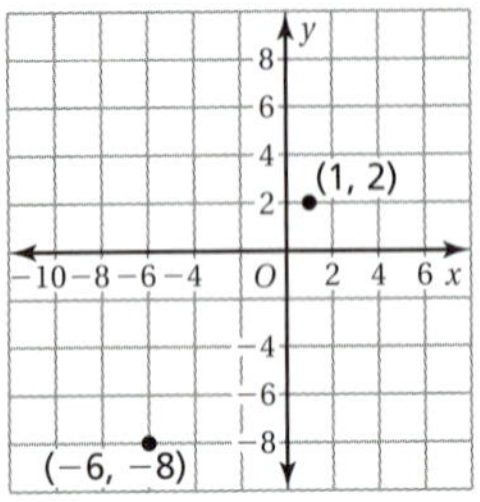

45. B

Section 2.1 Graphing Linear Equations

(pages 52 and 53)

1. a line

3. *Sample answer:*

x	0	1
$y = 3x - 1$	−1	2

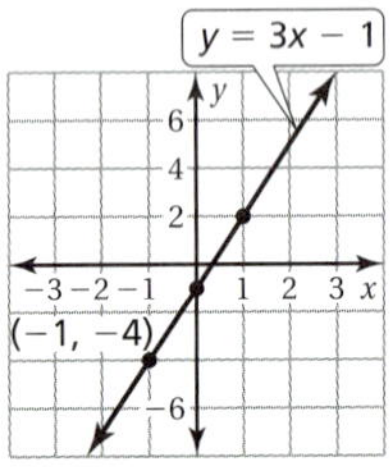

5.

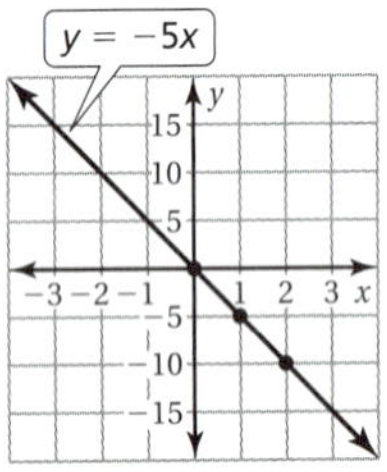

7.

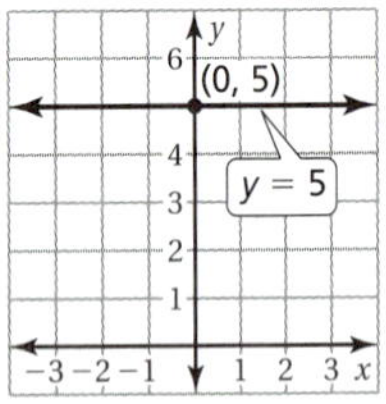

9.

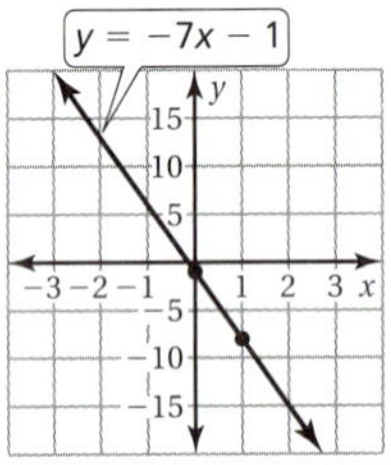

11.

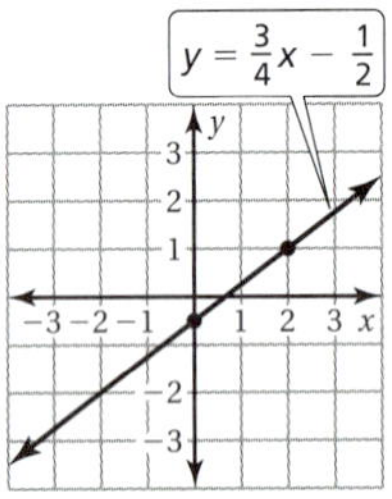

13.

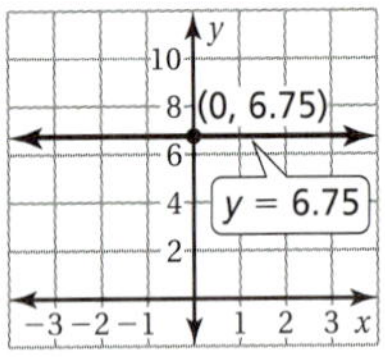

15.

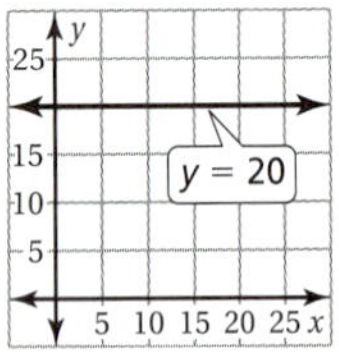

17. $y = 3x + 1$

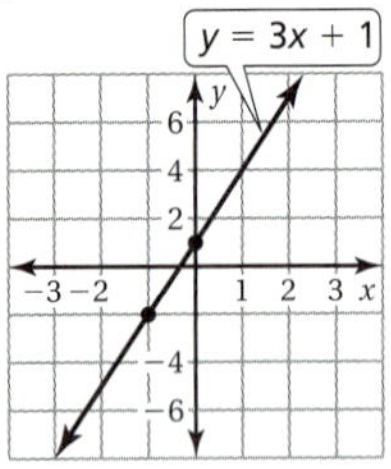

19. $y = 12x - 9$

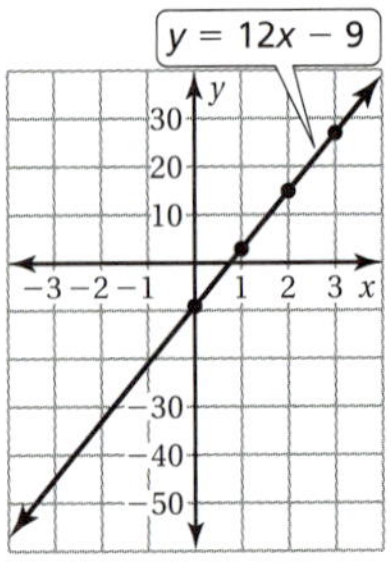

21. **a.** $y = 100 + 12.5x$

b. 6 mo

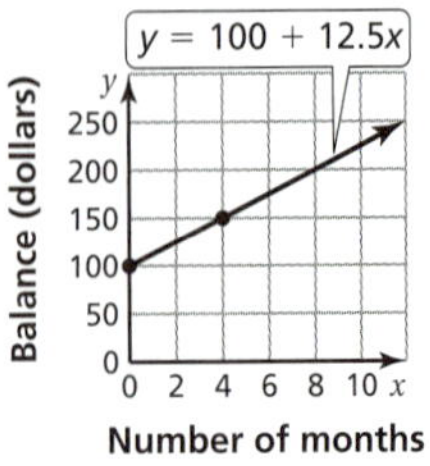

23. **a.** $y = 2x$

b. *Sample answer:* If you are 13 years old, the sea level has risen 26 millimeters since you were born.

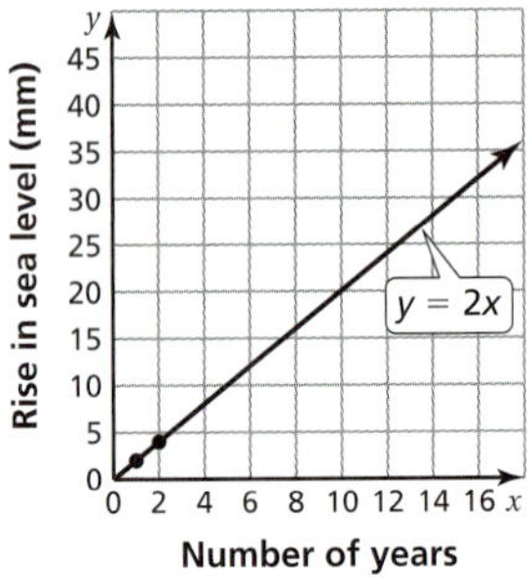

25. (5, 3) **27.** (2, −2) **29.** B

Section 2.2 Slope of a Line *(pages 59–61)*

1. **a.** B and C

b. A

c. no; All of the slopes are different.

3. The line is horizontal.

5.

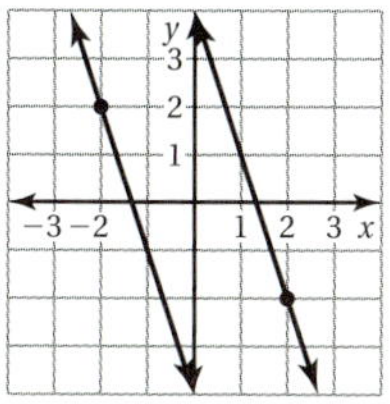

The lines are parallel.

7. $\frac{3}{4}$

9. $-\frac{3}{5}$

11. 0

13. The 2 should be -2 because it goes down.

Slope $= -\frac{2}{3}$

15. 4

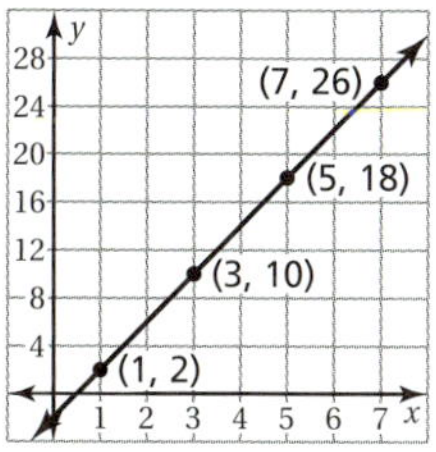

17. $-\frac{3}{4}$

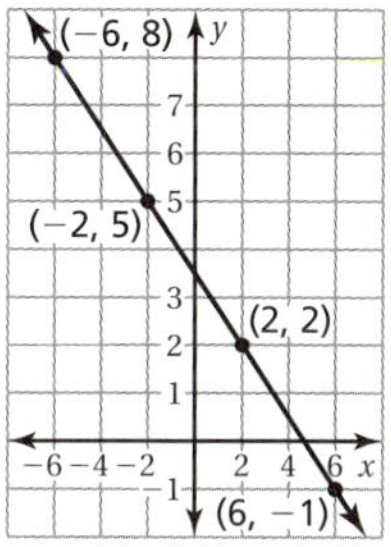

19. $\frac{1}{3}$

21. red and green; They both have a slope of $\frac{4}{3}$.

23. no; Opposite sides have different slopes.

25. **a.** $\frac{3}{40}$

b. The cost increases by \$3 for every 40 miles you drive, or the cost increases by \$0.075 for every mile you drive.

27. You can draw the slide in a coordinate plane and let the x-axis be the ground to find the slope.

29.

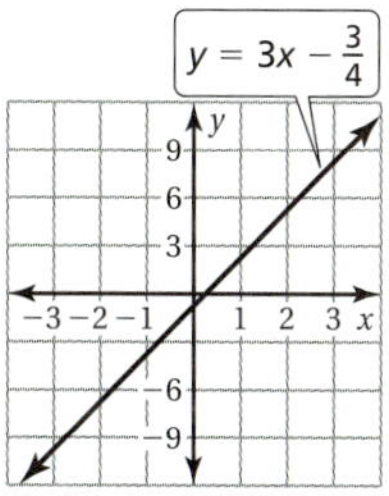

31. B

Section 2.3 Graphing Linear Equations in Slope-Intercept Form *(pages 66 and 67)*

1. Find the x-coordinate of the point where the graph crosses the x-axis.

3. *Sample answer:* The amount of gasoline y (in gallons) left in your tank after you travel x miles is $y = -\frac{1}{20}x + 20$. The slope of $-\frac{1}{20}$ means the car uses 1 gallon of gas for every 20 miles driven. The y-intercept of 20 means there is originally 20 gallons of gas in the tank.

5. A; slope: $\frac{1}{3}$; y-intercept: -2

7. slope: 4; y-intercept: -5

Section 2.3 Graphing Linear Equations in Slope-Intercept Form (pages 66 and 67) (continued)

9. slope: $-\frac{4}{5}$; y-intercept: -2

11. slope: $\frac{4}{3}$; y-intercept: -1

13. slope: -2; y-intercept: 3.5

15. slope: 1.5; y-intercept: 11

17. a.

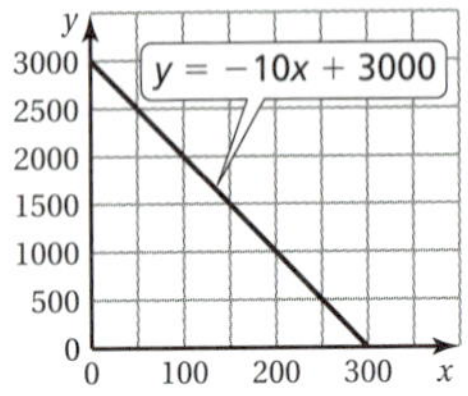

b. The x-intercept of 300 means the skydiver lands on the ground after 300 seconds. The slope of -10 means that the skydiver falls to the ground at a rate of 10 feet per second.

19.

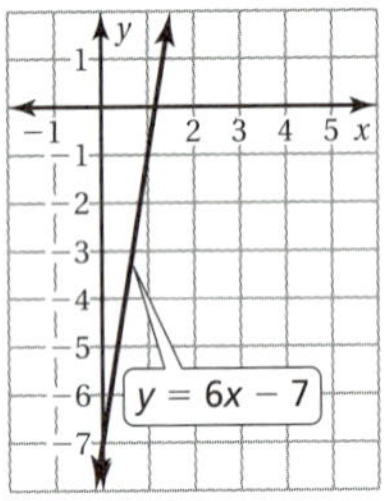

x-intercept: $\frac{7}{6}$

21.

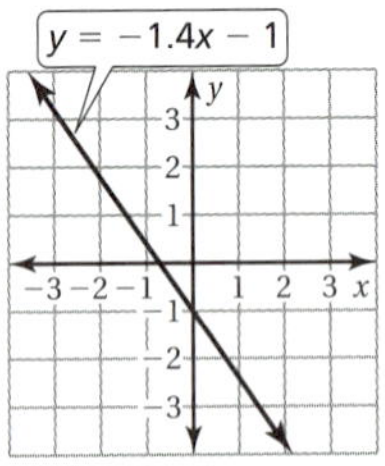

x-intercept: $-\frac{5}{7}$

23.

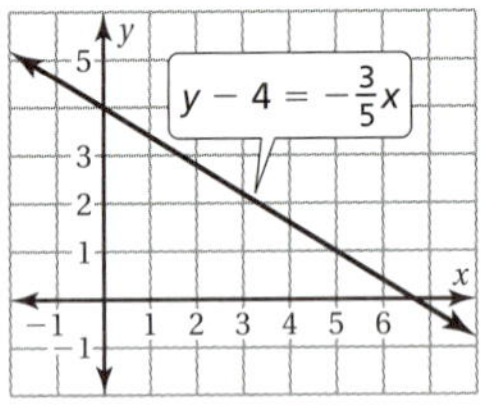

x-intercept: $\frac{20}{3}$

25. $y = 0.75x + 5$

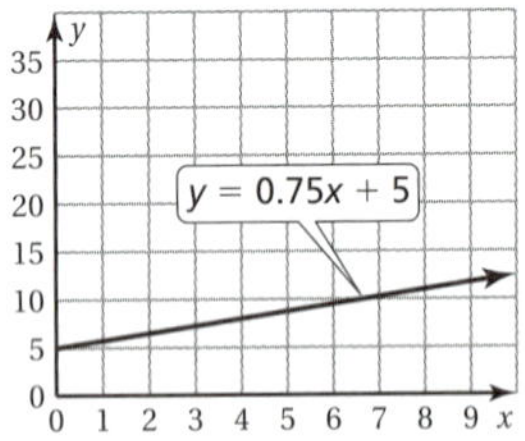

27. $y = 0.15x + 35$

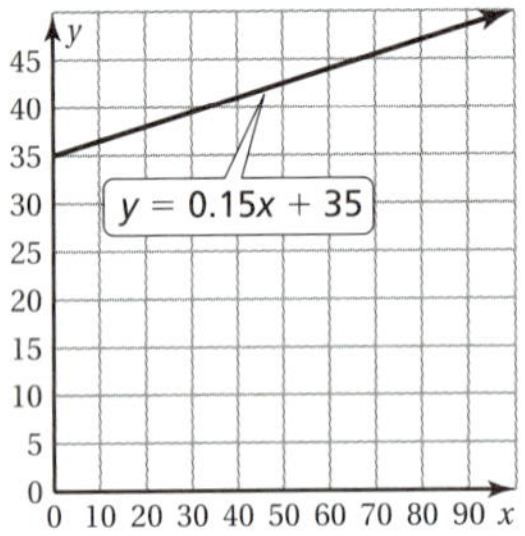

29. $y = 2x + 3$

31. $y = \frac{2}{3}x - 2$

33. B

Section 2.4 Graphing Linear Equations in Standard Form (pages 72 and 73)

1. no; The equation is in slope-intercept form.

3. x = pounds of peaches

y = pounds of apples

$y = -\frac{4}{3}x + 10$

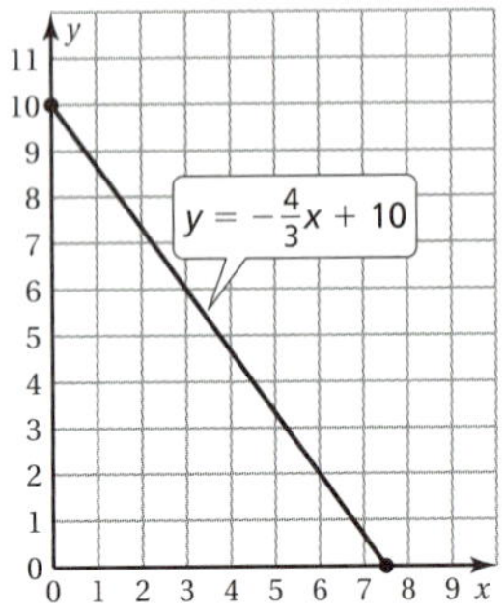

5. $y = -2x + 17$

7. $y = \frac{1}{2}x + 10$

9.

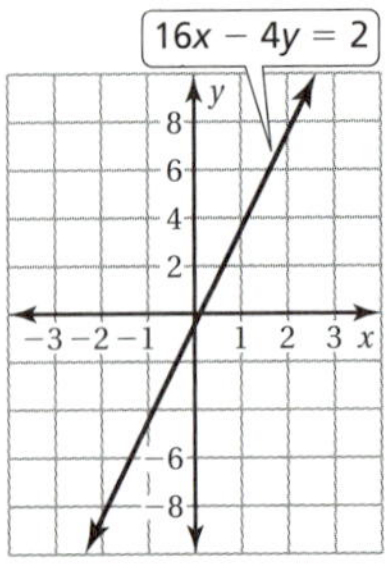

11. x-intercept: -6

y-intercept: 3

13. x-intercept: none

y-intercept: -3

15. a. $y - 25x = 65$

b. $390

17. 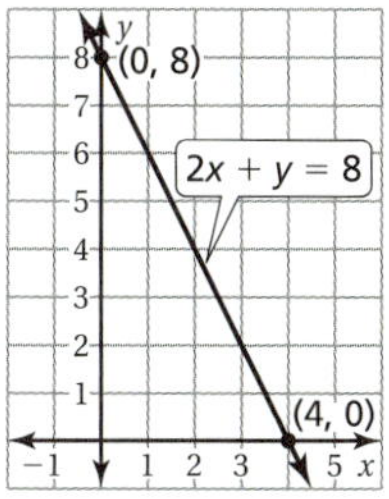

19. x-intercept: 9

y-intercept: 7

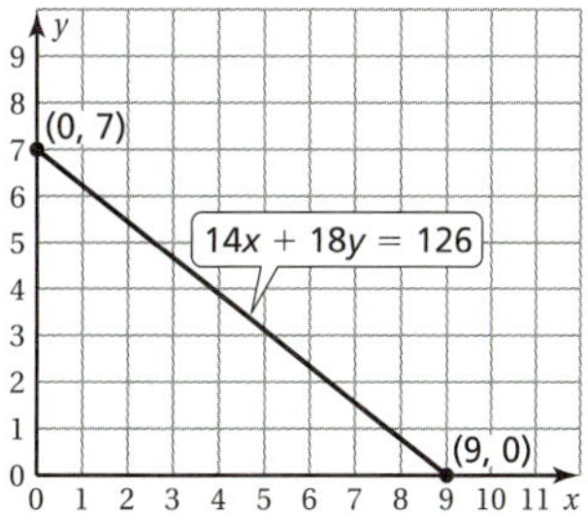

21. a. $9.45x + 7.65y = 160.65$

b. 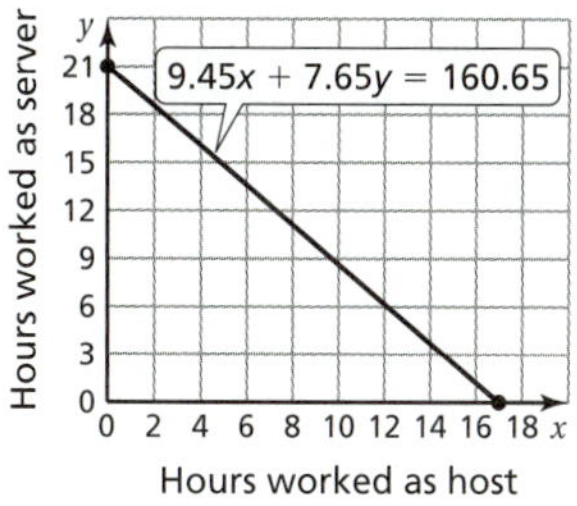

23. a. $y = 40x + 70$

b. x-intercept: $-\frac{7}{4}$; It will not be on the graph because you cannot have a negative time.

c. 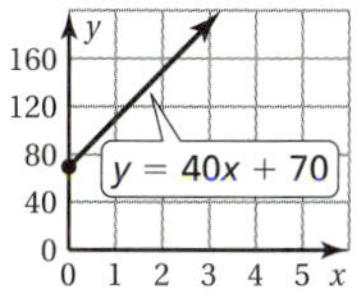

25.

x	-2	-1	0	1	2
$-5 - 3x$	1	-2	-5	-8	-11

Section 2.5

Systems of Linear Equations
(pages 80 and 81)

1. yes; The equations are linear and in the same variables.

3.

x	0	1	2	3	4	5	6
C	150	165	180	195	210	225	240
R	0	45	90	135	180	225	270

(5, 225)

5. (2.5, 6.5)

7. (3, -1)

9. a. $R = 35x$ **b.** 100 rides

11. (-5, 1)

13. (12, 15)

15. (8, 1)

17. a. 6 h **b.** 49 mi

19. yes

21. no

Section 2.6

Special Systems of Linear Equations
(pages 86 and 87)

1. The graph of the system with no solution has two parallel lines, and the graph of a system with infinitely many solutions is one line.

3. one solution; because the lines are not parallel and will not be the same equation

5. no solution

7. infinitely many solutions; all points on the line $y = -\frac{1}{6}x + 5$

9. one solution; (2, -3)

11. no solution

13. no; because they are running at the same speed and your pig had a head start

15. no solution

17. a. 6 h

b. You both work the same number of hours.

Section 2.6 Special Systems of Linear Equations
(pages 86 and 87) (continued)

19. **a.** *Sample answer:* $y = -7$

b. *Sample answer:* $y = 3x$

c. *Sample answer:* $2y - 6x = -2$

21. $x = -3$

23. B

Section 2.7 Solving Equations by Graphing
(pages 92 and 93)

1. algebraic method; Graphing fractions is harder than solving the equation.

3. $x = 6$

5. $x = 6$

7. $x = 3$

9. yes; Because a solution of $3x + 2 = 4x$ exists ($x = 2$).

11. $x = 2$

13. The two lines are parallel, which means there is no solution. Using an algebraic method, you obtain $-5 = 8$, which is not true and means that there is no solution.

15. Organize the home and away games for last year and this year in a table before solving.

17. 4

19. -3

21. A

Section 3.1 Writing Equations in Slope-Intercept Form
(pages 110 and 111)

1. *Sample answer:* Find the ratio of the rise to the run between the intercepts.

3. $y = 3x + 2$; $y = 3x - 10$; $y = 5$; $y = -1$

5. $y = x + 4$

7. $y = \frac{1}{4}x + 1$

9. $y = \frac{1}{3}x - 3$

11. The x-intercept was used instead of the y-intercept. $y = \frac{1}{2}x - 2$

13. $y = 5$

15. $y = -2$

17. **a–b.**

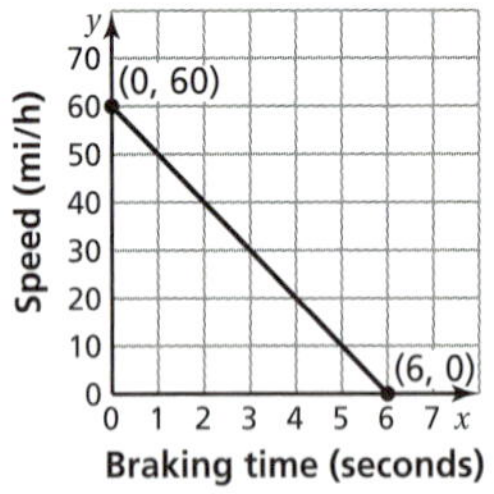

(0, 60) represents the speed of the automobile before braking. (6, 0) represents the amount of time it takes to stop. The line represents the speed y of the automobile after x seconds of braking.

c. $y = -10x + 60$

19. Be sure to check that your rate of growth will not lead to a 0-year-old tree with a negative height.

21–23.

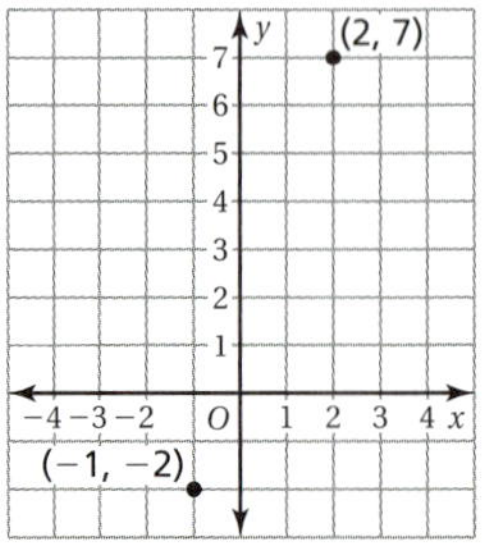

Section 3.2 Writing Equations Using a Slope and a Point
(pages 116 and 117)

1. *Sample answer:* slope and a point

3. $y = \frac{1}{2}x + 1$ **5.** $y = -3x + 8$ **7.** $y = \frac{3}{4}x + 5$

9. $y = -\frac{1}{7}x - 4$ **11.** $y = -2x - 6$ **13.** $V = \frac{2}{25}T + 22$

15. The rate of change is 0.25 degree per chirp.

17. **a.** $y = -0.03x + 2.9$

b. 2 g/cm^2

c. *Sample answer:* Eventually $y = 0$, which means the astronaut's bones will be very weak.

19. B

Section 3.3 Writing Equations Using Two Points
(pages 122 and 123)

1. Plot both points and draw the line that passes through them. Use the graph to find the slope and y-intercept. Then write the equation in slope-intercept form.

3. slope $= -1$; y-intercept: 0; $y = -x$

5. slope $= \frac{1}{3}$; y-intercept: -2; $y = \frac{1}{3}x - 2$

7. $y = 2x$ **9.** $y = \frac{1}{4}x$ **11.** $y = x + 1$ **13.** $y = \frac{3}{2}x - 10$

15. They switched the slope and y-intercept in the equation. $y = 2x - 4$

17. **a.**

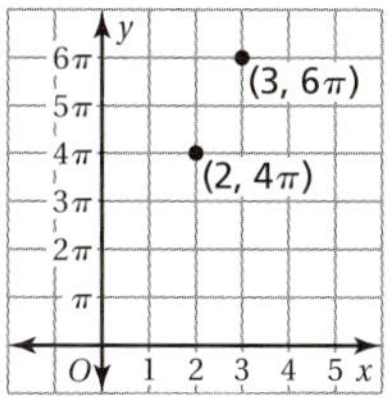

b. $y = 2\pi x$

19. **a.** $y = -2000x + 21{,}000$

b.

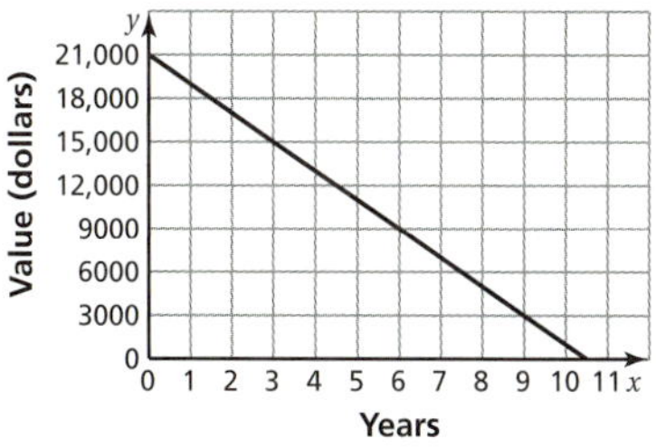

c. \$21,000; the original price of the car

21. **a.** $y = 14x - 108.5$

b. 4 m

23. 175 **25.** D

Section 3.4 Solving Real-Life Problems
(pages 130 and 131)

1. The y-intercept is -6 because the line crosses the y-axis at the point $(0, -6)$. The x-intercept is 2 because the line crosses the x-axis at the point $(2, 0)$. You can use these two points to find the slope.

$$\text{Slope} = \frac{\text{change in } y}{\text{change in } x} = \frac{6}{2} = 3$$

3. *Sample answer:* the rate at which something is happening

5. *Sample answer:* On a visit to Mexico, you spend 45 pesos every week. After 4 weeks, you have no pesos left.

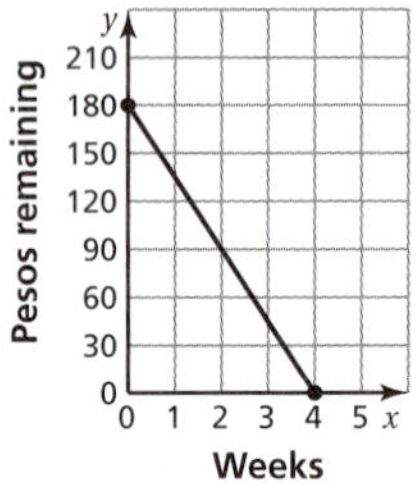

7. **a.** slope: -3.6; y-intercept: 59 **b.** $y = -3.6x + 59$

c. 59°F

9. **a.** Antananarivo: 19°S, 47°E; Denver: 39°N, 105°W; Brasilia: 16°S, 48°W; London: 51°N, 0°W; Beijing: 40°N, 116°E

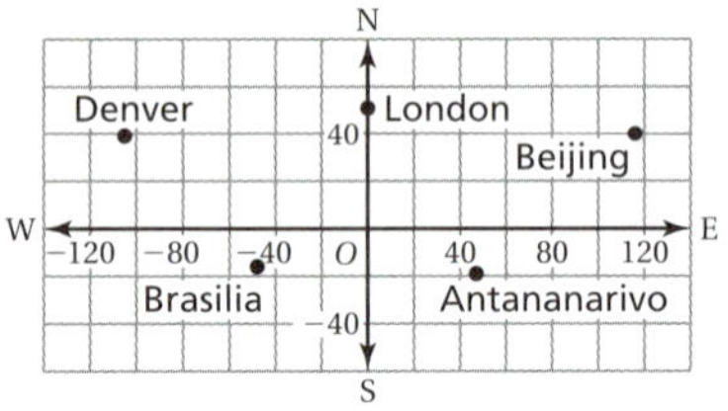

b. $y = \frac{1}{221}x + \frac{8724}{221}$

c. a place that is on the prime meridian

11. infinitely many solutions **13.** no solution

Section 3.5 Writing Systems of Linear Equations
(pages 136 and 137)

1. because its graph is a line

3. You can use a table to see when the two equations are equal. You can use a graph to see whether or not the two lines intersect. You can use algebra and set the equations equal to each other to see when they have the same value.

5. **a.** $x + y = 12$

$3x + 2y = 32$

b.

x	0	1	2	3	4	5	6	7	8
$y = 12 - x$	12	11	10	9	8	7	6	5	4
$y = 16 - \frac{3}{2}x$	16	14.5	13	11.5	10	8.5	7	5.5	4

8 lilies and 4 tulips

c.

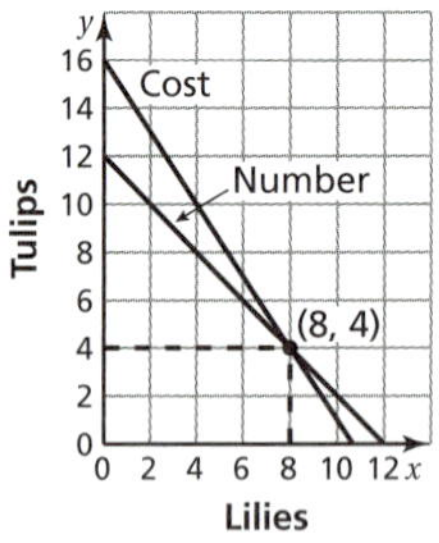

8 lilies and 4 tulips

d. $12 - x = 16 - \frac{3}{2}x$;

$x = 8, y = 4$;

8 lilies and 4 tulips

7. **a.** no; You need to know how many more dimes there are than nickels or how many coins there are total.

b. *Sample answer:* 9 dimes and 1 nickel

9. no; A linear system must have either one, none, or infinitely many solutions. Lines cannot intersect at exactly two points.

11. Each equation is the same. So, the graph of the system is the same line.

13. $(1, 0), (-2, 3), (-6, 1)$ **15.** $y = \frac{1}{4}x - 2$ **17.** B

Section 4.1 Domain and Range of a Function
(pages 152 and 153)

1. no; The equation is not solved for y.

3. **a.** $y = 6 - 2x$ **b.** domain: 0, 1, 2, 3; range: 6, 4, 2, 0

c. $x = 6$ is not in the domain because it would make y negative, and it is not possible to buy a negative number of headbands.

5. domain: −2, −1, 0, 1, 2; range: −2, 0, 2

7. The domain and range are switched. The domain is −3, −1, 1, and 3. The range is −2, 0, 2, and 4.

9.

x	−1	0	1	2
y	−4	2	8	14

domain: −1, 0, 1, 2

range: −4, 2, 8, 14

11.

x	−1	0	1	2
y	1.5	3	4.5	6

domain: −1, 0, 1, 2

range: 1.5, 3, 4.5, 6

13. Rewrite the percent as a fraction or decimal before writing an equation.

15.

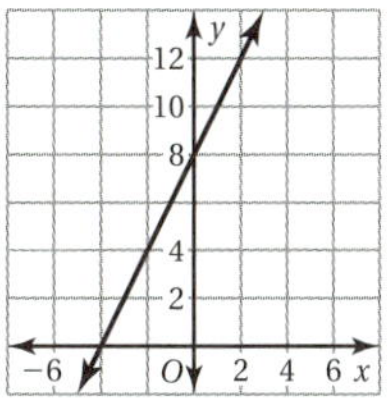

17.

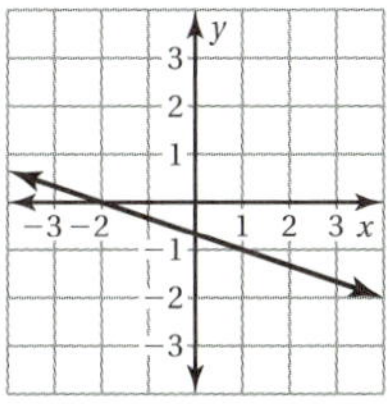

19. D

Section 4.2 Discrete and Continuous Domains
(pages 158 and 159)

1. A discrete domain consists of only certain numbers in an interval, whereas a continuous domain consists of all numbers in an interval.

3. domain: $x \geq 0$ and $x \leq 6$

range: $y \geq 0$ and $y \leq 6$;

continuous

5.

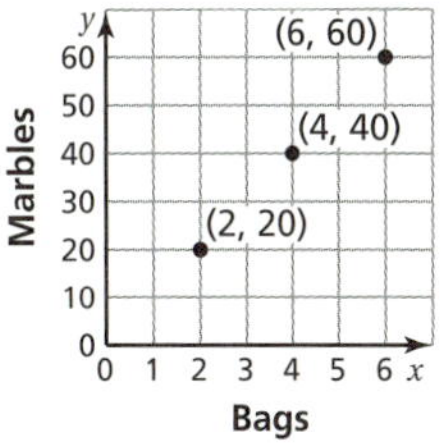

discrete

7.

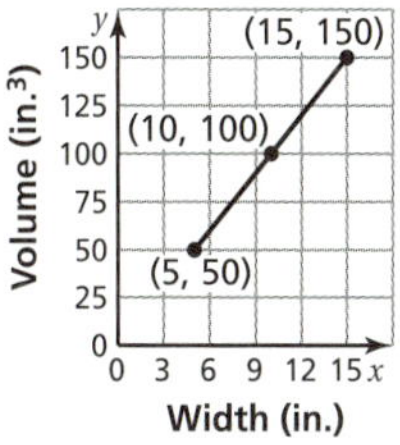

continuous

9. The domain is discrete because only certain numbers are inputs.

11. The function with an input of length has a continuous domain because you can use any length, but you cannot have half a shirt.

13.

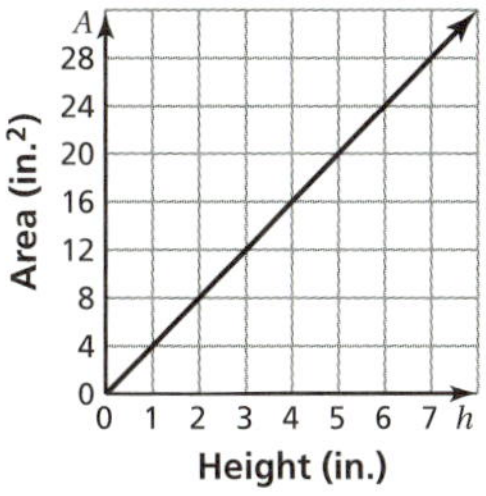

continuous

15. Before writing a function, draw one possible arrangement to understand the problem.

17. $-\frac{5}{2}$

19. C

Section 4.3 Linear Function Patterns
(pages 166 and 167)

1. words, equation, table, graph

3. $y = \pi x$; x is the diameter; y is the circumference.

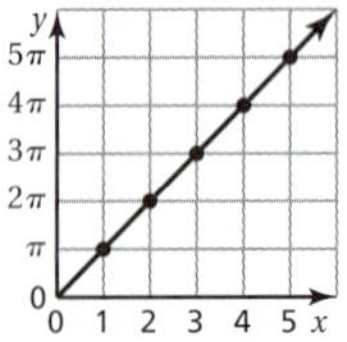

5. $y = \frac{4}{3}x + 2$

7. $y = 3$

9. $y = -\frac{1}{4}x$

11. a.

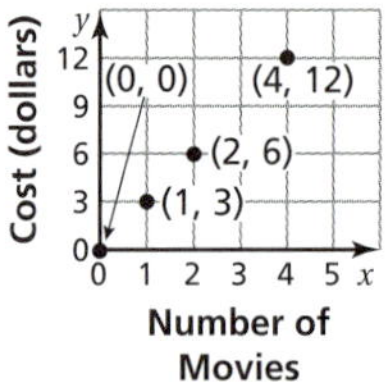

discrete

b. $y = 3x$

c. \$9

13. Substitute 8 for t in the equation.

15. 5%

17. B

Section 4.4 Comparing Linear and Nonlinear Functions
(pages 172 and 173)

1. A linear function has a constant rate of change. A nonlinear function does not have a constant rate of change.

3.

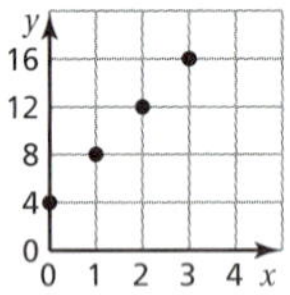

linear

5.

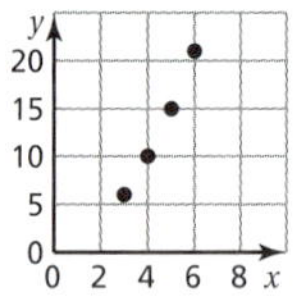

nonlinear

7. linear; The graph is a line.

9. linear; As x increases by 6, y increases by 4.

11. nonlinear; As x increases by 1, V increases by different amounts.

13. linear; The equation can be written in slope-intercept form.

15. Because you want the table to represent a linear function and 3 is half-way between 2 and 4, the missing value is half-way between 2.80 and 5.60.

17. nonlinear; The graph is not a line.

19. linear

21. straight

23. right

Section 5.1 Classifying Angles
(pages 188 and 189)

1. The sum of the measures of two complementary angles is 90°. The sum of the measures of two supplementary angles is 180°.

3. sometimes; Either x or y may be obtuse.

5. never; Because x and y must both be less than 90° and greater than 0°.

7. complementary

9. supplementary

11. neither

13. 128

15. Vertical angles are congruent. The value of x is 35.

17. 37

19. 20

21. **a.** $\angle CBD$ and $\angle DBE$; $\angle ABF$ and $\angle FBE$

b. $\angle ABE$ and $\angle CBE$; $\angle ABD$ and $\angle CBD$; $\angle CBF$ and $\angle ABF$

23. 54°

25. $7x + y + 90 = 180$; $5x + 2y = 90$; $x = 10$; $y = 20$

27. 29.3

29. B

Section 5.2 Angles and Sides of Triangles
(pages 194 and 195)

1. An equilateral triangle has three congruent sides. An isosceles triangle has at least two congruent sides. So, an equilateral triangle is a specific type of isosceles triangle.

3. right isosceles triangle

5. obtuse isosceles triangle

7. 94; obtuse triangle

9. 67.5; acute isosceles triangle

11. 24; obtuse isosceles triangle

13. **a.** 70 **b.** acute isosceles triangle

15. no; 39.5°

17. yes

19. If two angle measures of a triangle were greater than or equal to 90°, the sum of those two angle measures would be greater than or equal to 180°. The sum of the three angle measures would be greater than 180°, which is not possible.

21. $x + 2x + 2x + 8 + 5 = 48$; 7

23. $4x - 4 + 3\pi = 25.42$ or $2x - 4 = 6$; 5

Section 5.3 Angles of Polygons
(pages 201–203)

1. 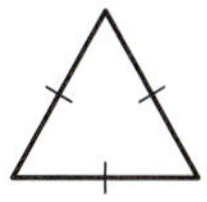

3. What is the measure of an angle of a regular pentagon?; 108°; 540°

5. 1260°

7. 720°

9. 1080°

11. no; The angle measures given add up to 535°, but the sum of the angle measures of a pentagon is 540°.

13. 135

15. 140°

17. 140°

19. The sum of the angle measures should have been divided by the number of angles, 20. $3240° \div 20 = 162°$; The measure of each angle is 162°.

21. 24 sides

23. convex; No line segment connecting two vertices lies outside the polygon.

25. no; All of the angles would not be congruent.

27. 135°

29. 120°

31. You can determine if it is a linear function by writing an equation or by graphing the points.

33. 9

35. 3

37. D

Section 5.4 Using Similar Triangles
(pages 210 and 211)

1. Write a proportion that uses the missing measurement because the ratios of corresponding side lengths are equal.

3. Student should draw a triangle with the same angle measures as the textbook. The ratio of the corresponding side lengths, $\frac{\text{student's triangle length}}{\text{book's triangle length}}$, should be greater than one.

5. yes; The triangles have the same angle measures, 107°, 39°, and 34°.

7. no; The triangles do not have the same angle measures.

9. The numerators of the fractions should be from the same triangle.

$$\frac{18}{16} = \frac{x}{8}$$
$$16x = 144$$
$$x = 9$$

11. 65

13. no; Each side increases by 50%, so each side is multiplied by a factor of $\frac{3}{2}$. The area is $\frac{3}{2}\left(\frac{3}{2}\right) = \frac{9}{4}$ or 225% of the original area, which is a 125% increase.

15. When two triangles are similar, the ratios of corresponding sides are equal.

17. linear; The equation can be rewritten in slope-intercept form.

19. nonlinear; The equation cannot be rewritten in slope-intercept form.

Section 5.5 Parallel Lines and Transversals
(pages 217–219)

1. *Sample answer:*

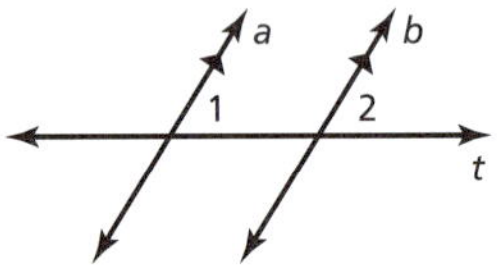

3. m and n

5. 8

7. $\angle 1 = 107°$, $\angle 2 = 73°$

9. $\angle 5 = 49°$, $\angle 6 = 131°$

11. 60°; Corresponding angles are congruent.

13. $\angle 1$, $\angle 3$, $\angle 5$, and $\angle 7$ are congruent. $\angle 2$, $\angle 4$, $\angle 6$, and $\angle 8$ are congruent.

15. $\angle 6 = 61°$; $\angle 6$ and the given angle are vertical angles.
$\angle 5 = 119°$ and $\angle 7 = 119°$; $\angle 5$ and $\angle 7$ are supplementary to the given angle.
$\angle 1 = 61°$; $\angle 1$ and the given angle are corresponding angles.
$\angle 3 = 61°$; $\angle 1$ and $\angle 3$ are vertical angles.
$\angle 2 = 119°$ and $\angle 4 = 119°$; $\angle 2$ and $\angle 4$ are supplementary to $\angle 1$.

17. $\angle 2 = 90°$; $\angle 2$ and the given angle are vertical angles.
$\angle 1 = 90°$ and $\angle 3 = 90°$; $\angle 1$ and $\angle 3$ are supplementary to the given angle.
$\angle 4 = 90°$; $\angle 4$ and the given angle are corresponding angles.
$\angle 6 = 90°$; $\angle 4$ and $\angle 6$ are vertical angles.
$\angle 5 = 90°$ and $\angle 7 = 90°$; $\angle 5$ and $\angle 7$ are supplementary to $\angle 4$.

19. 132°; *Sample answer:* $\angle 2$ and $\angle 4$ are alternate interior angles and $\angle 4$ and $\angle 3$ are supplementary.

21. 120°; *Sample answer:* ∠6 and ∠8 are alternate exterior angles.

23. 61.3°; *Sample answer:* ∠3 and ∠1 are alternate interior angles and ∠1 and ∠2 are supplementary.

25. They are all right angles because perpendicular lines form 90° angles.

27. 130

29. **a.** no; They look like they are spreading apart. **b.** Check students' work.

31. 13 **33.** 51 **35.** B

Section 6.1 Finding Square Roots
(pages 234 and 235)

1. no; There is no integer whose square is 26.

3. $\sqrt{256}$ represents the positive square root because there is not a – or a ± in front.

5. 1.3 km **7.** 3 and −3 **9.** 2 and −2

11. 25 **13.** $\frac{1}{31}$ and $-\frac{1}{31}$ **15.** 2.2 and −2.2

17. The positive and negative square roots should have been given.

$\pm\sqrt{\frac{1}{4}} = \frac{1}{2}$ and $-\frac{1}{2}$

19. 9 **21.** 25 **23.** 40

25. because a negative radius does not make sense

27. = **29.** 9 ft **31.** 8 m/sec **33.** 2.5 ft

35. 25 **37.** 144 **39.** B

Section 6.2 The Pythagorean Theorem
(pages 240 and 241)

1. The hypotenuse is the longest side and the legs are the other two sides.

3. 24 cm **5.** 9 in. **7.** 12 ft

9. The length of the hypotenuse was substituted for the wrong variable.

$$\begin{aligned} a^2 + b^2 &= c^2 \\ 7^2 + b^2 &= 25^2 \\ 49 + b^2 &= 625 \\ b^2 &= 576 \\ b &= 24 \end{aligned}$$

11. 16 cm

13. 10 ft

15. 8.4 cm

17. **a.** *Sample answer:*

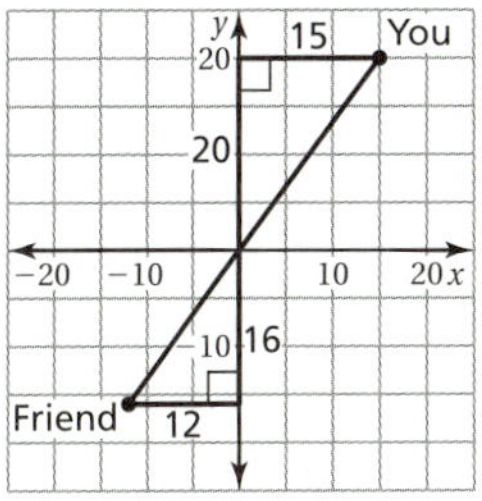

b. 45 ft

19. 6 and −6

21. 13

23. C

Section 6.3 Approximating Square Roots
(pages 249–251)

1. A rational number can be written as the ratio of two integers. An irrational number cannot be written as the ratio of two integers.

3. all rational and irrational numbers; *Sample answer:* $-2, \frac{1}{8}, \sqrt{7}$

5. yes

7. no

9. rational; $3.\overline{6}$ is a repeating decimal.

11. irrational; 7 is not a perfect square.

13. rational; $-3\frac{8}{9}$ can be written as the ratio of two integers.

15. 144 is a perfect square. So, $\sqrt{144}$ is rational.

17. **a.** natural number **b.** irrational number **c.** irrational number

19. 26

21. -10

23. -13

25. 10; 10 is to the right of $\sqrt{20}$.

27. $\sqrt{133}$; $\sqrt{133}$ is to the right of $10\frac{3}{4}$.

29. -0.25; -0.25 is to the right of $-\sqrt{0.25}$.

31. 8 ft

33. *Sample answer:* $a = 82$, $b = 97$

35. 1.1

37. 30.1 m/sec

39. Falling objects do not fall at a linear rate. Their speed increases with each second they are falling.

41. $-3x + 3y$

43. $40k - 9$

Section 6.4 Simplifying Square Roots
(pages 256 and 257)

1. *Sample answer:* The square root is like a variable. So, you add or subtract the number in front to simplify.

3. about 1.62; yes

5. about 1.11; no

7. $\frac{\sqrt{7}+1}{3}$

9. $6\sqrt{3}$

11. $2\sqrt{5}$

13. $-7.7\sqrt{15}$

15. You do not add the radicands. $4\sqrt{5} + 3\sqrt{5} = 7\sqrt{5}$

17. $10\sqrt{2}$

19. $4\sqrt{3}$

21. $\frac{\sqrt{23}}{8}$

23. $\frac{\sqrt{17}}{7}$

25. $10\sqrt{2}$ in.

27. $6\sqrt{6}$

29. 210 ft^3

31. **a.** $88\sqrt{2}$ ft **b.** 680 ft^2

33. Remember to take the square root of each side when solving for r.

35. 24 in.

37. C

Section 6.5 Using the Pythagorean Theorem
(pages 262 and 263)

1. *Sample answer:* You can plot a point at the origin and then draw lengths that represent the legs. Then, you can use the Pythagorean Theorem to find the hypotenuse of the triangle.

3. 27.7 m
5. 11.3 yd
7. 7.2 units
9. 27.5 ft
11. 15.1 m
13. yes
15. no
17. yes
19. 12.8 ft
21. a. *Sample answer:* 5 in., 7 in., 3 in.
    b. *Sample answer:* $BC \approx 8.6$ in.; $AB \approx 9.1$ in.
    c. Check students' work.
23. mean: 13; median: 12.5; mode: 12
25. mean: 58; median: 59; mode: 59

Section 7.1 Measures of Central Tendency
(pages 278 and 279)

1. no; The definition of an outlier means that it is not in the center of the data.
3. If the outlier is greater than the mean, removing it will decrease the mean. If the outlier is less than the mean, removing it will increase the mean.
5. mean: 1; median: 1; mode: -1
7. mean: $1\frac{29}{30}$ h; median: 2 h; modes: $1\frac{2}{3}$ h and 2 h
9. They calculated the mean, not the median. Test scores: 80, 80, 90, 90, 90, 98

   $\text{Median} = \frac{90 + 90}{2} = \frac{180}{2} = 90$

11. 4
13. 16
15. a. 105°F b. mean
17. The mean and median both decrease by \$0.05. There is still no mode.
19. $-8, -5, -3, 1, 4, 7$
21. B

Section 7.2 Box-and-Whisker Plots
(pages 284 and 285)

1. 25%; 50%
3. The length gives the range of the data set. This tells how much the data vary.
5. 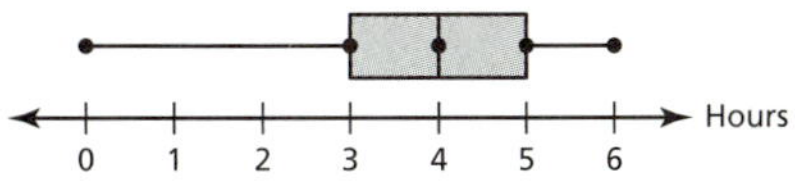

7. 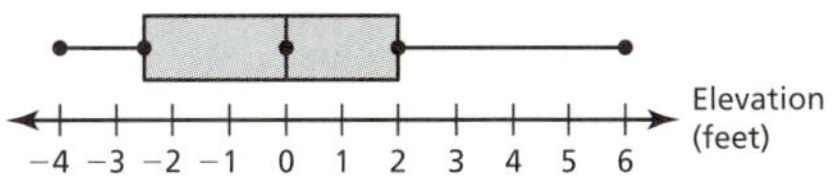

9. 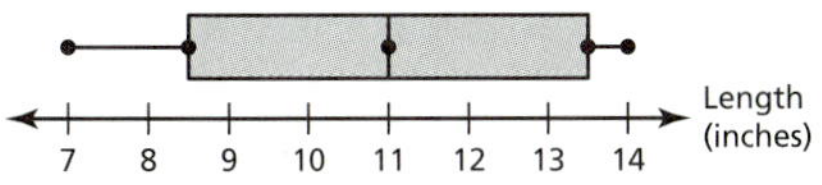

range = 7

Section 7.2 Box-and-Whisker Plots *(continued)*

(pages 284 and 285)

11. a. 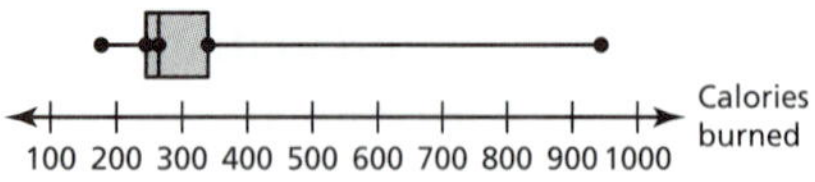

b. 944 calories

c. 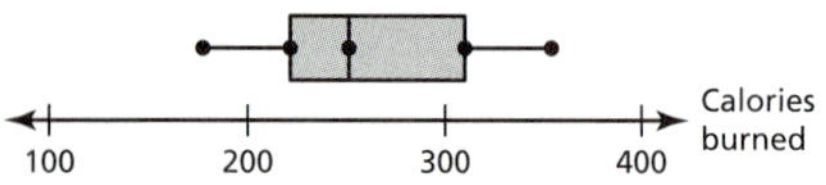

d. The outlier makes the right whisker longer, increases the length of the box, increases the third quartile, and increases the median. In this case, the first quartile and the left whisker were not affected.

13. *Sample answer:* 0, 5, 10, 10, 10, 15, 20

15. *Sample answer:* 1, 7, 9, 10, 11, 11, 12

17. $y = 3x + 2$

19. $y = -\frac{1}{4}x$

21. B

Section 7.3 Scatter Plots and Lines of Best Fit

(pages 293–295)

1. They must be ordered pairs so there are equal amounts of x- and y-values.

3. a–b. 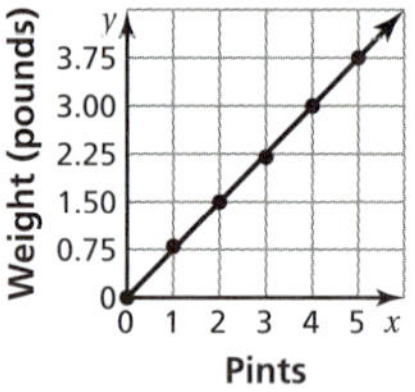

c. *Sample answer:* $y = 0.75x$

d. *Sample answer:* 7.5 lb

e. *Sample answer:* \$16.88

5. a. 3.5 h **b.** \$85

c. There is a positive relationship between hours worked and earnings.

7. positive relationship

9. negative relationship

11. a–b. 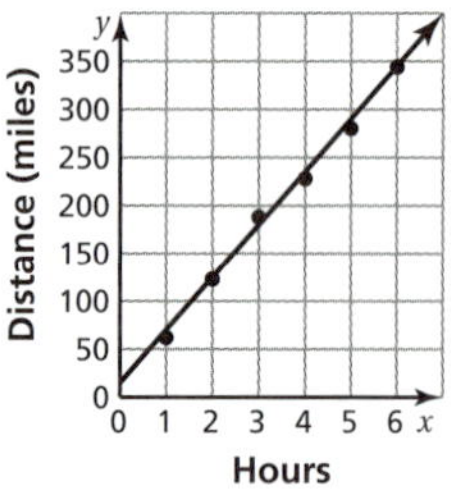

c. *Sample answer:* $y = 55x + 15$

d. *Sample answer:* 400 mi

13. a. positive relationship

b. The more time spent studying, the better the test score.

15. The slope of the line of best fit should be close to 1.

17. 2

19. -4

Section 7.4 Choosing a Data Display
(pages 300 and 301)

1. yes; Different displays may show different aspects of the data.

3. *Sample answer:*

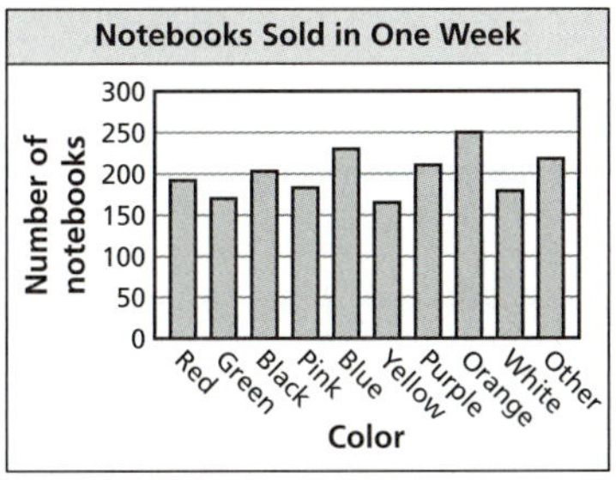

A bar graph shows the data in different color categories.

5. *Sample answer:* Line graph: shows changes over time.

7. *Sample answer:* Line graph: shows changes changes over time.

9. The pictures of the bikes are larger on Monday, which makes it seem like the distance is the same each day.

11. The intervals are not the same size.

13. *Sample answer:* bar graph; Each bar can represent a different vegetable.

15. *Sample answer:* line plot

17. Does one display better show the differences in digits?

19. $8x = 24$

Section 8.1 Writing and Graphing Inequalities
(pages 316 and 317)

1. An open circle would be used because 250 is not a solution.

3. no; $x \geq -9$ is all values of x greater than or equal to -9. $-9 \geq x$ is all values of x less than or equal to -9.

5. $x < -3$; all values of x less than -3

7. $y + 5.2 < 23$

9. $k - 8.3 > 48$

11. yes

13. yes

15. no

17. 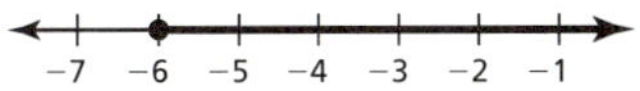

19. 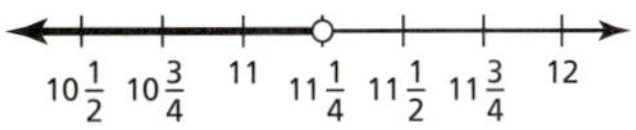

21. $x \geq 21$

23. yes

25. **a.** $a \geq 10$; 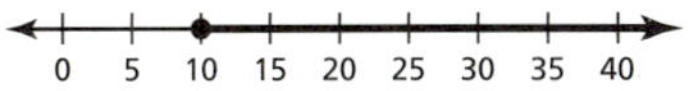

$s \geq 200$; 

$t \geq 10$; 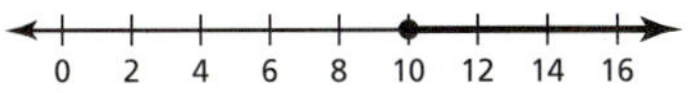

b. yes; You satisfy the swimming requirement of the course because $10(25) = 250$ and $250 \geq 200$.

27. **a.** $m < n$; $n \leq p$ **b.** $m < p$

c. no; Because n is no more than p and m is less than n, m cannot be equal to p.

29. -1.7

31. D

Section 8.2 Solving Inequalities Using Addition or Subtraction *(pages 322 and 323)*

1. no; The solution of $r - 5 \leq 8$ is $r \leq 13$ and the solution of $8 \leq r - 5$ is $r \geq 13$.

3. *Sample answer:* $A = 350$, $C = 275$, $Y = 3105$, $T = 50$, $N = 2$

5. *Sample answer:* $A = 400$, $C = 380$, $Y = 6510$, $T = 83$, $N = 0$

7. $t > 4$; 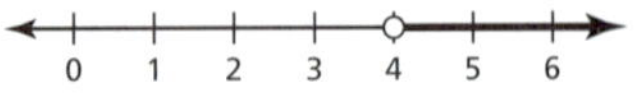

9. $a > -8$; 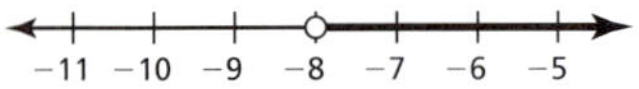

11. $-\frac{3}{5} > d$; 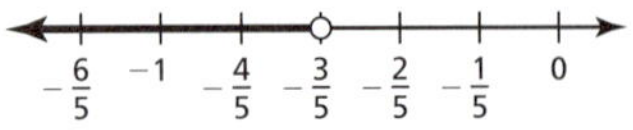

13. $m \leq 1$; 

15. $h < -1.5$; 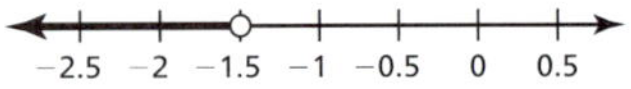

17. $9.5 \geq u$; 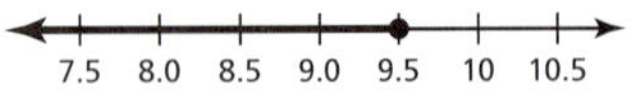

19. **a.** $100 + V \leq 700$; $V \leq 600$ in.^{3} **b.** $V \leq \frac{700}{3}$ in.^{3}

21. $x + 2 > 10$; $x > 8$

23. 5

25. **a.** $4500 + x \geq 12{,}000$; $x \geq 7500$ points

b. This changes the number added to x by 60%, so the inequality becomes $7200 + x \geq 12{,}000$. So, you need less points to advance to the next level.

27. $2\pi h + 2\pi \leq 15\pi$; $h \leq 6.5$ mm

29. 10

31. 12

Section 8.3 Solving Inequalities Using Multiplication or Division *(pages 331–333)*

1. Multiply each side of the inequality by 6.

3. *Sample answer:* $-3x < 6$

5. $x \geq -1$

7. $x \leq -3$

9. $x \leq \frac{3}{2}$

11. $c \leq -36$; 

13. $x < -28$; 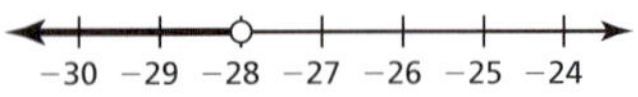

15. $k > 2$; 

17. $y \leq -4$; 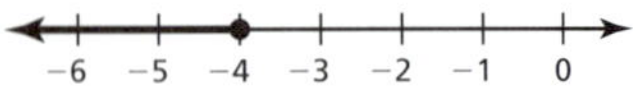

19. The inequality sign should not have been reversed.

$$\frac{x}{2} < -5$$

$$2 \cdot \frac{x}{2} < 2 \cdot (-5)$$

$$x < -10$$

21. $\frac{x}{8} < -2$; $x < -16$

23. $5x > 20$; $x > 4$

25. $0.25x \leq 3.65$; $x \leq 14.6$; You can make at most 14 copies.

27. $n \geq -5$; 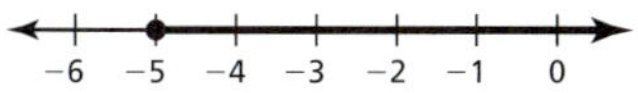

29. $h \leq -42$; 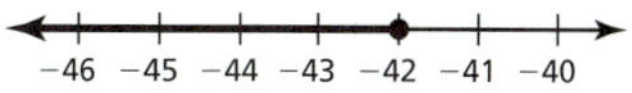

31. $y > \frac{11}{2}$; 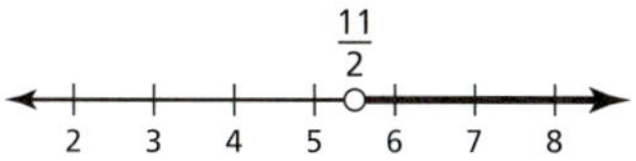

33. $m > -12$; 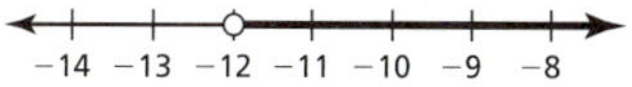

35. $b > 4$; 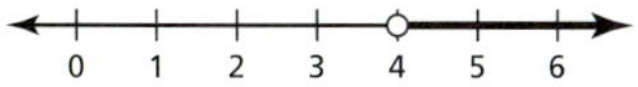

37. no; You need to solve the inequality for x. The solution is $x < 0$. Therefore, numbers greater than 0 are not solutions.

39. $12x \geq 102$; $x \geq 8.5$ cm

41. $\frac{x}{4} < 80$; $x < \$320$

43. *Answer should include, but is not limited to:* Using the correct number of months that the CD has been out. In part (d), an acceptable answer could be never because the top selling CD could have a higher monthly average.

45. $n \geq -6$ and $n \leq -4$;

47. $m < 20$;

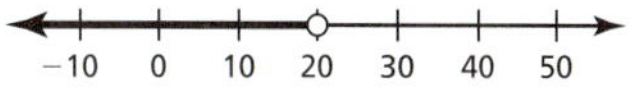

49. $8\frac{1}{4}$

51. 84

Section 8.4 Solving Multi-Step Inequalities
(pages 338 and 339)

1. *Sample answer:* They use the same techniques, but when solving an inequality, you must be careful to reverse the inequality symbol when you multiply or divide by a negative number.

3. $k > 0$ and $k \leq 16$ units

5. $b \geq 1$; 

7. $m \geq -15$; 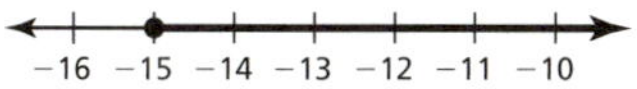

9. $p < -1$; 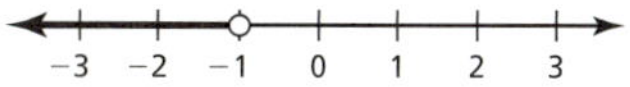

11. They did not perform the operations in proper order.

$$\frac{x}{4} + 6 \geq 3$$
$$\frac{x}{4} \geq -3$$
$$x \geq -12$$

13. $y \leq 13$; 

15. $u < -17$; 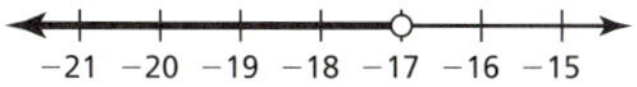

17. $z > -0.9$; 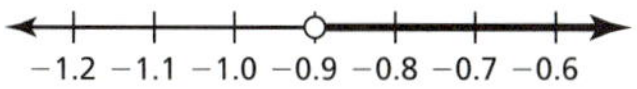

19. $x \leq 6$; 

21. $\frac{3}{16}x + 2 \leq 11$; $x > 0$ and $x \leq 48$ lines

23. Remember to add the height of the truck to find the height the ladder can reach.

25. $r \geq 3$ units

27. 625π in.^{2}

29. A

Section 9.1 Exponents
(pages 354 and 355)

1. An exponent describes the number of times the base is used as a factor. A power is the entire expression (base and exponent). A power tells you the value of the factor and the number of factors. No, the two cannot be used interchangeably.

3. 3^4 **5.** $\left(-\frac{1}{2}\right)^3$ **7.** π^3x^4 **9.** 8^4b^3

11. 25 **13.** 1 **15.** $\frac{1}{144}$

17. The exponent 3 describes how many times the base 6 should be used as a factor. Three should not appear as a factor in the product. $6^3 = 6 \cdot 6 \cdot 6 = 216$

19. $-\left(\frac{1}{4}\right)^4$ **21.** 29 **23.** 5 **25.** 66

27.

h	1	2	3	4	5
$2^h - 1$	1	3	7	15	31
2^{h-1}	1	2	4	8	16

$2^h - 1$; The option $2^h - 1$ pays you more money when $h > 1$.

29. Remember to add the black keys when finding how many notes you travel.

31. Associative Property of Multiplication

33. B

Section 9.2 Product of Powers Property
(pages 360 and 361)

1. When multiplying powers with the same base

3. 3^4 **5.** $(-4)^{12}$ **7.** h^7

9. $\left(-\frac{5}{7}\right)^{17}$ **11.** 5^{12} **13.** 3.8^{12}

15. The bases should not be multiplied. $5^2 \cdot 5^9 = 5^{2+9} = 5^{11}$

17. $216g^3$ **19.** $\frac{1}{25}k^2$ **21.** $r^{12}t^{12}$

23. no; $3^2 + 3^3 = 9 + 27 = 36$ and $3^5 = 243$

25. 496 **27.** 78,125

29. **a.** $16\pi \approx 50.24$ in.^{3}

b. $192\pi \approx 602.88$ in.^{3} Squaring each of the dimensions causes the volume to be 12 times larger.

31. Use the Commutative and Associative Properties of Multiplication to group the powers.

33. 4 **35.** 3 **37.** B

Section 9.3 Quotient of Powers Property
(pages 366 and 367)

1. To divide powers means to divide out the common factors of the numerator and denominator. To divide powers with the same base, write the power with the common base and an exponent found by subtracting the exponent in the denominator from the exponent in the numerator.

3. 6^6 **5.** $(-3)^3$ **7.** 5^6

9. $(-17)^3$ **11.** $(-6.4)^2$ **13.** b^{13}

15. You should subtract the exponents instead of dividing them. $\frac{6^{15}}{6^5} = 6^{15-5} = 6^{10}$

17. 2^9 **19.** π^8 **21.** k^{14}

23. $64x$ **25.** $125a^3b^2$ **27.** x^7y^6

29. You are checking to see if there is a constant rate of change in the prices, not if it is a linear function.

31. 10^{13} galaxies **33.** -9

35. 61 **37.** B

Section 9.4 Zero and Negative Exponents
(pages 374 and 375)

1. no; Any nonzero base raised to the zero power is always 1.

3. $5^{-5}, 5^0, 5^4$

5.

n	4	3	2	1
$\frac{5^n}{5^2}$	$5^2 = 25$	$5^1 = 5$	$5^0 = 1$	$5^{-1} = \frac{1}{5}$

7. One-fifth of 5^1; $5^0 = \frac{1}{5}(5^1) = 1$

9. $\frac{1}{36}$ **11.** $\frac{1}{16}$ **13.** 1 **15.** $\frac{1}{125}$

17. The negative sign goes with the exponent, not the base. $(4)^{-3} = \frac{1}{4^3} = \frac{1}{64}$

19. 2^0; 10^0 **21.** $\frac{a^7}{64}$ **23.** $5b$

25. 12 **27.** $\frac{w^6}{9}$

29. 10,000 micrometers **31.** 1,000,000 micrometers

33. Convert the blood donation to cubic millimeters before answering the parts.

35. If $a = 0$, then $0^n = 0$. Because you can not divide by 0, the expression $\frac{1}{0}$ is undefined.

37. 10^3 **39.** D

Section 9.5 Reading Scientific Notation
(pages 380 and 381)

1. Scientific notation uses a factor of at least one but less than 10 multiplied by a power of 10. A number in standard form is written out with all the zeros and place values included.

3. 0.00015 m

5. 20,000 mm^3

7. yes; The factor is at least 1 and less than 10. The power of 10 has an integer exponent.

9. no; The factor is greater than 10.

11. yes; The factor is at least 1 and less than 10. The power of 10 has an integer exponent.

13. no; The factor is less than 1.

15. 70,000,000

17. 500

19. 0.000044

21. 1,660,000,000

23. 9,725,000

25. a. 810,000,000 platelets
  b. 1,350,000,000,000 platelets

27. a. Bellatrix
  b. Betelgeuse

29. 5×10^{12} km^2

31. Be sure to convert some of the speeds so that they all have the same units.

33. 10^7

35. $\frac{1}{10^{16}}$

Section 9.6 Writing Scientific Notation
(pages 386 and 387)

1. If the number is greater than or equal to 10, the exponent will be positive. If the number is less than 1 and greater than 0, the exponent will be negative.

3. 2.1×10^{-3}

5. 3.21×10^8

7. 4×10^{-5}

9. 4.56×10^{10}

11. 8.4×10^5

13. 72.5 is not less than 10. The decimal point needs to move one more place to the left.
 7.25×10^7

15. 9×10^{-10}

17. 1.6×10^8

19. 2.88×10^{-7}

21. 4.01×10^7 m

23. 5.612×10^{14} cm^2

25. 9.75×10^9 N•m per sec

27. *Answer should include, but is not limited to:* Make sure calculations using scientific notation are done correctly.

29. a. 2.65×10^8
  b. 2.2×10^{-4}

31. 200

Section B.1 Simple and Compound Interest

(pages A14 and A15)

1. simple interest

3. a.

Simple Interest			
t	Principal	Annual Interest	Balance at End of Year
3	\$500	\$15	\$545
4	\$500	\$15	\$560
5	\$500	\$15	\$575

Compound Interest			
t	Principal and Interest	Annual Interest	Balance at End of Year
3	\$530.45	\$15.91	\$546.36
4	\$546.36	\$16.39	\$562.75
5	\$562.75	\$16.88	\$579.63

b. compound interest

5. \$1405.09

7. The table shows compound interest.

t	Principal	Annual Interest	Balance at End of Year
1	\$700.00	\$21	\$721.00
2	\$700.00	\$21	\$742.00

9. a. \$300 **b.** \$13

11. One account will not always be greater than the other account.

13. $3.375x^6$

15. $29.16d^{18}$

Section B.2 Formula for Compound Interest

(pages A20 and A21)

1. Compound interest earns more than simple interest for a given interest rate because it is interest earned on the principal and on the previously earned interest.

3.

Annual Interest Rate	Balance after 5 Years
3.0%	\$579.64
3.5%	\$593.84
4.0%	\$608.33
4.5%	\$623.09

5. a. \$688.51

b. \$88.51

7. The interest rate should be in decimal form.

$$\begin{aligned} B &= P(1 + r)^t \\ &= 2000(1 + 0.04)^2 \\ &= 2000(1.04)^2 \\ &= \$2163.20 \end{aligned}$$

9. Account B; Its balance is greater after 8 years.

11. Compounding interest four times a year is compounding it quarterly.

13. 27,915

15. 0.4

17. D

Section B.3 Installment Loans
(pages A26 and A27)

1. The principal of a loan is less than the total amount due because interest is included in the amount due.

3.

Balance Before Payment	Monthly Payment	Monthly Interest	Balance After Payment
$1135.00	$233.86	$11.35	$912.49
$912.49	$233.86	$9.12	$687.75
$687.75	$233.86	$6.88	$460.77
$460.77	$233.86	$4.61	$231.52
$231.52	$233.84	$2.32	$0.00

$34.28

5. $28.74

7. $60.88

9.

Interest Paid		
Term \ Annual Interest Rate	5%	7%
20	$70,068.00	$103,286.40
25	$90,453.00	$134,442.00
30	$111,908.40	$167,409.60

11. about 24.9%

13. For Loan A, you will use the amount of money in your savings account after 3 years to pay off the loan.

15. -28.06

Section B.4 Checking Accounts
(pages A32 and A33)

1. date, check number (if one is written), transaction description, credit or debit amount

3. a.

Date	Check #	Transaction	Credit	Debit	Balance
		Balance Forward			300.00
3/4/12	506	Dinner		45.32	254.68
3/7/12		ATM Withdrawal		80.00	174.68
3/8/12	507	Electric Company		39.76	134.92
3/14/12		Deposit	150.00		284.92
3/18/12	508	Credit Card Payment		129.50	155.42

b. yes

5. more than 20 withdrawals each month

7. You should balance your checkbook after every transaction so that you do not write any bad checks.

9. 7.9×10^{-8}

11. 3.6×10^{11}

13. C

Section B.5 Credit Cards
(pages A40 and A41)

1. *Sample answer:* You can use both cards to make a purchase, instead of using cash or a check. *Sample answer:* A credit card is similar to a loan in that payment is postponed and interest is charged.

3. \$2.70   **5.** \$2.42

7. **a.** \$3.33 **b.** \$107.76 **c.** \$102.47   **9.** **a.** \$3.94 **b.** \$74.81 **c.** \$71.06

11. To find the percent the store keeps, divide the amount the store keeps by the total amount with the sales tax.

13. \$740.98; \$240.00; \$59.02; The amount you paid is more than four times the decrease in the amount due. You end up decreasing your balance by about \$5 each month because of the interest. You should pay more than the minimum so that more money goes toward decreasing the balance, rather than paying mostly interest.

15. 12   **17.** $6\sqrt{19}$

Section B.6 Payroll Deductions
(pages A46 and A47)

1. never; At most, net pay can equal gross pay if there are no deductions, but it will never be greater than gross pay.

3. \$1173.75   **5.** \$612.08   **7.** every 2 weeks

9. \$17,468.10   **11.** \$228.15

13–15. Make sure to look back at page A45 and divide by the correct amounts.

17. *Sample answer:* Colorado: 4.63%; Florida: none; Massachusetts: 5.3%; Florida has the lowest income tax and Massachusetts has the highest income tax.

19. 3   **21.** 343   **23.** D

Student Index

This student-friendly index will help you find vocabulary, key ideas, and concepts. It is easily accessible and designed to be a reference for you whether you are looking for a definition, real-life application, or help with avoiding common errors.

A

D

E

F

G

Student Index

Q

R

Student Index

T

U

V

W

Additional Answers

Chapter 1

Section 1.2

Record and Practice Journal

2. indigo: 45°, 45°, 90°; violet: 60°, 60°, 60°; orange: 75°, 65°, 40°; yellow: 25°, 60°, 95°; blue: 75°, 75°, 30°; green: 15°, 135°, 30°

Chapter 2

Section 2.1

On Your Own

1.

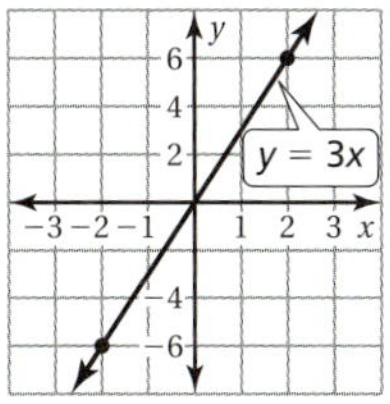

2.

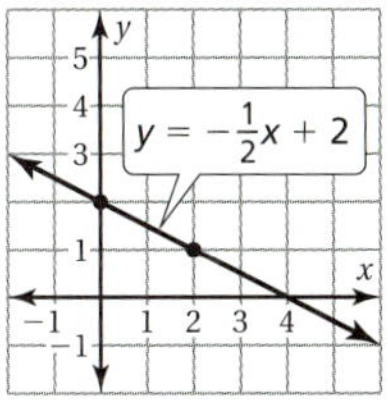

3.

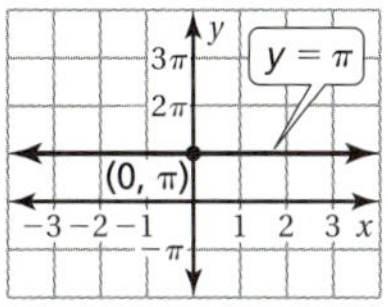

4.

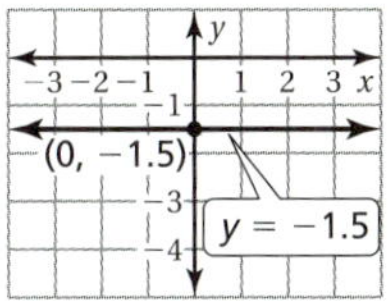

Practice and Problem Solving

6.

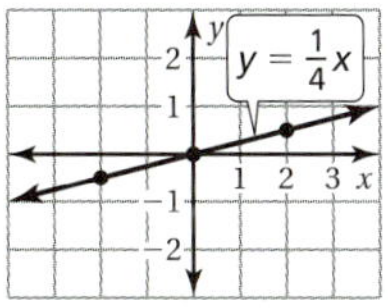

7.

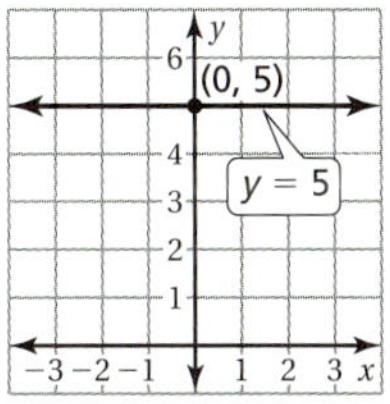

8.

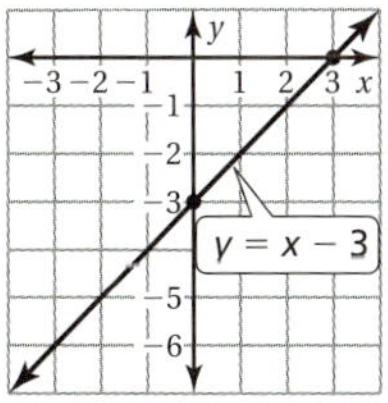

9.

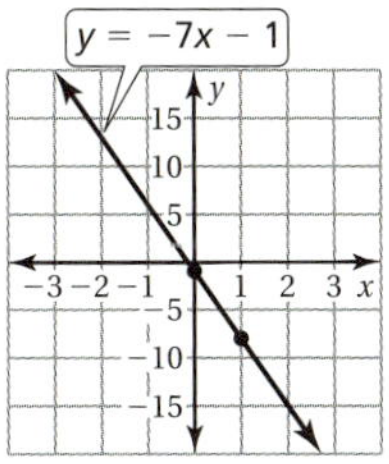

10.

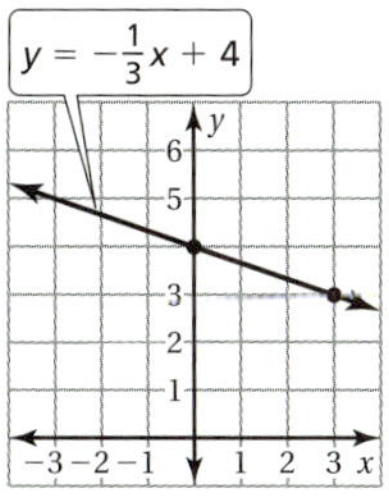

11.

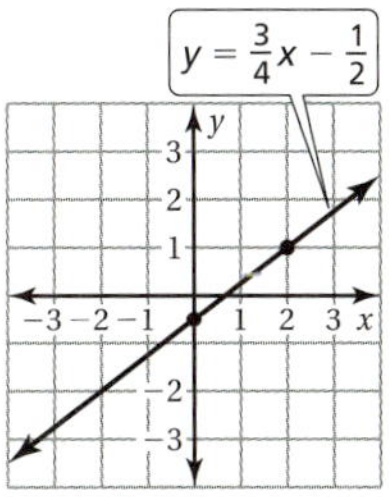

12.

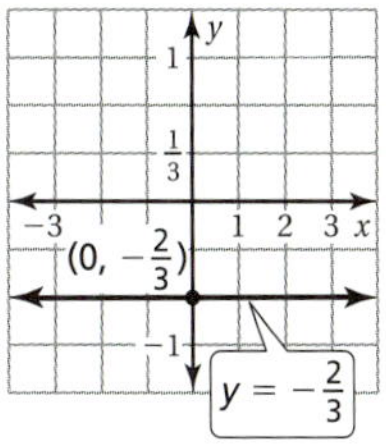

13.

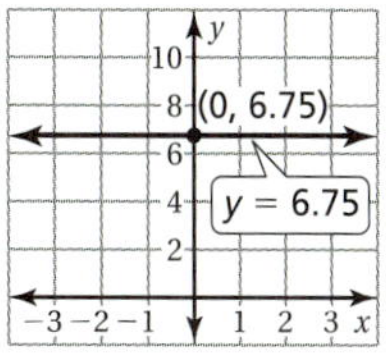

14. The equation $x = 4$ is graphed, not $y = 4$.

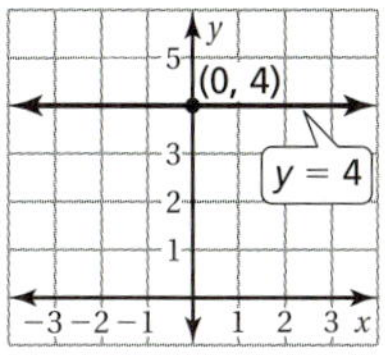

15.

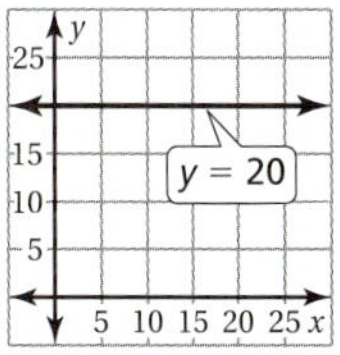

16. **a.**

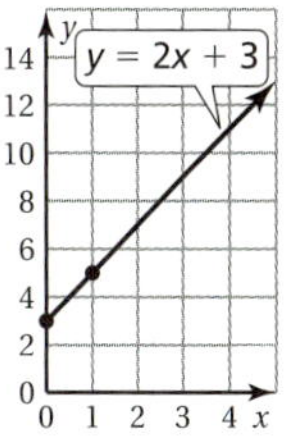

b. about \$5

c. \$5.25

18. $y = -\frac{5}{2}x + 2$

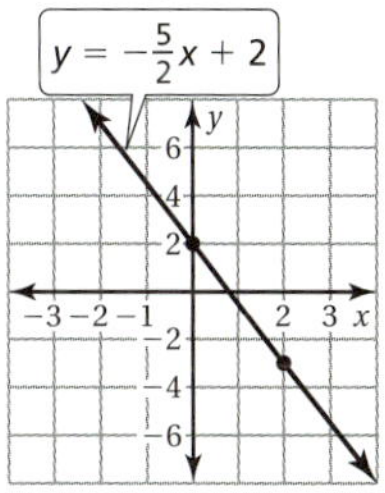

19. $y = 12x - 9$

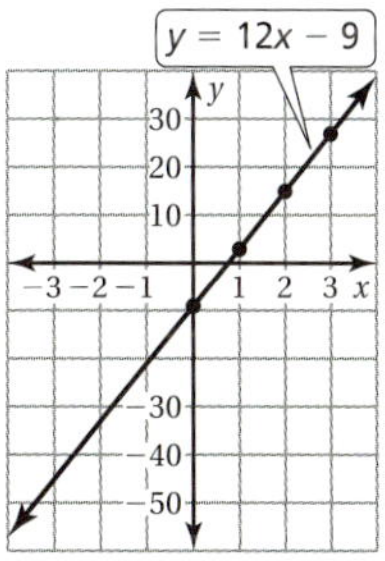

20. $y = -2x + 3$

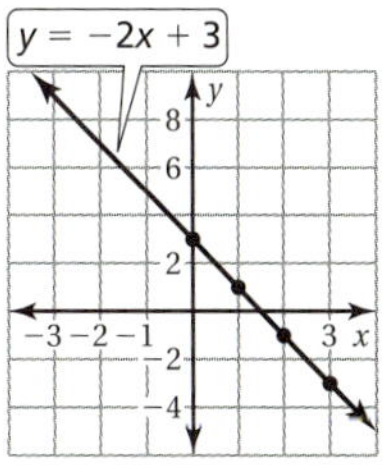

21. a. $y = 100 + 12.5x$

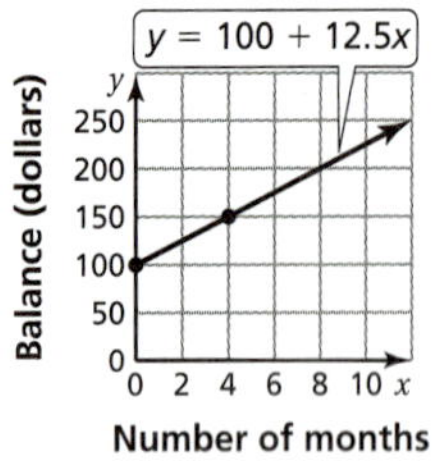

b. 6 mo

23. a. $y = 2x$

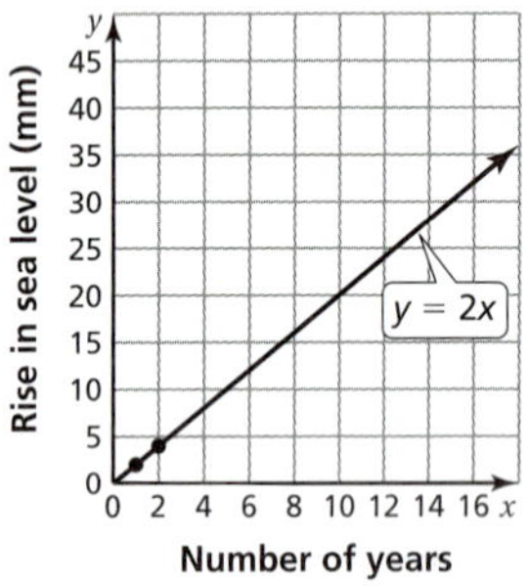

b. *Sample answer:* If you are 13 years old, the sea level has risen 26 millimeters since you were born.

24. *Sample answer:*

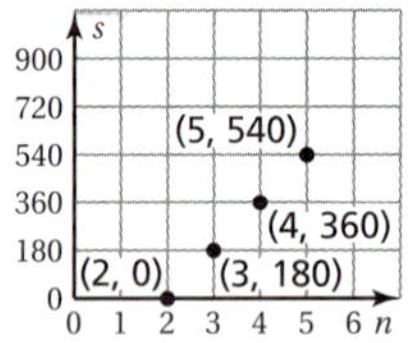

Yes, the points lie on a line.

Section 2.2

Practice and Problem Solving

18. $-\frac{7}{6}$

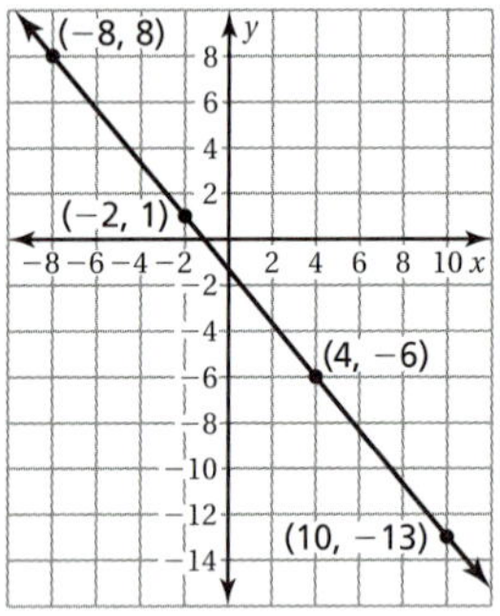

19. $\frac{1}{3}$

20. *Sample answer:*

a. Yes, it follows the guidelines.

b.

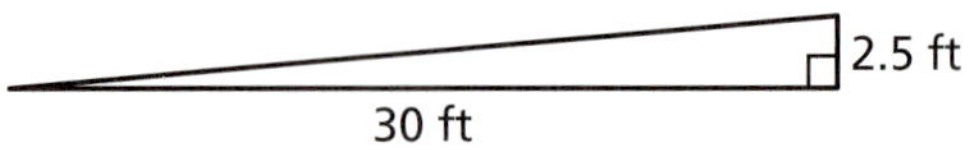

21. red and green; They both have a slope of $\frac{4}{3}$.

22. blue and red; They both have a slope of -3.

23. no; Opposite sides have different slopes.

24. yes; The opposite sides have equal slopes and lengths.

25. a. $\frac{3}{40}$

b. The cost increases by \$3 for every 40 miles you drive, or the cost increases by \$0.075 for every mile you drive.

26. The boat ramp, because it has a 16.67% grade.

Section 2.3

On Your Own

3.

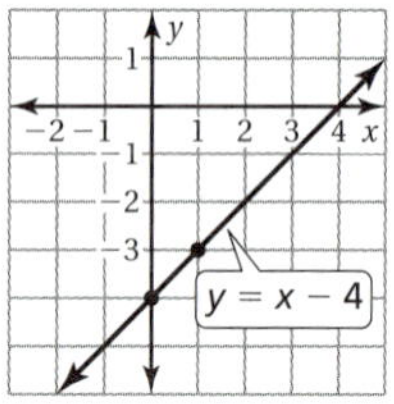

x-intercept: 4

4.

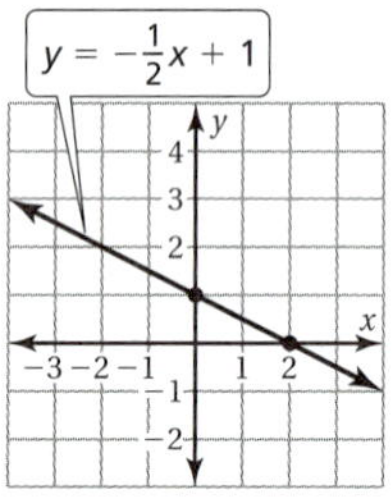

x-intercept: 2

5. The y-intercept means that the taxi has an initial fee of \$1.50. The slope means the taxi charges \$2 per mile.

Practice and Problem Solving

16. The y-intercept should be -3.
$y = 4x - 3$
The slope is 4 and the y-intercept is -3.

17. a.

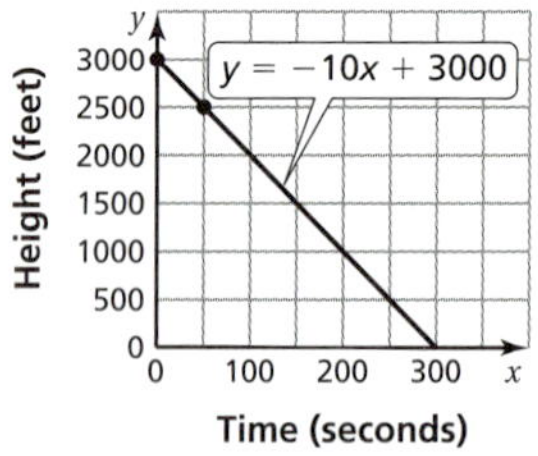

b. The x-intercept of 300 means the skydiver lands on the ground after 300 seconds. The slope of -10 means that the skydiver falls to the ground at a rate of 10 feet per second.

19. 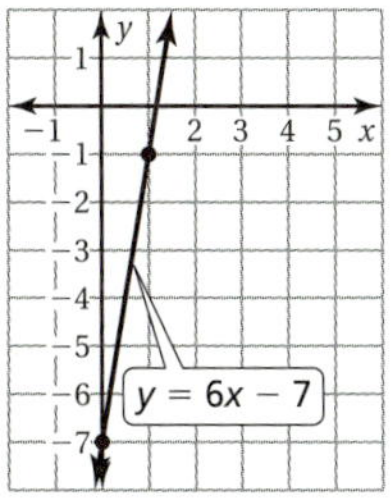

x-intercept: $\frac{7}{6}$

20. 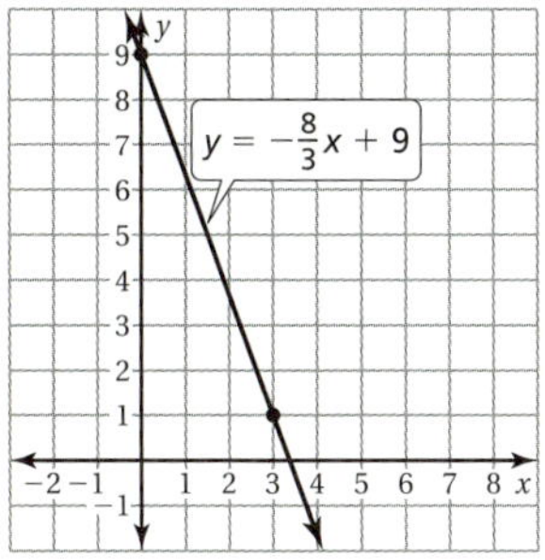

x-intercept: $\frac{27}{8}$

21. 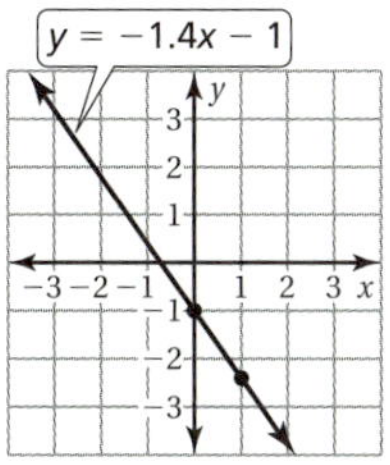

x-intercept: $-\frac{5}{7}$

22. 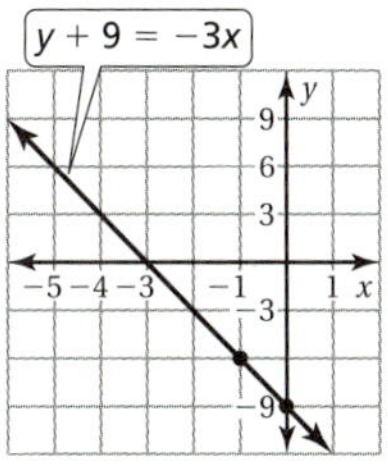

x-intercept: -3

23. 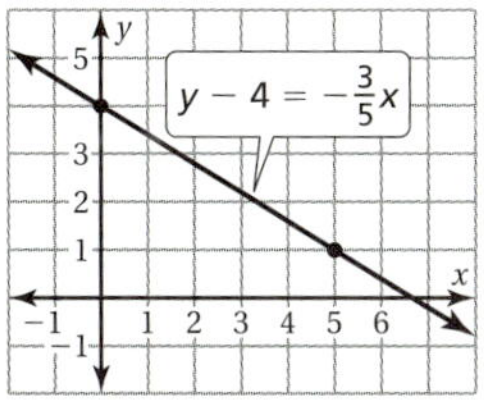

x-intercept: $\frac{20}{3}$

24. a. 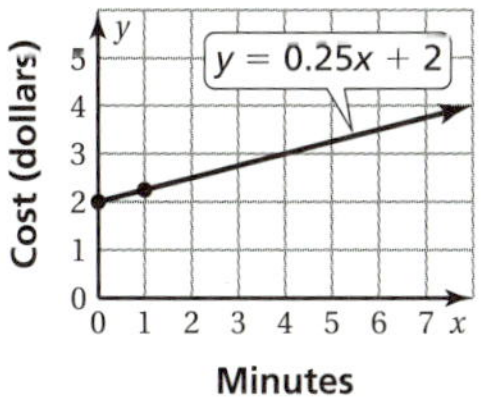

b. The slope of 0.25 means that it costs \$0.25 for each minute spent making a long distance call. The y-intercept of 2 means that there is an initial fee of \$2.

25. $y = 0.75x + 5$

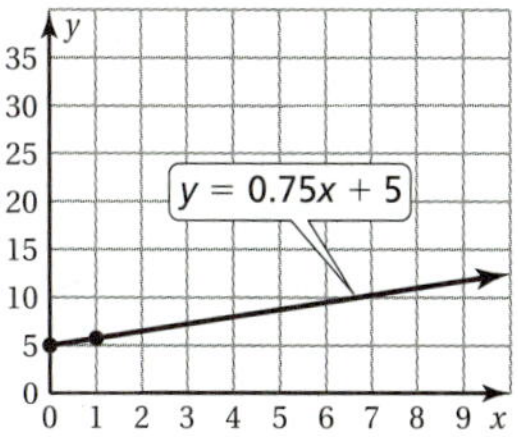

26. $y = 5x - 40$

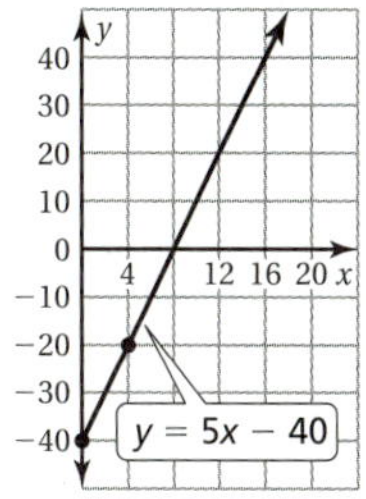

27. $y = 0.15x + 35$

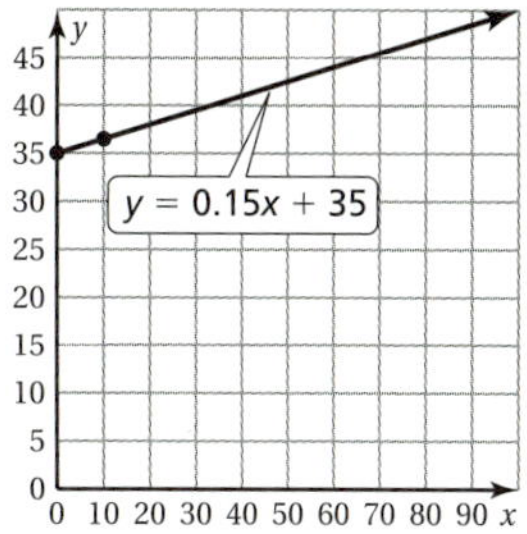

Section 2.4

On Your Own

3. 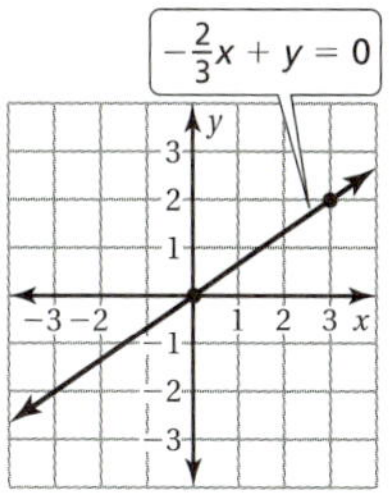

4. 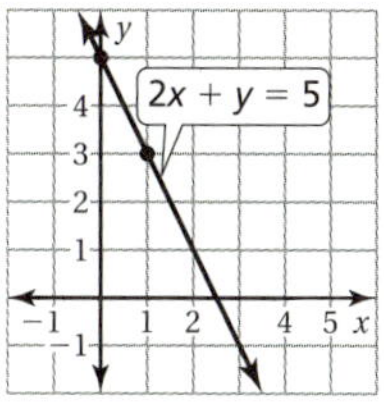

5. 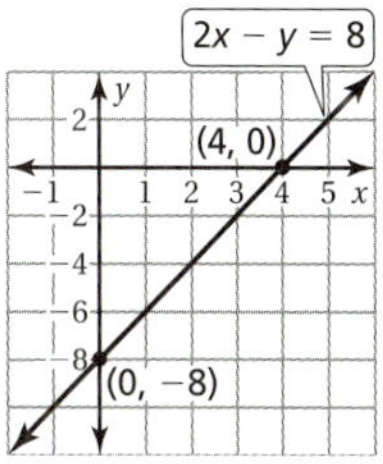

6. 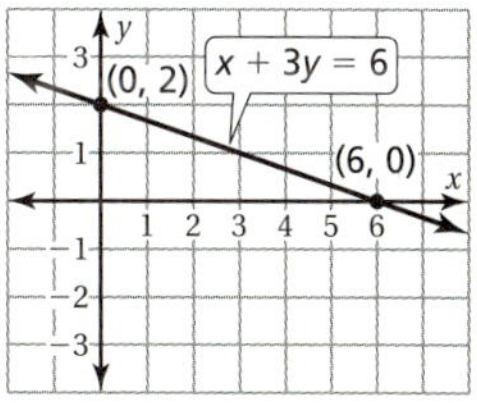

7. 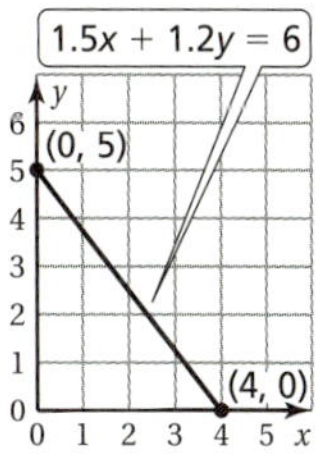

The x-intercept shows that you can buy 4 pounds of apples if you don't buy any oranges. The y-intercept shows that you can buy 5 pounds of oranges if you don't buy any apples.

Additional Answers

Practice and Problem Solving

9.

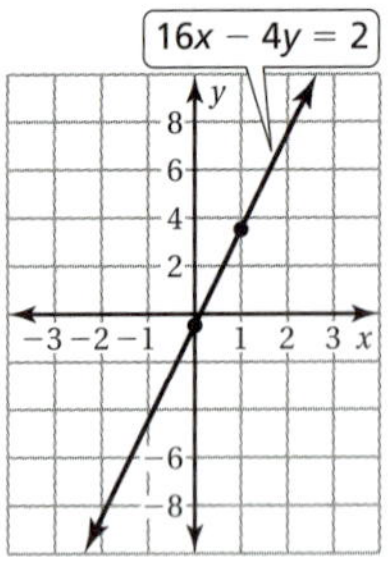

10.

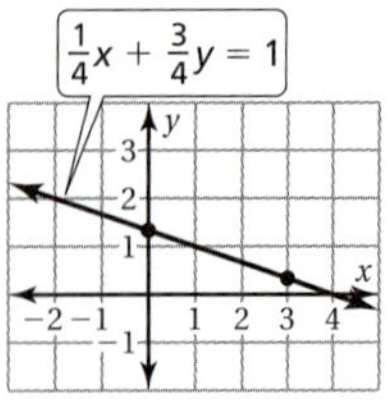

11. x-intercept: -6
y-intercept: 3

12. x-intercept: -4
y-intercept: -5

13. x-intercept: none
y-intercept: -3

14. They should have let $y = 0$, not $x = 0$.

$$-2x + 3y = 12$$
$$-2x + 3(0) = 12$$
$$-2x = 12$$
$$x = -6$$

15. **a.** $y - 25x = 65$

b. \$390

16.

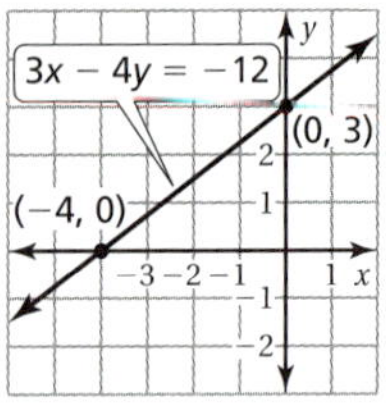

17.

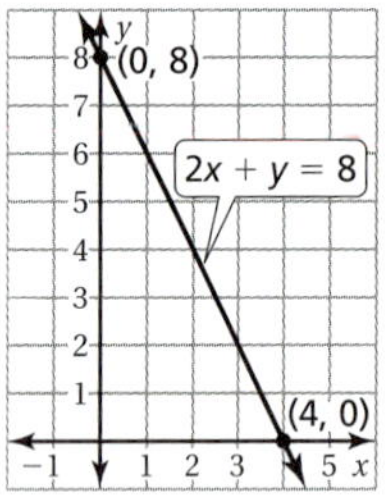

18.

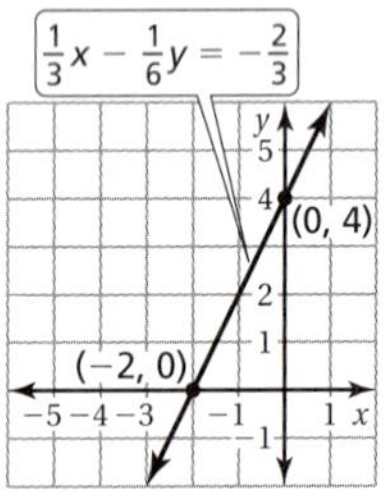

19. x-intercept: 9
y-intercept: 7

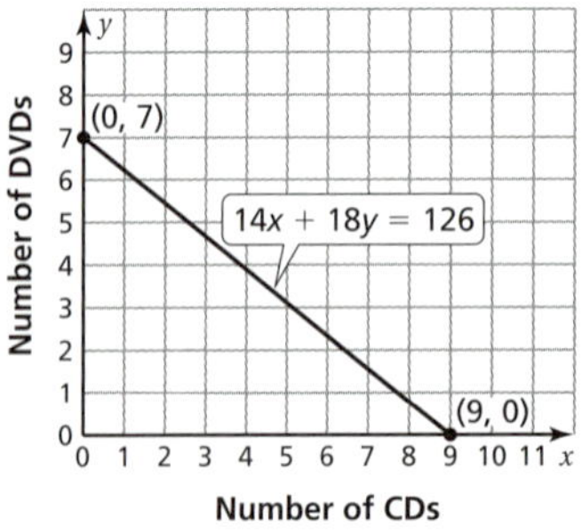

21. **a.** $9.45x + 7.65y = 160.65$

b.

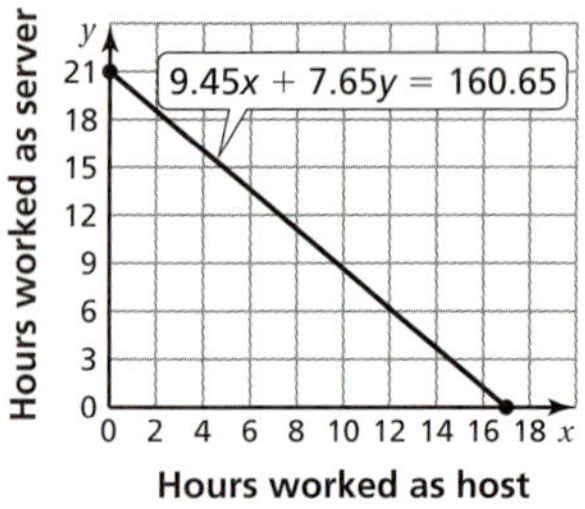

22. no; For example, $y = 5$ does not have an x-intercept, neither do any horizontal lines except $y = 0$.

23. **a.** $y = 40x + 70$

b. x-intercept: $-\frac{7}{4}$; It will not be on the graph because you cannot have a negative time.

c.

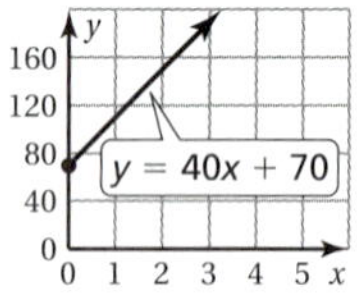

2.1–2.4 Quiz

1.

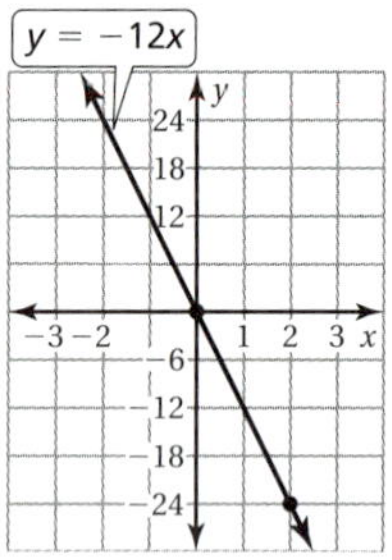

2.

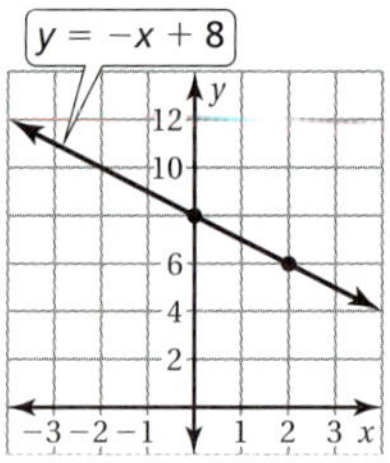

3.

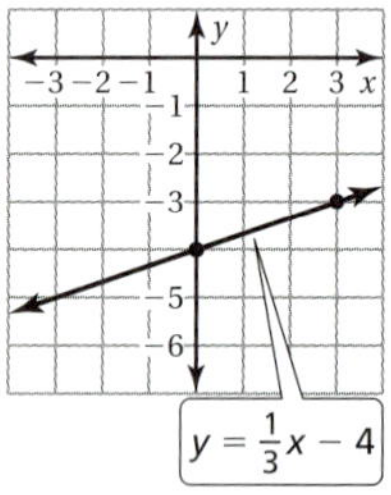

4.

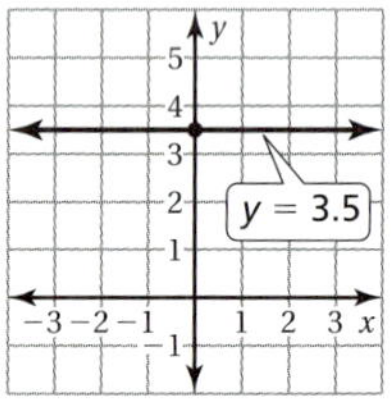

12.

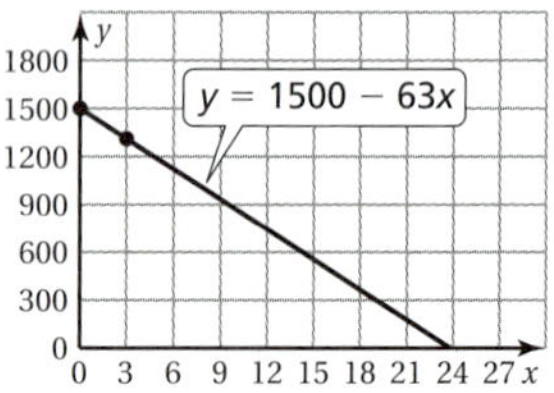

13. 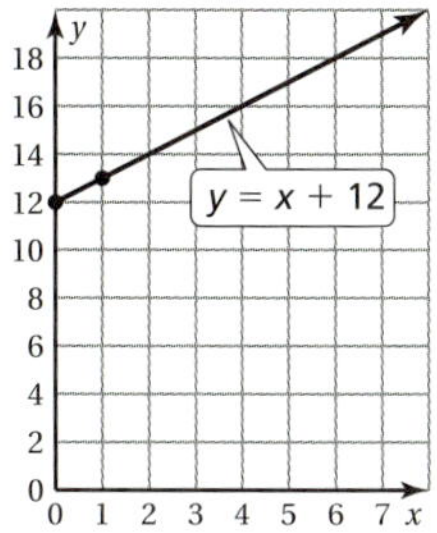

14. a. 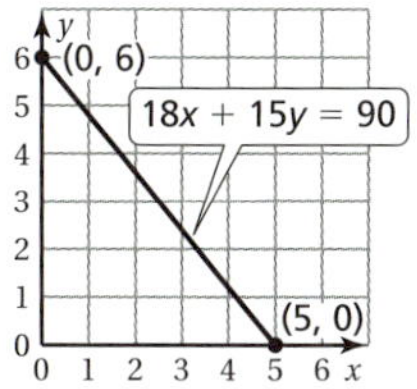

b. The x-intercept, 5, shows that you can buy 5 gallons of blue paint if you do not buy any white paint. The y-intercept, 6, shows that you can buy 6 gallons of white paint if you do not buy any blue paint.

Section 2.5

Practice and Problem Solving

3.

x	0	1	2	3	4	5	6
C	150	165	180	195	210	225	240
R	0	45	90	135	180	225	270

(5, 225)

4.

x	0	1	2	3	4	5	6
C	80	104	128	152	176	200	224
R	0	44	88	132	176	220	264

(4, 176)

Section 2.7

Record and Practice Journal

5. You can graph a linear equation and find the value of y for a given value of x. You can graph a system of linear equations and the point of intersection is the solution of the system. You can take an equation that has variables on both sides and use the left side of the equation to write one linear equation and use the right side to write another linear equation. Then you can graph the equations. The solution is the x-coordinate of the point of intersection.

2.5–2.7 Quiz

3.

x	0	1	2	3	4	5	6
C	180	190	200	210	220	230	240
R	0	46	92	138	184	230	276

(5, 230)

Chapter 2 Test

7. 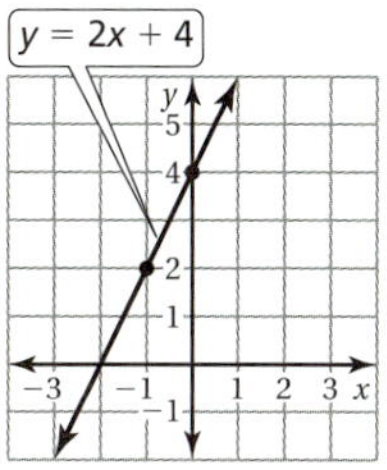

8. 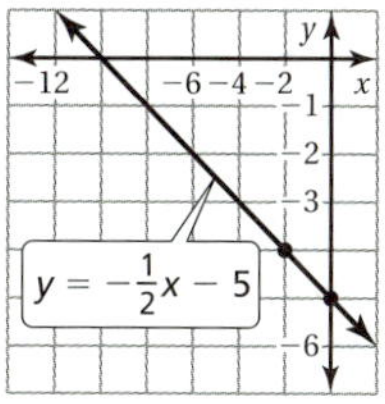

9. 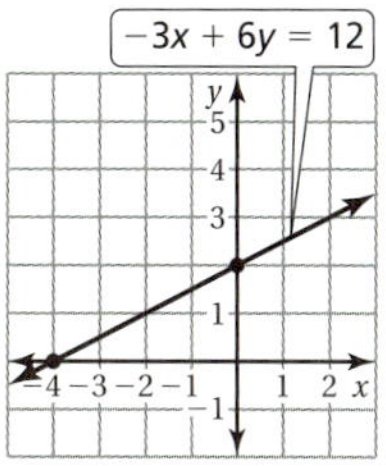

Chapter 3

Section 3.1

Record and Practice Journal

1.

a. top line: slope: $\frac{1}{2}$; y-intercept: 4; $y = \frac{1}{2}x + 4$

middle line: slope: $\frac{1}{2}$; y-intercept: 1; $y = \frac{1}{2}x + 1$

bottom line: slope: $\frac{1}{2}$; y-intercept: -2;

$y = \frac{1}{2}x - 2$

The lines are parallel.

b. right line: slope: -2; y-intercept: 3; $y = -2x + 3$

middle line: slope: -2; y-intercept: -1;

$y = -2x - 1$

left line: slope: -2; y-intercept: -5;

$y = -2x - 5$

The lines are parallel.

c. line passing through (3, 2): slope: $-\frac{1}{3}$;

y-intercept: 3; $y = -\frac{1}{3}x + 3$

line passing through (3, 7): slope: $\frac{4}{3}$;

y-intercept: 3; $y = \frac{4}{3}x + 3$

line passing through (6, 4): slope: $\frac{1}{6}$;

y-intercept: 3; $y = \frac{1}{6}x + 3$

The lines have the same y-intercept.

d. line passing through (1, 2): slope: 2;

y-intercept: 0; $y = 2x$

line passing through (1, −1): slope: −1;

y-intercept: 0; $y = -x$

line passing through (3, 1): slope: $\frac{1}{3}$;

y-intercept: 0; $y = \frac{1}{3}x$

The lines have the same y-intercept.

2. a. $y = 5$ $\quad y = x + 1$

$y = -2$ $\quad y = x + 5$

b. $y = 5$ $\quad y = x + 1$

$y = -2$

$y = x + 5$

Practice and Problem Solving

17. a–b.

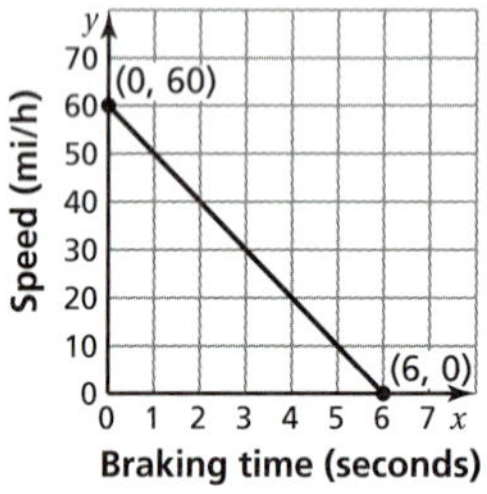

(0, 60) represents the speed of the automobile before braking. (6, 0) represents the amount of time it takes to stop. The line represents the speed y of the automobile after x seconds of braking.

c. $y = -10x + 60$

Section 3.3

Practice and Problem Solving

17. a.

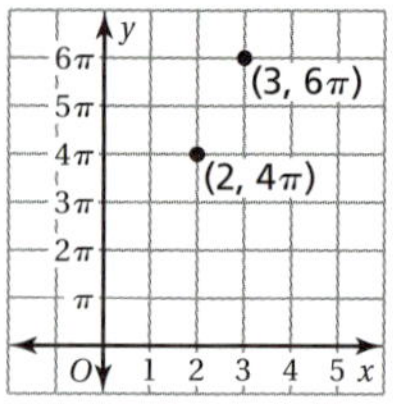

b. $y = 2\pi x$

19. a. $y = -2000x + 21{,}000$

b.

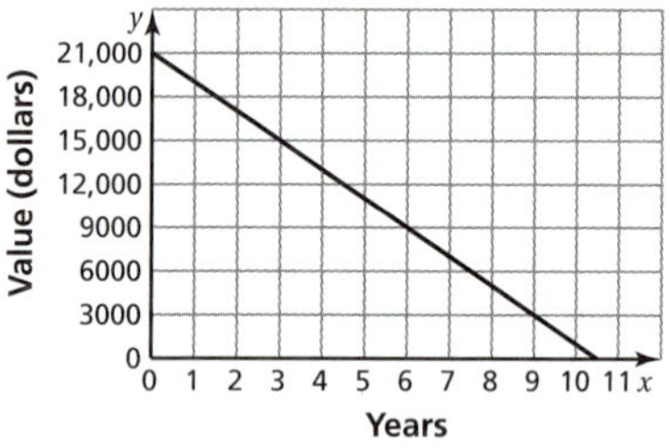

c. \$21,000; the original price of the car

Section 3.4

On Your Own

1. a.

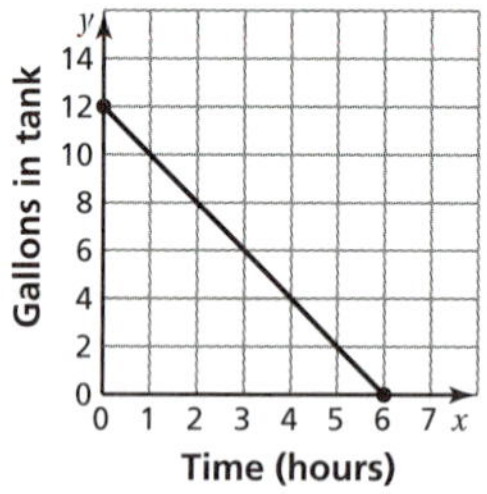

b. The x-intercept is 6. So, you can drive 6 hours before the tank is empty. The y-intercept is 12. So, there are 12 gallons in the tank before you start driving.

c. 3.5 or $3\frac{1}{2}$ h

Practice and Problem Solving

4. *Sample answer:* A gasoline tank initially has 16 gallons of gas. After 10 hours, the tank is empty.

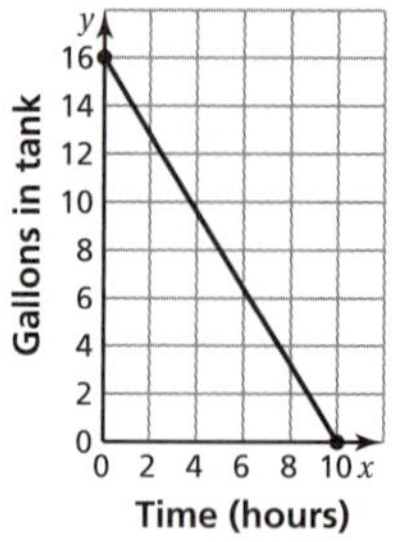

5. *Sample answer:* On a visit to Mexico, you spend 45 pesos every week. After 4 weeks, you have no pesos left.

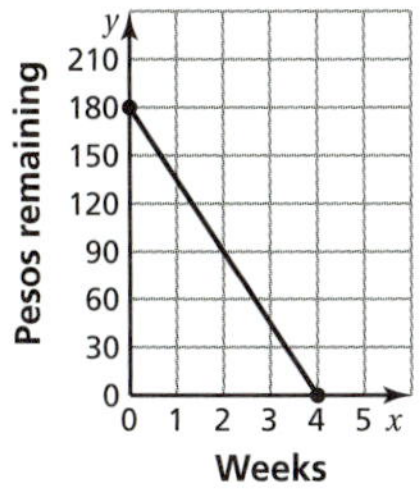

8. a. The x-intercept is 6. So, it takes 6 hours for your family to drive from Cincinnati to St. Louis. The y-intercept is 360. So, it is 360 miles from Cincinnati to St. Louis.

b. -60; Your distance from Cincinnati decreases at a rate of 60 miles per hour

c. $y = -60x + 360$; Both intercepts would be less and the slope would be the same.

Record and Practice Journal Practice

1.

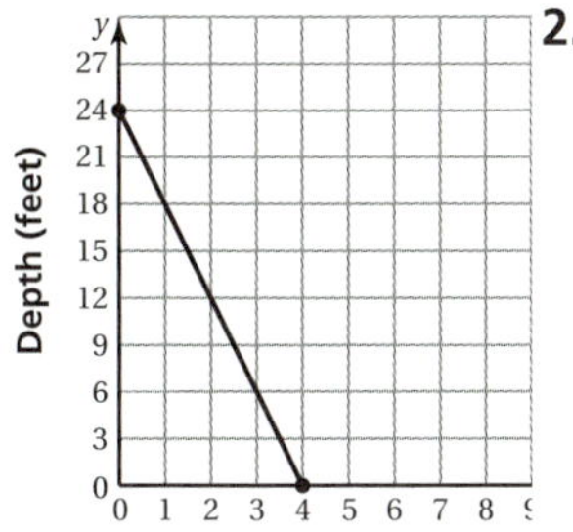

2.

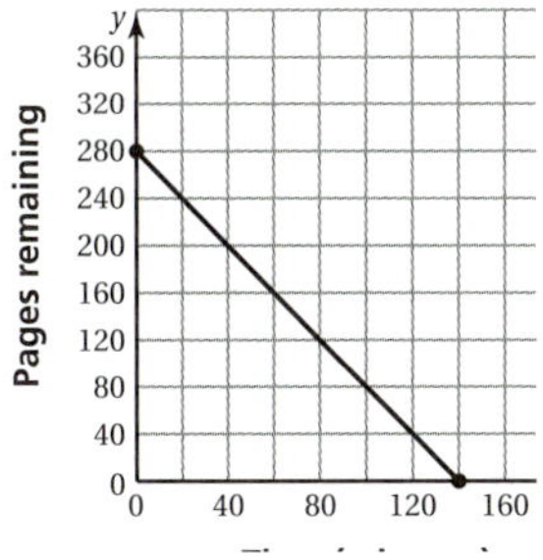

Section 3.5

Practice and Problem Solving

5. a. $x + y = 12$
$3x + 2y = 32$

b.

x	0	1	2	3	4	5	6	7	8
$y = 12 - x$	12	11	10	9	8	7	6	5	4
$y = 16 - \frac{3}{2}x$	16	14.5	13	11.5	10	8.5	7	5.5	4

8 lilies and 4 tulips

c.

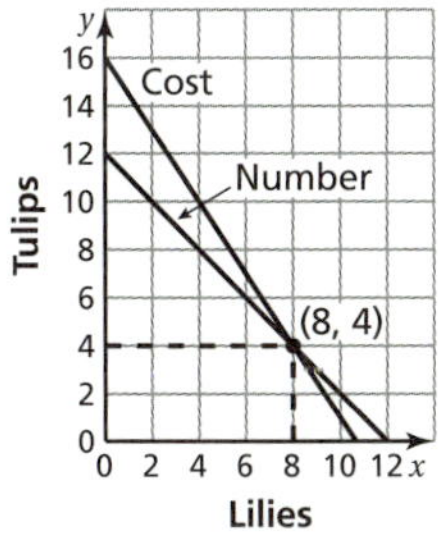

8 lilies and 4 tulips

d. $12 - x = 16 - \frac{3}{2}x$; $x = 8$, $y = 4$;
8 lilies and 4 tulips

6. a. $x + y = 63$
$x = y + 11$

b.

x	30	31	32	33	34	35	36	37
$y = 63 - x$	33	32	31	30	29	28	27	26
$y = x - 11$	19	20	21	22	23	24	25	26

37 boys and 26 girls

c.

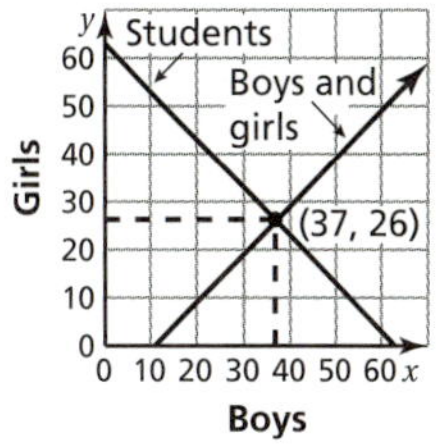

37 boys and 26 girls

d. $63 - x = x - 11$; $x = 37$, $y = 26$;
37 boys and 26 girls

3.4–3.5 Quiz

5. $2\ell + 2w = 36$; $\ell = 2w$;
$w = 6$ ft, $\ell = 12$ ft

6. $x - y = 8$; $x = 2y - 1$;
$x = 17$, $y = 9$

7. $x + y = 23$; $x = y + 5$;
$x = 14$ cats, $y = 9$ dogs

Chapter 4

Record and Practice Journal Fair Game Review

7.

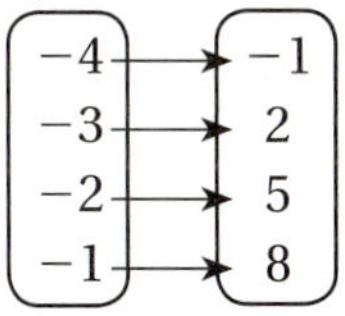

As the input increases by 1, the output increases by 3.

8.

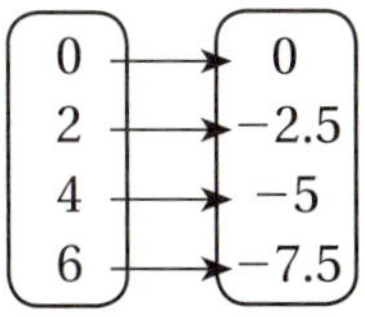

As the input increases by 2, the output decreases by 2.5.

9. Input Output

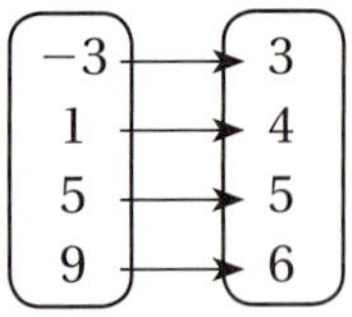

As the input increases by 4, the output increases by 1.

10. Input Output

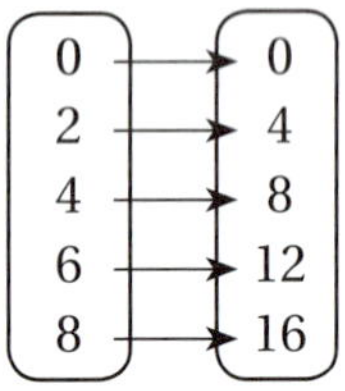

Section 4.1

Record and Practice Journal

4. b–c. women:

x (Domain)	$5\frac{1}{2}$	6	$6\frac{1}{2}$	7	$7\frac{1}{2}$	8	$8\frac{1}{2}$
y (Range)	8.8	9	9.2	9.3	9.5	9.7	9.8

x (Domain)	9	$9\frac{1}{2}$	10	$10\frac{1}{2}$	11	$11\frac{1}{2}$	12
y (Range)	10	10.2	10.3	10.5	10.7	10.8	11

4. b–c. men:

x (Domain)	$5\frac{1}{2}$	6	$6\frac{1}{2}$	7	$7\frac{1}{2}$	8	$8\frac{1}{2}$
y (Range)	9.1	9.3	9.5	9.6	9.8	10	10.1

x (Domain)	9	$9\frac{1}{2}$	10	$10\frac{1}{2}$	11	$11\frac{1}{2}$	12
y (Range)	10.3	10.5	10.6	10.8	11	11.1	11.3

On Your Own

3.

x	−1	0	1	2
y	−5	−3	−1	1

domain: −1, 0, 1, 2
range: −5, −3, −1, 1

4.

x	0	1	2	3
y	−3	−4	−5	−6

domain: 0, 1, 2, 3
range: −3, −4, −5, −6

Practice and Problem Solving

14. a. The domain is all real numbers because you can find the absolute value of any number. The range is all real numbers greater than or equal to 0 because the least an absolute value can be is 0.

b. The domain is all real numbers because you can find the absolute value of any number. The range is all real numbers less than or equal to 0 because the negative sign will make every *y*-value be 0 or negative.

c. The domain is all real numbers because you can find the absolute value of any number. The range is all real numbers greater than or equal to −6 because the least an absolute value can be is 0 and you subtract 6 from that.

d. The domain is all real numbers because you can find the absolute value of any number. The range is all real numbers less than or equal to 4 because the negative sign will make the greatest absolute value be 0 and you add 4 to that.

Fair Game Review

15.

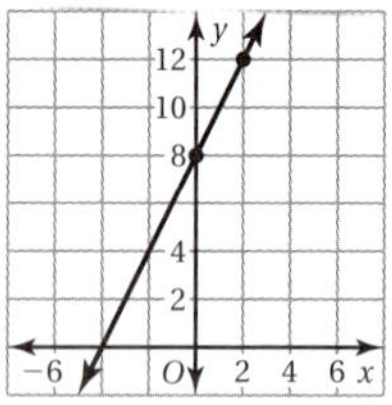

16.

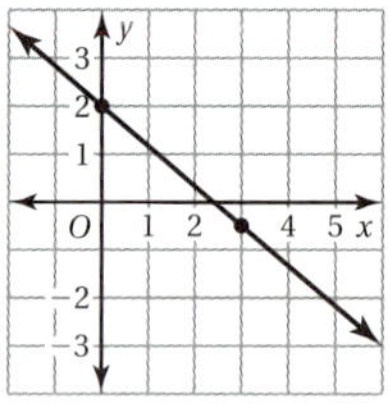

17.

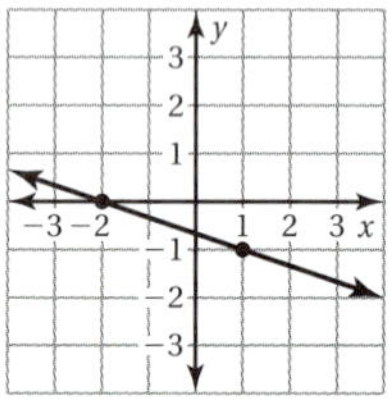

18.

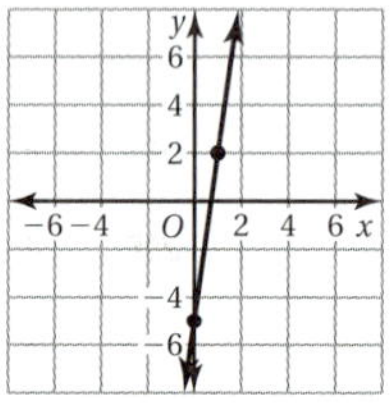

Record and Practice Journal Practice

5. b.

x	1	2	4	8	10
y	18,000	36,000	72,000	144,000	180,000

Section 4.2

On Your Own

2.

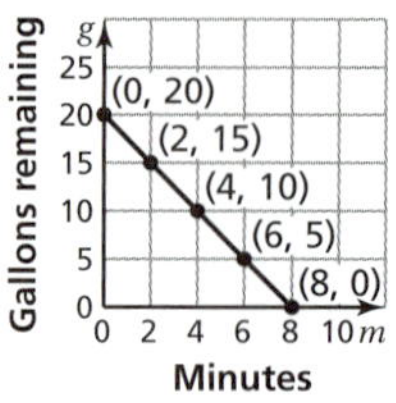

continuous

Practice and Problem Solving

6.

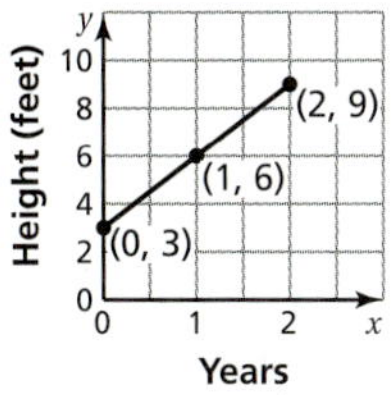

continuous

7.

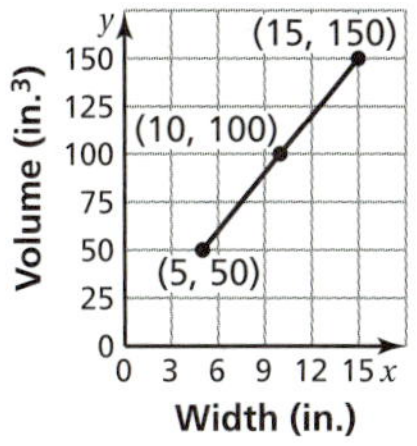

continuous

8.

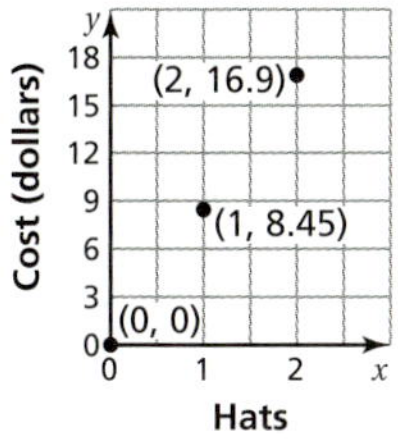

discrete

12.

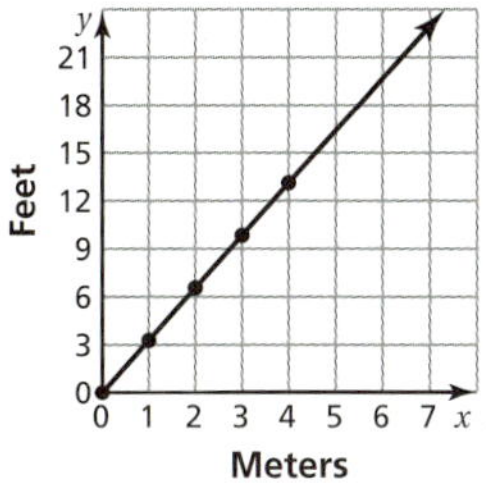

continuous

13.

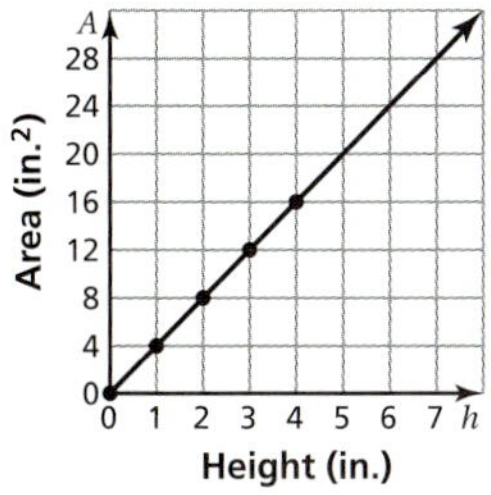

continuous

4.1–4.2 Quiz

7.

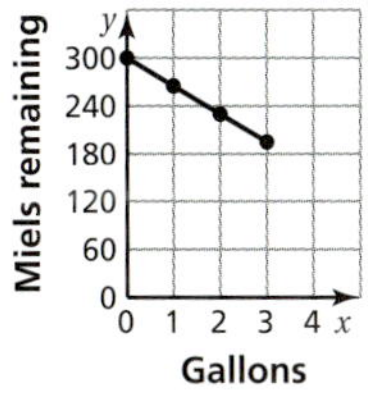

continuous

8.

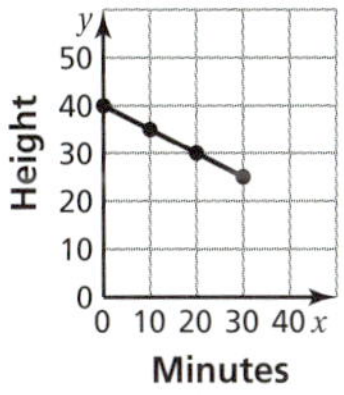

continuous

9.

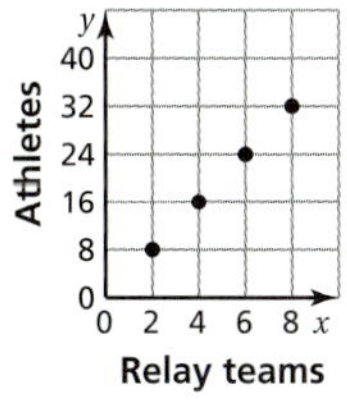

discrete

10.

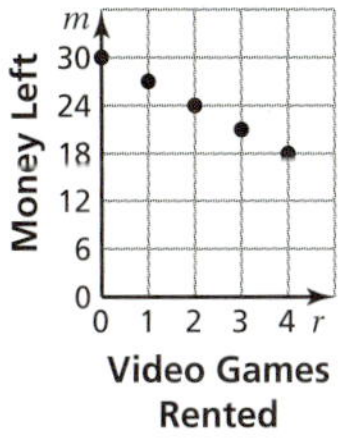

discrete

11. a. $y = 0.6x$

b.

x	100	120	140	160
y	60	72	84	96

Section 4.3

Practice and Problem Solving

14. a.

Temperature (°F), *t*	94	95	96	97	98
Heat Index (°F), *H*	122	126	130	134	138

b. $H = 4t - 254$

c. 146°F

Chapter 4 Test

3.

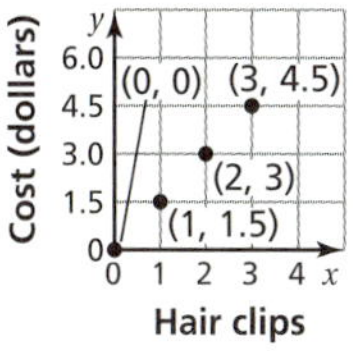

discrete

4.

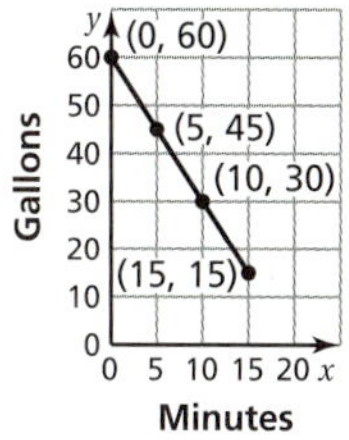

continuous

Chapter 5

Section 5.1

Record and Practice Journal

1. a. linear; $y = 90 - x$; greater than 0 and less than 90.

b. linear; $y = 180 - x$; greater than 0 and less than 180.

3. b. $\angle A$ and $\angle ABE$; $\angle A$ and $\angle C$; $\angle A$ and $\angle BED$; $\angle A$ and $\angle BEF$; $\angle A$ and $\angle F$; $\angle ABE$ and $\angle C$; $\angle ABE$ and $\angle BED$; $\angle ABE$ and $\angle BEF$; $\angle ABE$ and $\angle F$; $\angle C$ and $\angle BED$; $\angle C$ and $\angle BEF$; $\angle C$ and $\angle F$; $\angle BED$ and $\angle BEF$; $\angle BED$ and $\angle F$; $\angle BEF$ and $\angle F$; $\angle A$ and $\angle CDE$; $\angle A$ and $\angle CBE$; $\angle ABE$ and $\angle CBE$; $\angle ABE$ and $\angle CDE$; $\angle BEF$ and $\angle CBE$; $\angle BEF$ and $\angle CDE$; $\angle F$ and $\angle CBE$; $\angle F$ and $\angle CDE$; $\angle CBE$ and $\angle C$; $\angle CBE$ and $\angle CDE$; $\angle CBE$ and $\angle BED$; $\angle C$ and $\angle CDE$; $\angle CDE$ and $\angle BED$

Section 5.2

Practice and Problem Solving

19. If two angle measures of a triangle were greater than or equal to 90°, the sum of those two angle measures would be greater than or equal to 180°. The sum of the three angle measures would be greater than 180°, which is not possible.

Section 5.3

Practice and Problem Solving

26. *Sample answer:* rhombus

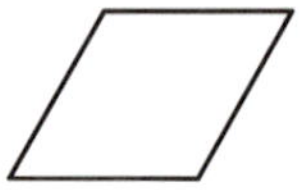

32. a. *Sample answer:*

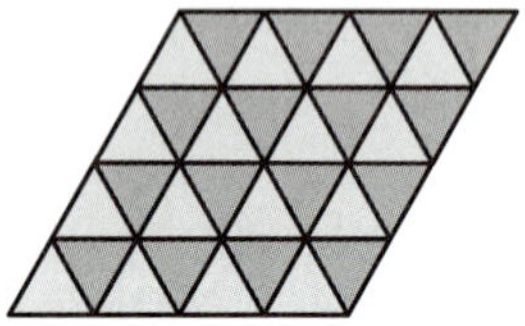

b. *Sample answer:* square, hexagon

c. *Sample answer:*

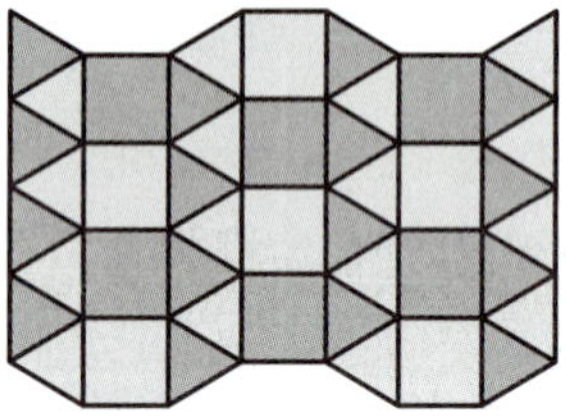

Section 5.4

Practice and Problem Solving

14. *Sample answer:* 10 ft

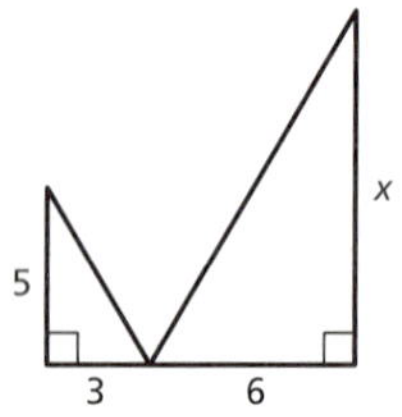

Section 5.5

Practice and Problem Solving

17. $\angle 2 = 90°$; $\angle 2$ and the given angle are vertical angles.
$\angle 1 = 90°$ and $\angle 3 = 90°$; $\angle 1$ and $\angle 3$ are supplementary to the given angle.
$\angle 4 = 90°$; $\angle 4$ and the given angle are corresponding angles.
$\angle 6 = 90°$; $\angle 4$ and $\angle 6$ are vertical angles.
$\angle 5 = 90°$ and $\angle 7 = 90°$; $\angle 5$ and $\angle 7$ are supplementary to $\angle 4$.

18. 56°; *Sample answer:* $\angle 1$ and $\angle 8$ are corresponding angles and $\angle 8$ and $\angle 4$ are supplementary.

19. 132°; *Sample answer:* $\angle 2$ and $\angle 4$ are alternate interior angles and $\angle 4$ and $\angle 3$ are supplementary.

20. 55°; *Sample answer:* $\angle 4$ and $\angle 2$ are alternate interior angles.

21. 120°; *Sample answer:* $\angle 6$ and $\angle 8$ are alternate exterior angles.

22. 129.5°; *Sample answer:* $\angle 7$ and $\angle 5$ are alternate exterior angles and $\angle 5$ and $\angle 6$ are supplementary.

23. 61.3°; *Sample answer:* $\angle 3$ and $\angle 1$ are alternate interior angles and $\angle 1$ and $\angle 2$ are supplementary.

24. 40°

25. They are all right angles because perpendicular lines form 90° angles.

26. *Sample answer:* $\angle 1$ and $\angle 7$ are congruent because they are alternate exterior angles; $\angle 1$ and $\angle 5$ are corresponding angles and $\angle 5$ and $\angle 7$ are vertical angles. So, $\angle 1$ and $\angle 7$ are congruent.

5.4–5.5 Quiz

11. 119°; $\angle 5$ and $\angle 3$ are alternate interior angles.

12. 60°; $\angle 4$ and $\angle 6$ are alternate exterior angles.

13. $\angle 1 = 108°$, $\angle 2 = 108°$; Because of alternate interior angles, the angle below $\angle 1$ is 72°. This angle is supplementary to both $\angle 1$ and $\angle 2$.

14. Yes, if you have

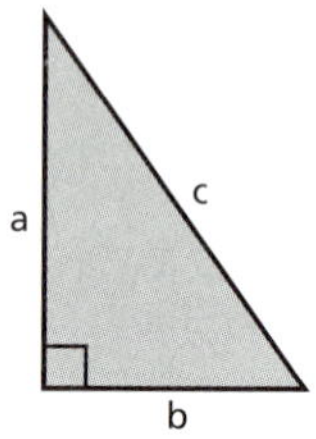

the perimeter is $a + b + c$, and the perimeter of

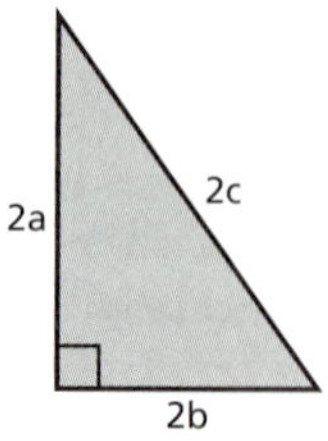

is $2a + 2b + 2c = 2(a + b + c)$.

Chapter 6

Section 6.5

Practice and Problem Solving

22. yes; *Sample answer:* Plot the points and connect to form a triangle. Then draw three right triangles outside the original triangle so that the hypotenuses are the side lengths of the original triangle. Then use the Pythagorean Theorem to find that the side lengths are $\sqrt{18}$, $\sqrt{50}$, and $\sqrt{68}$.

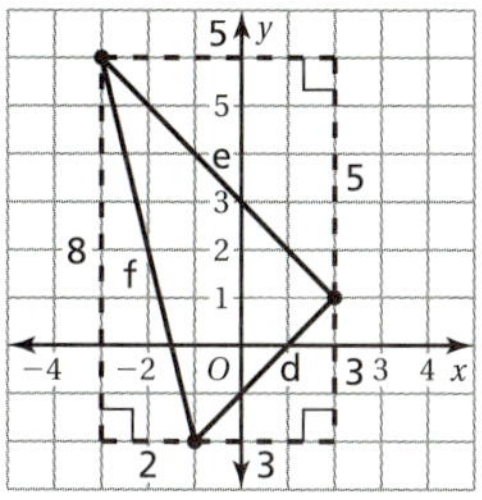

Chapter 7

Record and Practice Journal Fair Game Review

6.

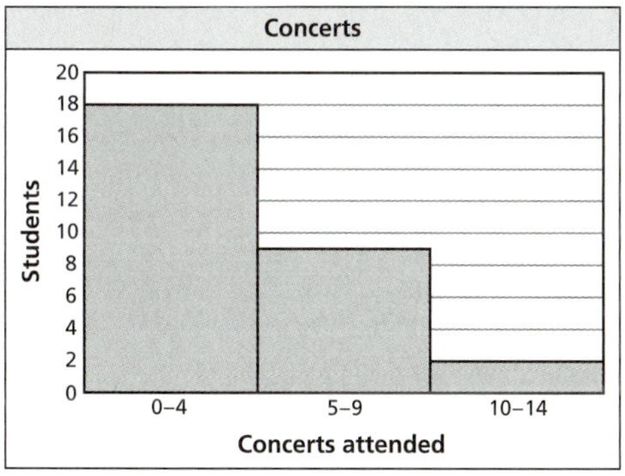

7.

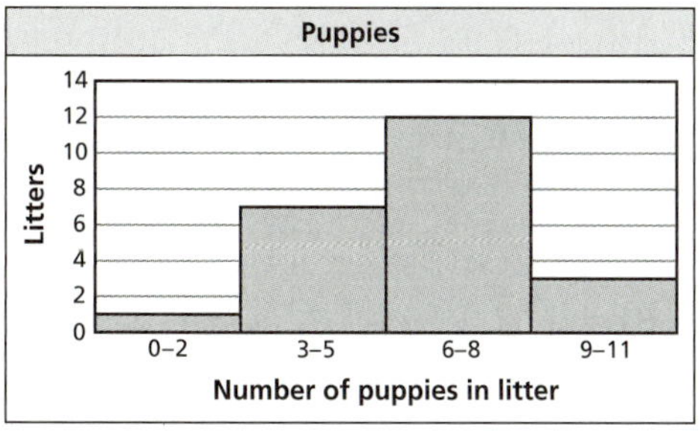

8.

9.

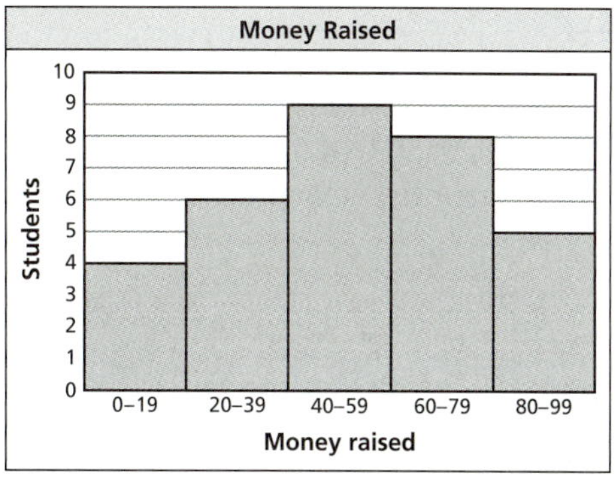

10.

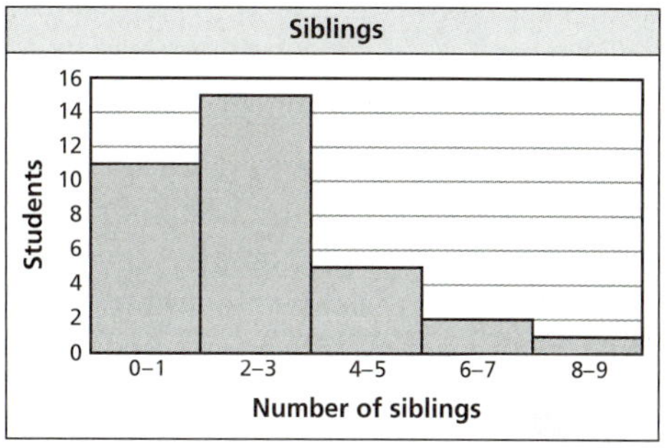

Section 7.2

On Your Own

1.

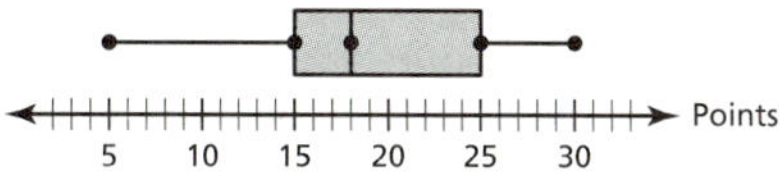

Practice and Problem Solving

5.

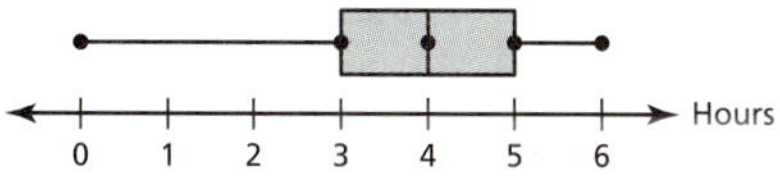

6.

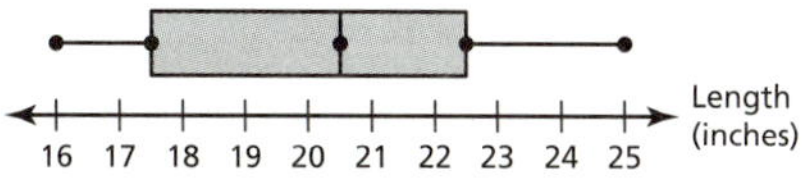

7.

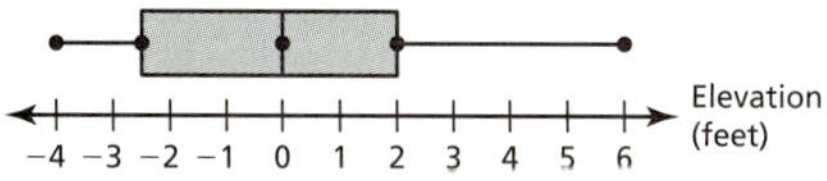

8. The median should be 5.5.

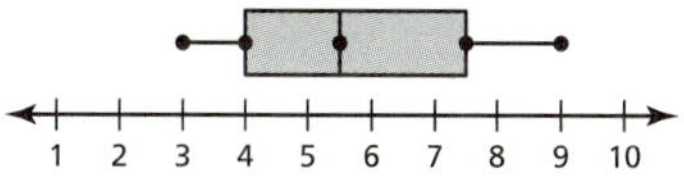

9.

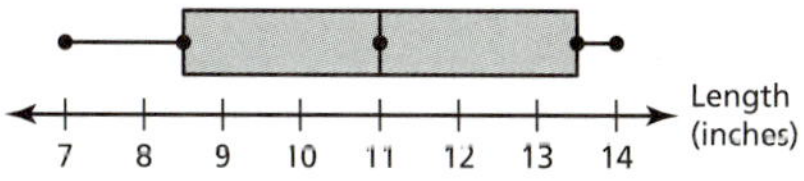

range = 7

10. 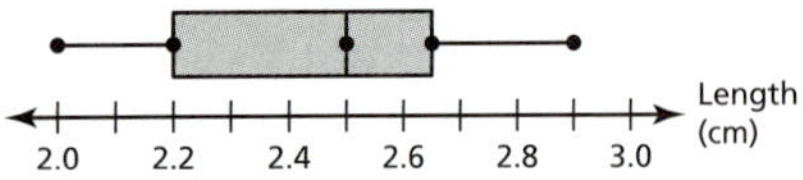

Sample answer: 50% of the inchworms are between 2.2 centimeters and 2.65 centimeters long.

11. a. 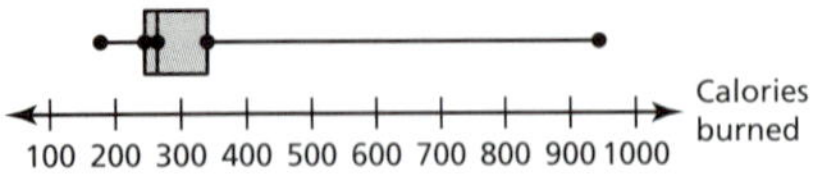

b. 944 calories

c. 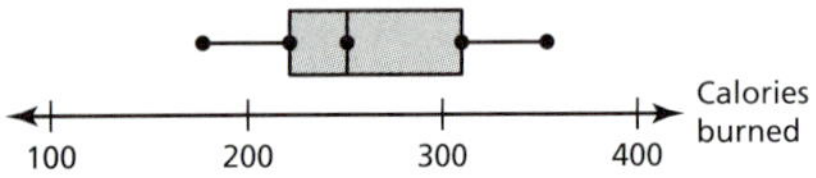

d. The outlier makes the right whisker longer, increases the length of the box, increases the third quartile, and increases the median. In this case, the first quartile and the left whisker were not affected.

7.1–7.2 Quiz

5. 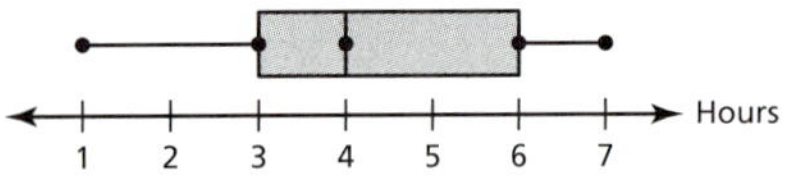

6. 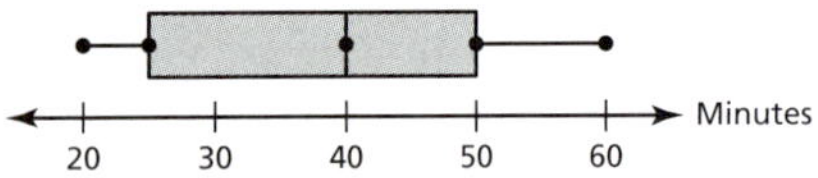

7. 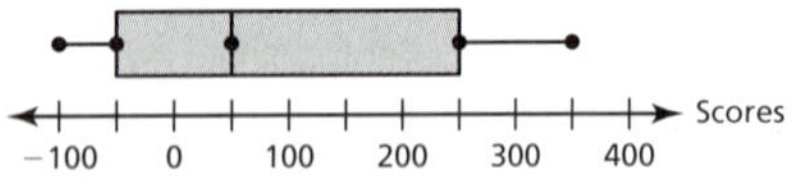

9. 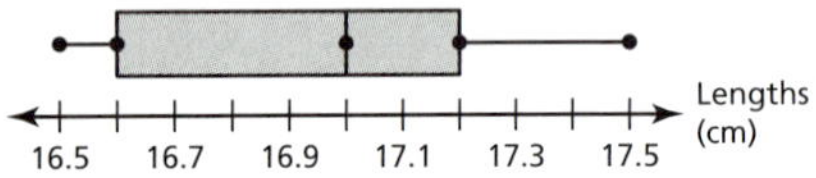

Most green anoles are between 16.6 and 17.2 centimeters long.

Section 7.3

On Your Own

2. 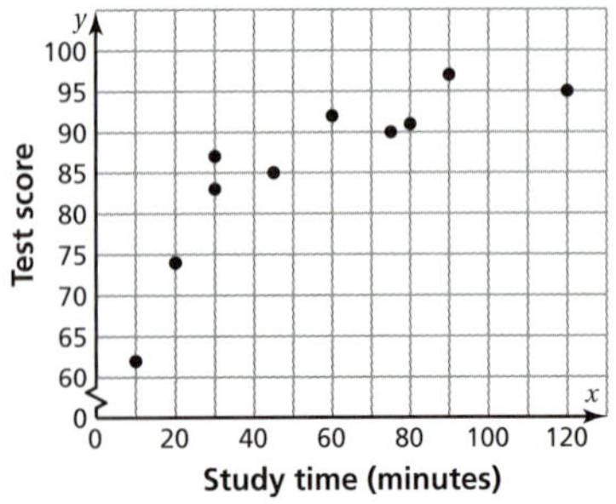

positive relationship

Practice and Problem Solving

16. a. (24.00, 40,000)

b. Because the outlier is below the other values, it will increase the steepness of the line of best fit.

c. 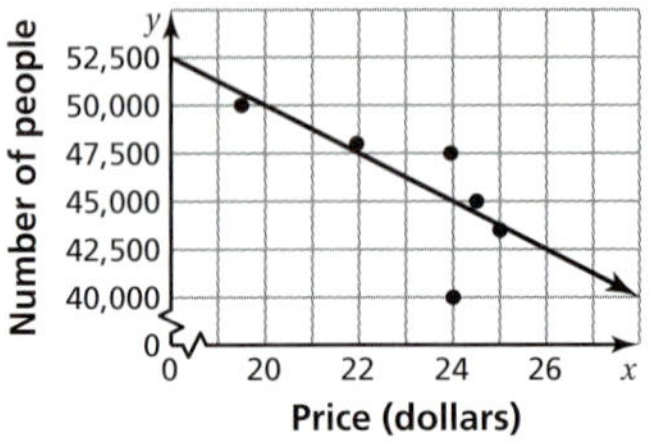

d. *Sample answer:* 41,000 people

Record and Practice Journal Practice

5. b. 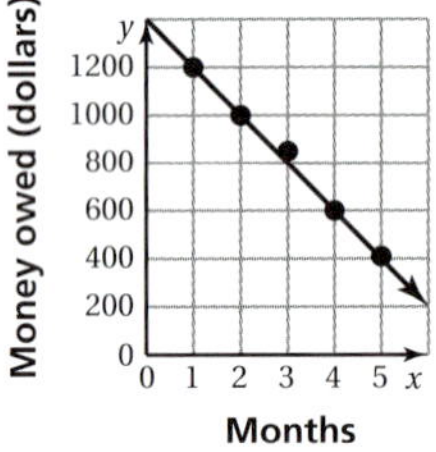

Section 7.4

Record and Practice Journal

1. c. *Sample answer:*

Raccoon Road Kill Weights

Stem	Leaf
9	4 5
10	
11	0
12	4 9
13	4 6 9
14	0 5 8 8
15	2 7
16	8
17	0 2 3 5
18	5 5 6 7
19	0 1 4
20	4
21	3 5 5 5
22	
23	
24	
25	4

Key: 9 | 4 = 9.4 pounds

The stem-and-leaf plot shows how the raccoon weights are distributed.

Practice and Problem Solving

10. The break in the scale for the vertical axis makes it appear as though there is a greater difference in sales between months.

11. The intervals are not the same size.

12. The width of the bars are different, so it looks like some months have more rainfall.

Record and Practice Journal Practice

4. Because the rain icon is larger than the sun icon, it makes it look as if there were equal amounts of sunny and rainy days when there was not.

7.3–7.4 Quiz

8. a–b.

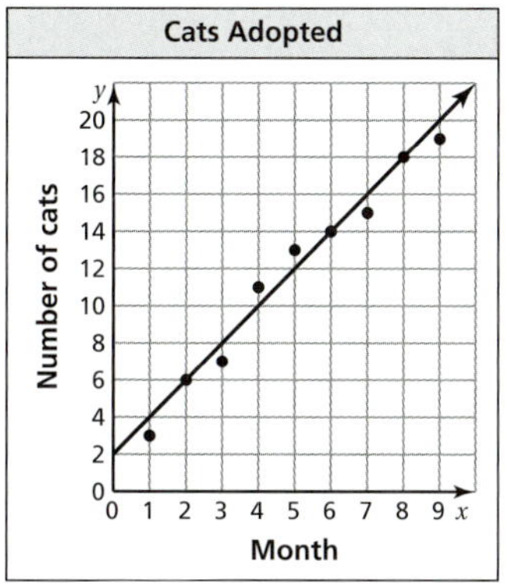

c. *Sample answer:* $y = 2x + 2$

d. *Sample answer:* 22 cats

Chapter 7 Review

4.

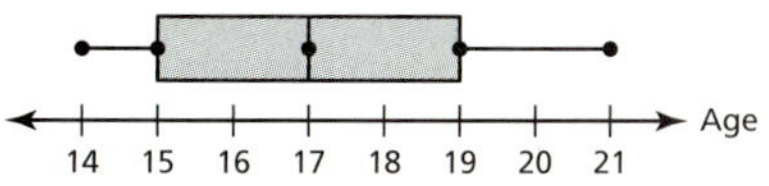

5.

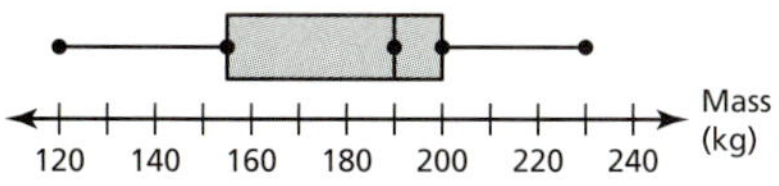

Chapter 7 Test

3.

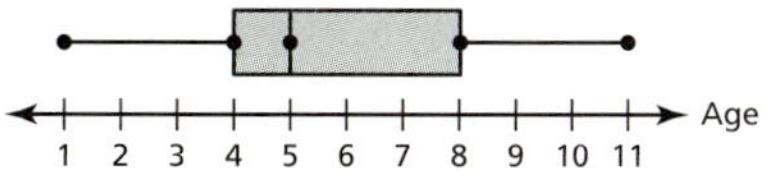

4.

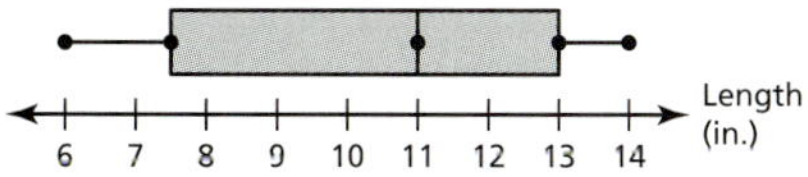

5.

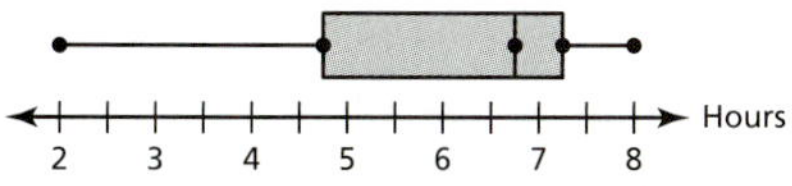

9.

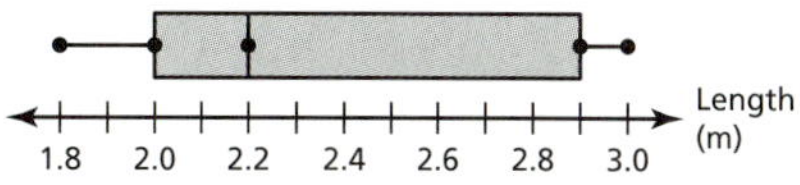

Half of the alligators are between 2 and 2.9 meters long.

11. a–b.

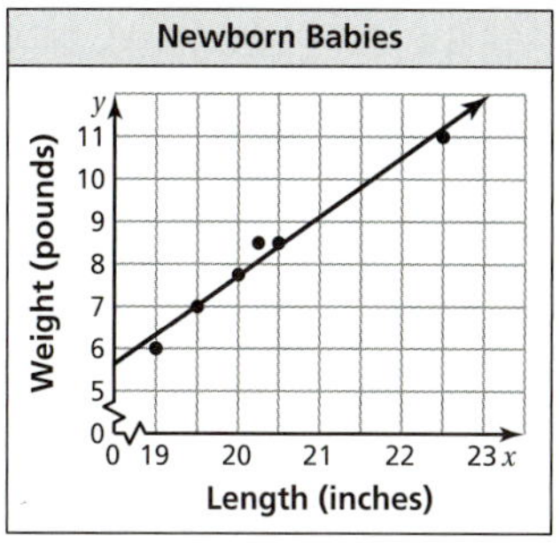

c. *Sample answer:* $y = 1.5x - 22.25$

d. *Sample answer:* 7.38 pounds

Chapter 8

Record and Practice Journal Fair Game Review

8.

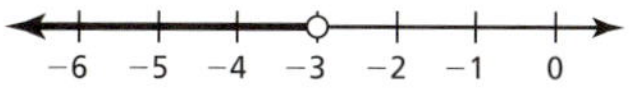

9.

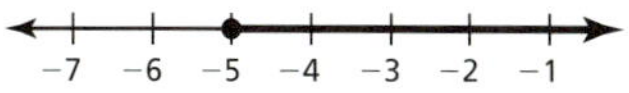

10.

11.

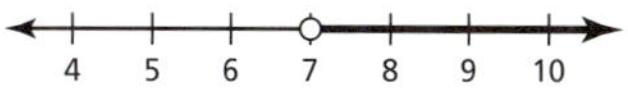

12.

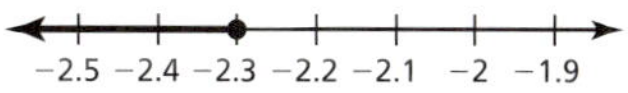

13.

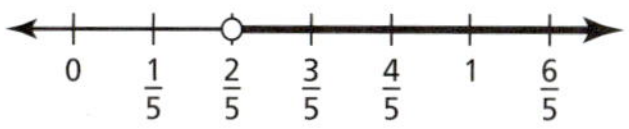

14.

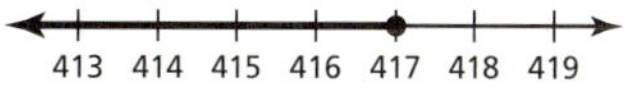

Section 8.1

Practice and Problem Solving

25. a. $a \geq 10$;

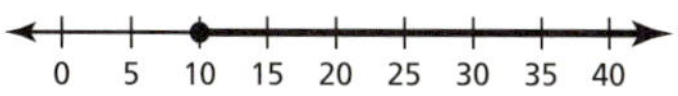

$s \geq 200$;

$t \geq 10$;

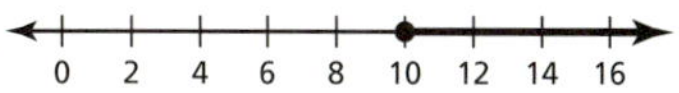

b. yes; You satisfy the swimming requirement of the course because $10(25) = 250$ and $250 \geq 200$.

Section 8.2

On Your Own

5. $\frac{1}{6} \le z$; 

6. $p > -3$; 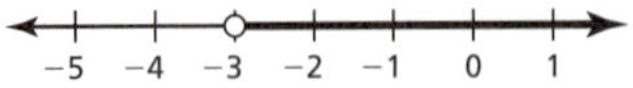

Practice and Problem Solving

10. $5\frac{1}{2} < v$; 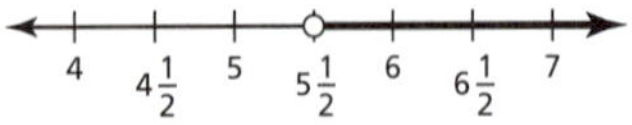

11. $-\frac{3}{5} > d$; 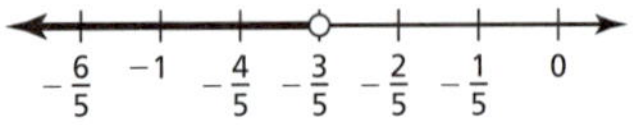

12. $-\frac{1}{3} \le g$; 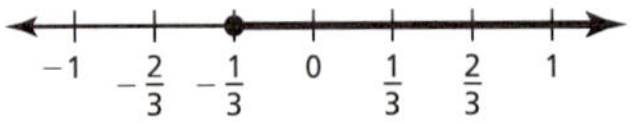

13. $m \le 1$; 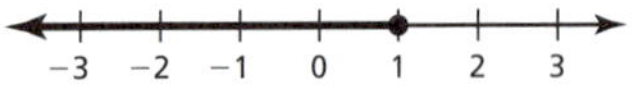

14. $1.4 \le k$; 

15. $h < -1.5$; 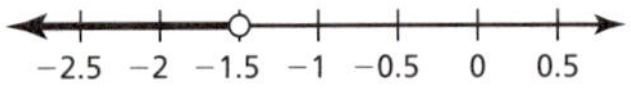

16. $-\pi > s$; 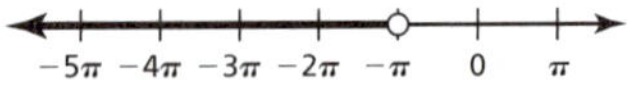

17. $9.5 \ge u$; 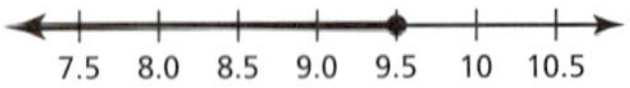

25. a. $4500 + x \ge 12{,}000$; $x \ge 7500$ points

b. This changes the number added to x by 60%, so the inequality becomes $7200 + x \ge 12{,}000$. So, you need less points to advance to the next level.

8.1–8.2 Quiz

17. a. $s \ge 100$;

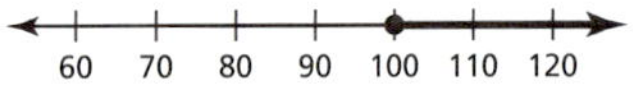

$t \ge 5$;

$u \ge 10$;

b. yes; Because 100 yards is equal to 300 feet and $350 \ge 300$.

Section 8.3

Practice and Problem Solving

12. $m < 10$; 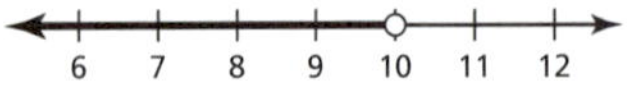

13. $x < -28$; 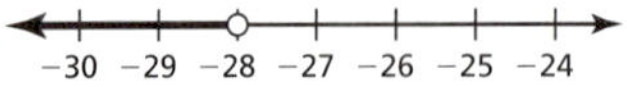

14. $w \ge -13$; 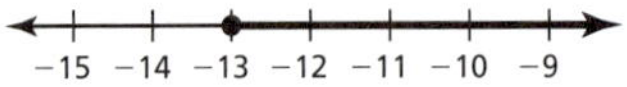

15. $k > 2$; 

16. $x \le -\frac{3}{8}$; 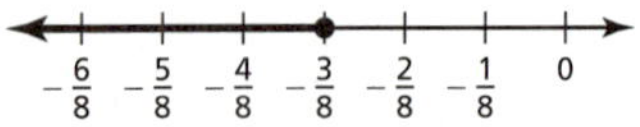

17. $y \le -4$; 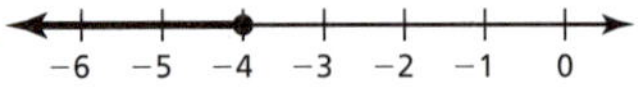

18. $b < 34.68$; 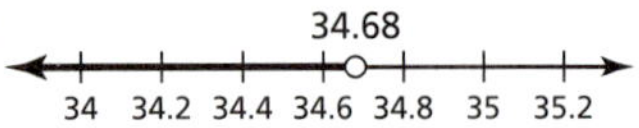

29. $h \le -42$; 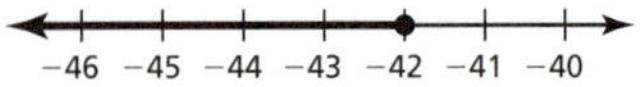

30. $x < 24$; 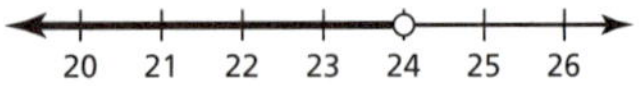

31. $y > \frac{11}{2}$; 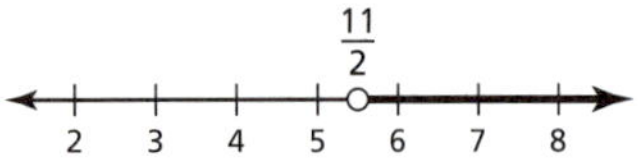

32. $d \le -8$; 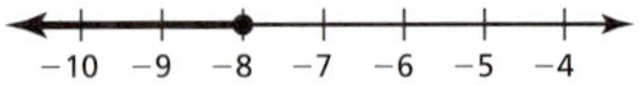

33. $m > -12$; 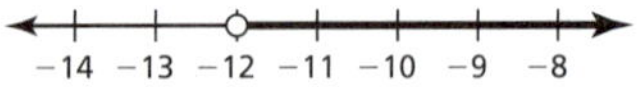

34. $k \ge -9$; 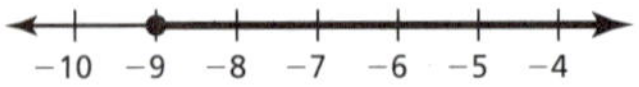

35. $b > 4$; 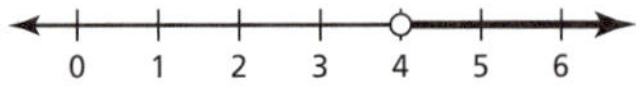

42. *Sample answer:* Consider the inequality $5 > 3$. If you multiply or divide each side by -1 without reversing the direction of the inequality symbol, you obtain $-5 > -3$, which is not true. So, whenever you multiply or divide an inequality by a negative number, you must reverse the direction of the inequality symbol to obtain a true statement.

43. *Answer should include, but is not limited to:* Make sure students use the correct number of months that the CD has been out. In part (d), an acceptable answer could be never because the top selling CD could have a higher monthly average.

Section 8.4

Practice and Problem Solving

12. $g \le 1$; 

13. $y \le 13$; 

14. $h \le -3$; 

15. $u < -17$; 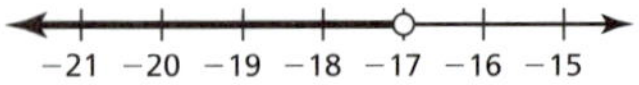

16. $n < 4.7$; 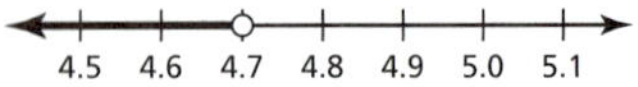

17. $z > -0.9$; 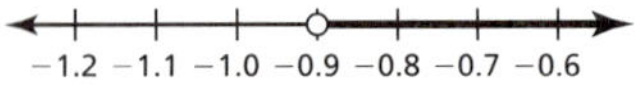

18. $20x + 100 \leq 320; x \leq 11$
\$20 bills

24. $3.5x + 350 \geq 500; x \geq 42\frac{6}{7}$; at least 43 more cars, so at least 143 cars total

25. $r \geq 3$ units

8.3–8.4 Quiz

9. $j < -25$;

10. $w > -1$;

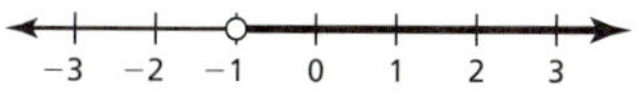

11. $0.5x \geq 200; x \geq 400$ flowers

12. $10x + 20 \leq 100; x \leq 8$; at most 8 guests

Chapter 8 Review

16. $q < 3$;

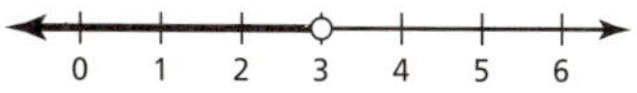

17. $p \geq -40$;

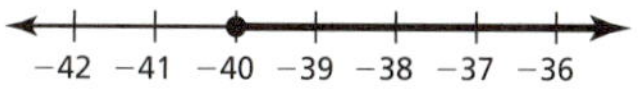

18. $k \geq 1.4$;

Chapter 8 Test

10. $v \geq -2.1$;

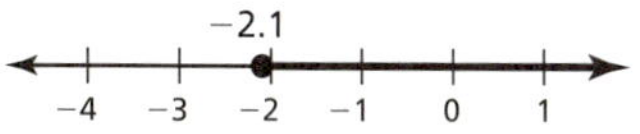

11. $t > 1$;

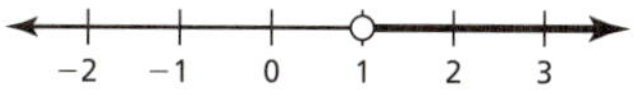

Chapter 9

Section 9.1

Practice and Problem Solving

27.

h	1	2	3	4	5
$2^h - 1$	1	3	7	15	31
2^{h-1}	1	2	4	8	16

$2^h - 1$; The option $2^h - 1$ pays you more money when $h > 1$.

Section 9.6

Record and Practice Journal

4. *Sample answer:* Move the decimal point left or right so the number is at least one but less than 10. Then multiply by ten raised to the number of times you moved the decimal. If you moved the decimal point to the left, the exponent will be positive. If you moved the decimal point to the right, the exponent should be negative.

Appendix B

Section B.1

Record and Practice Journal

3. *Sample answer:* If an account earns simple interest, you can find the balance B by using the equation $B = P(1 + rt)$ where P is the principal, r is the annual interest rate (in decimal form), and t is the time (in years). If an account earns compound interest, you can find the balance by multiplying the previous end-of-year balance by the annual interest rate and adding the result to the previous end-of-year balance.

Practice and Problem Solving

3. a.

Simple Interest			
t	Principal	Annual Interest	Balance at End of Year
3	\$500	\$15	\$545
4	\$500	\$15	\$560
5	\$500	\$15	\$575

Compound Interest			
t	Principal and Interest	Annual Interest	Balance at End of Year
3	\$530.45	\$15.91	\$546.36
4	\$546.36	\$16.39	\$562.75
5	\$562.75	\$16.88	\$579.63

b. compound interest

Section B.2

Record and Practice Journal

1. a. You can use the Distributive Property to rewrite the balance after the first year.

b. For year 2, the annual interest is the interest rate multiplied by the balance from year 1.

c. You can use the Distributive Property and Properties of Exponents to rewrite the balance after the second year.

d. The balance after the fifth year is a continuation of the pattern developed in parts (a) through (c).

Section B.3

Record and Practice Journal

2. a.

Interest You Pay					
Term \ Interest Rate	6%	8%	10%	12%	14%
36 mo	$1713.24	$2305.80	$2909.16	$3522.96	$4147.20
48 mo	$2291.04	$3092.64	$3913.44	$4752.48	$5610.24
60 mo	$2879.40	$3898.80	$4947.00	$6024.00	$7129.80
72 mo	$3478.32	$4723.20	$6009.84	$7336.80	$8704.80

b.

Interest You Pay			
Term \ Interest Rate	4%	6%	8%
20 yr	$81,782.40	$129,499.20	$181,341.60
30 yr	$129,366.00	$208,508.40	$295,480.80
40 yr	$181,099.20	$295,382.40	$420,748.80
50 yr	$236,562.00	$388,518.00	$553,614.00

Interest You Pay		
Term \ Interest Rate	10%	12%
20 yr	$236,889.60	$295,670.40
30 yr	$388,666.80	$486,540.00
40 yr	$553,660.80	$691,344.00
50 yr	$726,234.00	$902,766.00

Record and Practice Journal Practice

3.

Interest Paid on $150,000 Mortgage		
Term \ Interest Rate	4%	6%
20 yr	$68,152.80	$107,916.00
25 yr	$87,528.00	$139,935.00
30 yr	$107,803.20	$173,758.80

4.

Interest Paid on $180,000 Mortgage		
Term \ Interest Rate	5%	7%
20 yr	$105,100.80	$154,929.60
25 yr	$135,678.00	$201,660.00
30 yr	$167,860.80	$251,114.40

Section B.4

Record and Practice Journal

1. b–c.

Date	Check #	Transaction	Credit	Debit	Balance
		Balance Forward			100.00
5/11/12	996	Grocery Store		29.55	70.45
5/11/12		Deposit Paycheck	482.75		553.20
5/16/12		ATM Withdrawal		100.00	453.20
5/17/12	997	Gym Membership		22.35	430.85
5/17/12	998	Cell Phone Company		58.00	372.85
5/17/12	999	Car Payment		82.66	290.19
5/18/12		Deposit Paycheck	501.50		791.69
5/18/12	1000	Birthday Gift		41.28	750.41
5/20/12	1001	Green Jean Shop		146.77	603.64

2.

Date	Check #	Transaction	Credit	Debit	Balance
5/17/12	998	Cell Phone Company		58.00	372.85
5/17/12	999	Car Payment		82.66	290.19
5/18/12		Deposit Paycheck	501.50		791.69
5/18/12	1000	Birthday Gift		41.28	750.41
5/20/12	1001	Green Jean Shop		146.77	603.64
5/16/12		ATM Fee		2.50	601.14
5/24/12		Interest	1.82		602.96

Your balance now agrees with the bank statement.

On Your Own

1.

Date	Check #	Transaction	Credit	Debit	Balance
9/30/12	1317	Auto Shop		74.95	765.86
10/1/12	1318	Plumber		120.85	645.01

The ending balance is $645.01.

B.1–B.4 Quiz

1. a. You: $B = 1200 + 30t$
Friend: $B = 1000 + 50t$

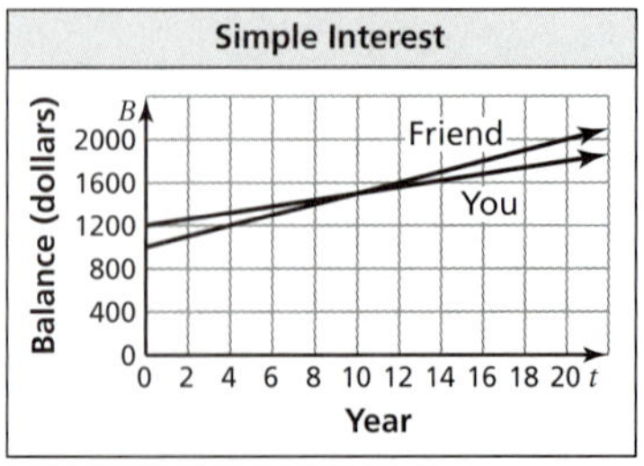

b. Yes, the graphs intersect when $t = 10$. So, after 10 years, the accounts are equal.

Section B.6

Record and Practice Journal

1. d. The gross pay is the amount of money earned during a pay period. The net pay is the amount of money that an employee receives after all the deductions have been made from the gross pay.

Practice and Problem Solving

17. *Sample answer:* Colorado: 4.63%; Florida: none; Massachusetts: 5.3%; Florida has the lowest income tax and Massachusetts has the highest income tax.

Conversions

U.S. Customary
1 foot = 12 inches
1 yard = 3 feet
1 mile = 5280 feet
1 acre ≈ 43,560 square feet
1 cup = 8 fluid ounces
1 pint = 2 cups
1 quart = 2 pints
1 gallon = 4 quarts
1 gallon = 231 cubic inches
1 pound = 16 ounces
1 ton = 2000 pounds
1 cubic foot ≈ 7.5 gallons

U.S. Customary to Metric
1 inch ≈ 2.54 centimeters
1 foot ≈ 0.3 meter
1 mile ≈ 1.6 kilometers
1 quart ≈ 0.95 liter
1 gallon ≈ 3.79 liters
1 cup ≈ 237 milliliters
1 pound ≈ 0.45 kilogram
1 ounce ≈ 28 grams
1 gallon ≈ 3785 cubic centimeters

Time
1 minute = 60 seconds
1 hour = 60 minutes
1 hour = 3600 seconds
1 year = 52 weeks

Temperature
$$C = \frac{5}{9}(F - 32)$$

$$F = \frac{9}{5}C + 32$$

Metric
1 centimeter = 10 millimeters
1 meter = 100 centimeters
1 kilometer = 1000 meters
1 liter = 1000 milliliters
1 kiloliter = 1000 liters
1 milliliter = 1 cubic centimeter
1 liter = 1000 cubic centimeters
1 cubic millimeter = 0.001 milliliter
1 gram = 1000 milligrams
1 kilogram = 1000 grams

Metric to U.S. Customary
1 centimeter ≈ 0.39 inch
1 meter ≈ 3.28 feet
1 kilometer ≈ 0.6 mile
1 liter ≈ 1.06 quarts
1 liter ≈ 0.26 gallon
1 kilogram ≈ 2.2 pounds
1 gram ≈ 0.035 ounce
1 cubic meter ≈ 264 gallon

Equations of Lines

Slope-intercept form

$$y = mx + b$$

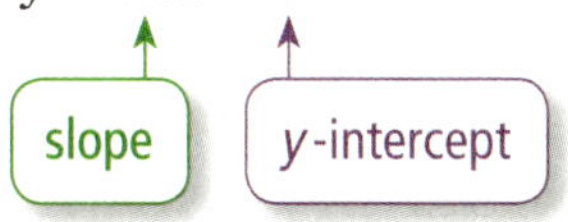

Standard form

$$ax + by = c, a, b \neq 0$$

Pythagorean Theorem

$$a^2 + b^2 = c^2$$

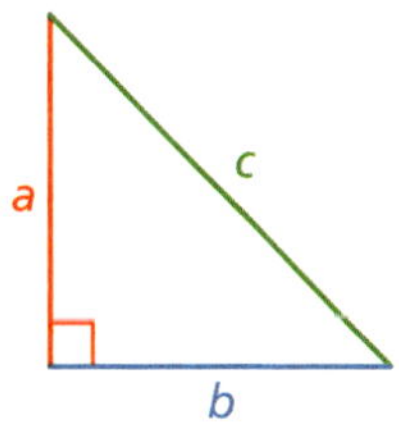

Rules of Exponents

Product of Powers Property	$a^m \cdot a^n = a^{m+n}$
Quotient of Powers Property	$\frac{a^m}{a^n} = a^{m-n}$, where $a \neq 0$
Zero Exponents	$a^0 = 1$, where $a \neq 0$
Negative Exponents	$a^{-n} = \frac{1}{a^n}$, where $a \neq 0$

NCTM Curriculum Focal Points

K

FOCAL POINT Represent, compare, and order whole numbers and join and separate sets.

FOCAL POINT Describe shapes and space.

FOCAL POINT Order objects by measurable attributes.

Connections

Algebra • Data Analysis • Geometry

1

FOCAL POINT Develop understandings of addition and subtraction and strategies for basic addition facts and related subtraction facts.

FOCAL POINT Develop an understanding of whole number relationships, including grouping in tens and ones.

FOCAL POINT Compose and decompose geometric shapes.

Connections

Algebra • Data Analysis • Measurement • Number and Operations

2

FOCAL POINT Develop an understanding of the base-ten numeration system and place-value concepts.

FOCAL POINT Develop quick recall of addition facts and related subtraction facts and fluency with multi-digit addition and subtraction.

FOCAL POINT Develop an understanding of linear measurement and facility in measuring lengths.

Connections

Algebra • Geometry • Measurement • Number and Operations